TRACE ELEMENTS
in the
ENVIRONMENT

Biogeochemistry, Biotechnology, and Bioremediation

Edited by
M.N.V. Prasad
Kenneth S. Sajwan
Ravi Naidu

CRC Press
Taylor & Francis Group
Boca Raton London New York

CRC Press is an imprint of the
Taylor & Francis Group, an **informa** business
A TAYLOR & FRANCIS BOOK

CRC Press
Taylor & Francis Group
6000 Broken Sound Parkway NW, Suite 300
Boca Raton, FL 33487-2742

First issued in paperback 2019

© 2006 by Taylor & Francis Group, LLC
CRC Press is an imprint of Taylor & Francis Group, an Informa business

No claim to original U.S. Government works

ISBN-13: 978-1-56670-685-8 (hbk)
ISBN-13: 978-0-367-39196-6 (pbk)

Library of Congress Card Number 2005041838

Library of Congress Cataloging-in-Publication Data

Trace elements in the environment: biogeochemistry, biotechnology, and bioremediation / edited by M.N.V. Prasad, Ravi Naidu, Kenneth S. Sajwan.
     p. cm.
   Includes bibliographical references and index.
   ISBN 1-56670-685-8
   1. Trace elements--Environmental aspects. I. Prasad, M. N. V. (Majeti Narasimha Vara), 1953- II. Sajwan, Kenneth S. III. Naidu, R.

QH545.T7T73 2005
628.5--dc22
                                                                                          2005041838

Visit the Taylor & Francis Web site at
http://www.taylorandfrancis.com

and the CRC Press Web site at
http://www.crcpress.com

Preface

Plant and soil form an integrated system. Technogenic contamination of soils with potentially toxic trace elements (PTE) are reflected in the functioning of plants and soil biota. Soil contamination by PTE has several implications for human health, as well as for the biosphere. Trace element "biogeogenic cycling" in the environment is an integral function of the ecosystem (aquatic, terrestrial, and atmospheric). Therefore, the aim of this collective work is to deal with the trace elements in the holistic environment, considering advancements in the state-of-the-art analytical techniques, molecular biology, and contemporary biotechnology that enhance our knowledge of the behavior of trace elements in the biogeosphere and organismal levels, i.e., at the cellular and molecular levels. Various chapters of this book provide the background with appropriate examples to understanding the trace elements in the biogeosphere on bioavailability, biogeochemistry, biotechnology, bioremediation, and risk assessment.

Trace element behavior and fate depend on their chemistry in soil inorganic and organic phases; their bioavailability depends on a variety of factors concerning the ambient environment, soil, and/or sludge. Trace element enrichment in soil, water, and air may result from natural sources and/or anthropogenic activities such as smelting, mining, agricultural, and waste disposal technologies. For example, coal fly ash application to soils and its effect on boron and other trace element availability to plants; bioavailability of trace elements in relation to root modification in the rhizosphere; and availability through sewage sludge are some important issues discussed in this book. To better explore adaptive physiology of plants exposed to elevated doses of trace elements, knowledge of the behavior of the essential and nonessential elements, aspects related to biogeogenic cycling, accumulation, and exclusion mechanisms by target organisms is a must.

It is generally accepted that the rhizosphere plays an important role in the bioavailability of trace elements. The mechanisms involved in chemical modifications in the rhizosphere, as well as on uptake of trace elements, differ among plant species and soil conditions. The ability to manipulate siderophore production in the rhizosphere to improve plant trace element nutrition will remain a significant challenge for the future to investigate. The importance of mycorrhizal symbiosis for the establishment of a sustainable plant cover on soils with PTE is therefore obvious. Microbial genomics is an integrated tool for developing biosensors for toxic trace elements in the environment, arbuscular mycorrhizal fungi, and the role of arbuscular mycorrhiza and associated microorganisms are increasingly considered in phytoremediation of heavy metal polluted sites. Plant metallothionein genes; genetic engineering for the cleanup of toxic trace elements; and "metallomics," a multidisciplinary metal-assisted functional biogeochemistry — its scope and limitations as the crux of biotechnology and its role in dealing with the PTE in the environment are some of the themes reviewed in different chapters.

Self-cleaning of soils does not take place or, rather, takes place extremely slowly. The toxic metals in top soil, thus get accumulated in plants. Plants can remediate metal pollutants mainly in two ways: (1) phytostabilization, in which plants convert pollutants to a less bioavailable form and/or prevent pollutants' dispersal by wind erosion or leaching; and (2) phytoextraction, in which plants accumulate pollutants in their harvestable tissues, thus decreasing the concentration of the pollutants in the soil. Plants that accumulate and/or exclude toxic trace elements; tolerant plants and biodiversity prospecting to promote phytotechnologies for environmental cleanup; phytomanagement of abandoned mines and biogeochemical prospecting; phytoremediation of contaminated soil with cereal crops; and the role of fertilizers and bacteria in biavailability of metals are reviewed.

Phytotechnologies using trees; stabilization, remediation, and integrated management of metal-contaminated ecosystems by grasses; applications of weeds more adapted to unfavorable soil conditions such as low moisture; presence of toxic metals easily acclimatized to local situation that would act as sentinels for monitoring trace element pollution; detoxification and defense mechanisms in metal-exposed plants; biogeochemical cycling of trace elements by aquatic and wetland plants and its relevance to phytoremediation; plants that hyperaccumulate PTE and biodiversity prospecting for phytoremediation; phytomanagement of radioactively contaminated sites; phytoextraction of Cd and Zn by willows — advantages and limitations; adaptive physiology; and rhizosphere biotechnology are covered in the sections on biotechnology and bioremediation.

Bacterial biosorption of trace elements; processes and applications of electroremediation of heavy metal-contaminated soils; and application of novel nanoporous sorbents for the removal of heavy metals, metalloids, and radionuclides are some of the emerging areas of research that have been included in this book.

The increasing level of trace elements in the tissues of plants and animals due to bioaccumulation and trophic transfer has adverse effects on ecological and human health. Therefore, the risk assessment, pathways, and trace element toxicity of sewage sludge-amended soils and usage in agroforestry; trophic transfer of trace metals and associated human health issues; and PTE accumulation, movement, and remediation in soils receiving animal manure are also covered.

Editors

M.N.V. Prasad is a professor of environmental biology at the University of Hyderabad, India. The author, coauthor, editor, or coeditor for 6 books; Dr. Prasad has published more than 140 research papers in the broad area of environmental botany and heavy metal stress in plants. From 2000 to 2004, Dr. Prasad served as head of the Department of Plant Sciences, School of Life Sciences; during 2001-2003, he functioned as a coordinator for the M.Sc. Biotechnology programme of the School of Life Sciences (sponsored by the Department of Biotechnology, Government of India); also as coordinator since 1995, and post graduate diploma in Environmental Education and Management, Centre for Distance Education. He is an elected fellow of the Linnean Society of London, England, and the National Institute of Ecology, New Delhi, India, and a member of the International Allelopathy Society; the National Institute of Ecology; the Bioenergy Society of India; and the Indian Network for Soil Contamination Research. He received B.Sc. (1973) and M.Sc. (1975) degrees from Andhra University, India, and a Ph.D. degree (1979) in botany from the University of Lucknow, India.

Academic distinctions:

- 2003: Academy of Finland, visiting scientist, Department of Biology/Botany, University of Oulu, Finland
- 2000: Swedish International Development Cooperation Agency (SIDA) visiting fellowship, Institute of Botany Stockholm University
- 1998: Elected fellow, Linnean Society of London
- 1998: Fundacao para a Ciencia e a Technologia (FCT), Portugal visiting professorship at the Departamento de Botanica, Universidade de Coimbra, Coimbra, Portugal
- 1996: Indo–Polish Cultural Exchange Program, visiting fellowship, Institution of Molecular Biology, Jagiellonian University, Krakow, Poland
- 1996: Elected fellow, National Institute of Ecology, New Delhi
- 1994: Natural and Engineering Research Council (NSERC), Canada, Foreign Researcher Award

Professor Ravi Naidu is foundation professor and the inaugural director of the Australian Research Centre for Environmental Risk Assessment and Remediation. He has researched environmental contaminants, bioavailability, and remediation for over 20 years. Naidu has co-authored over 300 technical publications and co-edited 8 books in the field of soil and environmental sciences including remediation of contaminated sites. Since 1994, he has worked with scientists from the Asia region on environmental contamination, including the recent arsenic poisoning of people in Bangladesh, India, and China. In recognition of his contribution to environmental research, he was awarded the Gold Medal in environmental science in 1998 by Tamil Nadu Agricultural University, elected to the Fellow of Soil Science Society of America in 2000, and also elected Fellow of the Soil Society of New Zealand in 2004. Naidu is the chair of the Standards Australia–New Zealand Technical Committee on Sampling and Analyses of Contaminated Soils; chair of the International Committee on Bioavailability and Risk Assessment; chair of the International Union of Soil Sciences' Commission for Soil Degradation Control, Remediation, and Reclamation; president of the International Society on Trace Element Biogeochemistry; and sitting member of the Victorian EPA Contaminated Sites Auditor Panel.

Kenneth S. Sajwan is a professor and director of the environmental science program in the department of natural sciences and mathematics at Georgia's Savannah State University. Dr. Sajwan earned a B.S. degree in agriculture and animal husbandry, an M.S. degree in agronomy, and Ph.D. degrees in science from the Indian Institute of Technology, Kharagpur and in soil chemistry and plant nutrition from Colorado State University. Dr. Sajwan joined the faculty of Savannah State University in 1992 as an associate professor and was promoted to full professor in 1996. Prior to joining the faculty of Savannah State University, he was an assistant professor at the University of Georgia's Savannah River Ecology Laboratory in Aiken, South Carolina. His previous work experience includes a World Bank consultancy to Colombia, South America, and research associateships at the universities of Wisconsin and Kentucky. Dr. Sajwan holds adjunct professorship appointments at Alabama Agricultural & Mechanical University and the University of South Carolina at Aiken, and a faculty affiliate appointment at the Institute of Ecology at the University of Georgia.

Dr. Sajwan has been recognized as a devoted and talented teacher and his accomplishments are reflected in his ability to motivate, challenge, and inspire his students to excel in the classroom and beyond. Dr. Sajwan has received several awards for his outstanding contribution to teaching and research. He is a recipient of the Richard Nicholson National Award for Excellence in Science Teaching (2005); the National Science Teacher's Association Distinguished Science Teacher Award (2004); the Ernest L. Boyer International Award for Teaching, Learning, and Technology (2003); the Board of Regents' University System of Georgia Teaching Excellence Award (2002); and the White House Millennium Award for Teaching and Research Excellence (2001). In addition, he is the recipient of the Board of Regents' University System of Georgia Distinguished Professor of Teaching and Learning Award for the 1998-1999 academic year at Savannah State University and is the recipient of the 1999 International Award for Innovative Excellence in Teaching, Learning, and Technology.

Dr. Sajwan has edited four books, *Coal Combustion Products and Environmental Issues, Chemistry of Trace Elements in Fly Ash, Biogeochemistry of Trace Elements in Coal and Coal Combustion Byproducts,* and *Trace Elements in Coal and Coal Combustion Residues.* In addition, he has published two laboratory manuals and over 100 articles in peer reviewed journals, serials, conference proceedings, and symposia. Dr. Sajwan's primary research areas of interest include biogeochemistry of trace elements, environmental chemistry, ecotoxicology, and chemical equilibria in soils.

Contributors

J. Afolabi
Department of Natural Sciences and Mathematics
Savannah State University
Savannah, Georgia

Clark Alexander
Department of Chemistry
Murray State University
Murray, Kentucky

A.K. Alva
Department of Natural Sciences and Mathematics
Savannah State University
Savannah, Georgia

V. Antoniadis
Institute of Soil Mapping and Classification
National Agricultural Research Foundation
Larissa, Greece

P. Aravind
Department of Plant Sciences
University of Hyderabad
Hyderabad, India

C.R. Babu
Center for Environmental Management of Degraded Ecosystems
School of Environmental Studies, University of Delhi
Delhi, India

J.M. Barea
Departamento Microbiologia del Suelo y Sistemas Simbioticos
Estación Experimental del Zaidin
Granada, Spain

D.I. Bashmakov
Department of Botany and Plant Physiology
Mordovian N.P. Ogariov State University
Saransk, Russia

Nanthi Bolan
Institute of Natural Resources
Massey University
Palmerston North, New Zealand

Ranadhir Chakraborty
Department of Botany, Microbiology Laboratory
University of North Bengal
West Bengal, India

I-Lun Chien
Department of Chemistry
Murray State University
Murray, Kentucky

Brent Clothier
HortResearch
Palmerston North, New Zealand

Jan Colpaert
Environmental Biology
Limburgs University Centre
Diepenbeek, Belgium

Nicholas G. Danalatos
Department of Environmental Studies
University of Aegean
Mytilini, Greece

Rupali Datta
Earth and Environmental Science Department
The University of Texas at San Antonio
San Antonio, Texas

Ma. del Carmen Angeles González Chávez
Natural Resources Institute
Colegio de Postgraduados
Montecillo, México

N.M. Dickinson
School of Biological and Earth Sciences
Liverpool John Moores University
Liverpool, England

H. Freitas
Departamento de Botânica
Universidade de Coimbra
Coimbra, Portugal

Glen E. Fryxell
Pacific Northwest National Laboratory
Richland, Washington

V. Gianinazzi–Pearson
Plante-Microbe-Environment
Dijon, France

Maria Greger
Department of Botany
Stockholm University
Stockholm, Sweden

B.H. Hulin
Department of Renewal Resources
University of Wyoming
Laramie, Wyoming

V.K. Jha
Institute of Life Sciences
Orissa, India

A. Jurkiewicz
Institute of Botany of the Jagiellonian University
Krakow, Poland

So-Young Kang
Department of Environmental Science and Engineering
Gwangju Institute of Science and Technology (GIST)
Gwangju, South Korea

Catherine Keller
CEREGE
Universitè Aix-Marseille III
Aix-en-Provence, France

Kyoung-Woong Kim
Department of Environmental Science and Engineering
Gwangju Institute of Science and Technology (GIST)
Gwangju, South Korea

N. Kundu
Department of Geology and Geophysics
Indian Institute of Technology
West Bengal, India

Corinne Leyval
Laboratoire des Interactions Microorganismes-Minèraux-Matiere
Organique dans les Sols
Vandoeuvre-lès-Nancy, France

Yuehe Lin
Pacific Northwest National Laboratory
Richland, Washington

G. Lingua
Dipartimento di Scienze dell'Ambiente e della Vita
Università del e gustes Piemonte Orientale Amedeo Avogadro
Alessandria, Italy

Bommanna G. Loganathan
Department of Chemistry
Murray State University
Murray, Kentucky

Amit Love
Center for Environmental Management of Degraded Ecosystems
School of Environmental Studies, University of Delhi
Delhi, India

A.S. Lukatkin
Department of Botany and Plant Physiology
Mordovian N.P. Ogariov State University
Saransk, Russia

Santiago Mahimairaja
Institute of Natural Resources
Massey University
Palmerston North, New Zealand

Theodora Matsi
Soil Science Laboratory
Aristotle University of Thessaloniki
Thessaloniki, Greece

Shas V. Mattigod
Pacific Northwest National Laboratory
Richland, Washington

Ioannis K. Mitsios
Department of Agriculture, Crop Production and Rural Environment
School of Agricultural Sciences, University of Thessaly
Magnesia, Greece

Jeffrey M. Novak
USDA–ARS
Coastal Plains Research Center
Florence, South Carolina

M.K. Panigrahi
Department of Geology and Geophysics
Indian Institute of Technology
West Bengal, India

S. Paramasivam
Department of Natural Sciences and Mathematics
Savannah State University
Savannah, Georgia

Kent E. Parker
Pacific Northwest National Laboratory
Richland, Washington

M.N.V. Prasad
Department of Plant Sciences, School of Life Sciences
University of Hyderabad
Hyderabad, India

J. Pratas
Departamento de Ciências da Terra
Universidade de Coimbra
Coimbra, Portugal

I.D. Pulford
Environmental, Agricultural, and Analytical Chemistry
Chemistry Department, University of Glasgow
Glasgow, Scotland

K.J. Reddy
Department of Renewal Resources
University of Wyoming
Laramie, Wyoming

Alexandra B. Ribeiro
Departamento de Ciências e Engenharia do Ambiente, Faculdade de Ciências e Tecnologia
Universidade Nova de Lisboa
Caparica, Portugal

Brett Robinson
Swiss Federal Institute of Technology
Zurich Institute of Terrestrial Ecology
Schlieren, Switzerland

Jose M. Rodríguez–Maroto
Department of Chemical Engineering, Faculty of Sciences
University of Malaga
Malaga, Spain

Pradosh Roy
Department of Microbiology
Bose Institute
Kolkata, India

Shivendra Sahi
Biotechnology Center, Department of Biology
Western Kentucky University
Bowling Green, Kentucky

B.B. Sahu
Institute of Life Sciences
Orissa, India

Kenneth S. Sajwan
Department of Natural Sciences and Mathematics
Savannah State University
Savannah, Georgia

V. Samaras
Institute of Soil Mapping and Classification
National Agricultural Research Foundation
Larissa, Greece

Dibyendu Sarkar
Earth and Environmental Science Department
The University of Texas at San Antonio
San Antonio, Texas

K. Chandra Sekhar
DRDO, Ministry of Defense
Government of India
Defense of Metallurgical Research Laboratory
Hyderabad, India

J. Sgouras
National Agricultural Research Foundation
Institute of Soil Classification and Mapping
Larissa, Greece

Nilesh Sharma
Biotechnology Center, Department of Biology
Western Kentucky University
Bowling Green, Kentucky

B.P. Shaw
Institute of Life Sciences
Orissa, India

Irina Shtangeeva
St. Petersburg University
St. Petersburg, Russia

Karamat R. Sistani
USDA–ARS
Animal Waste Management Research Unit
Bowling Green, Kentucky

Q.D. Skinner
Department of Renewal Resources
University of Wyoming
Laramie, Wyoming

S. Stamatiadis
Gaia Environmental Research and Education Center
Goulandris Natural History Museum
Athens, Greece

S. Tripathy
Department of Geology and Geophysics
Indian Institute of Technology
West Bengal, India

Christos Tsadilas
National Agricultural Research Foundation
Institute of Soil Classification and Mapping
Larissa, Greece

E. Tsantila
Gaia Environmental Research and Education Center
Goulandris Natural History Museum
Athens, Greece

K. Turnau
Institute of Botany
The Jagiellonian University
Krakow, Poland

H. Vandenhove
SCK-CEN, Radiation Protection Research Department
Boeretang, Belgium

Jaco Vangronsveld
Environmental Biology
Limburgs University Center
Diepenbeek, Belgium

Table of Contents

Section I

Bioavailability

1 Coal Fly Ash Application to Soils and its Effect on Boron Availability to Plants

Theodora Matsi and Christos Tsadilas

CONTENTS

1.1 INTRODUCTION

Fly ash, a by-product of coal combustion, is produced in large quantities in many countries and is partially disposed of in soils. Physically, fly ash consists mainly of silt-sized particles, is characterized by moderate to extremely high water-holding capacity, and possesses cementing properties. Chemically, fly ash is a ferro–alumino–silicate material and contains substantial amounts of macro- and micronutrients for plants (except C and N). Fly ash can be a strongly acidic or a strongly alkaline material, depending on its composition. Consequently, its application to soils can enhance soil fertility, improve soil's physical properties, and raise the pH of acid soils, if it is alkaline in reaction.

However, fly ash may cause undesirable environmental problems, i.e., unacceptable soil pH values, salinization, and B toxicity [1,2]. In fact, certain trace elements found in different fly ashes are considered to be potentially toxic in plants and animals [2], although, according to Page et al., [3] concentrations of trace elements in fly ash and soils are generally comparable, except those of B, Mo, and Se. Concentrations of these three elements in fly ash greatly exceed those generally found in soils.

Boron, an essential micronutrient for plant growth, is toxic at levels even slightly higher than the optimum for normal growth. Boron concentration in the soil solution plays an important role in plant nutrition because it is the determining factor of plant response [4]. Boron concentration in the soil solution, however, is buffered by B adsorbed onto soil particles [5]. Thus, knowledge of factors affecting B adsorption on and release from soil solids is important to understanding or predicting plant response when B-rich fly ash is applied to soils.

Application of fly ash to soil generally increases trace elements' concentrations to a degree that depends on fly ash properties, rate of application, and soil physicochemical characteristics. Among the potentially toxic trace elements in fly ash, however, only B has been associated with significant reductions in plant production under field conditions; fly ash B is water soluble and consequently may be readily absorbed by plants grown on soils amended with fly ash. In fact, many researchers consider B a major limiting factor for the beneficial utilization of fly ashes in croplands, especially unweathered fly ash [1]. In many cases, plant tolerance to fly ash is associated with B tolerance. Suggested measures to obviate detrimental effects to plants due to the agronomic use of fly ash include weathering of fly ash, followed by adequate drainage [1,2] or selection of species tolerant to B toxicity [2,6].

To evaluate the risk of B phytotoxicity in connection with the agronomic use of fly ash, knowledge relevant to B status in fly ash and soils, as well as to the factors controlling B availability, is needed. The objectives of this chapter are to compile and evaluate existing research and knowledge concerning: (1) concentrations and chemical behavior of B in fly ashes; (2) soil factors and constituents influencing the fate of B after fly ash application to soils; and (3) rates of fly ash application to soils that could result in B phytotoxicity or correction of B deficiency, in relation to properties of ashes and soils.

1.2 BORON IN FLY ASH

1.2.1 Boron Concentration in Fly Ash, as Affected by its Origin, Particle Size Distribution, and Degree of Weathering

Boron is among the trace elements with which fly ash is enriched [3], in comparison to other by-products of coal combustion. Goodarzi and Swaine [7] reported that, during the combustion of Netherlands coals (with an average B content of 35 mg kg^{-1}), B was highly concentrated in flue gases after desulfurization (average 1473 µg B m^{-3}) and relatively less concentrated in bottom ash (135 mg B kg^{-1}). Scrubbing by electrostatic precipitators reduced B content of flue gases to 237 µg m^{-3}, while the rest of B remained in fly ash. Furthermore, the same researchers reported that Australian bituminous coals have produced bottom ash with B concentrations of 40 to 80 mg kg^{-1} and fly ash containing 20 to 400 mg B kg^{-1}. Llorens et al. [8] studied the composition of Spanish fly and bottom ashes derived from a mixture of 95% subbituminous and 5% bituminous coal. They found that total B concentration in fly ash was almost 3.7 times higher than that in bottom ash and that water-extractable B of fly ash was 10.6 times greater than that of bottom ash.

Boron concentration in fly ash varies greatly. Table 1.1 summarizes the values of B content of fresh or weathered fly ashes derived from different types of coal, in various countries. It is reported that B concentration in fly ash is generally higher than that of other trace elements, ranging from 48 to 618 mg kg^{-1} [25]. Certain extremely high values of B content in fly ashes are reported in the literature. Cox et al. [29] and James et al. [27] found that B concentration in the ashes studied was at levels up to about 2000 mg kg^{-1}; Roy et al. [26] reported that B content in a western U.S. fly ash was 5000 mg kg^{-1}.

Among the trace elements contained in fly ash, B is the most soluble in water. Pagenkopf and Connolly [30] reported that more than 50% of the total B in fly ashes was water soluble. In a study involving 18 alkaline fly ashes in the U.S., James et al. [27] found that seven western fly ashes had

TABLE 1.1
Boron Concentrations in Fly Ashes

Country	Coal type	Fly ash (weathering / reaction)	B concentration (mg kg⁻¹) total	Water soluble	Ref.
Australia (Queensland, N. South Wales, S. Australia, W. Australia, Victoria)	Bituminous black or lignite brown	No / alkaline		1.1–41.6	9
		Yes / alkaline		0.1–4.2	
Australia (Queensland, N. South Wales, S. Australia)		No / alkaline		3.0–32.9	10
Australia	Bituminous	No / acid, alkaline	20–400		7
Australia				1.15–2.86 mg l⁻¹	11
Canada				42 mg l⁻¹	12
Greece	Lignite	Yes / alkaline		2.8–3.1	13
Greece	Lignite	No / alkaline	52		14
Greece	Lignite	No / alkaline	0.28–0.57		15
India		No	123		16
		Yes	150		
Slovenia	Brown coal	No	≅ 1		17
South Africa			23–600		18
Spain		No	61–339		19
UK			3–150 (mean 43)		20
UK				3–250	21
U.S. (9 States)	Bituminous or lignite	No	234–618		22, 23
U.S.			234–415		24
U.S. (Kentucky, Ohio, Virginia, W. Virginia)					
U.S. (Southeastern)		Acid		22	25
		Alkaline		50	
U.S. (Western)		No	10–618	24 (mean)	3
U.S. (Illinois)		No	870–1600		26
U.S. (Western)		Alkaline	800 (mean)		27
U.S. (Midwestern)		Alkaline	675 (mean)		27

TABLE 1.1
Boron Concentrations in Fly Ashes (continued)

Country	Coal type	Fly ash (weathering / reaction)	B concentration (mg kg⁻¹) total Water soluble	Ref.
U.S. (Midwestern) (Illinois, Indiana, Kansas, Wisconsin, Michigan, Kentucky, Missouri, Iowa)		Alkaline	172–1480	27
U.S. (Alabama, Colorado, Dalaware, Georgia, Iowa, Kentucky, Maryland, Massachusetts, Michigan, Minnesota, Montana, N. Hampshire, N. Mexico, N. York, N. Carolina, Ohio, S. Carolina, S. Dakota, Utah, W. Virginia, Wisconsin)			10–900	28

an average total B concentration of 800 mg kg^{-1}, with 37% water leachable. The remaining 11 midwestern ashes had 675 mg kg^{-1} total B, 47% of which was water leachable.

According to Carlson and Adriano [2], concentrations of water-soluble B in fly ash may exceed the value of 250 mg kg^{-1} and 17 to 64% of the total B in fly ash is immediately soluble in water. Churey et al. [31] studied the solubility of As, B, Mo, and Se in aqueous equilibrates of many ashes. Boron was found to be more soluble than the other elements. On a weight basis, the most soluble trace elements of nine Illinois basin fly ashes were found to be B and Cd; the percentages of soluble matrix B ranged from 24 to 56% with an average of 44 ± 10% [26]. Based on a study of the mobility of water-soluble major and trace elements in four alkaline Spanish fly ashes, Querol et al. [32] reported that the leaching rate of B was the highest among 19 trace elements studied. Up to 78% of the B present in the fly ash samples was water soluble.

Boron concentration in fly ashes depends on a variety of factors, such as the nature of the element; the parent coal; the combustion conditions; the efficiency and type of emission control devices; and the degree of weathering [2]. The manner in which these factors affect B concentration in fly ash is discussed in the following paragraphs.

Certain elements during combustion are volatilized and then condense onto the surface of the ash particles as the flue gas cools. Boron has been classified among the trace elements that are partly or fully volatilized and partitioned between the fly ash and the combustion flue gas in the upper part of the combustion system [8]. Trace elements condense mainly on smaller particles because of greater surface area, as stated by El-Mogazi et al. [33]. Similarly, Elseewi et al. [34] and Pougnet et al. [18] reported that B concentration in fly ash is inversely proportional to the size of fly ash particles, meaning that B concentration increases as particle size decreases.

Boron concentration of two fly ashes was found to increase from 8 to 18 mg kg^{-1} and from 49 to 180 mg kg^{-1} as the fly ash particle size decreased from 51.7 to 7.7 μm and from 26.7 to 2.5 μm, respectively [35]. Hansen et al. [36] reported that total B in coal fly ash increased from 36.7 to 48.5 μmol g^{-1} as the particle size decreased from 20 to 2.2 μm. Boron was found to be more concentrated in the <53-μm fraction of fly ashes [3,37]. In addition, Halligan and Pagenkopf [38] reported that most of the available B of a fly ash was associated with the finest ash particles (<45 μm) and Hollis et al. [39] found that water-soluble B from a fly ash was concentrated in the smaller size fractions and only a small amount was dissolved from particles > 20 μm.

Boron concentration of fly ash depends also on B content of the coal burned. According to Pagenkopf and Connolly [30], the amount of B found in U.S. coal ashes varied from 5 to 200 mg kg^{-1}, depending upon the mine site. Ashes of eastern and midwestern (bituminous and subbituminous) coal deposits of the U.S. usually have lower B concentration than those from the western states (derived from subbituminous coals and lignite). Similarly, Adriano et al. [1] reported that total B content of fly ashes derived from bituminous and subbituminous coal and lignite were 36, 50, and 500 mg kg^{-1}, respectively.

The efficiency and type of emission control devices may affect the amount of the finest fly ash particles released into the atmosphere and, consequently, the amount of B released bound on these particles. Page et al. [3] reported that the estimates of the daily atmospheric emission of fly ash from two U.S. midwestern power plants, equipped with electrostatic precipitators, were 4 to 5 Mg day^{-1}; from another plant equipped with mechanical dust collector, it was 120 Mg day^{-1}. The latter station also had the greatest B emission. Gladney et al. [40], Llorens et al. [8], and Pagenkopf and Connolly [30] supported the idea that, except for B condensation on fly ash retained by the electrostatic precipitators, a significant amount of B probably is released to the atmosphere bound on uncaptured particles of fly ash. Specifically, Pagenkopf and Connolly [30] reported that as much as 71% of the total B present in coal may be lost to the atmosphere upon combustion.

Conversely, Goodarzi and Swaine [7] suggested that, because B is partly volatile, small amounts of B are expected to be emitted into the atmosphere through fine fly ash particles. These researchers studied the deposition of B around a power station located near Sydney, Australia, for 3 years. Their results showed that B deposition on land, over the whole period, ranged from <0.01 to 13 mg B m^{-2} and that most B was deposited near the power station, within a radius of about 5 km, with a tendency to decrease with distance from the station.

Weathering of fly ash reduces soluble B [2], thus minimizing the risk of B phytotoxicity to plants grown on ash deposits or in soils amended with fly ash. However, the beneficial impact of fly ash weathering on plant growth, with respect to B, depends upon the initial B content of fly ash and the rate of B leaching. Weathering of fly ash under laboratory or natural conditions and its effect on plant growth are discussed in more detail in the next sections.

1.2.2 FLY ASH CONSTITUENTS ASSOCIATED WITH B AND MECHANISMS OF B RELEASE AND RETENTION

Fly ash consists mainly of alumino–silicate materials, 70 to 90% of which are glassy spheres; the remainder consists of quartz, mullite, hematite, magnetite, calcite, gypsum, and anhydrite [1,41]. Furthermore, tourmaline, a mineral containing B, was identified and the presence of B salts such as borax, rhodium boride, boron arsenite, and boron phosphate was suggested [1].

Section 1.2.1 mentioned that a significant proportion of total B in fly ash seems to be readily water soluble. It was found that at least two independent B chemical species must be present in fly ash, i.e., soluble and insoluble; the first one must not be present in bottom ash [29,34]. Narukawa et al. [35] conducted fractionation experiments and concluded that most of the B chemical species in fly ash were water soluble. In order to explain the high mobility of fly ash B, James et al. [27] suggested that the predominant forms of B in fly ash are probably very soluble borates and less soluble borosilicates, but Pagenkopf and Conolly [30] claimed that the B-soluble species in fly ash are borates and boric acid. In addition, Gangloff et al. [42] attributed the rapid leaching of B observed from fly ash–soil mixtures to the presence of a moderately soluble borate salt residing on fly ash surface.

Several researchers conducted experiments to determine the components of fly ash connected to highly or less soluble B species. Elseewi et al. [34] reported that the low solubility B fractions may be associated with Fe, Al, and Mg oxides, which exhibit high affinity for B existing in fly ash at relatively high concentrations. Narukawa et al. [35] reported that the majority of B chemical species (including the water-soluble ones) were associated with Fe and Mn oxides in acidic fly ashes and with carbonates in alkaline fly ashes. Warren et al. [43] studied the release of B from an alkaline fly ash in relation to the pH of ash aqueous suspensions containing HClO$_4$. They observed that B dissolution increased with decreasing pH of the suspension and concluded that B released from the fly ash at pH < 6.0 seemed to be associated with Ca and Mg species in the ash, which have high affinity to B and not with alumino–silicate or silicate glass material. In addition, Plank et al. [44] suggested that the less soluble B species probably are connected to the vitreous phase of fly ash.

The association of soluble B with Ca or Fe species and of less soluble B with Al–Si material in fly ash could be supported by the scheme of secondary minerals formation during acid weathering of fly ash proposed by Warren and Dudas [41]. They reported that CaO is transformed to less soluble minerals, such as calcite and possibly aragonite and gypsum, and Fe compounds are dissolved; followed by precipitation as an amorphous oxyhydroxide; Al–Si material is solubilized; and followed by precipitation of an amorphous clay material identified as protoimogolite.

To explain the high solubility of fly ash B under acidic conditions, Dreesen et al. [45] suggested that B is among the trace elements that are predominant at the surface of fly ash particles. Moreover, Elseewi et al. [34] reported that the water-soluble B fraction, which is considered plant available, is probably associated with the outer surfaces of fly ash spheres; the less soluble fraction is associated with the inner matrix of fly ash and in compounds of low solubility. Based on data obtained from fly ash leaching experiments, Querol et al. [32] concluded that the leaching trends of trace elements, including B, were consistent with the dissolution of small particles or coatings on the surface of the ash solid phases rather than with the dissolution of a homogeneous glass phase.

Dudas and Warren [46] proposed a submicroscopic model for solid and thick-walled glassy particles of fly ash, which comprise the bulk of any coal fly ash. The model is consistent with the results of fly ash weathering and leaching experiments. It explains the behavior of fly ash elements on the basis of their partitioning into a reactive exterior glass surface and subsurface and into an interior, less reactive glass matrix, which weathers at a slow rate. Elements readily leached from ash, like B, are mainly associated with the exterior reactive surface of ash particles. The partitioning of elements is related to the forms of each element in the coal and to combustion conditions.

These researchers stated that B occurs in solid form as hydrous borates in shale fragments mixed with coal. During combustion, the hydrous borates would be transformed to particles of simple oxides such as B_2O_3, which then could fuse with the reactive exterior of ash particles. Moreover, it was proposed that most of the elements contained in the carbonaceous material can exhibit similar behavior; according to Boyd [47], a significant fraction of B that occurred in coal was organically bound.

Although the majority of studies are focused on B release from fly ash, B retention by fly ash constituents has been also investigated. Zhang and Reardon [48] attributed the reduction of B concentration in fly ash leachates to ettringite formation during leaching. Also, B concentration in the leachates of a lime-treated fly ash decreased markedly, and this decrease was attributed to hydrocalumite identified in the leached fly ash.

Pagenkopf and Connolly [30] proposed three possible physicochemical reactions that may influence B release from coal fly ash:

- B adsorption by hydrous oxides
- Solubilization of metal borates
- Formation of surface coatings that occlude B within the ash particles

According to these researchers, adsorption of B by the alumino–silicate matrix seems to be the major mechanism for B retention by ash particles. In addition, Hollis et al. [39] reported that coprecipitation of B with $CaCO_3$ is also a possible mechanism of B retention in alkaline fly ashes.

1.2.3 FACTORS AFFECTING B RELEASE FROM AND RETENTION BY FLY ASH

Total B content of fly ash; ash particle size (see Section 1.2.1); contact time and ratio of ash to leachates; and pH are the main factors influencing B release from and retention by fly ash particles. They are discussed in more detail in the next paragraphs.

Water-soluble concentration of B was highly correlated with its total concentration in soft coal fly ash [31]. Furr et al. [28] reported that B absorption by cabbage (*Brassica oleracea*) grown on soil amended with U.S. fly ashes at a rate of 7% w/w was strongly correlated with the ash total B content. On the other hand, Pougnet et al. [18] reported that leached B from 31 strongly alkaline coal fly ashes equilibrated with water at a ratio 1:100 for 20 h ranged from 11 to 39% of the total B in fly ashes. No relationship was observed between the total B content of the fly ashes and the percentages of B leached.

The rate of B release from fly ash seems to decrease with time. Jones and Lewis [49] and Halligan and Pagenkopf [38] observed that the amount of B released from coal fly ashes exhibited a rapid initial increase; the rate of release decreased with time and eventually reached a plateau. Townsend and Hodgson [20] studied the change of B concentration in fly ash–water extracts over a period of 9 h. Their data showed that B in water extracts was sharply increased during the first 2 to 4 h and then decreased. Hollis et al. [39] found that the water-soluble B concentration in extracts derived from an alkaline fly ash decreased from 3.7 to 0.7 mmol dm^{-3} with increasing contact time from 0.5 to 96 h.

It was observed that the rate of B release increased as the dilution of ash–water suspensions increased [34,38]. This increase was attributed to the enhanced migration of B from the inner to the outer surfaces of fly ash particles, due to dilution.

Many researchers have studied the solubilization of B from fly ash under different pH values [3,10,29,34,35,37,39,45,50–53]. The extractability of B from fly ash was found to be strongly and inversely related to pH (described in detail in the next paragraphs).

Boron in a strongly alkaline fly ash was extracted with water and different acid and alkaline solutions. The higher percentages of extractable B were found in strongly acidic pH media. Specifically, 94% of fly ash B was extracted with 0.1 M citric acid (pH = 3.6), whereas only 1.5% was extracted with water (pH = 11.9) [45]. In an experiment investigating the release of several trace elements from a British fly ash by using several extraction techniques, it was found that B concentration in the extractant solution decreased as pH increased from 4.6 to 9.3 [50]. In addition, Cox et al. [29] reported that the leaching rate of fly ash B was higher into acid solutions; total soluble B did not depend on the pH over the range of 6 to 8, but it was decreased at strong alkaline pH.

Also, Page et al. [3] found that, although B concentration in strongly alkaline fly ash–water extracts (pH > 12) was very low (<0.6 mg l^{-1}), it increased significantly (65 mg l^{-1}) when the pH decreased to 6.5. Similar results were reported by Elseewi et al. [51] and Phung et al. [37]. In a relevant work, Elseewi et al. [34] used water and 0.01 N NaCl, both neutral and acidified, as B extracting solutions from fly ash. They found that the amounts of B extracted by the acidified NaCl solutions were substantially higher than those extracted by water or the neutral electrolyte. The maximum percentage of fly ash total B, extracted with water, was 27.4%; that extracted by the acidified NaCl was 41.4%.

In addition, Hollis et al. [39] found that, at pH 9, three extractions with water removed only 60% of the fly ash total B; however, at pH 6, two extractions removed all total B contained in the ash, with the highest amount of B removed in the first extraction. Narukawa et al. [35] found that the highest percentages of B leached from acidic and alkaline fly ashes did not depend on the initial pH of the ash and were obtained by using an acid solution (pH = 4), in comparison to water (pH similar to that of the ash) and to an alkaline solution (pH = 12).

Weathering, in conjunction with the final pH, seems to affect B availability in a different way in acidic and alkaline fly ashes. Thus, Khandkar et al. [52] reported that hot water-extractable B of three acidic fly ashes derived from Gondwana coals decreased after weathering; that of an alkaline fly ash derived from lignite coal increased. This was attributed to the fact that, during weathering, the pH of the acidic fly ashes changed to alkaline, but that of the alkaline fly ash decreased to the acid range. In a relevant study with two alkaline fly ashes, Kukier et al. [53] found that the amounts of hot water-extractable B increased with increasing acidity of the boiling solutions. In another similar study with five Australian strongly alkaline unweathered fly ashes, Aitken and Bell [10] reported that hot water-extractable B ranged from 3.0 to 32.9 mg kg^{-1} in the fresh ashes, but decreased to 0.3 to 12.2 mg kg^{-1} when the ashes were leached with deionized water, and to 0.3 to 4.7 mg kg^{-1} when ashes were adjusted to pH 6.5 and subsequently leached.

All the previously mentioned findings agree that the solubility of fly ash B is strongly and inversely related to the pH of the extracting solution. As the pH decreases, the dissolution of fly ash particles is enhanced, and the smaller particles are dissolved more quickly than the bigger, starting from the surface to the interior of the particle. Consequently, these findings are in agreement with those mentioned in Section 1.2.2 that a high amount of the B in fly ash is in available forms; is associated mainly with the smaller particles; and is located at the surface rather than in the interior of fly ash particles. Most of the relevant studies referred to pH values ≥ 4.0. Because the solubility of Al–Si matrix of fly ash is rather low at these pH values, it should be concluded that the readily available B from fly ash is probably associated with other than the alumino–silicate phase. This conclusion is also in agreement with the findings reported in the Section 1.2.2.

1.2.4 PLANT GROWTH ON FLY ASH WITH RESPECT TO ITS B CONTENT

The main factors limiting plant growth on fly ash deposits are excessive B; high pH; high salinity; and lack of N and, to some extent, P. In addition, indurated layers of ash produced by compaction and pozzolanic action may inhibit the normal growth of the root system. Agricultural and horticultural crops have been classified on the basis of tolerance to fly ash. According to Hodgson and Buckley [54], Chenopodiaceae are all highly tolerant to fly ash, and Leguminosae, Cruciferae, and Graminae species show considerable variation in their capacity to tolerate fly ash. It is reported that grouping species in relation to their ash tolerance is similar to grouping them on the basis of B requirements. Although the ability of plants to grow on fly ash indicates mainly their tolerance to B excess, the role of soluble salts, unstable crumb structure and macro- and micronutrient deficiencies should also be considered.

On the basis of data from British fly ashes, Hodgson and Townsend [55] proposed certain B concentration ranges for plant growth on fly ashes. Specifically, they proposed that concentration of hot water-extractable B at levels less than 4 mg kg^{-1} be considered as nontoxic; 4 to 10 mg kg^{-1} as slightly toxic; 11 to 20 mg kg^{-1} as moderately toxic; 21 to 30 mg kg^{-1} as toxic; and greater than 30 mg kg^{-1} as highly toxic. Also, Townsend and Gillham [21] suggested that hot water-extractable B levels in fly ashes higher than 20 mg kg^{-1} are probably toxic to most agricultural crops; more sensitive crops such as barley (*Hordeum vulgare*), peas (*Pisum sativum*), and beans (*Phaseolous vulgaris*) could show symptoms at values as low as 7 mg kg^{-1}. However, such data do not take into account the long-term B release characteristics of ash, especially unweathered fly ashes, and the genetic characteristics of plant species, concerning their behavior to B excess. For example, an Australian fly ash containing 3 mg kg^{-1} hot water-extractable B — a level considered nontoxic — resulted in B toxicity and reduced yields of French beans and Rhodes grass (*Chloris gavana*) [10].

As Section 1.2.1 mentioned, weathering of fly ash reduces B content. However, the time needed for reducing B concentration in fly ash at acceptable levels for plant growth, under natural conditions, varies widely. Some reports in the literature suggest a period from a few to several years of fly ash aging, depending on the initial B content of fly ash and the climate. Jones and Lewis [49], for example, reported that in a fly ash stock pile, B content of some recently deposited fly ash was 216 mg kg^{-1} and that of 25-year-old ashes was 4.3 mg kg^{-1}. Boron content of wild white clover (*Trifolium repens*) grown on ash decreased with increasing ash aging and became similar to that of the soil control in the 25-year-old ash.

According to Townsend and Gillham [21], weathering of fly ash with moderately high initial B content for 4 years could reduce B to acceptable levels. Burns and Collier [56] developed a simulation model to predict the period required to leach B from the top 30 cm of a coal ash deposit in the U.K. This model predicted that B concentration would decrease at acceptable levels for most crops after 5 to 15 years. According to Nass et al. [57], B concentration in grasses grown on 20- and 30-year-old fly ash deposits was lower than that obtained for plants grown on fresh fly ashes, but higher than the values considered normal for these specific species.

In general, plant species grown on fly ash deposits weathered for certain years attained better growth and absorbed B at levels probably high but nontoxic. The growth of species in relatively weathered or fresh fly ash was limited by B toxicity, among other factors. However, the opposite cannot be precluded, depending on the fly ash B content and the plant species.

Several cases of different species grown well on weathered fly ash deposits are reported in the literature. It was found that B concentration of white sweet clover grown on a coal fly ash containing 7.5 mg B kg^{-1} was slightly higher (51 mg kg^{-1}) than that grown on a soil containing 5.2 mg B kg^{-1} (45 mg kg^{-1}) [58]. In a study with grasses and legumes growing on four soil-capped ash landfills in New York, Weinstein et al. [59] found elevated levels of B in most of the plants, with generally higher levels in the legumes than in grasses. To explain this difference, they hypothesized that the deeper rooted legumes had probably penetrated the underlying ash deposit.

On the other hand, Woodbury et al. [60] found that B concentrations of several plant species, i.e., bird's foot trefoil (*Lotus corniculatus*); red clover (*Trifolium pratense*), timothy (*Phleum pratense*), orchard grass (*Dactylis glomerata*), velvet grass (*Halcus lanatus*); and several fescues (*Festuca* spp.) grown on a soil-capped ash landfill ranged at levels similar to the soil control (29 to 53 mg kg^{-1} in the legumes and 2 to 11 mg kg^{-1} in the grasses). Conversely, growth of cucumber (*Cucumis sativus*), which is considered a semitolerant crop to B toxicity, was suppressed in a weathered acidic fly ash, although the ash had a low B content (total B 28 mg kg^{-1} and water-extractable B 3.7 mg kg^{-1}). After 8 weeks, all plants exhibited certain toxicity symptoms and plant biomass was significantly less compared to the control soil. It was concluded that B toxicity was the primary cause for plant growth suppression because B concentrations in ash grown plants were tenfold of the control and exceeded the B toxicity threshold for cucumber (>300 mg kg^{-1}) [61].

In addition, several woody species have been tested in fly ash deposits. In certain cases, the deeper rooted system was proved to be the main factor preventing B accumulation at levels higher than those considered toxic, although the ash was enriched in B. For example, *Polpulus robusta* and *Picea sitchensis* grown on two ash deposits containing 14 and 100 mg B kg^{-1}, respectively, did not exhibit severe B toxicity symptoms. This was attributed to the fact that B concentration at depths below 30 to 40 cm was low [54]. Similarly, eight woody species — European black alder (*Alnus glutinosa*); sweet birch (*Betula lenta*); sycamore (*Platanus occidentalis*); sawtooth oak (*Quercus acutissima*); cherry olive (*Elaeagnus multiflora)*; autumn olive (*Elaeagnus umbellata*); silky dogwood (*Cornus amomum*); and gray dogwood (*Cornus racemosa*) — grown on a strongly acidic fly ash did not show B toxicity symptoms, although plant B levels were elevated [62].

Carlson and Adriano [63] reported similar results for sweet gum (*Liquidambar styraciflua*) and sycamore growing on a 20-year-old fly ash wet basin (pH = 5.6). Trees growing on the ash attained better growth, even though they had higher trace element concentrations in comparison to the control soil. As far as B is concerned, elevated concentrations were observed in trees growing on the ash (three times the control); however, in almost all cases, B levels in the foliage were below 100 mg kg^{-1}. The differences between fly ash-wet basin and soil observed in foliar B concentrations were attributed to differences in B substrate concentrations.

The deleterious effects of fly ash, especially the unweathered, on plant growth due to high B content can be avoided by weathering or by mixing the fly ash with an inert medium. Holliday et al. [64] conducted a pot experiment with oat (*Avena vulgare*) grown on an inert medium mixed with a strongly alkaline fresh fly ash (pH = 9.0) and on the same ash treated with acid (pH = 6.4), at rates up to 100%. They found that B toxicity symptoms were obvious in the plants grown in the inert medium with fresh ash ≥ 6% and acid-treated ash ≥ 25%. At these rates, B concentrations in the biomass were 260 and 310 mg kg^{-1} for the untreated and the treated fly ash, respectively.

Similar research was carried out by Townsend and Gillham [21] by means of a small-scale field experiment with cereals, grasses, legumes, and other crops, using initially fresh and weathered fly ash, amended or not with silt. Among the species studied, red clover; white clover and Lucerne; timothy; cocksfoot and ryegrass (*Lolium perenne*); and potatoes (*Solanum tuberosum*) have been grown well for a number of years, apart from failures in the early years on the new ash plots (with no silt), attributable to excess B. The improvement of all crops' performance with years was attributed to a decrease of B content of the ash by weathering.

Aitken and Bell [10] studied the uptake of B by French beans and Rhodes grass grown on alkaline fly ash–sand mixtures in a pot experiment. The five fly ashes used were untreated, leached, or adjusted to pH 6.5 and subsequently leached and mixed with sand at rates of 5 and 10% w/w. The untreated ashes resulted in lower yields than the leached and pH adjusted and subsequently leached for both species; this was attributed mainly to B toxicity because plant B uptake and hot water-extractable B were higher in the untreated ashes. For both levels of ash addition and both species, B concentrations and uptake decreased in the order: untreated ash > leached ash > pH-adjusted and leached ash.

Although B phytotoxicity is a possible risk for plants grown on unweathered fly ashes, especially those having high B content, the opposite is also true as stated by Nass et al. [57]. For 6 years, they studied the growth of a mixture of grasses (*Lolium perenne, Festuca rubra commutata,* and *Phleum pratense*) and white clover, in lysimeters filled with three fresh fly ashes that differed in their reaction (alkaline, neutral, and acidic). Boron concentrations in plant tissues were always below 33 mg kg^{-1} and only concentrations higher than 50 mg kg^{-1} would cause B toxicity.

Some cases of detrimental effects on seedling growth in fly ash, due to trace elements, are also reported in the literature. Excessive levels of certain trace elements, including B, impaired the growth of lettuce (*Lactuca sativa*) seedlings on a British neutral fly ash [50]. In addition, Shulka and Mishra [65] attributed the deleterious effects on the seedlings of corn (*Zea mays*) and soybean grown in nutrient solution amended with 2.5% Indian alkaline fly ash to B or heavy metals.

1.3 BORON AVAILABILITY TO PLANTS AS INFLUENCED BY FLY ASH APPLICATION TO SOILS

1.3.1 Soil B

Boron is a micronutrient, essential for plant growth, for which the range of deficiency, sufficiency, and toxicity is too narrow. According to Evans and Sparks [66], B concentrations in soil solution of <1 mg l^{-1} are considered deficient; 1 to 5 mg l^{-1} sufficient; and >5 mg l^{-1} toxic for the normal growth of most plants. Keren and Bingham [6] suggested that soil concentrations of hot water-extractable B (B availability index) exceeding the value of 5 mg kg^{-1} are considered phytotoxic to most crops. Boron concentration in the soil solution is very important for plant nutrition because plants respond only to B concentration in it [4]. However, B of the soil solution is in equilibrium to B adsorbed by soil particles. Plants obtain B from the soil solution and the adsorbed pool of B acts as a buffer against changes in solution B [5]. Thus, it is important to know the soil constituents absorbing B and the soil factors affecting B adsorption in order to predict or prevent drastic changes in soil B content caused by external sources (e.g., by soil application of fly ash).

Boron in the soil solution at the common range of soil pH exists as two chemical species, i.e., boric acid, H_3BO_3, and borate anion, $B(OH)_4^-$; the latter appears at the alkaline range of soil pH and has greater affinity to soil colloids. Both species are quite water soluble [6,67]. The soil constituents that adsorb B are Al and Fe oxyhydroxides, clay minerals, $CaCO_3$, and organic matter. Specifically, clay minerals' B adsorption capacity has been found to increase in the order: kaolinite < montmorillonite < illite. Various mechanisms have been proposed to explain B adsorption by Al and Fe oxyhydroxides, edges of clay minerals, and organic matter, among which ligand exchange seems to be the dominant. In addition, B coprecipitation with $CaCO_3$ has been reported as a B sorption mechanism. Several soil factors influence B retention by and release from soil constituents, including [5,67]:

- pH
- Texture
- B concentration and ionic composition of the soil solution
- Cation exchange capacity
- Exchangeable ions
- Other anions
- Soil moisture
- Wetting and drying
- Temperature

1.3.2 Soil Factors Affecting Availability of Fly Ash B to Plants

The same soil factors affecting the availability of soil B affect also the availability of B released from fly ash added to the soil. Of all factors mentioned in Section 1.3.1, soil pH is probably the most important factor, followed by soil texture. Boron adsorption by soil solids increases as the soil pH increases, reaching a maximum at the pH range of 8.5 to 10. Soils with low clay content adsorb less B, and B in sandy soils is leached more easily and rapidly than in clayey soils [5,67].

Fly ash application to soils results in an increase (alkaline ash) or a decrease (acidic ash) in pH. Through that, in conjunction to soil texture, it affects B adsorption or B leaching. Final acid or neutral values of soil pH and coarse texture promote rapid leaching of B from the soils, and alkaline values and fine texture promote B adsorption by soil colloids. Both processes may be utilized in removing excessive amounts of B from the soil solution to avoid risk of B phytotoxicity. In both cases, however, the final result depends significantly on fly ash B content and the rate of fly ash addition.

Several studies have showed that incorporating fly ash into a coarse textured soil and allowing for subsequent leaching would result in the removal of certain potentially phytotoxic elements from the root zone, thus minimizing threats to crops [37,42,68–70]. In most of these studies, the pH of the leachates or the ash–soil mixtures was lower than that connected to B adsorption maximum. In a leaching experiment in columns, with an acid sand and a calcareous sandy loam soil amended with 5% w/w strongly alkaline fly ash of high B content, it was found that B was leached from the fly ash-amended acid sand more readily compared to the calcareous soil. The pH of the leachates was similar to the initial pH of the soils during the whole leaching period. However, at the end of the experiment, less than 20% of the total B was removed from both soils [70].

In another experiment with an alkaline fly ash added to three soils — one acid sand and two calcareous soils (a loamy sand and a silt loam) — at rates up to 1%, water-soluble B increased in all cases. The calcareous soil with the highest pH (8.2) and the finest texture (silt loam) was the least affected by fly ash application [37]. In leaching experiments with a loamy sand soil, Ghodrati et al. [68] amended soil with two alkaline fly ashes at a rate of 30% w/w and observed that the initial B concentration in the leachates was high, but decreased to background levels after about 30 cm of water had passed through the soil. More than 92% of the B initially present in the ash-amended soil had leached from ash–soil mixtures after leaching with 150 cm of water. Similar experiments and results have been reported [69].

In another experiment, Gangloff et al. [42] added unweathered alkaline (pH = 8.4) fly ash, with low total B, in an acid (pH $\cong$ 5.0) loamy sand soil at a high rate ($\cong$ 662 Mg ha^{-1}). Their goal was to study the leachability of B under field conditions, with lysimeters installed at a depth of 120 cm. A peak B concentration of 4.8 mg l^{-1} in the leachates occurred after 35 to 40 cm of cumulative rainfall. At this point, the pH of the leachates was extremely acidic (pH < 4.0), although it was alkaline (pH $\cong$ 8.0) in the beginning. Approximately 82 cm of cumulative rainfall was needed to reduce B concentration in the leachates from ash-amended plots to initial levels ($\cong$1 mg l^{-1}).

At the same field, the same researchers [42] established microplots in order to characterize leachate properties of fly ash-amended soil under more controlled conditions. The same fly ash was applied at a higher rate (950 Mg ha^{-1}) and lysimeters were established at various depths to 120 cm. A peak B concentration of approximately 8.5 mg l^{-1} was observed in the leachates at 15 cm depth after 18 cm of irrigation water. Leachate samples from the lysimeters installed at depths of 15, 30, and 60 cm generally had B concentrations much higher than those collected at 120 cm. The pH of the amended microplots was less than 8.0 during the course of this study.

Although leaching of fly ash-amended soils coarse or moderately coarse in texture is expected to reduce soluble B and consequently plant available B, this is not always the case. A large proportion of B added in the form of fly ash might be leached but the remaining could be high enough to support elevated B concentrations in the plants grown in ash-amended soils. This might be due to high B content and high fly ash application rates. In a greenhouse study, a loamy sand, moderately acidic soil was amended with alkaline fly ash at rates up to 40% w/w (equivalent to 600 Mg ha^{-1}) and subjected to four leaching regimes (equivalent to 0, 25, 50, and 100 cm of rainfall). Of particular importance was the rapid leaching of B from ash–soil mixtures. After leaching, corn was grown in fly ash–soil mixtures. Leaching enhanced corn growth, but did not result in equivalent dry matter yield to that obtained from the control soil. This was attributed to the fact that B concentrations in corn grown in the leached ash-amended soils decreased compared to the unleached, but remained at levels that could be considered high [69].

Application of alkaline fly ash with a high B content to soils is expected to raise soil pH and, consequently, enhance B adsorption by soil constituents, thus resulting in low levels of B in the soil solution. This hypothesis was tested by Matsi and Keramidas [71], who studied the changes in B adsorption characteristics of three soils (two acid and one calcareous) upon addition of two Greek, alkaline fly ashes, by using adsorption and desorption experiments. The fly ashes were added to the soils at rates up to 5% w/w (equivalent to $\cong$100 Mg ha^{-1}). Because both fly ashes did not contain appreciable amounts of B (hot water-extractable B was 2.78 and 3.12 mg kg^{-1}), B was externally added to the fly ash–soil mixtures used in the desorption experiment. Fly ash application increased only B sorption capacity of all soils, but not the strength of B retention by soil solids. Within a desorption period of 120 h, 80 to 100% of the adsorbed B was released.

These findings suggested that, upon fly ash addition to soils, any increase in B sorption capacity is not necessarily accompanied by an increase in the strength of B retention by soil constituents; consequently, added B might remain in a loosely held condition onto soil solids and could be easily released into the soil solution. Martens et al. [22] reported that application of three alkaline fly ashes containing high levels of total B at various rates increased B uptake by alfalfa (*Medicago sativa*) grown on a silt loam soil, despite the increase in soil pH. They concluded that B in fly ash is sufficiently mobile to overcome any decrease in B availability resulting from the increase in soil pH. In a similar study with an alkaline fly ash with high content of total B and an acidic clay soil, Warren et al. [43] found that B adsorption increased because of increase in pH (from 4.5 to 5.9); however, even at the highest application rate (167 Mg ha^{-1} $\cong$ 8.35% w/w), adsorbed B accounted for only 10% of the total B.

Most of the findings in B adsorption studies in fly ash-amended soils suggest that, upon ash application to soils, B adsorption by soil particles plays a rather insignificant role in removing B excess from the soil solution. This occurred at all pH ranges common in soils, even at the alkaline pH range where B adsorption maximum occurs. Consequently, instead of relying on adsorption, leaching with adequate drainage is a more effective strategy in order to prevent B phytotoxicity due to fly ash addition in soils, especially in the cases of ashes with high B content.

1.3.3 PLANT GROWTH IN FLY ASH-AMENDED SOILS WITH RESPECT TO B

Using alkaline fly ash as a liming material to acid soils is probably the most common agronomic utilization of fly ash. In addition, cases of alkaline or acid fly ash application to soils differing in their physicochemical characteristics to improve soil fertility or soil physical properties are also frequently reported in the literature. In all cases, however, caution is needed with respect to B phytotoxicity risk. Although several studies reported in the previous sections of this chapter proved the high plant availability of fly ash B, this is further demonstrated by the enrichment ratio (ER). The ER of an element is calculated by dividing the elemental concentration of the plant grown in a treated soil with that grown in the untreated. The higher the ER for an element is, the greater is the plant adsorption of the particular element due to treatment.

Adriano et al. [1] determined ER for 27 elements for plants grown in soils and fly ash-amended soils. They reported that the ER values for most of the elements were $\cong$ 1.0 (meaning no effect on elemental absorption). However, several potentially hazardous elements, such as As, Ba, B, Mo, Se, Sr, and V, showed remarkable concentrating effects when plants were grown in fly ash-treated substrates (ER $\cong$ 3 to 8 and, specifically, for B $\cong$ 5.0). Tolle and Arthur [72] reported that plant uptake of trace elements due to fly ash addition reached plant toxic levels for only one element: B. Among ER values determined for 25 elements, exceptionally high values (ER > 15) were noted only for four elements, one of which was B.

1.3.3.1 Cases of B Phytotoxicity

It is postulated that B phytotoxicity may persist for a number of years after fly ash application to soils [44]. Cases of B toxicity caused by fly ash application to soils are summarized in Table1.2. Boron toxicity was evident mainly by yield reduction or poor plant growth in conjunction with elevated B concentrations in plant and soils. However, B toxicity symptoms were obvious in plants only in certain cases. Although different rates of fly ash application are reported, as can be seen in Table 1.2, the usual rates have not exceeded the value of 10% w/w (equivalent to $\cong$ 200 Mg ha^{-1}; Table 1.2); in most of the cases, fly ash was unweathered [10,34,53,72,73,79–81].

It will be evident from the following discussion that the risk of B phytotoxicity is mainly connected to the agronomic use of unweathered rather than weathered fly ash because adequate weathering is expected to reduce B content of ash significantly (see also Section 1.2.3 and Section 1.2.4). Phung et al. [82] tested the possibility of using a fresh, strongly alkaline fly ash as a liming material to a strongly acidic, silty clay soil. Ash application at the highest rates (5 and 10% w/w) increased soil B at levels that could be considered toxic to sensitive plants. Martens and Beahm [81] compared the effects of a weathered and a fresh ash applied to a strongly acidic silt loam soil. They found that addition of the weathered ash at rates up to 144 Mg ha^{-1} had no harmful effect, but fresh ash at rates of 96 to 144 Mg ha^{-1} resulted in B toxicity to corn plants.

In addition, in a glasshouse experiment with French beans and Rhodes grass, Aitken and Bell [10] used an Australian fly ash (untreated, leached, or adjusted to pH 6.5 and subsequently leached) as an amendment (0 to 70% w/w) for a sandy loam soil. They found that, for both species, heavy applications of untreated fly ash ($\geq$30% for beans and 70% for Rhodes grass) resulted in poor plant growth, primarily due to B toxicity. The risk of B phytotoxicity was reduced by leaching the fly ash and even more by pH adjustment and subsequent leaching, prior to soil addition. Boron concentrations in plant tissues of both species were above 100 mg kg^{-1} for almost all untreated ash treatments and below 100 mg kg^{-1} for almost all pH adjusted and leached ash treatments.

1.3.3.2 Cases of B Deficiency Correction

Although B phytotoxicity due to soil application of fly ash is the common case reported in the literature, beneficial effects of the agronomic use of fly ash, such as alleviation of B deficiencies for several plants, have also been reported and are summarized in Table 1.3. The usual rates of ash application to soils were similar to those reported in the previous section (up to 10% w/w $\cong$ 200 Mg ha^{-1}). According to Bradford [91], if a B concentration in plants equal to about 200 mg kg^{-1} is considered the threshold of toxicity for most crops, then B levels in plants shown in Table 1.3 were much lower and, in most cases, were lower than 100 mg kg^{-1} [13,22,25,44,83,86–88,90].

The availability of B from fly ash has been compared to common B fertilizers (i.e., borax) by means of pot and field experiments [22,23,25,87]. In a pot experiment, Martens et al. [22] found that application of three alkaline fly ashes containing high levels of total B (319, 415, and 618 mg kg^{-1}) at various rates increased B uptake by alfalfa grown in a silt loam soil, at levels similar to those resulted from the application of borax at equal rates. Similarly, Plank and Martens [25] used two U.S. fly ashes (acid and alkaline) with high B content as B sources for alfalfa in a 3-year field experiment and compared the results to those obtained by the use of borax. Fly ashes were applied at rates equivalent to 1.7 and 3.4 kg B ha^{-1} (based on their total B content) in a single dose in the beginning of the experiment; borax was applied in the beginning of each growing season. They found that yield and B uptake by alfalfa was significantly increased upon application of both fly

TABLE 1.2
Cases of Detrimental Effects of Fly Ash on Plant Growth with Respect to B

Plant species	Soil characteristics (reaction / texture)	Fly ash characteristics (reaction / B content / maximum rate)	Detrimental effects	Ref.
Pot experiments				
Barley		Relatively low / 50% v/v	B toxicity symptoms / elevated plant B (at ash rate $\geq$ 12.5% v/v)	64
Barley	Almost neutral / medium	Strongly alkaline / relatively high / 75% v/v	Yield reduction (at ash rate $\geq$ 50% v/v) / B toxicity symptoms / elevated plant B (at ash rate $\geq$ 6.25% v/v)	73
Barley, brittlebush (*Encelia farinose*)	Acid / medium; Calcareous / medium	Alkaline / high / 8% w/w	Probably B toxicity symptoms / elevated plant and soil B	51
Corn	Acid / medium	Alkaline / high / 10% w/w	B toxicity symptoms / elevated plant and soil B (at ash rate $\geq$ 0.6% w/w)	53
Corn	Acid / light; Acid / light	Alkaline / high / 40% w/w	B toxicity symptoms / elevated plant and soil B (at ash rate $\geq$ 20% w/w)	74
Corn	Strongly acid / medium	Strongly alkaline / high / 25% w/w	Elevated plant B (at ash rate $\geq$ 1% w/w)	75
Soybean		10% w/w	Elevated plant B	76
Tomato	Slightly acid	Alkaline / high / 90% v/v	Excessive plant B at ash rate (>50% v/v)	77
Wheat	Strongly acid / heavy	Alkaline / high / 167 Mg ha^{-1} ($\equiv$8.35% w/w)	Elevated soil B	43
Alfalfa, Bermuda grass, white clover	Acid; Calcareous	Alkaline / 8% w/w	White clover yield reduction / elevated plant and soil B	34
Alfalfa, barley, Bermuda grass (*Cynodon dactylon*), brittlebush, lettuce, Swiss chard, white clover	Acid; Calcareous	8% w/w	Lettuce yield reduction	3
French beans, Rhodes grass		70% w/w	Yield reduction for French beans (at ash rate $\geq$ 30% w/w) and Rhodes grass (at ash rate 70% w/w) / B toxicity symptoms	10
Alfalfa, corn, soybean, wheat	Strongly acid / heavy	Alkaline / high / 167 Mg ha^{-1} ($\equiv$8.35% w/w)	Yield reduction / B toxicity symptoms for corn and soybean / elevated plant and soil B (at ash rate > 110 Mg ha^{-1})	78
Field experiments				
Centipedegrass (*Eremochloa ophiroides*)	Acid / medium	Alkaline / relatively low / 1120 Mg ha^{-1}	Poor initial plant establishment / elevated plant and soil B	79, 80
Corn	Acid / medium	Alkaline / 144 Mg ha^{-1}	Toxicity symptoms (at ash rate $\geq$ 96 Mg ha^{-1})	81
Alfalfa, timothy, oat	Slightly acid / medium	Acid / high / 700 Mg ha^{-1}	Reduced yield (at ash rate $\geq$ 400 Mg ha^{-1}) / toxic plant B uptake	72

TABLE 1.3
Cases of Beneficial Effects of Fly Ash on Plant Growth with Respect to B

Plant species	Soil characteristics (reaction / texture)	Fly ash characteristics (reaction / B content / maximum rate)	Increases	Ref.
Pot experiments				
Alfalfa	Acid / medium	Alkaline / high / 5% w/w	Plant B uptake	22, 23
Alfalfa	Acid	Alkaline / 3.2% w/w (≡80 Mg ha⁻¹)	Growth / soil B / plant B	12
Ryegrass	Strongly alkaline	Acid / 8% w/w	Growth / plant B	83
Ryegrass	Acid / medium	Alkaline / low / 5% w/w	Growth / plant B uptake	13
Tomato (*Lycopersicon esculentum*)	Acid	Alkaline / high / 50% v/v	Yield / plant B / soil B	77
Vegetables, millet (*Echinochloa crusgalli*)	Acid / medium	Acid / 10% w/w	Plant B uptake	84
	Neutral / medium	Acid / 10% w/w	Plant B uptake	85
Field experiments				
Alfalfa	Slightly alkaline	Acid / high / 3.4 kg B ha⁻¹	Yield / plant B uptake	25
	Slightly alkaline	Alkaline / high / 3.4 kg B ha⁻¹		
Corn	Acid / medium	Alkaline / low / 144 Mg ha⁻¹	Soil B / plant B	44
	Acid / light			
Corn	Acid / medium	Alkaline / 144 Mg ha⁻¹		81
Couch grass (*Cynodon dactylon*)	Acid / light	Acid / low / 20% w/w	Root growth / plant B	86
Garlic (*Allium sativum*)	Acid	Alkaline / 825 g B ha⁻¹	Yield	87
Alfalfa, bird's foot trefoil, brome (*Bromus*), orchard grass, timothy	Acid / medium	Acid / low / 5% w/w (≡112.5 Mg ha⁻¹)	Plant B	88
Alfalfa, bird's foot trefoil, brome, corn, millet, orchard grass, sorghum (*Sorghum bicolor*), timothy	Acid	125 Mg ha⁻¹ (≡5% w/w)	Plant B	89
Beans, cabbage, carrots (*Daucus carota*), onions (*Illium cepa*), potatoes, tomatoes				
Corn, soybean, wheat	Acid	Alkaline / 50 Mg ha⁻¹	Growth / soil B / plant B	12
Scots pine	Strongly acid	Alkaline / 20 Mg ha⁻¹	Plant B	90

ashes and, in addition, the B-supplying power of the fly ashes was as efficient in providing B to plants as the annual applications of B as borax.

Although the use of alkaline fly ash as a liming agent for acid soils is the common case, the use of acidic fly ash as an alternative to gypsum for amelioration of soils with sodicity problems has also been tested. In a pot experiment with padi followed by wheat, Kumar and Singh [92] applied fly ash to a sandy loam soil with sodicity problems at rates up to 7.5% w/w. They found that available soil and plant B were significantly increased with different fly ash levels, but in all cases remained at levels considered acceptable.

1.4 CONCLUSIONS

Any general conclusion concerning fly ash and its use as a soil amendment must be drawn with caution due to the variability of physicochemical properties of ashes and amended soils. Thus, summarizing, fly ash is expected to be rich in B because B is volatilized during coal combustion and condenses onto fly ash particles. Boron availability from fly ash is usually high due to the concentration of significant amounts of B on the external surfaces of fly ash particles and to the solubility of B species releasing upon contact of fly ash with water. Boron release from fly ash is strongly and inversely related to pH. Leaching of fly ash in conjunction to pH decrease is expected to reduce total and available B significantly.

Upon fly ash application to soils, available B concentration is expected to increase, but the extent of this increase depends mainly on fly ash B content and rate of application, degree of ash weathering, soil texture, and final soil pH. Due to the enhanced availability of fly ash B, soil constituents are not expected to play a significant role in removing and thus substantially reducing B in soil solution after fly ash addition to soils, even in cases of alkaline pH range where B adsorption maximum occurs.

Leaching of fly ash-amended soil is a more effective way in decreasing excess B, due to fly ash application. Soil texture significantly affects the time needed for effective leaching. Coarsely textured soils are leached more readily than finely textured soils. From a practical approach, in order to avoid the risk of B phytotoxicity, the use of adequately weathered fly ash rather than unweathered and adequate leaching of amended soil in conjunction with the cultivation of plant species tolerant to B could be suggested. Of course, all the preceding points depend significantly on fly ash B content and rate of application. As far as ash application rate is concerned, in general, fly ashes with high B content must be incorporated into the soils at low rates and vice versa.

ACKNOWLEDGMENTS

The help of Professor V.Z. Keramidas in editing this chapter is greatly appreciated.

REFERENCES

1. Adriano, D.C. et al., Utilization and disposal of fly ash and other coal residues in terrestrial ecosystems: a review, *J. Environ. Qual.*, 9, 333, 1980.
2. Carlson, C.L. and Adriano, D.C., Environmental impacts of coal combustion residues, *J. Environ. Qual.*, 22, 227, 1993.
3. Page, A.L., Elseewi, A.A., and Straughan, I.R., Physical and chemical properties of fly ash from coal-fired power plants with reference to environmental impacts, *Res. Rev.*, 71, 83, 1979.
4. Keren, R., Bingham, F.T., and Rhoades, J.D., Plant uptake of boron as affected by boron distribution between liquid and solid phases in soils, *Soil Sci. Soc. Am. J.*, 49, 297, 1985.
5. Gupta, U.C. et al., Boron toxicity and deficiency: a review, *Can. J. Soil Sci.*, 65, 381, 1985.
6. Keren, R. and Bingham, F.T., Boron in water, soils and plants, *Adv. Soil Sci.*, 1, 229, 1985.

7. Goodarzi, F. and Swaine, D.J., Behavior of boron in coal during natural and industrial combustion processes, *Energy Sources*, 15, 609, 1993.

8. Llorens, J.F., Fernandez–Turiel, J.L., and Querol, X., The fate of trace elements in a large coal-fired power plant, *Environ. Geol.*, 40, 409, 2001.

9. Aitken, R.L., Campbell, D.J., and Bell, L.C., Properties of Australian fly ashes relevant to their agronomic utilization, *Aust. J. Soil Res.*, 22, 443, 1984.

10. Aitken, R. and Bell, L., Plant uptake and phytotoxicity of boron in Australian fly ashes, *Plant Soil*, 84, 245, 1985.

11. Pathan, S.M., Aylmore, L.A.G., and Colmer, T.D., Properties of several fly ash materials in relation to use as soil amendments, *J. Environ. Qual.*, 32, 687, 2003.

12. Cline, J.A., Bijl, M., and Torrenueva, A., Coal fly ash as a soil conditioner for field crops in Southern Ontario, *J. Environ. Qual.*, 29, 1982, 2000.

13. Matsi, T. and Keramidas, V.Z., Fly ash application on two acid soils and its effect on soil salinity, pH, B, P and on ryegrass growth and composition, *Environ. Pollut.*, 104, 107, 1999.

14. Iordanidis, A. et al., A correlation study of trace elements in lignite and fly ash generated in a power station, *Intern J. Environ. Anal. Chem.*, 79, 133, 2001.

15. Georgakopoulos, A. et al., Environmentally important elements in fly ashes and their leachates of the power stations of Greece, *Energy Sources*, 24, 83, 2002.

16. Sikka, R. and Kansal, B.D., Characterization of thermal power-plant fly ash for agronomic purposes and to identify pollution hazards, *Bioresource Technol.*, 50, 269, 1994.

17. Grilc, V. and Petkonsek, A., Stabilization of boron-containing mineral sludge with various solidification agents, *Waste Manage. Res.*, 15, 73, 1997.

18. Pougnet, M.A.B., Wyrley–Birch, J.M., and Orren, M.J., The boron and lithium content of South African coals and coal ashes, *Intern. J. Environ. Anal. Chem.*, 38, 539, 1990.

19. Querol, X. et al., Physicochemical characterization of Spanish fly ashes, *Energy Sources*, 21, 883, 1999.

20. Townsend, W.N. and Hodgson, D.R., Edafological problems associated with deposits of pulverized fuel ash, in *Ecology and Reclamation of Devastated Land*, Huntick, R.J. and Davis, G., Eds., Gordon and Breach, New York, 1973, vol. 1, 45.

21. Townsend, W.N. and Gillham, E.W.F., Pulverised fuel ash as a medium for plant growth, in *The Ecology of Resource Degradation and Renewal*, Chadwick, M.J. and Goodman, G.T., Eds., Blackwell Scientific Publications, Oxford, 1975, 287.

22. Martens, D.C. et al., Fly ash as a fertilizer, in *Proc. 2nd Ash Utilization Symp.*, Pittsburgh, 1970, 310.

23. Mulford, F.R. and Martens, D.C., Response of alfalfa to boron in fly ash, *Soil Sci. Soc. Am. Proc.*, 35, 296, 1971.

24. Plank, C.O. and Martens, D.C., Amelioration of soils with fly ash, *J. Soil Water Conserv.*, 177, 1973.

25. Plank, C.O. and Martens, D.C., Boron availability as influenced by application of fly ash to soil, *Soil Sci. Soc. Am. Proc.*, 38, 974, 1974.

26. Roy, W.R. et al., Illinois basin coal fly ashes. I. Chemical characterization and solubility, *Environ. Sci. Technol.*, 18, 734, 1984.

27. James, W.D. et al., Water-leachable boron from coal ashes, *Environ. Sci. Technol.*, 16, 195, 1982.

28. Furr, A.K. et al., National survey of elements and radioactivity in fly ashes. Absorption of elements by cabbage in fly ash–soil mixtures, *Environ. Sci. Technol.*, 11, 1194, 1977.

29. Cox, J.A. et al., Leaching of boron coal ash, *Environ. Sci. Technol.*, 12, 722, 1978.

30. Pagenkopf, G.K. and Connolly, J.M., Retention of boron by coal ash, *Environ. Sci. Technol.*, 16, 609, 1982.

31. Churey, D.J. et al., Element concentrations in aqueous equilibrates of coal and lignite fly ashes, *J. Agric. Food Chem.*, 27, 910, 1979.

32. Querol, X. et al., Extraction of soluble major and trace elements from fly ash in open and closed leaching systems, *Fuel*, 80, 801, 2001.

33. El-Mogazi, D., Lisk, D.J., and Weinstein, L.H., A review of physical, chemical and biological properties of fly ash and effects on agricultural ecosystems, *Sci. Total Environ.*, 74, 1, 1988.

34. Elseewi, A.A. et al., Boron enrichment of plants and soils treated with coal ash, *J. Plant Nutr.*, 3, 409, 1981.

35. Narukawa, T. et al., Investigation into the relationship between major and minor element contents and particle size and leachability of boron in fly ash from coal fuel thermal power plants, *J. Environ. Monit.*, 5, 831, 2003.

36. Hansen, L.D. et al., Chemical speciation of elements in stack-collected, respirable size, coal fly ash, *Environ. Sci. Technol.*, 18, 181, 1984.

37. Phung, H.T. et al., Trace elements in fly ash and their release in water and treated soils, *J. Environ. Qual.*, 8, 171, 1979.

38. Halligan, A.S. and Pagenkopf, G.K., Factors influencing the release of boron from coal ash materials, *Environ. Sci. Technol.*, 14, 995, 1980.

39. Hollis, J.F., Keren, R., and Gal, M., Boron release and sorption by fly ash as affected by pH and particle size, *J. Environ. Qual.*, 17, 181, 1988.

40. Gladney, E.S. et al., Observations on boron release from coal-fired power plants, *Environ. Sci. Technol.*, 12, 1084, 1978.

41. Warren, C.J. and Dudas, M.J., Formation of secondary minerals in artificially weathered fly ash, *J. Environ. Qual.*, 14, 405, 1985.

42. Gangloff, W.J. et al., Field study: influence of fly ash on leachate composition in an excessively drained soil, *J. Environ. Qual.*, 26, 714, 1997.

43. Warren, C.J., Evans, L.J., and Sheard, R.W., Release of some trace elements from sluiced fly ash on acidic soils with particular reference to boron, *Water Manage. Res.*, 11, 3, 1993.

44. Plank, C.O., Martens, D.C., and Hallock, D.L., Effect of soil application of fly ash on chemical composition and yield of corn (*Zea mays* L.) and on chemical composition of displaced solutions, *Plant Soil*, 42, 465, 1975.

45. Dreesen, D.R. et al., Comparison of levels of trace elements extracted from fly ash and levels found in effluent waters from a coal-fired power plant, *Environ. Sci. Technol.*, 11, 1017, 1977.

46. Dudas, M.J. and Warren, C.J., Submicroscopic model of fly ash particles, *Geoderma*, 40, 101, 1987.

47. Boyd, R.J., The partitioning behavior of boron from tourmaline during ashing of coal, *Int. J. Coal Geol.*, 53, 43, 2002.

48. Zhang, M. and Reardon, E.J., Removal of B, Cr, Mo, and Se from wastewater by incorporation into hydrocalumite and ettringite, *Environ. Sci. Technol.*, 37, 2947, 2003.

49. Jones, L.H. and Lewis, A.V., Weathering of fly ash, *Nature*, 185, 404, 1960.

50. Collier, G.F. and Greenwood, D.J., Potential phytotoxic components of pulverized fuel ash, *J. Sci. Food Agric.*, 28, 137, 1977.

51. Elseewi, A.A., Straughan, I.R., and Page, A.L., Sequential cropping of fly ash-amended soils: effects on soil chemical properties and yield and elemental composition of plants, *Sci. Total Environ.*, 15, 247, 1980.

52. Khandkar, U.R. et al., Edaphological characteristics of unweathered and weathered fly ashes from Gondwana and lignite coal, *Environ. Pollut.*, 79, 297, 1993.

53. Kukier, U., Sumner, M.E., and Miller, W.P., Boron release from fly ash and its uptake by corn, *J. Environ. Qual.*, 25, 596, 1994.

54. Hodgson, D.R. and Buckley, G.P., A practical approach towards the establishment of trees and shrubs on pulverized fuel ash, in *The Ecology of Resource Degradation and Renewal*, Chadwick M.J. and Goodman G.T., Eds., Blackwell Scientific Publications, Oxford, 1975, 305.

55. Hodgson, D.R. and Townsend, W.N., The amelioration and revegetation of pulverized fuel ash, in *Ecology and Reclamation of Devastated Land*, Hutnik, R.J. and Davis G., Eds., Gordon and Breach, New York, 1973, vol. 2, 247.

56. Burns, I.G. and Collier, G.F., A simulation model for the leaching of borate in soils and precipitated fuel ash, *J. Sci. Food Agric.*, 31, 743, 1980.

57. Nass, M.M. et al., Long-term supply and uptake by plants of elements from coal fly ash, *Commun. Soil Sci. Plant Anal.*, 24, 899, 1993.

58. Furr, A.K. et al., Elemental content of tissues and excreta of lambs, goats, and kids fed white sweet clover growing on fly ash, *J. Agric. Food Chem.*, 26, 847, 1978.

59. Weinstein, L.H. et al., Elemental analysis of grasses and legumes growing on soil covering coal fly ash landfill sites, *J. Food Safety*, 9, 291, 1989.

60. Woodbury, P.B. et al., Assessing trace element uptake by vegetation on a coal fly ash landfill, *Water Air Soil Pollut.*, 111, 271, 1999.

61. Dosskey, M.G. and Adriano, D.C., Trace element toxicity in a mycorrhizal cucumber grown on weathered coal fly ash, *Soil Biol. Biochem.*, 25, 1547, 1993.

62. Scanlon, D.H. and Duggan. J.C., Growth and elemental uptake of woody plants on fly ash, *Environ. Sci. Technol.*, 13, 311, 1979.

63. Carlson, C.L. and Adriano, D.C., Growth and elemental content of two species growing on abandoned coal fly ash basins, *J. Environ. Qual.*, 20, 581, 1991.

64. Holliday, R. et al., Plant growth on fly ash, *Nature*, 12, 1079, 1958.

65. Shulka, K.N. and Mishra, L.C., Effect of fly ash extract on growth and development of corn and soybean seedlings, *Water Air Soil Pollut.*, 27, 155, 1986.

66. Evans, C.M. and Sparks, D.L., On the chemistry and mineralogy of boron in pure and mixed systems. A review, *Commun. Soil Sci. Plant Anal.*, 14, 827, 1983.

67. Goldberg, S., Chemistry and mineralogy of boron in soils, in *Boron and its Role in Crop Production*, Gupta U.C., Ed., CRC Press Inc, Boca Raton, FL, 1993, chap. 2.

68. Ghodrati, M., Sims, J.T., and Vasilas, B.L., Evaluation of fly ash as a soil amendment for the Atlantic Coastal Plain: I. Soil hydraulic properties and elemental leaching, *Water Air Soil Pollut.*, 81, 349, 1995.

69. Ghodrati, M. et al., Enhancing the benefits of fly ash as a soil amendment by preleaching, *Soil Sci.*, 150, 244, 1995.

70. Phung, H.T. et al., The practice of leaching boron and soluble salts from fly ash-amended soils, *Water Air Soil Pollut.*, 12, 247, 1979.

71. Matsi, T. and Keramidas, V.Z., Alkaline fly ash effects on boron sorption and desorption in soils, *Soil Sci. Soc. Am. J.*, 65, 1101, 2001.

72. Tolle, D.A. and Arthur, M.F., Microcosm/field comparison of trace element uptake in crops grown in fly ash-amended soil, *Sci. Total Environ.*, 31, 243, 1983.

73. Salé L.Y., Naeth, M.A., and Chanasyk, D.S., Growth response of barley on unweathered fly ash–amended soil, *J. Environ. Qual.*, 25, 684, 1996.

74. Sims, J.T., Vasilas, B.L., and Ghodrati, M., Evaluation of fly ash as a soil amendment for the Atlantic Coastal Plain: II. Soil chemical properties and crop growth, *Water Air Soil Pollut.*, 81, 363, 1995.

75. Clark, R.B. et al., Boron accumulation by maize grown in acidic soil amended with coal combustion products, *Fuel*, 78, 179, 1999.

76. Romney, E.M., Wallace, A., and Alexander, G.V., Boron in vegetation in relationship to a coal–burning power plant, *Commun. Soil Sci. Plant Anal.*, 8, 803, 1977.

77. Khan, M.R. and Khan, M.W., The effect of fly ash on plant growth and yield of tomato, *Environ. Pollut.*, 92, 105, 1996.

78. Warren, C.J., Some limitations of sluiced fly ash as a liming material for acidic soils, *Waste Manage. Res.*, 10, 317, 1992.

79. Adriano, D.C. and Weber, J.T., Influence of fly ash on soil physical properties and turfgrass establishment, *J. Environ. Qual.*, 30, 596, 2001.

80. Adriano, D.C. et al., Effects of high rates of coal fly ash on soil, turfgrass, and groundwater quality, *Water Air Soil Pollut*, 139, 365, 2002.

81. Martens, D.C. and Beahm, B.R., Growth of plants in fly ash amended soils, in *Proc. 4th Int. Ash Utilization Symposium*, Faber, J.H. et al., Eds., Morgantown Energy Research Center, Morgantown, WV, 1976, 657.

82. Phung, H.T., Lund, L.J., and Page, A.L., Potential use of fly ash as a liming material, in Adriano, D.C. and Brisdin I.L., Eds., *Environmental Chemistry and Cycling Process*, U.S. Dep. Commerce, Springfield, Va., 1978, 504.

83. Wright, R.J. et al., Influence of soil-applied coal combustion by-products on growth and elemental composition of annual ryegrass, *Environ. Geochem. Health*, 20, 11, 1998.

84. Furr, A.K. et al., Multielement uptake by vegetables and millet grown in pots of fly ash-amended soil, *J. Agric. Food Chem.*, 24, 885, 1976.

85. Furr, A.K. et al., Elemental content of apple, millet and vegetables grown in pots of neutral soil amended with fly ash, *J. Agric. Food Chem.*, 27, 135, 1979.

86. Pathan, S.M., Aylmore, L.A.G., and T.D. Colmer, T.D., Soil properties and turf growth on a sandy soil amended with fly ash, *Plant Soil*, 256, 103, 2003.

87. Chermsiri, C. et al., Effect of boron sources on garlic (*Allium sativum* L.) productivity, *Biol. Fertil. Soils*, 20, 125, 1995.

88. Gutenmann, W.H. et al., Arsenic, boron, molybdenum, and selenium in successive cuttings of forage crops field grown on fly ash amended soil, *J. Agric. Food Chem.*, 27, 1393, 1979.

89. Furr, A.K. et al., Elemental content of vegetables, grains and forages field-grown on fly ash amended soil, *J. Agric. Food Chem.*, 26, 357, 1978.

90. Hytönen, J., Effects of wood, peat and coal ash fertilization on Scots pine foliar nutrient concentrations and growth on afforested former agricultural peat soils, *Silva Fenn.*, 219, 2003.

91. Bradford, G.R., Boron, in *Diagnostic criteria for plants and soils*, Chapman, H.D., Ed., Univ. of California, Riverside, 1966, chap. 4.

92. Kumar, D. and Singh, B., The use of coal fly ash in sodic soil reclamation, *Land Degrad. Develop.*, 14, 285, 2003.

2 Bioavailability of Trace Elements in Relation to Root Modification in the Rhizosphere

Ioannis K. Mitsios and Nicholas G. Danalatos

CONTENTS

2.1 INTRODUCTION

The immediate vicinity of plant roots is of particular importance for plant nutrient turnover and bioavailability. The part of the soil directly influenced by the roots is called "rhizosphere." Because the root affects the adjacent soil, it is obviously very interesting to examine the spread out zone plainly exploited for a particular solute. If this is wide, there may be no point in emphasizing effects close to the roots; however, if it is narrow, predictions based on the behavior of the bulk soil may be made.

The effect of the root on the adjacent soil medium is mainly the release of organic and inorganic material into the soil. Predicting the bioavailability of trace elements to plants is a major agricultural and environmental issue. The prime entry of toxic elements in the food chain comprises the plants and the animals. Trace elements occur naturally in rocks and soils, but principally in forms that are nonavailable to living organisms. Large quantities of trace elements are released into the environments by anthropogenic activities — for example, industrial processes, manufacturing, and agricultural amendments.

Several metals and metalloids, such as Mn, Cu, Zn, Ni, Mo, and B, play an important role in the soil–plant system. The bioavailability of these elements is therefore of fundamental importance. A small available amount of these elements can cause deficiencies in plants. However, in the case of pollution, a large available amount of these elements and other similar nonessential elements (such as Pb, Cr, Ae, etc.) can be detrimental to plant growth. Recently, many workers [1,2] have described a range of factors and mechanisms for nutrients that implied in the transfer of elements from the soil into plant roots.

Not only the uptake of trace elements but also the acquisition of these elements encompasses chemical processes that occur at the soil–root interface and that can influence the dynamics of trace elements in the rhizosphere [2,3]. The problem arises from the fact that modification in the rhizosphere may be physical [4] and chemical through several other processes [2,5–8] or microbiological. Therefore, understanding all the mechanisms implied in the process of acquisition of elements by plant roots is a prerequisite for assessing their bioavailability. Nye and Tinker [9] reported that microbial biomass and activity occur in the rhizosphere. This depends on the flux of C from root exudation. Root exudates include organic substances, some of which are very reactive compounds that can affect the chemical characteristics of the rhizosphere.

This chapter reviews the chemical processes in the rhizosphere that affect the uptake of trace elements, their bioavailability to plants, and plant nutrition. Also, in this chapter, emphasis is given to the role of rhizosphere microorganisms in the nutrient supply to plants.

2.2   TRACE ELEMENTS IN THE SOILS

Trace elements accumulate locally in soils due to weathering of rock minerals. Because trace elements are essential for plants, animals, and human beings, it is necessary to ensure their adequate levels in agricultural products. Apart from trace elements originating in parent materials and entering the soil through chemical weathering processes, soil toxic trace elements have many anthropogenic sources. Campbell et al. [10] compared natural and anthropogenic quantities of trace metals emitted to the atmosphere and showed that around 15 times more Cd; 100 times more Pb; 13 times more Cu; and 21 times more Zn are emitted by man's activities than by natural processes.

Trace elements and their total concentrations can vary widely in different soils derived from different parent materials. Levinson [11] and Alloway [12] proposed that the more basic igneous parent materials contribute highest quantities of Cr, Mn, Co, and Ni to soils; among the sedimentary parent materials, shales potentially contribute the highest quantities of Cr, Co, Ni, Zn, and Pb. Mineral weathering rates determine the release of the elements into soil — initially, in the form of simple or complex inorganic ligands. This process depends on mineral solubility under prevailing soil pH and Eh conditions. In soils, Mn, Ni, and Cr are present in highest quantities, and Cd and Hg are present in the smallest amounts.

The concentration of trace elements such as Mn, Ni, and Cr in the soil profile depends on parent material when the parent material is the main source in the soil. Brooks [13] found that the concentration of Ni in soils developed on serpentine was as high as 100 to 7000 μg Ni g^{-1}. Because many minerals are resistant to weathering, trace elements are insoluble; higher concentrations can be found in tropical soils where the weathering process is intensive.

Large quantities of Cd are applied to agricultural soils in phosphate fertilizers; the highest concentration occurs in topsoil. Topsoil concentration is maintained by vegetation recycling, and Cd and Pb show very slow movement down the profile [14]. Colbourn and Thornton [15] calculated the relative topsoil enhancement (RTE) as the ratio of Pb concentration in topsoil (<15 cm) to the Pb concentration in subsoil (>15 cm). They reported RTE values of 1.2 to 2.0 in remote agricultural areas and values of 4 to 20 in areas contaminated by mining.

2.3 ROOT MODIFICATION OF THE RHIZOSPHERE AND BIOAVAILABILITY OF TRACE ELEMENTS

2.3.1 PH CHANGES IN THE RHIZOSPHERE AND BIOAVAILABILITY OF TRACE ELEMENTS

Without doubt, most plants growing under natural conditions cause pH to rise in the rhizosphere. As pointed out by Cunningham [16], 62 plant species contain an average of 250 meq of absorbed cations per 100 g of oven-dry shoots and 360 meq anions per 100 g of oven-dry shoots. In general, roots release HCO_3 or OH to maintain electrical neutrality in the rhizosphere: an excess of 1100 meq of HCO_3 per 100 g dry matter, rather than H^+ ion, and the soil near roots become more alkaline instead of more acid.

Some other workers have showed that, when plants are supplied with NH_4^+ rather than NO_3, the pH in the root zone falls [17–19]. Nitrate normally constitutes more than half the total absorbed anions (180 meq per 100 g dry matter). If this amount of nitrogen is taken up as NH_4^+ instead of NO_3^-, cations are 430 meq per 100 g dry matter and anions are 180 me per 100 g dry matter; this yields a net release of 250 meq/100 g dry matter. For legumes, where nitrogen is fixed symbiotically and little is taken up as NO_3^- or NH_4^+, the net effect is 70 meq H^+ released per gram of dry matter. The changes in the rhizosphere are at a maximum when the pH in soil is about 5.3 because, at this value of pH, diffusion of acidity is low [20]. The thickness of the zone influenced by the root decreases to about 1 mm. It is from this thin zone that most of the micronutrients and phosphate uptake probably occurs and in which a large microbial population exists [21].

Nye [21] proposed that at soil pH 8 changes in the pH are small because HCO_3^- concentration is high in the soil solution. The gradients in the rhizosphere will have a more significant effect on pH in alkaline soils. The release of H^+ or HCO_3^- and the production of CO_2 and the variation of soil pH depend upon the plant root. At pH values above 7 for normal inputs of H^+ or HCO_3^- by the root, the soil may have a relatively high acidity diffusion coefficient and thus pH gradients are predicted to be very small. However, the production of CO_2 by the root may have a more significant effect. The solubility of CO_2 is greater than O_2 in soil water; this means that small changes in partial pressure of CO_2 lead to relatively large concentration gradients and therefore rapid diffusion [21].

Increased buffer capacity increases the time necessary to establish a given profile, and increased water content reduces the pH changes in the rhizosphere because the ions can diffuse more rapidly. Mitsios and Powell [1] measured changes around single onion roots giving average pH values in small volumes of soil. The preceding processes result in severe pH changes in the rhizosphere, which are directly involved in the dissolution of minerals [2] such as silicates [22], carbonates [23], and phosphates in the rhizosphere [24]. In the rhizosphere, pH can also affect phosphate uptake.

If plants can induce the release of Ca from Ca-carbonates and phosphates due to release of protons by roots, this process is also likely to induce a release of trace elements from them. Hinsinger and Gilkes [24] showed an increase in Ca and P concentration in the rhizosphere of ryegrass and subclover due to release of protons by their roots. It is well known that phosphorus fertilizers are a source of input of Cd in agricultural soils due to substitution of Cd that occurs in phosphate rocks used for manufacturing phosphorus fertilizers [25].

The released proton by plant roots can cause an increased dissolution of goethite. It is known that most plant species (all but grasses) have been described as strategy I plants that respond to Fe deficiency. Marschner and Roemheld [26] concluded that the response to Fe deficiency is due to increased acidifying and reducing capacity of their roots.

Fenn and Assadian [27] showed that, in the rhizosphere of *Cynodon dactylon*, pH changes and dissolution of carbonates could mobilize Pb, Cu, and Mn in the rhizosphere and accumulate them in the leaves. Youssef and Chino [28,29] found that the mobility of Zn and Cu increased in the soil surrounding the roots because of the acidification of the rhizosphere. Neng–Chang and Huai–Man [30] studied the chemical behavior of Cd in wheat rhizosphere and concluded that the mobility of

Cd increased in the soil–root interface due to acidification of the rhizosphere. The form of nitrogen taken up by the plants was the main factor responsible for acidification process.

Some works has been done to investigate pH changes and the mobility of trace elements in the rhizosphere of hyperaccumulator species. Bernal and McGrath [31] studied the effects of pH and heavy metal concentrations in solution culture on the proton release growth and elemental composition of *Alyssum murale* and *Raphanus sativus*.

Bernal et al. [32] compared redox potential and pH changes in the rhizosphere of the Ni hyperaccumulator *Alyssum murale* and the nonhyperaccumulator *Raphanus sativus*. These workers concluded that the form of N taken up by the plants was the main factor responsible for pH changes and that the plants were able to reduce system more effectively than the hyperaccumulator. These results indicate that the hyperaccumulator mechanisms may be due to other rhizosphere processes, such as the release of chelating agents, or to differences in the number and affinity of metal root transporters. McGrath et al. [33] studied the heavy metals uptake and chemical changes in the rhizosphere of *Thlaspi caerulescens* and *Thlaspi ochroleucum* grown in contaminated soils. Knight et al. [34] investigated the Zn and Cd uptake by the hyperaccumulator *Thlaspi caerulescens* in contaminated soils and its effects on the concentration and chemical speciation of metals in soil solution. They found that the decrease in the mobile Zn fraction could explain only less than 10% of the total Zn uptake by the plants. The mobile fraction of Zn was depleted by *Thlaspi caerulescens* more than the closely related, but nonaccumulating *Thlaspi ochroleucum*.

In the rhizosphere of *Thlaspi caerulescens*, no significant differences in pH were observed. To explain these results, Knight et al. [34] suggested two possible mechanisms: *T. caerulencens* is able to mobilize Zn from the soil, or the soil studied had a large capacity to buffer the concentration of Zn in soil solution. Recently, Hamon and McLaughlin [35] showed that there is no difference in specific activity of Cd or Zn taken up by *T. caerulescens* or wheat. This indicates that the hyperaccumulator plant was able to access the same pools of metals available to the wheat plants. However, the Zn added in biosolids was highly labile, and the *T. caerulescens* in this experiment acts more as a Cd-tolerant species than as hyperaccumulator for Cd. These results show that hyperaccumulator plants seem to take up from the same phytoavailable metal pools, from which the other plants can take up metals when this pool is large enough. Mechanisms of hyperaccumulators, such as root exudation, may exist in the rhizosphere that can support metal uptake from less accessible pools.

Moving to the field conditions, where metals are returned to the soil from several sources, the soil is an easily available metal pool.

2.3.2 Concentration Changes of Ions in the Rhizosphere

The root may alter simple predictions of solute movements by release of H^+ or HCO_3^- ions; evolution of CO_2 from respiration; creation of changes in concentration of other ions and solutes; and excretion of organic substances. Change in pH-associated bicarbonate ion concentration is only one example of the disturbance near the root, which may influence diffusion of other ions. In fact, multiple ion diffusion is present and an accurate solution has not been attempted [9].

Nitrate and chlorite move towards the root from a distance. These anions are accompanied by cations — for example, calcium, because calcium is the dominant cation in a normal soil. Measurements of the diffusion coefficient of an ion require knowledge of the concentration and concentration gradients of all other ions [37]. The soil moisture level has a marked effect on the absorption of solutes by whole plants. The detailed interpretation of these effects is extremely complicated because it involves the transport of soil solutes and plant physiological responses.

Some of the effects of a single root are as follows:

- Plant effects:
  - The root absorbing power may be reduced by decreased water potential within the plant affecting plant growth.

- The contact between root and soil may be reduced by shrinkage of soil or root.
- Soil effects:
  - The diffusion coefficient will decrease because the moisture content and the impedance factors are reduced. The impedance factor takes account primarily of the tortuous pathway followed by the solute through the pores. This has the effect of increasing the path length to be traversed and of reducing the concentration along this path length. It may also include the effect on the increase in the viscosity of water near charged surfaces, which will affect the mobility of all solutes, though it is unlikely to be significant except in dry soils.
  - Convection to a root may be reduced by decrease in rate of transpiration.
  - In drier soils, the diffusion coefficient near the root may sharply decrease because the water level decreases sharply near the surface if transpiration is still appreciable.
  - The solution concentration of nonadsorbed solutes, e.g., chloride and nitrate, will increase.
  - The concentration in solution of exchangeable cations will increase. Because the anion concentration rises, the concentration of cations will rise. If Ca and Mg are the dominant cations, their concentration will increase approximately directly with the total anion concentration. The monovalent cations, e.g., potassium, will increase so that the reduced activity ratio is maintained:

$$\text{i.e.} \frac{(\text{K})^2 \, \text{dry}}{(\text{K})^2 \, \text{wet}} \approx \frac{(\text{Ca}) \, \text{dry}}{(\text{Ca}) \, \text{wet}} \tag{2.1}$$

  - The concentration of adsorbed anions, e.g., phosphate, will tend to decrease because the activity product $(\text{Ca}) (\text{H}_2\text{PO}_4)_2$ tends to be constant and (Ca) increases according to the preceding bulleted statement.
  - The processes of mass flow and diffusion occur together and the resulting concentration profile is not the result of two independent processes. Thus, it is not possible to state that a given proportion of the total solute absorbed has "arrived" or is "taken up" by mass flow and the remaining proportion by diffusion [9]. However, it is correct to state that a given amount of solute has been brought into a zone of disturbance around the root by mass flow, assuming that the radial inflow of solution from outside it does not vary with distance from the root.

Lorenz et al. [38] found that the amount of Ca and Mg transferred toward the roots by mass flow was three- to sixfold the actual rates of uptake of these nutrients by radish roots, depending on the method used for measuring soil solution concentrations. For the elements that occur at low concentrations in the soil solution, mass flow will account for only a portion of the actual flux taken up by plant roots. This is the case for P and K among the major nutrients and of trace elements such as Zn and Cd [38].

McGrath et al. [33] found that more than 90% of the Zn taken up by the hyperaccumulator plants of the *Thlaspi* genus was originating from the nonexchangeable pool of soil Zn. Nye and Tinker [9] proposed that no simple algebraic expression can describe the spread of the disturbance zone of depletion, and the concentration is in fact asymptotic to the distance axis. In practice, measurements of concentrations are not accurate enough to detect it. The spread of the disturbance zone of phosphate and potassium depends mainly on the diffusion coefficient and time rather than root-absorbing power. When the concentration of phosphate in the soil solution in the rhizosphere is less than about 10^{-6} m, the value of root-absorbing power for young roots is high, and their uptake rate depends very largely on the rate of diffusion through the soil.

In addition to the depletion of trace elements in the rhizosphere, which has a direct effect on adsorption/desorption and precipitation-dissolution equilibria, the accumulation and depletion of major elements are also likely to interfere with the kinetics of trace elements and their bioavailability to plants.

2.3.3 ROOT ORGANIC EXUDATES, TRACE ELEMENT MOBILIZATION IN THE RHIZOSPHERE, AND THEIR BIOAVAILABILITY

The plant can play an important role in metal bioavailability through mechanisms causing release of root organic exudates. Among the root exudates released in the rhizosphere, some compounds can form strong complexes or chelates with a range of metals. This is particularly the case of aliphatic and phenolic acids, on the one hand, and of phytosiderophone, on the other hand.

Plants growing under conditions in which Fe and/or Zn are deficient can actively increase the availability of these metals, releasing specific organic compounds. Different strategies have been proposed by Marschner et al. [39] and Roemheld [40]. Among them, strategy I is active in dicotyledonous and monocotyledonous species with the exception of graminaceous species. This strategy is based on a three-component system constituting a plasma membrane-bound inducible reductase; an enhanced excretion of protons; and the release of reducing and chelating agents [26]. Strategy II (active in graminaceous plants) is based on the release of phytosiderophores in the rhizosphere and specific uptake system on the root surface.

The nature and the rate of release of phytosiderophores differ among plant species and even cultivars [26,41,42]. Treeby et al. [43] and Tagaki et al. [44] proposed that phytosiderophores form chelates not only with Fe but also with Zn, Cu, and Mn.

Wiren et al. [45] found evidence that Zn can be taken up in grasses in the form of nondissociated Zn-phytosiderophores. Bienfait [46] proposed a third strategy concerning the capacity of microorganisms to release siderophores and the possibility for the plants to take up these compounds. Also, Crowley et al. [47,48] proposed the existence of a microbial siderophore Fe transport system in oat and maize.

Hofflandy et al. [49] observed exudation of organic acids in P-deficient plants. Lundstroem [50] suggested the significance of organic acids for weathering and the podzolization process. Gahoonia and Nielsen [51] proposed mechanisms for controlling the pH at the soil–root interface. It is known that mineral weathering and dissolution of P-containing minerals can increase the mobility of metals. Plant uptake of trace elements depends on the metal availability in the rhizosphere; exudation of phytosiderophores and organic acids and changes in pH and redox potential are considered key factors controlling metal mobility in the rhizosphere.

The exudation of organic acids may also be important for increasing nutrient availability. Moghimi et al. [53] were able to isolate a-ketogluconate from the rhizosphere of wheat roots in quantities that could solubilize considerable amounts of phosphate from hydroaxyapatite. This appears to be a direct effect of organic acids on phosphate availability by lowering rhizosphere pH. However, organic acids may increase phosphate availability by desorbing phosphate from the surface of sesquioxides by anion exchange and by increasing phosphate mobilization by chelation with Fe or Al phosphate or both.

Gardner et al. [54] found that the proteoid roots of white lupine release citrate that, as they suggest, is responsible for increasing phosphate availability to the plant. The authors concluded that because the plant of white lupine has the coarse proteoid root system of low surface area, the mechanism by which citrate increases phosphate availability is to increase the rate at which phosphate reaches the root surface. These workers postulate that citrate exuded from the roots reacts in the soil to form ferric hydroxyphosphate polymers, which diffuse to the root surface. In this work, Fe^{3+} is reduced to Fe^{2+}, citrate is released, and the phosphate is taken up by the root. Citrate thus acts as a shuttle mobilizing Fe phosphate in the acquisition of P from the rhizosphere.

Jauregui and Reisenauer [55] have proposed that the MnO_2 is reduced by exudated malate. Chelation of the Mn_2^+ produced prevents reoxidation and increases the mobility of Mn_2^+ in the rhizosphere.

Mench and Martin [52] studied the mobilization of Cd and other metals from two soils by root exudates of *Zea mays* L., *Nicotiana tabacum* L., and *Nicotiana rustica* L. They found that root exudates of *Nicotiana tabacum* were able to extract more Cd from soil than those of *Nicotiana rustica* and *Zea mays*.

2.3.3.1 Function of Siderophores in the Plant Rhizosphere

Siderophores are iron chelating agents secreted by microorganisms and graminaceous plants in response to iron deficiency. The nature and the rate of release of siderophores differ among plant species and even cultivars [26,41,42].

These compounds are important for iron nutrition and are also speculated to function in the ecology of microorganisms in the plant rhizosphere. Under aerated conditions, at neutral to alkaline pH, inorganic iron is extremely insoluble. In such conditions, plants and microorganisms rely absolutely on iron uptake from organic complexes or iron that has been solubilized by siderophores and organic compounds from root exudates. Organic acids secreted by plant roots dissolve iron as a specific response to iron deficiency [26]. Graminaceous plants initially release organic acids in the rhizosphere, but as the plant becomes more iron stressed, these are followed by increased production of highly efficient chelators, called phytosiderophores, secreted in localized zones behind the root tips [56].

Recently, exchange of metals between siderophores and phytosiderophores has been proposed as a primary mechanism for plant use of microbial siderophores [57–58]. It has also been shown that microbial siderophores may strip iron from phytosiderophores [59]. The partitioning of metals between different types of siderophores and other iron complexes depends on the stability constants as well as the concentration of each chelator and the ability of the chelators to attack the surface of iron minerals and undergo exchange.

The competition for iron between plant and microorganisms involves very complex interactions that depend on a number of factors. For example, differences in the level of siderophore production by all the competing microorganisms; the chemical stabilities of various siderophores and other chelators with iron; their resistance to degradation; and the ability of different siderophores in the soil solution may interact through ligand exchange.

2.3.4 Bioavailability of Trace Elements and Oxidation Reduction Processes in the Rhizosphere

2.3.4.1 Oxidation, Reduction Processes, and pH in Aerobic Conditions

It is well known that the pH of aerobic soils depends on the nature of the parent material, the degree of weathering and leaching, and the effects of any additions of amendments such as agricultural lime or fertilizers. In aerobic soils, the pH is not affected very much by the oxidation and reduction processes involved in aerobic respiration. For the glucose – pyruvic acid and oxygen – water couples protons and electrons are produced in equal numbers and are utilized in equal numbers by oxygen. This appears to be so in all oxidation steps in the breakdown of cellulose. There is no net gain or loss of protons during aerobic respiration. Although some of the steps in the respiration chain linking pyruvic acid to O_2 may not involve equal numbers of protons and electrons, the steps are cyclic and the overall reaction is balanced. However, CO_2 is produced by aerobic respiration, and an increase in partial pressure of CO_2 reduces pH via the H_2CO_3–HCO_3^-–CO_3^- system in the soil water [2]. The effects are mostly in calcareous and sodic soils. The pH of calcareous and sodic soils is close to 8 and 9, respectively; in equilibrium with the atmosphere, it is close to 7.5 and 8, respectively, when the relative pressure of CO_2 is 1%.

During the respiration process, the production of organic acids may have a slight acidifying effect. The oxidation of NH_4^+ to NO_3^- also causes a drop in pH.

2.3.4.2 Redox Processes and pH in Anaerobic Conditions

In anaerobic conditions, the couple MnO_2–Mn^{2+} causes a rise in pH, but the NO_3^-–NO_2^- and CO_2–CH_4 couples produce no change. In many soils, the $Fe(OH)_3$–Fe_2^+ couple tends to predominate, thus giving a rise in pH. The production of CO_2 tends to cause a decrease in pH; the resultant effect is that, in acid soils where the CO_2 effect is small, pH rises. In alkaline soils, where the CO_2 effect is large, pH drops.

2.3.4.3 The Significance of Redox in Rhizosphere

The supply of manganese to plants is as Mn^{2+}. Solid manganese compounds have very low solubility. When the pH of the soil is high, the plants may suffer from deficiencies and when the pH is low, toxicities may appear. Waterlogged soils would be expected to have adequate and perhaps toxic amounts of manganese. Most flooded soils contain sufficient water-soluble manganese for the growth of rice. Paddy rice has a high demand and a high tolerance for manganese.

Boron, cobalt, copper, molybdenum, and zinc are not involved in oxidation-reduction reactions in soils. However, the changes in pH in iron and manganese compounds when soils become anaerobic may affect the mobility and availability of the nutrients. Thus, the concentrations of Cu, Co, and Zn may increase in soil solution due to release from their association with ferric and manganese hydroxides when these are reduced. The organic compounds may chelate the nutrients and also increase their concentration. The net result is to increase the availability of Co and Cu but to decrease that of Zn.

In waterlogged or submerged soils, roots do not grow except for the case of flood-tolerant plants such as rice. The most obvious factor for the lack of growth is the lack of O_2 at the root surface. The concentration O_2 required at the root surface varies depending on the age and type of plant root. The bacteria seem to need about $4 \times 10^{-6}\ M\ O_2$ at their surface, although the concentration required at the site of enzyme activity may be only $2.5 \times 10^{-8}\ M$. Because of the longer diffusion path in plant roots, the concentration required at their surface may be larger than $4 \times 10^{-6}\ M$.

The leakage of O_2 from the roots, as a consequence of its transfer from the shoots through the aerenchyma, provides an adequate supply of O_2 for rice root respiration. It also enables rice plants to alleviate metal toxicities that can occur due to the ambient reducing conditions of the bulk soil and much increased solubility of Fe and Mn oxides [60]. Flessa and Fischer [61] and Begg et al. [62] found that leakage of O_2 leads to a substantial reoxidation of the rhizosphere, which can result in a precipitation of Fe and Mn oxides at the root surface or even in the root cell wall [63].

The reduction of Fe is a major mechanism involved in the acquisition of Fe by most plants, as indicated previously [26,40]. Strategy I species, in particular, have been defined as plants exhibiting an enhanced reduction activity as a response to Fe deficiency [26].

2.3.5 Bioavailability of Trace Elements as Related to Root–Microorganism Interactions in the Rhizosphere

In natural conditions, all normal roots support the microorganisms on their external surfaces and in a thin sheath of soil immediately adjacent to the rhizosphere [64,65]. Pathogens, which cause specific diseases in roots by invading the roots, are excluded from this discussion, which is largely confined to the effects of the organisms on the nutrient supply to apparently normal plants. Organisms can cause changes in the phase equilibria of soil nutrients. These become more easily absorbed by plants or become more readily transported to the roots — e.g., complex formation or redox changes or pH changes. They also can cause changes in chemical composition of the soil such as mineralization of organic matter or decomposition of soil minerals. In symbiotic processes, nutrients are transferred directly to the plant from their organisms.

Root microorganisms in the rhizosphere play an effective role in the bioavailability of trace elements. It is well known that root exudates can cause balance between Mn-oxidizing and Mn-reducing bacteria. A large number of interactions among roots, microorganisms, and the trace elements in the rhizosphere cannot be discussed in detail in this chapter. The attention here will focus on the bioavailability of trace elements as influenced by mycorrhizae. Mycorrhizal fungi occur in soils in close association with plant roots that permit plants to acquire nutrients from the soil. These fungi may be divided into two groups: the ectotrophic and the endotrophic mycorrhizae.

The ectotrophic mycorrizal fungi or ectomycorrhizae (ECM) alter the root very clearly, even in its physical form. A mass of fungal hyphae form a sheath around the root proper; hyphae extend from this into the soil and into the intercellular spaces in the cortex. These fungi, which belong to the Basidiomycetes, depend on carbohydrates supplied by the root. The thick sheath of hyphae covering the roots favors the uptake of water and inorganic nutrients — especially phosphate because it effectively enlarges the surface area of the root in direct contact with the soil. The thin mycorrhizal hyphae, with a diameter of 2 to 4 μm, are able to penetrate soil pores not accessible to the root hairs with a diameter of about five times greater than the hyphal diameter [66].

Ectotrophic mycorrhizae are mainly found on the roots of trees and shrubs. Numerous investigations have shown that they promote the growth of trees when grown on soils low in available phosphate. It has been reported that seedlings of *Pinus* and *Picea* planted on newly drained organic soils only grew when the roots were infected with appropriate mycorrhizal fungus. Endotrophic mycorrhizae form no definite sheath around the root. The most important group is known as the arbuscular mycorrhizae (AM). These are formed by most herbaceous plants (including agricultural plants) and with the majority of trees in Mediterranean and tropical ecosystems (soils with neutral pH and low organic matter content).

The rhizosphere of mycorrhizal plants may be defined as mycorrhizosphere. The rhizosphere is modified in several respects, depending on the presence or absence of mycorrhizae. Toxicity of metals in soil depends on their bioavailability — defined as their ability to be transferred from a soil to a living organism. This is a function not only of their total concentration, but also of physicochemical (pH, Eh, organic matter, clay content, etc.) and biological factors [67].

Soil microorganisms including mycorrhizal fungi are affected by the presence of high metal concentrations in soil [68]. The organisms influence the availability of trace elements in soil directly, through alterations of pH, Eh, biosorption, or uptake, or indirectly in the rhizosphere through their effect on plant growth, root exudation, and resulting rhizosphere chemistry.

2.3.5.1   Bioavailability of Trace Elements as Related to Mycorrhizal Fungi

Mycorrhizal fungi can directly affect bioavailability of trace elements by an effect on the free metals in soil solution (immobilization by adsorption, absorption, and accumulation) and indirectly by modifying root exudation or by affecting solubilization of metal-bearing minerals. Mycorrhizae have the capacity to protect their host plants when the metal uptake is excessive. Metal uptake decreases from ericoid to ectomycorrhizae (CECM) and arbuscular mycorrhizae (AM) fungi. This is the result of pH changes in soil and thus metal availability, but it is also related to morphology and biomass of the fungal structure of mycorrhizae.

The ectomycorrhizae are exceptional in this respect because they form a dense and thick sheet of fungal tissue (the mantle) that covers the root surface completely. It seems that any ion entering the root must pass the fungal mantle. In other words, ectomycorrhizae have the ability to "filter" ions that enter the plant, which the other types of mycorrhizae do not have. Mycorrhizal fungi are part of the rhizosphere, so their metal sorption capacity is a fundamental issue to researchers concerning the fate of metals in the rhizosphere. On one hand, the contact between metal ions and the hyphae in soil is the first interaction in mycorrhizal metal transport and the processes that take place at the hyphae surface affect the fate of trace elements in the rhizosphere. On the other hand, sorption of metals on mycorrhizal mycelium may be extensive and limit the amount of metals taken up by the fungi.

In a recent work on cation exchange capacity and Pb sorption in ectomycorrhizal fungi, Marschner et al. [69] found that the electron-dense lead deposits on the surface of ectomycorrhizal fungus with the highest Pb sorption capacity contained molar equivalents of P. The sorption characteristics of a fungus may differ between fungi *in vitro* and in symbiosis with a plant; Marschner et al. have found higher sorption [64]. Colpaert and Assche [70] showed heavy metal uptake and accumulation by ectomycorrhizal fungi, using axenic cultures, where the metals were added as soluble salts. Under these conditions, the fungal-soil concentration ratios were around 200 and 80 for Cd, and 40 and 30 for Zn for nontolerant and metal-tolerant isolates of *Suillus bovines*, respectively.

Gast et al. [71] studied the heavy metals in mushrooms and their relationship with soil characteristics and they found large differences between metals with very high accumulation for Cd, exclusion for Pb, and a narrower range of concentrations for Zn and Cu. These workers suggested a regulation of uptake for essential elements and concluded that species differences — not soil factors — are the primary determinants of metal levels in fungi.

Ectomycorrhizal fungi can increase the bioavailability of heavy metals in the rhizosphere by solubilizing minerals containing metals such as rock phosphates [72]. Mycorrhizal fungi associated with plant roots in symbiosis affect plant root exudation quantitatively and modify the composition of root exudates containing carbohydrate, amino acids, and aliphatic acids [73]. Leyval and Berthelin [74] proposed that the modification of the composition of root exudates by mycorrhizal fungi influence the bioweathering of minerals in the rhizosphere and the availability of metals in the mycorrhizosphere.

2.4 CONCLUSIONS

It is generally accepted that the rhizosphere plays an important role in the bioavailability of trace elements. A range of mechanisms exists, such as excretion of H^+ or HCO_3^-; respiration and release of CO_2; exudation; desorption and adsorption of trace elements; dissolution–precipitation; redox reaction; and chelation, by which plant roots can modify the chemical conditions in the rhizosphere. These factors can affect dramatically the behavior of root in the rhizosphere and substantially the bioavailability of trace elements. The mechanisms involved in chemical modifications in the rhizosphere as well as on uptake of trace elements differ among plant species and soil conditions.

The extent to which siderophores alter the ecology of the rhizosphere is a question that remains to be investigated. Scientists need to have a better understanding of how siderophores can function in the rhizosphere and under different soil conditions rather than in laboratory experiments. Research efforts must be focused on how these compounds influence heavy metal transport and bioavailability in soils. In the future, we hope these compounds may have application and use siderophores as iron fertilizers. The ability to manipulate siderophore production in the rhizosphere to improve plant trace metal nutrition will remain a significant challenge.

REFERENCES

1. Mitsios, I.K. and Rowell, D.L., Plant uptake of exchangeable and nonexchangeable potassium. I. Measurement and modeling for onion roots in a chalky boulder clay soil, *J. Soil Sci.*, 38, 53, 1987.
2. Hinsinger, P., How do plant roots acquire mineral nutrients? Chemical processes involved in the rhizosphere, *Adv. Agron.*, 64, 225, 1998.
3. Clarkson, D.T., Factors affecting mineral nutrient acquisition by plants, *Annu. Rev. Plant Physiol.* 36, 77, 1985.

4. Gregory, P.J. and Hinsinger, P., New approaches to studying chemical and physical changes in the rhizosphere: an overview, *Plant Soil*, 211, 1, 1999.
5. Marschner, H., *Mineral Nutrition of Higher Plants*, 2nd ed., Academic Press, London, 1995.
6. Marschner, H., Roemheld, V., Horst, W.J. and Martin, R., Root-induced changes in the rhizosphere: importance for the mineral nutrition of plants, *Z. Pflanzenernaehr. Bodenkd.*, 149, 441, 1986.
7. Gobran, G.R., Clegg, S. and Courchense, F., The rhizosphere and trace elements acquisition in soils, in *Fate and Transport of Heavy Metals in the Vadose Zone*, Selim, H.M., Iskandar, A., Eds., CRC Press, Boca Raton, FL, 1999, 225.
8. Darrah, P.R., The rhizosphere and plant nutrition: a quantitative approach, *Plant Soil* 155/156, 1, 1993.
9. Nye, P.H. and Tinker P.B., *Solute Movement in the Soil–Root System*, Blackwell Scientific Publications, Oxford, 1977.
10. Campbell, P.G.C., Stokes, P.M. and Galloway, J.N., The effect of atmospheric deposition on the geochemical cycline and biological availability of metals, in *Heavy Metals in the Environment* 2, Heidelberg International Conference. CEP Consultants, Edinburgh, 1983, 760.
11. Levinson, A.A., *Introduction to Exploration Geochemistry*, Applied Publishing, Calgary, 1974.
12. Alloway, B.J., The origin of heavy metals in soils, in *Heavy Metals in Soils*, Alloway, B.J., Ed., Blackie, London, Wiley, New York, 1990, 29.
13. Brooks, R.R., *Serpentine and its Vegetation*, Dioscorides Press, Portland, 1987.
14. Biddappa, C.C., Chino, M. and Kumazawa, K., Migration of heavy metals in two Japanese soils, *Science*, 173, 233, 1982.
15. Colbourn, P. and Thornton, I., Lead pollution in agricultural soils, *J. Soil Sci.*, 29, 513, 1978.
16. Cunningham, R.K., Cation anion relationships in crop nutrition: III. Relationships between the ratios of sum of the cations: some of the anions and nitrogen concentration in several plant species, *J. Agric. Sci.*, 63, 109, 1964.
17. Miller, M.H., Effects of nitrogen on phosphorus absorption by plants, in *The Plant Root and its Environment*, Carson, E.W., Ed., University of Virginia Press, Charlottesville, 1974, 643.
18. Miller, M.H., Mamaril, C.P. and Blair, G.J., Ammonium effects on phosphorus absorption through pH changes and phosphorus precipitation at the soil–root interface, *J. Agron.*, 62, 1970, 524.
19. RileyDand Barber, S.A., Effect of ammonium and nitrate fertilization on phosphorus uptake as related to root-induced pH changes at the root–soil interface, *Soil Sci. Soc. Am. Proc.*, 35, 301, 1971.
20. Rowell, D.L., Soil acidity and alkalinity, in *Russell's Soil Conditions and Plant Growth*, 11th ed., Wild, A., Ed., Longman Scientific and Technical, New York, 1988, 844.
21. Nye, P.H., pH changes across the rhizosphere induced by roots, *Plant Soil* 61, 7, 1981.
22. Hinsinger, P., Elass, B. and Robert, M., Root-induced irreversible transformation of a trioctahedral mica in the rhizosphere of rape, *J. Soil Sci.*, 44, 535, 1993.
23. Bertnand, I., Hinsinger, P., Jaillard, B. and Arvieu, J.C., Dynamics of phosphorus in the rhizosphere of maize and rape grown on synthetic, phosphated calcite and goethite, *Plant Soil*, 211, 111, 1999.
24. Hinsinger, P. and Gilkes, R.J., Mobilization of phosphate from phosphate rock and alumina-sorbed phosphate by the roots of ryegrass and clover as related to rhizosphere pH, *Eur. J. Soil Sci.*, 47, 533, 1996.
25. McLaughlin, M., Tiller, K.G., Naidu, R. and Stevens, D.P., Review: the behavior and environmental impact of contaminants in fertilizers, *Aust. J. Soil Res.*, 34, 1, 1996.
26. Marschner, H. and Roemheld, V., Strategies of plants for acquisition of iron, *Plant Soil*, 165, 261, 1994.
27. Fenn, L.B. and Assadian, N., Can rhizosphere chemical changes enhance heavy metal absorption by plants growing in calcareous soil? In *Proc. 5th Int. Conf. Biogeochem. Trace Element*, Vienna, 1, 11–15, Wenzel, W.W., Adriano, D.C., Alloway, B., Donaidu, H.E. and Pierzynski, G.M., Eds., International Society for Trace Element Biochemistry, 1999, 154.
28. Youssef, R.A. and Chino, M., Root induced changes in the rhizosphere of plants. I. Changes in relation to the bulk soil, *Soil Sci. Plant Nutr.*, 35, 461, 1989.
29. Youssef, R.A. and Chino, M., Root-induced changes in the rhizosphere of plants. II. Distribution of heavy metals across the rhizosphere in soils, *Soil Sci. Plant Nutr.*, 35, 609, 1989b.
30. Neng–Chang, C. and Huai–Man, C., Chemical behavior of cadmium in wheat rhizosphere, *Pedosphere*, 2, 363, 1992.

31. Bernal, M.P. and McGrath, S.P., Effects of pH and heavy metal concentrations in solution culture on the proton release, growth and elemental composition of *Alyssum murale* and *Raphanus sativus* L., *Plant Soil*, 166, 83, 1994.

32. Bernal, M.P., McGrath, S.P., Miller, A.J. and Baker, A.J.M., Comparison of the chemical changes in the rhizosphere of the nickel hyperaccumulator *Alyssum murale* with the nonaccumulator *Raphanus sativus*, *Plant Soil*, 164, 251, 1994.

33. McGrath, S.P., Shen, Z.G. and Zhao, F.J., heavy metals uptake and chemical changes in the rhizosphere of *Thlaspi caerulescens* and *Thlaspi ochroleucum* grown in contaminated soils, *Plant Soil*, 188, 153, 1997.

34. Knight, B., Zhao, F.J., McGrath, S.P. and Shen, Z.G., Zink and cadmium uptake by the hyperaccumulator *Thlaspi caerulescens* in contaminated soils and its effects on the concentration and chemical speciation of metals in soil solution, *Plant Soil*, 197, 71, 1997.

35. Hamon, R.E. and McLaughlin, J.M., Use of the hyperaccumulator *Thlaspi caerulescens* for bioavailable contaminant stripping, in *Proc. 5th Int. Conf. Biochem. Trace Elements*, Vienna, 2, 11–15, Doner, H.E., Keller, C., Lepp, N.W., Mench, M., Nuidu, R. and Pierzynski, G.M., Eds., International Society for Trace Element Biochemistry, 1999, 908.

36. Wenzel, N.W. and Jockwer, F., Accumulation of heavy metals in plant grown on mineralized soils of the Austrian Alps, *Environ. Pollut.*, 104, 145, 1999.

37. Nye, P.H., The measurement and mechanism of ion diffusion in soil: I. The relation between self-diffusion and bulk diffusion, *J. Soil Sci.*, 17, 16, 1996.

38. Lorenz, S.E., Hamon, R.E. and McGrath, S.P., Differences between soil solutions obtained from rhizosphere and nonrhizosphere soils by water displacement and soil centrification, *Eur. J. Soil Sci.*, 45, 431, 1994.

39. Marschner, H., Roemheld, V. and Kissel, M., Different strategies in higher plants in mobilization and uptake of ions, *J. Plant Nutr.*, 9, 695, 1986.

40. Roemheld, V., Existence of two different strategies in the acquisition of iron in higher plants, in *Iron Transport in Microbes, Plant and Animals,* Winkelman, S.V., Ed., Verlag Weinheim, GE, 1987, 353.

41. Mench, M.J. and Fargues, S., Metal uptake by iron-efficient and -inefficient oats, *Plant Soil*, 165, 227, 1994.

42. Brown, J.C., Von Jolley, D. and Lytle, M., Comparative evaluation of iron solubilizing substances (phytosiderophores) released by oats and corn: iron efficient and inefficient plants, *Plant Soil*, 130, 157, 1991.

43. Treeby, M., Marschner, H. and Roemheld, V., Mobilization of iron and other micronutrients from a calcareous soil by plant-borne microbial and synthetic metal chelators, *Plant Soil*, 114, 217, 1989.

44. Tagaki, S., Nomoto, K. and Takemoto, T., Physiological aspect of mugeneic acid, a possible phytosiderophores of graminaceous plants, *J. Plant Nutr.*, 7, 469, 1984.

45. Wiren, N., Marschner, H. and Roemheld, V., Roots of iron-efficient maize also absorb phytosiderophore-chelated, Zn, *Plant Physiol.*, 111, 1119, 1996.

46. Bienfait, F., Prevention of stress in iron metabolism of plants, *Acta Bot. Neerl.*, 38, 105, 1989.

47. Crowley, D.E., Reid, C.P.P. and Szaniszlo, P.J., Utilization of microbial siderophores in iron acquisition by oat, *Plant Physiol.*, 87, 680, 1988.

48. Crowley, D.E., Wang, Y.C., Reid, C.P.P. and Szaniszlo, P.J., Mechanisms of iron acquisition from siderophores by microorganisms and plants, *Plant Soil*, 113, 155, 1991.

49. Hofflandy, E., Findenegg, G.R. and Nelemans, J.A., Solubilization of rock phosphate by rape. I. Evaluation of the role of uptake pattern, *Plant Soil*, 113, 155, 1989.

50. Lundstroem, U.S., Significance of organic acids for weathering and podzolization process, *Environ. Int.*, 20, 21, 1994.

51. Gahoonia, T.S. and Nielsen, N.E., Control of pH at the soil–root interface, *Plant Soil*, 140, 49, 1992.

52. Mench, M. and Martin, E., Mobilization of cadmium and other metals from two soils by root exudates of *Zea mays* L., *Nicotiana tabacum* L. and *Nicotiana rustica* L., *Plant Soil*, 132, 187, 1991.

53. Moghimi, A., Tate, M.E. and Oades, J.M., Phosphate dissolution by rhizosphere products. II. Characterization of rhizosphere products especially α-ketogluconic acid. *Soil Bio. Biochem.*, 19, 283, 1978.

54. Gardner, W.K., Barber, D.A. and Parbery, D.G., The acquisition of phosphorus by *Lupinus albus* L., *Plant Soil*, 70, 107, 1983.

55. Jauregui, M.A. and Reisenauer, H.M., Dissolution of oxides of manganese and iron by root exudates components, *Soil Sci. Soc. Am. J.*, 46, 314, 1982.

56. Ma, J.F. and Nomoto, K., Effective regulation of iron potential role in graminaceous plants: the role of mugineic acids as phytosiropheres, *Physiol. Planta*, 97, 609, 1996.

57. Yehuda, Z., Shenker, M., Roemheld, V., Marschner, P., Hadar, Y. and Chen, Y., The role of ligand exchange in the uptake of iron from microbial siderophores by gramineous plants, *Plant Physiol.*, 112, 1273, 1996.

58. Shenker, M., Hadar, Y. and Chen, Y., Stability constants of the fungal siderophore rhizoferrin with various microelements and calcium, *Soil Sci. Soc. Am. J.*, 60, 1140, 1996.

59. Jurkevitch, E., Hadar, Y., Chen, Y, Chino, M. and Mori, S., Indirect utilization of the phytosiderophore mugineic acid as an iron source to rhizosphere fluorescent, *Pseudomonas Biometals*, 6, 119, 1993.

60. Lindsay, W.L. *Chemical Equilibria in Soils*, John Wiley & Sons, New York, 1979.

61. Flessa, H. and Fischer, W.R., Plant-induced changes in the redox potentials of rice rhizosphere, *Plant Soil*, 143, 55, 1992.

62. Begg, C.B.M., Kirk, G.J.D., MacKense, A.F. and Neue, H., Root-induced iron oxidation and pH changes in the lowland rice rhizosphere, *New Phytol.*, 128, 469, 1994.

63. Chen, C.C., Dixon, J.B. and Turner, F.T., Iron coatings on rice roots: mineralogy and quantity influencing factors, *Soil Sci. Soc. Am.*, 44, 635, 1980.

64. Clark, F.E., Soil microorganisms and plant roots, *Adv. Agron.*, 1, 242, 1949.

65. Clark, F.E. and Paul, E.A., The microflora of grassland, *Adv. Agron.*, 22, 375, 1970.

66. Schleghte, G., Nutrient uptake of plants and mycorrhiza. I. Ectotrophic mycorrhiza. *Kalibriefe, (Buentehof) Fachreb.*, 2, 6e Folge, 1976.

67. Berthelin, J. Munier–Lamy, C. and Leyval, C., Effect of microorganisms on mobility of heavy metals in soils, in *Environmental Impacts of Soil Component Interractions: Vol. 2. Metals, Other Inorganics, and Microbial Activities*, Huang, P.M. et al., Eds., Lewis Publishers, Boca Raton, 1995.

68. Gadd, G.M., Interractions of fungi with toxic metals, *New Phytol.*, 124, 25, 1993.

69. Marschner, P., Jentschke, G. and Godbold, D.L., Cation exchange capacity and lead sorption in ectomycorrhizal fungi, *Plant Soil*, 205, 93, 1998.

70. Colpaert, J.V. and Assche, J.A., The effect of cadmium and the cadmium–zink interaction on the axenic growth of ectomycorrhizal fungi, *Plant Soil*, 145, 237, 1992.

71. Gast, C.H., Jansen, E. Bierling, J. and Haanstra, L., Heavy metals in mushrooms and their relationship with soil characteristics, *Chemosphere*, 17, 789, 1988.

72. Leyval, C., Surtiningsih, T. and Berthelin, J., Mobilization of Cd from rock phosphates by rhizospheric microorganisms (phosphate dissolving bacteria and ectomycorrhizal fungi), *Phosphorus, Sulfur Silicon*, 77, 133, 1993.

73. Laheure, F., Leyval, C. and Berthelin, J., Root exudates of maize, pine beech seedlings influenced by mycorrhizal and bacterial inoculations, *Symbiosis*, 9, 111, 1990.

74. Leyval, C. and Berthelin, J., Weathering of phlogopite in pine (*Pinus silvestris* L.) rhizosphere inoculated with an ectomycorrhizal fungus and an acid-producing bacterium, *Soil Sci. Soc. Am. J.*, 55, 1009, 1991.

3 Availability of Heavy Metals Applied to Soil through Sewage Sludge

V. Antoniadis, C.D. Tsadilas, V. Samaras, and J. Sgouras

CONTENTS

3.1 INTRODUCTION

Sewage sludge is the residue product of the treatment of municipal or industrial waste water. Usually, it only contains 10 to 30% solids. The disposal management options include disposal to landfill sites, application to land (as a soil amendment or on dedicated waste land), or incineration [1]. Sewage sludge contains high amounts of organic matter and significant quantities of macro- and micronutrients [2]. This makes it an efficient organic fertilizer because it is cheap and produced in great amounts all year. The fertilizer value of sludge is considerable, especially in warm or semiarid regions, where common agricultural practices (e.g., intense cultivation, plowing surface soil) tend to reduce the average soil organic carbon content. In addition, sewage sludge may improve soil physical properties because it promotes the development of good structure and macroporosity and prevents soil erosion [3].

However, sludge also contains various contaminants, such as inorganic potentially toxic elements, and persistent organic compounds. It is believed that sewage sludge is source as well as sink of heavy metals; however, because, along with the metals, it also provides adsorption sites in the humic phases, a great controversy still exists concerning the availability of heavy metals long after the sewage sludge application has ceased. A number of researchers claim that "all evidence available indicates that the specific metal adsorption capacity added with sludge will persist as long as the heavy metals of concern persist in the soil" (Chaney and Ryan [4], as quoted by McBride [5]). On the other hand, some concerns are that heavy metal availability to plants may increase over time.

The preceding exhibits the need to address and study the chemical behavior of heavy metals in relation to various organic phases carefully. The reason is that, although organic matter content as a whole will decrease over time, important fractions of it may increase, in absolute terms or in proportion to the total organic matter content. These fractions may lead to an increase in heavy metal availability over time. However, the chemistry and evolution of humic substances resulting from organic matter decomposition is not yet clear for a number of reasons.

One reason is that long-term experiments are scarce, so results are not always safe to extrapolate and generalize outside the particular conditions of specific experiments. A second reason may be that separation techniques are costly in time and money, and the protocols are not necessarily identical among the various university laboratories. This leads to production of a variety of results that are difficult to combine and interpret. For example, high molecular weight humic acids bind metals strongly enough to keep them away from available pools, and dissolved low molecular weight substances tend to chelate metals and to inhibit their adsorption from solid surfaces, thus increasing metal availability [6]. The substance that will dominate the organic phase of the soil–sludge mixtures will depend on a variety of reasons and conditions that are not always easy to recognize and predict.

In addition, inorganic phases in soil may also play an important role in the complexation reactions with heavy metals. The solubility and therefore the mobility of heavy metals will depend on adsorption, precipitation, and complexation mechanisms; these will, in turn, be affected by heavy metal concentration, pH, colloidal phase content, etc. Thus, it is of great importance to be able to assess heavy metal availability risks because metals may be taken up by plant roots or leached down the soil profile months or years after sewage sludge has been applied. The possible adverse environmental effects of sludge-borne heavy metals need to be understood and managed.

The objectives of this review are to discuss the chemical behavior and environmental fate of heavy metals after sewage sludge has been applied to soils. In the first section, the role of sludge- and soil-borne organic substances found or evolving in the soil matrix is considered. Also, the inorganic phase equilibria are discussed. The next section discusses the availability and environmental fate of heavy metals. The possible risks of heavy metals being introduced to the human food chain or leached down the soil profile, as well as how this behavior may change over time, are examined. In each section, results from various published data are discussed and assessed in the light of the authors' findings.

3.2 THE COMPOSITION OF SEWAGE SLUDGE

3.2.1 Organic Phase

3.2.1.1 Origination, Composition, and Metal-Chelating Properties

Sewage sludge contains a load of organic and inorganic phases. The organics will eventually be assimilated into the soil via the cycling of nutrients as N, P, and S, while metals will accumulate in the surface soil [7]. The more easily decomposable groups of sludge-borne organic matter (protein, cellulose, and hemicellulose) will be rapidly acted upon by soil microorganisms, once sludge is applied to soil, because they are utilized by them as a source of C and N. The resistant groups (lignin, waxes, and tanin) will remain "as is" for a long time and, along with newly formed substances, will define the humic substances of soil [8].

Sludge-borne organic matter will improve physical and chemical soil conditions. In physical terms, organic matter improves the stability of soil aggregates, thus resulting in better aeration, and increases water retention capacity of the soil. In chemical terms, soil solid phase per unit mass increases significantly, due to the large specific surface of the humic substances. Thus, it contributes to the increase of the adsorption sites, improves the nutrient retention of soil, and reduces the losses by leaching. Further decomposition of humic substances will produce low molecular weight organic substances (LMWOS) or dissolved organic carbon (DOC); these are highly active in the soil environment and bear significant binding abilities for heavy metals. These LMWOS comprise free amino acids, sugars, peptides, aliphatic acids, and their polymers, with molecular weights usually not exceeding 5000 to 10,000 g mol^{-1} [6].

Humic substances have pH-dependent behavior, according to the reaction:

$$R–COO^- + H^+ = R–COOH \qquad (3.1)$$

where R is the carbon chain of organic matter.

The deprotonated form ($R–COO^-$) dominates at neutral to high pH, and the $R–COOH$ form exists at acidic pHs. Spark et al. [9] reported that humic substances are insoluble in the soil solution at acidic pH. As the pH increases, humic phases dissolve in the solution and the deprotonation of the functional groups increases the repulsion forces among the organic molecules. This leads to a more open and linear form at neutral pH, compared to the "spherocolloidal" form at lower pH. On the other hand, water-soluble organic carbon (DOC) and fulvic acids have lower molecular weight, bear more active groups per unit mass than the humic acids, and are thus soluble to the whole range of pH values — from acidic to alkaline. Organic ligands on humic and fulvic acids behave as soft Lewis bases; metals are soft acids and thus they tend to form organometallic complexes. The complexation reaction follows the formula:

$$R–L^{1-} + M^{m+} = R–L–M^{m-1} \qquad (3.2)$$

where L is the active group; M is the metal; and the valencies are represented as superscripts.

The main complexing functional groups are carboxyls, phenols, and alcohols. Bonding can be covalent, where the metal and counter-ion contribute one electron each, or coordinate, where the ligand provides both electrons. Theoretically, any organic molecule containing trivalent N or P or divalent O or S can act as a ligand [10].

Heavy metals added in soil with sewage sludge are greatly influenced by the presence of humic substances, of native soil and sludge borne. Heavy metal availability depends on the binding capacity of humus, especially for elements like Cu and Pb. Addition of humic substances with sewage sludge application increases the cation exchange capacity of the soil, thus enhancing the ability of soil to retain greater quantities of heavy metals. As stated, this ability is pH dependent. At pH 6 to 8, humic acids develop the greatest surface area as they become more linear; subsequently, heavy metal retention reaches a peak [9]. Spark et al [9] also concluded that humic acids have greater affinity for metals than inorganic-hydroxide forms and thus control metal solubility at this range of pH.

On the other hand, fulvic acids tend to behave in a different way than humic acids do because they can be taken up by plants directly, along with the metals that are bound onto them [11]. Dissolved organic carbon (DOC) is often reported to reduce metal adsorption soil solid constituents by competing more effectively for the free metal ion and forming organometallic complexes or by being preferentially adsorbed onto the solid phases [12]. This behavior is more effective at near

TABLE 3.1
Relative Affinity for Metals of Humic and Fulvic Acids

Material	Relative affinity	Ref.
FA (pH 5)	Cu > Pb > Zn	14
HA (pH 4.5)	Pb > Cu > Zn > Ni > Cd = Cr	15
HA (pH 6.5)	Pb > Cu = Zn > Cd > Ni > Cr	15
HA (pH 4–7)	Zn > Cu > Pb	14
HA (pH 4–8)	Cu > Pb >> Cd > Zn	14
HA (pH 3.5)	Cu > Ni > Co > Pb > Zn	16
HA (pH 5)	Ni > Co > Cu > Cu > Pb > Zn	16
HA (pH 4–7)	Hg = Pb = Cr > Zn >Ni = Zn	17
HA (pH 5.8)	Hg = Pb = Cr = Cu > Cd > Zn > Ni	17

neutral pH. This is probably because hydroxide, carbonate, and phosphate phases control heavy metal solubility at higher pHs, and, at acidic pHs, free metal ions are predominant. Metal–ligand complex stability generally decreases with a reduction in pH, reflecting the role of R–COO⁻ in metal complexation [13].

However, not only do the organic phases bear different binding abilities for metals, but also different metals have different affinities for humic substances. Metals like Cd, Ni, and Zn are generally reported to be more mobile in the soil environment than Pb, Cu, and Co. Table 3.1 shows the relative affinities of different metals onto humic and fulvic acids at various pH values, as reported by various researchers [14–17].

In addition to this, various researchers have differentiated humic behavior in accordance with the hydrophobic (Ho) or hydrophilic (Hi) nature of humic substances. According to Han and Thompson [18], in the DOC of molecular weight (MW) < 3500 g mol⁻¹, 56% is Hi and 44% is Ho; the total concentration is 860 mg DOC kg⁻¹ organic matter. A MW > 14,000 g mol⁻¹ 83% is Hi and only 17% is Ho (with a total concentration of 580 mg DOC kg⁻¹ organic matter).

Explaining the binding of metals by organic matter has two main approaches. The first one is the discrete ligand model (DLM) and the second is the continuous distribution model (CDM). The DLM approach suggests that only a few ligands on the humic substance are required to fit experimental data. Given the complexity of humic phases, these ligands (or ligand classes) must be less than the actual total number presented on these molecules. For an individual adsorption site, v_i, the following formula is valid:

$$v_i = \frac{K_i[M]}{1 + K_i[M]} \tag{3.3}$$

where K_i is the stability constant of the binding of metal M onto the given ligand class v_i and in brackets are the metal concentrations.

Although the DLM represents metal binding quite successfully, the CDM approach allows a large number of sites to be involved in binding metals, and thus it represents real conditions more closely. In this model, no discrete Ki defined, but rather a continuum of K values. This model assumes an irregular frequency distribution of the functional groups. The formula describing the macroscopic free ligand concentration, L_T^f, is:

$$[L_T^f] = \int_0^\infty \frac{L(K)dK}{1 + K[M]} \tag{3.4}$$

which is the analogue of Equation 3.3. Kaschl et al. [19] suggested that the CDM was rather valid, in their work with the binding abilities of organic fractions for Cd. They differentiated further the Hi and Ho fractions to acid (A), neutral (N), and basic (B) and calculated the relative and absolute complexing capacities of these hydrophilic and hydrophobic fractions, as given in Table 3.2.

3.2.1.2 Sludge-Borne and Soil-Organic Matter Properties after Sewage Sludge Is Applied to Soil

When sewage sludge is added to soil, its organic matter decomposes rapidly. However, not only the sludge-borne organic matter, but also soil organic matter exhibits an increased rate of decomposition, even if soil organic matter decomposition had reached equilibrium before sludge was

TABLE 3.2
Binding Capacities and pK Values of Hydrophobic (Ho) and Hydrophilic (Hi) Acids (A), Neutrals (N) and Black (B) for Cd in Dissolved Organic Carbon (DOC) Extracted from Municipal Solid Waste Compost

	% of DOC	Complexing capacity (c.c.) in μmol Cd g^{-1} DOC	% of total c.c. of DOC	pK of highest c.c. groups	pK of most abundant groups
HoA	22	1100	38	7.74	5.51
HoN	17	400	14	7.69	6.21
HoB	1 (total Ho = 40%)	—	— (total Ho = 40%)	—	—
HiA	26	800	26	7.02	5.43
HiN	21	600	21	6.93	5.35
HiB	13 (total Ho = 60%)	400	13 (total Ho = 60%)	8.11	5.73

Notes:

HoA = Hydrophobic acids, representing the "young" fulvic acids (FA), the most soluble part of DOC, comprising polyphenols, and humic-associated carbohydrates.

HoN = Hydrophobic neutrals, apolar, closely related to humic acids (HA), the least soluble DOC, comprising fats, waxes, oils, resins, amides, phosphate esters, chlorinated hydrocarbons, high molecular weight alcohols, esteres, ketones, and aldehydes.

HoB = Hydrophobic bases, the least significant and least extractable fraction in DOC, comprising complex polynuclear amides, nucleic acids, quinones, porphyrins, aromatic amines, and ethers.

HiA = Hydrophilic acids, comprising small, highly oxidized organic compounds, polyxydroxy phenols, and uronic acids high in inorganic salts.

HiN = Hydrophylic neutrals, comprising oligo- and polysaccharides, carbohydrates, polyfunctional alcohols, or phosphate salts.

HiB = Hydrophilic bases, comprising proteins, peptides and amino–sugar polymers, low molecular weight amines and pyridine.

Sources:

Data rearranged from Kaschl, A. et al., *J. Environ. Qual.*, 31, 1885–1892, 2002.

Information on hydrophilic and hydrophobic compounds obtained from Kaschl, A. et al., *J. Environ. Qual.*, 31, 1885–1892, 2002; Smith, S.R., *Agricultural Recycling of Sewage Sludge and the Environment*, CAB International, Wallingford, 1996; Han, N. and Thompson, M.L., *J. Environ. Qual.*, 28, 652–658, 1999; and Chefetz, B. et al. *Soil Sci. Soc. Am. J.*, 62, 326–332, 1998.

added [16]. Although decomposition is a complex phenomenon, it is usually described by a first-order kinetics equation, as follows:

$$C(t) = C_0 e^{-k(t-t_0)} \qquad (3.5)$$

where $C(t)$ is the residual organic carbon remaining in the soil after time t; C_0 is the initial organic carbon content at time t_0; and k is the decomposition rate. The k factor rarely stays constant, although it is considered so in the preceding equation. Decomposition depends on a variety of factors, which fall into four main categories [3]:

- Sludge characteristics. Decomposition depends on the source, the stabilization processes, and the composition. Decomposition is generally higher when C/N ratio is low (especially below 10) and when the sludge is dry rather than liquid.
- Soil parameters. Soil properties such as texture and pH play a key role. Decomposition is generally higher in a sandy and slower in a clayey soil. pH values need to favor microbe activity; thus, acidic values will mean less activity and subsequently poor organic matter decomposition.
- Environmental conditions. Ambient temperature and moisture will dramatically affect decomposition rate. The higher the temperature is and the closer the moisture is to the field capacity, the faster the decomposition is.
- Application conditions. Decomposition depends on application rate and method. For the application method, it is agreed that mineralization is faster when sewage sludge is surface applied than when it is incorporated into the soil. For the application rate, the greater the quantity of sludge is per hectare, the faster the decomposition is [21]. However, there is a disagreement on that [7,22].

Although the preceding are related to the individual decomposition rates of a given sewage sludge in a given soil under certain ambient environmental conditions, it is evident that k reduces with time (Figure 3.1). It can be observed that k, irrespective of conditions, will be higher when sewage sludge is first applied to soil and will significantly reduce as time proceeds. However, as stated before, different groups of sludge-borne organic matter will behave differently. Hydrophobic

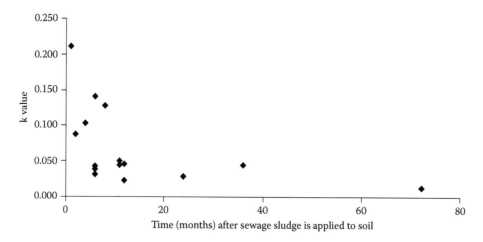

FIGURE 3.1 Mean k values (rate of organic matter decomposition) after sewage sludge is applied to soils. These k values represent a variety of experimental data. (Data modified from Metzger, L. and Yaron, B., *Adv. Soil Sci.*, 7, 141–162, 1987; and Antoniadis, V. and Alloway, B.J., *Water Air Soil Pollut.*, 132, 201–214, 2001.)

compounds increase remarkably in soil organic carbon after sewage sludge application and, as a result, they induce the formation of hydrophobic aggregates. Thus, Hi/Ho ratio will decrease as sewage sludge application rate increases [8].

Nevertheless, as sewage sludge application increases, DOC also increases in absolute as well as relative terms compared to the total organic carbon added to soil. Experimental data seem to agree on the fact that humic and fulvic acids in soil–sludge mixtures evolve, increasing their concentrations months after sewage sludge application. Heavy metals are preferentially associated with hydrophilic phases, so one could state that metal availability will tend to decrease with time after the termination of sludge application. However, this trend may be different if DOC increases more than the humic acids, and some evidence suggests that this may take place. Metzger and Yaron [3] reported that the ratio of humic over fulvic acids was reduced following sludge application, and Han and Thompson [18] found that, in the surface soil (0 to 5 cm), Hi/Ho ratio increased. These data point to the risk of heavy metals exhibiting enhanced availability even months after the termination of sewage sludge application.

3.2.2 Inorganic Phase

3.2.2.1 Fertilizer Value of Sewage Sludge

Sewage sludge is increasingly used as a fertilizer because land application is considered a more environmentally friendly disposal method and because of its great fertilizer value (Table 3.3). Nowadays, sewage sludge can be produced for agricultural use as dry granular, palletized, or fortified organic-based materials, which have more specialized uses in agriculture and horticulture compared with the conventional forms [23]. The technology for applying sludge is advanced and includes surface spreading and injection of the material into the soil, a practice that helps reduce odor problems and facilitates proper incorporation into the soil. The optimum dose of application is difficult to determine because of restrictions that depend on soil parameters, such as pH, clay content, and contaminant and nutrient content in sludge [24]. Usually, if no other restricting factors are present, the rate of application should not exceed 10 t ha^{-1} yr^{-1}, equivalent to 250 kg N ha^{-1} [25].

3.2.2.2 Heavy Metal Loadings in Sewage Sludge and in Soils where Sludge Is Added

Heavy metals occur in sewage sludge through domestic, run-off, and industrial inputs, with the industrial input contributing only a small percentage of the total in specific elements, such as Pb and Zn [1]. Nowadays, heavy metal loads in sewage sludge have been significantly reduced as a result of improved effluent control and the use of cleaner technologies from industries due to tighter state legislations. As a result, the metal loadings in sewage sludge and top soils are being reduced (e.g., reports by Chaudri et al. [26] for England and Wales concerning Cd; Barbarick et al. [27], reporting a period from 1982 to 1992; Sloan et al. [28]; and Berti and Jacobs [29]).

TABLE 3.3
Typical Concentrations of Nutrients and Organic Carbon in Sewage Sludges

	Organic carbon (%)	Total N (%)	NH$_4$–N (mg kg^{-1})	NO$_3$–N (mg kg^{-1})	Total P (%)	Total S (%)
Range	16–40	0.5–17.6	800–5600	79–160	0.5–14.3	0.6–1.5

Sources: Data obtained from O'Riordan, E.G. et al., *Irish J. Agri. Res.*, 25, 223–229; Gerba, C.P., in *Pollution Science*, Pepper, L., Gerba, C.P., and Brusseau, M.L., Eds., Academic Press, San Diego, 1996, 301–319; and Smith, S.R., *Agricultural Recycling of Sewage Sludge and the Environment*, CAB International, Wallingford, 1996.

Nevertheless, it is inevitable that sewage sludge application to soil will increase heavy metal concentrations well beyond their background concentrations. This is the reason that governments deal with this prospect, setting upper limits of heavy metals in soils and sludges to be added to soils. Heavy metal concentration limits in sewage sludge for the U.S. and the E.U. are shown in Table 3.4. Background metal concentrations in soils and maximum permissible loading rates are shown in Table 3.5. It is evident that metal-loading standards differ greatly between the U.S. and the E.U., and this stresses that sludge-induced metal contamination is still quite controversial and a matter not yet fully resolved [20].

TABLE 3.4
Concentrations of Heavy Metals in Sewage Sludges in the U.S. and the E.U.

	Permissible limits in the U.S. (mg kg⁻¹)[a]	Cumulative pollutant loading rate limits in the U.S. (kg ha⁻¹)[a]	Permissible limits in the E.U. (mg kg⁻¹)[b]	Typical metal concentrations in sludges applied in agricultural land in 1991 (50 percentile)[c]
As	75	41	—	3.2
Cd	85	39	20–40	3.2
Cu	4300	1500	1000–1750	473
Pb	840	300	750–1200	217
Hg	57	17	16–25	3.2
Mo	75	—	—	1.0
Ni	420	420	300–400	37
Se	100	100	—	0.28
Zn	7500	2800	2500–4000	889

[a] Harrison, E.Z. et al., *Int. J. Environ. Pollut.*, 11, 1–36, 1999.
[b] Alloway, B.J., in *Heavy Metals in Soils*, Alloway, B.J., Ed., Blackie Academic and Professional, London, 1995, 38–57.
[c] Smith, S.R., *Agricultural Recycling of Sewage Sludge and the Environment*, CAB International, Wallingford, 1996.

TABLE 3.5
Typical and Upper Permissible Heavy Metal Concentrations in Soils

	Heavy metal concentrations in soils of England and Wales (National Soil Inventory)[a]	Heavy metal concentrations in agricultural soils in the U.K. in 1991[b]	U.K. regulation limits[b]	Annual pollutant loading rate (kg ha⁻¹ yr⁻¹)[c]
As	—	—	—	2
Cd	0.7	0.55	3	1.9
Cu	18.1	17	80	75
Pb	40.0	33	300	15
Hg	0.1	0.13	1	0.85
Mo	—	—	—	—
Ni	22.6	14	50	21
Se	—	—	—	5
Zn	82.0	57	200	140

[a] McGrath, S.P. and Loveland, P.J., *The Soil Geochemical Atlas of England and Wales*, Blackie Academic and Professional, London, 1992.
[b] Alloway, B.J., in *Heavy Metals in Soils*, Alloway, B.J., Ed., Blackie Academic and Professional, London, 1995, 38–57.
[c] Harrison, E.Z. et al., *Int. J. Environ. Pollut.*, 11, 1–36, 1999.

3.2.2.3 Heavy Metal Chemistry in Soils and Heavy Metal Properties after Termination of Sewage Sludge Application

After sewage sludge-borne heavy metals are introduced into the soil, a series of mechanisms tend to bind metals to the organic (soil and sludge-borne) and the inorganic constituents. A percentage of the metals will be reversibly adsorbed onto clay surfaces; some will precipitate out of the solution with other inorganic phases (carbonates, phosphates, etc.), especially at alkaline pH values. Some will be specifically adsorbed onto solid surfaces, and the organic colloids (fulvic and humic acids, as discussed earlier) compete for metal binding effectively. A small portion of the added metals will be soluble into the soil solution in pure ionic form and thus readily available for plant uptake or easily leached out of the root zone and probably into the groundwater.

These processes are usually pH dependent, with the metals bound more strongly onto the inorganic phases with an increase in pH value and heavy metals becoming more available in acidic pHs. In his early work, Harmsen [30] suggested two mechanisms to explain this metal behavior:

- Selective adsorption of metals on sites previously occupied by H^+ (M/H exchange, M = metal)
- Selective adsorption of hydrolysis products of metals onto sites previously occupied by H^+ (M(OH)/H exchange)

In a more recent review, Naidu et al. [31] agreed with the significance of the preceding mechanisms and they added that this process is related to the increase of the adsorption density of clay mineral margins. Evans [17] further explained that pH controls the two main retention mechanisms, i.e., adsorption and precipitation, because metal solubility depends on the solubility product of the solid phase (precipitate) containing the metal. The most important precipitation reaction, e.g., hydrolysis, can be described by the following reaction:

$$M(OH)_m + mH^+ = M^{m+} + mH_2O \tag{3.6}$$

The reaction constant (K) is then given by the equation:

$$K = \frac{(M^{m+})}{(H^+)^m} \rightarrow pM = mpH + pK \tag{3.7}$$

where M is the metal and m+ the positive valency that it bears.

The preceding equations indicate that, as pH increases, pM also increases and, subsequently, metal concentration decreases as a result of precipitation. This means that metal solubility depends on the solubility product of metal hydroxides. Specific adsorption of metal hydroxides increases with decreasing pK values. For elements with equal pK, the larger ionic radius plays a decisive role. The order of specific adsorption for the most common heavy metals given by Alloway [25] is as follows:

Cd (pK 10.1) < Ni (pK 9.9) < Co (pK 9.7) < Zn (pK 9.0) << Cu (pK 7.7)
< Pb (pK 7.7) (Cu has larger ionic radius than Pb) < Hg (pK 3.4)

At neutral pHs, the cation exchange (or retention) capacity of soils is also higher than that of acidic pHs, especially in soils with high content in variable charge constituents (such as Al and Fe hydroxides, 1:1 clays and clay edges, and humic substances); this explains the stronger binding capacity of soil for heavy metals when pH is high. In addition, metal diffusion into the inorganic lattice of phases such as Mn hydroxides and some 2:1 clays (illites) may also be possible. In that case, metals become almost irreversibly unavailable, with smaller ionic radius metals having greater affinity for such a process.

Many researchers have used the "sequential extraction" method to determine the relative affinity of heavy metals for the various organic and inorganic soil constituents (e.g., Mbila et al. [32]; Berti and Jacobs [33]; Sposito et al. [34]; Tsadilas et al. [35]; and Yong et al. [36]). It is agreed that organically bound forms are important in complexing metals and diminishing their environmental availability (mean values of the preceding works indicate contribution of metal retention by organic phases as follows: Cd 26%; Cu 55%; Pb 22%; Ni 17%; and Zn 24%). However, as explained earlier, organic phases are prone to intense mineralization and rapid alteration once sewage sludge has been introduced into the soil. With the humus decomposition, one significant part that contributes to metal retention approximately equal to the value mentioned earlier is ruled out. This means that, over time, the chemistry of metal retention will be altered. Whether soil will bear the same capacity to control heavy metal availability over time is a matter of great dispute within the scientific community. This will be discussed next.

3.3 AVAILABILITY OF HEAVY METALS AND THEIR FATE OVER TIME

3.3.1 Factors Affecting Heavy Metal Availability

In the complex soil environment in which heavy metals are introduced when sewage sludge is applied to agricultural land, their bioavailability to crop plants depends on various factors. These are well documented in papers reporting original experimental data and in review papers. A synopsis of these factors will be now discussed.

3.3.1.1 pH Value

Soil pH is probably the most widely recognized factor affecting heavy metal availability. With the exception of molybdenum, arsenic, and selenium, heavy metal retention increases with a pH increase and thus availability decreases. This is recorded in experiments measuring plant uptake or quantifying metal adsorption onto solid phases and metal extractability. In the first type of experiment, pH values of acidic soils increase with the effect of lime-stabilized sludge and are found to have depressed metal uptake by plants (e.g., Basta and Sloan [37]), or lime is applied to natural and metal-contaminated soils to improve soil pH and similar uptake effects are found (e.g., Oliver et al. [38]). In the second type of experiment, several solid phases, pure (oxides, clays, etc.) or natural soils, are used to evaluate the adsorption strength of heavy metals in a variety of pH values (e.g., Salam and Helmke [39]; Elliott et al. [14]; Yong and Phadungchewit [40]; Gray et al. [41]) (see also Figure 3.2).

These trends do not affect different heavy metals in a similar way. It is found that "mobile" elements such as Ni, Cd, and Zn are much more sensitive in changing their availability status with pH than other elements, such as Pb and Cu, which are more strongly retained by humus [1]. However, some data suggest that the positive pH effect on metal availability is not permanent, especially in high buffer capacity soils, because soils tend to return to initial pH over time. Ca^{2+} ions introduced to soil through liming may even compete with other metals for the limited adsorption sites, causing adverse effects [25]. Also, at certain periods of time, metal availability may have a dramatic peak, due to a temporary pH depression, even if pH values before and after that period are well controlled. Such periods may occur during early summer when the intense increase in soil microbial biomass may cause a rapid sludge-borne organic matter decomposition, which, in turn, tends to reduce soil pH due to the release of weak organic acids.

3.3.1.2 Organic Matter

This is a factor almost as important as pH. However, the role of different organic phases has been already discussed and will be stressed in subsequent chapters as well.

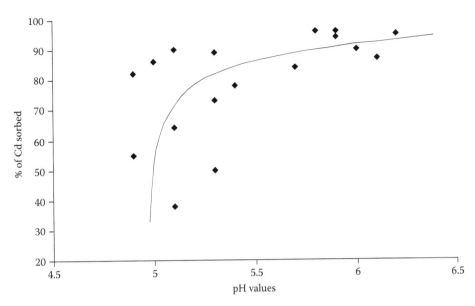

FIGURE 3.2 Effect of soil pH on percentage of Cd sorbed onto soils. Data concern five New Zealand soils that received 2 mg Cd kg^{-1} as $Cd(NO_3)_2$. (Data obtained from Gray, C.W. et al., *Austr. J. Soil Res.*, 36, 199–216, 1998.)

3.3.1.3 Redox Potential

Redox effects are particularly important in soils of aquic moisture regime (low lying soils in relief depressions, poorly drained, with water table normally into the root zone) and paddy soils intentionally flooded for cultivation reasons (rise fields). Generally, low redox values (reducing conditions) result in low metal availability, due to the formation of HS$^-$ and the subsequent precipitation of sulfide–metal solids out of the solution [42]. However, in soils in which Fe and Mn hydrous oxides are predominant, this trend may be altered. Prolonged reducing conditions may cause total depletion of hydrous oxides. Even if oxic conditions prevail again, the contribution of these phases to the sorptive capacity of the soils is not recovered. It is found that, when cultivated, such soils (developing prolonged anoxic conditions) exhibit an impressive increase in heavy metal uptake by test plants compared to other soils with no history in waterlogged regimes [43].

3.3.1.4 Competition Effects among Metals

The competition or synergism between heavy metals and/or other ions present in the soil system has a very significant effect on metal mobility in the solution and in the root zone. An overview of antagonistic and synergistic behavior of the elements is given in Figure 3.3. A lot of research effort has been put into the investigation of the interactions between Cd and Zn and the findings appear to be contradictory. Works supporting synergism suggest a mechanism by which the two metals compete for the same adsorption sites, and thus the soil retention capacity of the soil is reduced for any given metal when the other increases in concentration in the solution. Thus, when both elements are present in the soil solution, they exhibit enhanced bioavailability and reduced retention [44].

On the other hand, when plant uptake is concerned, antagonism is the usual effect because it is found that plant roots favor the uptake of the metal found in higher relevant concentrations over the other (e.g., McLaughlin et al. [45]). A well recognized competition in plant uptake occurs between micronutrients (such as Ca, P, and S) and heavy metals. This is due to favorable plant selectivity for the nutrients over heavy metals and to precipitation of insoluble $Cd(PO_4)_2$, especially

Adjacent to roots

	Cu	Zn	Cd	Pb	As	Se	Cr	Mo	Mn	Fe	Co	Ni
Cu	■		γ					α	α			
Zn	α	■	β						α	α		
Cd		α	■									
Pb		δ	β	■								
As		δ			■				α			
Se	α	α	α			■						
Cr	δ						■		α	α		
Mo	α							■	α	α		
Mn	γ	α	γ		α	α	α	γ	■			
Fe	α	α	γ				α	α	α	■	α	
Co									α	α	■	
Ni	γ	γ	γ						δ			■

Plants

FIGURE 3.3 Interactions of heavy metals within plants and at the root surface. 'α: antagonism, β :synergism, γ: antagonism and/or synergism, δ: possible antagonism.' (Redrawn from Kabata–Pendias, A. and Pendias, H., *Trace Elements in Soils and Plants*, 2nd ed., CRC Press, Boca Raton, FL, 1992, 1–87; 131–141. With permission.)

at neutral to alkaline pH, that depletes Cd of the soil solution [46]. Also, Pb mimics the physiological behavior of Ca, and thus it may inhibit several enzymes in plants when Ca is not readily available and the plant takes up Pb rather than Ca [47].

3.3.1.5 The Effect of Carbonates

At high pH values, calcite ($CaCO_3$) sorbs Cd as $CdCO_3$ and reduces its availability. The same effect is evident for other metals too. This provides a further explanation as to why liming of acidic soils is effective for reducing metal availability to crop plants. Above pH 7.3, the order of carbonate solids precipitating is [17]:

$$Mg > Ba > Ca > Sr > Zn > Fe > Cd$$

Papadopoulos and Powell [48] suggested for this sequence that the affinity of calcite for metals is related to their ionic radii. This is more evident for Cd^{2+}, which, at high pH, substitutes for Ca^{2+} by chemisorption because the Cd radius fits better for this process.

3.3.1.6 Chloride Ions

A factor especially important in arid soils is the effect of chloride ions, which are found to enhance heavy metal availability, particularly that of Cd [49]. Chloride complexes with Cd ($CdCl^-$, $CdCl_2^0$) may be very stable and inhibit Cd from being adsorbed onto soil solids [50]. Chloride ions can be found in soils because of environmental reasons (soil salinity due to high evapotranspiration), poor irrigation practices, or sewage sludge application to soils prone to salinity.

3.3.2 HEAVY METAL ACCUMULATION IN CROP PLANTS

All the preceding factors governing heavy metal availability may affect plant uptake of heavy metals. However, it should be stressed that other parameters related to plant physiology (such as plant genotype) are not studied here, but are of great importance as well. One of the greatest

TABLE 3.6
Soil–Plant Transfer Coefficients (TC) of Heavy Metals

Element	TC estimations[a]	Pot experiments[b]	Field experiments[c]
Cd	1–10	1	0.94
Cr	0.01–0.1	—	0.0005
Cu	0.1–10	0.25	0.21
Ni	0.1–1	1.29	0.06
Pb	0.01–0.1	0.09	0.02
Zn	1–10	0.88	1.05

[a] Obtained from Alloway, B.J., in *Heavy Metals in Soils*, Alloway, B.J., Ed., Blackie Academic and Professional, London, 1995, 38–57.

[b] Data from Antoniadis, V. and Alloway, B.J., *Water Air Soil Pollut.*, 132, 201–214, 2001; Antoniadis, V. and Alloway, B.J., *Environ. Pollut.*, 117, 515–521, 2002; Tsadilas, C.D. et al. *Commun. Soil Sci. Plant Anal.*, 26, 2603–2619, 1995; Hooda, P.S. and Alloway, B.J., *J. Soil Sci.*, 44, 97–110, 1993; and Jackson, A.P. and Alloway, B.J., *Plant Soil*, 132, 179–186, 1991.

[c] Data from Antoniadis, V. et al., in *Proc. 8th Conf. Hellenic Soil Sci. Soc.*, Kavala, 2000, 459–469; Berti, W.R. and Jacobs, L.W., *J. Environ. Qual.*, 25, 1025–1032, 1996; Bergkvist, P. et al., *Agri. Ecosyst. Environ.*, 97, 167–179, 2003; Brown, S.L. et al., *J. Environ. Qual.*, 127, 1071–1078, 1998; Chang et al., *J. Environ. Qual.*, 12, 391–397, 1983; and Schaecke, W., Tanneberg, H., and Schilling, G., *J. Plant Nutr. Soil Sci.*, 165, 609–617, 2002.

concerns is that after sewage sludge has been applied to soils, heavy metals borne in it may accumulate in plants and, subsequently, enter the human food chain or have toxic effects on plants and/or animals grazing on them. Heavy metal concentrations in plants depend on the concentrations (total and available) of the metals in the soil in which the plants are grown. The transfer coefficient (TC) between soil and plant gives a measure of metal mobility. Table 3.6 gives the TC of the main heavy metals as obtained by Alloway [25] and values of the same coefficient from several other works.

Although TCs are not supposed to be precise in predicting metal uptake by plants (because only orders of magnitude are indicated), it is evident that Cd, Ni, and Zn are among the most mobile elements and that the ones strongly sorbed onto humic substances (including Cr and Pb) are less available. From Table 3.6, it is evident that heavy metals exhibit behavior inside a green-house in pot studies different from that in field experiments; the latter usually gives smaller TC values. An explanation of this trend was given by De Vries and Tiller [51], who attributed the higher metal concentrations in pot experiments to the more favorable conditions (temperature, humidity, and light distribution) usually found inside a greenhouse.

Although McBride [5] agreed with that, he added that differences in plant rooting patterns are more critical. Most of the crops' rooting systems in the field go well beyond the sludge incorporation zone and thus heavy metal uptake is more conservative. In pot experiments, plant roots are "forced" to remain where sludge is thoroughly incorporated with the soil. Some researchers, however, have found no relationship between metal concentration in soil and in the plant. Kuo [52] and Del Castilho and Chardon [53] found no apparent relationship between DTPA-extractable Cd (probably the most widely used availability index) and plant Cd in five soils with a pH of 5 to 6. The same was also found by Barbarick and Workman [54] and O'Connor [55].

Although Cd and Pb are nonessential elements with no biological function within the plant, Ni and Zn are essential micronutrients for plant growth. However, Ni and Zn can cause toxicities when they are present in plants in excess. Zinc can be relatively easily translocated from roots to shoots in a plant and, in high concentrations, tends to accumulate in mature leaves although heavy metals accumulate mostly in roots. This is also shown in Figure 3.4. Other heavy metals, however,

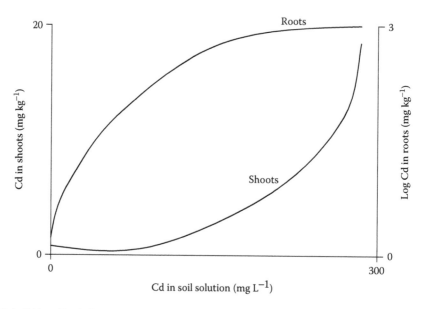

FIGURE 3.4 Cd in soil solution, roots, and shoots. (Redrawn from Kabata–Pendias, A. and Pendias, H., *Trace Elements in Soils and Plants*, 2nd ed., CRC Press, Boca Raton, FL, 1992, 1–87; 131–141. With permission.)

stay in the roots and move slowly upwards [56]. Kabata–Pendias and Pendias have summarized the several ways by which plants can tolerate high heavy metal concentrations [57]:

- Selective uptake of ions
- Decreased permeability through cell membranes
- Immobilization of ions in roots, foliage, or seeds
- Removal of ions from metabolism by deposition in insoluble forms
- Alteration in metabolic patterns: increased enzyme system that is inhibited, or increased antagonistic metabolite, or reduced metabolic pathway by-passing an inhibited site
- Adaptation to toxic metal replacement of a physiological metal in an enzyme
- Release of ions to plants by leaching from foliage and excretion from roots

However, it is evident that different plant species and different cultivars in the same species have different capabilities in tolerating heavy metals [58]. That is, different plants, grown in the same soil contaminated with heavy metals, may accumulate variable quantities of these metals. The response of some plant species to heavy metal exposure is shown in Table 3.7.

3.3.3 CONCERNS OF HEAVY METAL LEACHING OUT OF SOIL AND INTO GROUNDWATER

The movement of heavy metals down the soil profile is of a great importance because it involves the risk of groundwater contamination and deterioration of drinking water quality. The literature provides some extreme examples of cases of massive heavy metal movement up to 3 m depth [59]. After heavy metals have been introduced with sewage sludge application to land, downward movement will be a possibility when some, or any combination, of the following is evident:

- High sludge application rates
- Enhanced heavy metal load in sludge

TABLE 3.7
Relative Accumulation of Heavy Metals by Different Crops

High plant uptake	Low plant uptake
Lettuce	Potato
Spinach	Maize
Celery	Peas
Kale	Leek
Ryegrass	Onion
Sugar beet	Tomato
Turnip	Berry fruits

Sources: From Alloway, B.J., in *Heavy Metals in Soils*, Alloway, B.J., Ed., Blackie Academic and Professional, London, 1995, 38–57; and Kloke, A., Saurbeck, D.R., and Vetter, H., in *Changing Metal Cycles and Human Health*, Nriagu, J., Ed., Springer–Verlag, Berlin, 1994.

- Soils with low sorptive capacities (low organic matter and clay contents, acidic pH conditions)
- High rainfall or irrigation water rates

Many research efforts have been put into performing laboratory experiments with soil columns or monitoring metal behavior in soil profiles in areas with history in sludge application. As Camobreco et al. [60] pointed out, measuring metal movement in repacked soil columns is different from measuring it in the field under "real" conditions. The reason is that metal movement more often occurs through soil macropores or cracks. This leads to the preferential flow of the water and, subsequently, to the metals that it carries; this fragile soil structure is destroyed when soil is sampled and taken to the laboratory to be set up for a column experiment [61]. Thus, metal mobility in the profile is very often severely underestimated, and results seem very promising when conclusions rely only on laboratory trials. However, column experiments are useful because they provide a picture based on carefully controlled parameters.

Although it is often assumed that downward movement of heavy metals is not a significant process, there is a disagreement in the literature on that matter. For example, Davis et al. [62] applied 40 t ha^{-1} sludge and found heavy metal movement only down to 10 cm depth, after a residual time of 3 years. El-Hassanin et al. [63] found that Cd, Zn, and Pb only moved in the surface soil layer; similarly, Higgins [64] found that no heavy metal movement to the B horizon was evident. Chang et al. [65] reported that, 4 years after the termination of sewage sludge application, heavy metals were deposited only in the 0- to 15-cm upper soil layer, and Harmsen [30] found that Cd and Pb did not accumulate below 40 cm of the surface in a heavily contaminated Zn smelter area. Dowdy et al. [66] also reported that heavy metals did not move significantly below the incorporation zone (or plow layer). The same was concluded by Williams et al. [67], Emmerich et al. [68], and Miner et al. [69].

In contrast to this, data have also been published suggesting that metals can move below the incorporation zone, under certain conditions; such works many times concern long periods of sludge application. Works of this kind are very important and should be taken seriously under consideration before sludge utilization to agricultural soils is to be generalized. Antoniadis and Alloway [70] studied an area in central England where sewage sludge had been applied in a farm for decades at normal rates. They found that Ni was significantly higher than the control even at a depth of 80 cm; Zn had been moved to 60 cm and Cd to 40 cm. Lead was the only metal found higher than

TABLE 3.8

Heavy Metal Distribution in the Profile of a Soil Receiving Sewage Sludge for Many Decades at "Normal" Rates[a] and in a Nonsludged Control Soil

Soil depth (cm)	Cd		Ni		Pb		Zn	
	Farm	Control	Farm	Control	Farm	Control	Farm	Control
0–20	0.95	0.85	20.3	13.8	70.0	30.3	120.3	80.4
20–40	0.82	0.78	17.2	10.6	32.3	30.6	107.1	72.7
40–60	0.63	0.63	16.5	5.6	28.5	29.7	85.2	68.2
60–80	0.51	0.55	14.0	4.7	27.4	27.9	75.5	65.0
$LSD_{0.05}$	0.13		4.3		9.5		20.5	

[a] c. 3–5 t ha^{-1} at site S2.

Source: Data obtained from Antoniadis, V. and Alloway, B.J., *Commun. Soil Sci. Plant Anal.*, 34, 1225–1231, 2003.

the control only in the topsoil (see also Table 3.8). At a site where raw sewage wastewater had been applied for 5 decades, Schirado et al. [71] found a rather uniform distribution of Cd, Zn, and Ni in the soil profile up to a depth of 150 cm; this was attributed to the movement of the metals. Barbarick et al. [27] applied sewage sludge for 11 years at a rate of 27 t ha^{-1} yr^{-1}. Nine years after the commencement of their experiment, Zn had moved down to 125 cm depth, although other metals (Pb, Cd, and Ni) did not move significantly. Sloan et al. [72] found significant movement of all metals studied (Cd, Cr, Cu, Ni, Pb, and Zn), down to a depth of 30 cm, after 16 years of continuous sludge application.

The literature also offers examples of data exhibiting that it is not necessary to have many favorable factors for metal mobilization in order to measure significant movement. Using repacked columns of a neutral pH soil and single sludge application rates of 50 t ha^{-1} of a sludge relatively low in heavy metals (Cd 2.41; Ni 38.08; Pb 317; and Zn 645 mg kg^{-1}), Antoniadis and Alloway [73] found that Cd, Ni, and Zn moved to 8 cm depth below the incorporation zone. Also, some work that has been performed on sources of heavy metals different from sewage sludge reports significant metal movement through the soil profile. Li and Shuman [74] contaminated coarse soils with low sorptive capacities with flue dust and found movement of Zn down to a depth of 75 cm and of Cd and Pb to a depth of 30 cm. Similarly, Stevenson and Welch [75] found that Pb moved to a depth of 75 cm after inorganic Pb salt applications of 3.2 t ha^{-1} to soil.

In recent years, much work has concentrated on the effect of low molecular weight humic substances, often identified as dissolved organic carbon (DOC), in metal movement. As mentioned previously, DOC may preferentially sorb heavy metal, reducing metal adsorption onto solid phases enhancing and thus metal availability and mobility. Because DOC elution may be associated with the decomposition of sewage sludge-borne organic matter, several workers have hypothesized that DOC may facilitate heavy metal transport down the soil profile [76,77].

Sawhney et al. [78] mixed soil and compost containing heavy metals at rates of 0, 15, 50, and 100% compost, and they measured the leachate after applying 20 mm of water per day for 70 days. They found that all metals (Cd, Pb, Cr, Cu, and Zn) had moved into the leachate in the first half of their experiment. However, in the second half, the soluble pools (presumably these included DOC elution) were "exhausted." Bolton and Evans [79] measured sludge landfill leachates and applied a computer model to calculate the metal speciation; they found that metals were predominantly complexed with Cl$^-$ ions and fulvic acids. Zhu and Alva [80] found a positive correlation between metals (Zn and Cu) and DOC in the leachates eluted in their leaching experiment.

From the preceding, it is evident that heavy metal leaching from sludge-treated soils is not always exhibited and that a clear and positive correlation between metal leaching and actual groundwater contamination has not been recorded. However, the existing evidence discussed should be considered seriously and certain soil parameters should be studied further before sewage sludge application is advised, in order to avoid any unnecessary future risks of groundwater contamination. For example, Antoniadis and Alloway [73] found small but significant concentrations of heavy metals in the leachates eluted from their soil columns, although metals in these columns had moved only slightly. Richards et al. [81] concluded that preferential flow and metal complexation with soluble organic phases may facilitate metal leaching that leaves no traces of metal subsurface readsorption, and the same was recognized by McBride et al. [82].

3.3.4 Approaches to the Time Factor or the Residual Effects of Sludge-Borne Heavy Metals

After sludge has been applied to soils, the heavy metals introduced to the soil environment interact with other soil components for a long time. Thus, it is widely recognized that long-term experiments should be used for assessing heavy metal bioavailability. Alloway and Jackson [83] stated that it is difficult to assess whether heavy metals remain highly available to plants many years after sewage sludge application has ceased. The literature provides many examples of research findings suggesting that metal availability decreased or did not change over time. Chang et al. [84] studied a residual period of 10 years and concluded that metal availability to plants tends to reach a plateau a few years after the last sludge application. Over a residual time of 10 years, Hyun et al. [7] assessed the hypothesis that organic matter decomposition may adversely affect metal availability of Cd; they found that, although the former decreased by 43% over that period of time, Cd availability did not significantly increase.

Similarly, Canet et al. [85] applied sewage sludge in equivalent doses of 400, 800, and 1200 kg N ha^{-1} (approximately 13, 26, and 39 t sludge ha^{-1}) and 7 years later found no significant increase in metal uptake by Swiss chard and lettuce. Krauss and Diez [86] found that test crops, such as ryegrass, barley, maize, and potato, remained unaffected 12 years after the cessation of a period of intense sludge application (40 mg Cd and 808 mg Pb kg^{-1} had been applied to soil). Moreover, Bidwell and Dowdy [87] reported a dramatic decrease in heavy metal availability after the termination of sewage sludge application. Chaney and Ryan [4] found that metal availability decreased after the termination of sewage sludge application and maintained that metal availability is more likely to be kept to a minimum in sludge-amended soils due to the enhanced specific adsorption capacity of soils after the addition of sludge-borne organic matter.

However, McBride [5] criticized this work by saying that soil organic matter will not possess the same ability to sorb heavy metal over time; this capacity is likely to decrease with maturing, probably due to organic matter decomposition. He also stated that research works tend to underestimate metal dynamics with sludge application; metal availability is measured by plant shoots' uptake, but "uptake may be suppressed by changes in the translocating efficiency of the plant" which may change over time. He also gave reasons as to why heavy metal availability should not be expected to increase after the sewage sludge application. First, the microbial biomass may be less effective in decomposing organic matter due to possible toxic effects on it and, second, sludge particles will reach an equilibrium with soil matrix after some time of sludge application. All these may lead to a delayed decomposition of organic matter and reduced heavy metal availability.

In a long-term experiment, Chang et al. [84] discussed two hypotheses likely to be valid after the termination of sewage sludge application. The first is the "plateau effect" or "sludge protection," in which heavy metal accumulation in plants reaches a maximum and will not substantially increase, even if heavy metal loadings in sludge increase. In the second, the "time bomb," heavy metal concentrations of plant tissues remain unchanged or rise slowly soon after the termination of sewage sludge; however, as organic matter is decomposed in the soil, the heavy metal concentrations of the plant tissue rise rapidly. They concluded that data point to the first hypothesis, but they admitted that this conclusion could not be definite and that more data would be needed.

On the same subject, Petruzzelli et al. [88] found that Zn and Cd concentrations in plants grown in sludge-treated soils were higher in concentration 4 years after the last sludge application. Sloan et al. [28] found that Cd was still highly recovered from the soil 15 years after sludge application, and McBride et al. [82] found that Cd and Zn were still more available to plants in sludge-treated plots in their field trial than the counterpart control plots. Moreover, in the field experiment of Heckman et al. [89], up to 112 t ha^{-1} of sludge was applied; after a period of 5 to 9 years, the concentrations of Cd and Zn in soybean were still significantly higher than those in controls. Gardiner et al. [90] applied sewage sludge for 5 years at a rate of 20 t ha^{-1} yr^{-1} and found an impressive sixfold increase in Cd concentration in Swiss chard. After that time, the increase was slower. However, apart from Cd other metals, including Zn, Cu, and Ni, did not increase as much.

Other research works that reported enhanced heavy metal availability to plants in soils treated with sewage sludge include Mulchi et al. [91], who found that, after nearly 10 years of continuous sludge application to soil, concentrations of Zn, Cd, Mn, Ni, and Cu correlated very strongly in tobacco with "total" metal concentrations in the soil (R^2 values of 0.65 to 0.95), and Chang et al. [92], who found that Cd concentrations in plant tissues (radish and Swiss chard) tended to increase yearly with each sludge application. Chaudri et al. [93] measured higher Cd concentrations in wheat grains in the treated plots than in the controls 31 years after a single sludge dose of 125 t ha^{-1}. Moreover, McGrath et al. [94] found that, after 23 years of the cessation of sewage sludge application, Cd and Zn extractability and availability at high sludge rates (which introduced 2158 kg Zn ha^{-1} and 70.2 kg Cd ha^{-1} into the soil) did not equal that of the control. Other experiments recording shorter periods also found that heavy metal availability to plants did not decrease to equal that of the control after the termination of sludge application (Obrador et al. [95] and Antoniadis et al. [96] reported a residual period of 1 year).

An important consideration to bear in mind concerning the residual effect of heavy metals is the matter of temperature. In areas of higher temperature regimes, organic matter decomposition is faster, and the protective role of humic substances over metal availability is less important than in temperate countries. Antoniadis and Alloway [97] tested the behavior of heavy metals in two different temperature regimes (namely 15 and 25°C) in a pot experiment for 2 years. They found that, over time, heavy metal availability to plants and extractability with 0.01 M CaCl$_2$ were higher at 25°C than at 15°C.

White et al. [98] applied sludge at a rate of 45 and 90 t ha^{-1} in a land with warm temperatures and measured the extractability of metals with DTPA over a period of 9 years. They found that DTPA-extractable Cd, Pb, and Zn increased significantly (Cd and Zn had nearly a tenfold increase) until the fourth year; then they subsequently decreased. However, metal concentrations were significantly higher at the end of the experiment than at the beginning. Although they failed to explain the reasons for the peaks in metal concentrations in the fourth year, their work provided evidence that heavy metal extractability may be significantly enhanced, especially in warm environments.

3.4 CONCLUSIONS: IS LAND APPLICATION OF SEWAGE SLUDGE SAFE?

From the preceding discussion, it becomes evident that the answer to the question of whether application of sludge to agricultural lands is a safe practice may not be a simple "yes" or "no." Heavy metal behavior and fate depend on their chemistry in soil inorganic and organic phases; their bioavailability depends on a variety of factors concerning the ambient environment, soil, and sludge. For some, the fact that the research evidence of the last three decades — even in cases of sludge application far exceeding regulation limits — does not clearly point to any major contamination risk for crops, animals, and humans is sufficient to assure that "high-quality biosolids may be used beneficially in sustainable agriculture" [99] if appropriate regulations are followed.

Others adopt a different perspective and consider that the lack of evidence of any significant contamination risk is not sufficient to render the matter settled. The fact that there is no clear evidence of absolute safety should lead to more caution and impose the need for more research. Otherwise, as McBride [5] put it, "the long-term consequences of the application of metal-laden sewage sludges…are still unknown."

It is undisputable, however, that heavy metals deposited in the soils with sludge application will stay there for a long time. Evidence so far bears promising elements in support of the use of sludge because metals are strongly bound onto organic phases, especially when the sludge application follows the relevant regulations. However, this does not rule out any future risks because of serious predictions for changes in climatic conditions that will alter heavy metal chemistry due to possible alteration of organic matter decomposition patterns and heavy metal behavior when concentrations approach a crucial threshold, after which their mobility may be less beneficial for the environment [100]. Ongoing research efforts may provide a more substantial answer to the matter.

REFERENCES

1. Smith, S.R., *Agricultural Recycling of Sewage Sludge and the Environment*, CAB International, Wallingford, 1996.
2. Schultz, R. and Romheld, V., Recycling of municipal and industrial organic wastes in agriculture: benefits, limitations and means of improvement, *Soil Sci. Plant Nutr.*, 43, 1051–1056, 1997.
3. Metzger, L. and Yaron, B., Influence of sludge organic matter on soil physical properties, *Adv. Soil Sci.*, 7, 141–162, 1987.
4. Chaney, R.L. and Ryan, J.A., Heavy metals and toxic organic pollutants in MSW-compost: research results on phytoavailability, bioavailability, fate, etc., in *Science and Engineering of Composting: Design, Environmental, Microbiological and Utilization Aspects*, Hoitink, H.A.J. and Keener, H.M., Eds., Renaissance Publications, Worthington, OH, 1993, 451–506.
5. McBride, M.B., Toxic metal accumulation from agricultural use of sludge: are USEPA regulations protective? *J. Environ. Qual.*, 24, 5–18, 1995.
6. Antoniadis, V. and Alloway, B.J., The role of DOC in the mobility of Cd, Ni and Zn in sewage sludge-amended soils, *Environ. Pollut.*, 117, 515–521, 2002.
7. Huyn, H.-N. et al., Cadmium solubility and phytoavailability in sludge-treated soil: effects of soil organic carbon, *J. Environ. Qual.*, 27, 329–334, 1998.
8. Chefetz, B. et al., Characterization of dissolved organic matter extracted from composted municipal solid waste, *Soil Sci. Soc. Am. J.*, 62, 326–332, 1998.
9. Spark, K.M., Wells, J.D., and Johnson, B.B., The interaction of a humic acid with heavy metals, *Austr. J. Soil Res.*, 35, 89–101, 1997.
10. Livens, F.R., Chemical reactions of metals with humic material, *Environ. Pollut.*, 70, 183–208, 1991.
11. Hamon, R.E. et al., Changes in trace metal species and other components of the rhizosphere during growth of radish, *Plant Environ.*, 18, 749–756, 1995.
12. Giusquiani, P.L. et al., 1998. Fate of pig sludge liquid fraction in calcareous soil: agricultural and environmental implications, *J. Environ. Qual.*, 27, 364–371, 1998.
13. Naidu, R. and Harter, R.D., Effect of different organic ligands on cadmium sorption by and extractability from soils, *Soil Sci. Soc. Am. J.*, 62, 644–650, 1998.
14. Elliott, H.A., Liberati, M.R., and Huang, L.P., Competitive adsorption of heavy metals by soils, *J. Environ. Qual.*, 15, 214–219, 1986.
15. Gao, S.A. et al., Simultaneous sorption of Cd, Cu, Ni, Zn, Pb and Cr on soils treated with sewage sludge supernatant, *Water Air Soil Pollut.*, 93, 331–345, 1997.
16. Linehan, D.J., Organic matter and trace metals in soils, in *Soil Organic Matter and Biological Activity*, Vaugham, D. and Malcolm, R.E., Eds., Martinus Nijhoff/Dr W. Junk Publishers, Dordrecht, 1986, 403–422.
17. Evans, L.J., Chemistry of metal retention by soils, *Environ. Sci. Technol.*, 23, 1046–1056, 1989.

18. Han, N. and Thompson, M.L., Soluble organic carbon in biosolids-amended Mollisol, *J. Environ. Qual.*, 28, 652–658, 1999.
19. Kaschl, A., Romheld, V., and Chen, Y., Cadmium binding by fractions of dissolved organic matter and humic substances from municipal solid waste compost, *J. Environ. Qual.*, 31, 1885–1892, 2002.
20. Brady, N.C. and Weil, R.R., *The Nature and Properties of Soils*, 13th ed., Prentice Hall, Upper Saddle River, NJ, 2002.
21. Giusquiani, P.L. et al., Urban waste compost: effects on physical, chemical, and biochemical soil properties, *J. Environ. Qual.*, 24, 175–182, 1995.
22. Terry, R.E., Nelson, D.W., and Sommers, L.E., Carbon cycling during sewage sludge decomposition in soils, *Soil Sci. Soc. Am. J.*, 43, 494–499, 1979.
23. Smith, S.R. and Hadley, P., Nitrogen-fertilizer value of activated sewage sludge derived protein — effect of environment and nitrification inhibitor on NO_3 release, soil microbial activity and yield of summer cabage, *Fertilizer Res.*, 33, 47–57, 1992.
24. Davis, R.D., The impact of the EU and UK environmental pressures on the future of sludge treatment and disposal, *J. Chartered Instit. Water Environ. Manage.*, 10, 65–69, 1996.
25. Alloway, B.J., Soil processes and the behavior of metals, in *Heavy Metals in Soils*, Alloway, B.J., Ed., Blackie Academic and Professional, London, 1995, 38–57.
26. Chaudri, A.M. et al., The cadmium content of British wheat grain, *J. Environ. Qual.*, 24, 850–855, 1995.
27. Barbarick, K.A., Ippolito, J.A., and Westfall, D.G., Extractable trace elements in the soil profile after years of biosolids application, *J. Environ. Qual.*, 27, 801–805, 1998.
28. Sloan, J.J. et al., Long-term effects of biosolids applications on heavy metal bioavailability in agricultural soils, *J. Environ. Qual.*, 26, 966–974, 1997.
29. Berti, W.R. and Jacobs, L.W., Distribution of trace elements in soil from repeated sewage sludge application, *J. Environ. Qual.*, 27, 1280–1286, 1998.
30. Harmsen, K., *Behaviour of Heavy Metals in Soils*, Centre for Agricultural Publishing and Documentation, Wageningen, 1977, 5–171.
31. Naidu, R. et al., Cadmium sorption and transport in variable charge soils: a review, *J. Environ. Qual.*, 26, 602–617, 1997.
32. Mbila, M.O. et al., Distribution and movement of sludge-derived trace metals in selected Nigerian soils, *J. Environ. Qual.*, 30, 1667–1674, 2001.
33. Berti, W.R. and Jacobs, L.W., Chemistry and phytotoxicity of soil trace elements from repeated sewage sludge applications, *J. Environ. Qual.*, 25, 1025–1032, 1996.
34. Sposito, G., Lung, L.J., and Chang, A.C., Trace metal chemistry in arid-zone field soils amended with sewage sludge: I. Fractionation of Ni, Cu, Zn, Cd and Pb in solid phases, *Soil Sci. Soc. Am. J.*, 46, 260–264, 1982.
35. Tsadilas, C.D. et al., Influence of sewage sludge application on soil properties and on distribution and availability of heavy metal fractions, *Commun. Soil Sci. Plant Anal.*, 26, 2603–2619, 1995.
36. Yong, R.N., Galvez–Cloutier, R., and Phadungchewit, Y., Selective sequential extraction analysis of heavy-metal retention in soil, *Can. Geotech. J.*, 30, 834–847, 1993.
37. Basta, N.T. and Sloan, J.J., Bioavailabilityof heavy metals in strongly acidic soils treated with exceptional quality biosolids, *J. Environ. Qual.*, 28, 633–638, 1999.
38. Oliver, D.P. et al., Effect of soil pH and applied cadmium on cadmium concentration in wheat grain, *Austr. J. Soil Res.*, 36, 571–583, 1998.
39. Salam, A.K. and Helmke, P.A., The pH dependence of free ionic activities and total dissolved concentrations of copper and cadmium in soil solution, *Geoderma*, 83, 281–291, 1998.
40. Yong, R.N. and Phadungchewit, Y., pH influence on selectivity and retention of heavy metals in some clay soils, *Can. Geotech. J.*, 30, 821–833, 1993.
41. Gray, C.W. et al., Sorption and desorption of cadmium from some New Zealand soils: effect of pH and contact time, *Austr. J. Soil Res.*, 36, 199–216, 1998.
42. Xiong, L.M. and Lu, R.K., Effect of liming on plant accumulation of cadmium under upland or flooded conditions, *Environ. Pollut.*, 79, 199–203, 1993.
43. Alloway, B.J., The mobilization of the trace elements in soils, in *Contaminated Soils*, Prost, R., Ed., *3rd International Conference on the Biogeochemistry of Trace Elements*, 15–19 May 1995, Paris, 1997, 133–145.

44. Antoniadis, V. and McKinley, J.D., Measuring heavy metal migration rates in a low permeability soil, *Environ. Chem. Letters,* 1, 103–106, 2003.

45. McLauhglin, M.J. et al., Review: the behavior and environmental impact of contaminants in fertilizers, *Austr. J. Soil Res.* 34, 1–54, 1996.

46. Jing, J. and Logan, T.J., Effects of sewage sludge cadmium concentration on chemical extractability and plant uptake, *J. Environ. Qual.,* 21, 73–81, 1992.

47. Bunce, N.J., *Introduction to Environmental Chemistry,* Wherz Publishing Ltd, Winnipeg, 1993, 341–522.

48. Papadopoulos, P. and Rowel, D.L., The reactions of cadmium with calcium carbonate surfaces, *J. Soil Sci.,* 39, 23–36, 1988.

49. Sommers, E. and McLaughlin, M.J., Chloride increases cadmium uptake in Swiss chard in a resin-buffered nutrient solution, *Soil Sci. Soc. Am. J.,* 60, 1443–1447, 1996.

50. Smolders, E. et al., Effect of soil solution chlorine on cadmium availability to Swiss chard, *J. Environ. Qual.,* 27, 426–432, 1998.

51. De Vries, M.P.C. and Tiller, K.G., Sewage sludge as a soil amendment, with special reference to Cd, Cu, Mn, Pb and Zn — comparison of result from experiments conducted inside and outside a glasshouse, *Environ. Pollut.,* 16, 321–240, 1978.

52. Kuo, S., Cadmium buffering capacity and accumulation in Swiss chard in some sludge-amended soils, *J. Soil Sci. Soc. Am.,* 54, 86–91, 1990.

53. Del Castilho, P. and Chardon, W.J., Uptake of soil cadmium by three field crops and its prediction by a pH-dependent Freundlich sorption model, *Plant Soil,* 171, 263–266, 1995.

54. Barbarick, K.A. and Workman, S.M., Ammonium bicarbonate-DTPA and DTPA extractions of sludge-amended soils, *J. Environ. Qual.,* 16, 125–130, 1987.

55. O'Connor, G.A., Use and misuse of the DTPA soil test, *J. Environ. Qual.,* 17, 715–718, 1988.

56. Wang, P. et al., Fractions and availability of nickel in Loessial soil amended with sewage or sewage sludge, *J. Environ. Qual.,* 26, 795–801, 1997.

57. Kabata–Pendias, A. and Pendias, H., *Trace Elements in Soils and Plants,* 2nd ed., CRC Press, Boca Raton, FL, 1992, 1–87, 131–141.

58. Young, X. et al., Influx transport and accumulation of cadmium in plant species grown at different Cd^{2+} activities, *J. Environ. Sci. Health Part B — Pesticides, Food Contamination Agri. Wastes,* 30, 569–583, 1995.

59. Lund, L.J., Page, A.L., and Nelson, C.O., Movement of heavy metals below sewage disposal ponds, *J. Environ. Qual.,* 5, 330–334, 1976.

60. Camobreco, V.J. et al, Movement of heavy metals through undisturbed and homogenized soil columns, *Soil Sci.,* 161, 740–750, 1996.

61. Dowdy, R.H. and Volk, V.V., Movement of heavy metals in soils, in *Chemical Mobility and Reactivity in Soil Systems,* Soil Science Society of America and American Society of Agronomy, Madison, WI, 1983, 229–240.

62. Davis, R.D. et al., Distribution of metals in grassland soils following surface applications of sewage sludge, *Environ. Pollut.,* 49, 99–115, 1988.

63. El-Hassanin, A.S., Labib, T.M., and Dobal A.T., Potential Pb, Cd, Zn, and B contamination of sandy soils after different irrigation periods with sewage effluent, *Water Air Soil Pollut.,* 66, 239–249, 1993.

64. Higgins, A.J., Environmental constrains of land application of sewage sludge, *Am. Soc. Agri. Eng.,* 27, 407–414, 1984.

65. Chang, A.C., Page, A.L., and Bingham, F.T., Heavy metal absorption by winter wheat following termination of cropland sludge applications, *J. Environ. Qual.,* 11, 705–708, 1982.

66. Dowdy, R.H. et al., Trace metal movement in an aeric Ochraqualf following 14 years of annual sludge application, *J. Environ. Qual.,* 20, 119–123, 1991.

67. Williams, D.E. et al., Trace element accumulation, movement and distribution in the soil profile from massive applications of sewage sludge, *Soil Sci.,* 129, 119–132, 1980.

68. Emmerich, W.E. et al., Movement of heavy metals in sewage sludge-treated soils, *J. Environ. Qual.,* 11, 174–178, 1982.

69. Miner, G.S., Gutierrez, R., and King, L.D., Soil factors affecting plant concentrations of cadmium, copper and zinc on sludge-amended soils, *J. Environ. Qual.,* 26, 989–994, 1997.

70. Antoniadis, V. and Alloway, B.J., Evidence of heavy metal movement down the profile of a heavily sludged soil, *Commun. Soil Sci. Plant Anal.*, 34, 1225–1231, 2003.

71. Schirado, T. et al., Evidence for movement of heavy metals in a soil irrigated with untreated wastewater, *J. Environ. Qual.*, 15, 9–12, 1986.

72. Sloan, J.J., Dowdy, R.H., and Dolan, M.S., Recovery of biosolids-applied heavy metals sixteen years after application, *J. Environ. Qual.*, 27, 1312–1317, 1998.

73. Antoniadis, V. and Alloway, B.J., Leaching of Cd, Ni and Zn down the profile of a sewage sludge-treated soil, *Commun. Soil Sci. Plant Anal.*, 33, 273–286, 2002.

74. Li, Z.B. and Shuman, L.M., Heavy metal movement in metal contaminated soil profile, *Soil Sci.*, 161, 656–666, 1996.

75. Stevenson, F.J. and Welch, L.F., Migration of applied lead in a field soil, *Environ. Sci. Technol.*, 13, 1255–1259, 1979.

76. Jardine, P.M. et al., Comparison of models for describing the transport of dissolved organic carbon in aquifer columns, *Soil Sci. Soc. Am. J.*, 56, 393–401, 1992.

77. Fotovat, A., Naidu, R., and Oades, J.M., The effect of major cations and ionic strength on desorption of native heavy metals in acidic and sodic soils, in *1st Int. Conf. Contaminants Soil Environ.*, Adelaide, 1996, 193–194.

78. Sawhney, B.L., Bugbee, G.J., and Stilwell, D.E., Leachability of heavy metals from growth media containing source-separated municipal solid waste compost, *J. Environ. Qual.*, 23, 718–722, 1994.

79. Bolton, K.A. and Evans, L.J., Elemental composition and speciation of some landfill leachates with particular reference to cadmium, *Water Air Water Pollut.*, 60, 43–53, 1991.

80. Zhu, B. and Alva, A.K., Trace-metal and cation-transport in a sandy soil with various amendments, *Soil Sci. Am. J.*, 57, 723–727, 1993.

81. Richards, B.K. et al., Metal mobility at an old, heavily loaded sludge application site, *Environ. Pollut.*, 99, 365–377, 1998.

82. McBride, M.B. et al., Mobility and solubility of toxic metals and nutrients in soil 15 years after sludge application, *Soil Sci.*, 162, 487–500, 1997.

83. Alloway, B.J. and Jackson, A.P., The behavior of heavy metals in sewage sludge-amended soils, *Sci. Total Environ.*, 100, 151–176, 1991.

84. Chang, A.C., Hyun, H.-N., and Page, A.L., Cadmium uptake for Swiss chard on composted sewage sludge treated field plots: plateau or time bomb?, *J. Environ. Qual.*, 26, 11–19, 1997.

85. Canet, R. et al., Sequential fractionation and plant availability of heavy metals as affected by sewage sludge applications to soils, *Commun. Soil Sci. Plant Anal.*, 29, 697–716, 1998.

86. Krauss, M. and Diez, T., Uptake of heavy metals by plants from highly contaminated soils. *Agribiological Res. — Zeitschrift Agrarbiologie Agrikulturechemie Okologie*, 50, 340–349, 1997.

87. Bidwell, A.M. and Dowdy, R.H., Cadmium and zinc availability to corn following termination of sewage sludge applications, *J. Environ. Qual.*, 16, 438–442, 1987.

88. Petruzzelli, G., Lubrano, L., and Guidi, G., Uptake by corn and chemical extractability of heavy metals from a 4-year compost treated soil, *Plant Soil*, 116, 23–27, 1989.

89. Heckman, J.R., Angle, J.S., and Chaney, R.L., Residual effects of sewage sludge on soybean: I. Accumulation of heavy metals, *J. Environ. Qual.*, 16, 113–117, 1987.

90. Gardiner, D.T. et al., Effects of repeated sewage sludge applications on plant accumulation of heavy metals, *Agri. Ecosyst. Environ.*, 55, 1–6, 1995.

91. Mulchi, C.L. et al., Residual heavy metal concentrations in sludge-amended coastal plain soils: 2. Predicting metal concentrations in tobacco from soil test information, *Commun. Soil Sci. Plant Anal.*, 23, 1053–1069, 1992.

92. Chang, A.C., Page, A.L., and Wernicke, J.E., Long-term sludge application on cadmium and zinc accumulation in Swiss chard and radish, *J. Environ. Qual.*, 16, 217–221, 1987.

93. Chaudri, A.M. et al., Cadmium content of wheat grain from a long-term field experiment with sewage sludge, *J. Environ. Qual.*, 30, 1575–1580, 2001.

94. McGrath, S.P. et al., Long-term changes in the extractability and bioavailability of zinc and cadmium after sludge application, *J. Environ. Qual.*, 29, 875–883, 2000.

95. Obrador, A. et al., Metal mobility and potential bioavailability in organic matter-rich soil-sludge mixtures: effect of soil type and contact time, *Sci. Total Environ.*, 206, 117–126, 1997.

96. Antoniadis, V. et al., Bioavailability of Pb and Zn in soils amended with sewage sludge: comparative tests in England and Greece, in *Proc. 8th Conf. Hellenic Soil Sci. Soc.*, Kavala, 2000, 459–469.

97. Antoniadis, V. and Alloway, B.J., Availability of Cd, Ni and Zn to ryegrass in sewage sludge-treated soils at different temperatures, *Water Air Soil Pollut.*, 132, 201–214, 2001.

98. White, C.S., Lotfin, S.R., and Aguilar, R., Application of biosolids to degraded semiarid rangeland: nine-year responses, *J. Environ. Qual.*, 26, 1663–1671, 1997.

99. Chaney, R.L., Trace metal movement: Soil-plant systems and bioavailability of biosolids-applied metals, in *Sewage Sludge: Land Utilization and the Environment*, Clapp, C.E., Larson, W.E., and Dowdy, R.H., Eds., SSSA Miscellaneous Publications, Madison, WI, 1994, 27–32.

100. Harrison, E.Z., McBride, M.B., and Bouldin, D.R., Land application of sewage sludges: an appraisal of the U.S. regulations, *Int. J. Environ. Pollut.*, 11, 1–36, 1999.

4 Influence of Fly Ash Application on Heavy Metal Forms and Their Availability

Christos Tsadilas, E. Tsantila, S. Stamatiadis, V. Antoniadis, and V. Samaras

CONTENTS

4.1 INTRODUCTION

Despite continuing efforts to find alternative and environmentally safe energy sources, coal remains a major fuel used for production of electricity worldwide. In the world today, about 25% of the energy produced, or 38% of the total electricity consumed, is generated from coal. Although a declining trend of coal use was recorded recently [1], the forecast is that coal consumption for energy production will increase. This is because the energy requirements continue to increase in response to a steadily increasing world population and coal remains an abundant and relatively cheap natural energy source.

Countries substantially dependent on coal for electric power generation include the U.S. (56%), China (80%), and India (66%) [2]. Europe (EU-15) on average produces 25% of its electricity from coal, although in some member countries this percentage is over 50%. For example, electricity power in Poland is almost fully coal based and in countries like Denmark and Germany over 50% of the electricity consumed is generated from coal [2]. Greece produces the second largest amount of coal in Europe and the sixth largest in the world. The available coal deposits are estimated to be about 3.2 billion tons; this is considered adequate for the next 45 years [3]. Over 73% of the power requirements of Greece come from lignite and its annual consumption is estimated at about 64 Mt [4].

During combustion in a power plant, coal generates great quantities of residues known as coal combustion products (CCP). They include the noncombustible mineral components of coal that are partitioned into fly ash (FA), bottom ash (BA), boiler slag (BS), and flue gas desulfurization residue (FGD). Fly ash generally consists of fine particles that are not easily disposed of. Most of this fraction of CCPs is captured by pollution control devices before release to the atmosphere. Mechanical collectors remove the coarser and predominantly sand-sized particles, and the finer silt-sized particles are removed by electrostatic precipitators. The amount of fly ash produced each year is enormous. For example, the total fly ash production in the U.S. was 63 million tons in 1998. The respective amount of fly ash produced annually in Greece is estimated to be nearly 13 million tons [5].

Knowledge of fly ash properties may provide insights for determining appropriate methods of treatment. The properties of fly ash are determined by the composition of the parent material (coal) and the conditions during coal combustion. Efficiency of emission control devices as well as storage and handling affect at large fly ash properties. Adriano et al. [7] and El-Mogazi et al. [8] provided excellent reviews of fly ash properties. They concluded that morphology of fly ash helps in understanding its physical properties and leaching behavior.

Fly ash is usually considered an amorphous ferro–aluminum silicate mineral [7] because a typical fly ash aggregate consists of spherical particles embedded in an amorphous matrix. Chang et al. [9] determined that fly ash is composed of approximately 63% silt-sized particles in the range of 2 to 50 μm; 33% of the particles had a diameter > 50 μm and the remainder had a diameter < 2 μm. The surface area varies between 0.46 and 1.27 m^2 g^{-1} and this influences the sorptivity of nutrients [10]. Chemically, fly ash contains almost all existing elements in nature. However, the major matrix elements are Si, Al, and Fe, as well as Ca, K, Na, and Ti.

Most of the fly ash produced is disposed of by landfilling, stockpiling, and storage in settling ponds. The remainder is used as cement, concrete, or ground material; raw feed for cement clinker; structural fill; road base; mineral filler; snow and ice control; mining applications; agricultural application; and so on. In 2000, from a total of 57.14 million tons of fly ash produced in the U.S., about 68% was disposed of; the remaining was recycled through the previously mentioned uses [6].

Nevertheless, fly ash has become an increasing disposal problem in many countries of the world. A sound alternative to fly ash recycling is its agronomic utilization. Research so far has shown many advantages of fly ash utilization as soil amendment, which brings about improvement of soil texture [11] and bulk density [12]; increase of water-holding capacity through increased porosity [12]; increase of soil pH in acid soils [13,14]; and increase of soil salinity, especially with unweathered fly ash, and of nutrient concentrations [13]. All these beneficial effects usually result in increased crop yield as has been reported by several workers [13,15–17]. However, fly ash addition to soils may also have adverse effects on crop growth, usually due to B, salt, and heavy metal toxicity. El-Mogazi et al. [8] summarized these adverse effects on several agricultural crops. In most cases, B and salt toxicity can be reduced if fly ash is preleached [18].

From an environmental point of view, the effect of fly ash on the concentration of heavy metals in soil is of special interest. Heavy metals of environmental significance in fly ash include As, Cd, Pb, Mo, Ni, Se, and Zn [8]. Relevant research has shown that application of fly ash to soil at various rates increases heavy metal content of some crops while others remain unaffected. In a pot experiment, Furr et al. [19] investigated the influence of fly ash application on the uptake of 42 elements by several crops. They found that fly ash application increased As, Mo, and Ni concentration in the edible parts of beans; As in cabbage; and As and Mo in carrots. However, fly ash application did not increase heavy metal concentration in onions, potatoes, and tomatoes.

In a similar study, Furr et al. [20] found that application of 50 tons/acre of fly ash increased As and Se concentration in several crops, including the edible parts of vegetables like beans, cabbage, carrots, potatoes, and tomatoes. Increased heavy metal uptake was also found for apple, millet, and vegetables [21] for the elements B, Co, Cu, and Mo and corn grain [22].

The main alternatives to the agronomic utilization of fly ash in Greece are cement additives and disposal in landfills. Research on the implications of application of domestic fly ash to soil is limited, although an increasing interest was recently expressed [13,23]. This chapter presents the results of an experiment on the influence of fly ash and sewage sludge on soil quality and on wheat growth and heavy metal uptake.

4.2 MATERIALS AND METHODS USED

The influence of fly ash on soil properties and wheat characteristics was investigated in a greenhouse pot experiment in which an acidic soil classified as Typic Haploxeralf was used, and in a field experiment in an acid soil from Central Greece, classified also as Typic Haploxeralf. Some chemical properties of the soils used are presented in Table 4.1.

Fly ash was collected from the electrostatic precipitator of a lignite-fired electric power plant in Northern Greece and aged for 3 months by maintaining in open air and leaching periodically with deionized water. The material was strongly alkaline (pH 12.1); its heavy metal composition is shown in Table 4.2. Sewage sludge was the same for both experiments and it was collected from the waste water treatment plant of the city of Tirnavos (central Greece). Its composition determined according to Leschber et al. [24] is shown in Table 4.3.

In the greenhouse experiment, a completely randomized block design was composed of seven treatments with four replicates each (Table 4.4). The application rates of fly ash (fa1, fa2) were determined in a preliminary incubation experiment in which addition of 30 g fly ash/pot resulted in a

TABLE 4.1
Selected Properties of the Soils Used in the Experiments

Properties	Pot experiment soil	Field experiment soil
pH (water 1:1)	4.85	5.01
Clay content, %	15	13
Organic matter content, %	1.8	0.61
Cation exchange capacity, cmol(+) kg^{-1}	13	9.50
Base saturation, %	38	55
Electrical conductivity (1:5), mmhos cm^{-1}	0.11	0.05
Total Zn, mg kg^{-1}	42.3	29.86
Total Mn, mg kg^{-1}	275	254
Total Cu, mg kg^{-1}	38.7	24.20
Total Ni, mg kg^{-1}	34.65	12.23
Total Pb, mg kg^{-1}	49.52	11.93
Total Mo, mg kg^{-1}	—	11.43
Total Se, mg kg^{-1}	—	0.33
Total Cd, mg kg^{-1}	—	0.62
Total Co, mg kg^{-1}	—	5.82
Total As, mg kg^{-1}	—	1.68

TABLE 4.2
Heavy Metal Concentration (mg kg^{-1}) Extracted by DTPA and HNO$_3$ of the Fly Ash Used in the Experiments

Extraction method	As	Cd	Co	Cu	Mn	Mo	Ni	Pb	Se	Zn
DTPA	nd	nd	nd	0.8	nd	nd	0.5	0.5	nd	nd
HNO$_3$	17.2	3.2	10.7	28.1	135.1	35.9	170.6	36.5	3	18.5

nd: not detected.

TABLE 4.3
Some Characteristics of the Sewage Sludge Used in the Experiments

pH (H$_2$O 1:1)	CaCO$_3$ (%)	EC, mmhos/cm	Organic matter content (%)	Total N (%)	Zn (mg kg^{-1})
6.5	1.1	2.87	34.0	4.45	1320
Cu, mg kg^{-1}	Ni, mg kg^{-1}	Pb, mg kg^{-1}	Cd, mg kg^{-1}		
287	46.4	183.0	1.49		

TABLE 4.4
Treatments of the Pot Experiment

Codes	Treatments
c (control)	Soil without amendments
if	Soil plus inorganic fertilizers at 150 kg N/ha and 80 kg P$_2$O$_5$/ha
fa1	Soil plus 30 g fly ash/pot only
fa2	Soil plus 90 g fly ash/pot only
fass1	Soil plus 2.65 g sewage sludge/pot plus 30 g fly ash/pot
fass2	Soil plus 5.25 g sewage sludge/pot plus 30 g fly ash/pot
fass3	Soil plus 10.5 g sewage sludge/pot plus 30 g fly ash/pot

TABLE 4.5
Treatments of the Field Experiment

Codes	Treatments
C (control)	Soil without amendments and fertilizers
IF	Without amendments but 160 kg N/ha and 80 kg P$_2$O$_5$/ha
FA1	22.5 tons fly ash/ha plus the same amount of N and P
FA2	67.5 tons fly ash/ha plus the same amount of N and P
FASS1	22.5 tons fly ash/ha plus 6.3 tons/ha sewage sludge
FASS2	22.5 tons fly ash/ha plus 12.6 tons/ha sewage sludge

pH of 6.5 and addition of 90 g/pot further increased soil pH to 7.5. The mixtures of fly ash and sewage sludge with 2.5 kg of soil were wetted to field capacity and equilibrated in the pots for 1 month. Fifteen seeds/pot of durum wheat (*Triticum aestivum* L.) were planted in the pots and grown for 2 months in a nonheated greenhouse. The ground plant parts were then harvested, dried, and weighed. At the same time, soil samples from all pots were collected, air-dried, crushed, and sieved with a 2-mm sieve. Soil samples were analyzed for pH, and DTPA-extractable and total heavy metal content.

A 2-year field experiment was conducted near the city of Larissa in central Greece. The selected treatments in a completely randomized block design are presented in Table 4.5. Fly ash rates were selected after a preliminary incubation experiment. Each treatment was replicated four times in experimental plots of 16 m^2 each.

The amendments were applied in October and the field was planted with durum wheat (*Triticum aestivum* L.) in November. At the harvesting time in June, surface composite soil samples were taken in order to determine heavy metal content (As, Ni, Se, Mo, Cd, Co, Zn, and Pb). Heavy metal determination included extraction with DTPA [25]; digestion with 4 M HNO$_3$ at 80°C overnight [26]; and fractionation to exchangeable, organic matter and carbonate bound, and clay-associated forms according to Emmerich et al. [27]. Soil pH was measured in a water:soil 1:1 suspension according to McLean [28]. Plant samples were analyzed for heavy metal content after dry ashing at 500°C and dialysis with dilute HNO_3. The same heavy metals were measured in wheat grain after the appropriate preparation. Heavy metal content in the extraction solutions was determined by atomic absorption spectrometry or inductively coupled plasma (ICP).

In addition to these analyses, soil samples taken from the field plots were sequentially extracted to determine the distribution of Cu, Ni, and Zn in the exchangeable, organic, carbonate, and residual pools. The exchangeable-associated metals were extracted with KNO_3; the organic-associated forms with NaOH; the carbonate-bound metals with EDTA; and the residual metal concentrations with 4 N HNO_3 digestion [27].

Grain yield and concentration of heavy metals in soil, plant, and wheat grain were evaluated using analyses of variance (ANOVA) procedures. To separate treatment means, the least significant difference (LSD) at $P < 0.05$ test was used. Regression analysis was performed when necessary. When differences between treatments are significant, they are indicated with 1 asterisk for a probability level of 95%, with 2 asterisks for a probability level of 99%, and with 3 asterisks for a probability of 99.9%.

4.3 RESULTS AND DISCUSSION

4.3.1 POT EXPERIMENT

Fly ash application increased soil pH significantly. Soil pH progressively increased from 5.02 in the control to 6.62 in the low fly ash treatment to 7.48 in the high fly ash treatment (Table 4.6). The addition of sewage sludge to fly ash did not further increase soil pH because pH in the sewage sludge treatments was not greater than that in the fa1 treatment. This finding was expected because the initial pH of sludge was 6.5 (Table 4.1). The application of the inorganic fertilizer, however, decreased soil pH by about 0.2 units, probably due to nitrification of ammonium.

Fly ash and sewage sludge application significantly increased total heavy metal concentration in soil (Table 4.6). The addition of sludge also increased total metal concentration in soil. In contrast to total concentrations, the concentration of all DTPA-extracted metals was higher in the control treatment. The lowest DTPA (or available) concentrations were found in the high fly ash treatment (fa2) despite the increased total concentration of these metals. The reduced metal availability in the soil appears to be caused by the great pH increase due to fly ash addition.

The addition of sludge to fly ash progressively increased the available forms of the metals to levels similar to that of the control (Table 4.6). The increased metal extractability in the presence

TABLE 4.6
Effects of Fly Ash and Sewage Sludge Application on Soil pH, DTPA-Extractable and HNO_3-Digested Heavy Metals[a] in the Pot Experiment

		Treatment						
		CO	IF	FA1	FA2	FASS1	FASS2	FASS3
pH		5.02b[b]	4.81a	6.62e	7.48f	6.64d,e	6.39c	6.43c,d
DTPA	Cu	11.41c	9.36b	9.19b	7.84a	9.35b	9.49b	11.30c
	Pb	0.78b	0.42a	0.42a	0.39a	0.38a	0.80b	0.65a,b
	Cd	0.14d	0.10b	0.10b	0.07a	0.09b	0.09b	0.12c
	Zn	2.44d	1.71c	1.26a,b	1.09a	1.80c	1.64bc	2.65d
HNO_3	Cu	34.95a	75.31c	64.06b	74.67c	73.87c	76.69c	79.06c
	Pb	4.16a	7.87c	6.93bc	7.33b,c	7.31b,c	6.94b,c	6.62b
	Ni	88.95a	105.37c,d	91.03b	110.12c,d	106.23c	104.22c	118.12c
	Zn	63.95a	89.68c	74.84b	102.92c	98.12c	99.06c	95.94c

[a] mg kg^{-1}.
[b] Means within rows followed by different letter(s) are significantly different at probability level $P < 0.05$.

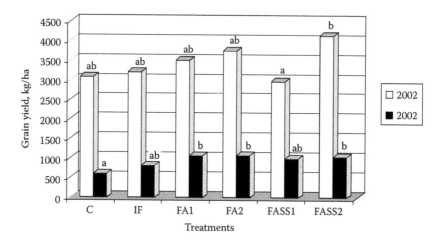

FIGURE 4.1 Influence of fly ash and sewage sludge application on wheat yield. (Bar means within years followed by different letter(s) are significantly different at $P < 0.05$.)

of sludge cannot be explained in terms of pH changes because soil pH did not increase after sludge addition. However, the increase in heavy metal availability after increased sewage sludge application to soil has been reported by many workers (for example, Antoniadis and Alloway [30]). The reduced availability of heavy metals due to fly ash application is a desirable property of this material that may prevent excessive uptake of heavy metals by crops. However, close monitoring is required in order to maintain an elevated soil pH (6.6 to 7.5 according to the data of this experiment), thus avoiding remobilization of heavy metals.

4.3.2 Field Experiment

Caution should be taken in the interpretation of the results because pot experiments have restricted value and may not be representative of field conditions. This is the reason that, in addition to the pot experiment, a field experiment was conducted, the results of which are presented below. A summary of the field treatments is shown in Table 4.5.

4.3.2.1 Influence of Fly Ash on Wheat Grain Yield

Fly ash application had no significant effect on grain yield the first year of the experiment (Figure 4.1). Significant difference in wheat yield occurred only between the two sludge-containing treatments. Yield was greatly reduced in the second year and this was attributed to adverse weather conditions. Even so, fly ash increased wheat yield when applied alone or in combination with sewage sludge (Figure 4.1). The only exception was the plant yield of the FASS1 treatment, in which it was significantly lower than in the rest of the treatments, even less than the control. This, however, should not be attributed to any toxicity effect because the trend was not consistent at FASS2 and was not recorded in the subsequent year.

The increased yield appears to be caused by a corresponding increase in soil pH due to fly ash application (Figure 4.2). Fly ash application increased soil pH from 4.61 in the control and 4.62 in the treatment with inorganic fertilizer to 5.32 at FA2 and 6.56 at FASS2. In contrast to the pot experiment, the addition of sludge to fly ash increased soil pH.

4.3.2.2 Influence of Fly Ash on Heavy Metal Content of Soil and Wheat

4.3.2.2.1 Total Heavy Metal Content in Soil

The total concentration of heavy metals in soil generally remained unchanged after application of fly ash or in combination with sewage sludge. This is indicated in Table 4.7 for the metals Cd, Co,

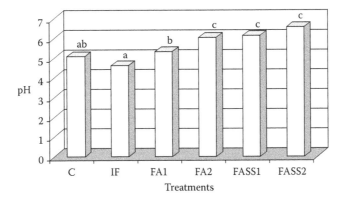

FIGURE 4.2 Influence of fly ash and sewage sludge on soil pH. (Different letters on bars show statistically significant differences at probability level $P < 0.05$.)

TABLE 4.7
Influence of Fly Ash and Sewage Sludge Application on Soil Heavy Metal Concentrations in the Field Experiment

Treatments	Metal concentration (mg kg^{-1})					
	Cd_{HNO3}	Cd_{DTPA}	Co_{HNO3}	Co_{DTPA}	As_{HNO3}	As_{DTPA}
C	0.575a[a]	nd	5.72a	0.175b,c	1.7a	0.12a
IF	0.525a	nd	5.77a	0.3d	1.8a	0.17a,b
FA1	0.575a	nd	5.82a	0.3c	1.85a	0.17a,b
FA2	0.6a	nd	5.75a	0.13b	1.7a	0.22b
FASS1	1.8a	nd	5.42a	0.10a	1.6a	0.3c
FASS2	0.6a	nd	5.45a	0.10a	1.77a	0.25b
	Mn_{HNO3}	Mn_{DTPA}	Mo_{HNO3}	Mo_{DTPA}	Ni_{HNO3}	Ni_{DTPA}
C	254a	20.78b	11.4b	0.125c	12.27a	0.95c
IF	248a	22.45b	9.22b	0.094b,c	11.7a	1.12d
FA1	250a	20.63b	2.55a	0.025b	12.37a	1cd
FA2	249a	13.38a	2.52a	0.028b	12.92a	0.8b
FASS1	242a	12.75a	2.37a	0.0002a	11.57a	0.65a,b
FASS2	245a	9.35a	9.68b	0.006a	12.1a	0.52a
	Zn_{HNO3}	Zn_{DTPA}	Se_{HNO3}	Se_{DTPA}	Pb_{HNO3}	Pb_{DTPA}
C	26.87a	0.62a	0.32c	nd	12.07a	0.57a,b
IF	24.7a	0.7a	0.32b,c	nd	11.42a	0.9b
FA1	16.75a	0.52a	0.25a	nd	11.92a	0.62a,b
FA2	16.65a	0.65a	0.3a,b,c	nd	11.9a	0.47a
FASS1	18.3a	0.87a	0.27a,b,c	nd	12.2a	0.5a
FASS2	26.4a	3b	0.35c	nd	12.6a	0.52a

[a] Values in the same columns followed by different letter(s) differ significantly at the probability level $p < 0.05$.

As, Mn, Ni, Zn, and Se and in Figure 4.3 for elements such as Mn, Zn, and Co. Total Mo concentration decreased in the treatments with fly ash compared to the control and inorganic fertilizer treatment. Only the high sludge application (FASS2) caused Mo concentration to increase to the level of the control treatment, possibly indicating that sewage sludge was enriched in Mo.

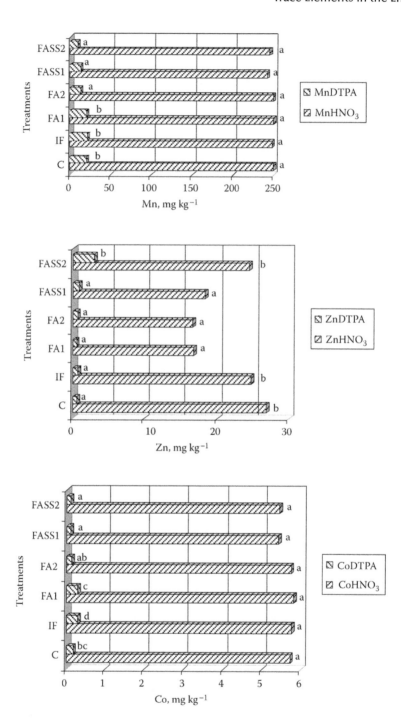

FIGURE 4.3 Influence of fly ash and sewage sludge application on total and DTPA extractable Mn (top), Zn (middle), and Co (bottom) in the field experiment. (Different letters on bars representing the same parameter indicate significant difference at probability level $P < 0.05$.)

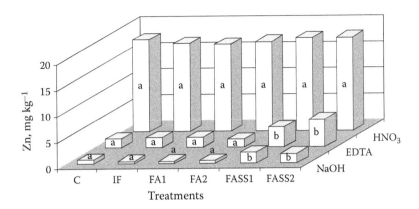

FIGURE 4.4 Influence of fly ash and sewage sludge on soil Zn fractions in the field experiment. (Different letters on bars representing the same fraction indicate significant difference at probability level $P < 0.05$.)

However, in the case of Ni, its concentration was higher in the treatment with the higher fly ash rate (FA2), probably due to the high Ni concentration in fly ash (170.6 mg kg^{-1}).

4.3.2.2.2 "Available" Heavy Metal Contents in Soil

The situation was different with the "available" forms of heavy metals, i.e., those extracted with DTPA. The "available" forms of heavy metals such as Co, Mn, Mo, Ni, and Pb decreased upon application of fly ash (Figure 4.3 and Table 4.7). Cadmium and Se concentrations extracted with DTPA were negligible. DTPA-extractable Zn and As concentration was higher in the sludge treatments; this should be attributed to the positive relationship of Zn and As with sewage sludge-borne organic matter. The fact that heavy metal availability decreased with fly ash application would be attributed to the pH effect as the significant negative correlations between soil pH and DTPA-extractable Co ($r = -0.74***$), Ni ($r = -0.74***$), and Pb ($r = -0.52**$) (data not shown) indicate. These correlations clearly show the effect of soil pH on metal availability, leading to the conclusion that fly ash application is a safe practice from the standpoint of reducing the availability of most metals in acid soils.

4.3.2.2.3 Heavy Metal Distribution in Soil

Three of the metals studied (Cu, Ni, and Zn) were fractionated into exchangeable-, organic- and carbonate-associated fractions. The changes in these fractions as a result of fly ash and sewage sludge application are presented in Figure 4.4 for Zn and in Figure 4.5 for Ni (data for Cu are not shown). The exchangeable fraction of Zn (KNO$_3$ extractable) was not affected by fly ash application. However, in the case of Ni, this fraction was negatively influenced by fly ash application. The organic-associated fraction (NaOH extractable) and the carbonate-bound fraction (EDTA extractable) were increased for Zn and Ni, but these fractions remained unaffected in the case of Cu.

The residual fraction (HNO$_3$ digested) remained unchanged in Zn and significantly increased in Ni. The implication of soil pH in determining the decrease in the exchangeable form is demonstrated by the significant correlations in Figure 4.6 for the exchangeable-, organic-, and carbonate-bound fractions of Ni and Cu. The data, therefore, indicate that fly ash provides metal forms that tend to diminish availability to plants.

4.3.2.2.4 Heavy Metal Concentrations in Wheat Grain

Heavy metal concentration was also measured in wheat grain. Cadmium and Co concentrations were at undetectable levels. Nickel and Se concentration in the treatments including fly ash and sewage sludge applications ranged between 3.52 and 7.20 and 1.13 and 2.14 mg kg^{-1}, respectively; this was not significantly different from that in the control and the inorganic fertilizer treatment

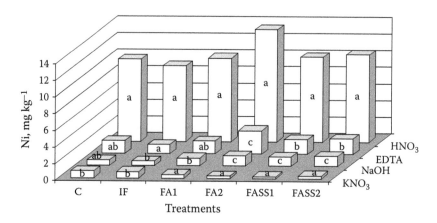

FIGURE 4.5 Fly ash and sewage sludge influence on Ni fractions in the field experiment. (Different letters on bars representing the same fraction indicate significant difference at probability level $P < 0.05$.)

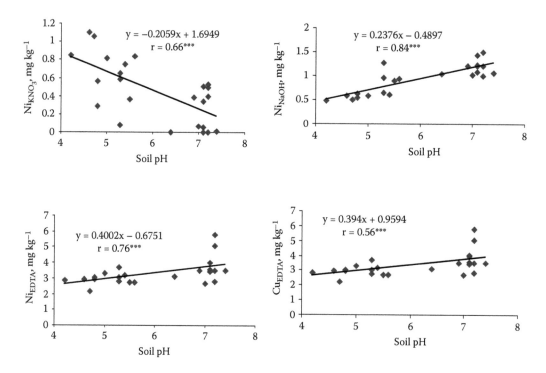

FIGURE 4.6 Relationship between soil pH and exchangeable Ni (top left), organic matter associated (top right), and carbonate bound fractions of Ni and Cu (bottom left and right, respectively).

(data not shown). Copper concentration in wheat seeds was significantly increased in all treatments containing fly ash and sewage sludge (Figure 4.7).

Molybdenum showed an unexpected trend in that its concentration was higher in the control (C) and the inorganic fertilizer treatment (IF) in which soil pH was lower, although one would expect the opposite behavior (Figure 4.7). Zinc, on the other hand, increased significantly from 20 mg kg^{-1} (in the control) to around 33 to 35 mg kg^{-1} (at FASS1 and FASS2 treatments) (Figure 4.8). Finally, Mn showed an unclear trend (Figure 4.8). Therefore, heavy metal accumulation in wheat

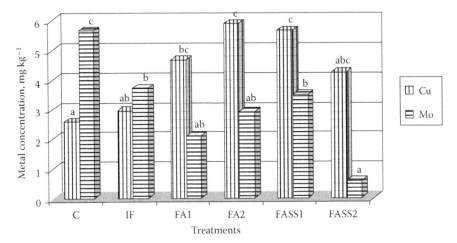

FIGURE 4.7 Influence of fly ash on the concentration of Cu and Mo in wheat grain. (Different letters on bars symbolizing the same parameter show significant difference at probability level $P < 0.05$.)

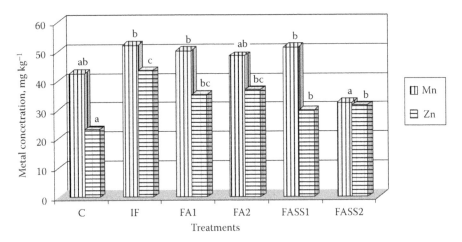

FIGURE 4.8 Influence of fly ash on the concentration of Mn and Zn in wheat grain. (Different letters on bars symbolizing the same parameter show significant difference at probability level $P < 0.05$.)

grain had no clear effect in all cases studied except for Zn and Cu, which clearly exhibited an increasing trend with increase in fly ash application rate.

4.4 CONCLUSIONS

The results of the present study showed that, although fly ash application to the soil along with sewage sludge increased total heavy metal content in soil (pot experiment), in general it decreased their availability to wheat plants, obviously due to pH increase. More specifically, total concentration of exchangeable fractions of Ni (readily available) in soil decreased with fly ash application (Cu- and Zn-exchangeable fractions were not detected). This was consistent with the decrease in DTPA-extractable heavy metals with fly ash addition. However, sewage sludge application increased DTPA-extractable As and Zn. Wheat grain concentration of heavy metals did not increase significantly except for Cu and Zn; however, both are essential to plant growth. Other potentially toxic elements such as Cd and Pb were completely excluded from plant uptake.

REFERENCES

1. Dunn, S., Decarbonizing the energy economy, in *State of the World 2001*, Lind Strake, Ed., The World Watch Institute, W.W. Norton Comp. New York, 2001, 83–102.
2. Twardovska, I., Szczepanska, J., and Stefaniak, S., Occurrence and mobilization potential of trace elements from disposed coal combustion fly ash, in *Chemistry of Trace Elements in Fly Ash*, Sajwan, K.S., Alva, A.K., and Keefer, R.F., Eds., Kluwer Academic/Plenum Publishers, New York, 2003, 13–24.
3. Public Power Corporation, S.A. http://www.dei.gr.
4. Arditsoglou, A. et al, Size distribution of trace elements and polycyclic aromatic hydrocarbons in fly ash generated in Greek lignite-fired power plants, *Sci. Total Environ.*, 323, 153–167, 2004.
5. ACAA, *1998 Coal Combustion Product (CCP): Production and Use (Short Tons)*, American Coal Ash Association, Alexandria, VA, 1999.
6. Punshon, T., Seaman, J.C., and Sajwan, K.S., The production and use of coal combustion products, in *Chemistry of Trace Elements in Fly Ash*, Sajwan, K.S., Alva, A.K., and Keefer, R.F., Eds., Kluwer Academic/Plenum Publishers, New York, 2003, 1–11.
7. Adriano, D.C. et al., Utilization and disposal of fly ash and other coal residues in terrestrial ecosystems: a review, *J. Environ. Qual.*, 9, 333–344, 1980.
8. El-Mogazi, D., Lisk, D.J., and Weinstein, L.H., A review of physical, chemical, and biological properties of fly ash and effects on agricultural ecosystems, *Sci. Total Environ.*, 74, 1–37, 1988.
9. Chang, A.C. et al., Physical properties of fly ash amended soils. *J. Environ. Qual.*, 6, 267–270, 1979.
10. Schure, M.R. et al., Surface area and porosity of coal fly ash, *Environ. Sci. Technol.*, 19, 82–86, 1985.
11. Fail, J.L. and Wochock, Z.S., Soybean growth on fly ash-amended strip mine soils, *Plant Soil*, 48, 472–484, 1977.
12. Chang, A.C. et al., Physical properties of fly ash-amended soils, *J. Environ. Qual.*, 6, 267–270, 1977.
13. Matsi, T. and Keramidas, V.Z., Fly ash application on two acid soils and its effect on soil salinity, pH, B, P and on ryegrass growth and composition, *Environ. Pollut.*, 104, 107–112, 1999.
14. Elseewi, A.A., Straughan, I.R., and Page, A.L, Sequential cropping of fly ash-amended soil: effects on soil chemical properties and yield and elemental composition of plants, *Sci. Total Environ.*, 15, 247–259, 1980.
15. Elseewi, A.A., Bigham, F.T., and Page, A.L., Availability of sulfur in fly ash to plants, *J. Environ. Qual.*, 7, 69–73, 1978.
16. Martens, D.C., Availability of plant nutrients in fly ash, *Compost Sci.*, 12, 15–19, 1971.
17. Doran, J.W. and Martens, D.C., Molybdenum availability as influenced by application of fly ash to soil, *J. Environ. Qual.*, 1, 186–189, 1972.
18. Adriano, D.C. et al., Cadmium availability to Sudan grass grown on soils amended with sewage sludge and fly ash, *J. Environ. Qual.*, 11, 179–203, 1982.
19. Furr, A.K. et al., Multielement uptake by vegetable and millet grown in pots on fly ash amended soil, *J. Agri. Food Chem.*, 24, 885–888, 1976.
20. Furr, A.K. et al., Elemental content of vegetables, grains and forages field-grown on fly ash-amended soil, *J. Agric. Food Chem.*, 26, 357–359, 1978.
21. Furr, A.K. et al., Elemental content of apple, millet, and vegetables grown in pots of neutral soil amended with fly ash, *J. Agric. Food Chem.*, 27, 135–138, 1979.
22. Combs, G.F., Jr., Barrows, S.A., and Swader, F.N., Biologic availability of selenium in corn grain produced on soil amended with fly ash, *J. Agric. Food Chem.*, 28, 406–409, 1980.
23. Matsi, T., Effect of application of fly ash, derived from lignite fired plants, on some physicochemical characteristics of three soils and on nutrient uptake by plants, Ph.D. Thesis, Aristotle University of Thessaloniki, Thessaloniki, 1997.
24. Leschber, R., Davis, R.D., and L'Hermite, P., *Chemical Methods for Assessing Bioavailable Metals in Sludges and Soils*, Elsevier Applied Science Publishers, London, 1984, 95.
25. Baker, D.E. and Amacher, M.C., Nickel, copper, zinc and cadmium, in *Methods of Soil Analysis — Part 2: Chemical and Microbiological Properties*, Page, A.L., Miller, R.H., and Keeney, R., Eds., 2nd ed., Agronomy No 9, ASA, SSSA, Madison, 1982, 323–336.

26. Sposito, G., Lund, L.J., and Chang, A.C., Trace metals chemistry in arid-zone field soils amended with sewage sludge. I. Fractionation of Ni, Cu, Zn, Cd, and Pb in solid phases, *Soil Sci. Soc. Am. J.*, 46, 260–264, 1982.

27. Emmerich, W.E. et al., Solid phase forms on heavy metals in sewage sludge treated soils, *J. Environ. Qual.*, 11, 178–191, 1982.

28. McLean, E.O., Soil pH and lime requirements. *Methods of Soil Analysis — Part 2: Chemical and Microbiological Properties*, Page, A.L., Miller, R.H., and Keeney, R., Eds., 2nd ed., Agronomy No 9, ASA, SSSA, Madison, 1982, 199–224.

29. Tsadilas, C.D. et al., Influence of sewage sludge application on soil properties and on distribution and availability of heavy metal fractions, *Commun. Soil Sci. Plant Anal.*, 26, 2603–2619, 1995.

30. Antoniadis, V. and Alloway, B.J., Availability of Cd, Ni and Zn to ryegrass in sewage sludge-treated soils at different temperatures, *Water Air Soil Pollut.*, 132, 201–214, 2001.

31. Antoniadis, V. and Alloway, B.J., The role of DOC in the mobility of Cd, Ni and Zn in sewage sludge-amended soils, *Environ. Pollut.*, 117, 515–521, 2002.

5 Arsenic Concentration and Bioavailability in Soils as a Function of Soil Properties: a Florida Case Study

Dibyendu Sarkar and Rupali Datta

CONTENTS

ABSTRACT

Background arsenic (As) concentrations are important measures of defining the level of contamination in soils and for setting up soil cleanup goals. Twelve surface soil samples from each of three ecological zones of Florida, namely, salt marsh, freshwater marsh, and pine flatwoods, were characterized to determine which soil properties most influence As retention and availability. Soils were analyzed for total As, phytoavailable As, and bioavailable As concentrations. The most important soil properties that control As biogeochemistry in soils were Fe, Al, Ca, and P contents; cation exchange capacity; and soil organic matter. Arsenic concentration was significantly higher in the salt marsh soils compared to the freshwater marsh and pine flatwoods soils; the latter two were not significantly different from one another. Phytoavailable As contributed towards 2 to 16% of total soil As. The amount of As bioavailable to the human gastrointestinal system was generally higher than As available to plants. Although total As concentration was greatest in salt marsh soils, percent bioavailability was the lowest. In contrast, pine flatwoods' soils had the highest percentage of bioavailable As in spite of having the lowest total As concentrations. Results demonstrate that total As content is not a good indicator of soil contamination level as far as risk to human health is concerned.

5.1 INTRODUCTION

Ecologists have classified natural vegetation in Florida into 26 ecological zones such as North and South Florida coastal strands; sand pine scrub; North and South Florida flatwoods; oak hammock; scrub cypress; salt marsh; mangrove swamp; sawgrass marsh; freshwater marsh; etc [1]. The type of vegetative community in an ecological zone is likely to depend on the type of soil that supports it. Three ecological zones were chosen to study background As concentrations and to identify the soil properties that influence soil As concentration — namely, salt marsh, freshwater marsh, and pine flatwoods.

Salt marshes of Florida are coastal ecosystems with communities of nonwoody, salt-tolerant plants occupying intertidal zones that exhibit characteristics of terrestrial and marine ecosystems [2]. The salt marsh community is located on the Atlantic coast, extending southwest towards the Gulf of Mexico and inland along tidal rivers (Figure 5.1a). This community appears as an open expanse of grasses, sedges, and rushes. Vegetation occurs in distinct zones as a result of the fluctuating water levels and salinity concentrations due to tidal action. Major plant species include black needlerush and seashore saltgrass [2]. Low, regularly flooded marshes are dominated by smooth cordgrass; high marsh supports salt myrtle, marshhay cordgrass, marshelder, saltwort, and sea oxeye. Soils are generally poorly drained, sandy clay loams underlain by loam sands or organic soils underlain by loam sands. Salt marshes are considered nutrient sinks where plants accumulate nutrients rapidly during the growing season [2].

Freshwater marshes are not uniformly distributed throughout the state. The greatest expanse is the Everglades in South Florida (Figure 5.1b). Several other marshes are associated with river floodplains and occur throughout Florida, appearing as an open expanse of grasses, sedges, and rushes, as well as other herbaceous plants in areas where soils are usually saturated or covered with surface water for 2 or more months during the year [3]. Plants that characterize this community include grasses (beakrushes, blue maidencane, bulrushes, common reed, etc.); herbaceous plants (arrow head, blue flag, cattail, fire flag, etc.); and shrubs (St. Johns wort, primrose willow, elderberry, etc.). Soils are generally dominated by peat, marl, and sand with high organic matter and nitrogen, but low phosphorus and potassium levels [3].

Pine flatwoods comprise the most widespread biological community in Florida, constituting about 30 to 50% of Florida's uplands ranging from northern to southern Florida (Figure 5.1c). It is easily identified by its flat topography and its typically scattered pine trees with an understory of saw palmetto and grasses [4]. The dominant trees characteristic of flatwoods are longleaf pine, slash pine, South Florida slash pine, and pond pine. The understory shrub layer includes, in addition

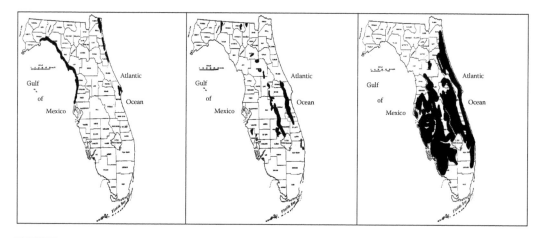

FIGURE 5.1 Ecological zones of Florida: (a) salt marsh; (b) freshwater marsh; and (c) pine flatwoods. (From Myers, R. and J. Ewel (Eds.), 1990, *Ecosystems of Florida*. University of Central Florida Press, Orlando, FL.)

to saw palmetto, gallberry, fetterbush, staggerbush, wax myrtle, etc. Soils that characterize pine flatwoods are poorly drained acidic fine sands with low amounts of nutrients and low organic matter content [4].

The U.S. Environmental Protection Agency (USEPA) has classified inorganic As as a Group A human carcinogen [5]. Arsenic is typically found in soils at background concentrations ranging from 0.1 to 40 mg/kg [6]. The main source of As in uncontaminated soils is the parent material from which the soils are derived [7]. Other sources of As in soils include atmospheric deposition [8] and bioconcentration by plants, aquatic organisms, and lower invertebrates [9–11]. An elevated health risk is associated with long-term human exposure to As in soils. Several studies have been conducted to establish background As concentrations in uncontaminated soils of Florida, with widely varying results. A level of 8 mg/kg was reported in agricultural soils [8,12] and a geometric mean (exponentiated mean of a log-transformed distribution) was reported of 1.1 mg/kg in surface soil based on a study of 40 mineral soils of Florida. Scarlatos and Scarlatos [13] reported As concentrations ranging from 1.1 to 54.3 mg/kg in 115 surface soil samples collected from Florida.

The residential soil cleanup goal set by the Florida Department of Environmental Protection (FDEP) is 0.80 mg/kg [14]. The maximum allowable level of As for oral intake is 0.3 µg/kg/d [15]. The importance of considering dietary intake of As through the food chain via uptake from contaminated soils [16], as well as soil ingestion from incidental hand-to-mouth activity by children playing in backyards, has been repeatedly emphasized in recent studies assessing public health risks associated with long-term exposure to low-level As-contaminated systems [17,18]. Adriano [19] summarized several soil properties that are most likely to influence soil availability of As: pH, texture (clay content); amorphous Fe–Al oxides; organic matter content; sulfur content; phosphate concentration; and soil redox conditions.

The most important parameter for accurate assessment of the risk posed by As to human health is its bioavailability. Bioavailability refers to the extent of absorption of a chemical into the bloodstream from the gastrointestinal tract, lungs, or skin [20]. It is also essential to consider phytoavailability (potential availability for plant uptake) because dietary intake of As can occur through the food chain via plant uptake from contaminated soils. Bioavailability and phytoavailability depend on soil characteristics; thus, the current study was geared towards examining As phytoavailability and bioavailability using 12 surface soil samples from each of the three aforementioned ecological zones of Florida. The main objectives of this study were:

- To identify and characterize the soil properties most likely to influence As concentration in surface soils collected from three ecological zones of Florida
- To measure total As concentration as well as the phytoavailable and bioavailable fractions of As in those soils
- To investigate whether the soils from the three ecological zones are significantly different in terms of total, phytoavailable, and bioavailable As

5.2 MATERIALS AND METHODS

5.2.1 Soil Sampling, Preparation, and Characterization

Surface soil samples from three ecological zones in Florida — namely, salt marsh, freshwater marsh and pine flatwoods — were used for this study. Twelve soil samples from each zone were obtained. A detailed description of the soil samples including their soil order, series, horizon, and type is listed in Table 5.1. Soil samples were air-dried and passed through a 2-mm sieve. Soils were characterized for pH, electrical conductivity, and water content using standard protocols; organic matter content was determined using the loss-on-ignition method [21]. Total C and N were determined using an

TABLE 5.1
Description of Soil Samples

Ecological zone	Sample ID	Soil order	Soil series	Horizon	Soil type
Salt marsh	SM-1	Entisol	Turnbull	0A	Muck
	SM-2	Entisol	McKee	A	Variant silt
	SM-3	Inceptisol	Hansboro	0E	Variant muck
	SM-4	Entisol	Hallandale	A	Fine sand
	SM-5	Entisol	Kesson	A	Fine sand
	SM-6	Entisol	Peckish	A1	Muck fine sand
	SM-7	Spodosol	Estero	0A	Muck
	SM-8	Inceptisol	Weckiwachee	0A1	Muck
	SM-9	Inceptisol	Durbin	0A1	Muck
	SM-10	Inceptisol	Durbin	0A1	Muck
	SM-11	Inceptisol	Tisonia	0A1	Variant muck
	SM-12	Spodosol	Lynn Haven	0A1	Fine sand
Freshwater marsh	FW-1	Inceptisol	Samsula	0A1	Loamy sand
	FW-2	Inceptisol	Brighton	0A	Muck
	FW-3	Entisol	Placid	A1	Fine sand
	FW-4	Alfisol	Pineda	A1	Fine sand
	FW-5	Alfisol	Winder	A	Sand
	FW-6	Inceptisol	Samsula	0A1	Muck
	FW-7	Entisol	Sellers	A1	Loamy sand
	FW-8	Inceptisol	Shenks	0A1	Muck
	FW-9	Alfisol	Winder	A	Sand
	FW-10	Inceptisol	Terra Ceia	0A1	Muck
	FW-11	Mollisol	Hendry	A	Sandy loam
	FW-12	Inceptisol	Kaliga	0A1	Muck
Pine flatwoods	PF-1	Entisol	Osceolla	A	Sand
	PF-2	Spodosol	Myakka	A	Fine sand
	PF-3	Spodosol	Eaugallie	A1	Fine sand
	PF-4	Spodosol	Smyrna	A	Fine sand
	PF-5	Spodosol	Smyrna	A	Fine sand
	PF-6	Spodosol	Immokalee	A	Fine sand
	PF-7	Spodosol	Immokalee	A	Sand
	PF-8	Spodosol	Eaugallie	A	Sand
	PF-9	Spodosol	Eaugallie	A	Fine sand
	PF-10	Spodosol	Immokalee	AP	Sand
	PF-11	Spodosol	Mandarin	A	Fine sand
	PF-12	Spodosol	Leon	A	Fine sand

elemental analyzer (Perkin Elmer 2400 Series II). Total Ca, Mg, Fe, Al, P, and As were extracted following USEPA method 3050B [22]. Phytoavailable As was obtained by shaking 1 g of soil with 50 ml of 1 M NH$_4$Cl solution for 30 min. Particle size and cation exchange capacity were not measured; values were obtained from the soil characterization database of Florida Cooperative Soil Survey (1967–1989).

Phosphorus was measured colorimetrically using the molybdate-ascorbic acid method [23] using Varian Cary 50 Spectrophotometer. Iron was also determined colorimetrically according to Olson and Ellis [24] as a complex with 1,10-phenanthroline reagent. Ca, Mg, and Al were analyzed using flame atomic absorption spectrometry and As was determined using graphite furnace atomic absorption spectrometry (GFAAS).

5.2.2 DETERMINATION OF BIOAVAILABLE ARSENIC

Bioavailable As was estimated according to the *in vitro* gastrointestinal method of Sarkar and Datta [25]. The reactions were carried out in 250-ml beakers in a 37°C water bath to simulate body temperature. Anaerobic conditions were maintained by passing argon gas through the solutions. Constant mixing of the solution was maintained using a stirrer to simulate gastric mixing. The extractant used consisted of 0.15 M NaCl and 1% porcine pepsin. One gram of soil sample was added to 150 ml of gastric solution, and the pH of the solution was adjusted to 1.8 using 1 N HCl. The solution was incubated for 1 h.

The solution was then modified for the intestinal phase by adjusting the pH to 7.0 using a saturated solution of $NaHCO_3$, followed by the addition of 525 mg of porcine bile extract and 52.5 mg of porcine pancreatin (Sigma Chemical Co., St. Louis, MO). In order to simulate absorption through the intestinal lining, a 40-cm^2 filter paper strip coated with Fe oxide was used. The Fe oxide strip was placed in a square bag (sides 6.5 cm) made of nylon membrane filter of 8-μm pore size. The bag was tied with a string and suspended in the reaction vessel. The solution was incubated for 1 h, at the end of which 10 ml of solution was collected, centrifuged at 5000 rpm for 30 min, and analyzed by GFAAS. Arsenic adsorbed by the Fe oxide strip was desorbed by shaking it vigorously in 80 ml 1 N HNO_3 for 1 h.

5.3 RESULTS AND DISCUSSION

5.3.1 CHARACTERIZATION OF SOILS

Soil physicochemical properties that are most likely to influence As concentrations in the soil samples collected from salt marsh, freshwater marsh, and pine flatwoods are presented in Table 5.2. The properties are described in terms of range, arithmetic mean (AM), geometric mean (GM), and their respective standard deviations (ASD and GSD). The majority of the procedures available in the literature on environmental statistics for computing AM assume that contaminant concentrations are approximately normally distributed [26]. The upper confidence level of the mean moves closer to the true mean as sample size increases; the 95th percentile of the distribution remains at the upper end of the distribution [27].

However, the distributions of elemental concentration in most soils are positively skewed and usually follow a log-normal distribution [16]. When dealing with a skewed distribution, the GM is a better maximum likelihood estimator of the central tendency than AM [16] is. The calculated baseline concentrations, assuming log-normality of the elemental distribution, better represent the natural level of chemicals in soils because the distorting effects of a few high values are minimized [16]. Table 5.2 shows that the surface soils studied represented a wide range of properties that could potentially affect arsenic retention and, thus, its availability to plants and humans.

5.3.2 TOTAL ARSENIC CONCENTRATIONS IN SOILS

Total As concentrations in the soil samples studied are given in Table 5.3. The salt marsh soils had the highest concentration of total As (GM 12.22 mg/kg), with one sample as high as 89.84 mg/kg. The pine flatwoods samples exhibited the lowest concentration of total As (GM 0.38 mg/kg), and the freshwater marsh samples were intermediate between the other two ecological zones (GM 2.41 mg/kg). A comparison of means using the least square difference (LSD) method revealed that mean total As concentration in the salt marsh was significantly higher than in the other two ecological zones at the 99% confidence interval ($\alpha < 0.01$). The mean total As concentrations in the freshwater marsh and the pine flatwoods were not significantly different from each other. All 12 of the salt marsh soils, 10 out of 12 freshwater marsh soils, and 1 out of 12 pine flatwoods soils exceeded the residential soil cleanup goal of 0.80 mg/kg set by the FDEP [14].

TABLE 5.2
Soil Characterization

Zone	No. of samples	Data	Soil properties											
			pH	Clay (%)	EC (µS/cm)	Fe (mg/kg)	Al (mg/kg)	Ca (mg/kg)	Mg (mg/kg)	S (mg/kg)	C (%)	SOM (%)	CEC (cmol/kg)	P (mg/kg)
Salt marsh	12	Hi	8.09	3.3	68000	78.66	10.07	14643	57.04	6400	27.69	58.13	466.4	591.4
		Lo	4.41	0.0	925	ND	1.38	2012	29.80	100	0.10	0.69	12.79	120.7
		AM	5.33	0.72	24084	20.56	6.75	8335	44.71	3280	13.75	24.30	178.1	321.5
		ASD	1.16	1.26	23230	25.55	3.54	4594	7.38	2622	9.91	18.99	132.9	140.5
		GM	5.22	NC	13058	8.36	5.52	7024	44.10	1385	5.53	12.72	118.1	289.8
		GSD	1.10	NC	1.78	2.00	1.38	1.34	1.08	2.24	2.55	2.02	1.64	1.25
FW marsh	12	Hi	6.61	3.9	1110	54.40	8.06	13414	41.72	6300	44.76	52.31	218.1	506.1
		Lo	3.95	0.0	99	ND	ND	1365	16.49	ND	0.48	0.41	1.39	120.4
		AM	5.03	0.82	513.1	10.70	4.44	31.88	27.81	2770	17.06	10.94	76.76	220.5
		ASD	0.87	1.29	291.3	17.98	2.78	3503	7.61	2482	18.66	18.35	71.27	117.8
		GM	4.96	NC	435.9	4.43	3.33	2376	26.85	1596	8.08	4.61	37.13	198.4
		GSD	1.08	NC	1.32	1.82	1.49	1.34	1.13	1.77	1.88	1.86	1.95	1.22
Pine FW	12	Hi	7.48	3.6	235	0.49	5.03	1794	203.9	1400	8.14	8.26	18.34	182.0
		Lo	3.99	0.4	73	ND	ND	240.9	12.86	ND	0.80	0.27	3.23	110.1
		AM	4.62	1.30	177.6	0.25	1.81	1215	41.64	466.7	2.70	2.91	10.83	140.5
		ASD	0.94	0.96	56.7	0.34	1.55	390	51.56	457.9	2.82	2.33	6.26	23.77
		GM	4.55	1.03	168.1	0.07	1.28	1116	30.83	324.3	1.80	2.10	8.94	138.8
		GSD	1.08	1.36	1.17	3.31	1.51	1.25	1.33	1.46	1.48	1.49	1.35	1.07
Total	36	Hi	8.09	3.9	68000	78.66	10.07	14643	57.04	6400	44.76	58.13	466.4	591.4
		Lo	3.95	0.0	73	ND	ND	240.9	12.86	ND	0.10	0.27	1.39	110.1

TABLE 5.3
Arsenic Concentrations in Soils of Florida Ecological Zones

Arsenic pools	Ecological zone	No. of samples	Range (mg/kg)	AM ± ASD (mg/kg)[a]	GM ± GSD (mg/kg)[b]
Total	Salt marsh	12	1.10–89.8	21.8 ± 24.8 *a*	12.2 ± 1.73
	Freshwater marsh	12	0.1–9.38	3.98 ± 3.15 *b*	2.41 ± 1.76
	Pine flatwoods	12	0.20–1.21	0.42 ± 0.26 *b*	0.38 ± 1.21
Bioavailable	Salt marsh	12	ND–16.4	3.91 ± 5.78 *a*	1.40 ± 2.06
	Freshwater marsh	12	ND–1.29	0.36 ± 0.45 *b*	0.15 ± 1.87
	Pine flatwoods	12	ND–0.10	0.05 ± 0.03 *b*	0.04 ± 1.33
Phytoavailable	Salt marsh	12	ND–1.92	0.67 ± 0.72 *a*	0.26 ± 2.16
	Freshwater marsh	12	ND–0.19	0.06 ± 0.07 *b*	0.04 ± 1.68
	Pine flatwoods	12	ND–0.08	0.05 ± 0.02 *b*	0.04 ±1.19

[a] AM ± ASD = arithmetic mean ± standard deviation.
[b] GM ± GSD = geometric mean ± standard deviation.
Note: Different letters in the same column (specific to arsenic pool) indicate significantly different at α 0.1 using the LSD test.

Out of the 12 soil samples evaluated for the salt marsh ecological zone, 5 belonged to the Entisol soil order, 5 to the Inceptisol soil order, and 2 to the Spodosol soil order (Table 5.1). However, despite being officially classified as Inceptisols, all the five soil samples in that order had typical variant muck texture characteristic of Histosols. Histosols are often referred to as peat or muck. They contain at least 20 to 30% organic matter by weight and are more than 40 cm thick. Histosols typically form in areas of poor drainage, where soils are frequently wet. Such conditions retard the decomposition of plant and animal remains, which accumulate over time [28].

Entisols are soils that were formed relatively recently and are thus poorly developed, with only A horizon. Inceptisols are soils that exhibit minimal horizon development. They are more developed than Entisols, but still lack the features characteristic of other soil orders. Inceptisols are widely distributed and occur under a wide range of environmental settings. They are often found on fairly steep slopes, young geomorphic surfaces, and resistant parent materials [28]. Spodosols are acid soils characterized by an accumulation of aluminum and iron oxides beneath the surface. They typically have a light-colored E horizon (layer) overlying a reddish-brown spodic horizon. Spodosols often occur under coniferous forests in moist climates.

A qualitative evaluation of soil order impact on As concentrations revealed that total As varied as Inceptisol (Histosol) > Entisol > Spodosol. Inceptisols had the highest values of soil organic matter, CEC, total Fe, total Ca, and total P; Spodosols had the lowest. Similar findings were reported by Chen et al. [7], who observed that the highest mean As concentrations were associated with Histosols.

Out of the 12 samples evaluated for the freshwater marsh ecological zone, 6 were Inceptisols, 2 were Entisols, 3 were Alfisols, and 1 was a Mollisol (Table 5.1). Alfisols are forest soils that have relatively high native fertility. These soils are well developed and contain a subsurface horizon in which clays have accumulated. Alfisols are mostly found in temperate humid and subhumid regions of the world. Formed in grasslands, Mollisols occur primarily in middle latitudes. They are famous for their thick, dark surface horizon, which results from the addition of organic materials derived from plant roots [28]. The Mollisol sample was discounted even in a qualitative analysis because one sample cannot be considered to be representative of an entire soil order.

Total As varied with soil order as follows: Inceptisol > Alfisol > Entisols. Eleven out of twelve pine flatwood samples were Spodosols and one was an Entisol. Apparently, the Spodosols had the lowest As concentration corresponding to the lowest CEC, total Fe, Ca, P, and soil organic matter among the soil orders discussed.

5.3.3 Influence of Soil Properties on Arsenic Concentrations

Soil pH generally has an impact on soil concentration of As, but such impacts depend on the nature of the variable charge mineral surfaces common in the soils [19]. Generally, sorption of As decreases with increasing pH [19]. This can be attributed to the negative surface charge on the adsorptive surface at higher pH, as well as the negative charge of As oxyanions [29]. Although the range of pH of the soils in the current study was quite large (3.95 to 8.09), the majority of the samples clustered in a small group ranging between 4.55 and 5.22 (Table 5.2). Not much difference is anticipated in the expected behavior of As oxyanions in small pH range, so soil pH was not expected to play a major role in explaining the variability in total As concentrations in the soils studied. A correlation coefficient of 0.001 (not significant at α 0.1) was obtained when total soil As data were regressed against the soil pH values (Figure 5.2a).

Sorption of As varies among soils and appears to be related to the oxide content of the soil [8,30,31]. Like phosphate, As is strongly adsorbed by amorphous iron and aluminum oxides [19]. Thus, presence of amorphous Fe and Al-oxides in soils is likely to enhance As retention. In the current suite of samples, total Fe concentrations varied between undetectable and 78.7 mg/kg (Table 5.2). The total Al content varied between below detection level and 10.1 mg/kg. Fe and Al were in much higher concentration in the salt marsh soils (GM 8.36 and 5.52, respectively) compared to the pine flatwoods soils (GM 0.07 and 1.28, respectively). Apparently, the salt marsh soils have greater As retention capacity than the pine flatwoods soils.

Total As concentration correlated significantly (at α 0.1) with total Fe + Al, and yielded a correlation coefficient of 0.57 (Figure 5.2b), indicating the role of Fe and Al oxides in retaining As in the surface soils. Calcium and magnesium in oxidized soils systems have the potential to precipitate As as Ca or Mg arsenate. Therefore, high Ca and/or Mg concentrations might imply greater retention of arsenic; as Figure 5.2c demonstrates, total soil As correlated significantly with total Ca + Mg concentrations. Table 5.2 demonstrates that Ca concentrations varied greatly among the soil samples studied (241 mg/kg to 14,643 mg/kg) with the highest concentration in salt marsh soils (7024 mg/kg). Magnesium contents were negligible compared to Ca and varied between 12.9 mg/kg and 57.1 mg/kg.

In general, As mobility and bioavailability are greater in sandy soils than in clayey soils. Woolson [32] reported that As phytotoxicity to horticultural crops was highest on a loamy sand soil and lowest on a silty clay loam. Others have also noted this inverse relationship between clay content and trend in bioaccumulation [33–35]. The main reason for this phenomenon is that hydrous Fe, Mn, and Al oxides vary directly with the clay content of the soil [19] because they primarily occur as coatings on the clay minerals or in the clay size fractions of the soils. Thus, the water-soluble fraction of As is generally higher in soils with low clay content than in those with high clay content [19].

However, certain clay minerals have permanent negative charge due to isomorphic substitution of a lower valence cation for a higher valence cation in the tetrahedral or octahedral layer [36]. Given that inorganic As predominantly occurs as oxyanions (arsenate or arsenite) in soil solutions, it is possible that negatively charged As ions are repelled by the negatively charged surface sites and thereby become more available for uptake by plants. Percent clay in the studied suite of soil samples under consideration varied between 0 and 3.9 (Table 5.2), representing a very small cluster. Therefore, As concentration in these soils was not expected to be influenced by clay contents, as demonstrated by their poor correlation with total soil As concentrations (Figure 5.2d).

Soil organic matter (SOM) is primarily composed of humic and fulvic acids [37]. Because of their acid-base, sorptive, and complexing properties, humic substances have a strong effect on the properties of water [38] and play an important role in the mobility of As in environmental waters. The adsorption of As by humic acids is high — in the pH range of 5 to 7 and when the humic acids have high ash and calcium content [39]. Humic acids can contribute more to the retention of

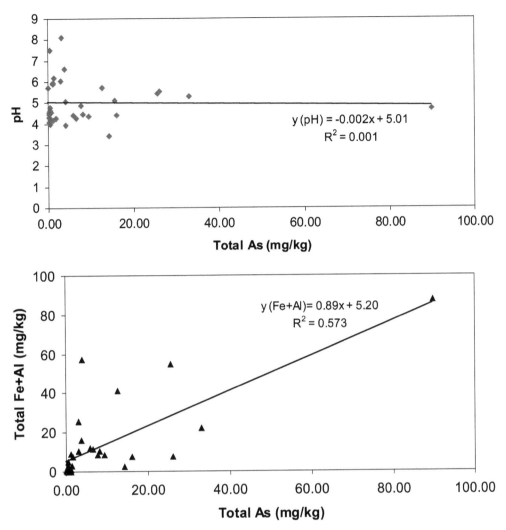

FIGURE 5.2 (A) Arsenic concentration in Florida soils as a function of soil pH ($n = 36$). (B) Arsenic concentration in Florida soils as a function of iron and aluminum concentrations ($n = 36$). (C) Arsenic concentration in Florida soils as a function of calcium and magnesium concentrations ($n = 36$). (D) Arsenic concentration in Florida soils as a function of clay content ($n = 36$). (E) Arsenic concentration in Florida soils as a function of soil organic matter content ($n = 36$). (F) Arsenic concentration in Florida soils as a function of cation exchange capacity ($n = 36$). (G) Arsenic concentration in Florida soils as a function of soil salinity measured as electrical conductivity ($n = 36$). (H) Arsenic concentration in Florida soils as a function of sulfur concentration ($n = 36$). (I) Arsenic concentration in Florida soils as a function of phosphorus concentrations ($n = 36$).

As in acidic environments than do clays and some metal oxides. The major retention sites on the humic acids at low pH systems are the amine groups [40].

Given the ideal pH conditions and a relatively high concentration of SOM (12.7%; Table 5.2), the soils from the salt marsh are likely to retain higher amounts of As than do freshwater marsh (4.61%) and pine flatwoods (2.1%). However, it should also be remembered that, although humic acids have strong metal retention capabilities, fulvic acids form soluble complexes of metals in waters; they probably keep some of these metal ions in solution and are particularly involved in iron solubilization and transport [38]. Therefore, the presence of fulvic acids can increase the mobility and the potential for leaching of As. This is because fulvic acids can be adsorbed onto

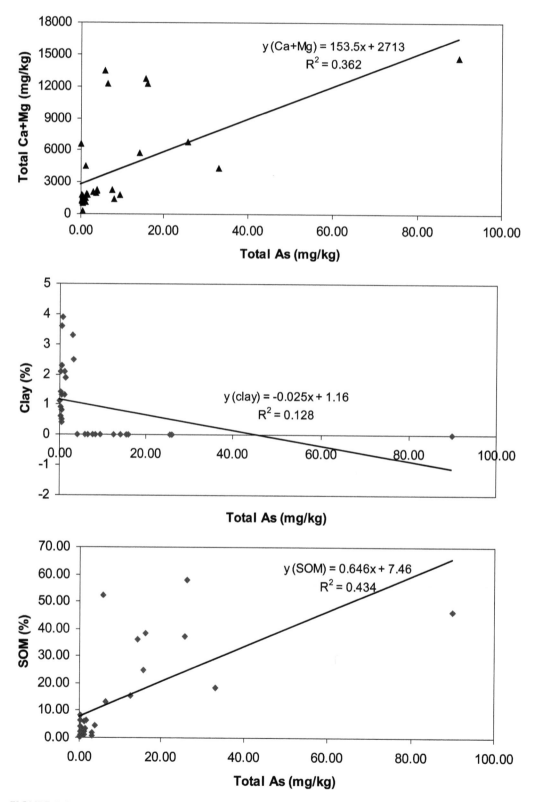

FIGURE 5.2 (continued)

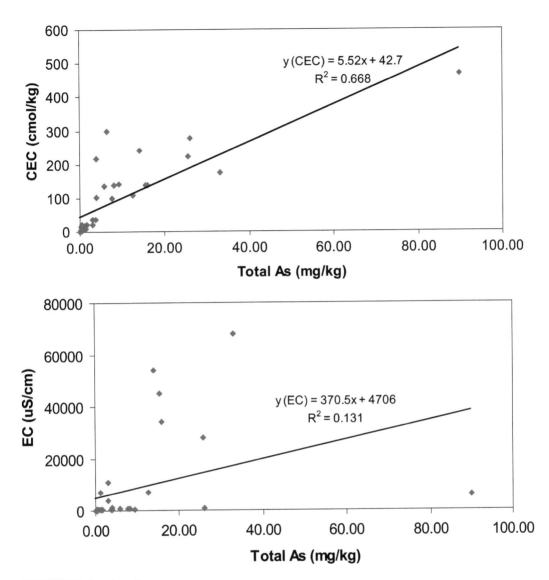

FIGURE 5.2 (continued)

the hydrous oxides of sediments by coulombic attraction, which leads to the formation of predominantly negatively charged surfaces due to the deprotonation of the functional groups [39].

In addition, fulvic acids can directly react with As, resulting in a decrease in adsorption of the corresponding As complex [41]. In this study, total soil As concentrations (pooled across the ecological zones) demonstrated a strong positive correlation with the SOM content with an R^2 value of 0.43, significant at α 0.1 (Figure 5.2e). Given the fact that the salt marsh samples contained significantly more As than freshwater marsh and pine flatwoods did (which clustered in the lower left hand corner of Figure 5.2e), it is logical to attribute this strong correlation to the As-binding capacity of SOM in salt marsh.

The quantity of reversibly adsorbed cations per unit weight of adsorbent (e.g., cmol(+)/kg) is called the cation exchange capacity or CEC [19]. The higher the CEC is, the greater is the amount of positive charge on the surface and the higher is the potential of the As oxyanions to form electrostatic bonds with the positively charged surface sites. Chen et al [7] showed a high degree of correlation between CEC and As in surface soils ($r = 0.97$) based on soil order. This is possibly

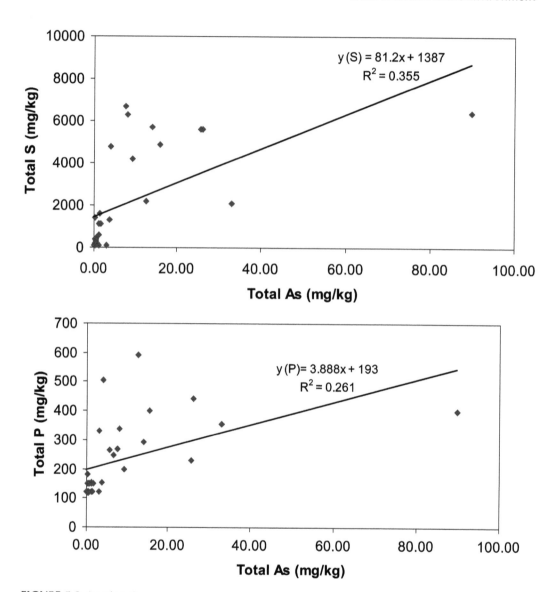

FIGURE 5.2 (continued)

because CEC was affected by the contents of clays, organic carbon, total Fe, Al, and probably P in soils. In the current study, the CEC of soils varied between 1.39 cmol/kg and 466 cmol/kg. The largest mean CEC was recorded in the salt marsh samples (118 cmol/kg), followed by freshwater marsh (37.1 cmol/kg) and pine flatwoods (8.94 cmol/kg), implying that As retention is likely to be greater in the salt marsh soils. A strong relationship between total soil As concentration and CEC was observed (R^2 value of 0.43, significant at α 0.1), as evident from Figure 5.2f.

Electrical conductivity is an expression of ionic strength of soil solution. Ions that undergo specific, inner-sphere complexation with surface functional groups are generally not affected by changes in solution ionic strength [42]. As has been reported to be predominantly adsorbed on metal hydroxides and other mineral surfaces via inner-sphere surface complexation mechanism via ligand exchange [43]; however, in certain cases, it may also undergo outer-sphere complexation if the bonding between As and the surface functional group is primarily electrostatic [44]. In such circumstances, soil EC is likely to have a major impact on As retention by soils.

Based on the strong correlation between soil As and CEC (which implied an electrostatic phenomenon), this possibility seemed plausible in the case of the soil samples studied. The EC of the soils exhibited an enormous range, varying between 73 and 68,000 µS/cm. The mean EC of the salt marsh samples (13,058 µS/cm) was significantly higher than that of freshwater marsh (435.9 µS/cm) and pine flatwoods (168.1 µS/cm). However, no apparent correlation was observed between total soil As and soil salinity; the R^2 value of 0.13 was not significant at α 0.1 (Figure 5.2g).

Sulfides also play an important role in retaining and remobilizing As from contaminated soils and sediments [39]. In a contaminated fluvial system, where surface sediments are oxidized while the lower buried layers are reduced, the concentration and partitioning of As are controlled by the redox interface. When reduction occurs, the adsorbed and coprecipitated As is released to the porewater [46]. Sulfides in the reduced environment will scavenge the As with the formation of arsenic sulfide precipitates [45,47]. Although a reducing environment is quite unlikely for the soil samples investigated in the current study, (which were collected primarily from the surface horizons), the marshlands of Florida were formed under reducing conditions [48]; thus, a significant portion of total As is likely to be associated with S in those soils.

As evidenced from the data presented in Table 5.2, freshwater and saltwater marsh samples had high mean S content (1596 mg/kg and 1385 mg/kg, respectively) compared to pine flatwoods samples, in which the S content varied between undetectable to 1400 mg/kg with a mean value of 324.3 mg/kg. Total soil As correlated positively with S content in the soil samples studied with a correlation coefficient of 0.36 significant at α 0.1 (Figure 5.2h), indicating the role of sulfides in retaining As in the surface soils, particularly those from the marsh ecosystems.

Phosphorus and arsenic exhibit very similar chemical properties. P and As are group VA elements and have very similar atomic radius (1.33 vs. 1.23); bonding radius (1.2 vs. 1.06); ionization potential (9.81 vs. 10.48 V); and electronegativity (–2.18 vs. –2.19). Both form oxyanions in +5 oxidation state; these are their most common environmental species. Phosphates are stable over a large range of pH and pE, and As can exist in the +3 oxidation state and easily forms links with sulfur and carbon [49]. Phosphates and arsenates undergo similar types of retention in soil minerals via (primarily) inner-sphere surface complexation. Thus, phosphates have demonstrated strong abilities to compete with As for sorption sites in environmentally important pH ranges [50].

Therefore, it is logical to assume that As, like P, will be similarly and significantly retained by positively charged, high specific surface soil components, such as amorphous Fe/Al oxides; presence of high concentration of P in soils could be indicative of As enrichment. Elevated As concentrations in sediments from the south Atlantic and along the Gulf Coast of Florida have been reported [51,52]; P deposits and soil pesticide residues were the hypothesized main sources of elevated As [52]. This is possible because phosphate rocks have a relatively high As concentration (6.6 to 121 mg/kg) compared with average (1.8 to 6.6 mg/kg) nonphosphorite rocks [53].

Chen et al. [7] identified soil P content as one of the most important soil properties that influence soil As concentration. In the studied soil samples, P concentration varied between 110.1 and 594.4 mg/kg, with the highest concentration in salt marsh soils (289.8 mg/kg), followed by freshwater marsh (198.4 mg/kg) and pine flatwoods (138.8 mg/kg) soils (Table 5.2). Total soil As correlated positively and significantly (at α 0.1) with total soil P, yielding a correlation coefficient of 0.26 (Figure 5.2i).

5.3.4 Arsenic Availability in Soils

Bioavailable and phytoavailable As fractions followed the same general trend as total soil As concentrations (Table 5.3). Phytoavailable As fraction, geared towards estimating the approximate amount of As in soils likely to be readily available for uptake by plants, represented only a small percentage of total As with the geometric mean concentration varying between 0.26 mg/kg in the salt marsh and 0.04 mg/kg in the freshwater marsh and the pine flatwoods.

The extractant used for estimating phytoavailable As was a neutral salt solution (ammonium chloride) designed to extract the soluble and exchangeable fractions of soil As. Apparently, only a small percentage of the total As in soils of all three ecological zones was soluble or exchangeable. This result is expected because the soil samples investigated in this study were almost exclusively collected from the surface horizons, and the soluble/exchangeable fractions were already exhausted as a result of plant uptake or they might have leached through the soil profile. This may be particularly true for the Spodosols, which are characterized by highly leachable soil types. O'Connor and Sarkar [54] studied leaching of phosphate in a typical Florida Spodosol and reported rapid vertical migration of P through the soil profile until it reached the characteristic hardpan, i.e., the spodic horizon, when the transport became lateral, controlled by subsurface topography.

The *in vitro* bioavailable fraction, geared towards estimating the approximate amount of As in soils likely to be available to human intestinal system in case of incidental hand-to-mouth exposure [55,56], also represented a small percentage of total As, although the concentrations were generally higher than the phytoavailable As fractions. Apparently, this fraction was extracting not only the soluble and exchangeable As, but also the reversibly adsorbed As fraction such as those possibly complexed by the SOM. Still, a significant amount of total soil As remained potentially unavailable to the human gastrointestinal system.

Figure 5.3 represents the bioavailable and phytoavailable As fractions as percentages of total soil As. The phytoavailable fraction varied between 1.8% in salt marsh soils and 13.5% in the pine flatwoods soils. Comparison of means using the LSD method revealed that, at the α 0.1 level, the mean percent total As phytoavailable in the pine flatwoods soils was significantly higher than that in the freshwater marsh and salt marsh soils. Although such a clear statistical differentiation among the ecological zones was not obtained, the mean percent of bioavailable As varied in the order: pine flatwoods (13.15%) > freshwater marsh (9.67%) > salt marsh (7.62%). Apparently, although

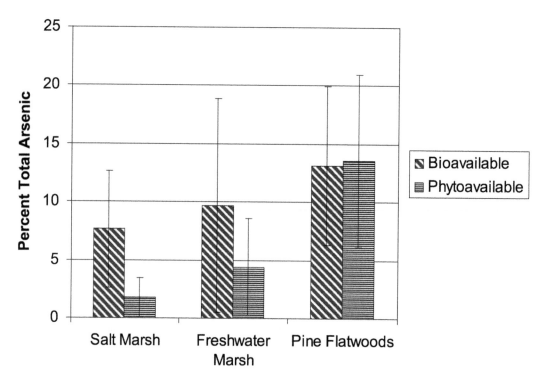

FIGURE 5.3 Percent of total arsenic that is bioavailable or phytoavailable. Data represent mean values and standard deviations.

the salt marsh soil samples had much higher concentrations of total As than the other ecological zones did, their availability to plants or the human intestinal system was the lowest.

On the other hand, although pine flatwoods had the lowest mean total As concentration, a relatively higher percentage was extracted by the solutions designed to extract the bioavailable and phytoavailable fractions. This is probably because As in the salt marsh soils is much more strongly retained by soil components by irreversible adsorption onto Fe/Al oxides and/or in the form of Ca/Mg-arsenical precipitates, as indicated by the data presented in Table 5.2. Thus, total As is not a good indicator of the level of contamination in soils from a human health risk assessment perspective.

5.4 CONCLUSIONS

- The soil properties most likely to influence As concentration in the Florida surface soils studied are the concentrations of Fe, Al, Ca, P, S, SOM, and CEC.
- Mean As concentration was significantly higher in the salt marsh soils compared to that in the freshwater marsh and pine flatwoods soils; the later two were not significantly different from each other.
- Only a small portion of the total As was phytoavailable.
- The bioavailable portion of As was generally higher than that of phytoavailable As in freshwater marsh and pine flatwoods soils.
- Although total As concentration was highest in salt marsh soils, percent bioavailability of As was the lowest. In contrast, pine flatwoods soil had the highest percentage of bioavailable As despite having the lowest total As concentration.
- Total As concentration is not a good indicator of soil contamination level as far as risk to human health is concerned.

ACKNOWLEDGMENTS

The authors would like to acknowledge the Center for Water Research, UTSA, for providing the support of a technical assistant. Thanks are also due to Mr. Larry Schwandes of the University of Florida for providing the soils used in this study and to the students of UTSA enrolled in the course GEO 6203 of spring 2002, who were offered by the primary author for collecting some of the data presented in this chapter.

REFERENCES

1. USDA Soil Conservation Service. 1984. 26 ecological communities of Florida. U.S. Govt. Printing Office, Washington, D.C.
2. Montague C.L. and R.G. Weigert. 1990. Salt marshes. In R.L. Myers and J.J. Ewel (Eds.) *Ecosystems of Florida*. University of Central Florida Press, Orlando, FL.
3. Kushlan, J. 1990. Freshwater marshes. In R. Myers and J. Ewel (Eds.) *Ecosystems of Florida*. University of Central Florida Press, Orlando, FL.
4. Abrahamson, W.G. and D.C. Hartnett. 1990. Pine flatwoods and dry prairies. In R.L. Myers and J.J. Ewel (Eds.) *Ecosystems of Florida*. University of Central Florida Press, Orlando, FL.
5. Southworth, R.M. 1995. Land application pollutant limit for arsenic. Part 503. USEPA, Washington, D.C.
6. Bowen, H.J.M. 1979. *Environmental Chemistry of the Elements*. Academic Press, New York.
7. Chen, M., L.Q. Ma and W.G. Harris. 2002. *Arsenic concentrations in Florida surface soils: Influence of soil type and properties. Soil Sci. Soc. Am. J.* 66:632–640.

8. Smith, S., R. Naidu and A.M. Alston. 1998. Arsenic in the soil environment: A review. In D.L. Sparks (Ed.) *Advanced Agronomy.* 64: 149–195. Academic Press, San Diego, CA.
9. Cullen, W.R. and K.J. Reimer. 1989. Arsenic speciation in the environment. *Chem Rev.* 89:713–764.
10. Otte, M.L., J. Rozema, M.A. Beek, B.J. Kater and R.A. Broekman. 1990. Trace metals in Gulf of Mexico oysters. *Sci. Total Environ.* 97/98:551–593.
11. Strom, R.N., R.S. Braman, W.C. Japp, P. Dolan, K.B. Donelly and D.F. Martin. 1992. Analysis of selected trace metals and pesticides offshore of the Florida Keys. *Florida Scientist.* 55:1–13.
12. Ma, L.Q., F. Tan, and W.G. Harris. 1997. Concentrations and distributions of eleven elements in Florida Soils. *J. Environ. Qual.* 26:769–775.
13. Scarlatos, P.D. and D. Scarlatos. 1997. Ecological impact of arsenic and other trace metals from application of recovered screen materials on Florida soils. FCSHWM Special Waste Publication No. 97-5, Florida Atlantic Univ., Boca Raton, FL.
14. Tonner–Navarro, L., N.C. Halmes and S.M. Roberts. 1998. Development of soil cleanup target levels (SCTLs) for Chapter 62-785, Florida Administration Code, Gainesville, FL.
15. USEPA. 1998. Integrated risk information system (IRIS): arsenic, inorganic. CASRN 7440-38-2. USEPA, Cincinnati, OH.
16. Chen, M., L.Q. Ma, and W.G. Harris. 1999. Baseline concentrations of 15 trace elements in Florida surface soils. *J. Environ. Qual.* 28:1173–1181.
17. Clausing, P., B. Brunekreff, and J.H. van Wijnen. 1987. A method for estimating soil ingestion by children. *Int. Arch. Occup. Env. Med.* 59:73–82.
18. Calabrese, E., R. Barnes, and E.J. Stanek. 1989. How much soil do young children ingest: An epidemiologic study. *Reg. Toxicol. Pharmacol.* 10:123–137.
19. Adriano, D.C. 2001. Trace elements in terrestrial environments: biogeochemistry, bioavailability, and risks of metals. Springer–Verlag, New York.
20. Halmes, N.C.H. and S.M. Roberts. 1997. Technical Report 97-02. Center for Environmental and Human Toxicology. University of Florida.
21. Klute, A. 1996. *Methods of Soil Analysis: Part 1: Physical and Mineralogical Methods.* SSSA Publications, Madison, WI.
22. USEPA. 1996. Test methods for evaluating solid waste, SW 846, 3rd ed. Office of Solid Waste and Emergency Response, Washington, D.C.
23. Watanabe, F.S and S.R. Olsen. 1965. Test of an ascorbic acid method for determining phosphorous in water and $NaHCO_3$ extracts from soil. *Soil Sci. Soc. Am. Proc.* 29:677–678.
24. Olson, R.V and R. Ellis. 1982. *Methods of Soil Analysis. Part 2. Agronomy Monograph.* 9. ASA and SSSA, Madison, WI. 301–312.
25. Sarkar D. and R. Datta. 2003. A modified *in-vitro* method to assess bioavailable arsenic in pesticide-applied soils. *Environ. Pollut.* 126:363–366.
26. Chen, M. and L.Q. Ma. 2001. Taxonomic and geographic distribution of total P in Florida surface soils. *Soil Sci. Soc. Am. J.* 65:1539–1547.
27. Myers, J.C. 1997. *Geostatical Error Management, Quantifying Uncertainty of Environmental Sampling and Mapping.* Van Nostrand Reinhold, New York.
28. Geobopological Society. 1998. The 12 soil orders.
29. Wasay, S.A., W. Parker, P.J. Van Geel, S. Barrington, and S. Tokunaga. 2000. Arsenic pollution of a loam soil: Retention, form and decontamination. *J. Soil Contamination.* 9:51–64.
30. Jain, A., K.P. Raven, and R.H. Loeppert. 1999. Arsenite and arsenate adsorption on ferrihydrite: Surface charge reduction an net OH release Stoichiometry. *Environ. Sci. Technol.* 33:1179–1184.
31. Roussel, C., H. Brill, and A. Fernandez. 2000. Arsenic speciation: involvement in the evaluation of environmental impact caused by mine wastes. *J. Environ. Qual.* 29:182–188.
32. Woolson, E.A. 1973. Arsenic phytotoxicity and uptake in six vegetable crops. *Weed Sci.* 21:524–527.
33. Reed, J.F. and M.B. Sturgis. 1936. Toxicity from arsenic compounds to rice on flooded soils. *Am. Soc. Agron.* 28:432–436.
34. Crafts, A.S. and R.S. Rosenfels. 1939. Toxicity studies with arsenic in eighty California soils. *Hilgardia.* 12:177–200.
35. Jacobs, L.W. and D.R. Keeney. 1970. Arsenic-phosphorous interactions on corn. *Commun. Soil Sci. Plant Anal.* 1:85–94.

36. Evangelou, V.P. 1998. *Environmental Soil and Water Chemistry: Principles and Applications.* John Wiley & Sons, New York.
37. Sposito, G. 1984. *The Surface Chemistry of Soils.* Oxford University Press, New York.
38. Manahan, S.E. 1991. *Environmental Chemistry,* 5th ed. Lewis Publishers, Chelsea, MI.
39. Mok, W.M. and C.M. Wei 1994. In J.O. Nrigau (Ed.) *Arsenic in the Environment Part I: Cycling and Characterization.* John Wiley & Sons, New York.
40. Thanabalasingam, P. and W.F. Pickering 1986. Arsenic sorption by humic acids. *Environ. Pollut.* Ser. B 12:233–246.
41. Xu H., B. Allard and A. Grimvall. 1988. Influence of pH and organic substance on the adsorption of As(V) on geologic materials. *Water, Air, Soil Pollut.* 40:293-305.
42. Sarkar, D., M.E. Essington, and K.C. Misra. 1999. Adsorption of Mercury (II) by Variable Charge Surface of Quartz and Gibbsite. *Soil Sci. Soc. Am. J.* 63:1626–1636.
43. Huang, P.M. 1980. Adsorption process in soil. In O. Hutzinger (Ed.), *Handbook of Environmental Chemistry,* Springer–Verlag, Berlin and New York, vol. 2A, 47–58.
44. Prasad, G. 1994. In J.O. Nrigau (Ed.) *Arsenic in the Environment Part I: Cycling and Characterization.* John Wiley & Sons, New York.
45. Aggett, J. and G.A. O'Brien. 1985. Detailed model for the mobility of arsenic in lacustrine sediments based on measurements in Lake Ohakuri. *Environ. Sci. Technol.* 19:231–238.
46. Moore, J.N., W.H. Kicklin, and C. Johns. 1988 Partitioning of arsenic and metals in reducing sulfidic sediments. *Environ. Sci. Technol.* 22:432–437.
47. Boyle, R.W. and I.R. Jonasson. 1973. *J. Geochem. Explor.* 2:252–296.
48. Myers, R. and J. Ewel (Eds.). 1990. *Ecosystems of Florida.* University of Central Florida Press, Orlando, FL.
49. O'Neill, P. 1995. Arsenic. In *Heavy Metals in Soils.* Alloway, B.J. (Ed.), Blackie Academic and Professional, Glasgow.
50. Matera, V. and I. Le Hecho. 2001. In *Heavy Metal Release in Soils.* Selim, H.M. and D.L. Sparks (Eds.), Lewis Publishers, Boca Raton, FL.
51. Presley, B.J., R.J. Taylor, and P.N. Boothe. 1990. Trace metals in Gulf of Mexico oysters. *Sci. Total Environ.* 97/98:551–593.
52. Valette–Silver, N.J. G.F. Reidel, E.A. Crecelius, H. Windom, R.G. Smith and S.S. Dolvin. 1999. Elevated arsenic concentrations in bivalves from the southeast coasts of the USA. *Mar. Environ. Res.* 48:311–333.
53. Van Kauwenbergh, S.J. 1997. Cadmium and other minor elements in world resources of phosphate rock. Paper presented at the Fertilizer Soc. London. Int. Fert. Soc., York, U.K.
54. O'Connor, G.A. and D. Sarkar. 1999. Fate of applied residuals-bound phosphorus. Final Report, Contract WM 661, Florida Department of Environmental Protection, Tallahassee, FL.
55. Rodriguez, R.R., N.T Basta, S.W Casteel, and L. Pace. 1999. An *in-vitro* gastro-intestinal method to estimate bioavailable arsenic in contaminated soil and solid media. *Environ. Sci. Technol.* 33:642–649.
56. Ruby, M.V., A. Davis, T.E. Link, R. Schoof, R.L. Chaney, G.B. Freeman and P. Bergstrom. 1993. Estimation of lead and arsenic bioavailability using a physiologically based extraction test. *Environ. Sci. Technol.* 27:2870–2877.

Section II

Biogeochemistry

6 Solubility, Mobility, and Bioaccumulation of Trace Elements: Abiotic Processes in the Rhizosphere

Brett Robinson, Nanthi Bolan, Santiago Mahimairaja, and Brent Clothier

CONTENTS

6.1 INTRODUCTION

Trace elements are elements that are present in small concentrations (<1000 mg/kg) in living organisms. This definition includes all the elements except H, C, N, O, Na, Mg, P, S, Cl, K, and Ca. Despite their presence as only traces, they nonetheless affect biological processes positively as well as deleteriously.

Trace elements in soils originate from natural and anthropogenic sources. Consequently, their concentrations can vary considerably. The trace element loading in soil is a function of the parent material plus subsequent atmospheric or water-borne deposition. Elevated levels of trace elements in soil can adversely affect the soil's fertility and may represent an ecological and human health risk if they enter the food chain or leach into receiving waters. However, the impact of trace metals on soil and the surrounding environment often cannot be predicted simply by measuring the total concentration. This is because only the soluble and mobile fraction has the potential to leach or to be taken up by plants and enter the food chain.

The mobility, solubility, and bioaccumulation of trace elements depend on a plethora of soil, microbial, and plant factors, as well as the properties of the trace element. Chemical solubility is a prerequisite for physical mobility and bioavailability. Bioaccumulation can result from trace element transport in the rhizosphere and absorption onto soil organisms or passage through plants' plasmalemma at the root: soil interface. These processes can be augmented or retarded through soil amendments or modification of the soil's vegetation. The goal of such engineering approaches may be to diminish trace element mobility so that it poses minimum risk to the surrounding environment, or to promote solubility so that trace elements are taken up by plants or leached out of the rhizosphere.

Enhancing trace element solubility may be a tool for the remediation of contaminated soils through the use of plants to remove the trace element, phytoremediation, or by leaching contaminants out of the root zone. Alternatively, plant and animal productivity may be improved by increasing the trace element solubility in a soil that is deficient in one or more essential elements.

This chapter discusses the physicochemical processes that affect the fate of trace elements in the rhizosphere with a view to the remediation of contaminated soils.

6.2 TRACE ELEMENT SOLUBILITY IN THE RHIZOSPHERE

The key abiotic mechanisms that control solubility will now be discussed and how these processes can be represented analytically will be outlined. For any trace element, only some fraction of the total concentration will be in soil solution, with the remainder bound to the soil matrix. Mass balance of this distribution gives

$$M = \theta C + \rho S ,$$

(6.1)

where

M is the total concentration (mg/kg)
θ is the volumetric water content (m^3/m^3)
C is the trace element concentration in the soil solution (mg/L)
ρ is the bulk density of the soil (t/m^3)
S is the concentration bound to the soil matrix (mg/kg)

The solubility of a trace element in the rhizosphere is often described in a simple way by a distribution coefficient (K_d) where

$$K_d = \frac{S}{C} .$$

(6.2)

In some cases, however, as the total concentration of trace elements in a soil increases, unlike the representation of Equation 6.2, the soil's ability to adsorb these further loadings decreases, due to a saturation of the chemical-binding sites in the soil. The observations of sorption of most trace elements can be described using a Langmuir (Equation 6.3) or a Freundlich (Equation 6.4) isotherm, which accounts for this nonlinearity in sorption.

Langmuir:
$$S = \frac{KQA_rC}{(1+KC)}$$

(6.3)

Freundlich:
$$S = KC^n ,$$

(6.4)

where

K is the adsorption constant
Q is number of sorption sites (mol/m^3)
A_r is the atomic mass of the trace element (g/mol)
n is the Freundlich exponent

If $n = 1$, then Equation 6.4 collapses to the linear model of Equation 6.1.

In the Langmuir case, when Q is finite, as in the case of soils and sediments, the value of S approaches QA_r as the concentration of trace element in soil solution increases. Figure 6.1 shows the adsorption of Cd by a silt loam at a range of Cd concentrations in soil solution. In this case, the isotherm is described by the Langmuir equation (Equation 6.3) with values of Q and K equal to 28.7 and 0.14, respectively.

The solubility of trace elements in the rhizosphere is a function of the soil's chemical and physical properties. Most trace element ions carry a positive charge and can therefore be retained by the negative binding sites of the soil's matrix. The soil's cation exchange capacity (CEC), which indicates the number of negative charges per unit mass, provides an indication as to the soil's potential to retain positively charged ions. The negatively charged binding sites for trace elements occur on organic matter, clays, and the oxides of Fe, Mn, and Al, which make up the soil's matrix.

Trace elements that carry a negative charge, such as F$^-$, Br$^-$, and the oxyanions AsO_2^- and CrO_4^{2-}, can bind electrostatically to positively charged sites in the soil matrix, as occurs in variably charged soils. This is measured by the soil's anion exchange capacity (AEC). In many temperate soils, the AEC is so small as to be insignificant. Therefore, many negatively charged ions, such as Br$^-$, can move freely with soil moisture unaffected by exchange. They can be used as chemical "tracers" of water movement through soil.

However, some highly weathered soils, and those that contain significant quantities of such volcanic minerals as allophane and imogolite can have a significant AEC. The AEC depends on pH. Some trace element anions, such as arsenate and selenate, also form specific chemical bonds with soil components. This results in their adsorption exceeding the AEC of the soil. The strength of the

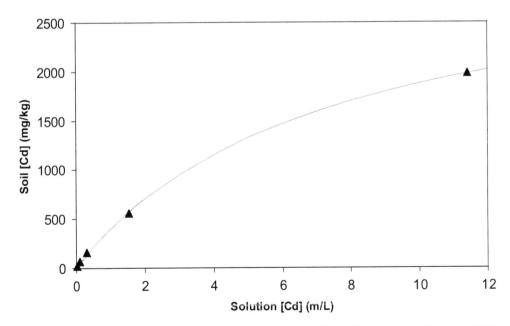

FIGURE 6.1 The effect of Cd^{2+} concentration on soil adsorption. The soil used was a silt loam pH 5.7, an organic matter content of 6.3%.

bond between the binding site of the soil and the trace element is a function of the size and charge on the trace element ion or complex. Smaller ions, with a higher charge, form the strongest bonds.

Although a small percentage of a soil's clay fraction carries a permanent negative charge, the charge carried by organic matter and, as mentioned some variable-charge clay minerals is pH dependent. Therefore, pH also profoundly affects the binding of trace elements in the rhizosphere. For positively charged ions such as Cd^{2+}, soil acidification invariably results in increased trace element solubility due to increased competition from H^+ ions at the negatively charged binding sites [1] (Figure 6.2). Conversely, soil adsorption of some trace elements such as Zn^{2+} can lower soil pH by releasing H^+ ions from bound surfaces [2]. Negatively charged trace elements, or trace element complexes, tend to be more soluble at a higher pH. The adsorption of trace elements onto variable charge minerals such as Fe, Al, and Mn oxides is also pH dependent. As these materials assume more negative charge under alkaline conditions, their capacity to absorb trace elements, which is generally positively charged, increases [3].

Trace elements may be displaced from exchange sites by other ions attracted from the soil solution. The extent of this competition for binding sites depends on the type and concentration of the trace element as well as that of the competing ion. As a general rule, trace elements such as Cd^{2+}, which has an atomic radius (r) of 0.97Å, can be displaced by other ions of a similar size and charge in soil solution, such as Ca^{2+} ($r = 0.99$Å). Therefore, soil amendments such as phosphates that are designed to immobilize heavy metals may actually promote the solubility of some co-contaminants such as As.

Trace element adsorption onto charged exchange sites is not the only mechanism governing trace element solubility in the rhizosphere. The extent, soluble–insoluble partitioning, and mobility of soil organic matter play an important role in the solubility and environmental fate of trace elements. Metal complexation by organic matter can promote or reduce metal solubility, depending on the solubility of the organic ligand.

Although the exact composition of dissolved organic matter is variable and complex, a large portion of this mobile material is composed of fulvic and humic acids. Minor components can also include macromolecular hydrophilic acids, carbohydrates, and carboxylic and amino acids [4]. Dissolved organic matter has been demonstrated to promote heavy-metal solubility [5] and mobility,

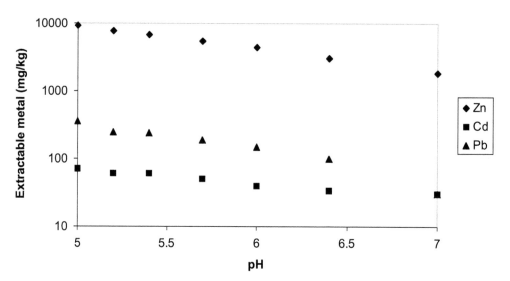

FIGURE 6.2 Metal extracted using 1 M ammonium acetate buffered at various pHs. Note the log scale on the y-axis. The total concentrations of Zn, Cd, and Pb in the soil were 40,416, 360, and 6209 mg/kg, respectively.

as will be discussed. In particular, Cu^{2+}, Hg^{2+}, and Pb^{2+} are strongly bound by humic acid [6], so their solubility and transport is thus promoted by increasing concentrations of dissolved organic matter. Wu et al. [7] demonstrated that, in a kaolinite soil, the Cu–humate complexes are mobile in acid and alkaline conditions, but not in neutral conditions, where they are sorbed. Almås et al. [8] found that the application of organic matter to soil enhanced the solubility of Zn and Cd due to the formation of complexes with dissolved organic acids.

A continual source of soluble organic matter in the rhizosphere is provided by plant exudates. These include metabolic products of fungi and microorganisms, the degradation of leaf litter and insoluble organic matter. The addition of organic fertilizers, effluents, and sludges can also provide sources. Fischer [4] demonstrated that amino acids derived from these solubilized heavy metals that have been bound to typical soil components.

The solubility of trace elements in the rhizosphere can also change over time. The soil's organic matter content is naturally dynamic. Furthermore, it may be augmented through the establishment of vegetation or the addition of composts or sewage sludge. Conversely, continual oxidation of organic matter occurs, particularly in tropical soils. Compost or sewage sludge amendment can offset this decline. The solubility trace elements that are strongly adsorbed by humic and fulvic acids will be affected by a change in the soil's organic matter content.

Over time, the solubilized trace elements can migrate and diffuse into the less accessible and immobile soil–water regions. There, they may become bound to the soil's solid phase. Consequently, the time needed for a contaminant to be redissolved or again be transferred from the stagnant- to the mobile-water phase may increase with aging of the compound in soil [9].

6.3 TRACE ELEMENT SPECIATION IN THE RHIZOSPHERE

The chemical speciation of trace elements in the rhizosphere profoundly affects their solubility, mobility, and toxicity. In soils, Cr that is present in the +6 oxidation state [Cr(VI)] is more mobile, more readily bioaccumulated, and 100 to 1000 times more toxic when present in the +3 oxidation stare [Cr(III)] [6].

In the rhizosphere, the various phases of trace elements are in dynamic equilibrium. Therefore, as trace elements are removed from soil solution by leaching or plant uptake, more will be desorbed from soil particles into solution. The rate at which this occurs depends on the distribution (see earlier equations) as well as the speciation of the trace element in the soil matrix.

Trace elements in soil solution and those weakly bound to exchange sites on the surface of the soil matrix are the most mobile and biologically active. Carbonate or sulphide precipitates, which may be occluded in Fe, Mn, or Al oxides, are immobile and biologically inactive. However, they may readily become mobilized by a shift in equilibrium caused by a change in soil pH and/or the redox conditions. The trace elements of the crystalline lattices of clays are generally inert. Other trace elements may be bound by more than one soil component. Alcacio et al. [10] demonstrated three possible binding configurations of Cu^{2+} on complexes of oxide minerals and organic matter. These are the Cu^{2+} bound to mineral surfaces and the Cu^{2+} bound to the organic matter that is absorbed onto oxides, along with Cu^{2+} bridges between oxides and organic matter.

The speciation of trace elements in the rhizosphere may be measured directly using analytical techniques, or it can be calculated using other parameters. The analytical procedures used to determine the speciation of trace elements in soils are usually specific for the solid phase or the solution phase. For example, sequential extraction procedures can be used to determine the distribution of trace elements in various soil fractions: soluble; exchangeable; sulfide/carbonate bound; organically bound; oxide bound; and residual or lattice mineral bound. Given the limitations and costs of analytical procedures, rather than measuring speciation and distribution of trace elements in soils, it is often calculated using speciation models such as GEOCHEM [11] and MINTEQ2 [12]. These are based on theoretical thermodynamics. Despite the fundamental principles that drive these models, their simplicity makes it problematic to model accurately trace element speciation and distribution in soils that contain a plethora of organic and inorganic matrices.

6.4 TRACE ELEMENT MOBILITY

Water is the vehicle for transport of solutes, including trace elements, through soil. Although solubility is a prerequisite of mobility, various rate-limited or kinetic geochemical and hydrological processes in the rhizosphere affect the transport of trace elements. The transport of dissolved trace elements in soils depends primarily on their concentration gradient spatially, and the mass flow of water. The latter is a function of the soil's matric water potential, its porosity, and long-distance preferential processes of transport through macropores. Nonequilibrium chemical or physical reactions may also occur. Physical nonequilibrium can result from nonuniform water flows and preferential transport.

The long-distance transport of trace elements in soils can be described using the convection dispersion equation (CDE), which describes the movement of solutes during transient flow [13]: In one-dimensional form, this is

$$\frac{\partial M}{\partial t} = \frac{\partial}{\partial z}\left[(\theta D_s)\frac{\partial C}{\partial z}\right] - \partial\frac{(q_w C)}{\partial z} - S_m(z) ,$$ (6.5)

- where
- M is the total trace element concentration (mg/kg)
- t is time (s)
- z is depth (m)
- θ is the volumetric water content (m^3/m^3)
- C is the local trace element concentration in the soil solution (mg/L)
- D_s is the solution diffusion coefficient (m^2/s)
- q_w is the soil water flux (m/s)
- $S_m(z)$ is the solute uptake, or release, by plant roots as a function of depth

Here, M is the total concentration, which is partitioned according to Equation 6.1.

Dissolved trace elements will be more mobile in sandy soils because the diffusive and connective water fluxes tend to be higher than in loams or clays. Mobility is reduced by plant uptake or sorption onto plant roots that are a sink for water and solutes. Mobility will be enhanced by plant release or desorption.

6.5 BIOACCUMULATION OF TRACE ELEMENTS

Some soil microorganisms accumulate trace elements, so-called bioaccumulation, from substrates that have a low total concentration. Bacteria and fungi can bioaccumulate trace elements. This uptake is via two processes: (1) sorption of metals by microbial biomass and its byproducts; and (2) the physiological uptake of trace elements by microorganisms through metabolically active and passive processes.

The uptake of trace elements by organisms is a function of biological species as well as the trace element's solubility and mobility. Soil fauna and microorganisms behave similarly to soil organic matter in that they possess binding sites for some trace elements. Robinson et al. [14] found that rhizobacteria such as *Pseudomonas fluorescens* from New Zealand pasturelands accumulated Cd to levels about 100 times that of the ambient solution in which they were grown. As with soil organic matter, the adsorption of Cd by these microorganisms decreased at lower solution pHs.

Motile soil animals such as worms and rotifers affect the transport and distribution of trace elements in soils. However, these biotic factors are beyond the scope of this chapter.

Plant roots differ from soil fauna and microorganisms in that they are a sink for soil water and by absorption they have the capacity to remove some trace elements from the rhizosphere. Soil solution is drawn from the root zone into the plants' roots and then via the stems to the leaves; it is lost to the atmosphere via transpiration. Any trace element taken up in the soil solution and entering the roots will accumulate in the roots or the shoots of the plant. High trace-element concentrations in the roots can result from water uptake inducing migration of the trace elements, via mass flow, to the root surface where they are precipitated [15]. In the aboveground portions, the highest concentrations are often found in the leaves because they are the major water sink prior to evaporation of the water. Metal translocation from the roots to the shoots is driven by water uptake [16,17].

The total amount of metal that accumulates in the plant does not necessarily equal the cumulative product of the soil-solution metal concentration times the volume of water transpired by the plant, as might be predicted for passive uptake. For a metal to be translocated to the aerial parts of a plant, it must enter the root via the symplastic or apoplastic [18] pathways, where some active or passive filtering may occur.

The fraction of dissolved trace elements that passes into the root xylem may be described by a root absorption factor (ϕ), a dimensionless lumped parameter that represents the root xylem/soil solution metal concentration quotient [19].

$$\phi = \frac{[C]_r}{[C]} \,, \tag{6.6}$$

where $[C]_r$ is the soluble metal concentration (mg/L) in the root xylem and $[C]$ is the soluble metal concentration (mg/L) in the soil solution.

The quotient ϕ is a lumped parameter incorporating many complex and often poorly understood biogeochemical factors that influence the passage of metals from soil into roots. Rhizobiological activity, root exudates, temperature, moisture, pH, and the concentration of competing ions will all affect ϕ. Also, ϕ will change depending on the trace element concentration in the soil solution. This would be particularly pronounced for essential elements that are subject to active uptake or root exclusion [20].

The speciation of the trace element in the rhizosphere is important in determining ϕ. Trace elements complexed with large organic molecules cannot easily pass through the roots' plasmalemma. This has implications for engineering plant uptake, which is discussed later. Free trace element ions are the most readily adsorbed. However, there is increasing evidence that some complexes, such as chloro–complexes can also be taken up by plant roots, although at lower efficiency [21].

6.6 BIOACCUMULATION AS AFFECTED BY Φ

Baker [22] divided plant species into three groups according to their above-ground metal concentrations in relation to the metal concentration in the soil. These three groups may be delineated using ϕ:

- For nonessential elements such as cadmium, nickel, and arsenic, plants with a very low ϕ are termed "excluders." Most plants that occur naturally on metalliferous soils are recognized as excluders.
- Plants that have a relatively constant ϕ over a wide range of metal solution concentrations are known as "indicators." In this case, the concentration in the plant has a near-linear relationship to the soluble metal concentration in soil solution. Plants that do not occur

naturally on metalliferous soils usually behave as "indicators" when grown in the presence of nonessential elements.

- The third category of plants are those that tolerate very high concentrations of metal in their aerial parts or have an active uptake mechanism even for nonessential metals (high ϕ). These plants are known as "hyperaccumulators" [23].

For excluders and hyperaccumulators, ϕ might be constant over just a narrow concentration range. There can be a sudden increase in plant metal concentration at high soil–solution concentrations. At this point, control mechanisms break down, and metal "floods" into the plant. This may be an overload of the regulatory mechanism or a break down of the plasma membrane at the apoplast/symplast interface. When this occurs, the plants show toxicity symptoms and biomass production is reduced.

The bioaccumulation of trace elements plays a major role in determining their environmental fate. This determines whether the element remains in the rhizosphere, leaches, or is removed by plants. Adjusting trace element mobility or bioaccumulation can be used as a tool to remediate contaminated soils or improve the fertility of soils that have low bioavailability of essential nutrients.

6.7 ENGINEERING TRACE ELEMENT SOLUBILITY, MOBILITY, AND BIOACCUMULATION FOR IMPROVED FERTILITY OR ENVIRONMENTAL PROTECTION

Soil amendments can be used to promote or reduce trace element solubility and bioaccumulation in contaminated soils. Amendments that induce solubility can be used to cleanse polluted sites via leaching the contaminating trace elements or by facilitating their uptake by plants. Subsequent removal of the plants would also remove the contaminant from the site. Alternatively, soil amendments that immobilize trace elements can lessen their impact on soil-borne organisms and reduce exposure pathways.

Amendments may also serve to enhance the fertility of a soil in which the bioavailability of one or more trace elements is limiting agricultural production. Essential trace elements such as Fe, Co, and Se may be present in soils at relatively high total concentrations, yet be unavailable to organisms due to physicochemical conditions in the soil.

6.8 TRACE ELEMENT SOLUBILIZATION

Solubilizing trace elements for the purposes of *ex situ* soil washing has been used widely for the remediation of contaminated soils in Europe [24]. Tokunaga and Hakuta [25] evaluated an acid-washing process to extract the As(V) from a soil contaminated at 2830 mg/kg by As. Phosphoric acid proved to be a promising extractant, attaining 99.9% As extraction at 9.4% acid concentration. The success of soil washing largely depends on the speciation of the trace elements because it is based on the desorption or dissolution of trace elements from the soil inorganic and organic matrix during washing with acids or chelating agents. Although soil washing is suitable for off-site treatment, it can also be used for on-site remediation using mobile equipment.

6.9 INDUCED BIOACCUMULATION

Plants can be induced to take up trace elements by the use of amendments. Chelating agents, such as ethylenediaminetetraacetic acid (EDTA), diethylenetriaminepentaacetic acid (DTPA), and nitrilotriacetic acid (NTA), and organic acids such as citric and oxalic acids are commonly proposed. These agents have been proven effective in enhancing the solubility of Pb, Cd, Cu, Zn, and other trace element cations [26–29].

Addition of thiosulphate and thiocynate salts to mine spoil induced plants to accumulate Hg [30] and, auspiciously, Au [31]. Chloride anions increased the Cd solubility in soils due to the formation of relatively stable chloride ion complexes ($CdCl^+$ and $CdCl_2$) [32]. Similarly, the addition of chloride to soils has also been demonstrated to enhance the uptake of Cd by plants [33].

Solubilization, however, does not necessarily induce bioaccumulation. For example, Cu^{2+} is solubilized by dissolved organic matter, but the resulting complex is not taken up by plants due to its inability to pass through the plasmalemma [34]. Experiments using the nickel hyperaccumulator *Berkheya coddii* have shown that the chelating agents cause a decrease in nickel uptake, despite enhancing the nickel solubility in the soil (Figure 6.3A). However, increasing the concentration of EDTA and soluble metal in the substrate can induce plant uptake. Figure 6.3B shows that when 4 g of EDTA per kg of soil is added, *Arrhenatherum elatius* can be induced to take up nearly 1000 mg/kg Ni on a dry matter basis. Depending on the metal species, induced uptake can cause plant death. Disruption of the plasmalemma possibly allows the Ni–EDTA complex to enter the xylem.

The strategy used for chelate-enhanced phytoremediation is to apply chelate to a mature crop growing on a contaminated soil. As well, a pesticide can be used to disrupt root membranes allowing the complexed metal to pass directly into the root xylem [35].

Blaylock [35] showed an impressive decrease in soil-lead concentration over 2 years at two sites in the U.S. using a combination of *Brassica juncea* and EDTA to induce accumulation. Unfortunately, the mass balance of lead was not shown. It is therefore uncertain just how much lead the plants removed and how much deleteriously leached through the soil profile to contaminated receiving environments.

Environmental concerns have been raised over the use of induced bioaccumulation due to the possibility that some of the metals might leach through the soil profile, possibly entering groundwater [36]. Processes such as preferential flow may exacerbate metal leaching [37]. Soil amendments may also persist in the environment, creating additional problems. The addition of chelating agents is likely to induce the solubilization of other than target metals — which may be phytotoxic — such as Al and Mn. Chelators such as EDTA can act as chemical plows, redistributing surface contamination down the soil profile. Concentration near the soil surface is reduced, thereby reducing exposure pathways, but the total amount of contaminant is not affected. Induced bioaccumulation should thus only be used on hydraulically isolated treatment sites where the connection to receiving waters has been "broken."

6.10 *IN SITU* IMMOBILIZATION

Reducing trace element solubility, mobility, and bioaccumulation in contaminated soils is an effective, low-cost means of remediation that does not require drastic disturbance of the site. Naturally occurring or artificial soil amendments such as liming material, phosphate, zeolite, bentonite, clay, Fe metal, Fe and Mn oxides, and organic matter, may be used to mitigate the toxic effects of trace elements [38]. These amendments reduce trace element mobility by promoting the formation of insoluble precipitates or by enhancing the soil's capacity to bind the trace element. The latter can be achieved directly through the addition of adsorbent material or indirectly by adjusting the soil's pH–Eh conditions to promote trace element absorption onto the soil's matrix.

Chemical immobilization using phosphate amendments, such as mineral apatite, synthetic hydroxyapatite, and phosphate salts, has proven effective in reducing heavy-metal solubility by the formation of metal–phosphate complexes [39] and by increasing the number of negatively charged exchange sites [40]. These additions reduce metal bioavailability not only to plants, but also for humans who may have ingested contaminated soil [41]. The solubility of lead in soil can be greatly reduced by the formation of chloropyromorphite [$Pb_5(PO_4)_3Cl$]. Several microcosm studies have shown that chloropyromorphite can be formed through the addition of hydroxyapatite [$Ca_{10}(PO_4)_6(OH)_2$] [42,43]. Brown et al. [44] demonstrated that phosphate fertilizers could poten-

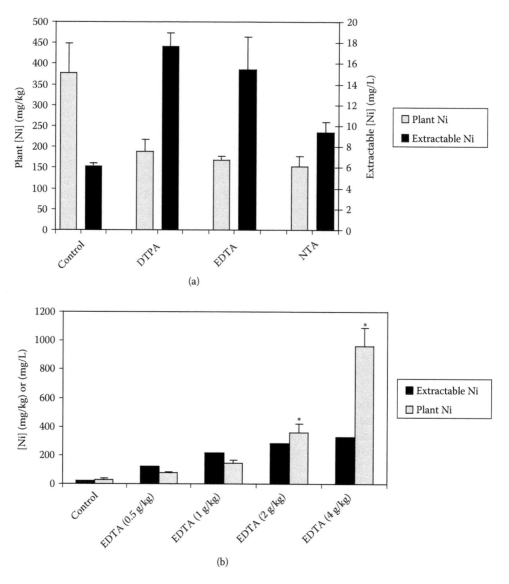

(a)

(b)

FIGURE 6.3 (A) The effect of chelating agents added at 2 g/kg soil, on shoot Ni concentration (dry matter) by the hyperaccumulator *Berkheya coddii*. (B) The effect of increasing EDTA addition on the shoot Ni concentration (dry matter) of *Arrhenatherum elatius*. Extractable Ni was determined using a 1 M NH$_4$OAc extractant. * denotes plant death.

tially be used in combination with organic matter and Fe-rich material, as soil amendments to reduce plant availability of Cd, Zn, and Pb.

Phosphate amendments such as hydroxyapatite are effective in reducing the solubility of Pb, Cd, Zn, Al, Ba, Co, Mn, Ni, and U. However, phosphate has been shown to promote the solubility of As and Cr [45], possibly through reduced sorption of the oxyanions due to an increase in pH and competition from PO$_4^{3-}$.

A variety of inorganic and organic amendments have been used to reduce Cr(VI) to the less mobile, less toxic Cr(III) species. Fe(II)-bearing minerals form effective reductants. Surface-bound organic matter does this also. The latter is catalyzed by soil mineral surfaces, and Cr(III) binds tightly to surface species or is precipitated as Cr(OH)$_3$ [46]. Similarly, Bolan et al. [47] have shown that organic amendments, such as animal and poultry manures rich in dissolved organic carbon, are very effective in reducing Cr(VI) to Cr(III).

Lombi et al. [48] demonstrated that the 2% addition of bauxite residue, "red mud," to contaminated soils reduced the solubility of Cd, Pb, Ni, and Zn, but not Cu. The remedial action of this material was attributed to a rise in soil pH and adsorption of the metals onto oxides of Fe and Mn.

Liming has been demonstrated to be effective in reducing the mobility of trace element cations in variable-charge soils by increasing the negative charge on oxides, clays, and organic matter. The effectiveness of raising the pH on metal immobilization also depends on the liming agent. Bolan et al. [49] found that $Ca(OH)_2$ was less effective than KOH in immobilizing Cd^{2+} due to competition between Ca^{2+} and Cd^{2+} for adsorption sites.

Establishing vegetation on a contaminated site can reduce the solubility and mobility of trace elements. Plant growth reduces trace element mobility in the substrate through the addition of organic matter, creation of an aerobic environment, root uptake, and returning rainfall to the atmosphere by evapotranspiration [50]. Römkens et al. [51] showed that Cu solubility was lowered significantly in the root zone of *Agrostis capillaries* (var Parys Mountain). Turpeinen et al. [52] demonstrated that Pb solubility was reduced by up to 93% by pine seedlings. The use of vegetation for the remediation of contaminated sites, phytoremediation, has some advantage over other *in situ* immobilization techniques. Once established, the physicochemical change induced by vegetation is permanent (Figure 6.4).

FIGURE 6.4 Vegetation established on the Tui mine tailings, Te Aroha, New Zealand. The vegetation has reduced metal mobility by adding organic matter to the substrate and returning some rainfall to the atmosphere via evapotranspiration.

6.11 CONCLUSIONS

The effect of soil-borne trace elements on ecosystems and human health is a function of their solubility and mobility. These properties are often disproportionate to the trace element's total concentration. Rather, the speciation of the trace element and the physicochemical properties of the ambient soil are of overriding importance in determining bioavailability. Solubility is primarily a function of soil pH and organic matter content. Mobility depends on solubility and water transport processes.

Trace element bioaccumulation occurs via adsorption onto roots and soil microorganisms as well as absorption into roots and translocation to the aerial portions plants. The latter is affected by the permissivity of the roots' plasmalemma to the dissolved trace element species.

An innovative low-cost strategy for the remediation of trace element contaminated soils is *in situ* immobilization using chemical amendments, modified vegetation, or a combination of the two. Remediation technologies that rely on amendments that promote trace element solubility will be limited to sites where the risk of contaminant leaching to groundwater has been eliminated.

Future development of whole-system models that calculate the leaching and plant uptake of trace elements in the rhizosphere will facilitate the design and implementation of remediation technologies designed to isolate contaminated soils from the surrounding biota.

REFERENCES

1. Naidu, R. et al. Ionic-strength and pH effects on the sorption of cadmium and the surface charge of soils. *Eur. J. Soil Sci.* 45, 419, 1994.
2. Yu, S. et al. Adsorption–desorption behavior of copper at contaminated levels in red soils from China. *J. Environ. Qual.* 31, 1129, 2002.
3. Atanassova, I.D. Adsorption and desorption of Cu at high equilibrium concentrations by soil and clay samples from Bulgaria. *Environ. Pollut.* 87, 17, 1995.
4. Fischer, K. Removal of heavy metals from soil components and soil by natural chelating agents. Part 1: Displacement from clay minerals and peat by L-cysteine and L-penicillamine. *Water, Air, Soil Pollut.* 137, 267, 2002.
5. Weng, L.P. et al. Complexation with dissolved organic matter and solubility control of heavy metals in a sandy soil. *Environ. Sci. Technol.* 36, 4804, 2002.
6. Kerndorff, H. and Schnitzer, M. Sorption of metals on humic acids. *Geochim. Cosmochim. Acta.* 44,1701, 1980.
7. Wu, J., West, L.J., and Stewart, D.I. Copper (II) humate mobility in kaolinite soil. *Eng. Geol.* 60, 275, 2001.
8. Almås, S., Sing, B.R., and Salbu, B. Mobility of cadmium-109 and zinc-65 in soil influenced by equilibrium time, temperature, and organic matter. *J. Environ. Qual.* 28, 1742, 1999.
9. Seuntjens, P. et al. Aging effects on cadmium transport in undisturbed contaminated sandy soil columns. *J. Environ. Qual.* 30, 1040, 2001.
10. Alcacio, T.E. et al. Molecular scale characteristics of Cu(II) bonding in goethite-humate complexes. *Geochim. Cosmochim. Acta* 65, 1355, 2001.
11. Mattigod, S.V. and Sposito, G. GEOCHEM: a computer program for the calculation of chemical equilibria in soils solutions and other water systems. Department of Soils and Environmental Sciences, University of California, Riverside, 1979.
12. Allison, J.D., Brown, D.S., and Novo–Gardac, K.J. MINTEQA2/PRODEFA2, A geochemical assessment model for environmental system (EPA/600/3-91/021). U.S. Environmental Protection Agency, Athens, GA, 1991.
13. Vogeler, I. et al. Contaminant transport in the root zone. in *Trace Elements in Soil: Bioavailability, Flux and Transfer.* I.K. Iskandar and M.B. Kirkham, Eds. Lewis Publishers. Boca Raton, FL, 2001 chap. 9.
14. Robinson, B.H. et al. Cadmium adsorption by rhizobacteria: implications for New Zealand pastureland. *Agric. Ecosyst. Environ.* 87, 315, 2001.

15. Zhao, F.J. et al. Zinc hyperaccumulation and cellular distribution in *Arabidopsis halleri. Plant Cell Environ.* 23, 507, 2000.

16. Salt, D.E. et al. Mechanisms for cadmium mobility and accumulation in Indian mustard. *Plant Physiol.* 109, 1427, 1995.

17. Hinchman, R.R., Negri, C.M., and Gatliff, E.G. Phytoremediation: using green plants to clean up contaminated soil, groundwater and wastewater, in Proc. Int. Topical Meeting Nucl. Hazardous Waste Manage., Spectrum 96, Seattle, WA, 1996.

18. Marschner, H. Mineral Nutrition of Higher Plants. Academic Press. London, 1985, 65.

19. Robinson, B.H. et al. Phytoextraction: an assessment of biogeochemical and economic viability. *Plant Soil* 249, 117, 2003.

20. Salisbury, F.B. and Ross, C.W. *Plant Physiology*, 4th ed., Wadsworth, CA, 1992, 154.

21. Smolders, E. and McLaughlin, M.J. Chloride increases cadmium uptake in Swiss chard in a resin buffered nutrient solution. *Soil Sci. Soc. Am. J.* 60, 1443, 1996.

22. Baker, A.J.M. Accumulators and excluders — strategies in the response of plants to heavy metals. *J. Plant Nutr.* 3, 643, 1981.

23. Brooks, R.R. et al. Detection of nickeliferous rocks by analysis of herbarium specimens of indicator plants. *J. Geochem. Explor.* 7, 49, 1977.

24. Tuin, B.J.W. and Tels, M. Continuous treatment of heavy metal contaminated clay soils by extraction in stirred tanks and counter current column. *Environ. Technol.* 12, 178, 1991.

25. Tokunaga, S. and Hakuta, T. Acid washing and stabilization of an artificial arsenic-contaminated soil. *Chemosphere* 46, 31, 2002.

26. Huang, J.W. and Cunningham, S.D. Lead phytoextraction: species variation in lead uptake and translocation. *New Phytol.* 134, 75, 1996.

27. Blaylock, M.J. et al. Enhanced accumulation of Pb in Indian mustard by soil-applied chelating agents. *Environ. Sci. Technol.* 31, 860, 1997.

28. Robinson, B.H., Brooks, R.R., and Clothier, B.E. Soil amendments affecting nickel and cobalt uptake by Berkheya coddii: potential use for phytomining and phytoremediation. *Ann. Bot.* 84, 689, 1999.

29. Thayalakumaran, T. et al. Plant uptake and leaching of copper during EDTA-enhanced phytoremediation of repacked and undisturbed soil. *Plant Soil* 254, 415, 2003.

30. Moreno, F. et al. Phytoremediation of mercury-contaminated mine tailings by induced plant-mercury accumulation. *Environ. Pract.* 6, 165, 2004.

31. Anderson, C.W.N. et al. Induced hyperaccumulation of gold in plants. *Nature* 395, 553, 1998.

32. Weggler, K., McLaughlin, M.J. and Graham, R.D. 2004. Effect of chloride in soil solution on the plant availability of biosolid-borne cadmium. *J. Environ. Qual.* 33, 496, 2004.

33. McLaughlin, M.J., et al. Increased soil salinity causes elevated cadmium concentrations in field grown potato tubers. *J. Environ. Qual.* 23, 1013,1994.

34. Bolan, N.S. and Duraisamy, V.P. Role of inorganic and organic soil amendments on immobilization and phytoavailability of heavy metals: a review involving specific case studies. *Aust. J. Soil. Res.* 41, 533, 2003.

35. Blaylock, M.J. Field demonstrations of phytoremediation of lead contaminated soils, in *Phytoremediation of Contaminated Soil and Water*, Terry, N. and Bañuelos, G., Eds. Lewis publishers, Boca Raton, FL, 2000, chap 1.

36. Lombi, E. et al. Phytoremediation of heavy-metal contaminated soils: natural hyperaccumulation versus chemically enhanced phytoextraction. *J. Environ. Qual.* 30, 1919, 2001.

37. Bundt, M. et al. Impact of preferential flow on radionuclide distribution in soil. *Environ. Sci. Technol.* 34, 3895, 2000.

38. Cheng, S. and Hseu, Z. *In situ* immobilization of cadmium and lead by different amendments in two contaminated soils. *Water, Air, Soil Pollut.* 140, 73, 2002.

39. McGowen, S.L., Basta, N.T., and Brown, G.O. Use of diammonium phosphate to reduce heavy metal solubility and transport in smelter-contaminated soil. *J. Environ. Qual.* 30, 493, 2001.

40. Bolan, N.S. et al. The effects of anion sorption on sorption and leaching of cadmium. *Aust. J. Soil. Res.* 37, 445, 1999.

41. Zhang, P., Ryan, J.A., and Yang, J. *In vitro* soil Pb solubility in the presence of hydroxyapatite. *Environ. Sci. Technol.* 32, 2763, 1998.

42. Ma, Q.Y., Traina, S.J., and Logan T.J. *In situ* lead immobilization by apatite. *Environ. Sci. Technol.* 27, 1803–1810, 1993.
43. Ryan, J.A. et al. Formation of chloropyromorphite in a lead-contaminated soil amended with hydroxyapatite. *Environ. Sci. Technol.* 35, 3798, 2001.
44. Brown, S. et al. *In situ* soil treatments to reduce the phyto- and bioavailability of lead, zinc, and cadmium. *J. Environ. Qual.* 33, 522, 2004.
45. Seaman, J.C., Arey, J.S., and Bertsch, P.M. Immobilization of nickel and other metals in contaminated sediments by hydroxyapatite addition. *J. Environ. Qual.* 30, 460, 2001.
46. Jardine, P.M. et al. Fate and transport of hexavalent chromium in undisturbed heterogenous soil. *Environ. Sci. Technol.* 33, 2939, 1999.
47. Bolan, N.S. et al. Reduction and phytoavailability of Cr(VI) as influenced by organic manure compost. *J. Environ. Qual.* 32, 120, 2003
48. Lombi, E. et al. *In situ* fixation of metals in soils using bauxite residue: chemical assessment. *Environ. Pollut.* 118, 435, 2002.
49. Bolan, N.S. et al. Immobilization and phytoavailability of cadmium in variable charge soils. II. Effect of lime addition. *Plant Soil* 251, 187, 2003.
50. Robinson, B.H. et al. Phytoremediation: using plants as biopumps to improve degraded environments. *Aust. J. Soil. Res.* 41, 599, 2003.
51. Römkens, P.F.A.M., Bouwman, L.A., and Boon, G.T. Effect of plant growth on copper solubility and speciation in soil solution samples. *Environ. Pollut.* 106, 315, 1999.
52. Turpeinen, R., Salminen, J., and Kairesalo, T. Mobility and bioavailability of lead in contaminated boreal forest soil. *Environ. Sci. Technol.* 34, 5152, 2000.

7 Appraisal of Fluoride Contamination of Groundwater through Multivariate Analysis: Case Study

S. Tripathy, M.K. Panigrahi, and N. Kundu

CONTENTS

ABSTRACT

Groundwater in the surrounding area of a hot spring in the Nayagarh district of Orissa, India, is enriched in fluoride. Although the hot-spring water has higher pH and bicarbonate concentration but is depleted in Mg, the concentrations of other species are more or less similar to the uncontaminated groundwater. The temperature and pH of the hot spring water mostly rules out any abnormal input of heat and addition of species from deeper crustal origin other than the expected exchange of components between the circulating meteoric water and the surrounding rock. Principal component analysis on measured parameters in the hot spring water, contaminated and uncontaminated groundwater from the area, confirmed mixing of water from the hot spring with that in the subsurface, causing fluoride enrichment in the latter. Furthermore, R-mode factor analysis on selected hydrochemical parameters explains the interrelationships among pH, fluoride, and bicarbonate concentrations; the scores of the two principal factors discriminate the water types. The hydrochemical parameters were classified into two groups by cluster analysis. The first group comprised pH, F^-, Na, and Fe ascribed to the process operating within the hot spring and the second, consisting of Ca, Mg, and HCO_3^-, is related to the host rock.

7.1 INTRODUCTION

The hydrosphere chemistry in general is influenced by a multitude of anthropogenic factors besides natural causes — namely, mineralogy and chemistry of the country rock through which water flows and the interactions among them in a given environment. The residence time of water in any reservoir (host aquifer rock), the soil through which the water seeps, and phenomena like melting of ice and discharge of geothermal fluids are additional factors affecting the constitution of hydrochemical species of any water.

Various anthropogenic activities generate many chemical species that mix and interact with the surface and groundwater reservoirs through a series of processes. The general chemical makeup of the water, particularly anomalous concentrations of chemical species that pose serious threats to life, are the results of such processes. It is generally not possible to work out the cause and effect relationship in all cases deterministically. In an attempt to understand the process, many chemical parameters in the water are analyzed, landing up in a multivariate system; the objective then is to transform the dataset to an easily understandable form.

Several multivariate statistical techniques have been adopted for this purpose and results are quite encouraging [1–10]. Among the multivariate techniques, principal component analysis, factor analysis, and cluster analysis have been quite effective in understanding the processes controlling the chemistry of natural waters [1,11–15] and contaminated waters, in particular [16].

Fluoride enrichment of surface and groundwater is a frequent phenomenon and health problems arising out of excess intake of the anion are quite widespread as indicated by the statistics on the regions and populations facing the problem all over the world [17–20]. Fluoride enrichment of water and resultant health problems are reported from many parts of India [21–23].

7.2 STUDY AREA

The study area comprises a cluster of villages (Figure 7.1) in the Nayagarh district of Orissa, India, where enrichment of fluoride in ground and surface water was reported earlier [24]. The population of the area is affected by dental and skeletal fluorisis due to intake of fluoride-enriched water. The area spreads over approximately 100 km^2 (Figure 7.1) and experiences subtropical climate with maximum temperature rising up to 42°C during the month of May. The southwest monsoon is the principal source of rainfall; it breaks in the middle of June and continues till early October. The average annual rainfall recorded over the period of the last 12 years is 1145.35 mm. The area is a part of the Eastern Ghats geological province, comprising mainly granite gneisses and more Al-rich quartzofeldspathic gneisses (khondalites) with Fe–Mg-rich charnockites and mafic granulites. Among the structural elements, joints/fractures and foliations (trending NE–SW with gentle easterly dips) form the prominent conduits for subsurface fluid movement.

In the area, groundwater generally occurs under unconfined conditions. The area is also characterized by the presence of a hot spring, one among many such occurrences in the Eastern Ghats province of Orissa over a linear tract of 200 km. The surface temperature of the hot spring water varies from 40 to 60°C. The hot spring is located near the topographic low of the area with a steeper slope in the northern, western, and southwestern parts compared to that of the eastern and southeastern parts. The soil cover varies in thickness from 2 to 4 m and is thinner over elevated ground. The fluoride-enriched zones in the study were earlier delineated by employing geoelectrical resistivity sounding in conjunction with hydrogeochemical study [13]. The present work has attempted a more detailed analysis of the hydrogeochemical parameters employing multivariate geostatistics to signify the observed variability in the data physically in terms of natural processes giving rise to fluoride enrichment in water.

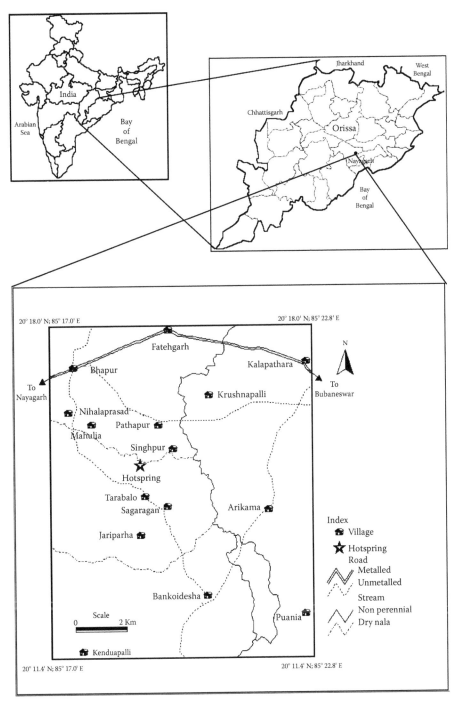

FIGURE 7.1 Location map of the study area around the hot spring in the Nayagarh district of Orissa, India.

7.3 MATERIALS AND METHODS

Water samples in duplicate from various potable sources comprising bore wells and open wells in 55 villages in the study area were collected during the postmonsoon period. In addition, premonsoon water samples collected from selected locations were used to study the temporal variations of different parameters. In addition, water samples from the hot spring were also collected in the pre- and postmonsoon periods. One set of samples was filtered and acidified with 2 ml of 35% HNO_3 and stored in air-tight polyethylene bottles for use in the determination of pertinent cations; the unfiltered and nonacidified ones were used to determine the concentrations of anions, pH, electrical conductivity (EC), and total dissolved solids (TDS).

Water samples collected from the hot spring and potable sources were analyzed for fluoride, chloride, EC, and pH by Orion ion selective electrode model 1260. The instrument was calibrated using E-MERCK buffers of pH 4, 7, and 9, respectively. For fluoride measurement, the instrument was calibrated using 0.1 M standard sodium fluoride (NaF) solution; for chloride, 1 M chloride solution was used. Blank solutions made up from MILLI-Q water (resistivity: 10 to 15 $M\Omega cm$; conductivity: <0.2 $\mu s/cm$; TOC: <30 ppb) were also analyzed. To each 50-ml sample, including the blank before analysis for fluoride, 2 ml of total ionic strength adjustment buffer (TISAB IV) was added. Similarly, 2 ml of ionic strength adjuster (ISA) was added to 50 ml of each sample and the blank, during chloride measurement. The buffer solutions helped in minimizing complex formation and buffered the solution pH to 5.2. The samples were continuously stirred using a magnetic stirrer during measurement.

The concentrations of Na, Ca, K, Mg, Al, and Fe were determined by analyzing the acidified water samples employing the Perkin–Elmer Model 3300 inductively coupled plasma optical emission spectroscopy (ICP-OES) at the Department of Earth Sciences, University of Western Ontario, Canada. Standard wet chemical methods [25] were employed in the determination of carbonate and bicarbonate concentrations in water samples. Sulphate and silica were determined by CINTRA-5 UV-VIS spectrophotometer. Results of analyses are presented in Table 7.1.

7.4 SALIENT HYDROCHEMICAL FEATURES

The pH of hot spring water ranges from 7.9 to 8.9, with a mean of 8.2, and the fluoride concentration varies from 12 to 14.2 ppm, with a mean of 13 ppm; the temperature of the surface discharge varies from 40 to 60°C, with a mean of 53°C. The thermal waters are soft and more alkaline in nature than the groundwater. The hot spring water is mostly depleted in Mg and low in Ca and K; however, Na content is considerably higher and the bicarbonate content is marginally higher than the groundwater. The inverse relationship of fluoride concentration in water with Ca and Mg is reported quite often [26,27].

The subsurface temperature of the hot spring is about 100 to 125°C on the basis of Na–K–Mg triangular plot and its chemistry indicates partial equilibrium with rocks of average continental crust [28]. Because the area is a part of a stable continental shield with a normal geothermal gradient, input of additional heat for sustaining a geothermal system at this temperature is required. Crustal radiogenic heat is a strong possibility. The temperature, along with the alkaline pH of the hot spring, also would rule out any additional input of volatile component from the deeper part and the chemistry of the hot spring is a result of exchange of components with the near-surface rocks.

Water samples with more than 1.5 ppm fluoride fall into "sodium" or "potassium" type in cation facies and include samples from the hot spring. About 60% of these are the chloride type in anion facies. Among the samples containing less than 1.5 ppm fluoride, 50% fall in "no dominant type," and approximately 25% are sodium type or potassium type; the remainder is "calcium" type. Concurrently, nearly half of these samples fall into the "bicarbonate" type and the rest into the "chloride" type in anionic facies [12].

Spatial distribution of fluoride concentration of groundwater (Figure 7.2) was obtained from the analytical data using SURFER 6.04. The semicircular area around the hot spring, including the villages of Singhpur and Sagargaon, shows high concentrations of fluoride in groundwater above the permissible limit; in other words, a delineation of the contaminated zone in the area is revealed.

TABLE 7.1
Results of Analysis of the Ground Water and Hot Spring Water of the Study Area

No.	pH	EC	TDS	TH	HCO_3^-	CO_3^{2-}	F^-	Cl^-	SO_4^{2-}	Ca^{2+}	Mg^{2+}	Na^+	K^+	Al^{3-}	Fe^{3-}
1	7.7	428	257	86	162	7.8	0.28	28.5	42.0	15	11.7	97.8	1.3	0.96	0.97
2	7.5	1037	622	738	198	2.3	0.38	181	0.8	183	68.3	175.2	2.1	0.52	4.69
3	7.6	1341	805	96	301	4.7	1.06	202	4.3	1.9	22.2	143.5	1.7	0.46	0.02
4	7.7	1382	829	95	301	20.3	0.85	219	12.3	1.8	22.1	109.6	1.4	0.72	0.47
5	7.6	2273	1364	168	368	12.5	0.65	464	11.6	2.3	39.4	114.0	10.2	0.82	2.65
6	7.7	1226	736	220	257	7.8	0.64	204	12.9	76.5	7.0	76.9	3.9	0.28	0.51
7	7.5	2000	1547	44	285	6.2	0.55	543	16.5	4.4	7.9	10.7	15.8	0.16	3.93
8	7.5	550	352	249	160	0.9	0.67	26.5	0.1	49.1	18.0	19.0	1.6	0.40	0.37
9	7.2	2054	1232	197	273	9.4	0.47	385	0.5	2.3	46.5	18.6	11.0	0.92	0.98
10	7.7	2000	1800	48	371	14.0	0.33	712	12.3	6.8	7.6	100.8	56.2	0.71	0.72
11	7.9	2634	1580	35	582	17.9	1.11	411	17.4	2.2	7.3	96.7	94.3	0.14	0.32
12	7.6	620	397	128	130	0.3	0.39	9.2	1.4	35.3	9.6	15.6	2.0	0.55	1.10
13	7.6	617	370	468	222	7.8	0.49	37.9	8.9	46.5	85.6	17.8	3.2	0.99	0.41
14	6.8	1453	930	515	170	0.6	0.53	228	26.8	159.6	28.6	35.5	3.2	0.86	0.45
15	7.1	415	249	92	105	0.0	0.21	64.3	14.1	14	13.9	35.9	2.8	0.77	29.2
16	7.9	976	586	87	523	15.6	0.69	32.3	9.5	1.4	20.4	53.6	1.2	0.59	0.24
17	7.7	106	68	268	150	0.5	1.15	16.3	4.8	46.4	37.0	28.9	2.8	0.37	0.22
18	7.0	271	163	58	86	0.0	0.27	27.7	8.1	10	8.0	16.6	5.2	0.56	1.38
19	7.7	2481	1489	233	514	15.6	1.60	441	12.3	2.2	55.2	108.6	17.0	0.38	0.23
20	7.7	216	138	71	120	0.3	0.28	36.3	0.1	22.7	3.6	13.9	3.0	0.35	1.56
21	6.8	1259	755	80	127	0.0	0.36	303	8.3	1.9	18.3	14.7	2.2	0.67	17.70
22	7.8	945	567	484	377	10.9	0.38	69.9	13.4	93.5	60.9	190.9	28.5	0.12	1.36
23	7.3	254	152	31	67	3.1	0.17	35	19.3	9	2.1	91.3	17.4	0.38	0.31
24	7.7	886	532	486	209	1.6	1.17	157	1.4	47.5	89.2	86.4	1.6	0.94	1.97
25	8.0	844	506	433	320	12.5	1.46	45.1	14.1	35.5	83.6	80.6	1.9	0.21	0.42
26	7.9	1114	668	884	428	14.0	0.62	92.3	1.3	342	7.3	33.1	1.5	0.25	0.30
27	7.8	1572	943	454	301	7.8	0.41	252	6.0	169.5	7.5	15.4	1.7	0.73	0.21
28	7.6	100	64	432	110	0.3	0.29	247	12.9	144.4	17.6	38.5	2.9	0.39	0.70
29	7.6	1127	676	34	374	14.0	0.43	137	116.5	2.2	6.9	27.1	2.6	0.70	0.41
30	7.4	332	199	46	111	4.7	0.61	18.9	13.6	6	7.6	66.5	1.1	0.16	1.07
31	6.6	159	102	27	58	0.0	0.16	16.7	4.8	6.3	2.9	14.1	1.5	0.04	15.8
32	7.7	452	271	126	216	7.8	1.07	5.58	34.8	19.5	18.9	67.5	1.1	0.46	0.27
33	7.5	488	293	161	243	7.8	1.18	5.11	15.3	23.5	24.8	42.9	1.0	0.53	0.41
34	7.2	106	68	40	84	0.1	0.25	7.52	1.5	11.5	2.7	10.8	2.0	0.46	0.57
35	7.3	327	196	89	106	3.1	0.36	32.6	50.7	10.5	15.2	43.5	2.4	0.89	0.32
36	7.7	456	274	163	198	8.6	0.48	15.4	0.1	24	25.0	16.4	2.0	0.46	0.34
37	7.1	862	552	768	86	0.1	0.30	552	34.2	190.1	71.3	117.7	2.6	0.40	0.22
38	7.6	2719	1631	271	374	9.4	0.48	562	21.3	2.4	64.3	160.8	1.9	0.28	0.99
39	8.6	427	256	147	193	0.8	1.19	17.8	21.5	31.7	16.5	19.0	0.9	0.21	0.59
40	7.9	1363	818	149	336	7.8	1.10	198	1.5	2.1	35.0	102.3	0.9	0.65	0.29
41	7.6	1427	856	37	374	12.5	0.32	226	15.2	2.4	7.5	32.6	1.3	0.35	0.36
42	6.6	377	241	92	100	0.0	0.30	47.9	15.6	24.2	7.7	19.5	2.0	0.07	1.44
43	7.8	650	390	239	263	3.1	2.51	37.1	7.5	75.6	12.2	203.2	1.1	0.41	0.24

TABLE 7.1
Results of Analysis of the Ground Water and Hot Spring Water of the Study Area (continued)

No.	pH	EC	TDS	TH	HCO_3^-	CO_3^{2-}	F^-	Cl^-	SO_4^{2-}	Ca^{2+}	Mg^{2+}	Na^+	K^+	Al^{3-}	Fe^{3-}
44	7.4	201	129	255	137	0.2	0.62	44.6	6.0	62.4	23.9	82.9	2.4	0.72	0.24
45	7.9	692	415	1081	330	18.7	0.52	17.2	22.8	356.5	46.3	15.2	1.1	0.23	0.11
46	8.0	1835	1101	836	416	17.2	0.91	301	1.5	322.5	7.4	24.4	1.8	0.35	0.75
47	7.1	262	157	61	89	0.0	0.21	26.1	7.5	9.4	9.1	18.6	0.9	0.89	9.54
48	7.6	1678	1007	36	200	4.7	0.61	364	31.4	2.6	7.3	31.2	10.8	0.61	0.29
49	7.7	905	579	123	366	1.1	3.06	42.1	5.0	37.5	7.0	124.0	1.9	0.07	0.20
50	7.7	85	54	126	366	1.1	2.71	45.3	4.8	38.5	7.2	130.6	2.1	0.42	0.30
51	8.2	914	548	50	397	9.4	3.12	37.7	5.8	1.1	11.5	132.9	1.1	0.33	0.34
52	7.8	416	250	163	168	3.1	0.56	17.1	18.3	50	9.3	29.5	1.9	0.93	0.07
53	7.7	1637	948	51	232	0.7	4.64	33.3	2.2	14.1	3.7	137.6	1.6	0.05	0.90
54	7.9	96	61	132	285	1.4	8.07	184	16.5	30.1	13.8	152.3	1.7	0.43	0.20
55	8.1	652	417	54	315	2.4	5.35	37.1	1.0	15.2	3.9	144.4	1.7	0.68	1.80
56	7.8	1454	872	156	374	9.4	1.39	203	16.5	2	36.7	153.9	1.0	0.55	0.20
57	7.4	1537	922	34	270	7.8	0.32	272	33.7	2.2	6.9	29.9	3.3	1.32	0.52
58	8.0	829	531	13	167	1.5	12	180	16.1	5.1	0.1	125.5	3.3	0.11	0.15
59	8.9	853	546	15	382	28.5	12.7	192	17.0	5.9	0.2	132.4	4.4	0.13	0.14
60	7.9	1524	975	13	409	3.0	13.1	176	16.6	5	0.1	130.9	4.1	0.36	0.21
61	7.9	485	310	9	329	1.6	14.2	173	22.8	3.5	0.2	128.2	3.2	0.58	0.22
62	8.1	621	397	14	311	3.6	13.1	171	17.4	5.2	0.1	132.8	3.5	0.46	0.13

Note: Analyses # 58 to 62 are for the hot-spring water.

7.5 MULTIVARIATE ANALYSIS OF HYDROCHEMICAL DATA

The various hydrochemical parameters analyzed for all the types of water from the area represent a set of multivariate data. Three different types of multivariate statistical analyses were attempted in order to reduce the dimensionality of the dataset for meaningful interpretation in terms of the processes responsible for the variability of the data. Principal component analysis was attempted in order to resolve the variance into mutually orthogonal axes for identification of different fluid components in the study area. Results of similar analysis on a fewer number of analytical data were reported earlier [12].

Because of the three distinct components — thermal spring, groundwater with high fluoride concentration, and groundwater with low fluoride concentration, it is necessary to determine the actual number of end member components. The permissible limit of 1.5 ppm was taken as the dividing line between the contaminated and uncontaminated groundwater. The present statistical analysis is based on 49 water samples from the uncontaminated zone, 7 samples from the inner zone of contamination, and 5 samples from the thermal spring, as shown in Figure 7.2. A robust principal component analysis program (ROPCA) was used for the Q-mode principal component analysis [29] and the first two components were chosen for plotting the principal component scores.

The data collected during the study are well posed, as revealed from the first two eigenvalues extracted from the dispersion matrix accounting for more than 55% of the total variance. Principal factor analysis was also attempted on the data presented in Table 7.1. Eigenvalues were extracted from the dataset and the corresponding factor loadings for variables and factor scores for each case were calculated using STATISTICA 4.0. Only nine variables with significant correlation

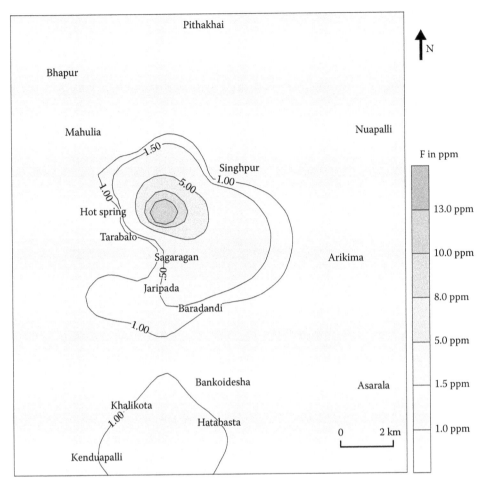

FIGURE 7.2 Spatial distributions of fluoride concentrations in ground water in the study area (see text for details).

coefficient ($r > 0.5$) were considered for the analysis. Cluster analysis of 15 chemical parameters was also done on the same dataset using Ward's method and Euclidean distance measure for the clustering.

7.6 RESULTS AND DISCUSSION

7.6.1 PCA

Principal component analysis has been quite useful in deducing the end-member components of groundwater [1]. Plots of the first two principal components in the present case as shown in Figure 7.3 reveal an interesting mixing trend between the hot spring water and the uncontaminated groundwater, resulting in enriched fluoride content in the latter. The colinearity of the plots necessarily rules out the involvement of any other water type. The cause of fluoride enrichment in the groundwater is thus discernible.

7.6.2 FACTOR ANALYSIS

The parameters presented in Table 7.1 lack good correlation and therefore the resolvability of the total variance of the system is not satisfactory. Out of the various combinations that were tried,

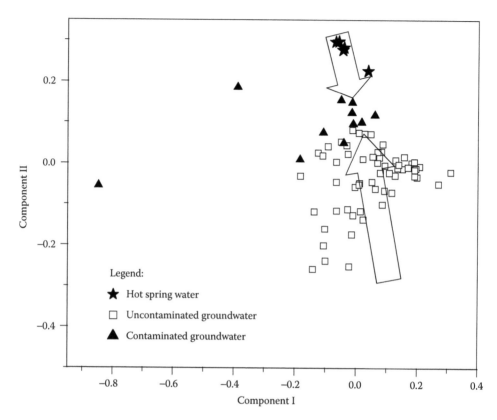

FIGURE 7.3 Plots of the first two principal component scores computed on the hydrochemical parameters of ground water and hot-spring water of the study area. The arrows indicate the mixing trend.

TABLE 7.2
Eigenvalues Extracted by Principal Factor Analysis

	Eigenvalue	% Total variance	Cumulative eigenvalue	Cumulative %
1	2.619194	29.10216	2.619194	29.10216
2	2.321811	25.79790	4.941005	54.90006
3	1.125874	12.50971	6.066879	67.40977

only the combination of the nine parameters (pH; total hardness (TH); HCO_3; F; Ca; Mg; Na; Al; and Fe) yielded satisfactory results. The three eigenvalues extracted account for 67% of the total variance (Table 7.2).

The first three principal factors extracted from the dataset and the raw loadings of the parameters on these factors are presented in Table 7.3. Only factor loadings exceeding absolute value of 0.6 are considered. As seen from the table, pH, HCO_3, F, and Na are significantly loaded on the first factor, and TH, Ca, and Mg are loaded significantly on the second factor. The third factor is apparently insignificant, so the system can be visualized as being controlled by two factors. The first factor indicates strong mutual influences of F, HCO_3, Na, and pH.

As explained earlier on the basis of the equilibrium governing calcite, fluorite, bicarbonate, and fluoride ions, at a constant pH, an increase in the bicarbonate concentration would increase the fluoride concentration in the water. Figure 7.4(a–c) shows the plots of fluoride concentration

TABLE 7.3
Factor Loadings (Unrotated) on Three Principal Factors

	Factor 1	Factor 2	Factor 3
PH	-0.825^a	0.156	0.016
TH	0.116	0.972^a	-0.105
HCO3	-0.7010^a	0.229	0.170
F	-0.703^a	-0.314	-0.150
CA	0.080	0.860^a	-0.403
MG	0.118	0.619^a	0.600^a
NA	-0.690^a	0.019	0.287
AL	0.381	-0.066	0.669
FE	0.546	-0.270	-0.103
Expl. var	2.619	2.322	1.126
Prp. totl	0.291	0.258	0.125

^{a} Significant loadings (>0.6).

vs. pH, HCO_3, and Na and reveals an interesting commonality — i.e., there is a high in the fluoride concentration at restricted ranges of variation of the other three parameters. In other words, the hot spring and the contaminated groundwater correspond to a restricted range of pH (7.5 to 8.5), HCO_3^- (200 to 400 ppm), and Na (between 100 to 150 ppm). This possibly needs to be resolved further from speciation study of the water and aqueous equilibria considerations. The principal factor analysis only indicates the interplay of the two factors. The plot of the first two principal factors' scores for all the samples (Figure 7.5) very clearly separates the three types of water in the area.

7.6.3 Cluster Analysis

R-mode cluster analysis was carried out on the same set of variables used for factor analysis and the tree diagram (Figure 7.6) using Ward's method, and Euclidian distance measures are obtained. The figure shows three different clusters of chemical species in the water: pH, F⁻, Al, Fe, and Mg make up one; Ca and Na constitute the second; and HCO_3^- and TH comprise the third. It may be noted that, though pH, Na, and HCO_3^- were associated with the first factor obtained in the factor analysis, these species are now separated into different clusters. However, pH and F⁻ fall into the same cluster and are in accordance with their strong association with the first factor and therefore establish the strong control of pH on the fluoride.

As was noted earlier [12,13,30], the hot spring represents a hydrothermal convection cell operating by percolation of meteoric water through the fracture network in the rock — acquiring higher temperature at deeper levels and the chemical species by interaction with the rock and, finally, venting the hot water through the major fractures/fissures. Such evolved waters in the up-flow zones are characterized by higher pH [17] mostly in the stability fields of HCO_3^-/CO_3^{-2} and K-feldspar. The close relationship of fluoride, bicarbonate, and pH is thus explainable.

However, the presence of a separate cluster for bicarbonates suggests that its concentration in the water is not related to the hot spring activity; rather, bicarbonate, along with Ca, Mg, and Na, is basically controlled by the chemical and mineralogical composition of the surrounding country rocks. It is generally observed that pH of water in land-based geothermal systems is mostly independent from the dominance of anionic species [31]. In summary, the cluster analysis brings out a classification of the chemical species in the water in the area; pH, fluoride, and, to some extent, Fe and Al are associated with the hot spring; Ca, Mg, and HCO_3^- are controlled by the surrounding rocks.

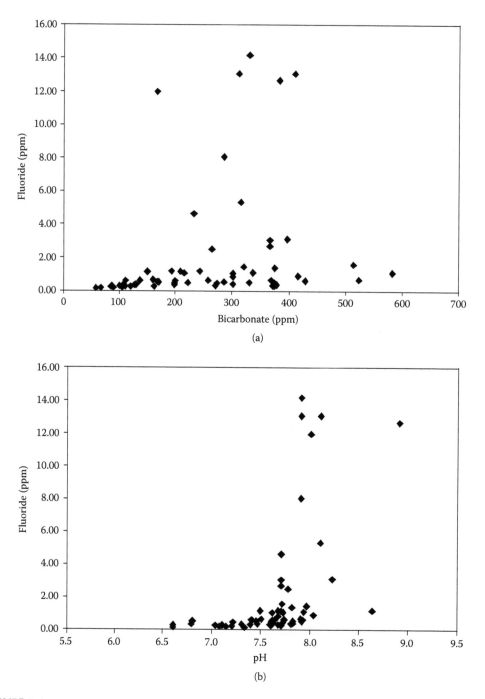

FIGURE 7.4 Variation of fluoride concentration with (a) HCO_3^-; (b) pH; and (c) Na. These parameters constitute the first and the most important factor that underlies the data examined (see text).

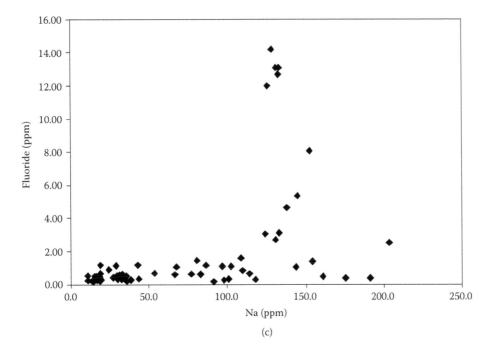

(c)

FIGURE 7.4 (continued)

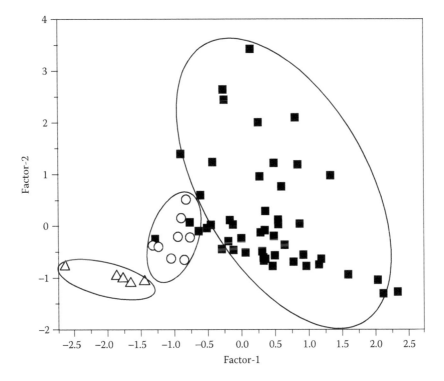

FIGURE 7.5 Plots of scores of the first two factors extracted from a subset of the data (see text for details). The three types of waters are well discriminated on the plot with their respective fields shown.

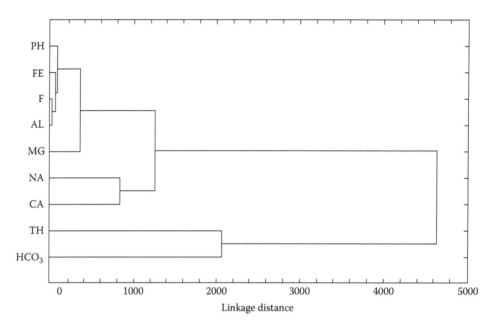

FIGURE 7.6 Dendogram of the ten important chemical species considered for cluster analysis using Ward's method and Euclidean distance measures.

7.7 CONCLUSION

The fluoride-enriched groundwater in the study area is a result of mixing the hot-spring water and normal groundwater as revealed from the principal component analysis. Two major factors that control the water chemistry are identified; the one associated with F^-, HCO_3^-, and pH is the major one. Furthermore, cluster analysis classified the hydrochemical species into two identifiable categories: fluoride and pH with or without Na, Al, and Fe are related to the hot-spring activity; calcium, magnesium, and bicarbonate contents in the water are mainly governed by the chemistry of the rocks.

ACKNOWLEDGMENTS

Instrument support for this work was mainly available through a project sanctioned to ST from India Canada Environment Facility. N. Kundu was supported by a Research Fellowship from the Ministry of HRD, Government of India.

REFERENCES

1. Laaskoharju, M., Skarman, C., and Skarman, E. (1999) Multivariate mixing and mass balance (M3) calculations, a new tool for decoding hydrogeochemical information, *Appl. Geochem.*, 14: 861–871.
2. Evans, C.D., Davies, T.D., Wigington, P.J., Jr., Tranter, M., and Krester, W.A. (1996) Use of factor analysis to investigate processes controlling the chemical composition of four streams in the Adirondack Mountains, New York. *J. Hydrol.*, 185: 297–316.
3. Meng, S.X. and Maynard, J.B. (2001) Use of statistical analysis to formulate conceptual models of geochemical behavior: water chemical data from the Botucatu aquifer in Sao Paulo state, Brazil. *J. Hydol.*, 250: 78–97.

4. Liu, C.W., Lin, K.H., and Kuo, Y.M. (2003) Application of factor analysis in the assessment of groundwater quality in a blackfoot disease area in Taiwan. *Sci. Total Env.*, 313: 77–89.

5. Briz–Kishore, B.H. and Murali, G. (1992) Factor analysis for revealing hydrochemical characteristics of a water shed, *Environ. Geol.*, 19: 3–9.

6. Dawdy, D.R. and Feth, J.H. (1967) Application of factor analysis in study of chemistry of groundwater quality, Kojave River Valley, California, *Water Resour. Res.*, 3: 505–510.

7. Jayakumar, R. and Siraz, L. (1996) Factor analysis in hydrogeochemistry of coastal aquifers — a preliminary study, *Environ. Geol.*, 31(3/4): 174–177.

8. Razack, M. and Dazy, J. (1990); Hydrochemical characterization of groundwater mixing in sedimentary and metamorphic reservoirs with combined use to piper's principal and factor analysis, *J. Hydrol.*, 114: 371–393.

9. Seyhan, E.V., Van De Caried, A.A., and Engelen, G.B. (1985) Multivariate analysis and interpretation of the hydrochemistry of a dolomite reef aquifer, Northern Italy, *Water Resour. Res.*, 21: 1010–1024.

10. Subbarao, C., Subbarao, N.V., and Chandu, S.N. (1996) Characterization of groundwater contamination using factor analysis, *Environ. Geol.*, 28(4): 175–180.

11. Lewis–Beck, M.S. (1994) Factor analysis and related techniques. *Sage Inc*, Jurong, Singapore.

12. Kundu, N., Panigrahi, M.K., Tripathy, S., Munshi, S., Powell, M.A. and Hart, B.R. (2001) Geochemical appraisal of fluoride contamination of groundwater in Nayagargh District of Orissa, India, *Environ. Geol.* 41(3–4): 451–460.

13. Kundu, N., Panigrahi, M.K., Sharma, S.P., and Tripathy, S. (2002) Delineation of fluoride contaminated groundwater zones around a hot spring using resistivity sounding in Nayagarh district of Orissa, India, *Environ. Geol.*, 43: 228–235.

14. Ruiz, F., Gomis, V., and Blasco, P. (1990) Application of factor analysis to the hydrogeochemical study of a coastal aquifer, *J. Hydrol.*, 119: 169–177.

15. Lawrence, F.W. and Upchurch, S.B. (1983) Identification of recharge areas using geochemical factor analysis, *Groundwater*, 20: 680–687.

16. Ashley, R.P. and Lloyd, J.W. (1978); An example of the use of factor analysis and cluster analysis in groundwater chemistry interpretation, *J. Hydrol.*, 39: 355–364.

17. World Health Organization (1984) *Guidelines for Drinking Water Quality*, Vol. 2, *Health Criteria and Other Supporting Information*, WHO, Geneva.

18. Apambire, W.M., Boyle, D.R., and Michel, F.A. (1997) Geochemistry, genesis, and health implications of fluoriferous groundwaters in the upper regions of Ghana, *Environ. Geol.*, 35(1): 13–24.

19. Fuhong, R. and Shuqin, J. (1988) Distribution and formation of high-fluorine groundwater in China, *Environ. Geol. Water. Sci.*, 12(1): 3-10, 1–20.

20. Gaciri, S.J. and Davis, T.C. (1993) The occurrence and geochemistry of fluoride in some natural waters of Kenya, *J. Hydrol.*, 43: 395–412.

21. Chadha, D.K. and Tamta, S.R. (1999) Occurrence and origin of groundwater fluoride in phreatic zone of Unnao district, Uttar Pradesh, *J. Appl. Geochem.*, 1: 21–26.

22. Handa, B.K. (1975) Geochemistry and genesis of fluoride containing groundwaters in India, *Groundwater*, 13: 275–281.

23. Maithani, P.B., Gurjar, R., Banerjee, R., Balaji, B.K., Ramchandran, S., and Singh, R. (1998) Anomalous fluoride in groundwater from Western part of Sirohi district, Rajasthan and its crippling effects on human health, *Curr. Sci.*, 74(9): 773–777.

24. Das, S., Meheta, B.C., Samanta, S.K., Das, P.K., and Srivastava, S.K. (2000) Fluoride hazards in groundwater of Orissa, India, *Indian J. Environ. Health*, 1: 40–46.

25. APHA (1971) *Standard Methods of Examination of Water and Waste Water*, 13th ed., American Public Health Association, New York, 769.

26. Gaciri, S.J. and Davis, T.C. (1993) The occurrence and geochemistry of fluoride in some natural waters of Kenya, *J. Hydrol.*, 43: 398–412.

27. Maina, J.W. and Gaciri, S.J. (1984) Contributions to the hydrogeochemistry of the area to the immediate north of Nairobi conservation area, Kenya, *J. Afr. Earth Sci.*, 2: 227–232.

28. Giggenbach, W.F. (1988) Geothermal solute equilibria: derivation of Na-K-Mg-Ca geoindicators. *Geochim. Cosmochim. Acta*, 52: 2749–2765.

29. Zhou, D. (1989); ROPCA–FORTRAN program for robust principal component analysis, *Computer Geosci.*, 15: 59–78.

30. Kundu, N. (2004) Fluoride contamination in water and soil vis-à-vis hot spring activity in parts of Nayagarh district, Orissa, India: a geochemical appraisal (unpublished), Ph.D. thesis, IIT, Kharagpur, 104.

31. Ellis, A.J. (1979) Explored geothermal systems. In: Barnes, H.L. (Ed.) *Geochemistry of Hydrothermal Ore Deposits,* 2nd ed., New York, John Wiley & Sons, 632–683.

8 Geochemical Processes Governing Trace Elements in CBNG-Produced Water

K.J. Reddy, Q.D. Skinner, and B.H. Hulin

CONTENTS

ABSTRACT

Trace elements in natural waters are a result of natural and anthropogenic processes. The toxicity and deficiency of trace elements to humans, plants, animal, and aquatic life depend on geochemical processes. These processes include adsorption–desorption and dissolution–precipitation processes. The objective of this chapter is to discuss geochemical processes controlling trace elements [e.g., arsenic (As), selenium (Se), iron (Fe), aluminum (Al), barium (Ba), and fluoride (F)] in coalbed methane natural gas (CBNG)-produced water in semiarid environments. The chapter presents

- Background information
- The CBNG extraction process and quality of produced water
- Geochemical processes of trace elements in CBNG-produced water
- Potential impacts of CBNG-produced water on rangeland and riparian zone plants, soils, and sediments

Results show that cationic trace elements in produced water precipitate as oxides and hydroxides when they interact with semiarid soils or channel sediment. However, anionic trace elements in produced water become more soluble and mobile in semiarid environments. Findings of this study will be useful in properly managing CBNG-produced water in western ecological environments.

8.1 INTRODUCTION

Trace elements in excess of small concentrations can be toxic to human, plants, animals, and aquatic life. The toxicity of trace elements in natural waters depends on various geochemical processes, including mineral dissolution; ion desorption; complexation; speciation; adsorption; precipitation; and transport [1]. As rain water infiltrates through watershed soils, trace elements are released into surface water as a result of mineral dissolution and desorption (Figure 8.1). Once trace elements are in water, they undergo rapid hydrolysis, ion complexation, and ion-pair formation. Also, trace elements in water are adsorbed by the colloidal particles and are precipitated into sediment as solid

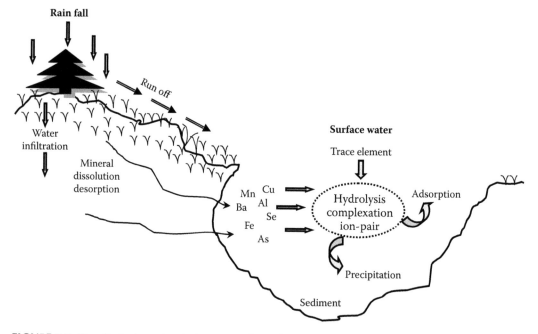

FIGURE 8.1 Hypothetical geochemical processes in surface water systems.

phases. These processes in turn control availability, toxicity, and transport of trace elements in water. Thus, understanding geochemical processes of trace elements in natural waters is key to predicting toxicity issues, which have gained interest in recent years because of human health problems.

In numerous cases, trace element contamination of drinking water supplies has led to serious human health problems. For example, arsenic contamination of drinking water supplies was linked to mass poisoning of people in Bangladesh and India [2]. Some estimates suggest that several million people are potentially at risk from drinking arsenic-contaminated water [3]. Although trace element contamination of surface water may be a natural process of weathering and erosion, anthropogenic sources exist as well. These include

- Energy production processes (e.g., coal- and natural gas-burning power plants, oil refineries)
- Mineral mining (e.g., coal, uranium)
- Oil and natural gas extraction
- Agriculture production (e.g., intensive application of irrigation, intensive application of fertilizers, pesticides, and herbicides)
- Livestock production (e.g., confined and unconfined animal feeding operations)
- Others (e.g., urbanization, solid waste disposal, deforestation)

The focus of this chapter is coalbed methane, which is a major source of natural gas. Worldwide use of different energy sources is shown in Figure 8.2 [4]. These results suggest a significant increase in use of natural gas over a period of approximately 20 years because extraction of natural gas is less expensive compared to other energy sources and is a clean-burning fossil fuel.

Coalbed methane natural gas (CBNG) extraction from coal deposits of Wyoming, Montana, Colorado, and New Mexico is occurring rapidly to meet energy needs of the U.S. For example, the natural gas industry produced 10 Bcf (billion cubic feet) of methane from 282 wells in 1984. By 1998, this had increased to 1200 Bcf from 7500 wells [5]. It is estimated that Wyoming coal deposits contain approximately 31.7 Tcf (trillion cubic feet) of recoverable natural gas. Within Wyoming, the Powder River Basin (PRB) contains the majority of natural gas and is in the forefront of development in North America [6,7]. Extraction of methane from a confined coal seam aquifer

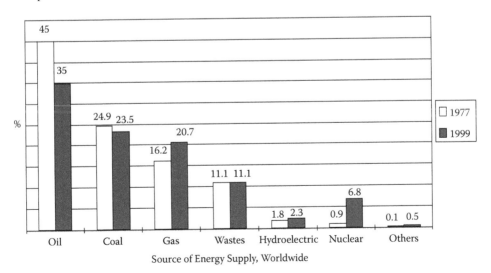

FIGURE 8.2 Worldwide use of different energy sources. (From Kluger, J. and A. Dorfman. 2002. *Time* Special Report. A6).

is facilitated by pumping large volumes of groundwater. The pumped groundwater is called "product water" or "produced water"; it is discharged into disposal ponds and stream channels, and reinjected into an aquifer. However, the chemistry of produced water (salinity, sodicity, and trace elements) varies with aquifer geology and watershed physical and chemical properties.

The objective of this chapter is to discuss geochemical processes governing trace elements in produced water. Also, salinity and sodicity of produced water are discussed because these parameters influence geochemical processes of trace elements in produced water. Information is organized as sections in a sequential order to help readers understand CBNG concepts.

8.2 BACKGROUND INFORMATION

It is important to discuss geology, groundwater, surface water, soils, and plant communities of the PRB because quality of produced water depends on the aquifer geology. Also, discharge of produced water (groundwater) from natural gas development increases overall flow of receiving tributaries, which drain through local soils and plant communities. Subsequently, these processes influence the quality of CBNG-produced water.

8.2.1 GEOLOGY AND GROUNDWATER

The Powder River Basin (Figure 8.3), which is a semiarid basin with average annual precipitation ranging from 30 to 60 cm, is located in northeastern Wyoming within Campbell County and small portions of Converse, Johnson, and Sheridan Counties. This basin (bounded by the Black Hills on the east, the Hartville Uplift to the south, the Big Horn Mountains on the west, and the Yellowstone

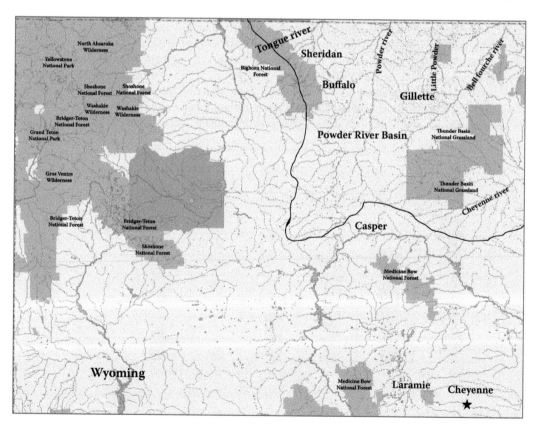

FIGURE 8.3 Study area: the Powder River Basin showing different watersheds.

River to the north) is generally high plains with elevations from 1640 to 1800 m above sea level. Eastern portions of the basin are typically high plains, but the central and western portions contain the Powder River Breaks and have a more diversified landscape setting and associated flora. More details regarding the geology and flora of the PRB are published elsewhere [8–10].

The aquifer geology of the PRB primarily comprises five distinct coalbed formations. These include: (1) Alluvium; (2) Wasatch; (3) Wyodak; (4) Upper Fort Union; and (5) Lower Fort Union. Methane is primarily extracted from confined deeper aquifers of Wasatch and Fort Union formations. Geologic studies of PRB suggest that methane is formed deep in confined coalbed aquifers through biogeophysical processes and remains trapped by water pressure.

It is not clearly understood whether methane contained in these aquifers is in a free state, adsorbed to the interior pore surfaces, or dissolved in aquifer water. Recovery of the methane is facilitated by depressurizing and pumping water from the aquifer (Figure 8.4). The quality of aquifer water varies from basin to basin. For example, aquifer water from deeper formations such as Fort Union is generally salty and dominated by sodium and bicarbonate. A shallow aquifer such as Wasatch formation contains low salt, sodium, and bicarbonate. The groundwater uses within the PRB include domestic, industrial, irrigation, municipal, and livestock watering.

8.2.2 Surface Water

The PRB consists of five perennial rivers, Cheyenne River, Belle Fourche River, Little Powder River, Powder River, and Tongue River, which are all tributaries of the Missouri River. Tributaries

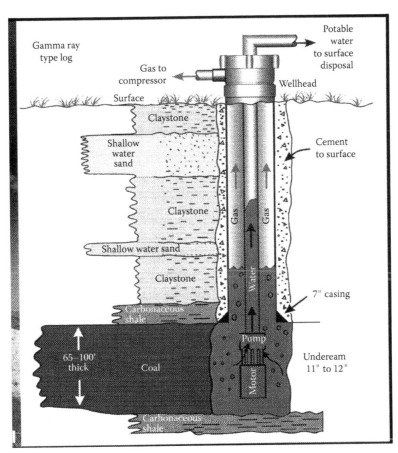

FIGURE 8.4 Coalbed methane natural gas extraction well.

of these perennial rivers are ephemeral with flow occurring in response to snowmelt and storm event. The Cheyenne River flows to the east or southeast part of the basin. The Belle Fourche River flows to the northeast of the basin and Little Powder River flows north of the basin. The Powder River flows northwest of the basin. The Tongue River flows through the northeast of the PRB. Surface water found in the PRB, except for Tongue River, is generally dominated by sodium sulfate, followed by calcium, magnesium, and alkalinity. These surface waters also contain higher total dissolved solids. The surface water uses within the basin include livestock watering, agriculture, and domestic.

8.2.3 Soils

Powder River Basin rangeland soils are loamy, sandy clay loam and clay loams underlaid with gravels and sands. Soils of PRB vary from basin to basin; the landscape is dominated by short hills and deep erosion gullies. The soils carry a chemistry rich in natural sodium and calcium carbonate. Cheyenne River Basin watershed soils are dominated by Ustic Haplargids (clay loam). The upper 0 to 11 cm of these soils contain a neutral pH of 7.2 with 1% organic carbon. These soils are well drained with slow permeability and have low to high runoff. The BFR soils are Ustic Calciargids (fine loamy). These soils also are well drained with moderate permeability and have low or medium runoff. The upper 0 to 11 cm of these soils have a neutral pH of 7.2 with 1.2% organic carbon.

The LPR watersheds soils are Ustic Torriorthents (loamy). The upper 0 to 11 cm contain 0.85% organic carbon with a pH of 8.0. These soils are well drained with moderate permeability and have medium or high runoff depending upon slope. Soils in PR are loams, silty clays, fine sandy loams, and sandy loams, with shale and sandstone outcrops. Soils in the study area are classified as moderately to well drained [11,12].

8.2.4 Plant Communities

The PRB lies on the western edge of the U.S. Northern Mixed Grass Prairie [13]. Lower elevation grassland is interspersed with shrub species of sagebrush (*Artemisia* L.). Scattered ponderosa pine (*Pinus ponderosa* Laws. and Laws.) and juniper (*Juniperus* L.) forested areas may occupy higher elevation hills and rugged drainage basin ridge areas. Scattered cottonwood (*Populus* L.) trees and willows (*Salix* L.) may occupy stream channel flood plains of larger stream systems [14]. Linear meadow land parallels larger streams where water supplies have been developed for irrigation. Some dryland wheat (*Triticum avestivum* L.) and irrigated barley (*Hordeum vulgar* L.), oats (*Avena sativa* L.), and alfalfa (*Medicago* L.) are produced within this region's rangeland.

Basin grassland is composed of cool- and warm-season grass species. Cool season grasses grow in response to soil moisture stored during fall and early spring snow melt and rain. Warm season grasses respond to late spring and summer precipitation. Stream flow that may occur because of snow melt, rain, or groundwater springs may support riparian grasses along channel flood plains. Existing grasslands are supported by soil water that is concentrated in the upper part of the soil profile and seldom percolates below the root zone. Therefore, plant nutrients and salts are generally concentrated near the soil surface and root zone. Although the near surface soils may be nutrient limited, nutrients are regenerated because precipitation is insufficient to leach them below the root zone and to underlying aquifers. Limited grass production of the PRB is therefore nutritious relative to that of higher rainfall areas. Because rangeland of the PRB is composed of cool- and warm-season grass species, the overall nutritional value of this grassland is prolonged later in dry summer months by the later warm-season grass production. Dominant grass species of the PRB by plant community are presented in Table 8.1 [15].

TABLE 8.1
Dominate grass species of the Norther Mixed Grass Prairie within the Powder River Basin, Wyoming from the Grasses of Wyoming.

Upland rangeland	Riparian Zones	Meadows
Poa secunda Presl.[a]	*Stipa viridula* Trin.[a]	*Elytrigia intermedia*
Poa cambyi Howell[a]	*Pascopyron smithii*	*Bromus inermis* Leyss.[c]
Stipa comata Trin. & Rupr.[a]	*Elymus canadensis* L.[a]	*Alopercurus pratensis* L.[a]
Oryzopsis hymenoides Ricker	*Leymus cinereus* A. Love[a]	*Alopercurus arundinaceus* Poir.[c]
Koeleria cristata Pers.[a]	*Spartina gracilis* Trin.[b]	*Alopercurus carolinius* Walt.[a]
Pascopyron smithii A. Love[a]	*Spartina pectinata* Link[b]	*Phleum pratense* L.[a]
Elymus lanceolatus Gould	*Sporobolus airoides* Torr.[b]	*Agrostis alba* L.[a]
Elymus trachycaulus Gould ex Shinners[a]	*Distichlis spicata* Green[b]	*Dactylis glomerata* L.[a]
Bouteloua gracilis Lag. Ex Steud.[b]	*Hordeum jubatum* L.[a]	*Phalaris arundinaceae* L.[a]
Pseudoroegneria spicata A. Love1	*Elymus lanceolatus*	*Pascopyron smithii*
Calamovilfa longifolia Scrib.[b]	*Sporobolus cryptandrus*	*Elymus lanceolatus*
Sporobolus cryptandrus Gray[b]	*Puccinellia nuttalliana* Hitchc.[b]	*Elytrigia intermedia*
Sporobolus heterolepis Gray[b]	*lytrigia intermedia* Nevski[a]	*Bromus tectorum*
Andropogon hallia Hack.[b]	*Bromus tectorum*	*Bromus japonicus*
Bromus tectorum L.[c]	*Bromus japonicus*	
Bromus japonicus Thunb.[c]		

[a] cool season
[b] warm season
[c] introduced

8.3 CBNG EXTRACTION PROCESS AND QUALITY OF PRODUCED WATER

The previous section reviewed background information of the PRB. This section discusses the CBNG extraction process and disposal methods for produced water and chemistry of major elements in CBNG-produced water — particularly, salinity, sodium adsorption ratio (SAR), and sodicity. Chemistry of trace elements in produced water is discussed in detail later in the section on geochemical processes.

It is estimated that a single CBNG well in the PRB may produce from 8 to 80 L of produced water per minute, but this amount varies with aquifer pumped and the density of wells (Figure 8.4). At present, more than 16,000 wells are under production in the PRB and this number is expected to increase to at least 30,000. Approximately 2 trillion L of produced water eventually will be produced over a period of 15 to 20 years from CBNG extraction in Wyoming [6]. Commonly, several CBNG extraction wells are placed together in a manifold system discharging to a single point and releasing into constructed unlined disposal ponds and/or into stream channel systems (Figure 8.5).

The chemistry of produced water (salinity, sodicity, and trace elements) in the PRB varies with geology and depth of the coal formation. Only a few studies have examined the changes in the chemistry of produced water in the PRB [11,16]. Rice et al. [16] examined the chemistry of discharge water at wellhead across the PRB, Wyoming, and reported that moving from north and west toward deeper coal seams within the PRB produces saline and alkaline water. The studies of McBeth et al. [11] examined the chemistry of produced water at wellhead and in associated disposal ponds of the PRB. The study area consisted of the Cheyenne River Watershed, Belle Fourche Watershed, and Little Powder Watershed within the PRB. Results suggested an increase in SAR, EC, TDS, and pH of discharge water in associated disposal ponds moving from the south to the north of the PRB.

Pond disposal
(a)

Channel disposal
(b)

FIGURE 8.5 Disposal of coalbed methane natural gas-produced water.

Patz et al. [17] studied the chemistry of produced water released to the upland channel systems that support ephemeral streams in the PRB. Results suggest that produced water consisted of higher concentrations of sodium and alkalinity compared to other components. The pH of produced water increased in the downstream channel water; however, the pH of sediment water interface within the top 2 cm decreased due to the precipitation of calcite. Subsequently, precipitation of calcium carbonate in the downstream channel decreased calcium concentration and increased SAR. This literature review suggested that salinity and sodicity of produced water increase within the PRB moving from south to northwest. Often, produced water is dominated by sodium and bicarbonate.

8.4 GEOCHEMICAL PROCESSES OF TRACE ELEMENTS IN CBNG-PRODUCED WATER

This section presents information about the chemistry of trace elements in produced water interacting with channel sediment of Dead Horse Creek from the PRB and geochemical processes governing trace elements in produced water under semiarid environments.

8.4.1 STUDY AREA

This study was conducted using sediments collected from the channel system of Dead Horse Creek, a tributary to the Powder River in Johnson and Campbell Counties, Wyoming [18]. Dead Horse Creek is an intermittent and fifth-order stream, map scale 7.5″ [19,20] located in the grassland and sagebrush Mixed Grass Prairie [13]. The drainage basin has a southeast to northwest orientation with headwaters just west of Gillette, Wyoming (Figure 8.3). This region receives 25 to 45 cm of annual precipitation, mostly as winter snow. High-intensity and short-duration thunder storms are common in summer. Dominant upland soils are predominantly classified as Ustic Torriorthents (loamy). Sediment samples collected from channel alluvium were approximately 40% clay. Major coal formations in this region include the Tertiary White River formation and the Tertiary Tongue River member of the Fort Union formation [11].

8.4.2 SEDIMENT SAMPLING

Three individual sediment samples were collected along Dead Horse Creek. Site 1 was located on a headwater low gradient channel reach along a third-order tributary of Dead Horse Creek. Site 2 was located on a low gradient channel reach along a fourth-order tributary of Dead Horse Creek just below the confluence of the site 1 channel; site 3 was located on a low gradient channel reach along the main stem of Dead Horse Creek just before it joined the Powder River.

Slit trenches, approximately 1ft deep and 1 ft wide, were dug, perpendicular to flow, across the creek channel bottom. Samples were loaded and transported in a plastic liner to the University of Wyoming. The sediment samples were spread out on large plastic sheets and allowed to air dry and then crushed using a 20-lb rubber-coated sledge hammer. Larger plant parts were removed by hand and samples were then passed through a no. 10 mesh screen to remove smaller plant and larger sediment materials. Each screened sample was stored in a separate plastic container for future use in column studies.

8.4.3 WATER SAMPLING

One 400-gal sample of CBNG-produced water was collected from a single extraction well located in the Dead Horse Creek watershed after the well was flushed to obtain a representative aquifer sample. The sample was pumped directly from the well head to a clean, plastic, truck-mounted tank and transferred to a clean and rinsed plastic storage tank in the Environmental Simulation Laboratory, College of Agriculture at the University of Wyoming. Distilled water was obtained, when needed, from a central distillery within the College of Agriculture. This supply was stored in a smaller clean and rinsed 300-gal storage tank located next to the produced water.

8.4.4 COLUMN FACILITY

Using 15.2 cm diameter PVC pipe, 48 1-m columns were constructed (Figure 8.6). The PVC pipe was inserted and sealed with silicone into 0.50-cm bottom plates. Five inlets or outlets evenly spaced in the column walls were inserted along one side of the column to within 5 cm of the top. One inlet or outlet was located 5 cm down and directly across from the top outlet on the opposite side. Constructed columns were then placed on three benches under lights within the column facility.

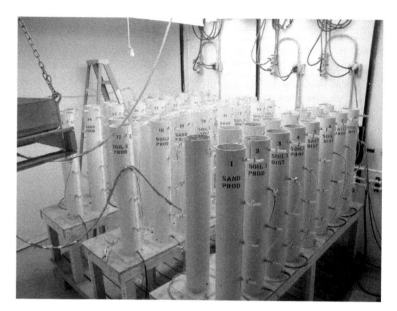

FIGURE 8.6 Design of column studies.

Feed and drain lines using tubing were then attached to appropriate inlets or outlets of each column and produced and distilled water storage tanks as planned for individual experiments. Control valves within the tubing apparatus were located at the end of each bench and between individual columns to control flow when appropriate.

8.4.5 COLUMN EXPERIMENTAL DESIGN

The 48 columns were allocated to six blocks along bench space and under the light system, where eight combinations of two water types (produced water and distilled water) and four sediment types (sediment samples one through three and a sand control) would be positioned within each block. Position of one of each of the eight possible combinations of water and sediment in each block was randomly selected within each block using random number tables. This provided one repetition for each of eight possible combinations in each of six blocks. This design provided each column combination of water and sediment equal exposure to possible variation in light quality produced by the light apparatus.

Columns were then placed on bench space and labeled according to bench positions within blocks and for each water and sediment combination. They loaded in approximately 15-cm lifts by placing each sediment in each assigned water and letting the sediment settle to within 2.54 cm of the water surface. This procedure ensured complete saturation throughout the filling process and simulated settling of sediment in surface runoff. Sediments in the column were filled to the bottom of top inlets and were allowed to settle or swell for 6 days. In columns where swelling occurred, sediment was removed to the depth of 5 cm. Each assigned column water was then added to a depth of 5 cm above sediment level for 5- and then 10-day contact experiments.

8.4.6 LIGHTS

Three light fixtures, 2.4 m long, comprised a combination of high-output fluorescent lights and 60-W incandescent bulbs attached to automatic timers for control of light intensity and duration. Each fixture provided 570 to 600 $\mu o\lambda\,\sigma^1\,m^{-2}$ light intensity within 15 to 30 cm of the light fixture. Full light intensity was provided for 14 h each day for all column experiments. Light fixture fans plus one oscillating fan provided air circulation for cooling the atmosphere above the columns and between the lighting apparatuses.

8.4.7 ANALYSIS

Differences in water quality data of sediment and water combinations were determined using one-way analysis of variance (ANOVA) procedures using a SAS [34] package. An alpha level of 0.05 was used to determine if differences existed at a 95% confidence interval. The Levene test was used to determine homogeneity of variance. The Tukey HSD post hoc test allowed for a pairwise comparison for all designated combinations of data.

8.4.8 EXPERIMENTS

8.4.8.1 Saturated Paste Experiments

Standard saturated paste procedures were used for the first experiment to determine whether drainage net sampling site locations and respective sediment supplies would produce different sediment water quality. This procedure requires using distilled water. Once distilled water was used, this procedure was modified by substituting distilled water with the produced water and conducting a second test. The objective of these two tests was to provide data sets that could be used to determine whether it would be justifiable to combine the results of the three channel sediments into one sediment source. No or minor differences in sediment chemistry by site location and respective sediment samples meant that sample numbers could be increased for the analysis of remaining experiments. Sodium (Na), magnesium (Mg), calcium (Ca), potassium (K), sulfate (SO$_4$), pH, and electrical conductivity (EC) were used to determine any differences in sediment water quality. It was determined that no significant differences existed between sample site locations.

8.4.8.2 Surface Ponding Experiments

These column experiments were designed to evaluate change in surface water chemistry after setting on the surface of the saturated column sediments for 5 and then 10 days. The columns were filled 5 cm deep above the sediment surface with produced water or distilled water and stirred twice daily to simulate turbulence similar to that caused by channel runoff or wind action for the 5-day test. All surface water was then drained, labeled, preserved, and stored while the columns were again filled to 5 cm deep above the sediment for the 10-day test. Change in chemistry for the 5-day test was analyzed during the 10-day test. Following the 10-day test, this water was collected, labeled, preserved, and analyzed for change in water chemistry.

All column water experiments were analyzed for major and trace elements. Trace elements analyzed were manganese (Mn), iron (Fe), arsenic (As), selenium (Se), fluoride (F), and barium (Ba). All samples were filtered through a 0.45-μmpore filter. Each filtered sample was divided into two subsamples. One sample was acidified to pH of 2.0 with HNO$_4^-$ and the other remained unacidified. Acidified samples were analyzed for Mn, Fe, As, Se, and Ba using inductively coupled plasma mass spectrometry (ICP-MS). Unacidified samples were analyzed for pH, alkalinity, and major elements. All analyses were performed following standard procedures [21].

8.4.8.3 Groundwater Experiments

After the surface water experiments, 500 ml of assigned water type was added to determine change in sediment water chemistry (groundwater) of all columns. The groundwater was collected from the second outlet from the top of the columns in individually sealed plastic containers. The second outlet was used for extractions because of the high clay content of sediments used and expectations that this depth would also produce the highest root biomass when plants were established in the columns. Once a sufficient amount of groundwater was collected, samples were analyzed for changes in trace element chemistry following procedures described earlier for column experiments.

Nebraska sedge (*Carex nebraskensis* Dewey), water sedge (*Carex aquatilis* Wahl.), and Baltic rush (*Juncus balticus* Wild) nursery stock grown in individual containers in New Mexico were transplanted to columns immediately following the completion of the first groundwater experiment. Within each of the eight water and sediment combinations, 16 plants of each species were randomly assigned. This resulted in 2 plants of each species for each combination; 8 plants of each species for each water type; 4 plants of each species for sand; and 12 plants of each species for the three sediments combined before transplanting occurred. Plants were watered daily for 2 months with the original water assignment so that column sediments were always flooded.

When plant roots and new stems were established, 500 ml of prescribed water was added to the sediment surface and the groundwater experiment was replicated to determine change in sediment water chemistry (groundwater) of all columns after establishing plants. Collection of groundwater and analysis of chemistry for this experiment remained as described previously.

8.4.8.4 CBNG-Produced Water Tank

Produced water tank samples were analyzed prior to the first saturated paste experiments, following the analysis of the surface water experiments, and prior to the final ground water test. Produced water samples were analyzed for the same chemical constituents as those for surface- and ground-water.

8.5 RESULTS AND DISCUSSION

8.5.1 CBNG-Produced Water Samples

The trace element water quality for the produced-water holding tank changed for some elements. Manganese and Fe decreased by approximately 50% from 4.0 to 1.7 mg/L and 290.3 to 115.2 mg/L, respectively. All other parameters remained unchanged. The pH increased from 7.89 to 8.49 and remained at 8.49 for the third sampling.

8.5.2 Surface Ponding

The 5- and 10-day tests (Table 8.2) were designed to evaluate change in trace element ponded water quality with prolonged exposure to surface sediments. Presented data of distilled water vs. produced water over sand illustrates the influence of water type on trace element water quality. Sediment vs. sand data comparisons isolate probable influence of channel sediment on water quality.

Arsenic, Ba, and pH in ponded distilled vs. produced water over sand-filled columns were significantly different and the remaining six parameters were not. Arsenic, Ba, and Cl increased from 1.0 to 3.7 µg/L, 5.6 to 239.6 µg/L, and 4.4 to 15.9 mg/L, respectively (Table 8.2). The pH increased from 6.27 for distilled water to 9.15 for produced water. Mn, Fe, Se, and F, remained constant. Mn, Fe, Se, and Ba in ponded distilled water over sand vs. ponded distilled water over

TABLE 8.2
Trace element release from Dead Horse Creek channel sediment with and without produced water for 5 and 10-day surface water.

	pH	Mn mg/L	Fe mg/L	As ug/L	Se ug/L	Ba ug/L	F mg/L
Produced Water/Sediment							
(5-day Surface water) Mean	7.82c	1.01ab	1.139b	4.05a	4.49b	99.36c	1.58a
(10-day Surface water) Mean	7.93c	1.01ab	1.139b	4.05a	4.49b	99.36c	1.58a
Distilled Water/Sediment							
(5-day Surface water) Mean	7.44c	1.743b	1.06b	2.14ab	3.36b	39.6d	2.95d
(10-day Surface water) Mean	8.00a	0.009a	1.22a	2.72a	3.04a	46.7a	1.83a
Produced Water/Sand							
(5-day Surface water) Mean	9.15a	0.0003a	0.039a	3.7a	1.20a	239.6a	1.15a
(10-day Surface water) Mean	9.13a	0.0006a	0.025a	1.8a	0.80a	193.7a	1.05a
Distilled Water/Sand							
(5-day Surface water) Mean	6.27b	0.0005a	0.046a	1.0b	0.50a	5.56b	0.01a
(10-day Surface water) Mean	8.74b	0.0001a	0.028a	0.90a	0.30a	5.14b	0.0001a
Produced Water							
Mean	8.49	0.0017	0.115	0.50	0.70	593.1	0.93

sediment increased from 0.5 to 1753.8 μg/L, 46.3 to 1066.6 μg/L, 0.5 to 2.8 μg/L, and 5.56 to 39.6 μg/L, respectively (Table 8.2). The pH varied between 6.27 for sand water to 7.44 for sediment water. Other measured parameters were not significantly different; however, As (1.0 to 2.14 μg/L) and F (0.01 to 2.95 mg/L) followed a trend of increasing when distilled water was over sediment instead of sand.

The results of the 10-day test (Table 8.2) are similar to that of the 5-day test, with only As changing. Arsenic went from 1.0 to 3.7 μg/L and significantly different, to not different (0.9 to 1.8 μg/L) in ponded distilled vs. water over sand-filled columns. Barium increased from 5.14 to 193.8 μg/L. The pH increased from 8.74 for distilled water to 9.31 for produced water. Manganese, Fe, Se, and F were not significantly different, but trends in the data suggest that they could increase with produced water over sand.

When comparing distilled water over sand with sediment, Mn, Fe, and Se increased from 1.0 to 913.0 μg/L, 28.9 to 1225.0 μg/L, and 0.3 to 2.7 μg/L, respectively (Table 8.2). All other parameters were not significantly different. However, As, Se, Ba, and F did follow a trend in increasing from 0.9 to 2.7 μg/L, 0.3 to 2.7 μg/L, 5.1 to 46.7 μg/L, and 0.0 to 1.83 mg/L, respectively. The pH dropped from 8.74 for sand to 8.00 for sediment. Analysis of the data from sand-filled columns of both tests would suggest that application of produced water may increase the concentrations of As and Ba. The data from distilled water columns would indicate that application of rain or produced water applied to the sediment in the Dead Horse Creek drainage will likely increase the concentrations of Mn, Fe, Se, and Ba.

8.5.3 GROUNDWATER

The groundwater sampled from all columns represented over 2 months of contact with sediments and should be as close as possible to equilibrium conditions between water types and sediment. These data suggest that, given time, distilled water in the column sediments will approach the measured chemistry of those subjected to produced water and that both will be higher than observed for surface contact only.

8.5.4 SURFACE- AND GROUNDWATER

The difference in surface- and groundwater trace element chemistry with distilled water and then CBNG-produced water are shown in Table 8.3. Significant differences are present for Se and Ba between the surface 5- and 10-day and groundwater data. The Se concentration increased from surface pond water (2.1 and 2.7 µg/L) to groundwater (8.4 µg/L) and Ba decreased from surface- (39.6 and 46.7 µg/L) to groundwater (11.6 µg/L). No significant difference in concentrations was found for all other chemical parameters analyzed. However, trend in the data suggests that Mn decreased from surface- (1753 and 913 µg/L) to groundwater (236.3 µg/L) and Ba decreased from surface- (39.60 and 46.7 µg/L) to groundwater (11.6 µg/L). In contrast, As increased from surface- (2.1 and 2.7 µg/L) to groundwater (75.9) and Se increased from surface- (2.8 and 2.7 µg/L) to groundwater (8.4 µg/L). The pH increased from surface- (7.44) to groundwater (8.59). Fe and F do not appear to be different.

The difference in surface- and groundwater trace element chemistry with produced water was similar to those found for distilled water. Se concentration increased from surface pond water (2.9 and 3.6 µg/L) to groundwater (7.7 µg/L) and Ba decreased from surface- (99.36 and 57.17 µg/L) to groundwater (18.91 µg/L). These differences were significant, as was pH surface- (7.82 and 7.93) to groundwater (8.89). The trends in changes of all other chemical parameter concentrations were identical to those observed for distilled and ponded surface water.

A recent study by McBeth et al. [22] indicated that Mn and Ba concentrations will decrease with time when stored in retention ponds. This same trend was noted in the distilled- and produced-water experiments. Mn was observed to drop (1010 to 917 for produced and 1753 to 913 µg/L for distilled) for both surface water tests and in the groundwater test (134 for produced water and 236 µg/L for distilled water). The Ba concentration fell from the surface water test (99.4 to 57.2 for produced water) to the groundwater test (18.9 µg/L). Though the Ba concentration remained similar

---

TABLE 8.3
Comparison of Trace Element Release from Dead Horse Creek Channel Sediment for 5- and 10-Day Surface Water vs. Groundwater

	pH	Mn (mg/L)	Fe (mg/L)	As (µg/L)	Se (µg/L)	Ba (µg/L)	F (mg/L)
Produced water/sediment							
5-Day surface water (mean)	7.82c	1.01ab	1.139b	4.05a	4.49b	99.36c	1.58a
10-Day surface water (mean)	7.93c	1.01ab	1.139b	4.05a	4.49b	99.36c	1.58a
Groundwater	8.89a	0.134a	1.06a	62.2a	7.66b	18.91c	0.86a
Distilled water/sediment							
5-Day surface water (mean)	7.44c	1.743b	1.06b	2.14ab	3.36b	39.6d	2.95d
10-Day surface water (mean)	8.00a	0.009a	1.22a	2.72a	3.04a	46.7a	1.83a
Groundwater	8.59a	0.236a	1.36b	75.9a	8.38a	11.55b	0.84a
Produced water							
Mean	8.49	0.0017	0.115	0.50	0.70	593.1	0.93

for the 5- and 10-day surface water test (39.6 to 46.7 µg/L), it did drop in the groundwater test (11.6 µg/L). Overall, results observed in this study are consistent with results reported by McBeth et al. [22]. A detailed discussion regarding geochemical processes controlling trace elements in semiarid environments is presented later in this section.

In summary, trace element analyses of column studies suggest that sediments of Dead Horse Creek drainage will alter the quality of produced water and precipitation surface- and groundwater runoff. Though produced water discharges can slightly alter the water quality of runoff, the sediment appears to be a stronger contributor to the overall water quality of stream channel runoff. Precipitation events will likely alter water quality of stream runoff by dissolving and leaching chemicals and thus increasing the chemical concentrations of downstream flows. Trace element availability seems to be predominantly affected by sediment interaction, with Mn and Ba decreasing and As and Se increasing from surface water retention to groundwater retention. Iron and F appear to be less affected by this difference in interaction.

Plants did not alter the sediment trace element water chemistry but did increase the rate of infiltration. Release of produced water may increase the pH of stream channel runoff and the mobility of anions, while decreasing the solubility of cations in soils. Overall results suggest that most constituents in produced and distilled water runoff were below the established criteria required for human consumption, except for Mn and Fe. Further research to determine the potential balance of produced water discharge and precipitation events (flushes) on downstream water quality will be useful in determining potential beneficial uses for produced water in the Powder River Basin.

8.5.5 Geochemical Processes

This section reviews geochemical processes controlling trace elements in produced water. McBeth et al. [22] examined the chemistry of trace elements in produced water at discharge points and in associated holding ponds across the PRB, Wyoming. Produced water samples from discharge points and associated holding ponds were collected from three different watersheds within the PRB (CHR, BFR, and LPR) and were analyzed for pH, Al, As, B, Ba, Cr, Cu, F, Fe, Mn, Mo, Se, and Zn. The chemistry of trace element concentrations was modeled with the MINTEQA2 geochemical equilibrium model.

Results of this study show that pH of produced water for three watersheds increased in holding ponds. For example, the pH of produced water increased from 7.21 to 8.26 for LPR watershed. Among three watersheds, produced water from CHR watershed exhibited relatively less change in trace element concentrations in holding ponds. Concentration of dissolved Al, Fe, As, Se, and F in produced water increased in BFR watershed holding ponds. For example, concentration of dissolved Fe increased from 113 to 135 µg/L. Boron, Cu, and Zn concentrations of produced water did not change in BFR watershed holding ponds.

However, concentration of dissolved Ba, Mn, and Cr in produced water decreased in BFR watershed holding ponds. For instance, Ba and Cr concentrations decreased from 445 to 386 µg/L and from 43.6 to 25.1 µg/L, respectively. In the LPR watershed, Al, Fe, As, Se, and F concentrations of produced water increased substantially in holding ponds. For example, Fe concentration increased from 192 to 312 µg/L. However, concentration of dissolved Ba, Mn, Cr, and Zn decreased in holding ponds.

Geochemical modeling calculations suggested that observed increase of Al and Fe concentrations in holding ponds was due to increase in concentration of $Al(OH)_4^-$ and $Fe(OH)_4^-$ species in water, which were responsible for pH increases [23]. Decreases in Ba, Mn, Cr, and Zn concentrations were attributed to the increase in pH, resulting in precipitates of $BaSO_4$ (barite), $MnCO_3$ (rhodochrosite), $Cr(OH)_2$ (chromium hydroxide), and $ZnCO_3$ (smithsonite) in pond waters, respectively. Several studies suggested that solubility of As, Se, and F in alkaline environments is controlled by the adsorption and desorption processes and increases as pH increases. Because As, Se, and F are anionic in nature and most of the soil mineral phases contain negative surface charge, these elements become soluble and mobile in alkaline watershed soils. The results found in this study are consistent with those in previous studies [24–27]. For example, concentration of dissolved As, Se, and F concentrations increased substantially in LPR watershed pond waters, which also had the highest increase in the pH.

Patz et al. [17] examined chemistry of trace elements in produced water interacting with Burger Draw and Sue Draw stream channels in the PRB, Wyoming. Significant change occurred for Fe, Mn, and As in flow of Burger Draw. As expected, Fe precipitates at once after discharge from wells (887.90 to 738.15 μg/L) as it transforms from a reduced to an oxidized state in Burger Draw (108.00 μg/L) or Sue Draw (3006.15 μg/L to 92.60 μg/L). Further reduction in Fe concentration in downstream flow did not appear to occur. Manganese reacted like Fe and decrease in concentration was low after initial contact with the atmosphere and the channel system. Arsenic and Se appeared to increase with downstream flow, but As was significantly higher after reservoir storage in Sue Draw (2.58 μg/L above and 4.48 μg/L below storage). McBeth et al. [22] and Hulin [18] report similar results.

In general, Al and Fe become less soluble and mobile in natural waters of the western alkaline environments. However, produced water may contain higher concentrations of Al and Fe in disposal pond water and these increases may be explained by the presence of Al and Fe hydroxyl species (e.g., $Fe(OH)_4^-$) in waters with a high pH [22]. Other studies also suggest that when anionic species of Al and Fe dominate in water, the solubility increases with increasing pH. Even though Al and Fe concentrations remained relatively unchanged, they may decrease with time in alkaline environments due to the likely precipitation of aluminum and iron hydroxides [28].

The decreases in Mn and Ba concentrations are likely due to the precipitation of oxides, hydroxides, and carbonates of these metals, common in the calcite-rich, alkaline environment of the PRB. Overall, the preceding review suggested that, in alkaline environments, cationic trace elements (e.g., Al, Fe, Mn, and Ba) become less soluble and mobile due to precipitation of oxides and hydroxides. However, anionic trace elements (e.g., As, Se, and F) become more soluble and mobile as a result of mineral dissolution and desorption processes.

8.6 POTENTIAL IMPACTS OF CBNG-PRODUCED WATER TO RANGELAND PLANTS, RIPARIAN PLANTS, SOILS, AND SEDIMENTS

In this final section, potential impacts of produced water to rangeland plants, soils, and sediment are summarized. Different management alternatives are proposed for produced water within the PRB, Wyoming. These options include surface land application (e.g., irrigation), aquaculture, and livestock watering. Nonetheless, any management option for produced water within the semiarid areas requires a careful evaluation of its salinity (e.g., EC, TDS), sodicity (e.g., SAR), and trace elements.

Common and observed issues associated with plant species and community impacts while developing CBNG in the PRB may be categorized as follows:

- Produced water and toxicity to native plants
- Native plant toxicity when produced water is applied to rangeland soils and to stored sediment in stream channel systems
- Change in plant species composition after produced water is released as perennial channel flow
- Vegetation response and ability to resist accelerated channel erosion after produced water is released as perennial channel flow
- Land application of produced water and change in soil chemistry and structure

- Change in land and stream channel characteristics as land application and channel release ends
- Role of plants in accumulation and turnover of trace elements

In the PRB, toxicity to plants was addressed in a preliminary laboratory study using produced water, nutrient solution, and perlite as a plant support system. Western wheatgrass, a dominate upland and riparian zone grass, was fed 100% produced water; 50% produced water and 50% nutrient solution water; and nutrient water alone. No plants died and no differences in total biomass were observed between 100% nutrient and 50% produced water and 50% nutrient solution; however, growth was reduced when no nutrients were supplied to 100% produced water (Figure 8.7). Plant nutrients in the tested produced water appear to limit growth response, and toxicity of the water did not appear to be an important issue [29].

Application of produced water to stored sediment of stream channel systems and plant toxicity was surveyed by Hulin [18] and Jackson [30] in a laboratory study using column techniques. Of three common wetland plants — Nebraska sedge (*Carex nebraskensis* Dewey); water sedge (*Carex aquatilis* Wahl.); and Baltic rush (*Juncus balticus* Wild) — grown in columns, only Baltic rush survived. These data suggest that when a produced water like that used for their experiments is applied to stored channel sediment, plant toxicity is an issue and that some native plant species cannot tolerate produced water and sediment chemistry under saturated conditions (Figure 8.8).

Change in plant species composition after produced water was released as perennial channel flow was observed by Patz [31]. Approximately 3000 containerized plants of each of the three

FIGURE 8.7 (A) Western wheatgrass grown in perlite and watered with nutrient solution; (B) 50% nutrient solution and product water; (C) product water only.

FIGURE 8.8 (A) All three plant species could survive in a control with soil and distilled water plus nutrients; (B) only Baltic rush was able to survive in soil plus product water; (B) Nebraska and water sedge could not survive product water and soil utilized for experiments.

FIGURE 8.9 Wetland plants adapted to saline conditions occupied stream channel area where saturated conditions exist after perennial release of product water. These plants replaced those species occupying the ephemeral channel area.

wetland species used in Hulin and Jackson's study were transplanted in and along three study sites of a single stream discharging produced water. Only Baltic rush survived and was confined to a narrow zone next to the water edge and bank. Rangeland plant species that had occupied the ephemeral channel area before release of produced water were being replaced by the riparian grass or grasslike species: inland salt grass (*Distichlis spicata* L.); foxtail barley (*Hordium jubatum* L.); Nutall's alkali grass (*Puccinellia nuttalliana* (Schult.) Hitchc.); Western wheatgrass (*Pascopyron smithii* Rydb.); American bullrush (*Scirpus americanus* Pers.); and maritime bullrush (*Scirpus maritumus* L. var. paludosus (A. nels.) Kukenth). These replacement species are known to be salt tolerant, whereas the grasses being replaced are not. This preliminary study suggests that upland rangeland grasses may be replaced by riparian zone and more salt-tolerant species when produced water is released as perennial flow, stored in ponds, or detained behind water spreaders or debris dams for extended lengths of time (Figure 8.9).

The change in plant composition observed by Patz [31] in addition to follow-up photo monitoring confirms that the vegetation response to perennial release of channel flow creates a stable riparian zone capable of resisting accelerated channel erosion. Plant height, cover, and stem density appear equivalent to riparian zones found throughout the PRB [35]. Three extreme flood events caused by summer convective storms have not caused low flow channel conditions to change. Figure 8.10 illustrates stability of riparian vegetation near Patz's 2002 study sites.

Land application of produced water and change in plant and soil characteristics in the PRB are yet to be confirmed within the scientific literature even though potential impacts of salinity and sodicity are well established. Figure 8.11 illustrates limited observations of soil conditions in which produced water is applied through sprinkler irrigation. Application of saline and sodic water for irrigation is not a problem in humid areas compared to arid and semiarid areas because rainfall is a major source of salt-free water. However, use of salty and sodic water for irrigation in arid and semiarid areas, particularly containing clay minerals with poor drainage, accumulates salts, decreases infiltration, and increases runoff and erosion [32]. A recent study reported that water with SAR as low as 5 will have an adverse effect on the soil structure and infiltration rates [33].

Surface disposal of produced water in ponds and in stream channels is creating riparian zone and wetland habitat in addition to that which is natural and has been developed through irrigation using historic water supplies. The additional produced water supply is limited to the time that it takes to remove the coalbed gas supply. Once this supply is harvested, it must be assumed that

FIGURE 8.10 Wetland plants that occupied ephemeral stream channel area after perennial release of product water continue to create stable conditions for reducing erosion and transport of sediment.

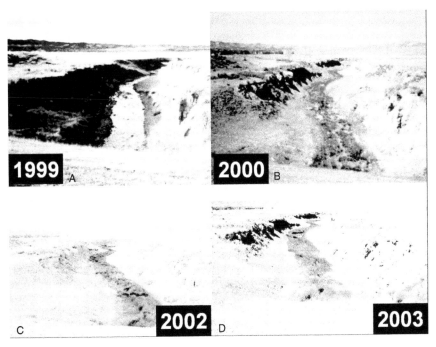

FIGURE 8.11 (A and B) sprinkler irrigation used to apply product water on (C) seeded and (D) native rangeland appears to cause crusting and cracking of some soil types.

riparian zone and forage plant resources created because of surface disposal of produced water will suffer drought conditions. The question of if and how vegetation will respond to no produced water has not been answered. It can be assumed that sodicity and salinity of soil and sediment resources should have increased through the influence of CBNG surface water supplies. However, fate and transport of trace elements by plants remain unknown. The questions associated with assimilation, accumulation, and transformation of trace elements by plants that can and cannot tolerate produced water should be answered knowing that reclamation of riparian zones, wetlands, and native range soils will be needed in the future.

Release of produced water into established watershed can have an impact on the water quality of the watershed, but this impact may be similar to that of naturally occurring precipitation events. However, precipitation is sporadic in causing runoff, and release of produced water may be perennial for some given period. From the authors' data, continual release of produced water would be expected to alter the chemistry of channel flow to assume eventually the chemical composition of the produced water discharge and naturally occurring precipitation combined. These data would further suggest that, following completion of CBNG discharge, naturally occurring precipitation would likely bring the concentrations of trace elements deposited back towards original levels.

The travel time of stream flow from the head water tributaries of Dead Horse Creek to the Powder River is not known. However, the authors' data suggest that containment behind debris dams, water spreader dikes, and even in ponds of low gradient stream channel reaches may produce local and high contributions of chemistry associated with trace elements to flow moving towards the Powder River. It did not matter if the stored water was CBNG-produced water or distilled water. When the water types were stored as ground water for approximately 2 months, both types of sediment water assumed about the same water chemistry. Although the pond water types were lower in chemistry after 5 and 10 days, they approached the water quality of the ground water supplies. Therefore, the issue of travel time to the Powder River as channel flow vs. detention storage is an important consideration for developing regulatory and management options for disposing of or using this new and potential CBNG-produced water supply.

It is important to note that stream flow of water from precipitation represents a baseline for evaluating potential water quality issues associated with managing CBNG-produced water. If CBNG water is released into Powder River tributaries, then water from precipitation represents a blending of two different supplies. The data suggest that rain water may contribute substantial amounts of chemistry to the Powder River system and this load should be considered as it varies in time and space. Then, blending and water treatment alternatives for adding produced water sources that range in quality and lie within the time and space distribution of precipitation runoff may be feasible for helping achieve acceptable water quality in a downstream direction.

REFERENCES

1. Stumm, W. and J.J. Morgan. 1996. *Aquatic Chemistry: Chemical Equilibria and Rates in Natural Waters*. 3rd ed., John Wiley & Sons., New York.
2. National Research Council. 1999. *Arsenic in Drinking Water*. National Academy Press, Washington, D.C.
3. Ahmed, M.F. 2002. An overview of arsenic removal technologies in Bangladesh and India. In: Technologies for Arsenic Removal from Drinking Water, M.F. Ahmed, M. Ashraf, and Z. Adeel, Eds., Bangladesh University of Engineering and Technology, Dhaka and The United Nations University, Tokyo, Japan, 251–269.
4. Kluger, J. and A. Dorfman. 2002. The challenges we face: energy and climate. Green century. *Time* Special Report. A6.
5. Hill, D.G., C.R. Nelson, and C. Brandenburg. 2000. Changing perceptions regarding the size and production potential of North America coalbed methane resources. In: Strategic Research Institute. Minimize Hazards and Maximize Profits through the Recovery of Coalbed and Coal Mine Methane: a New Energy Resource for the 21st Century. Conference Proceedings; Denver, Colorado.
6. DeBruin, R.H., R.M. Lyman, R.W. Jones, and L.W. Cook. 2000. Coalbed methane in Wyoming. Information pamphlet #7. Wyoming State Geological Survey; Laramie, WY.

7. Reddy, K.J., Q.D. Skinner, and I.H. McBeth. 2001. Exploring Uses for Coalbed Methane Product Water. *Reflections*, vol. 11, No. 1. University of Wyoming, College of Agriculture. Laramie, WY.

8. Ellis, M.S., Stricker, G.D., Flores, R.M., and Bader, L.R. 1998. Sulfur and ash in Paleocene Wyodak–Anderson coal in the Powder River Basin, Wyoming and Montana: a nonsequitur to externalities beyond 2000, Proceedings of the 23rd International Technical Conference on Coal Utilization, Clearwater, FL.

9. Flores, R.M., and Bader, L.R. 1999. Fort Union coal in the Powder River Basin, Wyoming and Montana: a synthesis: U.S. Geological Survey Professional Paper 1625-A, Chapter PS.GRI-99/0131.

10. Knight, D.H. 1994. *Mountains and Plains: The Ecology of Wyoming Landscapes*. Yale University Press, New Haven, CT.

11. McBeth, I.H., K.J. Reddy, and Q.D. Skinner. 2003. Chemistry of coalbed methane product water in three Wyoming watersheds. *J. Am. Water Resources Assoc.* 39: 575–585.

12. National Resource and Conservation Service. 2002. Soils of Wyoming. www.statelab.iastate.edu/soils/sc/wwwforms/fozzy.uwyo.edu.

13. Gould, F.W. 1968. *Grass Systematics*, McGraw–Hill, Inc., New York, 382 pp.

14. Dorn, R.D. 1977. *Manual of the Vascular Plants of Wyoming*, vol. 1, Garland Publishing, Inc. New York and London, 801 pp.

15. Skinner, Q.D., A.A. Beetle, and G.P. Hallsten. 1999. *Grasses of Wyoming*, Res. J. 202: 432, Agricultural Experimental Station, University of Wyoming.

16. Rice, C.A., M.S. Ellis, and J.H. Bullock. 2000. Water coproduced with coalbed methane in the Powder River Basin, Wyoming: preliminary compositional data. Open-File Report 00-372. U.S. Department of the Interior, U.S. Geological Survey: Denver, Colorado.

17. Patz, M.J., K.J. Reddy, and Q.D. Skinner. 2004. Chemistry of coalbed methane discharge water interacting with semiarid ephemeral channels. *J. Am. Water Res. Assoc.* 40: 1247-1255.

18. Hulin, B.K., 2001. Wyoming stream channel sediment and CBM product water. M.S. Thesis, department of Renewable Resources, University of Wyoming, Laramie, WY.

19. Horton, R.E. 1945. Erosional development of streams and their drainage basins, hydrologic approach to quantitative morphology, *Bull. Geol. Soc. Am.*, 56: 275–370.

20. Strahler, A.N. 1957. Dynamic basis of geomorphology, *Bull. Geol. Soc. Am.* 6: 913–920.

21. American Public Health Association. 1992. Standard methods for the examination of water and wastewater. American Public Health Association, Washington, D.C.

22. McBeth, I.H., K.J. Reddy, and Q.D. Skinner. 2003. Chemistry of trace elements in coalbed methane product water. *Water Res.* 37: 884–890.

23. Wesolowski, D.J. and Palmer, D.A. 1994. Aluminum speciation and equilibria in aqueous solutions: V. Gibbsite solubility at 500C and pH 3-9 in 0.1 molal NaCl solution. *Geochemica Cosmochimica Acta* 58: 2947–2970.

24. Niragu, J.O. and Pacyna, 1988. Quantitative assessment of worldwide contamination of air, water, and soils by trace metals. *Nature (London).* 333: 134–139.

25. Balistrieri, L. and Chao, T.T. 1987. Selenium adsorption by geothite. *Soil Sci. Soc. Am.* 51: 1145-1151.

26. Mok, W.M. and Wai, C.M. 1994. Mobilization of arsenic in contaminated river waters. In: *Arsenic in the Environment: Part I: Cycling and Characterization.* O.J. Nriagu (Ed.), John Wiley & Sons, Inc., New York.

27. Reddy, K.J. 1999. Selenium speciation in soil water: experimental and model predictions. In H.M. Selim and I.K. Iskandar (Eds.) *Fate and Transport of Metals in the Vadose Zone*, Ann Arbor Press, Inc., Chelsea, MI, 147–156, chap. 7.

28. Drever, J.I. 1997. *The Geochemistry of Natural Waters: Surface and Groundwater Environments*. 3rd ed., Prentice Hall, Englewood Cliffs, NJ.

29. Skinner, Q.D. K.K. Crane, J.G. Hiller, and J.D. Rodgers. 2000. *Wyoming Watersheds and Riparian Zones*, B-1085, Agriculture Resource Center University of Wyoming, Laramie, WY, 112.

30. Jackson, R.E. 2003. Wetland plant response to coalbed methane product water. M.S. thesis, Department of Renewable Resources, University of Wyoming, Laramie, WY.

31. Patz, M.J. 2002. Coalbed methane product water chemistry on Burger Draw, Wyoming. M.S. Thesis, Department of Renewable Resources, University of Wyoming, Laramie, WY.

32. Hanson, B., S.R. Grattan, and A. Fulton. 1993. *Agricultural Salinity and Drainage: a Handbook for Water Managers*. University of California Irrigation Program, University of California, Davis.

33. Mace, J.E. and C. Amerhein. 2001. Leaching and reclamation of a soil irrigated with moderate SAR waters. *Soil Sci. Soc. Am. J.* 65: 199–204.

34. SAS Institute, Inc. 1999. *Selected SAS Documentation for ENTO/STAT 5080: Statistical Methods for the Agricultural and Natural Resources Sciences*, 1st ed. SAS Publishing; North Carolina.

35. Skinner, Q.D. 2004. Unpublished data of current plant succession study associated with channel discharge of CBM product water.

9 Temporal Trends of Inorganic Elements in Kentucky Lake Sediments

Bommanna G. Loganathan, Clark Alexander, I-Lun Chien, and Kenneth S. Sajwan

CONTENTS

9.1 INTRODUCTION

Sediments are the ultimate reservoir for the numerous potential chemical contaminants that may be contained in effluents originating from urban, agricultural, and industrial lands and recreational activities [1–3]. Contaminated sediments in streams, rivers, lakes, and estuaries have the potential to pose ecological and human health risks [1,3]. Metals are a ubiquitous class of contaminants in aquatic sediments. Concentrations of certain metals such as cadmium, copper, nickel, lead, zinc, and other metals often are elevated above background levels in sediments that have been affected by human activity [4]. Earlier studies have correlated elevated concentrations of certain inorganic elements in sediments of rivers, estuaries, and coastal regions with increased industrial growth, agricultural operations, land use, etc. [5,6].

The sediment record, as revealed in sediment cores, has been used by many groups of investigators to reconstruct the history of contaminant input to reservoirs, lakes, rivers, and oceans throughout the world [7–9]. The basic assumptions are that contaminant inputs equilibrate relatively rapidly with sediment inputs and that the sediment column represents a continuous sequence, of sedimentation and associated contaminant accumulation. Using radiochemical chronologies, it is possible, theoretically, to date sediments over a period corresponding to the half-lives of two radiotracers: ^{210}Pb and ^{137}Cs. Ages can be assigned to different depth intervals within by use of the radiotracers. Reservoirs exhibit sedimentation rates that range from 10 to 200 mm/year [7]. Greater sedimentation rates convey several advantages to reconstructing temporal trends.

Although a large volume of literature exists on the levels and long-term trends of inorganic elements in sediments of natural lakes and coastal and oceanic environments, very limited information is available on the temporal trends of inorganic elements in manmade reservoirs or lakes. Reservoirs, particularly large river impoundments, are human-constructed ecosystems created and used for multiple purposes, including generation of hydroelectric power, water supply, flood control, transportation, and recreation. Therefore, the need to better understand the influence of human activities on these manmade ecosystems is obvious.

Kentucky Lake is the largest manmade lake in the southeastern U.S. (Figure 9.1). It is the ultimate repository of substances entering this watershed from portions of seven southeastern states, which include a sizable fraction of U.S. chemical processing, agricultural chemical products, and electronic manufacturing. Kentucky Dam tailwater, which connects the Ohio River, receives discharges from industries in the Calvert City Industrial Complex. During the past two decades, mass mortalities of mussels have been encountered in these regional waters; the causes have not been fully elucidated. In some cases, the quality and quantity of mussels harvested for the pearl industry have been substantially reduced [10].

Furthermore, a recent Public Advisory Committee Report [11] identified diminished air quality due to high levels of release of air pollutants from a variety of industries (chemical, metallurgical, and paper mills) in this region. Earlier studies have detected several metals and organics in sediments and freshwater mussel tissues collected from Kentucky Lake and Kentucky Dam tailwater [12,13]. These observations indicated the historical contamination of inorganic and organic pollutants in Kentucky Lake.

To the authors' knowledge, no comprehensive study has been conducted to describe historical contamination and trends of inorganic elements in this lake. This study measured the concentrations of chromium (Cr), nickel (Ni), copper (Cu), zinc (Zn), arsenic (As), selenium (Se), silver (Ag), cadmium (Cd), tin (Sn), antimony (Sb), lead (Pb), and mercury (Hg). It reconstructed the history of inorganic element input into Kentucky Lake over a 50-year period using the sedimentary record, as revealed in sediment core collected from Ledbetter embayment of Kentucky Lake.

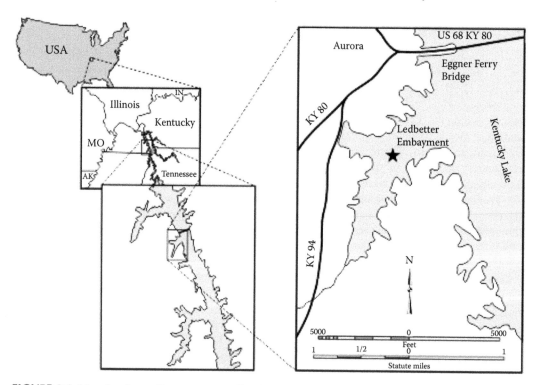

FIGURE 9.1 Map showing sediment core sampling location (* indicates sampling site).

9.2 SAMPLING AND ANALYSIS

The Ledbetter embayment of Kentucky Lake (Figure 9.1) was selected for this study. The sampling site (lat. 36° 45′ 02″ N; long. 88° 08′ 19″ W) in the Ledbetter embayment is situated apart from the mainstream of water current in Kentucky Lake so that the natural sedimentation process is not disturbed. The sediment core collected from this location should thus represent a continuous sequence of sedimentation and consequent accumulation of contaminants over the time frame of interest. A stainless steel inner liner (6 ft long and 2 in. in diameter) was used to collect the sediment core. The inner liner was inserted into a custom-made cast-iron sediment core sampler. It was pushed manually from a boat through the sediment under water to collect the sediment core. The sediment core collected was about 1 m long.

After collecting the sediment core, the inner liner was taken out of the core sampler and immediately transported to the laboratory. A stainless steel iron rod was used to push the sediment core out of the inner liner. The cores were cut into 5-cm sections and were collected in preweighed wide-mouth glass bottles. For chemical analysis, an aliquot of each core section was freeze-dried using a Freezone Freeze Dry System (model: 77535) for 60 h. Aliquots of each section were used for total organic carbon and radiochronological study.

9.2.1 CHEMICAL ANALYSIS

Sediment samples were digested with nitric, perchloric, and hydrofluoric acids. After digestion, the samples were analyzed for Cr, Ni, Cu, Zn, As, Se, Ag, Cd, Sn, Sb, and Pb. The samples were analyzed using Fisons/VG Model PQII+ inductively coupled plasma-mass spectrometry (ICP-MS). Indium was used as an internal standard because it has a major isotope at mass 115, which is approximately in the middle of the mass range (0 to 240 amu). Also, its natural occurrence in environmental samples is negligible. The concentration of the internal standard was in the range of 10 to 100 times greater than the other analytes in the sample.

All standards and samples were spiked with the internal standard of known concentrations. The samples were concentrated or diluted so that all analytes were bracketed by the calibration standards. Each sample batch of 20 samples included one or more certified reference standards, matrix spike and matrix spike duplicate, and continuing calibration standards. Total mercury in sediment was determined by isotope dilution-cold vapor (ICPMS) as described by Smith [14]. The samples are spiked with an enriched ^{201}Hg isotope prior to microwave digestion with HNO_3. An aliquot of the digest was reduced with sodium borohydride and the resulting mercury vapor was swept into the torch of the ICPMS.

9.2.2 ORGANIC CARBON AND NITROGEN

The sample preparation for the organic carbon and nitrogen analysis in sediments was performed following the procedure described by Wong et al. [15]. The analysis was carried out using a CHN analyzer (Perkin Elmer Series II CHNS/O Analyzer-2400 Series II with Perkin Elmer Auto Balance-AD4). Blanks (tin cups), conditioning reagents (sulfamic acid; Perkin Elmer-N241-0501), and standards (CHN standard; acetanilide; organic analytical standard; 0240-121; C = 71.09%, H = 6.71%, O = 11.86%, and N = 10.38%) were run to meet the analytical quality control and quality assurance criteria. After every ten samples, a duplicate, a blank, and a standard were run to verify the instrumental quality control criteria.

9.2.3 RADIOCHRONOLOGY

An aliquot of each of the 5-cm sections of the sediment core from the Ledbetter embayment site were collected in preweighed glass beakers and weighed and dried in an oven at 60°C for 72 h. The activity of ^{210}Pb and ^{137}Cs within each sample was determined concurrently using two low-background, planar intrinsic germanium detectors, a computer-based multichannel analyzer, and Maestro-II analysis software [5]. The sediment accumulation rate was calculated by producing a depth profile of total ^{210}Pb activity within the sediment core. The calculated ^{210}Pb accumulation rate was verified using ^{137}Cs as a complementary radiotracer.

9.3 TEMPORAL TRENDS OF INORGANIC ELEMENTS

Table 9.1 shows the concentrations of various inorganic elements detected in different sections of the sediment core. Cr, Ni, Cu, Zn, As, Se, Ag, Cd, Sn, Sb, Pb, and Hg were detected in all sections of the core sediment. Table 9.2 presents geochronology (ages) of different sections of core sediment up to 30 cm, percent organic carbon, and nitrogen contents.

Percent carbon and percent nitrogen ranged from 0.56 to 1.1 and 0.06 to 0.14, respectively. Among the various elements measured, Zn and Hg were found to be the highest (80 μg/g dry sediment) and lowest concentrations (0.02 μg/g dry sediment) detected in Ledbetter embayment core sediment, respectively. In general, Cr, Ni, Cu, Zn, and Pb concentrations in various sections of the core sediment were comparatively higher than As, Se, Ag, Cd, Sn, Sb, and Hg. Except for Pb, most elements measured showed elevated concentrations in the surface sediments and the concentrations decreased with depth up to 25 to 30 cm and then maintained a fairly steady state up to 110 cm.

Inorganic elements accumulate naturally in riverine sediments, associated with clays and other debris carried into the river. Because natural metal inputs are relatively constant, changes in metal concentrations are generally due to human activities [5]. Several investigators [16–19] have used aluminum as a proxy for the concentrations of aluminosilicate minerals. For natural sediments, the concentrations of metals covary with the concentrations of aluminum, thus providing a tool (Al) to normalize the results of metal analysis to identify natural and anthropogenic enrichments. In this study, Al was not determined; therefore, Al normalization could not be done. However, metals such as Cr, Ni, Cu, Sn, and As exhibited relatively higher concentrations in the surface sediments (0 to 5cm = years: 2000.6 to 1986.3) and the concentrations decreased with increasing depth (25 to 30 cm = years: 1929.2 to 1914.9). Then, a steady state was present in the deeper sections up to 110 cm (Table 9.1 and Figure 9.2). Age determination was not done for core sections below 30 cm.

The trends revealed relatively increased input of these chemicals from the early 1900s to 2000. However, Pb showed a gradual declining trend from the early 1970s to 2000 (Figure 9.2). Pb and Zn have been used by humans for a variety of purposes throughout the 19th and 20th centuries [20]. Point source inputs pf Pb and Zn to aquatic systems (streams, lakes, and reservoirs) include industrial effluents, municipal wastewater effluents, and fossil fuel combustion [21].

From 1950 to the 1970s, automobile use increased in response to economic and population growth, and the predominant source of Pb became automobile exhaust emission of tetraethyl Pb [22]. Population growth has continued to the present, but Pb concentrations have declined slowly, probably due to the removal of leaded gasoline. Temporal trend of Pb in Ledbetter embayment clearly reflected the use pattern. Similar trends were also reported by Callender and Rice [20]. Environmental Zn concentrations remained elevated, probably because of lack of specific regulatory actions regarding Zn [20].

Surface sediment (0 to 5 cm) metal concentrations represent current input of contaminants. Most of the new anthropogenic Pb and Zn additions to the environment may come from material sources. Pb is used in paper, plastics, and ceramics [23]. Zn is used in most commercial metal products, including brass, bronze, castings, and galvanized metal and is added during the manufacture of automobile tires, in the form of zinc oxide (ZnO), as an accelerator in the vulcanization process [24].

In addition, Zn is a common contaminant in agricultural and food wastes. Fossil fuel combustion is the main contributor to worldwide anthropogenic emissions of Zn [25]. Furthermore, atmospheric deposition of Pb, Zn, and other elements cannot be ruled out. Baker et al. [26] reported a number

TABLE 9.1

Inorganic Element Concentrations in Various Sections of a Sediment Core Collected at Ledbetter Embayment of Kentucky Lake, U.S.

Sample depth (cm)	Chromium (µg/g)	Nickel (µg/g)	Copper (µg/g)	Zinc (µg/g)	Arsenic (µg/g)	Selenium (µg/g)	Silver (µg/g)	Cadmium (µg/g)	Tin (µg/g)	Antimony (µg/g)	Mercury (µg/g)	Lead (µg/g)
0–5	46.9	20.4	19.2	79.9	3.80	0.46	0.20	0.25	1.33	0.70	0.11	21.3
5–10	46.1	19.4	18.1	75.6	3.43	0.64	0.27	0.26	1.55	0.71	0.14	22.8
10–15	45.7	18.0	17.4	70.7	3.11	0.51	0.20	0.24	1.18	0.53	0.17	23.2
15–20	43.1	17.4	16.1	68.6	2.99	0.54	0.19	0.21	1.01	0.50	0.21	22.5
20–25	40.2	16.4	15.7	62.6	3.04	0.46	0.22	0.17	1.37	0.59	0.11	20.1
25–30	80.2	13.7	13.4	48.9	2.71	0.41	0.20	0.10	1.02	0.51	0.05	17.1
30–35	50.9	11.9	11.8	41.0	2.43	0.28	0.17	0.07	0.74	0.34	0.02	15.6
35–40	33.0	12.2	11.8	40.1	2.71	0.45	0.16	0.07	0.61	0.33	0.02	15.6
40–45	30.6	11.2	10.7	37.1	2.40	0.43	0.14	0.06	0.53	0.30	0.02	14.6
45–50	29.6	11.7	10.4	36.9	2.55	0.29	0.18	0.06	0.64	0.31	0.02	14.6
50–55	34.9	13.7	12.8	44.1	2.98	0.61	0.15	0.06	0.57	0.29	0.02	16.5
55–60	28.0	12.2	10.6	38.7	2.46	0.39	0.13	0.05	0.54	0.31	0.02	14.4
60–65	27.2	12.7	10.2	38.4	2.16	0.42	0.09	0.05	0.23	0.21	0.02	14.3
65–70	76.8	12.7	10.8	39.7	2.18	0.40	0.14	0.05	0.37	0.30	0.02	15.0
70–75	25.5	11.5	9.8	36.3	2.17	0.37	0.13	0.04	0.33	0.20	0.02	13.2
75–80	23.1	9.8	8.6	35.3	2.00	0.28	0.13	0.04	0.43	0.29	0.02	11.3
80–85	24.9	11.1	9.5	36.8	2.07	0.53	0.10	0.04	0.30	0.28	0.02	12.2
85–90	29.5	11.5	11.1	40.4	2.37	0.63	0.17	0.04	0.67	0.29	0.02	13.6
90–95	29.2	11.2	11.1	39.1	2.45	0.52	0.17	0.03	0.63	0.36	0.02	13.7
95–100	30.5	10.9	10.7	39.9	2.53	0.43	0.13	0.03	0.48	0.32	0.02	14.2
100–105	31.0	10.1	9.9	35.0	2.32	0.41	0.10	0.02	0.31	0.23	0.02	13.1
105–110	30.1	9.5	8.7	33.6	2.24	0.39	0.09	0.02	0.23	0.20	0.02	12.5

TABLE 9.2
Total Organic Carbon, Nitrogen, and C/N Ratio in Selected Sections of a Sediment Core from Ledbetter Embayment, Kentucky Lake

Sample ID	Depth (cm)	Year	Percent carbon	Percent nitrogen	C/N ratio
LE 0–5	0–5	2000.6–1986.3	1.10	0.14	7.86
LE 5–10	5–10	1986.3–1972	0.84	0.09	9.33
LE 10–15	10–15	1972–1957.7	0.65	0.06	10.83
LE 15–20	15–20	1957.7–1943.4	0.63	0.06	10.5
LE 20–25	20–25	1943.4–1929.2	0.62	0.06	10.33
LE 25–30	25–30	1929.2–1914.9	0.56	0.06	9.33

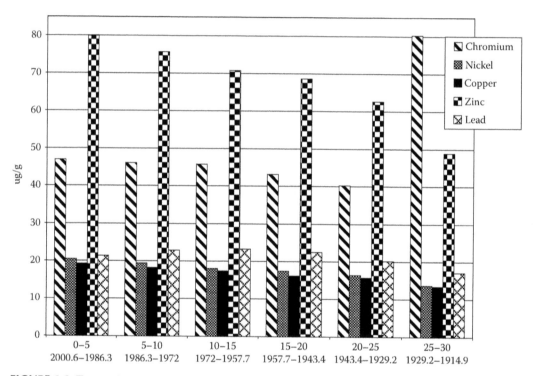

FIGURE 9.2 Temporal trends of selected inorganic elements in sections of a core sediment collected from Ledbetter embayment of Kentucky Lake, U.S.

of elements (Al, As, Br, Cd, Cu, Fe, Mn, Ni, Pb, S, Se, and V) including significant amount (over 10 μg/m³) of Zn in aerosol particles and over 1.5 μg/L Zn in precipitation samples collected at the three Chesapeake Bay Atmospheric Deposition Study sites in 1991 and 1992.

9.4 CONCLUSIONS

Very few studies have reported temporal trends of inorganic elements in an aquatic ecosystem constructed by humans. Vertical profile of inorganic elements in Ledbetter embayment of Kentucky Lake revealed increasing accumulation of Cr, Ni, Cu, Zn, As, Se, Ag, Cd, Sn, Sb, and Hg from the 1900s to the present. Pb has showed a gradual declining trend in recent years. In general, the profiles of the inorganic elements reflected production and usage trends of the respective chemicals.

Although the elemental concentrations of Ledbetter sediments do not seem to affect the living resources negatively, continued monitoring of theses elements is essential in order to prevent further contamination and harmful effects.

ACKNOWLEDGMENTS

This research was supported by the Special Research Program Award of Kentucky Academy of Science. The authors are thankful to Mr. Carl Woods for his help in preparation of this manuscript. Murray State University Center for Reservoir Research Publication Number for this article is: 103.

REFERENCES

1. Loganathan, B.G., Kannan, K., Sajwan, K.S. and Owen, D.A. Butyltin compounds in freshwater ecosystems. In: *Persistent, Bioaccumulative and Toxic Chemicals I: Fate and Exposure*. ACS Monograph Series vol. 772. Lipnick, R.L., Hermens, J., Jones, K. and Muir, D., Eds., American Chemical Society, Washington, D.C. 2001, chap. 10.
2. Loganathan, B.G., Kawano, M., Sajwan, K.S. and Owen, D.A. Extractable organohalogens (EOX) in sediment and mussel tissues from the Kentucky Lake and Kentucky Dam tailwater. *Toxicol. Environ. Chem.* 79, 233–242, 2001.
3. Apitz, S.E., Davis, J.W., Finkelstein, K., Hohreiter, D.W., Hoke, R., Jensen, R.H., Jersak, J., Kirtay, V.J., Mack, E.E., Magar, V.S., Moore, D., Reible, D. and Stahl, R.G. Jr. Assessing and managing contaminated sediments: part I. Developing an effective investigation and risk evaluation strategy. *Integrated Environ. Assessment Manage.* 1, 2–8, 2005.
4. Ankley, G.T., Di Toro, D.M., Hansen, D.J. and Berry, W.J. Assessing the ecological risk of metals in sediments. *Environ. Toxicol. Chem.* 15, 2053–2055, 1996.
5. Alexander, C.R., Smith, R., Loganathan, B.G. et al. Pollution history of the Savannah River estuary and comparisons with Baltic Sea pollution history. *Limnologica* 29, 267–273, 1999.
6. Macauley, J.M. et al. Annual statistical summary: EMAP — estuaries Louisiana Province — 1993. U.S. Environmental Protection Agency, Office of Research and Development, Environmental Research Laboratory, Gulf Breeze, FL. 1993, 95 pp.
7. Van Metre, P.C., Callender, E. and Fuller, C.C. Historical trends in organochlorine compounds in river basins identified using sediment cores from reservoirs. *Environ. Sci. Technol.* 31, 2339–2344, 1997.
8. Vallette–Silver, N.J. The use of sediment cores to reconstruct historical trends in contamination of estuarine and coastal sediments. *Estuaries* 16, 577–588, 1993.
9. Muller, G., Dominik, J., Reuther, R. et al. Sedimentary record of pollution in the western Baltic Sea. *Naturwissen* 67, 595–600, 1980.
10. Loganathan, B.G., Kannan, K., Senthilkumar, K. et al. Occurrence of butyltin residues in sediment and mussel tissues from the lowermost Tennessee River and Kentucky Lake, USA. *Chemosphere* 39, 2401–2408, 1999.
11. Kentucky Outlook 2000. A strategy for Kentucky's third century — executive summary. *The Kentucky Environment Protection Cabinet.* 105, 1997.
12. Loganathan, B.G., Neale, J.R., Sickel, J. et al. Persistent organochlorine concentrations in sediment and mussel tissues from the lowermost Tennessee River and Kentucky Lake, USA. *Organohalogen Compounds* 39, 121–124, 1998.
13. Seaford, K., Loganathan, B.G. and Owen, D.A. 2002. Seasonal variations of organic compounds and inorganic elements concentrations in surface sediments from Ledbetter embayment of Kentucky Lake. *J. Environ. Monit. Restoration* 1, 64–79, 2002.
14. Smith, R. Determination of mercury in environmental samples by isotope dilution/ICPMS. *Anal. Chem.* 65, 2485–2488, 1993.
15. Wong, J.M., Sweet, S.T., Brooks, J.M. et al. Total organic and carbonate carbon content of the sediment. National Status and Trends Program for Marine Environmental Quality. NOAA Technical Memorandum NOS ORCA 71, 1993.

16. Klinkhammer, G.P. and Bender, M.L. Trace metal distributions in the Hudson River estuary. *Estuarine, Coastal Shelf Sci.* 12, 629–643, 1981.

17. Schroop, S.J., Lewis, F.G., Windom, H.L. et al. Interpretation of metal concentrations in estuarine sediments of Florida using aluminum as a reference element. *Estuaries* 13, 227–235, 1990.

18. Trefry, J.H., Metz, S. and Trocine, R.P. The decline of lead transport by the Mississippi River. *Science* 230, 439–441, 1985.

19. Windom, H.L., Schroop, S.J., Calder, F.D. et al. Natural trace metal concentrations in estuarine and coastal marine sediments of the southeastern United States. *Environ. Sci. Technol.* 23, 314–320, 1989.

20. Callender, E. and Rice, K.C. The urban environmental gradient: Anthropogenic influences on the spatial and temporal distributions of lead and zinc in sediments. *Environ. Sci. Technol.* 24, 232–238, 2000.

21. Forstner, U. and Wittmann, G.T.W. *Metal Pollution in the Aquatic Environment.* Springer–Verlag, New York, 1979, 39–42.

22. U.S. Environmental Protection Agency. *National Air Pollution Emission Trends, 1900–1992,* EPA/454/R-933/032, Office of Air Quality Planning and Standards, U.S. Environmental Protection Agency, Washington, D.C. 1993.

23. Nriagu, J.O. *The Biogeochemistry of Lead in the Environment.* Elsevier, Amsterdam, 1978, Part A, 1–14.

24. Christensen, E.R. and Guinn, V.P. *Environmental Engineering Division, EE1,* 105, 165, 1979.

25. Nriagu, J.O. *Zinc in the Environment.* John Wiley & Sons, New York, 1980, Part 1, 113–159.

26. Baker, J.E., Poster, D.L., Clark, C.A. et al. Loadings of atmospheric trace elements and organic contaminants to the Chesapeake Bay, in *Atmospheric Deposition of Contaminants to the Great Lakes and Coastal Waters.* Baker, J.E., Ed. SETAC Press, Pensacola, FL, 1997, chap 9.

10 Chemical Association of Trace Elements in Soils Amended with Biosolids: Comparison of Two Biosolids

Kenneth S. Sajwan, S. Paramasivam, A.K. Alva, and J. Afolabi

CONTENTS

ABSTRACT

Understanding the chemical association of trace elements in soils amended with biosolids is very important because it determines their availability within and mobility beyond the rhizosphere. A sequential extraction method was used to determine the various chemical associations (exchangeable, sorbed, organic, carbonates, and sulfides) of Mn, Zn, Cu, Cd, Cr, Pb, and Ni at the end of sorghum–Sudan grass growth (65 days) in Candler fine sand (pH = 6.8) and in Ogeechee loamy sand (pH = 5.2) amended with sewage sludge (SS), which was obtained from two different sources at application rates of 0, 24.7, 49.4, 98.8, and 148.2 Mg ha^{-1}. Results of this study indicated that, irrespective of the soil type, Zn, Cd, Cr, Ni, Pb, and Mn in the labile fractions (exchangeable + sorbed) were almost nondetectable. Therefore, their availability to plants and mobility beyond the rhizosphere would be substantially low unless further transformations occurred from other fractions.

Results also indicated that the presence of substantial amounts of trace elements studied were in the sulfide (HNO_3) and organic (NaOH) fractions, irrespective of soil type. The exceptions were Pb, which was mainly present as carbonate (Na_2EDTA) fraction, and the remaining Pb equally as sulfide (HNO_3) and organic (NaOH) fractions. Furthermore, results indicated that the Cd was mainly present as carbonate (Na_2EDTA) fraction. Irrespective of soil type, source, and rate of SS application, summation of quantities of various fractions of all the trace elements studied through sequential extraction procedure were 1 to 7% lower than those of total quantities of these trace elements extracted with 4 M HNO_3 on fresh soil samples. They were also 5 to 20% lower than that of total trace elements determined on acid digestion described by the USEPA method. It was further evident that the application of SS shifted the solid phases containing the trace elements in soils away from those extractable with more severe reagents, such as 4 M HNO_3, to those extractable with milder reagents, such as dilute NaOH and Na_2EDTA.

10.1 INTRODUCTION

The agricultural use of biosolids has been a common practice in waste disposal in recent decades. Extensive research on the use of one of the biosolids (BS), SS, during the past three decades has helped to realize the beneficial effects of land application of biosolids. This resulted in renewed public interest in reusing and recycling of sewage sludge from wastewater treatment plants and a dramatic increase in land application of SS from 20 to 54% in the U.S. [1]. The traditional methods of organic waste disposal (landfilling, incineration, ocean dumping) are being restricted or outlawed due to air- and water-quality concerns.

Furthermore, wastewater treatment plants in cities around the world still encounter serious problems in disposing of SS; because of odor, high acidity, and levels of some heavy metals in excess of critical limits, public acceptance is lacking. This has caused exploration of alternative disposal methods. These endeavors have resulted in production of ash by incinerating dewatered activated sludge (ISS) and weathered ash (WISS) by dissolving and storing the incinerated by-product in ash ponds by wastewater treatment plants. Despite all this, the beneficial effects of amending soils with various organic amendments, including the use of sewage sludge, are still dramatically on the rise.

Most of these materials contain plant nutrients and organic matter and are beneficial in many ways to soils by improving their productivity. However, depending on origin and form, they contain variable amounts of trace elements such as Cu, Zn, Pb, Cr, Cd, and Ni, which tend to accumulate within the topsoil of the rhizosphere. Some of these elements can cause phytotoxicity and also pose toxicity problems to animals if any of these plants is used for grazing purpose. However, biosolids applications can also reduce trace element toxicity by redistributing them into various, less available forms [2]. Conversely, the slow mineralization of organic matter in biosolids could release various trace elements into more soluble forms [3]. The dynamics of sludge organic carbon appear, therefore, as a potential parameter driving the fate of trace elements in soil.

Research on the possible adverse effects of trace elements' accumulation in agricultural soils amended with municipal sewage sludge has begun to focus on the chemical forms of these elements in relation to their uptake by plants [4,5]. Later, scientists utilized a sequential extraction technique with various extractants to extract various chemical forms of these elements in the solid phase [2,6–9]. This chapter presents the results of a fractionation study of Mn, Zn, Cu, Cd, Cr, Pb, and Ni conducted at the end of a 65-day plant uptake experiment, and their chemical association in two soils amended with SS at varying rates from two different sources. This chapter also attempts to predict the potential fate of these trace elements with respect to mobility beyond the rhizosphere.

10.2 MATERIALS AND METHODS

10.2.1 Soils

Surface horizons (0 to 15 cm) of Candler fine sand (sandy, hyperthermic, uncoated, Typic Quartzipsamments) from a University of Florida experimental citrus grove in Lake Alfred, Florida, and Ogeechee loamy sand (silicious, thermic Typic Ochragult) from a fallow land in Savannah, Georgia, were used to study the effect of application of varying rates (0, 24.7, 49.4, 98.8, and 148.3 Mg ha^{-1}) of two SS materials on elemental uptake by sorghum–Sudan grass. Each of these treatments was replicated three times. At the end of 65 days of the plant growth experiment, various amended surface soils were sampled for follow-up fractionation study with the help of a sequential extraction procedure to identify the chemical association of residual trace elements, which could aid in predicting the potential fate of the trace elements in soil amended with these materials.

Selected properties of the surface samples of the two soils used in this study determined by standard methods are listed in Table 10.1. Characterization data of sewage sludge are also presented

TABLE 10.1
Selected Properties of Sewage Sludge and Soils Used in the Study

Properties[a]	Units	TSS[b]	PSS[c]	Candler[d]	Ogeechee[e]
pH (1:1 water:soil)		7.2	6.9	6.8	2
K	mg kg^{-1}	1785.0	1206.0	47.6	64.9
Na	mg kg^{-1}	1497.0	670.8	4.9	38.2
P	mg kg^{-1}	18540.0	13250.0	112.8	206.0
Ca	mg kg^{-1}	18890.0	15980.0	169.2	284.3
Mg	mg kg^{-1}	2571.0	1632.0	62.4	62.2
Mn	mg kg^{-1}	203.2	91.2	62.5	14.1
Fe	mg kg^{-1}	17210.0	10420.0	591.4	661.5
Cu	mg kg^{-1}	246.9	219.5	38.4	1.2
Zn	mg kg^{-1}	597.1	542.5	13.7	3.0
Pb	mg kg^{-1}	29.1	47.2	20.9	4.9
Cd	mg kg^{-1}	10.4	4.1	<0.5	<0.5
Ni	mg kg^{-1}	19.6	16.6	<1.0	<1.0
Cr	mg kg^{-1}	80.8	27.4	7.4	4.3
Al	mg kg^{-1}	16890.0	10170.0	1216.0	1383.0
B	mg kg^{-1}	97.8	107.6	1.1	<1.0
S	mg kg^{-1}	10150.0	8972.0	46.1	74.7
Sand	g kg^{-1}	—		967.0	860.0
Silt	g kg^{-1}	—		8.0	100.0
Clay	g kg^{-1}	—		25.0	40.0
Organic C	g kg^{-1}	—		5.1	11.6
CEC	cmol kg^{-1}			2.2	3.9
Texture				Fine sand	Loamy sand

[a] Total elemental compositions were determined on acid digest by ICP-OES.
[b] Collected from Travis field water quality control plant of the city of Savannah (70% domestic and 30% industrial).
[c] Collected from President Street water quality control plant of the city of Savannah (99% domestic and 1% industrial).
[d] Collected near Lake Alfred in Polk County, Florida.
[e] Collected near Savannah in Chatham County, Georgia.

in the same table. Detailed textural characterization indicated that soil properties were rather uniform in the field from which soil samples were collected and throughout the upper 15 cm of the soil profile. Because the soils in the pot were homogeneously mixed with varying rates of amendments, representative dried subsamples were collected from variously amended pots at the end of the 65-day plant uptake experiment for chemical association of trace elements study.

10.2.2 TRACE ELEMENT FRACTIONATION

The soil samples from each pot at the end of the 65-day plant uptake study were used for fractionation of chemical forms of Mn, Zn, Cu, Cd, Cr, Pb, and Ni. The total content of Mn, Zn, Cu, Cd, Cr, Pb, and Ni in the soil samples was determined on filtered extracts obtained from 2-g samples, which were digested with 12.5 mL 4 M HNO$_3$ at 80°C for 16 h. The trace elements in triplicate 2-g samples were fractionated by a modified sequential extraction procedure used by Sposito et al. [8] to quantify exchangeable, sorbed, organically bound, and precipitated forms (as carbonate and as sulfide).

Each 2-g soil sample was weighed into 50-mL centrifuge tubes, and 25 g each of the following reagents were sequentially added and shaken for the time specified for each extractant in the order listed:

- 0.5 M KNO$_3$, 16 h (exchangeable)
- Deionized water, 2 h (three times and combine all three fractions together; sorbed)
- 0.5 M NaOH, 16 h (organically bound)
- 0.05 M Na$_2$ EDTA, 6 h (carbonate from)
- 4 M HNO$_3$ at 80°C for 16 h (sulfide form)

At the end of each extraction period, the soil suspension was centrifuged, the supernatant was decanted into Whatman no. 42 filter paper, and then filtrate was collected. The centrifuge tube with soil was weighed once after the addition of the extractant and again after the solution was filtered, to estimate the quantity of the entrained solution. The quantity of each subsequent extractant was adjusted to account for the entrained solution from the previous extraction. The concentration of trace elements in the filtered solution was determined by Perkin Elmer optical emission spectroscopy (inductively coupled plasma optical emission spectroscopy) RL 3100. The amount of various trace elements in the entrained solution was carried over into subsequent extraction. The contribution of various trace elements in the entrained solution from the previous extraction was used to correct calculations of various trace elemental content in the subsequent extraction.

The concentrations of total trace elements were also determined by USEPA method no. 3050 [10]. This method (second method) involved digestion of 1.0 g soil in 10 mL of HNO$_3$. Then the solution was cooled and an additional 5 mL of concentrated HNO$_3$ was added and refluxed twice for 30 min each time. After the solution was cooled, 3 mL of 30% H$_2$O$_2$ was carefully added and refluxed in a block digester to obtain a clear solution. Then, 5 mL of concentrated HCl and 10 mL of deionized water were added and refluxed without boiling. The solution was then cooled and quantitatively transferred into 50 mL volumetric flask and volumized to 50 mL with deioinzed water. Total trace elemental concentrations in diluted digest was determined using Perkin Elmer optical emission spectroscopy (inductively-coupled plasma optical emission spectroscopy) RL 3100.

10.3 RESULTS AND DISCUSSION

10.3.1 TOTAL TRACE ELEMENT CONTENTS

The total elemental contents in soils amended with varying rates of two different sources of sewage sludge were determined by two different methods, as described under Section 10.2. However, total

TABLE 10.2
Effect of Soil Type and Rate of Raw Sewage Sludge Amendment on Total Content of Various Trace Elements at End of 65-day Sorghum–Sudan Grass Growth

Soil[a]	Source and amendment rates (Mg ha⁻¹)	Cu (mg kg⁻¹)[b]	Zn	Cd	Cr	Pb
	[PSS]					
Candler fine sand [FS]	0.0	16.92	8.10	0.06	1.92	2.46
	24.7	25.10	19.46	0.07	2.13	3.06
	49.4	31.20	31.15	0.13	2.69	9.26
	98.8	44.61	44.83	0.19	5.14	13.48
	148.3	56.25	58.88	0.20	6.27	16.09
Ogeechee loamy sand [GS]	0.0	8.78	8.89	0.07	6.32	11.71
	24.7	13.30	21.71	0.16	6.52	13.46
	49.4	23.14	36.51	0.20	9.01	15.84
	98.8	51.95	78.83	0.45	13.33	26.23
	148.3	63.17	121.96	0.51	14.71	42.38
	[TSS]					
Candler fine sand [FS]	0.0	12.92	7.21	0.06	2.54	5.13
	24.7	18.33	14.29	0.11	3.38	5.75
	49.4	22.67	46.69	0.36	5.83	6.97
	98.8	57.54	89.13	0.78	10.35	9.68
	148.3	75.00	141.25	1.29	15.40	11.88
Ogeechee loamy sand [GS]	0.0	4.33	3.38	0.05	5.43	5.88
	24.7	10.29	22.46	0.21	7.30	7.19
	49.4	15.00	42.88	0.46	9.66	8.31
	98.8	46.08	100.71	1.17	15.29	15.49
	148.3	59.67	142.08	1.56	20.27	18.75
LSD (p = 0.05)		3.12	5.84	0.10	0.43	1.53

[a] Labels next to the soil names in the parentheses [FS] and [GS] represent Florida soil and Georgia soil, respectively. [TSS] and [PSS] represent sewage sludge materials from Travis field and President Street water quality control plants in Savannah, Georgia.
[b] Numbers presented are means of three replicate sample analyses.

Note: LSD values were calculated for total trace element contents determined by HNO_3 acid digestion in soils amended with varying rates of two different sources of sewage sludge and collected at the end of 65 days of plant growth (U.S. Environmental Protection Agency, 1986, in: *Test Methods for Evaluating Solid Waste*. 3rd ed. USEPA SW-S846. U.S. Gov. Print. Office, Washington, D.C. 30501–30594USEPA, 1986).

content of Cu, Zn, Mn, Cr, and Pb in the soil samples-determined digestion step employed by USEPA method [10] is listed in Table 10.2. It is interesting to note that the trace element contents of the two soils, at any level of sewage sludge application, reflected the trace element composition of the sewage sludge. The order of trace element content in original sewage sludge material in TSS is Zn > Cu > Cr > Pb > Ni >Cd; this order in PSS material is Zn > Cu > Pb > Cr > Ni > Cd. The order of trace element content in unamended (preplant) Candler fine sand of Florida is Mn > Cu > Pb > Zn > Cr > Ni > Cd and, in unamended (preplant) Ogeechee loamy sand of Georgia, it is Mn > Pb > Cr > Zn > Cu > Ni > Cd (Table 10.1).

However, at the end of 65-d sorghum–Sudan grass growth, the order of total trace element in soils amended with varying rates of different sewage sludge material varied according to their source of SS. The order of total trace element contents amended with TSS in both soils was Zn > Cu > Cr > Pb > Cd; this order in PSS amended soils was Zn > Cu > Pb > Cr > Cd (Table 10.2). Interestingly, Ni was virtually nondetectable and therefore the concentrations of Ni were not

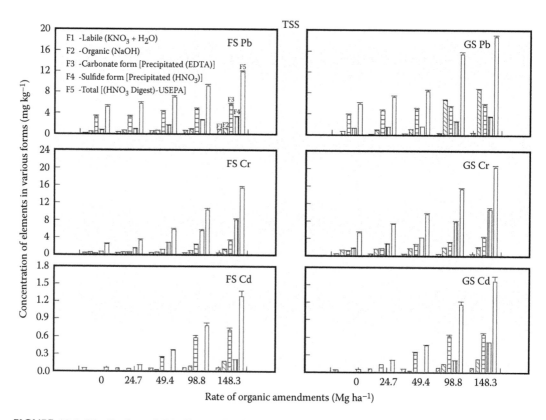

FIGURE 10.1 Distribution of Cd, Cr, and Pb into various chemical fractions at the end of 65-day sorghum–Sudan growth in sandy and loamy sand soils amended with various rates of industrial (70% domestic and 30% industrial — TSS) sewage sludge material obtained from a water quality-control plant in Savannah, Georgia.

presented in Table 10.2. Disappearance of Ni is attributed to greater uptake by sorghum–Sudan grass. A significant amount of Ni accumulation was observed in shoots and roots of sorghum–Sudan grass; Pb accumulation was observed mostly in roots. Accumulation of Ni in roots was greater than in shoots of sorghum–Sudan grass. Furthermore, in both soils, total trace element concentrations increased approximately proportionally to the application rates of different sewage sludge materials used to amend these soils.

10.3.2 DISTRIBUTION OF TRACE ELEMENTS INTO VARIOUS FRACTIONS

The chemical association of various residual trace elements in the soils amended with varying rates of different sources of sewage sludge at the end of sorghum–Sudan crop growth was determined by sequential extraction procedures. This procedure would help to identify the relative proportions of "exchangeable," "sorbed," "organic," "carbonate," or "sulfide" forms of these trace elements. Figure 10.1 through Figure 10.4 present distribution of various trace elements among these various fractions in soil amended with these two sewage sludge materials, along with total elemental contents determined by the USEPA method [10].

Theoretically, summation of various fractions of any trace elements should be equal to the total amount of that element extracted from fresh soil with 4 M HNO$_3$ digest. Total amount of various trace elements determined by 4 M HNO$_3$ digest is not presented here. The experimental precision at any step of the sequential extraction procedure was always better and the sum of extracted trace elements was always near, though usually systematically about 1 to 7% lower, than the total trace

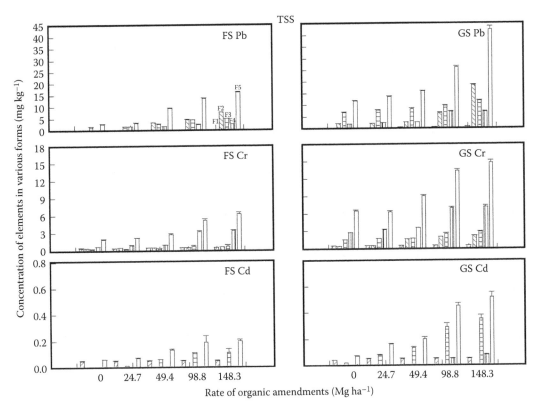

FIGURE 10.2 Distribution of Cd, Cr, and Pb into various chemical fractions at the end of 65-day sorghum–Sudan growth in sandy and loamy sand soils amended with various rates of domestic (99% domestic and 1% industrial — PSS) sewage sludge material obtained from water quality control plant in Savannah, Georgia.

elemental contents determined by 4 M HNO_3 digest on fresh samples. These deviations were perhaps due to some experimental errors. However, the summation values for various trace elements reported in this study were about 5 to 20% lower than that of total trace elements contents determined by HNO_3 + HCl and H_2O_2 digestion procedures [10] and reported in Table 10.2.

The overestimation of total trace element contents by the USEPA method compared to that of the other two procedures is basically due to recovery of trace elements from organic material through the oxidation process. Chemical association (fractionation) study data were presented only for Cd, Cr, Pb, Cu, Zn, and Mn. However, the trace element Ni was nondetectable for all the fractions studied and therefore was not reported.

10.3.2.1 Exchangeable Fraction

Exchangeable fractions of various trace elements are defined as the amount that could be extracted with 0.5 M KNO_3 during the first step of the sequential extraction procedure. In this study, irrespective of soil type, source, and rates of sewage sludge application, presence of very low to nondetectable amounts of Cd and Cr were evident as exchangeable fraction. However, presence of Mn in exchangeable fraction accounted for about 15% of the total contents in the soil at the end of 65 days of sorghum–Sudan growth. Other trace elements, such as Zn, Cu, and Pb in exchangeable form, accounted for only 1 to 4% of the total. Actual quantities of all the trace elements increased with increasing rates of sewage sludge amendments, irrespective of soil types or source of sewage sludge.

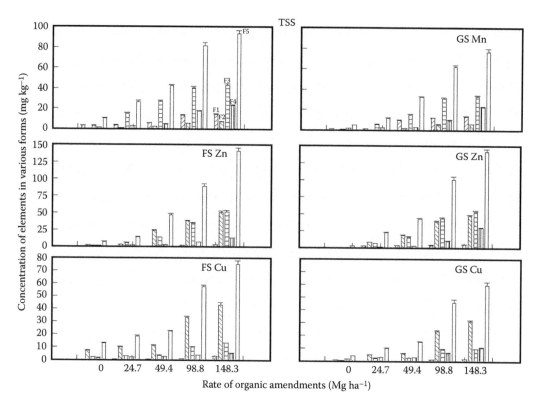

FIGURE 10.3 Distribution of Cu, Zn, and Mn into various chemical fractions at the end of 65-day sorghum–Sudan growth in sandy and loamy sand soils amended with various rates of industrial (70% domestic and 30% industrial — TSS) sewage sludge material obtained from water quality control plant in Savannah, Georgia.

10.3.2.2 Sorbed Fraction

Sorbed fractions of various trace elements are defined as the amount that could be extracted with deionized water as a second step in the sequential extraction procedure following the KNO_3 extraction step. Except for Cd and Cr, all the other trace elements studied (Zn, Cr, Pb, and Mn) in sorbed fraction were quantitatively smaller compared to that of the exchangeable form. The sorbed fraction of Cr accounted for about 2.6 to 22.0% of its respective total content; the sorbed Cd accounted for 5 to 85% of its respective total content.

10.3.2.3 Labile Fraction

Amounts of exchangeable and sorbed fractions were added together and presented as labile components of all these trace elements in Figures 10.1 through Figure 10.4. The labile fraction is the fraction that is readily available for plant uptake or leaching. Therefore, having more labile fraction of any of these trace elements would pose a serious concern because they can cause toxicity problems or groundwater contamination due to loading. The relationships among various fractions for various trace elements — how these various fractions are affected by soil type, sewage sludge source, and rates of amendments — are shown at a glance in Figure 10.1 through Figure 10.4. Overall, this labile fraction is substantially very low compared to other fractions.

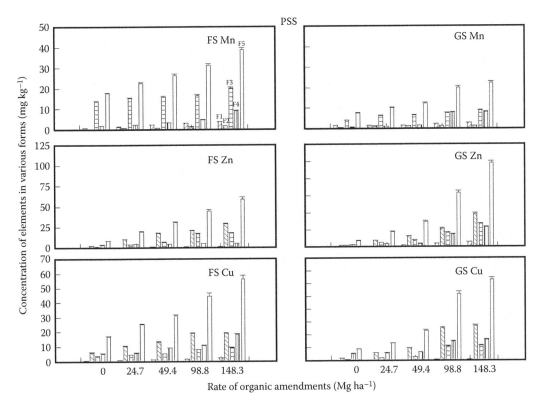

FIGURE 10.4 Distribution of Cu, Zn, and Mn into various chemical fractions at the end of 65-day sorghum–Sudan growth in sandy and loamy sand soils amended with various rates of domestic (99% domestic and 1% industrial — PSS) sewage sludge material obtained from water quality control plant in Savannah, Georgia.

10.3.2.4 Organic Fraction

Organically bound fractions of trace elements are defined as the amount that could be extracted from soil with 0.5 M NaOH for 16 h of equilibration under shaking conditions. This is the third step following, KNO_3 and water extraction steps in the sequential extraction procedure described by Sposito et.al. [8]. Among the trace elements studied for various chemical associations, residual Cu at the end of sorghum–Sudan crop growth in soils amended with varying rates of different sources of sewage sludge accounted for about 32 to 59% of the total Cu as in organic form. Similarly, Zn accounted for about 21 to 57% of the total Zn, and Pb accounted for about 7 to 48% of the total in organic form.

Furthermore, it was observed that the organic form of Cu almost doubled in actual quantity at low to moderate rates of sewage sludge amendment in Candler fine sand from Florida compared to that of Ogeechee loamy sand of Georgia for the corresponding rate of amendment application (Figure 10.3 and Figure 10.4). A somewhat similar trend was observed for Zn but not prominently as for Cu. The amount of organic fraction of Cu and Zn was not substantially affected by rate of amendment application at medium to higher rates, but a greater increase was observed at the highest rate of amendment application (148.3 Mg ha^{-1}). The organic fraction of Pb increased with increasing rates of amendment application and the increase in actual quantities was great at the highest rate of sewage sludge application. Organic fraction was very prominent in Ogeechee loamy sand of Georgia compared to that in Candler fine sand of Florida. In contrast, the organic form of Cr was higher at all application rates of sewage sludge in Ogeechee loamy sand, but not in Candler fine sand from Florida (Figure 10.1 and Figure 10.2).

10.3.2.5 Carbonate Fraction

The carbonate forms of trace elemental fractions are defined as the amount that could be extracted from soil with 0.05 M Na-EDTA for 6 h of equilibration under shaking conditions. This step is the fourth following the extraction of organic fraction with 0.5 M NaOH. All the trace elements studied in this study showed carbonate fraction in both soils amended with sewage sludge. However, among all the trace elements, a major portion of residual Pb, Cd, and Mn were present as carbonate, irrespective of source of sewage sludge and type of soil.

Among these three trace elements, about 50 to 60% of total residual Mn was present as carbonate form, irrespective of soil type and rate of sewage sludge application. The carbonate fraction of total Cd accounted for about 50 to 70% and the Pb carbonate fraction accounted for about 28 to 65% of the total Pb in both soils, irrespective of source of sewage sludge. Even though relative proportions of carbonate forms of Cu and Zn in soils amended with both sewage sludge materials and in both soils were somewhat comparable, the actual quantity of Cu present as carbonate form was greater in Candler fine sand from Florida and the actual quantity of Zn present as carbonate form was greater in Ogeechee loamy sand of Georgia. The carbonate form of Cr accounted only for 12 to 25% of the total Cr present in residual soil, irrespective of soil type and source of sewage sludge.

10.3.2.6 Sulfide Fraction

Trace elements in sulfide forms are generally extracted with the help of 4 M HNO$_3$ as the last step in the sequential extraction procedure described by Sposito et. al. [8]. All the trace elements studied in soils amended with sewage sludge indicated greater percentages present as sulfide form. In addition, irrespective of trace elements, the sulfide form of trace elements (Cu, Zn, Cr, Cd, Pb, and Mn) increased with increasing amount of sewage sludge application. In terms of actual quantity (Cr > Pb > Cd), quantities of Cd, Pb, and Cr were comparatively less when compared with other trace elements studied. Even though actual quantity of Cr was less, this represented about 29 to 65% of the total Cr content. Furthermore, about 5 to 30% of total Cu and Zn was present as sulfide form. Similarly, about 8 to 39% of total Mn was present as sulfide form.

10.4 GENERAL DISCUSSION AND CONCLUSION

The accumulation of Zn, Cu, Cr, Pb, Ni, Cd, and Mn in soils amended with sewage sludge was found to be governed by the content of the metals in amendment source and rate of sewage sludge application, and also the influence of plant growth and uptake. Soil properties, trace metal content, and rate of sewage sludge application appeared to play a role in determining the amounts of various fractions of trace elements. In general, it was evident that Ogeechee loamy sand of Georgia had greater specific trace element content at a particular SS application rate compared to that of Candler fine sand from Florida. Even though both SSs had substantially greater amounts of Ni (Table 10.1) prior to the application to these soils, no detectable amount of Ni was found in soils as any of those chemical fractions studied at the end of the 65-day sorghum–Sudan crop growth. Absence of Ni in residual soil was an indication of plant uptake; this was confirmed by the experimental data obtained by analyzing plant roots and shoots. However, the plant accumulation of Ni data is not presented in this chapter.

The actual quantity of labile Cu, Cr, Cd, and Pb in the soils, which was extractable with KNO$_3$ and H$_2$O, was very low, averaging between 0.0 to 3.0 mg L^{-1}, regardless of the types of soils and source and rate of SS application. However, for Zn and Mn, the KNO$_3$ extractable fraction was substantially greater compared to that of the H$_2$O extractable fraction. Combined (KNO$_3$ + H$_2$O) fractions for Zn accounted for between 2.5 and 10.0% of the total Zn; for Mn, they accounted for between 3 and 25% of the total Mn. The low amount of trace elements in the KNO$_3$ + H$_2$O fraction could signify a low availability of these elements to crop plants because the readily soluble form of a trace metal is often regarded as the most bioavailable form. However, the bioavailable forms are related to several properties and reactions within the soils [11,12] and several chemical extractants developed to represent plant absorption may extract trace elements from several fractions [13].

The application of sewage sludge to the field soils is expected to reduce the HNO_3 (sulfide) fraction percentages for all of the trace elements, regardless of the type of sewage sludge or soil, due to their acidic nature during their decomposition. This trend was generally observed with sewage sludge application for most of the trace elements studied in this experiment with the exception of Cr. The sulfide fraction of Cr represented about 28 to 62% of the total Cr content, which represents the slowly available fraction. It is also generally observed that the application of sewage sludge tended to increase percentages of NaOH (organic) fraction. Trace elements, such as Zn, Cu, and Pb, very closely followed this trend in the study, as expected. However, Cd and Mn did not follow this trend. In contrast, a significant portion (about 19 to 80%) of residual Cd and Mn was present as carbonate fraction, irrespective of soil type and source of sewage sludge.

Irrespective of soil type and source of sewage sludge material, the organic form of Cu for Cu and Zn was the predominant Cu form in residual soil amended with sewage sludge. However, the next greater percentage of Cu and Zn was precipitate form and the source of sewage sludge to a certain extent determined whether it was in sulfide form or carbonate form of precipitate. However, in general, a greater percentage of Zn was present as carbonate (Na_2EDTA) fraction irrespective of soil type or source of sewage sludge. In the case of Cu, a greater percentage of Cu was present as sulfide (HNO_3) in both soils receiving sewage sludge exclusively from a domestic source; the carbonate form of residual Cu was second dominant in both soils receiving sewage sludge from industrial and domestic mixture.

A major portion of Cr present as sulfide (HNO_3) fraction, irrespective of soil types, accounted for about 29 to 62% of total Cr. Relative proportions of various fractions of Pb were affected mainly by soil type; a quantitatively greater amount of Pb in any fraction was generally found in Ogeechee loamy sand compared to that of Candler fine sand of Florida. Among the various fractions studied, a quantitatively greater amount of Pb was present as organic (NaOH) fraction. Irrespective of soil type and source of sewage sludge precipitate form of Pb (carbonate or sulfide form) was the next dominant form Pb in residual soil. Irrespective of the soil type, and source of sewage sludge, the carbonate form of Cd and Mn was the dominant form present in residual soils. The various fractions of Mn were in the following order: $Na_2EDTA > HNO_3 > H_2O > NaOH > KNO_3$. For Cd, the order of fraction was $Na_2EDTA > H_2O > HNO_3 > NaOH > KNO_3$.

It is evident that the application of sewage sludge shifted the solid phases containing Cu, Zn, Cr, Pb, Cd, Ni, and Mn in the soils away from those extractable with more severe reagents, such as 4 M HNO_3, to those extractable with milder reagents such as dilute NaOH, NaEDTA, KNO_3, and H_2O. It was further evident that exchangeable (KNO_3) and sorbed (H_2O) fractions were substantially low and the major portion of the trace elements was in organic (NaOH) and carbonate (NaEDTA) and sulfide (HNO_3) fractions. This shift suggests that sewage sludge application would provide the trace elements in labile chemical forms that could be more readily available to crop plants with time. In the meantime, presence of substantially low amounts of exchangeable plus sorbed (labile) fractions of all the trace elements in sewage sludge-amended soils indicated very low mobility beyond the rhizosphere.

ACKNOWLEDGMENT

This research was supported by the U.S. Department of Energy and the Environmental Protection Agency Grant Award No. DE-FG09-02SR22248. Laboratory assistance rendered by Craig Young, Jeffrey Delise, and Marial Potts is greatly appreciated.

REFERENCES

1. Basta, N.T. (2000). Examples and case studies of beneficial reuse of municipal by-products. In: J.F. Power and W.A. Dick (Eds.) Land Application of Agricultural, Industrial, and Municipal By-Products. SSSA, Madison, WI. 481–504.
2. Sims, J.T. and Kline, J.S. (1991). Chemical fractionation and plant uptake of heavy metals in soils amended with co-composted sewage sludge. *J. Environ. Qual.* 20:387–395.
3. McBride, M.B. (1995). Toxic metal accumulation from agricultural use of sludge: are USEPA regulations protective? *J. Environ. Qual.* 24: 5–18.
4. Chaney, R.L. and Giordana, P.M. (1976). Microelements as related to plant deficiencies and toxicities. In: L.F. Elliot and F.J. Stevenson (Eds.) *Soils for Management of Organic Wastes and Wastewaters.* SSSA, Madison, WI. 234–279.
5. Mahler, R.J., Bingham, F.T., Sposito, G., and Page, A.L. (1980). Cadmium-enriched sewage sludge application to acid and calcareous soils: relation between treatment, Cd in saturation extracts, and Cd treatment. *J. Environ. Qual.* 9: 359–364.
6. Shuman, L.M. (1985). Fractionation method for soil microelements. *Soil Sci.* 140: 11–22.
7. Shuman, L.M. (1999). Organic waste amendments effect on zinc fractionation of two soils. *J. Environ. Qual.* 28:1442–1447.
8. Sposito, G., Lund, L.J., and Chang, A.C. (1982). Trace metal chemistry in arid-zone field soils amended with sewage sludge: I. Fractionation of Ni, Cu, Zn, Cd, and Pb in solid phases. *Soil Sci. Soc. Am. J.* 46:260–264.
9. Sposito, G., LeVesque, C.S., LeClaire, J.P., and Chang, A.C. (1983). Trace metal chemistry in arid-zone field soils amended with sewage sludge: III. Effect of time on the extraction of trace metals. *Soil Sci. Soc. Am. J.* 47:898–902.
10. U.S. Environmental Protection Agency. (1986). Acid digestion of sediments, sludge, and soils. In: *Test Methods for Evaluating Solid Waste.* 3rd ed. USEPA SW-S846. U.S. Gov. Print. Office, Washington, D.C. 30501–30594.
11. Fergusson, J.E. (1990). *The Heavy Elements. Chemistry, Environmental Impact and Health Effects.* Pergamon Press, Oxford._
12. Lee, B.D., Carter, B.J., Basta, N.T., and Weaver, B. (1997). Factors influencing heavy metal distribution in six Oklahoma benchmark soils. *Soil Sci. Soc. Am. J.* 61:218–223.
13. Tack, F.M.G. and Verloo, M.G. (1995). Chemical speciation and fractionation in soil and sediment heavy metal analysis: a review. *Int. J. Environ. Anal. Chem.* 59:225–238.

Section III

Biotechnology

11 Microbial Genomics as an Integrated Tool for Developing Biosensors for Toxic Trace Elements in the Environment

Ranadhir Chakraborty and Pradosh Roy

CONTENTS

11.1 INTRODUCTION

The process of the origin of life in the course of a long irreversible development of natural systems belongs to the great mysteries of nature. At present, we are only approaching its solution. There can be no doubt that the emergence and organization of living matter are connected with the properties of the atoms entering into its composition — primarily, with the properties of carbon. Thus, what brought about the origin of life consisted of those particular processes in space evolution of the substance of the solar system or, more precisely, in the nuclear synthesis (nucleosynthesis) that preceded the appearance of this system.

This nuclear synthesis led to the formation of the atoms of biophile elements H, C, N, and O in a relationship that proved to be favorable for the formation of complex organic compounds — the direct predecessors of life. When the primary gaseous nebula that was genetically associated with the early Sun was cooling, organic compounds appeared. They emerged mainly in the last stages of congelation, which was registered in the space rocks of the further periods: meteorites and particularly the carbonaceous chondrites. Thus, all the data on cosmochemistry of meteorites, asteroids, and comets prove that the formation of organic compounds in the solar system in the early stages of its development was a typical and mass phenomenon [1].

The biosphere represents a highly complicated organization of matter in which various inorganic forms of movement are related to living substances. Any form of life is connected with a particular environment; thus, the problem of the origin of life may well be regarded as precisely the problem of the emergence of the biosphere. It is not difficult to infer that the living organisms use, in the first place, the most accessible atoms, which are furthermore capable of forming stable and multiple chemical bonds.

It is common knowledge that carbon can give rise to the emergence of long chains, leading to the origin of innumerable polymers. Sulphur and phosphorus can also form multiple bonds. Sulphur enters into the composition of proteins, and phosphorus forms a constituent part of nucleic acids. The metabolism with the outer environment had become *a sine qua non* for the existence of every single organism. Presumably, the whole biosphere of this planet should be regarded as the part of its matter that undergoes the process of metabolism with the living matter. Any form of living organism consists of combinations of a few chemical elements (e.g., H, C, N, and O). The other elements enter into the composition of living organisms in relatively insignificant quantities, in spite of the exclusively important role that some of them have in performing physiological functions [1]. One may well think that all the elements of Mendeleev's table enter into the composition of the living matter of this planet, though in different quantities.

The simplest and earliest organized living forms, the bacteria, possess a selective ability of making use of the chemical elements from the environment according to their physiological needs. As with nutrient organic compounds, whatever was available in the environment provided selection for transport and metabolism. Most cell types found economy in accumulating trace elements for immediate and future needs through utilizing relatively unspecific uptake systems for a wide range of metal ions. Unlike carbon compounds, however, some inorganic nutrients at low levels are toxic when present at higher levels (Cu^{2+}, Co^{2+}, and Ni^{2+}, for example); other inorganic ions (e.g., Hg^{2+} and AsO_2^-) are always toxic and have no metabolic functions in bacteria [2].

With the emergence of intelligence, the highest form in the process of evolution from inanimate matter to human brain, the question that remains the most intriguing is the origin of life. Whatever complexity the process is, nature must drive to organize the system to evolve from simple molecules to multicellular organisms. Thus, the fundamental units in so many complex and diverse life systems seen today are the atoms of only little more than 100 elements. On the other hand, because every cell retains its history, all the ancestral evidences must be preserved in the living body. Thus, the present day molecules are able to guide understanding of the present biological phenomenon at the molecular level, as well as serve as the museum of all the prehistoric evolutionary processes. A biologist thus must find the answer to the fundamental question: selectivity of an essential metal atom (ion) encrypted in the life system, within the molecules.

This review hypothesizes that the toxicity potential of heavy metal ions has forced life in its early evolution to develop metal ion homoeostasis factors or metal resistance determinants. The amazing process of evolving resistance to heavy metal ions is not a result of successful accumulation of mistakes in replication of the genetic code, but is rather the outcome of designed creative processes. In the present context, it is apparent that the metabolic penalty for having uptake pumps more specific is greater than the genetic cost of having plasmid genes, which confers resistance to heavy metal ions, in the population that can spread when needed [2].

The new picture of creative cooperative evolution is based on the cybernetic capacity of the genome as suggested by Ben–Jacob [3]. This new picture of the genome as an adaptive cybernetic unit with self-awareness changed the neo-Darwinian paradigm in the life sciences because it has been shown that the genome is a dynamic entity capable of changing itself [4–7]. In this scenario, genome includes the chromosomes, all the extrachromosomal elements, and all the "chemical machinery" (like enzymes) involved in genomic activity and production of proteins. Every genome that has been sequenced to date has provided new insight into biological processes, activities, and potential of these species that was not evident before the availability of the genome sequence.

In the light of current perceptions of metabolic networks that have cropped up in the age of genomics is the one that microbial cells are not merely an assemblage of information but are constituted of atoms. The transporter determinants of genome as well as the extra cellular environment determine the atoms in a microbial cell. Therefore, if one wishes to understand how the cell is directed by the genome, one must know what atoms are present in the surroundings of the cell and how the cells react to complex mixtures of the chemical elements and compounds [8]. Genomics-based reconstruction of the cellular machinery, microbial remediation of complex environmental mixtures, and knowledge about microbial individuality have added dimension to the constellation of elements found in microbes [8]. Genome sequencing of prokaryotes of diverse origin occupying a wide range of environmental niches has resulted in shift of the primary focus of microbial genomics from the pathogen genomes to other genomes bearing potential for environmental applications.

Accessibility of bioinformatics tools enabled successful comparisons of different genomes to identify metabolic pathways and the analysis of transporter profiles across various species. For example, the genomic sequence of *Pseudomonas putida* KT2440 was used to survey the organism's possible mechanisms of uptake and resistance to and homoeostasis regulation of several metals and metalloids [9]. In the present context, it is relevant to cite that many cell types synthesize metal-binding proteins in response to the presence of specific metals and play important roles in metal homoeostasis and detoxification mechanisms. Due to a wide spectrum of selectivity, metal-binding proteins could be used as the basis of biosensors of varying specificity, depending on the required application.

Some microorganisms have developed genes for heavy metal resistance, which tend to be specific towards a particular metal instead of a general mechanism for all heavy metals. These systems can be used as the contaminant-sensing component of the biosensor by detecting the metal for which it is designed to detoxify or excrete. The contaminant-sensing component can find a route of biotechnology in combination with reporter genes to create biosensors that can identify toxic metals at very low levels.

11.2 GENESIS AND CHEMISTRY OF TOXIC TRACE METALS RELEVANT TO THEIR INTERACTION WITH LIFE PROCESSES

Elements were formed in stars billions of years ago for the metals on Earth. Stars mainly burn hydrogen to helium, and later helium to carbon with oxygen and nitrogen as intermediates; higher elements are just "accidentally" formed. Thus, the probability for any element to be formed decreases hyperlogarithmically with its atomic number. The heavier any element is, the lower is the probability to find it in the Earth's crust, and the lower is the probability that evolution has used it. The three exceptions to this rule are [1]:

- The element synthesis in the stars stops at iron; thus, iron and its neighbors in the periodic system are present in high amounts on Earth despite their relatively high atomic masses.
- Li, Be, and B, which are not heavy metals, are overstepped during element synthesis; thus, they are present in low amounts on Earth, despite their relatively low atomic masses.
- Some heavy metals with extraordinarily stable nuclei were formed in higher amounts than any other heavy metals.

Metals with a density beyond 5 g/CC are heavy metals. Therefore, the term "heavy metals" is ascribed to the transition elements from V (but not Sc and Ti) to the half metal As; from Zr (but not Y) to Sb; and from La to Po, the lanthanides and the actinides. There are 21 nonmetals, 16 light metals, and 53 heavy metals (including As) constituting the total of 90 naturally occurring elements [10]. The transition elements with incompletely filled d-orbitals are heavy metals. The ability of the heavy metal cations to form complex compounds, which may or may not be redox active, is provided by the d-orbitals.

The origin of life is reduced to that of the biosphere, which from its inception was a complicated self-regulating system. A great variety of geochemical functions of living matter emerged at least as a result of the fact that any most primitive cell, being in aqueous and marine medium, had the closest possible contact with all the chemical elements of Mendeleev's table. Because life depends on water, heavy metals interesting for any living cell must form soluble ions. All the 53 heavy metals do not have a positive or adverse biological function because of the nonavailability of some heavy metals to the living cell in the usual ecosystem [11]. To enable the bioavailability, heavy metals must be present at least in nanomolar concentration in a given ecosystem because a concentration of 1 nM means that, in a cell suspension of 10^9/ml, each cell may receive about 600 ions. Metal ions generally present in lower concentration may be used by a microorganism for very specific purposes; however, the lower the mean concentration of the metal ion is in an ecosystem, the lower is the probability that any species carries around genes to use or detoxify this specific heavy metal [11].

The living conditions in the world ocean were most favorable, so it is possible that the Earth's hydrosphere was distinguished by the constancy of the biomass throughout the whole period of existence. Weast [10] has differentiated heavy metals into four classes on the basis of their concentration in seawater:

- Frequent possible trace elements with concentrations between 100 nM and 1 µM are Fe, Zn, and Mo.
- Possible trace elements with concentration between 10 and 100 nM are Ni, Cu, As, N, Mn, Sn, and U.
- Rare possible trace elements would be Co, Ce, Ag, and Sb.
- Cd, Cr, W, Ga, Zr, Th, Hg, and Pb are just below the 1-nM level.

Other elements, e.g., Au with 55.8 pM in seawater, are not likely to become trace elements.

The relative solubility of heavy metals under physiological conditions dictates the difference in biological importance and the toxicity of heavy metals vis-a-vis affinity to sulfur and other interaction with macrobioelements [11]. Because of the low solubility of the tri- or tetravalent cation [10], Sn, Ce, Ga, Zr, and Th have no biological influence. Of the remaining 17 heavy metals, Fe, Mo, and Mn are important trace elements with low toxicity; Zn, Ni, Cu, V, Co, W, and Cr are toxic elements with high to moderate importance as trace elements; and As, Ag, Sb, Cd, Hg, Pb, and U have no beneficial function, but must be considered by cells as toxins [11].

In addition to playing an important catalytic role by complexing with biochemicals, heavy metal cations in sophisticated biochemical reactions, such as nitrogen fixation; water cleavage during oxygenic photosynthesis; respiration with oxygen or nitrate; one-electron catalysis, rearrangement of C–C bonds; hydrogen assimilation; cleavage of urea; transcription of genes into mRNA, etc., can also form nonspecific complex compounds at higher concentrations in the cells that lead to the toxic effects. The heavy metal cations like Hg^{2+}, Cd^{2+}, and Ag^+ are immensely toxic complex formers and are often perilous for any biological function. Even most important trace elements like Zn^{2+} or Ni^{2+} — and especially Cu^{2+} — are toxic at higher concentrations. Therefore, it was compelling to every life form to evolve some homoeostasis system for maintaining a tight control over intracellular concentration of heavy metal ions [11].

11.3 PHYSICAL PROPERTIES OF HEAVY METAL CATIONS AND OXYANIONS

The divalent cations Mn^{2+}, Fe^{2+}, Co^{2+}, Ni^{2+}, Cu^{2+}, and Zn^{2+} have ionic diameters in a range of 138 to 160 pm [10], a difference of 14%; all are, of course, double positively charged. Oxyanions like chromate, with four tetrahedrally arranged oxygen atoms, double negatively charged, differ mostly in the size of the central ion, so the structure of chromate resembles sulfate. The same is true for arsenate and phosphate. Thus, uptake systems for heavy metal ions must bind those ions tightly if they want to differentiate between a couple of similar ions. However, tight binding costs time and energy.

11.4 DUAL STRATEGY ADOPTION BY THE LIVING CELL FOR UPTAKE OF HEAVY METAL IONS AND THEIR COMPARISON

The uptake mechanisms that exist within a cell allow the entry of metal ions, including heavy metal ions that affect toxicity. There are two general uptake systems: the first one is driven by a chemi-

TABLE 11.1
Comparison between Two Basic Mechanisms of Uptake of Heavy Metal Ions

Nonspecific strategy	Specific strategy
Rapid	Slow
Constitutively expressed	Inducible; produced by the cell in times of need, starvation, or specific metabolic situation
Used by a variety of substrates	High substrate specificity
Usually driven only by the chemiosmoticgradient across the cytoplasmic	Uses ATP hydrolysis in many membrane cases as energy sources
Not expensive	Expensive
Uptake system:	Uptake system:
Nn-specific CorA (Ni^{2+}, Co^{2+}, Zn^{2+}, and Mn^{2+} uptake) [bacteria, archaea, and yeast]	P-type ATPases (Mg^{2+})
Pit-phosphate system (arsenate)	ABC transporters (Mn^{2+}, Zn^{2+}, and Ni^{2+})
Sulfate uptake system (chromate)	Slow and specific HoxN (Ni^{2+} and Co^{2+})

Sources: The table is constructed on the basis of the information assembled from MacDiarmid, C.W. and Gardner, R.C., *J. Biol. Chem.*, 273, 1727, 1998; and Tao, T. et al., *J. Bacteriol.*, 177, 2654, 1995.

osmotic gradient across the membrane results in an influx of a wider variety of heavy metals, and the second is driven by energy from ATP hydrolysis (Table 11.1).

11.5 INCORPORATION OF METALS IN BIOACTIVE MOLECULES IN THE PROCESS OF EVOLUTION

In the modern version of Mendeleev's table, transition metals are in the center and clearly share many attributes. They exist as (MnO_4^{2-}, WO_4^{2-} exceptions) charged cations in solution under biological conditions. Many of these ions can have similar ionic radii, so they tend to bind to similar classes of ligands. Indeed, ligating atoms that bind metal in a particular protein are found in four types of amino acid side chains: histidines, cysteine, glutamic, and aspartic acids. Less frequently, tyrosine and methionine coordination is found. Nature generally uses an array of two or five of these coordinating side chains in the majority of metalloenzymes.

Once it gathers these different metals, how does the cell know which metal to put into which enzyme-active site? A traditional view in the field of bioinorganic chemistry has been that metal selectivity is due to very sophisticated chelating properties of the individual apo-proteins. In this scenario, apo-proteins are thought to poise the exact orientation of these side chains to match the precise ionic radius and electronic preference of the functional metal ions, for example, Zn^{2+}, and to discriminate against all others, such as Cu^{2+} or Fe^{2+} ions. In some of these cases, very little difference is present in ionic radii. The proteins are thus viewed as highly specific chelating agents with finely tuned kinetic and thermodynamic properties that have been selected through evolution to bind only one type of transition metal ion. In this model, each apo-protein as it is produced in the cell simply selects a metal ion from the cytoplasm. In general, however, the field has moved to the point of thinking of membrane-bound proteins as transporters.

A breakthrough in the process of evolution of bioactive molecules from simple atoms must be the spontaneous association of small molecules to generate big molecules (i.e., polymers such as polypeptides and polynucleotides) that organized and resulted in the emergence of a giant molecule with structure and characteristics distinct from the inanimate matter in terms of its stability, spontaneous development, and awareness of environment. In that simplest form of the primitive life, it is possible that the minimum cellular processes were governed through inorganic catalysts.

In such reactions, the simplest redox reactions of the biological molecules were likely to be mediated and catalyzed by metal ions.

The primitive life processes evolved in Archean water were definitely anoxic and the environment was reducing the metal ions remained in the sparingly soluble state, limiting their concentration. When the reducing environment no longer existed upon exhaustion of readily available hydrogen, oxygen started appearing in the available form in the environment by splitting water by the photosynthetic apparatus. It is argued that metals available at that period at relatively higher concentration were selected by and encrypted in the bioactive molecules to perform the specific catalytic or other functions and thus became essential metals in the life processes. During transition from anoxic to oxic environment, some elements, like Cu, Fe, Ni, Zn, Mo, and Co, began to participate in biological functions.

The cells had to incorporate some of those metals in their life processes to scavenge the increasing oxygen availability. For instance, in catalase, cytochrome oxidase, or dismutase, metals such as copper, iron, or zinc are the cofactors and part of the active site of the enzyme is used in the production of oxygen or in the oxygen defense mechanisms. Among several thousand proteins expressed in a typical bacterium, about 30% are metalloproteins. The general order of prevalence of major transition metals in enzymes in *Escherichia coli* is Zn, Fe, Cu, Mo, Mn, Co, and Ni [8]. Although many of the metals thus became essential in the vital processes, the cell must face a difficult situation in uptake of the required metals only from the environment. The uptake system must be specific and regulated so that the system not only recognizes the specific metal ion but also senses any changes in the concentration of the metal in the surrounding environment. The microbial genome is the information bank and can be credited when needed in response to the signal from the environment.

11.5.1 Concepts of Heavy Metal Toxicity, Tolerance, and Resistance

Because nonspecific transporters are expressed constitutively, the cell hyperaccumulates heavy metal ions in the face of high concentration of any heavy metal in the environment. Once inside the cell, heavy metal cations like Hg^{2+}, Cd^{2+}, and Ag^+ with high atomic numbers tend to bind to –SH groups. The minimal inhibitory concentration of these metal ions is a function of the complex dissociation constants of the respective sulfides. By binding to SH groups, the metals may inhibit the activity of sensitive enzymes. Other heavy metal cations may interact with physiological ions (Cd^{2+} with Zn^{2+} or Ca^{2+}; Ni^{2+} and Co^{2+} with Fe^{2+}; Zn^{2+} with Mg^{2+}), thereby inhibiting the function of the respective physiological cation.

Heavy metal cations may bind to glutathione; the resulting bis-glutathione complexes tend to react with molecular oxygen to oxidized bis-glutathione GS-SG [17], the metal cation and H_2O_2. Because the oxidized bis-glutathione must be reduced again in an NADPH-dependent reaction and the metal cations immediately catch another two glutathione molecules, heavy metal cations cause considerable oxidative stress. Finally, heavy metal oxyanions interfere with the metabolism of the structurally related nonmetal (chromate with sulfate; arsenate with phosphate) and reduction of heavy metal oxyanion leads to the production of radicals, e.g., in the case of chromate. Therefore, the "gate" that was always "open" for Mg^{2+} or phosphate uptake turns out to be responsible for the heavy metal toxicity.

11.5.2 Emergence of Heavy Metal Tolerant Mutants: Misfit in Evolutionary Selection

Metal-tolerant mutants may arise out of mutations that affect the expression of the gene for the rapid and nonspecific transporters. In fact, *corA* and *pit* mutants with a tolerant phenotype towards cobalt and arsenate, respectively, have been isolated [18–20]. However, tolerant mutants are less fit than the wild type in the medium without toxic heavy metal ion and are thus rapidly overgrown by the revertant strain.

11.5.3 Evolution of Resistance Mechanism

The "open gate" paradox leads to a new evolutionary picture, where progress is not a result of successful accumulation of mistakes in replication of the genetic code, but is rather the outcome of designed creative processes. Progress happens when organisms are exposed to paradoxical environmental conditions — conflicting external constraints that force the organism to respond in contradicting manners [3].

In an attempt to ensure protection of sensitive cellular components, a cell may develop a metal resistance system. Several factors determine the extent of resistance in a microorganism: the type and number of mechanisms for metal uptake; the role that each metal plays in normal metabolism; and the presence of genes located on plasmids, chromosomes, or transposons that control metal resistance. Summing up all the information on metal resistance in microorganisms, the involvement of six mechanisms has been outlined [21]:

• Metal exclusion by permeability barrier
• Active transport of the metal away from the cell
• Intracellular sequestration of the metal by protein binding
• Extracellular sequestration
• Enzymatic detoxification of the metal to a less toxic form
• Reduction in metal sensitivity of cellular targets

The potential of toxicity of heavy metal has thus forced life in its early evolution to develop metal ion homoeostasis factors or metal resistance determinants [11]. Resistance determinants encode proteins with actions targeted directly against a heavy metal cation. Thus, resistance is different from tolerance: here, a protein with a main function not connected to the toxic heavy metal is changed to circumvent accumulation or action of the toxic metal.

The paradoxes are the gears of creativity, serving as the new principle on which the new paradigm is established. Heavy metal ions cannot be degraded or modified like toxic organic compounds and, paradoxically, cells obligately are required to maintain certain cytoplasmic concentrations of some trace metals in physiology; thus, microorganisms employ a number of mechanisms to establish proper equilibrium, including the uptake, chelation, and extrusion of metals [2]. In addition to the existing genetic mechanisms for heavy metal resistance affecting efflux [13,22], chelating, or conversion to a less toxic oxidation state by reduction, indications or possibilities of the involvement of diverse genes in metal (loid) homeostasis, tolerance, and resistance could be explored through data mining of genome databases [9].

11.5.4 Physicochemical Restriction in Detoxification Process vis-à-vis Choice for Getting Rid of Excess Heavy Metal Ions

The redox potential is an important criterion for detoxification by reduction. The redox potential of a given heavy metal should be between the hydrogen–proton couple (0 V) and the oxygen–hydrogen couple (1.229 V) [10], which is the physiological redox range for most aerobic cells. Thus, Hg^{2+} (0.851 V); chromate (1.350 V); arsenate (0.560 V); and Cu^{2+} (0.153 V) may be reduced by the cell, but Zn^{2+} (–0.762 V); Cd^{2+} (–0.4030 V); Co^{2+} (–0.28 V); and Ni^{2+} (–0.257 V) may not [10].

The second constraint arises in the context of scavenging the reduced product. The reduced product should be able to diffuse out of the cell or it might reoxidize itself. In fact, most reduction products are quite insoluble (Cr^{3+}) or even more toxic (AsO_2) than the educts. Logically, even if the cell decides to detoxify a compound like that by reduction, an efflux system should be present to push off the reduced products to the environment. Only in case of mercury do reducibility and a low vapor pressure of the metallic reduction product fit together; Hg is thus detoxified by reduction of Hg^{2+} to Hg^0 plus diffusional loss of the Hg^0. Thus, if the desirability of the reduction fails, the only alternatives left are complexation and efflux.

However, if a fast growing cell adopts complexation, the cost–benefit ratio rules out this strategy. This can be exemplified by assuming an aerobic cell that detoxifies Cd^{2+} by forming CdS. For this, sulfate must be taken up, which costs one ATP; PAPS must be formed (three ATPs) and reduced to sulfite (2 e^- lost, which may yield three ATPs during respiration); and, finally, sulfide (6 e^- = nine ATPs). This amounts up to 16 ATPs for the formation of one sulfide, which complexes one Cd^{2+}. If it is not only sulfide but also a thiole-group attached to cysteine or glutathione or derivatives thereof, or even a ribosomally synthesized protein like metallothionin, these costs are immense.

Because reduction is not possible or may not be sensible as a sole mechanism of detoxification, heavy metal ions must be detoxified by efflux, alone or in combination, in any organism growing rapidly in an environment contaminated with high concentrations of heavy metals. Heavy metal metabolism is therefore transport metabolism [11] and the protein families involved in heavy metal transport will be examined.

11.6 TOXIC METAL IONS AND MECHANISMS OF RESISTANCE

11.6.1 MERCURY

With the strongest toxicity, mercury does not have any beneficial function. The major form of mercury in the atmosphere is elemental mercury (Hg^0), which is volatile and is oxidized to mercuric ion (Hg^{2+}) photochemically. Hg^{2+}, the predominant form entering aquatic environments, readily adsorbs to particulate matter. The major microbial reaction observed is the methylation of mercury, yielding methylmercury, CH_3Hg^+. Because bacteria confront toxic Hg^{2+} and CH_3Hg^+, several methods of detoxifying mercury species exist. An NADPH-linked enzyme called mercuric reductase, which is related to glutathione reductase and other proteins [23], transfers two electrons to Hg^{2+}, reducing it to Hg^0, which is essentially nontoxic and leaves the cell through passive diffusion [2,24,25].

In Gram-negative bacteria, mercury resistance genes, called *mer* genes, are arranged in an operon and are under control of the regulatory protein MerR, which can function as a repressor and an activator [26–41]. In the absence of Hg^{2+}, MerR binds to the operator region of the *mer* operon and prevents transcription of *mer* genes. However, when Hg^{2+} is present, it forms a complex with MerR, which then functions as an activator of transcription of the *mer* operon [42,43]. MerD, the product of *merD*, also plays a regulatory role by preventing an overshot during induction [44].

The mercuric reductase is the product of the *merA* gene, whereas *merP* encodes a periplasmic Hg^{2+}-binding protein. This MerP binds Hg^{2+} and transfers it to a membrane protein, MerT, the product of merT, which transports Hg^{2+} into the cell for reduction by mercuric reductase [45,46]. Alternatively or in addition to MerTP, an alternative uptake route exists that involves the MerC protein [47,48]. The product of the reduction, Hg^0, which is volatile, escapes from the cell to the environment.

Methylmercury, as stated earlier, is soluble and can be concentrated in the aquatic food chain, primarily in fish, or further methylated by microorganisms to yield the volatile compound called dimethylmercury. Metabolically, methylation of mercury occurs by donation of methyl groups from $CH_3- B_{12}$. Methylation has been observed for arsenic, mercury, tin, lead, selenium, and tellurium [49]. Methylmercury and dimethylmercury bond to proteins and tend to accumulate in animal tissues, especially muscle. Organomercurials, which are always much more toxic than the Hg^{2+} is (e.g., methylmercury is about 100 times more toxic than Hg^0 or Hg^{2+}), may also be detoxified if the *mer* resistance determinant encodes a MerB organomercurial lyase in addition to the other proteins [35,38–40,50–54]. After cleavage by MerB, MerA reduces the resulting Hg^{2+}. The high toxicity of organomercurials and other methylated and alkylated heavy metal compounds makes it very unlikely that these kinds of chemical modifications of heavy metals are metal resistance mechanisms [11].

11.6.2 LEAD

Use of *lead* and its toxicity has been well known for a long time [55,56]. It is no transition element, but belongs to the element group IVa, C–Si–Ge–Sn–Pb. In seawater, it is even more rare than mercury [10]. Due to its low solubility, especially, lead phosphate is insoluble with a solubility product of 10^{-54}; its biologically available concentration is low. Molecular information on lead uptake is not available but Pb-tolerant bacteria have been isolated [57], and a process involving precipitation in intracellular lead phosphate granules in *Staphylococcus* has been reported [58,59].

11.6.3 ANTIMONITE

Antimonite enters the *E. coli* cell by the glycerol facilitator, GlpF [60]. It is detoxified by all systems giving resistance to arsenite by efflux [60,61].

11.6.4 CADMIUM

Uptake of cadmium is understood very little at the molecular level. In *Ralstonia* sp. CH34, cadmium is accumulated by the magnesium system(s) [22]. In contrast to cadmium transport by Mg^{2+} uptake systems in Gram-negative bacteria, the Mn^{2+} uptake system is primarily responsible for cadmium entry into the Gram-positive cells [62–64] and maybe also in *S. cerevisiae* [65].

Resistance to cadmium in bacteria is based on chemiosmotic antiporter efflux systems. These antiporter devices are three-component systems involved in drug or metal efflux and include proteins belonging to the RND (resistance, nodulation, cell division) family of integral membrane proteins. The best known archetype is the *czcCBA* system of *Ralstonia metallidurans*, which confers resistance to Cd, Zn, and Co. The cyanobacterium *Synechococcus* sp. contains metallothionein SmtA [66]. Amplification of the smt metallothionein determinant increases cadmium resistance [67] and deletion of it decreases resistance [68–70]. The SmtB regulator controls the actual metallothionein gene, smtA, which is also a metal-fist repressor [69,71,72]. This repressor also regulates a zinc-transporting P-type ATPase [73]; thus, in cyanobacteria, which are jammed with RND- and P-type transport systems, metal transport may play a more important role in cadmium resistance than the results concerning the metallothionein suggest.

In Gram-negative bacteria, cadmium seems to be detoxified by RND-driven system like Czc, which is mainly a zinc exporter [74,75], and Ncc, which is mainly a nickel exporter [76]. The CzcA protein of *R. metallidurans* is associated with the inner membrane and appears to have 12 transmembrane domains [77]. CzcC belongs to the outer membrane efflux protein (OEP) family that function as an auxiliary element for the export of metals. CzcB has the HylD (RND-associated membrane fusion proteins) domain in which the N-terminus is anchored to the inner membrane. Neither CzcB nor CzcC appears to be necessary for resistance to metals, but they may increase the efficiency and specificity of the main component, CzcA, of the system [77].

In Gram-positive bacteria, the first example for a cadmium-exporting P-type ATPase was the CadA pump from *Staphylococus aureus* [78,79]. This protein was the first member of a subfamily of heavy metal P-type ATPases, and all the copper (including Menke's and Wilson's), lead, and zinc transporters found later are related to this protein. Other cadmium resistances in Gram-positive bacteria were found to be mediated by CadA [65]. Regulator of cadA in most cases is the CadC metal-fist repressor [80–82]. Genomics of *P. putida* KT2440 revealed the presence of two CadA homologues, where putative *cadA2* gene is transcribed divergently under the control of the putative regulator CadR and putative cadA1 is located in the middle of the *czc* region [9]. The translated sequence of cadA2 gene (the putative N-terminal) bears resemblance to the N-terminal fragment of the P-type ATPase responsible for resistance to Cd and partial resistance to Zn (reported earlier in *P. putida* 06909 [83]) [9].

The mechanism by which an ABC transporter, YCF1p, transports the glutathione-bound cadmium (cadmium–bis-glutathionate complex) to the vacuole in yeast [84,85] may have a commonality in all eukaryotes because the multidrug resistance-associated protein from man can complement a YCF1 mutation with respect to cadmium resistance [86]. Another ABC transporter, HMT1p, is reported to transport cadmium–phytochelatin complexes and also bears resemblance to a transporter detected in *A. thaliana* to the vacuoles [86–90]. Despite reports on involvement of a cation diffusion facilitator, like ZRC1 transporter in cadmium extrusion, and binding of cadmium by metallothioneins, the most predominant cadmium-detoxification method in eukaryotes appears to be the vacuolar transport of glutathione/phytochelatin by ABC transporters [11].

11.6.5 SILVER

Since time immemorial, silver has been used in medicine as an antimicrobial agent [91]. Even today, silver is used in hospitals as an antiseptic on burned skin and in implanted catheters. It is not considered an essential trace metal because of its toxicity. In nature, silver-resistant bacteria have been evolved [92–101]. Although reduced accumulation of silver by silver-resistant bacteria was observed, binding of silver to other intracellular compounds like H_2S and phosphate also always occurred [11].

The copper-effluxing ATPase CopB from *E. hirae* was found to transport Ag as well as Cu [102]; the K_m for both substrates was identical. Genome analysis of *Pseudomonas putida* revealed that SilP and PacS are the prospective proteins involved in the transport of monovalent cations (Cu and/or Ag). The corresponding genomic segment is located in a large gene cluster probably involved in Ag resistance and/or Cu homoeostasis [9]. The Ag/Cu-related genomic segment could encode a P-type ATPase (putative SilP), a three polypeptide cation/proton antiporter (putative CusCBA), as well as metal-binding proteins (CopAB1); this could all be regulated by a two-component regulatory system (putative CopSR1) [9] resembling the *sil* gene of *Salmonella* plasmid pMG101 isolated from a hospital burn ward [103]. Summarily, silver resistance may be based on RND-driven transenvelope efflux in Gram-negative bacteria; efflux by P-type ATPases in Gram-positive organisms; and additional complexation by intracellular compounds.

11.6.6 MOLYBDENUM

In the form of molybdate capable of performing oxyanion catalysis without being as toxic as chromate, molybdenum is an important trace metal. The principal mechanism of molybdate uptake is mediated by an inducible ABC transporter [104–111]. Molybdate in bound to a specific molybdate cofactor [112,113], a pterin-mono- or dinucleotide in most enzymes; the exception is the nitrogenase enzyme where the specific iron-molybdenum cofactor does not involve a pterin, and Mo is bound to homocitrate, sulfur, and a histidine residue [114,115]. Genome analysis of *P. putida* KT2440 revealed another ABC metal transporter system for Mo uptake, modABC, where ModA, ModB, and ModC are putative proteins for periplasmic-binding protein, inner membrane permease, and the ATP-binding protein, respectively [9].

11.6.7 ARSENIC

Like silver, arsenic is predominantly toxic and devoid of any function as a trace element. Different forms of arsenic appear in any environmental sample, of which a fraction may be available to the biological system and are likely to be variable depending on the physicochemical status of that environmental niche. However, toxicity, solubility, and mobility can vary depending upon which species of arsenic is present, thus affecting the bioavailability of the arsenic contamination. Arsenate enters the bacterial cell via the rapid, nonspecific, and constitutive uptake systems for phosphate.

Mutation of this system leads to tolerance to arsenate. Both anions, phosphate and arsenate, may still be accumulated by the inducible, specific transport system Pst, but this system discriminates 100-fold better between both anions than Pit does [13,20,116,117]. However, tolerant mutant strains are always impaired in their phosphate metabolism and tend to revert readily.

In Gram-negative bacteria, the arsenic-resistance gene remains inactive with the absence of As (III) in the cell due to the binding of the ars operon repressor protein to the promoter region of the gene. The system is activated by As (III) binding to the repressor protein and freeing the promoter region for transcription. The freed promoter region is transcribed to produce various components of the mechanism such as ArsB, an arsenite-translocating protein that serves as a transmembrane efflux channel. This protein functions chemiosomotically, without any energy source, or by ATP hydrolysis when coupled with ArsA, an arsenite-specific ATPase [61,118–121].

The best studied example is the plasmid-encoded arsenic resistance from *E. coli* [122]. The ArsB protein in these systems is able to function alone [123]; therefore, arsenite efflux by the ArsAB complex is energized chemiosomotically and by ATP [124]. ArsA acts as a dimer with four ATP-binding sites, and homologs to this protein have been found in eubacteria, archebacteria, fungi, plants, and animals [121,125,126]. ArsC, the enzyme arsenate reductase, is also transcribed to reduce As(V) to As(III) because As(V) cannot pass through the ArsB/ArsA pump [127,128]. For the resistance determinant in *E. coli*, arsenate reduction by the ArsC protein is coupled to glutathione [129] via glutaredoxin [130,131]. For ArsC from *Staphylococus aureus*, the electron donor is thioredoxin [127].

ArsD is a regulatory protein for additional control over the expression of the system and ArsR is a transcriptional repressor. The mechanism varies slightly in Gram-positive bacteria, which lack ArsA and ArsD. Arsenical resistance is regulated mainly by the ArsR repressor, the first example of a "metal-fist" repressor in which an inducing metal — in this case As (III) — binds to the dimeric protein, thereby preventing binding to the operator region [117,132–137]. Related regulatory proteins regulate a Cd-P-type ATPase in *Staphyloccoccus aureus* and metallothionein synthesis in cyanobacteria.

In other arsenic resistance determinants, which are more expensive to run, an additional regulator exists, ArsD [138]. Although ArsR controls the basal level of expression of the ars operon, ArsD controls maximal expression, and both regulators compete for the same DNA-binding site [139]. Two copies of an *arsRBCH* operon are found in the chromosome of *P. putida* [9]. The *arsH*, originally identified in the *ars* cluster of a Tn*2502* transposon, appears to be necessary for arsenic resistance in *Yersinia enterocolitica* [140]. It has been suggested that ArsH might be a transcriptional activator because a plasmid containing the genes arsRBC without arsH in *Y. enterocolitica* did not cause an increase in arsenic resistance [140].

The protozoon *Leishmania*, in addition to having a P-glycoprotein-related transporter, an ABC-transporter responsible for arsenite resistance [141], also gets rid of the toxicity by conjugating As (III) with glutathione or trypathione let out by the glutathione conjugate transporter [142–146]. Homologous arsenite transporters related to ArsB were found in *S. cerevisiae*, the ARS3p protein [61,147], and also in man [148,149]. Thus, metabolism of arsenic resistance seems to follow the same pattern in all organisms again: uptake by the phosphate system; reduction; efflux by ArsB-related proteins or ABC-transporters; and maybe even additional energizing by ArsA-related ATPases.

11.6.8 ZINC

Zinc is an important trace element that is toxic at elevated concentrations. The other reality is that it must be transported into cells against concentration gradients. In the usual housekeeping cellular environment, some constitutively expressed broad ion range uptake systems are capable of satisfying the demand of macronutrients and trace elements of the bacterial cells [150]. Zinc is, however, important in forming complexes (such as Zn-fingers in DNA) and as a component in cellular enzymes [151–154]. Zinc is rapidly uptaken by Mg^{2+} transport systems, as shown in *Ralstonia* sp. CH34 [22]. Zn-uptake systems/transporters can be categorized into three groups.

The first group comprises CorA transporters found to be present in many bacteria, archaea, and yeast [12,14,15]. The second, narrowly distributed group comprising potential chemiosomotically driven transporters representing MgtE family is present in *Providencia stuartii* and few other Gram-negative and Gram-positive bacteria [155,156]. The third magnesium/zinc transporter is MgtA from *S. typhimurium*, a P-type ATPase that may transport zinc more efficiently than magnesium [156–159]. MgtA is regulated by magnesium starvation [16,160], and zinc may interfere with this process, which is at least partially dependent on the PhoPQ two-component regulatory system.

However, the MgtA P-type ATPase is not the inducible high-specificity uptake system for zinc. A periplasmic zinc-binding protein was found in *Haemophilus influenzae* to be important for zinc uptake [161], and ABC-transporters or the evidence for ABC-transporter involved in zinc uptake were found in *Streptococcus pneumoniae*, *Streptococcus gordonii,* and *E. coli* [162–164]. Proteins belonging to the Fur family were found to regulate iron/zinc uptake [164]. The first protein, Zur, is probably involved in the regulation of Zn uptake. Zur protein of *E. coli* binds two atoms of Zn per monomer of polypeptide. In *E. coli*, *zur* is located away from the *znuABC* genes, and acts as a repressor when it binds Zn [164]; in *P. putida*, zur is located within the znuACB1 genes [9]. The issue of iron regulons in pseudomonads involving not only the Fur proteins but also dedicated sigma factors, extra activators, and perhaps some small regulatory RNAs, is also related to metal homoeostasis [9].

Uptake of zinc in *S. cerevisiae* is also mediated by ZRT1p high-affinity and ZRT2p low-affinity transporters of the ZIP family having homologous counterpart in *Arabidopsis thaliana* (plants), protozoa, fungi, invertebrates, and vertebrates [165–169]. Two general efflux mechanisms are responsible for bacterial resistance to zinc. The first one is a P-type ATPase efflux system that transports zinc ions across the cytoplasmic membrane by energy from ATP hydrolysis. A chromosomal gene, *zntA*, was isolated from *E. coli* K12 and was found to be responsible for the ATPase that transports zinc and other cations across cell membranes' systems [170,171]. The related zinc transporter in the cyanobacterium *Synechocystis* is ZiaA [73]. In most cases, P-type ATPases mediating cadmium resistance can also efflux zinc.

The other mechanism involved in zinc efflux is an RND-driven transporter system that transports zinc across the cell wall of Gram-negative bacteria and is powered by a proton gradient and not ATP [22,74]. Czc from *Ralstonia* sp. is the first of these systems that has been cloned and sequenced [172,173]. The Czc determinant contains three structural genes coding for the three subunits of the membrane-bound efflux complex CzcCB2A [174,175]. The concept of "transenvelope transport" [176–180] has been recently developed as a model to describe common themes in certain efflux systems. The organization of the *czc*-like systems is generally conserved in all organisms studied, although regulatory proteins may be located in a variety of orientations with respect to the core *czcCBA* genes. CzcA is responsible for the translocation of metals across the plasma membrane; CzcCB would avoid their release into the periplasmic space and export them out of the cell.

Regulation of Czc is highly complicated and requires about six additional proteins [74,150,181]. Two or three genes of regulators, *czcN,* and Orf69 and *czcI*, are located upstream of the *czcCBA* structural genes. Transcription is initiated from three different promoters: *czcNp, czcIp,* and *czcCp*; possible RNA processing together with transcription initiation delivers an unstable *czcN-*, a *czcI-*, and a large *czcCBA*-message with a variety of rare large mRNAs such as *czcICBA* and *czcNICBA*. Downstream of *czcCBA*, three more regulator genes exist and are transcribed together as a *czcDRS* message. CzcRS forms a two-component regulatory system, which is not essential for Czc regulation. This system acts specifically on *czcNp* and seems to increase the sensitivity of the main Czc-regulator, probably an unknown sigma factor of the ECF (extracellular functions) sigma-70 factor family. The complexity of Czc regulation indicates the importance of this efflux system for metal homosostasis in *A. eutrophus*.

CzcD, a model cation diffusion facilitator (CDF) zinc transporter protein [13,182], finds its relative in *S. cerevisiae*, the CDF transporter, ZRC1 and COT1p, mediating resistance to zinc and cobalt [183,184]. ZRC1p is also involved in regulation of the glutathione level [185] and COT1p, which is basically a cobalt transporter, may substitute from ZRC1p [186,187]. Because COT1p transports its substrate across a mitochondrial membrane, both proteins could be involved in heavy metal metabolism of the yeast mitochondria. Thus, ZRC1p and COT1p might function in the efflux of surplus cations from the mitochondria [11]. Similar enough is the existence of four closely related mammalian CDF proteins: ZnT1, ZnT2, ZnT3, and ZnT4; these are responsible for Zn-efflux across the cytoplasmic membrane; Zn-transport into lysosomes; Zn-transport into synaptic vesicles; and Zn-secretion into milk, respectively [188–191].

11.6.9 Copper

Because copper is used by all living entities in cellular enzymes like cytochrome c oxidase, it is an important trace element. However, due to its radical character, copper is also very toxic and, because it is so widely used in mining, industry, and agriculture, high levels of copper may exist in some environments. Copper toxicity is based on the production of hydroperoxide radicals [192] and on interaction with the cell membrane [193]. As such, bacteria have evolved several types of mechanisms to resist toxicity due to high copper concentrations. Copper metabolism has been studied in *E. coli*; some species related to *Pseudomonas*; the Gram-positive bacterium *Enterococus hirae*; and in *S. cerevisiae*, which sheds some light on the copper metabolism of higher organisms [194–199].

The mechanism of a plasmid-encoded copper resistance [200] in *E. coli* is based on an efflux mechanism. The efflux proteins are expressed by plasmid-bound *pco* genes (structural genes, *pcoABCDE*; and the regulatory genes *pcoR* and *pcoS*), which in turn depend on the expression of chromosomal *cut* genes. Two cut genes, *cutC* and *cutF*, were shown to encode a copper-binding protein and an outer membrane lipoprotein [201–206]. In *Pseudomonas syringae*, resistance to copper via accumulation and compartmentalization in the periplasm and outer membrane is due to four proteins encoded on the plasmid-borne *cop* operon.

The two periplasmic blue copper proteins are CopA and CopC, and the inner and the outer membrane proteins are CopD and CopB, respectively. However, a mutant *cop* operon containing *copD* but lacking one or more of the other genes conferred hypersensitivity and hyperaccumulation of cellular copper, indicating a role for CopD in copper uptake by the cell [196,197,207–209]. As in *E. coli*, copper resistance in *Pseudomonas* is regulated by a two-component regulatory system composed of a membrane-bound histidine kinase and a soluble response regulator, which is phosphorylated by the kinase and switches on transcription of the *cop* genes [210,211].

The copper transport and resistance system in the Gram-positive bacterium *Enterococcus hirae* is the best understood system. The two genes, copA and copB, determining uptake and efflux P-type ATPases, respectively, are found in a single operon. Although CopA is probably responsible for copper uptake and copper nutrition, the 35% identical CopB is responsible for copper efflux and detoxification [167,212]. Copper and silver induce the system and they seem to transport silver besides copper [213]; obviously, the monovalent cations are being transported [102]. The first two genes in the cop operon, *copY* and *copZ*, determining a repressor and activator, respectively, constitute the regulatory protein pair [214,215].

P-type ATPases also seem to control copper flow in two pathogens: *Helicobacter pylori* [216] and *Listeria monocytogenes* [217]. Copper-transporting P-type ATPases have been found in cyanobacteria [218,219] and in yeast. The copper P-type ATPase does not transport copper across the cytoplasmic membrane. In yeast cells, the iron/copper-specific reductases FRE1p and FRE2p reduce Cu (II) to Cu (I) [220,221], which is transported into the cell by the CTR1p transporter [221–223]. A functional homologue of a novel protein in yeast with two related possible copper transporters (CTR2p, CTR3p) [167] is found in man [224]. Additionally, Cu (II) is accumulated by the CorA-related transporters ALR1p and ALR2p [221,223].

The metallothioneins of yeast, CUP1p and CRS5p, are probably copper-storage devices in yeast [225]. COX1p delivers copper into the mitochondria for the synthesis of cytochrome c oxidase [226–228]. A copper P-type ATPase, CCC2p, and a "copper chaperone," ATX1p, are also capable of transporting copper [229,230]. The protein factors, ACE1p, ACE2p, and MAC1p factors regulate the copper homoeostasis [220,231–234]. The progress in understanding of the copper homoeostasis in yeast has contributed significantly to copper homoeostasis in general [235–239]. In man, defects in function or expression of copper transporting P-type ATPases are responsible for two heredity diseases: Menke's and Wilson's diseases [240]; their functional homologues have been identified in mouse, rat, and *Caenorhabditis elegans* [241–243]. Thus, the copper-dependent cycling of the transporter may be true for all animals or maybe even for all eukaryotes.

11.6.10 NICKEL

Although nickel is a required nutrient, at high levels it can be toxic to microorganisms. It enters the bacterial and yeast cell mainly by the CorA system [12,158,244]. An additional nickel transporter identified in *Alcaligenes eutrophus* [245] as part of the hydrogenase gene cluster [246] was cloned [247] and expressed. The purified product HoxN [168] was found to be an archetype of a new class of transport proteins. Nickel is probably bound to hisitidine residues in the *Ralstonia eutropha* HoxN [248] and NixA from *Helicobacter pylori* [249]. Uptake of nickel (and cobalt in a related protein) [250,251] is probably driven by the chemiosmotic gradient. Surprisingly, nickel for hydro-genase formation in *E. coli* is supplied by an ABC-transporter [252–254], which involves a periplasmic nickel-binding protein. In Gram-negative bacteria, the *nikRABCDE* system is responsible for the uptake of nickel. In *E. coli*, NikA is the periplasmic-binding protein; NikB and NikC form a heterodimer inner membrane pore for the translocation of the metal; and NikD and NikE are the heterodimeric ATP-binding proteins.

The organization of these genes in *Pseudomonas putida* is different from that in *E. coli* but the same as in *Brucella suis*, where *nikR* is upstream of and transcribed divergently from *nikABCDE* [9]. Nickel detoxification may be done by sequestration and/or transport, as with most heavy metal cations. Metal resistance based on extracellular sequestration has been hypothesized only in bacteria, but has been found in several species of yeast and fungi [255–257]. Nickel uptake in *Saccharomyces cerevisiae* may be reduced by excreting large amounts of glutathione, which binds with great affinity to heavy metals [258]. Nickel is bound to polyphosphate in *Staphyloccus aureus* [259] and to free histidine in nickel hyperaccumulating plants [260]. In *S. cerevisiae*, nickel is disposed off into the vacuole and probably bound to histidine in there [256]. The transport into the vacuoles requires a proton-pumping ATPase [261]; thus, this kind of nickel transport may be also driven by a chemiosomotic gradient. Other yeasts and fungi probably detoxify nickel using similar mechanisms and by a mutation of the CorA uptake system [257,262].

The best known nickel resistance in bacteria is the nickel efflux driven by an RND transporter in *Ralstonia* sp. CH34 and related bacteria. Two systems, a nickel–cobalt resistance [*cnr*] [263] and a nickel–cobalt–cadmium resistance [*ncc*] [76] have been described that are closely related to the cobalt–zinc–cadmium resistance system *czc* from strain CH34. Regulation of *cnr/ncc* on one hand and *czc* on the other, however, is completely different. The nickel resistance systems are regulated by a sigma factor, a membrane-bound antisigma factor, and a periplasmic anti-antisigma factor, which probably senses the nickel. A P-type ATPase described to bind Ni^{2+}, Cu^{2+}, and Co^{2+}, ATPase 439, may be regarded as the first example of a nickel P-type ATPase in bacteria [11].

11.6.11 COBALT

Cobalt is required as a trace element in nitrile hydratases and cofactor B_{12}; however, in the ionized form, at higher concentration, it can directly inhibit microorganisms. It is accumulated in most bacteria by the CorA system [14,15,157,258]. Members of the CDF protein family have also been found to transport cobalt, like a protein in the Gram-positive bacterium *Staphylococcus aureus* [184] and COT1p protein from *S. cerevisiae* that transports cobalt across a mitochondrial membrane [186,187]. Excess Co^{2+} is generally pumped from Gram-negative bacteria in conjunction with nickel or zinc by a three-polypeptide membrane complex that is not an ATPase but functions as a divalent cation/2 H^+ antiporter [76,173,263]. A bacterium containing a nitrile hydratase, *Rhodococcus rhodochrous*, was reported to possess a system related to the nickel transporter HoxN from *R. eutropha* [250,251], thus implying that HoxN homolog might play a role to supply Co^{2+} for the production of this non-B_{12}-cobalt protein [11]. Thus, Co^{2+} is detoxified by RND-driven systems in Gram-negative bacteria and by CDF transporters in eukaryotes and Gram-positive bacteria.

11.6.12 CHROMIUM

Chromium does not have any essential function in microorganisms. Chromate enters the cell of *Alcaligenes eutrophus* and many other bacteria by the sulfate uptake system [13,22]. Changing the valence state by reducing Cr (VI) under aerobic and anaerobic conditions through electron-transport systems containing cytochromes may (though this is inconclusive) confer resistance to chromate. Chromate-resistant *Pseudomonas fluorescens* strain LB300 and other bacteria have been reported to reduce chromate [264–278]. The DNA sequences determining chromate resistance of the *Pseudomonas aeruginosa* and *Alcaligenes eutrophus* share homologous *chrA* genes encoding membrane. An additional upstream gene, *chrB*, that has been postulated to be responsible for the inducibility of the resistance in *A. eutrophus* is absent in *P. aeruginosa* [22,174,266,267].

11.6.13 VANADIUM

The heavy metal with the lowest atomic number is *Vanadium*. Vanadate, bearing structural similarity to phosphate, may be taken up by phosphate uptake systems and is known as an inhibitor of ATPases [279,280]. In addition to toxicity, there are few examples of the physiological role of vandate like vanadate-dependent nitrogenase in *Azotobacter chroococcum* for nitrogen fixation in absence of molybdate in the environment [281–293]. More functions as a trace element are obscure [294]; it also can be used as an electron acceptor for anaerobic respiration [295,296]. Vanadate resistance has been studied in an extremophile *Sulfolobus* and a eukaryote *S. cerevisiae* [297,298]; however, the detailed mechanism of vanadate resistance remains elusive.

11.7 EVALUATION OF THE UPTAKE AND EFFLUX CAPABILITIES OF ORGANISMS BASED ON WHOLE GENOME SEQUENCE ANALYSIS

By taking advantage of the availability of multiple complete genomes, some approaches offer new opportunities for predicting gene functions in each of these genomes. All these approaches rely on the same basic premise: the organization of the genetic information in each particular genome reflects a long history of mutations; gene duplications; gene rearrangements; gene function divergence; and gene acquisition and loss that has produced organisms uniquely adapted to their environment and capable of regulating their metabolism in accordance with the environmental conditions. This means that cross-genome similarities can be viewed as meaningful in the "evolutionary" sense and thus are potentially useful for functional analysis.

The most promising comparative methods specifically employ information derived from multiple genomes to achieve robustness and sensitivity that is not easily attainable with standard tools. It seems that they are indeed the tools for the "new genomics," whose impact will grow with the increase in the amount and diversity of genome information available. A disproportionate number of these examples are from the COG (clusters of orthologous groups) system.

- *Phylogenetic patterns (profiles).* The COG-type analysis applied to multiple genomes provides for the derivation of phylogenetic patterns, which are potentially useful in many aspects of genome analysis and annotation [299]. Similar concepts have been introduced by others in the form of phylogenetic profiles [300,301]. The phylogenetic pattern for each protein family (COG) is defined as the set of genomes in which the family is represented. A pattern search tool that allows the user to select COGs with a particular pattern accompanies the COG database. Predictably, genes that are functionally related tend to have the same phylogenetic pattern.
- *Profile scan.* Based on the classic Gribskov method of profile analysis [302,303], ProfileScan uses a method called pfscan to find similarities between a protein or nucleic acid query sequence and a profile library [304]. In this case, three profile libraries are available for searching. First is PROSITE, an ExPASy database that catalogs biologically significant sites through the use of motif and sequence profiles and patterns [305]. Second is pfam, which is a collection of protein domain families that differ from most such collections in one important aspect: the initial alignment of the protein domains is done by hand, rather than by depending on automated methods. As such, pfam contains slightly over 500 entries, but the entries are potentially of higher quality. The third profile set is referred to as the Gribskov collection.

All organisms for which complete genome sequences are available were analyzed for their content of cytoplasmic membrane transport proteins (work done in the Institute of Genomic Research). The transport systems present in each organism were classified according to (1) putative membrane topology; (2) protein family; (3) bioenergetics; and (4) substrate specificities. The overall transport capabilities of each organism were estimated and differences in their reliance on primary vs. secondary transport and the range of transporter substrate specificities in each organism were found to correlate generally with the respective ecological niches and metabolic capabilities of each organism. More than 80 distinct families of transport proteins were identified. The efficacy of a phylogenetic approach for predicting function was demonstrated.

Five major families of multidrug efflux transporters have been analyzed: the ATP-dependent ABC superfamily and four families of secondary transporters: MFS, RND, SMR, and MATE families. The ABC and MFS are large superfamilies including uptake and efflux proteins. *These two superfamilies are the two largest families of transporters known; they are extremely diverse and appear to be (almost) ubiquitous, suggesting a very ancient origin* [306]. The RND, SMR, and MATE families are smaller families, which to this date only include efflux systems; they are found in prokaryotes as well as eukaryotes. The major findings of these analyses and the complete complement of transports in each completely sequenced organism are available at http://www-biology.ucsd.edu/~ipaulsen/transport/.

11.8 TRANSPORTERS INVOLVED IN HEAVY METAL UPTAKE AND EFFLUX

Transporters of trace elements can be classified into four distinct groups [2,13,167,180,182,307–323]:

- *Electrochemical potential-driven transporters* utilize a carrier-mediated process to catalyze: transport of a single species by mediated diffusion or in a membrane potential-dependent manner (uniport); two or more species in opposite directions in a tightly coupled process without utilizing chemical free energy (antiport); or two or more species in the same direction in a coupled process (symport), again without using any form of energy other than the electrochemical potential gradient.

- *Primary active (P–P bond hydrolysis driven) transporters* utilize the free energy of the P–P bond hydrolysis to drive the movement of ions against their chemical or electro-chemical potential gradient. The transport protein may be transiently phosphorylated during the transport cycle but the substrate is not phosphorylated. These transporters occur universally in all domains of life.
- *Transporters whose mode of transport or energy coupling is unknown* are awaiting their final placement after their transport mode and energy coupling have been resolved. These families include at least one member for whom a transport function has been described, but the mode of transport or the energy coupling is not known.
- *Auxiliary transport proteins* in some way facilitate transport across one or more biological membranes, but they do not participate directly in the transmembrane translocation of a substrate. They may provide a function connected with energy coupling to transport; play a structural role in complex formation; serve a biogenic or stability function; or function in regulation.

Transporter families of the four groups involved in trace metal metabolism are summarized in Table 11.2.

11.9 INFLUX AND EFFLUX ARE COORDINATELY REGULATED

In the process of evolution of the ion-transport system, it was necessary to evolve specificity of the transporters. The occurrence of influx and efflux transporters for the same metal ion necessitated evolution of not only a specialized but also a highly regulated system. For a cell exposed to toxic metal ions, like Hg^{2+}, Cu^{2+}, Cd^{2+}, or oxyanions of arsenic and antimony, it is obligatory to maintain the cytosolic ion concentration below the toxic level. In general, the cells survive the toxic ion effect with the aid of a specific ion-efflux system. Thus, for a nonessential but toxic ion, the necessity and advantage of simultaneously operating an uptake- and an efflux-transport system associated in several metal ion resistance operons remains speculative. In many instances, however, cells had evolved a transport system for a particular essential ion that can be used by a structurally similar toxic ion to enter the bacterial cell.

One can also advocate that metabolic penalty for having uptake pumps more specific is greater than the genetic cost of having plasmid genes in the population that can spread when needed [2]. Thus, most of the toxic metal (or metalloid) resistance is encoded by mobilizable extrachromosomal replicons or transposons. Among the known resistance systems, mercury, cadmium (along with zinc, cobalt, and nickel), and arsenic (along with antimony) resistance genetic determinants have been well characterized. All these toxic metal ion resistance operon systems, except the ars operon, are common in having coordinated regulation with the specific gene(s) for the ion-uptake system associated with the plasmid encoded resistance operon or encoded by the genome [324].

Mutation in the resistance gene of the operon causes hypersensitivity if the associated uptake gene(s) remains unaltered. In the mercury resistance (*mer*) operon, at least three genes *merTPC* are cotranscribed with *merA* and specify Hg^{2+} uptake system. Thus, the mer operon, lacking merA gene, which encodes the mercuric reductase to convert mercuric ion to volatile mercury, resulted in more hypersensitivity to Hg^{2+} than the isogenic strains lacking mer operon [325]. This observa-tion, in fact, had helped to identify the mercuric ion uptake genes merTPC of the *mer* operon [47].

Similarly, a *copCD* clone of copper resistance *copABCD* [2] operon was phenotypically hyper-sensitive to copper [326]. Unlike mercuric ion, copper (or zinc) is an essential element but toxic at a higher level; thus, a homeostasis condition is maintained to ensure that the cells do not become copper (or zinc) depleted or copper (or zinc) surfeited [194]. Thus, complex and distinct operon systems having certain coordination between metabolism of these metals including uptake and resistance including efflux (and or binding) were evolved.

TABLE 11.2
Transporters Involved in Trace Metabolism

Family	Substrates	Size range[a]	Number of transmembrane segments[b]	Distribution	Examples
Electrochemical potential-driven transporters					
CDF	Cd^{2+}, Co^{2+}, Ni^{2+}	300–750	6	Archaea, bacteria, eucarya	Heavy metal uptake and efflux transporters of bacteria, eukaryoyic plasma membranes and mitochondria (CzcD of *Ralstonia eutropha*)
ZIP	Zn^{2+}, Fe^{2+}	376	8	Eucarya	Zn uptake transporter Zrt1 of *Saccharomyces cerevisiae*.
RND	Heavy metal ions, multiple drugs, organic solvents, etc.	800–1200	6, 12	Archaea, bacteria, eucarya	Drug efflux pump AcrA of *Escherichia coli*
TDT	Tellurite	300–350	10	Archaea, bacteria, eucarya	Arsenical resistance efflux pump of *Staphylococcus aureus*
ArsB	Arsenite, antimonite	400–900	12[c]	Archaea, bacteria, eucarya	Arsenical resistance efflux pump of *Staphylococcus aureus*
CHR	Chromate; sulfate (uptake or efflux)	c. 400	6,[c] 10	Archaea, bacteria	The chromate transporter ChrA of *Alcaligenes eutrophus*
NiCoT	Ni^{2+}, Co^{2+}, Ni^{2+}	300–400	8[c]	Bacteria	Ni^{2+} uptake permease HoxN of *Ralstonia eutropha*
Nramp	Divalent metal ions (uptake)	500–600	8–12[c]	Archaea, bacteria, eucarya	Divalent metal ion, H+ symporter Nram2 of *Homo sapiens*
ACR3	Arsenite	c.400	10	Archaea, bacteria, eucarya	Arsenical resistance-3 protein Acr3 of *S.cerevisie*
CadD	Cd, cations	150–200	5	Bacteria	Cadmium resistance protein CadD of *S. aureus*
Primary active transporters					
ABC	All kinds of inorganic and organic molecules from simple ions to macromolecules	1000–2000 (multidomain; usually mutisubunit)	(5[c]); (6[c]); 12_ variable	Archaea, bacteria, eucarya	MDR of *Homo sapiens*
P-ATPase	Cations (uptake and/or efflux)	600–1200 (sometimes mutisubunit)	(6–12); 8,[c] 10[c]	Archaea, bacteria, eucarya	KdpABC (K+ uptake) of *E. coli*
ArsAB	Arsenite, antimonite, (tellurite?)	c. 1100 (multidomain; two subunit)	12[c]	Archaea, bacteria, eucarya	Arsenite efflux pump ArsAB of *E. coli*
Transporters of unknown classification					
MerTP	Hg^{2+} (uptake)	c. 200	3[c]	Bacteria	Mercuric ion transporter MerTP encoded on the incJ plasmid pMERPH of *Shewanella putrifaciens*

TABLE 11.2
Transporters Involved in Trace Metabolism (continued)

Family	Substrates	Size range[a]	Number of transmembrane segments[b]	Distribution	Examples
MerC	Hg2+ (uptake)	c. 140	4	Bacteria	Hg uptake transporter MerC encoded on the incJ plasmid pMERPH of *Shewanella putrifaciens*
MerF	Hg2+	c. 80	2	Bacteria	MerF importer of *Pseudomonas aeruginosa* plasmid
FeT	Fe2+ (Co2+, Cd2+) (uptake)	c. 550	6	Eucarya (yeast)	Fe2+ transporter FeT4p of *S. cerevisiae*
Ctr1	Cu2+ (uptake)	150–200	3	Eucarya	Copper transporter Ctr2p of *S. cerevisiae*
PbrT	Pb2+	400–650	7	Bacteria	PbrT of *Ralstonia metallidurans*
MgtE	Mg2+, Co2+ (uptake)	300–500	4–5	Archaea, bacteria	Mg2+ transpoter MgtE of *Bacillus firmus*
Auxiliary transport proteins					
MFP	Proteins; polypeptides, signalling molecules, heavy metal ions, etc.	350–500	1	Bacteria	EmrA of *E. coli*

[a] Size range (in number of amino acid residue).
[b] Number of (putative) transmembrane alpha-helical segments.
[c] Number is established by substantial experimental data/x-ray crystallographic data.

Despite tremendous effort and progress made in the understanding of the molecular mechanism of arsenic resistance operon, including the characterization of structural genes and their products, nothing is known about the specific uptake system of this toxic metalloid [60,327,328]. In the early studies of phosphate transport system, it became apparent that the arsenate ion, being analogue of inorganic phosphate, acts as a substrate of the inorganic phosphate transport system, Pit and Pst (the low-affinity Pit system in particular) to enter the bacterial cell. As a result, mutation in phosphate transport may be associated with arsenate resistance phenotype. No other transport mechanism or arsenate-specific transporter is known in the microbes.

Arsenite or antimonite cannot use phosphate transport system; no investigation has been made to understand how these metalloids enter the cell. Moreover, because the solution chemistry of the metalloids is significantly complex, it adds further complications to investigate in the interactions of these toxic elements with microbes. However, elucidation in the molecular mechanism of arsenic or antimony transport (uptake and efflux) is essential [143]:

- Arsenite is a ubiquitous pollutant in the environment and a proven carcinogen.
- Antimony compounds are the only effective drug for treatment of leshmaniasis.
- Arsenic compounds are still used as an antitrypanosomal drug.
- Studies on the molecular biology of the arsenic–antimony resistance in microbes have proven highly beneficial in investigations of arsenic metabolism in higher organisms, including humans. Particularly, the ATP-dependent, efflux-mediated, multidrug resistance-associated protein is functionally similar to the ATP-coupled ArsAB pump.

A major difference in the *ars* and *mer* or *copABCD* (or *pcoABCD*) encoded resistance for arsenic and the mercury or copper, respectively, is the absence of any known arsenic (and antimony) uptake determinant in the *arsRDABC* or *arsRBC* operon. No coordinated regulation between arsenic uptake and efflux is known yet. Thus, the attempt to search the cellular transporters in the uptake mechanism of these metalloids in the bacterial genome should be rewarding.

Only a single antimonite-resistant mutant was identified from a pool of random transposon insertional mutants [60]. The insertion of the transposon was mapped in the glpF gene. Though the authors had selected for antimonite- and arsenite-resistant phenotype, only a single antimonite-resistant mutant (1 mM in LA) that showed wild type arsenite-sensitive phenotype was described. The results suggested that glpF gene coding for glycerol facilitator could be the major route of Sb (III) uptake.

The authors succeeded in selecting spontaneous resistant mutants of three kinds: arsenite-, arsenate-, and antimonite-resistant. The associated metal and antibiotic resistance phenotype of the two arsenite resistant mutants is similar to multidrug-resistant (MDR) bacterial and tumor cells [329]. Emergence of multidrug-resistant mutants in a bacterial population under drug selection pressure constitutes a serious concern in modern medicine; thus, a relevant field of study in adaptive mutation or evolutionary biology [7] has become pertinent. To understand the emergence of drug resistance in the protozoan-parasite *Leshmania*, arsenite-resistant *Leshmania* cells have been described since 1990 [330].

Arsenic drugs have been used since ancient times and most of those applications had been abandoned in the development of modern medicine. Nevertheless, several arsenical drugs of melarsen derivatives are still used to treat African sleeping sickness caused by a protozoan parasite *Trypanosoma brucei* [331]. The trypanolytic effect of melarsen oxide was abrogated by adenine or adenosine. This observation led to the conclusion that adenine transporters can be used by arsenic and thus parasites lacking P2 adenosine transport are resistant to melarsen due to loss of uptake. The other mechanism of arsenite-resistant development in a population of *Leshmania* exposed to drugs was due to amplification of drug resistance genes, such as *ltpgpA* encoding for P-glycoprotein [332–334].

11.10 EVOLUTION OF SPECIFIC GENETIC ELEMENTS TO SENSE AND RESPOND TO METALS IN THE ENVIRONMENT: A DOMAIN FOR DIVERSE METALS

11.10.1 Transcription Repressor Protein ArsR Binding with the Ars Operator–Promoter and ArsR Binding with Inducer Molecule

The *arsR* gene of R773 encodes the repressor protein of 117 amino acids. Transcriptional fusion with β-lactamase (*blaM*) demonstrated that the first 83 amino acid residues of ArsR protein were sufficient for repression [135]. Subsequently, experimental evidence also suggested that a core sequence of about 80 amino acids (9 to 89 residues) contained all the information for dimerization (the active form of ArsR), repression, and induced recognition [136]. ArsR binds to the operator sequence spanned from nt. −64 to −40 of the transcriptional start site identified by DNaseI protection assay [135]. More precisely, two short stretches, nt. −61 to −58 and −50 to −47, within the DNaseI protection region were suggested as the contact sites between the ArsR of R773 and the specific DNA sequence of the operon.

However, ArsR of pSX267 was shown to bind at −35 and −10 regions of the promoter [61]. This *in vitro* ArsR binding was substantiated by the constitutive expression with the clones having operator mutations or mutation in *arsR* of R773 [135]; deletion of *arsR* in pSX267 [61]; or mutation in *arsR* of pKW301. In order to perform the ArsR on the *ars* operon, it should have at least three domains specific for (1) metal binding; (2) DNA binding; and (3) dimerization, the active form of ArsR [136]. The ArsR of R773 is the first member of the ArsR metaloregulatory protein family [136,335]. The members of this family respond to a single or diverse metal-ion species, namely, arsenite, antimonite, bismuth (III), cadmium (II), and zinc (II). All these proteins possess a sequence [134,335] characteristic of DNA-binding helix–turn–helix (H–T–H) motif shown in Figure 11.1.

It was proposed that the N-terminal cysteine and at least one (generally two) histidine residues at C-terminal are the feature of this motif. These cysteines and histidines participate in metal binding, which thus disrupts DNA–ArsR binding [134]. Furthermore, ArsR of pSX267 was suggested to possess motifs of typical zinc-finger and leucine-zipper, identified in several DNA-binding proteins [336]. Arsenite, antimonite, or bismuth induces the transcription by releasing the repressor protein from the promoter, allowing the RNA polymerase to initiate the transcription. It is puzzling how these diverse oxyanions, arsenite, antimonite, and the cation bismuth (Bi^{3+}) could recognize the same metal-binding site, even though the efficiency of binding of these diverse metalloid species to the specific operator site is different [135,138].

11.11 PROBLEM OF TRANSPORTING TOXIC LEVELS OF THREE DIFFERENT DIVALENT CATIONS WITH A COMMON SYSTEM; THE NATURE OF THE MEANS REQUIRED TO COPE WITH THE DIFFICULTIES; AND THE TYPE OF GENETIC CHANGES: AN EXAMPLE OF SELF-AWARENESS — OUTCOME OF DESIGNED CREATIVE PROCESSES IN *RALSTONIA*

Ni^{2+} is accumulated by the rapid and nonspecific CorA (MIT) Mg^{2+} transport system. Highly specific nickel transporters are HoxN chemiosmotic transporters or ABC uptake transporters, which use a periplasmic nickel-binding protein, depending on the bacterial species. Characterized nickel-resistance systems are based on inducible, RND-driven transenvelope transporters. Moreover, a nickel efflux P-type ATPase may exist in *Helicobacter pylori*.

Zn^{2+} is accumulated by the rapid and nonspecific CorA (MIT) Mg^{2+} transport system in some bacterial species, and MgtE system in others. Inducible, high-affinity ABC transporters supply zinc in times of need. P-type ATPases may transport zinc in both directions: zinc uptake as a by-product

L C V C D L C M A L D Q S Q P K I S R H L A   ArsR R773 (Francestein *et al.*, 1994)

L C V C D F C T A L D E I P A Q D L P S S G   ArsR pKW301 (Suzuki *et al.*, 1998)

L C V C D L C T A L E Q S Q P K T S R H L A   ArsR R46 (Bruhn *et al.*, 1996)

 C V C D L C G A T S E S Q P K I S R H A   ArsR Tn 2502 (Neyt *et al.*, 1997)

L C V C E L M C A L A D S Q P K I S R H L A   ArsR P. aeruginosa (Cai *et al.*, 1998)

L C A C D L L E H F Q F S Q P T L S H H K   ArsR B. subtilis (Takemaru *et al.*, 1995)

L C A C D L L E H F Q F S Q P T L S H H K   ArsR pSX267 (Rosenstein *et al.*, 1992)

L C A C D L L E H F Q F S Q P T L S H H K   ArsR pI258 (Ji & Silver, 1992)

 C V C D I A N I L G V T I A N A S H H L R   CadC pI258 (SHi *et al.*, 1994)

 C V G D L A Q A I G V S E S A V S H Q L R   SmtB Synechococcus PCC7942 (Shi *et al.*, 1994)

FIGURE 11.1 Helix–turn–helix-DNA-binding motif of ArsR family of metalloregulatory proteins.

of Mg^{2+}-uptake again and efflux as detoxification. CDF transporters catalyze slow efflux, high-efficiency transenvelope efflux by inducible RND-driven transporters like Czc.

Magnesium and/or manganese uptake systems are responsible for the uptake of Cd^{2+}. Only in cyanobacteria, metallothionein-like proteins were characterized (Smt). Efflux is done in Gram-positive bacteria by P-type ATPases, in Gram-negative by RND-driven transenvelope transport, and maybe additionally to these systems by CDF transporters.

This part of the review will present the results of the conserved domain search with query sequences of resistance proteins derived from gene bank databases to support the findings of the earlier authors who have attempted to solve the evolutionary puzzles. Consider how a *Ralstonia* sp. strain CH34 (earlier known as *Alcaligenes eutrophus* CH34) solved the problem of transporting toxic levels of the three different divalent cations with a common system without causing starvation for the two, Co^{2+} and Zn^{2+}, that are essential trace nutrients.

It can well be assumed that regulation must occur not only at the level of transcription but also at the level of cation pump function. The kinetic properties of CzcABC complex assure that the toxic cation cadmium is always pumped out of the cell. Zinc, in contrast, is exported slowly at low concentration and rapidly at high concentration. Cobalt provides a problem with the *czc* system because it is a poor inducer of *czc* and is bound and transported with low affinity. One solution for problems with cobalt efflux for *Ralstonia* was to evolve related cobalt efflux systems that are not induced by zinc, such as the *cnr* resistance determinant.

The three subunits of CzcABC protein complex differ in structure and function, as first demonstrated by the analysis of *czc* deletion mutants [337]. CzcA and CzcB alone are able to catalyze a highly efficient Zn efflux. On the other hand, CzcC is needed to modify the substrate specificity of the complex to include cadmium and, for practical purposes, cobalt.

The CzcC was shown to be located in the membrane fraction, independent of the presence of CzcA and CzcB. The CzcC subunit does not contain apparent motifs for a metal-binding site; is membrane bound by itself; and does not need CzcA or CzcB as an anchor. Therefore, CzcC may function by altering the conformation of the CzcAB complex. Conserved domain search with CzcC query sequence produced significant alignments with COG1538 and pfam02321. COG1538 is TolC, an outer membrane protein involved in cell envelope biogenesis, outer membrane/intracellular trafficking, and secretion. The pfam02321 constitutes the outer membrane efflux protein (OEP). The OEP family form trimeric channels that allow export of a variety of substrates in Gram-negative bacteria. Each member of this family is composed of two repeats. The trimeric channel is composed of a 12-stranded, all-beta sheet barrel that spans the outer membrane, and a long all-helical barrel that spans the periplasm.

It has been demonstrated that after expression of *czcC* or *czcB* under the control of the phage T7 promoter, the cells containing the ^{35}S-labeled Czc proteins present in the crude extract were mainly found in the membrane fraction, and a part remained in the cytoplasmic fraction. Because it has been suggested that CzcC belongs to outer membrane-associated proteins, CzcC might be located at the periplasmic face of the outer membrane. Due to its high hydrophobicity [337], however, CzcC does not contain sufficient hydrophobic beta sheets or alpha helices to be an integral outer membrane protein [175]. Deletion of the *czcC* gene resulted in a loss of cadmium resistance and most of the cobalt resistance; however, most of the zinc resistance was retained [337]. On the other hand, deletion of the histidine-rich motifs of CzcB led to a decrease in zinc resistance, but resistance to cadmium and cobalt was only slightly affected. Therefore, cation transport through the Czc complex might use two different pathways: a CzcC-dependent pathway is used by cadmium, cobalt, and half of the zinc ions, and a CzcC-independent pathway that involves the histidine-rich motifs of CzcB is used by the other half of the zinc cations.

The predicted amino acid sequence of CzcB starts with a highly hydrophobic amino terminus followed by two small histidine-rich segments, which are homologous to each other. These two motifs are very good candidates for the required zinc-binding sites because they are absent in the otherwise related CnrB protein that does not recognize Zn^{2+}. Therefore, CzcB is a membrane-bound protein located on the cytoplasmic face of the membrane. CzcB shows some homology to membrane fusion proteins; it might space the periplasm and draw the outer membrane close to the CzcA antiporter, thereby connecting both membranes in a flexible fashion as has been proposed by HlyD. To the authors' query, position-specific scoring matrices (PSSMs) producing significant alignments were COG0845, AcrA, membrane fusion protein (cell envelope biogenesis, outer membrane); pfam00529, HlyD, HlyD family secretion protein; and COG1566, EmrA, multidrug resistance efflux pump (defense mechanisms).

CzcA alone may function as a cation–proton antiporter, although with low efficiency. From hydropathy analysis of the predicted sequence, four domains were predicted [150]: two hydrophobic regions composing the transport "tunnel" and two cytoplasmic hydrophilic domains. These regions are arranged in the following order: hydrophobic amino terminus; hydrophilic domain I; tunnel region I; hydrophilic domain II; tunnel region II; and hydrophilic carboxy terminus. This structure appears to be the result of an ancient gene duplication event shared by a family of proteins for which CzcA is the prototype [180].

A conserved domain search with the CzcA protein sequence produced significant alignments with gnl│CDD│13017 COG3696, putative silver efflux pump, inorganic ion transport, and metabolism; gln│CDD│15262, pfam00873, ACR_tran, AcrB/AcrD/AcrF family; and gln│CDD│10708, COG0841, AcrB, cation/multidrug efflux pump. The members of the protein family pfam00873 are integral membrane proteins. Some are involved in drug resistance. AcrB cooperates with a membrane fusion protein, AcrA, and an outer membrane channel, TolC. It is known that eight genes apparently are involved in bacterial silver resistance. Two of these, *silS* and *silR* encode a pair of cation-sensing sensor kinase (SilS) and transphosphorylated responder (SilR), homologous to PcoS/PcoR for copper resistance and CzcS/CzcR for cadmium, zinc, and cobalt resistances. As with *pco* system, silRS is followed by silE, which codes for a periplasmic Ag^+-binding protein, homologous to PcoE. However, the remainder of the silver resistance genes is transcribed in the opposite direction (unlike the situation in *pco*) and encodes a three-component CBA system, weakly homologous to CzcCBA, and a P-type ATPase [338].

CzcD protein is involved in the regulation of the Czc system. It is a membrane-bound protein with at least four transmembrane alpha helices and is a member of a subfamily of the cation-diffusion facilitator (CDF) protein family, which occurs in all three domains of life. CzcD protein sequence produced significant alignments with COG1230, Co/Zn/Cd efflux system component of inorganic ion transport and metabolism; pfm01545, Cation efflux family; and COG0053, MMT1, predicted Co/Zn/Cd cation transporters. Members of pfam01545 are integral membrane proteins that are found to increase tolerance to divalent metal cations such as cadmium, zinc, and cobalt. These proteins are thought to be efflux pumps that remove these ions from cells. The deletion of *czcD* in a *Ralstonia* sp. led to partially constitutive expression of the Czc system due to an increased transcription of the structural *czc CBA* genes in absence and presence of inducer. The *czcD* deletion could be fully complemented in trans by CzcD and two other CDF proteins from *Saccharomyces cerevisiae*, ZRC1p and COT1p. Thus CzcD appeared to repress the Czc system by an export of the inducing cations [338].

The ZRC1 protein is required in yeast for a zinc–cadmium resistance based on an unknown mechanism. Although multiple copies of the ZRC1 gene enable yeast cells to grow in the presence of high concentrations of zinc, disruption of the chromosomal ZRC1 locus renders the respective mutant strain more sensitive to Zn^{2+} [183]. The COT1 protein has a similar function as ZRC1 with respect to cobalt cation. Overexpression of COT1 protein increases cobalt tolerance and mutation of the COT1 gene makes the cells more sensitive to Co^{2+}. The COT1 protein is located in the mitochondrial membrane fraction and is involved in the uptake of Co^{2+} by mitochondria [187]. The sequence similarity of CzcD, ZRC1, and COT1 is so high that earlier authors were tempted to forward some working hypothesis about the function of these proteins: the CDF proteins are membrane-bound proteins involved in zinc, cobalt, and cadmium transport. That would lead to the assumption that CzcD functions as a sensor by a slow uptake of the zinc, cobalt, and cadmium cations.

Many of the regulatory systems by which cells sense and then respond to environment signals are called *two-component* systems. Such systems are characterized by having two different proteins: (1) a specific *sensor* protein located in the cell membrane; and (2) a cognate *response regulator* protein. The sensor protein has kinase activity and is often referred to as a sensor kinase. Sensor kinases detect a signal and in response phosphorylate themselves (autophosphorylation) at a specific histidine residue on their cytoplasmic surface. This phosphoryl group is then transmitted to another protein inside the cell, the response regulator.

The response regulator is a DNA-binding protein that regulates transcription. Such a two component regulatory system is involved in transcription control of heavy metal homoeostasis in *Ralstonia* [340]. A conserved domain search with a CzcS query sequence produced significant alignments with COG0642, BaeS, signal transduction histidine kinase (signal transduction mechanism); COG2205, KdpD, osmosensitive K^+ channel histidine kinase (signal transduction mechanism); and COG5002, VicK, signal transduction histidine kinase (signal transduction mechanism). Thus, it can be inferred that CzcS is a cation-sensing sensor kinase.

A conserved domain search with the CzcR query sequence revealed that it is a DNA-binding protein/transcriptional regulatory protein. PSSMs producing significant alignments (with high score bits) were COG0745, cd00156, smart0048, and pfam00072. The attributes of these notations are as follows: COG0745, 0mpR (transcription), cd00156, REC (homo dimer), smart00048 (receiver domain), pfam00072 (effector domain). COG0745, OmpR, response regulators consisting of a CheY-like receiver domain and a winged-helix DNA-binding domain (signal transduction mechanisms/transcription); cd00156, REC, signal receiver domain (originally thought to be unique to bacteria [CheY, OmpR, NtrC, and PhoB], now recently identified in eukaryotes ETR1 *Arabidopsis thaliana*; this domain receives the signal from the sensor partner in two-component systems; contains a phosphoacceptor site that is phosphorylated by histidine kinase homologs; usually found N-terminal to a DNA-binding effector domain; forms homodimers); smart00448, REC, cheY-homologous receiver domain; and pfam00072, response_reg, response regulator receiver domain. (This domain receives the signal from the sensor partner in bacterial two-component systems. It is usually found N-terminal to a DNA-binding effector domain.)

The mechanism used by the response regulator to control transcription depends on the system being described. In *E. coli*, the osmolarity of the environment controls which of two proteins, OmpC or OmpF, is synthesized as part of the outer membrane. The response regulator of this system is OmpR. When OmpR is phosphorylated, it acts as an *activator* of transcription of the *ompC* gene and a *repressor* of transcription of the *ompF* gene. In the Czc system, CzcD is the possible transducer, a transmembrane protein that can sense the cations. This CzcD should be in contact with CzcS. When CzcD binds to the cations, it probably changes conformation and causes a change in the autophosphorylation of CzcS. Phosphorylated CzcS (CzcS-P) then phosphorylates CzcR (forming CzcR-P), a *response regulator*, which is by its nature a transcriptional regulatory protein.

Mutagenesis, or the production of changes in the DNA sequence that affect the expression or structure of gene products, is one of the best methods for understanding gene function. Studying the phenotype of the mutants has become logically the subject of functional genomics. At the molecular level, for example, phenotype includes all temporal and spatial aspects of gene expression as well as related aspects of the expression, structure, function and spatial localization of proteins. In a recent study, comparison of the efflux systems of 63 sequenced prokaryotes with that of *Ralstonia metallidurans* indicated that heavy metal resistance is the result of multiple layers of resistance systems with overlapping substrate specificities, but unique functions [340]. The comparative genomics lead to the conclusion that the creation of the outstanding heavy metal resistance of *R. metallidurans* was because of gene multiplication by duplication and horizontal transfer, stepped up to the differentiation of function and, finally, combining the genes into highly efficient operons [340].

11.12 DEVELOPMENT OF PROMISING ANALYTICAL DEVICES FOR TRACE METAL DETECTION IN THE ENVIRONMENT: BIOSENSORS

The massive interest and commitment of resources in the public and private sectors flows from the generally held perception that genomics will be the single most fruitful approach to the acquisition of new information in basic and applied biology in the next several decades. Among many rewards in applied biology, the promise of facile new approaches for development of biosensors is one of the important inclusions.

11.12.1 PROTEIN-BASED BIOSENSORS

The low molecular weight proteins rich in cysteine residues, the metallothioneins, can bind heavy metal ions nonspecifically in metal-thiolate clusters. Metallothioneins, ubiquitously present in eukaryotes and prokaryotes, are synthesized in response to elevated concentrations of silver, bismuth, cadmium, cobalt, copper, mercury, nickel, or zinc metal ions. The cyanobacterium *Synechococcus* sp. synthesizes metallothionein SmtA in greater quantities when exposed to the elevated concentrations of Cd^{2+} or Zn^{2+}. The operon also encodes a divergently transcribed repressor, SmtB, which is a trans-acting repressor of expression from the *smt*A operator–promoter region. Metallothionein protein expression depends upon the interaction between the metal ions and the repressor protein that regulates the expression of metallothinein mRNA. Loss of the repressor gene, *smt*B, and subsequent unregulated transcription of *smt*A, has been shown to be advantageous to organisms constantly stressed with changing levels of cadmium, copper, lead, nickel, zinc, or arsenate [68].

Again, a mutation rendering inability to synthesize SmtA protein leads to the metal-sensitive phenotype. The purified recombinant SmtA-fusion protein, overexpressed in *E. coli* as a carboxy-terminal extension of glutathione S-transferase, was immobilized in different ways to a self-assembled thiol layer on a gold electrode placed as the working electrode in a potentiometric arrangement in a flow analysis system. This allowed detection of copper, cadmium, mercury, and zinc ions at femtomolar concentrations [341]. Similarly, the regulatory protein MerR encoded by the *mer* operon of Tn*501* in *Pseudomonas aeruginosa*, which controls the expression of itself as well as other *mer* gene products for mercury detoxification (by changing its conformation on binding to Hg^{2+} to align contacts at the promoter region of the operon-activating transcription of the *mer* genes) was also used as the recognition component of a biosensor. In both of these protein-based biosensors, GST-SmtA and MerR, a capacitive signal transducer was used to measure the conformational change following binding [341].

11.13 WHOLE-CELL-BASED BIOSENSORS FOR DETECTION OF BIOAVAILABLE HEAVY METALS

The quantitation of bioavailable metal is difficult with traditional analytical methods. However, the bioavailability of metals is an important factor in the determination of metal toxicity and therefore the detection of bioavailable metals is of interest. The new concept of analyzing bioavailability of heavy metals by creating microbial strains capable of sensing the environment will rely largely upon the functional genomics of a metal resistance genetic system. The greatest advantage is the ability of biosensors to detect the bioavailable fraction of the contaminant, as opposed to the total concentration. Such whole-cell bacterial biosensors will create a clearer picture by providing physiologically relevant data in response to a contaminant.

The essence of all metal resistance genetic systems is the specificity of genetic regulatory elements so that the corresponding metal controls the expression of the uptake or resistance gene products. For example, cobalt–zinc–cadmium resistance operon (*czc* operon) in *Ralstonia* sp. CH34 is regulated by a two-component regulatory system composed of the sensor histidine kinase CzcS and the response activator CzcR. Regulatory genes are arranged in an upstream as well as downsteam regulatory region. Genomics revealed the presence of the *czc*R and *czc*S together with *czc*D constituting the downstream regulatory region. Functional genomics with *czc*D::*lac*Z translation fusion and *czc*S::*lux* transcriptional fusion enabled the regulation of both genes by heavy metals to be understood [339].

These systems can be used as the contaminant-sensing component of the biosensor by detecting the substance for which it is designed to detoxify or excrete. The contaminant-sensing component is combined with the reporter genes to create biosensors that can identify toxic substances at very low levels. When the contaminant-sensing component detects the substance, it triggers the reporter gene. In the development of a mercury-specific biosensor, a hypersensitive clone was constructed using the regulatory sequence along with the mercury (Hg^{+2}) uptake genes *merTPC* of the mercury resistance operon. Such a clone was found responsive to Hg^{+2} with as low as 0.5 n*M* — several folds lower than the lowest concentration required to induce the operon without the *merTPC*.

A reporter gene encodes for a mechanism that produces a detectable cellular response. It determines the sensitivity and detection limits of the biosensor. Specific characteristics are needed for the reporter gene to be used in a biosensor. The gene must have an expression or activity that can be measured using a simple assay and reflects the amount of chemical or physical change. Also, the biosensor must be free of any gene expression or activity similar to the desired gene expression or activity being measured. The suitable reporter genes are (1) lacZ (β-galactosidase); (2) firefly (*Photinus pyralis*) luciferase, lucFF; and (3) bacterial luciferase, luxAB [342-43].

11.14 LUMINESCENCE-BASED BIOSENSORS

There are few successful reports of successful use of gene-fusion biosensor in the monitoring of a metal toxicity in field application. These strains express a sensitive reporter gene, luciferase, connected to a promoter element responding specifically to various heavy metals. Inserting a heavy metal responsive element directionally cloned in a suitable vector in front of the luciferase gene by using standard recombinant-DNA techniques created all the sensor plasmids. Using a nucleotide database, one can isolate a promoter/operator element generated by a polymerase chain reaction using specific oligonucleotide primers. The plasmids can then be expressed in different hosts, in order to obtain maximal and specific response to each of the metal tested. For instance, Biomet sensors (patented and recognized by OVAM) used a relevant soil bacterium, *Ralstonia metallidurans*, to modify it to deliver a light reaction when specific heavy metals go into the cell. The quantification of light emissions can easily be done with a microtiter plate luminometer. Freeze-dried sensor cells were also tested used like reagents. Another bioluminescent sensor (*mer-lux* fusion) for detection of mercury was successfully used to examine the feasibility of testing mercury concentration of natural water from a contaminated freshwater pond [343].

11.14.1 DEVELOPMENT OF AN *LacZ*-BASED ARSENIC BIOSENSOR [344]

Understanding the mechanism for arsenic resistance is necessary in order to develop an appropriate biosensor and to better understand its response. Therefore, to develop the biosensor, the arsenic-resistant gene and the reporter gene must be cloned and inserted onto one plasmid, which then is inserted into a host bacterium. The arsenic-sensing biosensor will then be triggered by arsenic, the analyte, entering the biosensor and activating the transcription of the resistant gene, which is to be followed by the transcription of the reporter gene. The entire resistant gene may not be needed, and the biosensor can only use the beginning components, such as the promoter, and should be able to recognize the arsenic and begin the transcription of the plasmid that contains the reporter gene.

The transcription of the reporter gene will produce proteins whose activity will be assessed in direct correlation to the amount of arsenic entering it. Bacterial arsenic resistance genetic operon *ars*RDABC or *ars*RBC present in diverse bacterial species have been well characterized [24,136]. Because the *ars* operator–promoter is inducible by arsenite (or arsenate in the presence of arsC),and inducibility is positively correlated with the concentration of the inducer, the activity of a reporter gene product, if cloned under this promoter, will reflect the availability of the inducer/analyte.

In the present study, an arsenic biosensor was created using the *lacZ* reporter gene. The promoterless *lacZ* gene, lacking a ribosome-binding sequence and the first eight nonessential amino acid codons, was coupled with the *ars* promoter along with *ars*R, *ars*D, and a part of *ars*A gene in the translational *lacZ* fusion vector, pMC1871. The recombinant plasmid pASH3, having truncated the *ars* operon, was found hypersensitive to arsenite and arsenate and the phenomenon was found suitable to develop a simple bioassay system for arsenic. However, it is not known whether the hypersensitivity to arsenite rendered by pASH3 is associated with active arsenite uptake like the mercury-hypersensitive clone. Thus far, the hypersensitivity of arsenite observed in *E. coli* cells conferred by pASH3 cannot be explained from the present state of knowledge of arsenic resistance and or uptake in microorganisms. However, the expression of *ars-lacZ* fusion was expected at a concentration lower than the values observed earlier with the complete *ars* operon *ars*RDABC.

The maximum specific activity was obtained with arsenite between 0.5 and 1 μM; however, β-galactosidase activity was found inducible even with 10 nM of arsenite. Therefore, the clone pASH3 could serve a biosensor to detect a very low concentration of arsenic, which may not be easily detectable by the standard chemical methods. When the arsenite-responsive regulation unit of the plasmid pI258 was used to express the firefly luciferase gene (*lucFF*), the lowest arsenite concentration required to induce the reporter gene was 100 nM and maximum induction was noted at 3.3 mM4 4 arsenite. In another method, expression of *arsD-lacZ* fusion was monitored by an electrochemical reaction to assay the arsenite-dependent β-galactosidase activity. The detection limit was 100 nM, but a 17-h induction period was required.

These two methods are apparently suitable for environmental samples of low bioavailable arsenic content; however, both require relatively costly instruments and may not be suitable in a field study. There are few reports of successful gene-fusion biosensors in the monitoring of a metal toxicity in field application. For instance, a luminescent bacterial biosensor was shown to be effective in the evaluation of arsenic bioavailability of chromated copper arsenate contamination [343]. Detection limit of arsenite by arsenic hypersensitive clone pASH3 was comparable with the earlier claims; the methodology is relatively simple as well. Moreover, the process can easily be improvised to an acceptable arsenic assay kit with a low-cost investment for monitoring a large number of samples for on-site analysis.

11.15 CONCLUSION

The creativity and imagination of the researcher are the two important variables that influence the potential application of genome data. It is prudent to follow a logical course when using genomic data. During the initial data acquisition step, in which genomic data are generated experimentally or retrieved from publicly available data sources, simultaneous evaluation of multiple data sets will ensure higher resolution and greater confidences while increasing the likelihood that the genomic elements of interest are represented. It is imperative to include an extensive mutagenic analysis of all the systems involved while studying the metal interactions at the genomic level. It would also be informative to study the regulation of all these genes by transcription profiling with DNA microarrays and proteomic analyses.

It is encouraging that the understanding of the genetic basis of metal resistance has opened up the ability to use these systems for environmental application. Although much still remains to be learned, the future development of a multianalytical biosensor will further improve the applicability of biosensors. It is probable that a multianalytical biosensor to test more than one trace element in the environment could be developed by genetically labeling the bacteria used in the biosensor. Continued research endeavors will produce best procedures and full applicability of whole-cell bacterial biosensors.

I cannot but wonder how this incredibly complex mechanism is still functioning at all. Thinking of life one clearly sees our science in the poorest and most primitive light. The properties of living matter are most probably predetermined by the fertilized cell in the same way as the prerequisite of life itself consists in the existence of an atom and consequently, the mystery of everything that existing can be found in its lowest stage of development.

A. Einstein

REFERENCES

1. Voitkevich, G.V., *Origin and Development of Life on Earth*, Mir Publishers, Moscow, 1990, 10.
2. Silver, S., Bacterial resistances to toxic metal ions — a review, *Gene*, 179, 9, 1996.
3. Ben–Jacob, E., Bacterial wisdom, Godel's theorem and creative genomic webs, *Physica. A.*, 248, 57, 1998.
4. Cairns, J., Overbaugh, J., and Miller, S., The origin of mutants, *Nature*, 335, 142, 1988.
5. Foster, P.L., Adaptive mutation: the uses of adversity, *Annu. Rev. Microbiol.*, 47, 467, 1993.
6. Hall, B.G., Adaptive evolution that requires multiple spontaneous mutations. I. Mutations involving an insertion sequence, *Genetics*, 120, 887, 1988.
7. Shapiro, J.A., Observations on the formation of clones containing arab — lacz cistron fusion, *Mol. Gen. Genet.*, 194, 79, 1984.

8. Wackett, L.P., Dodge, A.G., and Ellis, L.B.M., Microbial genomics and the periodic table, *Appl. Environ. Microbiol.*, 70, 647, 2004.

9. Canovas, D., Cases, I., and de Lorenzo, V., Heavy metal tolerance and metal homeostasis in *Pseudomonas putida* as revealed by complete genome, *Environ. Microbiol.*, 1, 2003.

10. Weast, R.C., *CRC Handbook of Chemistry and Physics*, 64th ed., CRC Press, Inc., Boca Raton, FL, 1984.

11. Nies, D.H., Microbial heavy-metal resistance, *Appl. Microbiol. Biotechnol.*, 51, 730, 1999.

12. MacDiarmid, C.W. and Gardner, R.C., Overexpression of the *Saccharomyces cerevisiae* magnesium transport system confers resistance to aluminum ion, *J. Biol. Chem.*, 273, 1727, 1998.

13. Nies, D.H. and Silver, S., Ion efflux systems involved in bacterial metal resistances, *J. Indust. Microbiol.*, 14, 186, 1995.

14. Smith, R.L. et al., Functional similarity between archaeal and bacteria CorA magnesium transporters, *J. Bacteriol.*, 180, 2788, 1998.

15. Smith, R.L. and Maguire, M.E., Distribution of the CorA Mg2+ transport system in Gram-negative bacteria, *J. Bacteriol.*, 177, 1638, 1995.

16. Tao, T. et al., Magnesium transport in *Salmonella typhimurium: mgtA* encodes a P-type ATPase and is regulated by Mg2+ in a manner similar of the *mgtB* P-type ATPase, *J. Bacteriol.*, 177, 2654, 1995.

17. Kachur, A.V., Koch, C.J., and Biaglow, J.E., Mechanism of copper-catalyzed oxidation of glutathione, *Free Radical Res.*, 28, 259, 1998.

18. Nelson, D.L. and Kennedy, E.P., Magnesium transport in *Escherichia coli*: inhibition by cobaltous ion, *J. Biol. Chem.*, 246, 3042, 1971.

19. Park, M.H., Wong, B.B., and Lusk, J.E., Mutants in three genes affecting transport of magnesium in *Escherichia coli*, *J. Bacteriol.*, 126, 1096, 1976.

20. Rosen, B.P., Bacterial resistance to heavy metals and metalloids, *J. Biol. Inorg. Chem.*, 1, 273, 1996.

21. Bruins, M.R., Kapil, S., and Oehme, F.W., Microbial resistance to metals in the environment, *Ecotoxic. Environ. Safety.*, 45, 198, 2000.

22. Nies, D.H. and Silver, S., Metal ion uptake by a plasmid-free metal-sensitive *Alcaligenes eutrophus* strain, *J. Bacteriol.*, 171, 4073, 1989.

23. Schiering, N. et al., Structure of the detoxification catalyst mercuric ion reductase from *Bacillus* sp. strain RC607, *Nature*, 352, 168, 1991.

24. Silver, S. and Phung, L.T., Bacterial heavy metal resistance: new surprises, *Annu. Rev. Microbiol.*, 50, 753, 1996.

25. Smith, T. et al., Bacterial oxidation of mercury metal vapor, Hg(0), *Appl. Environ. Microbiol.*, 64, 1328, 1998.

26. Barrineau, P. et al., The DNA sequence of the mercury resistance operon of the IncFII plasmid NR1. *J. Mol. Appl. Genet.*, 2, 601, 1984.

27. Brunker, P. et al., Regulation of the operon responsible for broad-spectrum mercury resistance in *Streptomyces lividans* 1326, *Mol. Gen. Genet.*, 251, 307, 1996.

28. Chu, L. et al., Regulation of the Staphylococcus aureus plasmid pI258 mercury resistance operon, *J. Bacteriol.*, 174, 7044, 1992.

29. Helmann, J.D. et al., Homologous metalloregulatory proteins from both Gram-positive and Gram-negative bacteria control transcription of mercury resistance operons, *J. Bacteriol.*, 171, 222, 1989.

30. Hobman, J.L. and Brown, N.L., Bacterial mercury-resistance genes, *Met. Ions. Biol. Syst.*, 34, 527, 1997.

31. Kholodii, G. et al., Molecular characterization of an aberrant mercury resistance transposable element from an environmental *Acinetobacter* strain, *Plasmid*, 30, 303, 1993.

32. Lee, I. W. et al., *In vivo* DNA-protein interactions at the divergent mercury resistance (mer) promoters. II. Repressor/activator (MerR)-RNA polymerase interaction with merOP mutants, *J. Biol. Chem.*, 268, 2632, 1993.

33. Livrelli, V., Lee, I.W., and Summers, A.O., *In vivo* DNA-protein interactions at the divergent mercury resistance (mer) promoters. I. Metalloregulatory protein MerR mutants, *J. Biol. Chem.*, 268, 2623, 1993.

34. Lund, P.A., Ford, S.J., and Brown, N.L., Transcriptional regulation of the mercury-resistance genes of transposon Tn501, *J. Gen. Microbiol.*, 132, 1986.

35. Nucifora, G. et al., Mercury operon regulation by the merR gene of the organomercurial resistance system of plasmid pDU1358, *J. Bacteriol.*, 171, 4241, 1989.

36. Osborn, A.M. et al., The mercury resistance operon of the IncJ plasmid pMERPH exhibits structural and regulatory divergence from other Gram-negative mer operons, *Microbiology*, 142, 337, 1996.

37. Peters, S.E. et al., Novel mercury resistance determinants carried by IncJ plasmids pMERPH and R391, *Mol. Gen. Genet.*, 228, 294, 1991.

38. Reniero, D., Galli, E., and Barbieri, P., Cloning and comparison of mercury- and organomercurial-resistance determinants from a *Pseudomonas stutzeri* plasmid, *Gene*, 166, 77, 1995.

39. Skinner, J.S., Ribot, E., and Laddaga, R.A., Transcriptional analysis of the *Staphylococcus aureus* plasmid pI258 mercury resistance determinant, *J. Bacteriol.*, 173, 5234, 1991.

40. Wang, Y. et al., Nucleotide sequence of a chromosomal mercury resistance determinant from a *Bacillus* sp. with broad-spectrum mercury resistance, *J. Bacteriol.*, 171, 83, 1989.

41. Yurieva, O. et al., Intercontinental spread of promiscuous mercury-resistance transposons in environmental bacteria, *Mol. Microbiol.*, 24, 321, 1997.

42. Caslake, L.F., Ashraf, S.I., and Summer, A.O., Mutations in the alpha and sigma 70 subunits of RNA polymerase affect expression of the *mer* operon, *J. Bacteriol.*, 179, 1787, 1997.

43. Summer, A.O., Untwist and shout — a heavy metal-responsive transcriptional regulator, *J. Bacteriol.*, 174, 3097, 1992.

44. Mukhopadhyay, D. et al., Purification and functional characterization of MerD. A coregulator of the mercury resistance operon in Gram-negative bacteria, *J. Biol. Chem.*, 266, 18538, 1991.

45. Hobman, J.L. and Brown, N.L., Overexpression of MerT, the mercuric ion transport protein of transposon Tn501, and genetic selection of mercury hypersensitivity mutations, *Mol. Gen. Genet.*, 250, 129, 1996.

46. Qian, H. et al., NMR solution structure of the oxidized form of MerP, a mercuric ion-binding protein involved in bacterial mercuric ion resistance, *Biochemistry*, 37, 9316, 1998.

47. Hamlett, N.V. et al., Roles of the Tn21 merT, merP, and merC gene products in mercury resistance and mercury binding, *J. Bacteriol.*, 174, 6377, 1992.

48. Sahlman, L., Wong, W., and Powlowski, J. A mercuric ion uptake role for the integral inner membrane protein, MerC, involved in bacterial mercuric ion resistance, *J. Biol. Chem.*, 272, 29518, 1997.

49. Fatoki, O.S., Biomethylation in the natural environment: a review, *S. Afr. J. Sci.*, 93, 366, 1997.

50. Babich, K. et al., Deletion mutant analysis of the *Staphylococcus aureus* plasmid pI258 mercury-resistance determinant, *Can. J. Microbiol.*, 37, 624, 1991.

51. Bhattacharyya, G., Chaudhuri, J., and Mandal, A., Elimination of mercury, cadmium and antibiotic resistance from *Acinetobacter lwoffi* and *Micrococcus* sp. at high temperature, *Folia Microbiol.*, 33, 213, 1988.

52. Liebert, C.A. et al., Phylogeny of mercury resistance (mer) operons of Gram-negative bacteria isolated from the fecal flora of primates, *Appl. Environ. Microbiol.*, 63, 1066, 1997.

53. Reniero, D. et al., Two aberrant mercury resistance transposons in the *Pseudomonas stutzeri* plasmid pPB, *Gene*, 208, 37, 1998.

54. Yu, H., Chu, L., and Misra, T. K., Intracellular inducer Hg2+ concentration is rate determining for the expression of the mercury-resistance operon in cells, *J. Bacteriol.*, 178, 2712, 1996.

55. Hong, S.M. et al., Greenland ice evidene of hemispheric lead pollution two millenia ago by Greek and Roman civilization, *Science*, 265, 1841, 1994.

56. Johnson, F.M., The genetic effects of environmental lead, *Mutat. Res-Rev. Mutat. Res.*, 410, 123, 1998.

57. Trajanovska, S., Britz, M.L., and Bhave, M., Detection of heavy metal ion resistance genes in Gram-positive and Gram-negative bacteria isolated from a lead-contaminated site, *Biodegradation*, 8, 113, 1997.

58. Levinson, H.S., and Mahler, I., Phosphatase activity and lead resistance in *Citrobacter freundii* and *Staphylococcus aureus*, *FEMS Microbiol. Lett*, 161, 135, 1998.

59. Levinson, H.S. et al., Lead resistance and sensitivity in *Staphylococcus aureus*, *FEMS Microbiol. Lett.*, 145, 421, 1996.

60. Sanders, O.I. et al., Antimonite is accumulated by the glycerol facilitator GlpF in *Escherichia coli*, *J. Bacteriol.*, 179, 3365, 1997.

61. Rosenstein, R. et al., Expression and regulation of the antimonite, arsenite, and arsenate resistance operon of *Staphylococcus xylosus* plasmid pSX267, *J. Bacteriol.*, 174, 3676, 1992.

62. Burke, B.E. and Pfister, R.M., Cadmium transport by a Cd2+-sensitive and a Cd2+-resistant strain of *Bacillus subtilis, Can. J. Microbiol.*, 32, 539, 1986.

63. Laddaga, R.A., Bessen, R., and Silver, S. Cadmium-resistant mutant of *Bacillus subtilis* 168 with reduced cadmium transport, *J. Bacteriol.*, 162, 1106, 1985.

64. Tynecka, Z. and Malm, A., Energetic basis of cadmium toxicity in *Staphylococcus aureus, Biometals,* 8, 197, 1995.

65. Liu, C.Q. et al., Genetic analysis of regions involved in replication and cadmium resistance of the plasmid pND302 from *Lactococcus lactis, Plasmid*, 38, 79, 1997.

66. Olafson, R.W., Abel, K., and Sim, R.S., Prokaryotic metallothionein: preliminary characterization of a blue-green algae heavy-metal binding protein, *Biochem. Biophys. Res. Commun.*, 89, 36, 1979.

67. Gupta, A. et al., Amplification and rearrangement of a prokaryotic metallothionein locus *smt* in *Synechococcus* PCC-6301 selected for tolerance to cadmium, *Proc. Royal Soc. London Ser. B – Biol. Sci.*, 248, 273, 1992.

68. Gupta, A. et al., Deletion within the metallothionein locus of cadmium-tolerant *Synechococcus* PCC 6301 involving a highly iterated palindrome (HIP1), *Mol. Microbiol.*, 7, 189, 1993.

69. Turner, J.S. et al., Zn2+-sensing by the cyanobacterial metallothionein repressor SmtB: different motifs mediate metal-induced protein-DNA dissociation, *Nucleic Acids Res.*, 24, 3714, 1996.

70. Turner, J.S., Robinson, N.J., and Gupta, A., Construction of Zn2+/Cd(2+)-tolerant cyanobacteria with a modified metallothionein divergon: further analysis of the function and regulation of smt, *J. Ind. Microbiol.*, 14, 259, 1995.

71. Huckle, J.W. et al., Isolation of a prokaryotic metallothionein locus and analysis of transcriptional control by trace metal ions, *Mol. Microbiol.*, 7, 177, 1993.

72. Morby, A.P. et al., SmtB is a metal-dependent repressor of the cyanobacterial metallothionein gene smtA: identification of a Zn inhibited DNA-protein complex, *Nucleic Acids Res.*, 21, 921, 1993.

73. Thelwell, C., Robinson, N.J., and TurnerCavet, J.S., An SmtB-like repressor from Synechocystis PCC 6803 regulates a zinc exporter, *Proc. Natl. Acad. Sci. USA.*, 95, 10728, 1998.

74. Nies, D.H., The cobalt, zinc, and cadmium efflux system CzcABC from *Alcaligenes eutrophus* functions as a cation-proton-antiporter in *Escherichia coli, J. Bacteriol.*, 177, 2707, 1995.

75. Nies, D.H. and Silver, S., Plasmid-determined inducible efflux is responsible for resistance to cadmium, zinc, and cobalt in *Alcaligenes eutrophus, J. Bacteriol.*, 171, 896, 1989.

76. Schmidt, T. and Schlegel, H.G., Combined nickel-cobalt-cadmium resistance encoded by the *ncc* locus of *Alcaligenes xylosoxidans* 31A, *J. Bacteriol.*, 176, 7045, 1994.

77. Taghavi, S. et al., *Alcaligenes eutrophus* as a model system for bacterial interactions with heavy metals in the environment, *Res. Microbiol.*, 148, 536, 1997.

78. Nucifora, G. et al., Cadmium resistance from *Staphylococcus aureus* plasmid pI258 cadA gene results from a cadmium-efflux ATPase, *Proc. Natl. Acad. Sci. USA.*, 86, 3544, 1989.

79. Silver, S., Misra, T.K., and Laddaga, R.A., DNA sequence analysis of bacterial toxic heavy metal resistances, *Biol. Trace Elem. Res.*, 21, 145, 1989.

80. Endo, G. and Silver, S., CadC, the transcriptional regulatory protein of the cadmium resistance system of *Staphylococcus aureus* plasmid pI258, *J. Bacteriol.*, 177, 4437, 1995.

81. Yoon, K.P., Misra, T.K., and Silver, S., Regulation of the cadA cadmium resistance determinant of *Staphylococcus aureus* plasmid pI258, *J. Bacteriol.*, 173, 7643, 1991.

82. Yoon, K.P. and Silver, S., A second gene in the *Staphylococcus aureus* cadA cadmium resistance determinant of plasmid pI258, *J. Bacteriol.*, 173, 7636, 1991.

83. Lee, S.W., Glickmann, E., and Cooksey, D.A., Chromosomal locus for cadmium resistance in *Pseudomonas putida* consisting of a cadmium-transporting ATPase and a MerR family response regulator, *Appl. Environ. Microbiol.*, 67,1437, 2001.

84. Li, Z.S. et al., A new pathway for vacuolar cadmium sequestration in *Saccharomyces cerevisiae*: YCF1-catalyzed transport of bis (glutathionato) cadmium, *Proc. Natl. Acad. Sci. U S A.*, 94, 42, 1997.

85. Li, Z.S. et al., The yeast cadmium factor protein (YCF1) is a vacuolar glutathione-S-conjugate pump, *J. Biol. Chem.*, 271, 6509,1996.

86. Tommasini, R. et al., The human multidrug resistance associated protein functionally complements the yeast cadmium resistance factor 1, *Proc. Natl. Acad. Sci. USA.*, 93, 6743, 1996.

87. Inouhe, M. et al., Resistance to cadmium ions and formation of a cadmium-binding complex in various wild-type yeasts, *Plant Cell Physiol.*, 37, 341, 1996.

88. Ortiz, D. et al., Heavy metal tolerance in the fission yeast requires an ATP-binding cassette-type vacular membrane transporter, *EMBO J.*, 11, 3491, 1992.

89. Ortiz, D.F. et al., Transport of metal-binding peptides by HMT1, a fission yeast ABC-type vacuolar membrane protein, *J. Biol. Chem.*, 270, 4721, 1995.

90. Wu, J.S., Sung, H.Y., and Juang, R.H., Transformation of cadmium-binding complexes during cadmium sequestration in fission yeast, *Biochem. Mol. Biol. Int.*, 36, 1169,1995.

91. Slawson, R.M. et al., Germanium and silver resistance, accumulation, and toxicity in microorganisms, *Plasmid*, 27, 72, 1992.

92. Deshpande, L.M. and Chopade, B.A., Plasmid mediated silver resistance in *Acinetobacter baumannii*, *Biometals*, 7, 49, 1994.

93. Haefeli, C., Franklin, C., and Hardy, K., Plasmid-determined silver resistance in *Pseudomonas stutzeri* isolated from a silver mine, *J. Bacteriol.*, 158, 389, 1984.

94. Kaur, P., Saxena, M., and Vadehra, D.V., Plasmid mediated resistance to silver ions in *Escherichia coli*, *Indian J. Med. Res.*, 82, 122, 1985.

95. Kaur, P. and Vadehra, D.V., Mechanism of resistance to silver ions in *Klebsiella pneumoniae*, *Antimicrob. Agents Chemother.*, 29, 165, 1986.

96. Slawson, R.M. et al., Silver resistance in *Pseudomonas stutzeri*, *Biometals*, 7, 30, 1994.

97. Starodub, M.E. and Trevors, J.T., Silver resistance in *Escherichia coli* R1, *J. Med. Microbiol.*, 29, 101, 1989.

98. Starodub, M.E. and Trevors, J.T., Mobilization of *Escherichia coli* R1 silver-resistance plasmid pJT1 by Tn5-Mob into *Escherichia coli* C600, *Biol. Met.*, 3, 24, 1990.

99. Starodub, M.E. and Trevors, J.T., Silver accumulation and resistance in *Escherichia coli* R1, *J. Inorg. Biochem.*, 39, 317, 1990.

100. Vasishta, R., Saxena, M., and Chhibber, S., Contribution of silver ion resistance to the pathogenicity of *Pseudomonas aeruginosa* with special reference to burn wound sepsis, *Folia Microbiol. Praha.*, 36, 498, 1991.

101. Yudkins, J., The effect of silver ions on some enzymes of *Bacterium coli*, *Enzymologia*, 2, 161, 1937.

102. Gupta, A. et al., Molecular basis for resistance to silver cations in *Salmonella*, *Nat. Med.*, 5, 183, 1999.

103. Solioz, M. and Odermatt, A., Copper and silver transport by CopA-ATPase in membrane vesicles of *Enterococcus hirae*, *J. Biol. Chem.*, 270, 9217, 1995.

104. Grunden, A.M. et al., Repression of the *Escherichia coli* modABCD (molybdate transport) operon by ModE, *J. Bacteriol.*, 178, 735, 1996.

105. Grunden, A.M. and Shanmugam, K.T., Molybdate transport and regulation in bacteria, *Arch. Microbiol.* 168, 345, 1997.

106. Imperial, J., Hadi, M., and Amy, N.K., Molybdate binding by ModA, the periplasmic component of the *Escherichia coli* mod molybdate transport system, *Biochim. Biophys. Acta.*, 1370, 337, 1998.

107. Maupin Furlow, J.A. et al., Genetic analysis of the modABCD (molybdate transport) operon of *Escherichia coli*, *J. Bacteriol.*, 177, 4851, 1995.

108. Menendez, C. et al., Molybdate-uptake genes and molybdopterin-biosynthesis genes on a bacterial plasmid — characterization of MoeA as a filament-forming protein with adenosinetriphosphatase activity, *Eur. J. Biochem.*, 250, 524, 1997.

109. Rech, S., Deppenmeier, U., and Gunsalus, R.P., Regulation of the molybdate transport operon, modABCD, of *Escherichia coli* in response to molybdate availability, *J. Bacteriol.*, 177, 1023, 1995.

110. Rosentel, J.K. et al., Molybdate and regulation of mod (molybdate transport), fdhF, and hyc (formate hydrogenlyase) operons in *Escherichia coli*, *J. Bacteriol.*, 177, 4857, 1995.

111. Walkenhorst, H.M., Hemschemeier, S.K., and Eichenlaub, R., Molecular analysis of the molybdate uptake operon, *modABCD*, of *Escherichia coli* and modR, a regulatory gene, *Microbiol. Res.*, 150, 347, 1995.

112. Romão, M.J. et al., Crystal structure of the xanthin-oxidase-related aldehyde oxidoreductase from *D. gigas*, *Science*, 270, 1170, 1995.

113. Schindelin, H. et al., Crystal structure of DMSO reductase: redox-linked changes in molybdopterin coordination, *Science*, 272, 1615, 1996.

114. Bolin, J.T. et al., *Molybdenum Enzymes, Cofactors and Model Systems*, American Chemical Society, Washington, D.C., 1993, 186.

115. Chan, M.K., Kim, J., and Rees, D.C., The nitrogenase FeMo-cofactor and P-cluster pair: 2.2 Å resolution structures, *Science*, 260, 792, 1993.
116. Rosen, B.P., Resistance mechanisms to arsenicals and antimonials, *J. Basic Clin. Physiol. Pharmacol.*, 6, 251, 1995.
117. Xu, C. et al., Zhou, T., Kuroda, M., and Rosen, B.P., Metalloid resistance mechanisms in prokaryotes, *J. Biochem. Tokyo.*, 123, 16, 1998.
118. Diorio, C. et al., *Escherichia coli* chromosomal ars operon homolog is functional in arsenic detoxification and is conserved in Gram-negative bacteria, *J. Bacteriol.*, 177, 2050, 1995.
119. Rosen, B.P., Biochemistry of arsenic detoxification, *FEBS Lett.*, 529, 86, 2002.
120. Wu, J., Tisa, L.S., and Rosen, B.P., Membrane topology of the ArsB protein, the membrane subunit of an anion-translocating ATPase, *J. Biol. Chem.*, 267, 12570, 1992.
121. Zhou, T. and Rosen, B.P., Tryptophan fluorescence reports nucleotide-induced conformational changes in a domain of the ArsA ATPase, *J. Biol. Chem.*, 272, 19731, 1997.
122. Chen, C.M. et al., Nucleotide sequence of the structural genes for an anion pump: the plasmid-encoded arsenical resistance operon, *J. Biol. Chem.*, 261, 15030, 1986.
123. Kuroda, M. et al., Alternate energy coupling of ArsB, the membrane subunit of the Ars anion-translocating ATPase, *J. Biol. Chem.*, 272, 326, 1997.
124. Dey, S. and Rosen, B.P., Dual mode of energy coupling by the oxyanion-translocating ArsB protein, *J. Bacteriol.*, 177, 385, 1995.
125. Li, J., Liu, S., and Rosen, B.P., Interaction of ATP-binding sites in the ArsA ATPase, the catalytic subunit of the Ars pump, *J. Biol. Chem.*, 271, 25247, 1996.
126. Li, J. and Rosen, B.P., Steric limitations in the interaction of the ATP-binding domains of the ArsA ATPase, *J. Biol. Chem.*, 273, 6796, 1998.
127. Ji, G.Y. et al., Arsenate reductase of *Staphylococcus aureus* plasmid pI258, *Biochemistry*, 33, 7294, 1994.
128. Ji, G.Y. and Silver, S., Reduction of arsenate to arsenite by the ArsC protein of the arsenic resistance operon of *Staphylococcus aureus* plasmid pI258, *Proc. Natl. Acad. Sci. USA.*, 89, 9474, 1992.
129. Oden, K.L., Gladysheva, T.B., and Rosen, B.P., Arsenate reduction mediated by the plasmid-encoded ArsC protein is coupled to glutathione, *Mol. Microbiol.*, 12, 301, 1994.
130. Gladysheva, T.B., Oden, K.L., and Rosen, B.P., Properties of the arsenate reductase of plasmid R773, *Biochemistry*, 33, 7288, 1994.
131. Liu, J. and Rosen, B.P., Ligand interactions of the ArsC arsenate reductase, *J. Biol. Chem.*, 272, 21084, 1997.
132. Rosen, B.P., Bhattacharjee, H., and Shi, W., Mechanisms of metalloregulation of an anion-translocating ATPase, *J. Bioenerg. Biomembr.*, 27, 85, 1995.
133. Shi, W. et al., The role of arsenic-thiol interactions in metalloregulation of the ars operon, *J. Biol. Chem.*, 271, 9291, 1996.
134. Shi, W., Wu, J., and Rosen, B.P., Identification of a putative metal-binding site in a new family of metalloregulatory proteins, *J. Biol. Chem.*, 269, 19826, 1994.
135. Wu, J. and Rosen, B.P., The arsD gene encodes a second trans-acting regulatory protein of the plasmid-encoded arsenical resistance operon, *Mol. Microbiol.*, 8, 615, 1993.
136. Xu, C. and Rosen, B.P., Dimerization is essential for DNA binding and repression by the ArsR metaloregulatory protein of *Escherichia coli*. *J. Biol. Chem.*, 272, 15734, 1997.
137. Xu, C., Shi, W., and Rosen, B.P., The chromosomal *arsR* gene of *Escherichia coli* encodes a trans-acting metalloregulatory protein, *J. Biol. Chem.*, 271, 2427, 1996.
138. Wu, J. and Rosen, B.P., Metaloregulated expression of the ars operon, *J.Biol. Chem.*, 268, 52, 1993.
139. Chen, Y. and Rosen, B.P., Metaloregulatory properties of the ArsD repressor, *J. Biol. Chem.*, 272, 14257, 1997.
140. Neyt, C. et al., Virulence and arsenic resistance in *Yersiniae*, *J. Bacteriol.*, 179, 612, 1997.
141. Papadopoulou, B. et al., Contribution of the Leishmania P-glycoprotein-related gene ltpgpA to oxyanion resistance, *J. Biol. Chem.*, 269, 11980, 1994.
142. Dey, S. et al., Interaction of the catalytic and the membrane subunits of an oxyanion-translocating ATPase, *Arch. Biochem. Biophys.*, 311, 418, 1994.
143. Dey, S. et al., An ATP-dependent As(III)-glutathione transport system in membrane vesicles of *Leishmania tarentolae*, *Proc. Natl. Acad. Sci. USA*, 93, 2192, 1996.

144. Legare, D. et al., Efflux systems and increased trypanothione levels in arsenite-resistant Leishmania, *Exp. Parasitol.*, 87, 275, 1997.
145. Mukhopadhyay, R. et al., Trypanothione overproduction and resistance to antimonials and arsenicals in Leishmania, *Proc. Natl. Acad. Sci. USA.*, 93, 10383, 1996.
146. Papadopoulou, B. et al., Gene disruption of the P-glycoprotein related gene pgpa of *Leishmania tarentolae*, *Biochem. Biophys. Res. Commun.*, 224, 772, 1996.
147. Wysocki, R., Bobrowicz, P., and Ulaszewski, S., The *Saccharomyces cerevisiae ACR3* gene encodes a putative membrane protein involved in arsenite transport, *J. Biol. Chem.*, 272, 30061, 1997.
148. KurdiHaidar, B. et al., Biochemical characterization of the human arsenite-stimulated ATPase (hASNA-I), *J. Biol. Chem.*, 273, 22173, 1998.
149. KurdiHaidar, B. et al., Dual cytoplasmic and nuclear distribution of the novel arsenite-stimulated human ATPase (hASNA-I), *J. Cell. Biochem.*, 71, 1, 1998.
150. Nies, D.H., CzcR and CzcD, gene products affecting regulation of resistance to cobalt, zinc and cadmium (*czc* system) in *Alcaligens eutrophus*, *J. Bacteriol.*, 174, 8102, 1992.
151. Daniels, M.J. et al., Coordination of Zn^{2+} (and Cd^{2+}) by prokaryotic metallothionein — involvement of His-imidazole, *J. Biol. Chem.*, 273, 22957, 1998.
152. DiazCruz, M.S. et al., Study of the zinc-binding properties of glutathione by differential pulse polarography and multivariate curve resolution, *J. Inorg. Biochem.* 70, 91, 1998.
153. Jiang, L.J., Maret, W., and Vallee, B.L., The ATP-metallothionein complex, *Proc. Natl. Acad. Sci. USA.*, 95, 9146, 1998.
154. Palmiter, R.D., The elusive function of metallothioneins, *Proc. Natl. Acad. Sci. USA.*, 95, 8428, 1998.
155. Smith, R.L., Thompson, L.J., and Maguire, M.E., Cloning and characterization of MgtE, a putative new class of Mg^{2+} transporters from *Bacillus firmus* OF4, *J. Bacteriol.*, 177, 1233, 1995.
156. Townsend, D.E. et al., Cloning of the *mgtE* Mg^{2+} transporter from *Providencia stuartii* and the distribution of *mgtE* in Gram-negative and Gram-positive bacteria, *J. Bacteriol.*, 177, 5350, 1995.
157. Snavely, M.D. et al., Magnesium transport in *Salmonella typhimurium*: $28Mg^{2+}$ transport by CorA, MgtA, and MgtB systems, *J. Bacteriol.*, 171, 4761, 1989.
158. Snavely, M.D. et al., Magnesium transport in *Salmonella typhimurium*: expression of cloned genes for three distinct Mg^{2+} transport systems, *J. Bacteriol.*, 171, 4752, 1989.
159. Snavely, M.D., Miller, C.G., and Maguire, M.E., The *mgtB* Mg^{2+} transport locus of *Salmonella typhimurium* encodes a P-type ATPase, *J. Biol. Chem.*, 266, 815, 1991.
160. Tao, T. et al., Magnesium transport in *Salmonella typhimurium*: biphasic magnesium and time dependence of the transcription of the *mgtA* and *mgtCB* loci, *Microbiology*, 144, 655, 1998.
161. Lu, D., Boyd, B., and Lingwood, C.A., Identification of the key protein for zinc uptake in *Hemophilus influenzae*, *J. Biol. Chem.*, 272, 29033, 1997.
162. Dintilhac, A. et al., Competence and virulence of *Streptococcus pneumoniae*: Adc and PsaA mutants exhibit a requirement for Zn and Mn resulting from inactivation of putative ABC metal permeases, *Mol. Microbiol.*, 25, 727, 1997.
163. Kolenbrander, P.E. et al., The adhesion-associated *sca* operon in *Streptococcus gordonii* encodes an inducible high-affinity ABC transporter for Mn^{2+} uptake, *J. Bacteriol.*, 180, 290, 1998.
164. Patzer, S.I. and Hantke, K., The ZnuABC high-affinity zinc uptake system and its regulator zur in *Escherichia coli*, *Mol. Microbiol.*, 28, 1199, 1998.
165. Fox, T.C. and Guerinot, M.L., Molecular biology of cation transport in plants, *Annu. Rev. Plant Physiol.*, 49, 669, 1998.
166. Grotz, N. et al., Identification of a family of zinc transporter genes from Arabidopsis that respond to zinc deficiency, *Proc. Natl. Acad. Sci. USA.*, 95, 7220, 1998.
167. Odermatt, A. et al., Suter, H., Krapf R., and Solioz, M., An ATPase operon involved in copper resistance by *Enterococcus hirae*, *Ann. NY. Acad. Sci.*, 671, 484, 1992.
167a. Paulsen, I.T. et al., Unified inventory of established and putative transporters encoded within the complete genome of *Saccharomyces cerevisiae*, *FEBS Lett.*, 430, 116, 1998.
168. Wolfram, L., Eitinger, T., and Friedrich, B., Construction and properties of a triprotein containing the high-affinity nickel transporter of *Alcaligenes eutrophus*, *FEBS Lett.*, 283, 109, 1991.
169. Zhao, H. and Eide, D., The *ZRT2* gene encodes the low affinity zinc transporter in *Saccharomyces cerevisiae*, *J. Biol. Chem.*, 271, 23203, 1996.

170. Beard, S.J. et al., Zinc (II) tolerance in *Escherichia coli* K-12: evidence that the *zntA* gene (o732) encodes a cation transport ATPase, *Mol. Microbiol.*, 25, 883, 1997.

171. Rensing, C., Mitra, B., and Rosen, B.P., The *zntA* gene of *Escherichia coli* encodes a Zn(II)-translocating P-type ATPase, *Proc. Natl. Acad. Sci. U S A.*, 24, 14326, 1997.

172. Mergeay, M. et al., *Alcaligenes eutrophus* CH34 is a facultative chemolithotroph with plasmid-bound resistance to heavy metals, *J. Bacteriol.*, 162, 328, 1985.

173. Nies, D. et al., Cloning of plasmid genes encoding resistance to cadmium, zinc, and cobalt in *Alcaligenes eutrophus* CH34, *J. Bacteriol.*, 169, 4865, 1987.

174. Nies, A., Nies, D.H., and Silver, S., Nucleotide sequence and expression of a plasmid-encoded chromate resistance determinant from *Alcaligenes eutrophus*, *J. Biol. Chem.*, 265, 5648, 1990.

175. Rensing, C., Pribyl, T., and Nies, D.H., New Functions for the three subunits of the CzcCBA cation-proton-antiporter, *J. Bacteriol.*, 22, 6871, 1997.

176. Nikaido, H., Multidrug efflux pumps of Gram-negative bacteria, *J. Bacteriol.*, 178, 5853, 1996.

177. Paulsen, I.T., Brown, M.H., and Skurray, R.D., Proton-dependend multidrug efflux systems, *Microbiol. Rev.*, 60, 575, 1996.

178. Saier, M.H. et al., The major facilitator superfamily, *J. Mol. Microbiol. Biotechnol.*, 1, 257, 1999.

179. Saier, M.H. et al., Two novel families of bacterial membrane proteins concerned with nodulation, cell division and transport, *Mol. Microbiol.*, 11, 841, 1994.

180. Saier, M.H.J., Computer-aided analyses of transport protein sequences: gleaning evidence concerning function, structure, biogenesis, and evolution, *Microbiol. Rev.*, 58, 71, 1994.

182. Paulsen, I.T. and Saier, M.H., Jr., A novel family of ubiquitous heavy metal ion transport proteins, *J. Membr. Biol.*, 156, 99, 1997.

183. Kamizomo, A. et al., Identification of a gene conferring resistance to zinc and cadmium ions in the yeast *Saccharomyces cerevisiae*, *Mol. Gen. Genet.*, 219, 161, 1989.

184. Xiong, A.M. and Jayaswal, R.K., Molecular characterization of a chromosomal determinant conferring resistance to zinc and cobalt ions in *Staphylococcus aureus*, *J. Bacteriol.*, 180, 4024, 1998.

185. Kobayashi, S. et al., Correlation of the *OSR/ZRC1* gene product and the intracellular glutathione levels in *Saccharomyces cerevisiae*, *Biotechnol. Appl. Biochem.*, 23, 3, 1996.

186. Conklin, D.S., Culbertson, M.R., and C.K., Interactions between gene products involved in divalent cation transport in *Sacchromyces cerevisiae*, *Mol. Gen. Genet.*, 244, 303, 1994.

187. Conklin, D.S. et al., COT1, a gene involved in cobalt accumulation in *Saccharomyces cerevisiae*, *Mol. Cell. Biol.*, 12, 3678, 1992.

188. Huang, L.P. and Gitschier, J., A novel gene involved in zinc transport is deficient in the lethal milk mouse, *Nature Genet.*, 17, 292, 1997.

189. Palmiter, R.D., Cole, T.B., and Findley, S.D., ZnT-2, a mammalian protein that confers resistance to zinc by facilitating vesicular sequestration, *EMBO J.*, 15, 1784, 1996.

190. Palmiter, R.D. et al., ZnT3, a putative transporter of zinc into synaptic vesicles, *Proc. Natl. Acad. Sci. USA.*, 93, 14934, 1996.

191. Palmiter, R.D. and Findley, S.D., Cloning and functional characterization of a mammalian zinc transporter that confers resistance to zinc, *EMBO J.*, 14, 639, 1995.

192. Rodriguez Montelongo, L. et al., Membrane-associated redox cycling of copper mediates hydroperoxide toxicity in Escherichia coli, *Biochim. Biophys. Acta.*, 1144, 77, 1993.

193. Suwalsky, M. et al., Cu2+ ions interact with cell membranes, *J. Inorg. Biochem.*, 70, 233, 1998.

194. Brown, N.L., Rouch, D.A., and Lee, B.T., Copper resistance determinants in bacteria, *Plasmid*, 27, 41, 1992.

195. Cervantes, C. and Gutierrez Corona, F., Copper resistance mechanisms in bacteria and fungi, *FEMS Microbiol. Rev.*, 14, 121, 1994.

196. Cooksey, D.A., Copper uptake and resistance in bacteria, *Mol. Microbiol.*, 7, 1, 1993.

197. Cooksey, D.A., Molecular mechanisms of copper resistance and accumulation in bacteria, *FEMS Microbiol. Rev.*, 14, 381, 1994.

198. Gordon, A.S., Howell, L.D., and Harwood, V., Response of diverse heterotrophic bacteria to elevated copper concentrations, *Can. J. Microbiol.*, 40, 408, 1994.

199. Trevors, J.T., Copper resistance in bacteria, *Microbiol. Sci.*, 4, 29, 1987.

200. Rouch, D. et al., Inducible plasmid-mediated copper resistance in *Escherichia coli*, *J. Gen. Microbiol.*, 1985, 131, 1985.

201. Brown, N.L. et al., Molecular genetics and transport analysis of the copper-resistance determinant (pco) from *Escherichia coli* plasmid pRJ1004, *Mol. Microbiol.*, 17, 1153, 1995.

202. Fong, S.T., Camakaris, J., and Lee, B.T.O., Molecular genetics of a chromosomal locus involved in copper tolerance in *Escherichia coli* K12, *Mol. Microbiol.*, 15, 1127, 1995.

203. Gupta, S.D. et al., Identification of *cutC* and *cutF (nlpF)* genes involved in copper tolerance in *Escherichia coli*, *J. Bacteriol.*, 177, 4207, 1995.

204. Gupta, S.D., Wu, H.C., and Rick, P.D., *Salmonella typhomurium* genetic locus which confers copper tolerance in copper sensitive mutants of *Escherichia coli*, *J. Bacteriol.*, 179, 4977, 1997.

205. Rogers, S.D. et al., Cloning and characterization of cutE, a gene involved in copper transport in *Escherichia coli*, *J. Bacteriol.*, 173, 6742, 1991.

206. Rouch, D.A. and Brown, N.L., Copper-inducible transcriptional regulation at two promoters in the *Escherichia coli* copper resistance determinant pco, *Microbiology*, 143, 1191, 1997.

207. Lee, Y.A. et al., Molecular cloning, chromosomal mapping, and sequence analysis of copper resistance genes from *Xanthomonas campestris pv. juglandis*: homology with small blue copper proteins and multicopper oxidase, *J. Bacteriol.*, 176, 173, 1994.

208. Lin, C.Z. and Olson, B.H., Occurrence of *cop*-like copper resistance genes among bacteria isolated from water distribution system, *Can. J. Microbiol.*, 41, 642, 1995.

209. Vargas, E. et al., Chromosome-encoded inducible copper resistance in *Pseudomonas* strains, *Antonie van Leeuwenhoek*, 68, 225, 1995.

210. Mills, S.D., Jasalavich, C.A., and Cooksey, D.A., A two-component regulatory system required for copper-inducible expression of the copper resistance operon of *Pseudomonas syringae*, *J. Bacteriol.*, 175, 1656, 1993.

211. Mills, S.D., Lim, C., and Cooksey, D.A., Purification and characterization of CopR, a transcriptional activator protein that binds to a conserved domain (*cop* box) in copper-inducible promoters of *Pseudomonas syringae*, *Mol. Gen. Genet.*, 244, 341, 1994.

212. Odermatt, A. et al., Primary structure of two P-type ATPases involved in copper homeostasis in *Enterococcus hirae*, *J. Biol. Chem.*, 268, 12775, 1993.

213. Odermatt, A., Krapf, R., and Solioz, M., Induction of the putative copper ATPases, CopA and CopB, of *Enterococcus hirae* by Ag+ and Cu2+, and Ag+ extrusion by CopB, *Biochem. Biophys. Res. Commun.*, 202, 44, 1994.

214. Odermatt, A. and Solioz, M., Two trans-acting metalloregulatory proteins controlling expression of the copper-ATPases of *Enterococcus hirae*, *J. Biol. Chem.*, 270, 4349, 1995.

215. Strausak, D. and Solioz, M., CopY is a copper inducible repressor of the *Enterococcus hirae* copper ATPases, *J. Biol. Chem.*, 272, 8932, 1997.

216. Ge, Z., Hiratsuka, K., and Taylor, D.E., Nucleotide sequence and mutational analysis indicate that two *Helicobacter pylori* genes encode a P-type ATPase and a cation-binding protein associated with copper transport, *Mol. Microbiol.*, 15, 97, 1995.

217. Francis, M.S. and Thomas, C.J., The *Listeria monocytogenes* gene *ctpA* encodes a putative P-type ATPase involved in copper transport, *Mol. Gen. Genet.*, 253, 484, 1997.

218. Kanamaru, K., Kashiwagi, S., and Mizuno, T., A copper-transporting P-type ATPase found in the thylakoid membrane of the cyanobacterium *Synechococcus* species PCC7942, *Mol. Microbiol.*, 13, 369, 1995.

219. Phung, L.T., Ajlani, G., and Haselkorn, R., P-type ATPase from the cyanobacterium *Synechococcus* 7942 related to the human Menkes and Wilson disease gene product, *Proc. Natl. Acad. Sci. USA.*, 91, 9651, 1994.

220. Georgatsou, E. et al., The yeast Fre1p/Fre2p cupric reductases facilitate copper uptake and are regulated by the copper modulated Mac1p activator, *J. Biol. Chem.*, 272, 13786, 1997.

221. Hassett, R. and Cosman, D.J., Evidence for Cu(II) reduction as a component of copper uptake by *Saccharomyces cerevisiae*, *J. Biol. Chem.*, 270, 128, 1995.

222. Dancis, A. et al., The *Saccharomyces cerevisiae* copper transport protein (CTR1p) — biochemical characterization, regulation by copper, and physiologic role in copper uptake, *J. Biol. Chem.*, 269, 25660, 1994.

223. Dancis, A. et al., Molecular characterization of a copper transport protein in *S. cerevisiae* — an unexpected role for copper in iron transport, *Cell*, 76, 393, 1994.

224. Zhou, B. and Gitschier, J., hCTR1: a human gene for copper uptake identified by complementation in yeast, *Proc. Natl. Acad. Sci. USA*, 94, 7481, 1997.

225. Presta, A. and Stillman, M. J., Incorporation of copper into the yeast *Saccharomyces cerevisiae*: identification of Cu(I) metallothionein in intact yeast cells, *J. Biol. Chem.*, 66, 231, 1997.

226. Amaravadi, R. et al., A role for the *Saccharomyces cerevisiae ATX1* gene in copper trafficking and iron transport, *Hum. Genet.*, 99, 329, 1997.

227. Beers, J., Glerum, D.M., and Tzagoloff, A., Purification, characterization, and localization of yeast Cox17p, a mitochondrial copper shuttle, *J. Biol. Chem.*, 272, 33191, 1997.

228. Glerum, D.M., Shtanko, A., and Tzagoloff, A. Characterization of *COX17*, a yeast gene involved in copper metabolism and assembly of cytochrome oxidase, *J. Biol. Chem.*, 271, 14504, 1996.

229. Lin, S.J. and Culotta, V.C., The *ATX1* gene of *Saccharomyces cerevisiae* encodes a small metal homeostasis factor that protects cells against reactive oxygen toxicity, *Proc. Natl. Acad. Sci. USA*, 92, 3784, 1995.

230. Yuan, D.S., Dancis, A., and Klausner, R.D., Restriction of copper export in *Saccharomyces cerevisiae* to a late Golgi or post-Golgi compartment in the secretory pathway, *J.Biol.Chem.*, 272, 25787, 1997.

231. Graden, J.A. and Winge, D.R., Copper mediated repression of the activation domain in the yeast MAC1p transcription factor, *Proc. Natl. Acad. Sci. USA*, 94, 5550, 1997.

232. Hottiger, T. et al., Physiological characterization of the yeast metallothionein (CUP1) promoter, and consequences of overexpressing its transcriptional activator, ACE1, *Yeast*, 10, 283, 1994.

233. Labbe, S., Zhu, Z.W., and Thiele, D.J., Copper specific transcriptional repression of yeast genes encoding critical components in the copper transport pathway, *J. Biol. Chem.*, 272, 15951, 1997.

234. Zhu, Z.W. et al., Copper differentially regulates the activity and degradation of yeast MAC1 transcription factor, *J. Biol. Chem.*, 273, 1277, 1998.

235. Askwith, C. and Kaplan, J., Iron and copper transport in yeast and its relevance to human disease, *Trends Biochem. Sci.*, 23, 135, 1998.

236. Dierick, H.A. et al., Immunocytochemical localization of the Menkes copper transport protein (ATP7A) to the trans-Golgi network, *Hum. Mol. Genet.*, 6, 409, 1997.

237. Francis, M.J. et al., A Golgi localization signal identified in the Menkes recombinant protein, *Hum. Mol. Genet.*, 7, 1245, 1998.

238. LaFontaine, S. et al., Functional analysis and intracellular localization of the human Menkes protein (MNK) stably expressed from a cDNA construct in Chinese hamster ovary cells (CHO-K1), *Hum. Mol. Genet.*, 7, 1293, 1998.

239. Vulpe, C.D. and Packman, S., Cellular copper transport, *Annu. Rev. Nutr.*, 15, 293, 1995.

240. Petris, M.J. et al., Ligand-regulated transport of the Menkes copper P-type ATPase efflux pump from the Golgi apparatus to the plasma membrane: a novel mechanism of regulated trafficking, *EMBO J.*, 15, 6084, 1996.

241. Koizumi, M. et al., A marked increase in free copper levels in the plasma and liver of LEC rats: an animal model for Wilson disease and liver cancer, *Free Radical Res.*, 28, 441, 1998.

242. Schilsky, M.L. et al., Spontaneous cholangiofibrosis in Long–Evans cinnamon rats: a rodent model for Wilson's disease, *Lab. Anim. Sci.*, 48, 156, 1998.

243. Yoshimizu, T. et al., Essential Cys-Pro-Cys motif of *Caenorhabditis elegans* copper transport ATPase, *Biosci. Biotechnol. Biochem.*, 62, 1258, 1998.

244. Hmiel, S.P. et al., Magnesium transport in *Salmonella typhimurium*: genetic characterization and cloning of three magnesium transport loci, *J. Bacteriol.*, 171, 4742, 1989.

245. Lohmeyer, M. and Friedrich, C.G., Nickel transport in *Alcaligenes eutrophus*, *Arch. Microbiol.*, 149, 81, 1987.

246. Eberz, G., Eitinger, T., and Friedrich, B., Genetic determinants of a nickel-specific transport system are part of the plasmid-encoded hydrogenase gene cluster in *Alcaligenes eutrophus*, *J. Bacteriol.*, 171,1340, 1989.

247. Eitinger, T. and Friedrich, B., Cloning, nucleotide sequence, and heterologous expression of a high-affinity nickel transport gene from *Alcaligenes eutrophus*, *J. Biol. Chem.*, 266, 3222, 1991.

248. Eitinger, T. et al., A Ni2+-binding motif is the basis of high affinity transport of the *Alcaligenes eutrophus* nickel permease, *J. Biol. Chem.*, 272, 17139, 1997.

249. Fulkerson, J.F., Garner, R.M., and Mobley, H.L., Conserved residues and motifs in the NixA protein of *Helicobacter pylori* are critical for the high affinity transport of nickel ions, *J. Biol. Chem.*, 273, 235, 1998.

250. Komeda, H., Kobayashi, M., and Shimizu, S., A novel gene cluster including the *Rhodococcus rhodochorus* J1 *nh1BA* genes encoding a low molecular mass nitrile hydratase (L-Nhase) induced by its reaction product, *J. Biol. Chem.*, 271, 15796, 1996.

251. Komeda, H., Kobayashi, M., and Shimizu, S., A novel transporter involved in cobalt uptake, *Proc. Natl. Acad. Sci. USA*, 94, 36, 1997.

252. de Pina, K. et al., Purification and characterization of the periplasmic nickel-binding protein NikA of *Escherichia coli* K12, *Eur. J. Biochem.*, 227, 857, 1995.

253. Navarro, C., Wu, L.F., and Mandrand Berthelot, M.A., The *nik* operon of *Escherichia coli* encodes a periplasmic-binding protein-dependent transport system for nickel, *Mol. Microbiol.*, 9, 1181, 1993.

254. Wu, L.F., Navarro C., and Mandrand Berthelot, M.A., The hydC region contains a multi-cistronic operon (nik) involved in nickel transport in *Escherichia coli*, *Gene*, 107, 37, 1991.

255. Joho, M. et al., Nickel resistance mechanisms in yeasts and other fungi, *J. Ind. Microbiol.*, 14, 164, 1995.

256. Joho, M. et al., The subcellular distribution of nickel in Ni-sensitive and Ni-resistant strains of *Saccharomyces cerevisiae*, *Microbios.*, 71, 149, 1992.

257. Joho, M. et al., Nickel resistance in yeast and other fungi, *J. Ind. Microbiol.*, 14, 64, 1995.

258. Murata, K. et al., Phenotypic character of the methylglycoxal resistance gene in *Saccharomyces cerevisiae* expression in *Escherichia coli* and application to breeding wild-type yeast strains, *Appl. Environ. Microbiol.*, 50, 1200, 1985.

259. Gonzalez, H. and Jensen, T.E., Nickel sequestering by polyphosphate bodies in *Staphylococcus aureus*. *Microbios.*, 93, 179, 1998.

260. Kramer, U. et al., Free hisitidine as a metal chelator in plants that accumulate nickel, *Nature*, 379, 635, 1996.

261. Nishimura, K., Igarashi, K., and Kakinuma, Y., Proton gradient-driven nickel uptake by vacuolar membrane vesicles of *Saccharomyces cerevisiae*, *J. Bacteriol.*, 180, 1962, 1998.

262. Ross, I.S., Reduced uptake of nickel by a nickel resistant strain of *Candida utilis*, *Microbios*, 83, 261, 1995.

263. Liesegang, H. et al., Characterization of the inducible nickel and cobalt resistance determinant *cnr* from pMOL28 of *Alcaligenes eutrophus* CH34, *J. Bacteriol.*, 175, 767, 1993.

264. Bopp, L.H. and Ehrlich H.L., Chromate resistance and reduction in *Pseudomonas fluorescens* strain LB300, *Arch. Microbiol.*, 150, 426, 1988.

265. Campos, J., Martinezpacheco, M., and Cervantes, C., Hexavalent chromium reduction by a chromate-resistant *Bacillus* sp. strain, *Antonie van Leeuwenhoek*, 68, 203, 1995.

266. Cervantes, C. et al., Cloning, nucleotide sequence, and expression of the chromate resistance determinant of *Pseudomonas aeruginosa* plasmid pUM505, *J. Bacteriol.*, 172, 287, 1990.

267. Cervantes, C. and Silver, S., Plasmid chromate resistance and chromium reduction, *Plasmid*, 27, 65, 1992.

268. Deleo, P.C. and Ehrlich, H.L., Reduction of hexavalent chromium by *Pseudomonas fluorescens* Lb300 in batch and continuous cultures, *Appl. Microbiol. Biotechnol.*, 40, 104, 1994.

269. Gopalan, R. and Veeramani, H., Studies on microbial chromate reduction by *Pseudomonas* sp in aerobic continuous suspended growth cultures. *Biotechnol. Bioengin.*, 43, 471, 1994.

270. Ishibashi, Y., Cervantes, C., and Silver, S., Chromium reduction in *Pseudomonas putida*, *Appl. Environ. Microbiol.*, 56, 2268, 1990.

271. Llovera, S. et al., Chromate reduction by resting cells of *Agrobacterium radiobacter*, *Appl. Environ. Microbiol.*, 59, 3516, 1993.

272. Ohtake, H. et al., Chromate-resistance in a chromate-reducing strain of *Enterobacter cloacae*, *FEMS Microbiol. Lett.*, 55, 85, 1990.

273. Peitzsch, N., Eberz, G., and Nies, D.H., *Alcaligenes eutrophus* as a bacterial chromate sensor, *Appl. Environ. Microbiol.*, 64, 453, 1998.

274. Schmieman, E.A. et al., Bacterial reduction of chromium, *Appl. Biochem. Biotech.*, 63, 855, 1997.

275. Schmieman, E.A. et al., Comparative kinetics of bacterial reduction of chromium, *J. Environ. Eng. Asce.*, 124, 449, 1998.

276. Shen, H. and Wang, Y.T., Modeling hexavalent chromium reduction in *Escherichia coli* 33456, *Biotechnol. Bioeng.*, 43, 293, 1994.

277. Turick, C.E. and Apel, W.A., A bioprocessing strategy that allows for the selection of Cr(VI)-reducing bacteria from soils, *J. Ind. Microbiol. Biotechnol.*, 18, 247, 1997.

278. Turick, C.E., Apel, W.A., and Carmiol, N.S., Isolation of hexavalent chromium-reducing anaerobes from hexavalent-chromium-contaminated and noncontaminated environments, *Appl. Microbiol. Biotechnol.*, 44, 683, 1996.

279. Mahanty, S.K. et al., Vanadate-resistant mutants of *Candida albicans* show alterations in phosphate uptake, *FEMS Microbiol. Lett.*, 68, 163, 1991.

280. Rehder, D., Structure and function of vanadium compounds in living organisms, *Biometals*, 5, 3, 1992.

281. Chatterjee, R., Ludden, P.W., and Shah, V.K., Characterization of VNFG, the delta subunit of the vnf-encoded apodinitrogenase from *Azotobacter vinelandii*: implications for its role in the formation of functional dinitrogenase 2, *J. Biol. Chem.*, 272, 3758, 1997.

282. Dilworth, M.J., Eady, R.R., and Eldridge, M.E., The vanadium nitrogenase of *Azotobacter chroococcum*: reduction of acetylene and ethylene to ethane, *Biochem. J.*, 249, 745, 1988.

283. Eady, R.R., The vanadium-containing nitrogenase of *Azotobacter*, *Biofactors*, 1, 111, 1988.

284. Eady, R.R., Vanadium nitrogenases of Azotobacter, *Met. Ions Biol. Syst.*, 31, 363, 1995.

285. Eady, R.R. et al., The vanadium nitrogenase of *Azotobacter chroococcum*: purification and properties of the Fe protein, *Biochem. J.*, 256, 189, 1988.

286. Eady, R.R. et al., The vanadium nitrogenase of *Azotobacter chroococcum*: purification and properties of the VFe protein, *Biochem. J.*, 244, 197, 1987.

287. Joerger, R.D. and Bishop, P.E., Bacterial alternative nitrogen fixation systems, *Crit. Rev. Microbiol.*, 16, 1, 1988.

288. Pau, R.N., Nitrogenases without molybdenum, *Trends. Biochem. Sci.*, 14, 183, 1989.

289. Raina, R. et al., Characterization of the gene for the Fe-protein of the vanadium dependent alternative nitrogenase of Azotobacter vinelandii and construction of a Tn5 mutant, *Mol. Gen. Genet.*, 214, 121, 1988.

290. Robson, R.L. Woodley, P.R., Pau, R.N., Eady, R.R., Structural genes for the vanadium nitrogenase from *Azotobacter chroococcum*, *EMBO. J.*, 8, 1217, 1989.

291. Smith, B.E. et al., The vanadium-iron protein of vanadium nitrogenase from *Azotobacter chroococcum* contains an iron-vanadium cofactor, *Biochem. J.*, 250, 299, 1988.

292. Thiel, T., Isolation and characterization of the VnfEN genes of the cyanobacterium *Anabaena variabilis*, *J. Bacteriol.*, 178, 4493, 1996.

293. Wolfinger, E.D. and Bishop, P.E., Nucleotide sequence and mutational analysis of the vnfENX region of *Azotobacter vinelandii*, *J. Bacteriol.*, 173, 7565, 1991.

294. Nielsen, F.H., Nutritional requirement for boron, silicon, vanadium, nickel, and arsenic — current knowledge and speculation, *FASEB J.*, 5, 26, 1991.

295. Lyalikova, N.N. and Yurkova, N.A., Role of microorganisms in vanadium concentration and dispersion, *Geomicrobiol. J.*, 10, 15, 1992.

296. Yurkova, N.A. and Lyalikova, N.N., New vanadate-reducing facultative chemolithotrophic bacteria, *Mikrobiologiya*, 59, 968, 1990.

297. Grogan, D.W., Phenotypic characterization of the archaebacterial genus *Sulfolobus*: comparison of five wild-type strains, *J. Bacteriol.*, 171, 6710, 1989.

298. Nakamura, T. et al., Cloning and characterization of the *Saccharomyces cerevisiae* SVS1 gene which encodes a serine- and threonine-rich protein required for vanadate resistance, *Gene*, 165, 25, 1995.

299. Tatusov, R.L., Koonin, E.V., and Lipman, D.J., A genomic perspective on protein families, *Science*, 278, 631, 1997.

300. Gaasterland, T. and Ragan, M.A., Microbial genescapes: phyletic and functional patterns of ORF distribution among prokaryotes, *Microb. Comp. Genomics*, 3, 199, 1998.

301. Pellegrini, M. et al., Assigning protein functions by comparative genome analysis: protein phylogenetic profiles, *Proc. Natl. Acad. Sci. USA*, 96, 4285, 1999.

302. Gribskov, M., McLachlan, A.D., and Eisenberg, D., Profile analysis: detection of distantly related proteins, *Proc. Natl. Acad. Sci. USA*, 84, 4355, 1987.

303. Gribskov, M. et al., Profile scanning for three-dimensional structural patterns in protein sequences, *Comput. Appl. Biosci.*, 4, 61, 1988.

304. Luthy, R., Xenarios, I., and Bucher, P., Improving the sensitivity of the sequence profile method, *Protein sci.*, 3, 139, 1994.
305. Hofmann, K. et al., The PROSITE database, its status in 1999, *Nucleic Acid Res.*, 27, 215, 1999.
306. Paulsen, I.T. et al., A family of Gram-negative bacterial outer membrane factors that function in the export of proteins, carbohydrates, drugs and heavy metals form Gram-negative bacteria, *FEMS Microbiol. Lett.*, 156, 1, 1997.
307. Blattner, F.R. et al., The complete genome sequence of *Escherichia coli* K-12 [comment], *Science*, 277, 1453, 1997.
308. Ching, M.H. et al., Substrate-induced dimerization of the ArsA protein, the catalytic component of an anion-translocating ATPase, *J. Biol. Chem.*, 266, 2327, 1991.
309. Dou, D. et al., Construction of a chimeric ArsA-ArsB protein for overexpression of the oxyanion-translocating ATPase, *J. Biol. Chem.*, 267, 25768, 1992.
310. Fagan, M.J. and Saier, M.H.J., P-typ ATPases of eukaryotes and bacteria: sequence comparisons and construction of phylogenetic trees, *J. Mol. Evol.*, 38, 57, 1994.
311. Fath, M.J. and Kolter, R., ABC-transporters: the bacterial exporters, *Microbiol. Rev.* 57, 995, 1993.
312. Inoue, Y., Kobayashi, S., and Kimura, A., Cloning and phenotypic characterization of a gene enhancing resistance against oxidative stress in *Saccharomyces cerevisiae*, *J. Ferment. Bioeng.*, 75, 327, 1993.
313. Karkaria, C.E. and Rosen, B.P., Trinitrophenyl-ATP binding to the ArsA protein: the catalytic subunit of an anion pump, *Arch. Biochem. Biophys.*, 288, 107, 1991.
314. Karkaria, C.E., Steiner, R.F., and Rosen, B.P., Ligand interactions in the ArsA protein, the catalytic component of an anion-translocating adenosine triphosphatase, *Biochemistry*, 30, 2625, 1991.
315. Liu, J. et al., Identification of an essential cysteinyl residue in the ArsC arsenate reductase of plasmid R773, *Biochemistry*, 34, 13472, 1995.
316. Lutsenko, S. and Cooper, M.J., Localization of the Wilson's disease protein product to mitochondria, *Proc. Natl. Acad. Sci. USA*, 95, 6004, 1998.
317. Ohta, Y., Shiraishi, N., and Nishikimi, M., Occurrence of two missense mutations in Cu-ATpase of the macular mouse, a Menkes disease model, *Biochem. Mol. Biol. Int.*, 43, 913, 1997.
318. Payne, A.S., Kelly, E.J., and Gitlin, J.D., Functional expression of the Wilson disease protein reveals mislocalization and impaired copper-dependent trafficking of the common H1069Q mutation, *Proc. Natl. Acad. Sci. USA*, 95, 10854, 1998.
319. Reddy, M.C.M. and Harris, E.D., Multiple transcripts coding for the Menkes gene: evidence for alternative splicing of Menkes mRNA, *Biochem. J.*, 334, 71, 1998.
320. Rensing, C., Mitra, B., and Rosen, B.P., Insertional inactivation of dsbA produces sensitivity to cadmium and zinc in *Escherichia coli*, *J. Bacteriol.*, 179, 2769, 1997.
321. Smith, D.L., Tao, T., and Maguire, M.E., Membrane topology of a P-type ATPase. The MgtB magnesium transport protein of *Salmonella typhimurium*, *J. Biol. Chem.*, 268, 22469, 1993.
322. Tam, R. and Saier, M.H.J., Structural, functional, and evolutionary relationships among extracellular solute-binding receptors of bacteria, *Microbiol. Rev.*, 57, 320, 1993.
323. Wolfram, L., Friedrich, B., and Eitinger, T., The *Alcaligenes eutrophus* protein HoxN mediates nickel transport in *Escherichia coli*, *J. Bacteriol.*, 177, 1840, 1995.
324. Silver, S. and Walderhaug, M., *Toxicology of Metals. Biochemical Aspects*, Berlin: Springer–Verlag, 1995, 435.
325. Nakahara, H. et al., Hypersensitivity to Hg2+ and hyperbinding activity associated with cloned fragments of the mercurial resistance operon of plasmid NR1, *J. Bacteriol.*, 140,161, 1979.
326. Cha, J.S. and Cooksey, D.A., Copper hypersensitivity and uptake in *Pseudomonas syringae* containing cloned components of the copper resistance operon, *Appl. Environ. Microbiol.*, 59, 1671, 1993.
327. Cai, J. and DuBow, M.S., Expression of the *Escherichia coli* chromosomal ars operon, *Can. J. Microbiol.*, 42, 662, 1996.
328. Kaur, P. and Rosen, B.P., Complementation between nucleotide-binding domains in an anion-translocating ATPase, *J. Bacteriol.*, 175, 351, 1993.
329. Gottesman, M.M. and Pastan, I., The multidrug transporter, a double-edged sword, *J. Biol. Chem.*, 263, 12163, 1988.
330. Haimeur, A. and Ouellette, M., Gene amplification in *Leishmania tarentolae* selected resistance to sodium stibogluconate, *Antimicrobial Agents Chemother.*, 42,1689, 1998.

331. Carter, N.S. and Fairlamb, A.H., Arsenical-resistant trypanosomes lack an unusual adenosine transporter, *Nature*, 361, 173, 1993.
332. Callahan, H.L. and Beverley S.M., Heavy metal resistance: a new role for P-glycoproteins in *Leishmania*, *J. Biol. Chem.*, 266, 18427, 1991.
333. Grondin, K. et al., Co-amplification of the gamma-glutamylcysteine synthetase gene *gsh1* and of the ABC transporter gene *pgpA* in arsenite-resistant *Leishmania tarentolae*, *EMBO J.*, 16, 3057, 1997.
334. Grondin, K., Papadopoulou, B., and Ouellette, M., Homologous recombination between direct repeat sequences yields P-glycoprotein containing circular amplicons in arsenite resistant *Leishmania*, *Nucl. Acid Res.*, 21, 1895, 1993.
335. Bairoch, A., A possible mechanism for metal-ion induced DNA-protein dissociation in a family of prokaryotic transcriptional regulators, *Nucl. Acid Res.*, 21, 2515, 1993.
336. Rosenstein, R., Nikoleit, K., and Gotz, F., Binding of ArsR, the repressor of the *Staphylococcus xylosus* (pSX267) arsenic resistance operon to a sequence with dyad symmetry within the *ars* promoter, *Mol. Gen. Genet.*, 242, 566, 1994.
337. Nies, D. H. et al., Expression and nucleotide sequence of a plasmid-determined divalent cation efflux system from *Alcaligenes eutrophus*, *Proc. Natl. Acad. Sci. USA*, 86, 7351, 1989.
338. Anton, A. et al., CzcD is a heavy metal ion transporter involved in regulation of heavy metal resistance in Ralstonia sp. strain CH34, *J. Bacteriol.*, 181, 6876, 1999.
339. van der Lelie, D. et al., Two component regulatory system involved in transcriptional control of heavy metal homoeostasis in *Alcaligenes eutrophus, Mol. Microbiol.*, 23, 493, 1997.
340. Nies, D.H, Efflux-mediated heavy metal resistance in prokaryotes, *FEMS Microbiol. Rev.*, 27, 313, 2003.
341. Bontidean, I. et al., Detection of heavy metal ions at femtomolar levels using protein-based biosensors, *Anal. Chem.*, 70, 4162, 1998.
342. Scott, D.L. et al., Genetically engineered bacteria: electrochemical sensing systems for antimonite and arsenite, *Anal. Chem.*, 69, 16, 1997.
343. Selifonova, O., Burlage, R., and Barkay, T., Bioluminescent sensors for detection of bioavailable Hg (II) in the environment, *Appl. Environ. Microbiol.*, 59, 3083, 1993.
344. Roy, P. et al., Microbe-based biosensor for detection of environmental pollutant arsenic in soil and water (unpublished 2005).

12 Arbuscular Mycorrhizal Fungi and Heavy Metals: Tolerance Mechanisms and Potential Use in Bioremediation

Ma. del Carmen Angeles González Chávez, Jaco Vangronsveld, Jan Colpaert, and Corinne Leyval

CONTENTS

12.1 INTRODUCTION

The arbuscular mycorrhizal (AM) symbiosis is commonly found in more than 80% of plant species. This association results in multiple benefits to the plants: improvement of plant growth; protection against root pathogens; adaptation to survive in extreme soil conditions; etc. Only recently have a role in soil conservation and a contribution to determine plant community structure in natural ecosystems been recognized for this symbiosis [1,2]. In contrast, its role in altered soil conditions, such as contaminated soils, is not completely understood yet.

Occurrence of arbuscular mycorrhizal fungi (AMF) is common in contaminated soils; however, it has been reported that the presence of high concentrations of potentially toxic elements has a negative effect on the diversity of these fungi [3,4]. Moreover, the present literature contains contradictory results concerning effects of PTEs on mycorrhizal plants.

Some PTEs are mainly accumulated in root systems and some authors have suggested that AMF might be involved in this accumulation; however, the mechanisms of retention and allocation of PTEs have been largely ignored. In ericoid and ectomycorrhizal fungi, the binding of PTEs to the external mycelium has been proposed as a tolerance mechanism [5–8] that reduces metal translocation to the shoots, but the validity of this hypothesis had not been proved for AMF. Novel information stressing AMF's possible importance in contaminated soils is available. A better understanding of the processes involved in dealing with high concentrations of PTEs on mycorrhizal plants may have strong implications for the use of these fungi in the bioremediation of contaminated soils.

12.2 AM SYMBIOSIS

The term "mycorrhiza" implies basically the association of fungi with roots. Indeed, mycorrhiza, not roots are the chief organs of nutrient uptake by the majority of land plants [9]. The different kinds of mycorrhizae have been described on the basis of their fungal associates into those aseptate endophytes in the class Glomeromycetes [10] and those formed by septate fungi in Ascomycetes and Basidiomycetes. This chapter will consider the association formed by members of the Glomeromycetes, which have been referred to as arbuscular mycorrhiza (AM), and were formally known as vesicular–arbuscular mycorrhiza. The plant hosts of this type of mycorrhiza may belong to Bryophyta, Pteridophyta, Gymnospermae, and the majority of families in the Angiospermae [9].

Arbuscular mycorrhizae are the most common type of mutualistic symbioses with plants. The association is formed between the roots of an enormously wide variety of host plants, which have true roots, and aseptate, obligate symbiotic fungi. The name of AM is derived from characteristic structures, the arbuscules, formed by the fungi within the cortical cells of the host plants (Figure 12.1a). Arbuscules are formed by all members of the phylum of the Glomeromycetes. In contrast, vesicles (Figure 12.1b), which occur intra- or intercellularly, are formed by all members of the families Glomaceae and Acaulosporaceae, but not by the members of the Gigasporaceae. Spores are considered as structures for reproduction or propagation (Figure 12.1c). An AM has three important components: the root; the fungal structures within the cells of the root; and the external mycelium (Figure 12.1d,e) functioning as a "bridge" between the root and the soil.

12.3 IMPORTANCE OF ARBUSCULAR MYCORRHIZA IN SOILS

The importance of the arbuscular mycorrhizal fungi is due to the multiple benefits to the plant hosts. One of the most recognized roles is related to plant nutrition [11]. These fungi play a central role in nutrient uptake in nutrient-deficient soils [12], especially for nutrients with low mobility, such as phosphate, Zn^{2+}, and Cu^{2+}. In nonmycorrhizal plants, nutrient absorption is confined to the outer cell layers of the root cortex and the rhizodermal cells with root hairs. In contrast, in mycorrhizal plants, higher absorption is realized by a much greater absorptive surface, created by the external mycelium of these fungi, which can extend beyond the nutrient depletion zone formed around the plant roots [13,14].

Other benefits of AMF are protection against root pathogens and greater tolerance to stress conditions such as salinity, extreme soil pH, and drought. Recently, it has been acknowledged that AMF can affect plant community structure [1,2]. Arbuscular mycorrhizae also participate in soil stability and conservation [15] through their physical and biological activities. The physical entan-

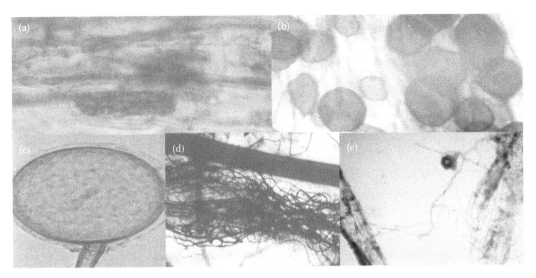

FIGURE 12.1 Fungal structures (a) arbuscules; (b) vesicles; (c) spores; and (d),(e) external mycelium.

glement of hyphae stabilizes soil aggregates. Active hyphae exudate glomalin, a glycoprotein produced abundantly by several if not all AMF, which acts as a biological cement and increases soil aggregation and resistance to soil erosion [15–17]. Additionally, Rillig et al. [18] have shown that this fungal glycoprotein significantly contributes to carbon sequestration in the soil. Nichols et al. [19] reported that glomalin, along with soil humic fraction and particulate organic matter, is a major contributor to soil organic C (up to 25%). They also concluded that this protein may be an important carbon storage pool. Glomalin production may be increased through sustainable agricultural practices.

Because the mycorrhizal condition is the normal state for the majority of the land plants, a lot of research has focused on nutritional and physiological studies in cultivated and natural ecosystems. In contrast, the importance of AM for PTE-contaminated soils has been poorly investigated. Much work has been done on MA fungi and their effect in PTE-polluted soils during the last 20 years; however, the importance of the role of mycorrhizal fungi in these soils — the fungal mechanisms dealing with PTEs — and their contribution to PTE tolerance in plants are still poorly known.

12.3.1 AMF in Contaminated Soils

Arbuscular mycorrhizae are an integral association consisting of two functioning partners: the fungus and the plant. Leyval et al. [20] stated that when interactions between fungi and PTEs are studied, two aspects should be considered: (1) the effect of PTEs on AMF; and (2) the effect of AMF on uptake and translocation of PTEs to the plant (Figure 12.2). In this relationship, fungi and plant PTE tolerance; soil and environmental factors influencing PTE availability; and ion PTEs' toxicity influence these effects.

12.3.1.1 Effect of PTEs on the Population of AMF

In recent years, more attention has been paid to understanding the effect of contaminants on soil biology. In general, PTEs seem to affect microbial species' richness, abundance, and diversity [21–23]. However, microorganisms are not uniformly affected.

Over 140 species of AMF had been described by the beginning of the 1990s [24]. In contaminated soils [25], different, identified, and, in some cases, unidentified species of AMF have been

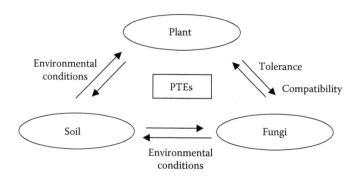

FIGURE 12.2 Interaction between potentially toxic elements (PTEs), arbuscular mycorrhizal fungi and plant. (Modified from Leyval, C. et al., *Mycorrhiza*, 7, 139, 1997.)

reported in studies investigating AMF population in contaminated sites [25]. Although identification of some AM species without validation with molecular techniques may be difficult and therefore questionable, at least one member of each of the seven genera has been found in these soils. *Glomus* is apparently the most common genus present in contaminated soils, but this should not disregard a wider distribution of species of other genera. However, predominance of one genus may be due, at least partially, to the techniques used to detect or identify these fungi (trap cultures, specific probes, etc.).

Leyval and Vandenkoornhuyse [26] showed that AMF diversity was changed by PTEs, even at concentrations below the European Community threshold values for agricultural soils. Fungal diversity was higher in a moderately contaminated soil than in a highly contaminated soil. Similar information was reported by del Val et al. [4]. These authors studied the diversity of AMF affected by the addition of different rates of sewage-amended sludge containing Pb, Cd, Cr, Cu, Ni, Hg, and Zn. The highest application rate of sludge (300 m^3 ha^{-1}year^{-1}) caused a significant decrease in the size and diversity of populations of AMF in soil, but diversity was increased in soils receiving intermediate rates of sludge application (100 m3 ha^{-1}year^{-1}).

Sambandan et al. [27] found a high level of diversity of AMF in soils (15 different species) contaminated by Zn, Cu, Pb, Ni, and Cd. Pawlowska et al. [28] reported different types of AMF in the undisturbed site of a calamine spoil (containing high concentrations of Cd, Pb, and Zn) in Poland. However, disturbance caused by surface soil removal proved to be an important factor determining frequency, distribution, and richness of AMF in the sites. They reported six different AMF on the undisturbed site and only two species on the disturbed site. *Glomus aggregatum, G. constrictum, G. pansihalos, Glomus* sp., *G. fasciculatum,* and *Entrophospora* sp. were found on the undisturbed soils; only the latter two fungi were found on the disturbed site. Apparently, *G. fasciculatum* and *Entrophosphora* sp. were less affected by disturbance.

Other authors have reported low population diversity of AMF in contaminated soils; some have observed only one to three species of AMF in the contaminated soils [29–31]. However, it is not clear whether this low diversity was really a consequence of PTE contamination or was due to other factors. PTEs tend to select species that sporulate easily in trap cultures. However, lack of sporulation does not necessarily mean absence of AMF from a soil. Other factors may be involved in situations showing low diversity of AMF in contaminated soils — for example, poor sporulation at the time of sampling; use of a restricted host range during propagation of AMF; and short propagation times to allow sporulation of those slow sporulating fungi [32]. Khan [33] reported that a soil of a Cr-contaminated site contained only spores of *Gigaspora* sp. in the rhizospheres of *Dalbergia sissoo, Acacia arabica,* and *Populus euroamericana*; however, their roots were also colonized by other mycorrhizal genera because vesicles were observed.

Analyzing Cd-rich slag and tailing piles (0.28 to 0.8 mg Cd-DTPA-extractable g^{-1}), González–Chávez [34] found species of *Scutellospora, Acaulospora, Entrophospora,* and *Glomus* colonizing abundantly soil and roots of different host plants (*Dalea* sp. *Crotalaria rotundifolia*, *Trifolium goniocarpum*, *Anagallis arvensis*, *Crusea longiflora*, *Sida rhombifolia*, and *Lopezia racemosa*). These materials contained amazing amounts of external mycelium (Up to 26.3 mg g^{-1} dry soil).

Molecular evaluation of AM fungal populations in colonized roots and soil is a promising tool for identifying naturally occurring fungi in field sites, such as these of contaminated soils. Molecular techniques such as those used by Turnau et al. [35] nested PCR in combination with taxon-specific primers for AMF fungal species (morphotypes) seem to be highly sensitive to detecting fungi in the field. These authors observed that five different AMF species were colonizing roots of *Fragaria vesca* growing on industrial waste sites located in Chrzanow (Poland). *Glomus* sp. HM-CL4 was detected with a mean frequency of 52%; *G. claroideum* and *Glomus* sp. MH-CL5 were 32 to 34% and *G. intraradices* and *Glomus mosseae* were present less frequently (5 to 7%). However, other AM species or genus may have been present and not identified because they were not amplified by the specific primers.

Cairney and Meharg [36] suggested that, to understand the functional significance of AMF in disturbed and contaminated systems, "keystone" species (species with an important role in the ecosystem) needed to be identified in the soil system in order to study the interaction processes with the contaminants. Thus, more assessments regarding fungal diversity on time and space in PTE contaminated soils are necessary. In addition, studies of the characterization of AMF species may be helpful for a better understanding of the participation of AMF in contaminated soils (see next section).

Arbuscular mycorrhiza colonization has also been observed in several metallophyte plants, which are plants commonly found on natural contaminated soils — for example: *Armeria maritima* spp. and *Viola calaminaria* [37,38] and in coexistent facultative metallophytes such as *Campanula rotundifolia* [39]. However, studies regarding the ecological role of AMF in these plants are just starting. From all these data on AM fungi in PTE polluted soils, it is rather difficult to relate the presence of AM fungi to the toxicity of PTE because physical and chemical soil characteristics and bioavailability of the PTEs were not always estimated; also, the effect of PTEs on AM fungi cannot be separated from their effect on the host plant [20].

12.3.1.2 Effect of AMF on Plant Uptake and Translocation of PTEs

It has been observed that AMF participate in the uptake of certain essential elements by plants growing on soils with nutrient deficiencies. Phosphorus uptake has been studied extensively. Apparently, the symbiosis is also important for the uptake and translocation of other essential and relatively immobile nutrients, such as Cu and Zn [9]. However, under high concentrations of PTEs in soil, there is disagreement in the literature about the participation of AMF to PTE uptake. Some authors have reported that AMF increase the uptake of PTEs [40–44]; others have reported a decreased uptake of PTEs and a reduction of toxic effects in plants with AMF [45–48].

The inconsistency of the AMF effect on PTE uptake may be a consequence of different factors involving fungi and plant species (nutritional conditions, PTE tolerance in both organisms) and the type of PTEs (degree of toxicity, levels, and speciation) in soil. In accordance with this, precautions should be taken in the selection of plants for studies of PTE uptake because differences in nutrient uptake between plants species and even varieties have been observed. Chaney [49] reported that plants growing on the same soil showed differences in uptake. For instance, spinach contained 10 times more Zn than did tall fescue, orchard grass 15-fold more Ni than did maize, and chard 5 times more Cu than did fescue. This wide difference in the uptake of PTEs among crop plants has been reported to be due to inherent root uptake (root distribution and depth) or to soil–plant interactions.

Streitwolf–Engel et al. [50] suggested investigating the effects of different AMF with more coexisting plant species. Additionally, Raju et al. [51] and Ravnskov and Jakobsen [52] mentioned that experiments demonstrating differential, and sometimes controversial, effects of AMF species have been carried out on single plant species and using AMF originating from different soils, which may never have naturally co-occurred with the plant under study.

Other factors affecting uptake of PTEs may be the source; amount of available metals; spatial distribution; and source and chemical speciation of the PTEs in the soils [49,53]. The source of PTEs may have a strong influence on AMF, and it may have an effect on the symbiosis. Overall, PTE cations are the most toxic inorganic metal forms, and their solubility generally increases with decreasing pH. As a result, mobility and bioavailability are enhanced and also toxicity to biological systems [54].

When PTEs are added as soluble salts, they generally show greater plant uptake and toxicity than when applied as sewage sludge or metal oxides [49]. Long incubation times are necessary to ensure that these are in equilibrium with the soil constituents and the less available bound forms complexed to organic matter and clays (which naturally takes place under field conditions). The types of PTEs have also been considered to be important for uptake and translocation by mycorrhizal plants because each PTE possesses specific chemical and physical characteristics and crop plants differ widely in uptake [49], translocation, and tolerance to PTEs [55].

12.4 MECHANISMS IN AMF TO TOLERATE PTEs

Unlike ericoid and ectomycorrhizal associations, in which the mechanisms of tolerance to PTE are more intensively studied, the participation of AMF in plant PTE tolerance is less clear. Consequently, the mechanisms involved have not been completely elucidated. An important reason is that a great proportion of AMF do not grow on conventional media; because these fungi are obligate symbionts, they cannot be cultured without a host root. This makes the studies more complicated. As a result, much of the information on the mechanisms involved has been inferred from plant response to PTEs and from observations on fungal structures in colonized roots. Thus, in some cases it is difficult to separate fungal vs. root participation.

Various mechanisms have been proposed for explaining the responses of plants to high PTE concentrations. Primary effects can be distinguished at the molecular, biochemical, and cellular level. Subsequently, effects at the physiological and organismal level can be observed [56].

Levitt [57] proposed three basic strategies for organisms dealing with high PTE concentrations: avoidance, detoxification, and biochemical tolerance. The avoidance mechanism involves PTE exclusion mechanisms, which operate at two levels: restriction of uptake and restriction of transport. The process of detoxification is essentially similar, but avoidance of toxicity results from subcellular PTE concentration or by binding. Biochemical tolerance reflects the presence of specialized metabolic pathways and enzymatic adaptations.

Research performed with microorganisms and plants has shown that the general PTE tolerance mechanisms are

- PTE-binding to cell wall
- Restricted influx through the plasma membrane
- Active PTE efflux
- Compartmentalization in the vacuole
- Chelation at the cell wall and cytoplasm

It is suggested that in plants [55,58] and in fungi [21], tolerance to PTEs may be a range of orchestrated responses and no universal pattern of response to PTEs exists because different species vary in their response to any particular metal. There is further variation in expression of tolerance to different PTEs within the same species or ecotypes.

12.4.1 BINDING TO CELL WALL

In related information with filamentous fungi, Gadd [21] reported that the cell wall is the first cellular site for interaction with PTEs. It is a protective barrier controlling the PTE uptake. Different components of wall structure and composition may create a variety of potential sites for sequestration at the cell wall level, including carboxyl, amine, hydroxyl, phosphate, and sulphydryl groups [59]. Polysaccharides like chitosan and chitin have been reported to participate in the binding of PTEs. However, other cell wall compounds may also participate — for example, melanins, glucans, and mannans [60–62]. Chitin and $\beta(1-3)$ glucans are the main wall components in Glomus species [9].

Binding of PTEs at the cell wall level has been reported on internal hyphae of AMF-colonized roots [63] and on external mycelium (EM) of these fungi [64] grown in contaminated soils. Using scanning and transmission electron microscopy equipped with an energy dispersion system and rhodizoniate histochemical staining, Turnau [63] was able to demonstrate localization of PTEs in colonized roots of *Euphorbia cyparissias* L. collected from Zn wastes. This author reported that 80% of the total intraradical mycelium showed a high content of PTEs. On the surface of this mycelium, amorphous or crystalloid deposits could be distinguished. High levels of S, P, and As were found in the former deposits, and Ca, Fe, Zn, and less As were found in the crystalloids. Si, Pb, Cu, and Zn were found in both deposits. PTEs were below detection limits within the walls of intraradical fungal hyphae. Additionally, small deposits containing PTEs were found on the inner layer of the spore wall of some of the morphotypes found at the site.

Gonzalez et al. [64] showed that the extraradical mycelium (EM) of AMF appears to provide an efficient surface for Cu sorption. These authors found that EM was able to adsorb Cu in a range from 3 to 14 mg g^{-1} of dry mycelium. A high Cu-sorption capacity was observed in the EM of two AMF originating from the same contaminated soil (*G. mosseae* BEG-132 and *G. claroideum* BEG-134) and also in *G. mosseae* BEG-25 coming from an agricultural soil.

Spores of AMF are structures that also are involved in the binding of PTEs. Sánchez Viveros et al. [65] reported that spores of *Glomus mosseae* BEG-132 sequestered Cu in a range from 470 to 680 µg g^{-1} of Cu (spore dry weight).

12.4.2 EXTRACELLULAR CHELATION

Gadd [21] mentioned that, at the cellular level, many organic fungal metabolites, which are efficient PTE chelators, may interact with PTE by complexation or precipitation. Some examples of these metabolites may be oxalic, citric acids [22,66], siderophores, and riboflavin [67].

Glomalin appears to be an efficient agent sequestering PTEs, not only by hyphae from the EM, but also by the colonized roots and through deposition of glomalin in the soil. In three different experiments, Gonzalez–Chavez et al. [68] tested whether glomalin sequesters different PTEs. Glomalin was extracted from polluted soils or purified from a *Gigaspora rosea* isolate or it was extracted from two AMF grown under three different levels of Cu. The results showed that:

- Glomalin sequestered Pb, Cu, and Cd in high concentrations in two polluted soils
- Glomalin produced by *Gigaspora rosea* was able to remove Cu, Zn, Co, and Ni from solution, but not Ca, K, and Mg
- The highest amount of Cu sequestered was 28 mg Cu g^{-1} glomalin

Binding and sequestration of PTEs by different fungal structures of AMF may result in the stabilization of these elements, reducing their availability and decreasing the toxicity risk to other soil microorganisms and plants growing in these sites. Table 12.1 shows the activity of glomalin and fungal structures sequestering different PTEs. When glomalin is involved in the sequestration, the importance of stabilization phenomena increases because of its copious production and recal-

TABLE 12.1
Sequestration of Potentially Toxic Elements (PTEs) by External Mycelium, Glomalin, and Spores of A MF

Fungal structure or bioproduct	PTE involved	Amount sequestered (mg/g)	Observations	Ref.
External mycelium	Cu	3–14	Excised mycelium exposed to 30 μM Cu solution for 120 min	64
Glomalin				
	Cu	1.5–4.3	Glomalin extracted from polluted soils	68
	Cd	0.2–0.08		
	Pb	0.6–1.12		
	Cu	7–28	Purified glomalin from an isolated of *Gigaspora rosea* exposed to 50 mM Cu solution. Precipitation occurred as soon as glomalin and Cu solution were mixed	
	Cu	1.13–1.6	Glomalin extracted from hyphae in active growth under three levels of Cu (0.8, 10, and 20 μM Cu)	
		0.3	Glomalin extracted from hyphae colonizing roots of *Sorghum vulgare*	
		0.4	Glomalin extracted from sand supporting external mycelium growth	
Spores	Cu	0.4–0.6	Spores of *Glomus mosseae* BEG-132	65

citrance in the soil [18]. Additionally, it is known that glomalin sequesters Cu not only by reversible reactions, such as ion exchange, but also by a strong and irreversible binding [68]. It is an important property of AMF and helps to understand their role in polluted soils.

12.4.3 CHELATION AT CYTOPLASM AND VACUOLE LEVEL

Chelation in the cytoplasm is an intracellular buffer system to control PTE toxicity through reduction of the concentration of cytotoxic-free PTE ions. In this system, metallothioneins (MT), metal-binding polypeptides, and/or MT-like proteins may be involved. In contrast to ectomycorrhizal fungi, the participation of these compounds in metal detoxification has not been confirmed in AMF [69,70]. However, evidence suggesting the presence of Cd-binding thiols in AMF as well as high concentrations of N and S in mycorrhizal roots exposed to Cd have been reported [71,72]. Galli et al. [72] studied the effect of Cu on the uptake, amount, and composition of Cu-binding peptides (Cu-BPs) of mycorrhizal maize plants inoculated with *Glomus intraradices*. They observed increased concentrations of the thiols cysteine, γ-glutamylcysteine, and glutathione up to an external Cu supply of 9 µg g^{-1}. However, the amount of thiols in Cu-BPs was not increased by mycorrhizal colonization in Cu-treated plants and no differences in Cu-uptake were detected between nonmycorrhizal and mycorrhizal plants.

Gonzalez [25] observed that the EM grown in a Cu-contaminated substrate presented higher tyrosinase activity (353 nmol mg^{-1} dry mycelium min^{-1}), in contrast to the tyrosinase activity in the EM propagated in noncontaminated substrate (153 nM mg^{-1} dry mycelium min^{-1}). The Cu substrate concentration (19 mg Cu g^{-1} substrate) induced a 2.3 times increase in tyrosinase activity in the EM of AMF tested. *Glomus caledonium* BEG-133 exhibited the highest tyrosinase activity, and *G. mosseae* BEG-132 and *G. claroideum* BEG-134 presented statistically similar activity. Comparisons using more species in different Cu levels are necessary in order to prove tyrosinase participation in Cu chelation and detoxification in AMF. This mechanism of chelation has been observed in other filamentous fungi, yeast, and ectomycorrhizal fungi [73–76].

An increase in PTE tolerance in plants has been suggested to be related to genes up-regulated in AMF-colonized roots. Rivera–Becerril et al. [77] reported that the expression of metallothionein genes increased in Cd-treated plants, but one gene (pcs) appeared to be activated specifically in Cd-treated mycorrhizal roots. In relation to these fungal PTE tolerance mechanisms, the use of visualization methods, such as transmission and scanning electron microscopy (TEM, SEM); energy-dispersive x-ray microanalysis (EDAX); and electron energy loss spectroscopy (EELS), to determine the extracellular and subcellular localization of elements has been very useful.

Gonzalez et al. [64] reported that crystal-like aggregates mainly comprising Fe were found on the mucilaginous outer hyphal wall layers of three AMF, when grown in Cu/As-contaminated soil. It was shown that, in the EM of *G. mosseae* BEG132 and *G. claroideum* BEG134, these aggregates contained significant amounts of Fe and Cu. This was not the case for *G. caledonium* BEG133; its aggregates only contained Fe.

In another comparative study, Turnau et al. [78] studied mycorrhizal roots of *Pteridium aquilinum* collected from Cd-treated experimental plots. They showed that hyphae of AMF uniformly colonizing cortical cells contained a much higher amount of PTEs than the cytoplasm of the host cells. They suggested that the ability of this plant to detoxify PTEs was due to its association with AMF. Most of the Cd was located in phosphate-rich material fungal vacuoles, which contained S, N, Al, Fe, Ti, and B. Thus, intracellular sequestration of PTE by fungal polyphosphate intracellularly into the fungus may contribute to decreasing its transfer to the plant.

Polyphosphates are produced widely by microorganisms forming an important intracellular storage of phosphate. However, these compounds may also be involved in the regulation of concentrations of PTE ion in cells [21]. Polyphosphate granules are maintaining ionic compartmentalization of PTEs and it has been observed that their biosynthesis accompanies vacuolar accumulation [79].

Gonzalez et al. [64] reported that the EM of *Glomus mosseae* BEG-132 growing in a contaminated substrate (As/Cu), contained intracellular Cu-rich bodies. In addition, traces of arsenic were also observed on the EDAX spectra of these bodies. The presence of arsenic was explained by the elevated concentrations in the soil in which the AMF were grown. Arsenate and phosphate are chemically very analogous, so this result suggests that arsenic may sequester Cu in the form of Cu-arsenate complexes in the cytoplasm of the EM of *G. mosseae* BEG132. The arsenic and Cu accumulation in *G. mosseae* BEG132 is interesting to study because detoxification of PTE within cells of several yeast species has been shown to be linked with polyphosphate granules located in the cytoplasm [79] and vacuoles [80].

12.4.4 OTHER POSSIBLE MECHANISMS IN AMF

Gonzalez et al. [81] reported that AMF isolated from a mine-spoil soil conferred enhanced tolerance to their host. AMF benefit the host on mine soils through a considerable reduction in arsenic uptake, particularly in the fraction of arsenic translocated to the shoot. The mechanisms remain unclear, but the authors suggested two alternative hypotheses:

- The fungi may have suppressed phosphate/arsenate uptake across the plasma membrane, which is the mechanism of arsenate tolerance observed in most higher plants investigated thus far [82].
- Enhanced efflux of arsenate, which is the mechanism used by an arsenate-tolerant ericoid mycorrhizal fungus (*Hymenoscyphus ericae*) and other microorganisms [83–85] may have taken place.

Elucidation of other mechanisms in AMF is required to gain knowledge about fungal participation on plant metal tolerance and their role in contaminated soils. Production of organic acids for metal chelation at the cell wall level; metallothioneins and polyphosphates for compartmentalization at the vacuolar level; and specific proteins may be potential mechanisms for PTE detoxification in AMF. Such mechanisms have already been observed in ectomycorrhizal fungi [69,70,86–88].

When considering other fungal mechanisms, indirect effects of AMF on plant nutrition; changes in root exudation; effects on rhizosphere microbial communities; soil structure; protection from stress environmental factors; etc. should be also mentioned.

12.5   CONTRIBUTION OF AMF IN PLANT TOLERANCE TO PTEs

The contribution of AMF to plant tolerance of PTE is poorly documented compared with that for ectomycorrhizal [7,89–92] and ericoid mycorrhizal plants [5,6]. Research has mainly focused on the effect of PTEs on the growth of arbuscular mycorrhizal plants compared with nonmycorrhizal plants and their effects on colonization.

Meharg and Cairney [36] suggested three possible roles of mycorrhizal fungi in plant tolerance to PTEs:

- Enhancing plant PTE tolerance. In this case, fungal tolerance is not necessarily an important characteristic to help plants establish and survive in contaminated soil conditions. This is true for ericoid mycorrhizal fungi [5,6]. Bradley et al. [6] demonstrated fungal protection in *Calluna vulgaris* by fungi isolated from contaminated and noncontaminated soils. Thus, fungal tolerance was not an important characteristic to confer tolerance to Zn and Cu in the ericoid association. It means tolerant fungi may be found in noncontaminated soils; therefore, the fact that a fungus is isolated from noncontaminated soils does not mean that it is not tolerant.
- Just satisfying their normal role in the association. In this case, plant benefits from the symbiosis may involve uptake of P and other essential plant nutrients and protection against abiotic and biotic stresses, without any enhanced tolerance to PTEs. In some cases, it is probable that plants may not require their fungal symbiont to achieve tolerance, but plant ability to grow on contaminated soils may be strongly enhanced by being mycorrhizal.
- Fungi enhance plant tolerance to PTEs. In the second and third points, mycorrhizal fungi, like their hosts, need to be PTE tolerant. Several authors also suggested that tolerant behavior of the fungi might be an important factor conferring plant tolerance. However, the efficiency of the protection differs among fungi, and it depends more on the compatibility between fungal isolates and the host plant rather than on fungal tolerance to PTEs [7,69,87,90,93].

Arbuscular mycorrhizal fungi have been thought to participate in plant tolerance because they can modify the host PTE uptake and offer plant benefit. Gonzalez et al. [81] observed that inoculation with AMF increased shoot growth and tiller production approximately twofold compared to noninoculated plants of an arsenate-tolerant genotype of *Holcus lanatus* (a grass species commonly accumulating As and different PTEs) when growing in polluted soils. Additionally, lower As shoot accumulation (threefold) was observed in mycorrhizal than nonmycorrhizal plants. Phosphate:arsenate ratio was also threefold higher in inoculated than in noninoculated plants. These authors discussed that the ability of this grass to grow on strongly contaminated soils was due to the mycorrhizal association. Additionally, they suggested that other benefits may result form the mycorrhizal plant status, which may improve fitness and health of *H. lanatus* grown in polluted soils.

In an experiment using mycorrhizal and nonmycorrhizal excised roots, it was shown that AMF, regardless of their arsenate tolerance, reduced arsenate influx in arsenate-tolerant and -nontolerant *Holcus lanatus* plants [81]. Nontolerant plants had higher levels of AM colonization than tolerant plants for tolerant and nontolerant AMF strains (*Glomus mosseae* BEG-132 and BEG-25, respectively). This suggests that AMF may play a greater role for nontolerant plants than for tolerant plants. In another experiment, the authors observed that AMF seem to down-regulate arsenate/phosphate transport. Higher P concentrations in colonized roots of tolerant plants in comparison to nonmycorrhizal roots were observed when plants were grown in contaminated soil for 16 weeks. The explanation given was that the higher P levels in the roots may be regulating the kinetics of phosphate:arsenate uptake [81].

Bethlenfalvay and Franson [94] also observed significantly lower concentrations of Mn in mycorrhizal plants than in nonmycorrhizal plants and the absence of symptoms of Mn toxicity. However, they concluded that the reduced toxicity of Mn was not a direct response of colonization of *Glomus mosseae*, but probably an indirect influence on plant functions. Nogueira and Harris [95] reported similar conclusions: higher P concentrations in mycorrhizal plants probably indirectly decreased the toxic effect of Mn.

12.5.1 IMPORTANCE OF THE EXTERNAL MYCELIUM OF AMF IN PLANT TOLERANCE TO PTEs

Participation of the EM of AMF, mediating almost all benefits that these fungi offer to their plant hosts, is a relevant fungal structure that should be taken into consideration. External mycelial networks link roots to the soil matrix. It is an important component in the symbiosis and participates extensively in nutrient cycling [96]. Although the EM has been neglected for many years in research of AMF, some authors have highlighted its importance [97–99].

Studies regarding the EM have shown that this fungal structure represents a fundamental component of the AM associations in noncontaminated soils. Olsson et al. [100] reported that EM may represent around 90% of the total mycelium in established arbuscular mycorrhiza. Interestingly, runner hyphae grow distantly from the colonized roots and can produce higher biomass than soil with high root densities [100].

Recently, the different methods of propagation of AMF using plants and compartmented pots, using mesh bags to keep roots in a compartment and hyphae in another compartment, have allowed studies regarding the EM and how it deals with PTEs. With compartmented pots, it is possible to obtain enough material for experimental purposes — as much as 100 mg of dry mycelium per 1-kg pot [101,102].

Excised external hyphae have the capacity to sequester Cd and Cu when exposed to solutions containing these elements [64,103]. Additionally, Gonzalez–Chavez et al. [68] showed the special ability of glomalin from hyphae of the EM to sequester Cu (1.6 mg Cu g^{-1} glomalin) when compared to glomalin extracted from roots (0.3 mg Cu g^{-1} glomalin) and sand (0.4 mg Cu g^{-1} glomalin). The hyphal capacity for sequestration increased significantly as the level of Cu increased in the growth substrate at 10 or 20 μM (1.60 and 1.63 mg Cu g^{-1} glomalin, respectively) vs. a 1.13 mg Cu g^{-1} glomalin without Cu addition.

12.6 CONSTITUTIVE AND ADAPTIVE METAL TOLERANCE IN AMF

Many plant species adapted to metalliferous soils are colonized by AMF in these environments [104]. For instance, Gildon and Tinker [29] and other researchers have reported abundant root colonization of some species, which are considered to belong as glacial relicts (10,000 to 15,000 years ago). *Viola calaminaria* is a good representative of this group [39]. Ernst [105] mentioned that evolution of the symbiosis demands coevolution of both organisms: AMF and plants. This suggests that AMF associated with plants necessarily take part in PTE tolerance and eventually plants and fungi colonize metalliferous soils. This also allows the coexistence of tolerant plants and other, less tolerant plants.

It has been suggested that AMF may evolve Zn, Cd, and As tolerance [3,29,81,106,107]. Tolerance to PTEs is typically a qualitative characteristic, which correlates with the prevailing levels of PTE availability in soil [108]. Apparently, many organisms have developed a variety of adaptive mechanisms for surviving in contaminated environments.

It was suggested that AMF isolated from contaminated soils would have a much higher PTE-binding capacity than AMF isolated from noncontaminated soils [3,106,107]; however, observations from Joner et al. [103] for Cd and Gonzalez et al. [64] for Cu show that these may have similar behavior independently of the nature of the site of isolation. Joner et al. [103] found that a Cd-tolerant *Glomus mosseae* strain possessed the highest Cd sorption capacity; two other AMF isolated from polluted soils (PS) had low sorption capabilities comparable to other strains from nonpolluted soils (NPS). Gonzalez et al. [64] observed that *G. mosseae* BEG-132 and *G. claroideum* BEG-134 (from PS) and *G. mosseae* BEG-25 (from NPS) had high Cu sorption capacity in comparison with *G. caledonium* BEG-132 (from PS) and *Gi. rosea* BEG 111 (from NPS). The fungi BEG-132, BEG-133, and BEG-134 were isolated from the same contaminated soil and they showed different responses to Cu. This suggests functional diversity of AMF.

Fungi differ in their mechanisms and capabilities to deal with PTEs, which may be related to the different strategies to control PTEs toxicity, i.e., using TEM, Gonzalez et al. [64] observed that *G. caledonium* BEG-133 appeared to avoid intracellular accumulation of As/Cu, but *G. mosseae* BEG-132 intracellularly accumulated high amounts of Cu and traces of As. Both these fungi were isolates from PS. Using spore germination test in polluted substrates, Gonzalez [25] reported that *Gigaspora rosea* BEG-111 spores were able to germinate at high concentrations of As and Cu, but retractile cytoplasm was observed in the germination tube. Sánchez–Viveros et al. [65] showed that spores of *G. claroideum* Zac-19 (isolated from NPS) germinating in polluted substrates presented negative chemotropism. This effect increased as the concentration of As/Cu increased in the substrate of germination.

Several studies with filamentous fungi and bacteria show that tolerance may be obtained by sequential exposition to PTEs [109]. However, irrefutable evidence showing that AMF present adaptive or constitutive tolerance is still lacking. Weissenhorn et al. [3] studied a fungal strain from noncontaminated soil and found that it was able to develop tolerance to Cd after 1 year of exposure to 40 mg of cadmium nitrate kg^{-1} of soil. However, when propagated under uncontaminated conditions, its tolerance was lost over a similar period. Malcová et al. [110] showed reduced Mn tolerance of AM fungal isolates due to exclusion of stress levels of Mn from the growth medium used to maintain their fungal cultures (2 years). They suggested that tolerance mechanisms are only advantageous when the selection pressure is present.

Sánchez–Viveros et al. [65] showed changes in tolerance (tested by spore germination) to As/Cu in spores of fungi propagated continuously or not in free-pollution substrate (Figure 12.3). In spores of *G. mosseae* BEG-132, when the fungus was propagated for 2 years in polluted substrate, higher germination percentages were observed at the three highest levels of As/Cu tested. In contrast, *G. caledonium* BEG-133 showed more stable responses at the different As/Cu concentrations. In relation to the fungus isolated from NP soil, *G. claroideum* Zac-19, spore germination was drastically affected by As/Cu concentrations. Germination was lower in spores propagated for 1 year in polluted substrate in comparison to free-exposition spores, except at the highest level of As/Cu concentration. These results indicate that the *G. mosseae* BEG-132 spores germinate properly under high pollutant concentrations if continuous propagation occurs in polluted substrate in comparison with spores propagated in unpolluted substrates or alternate discontinuous cycles. Conversely, for BEG-133, absence or not of pollution during propagation does not affect importantly the germination percentage.

As a strategy of PTE tolerance, AMF and other fungi may modify structurally [111,112]. Sánchez–Viveros et al. [65] observed that spore cell wall thickness of three fungal species increased after As/Cu exposition in polluted substrate (Table 12.2). The fungi tested were isolated from PS or NPS. Spore cell wall thickness of *G. mosseae* BEG-132 (from PS) significantly increased after a 2-year, continuous polluted-substrate propagation. Higher cell wall thickness was observed in *G. caledonium* BEG-133 (from PS) and *G. claroideum* Zac-19 (from NPS).

Some species seem to be able to adapt rapidly to changing soil conditions, such as contamination. For example, *G. mosseae, G. fasciculatum,* and *G. claroideum* are frequently found in

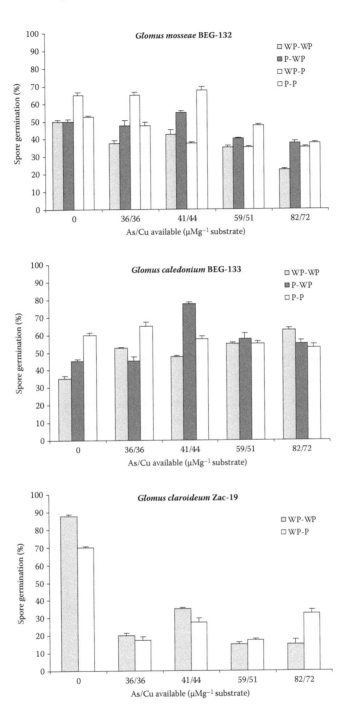

FIGURE 12.3 Spore germination of three arbuscular fungi after 2-year propagation in polluted (P) or not polluted substrate (WP). Germination tests were set on substrates with five As/Cu concentrations by open filter method (Brundett and Saito, 27, 85, 2005. *J. Soil Biochem.*,) At 25 days after germination. Treatments: WP–WP = free-pollution fungal propagation in two 1-year cycles; WP–P = free-pollution propagation first year, pollution second year; P–WP = pollution-propagation first year, free-pollution second year; P–P = pollution-propagation for two 1-year cycles. (Sánchez–Viveros, G. et al., *Rev. Int. Cont. Ambiental*, 20, 147, 2005. With permission.)

TABLE 12.2
Spore Thickness of Three Arbuscular Fungi Propagated in Polluted Substrate

Fungi	Propagation treatment	Spore thickness (μm)
G. mosseae BEG-132[a]	WP–WP	5.4b[c]
	P–P	5.9a
G. caledonium BEG-133[a]	WP–WP	3.9b
	P–P	6.0a
G. claroideum Zac-19[b]	WP–WP	5.4a
	WP–P	7.4b

[a]Fungi isolated from polluted soils and propagated for 2 years in P = polluted substrate and WP = free of pollution.
[b]Fungus isolated from nonpolluted soil propagated for one year in P = polluted substrate and WP = free of pollution.
[c]Means from 90 observations, values with the same letter within each fungus represent no significant difference (Tukey α = 0.05).

Source: Modified from Sánchez–Viveros, G. et al., *Rev. Int. Cont. Ambiental*, 2004, in press. With permission.

diverse terrestrial systems with high adaptation to different environmental and edaphic conditions.[4,26,113]. This may be the result of phenotypic plasticity as suggested by Weissenhorn et al. [3] and Meharg and Cairney [104].

To select fungi colonizing roots growing in contaminated soil and at the same time accumulating PTEs, Turnau et al. [35] used a molecular technique and histochemical staining with rhodizoniate simultaneously. They reported that only 5% of the roots showed PTE accumulation when stained with rhodizoniate. Interestingly, *G. mosseae* was the only fungus present in approximately 75% of these stained roots. Other fungi, such as *Glomus intraradices*, *G. claroideum*, and two *Glomus* species occurred with lower frequency in roots accumulating PTEs (25%). These observations in contaminated soils support the concept of functional biodiversity, which has been observed in natural ecosystems.

Sánchez–Viveros et al. [65] showed that some isolates of AMF may have an inherent ability to tolerate elevated PTE concentrations in soil. These authors observed that spores of *G. claroideum* (Zac-19), a fungus isolated from agricultural soils, propagated for 1 year in soils contaminated with As/Cu or in noncontaminated soils, were able to tolerate elevated concentrations of these pollutants, even if its spores presented a lower percentage of germination (20 to 30%) than in NP with 90%. In contrast, two AMF (from PS) grown on an As/Cu-contaminated soil presented high germination (50 to 70%), independently of the level of As/Cu present in the soil. This result shows that *G. claroideum* Zac-19 with low spore germination may be able to colonize plants, thus assuring its survival in polluted soils. For ectomycorrhizal fungi, Blaudez et al. [114] reported a strong interspecific variation in terms of PTE tolerance. In AMF, this variation may also be observed. Adaptive metal tolerance has been reported for a population of the ectomycorrhizal basidiomycete *Suillus luteus* originating from a metal-contaminated former zinc smelter site [115].

More research is necessary in order to gain knowledge about the stability of fungal tolerance during their cultivation in unpolluted substrates, including fungi isolates from polluted and nonpolluted soils. By now, it is still difficult to state the adaptive or constitutive tolerance in AMF. One problem in estimating arbuscular mycorrhizal tolerance to PTE using only the spore germination test is that it may not reflect the potential effect of the fungus on plant tolerance, as mentioned before. Even with a low germination percentage, AMF may colonize the plant.

Additionally, often no correlation exists between root rate colonization and fungal effect on the plant. It is possible to find significant effects on plant growth, nutrients, and survival at a low colonization rate. Thus, fungal tolerance within a plant should be investigated too; however, it is very difficult to evaluate. Another important remark is that qualifying a fungus to be "tolerant" to PTE means more tolerant to an element than another fungus, and tolerant to a certain degree in particular conditions ("bioavailability" is often not discussed). Thus, it is rather a relative tolerance.

It is relevant to emphasize that fungal species selection, imposed by any stress, does not always offer the optimal AMF symbionts for the function of the soils [116]. Selection of fungal species may have strong implications in the functioning of contaminated soils. Lokke et al. [117] suggested that a loss of AMF species diversity may increase the susceptibility of the host plant to suffering from environmental stresses.

12.7 USE OF AMF IN PHYTOREMEDIATION PRACTICES

Phytoremediation of soils polluted by EPTs is based on two processes: phytostabilization and phytoextraction [118]. In the first case, plants may reduce metal availability and risks of these elements in the soil. Phytostabilization considers the role of plant roots in order to control the uptake and inactive maintenance of EPTs in the soil, which decreases environmental and human risks of these toxic elements. This alternative demands the establishment of a closed vegetation cover, preferably within a short term [119]. Very high concentrations of PTEs in the soil and consequent plant toxicity are the most important limitations of this approach. The main aim of phytostabilization is to install a sustainable ecosystem, where plant biodiversity and functional stability are required [39]; thus, grasses and other herbaceous plants, shrubs, and trees may be useful for a successful remediation of soils using plants.

On the other hand, because PTEs are persistent elements, removal is the best way to ensure minimal ecological risk [39]. Phytoextraction demands decontamination of EPTs from the soil by use of high accumulating plants; especially, hyperaccumulating plants are required. These plants are able to take up and accumulate significant amounts of EPTs (more than 100 mg Cd kg^{-1} dry weight, 1000 mg Co, Cr, Cu, and Pb kg^{-1} dry weight, or 10,000 mg Ni and Zn kg^{-1} dry weight) in their aerial parts [118,120].

In these technologies, plant roots act directly for mechanical contaminant entrapment with consequent protection of the soil surface against wind and water erosion and a reduction of leaching. Microbial activity may indirectly act on these processes [121]. When using plants for remediation, participation of symbiotic microorganisms should not be ignored because these play an important role in nutrient cycling and in the behavior of PTEs in the soil. Plant–microorganism interactions have received little attention in remediation of soils polluted by PTEs.

Despite the role that AMF play in the establishment, survival, and productivity of plants [1,2], relatively few studies have focused on the use of these fungi in the bioremediation of soils polluted with PTEs. These organisms may be crucial for revegetation efforts following plant-based containment or removal technologies. For example, phytostabilization requires plant species adapted to excessive contaminant concentrations in the soil, well developed root systems, and closed coverage of the soil surface. A protective role against PTE toxicity for plants susceptible to colonization by AMF is an important alternative when remediation using plants is needed. Any mechanism reducing plant exposure to PTEs has potential as an applied remediation technology. The use of AMF may help accumulate PTEs in a nontoxic form within the plant roots and the external mycelium.

In addition to high toxic content of PTEs, polluted soils in general also have several limitations, such as low levels or low bioavailability of essential elements and low soil aggregation, which make establishment of plants and soil remediation more difficult. The use of AMF may help to alleviate some of these limitations. Evidence shows that these fungi colonize grasses and other plants, which are often growing in polluted and degraded soils, such as mine soils [81,104]. Some of these plants are "facultative mycorrhizal" in nonstressed conditions; this means that they are colonized by AMF, but the plants have low or no benefits from the mycorrhizal association. In contrast, under stressed conditions, these plants can profit from multiple nutritional and protection benefits from the association to AMF [81]. Some authors have suggested that AMF may help to alleviate other stresses found in PTE-polluted soils, such as drought, salinity, low availability of major nutrients, etc. [9, 104].

Although the role of AMF in remediation of polluted soils is uncertain, it could potentially help to establish a more diverse plant ecosystem [122]. Shetty et al. [123] reported that initial colonizers of PTE-polluted soils tend to be nonmycorrhizal, but the appearance of mycorrhizal plants derives successful restoration; an increase in plant community production; and the improvement of soil structure. Successful examples of the use of AMF in plant establishment on highly perturbated and polluted sites is starting to become a reality. For instance, Dodd et al. [113] showed that inoculation of different plants with AMF was a relevant strategy for soil reclamation of the chalk platforms deposited from the construction of the tunnel under the English Channel, between France and the U.K. These authors were able to promote plant establishment and revegetation in this area using the inoculation of native AMF and plants.

Many plants are obligate mycorrhizal, which means that they require the fungal symbionts to establish, grow, and survive under harsh physical and chemical soil conditions. Thus, phytoestablishment of PTEs using AMF may be an attractive, low-cost, and environmentally friendly alternative. However, some factors, such as selection of tolerant plants to EPTs and selection of the best fungal isolates for a given phytoremediation strategy, should be considered when using AMF in phytoremediation; when extremely high concentrations of PTEs are found, a combination with soil amendments in order to decrease bioavailability and toxicity of PTEs may be an important tool to realize successful soil remediation.

Selection of AMF for their use in phytoremediation practices may include fungal properties for root colonization; sequestration or accumulation of EPTs; uptake of essential nutrients; and soil stabilization properties.

Colpaert [39] suggested that inoculation of AMF suitable and adapted to polluted and climatic conditions is relevant to a successful soil remediation when native mycorrhizal fungi are absent. Plant inoculation of introduced plants in polluted soils and basic management agriculture practices increase the performance, development, and spread of AMF in these soils. For trees and shrubs colonized by AMF, inoculation in the nursery stage is the ideal method to introduce AMF in the soils, including polluted soils. When low AMF populations are present in polluted soils, proper management practices, such as multiple mycorrhizal hosts, low fertilization and pesticide levels, and no tillage, may help to increase fungal performance.

Small-scale demonstration experiments show that phytoremediation is an economical remediation alternative using different plants for slightly contaminated soils; however, for heavily contaminated soils, highly EPT-tolerant plants and hyperaccumulator plants should be used to revegetate [124]. Tolerant plants, which are mycorrhizal, have shown good performance in small-scale remediation experiments. Some examples include [39,119,125]:

- *Agrostis capillaris*
- *A. stolonifera*
- *Andropogon gerardii*
- *Festuca rubra*
- *F. arudinacea*
- *Holcus lanatus*
- *Deschampsia cespitosa*
- *D. flexuosa*
- *Dactylus glomerata*

Casuarina, *Acer*, *Salix*, and *Populus* spp. are also arbuscular mycorrhizal plants, which occur as pioneer species in early succession plant communities, and these are also used in phytoremediation of polluted soil.

Jasper [126] suggested that AMF in combination with other rhizospheric microorganisms, such as *Rhizobium* and *Frankia*, participating in N input to the soil diminish plant growth limitations and improve productivity of their hosts in polluted soils. *Rhizobium* and *Frankia* are important microbial components participating in N input to the soil.

Ernst [105] reported that few legumes may tolerate moderate PTE soil concentrations, for example, *Anthyllis vulneraria* and *Lotus corniculatus*. These plant species may be colonized by AMF and *Rhizobium*. Wu and Kruckeberg [127] reported *Lotus purshianus* and *Lupinus bicolor* as two tolerant species showing effective biological N fixation at a copper mine; however, *Lupinus* is a nonmycorrhizal plant. *Crotalaria* and *Coronilla* spp. may also be important species to recuperate polluted soils. Colpaert [39] suggested that leguminous species associated to *Rhizobium* and AMF may offer an advantage as pioneer plant colonizers in PTE-polluted soils.

It may be important to consider symbiotic microorganisms including AMF because they may assist on phytoremediation schemes. These rhizospheric microbes represent an alternative to clean or stabilize PTE-contaminated soils. Mycorrhizal plants may be of importance to build, cover, and avoid soil erosion, and at the same time stabilize PTEs in the soils.

In some plants, in an accumulator strategy, PTEs are actively concentrated within plant tissues over the full range of soil concentrations, implying a highly specialized physiology. The participation of AMF in this strategy has not been studied in detail. Davis Jr. et al. [128] showed that AMF enhance accumulation and tolerance to Cr in sunflower (*Helianthus annuus*). This plant has been reported to be able to accumulate high concentrations of PTEs, but, when colonized by AMF, higher Cr accumulation was observed without any evident toxic effect. In addition, greater mycorrhizal dependence, expressed by higher growth, was observed at higher Cr levels.

On the other hand, an extreme type of accumulation, described as hyperaccumulation, has also been identified in some plants. Many hyperaccumulators belong to the Chenopodiaceae and Brassicaceae families, which are nonmycorrhizal; however, the presence of AMF was recently demonstrated in members of Asteraceae that hyperaccumulate Ni [129]. Metallophytes in tropical and subtropical regions are potentially mycorrhizal plants, including species in the families Fabaceae, Lamiaceae, Asteraceae, and Poaceae. However, occurrence records on root colonization are lacking [39]. Knowledge about the role of AMF and soil biota in these kinds of plants is still lacking [37,38,130]. This topic needs more attention because zone bioavailability of EPTs is affected by microbial activities [131] and rhizospheric microorganisms represent an alternative to remediate soils, as mentioned by Ow [132].

Tonin et al. [37] reported that AMF from *Viola calaminaria* inoculated to clover increased eightfold and threefold the root concentration of Cd and Zn, respectively, without any significant difference in plant biomass and concentrations of metals in shoots. These AMF were efficient in sequestering metals at the root system. Hildebrandt et al. [38] reported that one *Glomus* isolated from this plant contributed to the accumulation of PTEs in plant roots in a nontoxic form.

The role of AMF in metallophyte or hyperaccumulator plants is unknown. Preliminary research showed that AMF may increase Zn accumulation in *V. calaminaria*; however, this effect depended on fungal species and Zn concentration. Fernandez–Fernandez et al. [133] found that *G. mosseae* BEG-132 (fungus from As/Cu-polluted soil) increased twofold higher Zn-shoot accumulation than *G. mosseae* BEG-25 (from NPS) and noninoculated plants of *V. calaminaria* grown in solutions containing 200 and 300 mg Zn L^{-1}. However, any significant difference was observed when plants were grown at 400 mg Zn L^{-1}.

This result opens another possibility to using AMF in phytoremediation practices into the phytoextraction alternative. All the former information shows the necessity of understanding the process involved in dealing with high concentrations of PTEs in mycorrhizal plants and fungal tolerance, which may have strong implications for the use of AMF in the biological remediation of polluted areas. More research on AMF effects in hyperaccumulators, metallophyte, and plants growing in highly polluted soils is necessary. It should involve more fungi, PTEs, and different kinds of soils.

Chaney et al. [134] have shown evidence of effective *in situ* inactivation of PTEs, such as Pb, using a combination of technologies. They used phosphate addition and Fe biosolids compost for a 3-year test. Reduction in soil Pb bioavailability was demonstrated. Several chemicals used for inactivation also may increase the fertility of the soil and eliminate PTE toxicity to plants and soil organisms. Growing a plant cover physically stabilizes the soil and its contaminants in place; this, in turn, minimizes soil erosion and the transport of PTEs through the soil. Incorporation of amendments and plants is a natural method for restoring soil ecology compared to other techniques. Application of metal-binding soil additives in order to decrease PTE availability may be necessary. These additives may be substances with high metal-adsorption capacity, such as beringite and zeolites [135]; manures and plant residues [136]; compost; and fertilizers [137]. A combination of these chemical and physical technologies, in addition to the biological ones involving AMF, is highly recommended for a more successful soil remediation.

12.8 FUTURE WORK

In order to use AMF into remediation technologies on a higher scale, it is still necessary to address different important aspects, such as to:

- Understand the ecology, physiology, and evolution of AMF involved in the sequestration and accumulation of PTEs and the mechanisms involved
- Learn about the biogeochemical processes involved in plant-based remediation of polluted soils
- Elucidate the role of AMF in the rhizosphere of high accumulators and hyperaccumulator plants
- Select tolerant AMF species by screening of different polluted conditions and metals (mines, sewages, and industrial sites
- Conduct more detailed studies regarding mechanisms, including genetic and molecular ones, involved in AM-mediated improvement of plant tolerance to PTE
- Increase the understanding of the extent of inter- and intraspecific variation of AMF
- Select successful plant–fungi combinations for suitable soil recuperation areas
- Compare different behavior of species alone and in consortia of native AMF in polluted soils to understand the function of the AMF species on this
- Conduct genetic and molecular studies regarding PTE tolerance in AMF, in general, because their mechanisms remain mostly unknown

12.9 CONCLUSIONS

A significant number of studies have shown the importance of mycorrhizal symbiosis for the establishment of a sustainable plant cover on soils with PTEs. However, among mycorrhizal fungi, considerable diversity in metal sensitivity is present and circumstantial evidence suggests that some mycorrhizal fungi have adapted to these adverse soil conditions. More studies are necessary to know whether PTE tolerance is indeed achieved at the population level. This includes the testing of large numbers of isolates, which can be difficult to realize. If particular tolerant isolates are used for applications, we need to better understand the stability of the trait over the long term in different conditions.

Our current knowledge of the PTE tolerance mechanisms in mycorrhizal fungi shows a very patchy picture. Several constitutive mechanisms in mycorrhizal fungi can avoid or reduce PTE toxicity. However, it could be more interesting to investigate the adaptive mechanisms that may have evolved in some taxa colonizing very toxic soils. To be able to do so, it is necessary to obtain sufficient numbers of isolates from polluted soils — isolates that differ considerably from normal ecotypes in metal sensitivity.

ACKNOWLEDGMENT

CGC thanks Dr. Ignacio Maldonado Mendoza for his critical comments to this paper. This work is part of the project SEMARNAT-CONACyT C0-01-2002-739.

REFERENCES

1. van der Heijden, M.G.A. et al., Different arbuscular mycorrhizal fungal species are potential determinants of plant community structure, *Ecology,* 79, 2082, 1998.
2. van der Heijden, M.G.A. et al., Mycorrhizal fungal diversity determines plant biodiversity ecosystem variability and productivity, *Nature,* 396, 69, 1998.
3. Weissenhorn, I. et al., Differential tolerance to Cd and Zn of arbuscular mycorrhizal (AM) fungal spores isolated from heavy metal polluted and unpolluted soils, *Plant Soil,* 167, 189, 1994.
4. del Val, C., Barea, J.M., and Azcon–Aguilar, C., Diversity of arbuscular mycorrhizal fungus populations in heavy metal-contaminated soils. *Appl. Environ. Microbiol.,* 65, 718, 1999.
5. Bradley, R., Burt, A.J., and Read, D.J., Mycorrhizal infection and resistance to heavy metal toxicity in *Calluna vulgaris, Nature,* 292, 335, 1981.
6. Bradley, R., Burt, A.J., and Read, D.J., The biology of mycorrhiza in the Ericaceae. VIII. The role of mycorrhizal infection in heavy metal resistance, *New Phytol.,* 91, 197, 1982.
7. Denny, H.J. and Wilkins, D.A., Zinc tolerance in *Betula* spp. IV. Mechanisms of ectomycorrhizal amelioration of Zn toxicity, *New Phytol.,* 106, 25, 1987.
8. Tam, P.C.F., Heavy metal tolerance by ectomycorrhizal fungi and metal amelioration by *Pisolithus tinctorius, Mycorrhiza,* 5, 181, 1995.
9. Smith, S.E. and Read, D. J., *Mycorrhizal Symbiosis,* 2nd. ed. Academic Press, London, 1997.
10. Schüler, A., Schwarzott, D., and Walker, C., A new fungal phylum, the Glomeromycota phylogeny and evolution, *Mycol. Res.,* 105, 1413, 2001.
11. Bethlenfalvay, G.J. and Schuepp, H., Arbuscular mycorrhizas in agrosystem stability, in *Impact of Arbuscular Mycorrhizas on Sustainable Agriculture and Natural Ecosystems,* Gianinazzi, S. and Schüepp, H., Eds., Birkhauser Verlag, Basel, Switzerland, 1994, 17.
12. George, E., Romheld, V., and Marschner, H., Contribution of mycorrhizal fungi to micronutrient uptake by plants, in *Biochemistry of Metal Micronutrients in the Rhizosphere.* Manthey, J.A., Crowley, D.E., and Luster, D.G., Eds., Lewis Publishers, London, 93, 1994.
13. Jakobsen, I., Research approaches to study the functioning of vesicular arbuscular mycorrhizas in the field, *Plant Soil,* 159, 141, 1994.
14. Munyanziza, E., Kehri, H.K., and Bagyaraj, D.J., Agricultural intensification, soil biodiversity and agro-ecosystem function in the tropics: the role of mycorrhiza in crops and trees, *Appl. Soil Ecol.,* 6, 77, 1997.
15. Wright, S.F. and Upadhyaya, A., A survey of soils for aggregate stability and glomain, a glycoprotein produced by hypha of arbuscular mycorrhizal fungi, *Plant Soil,* 198, 97, 1998.
16. Miller, R.M. and Jastrow, J.D., The role of mycorrhizal fungi in soil conservation. In *Proceedings of a Symposium on Mycorrhizae in Sustainable Agriculture,* Bethlenfalvay, G.J. and Linderman, R.G., Eds., ASA Special publication No. 54. Madison, WI, 1992.
17. Wright, S.F., Starr, J.L., and Paltineanu, I.C., Changes in aggregate stability and concentration of glomalin during tillage management transition, *Soil Sci. Soc. Am. J.,* 63, 1825, 1999.
18. Rillig, M.C. et al., Rise in carbon dioxide changes soil structure, *Nature,* 400, 628, 1999.
19. Nichols, K. et al., Carbon contributions and characteristics of humid acid, fulvic acid, particulate organic, and glomalin in diverse ecosystems, *11 Int. Humic Substances Soc.,* Boston, MA, 21–16 July 2002, 153.

20. Leyval, C., Turnau, K., and Haselwandter, K., Effect of heavy metal pollution on mycorrhizal colonization and function, physiological, ecological and applied aspects, *Mycorrhiza*, 7, 139, 1997.

21. Gadd, M.G., Interactions of fungi with toxic metals, *New Phytol.*, 124, 25, 1993.

22. Ross, S.M. and Kaye, K.J., The meaning of metal toxicity in soil–plant systems, in *Toxic Metals in Soil–Plant Systems.* Ross, S.M., Ed., John Wiley & Sons, Chichester, U.K., 1994.

23. McGrath, S.P., Chaudri, A.M., and Giller, K.E., Long-term effect of metals in sewage sludge on soils, microorganisms and plants, *J. Ind. Microbiol.*, 14, 94, 1995.

24. Schenck, N.C. and Perez, Y., *Manual for the Identification of VA Mycorrhizal Fungi.* 3rd ed. Synergistic Publications, Gainesville, FL, 1990.

25. Gonzalez, M.C., Arbuscular mycorrhizal fungi from As/Cu polluted soils, contribution to plant tolerance and importance of the external mycelium. Ph.D. thesis. Reading University, U.K., 2000.

26. Leyval, C. and Vandenkoornhuyse, Ph., Heavy metal pollution effect on the diversity of arbuscular mycorrhizal fungi in a old arable field experiment. In *Cost Action 8.21. Arbuscular Mycorrhizae in Sustainable Soil–Plant Systems,* Workshop and M.C. Meeting in Budapest, Abiotic stress alleviation by arbuscular mycorrhizal fungi, Hungary, 19-21 Sept. 1996.

27. Sambandan, K., Kannan, K., and Raman, N., Distribution of vesicular–arbuscular mycorrhizal fungi in heavy metal polluted soils of Tamil Nadu, *J. Environ. Biol., India*, 32, 159, 1992.

28. Pawlowska, T.E., Blaszkowski, J., and Ruhling, A., The mycorrhizal status of plants colonizing a calamine spoil mound in southern Poland, *Mycorrhiza*, 6, 499, 1996.

29. Gildon, A. and Tinker, P.B., A heavy metal-tolerant strain of a mycorrhizal fungus, *Trans. Br. Mycol. Soc.*, 77, 648, 1981.

30. Ietswaart, H.J., Griffioen, J.W.A., and Ernst, O.W.H., Seasonality of VAM infection in tree populations of *Agrostis capillaris* (Gramineae) on soil with or without heavy metal enrichment, *Plant Soil*, 139, 67, 1992.

31. Weissenhorn, I., Leyval, C., and Berthelin, J., Bioavailability of heavy metals and abundance of arbuscular mycorrhiza in a soil polluted by atmospheric deposition from a smelter, *Biol. Fert. Soils*, 19, 22, 1995.

32. Stuzt, J.C. and Morton, J.B., Successive pot cultures reveal high species richness of arbuscular endomycorrhizal fungi in arid ecosystems, *Can. J. Bot.*, 74, 1883, 1996.

33. Khan, A.G., Relationship between chromium biomagnification ratio, accumulation factor, and mycorrhizae in plants growing on tannery effluent-polluted soil, *Environ. Int.*, 26, 417, 2001.

34. Gonzalez–Chavez, M.C., unpublished data, 2004.

35. Turnau, K. et al., Identification of arbuscular mycorrhizal fungi in soils and roots of plants colonizing zinc wastes in southern Poland, *Mycorrhiza*, 10, 169, 2001.

36. Cairney, J.W.G. and Meharg, A.A., Influences of anthropogenic pollution on mycorrhizal fungal communities, *Environ. Pollut.*, 106, 169, 1999.

37. Tonin, C. et al., Assessment of arbuscular mycorrhizal fungi diversity in the rhizosphere of *Viola calaminaria* and effect of these fungi on heavy metal uptake by clover, *Mycorrhiza*, 10, 161, 2001.

38. Hildebrant, U., Kaldorf, M., and Bothe, H., The zinc violet and its colonization by arbuscular mycorrhizal fungi, *J. Plant Physiol.*, 154, 709, 1999.

39. Colpaert, J.V., Biological interactions: the significance of root-microbial symbioses for phytorestoration of metal-contaminated soils, in *Metal-Contaminated Soils:* in situ *Inactivation and Phytorestoration,* Vangronsveld, J. and Cunningham, S.D., Eds., Springer–Verlag, Berlin, 75, 1998.

40. El-Kherbawy, M.J.S. et al., Soil pH, rhizobia, and vesicular–arbuscular mycorrhiza inoculation effects on growth and heavy metal uptake of alfalfa (*Medicago sativa* L.), *Biol. Fert. Soils*, 8, 61, 1989.

41. Killham, K. and Firestone, M.K., Vesicular arbuscular mycorrhizal mediation of grass response to acid and heavy metal deposition, *Plant Soil*, 72, 39, 1983.

42. Giller, K.E. and McGrath. R., Pollution by toxic metals on agricultural soil, *Nature*, 335, 676, 1988.

43. McGrath, S., Effect of heavy metals from sewage sludge on soil microbes in agricultural ecosystems, in *Toxic Metals in Soil–Plant Systems,* Ross, S.M. Ed., John Wiley & Sons Ltd., New York, 1994.

44. Symeonidis, L., Tolerance of *Festuca rubra* L. to zinc in relation to mycorrhizal infection, *Biol. Metals*, 204, 1990.

45. Weissenhorn, I. et al., Arbuscular mycorrhizal contribution to heavy metal uptake by maize (*Zea mays* L.) in pot culture with contaminated soil, *Mycorrhiza*, 5, 245, 1995.

46. Heggo, A. and Angle, J. S., Effects of vesicular–arbuscular mycorrhizal fungi on heavy metal uptake by soybeans, *Soil Biol. Biochem.*, 22, 865, 1990.
47. Griffioen, W.A.J., Ietswaart, J.H., and Ernst, W.H.O., Mycorrhizal infection of an *Agrostis capillaris* population on a copper contaminated soil, *Plant Soil,* 158, 83, 1994.
48. Dueck, T.A. et al., Vesicular-arbuscular mycorrhizae decrease Zn-toxicity to grasses growing in Zn polluted soil, *Soil Biol. Biochem,* 18, 331, 1986.
49. Chaney, R.L., Plant uptake of inorganic waste constituents, in *Land Treatment of Hazardous Wastes,* Parr, J.M, Marsh, P.B., and Kla, J.M., Eds., Noyes Data Corporation, Park Ridge, NJ, 1983.
50. Streitwolf–Engel, R. et. al., Clonal growth traits of two *Prunella* species are determined by co-occurring arbuscular mycorrhizal fungi from a calcareous grassland, *J. Ecol.,* 85, 181, 1997.
51. Raju, P.S. et al., Effects of species of VA-mycorrhizal fungi on growth and mineral uptake of sorghum at different temperatures, *Plant Soil,* 121, 165, 1990.
52. Ravnskov, S. and Jakobsen, I., Functional compatibility in arbuscular mycorrhizas measured as hyphal P transport to the plant, *New Phytol.,* 129, 611, 1995.
53. Alloway, B.J., *Heavy Metals in Soils,* Chapman & Hall, London, 1995.
54. Roane, T.M., Pepper, I.L., and Miller, R.M., Microbial remediation of metals, in *Bioremediation Principles and Applications,* Biotech. Res. Series 6, Crawford, R.L., and Crawford, D., Eds., University Press, Cambridge, 1996.
55. Turner, A.P., The responses of plants to heavy metals, in toxic metals in soil–plant systems, Ross, S.M., Ed., John Wiley & Sons Ltd, New York, 1994.
56. Vangronsveld, J. and Clijster, H., Toxic effects of metals, in *Plants and the Chemical Elements,* Farago, M.E., Ed., VCH Verlagsgesellschaft, Weinheim, Germany, 1994, 150.
57. Levitt, J., *Responses of Plants to Environmental Stresses,* vol. 2, 2nd ed. Academic Press, New York, 1980.
58. Shaw, J.A., *Heavy Metal Tolerance in Plants: Evolutionary Aspects,* CRC Press, Boca Raton, FL, 1990.
59. Cervantes, C. and Gutierrez–Corona, F., Copper resistance mechanisms in bacteria and fungi, *FEMS Microbiol. Rev.,* 14, 121, 1994.
60. Gadd, G.M. and White, C., Heavy metals and radionucleotide accumulation and toxicity in fungi and yeast, in *Metal–Microbe Interaction.* Poole, R.M. and Gadd, G.M, Eds., IRL Press. Oxford, 1989.
61. Vasconcelos, M.T.S.D., Azenha, M.A.O., and Cabral, J.P.S., Comparison of availability of copper (II) complexes with organic ligands to bacterial cells and to chitin, *Environ. Toxicol. Chem,* 16, 2029, 1997.
62. Venkateswerlu, G., Yoder, M.J., and Stotzky, G., Morphological, ultrastructural, and chemical changes induced in *Cunninghamella blakesleeana* by copper and cobalt, *Appl. Microbiol. Biotech,* 31, 204, 1989.
63. Turnau, K., Heavy metal content and localization in mycorrhizal *Euphorbia cyparissias* from zinc wastes in southern Poland, *Acta Soc. Bot. Poloniae,* 67, 105, 1998.
64. Gonzalez, M.C. et al., Copper sorption and accumulation by the external mycelium of different *Glomus* spp. (arbuscular mycorrhizal fungi) isolated from the same polluted soil, *Plant Soil,* 240, 287, 2002.
65. Sánchez–Viveros, G. et al., Tolerancia adaptativa de tres hongos micorrizicos arbusculares al crecer en sustratos contaminados con As y Cu, *Rev. Int. Cont. Ambiental,* 20, 147 2005.
66. Murphy, R.J. and Levy, L.F., Production of copper oxalate by some copper-tolerant fungi, *Trans. Br. Mycol. Soc.,* 81, 165, 1983.
67. Gadd, M.G. and Edwards, S.W., Heavy metal-induced flavin production by *Debaryomyces hansenii* and possible connections with iron metabolism, *Trans. Br. Mycol. Soc.,* 87, 533, 1986.
68. Gonzalez–Chavez, M.C. et al., The role of glomalin, a protein produced by arbuscular mycorrhizal fungi, in sequestering potentially toxic elements, *Environ. Pollut.,* 2004, 130, 317.
69. Howe, R., Evans, R.L., and Ketteridge, S.W., Copper-binding proteins in ectomycorrhizal fungi, *New Phytol.,* 135, 123, 1997.
70. Morselt, A.F.W., Smiths, W.T.M., and Limonard, T., Histochemical demonstration of heavy metal tolerance in ectomycorrhizal fungi, *Plant Soil,* 96, 417, 1986.
71. Galli, U., Schuepp, H., and Brunold, C., Heavy metal binding by mycorrhizal fungi, *Physiol. Plantarum,* 92, 364, 1994.
72. Galli, U., Schuepp, H., and Brunold, Ch., Thiols of Cu treated maize plant inoculated with the arbuscular-mycorrhizal fungus *Glomus intraradices, Physiol. Plantarum,* 94, 247, 1995.

73. Gruhn, C.M. and Miller, K., Jr., Effect of copper on tyrosinase activity and polyamine content of some ectomycorrhizal fungi, *Mycol. Res.*, 95, 268, 1991.

74. Huber, M. and Lerch, K., The influence of copper on the induction of tyrosinase and laccase in *Neurospora crassa*, *FEBS Lett.*, 219, 335, 1987.

75. Lerch, C., Copper metallothionein, a copper-binding protein from *Neurospora crassa*, *Nature*, 284, 368, 1980.

76. Miranda, M. et al., Truffles, like black ones, are tyrosinase positive, *Plant Sci.*, 120, 29, 1988.

77. Rivera–Becerril, F. et al., Study of cadmium tolerance mechanisms in mycorrhizal and nonmycorrhizal pea genotypes, *Intercost Workshop on Bioremediation*, Sorrento, Spain, 2000, 140.

78. Turnau, K., Kottke, I., and Oberwinkler, F., Element localization in mycorrhizal roots of *Pteridium aquilinum* L. Kuhn collected from experimental plots treated with cadmium dust, *New Phytol.*, 123, 313, 1993.

79. Roomans, G.M., Localization of divalent cations in phosphate-rich cytoplasmic granules in yeast, *Physiol. Plantarum*, 48, 47, 1980.

80. Kunst, L. and Roomans, G.M., Intracellular localization of heavy metals in yeast *Saccharomyces cerevisiae* by x-ray microanalysis, *Scanning Electron Microscopy*, 1, 191, 1985.

81. Gonzalez, M.C. et al., Arbuscular mycorrhizal fungi confer enhanced arsenate resistance on *Holcus lanatus, New Phytol.*, 155, 163, 2002.

82. Meharg, A.A., Integrated tolerance mechanisms: constitutive and adaptative plant responses to elevated metal concentrations in the environment, *Plant Cell Environ.*, 17, 989, 1994.

83. Sharples, J.M. et al, Mechanisms of arsenate resistance in the ericoid mycorrhizal fungus, *Hymenoscyphus ericae, Plant Physiol.*, 124, 1327, 2000.

84. Sharples, J.M. et al., The symbiotic solution to arsenic contamination, *Nature,* 404, 951, 2000.

85. Rosen, B.P., Families of arsenic transporters, *Trends Microbiol.,* 7, 207, 1999.

86. Cromack, K., Jr. et al., Calcium oxalate accumulation and soil weathering in mats on the hypogeous fungus, *Hysterangium crassum, Soil Biol. Biochem,* 11, 463, 1979.

87. Hartley, J., Cairney, J.W.C., and Meharg, A.A., Do ectomicorrhizal fungi exhibit adaptive tolerance to potentially toxic metals in the environment? *Plant Soil*, 189, 303, 1997.

88. Turnau, K., Dexheimer, J., and Botton, B., Heavy metal sequestration and filtering effect in selected mycorrhizas from calamine dumps — EDAX microanalysis, *Proc. 10th Int. Conf. Heavy Metals Environ.*, Hamburg, 1995.

89. Adriaensen, K. et al., A zinc-adapted fungus protects pines from zinc stress, *New Phytol.*, 161, 549, 2004.

90. Colpaert, J.V. and Van Assche, J.A., Zinc toxicity in ectomycorrhizal *Pinus sylvestris, Plant Soil*, 143, 201, 1992.

91. Jones, M.D. and Hutchinson, T.C., The effect of mycorrhizal infection on the responses of *Betula papyrifera* to nickel and copper, *New Phytol.*, 102, 429, 1986.

92. Marschner, P., Jentschke, G., and Goldbold, D.L., Cation exchange capacity and lead sorption in ectomycorrhizal fungi, *Plant Soil*, 205, 93, 1998.

93. Godbold, D.L. et al., Ectomycorrhizas and amelioration of metal stress in forest trees, *Chemosphere,* 36, 757, 1998.

94. Bethenfalvay, G.J. and Franson, L., Manganese toxicity by mycorrhizae in soybean, *J. Plant Nutr.,* 12, 953, 1989.

95. Nogueira, A.V. and Harris, P.J., Influence of AMF on manganese toxicity, in *COST Action 8.21. Arbuscular Mycorrhizas in Sustainable Soil–Plant Systems.* Gianinazzi, S. and Schuepp, H., Eds., report activities, 1996, 103.

96. Olsson, P.A., Jakobsen, I., and Wallander, H., Foraging and resource allocation strategies of mycorrhizal fungi in a patchy environment, in *Mycorhizal Ecology,* van der Heijden, M.G.A. and Sanders, I.R., Eds., Springer–Verlag, London, 2002, 157.

97. Sylvia, D.M., Quantification of external hyphae of vesicular–arbuscular mycorrhiza fungi, in *Methods in Microbiology (Techniques for the Study of Mycorrhiza).* Norris J.R., Read, D.J., and Varma, A.K., Eds., Academic Press, London, 1992, vol. 24, 53.

98. Dodd, J.C., Approaches to the study of the extraradical mycelium of arbuscular mycorrhizal fungi, in *Impact of Arbuscular Mycorrhizas on Sustainable Agriculture and Natural Ecosystems.* Gianinazzi, S. and Schüepp, H., Eds., Birkhauser Verlag Basel, Switzerland, 147, 1994.

99. Dodd, J.C. et al., Mycelium of arbuscular mycorrhizal fungi (AMF) from different genera: form, function and detection, *New Phytol.,* 226, 1231, 2000.

100. Olsson, P.A., Baath, E., and Jakobsen, I., Phosphorus effect on the mycelium and storage structures of an arbuscular mycorrhizal fungus as studied in the soil and roots by analysis of fatty acid signatures, *Appl. Environ. Microbiol.,* 63, 3531, 1997.

101. Wright, S., personal communication, 2003.

102. Joner, E., personal communication, 2000.

103. Joner, E.J., Briones, R., and Leyval, C., Metal-binding capacity of arbuscular mycorrhizal mycelium, *Plant Soil,* 226, 227, 2000.

104. Meharg, A.A. and Cairney, J.W.G., Co-evolution of mycorrhizal symbionts and their hosts to metal contaminated environments, *Adv. Ecol. Res.,* 30, 70, 1999.

105. Ernst, W.H.O., Mine vegetation in Europe, in *Heavy Metal Tolerance in Plants: Evolutionary Aspects,* Shaw, A.J., Ed., CRC Press, Boca Raton, FL, 1990, 21.

106. Gildon, A. and Tinker, P.B., Interactions of vesicular–arbuscular mycorrhizal infections and heavy metals in plants. II. The effects of infection on uptake of copper, *New Phytol.,* 95, 263, 1983.

107. Weissenhorn, I. and Leyval, C., Root colonization of maize by a Cd-sensitive and a Cd-tolerant *Glomus mosseae* and cadmium uptake in sand culture, *Plant Soil,* 175, 233, 1995.

108. Verkleij, J.A.C. and Schat, H., Mechanisms of metal tolerance in higher plants, in *Heavy Metal Tolerance in Plants: Evolutionary Aspects,* Shaw, A.J., Ed., CRC Press, Boca Raton, FL, 1990, 180.

109. Valix, M., Tang, J.Y., and Malik, K., Heavy metal tolerance in fungi, *Miner. Eng.,* 14, 499, 2001.

110. Malcová, R., Rydová, J., and Vosátka, M., Metal-free cultivation of *Glomus* sp. BEG 140 isolated from Mn-contaminated soil reduces tolerance to Mn, *Mycorrhiza,* 13, 151, 2003.

111. Gardea–Torresdey, J.L. et al., Enhanced copper absorption and morphological alterations of cell of copper stressed *Mucor rouxxii, Environ. Toxicol. Chem.,* 16, 435, 1997.

112. Mullen, M.D. et al., Sorption of heavy metals by the soil fungi *Aspergillus niger* and *Mucor rouxxi, Soil Biol. Biochem.,* 24, 129, 1992.

113. Dodd, J.C. et al., The role of arbuscular mycorhizal fungi in plant community establishment at Shamphire Hoe, Kent, U.K. — the reclamation platform created during the building of the Channel Tunnel between France and U.K., *Biodiv. Conserv.,* 11, 39, 2001.

114. Blaudez, C. et al., Differential response of ectomycorrhizal fungi to heavy metals *in vitro, Mycol. Res.,* 104, 1366, 2000.

115. Colpaert, J.V. et al., Genetic variation and heavy metal tolerance in the ectomycorrhizal basidiomycete *Suillus luteus, New Phytol.,* 147, 367, 2000.

116. Johnson, N.C. and Pfleger, F.L., Vesicular–arbuscular mycorrhizal and cultural stress, in *Mycorrhizae in Sustainable Agriculture,* Bethenfalvay, G.J. and Linderman, R.G., Eds., ASA special publication No. 54, Madison, WI, 1992.

117. Lokke, H.J. et al., Critical loads of acidic deposition of forest soil: is the current approach adequate? *Ambio,* 25, 510, 1996.

118. Raskin, I., Phytoremediation: a novel strategy for the removal of toxic metals from the environment using plants, *Biotech.,* 13, 468, 1995.

119. Schat, H. and Verkleij, J.A.C., Biological interactions: the role for nonwoody plants in phytorestoration: possibilities to exploit adaptive heavy metal tolerance, in *Metal-Contaminated Soils:* in Situ *Inactivation and Phytorestoration,* Vangronsveld, J. and Cunningham, S.D., Eds., Berlin: Springer–Verlag, 1998, 51.

120. Cunningham, S.D. and Lee, C.R., Phytoremediation of contaminated soils, *Trends Biotech.,* 13, 393, 1995.

121. Wenzel, W.W. et al., Phytoremediation: a plant microbe-based remediation system, *Agronomy Monogr.* No. 37. *Bioremediation of Contaminated Soils,* ASSA, USA, 1999.

122. Oliveira, R. et al., Plants colonizing polluted sites: Integrating microbial aspects. *Intercost Workshop on Bioremediation,* Sorrento, Spain, 2000, 56.

123. Shetty, K.G. et al., Effects of mycorrhiza and other soil microbes on revegetation of heavy metal contaminated mine spoil, *Environ. Pollut.,* 86, 181, 1994.

124. Ernst, W.H.O., Bioavailability of heavy metals and decontamination of soils by plants, *Appl. Geochem.,* 11, 163, 1996.

125. Hetrick, B.A.D., Wilson, G., and Figge, D.A., The Influence of mycorrhizal symbiosis and fertilizer amendments on establishment of vegetation in heavy metal mine spoil, *Environ. Pollut.*, 86, 171, 1994.

126. Jasper, D.A., Management of mycorrhizas in vegetation, in *Management of Mycorrhizas in Agriculture, Horticulture and Forestry*, Robson, A.D., Abbott, L.K., and Malajczuk, N., Eds., Dordrecht. Kluwer, 1994, 211.

127. Wu, L. and Kruckeberg, A.L., Copper tolerance in two legume species from a copper mine habitat, *New Phytol.*, 99, 565, 1985.

128. Davis, F.T., Jr. et al., Mycorrhizal fungi enhance accumulation and tolerance of chromium in sunflower (*Helianthus annus*), *J. Plant Physiol.*, 158, 777, 2001.

129. Jeffries, P. et al., The contribution of arbuscular mycorrhizal fungi in sustainable maintenance of plant health and soil fertility, *Biol. Fert. Soils*, 37, 1, 2003.

130. Pawlowska, T.E. et al., Effects of metal phytoextraction practices on the indigenous community of arbuscular mycorrhizal fungi at a metal-contaminated landfill, *Appl. Environ. Microbiol.*, 66, 2526, 2000.

131. Wenzel, W.W., Biotic factors controlling pollutant bioavailability: rhizosphere processes, in *Proc. 2nd Int. Workshop*, Centro Stefano Franscini, Ascona, Switzerland, February 2–7, 2003, 107.

132. Ow, D.W., Heavy metal tolerance genes: prospective tools for bioremediation, *Resour., Conserv. Recycling*, 18, 135, 1996.

133. Fernandez–Fernandez, et al., unpublished data, 2004.

134. Chaney, F.L. et al., Evidence of effective in situ inactivation of soil Pb using phosphate or composted biosolids in the IINERT field test at Joplin, Missouri, in *Proc. 2nd Int. Workshop*, Centro Stefano Franscini, Ascona, Switzerland, February 2–7, 2003, 17.

135. Vangronsveld, J., Van Assche, F., and Clijsters, H., Reclamation of a bare industrial area contaminated by nonferrous metals: *in situ* metal immobilization and revegetation, *Environ. Pollut.*, 87, 51, 1995.

136. Amandi, A.A., Dickson, A., and Maate, G.O., Remediation of polluted soils. I. Effect of organic and inorganic nutrient supplements on the performance of maize (*Zea mays* L), *Water, Air Soil Pollut.*, 66, 59, 1993.

137. Rabinowitz, M.B., Modifying soil lead bioavailability by phosphate addition, *Bull. Environ. Contam. Toxicol.*, 51, 438, 1993.

13 Role of Arbuscular Mycorrhiza and Associated Microorganisms in Phytoremediation of Heavy Metal-Polluted Sites

K. Turnau, A. Jurkiewicz, G. Lingua, J.M. Barea, and V. Gianinazzi–Pearson

CONTENTS

13.1 INTRODUCTION

Various technologies exist that enable the detoxification/deactivation and removal of toxic compounds from the soil, mostly based on physicochemical extraction methods. They are costly and totally destroy soil microorganisms. The restitution of life on polluted sites or areas that were subjected to conventional technologies usually takes a very long time, is difficult, and often requires human intervention.

Phytoremediation is an alternative to physicochemical methods; it involves the use of plants in the process of decreasing the level of toxic compounds in soil, stabilizing the soil, and inhibiting erosion [1]. However, plants need appropriate below-ground ecosystems to establish diverse communities, especially on difficult sites [2–4]. Mycorrhizal fungi play a key role in increasing the volume of soil explored by the plant in search for nutrients and trace elements. Their activity directly or indirectly influences the microbial populations, qualitatively and quantitatively, including bacteria and fungi from the zone called the mycorrhizosphere.

The present chapter will focus on arbuscular mycorrhizal fungi (AMF). Under natural conditions, they are accompanied by bacteria such as legume symbiotic nodular bacteria, plant growth-promoting rhizobacteria (PGPR), and fungi, including other mycorrhizal symbionts and saprobic fungi. All of these organisms build specific consortia that influence the plant by means of interactions with abiotic [5] and biotic components of the soil [6] or stimulate plant growth through the production of vitamins and hormones [7,8]. Rebuilding/establishing such consortia is of utmost importance for the effectiveness of phytoremediation processes.

13.2 MYCORRHIZA AND ITS ROLE IN THE ENVIRONMENT

Mycorrhizal fungi are widespread and form symbiosis with a large majority of plant species on Earth [2]. A common trait of all mycorrhizal fungi is that their mycelium overgrows the soil surrounding the plant roots; the hyphal net stabilizes the soil and furthermore produces substances that bind or glue soil particles together [9]. The way in which the root and its surface are colonized depends on the type of mycorrhiza.

There are two main types of mycorrhiza: ecto- and endomycorrhiza. In the first case, the mycelium forms a more or less compacted fungal mantle on the surface of the root. Its protective properties depend on the species of the symbiotic fungus, the mantle's water-absorbing capacity, the production of pigments, and organic acids. The mycelium penetrates between cortical cells of the root, forming the so-called Hartig net, which is the site of exchange of compounds between the partners. Ectomycorrhiza is formed by several thousands of fungal species, more or less specific towards host plant species (usually trees from the temperate zone). These trees are obligate symbionts.

Endomycorrhiza is far more diverse. Its characteristic feature is the possibility to penetrate not only spaces between cells, but also the inside of live cortical cells, crossing the cell wall and then developing in touch with the plasma membrane of the plant cell. This type of symbiosis includes orchid, ericoid, and arbuscular mycorrhiza. In the first two types, the mycelium forms coils inside cortical cells, but in arbuscular mycorrhiza, characteristic tree-like structures termed arbuscules develop. Arbuscular mycorrhiza (AM) is the most widespread type of mycorrhiza, occurring in 80% of plant species; it is formed by about 120 species of fungi belonging to the Glomeromycota [10]. This symbiosis is believed to be phylogenetically the most ancient type of mycorrhiza.

Molecular and paleobotanical studies seem to support the hypothesis of a close relationship of the AMF with plants since they appeared on land [11,12]. This mycorrhiza plays a key role in the productivity, stability, and diversity of natural ecosystems. Natural soils with low levels or completely devoid of AMF propagules are rare. Several factors can influence the quantity (i.e., the number of propagules) and the quality (i.e., the composition in species) of AM fungi in the soil. The presence of heavy metals and/or other pollutants, the use of amendments to remediate pollution, and the kind of vegetation heavily affect the composition and abundance of the Glomalean fungi [13–15].

The disappearance of the propagules leads to serious consequences, such as the degradation of plant communities; decreased availability of essential elements; and loss of ecosystem stability. Among examples in which it is necessary to introduce AMF propagules during creation and rebuilding of plant communities are sites resulting from volcanic activity and cutting down of forests; industrial wastes; postmining open areas; excessively fertilized agricultural lands; and soils strongly polluted by toxic compounds such as heavy metals (HM) and xenobiotics [2]. In such situations, the introduction of mycorrhizal inoculum involving selected fungal strains adapted to survive in a given toxic environment and under given climatic conditions becomes a key tool in decreasing the toxicity of these compounds to the plants and in establishing a stable vegetation cover.

Fungi adapted to polluted soils should be a choice of preference for the production of inoculum for soil remediation. The number of spores in polluted areas can be affected by the presence of heavy metals, but different fungi show different sensitivity [15] and species-specific (or even strain-specific) behaviors can be observed. In order to reduce the costs of inoculation, the choice of plants is also a relevant point. Plants should be efficient in removing or stabilizing the pollutants, and they should also promote the establishing of strong mycorrhizal and microbial communities because different plant species can affect the species composition of the Glomalean community [15].

13.3 PHYTOREMEDIATION AND THE BEGINNING OF INTEREST IN MYCORRHIZA

At first, the necessity to include soil microorganisms in phytoremediation was neglected. People used compounds that increase the availability of toxic compounds, therefore stimulating the accumulation of metals in plants [16–18], as well as fertilizers to boost plant biomass production [19]. The most efficient varieties were selected; techniques involving genetic engineering were also used [20–23]. The plants' ability to produce organic compounds influencing the rhizosphere and increasing the availability of metals was also acknowledged [24–26].

Among plants that produce high amounts of organic acids in the rhizosphere, researchers' attention was drawn to the order *Lupinus*. Its cultivation can successfully replace the use of chemicals increasing the availability of soil metals. At last the fact that the activity of microorganisms is a factor strongly influencing the processes of mobilizing and immobilizing metals, by means of precipitating sulphides and hydrated iron oxides or their binding to polysaccharides, was noticed [24,27]. Elements such as Pb, Zn, and Cu can also bind to carbonates and oxalates produced by microorganisms [28]. Metals can as well bind to functional groups localized on the surface of the microorganisms' cell walls [29].

Biological methods of cleaning up contamination mainly use bacteria and saprobic fungi; the role of mycorrhizal fungi is still underestimated. Well-developed mycorrhiza can enhance plant survival in difficult areas because it increases the availability of biogens; reduces stress due to low water availability; increases the resistance to pathogens; stimulates the production of phytohormones; and generally improves the soil structure. These factors can significantly enhance bioremediation.

Among the usually considered bioremediation practices, special attention is due to phytostabilization, phytodegradation, and phytoextraction. Briefly, phytostabilization involves the immobilization of toxic compounds in the soil by means of plants that reduce soil erosion; leaking of contaminants into the ground waters; and their dispersion through wind erosion [30]. Phytodegradation includes various metabolic processes of plants and accompanying microorganisms leading to the breaking up of organic compounds such as polyaromatic hydrocarbons, pesticides, and explosives. Phytoextraction takes advantage of the ability of plants to hyperaccumulate metals. Plants are considered useful if they can take up over 1% of a given metal in the dry mass of their shoots. Such plants are grown on the given area and their above-ground parts are harvested, dried, and burned [31,32].

According to Gleba et al. [33], during phytoextraction the "giant underground networks formed by the roots of living plants function as solar driven pumps that extract and concentrate essential elements and compounds from soil and water." This observation has some important implications and consequences.

First, because heavy metals are taken up and transported in water solution, increased plant transpiration would increase metal translocation to the shoot. There is no doubt that mycorrhizal colonization affects the water relations of plants (Smith and Read [2] and references therein). A number of papers indicate that transpiration rates of mycorrhizal plants are significantly higher than those observed in nonmycorrhizal ones [34–39]. Mycorrhizal root systems are usually more branched [40] and therefore they present a larger absorbing surface even in the absence of changes in root biomass [34]. Also, the increased leaf area can be an important factor leading to increased transcription [41]; however, even comparing plants of the same size and root system length, the transpiration rates of mycorrhizal plants remain superior, due to the reduced stomatal resistance [38].

It has been noted that the high stomatal resistance observed in P-deficient nonmycorrhizal plants is a nutritional effect [38,39]. Nevertheless, stomatal behavior is affected by hormonal changes, which can depend on P nutrition as well as on mycorrhizal colonization [35,42–44]. Abscissic acid (ABA) is known to block transpiration and, consequently, metal accumulation in shoots [17]. According to Allen [35], ABA concentrations in leaves of *Bouteloua gracilis* decreased following mycorrhizal colonization by *Glomus fasciculatum*; on the other hand, Danneberg et al.

[43] report higher concentrations of ABA in leaves and roots of maize colonized by *Glomus* sp. (isolate T6) in comparison to nonmycorrhizal plants. Once more, the different combination of plant and fungus species might be important, as well as the growth conditions and the methods utilized for the measurements.

In the second place, roots and hyphae of mycorrhizal root systems explore an incredibly larger volume of soil in comparison with nonmycorrhizal root systems. Even if the contribution of the external hyphae to water uptake and its translocation to the roots has not been clarified [2], it was proved that the external mycelium may contribute to the uptake and translocation of some heavy metals (including Zn, Cu, and Cd [45–47]) to the host roots.

The preceding considerations show, once more, how important it is to gain deeper knowledge on the basic functioning of the mycorrhizal symbiosis for its exploitation in biotechnology and environmental applications.

13.4 MYCORRHIZA IN PHYTOSTABILIZATION

Mycorrhiza proved to be especially useful in phytostabilization. Although the first plants that colonize areas with increased heavy metal levels usually belong to nonmycorrhizal species [48,49], the development of a dense vegetation cover and improvement of the soil structure require the presence of symbiotic fungi. This is especially important for sites where postflotation material originating from zinc and lead ore processing is deposited. Such material is almost devoid of nitrogen and phosphorus compounds, has poor water-holding capacities, and is vulnerable to wind erosion [50]. Possible mechanisms of improved resistance of AMF-colonized plants to HMs include the enhancement of nutrient uptake, particularly phosphorus, and water supply [51,52], as well as metal sequestration through the production of binding substances or absorption of metals by microbial cells [47,53].

Recently, a gene called *GmarMT1*, encoding for a fungal metallothionein, has been identified in *Gigaspora margarita* (BEG 34) [54]. Metallothioneins (MT) are Cys-rich polypeptides able to chelate metal ions and important in the buffering of their intracellular concentration. Heterologous complementation in yeast revealed that the polypeptide encoded by *GmarMT1* confers increased tolerance to Cu and Cd. The gene expression in the symbiotic mycelia is up-regulated upon Cu exposure [54]. Spontaneous colonization of polluted substrate by arbuscular fungi takes a long time. However, it is possible to introduce propagules of selected strains of mycorrhizal fungi in the form of inoculum. Individual strains show a pronounced diversity in the effectiveness of metal binding and therefore also in reducing the toxicity of the substratum. Because the mycelium of certain strains of species, like *Glomus mosseae*, are tolerant to high heavy metal concentrations, they can bind a few times more metals than the mycelium of a saprobic species commonly used in bioremediation — *Rhizopus arrhizus* [55].

It was demonstrated for cadmium and zinc that, although these elements are detected in the mycelium developing inside plant roots, their accumulation in above-ground parts might be limited [46,56]. However, the analysis of tissue concentration and total shoot uptake of Cd, Zn, and Pb in *Plantago lanceolata*, grown in rhizoboxes on substratum collected from zinc wastes, inoculated with a number of arbuscular mycorrhizal fungal (AMF) strains has shown that metal uptake by the plant differs depending on AMF strain/species [57] (Figure 13.1).

The ability to bind and detoxify heavy metals in underground parts of plants might be of importance also for the stimulation of the growth of crops cultivated on polluted soils. Such plants were inoculated with a *Glomus intraradices* strain isolated from a metallophyte — *Viola calaminaria* [56,58]. AMF also eliminated the toxic effect of Cd on several pea genotypes [59]. It was noted that strains isolated from polluted areas are far more useful than strains originating from nonpolluted sites [60–62]. Differences in the effectiveness of metal detoxification and accumulation exist also among strains and species occurring on polluted sites [63]. This underlines the importance of the identification

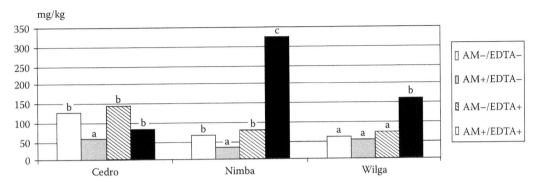

FIGURE 13.1 Pb content in shoots of three maize varieties, mycorrhizal/nonmycorrhizal and with/without EDTA treatment, as examples of different behavior patterns. Different letters above bars indicate statistically significant differences at $p < 0.05$. (Modified from Jurkiewicz, A. et al., *Acta Biol. Cracov. Bot.*, 46, 2004.)

and characterization of the strains, aimed at the selection of the most effective ones. The selected strains should also effectively compete with other fungi that might occur on the given site.

The inoculation of agricultural fields with mycorrhizal fungi in nonpolluted areas often does not improve the situation due to the presence of native strains, which are much better adapted to the given soil conditions, and therefore seems to be ineffective; however, in the case of polluted places that suffer from reduced propagule number and decreased range of fungal species, inoculation is far more effective. The recently developed molecular methods enable scientists to track the fate of the introduced strain in pot cultures as well as in field-collected material. Specific primers exist for a range of species and strains and allow detection of the presence of the introduced fungi in root samples stained to reveal the intraradical mycelium, using the nested PCR method [63–65]. In addition to analysis of the parameters of the fungal partner, it is also necessary to investigate the mechanisms that enable the plant to tolerate metals transferred from the mycelium. Plants show a range of reactions to heavy metals [66]; they can also regulate the degree of mycorrhizal colonization [67].

When selecting plant species for phytostabilization, special attention is usually given to grasses. Although C3 grasses are, under natural conditions, medium or poorly mycorrhizal, they become strongly colonized by AMF on industrial wastes [68–71] and in areas seriously polluted by heavy metals [72–74]. Introduction of the AMF inoculum simultaneously with mixtures of grasses adapted to given conditions is important for the grasses and may be the source of propagules for the establishment of trees such as *Acer* and *Populus* [75]. Similarly, the establishment of ectomycorrhizal tree plantations such as pine or birch (a common practice on industrial wastes) allows improvement of the structure of the soil and increases the soil organic matter content; this in turn creates better growth conditions for herbaceous plants and their symbiotic organisms [76].

Similarly to arbuscular mycorrhizae, ectomycorrhizae stimulate the growth of trees, protect them against pathogens, and may alleviate soil toxicity [77]. Ectomycorrhizal fungi appear on heavily polluted sites faster than arbuscular ones. This is because they form small spores in fruitbodies that usually expose the reproductive layer above the ground level, enabling easy, long-distance dispersal by wind. Individual species and strains of ectomycorrhizal fungi also show diversity in the effectiveness of protecting trees on polluted soils [53,78,79]. This phenomenon is clearly visible when comparing strains isolated from polluted and nonpolluted sites in laboratory conditions [80]. The mycelium can immobilize heavy metals and reduce their transfer to plant tissues, which is regarded as an important protection mechanism allowing trees to survive in polluted areas [53,74].

The phenomenon is explained by the binding of elements by pigments deposited on the surface of extraradical mycelium [60,81], within the hyphal wall [82,83], and in phosphate-rich vacuolar granules [81]. The biofiltration effect of the extraradical mycelium is the most pronounced. The fungal mantle can sometimes also play this role. Such a situation was described in *Rhizopogon roseolus* and *Suillus luteus* from zinc wastes in southern Poland [84,85]. In mycorrhizae of pines with the mentioned fungal species, a typical gradient of heavy metals, decreasing towards the inside of the mantle, was observed. The selection of mycorrhizal strains for the inoculation of tree seedlings planted in polluted areas is important for their establishment.

13.5 MYCORRHIZA IN PHYTODEGRADATION AND PHYTOEXTRACTION

Phytodegradation consists of accelerated degradation of polluting organic compounds, such as hydrocarbons, pesticides, or explosives, in the presence of plants. Existing technologies involve mostly saprobic bacteria and fungi [86–88]. Plants with an abundant root system also have a favorable effect on the degradation of polyaromatic hydrocarbons (PAH) [89,90]. The introduction of rhizosphere microorganisms into such cultures is an alternative to chemical compounds that increase the availability of toxic substances [89,91].

Arbuscular and ectomycorrhizal fungi can enhance phytodegradation [92]. Although the number of propagules of arbuscular fungi decreases with increasing concentration of xenobiotics [93], they can still stimulate plant growth by decreasing the stress related to low phosphorus availability [94] and water deficiency [95]; they can also boost the production of oxidation enzymes [96]. Ectomycorrhizal fungi can additionally produce enzymes taking part in preliminary or intermediate stages of xenobiotics' decomposition [52,97], which enables their further decomposition by other rhizosphere organisms [98–100]. In addition, soil polluted with organic compounds is usually rich in heavy metals. Although these metals cannot be degraded, development of the mycorrhizal mycelium can efficiently alter their availability and plant growth conditions.

Recently, attention has been paid to *Phragmites australis*, which is commonly used for constructed wetlands designed to treat organic effluents [101–103]. *P. australis* was so far believed to be nonmycorrhizal, but again it was recently found to form the symbiosis with enhanced frequency when the water level was reduced and during the flowering time [104]. The presence of potentially mycorrhizal plants in constructed wetlands might be important to enhance phytostabilization and improve the restoration of biodiversity in areas where the processes had ceased.

The least attention was paid to the use of mycorrhiza in phytoextraction. Recent years have brought an increase of interest in the hyperaccumulation of heavy metals by plants, due to the commercial potential of phytoremediation in cleaning up contaminated soil [19,105,106], and as a method to mine metals from low-grade ore bodies [107–110]. Although several reports have been issued on arbuscular mycorrhiza of plants occurring on heavy metal-rich soils such as serpentines [111,112] or strip mines [70,76], arbuscular mycorrhiza was only recently reported in a few hyperaccumulating species belonging to the Asteraceae family growing on nickel-enriched ultramafic soils in South Africa. All plants were found to be consistently colonized by AM fungi, including an abundant formation of arbuscules.

Among them, the most important for phytomining is *Berkheya coddii*, which is capable of accumulating up to 3.8% of Ni in dry biomass of leaves under natural conditions and produces a high yield exceeding that for most hyperaccumulators [113]. The species can also provide an excellent model for laboratory studies on mechanisms mediated by AMF fungi that allow for the phytoextraction process. It has been also shown to form well-developed mycorrhiza under greenhouse conditions. Preliminary results have shown that *B. coddii* inoculated with native fungi (*Gigaspora* sp. and *Glomus tenue*) had not only higher shoot biomass but also significantly increased Ni content (over two times) in comparison to noninoculated plants. This finding greatly contrasts with the conventional opinion that the presence of AMF reduces the uptake of trace elements if they occur in excessive amounts [53].

Mycorrhizal colonization was also reported in Zn and Pb hyperaccumulating *Thlaspi praecox* from the Alps; still, the colonization level of this plant is rather low and decreases with increasing content of heavy metals in the soil [114]. In addition, it is thus far not possible to obtain the formation of mycorrhiza of this group of plants under laboratory conditions; this makes the interpretation of field data hard to confirm.

Although mycorrhiza does not necessarily stimulate phytoextraction, its potential to increase the biomass of the plants, improve soil conditions, and protect the plants from pathogens offers important reasons to include this phenomenon in further research. Three possibilities to increase phytoextraction are being proposed: (1) transgenic plants; (2) hyperaccumulators or high biomass producing crops such as maize, especially for soils relatively less polluted [115,116], treated with chemical chelating substances such as EDTA or sulphur; and (3) stimulating the development of or introducing rhizosphere organisms that will increase the uptake of metals by the plants.

The biotechnological approach aims at producing genetically modified plants characterized by increased tolerance to toxic compounds, higher biomass, and high uptake of heavy metals. A number of transgenic plants have already been obtained by transferring appropriate genes from bacteria or yeasts (see, for example, references 117 through 119) or by generated somatic hybridization between plants such as *Brassica napus* and *Thlaspi caerulescens* [120]. Most transgenic plants have, thus far, only been tested under artificial conditions [121] and they still need further research before the application phase will start. Also, the transformation of AM fungi has been approached. The identification of genes with similar functions can be very important for understanding of the mechanisms of resistance and tolerance to heavy metals and for selection of the fungal strains most suitable for phytostabilization and phytoremediation.

However, although the transformation of many animal, plant, and fungal (e.g., *Saccharomyces*) species is now a relatively easy practice, standard protocols for the transformation of AM fungi are not available yet [122]. Beyond technical problems, the huge diversity of the fungal genome inside one single isolate [123] and the lack of knowledge about the factors controlling the expression of this diversity represent a major problem for an effective exploitation of this kind of approach. In addition, the below-ground environment in which the fungi live and their vegetative reproduction make control of the spread of the transformed fungi very difficult, suggesting a very cautious and careful introduction.

The use of synthetic chelates has been proposed because the amount of metals extracted from the soil by plants depends largely on the availability of the metals. In most soils, even highly polluted ones, only a relatively small percentage of the total metal pool is available to plants. These compounds mobilize metal ions and displace them into the soil solution. Among a variety of chelates tested by Huang et al. [116], EDTA was demonstrated to be the most effective in mobilizing Pb [124], showing that it also increased the availability of other metals such as Cd, Cu, Ni, and Zn.

Experiments carried out on maize show that this common crop can take up as much as 3000 mg kg^{-1} Pb in shoots when grown in laboratory conditions with EDTA (0.5 g kg^{-1} of soil) [116]. A study carried out on 15 commercially available Polish maize varieties inoculated with an AMF strain and treated with EDTA showed that most maize varieties cultivated on metal-rich substratum had higher shoot biomass; this clearly confirmed the role of mycorrhizal fungi in phytoremediation practices, although large differences between varieties have been observed [125].

The data show a large diversity in the effectiveness of phytoextraction among different varieties of the same species. This finding stresses the necessity to screen a large number of cultivars in order to select the best ones for further use. Moreover, the effect of individual varieties might vary when chelators such as EDTA are used (Figure 13.2). Although the use of chemical amendments should be considered carefully, heavy metal release into water flowing out from EDTA-treated pots was found to be substantially lower in the case of mycorrhizal plants than in nonmycorrhizal plants.

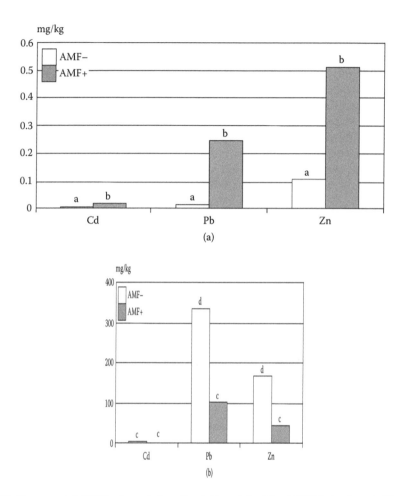

FIGURE 13.2 Heavy metal (HM) release into soil solution: (a) without EDTA treatment; (b) after EDTA treatment. Different letters above bars indicate statistically significant differences at $p < 0.05$. (Modified from Jurkiewicz, A. et al., *Acta Biol. Cracov. Bot.*, 46, 2004.)

This would suggest that mycorrhizal fungi increase the availability of metals to the plants but, at the same time, may decrease pollutant run-off into the ground water (Figure 13.3).

Supplementing soil with sulphur can reduce the soil pH and the application of amendments and fertilizers can cause variations in the abundance of species [14] and modifications in the colonization of roots, thus increasing the amount of vesicles [15]. Also, the use of chelating chemicals, like EDTA, should be applied with much care: it mobilizes the toxic substances, but with the risk of leaching deeper into the soil profile.

In the general literature, not much attention has been paid to the fact that a large diversity in the effectiveness of phytoextraction might exist among different varieties of the same species, as shown by the previously mentioned study. These findings stress the necessity to screen a large number of cultivars in order to select the best ones for further use.

13.6 INFLUENCE OF SOIL BACTERIA ON MYCORRHIZA EFFICIENCY IN POLLUTED ENVIRONMENTS

Rhizosphere bacteria are known to improve mycorrhiza formation and activity by means of a number of so-called mycorrhizosphere activities, which benefit plant growth and health [126].

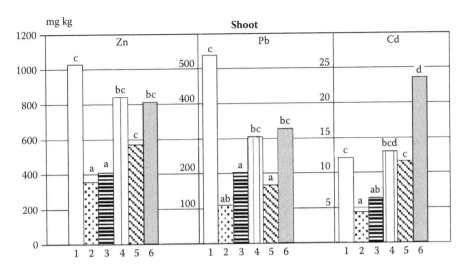

FIGURE 13.3 Heavy metal concentration in shoots of nonmycorrhizal and mycorrhizal *Plantago lanceolata* cultivated in rhizoboxes filled with zinc industrial wastes: 1: control plants (noninoculated); 2–6 — inoculated plants: 2: *Glomus clarum* isolated from zinc wastes; 3: *G. geosporum* from metal-polluted site; 4: *G. geosporum* from salt enriched site; 5: *G. claroideum* from zinc wastes; 6: *G. intraradices* from nonpolluted site. (From Orlowska, E. et al., *Polish Botanical Studies*, Polish Academy of Science [publisher] 2005.)

Therefore, it is to be expected that, with an appropriate selection of target bacteria, these could benefit the role of mycorrhiza in phytoremediation [127,128]. Selection procedures involve:

* Isolation of adapted bacteria from HM-contaminated soils
* Ecological compatibility with mycorrhizal fungi also adapted to HM contamination
* Functional compatibility of both types of microorganisms in terms of promoting phytoextraction and/or phytostabilization of metals from the polluted soil

Soil microbial diversity and activity are negatively affected by excessive concentration of HM [129]. However, indigenous bacterial populations must have adapted in a similar way to mycorrhizal fungi [14] and metal toxicity and evolved abilities that enable the bacteria to survive in polluted soils [127]. Adaptation of mycorrhizal fungi and associated bacteria to HM is considered a prerequisite for exploiting their potential role in phytoremediation [49].

The role of a tailored mycorrhizosphere in phytoremediation was investigated in a series of studies [130–134]. These studies consisted of:

* Isolation and characterization of microorganisms from a target HM contaminated site
* Development of several phytoremediation experiments
* Analysis of the mechanisms involved to account for the demonstrated phytoextraction and/or phytostabilization activities found

Under natural conditions, soil becomes contaminated with more than one metal; thus, it is difficult to determine which metals are responsible for the toxic effects observed [135]. Therefore, only long-term experiments using soils supplemented with a single metal salt can give the opportunity to study the individual toxic effects of each heavy metal on the beneficial microbes for a given time period [136]. In this context, a number of experiments are summarized in the present chapter, all of them using agricultural soil from Nagyhörcsök Experimental Station (Hungary). This soil was contaminated in 1991 with suspensions of 13 microelement salts applied separately. Each salt was applied at four levels (0, 30, 90, and 270 mg kg^{-1}) as described by Biró et al. [136].

Indigenous mycorrhizal fungi and bacterial strains were isolated from this HM-polluted soil 10 years after contamination. They were tested for their influence on plant growth and on the functioning of native mycorrhizal fungi in the face of Cd, Pb, or Ni toxicity.

The most efficient bacterial isolates were identified by means of 16S rDNA sequence analysis and confirmed to belong to the genus *Brevibacillus*. Particularly, *B. brevis* was the most abundant species [130,132]. *Glomus mosseae* was present in all the HM-polluted soil samples, so it was the target mycorrhizal fungus used for phytoremediation inoculation experiments. *G. mosseae* and *Brevibacillus* sp. strains from nonpolluted environments were also used as reference inocula. *Trifolium repens* L. was used as test plant and inoculated with a suspension of *Rhizobium leguminosarum* bv *trifoli*, also an HM-tolerant strain.

In the Cd-contaminated soil [132,133], a high level functional compatibility between both types of autochthonous microorganisms was demonstrated; this resulted in a biomass increase of 545% (shoots) and 456% (roots) and in the N and P content compared to nonmycorrhizal plants. Coinoculation of both microorganisms increased root biomass and symbiotic structures (nodules and AM colonization) to a highest extent, which may be responsible for the beneficial effect observed. The results suggest that bacterial inoculation improved the mycorrhizal benefit in phytostabilization.

Dual inoculation of the Cd-adapted autochthonous *Brevibacillus* sp. and the AM fungus lowered the Cd concentration in *Trifolium* plants. This effect can be due to the ability of the bacteria to accumulate great amounts of Cd. In spite of that, the total Cd content accumulated in plant shoots was higher in dually inoculated plants. This indicates a phytoextraction activity resulting from such a dual inoculation. Further studies [134] demonstrated that the inoculated Cd-adapted bacteria increased dehydrogenase, phosphatase, and β-gluconase activities in the mycorrhizosphere, indicating an improvement of microbial activities concerning plant development in the polluted test soil.

With respect to Pb-spiked soil, experiments following the same methodological approaches [130] showed that *B. agri* at all Pb-spiking levels assayed consistently enhanced plant growth and nutrient accumulation in mycorrhizal plants, as well as nodule numbers and mycorrhizal colonization. This suggests a phytostabilization activity. Auxin production by the test bacteria can account for the beneficial role of these bacteria on mycorrhizal plant development [126]. Dual inoculation increased Pb concentration in plant shoots at the highest level of Pb applied. However, the total content of this metal in plants was consistently enhanced at all levels of Pb, showing that bacterial inoculation enhanced phytoextraction activities in plants inoculated with mycorrhizal fungi.

Dual inoculation of an indigenous Ni-adapted mycorrhizal strain of *G. mosseae* and a Ni-adapted bacterium (*Brevibacillus* sp.) isolated from Ni-contaminated soil was also assayed with *Trifolium* plants growing in Ni-polluted soil [131]. Dual inoculation increased total plant content of this metal at all levels of Ni assayed. This indicates that the tailored mycorrhizosphere carries out phytoextraction of Ni from polluted soils.

The mechanisms by which the bacterial isolates tested enhanced phytoremediation activities in mycorrhizal plants can be therefore summarized as follows:

- Improving rooting, mycorrhiza formation, and functioning
- Enhancing microbial activities in the mycorrhizosphere
- Accumulating metals, thus avoiding their transfer to the aquifers

In conclusion, the dual inoculation of suitable symbiotic and saprophytic rhizosphere microorganisms isolated from HM-polluted soils seems to play an important role in the development and HM tolerance by plants and in bioremediation of HM-contaminated soils.

13.7 MYCORRHIZA AS INDICATOR OF SOIL TOXICITY AND REMEDIATION RATE

Mycorrhizal fungi can also be useful as indicators of soil toxicity [71,137] and the effectiveness of remediation [76]. The toxicity of heavy metals and other pollutants (xenobiotics, PAH) has been monitored using AMF spore germination [13], mycorrhizal colonization of roots analyzed by PCR technique [138], and mycorrhizal infectivity [61]. Recently, the toxicity of zinc wastes of different ages and resulting from different extraction technologies has been compared using various AMF strains and species. The activity of the alkaline phosphatase [139], a vital staining of mycorrhizal colonization, has been found to provide a more sensitive test than the estimation of the total mycorrhizal development [57,62]. Similarly to other indicator organisms such as plants, earthworms, algae, and fish, the disadvantage is that all of them, including AMF, are not specific to pollutants and also react to soil properties (P and N content, pH, etc.).

The appropriate selection of the control soil seems to be problematic. The examples studied thus far prove that they are sensitive indicators of the changes that occur during phytoremediation or during spontaneous succession. Concerning a wider range of soils, the most useful is the germination test, especially that the technique is presently standardized and may rely on the fungal strains supplied commercially (Leyval, C., personal communication, 2004). Further approaches have been recently done to establish new methods considering a wide range of features, such as abundance of intraradical and extraradical mycelium; formation of vesicles; distribution of lipid droplets; etc. They seem to react sensitively to pollutants, at least in the case of the most widely used *Glomus mosseae* strain, BEG12.

Some AMF can be more sensitive to pollutants than plants [72], although some species, especially those originating from strongly polluted places, are well adapted to survival under extremely harsh conditions and their disappearance might be caused only by the lack of the symbiotic plants. The selection of an appropriate fungal strain and plant varieties is therefore of utmost importance [76]. Among the plant species analyzed thus far, English plantain (*Plantago lanceolata* L.), strongly colonized by arbuscular fungi, deserves special attention as an indicator used for bioassays. It occurs in diverse habitats and is resistant to a wide range of stress factors; it is also easy to obtain clones of one plant, which eliminates genetic variability among individuals in their response to toxic compounds [140–142].

13.8 CONCLUSIONS

Central and Eastern Europe are regions where large industrial wastes, deposits of various kinds of waste materials, and places polluted by insufficiently secured unused plant protection products — as well as sites subjected to intense motorization and industrialization — are especially common. Despite the usually well-designed actions aiming at explaining the problem of pollution to the local community, one can still see the production of plants destined for human consumption on heavily polluted soils. Cheap and fast monitoring methods are needed here, followed by low-cost and effective phytoremediation techniques; mycorrhizal fungi should play the key role. It should be emphasized that research on this group of fungi should be conducted in a complex way and include other mycorhizosphere organisms with which the fungi interact [143].

The previously mentioned questions illustrate how broad and diverse are the possibilities to use natural phenomena in solving difficult problems of today's civilization. This will certainly stimulate the dynamic development of a range of scientific fields aimed at explaining the mechanisms of the mentioned phenomena and at optimizing their practical applications.

ACKNOWLEDGMENTS

This chapter was written within the framework of the EU Project GENOMYCA (QLK5-CT-2000-01319/NAS QLRT-CT-2001-02804).

REFERENCES

1. Adriano, D.C., Chlopecka, A. and Kaplan, D.I., Role of soil chemistry in soil remediation and ecosystem conservation, in *Soil Chemistry and Ecosystem Health*. Special Publication no. 52, Soil Science Society of America, Madison, WI, 1998, chap. 1.
2. Smith, S.E. and Read, D.J., *Mycorrhizal Symbiosis*, Academic Press, London, 1997, 605.
3. Linderman, R.G., Effects of mycorrhizas on plant tolerance to diseases, in *Arbuscular Mycorrhizas: Physiology and Function*, Kapulnik, Y. and Douds, D., Jr., Eds., Kluwer Academic Publishers, Dordrecht, The Netherlands, 2000, 345.
4. Turnau, K. and Haselwandter, K., Arbuscular mycorrhizal fungi, an essential component of soil microflora in ecosystem restoration, in *Mycorrhizal Technology in Agriculture. From Genes to Bioproducts*, Gianinazzi, S., Schepp, H., Barea, J.M. and Haselwandter, K., Eds., Birkhauser Verlag, Switzerland, 2002, 137.
5. Turnau, K. and Kottke, I., Fungal activity as determined by microscale methods with special emphasis on interactions with heavy metals, in *The Fungal Community*, Dighton, J., Ed., Marcel Dekker, 2005, Chapter 14.
6. Azcón–Aguilar, C. and Barea, J.M., Arbuscular mycorrhizas and biological control of soil-borne plant pathogens. An overview of the mechanisms involved, *Mycorrhiza*, 6, 457, 1996.
7. Barea, J.M., Mycorrhiza/bacteria interactions on plant growth promotion, in *Plant Growth-Promoting Rhizobacteria, Present Status and Future Prospects*, Ogoshi, A., Kobayashi, L., Homma, Y., Kodama, F., Kondon, N. and Akino, S., Eds., OECD, Paris, 1997,150.
8. Barea, J.M., Rhizosphere and mycorrhiza of field crops, in *Biological Resource Management: Connecting Science and Policy*, Toutant, J.P., Balazs, E., Galante, E., Lynch, J.M., Schepers, J.S., Werner, D. and Werry, P.A., Eds., OECD, INRA edition and Springer, 2000, 110.
9. Miller, R.M. and Jastrow, J.D., Mycorrhizal fungi influence soil structure, in *Arbuscular Mycorrhizas: Physiology and Function*, Kapulnik, Y. and Douds, D.D., Eds., Kluwer Academic Publishers, Netherlands, 2000, 3.
10. Schussler, A., Schwarzott, D. and Walker, C., A new fungal phylum, the Glomeromycota: phylogeny and evolution, *Mycol. Res.*, 105, 1413, 2001.
11. Remy, W., Taylor, T.N., Haas, H. and Kerp, H., Four-hundred-million-year-old vesicular-arbuscular mycorrhizae, *Proc. Natl. Acad. Sci.*, 91, 11841, 1994.
12. Simon, L., Bousquet, J., Lévesque, R.C. and Lalonde, M., Origin and diversification of endomycorrhizal fungi and coincidence with vascular land plants, *Nature*, 363, 67, 1993.
13. Weissenhorn, I. and Leyval, C., Spore germination of arbuscular mycorrhizal fungi in soils differing in heavy metal content and other parameters, *Eur. J. Soil Biol.*, 32, 165, 1996.
14. Del Val, C., Barea, J.M. and Azcon–Aguilar, C., Diversity of arbuscular mycorrhizal fungus populations in heavy metal contaminated soils, *Appl. Environ. Microb.*, 65, 718, 1999.
15. Pawlowska, T.E., Chaney, R.L., Chin, M. and Charvat, I., Effects of metal phytoextraction practices on the indigenous community of arbuscular mycorrhizal fungi at a metal-contaminated landfill, *Appl. Environ. Microbiol.*, 66, 2526, 2000.
16. Blaylock, M.J., Zakharova, O., Salt, D.E. and Raskin, I., Increasing heavy metal uptake through soil amendments. The key to effective phytoremediation, in *Agronomy Abstracts*, ASA, Madison, WI, 1995, 218.
17. Salt, D.E., Blaylock, N., Kumar, N., Dushenkov, V., Ensley, B.D., Chet, I. and Raskin, I., Phytoremediation: a novel strategy for the removal of toxic metals from the environment using plants, *Biotechnology*, 13, 468, 1995.
18. Chlopecka, A. and Adriano, D.C., Mimicked *in situ* stabilization of metals in a cropped soil. *Environ. Sci. Technol.*, 30, 3294, 1996.

19. Baker, A.J.M., Reeves, R.D. and Hajar, A.S.M., Heavy metal accumulation and tolerance in British populations of the metallophyte *Thlaspi caerulescens* J. and C. Presl. (Brassicaceae), *New Phytol.*, 127, 61, 1994.

20. Baker, A.J.M. and Brooks, R.R., Terrestrial higher plants which hyperaccumulate metallic elements: a review of their distribution, ecology and phytochemistry, *Biorecovery*, 1, 81, 1989.

21. Levebre, K.K., Miki, B.L. and Laliberte, J.F., Mammalian metallothionein functions in plants, *Biotechnology*, 5, 1053, 1987.

22. Misra, S. and Gedamu, L., Heavy metal tolerant *Brassica napus* L. and *Nicotiana tabacum* L. plants, *Theor. Appl. Genet.*, 78, 161, 1989.

23. Maiti, I.B., Wagner, G.J. and Hunt, A.G., Light inducible and tissue specific expression of a chimeric mouse metallothionein cDNA gene in tobacco, *Plant Sci.*, 76, 99, 1991.

24. Ernst, W.H.O., Bioavailability of heavy metals and decontamination of soils by plants, *Appl. Geochem.*, 11, 163, 1996.

25. Rao Gadde, R. and Laitinen, H.A., Studies of heavy metal adsorption by hydrous iron and manganese oxides, *Anal. Chem.*, 46, 2022, 1974.

26. Krishnamurti, G.S.R., Cieslinski, G., Huang, P.M., Van Rees, K.C.J., Kinetics of cadmium release from soils as influenced by organic acids: Implication in cadmium availability, *J. Environ. Qual.*, 26, 271, 1997.

27. Lodenius, M. and Autio, S., Effects of acidification on the mobilization of cadmium and mercury from soils, *Arch. Environ. Con. Tox.*, 18, 261, 1989.

28. Bloomfield, C., The translocation of metals in soils, in *The Chemistry of Soil Processes*, Greenland, D.J. and Hayes, M.H.B. Eds., John Wiley & Sons Ltd, Chichester, U.K., 1981, 50.

29. Fein, J.B., Daughney, C.J., Yee, N. and Davis, T.A., A chemical equilibrium model for metal absorption onto bacterial surfaces, *Geochem. Cosmochim. Acta*, 61, 3319, 1997.

30. Losi, M.E., Amrhein, C. and Frankenberger, W.T., Bioremediation of chromate contaminated groundwater by reduction and precipitation in surface soils, *J. Environ. Qual.*, 23, 1141, 1994.

31. Kumar, P., Duschenkov, V., Motto, H. and Raskin, I., Phytoextraction: the use of plants to remove heavy metals from soils, *Environ. Sci. Technol.*, 29, 1232, 1995.

32. Brooks, R.R., Plant hyperaccumulators of metals and their role in mineral exploration, archaeology, and land reclamation, in *Remediation of Metal-Contaminated Soils*, Iskandar, I.K. and Adriano, D.C., Eds., Science Reviews, Northwood, England, 1997, 123.

33. Gleba, D., Borisjuk, N.V., Borisjuk, L.G., Kneer, R., Poulev, A., Skarzhinskaya, M., Dushenkov, S., Logendra, S., Gleba, Y.Y. and Raskin, I., Use of plant roots for phytoremediation and molecular farming, *P. Natl. Acad. Sci. USA*, 96, 5973, 1999.

34. Allen, M.F., Sexton, J.C., Moore, T.S. and Christensen, M., Comparative water relations and photosynthesis of mycorrhizal and nonmycorrhizal *Bouteloua gracilis* (HBK) Lag ex Steud., *New Phytol.*, 88, 683, 1981.

35. Allen, M.F., Influence of vesicular–arbuscular mycorrhizae on water movement through *Bouteloua gracilis* (HBK) Lag ex Steud., *New Phytol.*, 91, 191, 1982.

36. Nelsen, C. and Safir, G.R., The water relations of well watered, mycorrhizal and nonmycorrhizal onion plants, *J. Am. Soc. Hortic. Sci.*, 107, 271, 1982.

37. Huang, R.S., Smith, W.K. and Yost, R.S., Influence of vesicular-arbuscular mycorrhiza on growth, water relations, and leaf orientation in *Leucaena leucocephala* (Lam.) De Wit., *New Phytol.*, 99, 229, 1985.

38. Koide, R., The effect of mycorrhizal infection and phosphorus status on sunflower hydraulic and stomata properties, *J. Exp. Bot.*, 36, 1087, 1985.

39. Fitter, A.H., Water relations of red clover, *Trifolium pratense* L., as affected by VA mycorrhizal infection and phosphorus supply before and during drought, *J. Exp. Bot.*, 39, 595, 1988.

40. Berta, G., Fusconi, A., Trotta, A. and Scannerini, S., Morphogenic modifications induce by the mycorrhizal fungus *Glomus* strain E3 in the root system of *Allium porrum* L., *New Phytol.*, 114, 207, 1990.

41. Kothari, S.K., Marschner, H. and George, E., Effect of VA mycorrhizal fungi and rhizosphere microorganisms on root and shoot morphology, growth and water relations in maize, *New Phytol.*, 116, 303, 1990.

42. Baas, R. and Kuiper, D., Effects of vesicular–arbuscular mycorrhizal infection and phosphate on *Plantago major* ssp. *pleiosperma* in relation to internal cytokinin concentration, *Physiol. Plantarum*, 76, 211, 1989.

43. Danneberg, G., Latus, C., Zimmer, W., Hundeshagen, B., Schneider–Poetch, H.J. and Bothe, H., Influence of vesicular–arbuscular mycorrhizae on phytohormone balances in maize (*Zea mays* L.), *J. Plant Physiol.*, 141, 33, 1992.

44. Torelli, A., Trotta, A., Acerbi, L., Arcidiacono, G., Berta, G. and Branca, C., IAA and ZR content in leek (*Allium porrum* L.) as influenced by P nutrition and arbuscular mycorrhizae, in relation to plant development, *Plant Soil*, 226, 29, 2000.

45. Burkert, B. and Robson, A.D., ^{65}Zn uptake in subterranean clover (*Trifolium subterraneum* L.) by three vesicular–arbuscular mycorrhizal fungi in a root-free sandy soil, *Soil Biol. Biochem.*, 26, 1117, 1994.

46. Joner, E.J. and Leyval, C., Uptake of ^{109}Cd by roots and hyphae of *Glomus mossae/Trifolium subterraneum* mycorrhiza from soil amended with high and low concentration of cadmium, *New Phytol.*, 135, 353, 1997.

47. Kaldorf, M., Kuhn, A.J., Schroeder, W.H., Hildebrandt, U. and Bothe, H., Selective element deposit in maize colonized by heavy metal tolerance conferring arbuscular mycorrhizal fungus, *J. Plant Physiol.*, 154, 718, 1999.

48. Shetty, K.G., Banks, M.K., Hetrick, B.A. and Schwab, A.P., Biological characterization of a southeast Kansas mining site, *Water Air Soil Pollut.*, 78, 169, 1994.

49. Shetty, K.G., Hetrick, B.A.D., Figge, D.A.H. and Schwab, A.P., Effects of mycorrhizae and other soil microbes on revegetation of heavy metal contaminated mine spoil, *Environ. Pollut.*, 86, 181, 1994.

50. Turnau, K., Heavy metal uptake and arbuscular mycorrhiza development of *Euphorbia cyparissias* on zinc wastes in South Poland, *Acta Soc. Bot. Pol.*, 67, 105, 1998.

51. El-Kherbawy, M., Angle, J.S., Heggo, A. and Chaney, R.L., Soil pH, rhizobia and vesicular-arbuscular mycorrhizae inoculation effects on growth and heavy metal uptake of alfalfa (*Medicago sativa* L.), *Biol. Fert. Soils*, 8, 61, 1989.

52. Meharg, A.A. and Cairney, J.W.G., Ectomycorrhizas — extending the capabilities of rhizosphere remediation? *Soil Biol. Biochem.*, 32, 1475, 2000.

53. Leyval, C., Turnau, K. and Haselwandter, K., Effect of heavy metal pollution on mycorrhizal colonization and function: physiological, ecological and applied aspects, *Mycorrhiza*, 7, 139, 1997.

54. Lanfranco, L., Bolchi, A., Cesale Ros, E., Ottonello, S. and Bonfante, P., Differential expression of a metallothionein gene during the presymbiotic vs. the symbiotic phase of an arbuscular mycorrhizal fungus, *Plant Physiol.*, 130, 58, 2002.

55. Joner, E.J., Briones, R. and Leyval, C., Metal-binding capacity of arbuscular mycorrhizal mycelium, *Plant Soil*, 226, 227, 2000.

56. Hildebrandt, U., Kaldorf, M. and Bothe, H., The zinc violet and its colonization by arbuscular mycorrhizal fungi, *J. Pl. Physiol.*, 154, 709, 1999.

57. Orlowska, E., Jurkiewicz, A., Anielska, T., Godzik, B. and Turnau, K., Influence of different arbuscular mycorrhizal fungal (AMF) strains on heavy metal uptake by *Plantago lanceolata* L., *Polish Botanical Studies*, Polish Academy of Science, publisher, 19, 65, 2006.

58. Tonin, C., Vandenkoornhuyse, P., Joner, E.J., Straczek, J. and Leyval, C., Assessment of arbuscular mycorrhizal fungi diversity in the rhizosphere of *Viola calaminaria* and effect of these fungi on heavy metal uptake by clover, *Mycorrhiza*, 10, 161, 2001.

59. Rivera–Becerril, F., Calantzis, C., Turnau, K., Caussanel, J-P., Belimov, A.A., Gianinazzi, S., Strasser, R.J. and Gianinazzi–Pearson, V.J., Cadmium accumulation and buffering of cadmium-induced stress by arbuscular mycorrhiza in three *Pisum sativum* L. genotypes, *Exp. Bot.*, 53(371), 1177, 2002.

60. Galli, U., Schuepp, H. and Brunold, C., Heavy metal binding by mycorrhizal fungi, *Physiol. Pl.* 92, 364, 1994.

61. Leyval, C., Singh, B.R. and Joner, E.J., Occurrence and infectivity of arbuscular mycorrhizal fungi in some Norwegian soils influenced by heavy metals and soil properties, *Water Air Soil Pollut.*, 83, 203, 1995.

62. Orlowska, E., Ryszka, P., Jurkiewicz, A. and Turnau, K., Effectiveness of arbuscular mycorrhizal fungal (AMF) strains in colonization of plants involved in phytostabilisation of zinc wastes, *Geoderma*, 2005, in press.

63. Turnau, K., Ryszka, P., Van Tuinen, D. and Gianinazzi-Pearson, V., Identification of arbuscular mycorrhizal fungi in soils and roots of plants colonizing zinc wastes in Southern Poland, *Mycorrhiza*, 10, 169, 2001.
64. Van Tuinen, D., Jacquot, E., Zhao, B., Gallotte, A. and Gianinazzi-Pearson, V., Characterization of root colonization profiles by a microcosm community of arbuscular mycorrhizal fungi using 25S rDNA-targeted nested PCR, *Mol. Ecol.*, 7, 879, 1998.
65. Jacquot–Plumey, E., van Tuinen, D., Chatagnier, O., Gianinazzi, S. and Gianinazzi–Pearson, V., 25S rDNA-based molecular monitoring of glomalean fungi in sewage sludge-treated field plots, *Environ. Microbiol.*, 3, 525, 2001.
66. Antosiewicz, D.M., Adaptation of plants to an environment polluted with heavy metals, *Acta Soc. Bot. Pol.*, 61, 281, 1992.
67. Koide, R.T. and Schreiner, R.P., Regulation of the vesicular-arbuscular mycorrhizal symbiosis, *Annu. Rev. Pl. Physiol. Pl. Mol. Biol.*, 43, 557, 1992.
68. Daft, M.J. and Nicolson, T.H., Arbuscular mycorrhizas in plants colonizing coal wastes in Scotland, *New Phytol.*, 73, 1129, 1974.
69. Daft, M.J., Hacskaylo, E. and Nicolson, T.H., Arbuscular mycorrhizas in plants colonizing coal spoils in Scotland and Pennsylvania, in *Endomycorrhizas*, Sanders, F.E., Mosse, B. and Tinker, P.B., Eds., Academic Press, London, 1975, 561.
70. Pawlowska, T.E., Baszkowski, J. and Rühling, A., The mycorrhizal status of plants colonizing a calamine spoil mound in southern Poland, *Mycorrhiza*, 6, 499, 1996.
71. Gucwa–Przepióra, E. and Turnau, K., Arbuscular mycorrhiza and plant succession in the zinc smelter spoil heap in Katowice–Wenowiec, *Acta Soc. Bot. Pol.*, 70/2, 153, 2001.
72. Weissenhorn, I. and Leyval, C., Root colonization of maize by a Cd-sensitive and a Cd-tolerant *Glomus mosseae* and cadmium uptake in sand culture, *Plant Soil*, 175, 233, 1995.
73. Noyd, R.K., Pfleger, F.L. and Norland, M.R., Field responses to added organic matter, arbuscular mycorrhizal fungi, and fertilizer in reclamation of turbonite iron ore tailing, *Plant Soil*, 179, 89, 1996.
74. Khan, A.G., Kuek, C., Chaudhry, T.M., Khoo, C.S. and Hayes, W.J., Role of plants, mycorrhizae and phytochelators in heavy metal contaminated land remediation, *Chemosphere*, 41, 197, 2000.
75. Vosatka, M., Investigation of VAM in *Sorbus aucuparia* and *Acer pseudoplatanus* stands on air-polluted localities and mine spoils in North Bohemia, *Agr. Ecosyst. Environ.*, 29, 443, 1989.
76. Orlowska, E., Zubek, Sz., Jurkiewicz, A., Szarek–Sukaszewska, G. and Turnau, K., Influence of restoration on arbuscular mycorrhiza of *Biscutella laevigata* L. (Brassicaceae) and *Plantago lanceolata* L. (Plantaginaceae) from calamine spoil mounds, *Mycorrhiza*, 12, 153, 2002.
77. Haselwandter, K. and Bowen, G.D., Mycorrhizal relations in trees for agroforestry and land rehabilitation, *Forest Ecol. Manage.*, 81, 1, 1996.
78. Hartley–Whitaker, J., Cairney, J.W.G. and Meharg, A.A., Sensitivity to Cd or Zn of host and symbiont of ectomycorrhizal *Pinus sylvestris* L. (Scots pine) seedlings, *Plant Soil*, 218, 31, 2000.
79. Blaudez, D., Jacob, C., Turnau, K., Colpaert, J.V., Ahonen–Jonnarth, U., Finlay, R., Botton, B. and Chalot, M., Differencial responses of ectomycorrhizal fungi to heavy metals *in vitro*, *Mycol. Res.*, 104, 1366, 2000.
80. Colpaert, J.V., Vandenkoornhuyse, P., Adriansen, K. and Vangronsveld, J., Genetic variation and heavy metal tolerance in the ectomycorrhizal basidiomycete *Suillus luteus*, *New Phytol.*, 147, 367, 2000.
81. Turnau, K., Kottke, I., Dexheimer, J. and Botton, B., Element distribution in *Pisolithus tinctorius* mycelium treated with cadmium dust, *Ann. Bot.*, 74, 137, 1994.
82. Galli, U., Meier, M. and Brunold, C., Effects of cadmium on nonmycorrhizal and mycorrhizal Norway spruce seedlings (*Picea abies* (L.) Karst.) and its ectomycorrhizal fungus *Laccaria laccata* (Scop. ex Fr.) Bk. and Br.: sulphate reduction, thiols and distribution of the heavy metals, *New Phytol.*, 125, 837, 1993.
83. Tam, P.C.F., Heavy metal tolerance by ectomycorrhizal fungi and metal amelioration by *Pisolithus tinctorius*, *Mycorrhiza*, 5, 181, 1995.
84. Turnau, K., Kottke, I. and Dexheimer, J., Toxic element filtering in *Rhizopogon roseolus/Pinus sylvestris* mycorrhizas collected from calamine dumps, *Mycol. Res.*, 100/1, 16, 1996.
85. Turnau, K., Przybylowicz, W.J. and Mesjasz–Przybylowicz, J., Heavy metal distribution in *Suillus luteus* mycorrhizas — as revealed by micro-PIXE analysis, *J. Nucl. Instruments*, 181, 649, 2001.

86. Wilson, N.G. and Bradley, G., Enhanced degradation of petrol (Slovene diesel) in an aqueous system by immobilized *Pseudomonas fluorescens*, *J. Appl. Bacteriol.*, 80(1), 99, 1996.

87. Bezalel, L., Hadar, Y. and Cerniglia, C.E., Enzymatic mechanisms involved in phenanthrene degradation by the white rot fungus *Pleurotus ostreatus*, *Appl. Environ. Microbiol.*, 63, 2495, 1997.

88. Schützendübel, A., Majcherczyk, A., Johannes, C. and Huttermann, A., Degradation of fluorene, anthracene, phenanthrene, fluoranthene, and pyrene lacks connection to the production of extracellular enzymes by *Pleurotus ostreatus* and *Bjerkandera adusta*, *Int. Biodeter. Biodegr.*, 43(3), 93, 1999.

89. Schwab, A.P. and Banks, M.K., Biologically mediated dissipation of polyaromatic hydrocarbons in the root zone, in *Bioremediation through Rhizosphere Technology*, Anderson, T.A. and Coats, J.R., Eds., American Chemical Society, Washington D.C., 1994, 132.

90. Reilley, K.A., Banks, M.K. and Schwab, A.P., Dissipation of polycyclic aromatichydrocarbons in the rhizosphere, *J. Environ. Qual.*, 25, 212, 1996.

91. Shann, J.R. and Boyle, J.J., Influence of plant species on *in situ* rhizosphere degradation, in *Bioremediation through Rhizosphere Technology*, Anderson, T.A. and Coats, J.R., Eds., American Chemical Society, Washington D.C., 1994, 70.

92. Leyval, C., Joner, E.J., del Val, C. and Haselwandter, K., Potential of arbuscular mycorrhizal fungi for bioremediation, in *Mycorrhizal Technology in Agriculture. From Genes to Bioproducts*, Gianinazzi, S., Schuepp, H., Barea, J.M. and Haselwandter K., Eds., Birkhauser Verlag, Switzerland, 2002, 175.

93. Leyval, C. and Binet, P., Effect of polyaromatic hydrocarbons in soil on arbuscular mycorrhizal plants, *J. Environ. Qual.*, 27, 402, 1998.

94. Joner, E.J. and Leyval, C., Influence of arbuscular mycorrhiza on clover and ryegrass grown together in a soil spiked with polycyclic aromatic hydrocarbons, *Mycorrhiza*, 10, 155, 2001.

95. Sanchez–Diaz, M. and Honrubia, M., Water relations and alleviation of drought stress in mycorrhizal plants, in *Impact of Arbuscular Mycorrhizas on Sustainable Agriculture and Natural Ecosystems*, Gianinazzi, S. and Schüepp, H., Eds., Birkhäuser, Basel, 1994, 167.

96. Salzer, P., Corbere, H. and Boller, T., Hydrogen peroxide accumulation in *Medicago truncatula* roots colonized by the arbuscular mycorrhiza-forming fungus *Glomus intraradices*, *Planta*, 208, 319, 1999.

97. Barr, D.P. and Aust, S.D., Mechanisms white rot fungi use to degrade pollutants, *Environ. Sci. Technol.*, 28, 79, 1994.

98. Donnelly, P.K. and Fletcher, J.S., Potential use of mycorrhizal fungi as bioremediation agents, in *Bioremediation through Rhizosphere Technology*, Anderson, T.A. and Coats, J. R., Eds., American Chemical Society, Washington D.C., 1994, 93.

99. Gilbert, E.S. and Crowley, D.E., Plant compounds that induce polychlorinated biphenyl biodegradation by *Arthrobacter* sp. strain B1B, *Appl. Environ. Microbiol.*, 63, 1933, 1997.

100. Green, N.A., Meharg, A.A., Till, C., Troke, J. and Nicholson, J.K., Degradation of 4-fluorobiphenyl by mycorrhizal fungi as determined by 19C radiolabelling analysis, *Appl. Environ. Microbiol.*, 65, 4021, 1999.

101. Dias, S.M., Tratamento de efluentes em zonas humidas construidas ou leitos de macrofitas, *Bol. Biotechnol.*, 60, 14, 1998.

102. Trautmann, N., Martin, J.H., Porter, K.S. and Hawk, K.C., Use of artificial wetlands treatment of municipal solid waste landfill leachate, in *Constructed Wetlands for Waste Water Treatment: Municipal, Industrial and Agricultural*, Hammer, D.A., Ed., Lewis, Chelsea, MI, 1989, 245.

103. Maehlum, T., Treatment of landfill leachate in on-site lagoons and constructed wetlands, *Water Sci. Technol.*, 32(3), 129, 1995.

104. Oliveira, R.S., Dodd, J.C. and Castro, P.M.L., The mycorrhizal status of *Phragmites australis* in several polluted soils and sediments of an industrialized region of Northern Portugal, *Mycorrhiza*, 10(5), 241, 2001.

105. Baker, A.J.M., McGrath, S.P., Reeves, R.D. and Smith, J.A.C., Metal hyperaccumulator plants: a review of the ecology and physiology of a biological resource for phytoremediation of metal polluted soils, in *Phytoremediation of Contaminated Soil and Water*, Terry, N. and Banuelos, G., Eds., CRC Press, Boca Raton, FL, 2000, 85.

106. Salt, D.E., Smith, R.D. and Raskin, I., Phytoremediation, *Annu. Rev. Plant Phys.*, 49, 643, 1998.

107. Nicks, L. and Chambers, M.F., A pioneering study of the potential of phytomining for nickel, in *Plants That Hyperaccumulate Heavy Metals — Their Role in Phytoremediation, Microbiology, Archeology, Mineral Exploration and Phytomining*, Brooks, R.R., Ed., CAB International, Wallingford, MA, 1998, 313.

108. Brooks, R.R. and Robinson, B.H., The potential use of hyperaccumulators and other plants for phytomining, in *Plants That Hyperaccumulate Heavy Metals — Their Role in Phytoremediation, Microbiology, Archeology, Mineral Exploration and Phytomining*, Brooks, R.R. Ed., CAB International, Cambridge, MA, 1998, 327.

109. Chaney, R.L., Mallik, M., Li, Y.M., Brown, S.L. and Brewer, E.P., Phytoremediation of soil metals, *Curr. Opin. Biotech.*, 8, 279, 1997.

110. Chaney, R.L., Lee, Y.M., Brown, S.L., Homer, F.A., Malik, M., Angle, J.S., Baker, A.J.M., Reeves, R.D. and Chin, M., Improving metal hyperaccumulator wild plants to develop commercial phytoextraction systems: approaches and progress, in *Phytoremediation of Contaminated Soil and Water*, Terry, N. and Banuelos, G., Eds., CRC Press, Boca Raton, FL, 2000, 129.

111. Goncalves, S.C., Goncalves, M.T., Freitas, H. and Martins Loucao, M.A., Mycorrhizae in a Portuguese serpentine community, in *The Ecology of Ultramafic and Metalliferous Areas,* Jaffre, T., Reeves, R.D. and Becquer, T., Eds., *Proc. Second Int.Conf. Serpentine Ecol.* in Noumea (1995), 1997, 87.

112. Hopkins, N.A., Mycorrhizae in a California serpentine grassland community, *Can. J. Bot.*, 65, 484, 1987.

113. Augustyniak, M., Mesjasz-Przybylowicz, J., Nakonieczny, M., Dybowska, M., Przybylowicz, W. and Migula, P., Food relations between *Chrysolina pardalina* and *Berkheya coddii*, a nickel hyperaccumulator from South African ultramafic outcrops, *Fresen. Environ. Bull.*, 11(2), 85, 2002.

114. Regvar, M., Vogel, K., Irgel, N., Wraber, T., Hildebrandt, U., Wilde, P. and Bothe, H., Colonization of pennycresses (*Thlaspi* spp.) of the *Brassicaceae* by arbuscular mycorrhizal fungi, *J. Plant Physiol.*, 160(6), 615, 2003.

115. Huang, J.W. and Cunningham, S.D., Lead phytoextraction: species variation in lead uptake and translocation, *New Phytol.*, 134, 75, 1996.

116. Huang, J.W., Chen, J., Berti, W.R. and Cunningham, S.D., Phytoremediation of lead-contaminated soils: role of synthetic chelates in lead phytoextraction, *Environ. Sci. Technol.*, 31, 800, 1997.

117. Rensing, C., Sun, Y., Mitra, B. and Rosen, B.P., Pb(II)-translocating P-type ATPases, *J. Biol. Chem.*, 273, 32614, 1998.

118. Karenlampi, S., Schat, H., Vangronsveld, J., Verkleij, J.A.C., Van Der Lelie, D., Mergeay, M. and Tervahauta, A.I., Genetic engineering in the improvement of plants for phytoremediation of metal polluted soils, *Environ. Pollut.*, 107, 225, 2000.

119. Riba, G. and Chupeau, Y., Genetically modified plants, *Cell. Mol. Biol.*, 47, 1319, 2001.

120. Brewer, E.P., Sauders, J.A., Angle, J.S., Chaney, R.L. and Mcintosh, M.S., Somatic hybridization between the zinc accumulator *Thlaspi caerulescens* and *Brassica napus*, *Theor. Appl. Genet.*, 99, 761, 1999.

121. Krämer, U. and Chardonnens, A.N., The use of transgenic plants in the bioremediation of soils contaminated with trace elements, *Appl. Microbiol. Biotechnol.*, 55, 661, 2001.

122. Harrier, L., Millam, S., Biolistic transformation of arbuscular mycorrhizal fungi: progress and perspectives, *Mol. Biotechnol.*, 18, 25, 2001.

123. Koch, A.M., Kuhn, G., Fontanillas, P., Fumagalli, L., Goudet, I. and Sanders, I.R., High genetic variability and low local diversity in a population of arbuscular mycorrhizal fungi, *Proc. Natl. Acad. Sci.*, 101, 2369, 2004.

124. Blaylock, M.J., Salt, D.E., Dushenkov, S., Zakharova, O., Gussman, C., Kapulnik, Y., Ensley, B.D. and Raskin, I., Enhanced accumulation of Pb in Indian mustard by soil-applied chelating agents, *Environ. Sci. Technol.*, 31, 860, 1997.

125. Jurkiewicz, A., Orlowska, E., Anielska, T., Godzik, B. and Turnau, K., The influence of mycorrhiza and EDTA application on heavy metal uptake by different maize varieties, *Acta Biol. Cracov. Bot.*, 46, 7, 2004.

126. Barea, J.M., Gryndler, M., Lemanceau, Ph., Schüepp, H. and Azcón, R., The rhizosphere of mycorrhizal plants, in *Mycorrhiza Technology in Agriculture. From Genes to Bioproducts*, Gianinazzi, S., Schüepp, H., Barea, J.M. and Haselwandter, K., Eds., Birkhäuser Verlag, Basel, Switzerland, 2002, 1.

127. Biró, B., Bayoumi, H.E.A.F., Balázsy, S. and Kecskés, M., Metal sensitivity of some symbiotic N$_2$-fixing bacteria and *Pseudomonas* strains, *Acta Biol. Hung.*, 46, 9, 1995.

128. Takács, T., Biró, B., Vörös., I., Arbuscular mycorrhizal effect on heavy metal uptake of ryegrass (*Lolium perenne* L.) in pot culture with polluted soil, in *Plant Nutrition Food Security and Sustainability of Agro-Ecosystems*, Horst, W.J. Ed., Kluwer Academic Publishers, The Netherlands, 2001, 480.

129. Giller, K., Witter, E. and McGrath, S., Toxicity of heavy metals to microorganisms and microbial processes in agricultural soils: a review, *Soil. Biol. Biochem.*, 30, 1389, 1998.

130. Vivas, A., Azcón, R., Biró, B., Barea, J.M. and Ruiz-Lozano, J.M., Influence of bacterial strains isolated from lead-polluted soil and their interactions with arbuscular mycorrhizae on the growth of *Trifolium pratense* L. under lead toxicity, *Can. J. Microbiol.*, 49, 577, 2003.

131. Vivas, A., Biró, B., Anton, A., Vörös, I., Barea, J.M. and Azcón, R., Possibility of phytoremediation by co-inoculated Ni-tolerant mycorrhiza-bacterium strains, in *Trace Elements in the Food Chain*, Simon, L. and Szilágyi György, M., Eds., Bessenyei Publisher, Nyíregyháza, Hungary, 2003, 76.

132. Vivas, A., Vörös, I., Biró, B., Barea, J.M., Ruiz-Lozano, J.M. and Azcón, R., Beneficial effects of indigenous Cd-tolerant and Cd-sensitive *Glomus mosseae* associated with a Cd-adapted strain of *Brevibacillus* sp. in improving plant tolerance to Cd contamination, *Appl. Soil Ecol.*, 24, 177, 2003.

133. Vivas, A., Vörös, I., Biró, B., Campos, E., Barea, J.M. and Azcón, R., Symbiotic efficiency of autochthonous arbuscular mycorrhizal fungus (*G. mosseae*) and *Brevibacillus* sp. isolated from cadmium polluted soil under increasing cadmium levels, *Environ. Pollut.*, 126, 179, 2003.

134. Vivas, A., Barea, J.M. and Azcón, R., Interactive effects of *Brevibacillus brevis* and *Glomus mosseae*, both from a Cd contaminated soil, on plant growth, physiological mycorrhizal characteristic and soil enzymatic activities in a Cd- spiked soil, *Environ. Pollut.* 134, 275, 2004.

135. Chaudri, A.M., McGrath, S.P. and Giller, K.E., Survival of the indigenous population of *Rhizobium leguminosarum* bv. *trifolii* in soil spiked with Cd, Zn, Cu and Ni salts, *Soil Biol. Biochem.*, 24, 625, 1992.

136. Biró, B., Köves–Péchy, K., Vörös, I. and Kádár, I., Toxicity of some field applied heavy metal salts to the rhizobial and fungal microsymbionts of alfalfa and red clover, *Agrokém. Talajtan*, 47, 265, 1998.

137. Weissenhorn, I., Leyval, C. and Berthelin, J., Cd-tolerant arbuscular mycorrhizal (AM) fungi from heavy metal polluted soils, *Plant Soil*, 157, 247, 1993.

138. Jacquot–Plumey, E., van Tuinen, D., Gianinazzi, S., Gianinazzi–Pearson, V., Monitoring species of arbuscular mycorrhizal fungi in planta and in soil by nested PCR: application to the study of the impact of sewage sludge, *Plant Soil*, 226,179, 2000.

139. Van Aarle, I.M., Olsson, P.A., and Söderström, B., Microscopic detection of phosphatase activity of saprophytic and arbuscular mycorrhizal fungi using a fluorogenic substrate, *Mycologia*, 93(1), 17, 2001.

140. Wu, L. and Antonovics, J., Experimental ecological genetics in *Plantago*. I. Induction of roots and shoots on leaves for large scale vegetative propagation and metal tolerance testing in *P. lanceolata*, *New Phytol.*, 75, 277, 1975.

141. Baroni, F., Boscagli, A., Protano, G. and Riccobono, F., Antimony accumulation in *Achillea ageratum*, *Plantago lanceolata* and *Silene vulgaris* growing in an old Sb-mining area, *Environ. Pollut.*, 109, 347, 2000.

142. Bakker, M.I., Vorenhout, M., Sijm, D.T.H.M. and Kollofel, C., Dry deposition of atmospheric polycyclic hydrocarbons in three *Plantago* species, *Environ. Toxicol. Chem.*, 18, 2289, 1999.

143. Jeffries, P., Gianinazzi, S., Perotto, S., Turnau, K. and Barea, J.M., The contribution of arbuscular mycorrhizal fungi in sustainable maintenance of plant health and soil fertility, *Biol. Fertil. Soils*, 37, 1, 2003.

14 Plant Metallothionein Genes and Genetic Engineering for the Cleanup of Toxic Trace Elements

M.N.V. Prasad

CONTENTS

14.1 INTRODUCTION

The use of plants beyond the necessity of food and fiber is the beginning of environmental biogeotechnology. Toxic metal pollution of the biosphere has accelerated rapidly since the onset of the Industrial Revolution and heavy metal toxicity poses major environmental and health problems. In this regard, plants that accumulate and/or exclude metals are increasingly considered for phytoremediation and phytostabilization. Lead is one of the most frequently encountered heavy metals in polluted environments. The primary sources of this metal include mining and smelting of metalliferous ores; burning of leaded gasoline; disposal of municipal sewage and industrial wastes enriched in Pb; and use of Pb-based paint [1].

The threat that heavy metals pose to human and animal health is aggravated by their long-term persistence in the environment. For instance, Pb is one of the more persistent metals and has been estimated to have a soil retention time of 150 to 5000 years. Also, the average biological half-life of cadmium has been estimated to be about 18 years. The use of biological materials to clean up heavy metal-contaminated soils has been targeted as an efficient and affordable form of bioremediation [2]. One affordable solution might be expressing metal-accumulating genes in nonaccumulating plants showing interesting skills for bioremediation in order to turn them into hyperaccumulators.

14.2 MOLECULAR/ADAPTIVE PHYSIOLOGY AND GENETICS OF METAL HYPERACCUMULATION IN PLANTS

Cloning and characterization of metallothionein (MT) gene families in plants has progressed considerably in the last decade (Table 14.1). MTs and phytochelatins in plants contain a high percentage of cysteine sulfhydryl groups, which bind and sequester heavy metal ions in very stable complexes. Phytochelatins bind Cu and Cd with high affinity and are induced by various metals [62,63]. Phytochelatins may play a role in plant Cd tolerance. Howden and Cobbett [64] have isolated *Arabidopsis* mutants with increased sensitivity to Cd while Cu tolerance was almost unchanged [65,66]. These *cad1* mutants were deficient in PC synthesis and showed greatly reduced levels of PC synthase activity.

MTs not only bind to metals but also regulate intracellular concentrations and detoxify lethal concentrations of metals [59,67]. Various MT genes, such as mouse MTI; human MTIA (alpha domain); human MTII; Chinese hamster MTII; yeast *CUP1*; and pea *PsMTA*, have been transferred to *Nicotiana* sp., *Brassica* sp., or *A. thaliana* [20–23,68–77]. As a result, varying degrees of constitutively enhanced Cd tolerance have been achieved compared with the control. Metal uptake was not markedly altered; in some cases, no differences were present and, in others, maximally 70% less or 60% more Cd was taken up by the shoots or leaves.

Only one study has been reported on a transgenic plant generated with MT of plant origin. When pea (*Pisum sativum*) MT-like gene *PsMTA* was expressed in *A. thaliana*, more Cu (several-fold in some plants) accumulated in the roots of transformed plants than in those of controls [22]. S. Karenlampi (Finland) and her associates have isolated an MT gene from metal-tolerant *Silene vulgaris* and transferred it into several metal-sensitive yeasts. Increases in Cd and Cu tolerance were observed in the modified yeasts. These studies suggest that the MT gene may be useful in improving metal tolerance of plants.

The hyperaccumulator *Thlaspi caerulescens* and the related nonaccumulator *T. arvense* differ in their transcriptional regulation of *ZNT1* (zinc transporter 1) capable of conferring uptake of Cd^{2+} and Zn^{2+} [78]. Expression of *ZNT1* and root zinc uptake rates is elevated in *T. caerulescens* when compared to *T. arvense*. Zinc-mediated down-regulation of *ZNT1* transcript levels in the hyperaccumulator occurs at about 50-fold higher external metal concentrations compared to the nonhyperaccumulator. In several nickel hyperaccumulators, metal exposure elicits a large and dose-dependent increase in the concentrations of free histidine, which can act as a specific chelator able to detoxify Ni^{2+} and which enhances the rate of nickel translocation from the rooting medium into the xylem for transport into the shoot via the transpiration stream.

In the shoots of hyperaccumulating plants, metal detoxification is achieved by metal chelation and subcellular compartmentalization into the vacuole and the apoplast [79–83] and by sequestration within specific tissues, e.g., in the epidermis or in trichomes. The plant detoxification systems remain to be characterized at the molecular level. The generation and analysis of crosses between hyperaccumulators and related nonhyperaccumulators will be one key tool in identifying the genes responsible for the metal hyperaccumulator phenotype.

Based on a preliminary genetic analysis of a number of F2 progeny from crosses between the cadmium- and zinc-tolerant zinc hyperaccumulator *Arabidopsis halleri* ssp. *halleri* (L.) and the closely related, nontolerant nonaccumulator *A. lyrata* ssp. *petraea* (L.), it was postulated that only a small number of major genes were involved in zinc hyperaccumulation and zinc tolerance in *A. halleri* [34,84].

14.3 PLANT METALLOTHIONEIN GENES AND GENETIC ENGINEERING FOR PHYTOREMEDIATION OF TOXIC METALS

The development of a phytoremediation technology for some trace elements requires the transfer of genes into plants across species borders. The molecular basis of trace element detoxification and hyperaccumulation in plants has been increasingly investigated [20,85–87].

TABLE 14.1
Cloning and Characterization of Metallothionein (MT) Gene Families in Plants Including cDNA Encoding for Metallothinoein-like proteins

Plant Name	Protein/Gene/Gene Family	Reference
Type 1		
Mimulus		3
Pea	*PsMTA*	4
Barley	*ids-1*	5
		6
Maize		7
		8
Wheat	*wali1*	9
White clover		10
Arabidopsis	*MT1/MT1a*	11–13
	MT1c	12
Brassica napus		14
Rice	*OsMT-1*	15,16
Cotton	*MT1-A*	17
Vicia faba (fava bean)	*MT1a*	18
	MT1b	
Red fescue		19
Brassica napus, Nicotiana tabacum	*MT2*	20
Nicotiana tabacum	*MT1*	21
Pisum sativum	*PsMT A*	22
Yeast	*CUP1*	23
Arabidopsis	*Glutathione-S-transferase (parB)*	24
Glycine max (soybean)	*Ferritin*	25
	ZAT (AtMTP1)	26
	Arabidopsis thaliana (CAX2)	27
	Nicotiana tabacum (NtCBP4)	28
	FRE1 and *FRE2*	29
	Arabidopsis thaliana (AtNramp1)	30
	Arabidopsis thaliana (AtNramp3)	31
	Arabidopsis thaliana (merA)	32
	Arabidopsis thaliana (merB)	33
	AHA2	34
Type 2		
Soyabean		35
Arabidopsis		36
	MT2/MT2a	11,12
	MT2b	12
Ricinus communis (castor bean)		37
Vicia faba (fava bean)		38
Kiwi fruit	*PLIWI504*	39
Coffee		40
Chinese cabbage		41
Sambucus		37
Nicitiana Tabacum (tobacco)		42
		43
Trifolium repens (white clover)		44

TABLE 14.1
Cloning and Characterization of Metallothionein (MT) Gene Families in Plants Including cDNA Encoding for Metallothionein-like Proteins (continued)

Plant Name	Protein/Gene/Gene Family	Reference
Brassica campestris		45
Rice	*OsMT-2*	46
Tomato (three sequences)	*LeMT$_A$*	47
	LeMT$_B$	48
		48
Brassica napus		49
Brassica juncea (five sequences)		50
Apricot		51
Common rice plant		52
Type 3		
Rice		53
		54
		16
Type 4		
Kiwi fruit	*Clone 503*	39
Apple		55
Papaya		56
Banana	*Clone3-6*	57
Rice(two sequences)		16
Sweet cherry		58
Others		
Arabidopsis	*MTIb*	12
Arabidopsis	*MT3*	59
Tomato		47
Brassica campestris		45
Douglas fir	*PM 2.1*	60
Strawberry	*FMET1*	61
Banana	*Clone 3-23*	57

Despite the difficulty in predicting the effects of microbial genes in a complex multicellular organism like a plant, the successful introduction of a modified bacterial mercuric ion-reductase gene into yellow poplar (*Liriodendron tulipifera*) and *Nicotiana tabacum* demonstrates that bacterial genes may be extremely valuable in phytoremediation. Two *Arabidopsis* mutants resistant to high levels of aluminum have been characterized [34]. The genes have yet to be cloned, but one of the mutants, on chromosome 1, secretes organic acids to bind Al in the soil before it enters the plant. The second mutant, mapped to chromosome 4, increases the flux of hydrogen outside the root, changing the pH, which transforms the Al^{3+} ions into aluminum hydroxides and aluminum precipitates. These forms are incapable of entering the plant via the roots.

The *Arabidopsis* transgenic plants with mer (mercury) operon have conferred tolerance to gold [86ñ88]. Implications for the glutathione, phytochelatin synthetase pathways in transgenics for remediations is increasingly gaining the attention of scientists [88,89]. Metallothioneins (low molecular weight proteins with high cysteine content and a high affinity for binding metal cations such as those of cadmium, copper, and zinc) from animal sources were introduced into plants in a transgenic approach, mainly to reduce metal accumulation in shoots by trapping the metal in the roots. Expression of a mammalian MT in *N. tabacum* L. under the control of a constitutive promoter was able to reduce the translocation of cadmium into the shoots. Following exposure to a low cadmium concentration (0.02 μM) in the rooting medium, leaf cadmium concentrations were 20% lower in the transgenics than in wild-type plants. However, under field conditions, a consistent difference between transgenic and control plants could not be observed in leaf cadmium content or plant growth.

These results demonstrate that trace element uptake observed on nonsoil substrates under glasshouse or growth chamber conditions may not be extrapolated to predict the performance of transgenic plants on soil substrates or under field conditions. Plants overexpressing mammalian MTs were reported to be unaffected by concentrations of 100 to 200 μM cadmium, whereas growth of *N. tabacum* control plants was severely inhibited at external cadmium concentrations of 10 μM [20].

After thorough examination of a number of wild plant species growing on soils highly contaminated by heavy metals in eastern Spain, *Nicotiana glauca* R. Graham (shrub tobacco) was selected for biotechnological modification because it showed the most appropriate properties for phytoremediation [90]. This plant has a wide geographic distribution, grows fast with a high biomass, and is repulsive to herbivores. Following *Agrobacterium*-mediated transformation, the induction and overexpression of a wheat gene encoding phytochelatin synthase (*TaPCS1*) in this particular plant greatly increased its tolerance to metals such as Pb and Cd; it developed seedling roots 160% longer than those of wild-type plants. In addition, seedlings of transformed plants grown in mining soils containing high levels of Pb (1572 ppm) accumulated double the concentration of this heavy metal than that in wild-type plants. These results indicate that the transformed *N. glauca* represents a highly promising new tool for use in phytoremediation efforts.

It has been suggested that phytoremediation would rapidly become commercially available if metal-removal properties of hyperaccumulator plants, such as *T. caerulescens*, could be transferred to high biomass-producing species, such as Indian mustard (*Brassica juncea*) or maize (*Zea mays*) [91]. Biotechnology has already been successfully employed to manipulate metal uptake and tolerance properties in several species. For example, in tobacco (*Nicotiana tabacum*), increased metal tolerance has been obtained by expressing the mammalian metallothionein, metal-binding proteins, and genes [68,71].

Possibly, the most spectacular application of biotechnology for environmental restoration has been the bioengineering of plants capable of volatilizing mercury from soil contaminated with methylmercury. Methylmercury is a strong neurotoxic agent that is biosynthesized in Hg-contaminated soils. To detoxify this toxin, transgenic plants (*Arabidopsis* and tobacco) were engineered to express bacterial genes *merB* and *merA*. In these modified plants, *merB* catalyzes the protonolysis of the carbon–mercury bond with the generation of Hg^{2+}, a less mobile mercury species. Subsequently, *merA* converts Hg(II) to Hg(0), a less toxic, volatile element, which is released into the atmosphere [87,92].

Although regulatory concerns restrict the use of plants modified with *merA* and *merB*, this research illustrates the tremendous potential of biotechnology for environmental restoration. In an effort to address regulatory concerns related to phytovolatilization of mercury, Bizili et al. [33] demonstrated that plants engineered to express *MerB* (an organomercurial lyase under the control of a plant promoter) may be used to degrade methylmercury and subsequently remove ionic mercury via extraction. Despite recent advances in biotechnology, little is known about the genetics of metal hyperaccumulation in plants. In particular, the heredity of relevant plant mechanisms, such as metal transport and storage [93] and metal tolerance [94], must be better understood. Recently, Chaney et al. [95] proposed the use of traditional breeding approaches for improving metal hyperaccumulator species and possibly incorporating significant traits, such as metal tolerance and uptake characteristics, into high biomass-producing plants. Experiments have been conducted with genetically engineered plants for enhanced uptake of metals.

14.3.1 MERCURIC ION REDUCTION AND RESISTANCE

Bacteria have the ability to reduce a number of heavy metals to less toxic forms. Mercury resistance in Gram-negative bacteria is encoded by an operon, which includes mercuric ion reductase gene (*mer*A) among them. *Mer*A is a soluble NADPH-dependent, FAD-containing disulfide oxidoreductase. This enzyme converts toxic Hg^{2+} to the less toxic metallic mercury (Hg^0). *Escherichia coli* cells expressing the *mer*A gene were shown to possess a weak reduction activity toward Au^{3+} and Ag^+ in addition to reduction of Hg^{2+} [96]. The *mer*A gene also weakly increased Hg^{2+} tolerance of *Saccharomyces cerevisiae* [97]. These studies suggested that the *mer*A gene might affect metal tolerance when expressed in plant.

Initial attempts to express the bacterial *mer*A gene from Tn21 in plants to produce Hg^{2+} resistance were unsuccessful in spite of the use of very efficient plant expression systems. No full-length *mer*A RNA or *mer*A-encoded protein was detected. The original bacterial *mer*A sequence is rich in CpG dinucleotide having a highly skewed codon usage, both of which are particularly unfavorable to efficient expression in plants because they are exposed to methylation and subsequent gene silencing [87,98]. Therefore, a mutagenized *mer*A sequence (*mer*Ape9) was constructed, modifying 9% of the coding region, and was transformed to *Arabidopsis thaliana*. The seeds germinated and the seedlings grew on medium containing up to 100 m*M* Hg, although the transgenic plants expressed only low levels of *mer*A mRNA. Transgenic seedlings evolved two to three times the amount of Hg^0 compared to control plants. Plants were also resistant to toxic levels of Au^{3+}.

Rugh et al. [97] give a good example of a successful modification of a bacterial metal tolerance gene for expression in plants. Recently, Rugh and coworkers [98] reported on the development of transgenic yellow poplar (*Liriodendron tulipifera*) for mercury phytoremediation using *mer*A gene (*mer*A18) modified even further to optimize the codon usage in the plant. Transgenic plants evolved ten times the amount of Hg^0 compared to control plants. Thus far, this system has not been tested in field conditions. This is a convincing indication that genetic engineering may improve a plant's capacity to phytoremediate metal-polluted soils.

Partial success has been reported in the literature. For example, in an effort to correct for the small size of hyperaccumulator plants, Brewer [99] generated somatic hybrids between *T. caerulescens* (a Zn hyperaccumulator) and *Brassica napus* (canola) followed by hybrid selection for Zn tolerance. High biomass hybrids with superior Zn tolerance were recovered. These authors have also advocated a coordinated effort to collect and preserve germ plasm of accumulator species. Initially, phytoremediation trials were performed using plants known to accumulate metals and/or to possess metal tolerance — *Silene vulgaris* (Moench) Garcke L.; Brassicaceae plants *Brassica oleracea* and *Raphanus sativus*; and metal hyperaccumulators like *Thlaspi caerulescens* and *Alyssum* L. spp. [100]. Metal hyperaccumulators were most efficient at metal removal in these field trials.

In order to clean up a moderately contaminated soil, 6 and 130 croppings would be needed for zinc and cadmium, respectively. In pot trials, a low rate of biomass production, common to most hyperaccumulators, was shown to limit zinc removal from a contaminated soil by *T. caerulescens*. High biomass nonaccumulator Brassica crops were more effective [101]. For phytoextraction to become a viable technology, dramatic improvements would be required in hyperaccumulator biomass yield or nonaccumulator metal accumulation [102]. Plants able to tolerate and accumulate several metals are required; polluted soils often contain high levels of several contaminant trace elements. Soils polluted with arsenic, cadmium, lead, or mercury provide major targets for remediation. To date, no plants that reproducibly hyperaccumulate lead or mercury have been identified. In most naturally occurring tolerant plants studied to date, tolerance to arsenic or lead appears to be based on exclusion from the plant [103–108].

The development of a phytoremediation technology for some trace elements is thus likely to require the transfer of genes into plants across species borders. Although little is known about the molecular basis of trace element detoxification and hyperaccumulation in plants, a number of trace element detoxification systems from bacteria and yeast have been characterized genetically and functionally at the molecular level for the detoxification of metals [108]. Despite the difficulty in predicting the effects of microbial genes in a complex multicellular organism like a plant, the successful introduction of a modified bacterial mercuric ion-reductase gene into yellow poplar (*Liriodendron tulipifera*) and *Nicotiana tabacum* demonstrates that bacterial genes may be extremely valuable in phytoremediation [92,98].

Furthermore, *ZIP* genes that confer Zn uptake activities in yeast have also recently been described. Moffat [34] reported the characterization of two *Arabidopsis* mutants that were resistant to high levels of aluminum. The genes have yet to be cloned, but one of the mutants, on chromosome 1, secretes organic acids to bind Al in the soil before it enters the plant. The second mutant, mapped to chromosome 4, increased the flux of hydrogen outside the root, changing the pH, which transformed the Al^{3+} ions into aluminum hydroxides and aluminum precipitates. These forms are incapable of entering the plant via the roots.

Rate-limiting steps in selenium assimilation and volatilization have been deduced in Indian mustard. ATP sulfurylase was determined to be involved in selenate reduction and, when overexpressed in Indian mustard, conferred Se accumulation, tolerance, and volatilization. The *Arabidopsis* transgenic plants with *mer* operon have conferred tolerance to gold. Schmöger et al. (2000) [87] demonstrated that phytochelatins are involved in detoxification of arsenic. This has important implications for the glutathione, phytochelatin synthetase metabolic pathway transgenics.

Vatamaniuk et al. (1999) [108] identified an *Arabidopsis* cDNA named *AtPCS1*. Expression of *AtPCS1* protein mediated an increase in Cd accumulation, pointing to a possible role in Cd chelation or sequestration. Clemens et al. 1999 [83] identified a wheat cDNA, *TaPCS1*, that increased Cd resistance in wild-type yeast. Just like *AtPCS1*, tl resistance mediated by *TaPCS1* was associated with an increase in Cd accumulation and was dependent on GSH. Both *AtPCS1* and *TaPCS1* metal tolerance is GSH dependent. For further evidence of the role of PCs, refer to chapter 16 in this book. Overexpression of glutathione synthetase and g-glutamyl cysteine synthetase enhances cadmium accumulation and tolerance in Indian mustard plants (Zhu et al., 1999 a, b) [109, 110]. These studies show that the manipulation of GSHa concentrations has significant potential for increasing the plant accumulation of metals (Meagher 2000) [111].

Because most metal hyperaccumulators are slow growing and have low biomass, bioengineering of nonaccumulators is essential for effective phytoremediation. Conventional breeding approaches have also been proposed to improve plants for metal extraction. However, the success of this approach is doubtful due to sexual incompatibility between the parent lines [93]. Biotechnology has the potential of overcoming this limitation. However, comprehensive knowledge of the genetic basis for hyperaccumulation is essential for effective use of biotechnology to design transgenic plants capable of efficient phytoremediation.

14.4 ZINC-TRANSPORTING GENES IN PLANTS

Zinc is a constituent of several enzymes: carbonic anhydrase dehydrogenases; aldolases; Cu/Zn superoxide dismutase; isomerases; transphosphory-lases; and RNA and DNA polymerases. Therefore, Zn deficiency results in malfunction or no function of these enzymes. Zn-metalloproteins (Zn-finger motif) are regulators of gene expression (DNA-binding transcription factors). In the absence of these, RNA polymerase cannot complete its function of transcribing genetic information from DNA into RNA. Zinc is a cofactor of more than 200 enzymes, such as oxidoreductases, hydrolases, transferases, lyases, isomerases, and ligases. Many of the metalloenzymes are involved in the synthesis of DNA and RNA and protein synthesis and metabolism.

Metal-transporting genes have been identified in *Arabidopsis* (Brassicaceae). Overexpression of an *Arabidopsis* zinc transporter cation diffusion facilitator (CDF) gene enhanced resistance to Zn accumulation. Transgenic plants showed increased Zn uptake and tolerance and antisense of this gene led to wild-type Zn tolerance in transgenic plants. Zinc transporters can be manipulated to increase selectivity and accumulation of metal ions. In Brassicaceae, about 21 species belonging to three genera (*Cochlearia, Arabidopsis,* and *Thlaspi)* are reported to be zinc hyperaccumulators. Zhao et al. [112] isolated and identified the gene *ZNT1* as one of the micronutrient transport genes with high sequence homology with other Zn transport genes isolated from yeast.

A family of zinc transporter genes that responds to zinc deficiency has also been identified in *Arabidopsis.* Zn hyperaccumulation in *Thlaspi caerulescens* is because of the *ZNT1* gene, which encodes a high-affinity Zn transporter. This gene is constitutively expressed at a much higher level in *T. caerulescens* than in *T. arvense,* where its expression is stimulated by Zn deficiency. In fact, plant Zn status is shown to alter the normal regulation of Zn transporter genes in *T. caerulescens.* An important aspect of Zn hyperaccumulation and tolerance in *T. caerulescens* is also the production of low molecular weight compounds involved in Zn detoxification in the cell (cytoplasm and vacuole) and in the long-distance transport of Zn in the xylem vessels.

Citrate was not shown to play an important role in Zn chelation and malate had constitutively high concentrations in the shoots of the accumulator *T. caerulescens* and the nonaccumulator *T. ochroleucum.* More recently, direct measurements of the *in vivo* speciation of Zn in *T. caerulescens* using the noninvasive technique of x-ray absorption have revealed that histidine is responsible for the transport of Zn within the cell, whereas organic acids (citrate and oxalate) chelate Zn during long-distance transport and storage. Another constitutive aspect of *T. caerulescens* is the high Zn requirement for maximum growth, compared to other species. This probably depends on the strong expression of the metal sequestration mechanism, which would subtract a large amount of intracellular Zn to normal physiological processes even when the Zn supply is low.

Several transporters implicated in the uptake of divalent nutrient cations like Ca^{2+}, Fe^{2+}, or Zn^{2+} appear to be able to transport other divalent cations. For example, heterologous expression in yeast of IRT1 from *A. thaliana,* an iron-repressed transporter in the *ZIP* family of metal transporters, suggests a broad-range specificity of transport for Cd^{2+}, Fe^{2+}, Mn^{2+}, Zn^{2+}, and possibly other divalent cations (TC 2.A.5.1).

The expression of IRT1 is strongly induced in plants under conditions of iron deficiency and is repressed in iron-replete plants. This correlates well with the finding that cadmium uptake is enhanced in iron-deficient pea seedlings. However, these transporters are tightly regulated at the transcriptional and post-transcriptional levels, and, to date, no reports on plants engineered to overexpress transporters of the *ZIP* family have been issued. Tobacco plants engineered to contain increased amounts of NtCBP4 protein (TC 1.A.1.5.1), a putative cyclic-nucleotide and calmodulin-regulated cation channel in the plasma membrane, displayed an increased sensitivity to lead, a 1.5- to 2.0-fold shoot accumulation of lead, and an increased nickel tolerance. Yeast cells expressing the wheat *LCT1* cDNA (TC 9.A.20.1.1), encoding a low-affinity cation transporter, were hypersensitive to Cd^{2+} and accumulated increased amounts of cadmium [84]. Plants overexpressing *AtNramp3* (TC 2.A.55), a member of the Nramp family of metal transporters, were hypersensitive to Cd^{2+}, but enhanced cadmium accumulation was not observed.

In phytoremediation, it must be considered that a transporter capable of transporting a specific contaminant metal cation is capable of transporting other competing cations, like Ca^{2+} or Zn^{2+}, under natural soil conditions if the latter ions are present in large excess. Therefore, it is desirable to better understand what governs the specificity of membrane transporters, in order to generate mutated transporters with altered specificities [84]. Understanding the regulation of *ZIP* family members in *T. caerulescens* and analyzing *Arabidopsis* mutants with altered metal responses will also help to identify novel target genes and strategies for the generation of plants with enhanced metal uptake [93].

To date, numerous examples have been demonstrated to have the potential for phytoremediation — for example, Pb, Ni, Zn, Al, Se, Au, and As. Arazi et al. [28] have described a tobacco plasma membrane calmodulin-binding transporter that confers Ni^{2+} tolerance and Pb^{2+} hypersensitivity. Zinc-transporter To investigate the *in vivo* role of this gene, transgenic plants with the ZAT coding sequence exhibited increased Zn resistance and accumulation in the roots at high Zn concentrations [114].

In maize, zinc accumulation was found to be genetically controlled and affected by additive genes [115]. Four genes were found to be the minimum segregation factors in the (high × low) crosses for Zn accumulation. Zinc deficiency also increases root exudation of amino acids, sugars, and phenolic substances at different degrees in different species [116].

Three wheat genotypes (*Triticum aestivum* and *T. turgidum*) differed in their root-growth response to low zinc levels [117]. The zinc-efficient genotype increased root and shoot dry matter and developed longer and thinner roots (a greater proportion of fine roots with diameter of 0.2 mm) compared with the less efficient genotype. Due to a larger root surface area, the efficiency of zinc uptake increased. In wheat, Zn can be remobilized from leaves under Zn deficiency. Also, spinach, potato, navy bean, tomato, sorghum, and maize show great variations in Zn efficiency [118,119].

Arabidopsis thaliana has multiple Zn transporters designated ZIP1, *ZIP2*, *ZIP3*, and *ZIP4*. Grotz et al. [119] demonstrated the specificities of each of the *ZIP* genes with their experiments. They tested other metal ions for their ability to inhibit Zn uptake mediated by these proteins. Zn uptake by *ZIP1* was not inhibited by a tenfold excess of Mn, Ni, Fe, and Co. Zn was the most potent competitor, demonstrating that *ZIP1* prefers Zn as its substrate over theses metal ions. Cd and Cu also inhibited Zn uptake, but to a lesser extent. This suggests that Cd and Cu may also be substrates for *ZIP1*.

The *ZIP* family members have 309 to 476 aa; this range is largely due to variation in the number of residues between transmembrane domains III and IV, a domain designated as "variable." The amino acid sequences of all the known *ZIP* family members were aligned, and a dendrogram describing their sequence similarities was generated [120]. *ZNT1* is 379 aa in length and shares the same structural features exhibited by the other members of the *ZIP* family, including eight putative transmembrane domains and a highly hydrophilic cytoplasmic region predicted to reside between transmembrane domains three and four. The putative cytoplasmic domain contains a series of histidine repeats, which may define a metal-binding region for the transporter. Zinc transporters can be manipulated to increase selectivity and accumulation of metal ions.

Pence et al. [113] reported on the cloning and characterization of a high-affinity Zn^{2+} transporter cDNA, *ZNT1*, from the Zn/Cd-hyperaccumulating plant, *Thlaspi caerulescens*. Through comparisons to a closely related, nonaccumulator species, *Thlaspi arvense*, the researchers determined that the elevated ability of *T. caerulescens* to take up Zn and Cd was due, in part, to an enhanced level of expression of Zn transporters. Previous physiological studies by the group indicated that the hyperaccumulating ability of *T. caerulescens* was linked to Zn transport at a number of sites in the plant. The researchers transformed a Zn transport-deficient strain of yeast, ZHY3, with a cDNA library from *T. caerulescens*. By screening for growth on low-Zn medium, they were able to isolate seven clones, five of which represented a 1.2-kb cDNA designated *ZNT1* (for Zn transporter) that restored the yeast's ability to grow under low-Zn conditions. The *ZNT1* gene displayed considerable identity to two *Arabidopsis* thaliana metal transporter genes, *ZIP4* (for transporting Zn) and *IRT1* (for transporting Fe).

For purposes of comparison, they then cloned the homolog of *ZNT1* (designated *ZNT1*-arvense) from the nonhyperaccumulator species *T. arvense*. Expression studies using northern blots of RNA isolated from the roots and shoots of both *Thlaspi* species revealed that *ZNT1* is expressed in *T. caerulescens* at extremely high levels. In contrast, expression of *ZNT1* in *T. arvense* could only be detected at a very low level in shoots and roots, and then only when the plants had been exposed to conditions of Zn deficiency. To further explore the role of Zn status on *ZNT1* expression, Pence et al. [113] exposed both species to a range of Zn concentrations. They found that when *T. caerulescens* was grown in a nutrient solution containing an excess of Zn (50 μM), the transcript level of *ZNT1* decreased, indicating that *ZNT1* is not expressed constitutively in *T. caerulescens*. *ZNT1* transcript levels in *T. arvense* appeared to be unaffected by exposure to excess Zn.

Transport studies in *T. caerulescens* show that *ZNT1* mediates high-affinity Zn transport. In many plant species, the induction of a high-affinity transporter is characteristic of a nutrient-deficiency response and would correlate with the expression pattern observed for *ZNT1* in *T. arvense*.

The authors speculate that the hyperaccumulation phenotype in *T. caerulescens* may then be due to a mutation in the plant's ability to sense or respond to Zn levels — that is, these plants may be functioning as if they are constantly experiencing Zn deficiency. They propose that this is likely the result of a change in global regulation linked to the plant's overall Zn status; this supports the concept that the Zn hyperaccumulation phenotype, at least in this species, is due to a change in the regulation and not the constitutive expression of a single gene.

Several mutants with altered response to heavy metals have been isolated from *A. thaliana*. Cadmium-hypersensitive mutants with defects in phytochelatin synthetase and possibly in g-glutamylcysteine synthetase and glutathione synthetase have been isolated by Howden et al. [65,66]. Chen and Goldsbrough [120] found an increased activity of g-glutamylcysteine synthetase in tomato cells selected for cadmium tolerance. Some of these genes may prove useful in modifying suitable target plants for phytoremediation, although there are doubts about the usefulness of genes involved in phytochelatin synthesis [122].

14.5   FERRITIN EXPRESSION IN RICE

Ferritin is an iron storage protein found in animals, plants, and bacteria. It comprises 24 subunits, which may surround in a micellar up to 4500 ferric atoms [123]. It provides iron for the synthesis of iron proteins such as ferredoxin and cytochromes. It also prevents damage from free radicals produced by iron/dioxygen interactions. Ferritin has been found to provide an iron source for treatment of anemia. It was thus proposed that increase of the ferritin content of cereals by genetic modification may help to solve the problem of dietary iron deficiency. To increase the Fe content of rice, Goto et al. [25] transferred soybean ferritin gene into the plant. Using the rice seed storage protein glutelin promoter (GluB-1), they could target the expression of ferritin in developing seeds. The Fe content in transformed seeds was threefold compared to that in control seeds.

14.6   GENETIC MANIPULATION OF ORGANIC ACID
BIOSYNTHESIS

It has been proposed that metal tolerance could be based on the organic acid-formed complexes. Ernst [123] observed high malate concentrations in Zn- and Cu-tolerant plants; also the content of citrate was increased. Hyperaccumulators are heavily loaded with these acids and acid anions might have some function in metal storage or plant internal metal transport. Free histidine has been found as a metal chelator in xylem exudates in plants that accumulate Ni and the amount of free histidine increases in Ni exposure [80]. By modifying histidine metabolism, it might be possible to increase the Ni-accumulating capacity of plants.

During the past few years, several metal transporters have been isolated from *Arabidopsis*: Zn transporters *ZIP1, 3, 4* [120]; Fe transporter IRT1 [125]; and Cu transporter *COPT1* [126, 127]. Several transporters, like *ZIP1, ZIP3* and *IRT1*, are expressed in response to metal deficiency. *IRT1* may also play a role in the uptake of other metals because Cd, Zn, Co, and Mn inhibited Fe uptake of *IRT1* [125]. Changing the regulation of the expression of these transporters may modify the uptake of metals to the cells or organelles in a useful way.

Most of the studies aimed at determining the role of organic acid excretion have been carried out by comparing different plant species or nonisogenic lines of the same plant species. Plant transformation allows the production of genetically identical plants that differ only in one or a few genes. Taking advantage of this technology, the author's research group produced transgenic plants with an enhanced capacity to synthesize and excrete citrate. It was reasoned that, by overproducing citrate (one of the most powerful cation chelators in the organic acid group), the actual relevance of organic acids in several aspects of the plant–soil relationship could be elucidated.

To produce citrate-overproducing (CSb) plants, the coding sequence of the bacterial citrate synthase gene was placed under control of the 35S CaMV promoter and the nopaline synthase 3% end sequence (35S–CSb). This construct was used to transform tobacco and papaya plants. To determine whether the expression of a citrate synthase in plant cells leads to an increase in their citrate content, total and root extracts of transgenic lines were examined by HPLC and compared to control plants. It was found that the tobacco lines expressing the 35S–CSb construct had up to tenfold higher levels of citrate in their root tissue. The amount of citrate exuded by the roots of these transgenic lines was also increased up to fourfold as compared to control plants [81].

De la Fuente et al. [81] characterized the novel transgenic CSb lines to determine whether these plants were tolerant to aluminum. Experimental evidence suggests that the citrate-overproducing plants could tolerate a tenfold higher Al concentration than control plants. A mitochondrial citrate synthase of *Arabidopsis thaliana* was introduced into carrot (*Daucus carota*) cells by *Agrobacterium tumefaciens*. Several transgenic carrot cell lines that produced the *Arabidopsis* CS polypeptides and had high CS activity were identified. The increase in CS expression resulted in an enhanced capacity of phosphate uptake from insoluble sources of P in these transgenic cells [81,88].

More recently, this research group has shown that transgenic *Arabidopsis* plants that express high levels of the carrot citrate synthase have an enhanced aluminum tolerance. It has been reported that organic acid excretion by lupin plants constitutes a drain of 5 to 25% of the total fixed C; however, this does not appear to affect dry matter production significantly. This fact has also been confirmed in transgenic tobacco plants that overproduce citrate, which grow efficiently even at high levels of P-fertilization [83]. The insights obtained from transgenic models highlight the potential of organic acid manipulation to generate novel crops more efficient in the use of soil P and well adapted for growth in marginal soils [89].

14.7 MOLECULAR GENETIC AND TRANSGENIC STRATEGIES FOR PHYTOREMEDIATION HYPERACUMULATION

Transgenic plants capable of tolerating high levels of accumulated cadmium and lead have been developed recently [128]. These plants take up heavy metals more rapidly than traditional bioremediation plants do, thus making them potential hyperaccumulators with application for phytoextraction and rhizofiltration in the field [129-131].

Observing that certain *Saccharomyces cerevisiae*, which possess the *YCF1*, or yeast cadmium factor 1 protein, are known to pump cadmium [Cd(II)] into vacuoles, Li et al. [129] tested whether *YCF1* would also confer resistance to lead [Pb(II)]. Also known as vacuolar glutathione S-conjugate transporter, *YCF1* belongs to the ATP-binding cassette superfamily [2,3]. Li's team confirmed that *YCF1* gene expression permitted *S. cerevisiae* to withstand the toxic effects of 3 mM lead (Pb II) and 0.1 mM cadmium (Cd II) concentrations in growth media. This protection against lead and cadmium toxicity was due to the uptake and storage of the heavy metals in yeast vacuoles.

Next, Li's group attempted to determine whether *YCF1* expression in plants produced the same results. *Arabidopsis thaliana* was investigated as a model for *YCF1* expression. First, the *YCF1* gene was created using RT-PCR from *YCF1* expressing yeast. For expression in *A. thaliana*, Li and colleagues subcloned the *YCF1* gene into two vectors: PBI121 and pCambia1302. To enhance expression in plants, the pCambia1302 vector was cloned with four copies of the CaMV 35S promoter. *Agrobacterium tumefaciens* was used for transformation of *A. thaliana*. Green fluorescent protein tagged to *YCF1*, used as an expression reporter, indicated the presence of *YCF1* protein in the vacuolar as well as in the plasma membrane of the transformed *Arabidopsis* cells.

Li and coworkers investigated the uptake and sequestering of lead and cadmium in the plants. Transformed *A. thaliana* was grown on gravel supplemented with half-concentration Murashige–Skoog agar medium containing 0.75 mM lead or 70 μM cadmium. After 3 weeks, the plant tissues were analyzed for metal uptake using atomic absorption spectroscopy. Li's findings showed that the transgenic plants were as effective as naturally occurring hyperaccumulators. Although the transgenic plants accumulated less than twofold higher concentrations of Cd and Pb compared to wild-type, this is likely much less than the hyperaccumulator plants (mentioned earlier). Plants that exclude heavy metals have been demonstrated by expressing bacterial heavy metal transporter in *Arabidopsis* that enhances resistance to and decreases uptake of heavy metals [131].

Genetic strategies and transgenic plant and microbe production and field trials will fetch phytoremediation field applications [20,104,114,132–134]. Mercury is a worldwide problem as a result of its many diverse uses in industry. Mercury has been used in bleaching operations (chlorine production, paper, textiles, etc.); as a catalyst; as a pigment for paints; for gold mining; and as a fungicide and antibacterial agent in seeds and bulbs. Elemental mercury, Hg(0), can be a problem because it is oxidized to Hg^{2+} by biological systems and subsequently is leached into wetlands, waterways, and estuaries. Additionally, mercury can accumulate in animals as methylmercury ($CH_3–Hg^+$), dimethylmercury ($(CH_3)_2–Hg$) or other organomercury salts. Organic mercury produced by some anaerobic bacteria is one to two orders of magnitude more toxic in some eukaryotes; it is more likely to biomagnify than ionic mercury and efficiently permeates biological membranes. Monomethyl–Hg is responsible for severe neurological degeneration in birds, cats, and humans.

Certain bacteria are capable of pumping metals out of their cells and/or oxidizing, reducing, or modifying the metal ions to less toxic species. One example is the *mer* operon, which contains genes that sense mercury (*mer*B); transport mercury (*mer*T); sequester mercury to the periplasmic space (*mer*P); and reduce mercury (*mer*A). *Mer*B is a subset of the *mer* operon and is capable of catalyzing the breakdown of various forms of organic mercury to Hg^{2+}. *Mer*B encodes an enzyme, organomercurial lyase, that catalyses the protonolysis of the carbon–mercury bond. One of the products of this reaction is ionic mercury [92,135]:

$$Hg^{2+}. \ R–CH_2–Hg^+ \ ----merB---->R–CH_3 + Hg(II)$$

$$Hg(II) + NADPH \ ----merA---->Hg(0) + NADP^+ + H^+$$

Hg(0) (elemental mercury) can be volatilized by the cell.

14.8 CONCLUSIONS

Tree crop improvement through biotechnology is important for the remediation of contaminated environments because trees have extensive root systems to ensure an efficient uptake of pollutants (cadmium, mercury, pesticides) from soil and provide the possibility of several cycles of decontamination with the same plants. Transgenic trees with increased tolerance to heavy metals might be a solution towards sustainable development. Therefore, if overexpressed in annual and perennial crops in combination with bacterial enzymes for GSH synthesis, specific glutathione-S-transferases and phytochelatin synthase might go a long way. The genetic manipulation of organic acid metabolism could be used to develop transgenic varieties more adapted to marginal soils and more efficient in the assimilation of nutrients.

Alkaline soils of semiarid climates, which have traditionally sustained traditional rain-fed cultivation — however low P or Fe availability may be (even when fertilizers have been used) — have strongly limited more productive crop management [85,87]. The knowledge of organic acid biosynthesis, its universal occurrence in plants, and the effectiveness of organic acid exudation in conferring Al tolerance and enhanced P uptake from sparingly soluble P compounds makes the organic acid pathway a promising target to develop transgenic varieties better adapted to grow in marginal soils. A major opportunity is to modify the quality and quantity of these organic compounds to target the rhizosphere; in this way, genetic manipulation can contribute to a better understanding of the specific carbon substrates preferred by beneficial microorganisms, such as N2-fixing soil bacteria and mycorrhizal fungi.

Currently, many different genes involved in organic acid biosynthesis have been cloned from several organisms. Through genetic engineering, it could be possible to overexpress these genes under the regulation of strong and tissue-specific plant promoters. The incorporation of transgenic crops into integrated plant management and land use strategies could represent a promissory option for agricultural expansion with a lower environmental cost. It is certainly a research priority to achieve a more sustainable agriculture for future generations.

REFERENCES

1. Shaw, B.P., Sahu, S.K., and Mishra R.K. Heavy metal induced oxidative damage in terrestrial plants, in *Heavy Metal Stress in Plants — from Biomolecules to Ecosystems,* Prasad, M.N.V. (Ed.). Springer–Verlag, Heidelberg, 2004, chap. 4.
2. Ross, S.M. *Toxic Metals in Soil Plant Systems*, John Wiley & Sons, Chichester, U.K., 469, 1994.
3. De Miranda, J.R. et al. Metallothionein genes from the flowering plant *Mimulus guttatus, FEBS Lett.,* 260, 277, 1990.
4. Evans, I.M. et al. A gene from pea (*Pisum sativum* L.) with homology to metallothionein genes, *FEBS Lett.,* 262, 29, 1990.
5. Okumura, N. et al. An iron deficiency-specific cDNA from barley roots having two homologous cysteine-rich MT domains, *Plant Mol. Biol.,* 17, 531, 1991.
6. Nakanishi, H. et al. A plant metallothionein-like gene from iron deficiency barley roots. GenBank accession no. D50641, 1995.
7. De Framond, A.J. A metallothionein-like gene from maize (*Zea mays*). Cloning and characterization, *FEBS Lett.,* 290, 103, 1991.
8. Chevalier, C. et al. Molecular cloning and characterization of six cDNAs expressed during glucose starvation in excised maize (*Zea mays* L.) root tips, *Plant Mol. Biol.,* 28, 473, 1995.
9. Snowden, K.C. and Gardner, R.C. Five genes induced by aluminum in wheat (*Triticum aestivum* L.) roots, *Plant Physiol.,* 103, 855, 1993.
10. Ellison, N.W. Sequence analysis of two cDNA clones encoding metallothionein-like proteins from white clover (*Trifolium repens* L.), GenBank accession no. Z26493, 1993.
11. Zhou, J. and Goldsbrough, P.B. Functional homologs of full gal metallothionein genes from Arabidopsis, *Plant Cell,* 6, 875, 1994.
12. Zhou, J. and Goldsbrough, P.B. Structure, organization and expression of the metallothionein gene family in Arabidopsis, *Mol. Gen. Genet.,* 248, 318, 1995.
13. Yeh, S-C., Hsieh, H-M., and Huang, P.C. Transcripts of metallothionein genes in *Arabidopsis thaliana.* DNA sequence, *J. Seq. Map,* 5, 141, 1995.
14. Buchanan–Wollaston, V. Isolation of cDNA clones for genes that are expressed during leaf senescence in *Brassica napus*. Identification of a gene encoding a senescence-specific metal lothionein-like protein, *Plant Physiol.,* 105, 839, 1994.
15. Hsieh, H-M., Liu, W-K., and Huang, P.C. A novel stress-inducible metallocell biochemistry and biophysics thionein-like gene from rice, *Plant Mol. Biol.,* 28, 381, 1995.
16. Lee, M.C., Kim, C.S., and Eun, M.Y. Characterization of metallothionein-like protein from rice, GenBank accession no. AFO17366, 1997.
17. Hudspeth, R.L. et al. Characterization and expression of metallothionein-like genes in cotton, *Plant Mol. Biol.,* 31, 701, 1996.
18. Foley, R.C., Liang, Z.M., and Singh, K.B. Analysis of type 1 metallothionein cDNAs in *Vicia faba, Plant Mol. Biol.,* 33, 583, 1997.
19. Ma, M. et al. Cloning and sequencing of the thetaliothionein-like cDNA from *Festuca rubra* cv. Merlin, GenBank accession no. U96646, 1997.
20. Misra, S. and Gedamu, L. Heavy metal tolerant transgenic *Brassica napus* L. and *Nicotiana tabacum* L. plants, *Theor. Appl. Genet.,* 78, 161, 1989.

21. Pan, A. Alpha-domain of human metallothionein IA can bind to metals in transgenic tobacco plants, *Molecular Gen. Genet.*, 242, 666, 1994.

22. Evans, K.M. et al. Expression of the pea metallothionein-like gene *PsMTA* in *Escherichia coli* and *Arabidopsis thaliana* and analysis of trace metal ion accumulation: implications for *PsMTA* function, *Plant Mol. Biol.*, 20, 1019, 1992.

23. Hasegawa, I. et al. Genetic improvement of heavy metal tolerance in plants by transfer of the yeast metallothionein gene (*CUP1*), *Plant Soil*, 196, 277, 1997.

24. Ezaki, B. et al. Expression of aluminum-induced genes in transgenic *Arabidopsis* plants can ameliorate aluminum stress and/or oxidative stress, *Plant Physiol.*, 122, 657, 2000.

25. Goto, F., Yoshihara, T., Shigemoto, N., and Toki Sand Takaiwa, F. Iron fortification of rice seed by the soybean ferritin gene, *Nat. Biotechnol.*, 17, 282, 1998.

26. Van der Zaal, B. J. et al. Overexpression of a novel *Arabidopsis* gene related to putative zinc-transporter genes from animals can lead to enhanced zinc resistance and accumulation, *Plant Physiol.*, 119, 1047–1055, 1999.

27. Hirschi, K.D. et al. Expression of *Arabidopsis* CAX2 in tobacco altered metal accumulation and increased manganese tolerance, *Plant Physiol.*, 124, 125, 1999.

28. Arazi, T. et al. A tobacco plasma membrane calmodulin-binding transporter confers Ni^{2+} tolerance and Pb^{2+} hypersensitivity in transgenic plants, *Plant J.*, 20, 171, 1999.

29. Samuelsen, A.I., Martin, R.C., Mok, D.W.S., and Machteld, C.M. Expression of the yeast FRE genes in transgenic tobacco, *Plant Physiol.*, 118, 51, 1998.

30. Curie, C. et al. Involvement of Nramp1 from *Arabidopsis thaliana* in iron transport, *Biochem. J.*, 347, 749, 2000.

31. Thomine, S. et al. Cadmium and iron transport by members of a plant metal transporter family in *Arabidopsis* with homology to Nramp genes, *Proc. Natl. Acad. Sci. USA*, 97, 4991, 2000.

32. Rugh, C.L., Bizily, S.P., and Meagher, R.B. Phytoreduction of environmental pollution, in *Phytoremediation of Toxic Metal Metals Using Plants to Clean up the Environment*, Raskin I. and Ensley, B.D., Eds. John Wiley & Sons, New York, 2000, 151–171.

33. Bizily, S.P. Phytoremediation of methylmercury pollution: *mer*B expression in *Arabidopsis thaliana* confers resistance to organomercurials, *Proc. Natl. Acad. Sci. USA*, 96, 6808, 1999.

34. Moffat, A.S. Engineering plants to cope with metals, *Science*, 285, 369, 1999.

35. Kawashima, I. et al. Isolation of a gene for a metallothionein-like protein from soybean, *Plant Cell Physiol.*, 32, 913, 1991.

36. Takahashi, K. GenBank accession no.X62818, 1991.

37. Weig, A. and Komor, E. Isolation of a class II metallothionein cDNA from *Ricinus communis* L., GenBank accession no. L02306, 1992.

38. Foley, R.C. and Singh, K.B. Isolation of a *Vicia faba* metallothionein-like gene, expression in foliar trichomes, *Plant Mol. Biol.*, 26, 435, 1994.

39. Ledger, S.E. and Gardner, R.C. Cloning and characterization of five cDNAs for genes differentially expressed during fruit development of kiwifruit (*Actinidia deliciosa* var. *deliciosa*), *Plant Mol. Biol.*, 25, 877, 1994.

40. Moisyadi, S. and Stiles, J.I. A cDNA encoding a metallothionein I-like protein from coffee leaves (*Coffea arabica*), *Plant Physiol.*, 107, 295, 1995.

41. Kim, H.U. et al. Nucleotide sequence of cDNA clone encoding a metallothionein-like protein from Chinese cabbage, *Plant Physiol.*, 108, 863, 1995.

42. LaRosa, P.C. and Smigocki, A.C. A plant metallothionein is modulated by cytokinin. GenBank accession no. U35225, 1995.

43. Choi, D. et al. Molecular cloning of a metallothionein-like gene from *Nicotiana glutinosa* L. and its induction by wounding and tobacco mosaic virus infection, *Plant Physiol.*, 112, 353, 1996.

44. Ellison, N.W. and White, D.W.R. Isolation of two cDNA clones encoding metallothionein-like proteins from *Trifolium repens* L., *Plant Physiol.*, 112, 446. GenBank accession no. Z26492, 1996.

45. Kitashiba, H. et al. Identification of genes expressed in the shoot apex of *Brassica campestris* during floral transition, *Sex. Plant Reprod.*, 9, 186, 1996.

46. Hsieh, H.M. et al. RNA expression patterns of a type 2 metallothionein-like gene from rice, *Plant Mol. Biol.*, 32, 525, 1996.

47. Giritch, A. et al. Cloning and characterization of metallothionein-like genes family from tomato, GenBank accession nos. Z68138, Z68309, Z68310, 1995, 1998.

48. Whitelaw, C.A. et al. The isolation and characterization of type II metallothionein-like genes from tomato (*Lycopersicon esculenturn* L.), *Plant Mol. Biol.*, 33, 504, 1997.

49. Buchanan–Wollaston, V. and Ainsworth, C. Leaf senescence in *Brassica napus*, cloning of senescence related genes by subtractive hybridization, *Plant Mol. Biol.*, 33, 821, 1997.

50. Schaefer, H.J., Haag–Kerwer, A., and Rausch, T. cDNA cloning and expression analysis of genes encoding GSH synthesis in roots of the heavy metal accumulator *Brassica juncea* L., evidence for Cd-induction of putative mitochondrial γ-glutamylcysteine synthetase isoform, GenBank accession nos. Y10849, Y10850, Y10851, Y10852, 1997.

51. Mbeguie, A., Mbeguie, D., Gomez, R-M., and Fils–Lycaon, B. Molecular cloning and nucleotide sequence of an abscisic acid-, ripening-induced (ASR)-like protein from apricot fruit (accession no. 093164). Gene expression during fruit ripening, *Plant Physiol.*, 115, 1288, 1997.

52. Davies, E.C. and Thomas, J.C. A metallothionein from a facultative halophyte confers copper tolerance, GenBank accession no. AFOO0935, 1997.

53. Lee, M.C. et al. Molecular cloning and characterization of metallothionein-like protein in rice, GenBank accession nos. Y08529, 077294, 1996.

54. Yu, L. et al. Characterization of a novel metallothionein-like protein gene with strong expression in the stem of rice, GenBank accession no. AB002820, 1997.

55. Reid, S.J. and Ross, G.S. Two cDNA clones encoding metallothionein-like proteins in apple are upregulated during cool storage, GenBank accession no. 061974, 1996.

56. Rosenfield, C.L., Kiss, E., and Hrazdina, G. MdACS-2 (Accession No. 073815) and MdACS-3 (Accession No. 073816), two new 1-aminocyclopropane-1-carboxylate synthase in ripening apple fruit, *Plant Physiol.*, 112, 1735. GenBank accession no. Y08322, 1996.

57. Clendennen, S.K. and May, G.D. Differential gene expression in ripening banana fruit, *Plant Physiol.*, 115, 463, 1997.

58. Wiersma, P.A., Wil, Z., and Wilson, S.M. A fruit-related metallothionein-like cDNA clone from sweet cherry (accession no. AF028013) corresponds to fruit genes from diverse species, *Plant Physiol.*, 116, 867, 1998.

59. Murphy, A. et al. Purification and immunological identification of metallothioneins 1 and 2 from *Arabidopsis thaliana*, *Plant Physiol.*, 113, 1293, 1997.

60. Chatthai, M. The isolation of a novel metallothionein-related cDNA expressed in somatic and zygotic embryos of Douglas fir, regulation by ABA, osmoticum, and metal ions, *Plant Mol. Biol.*, 34, 243, 1997.

61. Aguilar, M. et al. Isolation of a cDNA encoding metallothionein-like protein (Accession No. 081041) from strawberry fruit, *Plant Physiol.*, 113, 664, 1997.

62. Prasad, M.N.V. (Ed.). *Heavy Metal Stress in Plants: from Biomolecules to Ecosystems*. Springer–Verlag, Heidelberg. 2nd ed., 462, xiv, 2004.

63. Sanità Di Toppi, L., Gremigni, P., Pawlik Skowronska B., Prasad, M.N.V., and Cobbett C.S. Responses to heavy metals in plants — molecular approach, in *Abiotic Stresses in Plants*. Sanità Di Toppi, L. and Pawlik Skowronska, B. (Eds.). Kluwer Academic Publishers, Dordrecht, 133–156, 2003.

64. Howden, R. and Cobbett, C.S. Cadmium-sensitive mutants of *Arabidopsis thaliana*, *Plant Physiol.*, 99, 100, 1992.

65. Howden, R. et al. A cadmium-sensitive, glutathione-deficient mutant of *Arabidopsis thaliana*, *Plant Physiol.*, 107, 1067, 1995.

66. Howden, R. Cadmium-sensitive, cad1 mutants of *Arabidopsis thaliana* are phytochelatin deficient, *Plant Physiol.*, 107, 1059, 1995.

67. Murphy, A. and Taiz, L. Comparison of metallothionein gene expression and nonproteithiolsinten *Arabidopsis* ecotypes, *Plant Physiol.*, 109, 945, 1995.

68. Lefebvre, D. D. et al. Mammalian metallothioneins functions in plants, *BioTechnology*, 5, 1053, 1987.

69. Maiti, I.B., Hunt, A.G., and Wagner, G.J. Seed-transmissible expression of mammalian metallothionein in transgenic tobacco, *Biochem. Biophys. Res. Commun.*, 150, 640, 1988.

70. Maiti, I.B. Inheritance and expression of the mouse metallothionein gene in tobacco, *Plant Physiol.*, 91, 1020, 1989.

71. Maiti, I.B. Light-inducible and tissue-specific expression of a chimeric mouse metallothionein cDNA gene in tobacco, *Plant Sci.,* 76, 99, 1991.
72. Yeargan, R. et al. Tissue partitioning of cadmium in transgenic tobacco seedlings and field grown plants expressing the mouse metallothionein I gene, *Transgenic Res.,* 1, 261, 1992.
73. Brandle, J.E. Field performance and heavy metal concentrations of transgenic ue-cured tobacco expressing a mammalian metallothionein-β-glucuronidase gene fusion, *Genome,* 36, 255, 1993.
74. Pan, A. Construction of multiple copy of alpha-domain gene fragment of human liver metallothionein IA in tandem arrays and its expression in transgenic tobacco plants, *Prot. Eng.,* 6, 755, 1993.
75. Pan, A. Expression of mouse metallothionein-I gene confers cadmium resistance in transgenic tobacco plants, *Plant Mol. Biol.,* 24, 341, 1994.
76. Elmayan, T. and Tepfer, M. Synthesis of a bifunctional metallothionein β-glucuronidase fusion protein in transgenic tobacco plants as a means of reducing leaf cadmium levels, *Plant J.,* 6, 433, 1994.
77. Hattori, J., Labbe, H., and Miki, B.L. Construction and expression of a metallothionein-beta-glucuronidase gene fusion, *Genome,* 37, 508, 1994.
78. Kramer, U. Cadmium for all meals — plants with an unusual appetite, *New Phytol.,* 145, 1, 2000.
79. Chardonnens, A.N. et al. Properties of enhanced tonoplast zinc transport in naturally selected zinc-tolerant *Silene vulgaris, Plant Physiol.,* 120, 779, 1999.
80. Kramer, U. Free histidine as a metal chelator in plants that accumulate nickel, *Nature,* 373, 635, 1996.
81. De La Fuente, J. M. et al. Aluminum Tolerance in Transgenic Plants by Alteration of Citrate Synthesis, *Science,* 276, 1566-1568, 1997.
82. Vazquez, M.D. et al. Compartmentalization of zinc in roots and leaves of the zinc hyperaccumulator *Thlaspi caerulescens* J. & C. Presl., *Bot. Acta,* 107, 243, 1994.
83. Clemens, S. et al. Tolerance to toxic metals by a gene family of phytochelatin synthases from plants and yeast, *EMBO J.,* 18, 3325, 1999.
84. Zenk, M.H. Heavy metal detoxification in higher plants — a review, *Gene,* 179, 21, 1996.
85. Sanità Di Toppi, L., Prasad, M.N.V., and Ottonello, S. Metal chelating peptides and proteins in plants, in *Physiology and Biochemistry of Metal Toxicity and Tolerance in Plants,* Prasad, M.N.V. and Strzaka, K. (Eds.), Kluwer Academic Publishers, Dordrecht, 2002, 59–93.
86. Leustek, T. et al. Pathways and regulation of sulfur metabolism revealed through molecular and genetic studies, *Annu. Rev. Plant Physiol. Mol. Biol.,* 51, 141, 2000.
87. Schmöger, M.E.A., Oven, M., and Grill, E. Detoxification of arsenic by phytochelatins in plants, *Plant Physiol.,* 122, 793, 2000.
88. Hartley–Whitaker, J. et al., Phytochelatins are involved in differential arsenate tolerance in *Holcus lanatus, Plant Physiol.,* 126, 299, 2001.
89. Gisbert, C. et al. A plant genetically modified that accumulates Pb is especially promising for phytoremediation, *Biochem. Biophys. Res. Commun.,* 303, 440, 2003.
90. Brown, S.L., Zinc and cadmium uptake by *Thlaspi caerulescens* and *Silene vulgaris* grown on sludge-amended soils in relation to total soil metals and soil pH, *Environ. Sci. Technol.,* 29, 1581, 1995.
91. Heaton, A.C.P. et al. Phytoremediation of mercury- and methylmercury-polluted soils using genetically engineered plants, *J. Soil Contamination,* 7, 497, 1998.
92. Lasat, M.M. et al. Molecular physiology of zinc transport in the Zn hyperaccumulator *Thlaspi caerulescens, J. Exp. Bot.,* 51, 71, 2000.
93. Ortiz, D.F. et al. Transport of metal-binding peptides by *HMT1,* a fission yeast ABC-type vacuolar membrane protein, *J. Biol. Chem.,* 270, 4721, 1995.
94. Chaney, R.L. et al. Improving metal hyperaccumulator wild plants to develop commercial phytoextraction systems: approaches and progress, in Terry, N. and Bañuelos, G.S. (Eds.). *Phytoremediation of Contaminated Soil and Water.* CRC Press, Boca Raton, FL, 1999.
95. Summers, A.O. and Sugarman, L.I. Cell-free mercury(II) reducing activity in a plasmid-bearing strain of *Escherichia coli, J. Bacteriol.,* 119, 242, 1974.
96. Rensing, C., Expression of bacterial mercuric ion reductase in *Saccharomyces cerevisiae, J. Bacteriol.,* 174, 1288, 1992.
97. Rugh, C.L. et al. Development of transgenic yellow poplar for mercury phytoremediation, *Nat. Biotechnol.* 16, 925, 1998.
98. Brewer E. P. et al. Somatic hybridization between the zinc accumulator *Thlaspi caerulescens* and *Brassica napus. Theoretical and Applied Genetics,* 99, 761, 1999.

99. Brown, S.L. et al. Phytoremediation potential of *Thlaspi caerulescens* and bladder campion for zinc- and cadmium-contaminated soils, *J. Environ. Qual.*, 23, 1151, 1994.

100. Ebbs, S.D. Heavy metals in the environment. Phytoextraction of cadmium and zinc from a contaminated soil, *J. Environ. Qual.*, 26, 1424, 1997.

101. Chaney, R.L. et al. Improving metal hyperaccumulator wild plants to develop commercial phytoextraction systems: approaches and progress, in *Phytoremediation of Contaminated Soil and Water*, Terry, N., Banuelos, G., and Vangronsveld, J. (Eds.). CRC Press, Boca Raton, FL, 2000, 129.

102. Macnair, M.R., Tansley review no. 49. The genetics of metal tolerance in vascular plants, *New Phytol.*, 124, 541, 1993.

103. Stomp, A.M., Han, K.H., Wilbert, S., Gordon, M.P., and Cunningham, S.D. Genetic strategies for enhancing phytoremediation, *Ann. NY Acad. Sci.*, 721, 481, 1994.

104. Raskin, I. Plant genetic engineering may help with environmental cleanup, *Proc. Natl. Acad. Sci. USA*, 93 3164, 1996.

105. Barcelo, J. and Poschenrieder, C. Phytoremediation: principles and perspectives, *Contrib. Sci.*, 2, 333, 2003, Institit d'Estudis Catalans, Barcelona.

106. Arisi, A.C.M. et al. Responses to cadmium in leaves of transformed poplars overexpressing γ-glutamylcysteine synthetase, *Plant Physiol.*, 109, 143, 2000.

107. Raskin, I. and Ensley, B.P. *Phytoremediation of Toxic Metals — Using Plants to Clean up the Environment*, John Wiley & Sons, New York, 2000.

108. Vatamaniuk, O.K. et al. AtPCS1, a phytochelatin synthase from *Arabidopsis*: isolation and *in vitro* reconstitution, *Proc. Natl. Acad. Sci. USA*, 96, 7110, 1999.

109. Zhu, Y.L. et al. Overexpression of glutathione synthetase in Indian mustard enhances cadmium accumulation and tolerance, *Plant Physiol.*, 119, 73, 1999.

110. Zhu, Y.L. et al. Cadmium tolerance and accumulation in Indian mustard is enhanced by overexpressing γ-glutamyl cysteine synthetase, *Plant Physiol.*, 121, 1169, 1999.

111. Meagher, R.B. Phytoremediaiton of toxic elemental and organic pollutants, *Curr. Opin. Plant Biol.*, 3, 153, 2000.

112. Zhao, H. et al. The yeast *ZRT1* gene encodes the zinc transporter protein of a high affinity uptake system induced by zinc limitation, *Proc. Natl. Acad. Sci. USA*, 93, 2454, 1996.

113. Pence, N.S. et al. The molecular physiology of heavy metal transport in the Zn/Cd hyperaccumulator *Thlaspi caerulescens*, *Proc. Natl. Acad. Sci. USA*, 97, 4956, 2000.

114. El-Bendary, A.A. et al. Mode of inheritance of zinc accumulation in maize, *J. Plant Nutr.*, 16, 2043, 1993.

115. Zhang, F.S. Mobilization of iron and manganese by plant-borne and synthetic metal chelators, in *Plant Nutrition — from Genetic Engineering to Field Practice*, Barrow, N.J. (Ed.). Kluwer Academic Publishers, Dordrecht, 1993, 115–118.

116. Dong, B., Rengel, Z., and Graham, R.D. Root morphology of wheat genotypes differing in zinc efficiency, *J. Plant Nutr.*, 18, 2761, 1995.

117. Graham, R.D. and Rengel, Z. Genotypic variation in zinc uptake and utilization by plants, in *Zinc in Soils and Plants*, Robson A.D., (Ed.). Kluwer Academic Publishers, Dordrecht, 1993, 107–118.

118. Pearson, J.N. and Rengel, Z. Mechanisms of plant resistance to nutrient deficiency stresses, in *Mechanisms of Environmental Stress Resistance in Plants*, Basra, A.S. and Basra, R.K. (Eds.). Harwood Academic Publishers, The Netherlands, 1997, 213–240.

119. Grotz N. et al. Identification of a family of zinc transporter genes from *Arabidopsis thaliana* that respond to zinc deficiency, *Proc. Natl. Acad. Sci. USA*, 95, 7220, 1998.

120. Chen, J. and Goldsbrough, P.B. Increased activity of γ-glutamylcysteine synthetase in tomato cells selected for cadmium tolerance, *Plant Physiol.*, 106, 233, 1994.

121. De Knecht, J.A. et al. Synthesis and degradation of phytochelatins in cadmium-sensitive and cadmium-tolerant *Silene vulgaris*, *Plant Sci.*, 106, 9, 1995.

122. Theil, E.C. Ferritin: structure, gene regulation, and cellular function in animals, plants, and microorganisms, *Annu. Rev. Biochem.*, 56, 289, 1987.

123. Ernst, W.H.O. Physiological and biochemical aspects of metal tolerance, in *Effects of Air Pollutants on Plants* Mansfield, T.A., (Ed.), Cambridge University Press, Cambridge, 115, 1976.

124. Eide, D. et al. A novel, iron-regulated metal transporter from plants identified by functional expression in yeast, *Proc. Natl. Acad. Sci. USA*, 93, 5624, 1996.

125. Kampfenkel, K. et al. Molecular characterization of a putative *Arabidopsis thaliana* copper transporter and its yeast homologue, *J. Biol.Chem.*, 270, 28, 479, 1995.
126. Eng, B.H. et al. Sequence analyses and phylogenetic characterization of the ZIP family of metal ion transport proteins, *J. Membrane Biol.*, 166, 1, 1998.
127. Song, W-Y. et al. Engineering tolerance and accumulation of lead and cadmium in transgenic plants, *Nat. Biotechnol.*, 21, 914, 2003.
128. Lee, J. et al. Functional expression of a bacterial heavy metal transporter in Arabidopsis enhances resistance to and decreases uptake of heavy metals, *Plant Physiol.*, 133, 589, 2003.
129. Henry, J.R. An overview of the phytoremediation of lead and mercury. A report prepared for the U.S. Environmental Protection Agency Office of Solid Waste and Emergency Response Technology Innovation Office. http://www.clu-in.org/download/remed/henry.pdf, 2003.
130. Li, Z-S. et al. The yeast cadmium factor protein (YCF1) is a vacuolar glutathione s-conjugate pump, *J. Biological Chem.*, 271, 6509, 1996.
131. Arazi, T., Sunkar, R., Kaplan, B., and Fromm, H.A. Tobacco plasma membrane calmodulin-binding transporter confers Ni^{2+} tolerance and Pb^{2+} hypersensitivity in transgenic plants, *Plant J.*, 20, 171–182, 1999.
132. Cai, X.H., Bown, C., Adhiya, J., Traina, S.J., and Sayre, R.T. Growth and heavy metal binding properties of transgenic *Chlamydomonas* expressing a foreign metallothionein, *Int. J. Phytorem.*, 1, 53–65, 1999.
133. Palmer, E.F., Warwick, F., and Keller, W. Brassicaceae (Cruciferae) family, plant biotechnology and phytoremediation, *Int. J. Phytoremediation*, 3, 245–287, 2001.
134. Pilon–Smits, E.A.H. and Pilon, M. Breeding mercury-breathing plants for environmental cleanup, *Trends Plant Sci.*, 5, 235–236, 2000.
135. Rugh, C.L., Bizily, S.P., and Meagher, R.B. Phytoreduction of environmental mercury pollution, in *Phytoremediation of Toxic Metals — Using Plants to Clean up the Environment*, Raskin, I. and Ensley, B.D. (Eds.). John Wiley & Sons, Inc., New York, 151–170, 2000.

15 "Metallomics" — a Multidisciplinary Metal-Assisted Functional Biogeochemistry: Scope and Limitations

M.N.V. Prasad

CONTENTS

15.1 INTRODUCTION

"Metallomics," a metal-assisted functional biogeochemistry, is a new scientific field first proposed by Haraguchi [1] to integrate the research fields related to metal biomolecules and its metals. Metallomics is the scientific field of symbiosis with genomics and proteomics because syntheses and metabolic functions of genes (DNA and RNA) and proteins cannot function without the

TABLE 15.1

Typical Metalloenzymes (and Metalloproteins) and Their Biological Functions

Metalloenzyme (MWa/kDa)	Number of atoms	Biological function
Transferrin (66–68)	2Fe	Transportation of iron
Ferritin (473)	1Fe	Storage of iron
Catalase (225)	4Fe	Decomposition of H_2O_2
Nitrogenase (200–220)	24Fe, 2Mo	Nitrogen fixation
Chitochrome P-450 (50)	Fe	Metabolisms of steroids and drugs
Carbonic anhydrase(30)	1Zn	Catalyst of H_2CO_3 equilibrium
Carboxypeptidase (34)	1Zn	Hydrolysis of peptide bonds at carboxy terminal
Alcohol dehydrogenase (150)	4Zn	Dehydration of alcohol
Alkaline phosphatase (89)	3.5Zn	Hydrolysis of phosphate esters
DNA polymerase (109)	2Zn	DNA synthesis
RNA polymerase (370)	2Zn	RNA synthesis
Plastocyaneine (134)	1Cu	Electron transfer
Gluthathion peroxidase (76–92)	1Se	Decomposition of H_2O_2 and organic superoxides
Urease (480)	10Ni	Transformation of urease to ammonia

Source: Haraguchi, H., *J. Anal. At. Spectrum*, 19, 5, 2004. With permission.

coordination of various metal ions and metalloenzymes. In metallomics, metalloproteins, metalloenzymes, and other metal-containing biomolecules are defined as "metallomes" in a similar manner to genomes in genomics as well as proteomes in proteomics. Because the identification of metallomes and the elucidation of their biological or physiological functions in biological systems is the main research target of metallomics, chemical speciation for specific identification of bioactive metallomes is the crux of establishing metallomics as an integrated biometal science.

Hazardous or toxic elements such as Hg, Cd, Pb, Cr(VI), As, Sn, and Se have caused serious environmental pollution or toxicological problems; thus, such elements in the biological, environmental, and geochemical samples have been extensively determined for environmental management and/or protection from environmental hazards, Most of trace metals in biological fluids and organs bind with various proteins called "metalloproteins." Metalloproteins are called "metalloenzymes" when they work as the biological catalysts to regulate biological reactions and physiological functions in biological cells and organs.

Some typical metalloenzymes and metalloproteins are summarized in Table 15.1. Metalloenzymes contain the specific number of metal ions at the active sites in specific proteins and they work as biocatalysts for specific enzymatic reactions, including gene (DNA, RNA) synthesis; metabolism; antioxidation; and so forth. In addition, the bioavailability and toxicity of the elements also depend on their chemical forms. Thus, species analysis of metal-binding molecules in biological samples is an important subject in various scientific fields such as biochemistry; biology; medicine; pharmacy; nutrition; agriculture; environmental science; etc. Accordingly, in recent years, chemical speciation or elemental speciation has been extensively developed to elucidate the biological essentiality and toxicity of the elements on a molecular basis [1,2,10–17]. Highly sensitive analytical methods such as AAS (atomic absorption spectrometry); ICP-AES (inductively coupled plasma atomic emission spectrometry); ICP-MS (inductively coupled plasma mass spectrometry); XRF (x-ray fluorescence spectrometry); and NAA (neutron activation analysis) enable one to determine almost all elements in the major to ultratrace concentration ranges.

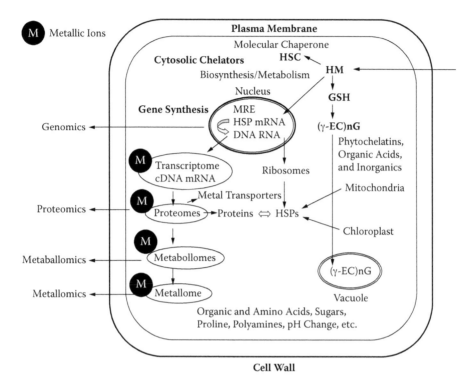

FIGURE 15.1 Basic principles of biological chemistry and proteomics for metabolic engineering of appropriate plants to control and transform hazardous pollutants. HM = heavy metal; M = metal ion; MRE = metal regulatory element; HSP = heat shock protein; HSC = heat shock cognate; ((-EC)nG – Glu – Cys- Gly aminoacids).

15.2 METALLOMICS AND METALLOMES

A schematic sketch of metallomics is proposed in Figure 15.1. On the left-hand side, academic technical terms such as genomics, proteomics, and metabollomics are shown along with metallomics to indicate their research areas in the biological system. Genomics deals with the scientific works on the genetic information of DNAs and RNAs encoded as the sequences of nucleic bases. DNA and RNA play an essential role in protein synthesis. Proteins are distributed inside and outside the cell, and they work as enzymes for synthesis and metabolism of various biomolecules of the cell. It is seen that a large number of proteins play essential roles in syntheses and metabolisms of many biological molecules to regulate and maintain the life system; protein science has been receiving great attention as postgenome science linked with genomics.

Many biological substances as well as metal ions are transported as raw materials inside the cell through the membrane. In general, material conversion is actively occurring inside the cell and also often in the cell membrane, and such material conversion and transportation in evolving specific transporters is termed "metabolism." Biological substances, which are usually small molecules such as amino acids, organic acids, and metal ions produced in metabolism, have recently been called "metabollomes" or "metabolites." Bioscience concerned with metallic elements and their applications has been studied independently in many scientific fields such as biochemistry; bioinorganic chemistry; nutritional science; pharmacy; medicine; toxicology; agriculture; and environmental science.

All such scientific fields have a deep interrelationship, with the common factor of metals, from the viewpoint of biological science. Therefore, it is desirable to promote it as an interdisciplinary field. Thus, Haraguchi [1] proposed the nomenclature of "metallomics" for biometal science. In the study of metallomics, elucidation of the physiological roles and functions of biomolecules binding with metallic ions in the biological systems should be the most important research target.

In recent years, genomics and proteomics have received great attention to appreciate various biological systems from the viewpoints of gene and protein sciences. Genomics and proteomics are indeed fundamentally important scientific fields because genes (DNAs and RNAs) contain the genetic information codes to synthesize various proteins. Genes and proteins cannot be synthesized without the assistance of metalloenzymes containing zinc and other metals. In this sense, metallomics may stand in the same position in scientific significance as genomics and proteomics. Thus, in metallomics, biological molecules bound with biometals are properly defined as "metallomes," corresponding to genomes and proteomes in genomics and proteomics, respectively. However, metallic ions such as alkali and alkaline earth metal ions, which exist mostly as free ions in biological fluids, should also be included in metallomes because they play many important roles in the occurrence of the physiological functions in the biological systems.

15.3   GLUTATHIONE AND ORGANIC ACIDS METABOLISM

Glutathione and organic acids metabolism plays a key role in metal tolerance in plants [2–5]. Glutathione is ubiquitous component cells from bacteria to plants and animals. In plants, it is the major low molecular mass thiol compound (28). Glutathione occurs in plants mainly as reduced GSH (95 %). Its synthesis is mediated by the enzymes glutamylcysteine synthetase (EC 6.3.2.2) and glutathione synthetase (EC 6.3.2.3). Glutathione metabolism is also connected with cysteine and sulphur metabolism in plants. Cysteine concentration limits glutathione biosynthesis. Low molecular thiol peptide phytochelatins (PCs), often called class III metallothioneins, are synthesized in plants from glutathione induced by heavy metal ions [6].

These peptides are synthesized from glutathione by means of α-glutamylcysteine transferase enzyme (EC 2.3.2.15), which is also called phytochelatin synthase (PCS), catalyzing transfer reaction of (α-Glu-Cys) group from a glutathione donor molecule to glutathione, an acceptor molecule. PCS is a cytosolic, constitutive enzyme and is activated by metal ions, namely, Cd^{2+}, Pb^{2+}, Ag^{1+}, Bi^{3+}, Zn^{2+}, Cu^{2+}, Hg^{2+}, and Au^{2+}. PCs thus synthesized chelate heavy metals and form complexes that are transported through cytosol in an ATP-dependent manner through tonoplast into vacuole. Thus, the toxic metals are swept away from cytosol. Some high molecular weight complexes (HMW) with S-2 can also be formed from these LMW complexes in vacuole [7].

Transgenic plants with modified genes of PCS and genes of glutathione synthesis enzymes, α-GCS and GS, and enzymes connected with sulphur metabolism, e.g., serineacetyltransferase, need special attention in order to achieve success in phytoremediation of metals in the environment. Plants under heavy metal stress produce free radicals and reactive oxygen species and must withstand the oxidative stress before acquiring tolerance to toxic metals. Glutathione is then used for the synthesis of PCs as well as for dithiol (GSSG) production. The ascorbate–glutathione pathway is involved in plant defense against oxidative stress. Organic acids play a major role in metal tolerance [8].

Organic acids play a role in metal chelation by forming complexes with metals, a process of metal detoxification. Chelation of metals with exuded organic acids in the rhizosphere and rhizospheric processes indeed form an important aspect of investigation for remediation. These metabolic pathways underscore the physiological, biochemical, and molecular bases for heavy metal tolerance [6].

15.4   METAL TRANSPORTERS AND INTERACTIONS IN MEMBRANES AT MOLECULAR LEVELS

Plants and humans require adequate amounts of micronutrients like iron and zinc, but accumulation of an excess or uptake of nonessential metals like cadmium or lead can be extremely harmful.

Proteins of the CDF (cation diffusion facilitator) family are involved in the homeostasis of Cd^{2+}, Co^{2+}, Fe^{2+}, and Zn^{2+} in microbes, animals, and plants [9]. Therefore, elucidation of the role of CDF proteins in *Arabidopsis thaliana* would be advantageous to the success of phytoremediation. Complementary DNAs are to be functionally expressed in appropriate mutants of *Saccharomyces cerevisiae* to test their function.

In a reverse genetics approach, several representative *Arabidopsis* CDFs will be used in RNA interference technology [10,11]. Regulation and localization of these CDFs need to be investigated by expressing promoter:GUS fusions and epitope-tagged fusion proteins in *A. thaliana*, and by development and use of specific antibodies. Very little information is available about protein–protein interactions of membrane. Such interactions might be vital for CDF function because their substrate metal cations are thought to be bound to metallochaperone proteins in the cytoplasm.

15.5 SPECIES-SELECTIVE ANALYSIS FOR METALS AND METALLOIDS IN PLANTS

The success of an analyst searching for a given organometallic moiety in a plant matrix depends on two factors. First, he or she must be sure to determine this and not another species (analytical selectivity). Second, the detection limit (sensitivity of the detector and noise level) of the instrumental setup should match the analyte's level in the sample.

Intrinsically, species-selective techniques, such as Mössbauer spectroscopy; x-ray photoelectron spectroscopy (XPS); electron spin resonance spectroscopy (ESR); or mass or tandem mass spectrometry (MS or MS/MS) usually fail at trace levels in the presence of a real-sample matrix. Al (70.4 MHz) NMR spectra were obtained on various intact samples of Al-accumulating plant tissues. None of the materials examined (*Hydrangea sepals* and leaves of three Theaceae species) contained detectable amounts of $Al(H_2O)_6^{3+}$, although the Al was present as hexacoordinated complexes and, in most instances, in at least two forms.

Selectivity in terms of species is typically achieved by on-line combination of a high-performance separation technique (chromatography or electrophoresis) with the parallel element-specific and molecule-specific detection. Nonspecific detectors [UV, flame ionization detector (FID)] suffer from a large background noise and poor sensitivity. Analysis by a coupled technique is often preceded by a more or less complex wet chemical sample preparation. The latter is mandatory for the sample to meet the conditions (imposed by the separation technique) in terms of form to be presented to the instrumental system. The tendency is to integrate this preliminary step into the whole experimental setup.

The coupled techniques available for elemental speciation analysis have been reviewed elsewhere (Table 15.2) [12]. The choice of the separation technique is determined by the physicochemical properties of the analyte (volatility, charge, polarity); that of the detection technique is determined by the analyte's level in the sample. This is the sample matrix (air, water, sediment, biomaterial) that dictates, on its turn, the choice of the sample preparation procedure. Separation techniques for speciation analysis have been comprehensively reviewed [12]. For volatile species or those convertible readily to volatile ones by means of derivation, gas chromatography (GC) is the method of choice. Species that do not fulfill the preceding requirement are separated by ion exchange or reversed-phase liquid chromatography. In particular, proteins and other biopolymers are separated by size-exclusion (gel-permeation) chromatography.

The physicochemical similarity of many proteins stimulates the use of electrophoretic techniques for efficient separations. In terms of detectors, plasma spectrometric techniques are favored over atomic absorption spectrometry (AAS) because of their much higher sensitivity. Microwave-induced plasma (MIP) atomic emission spectrometry is the choice for GC and a more energetic inductively coupled plasma (ICP) mass spectrometry is the choice for LC and capillary zone electrophoresis (CZE).

Element-specific detectors do not allow for the identification of the species eluted. This drawback gains in importance as the wider availability of more efficient separation techniques and more sensitive detectors makes the number of unidentified species grow. Therefore, the prerequisite for progress in speciation analysis is a wider application of sensitive on-line techniques for compound identification, i.e. mass or tandem mass spectrometry with soft (e.g., electrospray (ESI) or matrix-assisted laser desorption ionization (MALDI).

Thus far, the hyphenated techniques such as LC-ICP-MS, GC-ICP-MS, and LC-ICP-AES have been developed as the analytical methods for chemical speciation of trace metals in the biological samples. Such hyphenated methods are actually the most powerful techniques for chemical speciation. However, they are useful only for the identification of known and stable compounds such as methylmercury, methylated arsenics, butyltin compounds, and so on. From now on, however, it is obvious that the identification of biomolecules such as metalloproteins and metalloenzymes, as well as metal-binding nucleic acids and metabolites, will become more important in exploring biometal science in relation to their biological functions and metabolism. Then, the methods for the direct identification of biomolecules, such as ES-MS (electrospray mass spectrometry) and MALDITOFMS (matrix-assisted laser desorption ionization time-of-flight mass spectrometry), should be employed for study of chemical speciation.

Perhaps a system of LC doubly combined with ES-MS (or MALDI-TOFMS) and ICP-MS — i.e., LC-ES-MS-ICP-MS, which is a kind of tandem mass spectrometry and allows detection of organic molecules and trace metals simultaneously — may be the more ideal instrumentation for direct speciation of unknown organometallic compounds and metalloproteins to promote development and establishment of metallomics. It has been noted that the detection sensitivities for organic molecules obtained by organic mass spectrometry presently seem to be inferior to those for trace metals by ICP-MS. Therefore, the sensitivity matching in the hyphenated system, for example, using ES-MS and ICPMS, should be explored by improving the ionization efficiencies of bioorganic molecules in organic mass spectrometry to develop such a simultaneous detection system.

15.6 METABOLLOMICS

15.6.1 GLUTATHIONE METABOLISM AND PHYTOCHELATIN SYNTHESIS

Phytochelatin synthase (PCS) genes overexpressed in a PCS-positive background or transplanted into plants that lack an endogenous PCS homolog would serve as reliable biotechnological and molecular tools for heavy metal remediation in the environment. However, an integrated investigation of phytochelatin biosynthesis is warranted to understand the potential and limitations of a PCS-based metal detoxification as phytoremediation strategy [13]. Phytochelatins are considered to be the activated sulfate acceptors in the formation of a thiosulfate intermediate leading to sulfite formation upon reduction by thiosulfonate reductase [14]. This hypothesis, however, has been challenged by

- The fact that no plant thiosulfonate reductase has been identified thus far
- The recent demonstration that the main sulfite-forming pathway in plants relies on an enzyme adenosine 5′-phosphosulfate reductase (APS) that directly reduces activated sulfate using an intramolecular glutaredoxin domain [15]

15.7 CHEMICAL TRANSFORMATION

Selenate, the oxoanion of the element selenium, is taken up and metabolized by higher plants because of its chemical similarity to sulfate. Thus, growth on soils contaminated with selenate results in the formation of excess amounts of selenocysteine and selenomethionine. These are

TABLE 15.2
Research Subjects in Metallomics and Analytical Techniques Required in Metallomics Researches

Research subject	Analytical technique
Distributions of the elements in the biological fluids, cell, organs, etc.	Ultratrace analysis, all-elements analysis, one atom detection, one molecule detection
Chemical speciation of the elements in the biological samples and systems	Hyphenated methods (LC-ICP-MS, GC-ICP-MS, MALDI-MS, ES-MS)
Structural analysis of metallomes (metal-binding molecules)	X-ray diffraction analysis, EXAFS
Elucidation of reaction mechanisms of metallomes using model complexes (bioinorganic chemistry)	NMR, XPS, laser-Raman spectroscopy, DNA sequencer, amino acids sequencer, time-resolution and spatial-resolution fluorescence detection
Identification of metalloproteins and metalloenzymes	LC-ES-MS, LC-MALDI-MS, LC-ICP-MS
Metabolisms of biological molecules and metals (metabollomes, metabolites)	LC, GC, LC-MS, GC-MS, ES-MS, API-MS[a], biosensors
Medical diagnosis of health and disease related to trace metals on a multielement basis	ICP-AES, ICP-MS, graphite-furnace AAS, autoanalyzer, spectrophotometry
Design of inorganic drugs for chemotherapy	LC-MS, LC-ICP-MS, stable isotope tracers
Chemical evolution of living systems and organisms on the Earth	Isotope ratio measurement (chronological techniques, DNA sequencer
Other metal-assisted function biosciences in medicine, environmental science, food science, agriculture, toxicology, biogeochemistry, etc.	*In-situ* analysis, immunoassay, bioassay, food analysis, clinical analysis

[a] Atmospheric pressure chemical ionization mass spectrometry.

Source: Haraguchi, H., *J. Anal. At. Spectrum*, 19, 5, 2004. With permission.

incorporated into proteins of sensitive plants instead of cysteine and methionine and render the affected proteins nonfunctional.

In an approach to increase selenium assimilation by plants, the plastidic *A. thaliana* APS1 cDNA-encoding ATP sulfurylase was expressed in *B. juncea* under the control of a 35S promoter [16]. Transgenic plants exhibited a slightly increased tolerance to selenate when compared with wild-type controls and accumulated approximately twofold higher concentrations of selenium in their shoots. Enhanced sulfur assimilation in these transgenic plants resulted in an increase in glutathione concentrations by approximately 100 and 30% in shoots and roots, respectively; this suggested that ATP sulfurylase might also be an interesting target for the phytoremediation of other metals, especially cadmium [17].

Chemical transformation of a trace element into a less toxic, volatile compound is a very effective strategy for detoxification because the potentially harmful element is removed from the tissues. In mercury-contaminated soils and sediments, microbial activity results in the conversion of toxic Hg(II) into organomercurials — for example, the highly toxic methylmercury ($CH_3 Hg^+$). Mercury-resistant bacteria able to transform organomercurials and Hg(II) into significantly less toxic elemental mercury have been isolated. Methylmercury is converted to the less toxic Hg(II) by organomercurial lyase encoded by the gene *MerB*. A second enzyme, encoded by *MerA*, catalyzes the reduction of Hg(II) to elemental mercury, using NADPH as the electron donor. Under ambient conditions, elemental mercury enters the global biogeochemical cycle upon volatilization [18].

Mercury volatilization in plants has been established [16,18,19]. The nucleotide sequence of a bacterial *MerA* gene had to be modified to allow for high-level expression in plants. *Arabidopsis thaliana* expressing *Mer A* under the control of a constitutive cauliflower mosaic virus 35S promoter germinated and developed on agarose media containing 50 and 100 μM $HgCl_2$, concentrations that completely inhibited germination of wild-type seeds. The *MerA* plants showed a significantly higher tolerance to Hg^{2+} and volatalized Hg; their tolerance to methylmercury was unchanged. They were also more tolerant to Au^{3+}.

The *MerB* plants were significantly more tolerant to methylmercury and other organomercurials. They effectively converted highly toxic methylmercury to Hg^{2+}, which is 100 times less toxic to plants. To study the effects of both, the *MerA* and *MerB* plants were crossed and F1 generation was selfed. The F2–*MerA*, *MerB* double transgenics showed highest tolerance to organic mercury (10 micro molar = (uM)). *MerA* and *MerB* plants were shown to volatilize elemental mercury when supplied with organic mercury. Submicromolar concentrations of highly toxic organomercurials abolish germination of wild-type and *MerA*-expressing *A. thaliana*.

The combined expression of *MerA* and *MerB* in a high-biomass plant could be a promising step towards the generation of an improved mercury phytoremediator plant. Modified *MerA* genes were then introduced into higher biomass plants. Tobacco transformants expressing a modified *MerA* gene were able to develop and flower on soils containing up to 500 ppm Hg(II), but mercury removal from soil substrates has yet to be determined [18].

15.8 SULPHUR METABOLISM

Sulfur metabolism and HM detoxification are closely related processes [7] (see Chapter 16) (Figure 15.2). In addition, a possible homeostatic role of PCs towards essential HMs is clearly indicated by the fact that (1) low, but detectable, levels of PCs are present even in the absence of HM exposure; and (2) PC levels increase concomitantly with Cu and Zn depletion upon transfer of cell suspension cultures to a minimal micronutrient medium. In fact, it has been hypothesized that PCs are not simply an HM-detoxification system *sensu stricto*, especially in the presence of low concentrations of (essential) metals. Instead, under these conditions, PCs may primarily act as key components of metal homeostasis. The constitutive expression of PCS [20] might also be considered as an indication of a more general role of PCs not exclusively related to HM detoxification.

Further supporting this view is the strong protective effect of PCs against Cd-mediated inactivation of metal-sensitive enzymes and the ability of Zn– and Cu–PC complexes, mainly of the PC_2/PC_3 type, to reactivate metal-depleted or metal-poisoned metalloenzymes (although not more efficiently than the corresponding free salts). In plants, it is thus possible that PCs and MT-like proteins cooperatively act in the homeostasis of essential HMs. PCs have also been proposed as activated sulfate acceptors in the formation of a thiosulfate intermediate leading to sulfite formation upon reduction by thiosulfonate reductase. This hypothesis, however, has been challenged by the fact that no plant thiosulfonate reductase has been identified thus far and by the recent demonstration that the main sulfite-forming pathway in plants relies on an enzyme (adenosine 5′-phosphosulfate reductase) that directly reduces activated sulfate (APS) using an intramolecular glutaredoxin domain [15].

Although the metal detoxification and homeostatic roles of PCs are not mutually exclusive and may coexist at the whole plant level, the fact that *Arabidopsis* PC-deficient mutants (*cad1*) grow well in the presence of Cu and Zn micronutrient concentrations [21] suggests that the latter role, if real, is dispensable or easily replaceable by other metal-binding components, such as MTs. On the other hand, the idea of an exclusive metal detoxification function of PCs is somewhat weakened by the lack of correlation between the PC content and the HM sensitivity of metal-tolerant and nontolerant ecotypes of *Silene vulgaris* and *Datura innoxia* and by similar findings recently reported for the HM hyperaccumulator *Thlaspi caerulescens* and the closely related, nonaccumulator species *Thlaspi arvense* [22–24]. Neither PCS activity nor PC turnover upon transfer to a Cd-less medium differed between wild-type and HM-tolerant *Silene vulgaris* [15]. In addition, although PC_2 was the most abundant PC peptide in metal-tolerant plants, the more effective metal chelator, PC_3, prevailed in the nontolerant ecotype [26].

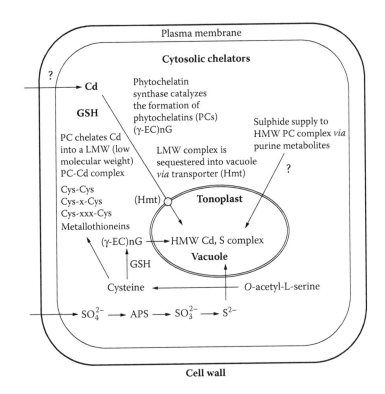

FIGURE 15.2 Sulfur metabolism and HM detoxification mechanisms — a comprehensive view.

Further to this point, a stronger correlation between Cu tolerance and the accumulation of MT mRNAs (r values ranging from 0.89 to 0.998) than with the total amount of intracellular nonprotein thiols, including PCs ($r = 0.77$), has been reported in *Arabidopsis* [27]. Therefore, it cannot be excluded that other systems, autonomously or in combination with PCs, may regulate HM homeostasis in the plant cell. For instance, Cu-MTs and Cu chaperones (termed Atx1, Lys7, and Cox 17) seem to be involved in Cu ion traffic in yeast cells [28] and analogous mechanisms might operate in higher plants [29].

Multiple connections exist between sulfur metabolism and heavy metal detoxification and homeostasis in plants. Metal ions are complexed in the cytosol with GSH and the derived PCs are transported into vacuole. The following functions have been implicated in Cd complexation and transport [30 and the references therein]:

- The ATP-binding cassette-type transport activity at the tonoplast
- The Cd^{2+}/H^+ antiport
- The vacuolar-type ATPase generating a proton gradient

Another notable cellular response depicted is that some metals interact with genes that have metal-regulating elements (MRE) at the promoter region, as found for MT genes of animals. For example, in the soybean, similar sequences have been found in the upstream region of heat shock gene coded for 17.5 kDa HSP (heat shock protein)[31]. Some heat shock genes and small HSPs have been induced by Cd ions in soybean [31,32]. These HSCs (heat shock cognates) thus may function as molecular chaperones [33,34].

Cui et al. [35] reported that the elemental sulphur and EDTA amendments increased the extractable fractions of soil Pb and Zn. EDTA was more effective than S. Shoot uptake of Pb and

Zn by Indian mustard and wheat was increased with S and EDTA amendments. Indian mustard shoots took up more Pb and Zn than winter wheat with or without S and EDTA amendment.

15.9 OVEREXPRESSION OF LCT1 (LOW-AFFINITY CATION TRANSPORTER) IN TOBACCO ENHANCED: THE PROTECTIVE ACTION OF CALCIUM AGAINST CADMIUM TOXICITY

Metal–metal interactions have been reported to have ameliorating functions [36–40]. For example, calcium involvement in zinc uptake and detoxification was studied in Zn^{2+}-tolerant and nontolerant populations of *Silene maritima*. Increasing calcium concentrations reduced Zn toxicity and led to a higher level of zinc accumulation by the roots of the tolerant plants; however, they decreased transport to the shoots of both types [41]. A higher calcium concentration in a medium was also reported to abolish the toxic effects of Cd^{2+} [42,43] and Pb^{2+} [44] on the activity of photosystem II. In addition, high Ca status and a high level of tolerance to Ca deficit accompanied enhanced Zn, Pb, Cu, and Al tolerance [45–47].

The regulation of heavy metal uptake and internal transport constitutes part of the basis of plant resistance to their toxicity. The mechanism accounting for the transport of heavy metals across membranes in plants and its regulation is far from understood, however. The general view is that nonessential metals usually cross plasmalemma and internal membranes through cation transporters with a broad substrate specificity or that they use pathways reserved for other ions [48,49]. For example, cadmium and lead were shown to be transported through pathways for calcium ions.

In plants, it has been suggested that putative tonoplast Ca^{2+}/H^+ antiporters encoded by CAX1 and CAX2 (calcium exchange protein) from *Arabidopsis* are involved in the transport of cadmium from the cytoplasm to the vacuole [50]. In turn, LCT1 cloned from wheat roots (however expressed in roots and leaves), a nonselective transmembrane transporter for Na^+, K^+ [51,52], and Ca^{2+}, also appeared to mediate Cd^{2+} transport to the cell [53]. Toxic metal ions such as Cd^{2+} or Pb^{2+} are known as very effective substituents of Ca^{2+}, e.g., in calmodulin, consequently interfering seriously with the role of this cation in a number of metabolic processes [54–56]. In this context, it seems possible that the regulation of heavy metal uptake/transport in a plant by the presence of calcium in various concentrations in the medium could in part contribute to an ameliorative effect of calcium on heavy metal toxicity.

To gain insight into the molecular mechanism of Ca^{2+}-dependent Cd^{2+} tolerance, tobacco was transformed with wheat cDNA LCT1, the first cloned plant influx system mediating the uptake of Ca^{2+} and Cd^{2+} ions into a cell [53]. Transformants were tested for the possible involvement of LCT1 in diminishing Cd toxicity with the enhancement of Ca^{2+} concentration in the medium. Antosiewicz and Hennig [57] also demonstrated that LCT1 is involved in calcium acquisition and in the alleviation of toxic effects of Cd^{2+} by enhanced external Ca^{2+} concentration.

Antosiewicz and Hennig [57] were the first to demonstrate the involvement of LCT1 in calcium acquisition and in the regulation of amelioration of Cd toxicity by calcium. Wheat cDNA LCT1 (low-affinity cation transporter gene), a nonspecific transporter for Ca^{2+}, Cd^{2+}, Na^+, and K^+, was overexpressed in tobacco. Transformants were tested for their sensitivity to a range of Ca^{2+} concentrations [0.01 to 10 mM Ca(NO3)2] with or without the presence of 0.05 mM Cd(NO3)2. LCT1-transformed plants expressed a phenotype distinct from controls only under conditions of low calcium (0.01 to 1 mM Ca^{2+}). They grew significantly better and had slightly higher shoot calcium concentration.

Transformants subjected to 0.05 mM Cd(NO3)2 in the presence of 1 mM Ca^{2+} displayed a substantially higher level of tolerance to cadmium and accumulated less Cd in roots. LCT1 increased calcium and decreased cadmium accumulation in transformed plants. LCT1, the putative plasmalemma nonspecific transporter for Ca^{2+}, Cd^{2+}, Na^+, and K^+ [51–53], was used for tobacco transformation in order to check whether the toxicity of cadmium administered in a range of Ca^{2+} concentrations to control and LCT1-transformed plants would be different. LCT1 improved plant performance at low external calcium; LCT1 expression in tobacco improved the growth of plants only within a limited range of calcium concentrations. LCT1 contributed to Ca^{2+}-dependent Cd^{2+} tolerance.

Numerous authors have described the phenomenon of calcium mitigating heavy metal toxicity [36,41–44]. The reported amelioration constitutes a part of the whole-plant defense system that includes uptake/transport and detoxification/sequestration. Based on the known broad spectrum of the role of Ca^{2+} in the regulation of metabolic processes [58], calcium might be an important factor in any of these components. It is not known whether the observed reduced toxicity might result from less cadmium uptake or from more efficient detoxification. For example, Choi et al. [59] demonstrated the contribution of calcium to the protection mechanism by immobilization of the metal as coprecipitates with calcium and phosphorous. Reduction of cadmium uptake and accumulation by calcium were also reported for several plant species [60–62] as was the opposite effect: inhibition of the accumulation of calcium by cadmium, leading to calcium deficiency [63].

15.10 OVEREXPRESSION OF ALFALFA ALDOSE/ALDEHYDE REDUCTASE CONFERS TOLERANCE TO CADMIUM STRESS

Wild-type (SR1) tobacco line and transgenic lines ALR1/5 and ALR1/9 overexpressing aldose/aldehyde reductase were less susceptible to cadmium-induced stress. Based on these observations, Hegedüs et al. [64] suggested that alfalfa aldose/aldehyde reductase overexpression may generally induce higher stress tolerance. These coworkers transformed tobacco (*Nicotiana tabacum* cv. Petit Havanna line SR1) by alfalfa aldose/aldehyde reductase cDNA attached to the viral constitutive promoter CaMv35S. The WT (SR1), a nonexpressing transgenic line (ALR1/7), and two transgenic lines (ALR1/5 and ALR1/9) previously shown by Western hybridization to overexpress the alfalfa ALR protein [21] were used in these experiments.

The ectopic synthesis of a novel alfalfa NADPH-dependent aldose/aldehyde reductase enzyme in transgenic tobacco plants provided substantial tolerance against oxidative stress, such as drought, paraquat, and UV-B [21,22]. Cd treatment caused a significant decrease in the chlorophyll content in all the genotypes tested. However, this decrease was less pronounced in the lines overproducing aldose/aldehyde reductase (72 and 74% for lines ALR1/9 and ALR1/5, respectively) than in the SR1 wild-type plants (52%). Similar tendencies were revealed for changes in the total carotenoid content of tobacco leaves.

The novel molecular methodologies, i.e., genomics and proteomics, have not been brought together in the past to characterize molecular mechanisms related to plant metal accumulation. Because the *Arabidopsis* genome has been completely sequenced, the full benefit of those data is available for identification of genes and proteins found in plants that hyperaccumulate metals. Various ecotypes of plants that hyperaccumulate metals are being currently investigated in various laboratories to identify metal responsive proteins. Proteomics provides a powerful additional tool for the identification of proteins induced or repressed under metal stress and also for comparison of various ecotypes. *Thlaspi* is the ideal plant because many of the proteins can be identified based on the homology to *Arabidopsis* or *Brassica* genes. On the other hand, it was evident that *Thlaspi* contains many proteins for which homology was not found from databases. These proteins may be of particular interest for further studies of metal tolerance, uptake, and accumulation.

15.11 HAIRY ROOTS OF HORSERADISH ARE AN IDEAL SYSTEM FOR INDUCTION OF PHYTOCHELATIN HOMOLOGS

When exposed to excess heavy metals, plants induce phytochelatins and related peptides (all designated as PCAs). Horseradish (*Armoracia rusticana*) was exposed for 3 days to cadmium (1

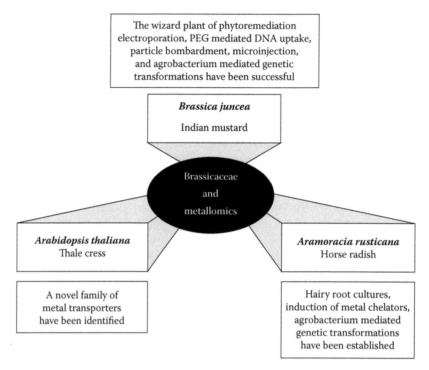

FIGURE 15.3 Biotechnology prospecting for phytoremediaiton of metals in the environment. In Brassicaceae *Arabidopsis thaliana*, *Brassica juncea*, and *Armoracia rusticana* have been extensively studied for metal sensitivity and resistance. In *A. thaliana*, a number of heavy metal accumulating and sequestering mutants have been identified. Brassicaceae are amenable to well-characterized biotechnological and molecular biological tools through which transgenic production can be achieved for field trials.

m*M*) along with reduced glutathione (2 m*M*); its hairy roots induced PCA. Brassicaceae are amenable for metallomics (Figure 15.3). The role of phytochelatins in sequestering toxic metals has been discussed in Chapter 16.

15.12 OVEREXPRESSION OF Γ-GLUTAMYLCYSTEINE SYNTHETASE IN INDIAN MUSTARD ENHANCED CADMIUM TOLERANCE AND ACCUMULATION

In an investigation of rate-limiting factors for glutathione and phytochelatin (PC) production and the importance of these compounds for heavy metal tolerance, Indian mustard (*Brassica juncea*) was genetically engineered to overexpress the *Escherichia coli* gene encoding g-glutamylcysteine synthetase (γ-ECS), targeted to the plastids. The γ-ECS transgenic seedlings showed increased tolerance to Cd and had higher concentrations of PCs, γ-GluCys, glutathione, and total nonprotein thiols compared with wild-type (WT) seedlings. When tested in a hydroponic system, γ-ECS mature plants accumulated more Cd than WT plants: shoot Cd concentrations were 40 to 90% higher. In spite of their higher tissue Cd concentration, the γ-ECS plants grew better in the presence of Cd than WT did. Thus, Zhu et al. [74] concluded that overexpression of γ-ECS increases biosynthesis of glutathione and PCs, which in turn enhances Cd tolerance and accumulation. Therefore, overexpression of g-ECS appears to be a promising strategy for the production of metal-tolerant plants for application in phytoremediation.

Nonprotein thiols (NPTs), which contain a high percentage of Cys sulfhydryl residues in plants, play a pivotal role in heavy metal detoxification. The reduced form of glutathione (γ-Glu-Cys-Gly,

GSH) is one of the most important components of NPT metabolism. GSH may play several roles in heavy metal tolerance and sequestration. It protects cells from oxidative stress damage, such as that caused by heavy metals in plants. PCs are heavy metal-binding peptides involved in heavy metal tolerance and sequestration [14]. PCs comprise a family of peptides with the general structure (γ-Glu-Cys)n-Gly, where $n = 2$ to 11 [30]. They were shown to be induced by heavy metals such as Cd in all plants tested [54], including Indian mustard [55]. The roles of GSH and PC synthesis in heavy metal tolerance were well illustrated in Cd-sensitive mutants of *Arabidopsis*. For example, the Cd-sensitive *cad2* mutant was defective in GSH and PC biosynthesis [21].

GSH is synthesized from its constituent amino acids in two sequential, ATP-dependent enzymatic reactions catalyzed by g-glutamylcysteine synthetase (γ-ECS) and glutathione synthetase (GS), respectively. PC synthase subsequently catalyzes the elongation of the (γ-Glu-Cys)n by transferring a g-GluCys group to GSH or to PCs [65]. Genes encoding PC synthase have been cloned from plants and yeast [67–69]. The rate-limiting step for GSH synthesis in the absence of heavy metals is believed to be the reaction catalyzed by γ-ECS because the activity of this enzyme is feedback regulated by GSH and dependent on Cys availability [70]. This view was supported by the observation that overexpression of the *Escherichia coli GSHI* gene (which encodes γ-ECS) in poplar resulted in increased foliar GSH levels [71,72].

In contrast, overexpression of GS did not lead to an increase in GSH levels in poplar [73] or in Indian mustard [74] in the absence of heavy metals. However, the Indian mustard GS overexpressing plants showed higher levels of GSH and PC2 relative to untransformed plants in the presence of heavy metals. These GS plants also showed enhanced heavy metal tolerance and accumulation [74]. It has been reported that overexpression of tomato γ-ECS could restore some degree of heavy metal tolerance to the *cad2 Arabidopsis* mutant. However, overexpression of this gene did not increase the Cd tolerance of wild-type (WT) *Arabidopsis* plants. *E. coli* γ-ECS enzyme was overexpressed in the chloroplasts of Indian mustard. The transgenic γ-ECS plants were compared with WT Indian mustard plants with respect to their Cd accumulation and tolerance, as well as their levels of heavy metal-binding peptides.

15.13 OVEREXPRESSION OF MTS AS A MEANS TO INCREASE CADMIUM TOLERANCE

Plants overexpressing mammalian MTs were reported to be unaffected by concentrations of 100 to 200 μM cadmium, but growth of *N. tabacum* control plants was severely inhibited at external cadmium concentrations of 10 μM [77]. Transformants of *Brassica oleracea* expressing the yeast metallothionein gene CUP1 tolerated 400 μM cadmium; however, wild-type plants were unable to grow at concentrations above 25 μM cadmium in a hydroponic medium. Transformants grown at 50 μM cadmium accumulated 10 to 70% higher concentrations of cadmium in their upper leaves than did nontransformed plants grown at 25 μM cadmium. This indicates that the enhanced tolerance observed in the transgenic plants was unlikely to be a consequence of excluding cadmium from the leaves.

The concept of using plants to clean up contaminated environments is not new. About 300 years ago, plants were proposed for use in the treatment of wastewater. At the end of the 19th century, *Thlaspi caerulescens* and *Viola calaminaria* were the first plant species documented to accumulate high levels of metals in leaves. Members of the genus *Astragalus* were capable of accumulating selenium up to 0.6% in dry shoot biomass. Despite subsequent reports claiming hyperaccumulators, the existence of plants hyperaccumulating metals other than Cr, Ni, Mn, Se, and Zn has been questioned and requires additional confirmation [78]. The idea of using plants to extract metals from contaminated soil and the first field trial on Zn and Cd phytoextraction were conducted by Baker et al. [79].

In the 1990s, extensive research was conducted to investigate the biology of metal phytoextraction. Despite significant success, understanding of the plant mechanisms that allow metal extraction is still emerging. In addition, relevant applied aspects, such as the effect of agronomic practices on metal removal by plants, are largely unknown. It is conceivable that maturation of phytoextraction into a commercial technology will ultimately depend on the elucidation of plant mechanisms and application of adequate agronomic practices. Natural occurrence of plant species capable of accumulating extraordinarily high metal levels makes the investigation of this process particularly interesting.

15.14 OVEREXPRESSION OF GLUTATHIONE SYNTHETASE IN INDIAN MUSTARD ENHANCED CADMIUM TOLERANCE

An important pathway by which plants detoxify heavy metals is through sequestration with heavy metal-binding peptides called phytochelatins or their precursor, glutathione. To identify limiting factors for heavy metal accumulation and tolerance and to develop transgenic plants with an increased capacity to accumulate and/or tolerate heavy metals, the *Escherichia coli* gshII gene encoding glutathione synthetase (GS) was overexpressed in the cytosol of Indian mustard (*Brassica juncea*). The transgenic GS plants accumulated significantly more Cd than the wild type: shoot Cd concentrations were up to 25% higher and total Cd accumulation per shoot was up to threefold higher. Moreover, the GS plants showed enhanced tolerance to Cd at the seedling and mature plant stages. Cd accumulation and tolerance were correlated with the gshII expression level. Cd-treated GS plants had higher concentrations of glutathione, phytochelatin, thiol, S, and Ca than wild-type plants. The conclusion was that, in the presence of Cd, the GS enzyme is rate limiting for the biosynthesis of glutathione and phytochelatins and that overexpression of GS offers a promising strategy for the production of plants with superior heavy metal phytoremediation capacity.

Heavy metal pollution of soils and waters, mainly caused by mining and the burning of fossil fuels, is a major environmental problem. Heavy metals, unlike organic pollutants, cannot be chemically degraded or biodegraded by microorganisms. An alternative biological approach used to deal with this problem is phytoremediation — i.e., the use of plants to clean up polluted waters and soils [78,80]. Heavy metals or metalloids can be removed from polluted sites by phytoextraction, which is the accumulation of the pollutants in the plant biomass [81]. Compared with other remediation technologies, phytoremediation is less expensive (1000-fold less expensive than excavation and reburial of soil [82]) and is particularly suitable for treatment of large volumes of substrate with low concentrations of heavy metals. However, the presence of heavy metals inhibits plant growth, limiting the application of phytoremediation. Therefore, one trait of great significance to phytoremediation is the ability of plants to tolerate the toxic metals extracted from the soil [75].

15.15 CONCLUSIONS

In spite of several achievements in metallomics, certain limitations still prevail. Phytoremediation technology in using genetically engineered plants has several advantages. There have been apprehensions about dispersal of these contaminants if plants are edible through the food chain. In this regard, industrial crops, e.g., energy and fiber crops, would be the best alternative land use option. Industrial crop (*Helianthus annuus, Brassica juncea, Armoracia rusticana, Arabidopsis halleri, Gossypium hirsutum, Eucalyptus, Amaranthus, Cannabis sativa,* and *Linum usitatissimum*)-based phytoremediation systems would contribute to sustainable development. The feasibility of genetic manipulation via *in vitro* culture techniques needs to be exploited for the success of phytoremediation for sustainable development (Figure 15.4). However, future research must focus on various tools of metallomics for maintaining the high quality of the environment (Figure 15.5).

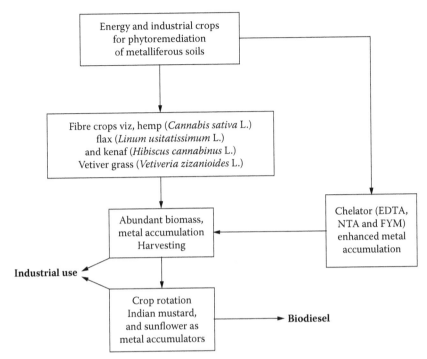

FIGURE 15.4 Industrial crops as potential tools of bioremediation for sustainable development. Sunflower for possible application in bioremediation of inorganic and organic pollutants.

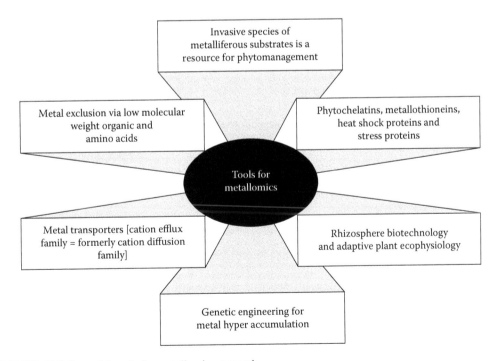

FIGURE 15.5 Potential tools for metallomics research.

REFERENCES

1. Haraguchi, H., Metallomics as integrated biometal science. *J. Anal. At. Spectrum*, 19, 5, 2004.
2. Arisi, A.C.M., Mocquot, B., Lagriffoul, A., Mench, M., Foyer, C.H., and Jouanin, L., Responses to cadmium in leaves of transformed poplars overexpressing γ-glutamylcysteine synthetase, *Physiol. Plant,* 109, 143–149, 2000.
3. Arisi, A.C.M., Noctor, G., Foyer, C.H., and Jouanin, L., Modification of thiol contents in poplars (*Populus tremula x P. alba*) overexpressing enzymes involved in glutathione synthesis, *Planta*, 1997, 362–372, 1997.
4. Clemens, S., Kim, E.J., Neumann, D., and Schroeder, J.I., Tolerance to toxic metals by a gene family of phytochelatin synthases from plants and yeast, *Embo J.,* 18, 3325–3333, 1999.
5. Huang, J.W., Blaylock, M.J., Kapulnik, Y., and Ensley, B.D., Phytoremediation of uranium-contaminated soils: role of organic acids in triggering uranium hyperaccumulation in plants, *Environ. Sci. Technol.,* 32, 2004–2008, 1998.
6. Krämer, U., Smith, R.D., Wenzel, W., Raskin, I., and Salt, D.E., The role of metal transport and metal tolerance in nickel hyperaccumulation by *Thlaspi Goesingense* Hálácsy, *Plant Physiol.,* 115, 1641–1650, 1997.
7. Lasat, M.M., Pence, N.S., Garvin, D.F., Ebbs, S.D., and Kochian, L.V., Molecular physiology of zinc transport in the Zn hyperaccumulator *Thlaspi Caerulescens, J. Exp. Bot.,* 51, 71–79, 2000.
8. Ma, J.F., Ryan, P., and Delhaize, E., Aluminum tolerance in plants and the complexing role of organic acids, *Trends Plant Sci.,* 1. 6, 273–278. 2001.
9. Persans, M.W., Salt, D.E., Kim, S.A., and Guerinot, M.L., Phylogenetic relationships within cation transporter families of *Arabidopsis, Plant Physiol.*, 126, 1646–1667, 2001.
10. Prasad, M.N.V. and Strzalka, K., *Physiology and Biochemistry of Metal Toxicity and Tolerance in Plants*, Kluwer Academic Publishers, Dordrecht, 460, 2002.
11. Schäfer, H.J., Haag–Kerwer, A., and Rausch, T., cDNA cloning and expression analysis of genes encoding gsh synthesis in roots of the heavy-metal accumulator *Brassica juncea* L.: evidence for Cd induction of a putative mitochondrial γ-glutamyl cysteine synthetase isoform, *Plant Mol. Biol.,* 37, 87–97, 1998.
12. Schaumlöffel, D., Szpunar, J., and Obiski, R., Species-selective analysis for metals and metalloids in plants, in *Heavy Metal Stress in Plants: from Biomolecules to Ecosystems,* 2nd ed., Prasad, M.N.V., Ed., Springer–Verlag, 2004, 409.
13. Sanità Di Toppi, L. et al., Metal chelating peptides and proteins in plants, in M.N.V. Prasad and K. Strzalka (Eds.), *Physiology and Biochemistry of Metal Toxicity and Tolerance in Plants*, Kluwer Academic Publishers, Dordrecht, 59–93, 2002.
14. Steffens, J.C., The heavy metal-binding peptides of plants, *Annu. Rev. Plant Physiol. Plant Mol. Biol.,* 41, 553, 1990.
15. Leustek, T. et al., Pathways and regulation of sulfur metabolism revealed through molecular and genetic studies, *Annu. Rev. Plant Physiol. Mol. Biol.,* 51, 141, 2000.
16. Pilon–Smits, E.A.H. et al., Overexpression of ATP sulfurylase in Indian mustard leads to increased selenate uptake, reduction and tolerance, *Plant Physiol.,* 119, 123, 1999.
17. Rauser, W.E., The role of thiols in plants under metal stress, in *Sulfur Nutrition and Sulfur Assimilation in Higher Plants*, C. Brunold, H. Rennenberg, and L.J. De Kok, Eds. Paul Haupt, Bern, Switzerland, 169–183, 2000.
18. Heaton, A. C. P. et al., Phytoremediation of mercury- and methylmercury-polluted soils using genetically engineered plants. *J. Soil Contam.,* 7, 497–510, 1998.
19. Rugh, C.L. et al., Development of transgenic yellow poplar for mercury phytoremediation, *Nat. Biotechnol.,* 16, 925, 1998.
20. Cobbet, C.S., Phytochelatins biosynthesis and function in heavy-metal detoxification, *Curr. Opin. Plant Biol.,* 3, 211, 2000.
21. Howden, R. et al., Cadmium-sensitive, cad1 mutants of *Arabidopsis thaliana* are phytochelatin deficient, *Plant Physiol.,* 107, 1059, 1995.
22. Ebbs, S.D. et al., Phytochelatins synthesis is not responsible for Cd tolerance in the Zn/Cd hyperaccumulator *Thlaspi caerulescens*, *Plant Biol.,* 2000 (The American Society of Plant Physiologists), Abstract no. 21004.

23. Sanità di Toppi, L. et al., Brassicaceae, in *Metals in the Environment: Analysis by Biodiversity,* Prasad, M.N.V., Ed., Marcel Dekker, New York, 2001, 219.

24. Sanità Di Toppi, L., Prasad, M.N.V., and Ottonello, S., Metal-chelating peptides and proteins in plants, in *Physiology and Biochemistry of Metal Toxicity and Tolerance in Plants,* Prasad, M.N.V. and Strzalka, K., Eds., Kluwer Academic Publishers, Dordrecht, 2002, 59.

25. de Knecht, J.A. et al., Synthesis and degradation of phytochelatins in cadmium-sensitive and cadmium-tolerant *Silene vulgaris, Plant Sci.,* 106, 9, 1995.

26. de Knecht, J.A. et al., Phytochelatins in cadmium-sensitive and cadmium-tolerant *Silene vulgaris, Plant Physiol.,* 104, 255, 1994.

27. Murphy, A. and Taiz, L., A new vertical mesh transfer technique for metal-tolerance studies in *Arabidopsis*: ecotypic variation and copper-sensitive mutants, *Plant Physiol.,* 108, 29, 1995.

28. Valentine, J.S. and Gralla, E.B., Delivering copper inside yeast and human cells, *Science,* 278, 817, 1997.

29. Sanità Di Toppi, L. et al., Responses to heavy metals in plants — molecular approach, in *Abiotic Stresses in Plants,* Sanità Di Toppi, L. and Pawlik Skowronska, B., Eds., Kluwer Academic Publishers, Dordrecht, 2003, 133.

30. Rauser, W.E., Phytochelatins and related peptides, *Plant Physiol.,* 109, 1141, 1995.

31. Czarenecka, E. et al., Comparative analysis of physical stress responses in soybean seedlings using cloned heat-shock cDNAs, *Plant Mol. Biol.,* 3, 45, 1984.

32. Neumann, D. et al., Heat-shock protein induced heavy-metaltolerance in higher plants, *Planta,* 194, 360, 1994.

33. Reddy, G.N. and Prasad, M.N.V., Tyrosine is not phosphorylated in cadmium-induced HSP cognate in maize (*Zea mays* L.) seedlings: role in chaperone function *Biochem. Arch.,* 9, 25, 1993.

34. Prasad, M.N.V., Trace metals, in *Plant Ecophysiology,* Prasad, M.N.V., Ed., John Wiley & Sons, New York, 1997, 207.

35. Cui, Y., Wang, O., Dong, Y., Li, H., and Christie, P., Enhanced uptake of soil Pb and Zn by Indian mustard and winter wheat following combined soil application of elemental sulphur and EDTA, *Plant Soil,* 261, 181–188, 2004.

36. Simon, E., Heavy metals in soils, vegetation development and heavy metal tolerance in plant populations from metalliferous area, *New Phytol.,* 81, 175, 1978.

37. Karataglis, S.S., Influence of the soil Ca on the tolerance of *Festuca rubra* populations against toxic metals, *Phyton,* 21, 103, 1981.

38. Aravind, P. and Prasad, M.N.V., Zinc alleviates cadmium induced toxicity in *Ceratophyllum demersum,* a fresh water macrophyte, *Plant Physiol. Biochem.,* 41, 4, 391, 2003.

39. Aravind, P. and Prasad, M.N.V., Carbonic anhydrase impairment in cadmium-treated *Ceratophyllum demersum* L. (free-floating freshwater macrophyte): toxicity reversal by zinc, *J. Anal.l Atomic Spectrophotometry,* 19, 52, 2004 (R. Soc. Chemists).

40. Aavind, P. and Prasad, M.N.V., Zinc protects chloroplasts and associated photochemical functions in cadmium exposed *Ceratophyllum demersum* L., a freshwater macrophyte, *Plant Sci.,* 166, 5, 1321, 2004.

41. Baker, A.J.M., The uptake of zinc and calcium from solutions culture by zinc-tolerant and nontolerant *Silene maritima* with. in relation to calcium supply, *New Phytol.,* 81, 321, 1978.

42. Skorzynska–Polit, E. et al., Calcium modifies Cd e.ect on runner bean plants, *Environ. Exp. Bot.,* 40, 275, 1988.

43. Skorzynska–Polit, E. and Baszynski, T., Does Cd^{2+} use Ca^{2+} channels to penetrate into chloroplasts? A preliminary study, *Acta Physiol. Plantarum,* 22, 171, 2000.

44. Rashid, A. and Popovic, R., Protective role of CaCl2 against Pb^{+2} inhibition in Photosystem II, *FEBS Lett.,* 271, 181, 1990.

45. Rengel, Z., Role of calcium in aluminum toxicity, *New Phytol.,* 121, 499, 1992.

46. Antosiewicz, D.M., Mineral status of dicotyledonous crop plants in relation to their constitutional tolerance to lead, *Environ. Exp. Bot.,* 33, 575, 1993.

47. Antosiewicz, D.M., The relationships between constitutional and inducible Pb-tolerance and tolerance to mineral deficits in *Biscutella laevigata* and *Silene inata, Environ. Exp. Bot.,* 35, 55, 1995.

48. Williams, L.E., Pittman, J.K., and Hall, J.L., Emerging mechanisms for heavy metal transport in plants, *Biochim. Biophys. Acta,* 1465, 104, 2000.

49. Clemens, S., Molecular mechanisms of plant metal tolerance and homeostasis, *Planta,* 212, 475, 2001.

50. Fox, T.C. and Guerinot, M.L., Molecular biology of cation transport in plants, *Annu. Rev. Plant Physiol. Plant Molecular Biol.,* 49, 669, 1998.

51. Schachtman, D.P. et al., Molecular and functional characterization of a novel low-affinity cation transporter (*LCT1*) in higher , *Proc. Natl. Acad. Sci. USA,* 94, 11079, 1997.

52. Amtmann, A. et al., The wheat cDNA *LCT1* generates hypersensitivity to sodium in a salt-sensitive yeast strain, *Plant Physiol.,* 126, 1061, 2001.

53. Clemens, S. et al., The plant cDNA LCT1 mediates the uptake of calcium and cadmium in yeast, *Proc. Natl. Acad. Sci. USA,* 95, 1243, 1998.

54. Habermann, E., Crowell, K., and Janicki, P., Lead and other metals can substitute for Ca^{+2} in calmodulin, *Arch. Toxicol.,* 54, 61, 1983.

55. Cheung, W.Y., Calmodulin: its potential role in cell proliferation and heavy metal toxicity, *Fed. Proc.,* 43, 2995, 1984.

56. Richardt, G., Federolf, G., and Habermann, H., A.nity of heavy metal ions to intracellular Ca^{+2}-binding proteins, *Biochem. Pharmacol.,* 35, 1331, 1986.

57. Antosiewicz, D.M. and Hennig, J., Overexpression of *LCT1* in tobacco enhances the protective action of calcium against cadmium toxicity, *Environ. Pollut.,* 129, 237, 2004.

58. Bush, D.S., Calcium regulation in plant cells and its role in signaling, *Annu. Rev. Plant Physiol. Plant Molecular Biol.,* 46, 95, 1995.

59. Choi, Y-E. et al., Detoxification of cadmium in tobacco plants: formation and active excretion of crystals containing cadmium and calcium through trichomes, *Planta,* 213, 45, 2001.

60. John, M.K., Interrelationship between plant cadmium and uptake of some other elements from culture solutions by oats and lettuce, *Environ. Pollut.,* 11, 85, 1976.

61. Hardimann, R.T. and Jacoby, B., Absorption and translocation of Cd in bush beans (*Phaseolus vulgaris*), *Physiol. Plantarum,* 61, 670, 1984.

62. Godbold, D.L., Cadmium uptake in Norway spruce (*Picea abies* L. Karst.) seedlings, *Tree Physiol.,* 9, 349, 1991.

63. Burzynski, M., The influence of lead and cadmium on the absorption and distribution of potassium, calcium, magnesium and iron in cucumber seedlings, *Acta Physiol. Plantarum,* 9, 229, 1987.

64. Hegedüs A. et al., Transgenic tobacco plants overproducing alfalfa aldose/aldehyde reductase show higher tolerance to low temperature and cadmium stress, *Plant Sci.,* 166, 1329, 2004.

65. Zenk, M.H., Heavy metal detoxification in higher plants: a review, *Gene,* 179, 21, 1996.

66. Speiser et al., *Brassica juncea* produces a phytochelatin cadmium sulfide complex, *Plant Physiol.,* 99, 817, 1992.

67. Vatamaniuk, O.K. et al., *AtPCS1*, a phytochelatin synthase from *Arabidopsis*: isolation and *in vitro* reconstitution, *Proc. Natl. Acad. Sci. USA,* 96, 7110, 1999.

68. Clemens S. et al., Tolerance to toxic metals by a gene family of phytochelatin synthases from plants and yeast, *EMBO J.,* 18, 3325, 1999.

69. Ha, S-B. et al., Phytochelatin synthase genes from *Arabidopsis* and the yeast *Schizosaccharomyces pombe, Plant Cell,* 11, 1153, 1999.

70. Noctor, G. et al., Glutathione: biosynthesis: metabolism and relationship to stress tolerance explored in transformed plants, *J. Exp. Bot.,* 49, 623, 1998.

71. Arisi, A.C.M. et al., Modification of thiol contents in poplars (*Populus tremula x P. alba*) overexpressing enzymes involved in glutathione synthesis, *Planta,* 203, 362, 1997.

72. Noctor, G. and Foyer, C.H., Ascorbate and glutathione: keeping active oxygen under control. Mustard produces a phytochelatin–cadmium–sulfide complex, *Plant Physiol.,* 99, 817, 1998.

73. Foyer, C.H. et al., Overexpression of glutathione reductase but not glutathione synthetase leads to increases in antioxidant capacity and resistance to photoinhibition in poplar trees, *Plant Physiol.,* 109, 1047, 1995.

74. Zhu, Y.L. et al., Cadmium tolerance and accumulation in Indian mustard is enhanced by overexpressing γ-glutamylcysteine synthetase, *Plant Physiol.,* 121, 1169, 1999.

75. Goldsbrough, P.B., Metal tolerance in plants: the role of phytochelatins and metallothioneins, in *Phytoremediation of Contaminated Soil and Water,* Terry, N. and Banuelos, G.S., Eds., CRC Press, Boca Raton, FL, 1998, 221.

76. Prasad, M.N.V. Ed., *Heavy Metal Stress in Plants: from Biomolecules to Ecosystems*, 2nd ed., Springer–Verlag, Heidelberg, 2004, 462, xiv.

77. Misra, S. and Gedamu, L., Heavy metal tolerant transgenic *Brassica napus* L. and *Nicotiana tabacum* L. plants, *Theor. Appl. Genet.*, 78, 161, 1989.

78. Salt, D.E. et al., Mechanism of cadmium mobility and accumulation in Indian mustard, *Plant Physiol.*, 109, 1427, 1995.

79. Baker, A.J.M. et al., The possibility if *in-situ* heavy metal decontamination of polluted soils using crops of metal-accumulating plants, *Resour. Conserv. Recycl.*, 11, 41, 1994.

80. Black, H., Absorbing possibilities: phytoremediation, *Environ. Health Perspect.*, 103, 1106, 1995.

81. Kumar, P.B.A.N. et al., Phytoextraction: the use of plants to remove heavy metals from soils, *Environ. Sci. Technol.*, 29, 1232, 1995.

82. Cunningham, S.D. and Ow, D.W., Promises and prospects of phytoremediation, *Plant Physiol.*, 110, 715, 1996.

16 Detoxification/Defense Mechanisms in Metal-Exposed Plants

B.P. Shaw, M.N.V. Prasad, V.K. Jha, and B.B. Sahu

CONTENTS

16.1 INTRODUCTION

Trace element "biogeogenic cycling" in the environment is an integral function of the ecosystem (aquatic, terrestrial, and atmospheric systems). Metal enrichments in these compartments may result from natural sources or from human activities, such as smelting, mining, processing, agricultural, and waste disposal technologies. Metals are present in the Earth's crust in various quantities [1]. Their relative abundance, however, differs greatly in regions over the globe, and the region at which a metal is found in high concentration serves as the source of the metal. Although a metal may be present in high concentration in a region, it does not pose any threat to the environment until the landmass of the region is used for agroindustry. This is because the metals present remain tightly bound to their Lewis components as sulfides, oxides, or carbonates, as the case may be [1], and the ore particles also remain tightly packed along with the particles of the soil, which makes them highly immobilized. It is only the mining of the ore, and subsequent uses of the extracted metals that lead to far and wide contamination of the environment. From the figures of the crustal abundance of various metals and their production per annum (Table 16.1), the magnitude of contamination or pollution by these metals as a result of anthropogenic activities may be imagined.

The concentration of a metal that existed in a region before the advent of industrial activity is termed its natural or background level. This is a result of release of the metal due to natural weathering of the metal-bearing formations in the area. The knowledge of natural contamination of a metal provides a true reference point for estimating the extent of pollution from the element and allows the contemporary situation to be seen in perspective — i.e., whether it is in excess from the point of view of its toxicity to organisms (Figure 16.1). However, the natural, or background levels of metals for some areas may be difficult to obtain because they may not exist due to human intervention; this is particularly true for lead, mercury, cadmium, and arsenic. In fact, although naturally occurring geochemical materials are the primary source of metals in the environment, not many examples are known for which the interaction between natural weathering processes and mineralized zones is completely devoid of a human contribution.

Anthropogenic activities lead to pollution of the three nonliving components of the environment — air, water, and soil — and the biosphere by metals [4]. The magnitude of pollution depends largely upon the nature and intensity of the activities; the most important among them are mining; industrial processing of ores and metals; and the use of metals and metal components, which affect the environment in a wide variety of ways [2,4,8–10]. However, this discussion will be restricted to terrestrial contamination and the adaptive process that plants undergo to face the challenge of the presence of high levels of natural or man-induced metals around them.

Excluders prevent metal uptake into roots and avoid translocation and accumulation in shoots [11]. They have a low potential for metal extraction, but they can be used to stabilize the soil to avoid further contamination spread due to erosion. Resistance of plants to heavy metal ions can be achieved by an avoidance mechanism, which includes mainly the mobilization of metal in root and in cell walls. Tolerance to heavy metals is based on the sequestration of heavy metal ions in vacuoles; on binding them by appropriate ligands like organic acids, proteins, and peptides; and on the presence of enzymes that can function at high levels of metallic ions (Figure 16.2 and Figure 16.3) [12].

The effective xylem loading of hyperaccumulators may be due to smaller sequestration of metals in the root vacuoles of hyperaccumulators [13]. Translocation of Ni from roots to shoots may involve specific ligands in some hyperaccumulator species. Kramer et al. [14] showed that spraying histidine on the leaves of the nonaccumulating *Alyssum montanum* greatly increased Ni tolerance and capacity for Ni transport to the shoots. The detoxification of heavy metals commences only when they enter the cells and occurs in the cell by the process of chelation, compartmentalization, or precipitation [15].

Metallothionein (MT)-II genes have been identified in plants [16,17]. Although detection of plant metallothioneins has been problematic, evidence suggests that they have the ability to bind heavy metals. Also, accumulation of heavy metals in plants has been shown to induce the production of phytochelatins (PCs), a family of thiol-rich peptides [18]. The synthesis of PCs has been documented to be induced by a variety of metals. However, PCs have been shown to be primarily involved in Cd and Cu tolerance [19]. A recent study suggested that PCs may also be involved in As detoxification [20].

The processes of heavy metal uptake, accumulation, distribution, and detoxification have been studied in a wide range of crop and herbaceous species [21]. The mechanisms involved in perennials have been partially investigated and reported to be considerably tolerant [22]. Several sequestration and detoxification strategies are reported to take place in plants exposed to elevated doses of toxic trace elements (Figure 16.4 and Figure 16.5) [23].

Complexation with phytochelatin peptides synthesized from glutathione has been identified as an important mechanism for detoxifying metals such as Cd, Pb, and Zn. Yet, phytochelatins do not appear to be the primary mechanism. Large increases in histidine levels and coordination of Ni with histidine have been reported in the xylem sap of *Alyssum lesbiacum*, suggesting that histidine is important for Ni tolerance and transport in hyperaccumulators [24].

TABLE 16.1
Worldwide Metal Production and Uses

Metal	Crustal abundance (mg kg^{-1})	Yearly production ($\times$ 1000 tonnes)	Major uses	Principal ores
Aluminum	83,000	16,200	Cable and wire for high-voltage electric transmission and various parts of autos, aircraft, and electrical equipment	Bauxite, Al$_2$O$_3$
Arsenic	1.80[a]	50	Making alloys for bullets and shot, storage batteries, herbicides, insecticides, and wood preservatives	Arsenide
Bismuth	0.20	4	Finds uses in phamaceuticals, electronics, cosmetics, and pigments, and as catalyst	Principally in flue dust as bismuthinite, Bi$_2$S$_3$, during smelting of Pb, Zn, or Cu
Chromium	110	10,800	Used in metal plating, making stainless-steel, wear-resistant and cutting-tool alloys, and as an anticorrosive	Chromite, FeOCr$_2$O$_3$
Cadmium	0.2	19	Used in electroplating, making Ni/Cd batteries, alloys, control rods in nuclear reactors, and pigments, and as stabilizer of polyvinyl chloride (PVC) plastic	Greenockite, CdS
Copper	63	8,700	Mainly used in making alloys and electrical products, the only wire used in windings in generators, motors, and transformers	As metal sulfides and oxides
Gold	0.0035	1.61	Used in jewelry and as the basis of currency	Calavarite (AuTe$_2$), Petzite [(Ag,Au)$_2$Te]
Iron	58,000	508,000	Most widely produced metal, usually as steel; also used in many alloys for special purposes	Hematite, Fe$_2$O$_3$, goethite, Fe$_2$O$_4$.H$_2$O, magnetite, Fe$_3$O$_4$
Lead	12	3,400	Making storage batteries, petrol additive, pigments, ammunition, cable sheathing	Galena, PbS
Manganese	1,300	22,000	Used as oxygen and sulfur scavenger in steel; manufacture of alloys, dry cells, chemicals	Found mainly as oxides
Mercury	0.089	6	Used as cathode in chlor-alkali cells, and also used in making paints, electrical apparatuses, fungicides	Cinnabar, HgS
Molybdenum	1.30	89	Used in making alloys, pigments, chemicals, lubricants, and as catalyst	Molybdenite, MoS$_2$, wulfenite, PbMoO$_4$
Nickel	89	800	Used in making coins, storage batteries, alloys, and as catalyst	Pentlandite [(Fe,Ni)$_9$S$_8$], Nicolite (NiAs)
Selenium	0.075	1.6[a]	Used in electronics, glass, pigments, photocopying	Mainly as clausthalite, PbSe, crokesite (Cu,Tl,Ag)$_2$Se
Silver	0.075	14	Finds uses mainly in making photographic materials and jewelry	Found with sulfide minerals

TABLE 16.1
Worldwide Metal Production and Uses (continued)

Metal	Crustal abundance (mg kg^{-1})	Yearly production (× 1000 tonnes)	Major uses	Principal ores
Tin	1.70	190	Used in coatings, solders; in making bearing alloys, bronze	Cassiterite, stannite
Titanium	6,400	4,200	Mainly used in making aircraft engines and their parts, also in making valves, pumps, paint pigments	As oxide, TiO$_2$
Vanadium	140	32	Used in making strong steel alloy	Primarily occurs as V(III) in igneous rocks
Zinc	94	7,200	Widely used in making brass (alloy), paint pigments; in galvanization	Found as sulfides, oxides, and silicates

Represents production during 1985.
Note: All production figures are of 1987.

Sources: Adapted from Manahan, S.E., *Environmental Chemistry*, Lewis Publishers, Boston, 1990, chap. 17; Ochiai, E.-I., *Bioinorganic Chemistry, an Introduction*, Allyn and Bacon, Inc., Boston, 1977, chap. 1; Fergusson, J.E., *The Heavy Elements: Chemistry, Environmental Impact and Health Effects*, Pergamon Press, New York, 1990, chap. 2; Evans, A.M., in *Introduction to Mineral Exploration*, Evans, A.M., Ed., Blackwell Science, Oxford, 1995, chap. 1; Chaterjee, K.K., *An Introduction to Mineral Economics*, Wiley Eastern Limited, Bombay, 1993, chap. 6; and Wedepohl, K.H., in *Metals and Their Compounds in the Environment: Occurrence, Analysis and Biological Relevance*, Merian, E., Ed., John Wiley & Sons, Inc., New York, 2000, chap. 1.

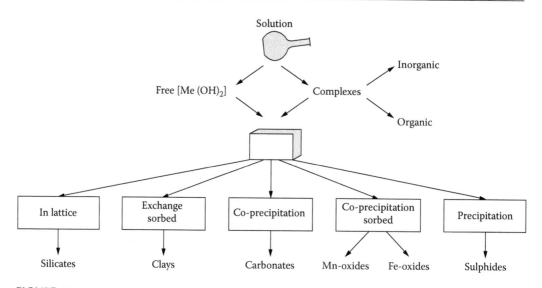

FIGURE 16.1 Fate of trace elements in the environment.

There are many indications that organic acids are involved in heavy metal tolerance, transport, and storage in plants, including for Al, Cd, Fe, Ni, and Zn. In plants that hyperaccumulate the metals as stated previously, the levels of citric, malic, malonic, and oxalic acids have been correlated with elevated concentrations of these metals in the biomass. Plant vacuoles are a major repository for organic acids, so an association between metals and organic acids suggests that metal detoxification occurs by vacuolar sequestration. However, other strategies for metal tolerance and

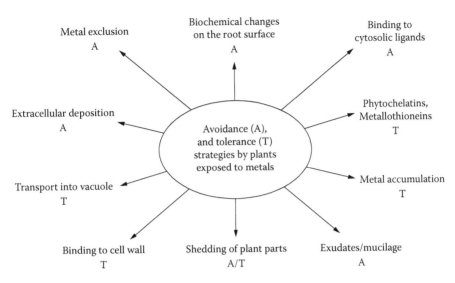

FIGURE 16.2 Avoidance and tolerance strategies adapted by plants exposed to elevated doses of trace elements.

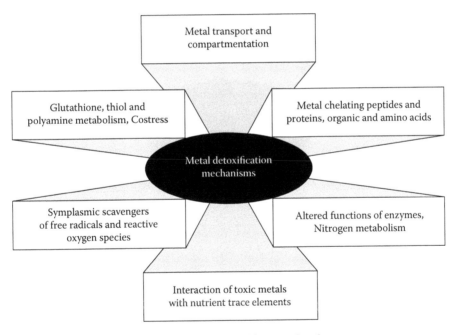

FIGURE 16.3 Metal detoxification mechanisms exhibited by vascular plants.

accumulation, such as binding to the cell wall or localization in the apoplast, may also be involved. The distribution of metals within plant tissues is therefore an important property that can act as an indirect indicator of detoxification mechanism. The distribution of metals between the apoplasm and symplasm of tissues, and between the cytosol and vacuole in cells, requires transmembrane transport; thus, the energizing and functioning of membrane processes may be of key significance in hyperaccumulation. The first, and still most common, parameter to characterize metal tolerance is the tolerance index, TI, which is calculated as:

$$TI = \text{response at elevated test metal concentration/response at control conditions}$$

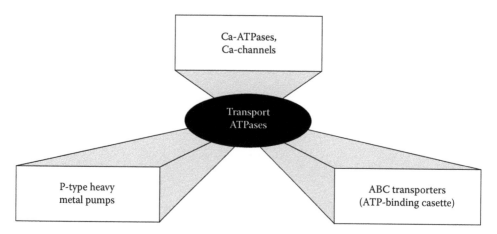

FIGURE 16.4 Role of plasmalemma in heavy metal tolerance involving transport ATPases.

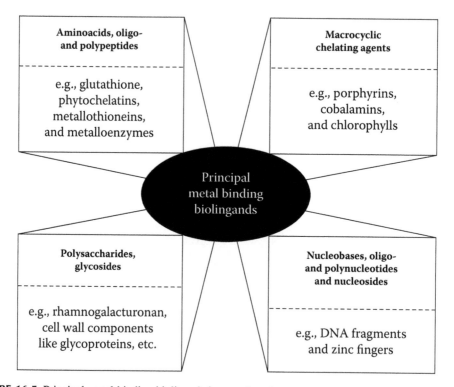

FIGURE 16.5 Principal metal-binding bioligands in vascular plants.

where response is a measurable character, e.g., increase in root length in the classical root elongation test. Alternatively, the effect index (EI) can be calculated as EI = 1 − TI. The response to control conditions and to the elevated metal concentration can be measured sequentially or in parallel [25].

16.2 HEAVY METAL CONTAMINATION OF SOIL AND ASSOCIATED AGRICULTURAL AND ENVIRONMENTAL PROBLEMS

Table 16.1 shows the yearly figures of production of heavy metals. Although the figures are of much environmental concern, these are of little importance as far as contamination of soil is concerned. This is because the use of heavy metals as industrial produce by mankind remains confined to cities and suburban areas, which may constitute less than 10 to 15% of the total inhabitable land mass. More importantly, the heavy metals used as industrial produce mostly find their way into aquatic environments through the drainage system and run-off water during the rainy season; from there, their return to the atmosphere and landmass through biogeochemical cycling is very slow [4].

Furthermore, it may also be noted that the use of heavy metals like Hg and As as components of pesticides in agriculture has been nearly discontinued, and the contamination of the land mass by these through agricultural practices is now only history. Also, although the use of fertilizers may result in contamination of the environment by various heavy metals present in them [26,27], this is unlikely to be of much significance because these are continuously removed from the soil along with each harvest.

Considering the network of the roads connecting one city to another, a significant source of heavy metals in the terrestrial environment, particularly of Pb, could be automobile exhaust [28–30]. Chamberlain [31] estimated that, since 1946, automobile-generated aerosol lead added about 3 µg g^{-1} lead to the topsoil in rural areas and <10 µg g^{-1} in urban areas in the U.K. However, this is said to be small compared to natural levels and the lead added from industrial discharges [4]. Furthermore, with increasing use of unleaded gasoline, the threat of contamination of the land mass may not remain of much importance. In fact, the threat of heavy metal pollution of the land mass, which may require some remedial measures such as phytoremediation, mainly results from two sources: (1) mining activities in the region for ore rich in one or more heavy metals; and (2) atmospheric emission by industries.

Mining of the earth for ore is the first step towards increasing contamination of the landmass by various heavy metals depending upon the type of the ore. The mining operation lets the ore particles loose; otherwise, they are bound tightly among each other, remaining virtually immobile. They become prone to be blown away by wind, thus contaminating a vast area around the mine, particularly in the windward direction. Additionally, the mining operation leaves stretches of mined lands devoid of vegetation because of their high metal contents. This problem of contamination of uncontaminated agricultural lands will further increase with increase in the area of mining and the mining operation. It is generally in practice to use only ores rich in metal for cost-effective extraction, but when the currently available stock of the metal-rich ores ends, the ores less rich in metal content may eventually be processed, leading to spatial increase in heavy metal-contaminated/polluted agricultural and other lands.

Processing of the ores for the extraction of metals is the second major step during which metals find their way into land mass; the metals escaping out of the chimneys of smelters are ultimately deposited in agricultural fields or other land, which may be far from the smelting unit. Atmospheric metal enrichment, leading subsequently to pollution of soil, is also associated with other higher temperature anthropogenic activities, like burning of fossil fuels, production of cements, etc. For illustration, the emission of a few heavy metals due to burning of coal is given in Table 16.2. Despite modern technological advances, smelting operations and fossil fuel burning in industries continue to be important sources of heavy metals in the terrestrial environment.

The environmental problem associated with Al needs special mention, although it is not a heavy metal when its specific gravity is taken into consideration. The two sources of metals to the terrestrial environment described earlier hold true for this metal also. In addition, Al as such occurs in high levels in soil, which may be appreciated from its high crustal abundance [1]. Wherever the soil pH is acidic, this causes serious agricultural losses. It has been estimated that approximately 40% of the world's cultivated lands and up to 70% of potentially arable lands are acidic [34], which speaks of the gravity of environmental problems and economical losses associated with Al contamination of soil. Al is most often found as oxide or silicate precipitates that are not toxic to plants. However, in acidic soil (pH < 5.0), Al speciates to soluble octahedral hexahydrate form, $Al(H_2O)_6^{3+}$ (commonly called Al^{3+} [35]) — the phytotoxic species responsible for agricultural losses.

TABLE 16.2
Emission[a] of a Few Heavy Metals from the Burning of Fossil Fuels

Heavy metals	Emission from coal burning	Emission from oil burning
Cobalt	700	30
Chromium	1400	50
Copper	2100	23
Nickel	2100	1600
Vanadium	3500	8200
Mercury	400	1600
Cadmium	140	2
Selenium	420	30
Arsenic	5000	10
Zinc	7000	40
Lead	3500	50

[a]Tons per year.

Sources: Adapted from Forstner, U. and Wittmann, G., *Metal Pollution in the Aquatic Environment*, Springer–Verlag, New York, 1979, chap. B; Bertine, K.K. and Goldber, E.D., *Environ. Sci. Technol.*, 11, 297, 1977; and Ruch, R.R. et al., *Environ. Geol. Notes*, 61, 1, 1973.

16.3 HEAVY METAL DETOXIFICATION MECHANISMS IN PLANTS

By virtue of their stationary status, plants, unlike animals, cannot migrate to avoid unfavorable fluctuation or changes in their environment; thus, they must change their metabolic activities suitably to allow them to cope with the changing environment or otherwise perish. The resulting changes in their metabolism are called "stress response," which may enable the plant to survive under the condition of stress. This may occur for a short time only, known as acclimation, or the changes induced may be good enough to support continuous growth of the plant, known as adaptation. The latter quality is being or may be exploited to find the solution to increasing heavy metal contamination of landmasses and forms the basis of "phytoremediation."

Cunningham and Ow [36] envisioned the working of phytoremediation as follows: "By growing plants over a number of years the aim is to either remove the pollutants from the contaminated matrix or to alter the chemical and physical nature of the contaminants within the soil so that it no longer presents a risk to human health and the environment." Thus, plants resistant to heavy metals can be used under the concept of phytoremediation in one or more of the following ways:

- To remove the metals from the soil
- To chelate the metals in the soil and bind the soil particles tightly so that their erosion by wind, as well as further contamination of the land in the windward direction, is prevented
- To make possible the use of the metal-contaminated land for agriculture

It is explicit that the plants to be used under the first category, i.e., for the removal of metals from soil, should be hyperaccumulators of the heavy metals contaminating the land; to be used under the second category, plants may or may not be hyperaccumulators, but should be resistant to the metals present in the soil and able to grow well, with good rooting systems. For plants to be used under the third category, in addition to being resistant to the metals contaminating the soil, it is necessary that they do not take up and accumulate them in their tissues; otherwise the agricultural products would be highly contaminated with the metals.

Heavy metal-resistant plants are available in nature in hyperaccumulator and excluder catego-ries. However, their use is limited because the character is not present in the desired species or the character requires additional desirable traits for effective phytoremediation. For example, a metal hyperaccumulator plant may not have the desired root structure, such as root depth and root density, and/or may not be fast growing with high biomass turnover, which could improve the metal accumulation or soil decontamination potential of the plant. Similarly, the metal-excluding character may not be present in the required crop species. Although breeding programs could be envisioned to put the desired traits together, knowledge of the biochemical and molecular bases of tolerance of plants to metals under both categories may allow scientists to take a biotechnological approach and introduce the required trait into the species of interest in a more comfortable and cost-effective manner.

Research on understanding the mechanism of metal tolerance dates back to as early as the 1950s, when only ecological and physiological differences between plants from metal-enriched and noncontaminated habitats were studied [37,38]. The investigation gained momentum only in the late 1960s when time- and cost-effective techniques for the analysis of metals — atomic absorption spectrophotometry — was developed [39]. However, during that period, the research was mainly concentrated upon the uptake of metals and their cellular compartmentalization [40,41].

From the 1970s, the physiological and genetic aspects of metal tolerance began to be studied using the rewarding approach of comparison of metal-tolerant and nontolerant cultivars of a species, or even isogenic line of a species, which differed as far as possible only in resistance to one or more metals [42,43]. Currently, understanding of tolerance of heavy metals in plants narrows basically to two categories: (1) resistance by exclusion of the metals; and (2) resistance by uptake, but subsequent sequestration of the metals to inactive form inside the cells. In addition, another concept is emerging in this field: heavy metal tolerance involving antioxidative machinery, which is worth discussing.

16.4 EXTRACELLULAR DETOXIFICATION OF METALS OTHER THAN ALUMINUM

The indication of possible involvement of plant exudates in ameliorating the toxic effect of metals other than Al mainly comes from work on algal systems, and that only on cyanobacteria. The indication mainly stems from the fact that many organic compounds, like amino acids [44,45]; mercaptans [46]; organic acids [47]; peptides [48]; and spent medium [46], supplied along with the heavy metals (Cu, Cd, Ag) alleviated their toxicity, and the fact that cyanobacteria produce weak acids, strong metal-complexing agents like hydroxamate siderophore [47]; amino acid-con-taining compounds [49]; catechol siderophore [50]; metallothioneins [51]; and unknown chelators [52,53].

However, the relationship between the excretion of organic compounds and their detoxification role has never been experimentally demonstrated [54]. Protective effects of plant exudates against heavy metal toxicity have not been reported. Recently, however, Arduini et al. [55] worked on Cd and Cu distribution in various Mediterranean tree seedlings and suggested that the well-developed root cap in plants has a protective role against metal uptake.

16.5 HEAVY METAL DETOXIFICATION THROUGH INTRACELLULAR SEQUESTRATION

Currently, it is accepted in principle that plants' resistance to heavy metals, unlike that to Al^{3+}, is achieved through their uptake and proper sequestration inside the cell, rather than by their exclusion that works for Al^{3+}. Plants' adaptation of a totally different mechanism for tolerance to heavy metals than for tolerance to Al^{3+} is probably because many essential metals in the heavy metal category must be taken up from the environment for various metabolic functions to continue [1], but their chemical properties match greatly with many of the nonessential heavy metals, thus making it difficult for plants to go for their selective uptake. Thus, the nonessential heavy metals are also taken up along with the essential ones when present in the environment.

Additionally, the essential heavy metals are also more or less as toxic to the organisms as the nonessential ones at a similar concentration. Thus, probably the only way for plants to counter the presence of elevated levels of essential or nonessential heavy metals in the environment is to sequester them properly out of the cytoplasm in inactive form. The problems associated with the presence of elevated levels of heavy metals in the environment rather may be viewed under broad perspectives: it is essential for plants to have mechanisms that (1) maintain the internal concentrations of essential metals between deficient and toxic limits; and (2) keep the nonessential metals below their toxicity threshold [18].

At present, approximately 400 plant species belonging to at least 45 families are hyperaccumulators of heavy metals to various degrees [56,57]. The field-collected samples of plants from metal-rich soils have been found to contain metals like Cu, Co, Cd, Mg, Ni, Se, or Zn up to levels that are 100 to 1000 times those normally accumulated by plants [56,57]. The concentration of some metals may reach as high as 1000 ppm or more on dry mass basis [58,59]. The exact mechanism involved in such hyperaccumulation is, however, still debatable. A few reports of involvement of organic acids in the hyperaccumulation process have been issued; however, general agreement in the scientific community is that the hyperaccumulation, and thus the resistance to the heavy metals accumulated, is facilitated by thiol-rich polypeptide, similar in function to that of metallothionein rich in amino acid cysteine, which was first discovered in horse kidney [60].

16.6 COMPLEX FORMATION WITH PHYTOCHELATINS

Polypeptides are designated "metallothionein" when they show several of the feature characteristics of equine renal metal-binding protein, like high metal content; high cysteine content with absence of aromatic amino acid and histidine; an abundance of Cys-x-Cys sequence, where x is an amino acid other than Cys; spectroscopic feature characteristics of metal thiolates; and metal thiolate clusters [61,62]. These have been subdivided into three classes:

- Class I: polypeptides with locations of cysteine closely related to those in equine renal metallothionein (MT)
- Class II: polypetides with location of cysteine only distantly related to those in equine renal metallothionein
- Class III: atypical nontranslationally synthesized metal thiolate polypeptides

The class III metallothioneins (MTs) are only known from the plant kingdom. They occur in the organisms as metal-binding complexes of various sizes, Mr (molecular mass) 3000 to 10,000, depending upon the ionic strength of the eluants; this suggests that these are accretions of multiple peptides of various lengths with the metal atom(s). The incorporation of varying amounts of sulfide or sulfite ions also contributes to the size heterogeneity of the class III MTs. Nevertheless, two broad categories of the complexes are generally recognized: low molecular weight (LMW) and high molecular weight (HMW). The grouping is based on the good resolution of the metal-binding complexes obtained in the extracts from fission yeast exposed to Cd in some of the earliest known experiments in the line [63,64].

Currently, several situations are known to exist in plants, ranging from good resolution of LMW and HMW complexes [65], partial resolution of the LMW complexes on the trailing shoulder of abundant HMW complexes [66], to no evidence for LMW complex [67]. Such variations are attributed to differences between organisms; type of nutrient medium for growth; concentration of the gel-filtration matrix; and column bed dimensions [18].

The indication of the existence of class III MTs in plants was first provided by Rauser and Curvetto [68] in roots of *Argostis* tolerant to Cu, and by Weigel and Jager [69] in bean exposed to Cd. The metal-binding complex was characterized to some extent only by Murasugi et al. [63,70] in extract from fission yeast, *Schizosaccharomyces pombe*, grown on Cd solution. Subsequently, these have been reported to be produced in cultured plant cells [71], algae [72], and virtually all higher plants tested [71,73]. These have also been reported to be induced by a variety of metals, including Cd, Cu, Zn, Pb, Hg, Bi, Ag, and Au [74–76]; some of these are soft metals on the Pearson's scale of softness. In addition, the complex is also induced by multiatomic anions like SeO_4^{2-}, SeO_3^{2-}, and AsO_4^{3-} [74]. Their induction depends not only on the type of the metal, but also on the plant species [77].

The information on composition and structure of phytochelatins (PCs) comes from the pioneering work of two groups: Kondo et al. [78,79] on fission yeast, *Schizosacchromyces pombe*, and Grill et al. [71] on cultured cells of *Rauvolfia serpentina*. The complexes produced by *S. pombe* have a structure identical to those of plants, consisting of heterogeneous population of polypeptides; each represents repeating units of γ-glutamyl-cysteine (γ-Glu-Cys) followed by a single C terminal glycine [γ-Glu-Cys)$_n$-Gly] with the number of repeating units (n) ranging from 2 to 11 [71,78,79]. Before their classification as class III MTs, various trivial names were given, depending upon one or more features associated with them — for example, as cadystin — as were found to be induced by cadmium in fission yeast [63,64,70,78,79]; phytochelatin (PC), representing the metal binding peptides of kingdom phyta [71]; γ-glutamyl peptides, after the presence of γ-glutamyl [80]; and poly(γ-glutamylcysteinyl)glycine, after their basic constituents [81].

However, none of these trivial names is appropriate because metals other than Cd also induce these polypeptides. Moreover, fungi are not considered to belong to kingdom phyta and diverse γ-glutamyl di- and tripeptides are known [61]. Nevertheless, the term "phytochelatin" is popularly used because fungi are always considered as plants. Besides, the term is meaningful — suggesting the chelating function of the molecule — and hence this will continue to be used in this chapter.

16.7 PHYTOCHELATINS: PRIMARY STRUCTURE AND CLASSES

The primary structure of phytochelatins (PCs) was basically derived from the Cd-binding chemical analysis of the Cd-binding complexes from various sources in the early phase of work in this field [71,72,78,79,81,82]. The polypeptides were found to be composed of three amino acids: Glu, Cys, and Gly, which occurred in the molar ratio: 2:2:1 and 3:3:1 in *S. pombe* [79] and *Datura innoxia* [81]; 3:3:1 and 4:4:1 in tomato [82]; 4:4:1 in *R. serpentine* [71]; and up to 5:5:1 in *Chlorella fusca* [72]. The primary structure in general is of nature (γ-Glu-Cys)$_n$-Gly, with n = 2 to 5. The value of n up to 11 has also been observed depending upon the plant species [73,74].

The primary structure of the polypeptides described previously is related to glutathione (GSH), γ-Glu-Cys-Gly. However, in some plants such as soybean (*Glycine max*), glutathione is replaced by homoglutathione with a nonprotein amino acid β-alanine (γ-Glu-Cys-β-Ala). In such plants, glycine at the carboxy terminal in the polypeptide is replaced by β-Ala and thus has been named homophytochelatin (h-PC) [83]. This has the general structure (γ-Glu-Cys)$_n$-β-Ala and has been observed in 36 species of legume; 13 species produce only h-PC and 23 species produce h-PC and PC, depending upon whether the plant produces only h-GSH or GSH and h-GSH [83].

The third family of the PC polypeptide was observed by Klapheck et al. [84] in certain species of Poaceae (rice, wheat, and oats) in which the terminal Gly is replaced by serine (Ser) showing the primary structure (γ-Glu-Cys)$_n$-Ser. These peptides are related to the tripeptide hydroxymethyl-glutathione (γ-Glu-Cys-Ser), so the polymer is termed hydroxymethyl-phytochelatin (hm-phytochelatin). In addition, the species studied also produced (γ-Glu-Cys)$_n$-Gly and (γ-Glu-Cys)$_n$ (desglycyl or desGly peptide).

The most recent addition in the family of phytochelatin polypeptides is that related to novel tripeptide γ-Glu-Cys-Glu, with the structure (γ-Glu-Cys)$_n$-Glu. This was identified in maize exposed to Cd [85]. Maize also produces in abundance another family of the polypeptides, the desGly-peptides [(γ-Glu-Cys)$_n$], which was first noticed as a minor constituent of Cd-binding complexes in the fission yeast [85]. The production of (γ-Glu-Cys)$_n$-Gly also occurred.

Thus, there are five families of polypeptides in class III MTs. They have the common features such as (1) Glu occupies the amino-terminal position; (2) Cys forms the next residue-forming peptide bond with the γ-carboxyl group of Glu; and (3) γ-Glu-Cys pairs are repeated two or more times with the subscript (n) specifying the exact number of repeats. The division of the polypeptides into five classes is only on the basis of the variation in the carboxy terminal amino acid. All the five classes of phytochelatins thus belong to one specific family of dipeptide γ-Glu-Cys, and the complement of (γ-Glu-Cys)$_n$ peptide from the five families largely varies according to the species.

16.8 PHYTOCHELATIN: SYNTHESIS

The γ-glutamyl linkages present in PCs suggest that these polypeptides are not primary gene products, i.e., they are not the translational products of mRNA and must be formed by ribosome-independent enzyme reactions. The similarity of PCs to GSH in containing γ-Glu-Cys moiety suggested that this could be involved in the synthesis of the polypeptides. A number of other observations also support the function of GSH as a precursor of PCs.

- The metal-induced synthesis of PCs is accompanied by a depletion of the GSH pool in cell cultures [74,86,87] and in plant tissue root [88–90].
- GSH is synthesized by the action of γ-glutamylcysteine (γ-Glu-Cys) synthetase (EC 6.3.2.2), which joins Glu with Cys followed by addition of Gly by GSH synthetase (EC 6.3.2.3). The activity of one or both of the enzymes increases upon exposure of the plants to Cd [88,91]. Furthermore, the plant cells incubated with buthione sulfoximie (BSO), a potent inhibitor of γ-Glu-Cys synthetase, are unable to synthesize PCs, and addition of GSH reestablishes PC synthesis [86–88].
- The mutants of *S. pombe* that lack γ-Glu-Cys synthetase or GSH synthetase do not synthesize PCs in response to Cd [92]. It has been demonstrated that, in cells of *Datura innoxia*, [^{35}S]GSH is rapidly incorporated into PCs after exposure to cadmium [93]. In the presence of BSO and GSH, PCs produced upon exposure of tomato cells to cadmium incorporate little [^{35}S]cysteine, indicating that these peptides are not synthesized by sequential addition of cysteine and glutamate to GSH [94].

At least three possible pathways of biosynthesis of PCs can be visualized:

- Transpeptidation, with GSH or the oligomeric PC peptide acting as an acceptor for the successive addition of γ-Glu-Cys moieties from GSH by transpeptidation reaction
- Dipeptide addition in which γ-Glu-Cys units, synthesized by the action of γ-Glu-Cys synthetase, are transferred to GSH and/or the oligomeric PC peptides
- Polymerization of the γ-Glu-Cys units to (γ-Glu-Cys)$_n$ oligomeric molecules that are transferred subsequently to Gly in a similar fashion known for GSH synthetase

Grill et al. [95] identified one activity in the extract from *Silene cucubalus* (= *vulgaris*) that conformed to the first possible pathway. *In vitro* experiments with 15-fold purified enzyme activity fraction with GSH as substrate showed induction of PC synthesis immediately after the addition of 0.1 mM Cd^{2+}. (γ-Glu-Cys)$_2$-Gly appeared without a noticeable lag phase. Heptapeptide ($n = 3$) was detected 15 min after the addition of Cd^{2+} and, after a further 20 min, nanopeptide ($n = 4$) was detected. (The reaction came to a halt after 100 min and resumed after a second addition of Cd^{2+}.)

When (γ-Glu-Cys)$_2$-Gly was present along with GSH, heptapeptide formation occurred immediately after the addition of Cd^{2+}. In the presence of only (γ-Glu-Cys)$_2$-Gly as substrate, the formation of first heptapeptides and then nanopeptides occurred. Simultaneously, the concentration of the pentapeptide (the substrate) decreased with concomitant release of GSH. These coworkers concluded that, in addition to adding γ-Glu-Cys unit to GSH, the enzyme, which was named γ-glutamylcysteine dipetidyl transpeptidase (trivial name phytochelatin synthetase or PC synthetase) could also add this to PC molecules. Also, the source of γ-Glu-Cys moiety can be GSH as well as the PC molecules.

Chen et al. [96] in addition to confirming the conclusion of Grill et al. [95] also showed that PCs could not be synthesized by the enzyme (PC synthetase) extract from tomato cells in the presence of γ-Glu-Cys alone or γ-Glu-Cys and Gly. They concluded that the enzyme identifies only two substrates, GSH and PCs, and probably has two binding sites, one specific for GSH and the other, less specific, for GSH and PCs. This conclusion was also in context of observations of Klapheck et al. [97]; they worked on pea (*Pisum sativum* L.), which produced GSH and h-GSH, and showed that the crude enzyme preparation from root produced (γ-Glu-Cys)$_2$-Gly in the presence of GSH. Given only γ-Glu-Cys-β-Ala (h-GSH) or γ-Glu-Cys-Ser (hm-GSH), however, the rate of production of their n_2 oligomers was much less.

However, in the presence of GSH and h-GSH or hm-GSH the synthesis of the respective β-alanyl or seryl n_2 oligomers was greatly increased. This led them to conclude that the enzyme has a γ-Glu-Cys donor-binding site specific for GSH and a less specific γ-Glu-Cys acceptor-binding site that is able to bind several tripeptides — namely, GSH, h-GSH, hm-GSH, and, of course, the PCs. Thus, the synthesis of PCs as well as of h-PCs and hm-PCs is possible only by a single enzyme. Whether the enzyme preparation of Chen et al. [96] and Klapheck et al. [97] could carry out PCs synthesis — even in the absence of GSH — with only added (γ-Glu-Cys)$_2$-Gly has not been checked. If so, then the possibility of existence of GSH specific binding site does not exist. Of course, it may be possible that the specificity of γ-Glu-Cys donor site of the enzyme lies in the recognition of Gly residue at the carboxy terminal end of the tripeptide, or the PCs.

The other two pathways of (γ-Glu-Cys)$_n$-Gly synthesis are indicated in the observation of Hayashi et al. [98]. The crude preparation of the enzyme differs from that described previously in two respects: (1) Cd is not necessary for its catalysis; and (2) some (γ-Glu-Cys)$_2$ appears in the reaction mixture along with PCs with GSH as substrate. Incubation of GSH with γ-Glu-Cys, (γ-Glu-Cys)$_2$, or (γ-Glu-Cys)$_3$ in the presence of the enzyme produces $n + 1$ oligomers of the (γ-Glu-Cys)$_n$ provided — i.e., PCs are produced by dipeptide addition, the second pathway. The preparation also polymerized γ-Glu-Cys into (γ-Glu-Cys)$_{2,3}$, suggesting a dipeptide transfer function of the enzyme. This is the only work that suggests a biosynthetic origin for (γ-Glu-Cys)$_n$. Furthermore, GSH synthetase added Gly to (γ-Glu-Cys)$_{2,3}$, giving $n = 2$ and $n = 3$ oligomers of the PCs, respectively. This allowed Hayashi et al. [98] to propose that polymerization of γ-Glu-Cys to (γ-Glu-Cys)$_n$, followed by GSH synthetase adding Gly, could be a third pathway for PC biosynthesis.

Klapheck et al. [97], however, believe the production of (γ-Glu-Cys)$_n$ to be a result of catabolic processes by the action of carboxypeptidase on (γ-Glu-Cys)$_n$-Gly removing the Gly moiety. By analogy to the reactions catalyzed by carboxypeptidase C, a proteolytic enzyme that also acts as dipetidyl transpeptidase [99], desGly-PCs may also arise from a hydrolytic activity of PC synthetase — i.e., cleavage of the Gly after binding of a PC molecule at the donor-binding site and transfer to water instead of to a γ-Glu-Cys acceptor [97].

The assumption of catabolic process in the formation of (γ-Glu-Cys)$_n$ is strengthened from the observation of production of (γ-Glu-Cys)$_n$-Glu in maize [85]. The tripeptide γ-Glu-Cys-Glu is found in maize only after the Cd-induced appearance of (γ-Glu-Cys)$_n$-Gly and (γ-Glu-Cys)$_n$, offering the possibility that the family of γ-Glu-Cys peptides with amino-terminal Glu are degradation products of other thiol peptides. Only action of γ-glutamyl transpeptidase for cleaving intramolecular γ-Glu linkages would be required [18]. Study on the Cd-sensitive mutant of *Arabidopsis thaliana* also supports only transpeptidation of γ-Glu-Cys moiety from GSH to GSH or (γ-Glu-Cys)$_n$-Gly as the probable pathway for PC synthesis; the mutant is deficient in GSH synthesis, producing significantly less PCs despite having PC synthetase activity similar to the wild type [66,100]. Nevertheless, the presence of desGly-PCs at the beginning of Cd incubation and at low Cd concentrations does indicate that *de novo* synthesis of these peptides is possible [84].

16.9 PHYTOCHELATINS: INDUCTION BY HEAVY METALS

Voluminous literature exists on the induction of PCs by heavy metals and their possible involvement in metal tolerance and has also been reviewed by many [18,39,61,101,102]. In fact, the phytochelatin response, or the synthesis of heavy metal-binding polypeptides, is one of the few examples in plant stress biology in which it can be readily demonstrated that the stress response (PC synthesis) is truly an adaptive stress response. Nevertheless, there are several exceptions. During the course of the stress response studies, attention has been focused not only on the rate of synthesis of the polypeptides, but also on the role of the precursors and the enzymes involved in their synthesis. Most of the information, however, comes from the work involving the heavy metal, cadmium, in response to which the induction of PC synthesis was first detected.

The argument in favor of the possible involvement of PCs in heavy metal tolerance mainly comes from its induction and accumulation by a wide range of plant species, including algae, in response to Cd, and also in response to a range of heavy metals [39,61,66,75,77,84,100–104]. In cell suspension cultures (of *Rauvolfia serpentina*), it has been observed that the tendency of metals to induce PCs decreases in the order [74]:

$$Hg \gg Cd, As, Fe > Cu, Ni > Sb, Au > Sn, Se, B > Pb, Zn$$

In the root culture of *Rubia tinctorum*, the PC induction by various heavy metals was in the order [76]:

$$Hg \gg Ag > Cd \gg As > Cu > Pd > Se > Ni > Pb > Zn > In > Ga$$

For the metals common to both cases, the order of induction is more or less similar, except for Ni and Se. The order, however, is based on the total metal concentration in the culture medium. For free ionic metals, the order may be different. This is evident from the work of Huang et al. [105], who applied the metal concentrations to the cell suspension culture depending upon their toxicity. Furthermore, Grill et al. [95] showed that the activation of the purified enzyme from cell suspension cultures of *Silene cucubalus* by Hg was only 27% of activation produced by equimolar concentration of Cd.

Further support in favor of the possible role of PCs in providing plants resistance to heavy metals comes from the study on metal (mostly Cd)-tolerant culture cell lines and strains of algae and mutants. The uptake of Cd by Cd-tolerant plant cell lines is somewhat greater than by the nontolerant ones prior to Cd becoming toxic [105–107]. The Cd-tolerant cells bind more than 80% of the cellular Cd as Cd–PC complex, but little binding of Cd occurs in the nontolerant cells, which grow poorly and die prematurely [87,105–107]. Furthermore, Gupta and Goldsbrough [108] observed that the tomato cell lines selected for resistance to various concentrations of Cd showed increased Cd and PC accumulation concomitant with increase in their tolerance level. At least 90% of the Cd in the most tolerant cell line was associated with Cd–PC complexes.

The evidence of PCs' protective function against heavy metal toxicity also comes from studies of the influence of the precursors of PCs, and the enzyme(s) involved in their synthesis, on the resistance of cell cultures or intact plants to the metals. Upon exposure of Cu-sensitive and Cu-tolerant *Silene cucubalus* (L.) to Cu, the loss of GSH pool was only observed in the former [109], suggesting that the maintenance of GSH pool for continued synthesis of phytochelatin is necessary for survival under metal stress. In a similar study on tomato cells, it was observed that the tolerance of CdR6-0 cells (cells selected for Cd tolerance) was associated with their enhanced capacity to synthesize GSH, nearly twofold higher than the unselected CdS cells, to maintain the production of PC [110]. It has also been demonstrated that the transgenic Indian mustard (*Brassica juncea*) overexpressing GSH synthetase contains greater amounts of GSH and phytochelatin, accumulates more Cd, and shows greater tolerance to Cd than the wild type [111]. Furthermore, the growth of the Cd-tolerant cells [87,105] or that of the nontolerant cells [94] remains unaffected in the presence of BSO alone, but is greatly inhibited in the presence of BSO together with Cd. The cells' growth is, however, restored in the presence of exogenous GSH and is accompanied by PC synthesis [86,94].

Howden et al. [66,100] used a genetic approach to establish the relationship between Cd-tolerance and phytochelatin synthesis and its accumulation. They could isolate an allelic series of Cd-sensitive mutants, *cad1* (*cad1-1*, *cad1-2*...), and a second Cd-hypersensitive mutant, *cad2*, affected at a different locus. They observed that the hypersensitivity of *cad1* mutants to Cd was associated with deficiency in their ability to accumulate PCs due to deficient PC synthetase activity, and that of *cad2* was associated with deficient GSH level resulting in deficient PC synthesis. Genetic studies using *S. pombe* have also shown that GSH-deficient mutants are also PC deficient and Cd hypersensitive [92,112].

In contrast to the preceding, many studies on naturally evolved heavy metal-tolerant varieties of plants, as well as on laboratory-selected tolerant cell lines, do not demonstrate a clear relationship between heavy metal resistance and PC production, thus creating doubts on the involvement of PCs in metal tolerance:

- Tolerant plants often do not produce more PCs than nontolerant ones [103,109,113].
- Although the level of PC in a Cu-sensitive ecotype of *Silene cucubalus* increases significantly at 0.5 μ*M* concentration of Cu, which is a nontoxic concentration for those plants, significant increase in the level of PC in the Cu-tolerant plants occurs only at 40 μ*M* or high Cu-concentrations, which are toxic for the ecotype [109].
- Distinctly Cu-tolerant (Marsberg) and nontolerant (Amsterdam) ecotypes of *Silene vulgaris* produce equal amount of PCs if they are grown at Cu-concentrations that cause equal degree of root growth inhibition, but such concentrations of the metal for tolerant plants are always greater than those for the nontolerant plants [113].
- The roots of Cd-tolerant plants of *S. vulgaris* exposed to a range of Cd concentrations accumulated greater amounts of the metal than the roots of Cd-sensitive plants, but contained significantly less PCs than the latter, particularly at the higher exposure concentrations [103].

16.10 HMW PC AND METAL TOLERANCE

Although the accumulation of PCs could be a major component of the heavy metal detoxification process, the increased tolerance to metals may involve other aspects of PC function. The first argument in favor of this came from Delhaize et al. [87], who observed that, although Cd-sensitive and Cd-tolerant cells of *Datura innoxia* synthesized the same amount of PCs during the initial 24-h exposure to 250 μ*M* Cd, the concentration was toxic to the Cd-sensitive cells only, as revealed by a cell viability study. However, they differed in their ability to form PC–Cd complexes: the sensitive cells formed complexes later than the tolerant cells. In addition, the complexes formed by the sensitive cells were of lower molecular weight than those of tolerant cells and did not bind all the Cd, unlike in the tolerant cells. Thus, the rapid formation of PC–Cd complexes sequestering most of the Cd within a short period could be a necessity for plants or cells showing tolerance to heavy metals. Evidence in support of this also comes from work on Cd-sensitive mutants of *Arabidopsis thaliana*, *CAD1*, which is deficient in its ability to sequester Cd [114].

Furthermore, Gupta and Goldsbrough [108] observed that the cell lines of tomato selected for their tolerance to various concentrations of Cd showed a trend towards accumulation of HMW PCs in addition to showing their enhanced synthesis. At least 90% of the Cd in the most tolerant cells was associated with PC complexes containing large amounts of SH. Thus, the size of PCs may be a determining factor in tolerance to heavy metals. It has been shown that PC_7 is more efficient than PC_2 in complexing Cd per mole of γ-Glu-Cys [115]. Moreover, exposure of maize seedlings to increasing concentrations of Cd results in the accumulation of longer PCs, with PC_4 the largest peptide accumulating PC [89].

Yet another way by which the metal-binding capacity of PCs (per mole of PC-SH) is increased is upon their association with acid labile sulfur (S^{2-}), which has been reported to increase the stability of Cd–PC complexes as well in *S. pombe* [80]. The relevance of the presence of S^{2-} in PC–metal complexes to metal detoxification is substantiated by the observation that the mutants of *S. pombe* that produce PC–Cd complexes without sulfide are hypersensitive to Cd [92]. In addition, Cd-tolerant *Silene vulgaris* plants exhibit a higher S:Cd ratio in the PC complexes than the Cd-sensitive plants [116]. It has also been observed that the HMW PC–metal complexes contain greater amount of S^{2-} than the LMW PC–metal complexes.

Two distinct peaks for HMW and LMW PC–metal complexes have not generally been observed in plants and have only been described for tomato [117] and a Se-tolerant variety of *Brassica juncea* [118]. Nevertheless, the two forms do exist even if they may not be distinctly separated. The acid labile sulfur associated with the two forms varies from species to species. HMW Cd-binding complexes in maize seedlings exposed to 3 μ*M* Cd show a S^{2-}:Cd molar ratio of 0.18; no acid labile sulfur occurred in the LMW complexes [119]. *Brassica juncea* grown in synthetic medium with 100 μ*M* Cd for 7 days produced HMW complexes with S^{2-}:Cd molar ratio of 1.0 and LMW complexes with ratio 0.42 [118]. Incompletely resolved complexes from roots of tomato exposed to 100 μ*M* Cd for 4 weeks had continuous S^{2-}:Cd molar ratios ranging from 0.15 to 0.41 for the HMW complexes and from 0.04 to 0.13 for the LMW complexes [117].

The yeasts *S. pombe* and *C. glabrata* grown in different media for 16 to 48 h and exposed to 500 or 1000 μ*M* Cd showed complexes with S^{2-}:Cd molar ratios of 0.11 to 0.55 [80,120]. How sulfide, Cd, and PC peptides interact within the complex is unclear for the cases in which the ratio is low. Cd–PC complexes with S^{2-}:Cd ratio greater than 0.4 appear as dense aggregates of 2-nm diameter particles called CdS crystallites. In yeast, each crystallite contains about 80 CdS units stabilized by a coating of about 30 peptides of glycyl [(γ-Glu-Cys)$_n$Gly] and desGly [(γ-Glu-Cys)$_n$] forms [120]. Reese et al. [117] showed the presence of such crystallites in plant (tomato) in PC–Cd formations with S^{2-}:Cd ratio of 0.41, but the crystallites were of less than 2-nm diameter coated with only (γ-Glu-Cys)$_n$Gly peptides.

The number of γ-Glu-Cys dipeptide repeats influences the stability of the complexes. Complexes formed with shorter peptides ($n = 1$ and $n = 2$) are more labile, and accretion of the crystallite to larger particles is more facile [117]. In yeasts *S. pombe* and *Candida glabrata*, although (γ-Glu-Cys)$_{2,3}$ peptides are present in C-binding complexes, (γ-Glu-Cys)$_{2-4}$-Gly peptides are usually more concentrated [120,121]. In tomato, the number of γ-Glu-Cys units varies from 3 to 6 with $n = 4$ the predominant peptides [117]. The Cd-binding complexes from several other sources are composed of (γ-Glu-Cys)$_n$-Gly, with $n = 3$ and $n = 4$ oligomers the most abundant [108,122]. In soybean (*Glycine max*), $n = 1,2,3,4$ oligomers of (γ-Glu-Cys)$_n$-β-Ala form the Cd-binding complexes [83]. The HMW complexes in maize are formed by the peptides from three families: (γ-Glu-Cys)$_n$-Gly; (γ-Glu-Cys)$_n$-Glu; and (γ-Glu-Cys)$_n$, of which (γ-Glu-Cys)$_n$ peptides remain present in highest concentrations, followed by (γ-Glu-Cys)$_n$-Gly and (γ-Glu-Cys)$_n$-Glu. The $n = 3$ oligomers of the three families form the highest constituent followed by an equally dominating concentration of $n = 4$ oligomers [119].

The preponderance of $n = 3$ and $n = 4$ oligomers in Cd-binding complexes from maize corroborates the increasing affinity of Cd for longer peptides [123], and their presence in HMW complex together with the acid labile sulfur speaks to the importance of HMW Cd complexes in metal (Cd) detoxification. Rauser and Meuwly [119] in their study showed that the concentration of $n = 3$ and $n = 4$ oligomers increased in maize with increase in the number of days of its exposure to Cd, and the HMW complexes sequestered 59% of the Cd after day 1, which increased from 88 to 92% by days 4 to 7.

16.11 CELLULAR COMPARTMENTALIZATION OF PC–METAL COMPLEXES AND METAL TOLERANCE

Another important aspect of PC-mediated tolerance of plants to heavy metals is probably the effective transportation of the metal to vacuoles for storage in which they could be playing an important role. Arguments in favor of this come from several observations. Vogeli–Lange and Wagner [124] isolated mesophyll protoplast from tobacco exposed to Cd and showed that the vacuoles contained $110 \pm 8\%$ of the protoplast Cd and $104 \pm 8\%$ of the protoplast PCs. These workers envisioned the synthesis of PCs in cytosol and transfer of Cd and the peptides, perhaps as complex, across the tonoplast into the vacuole, where the metal is chelated by the peptides and organic acids.

Working on tomato cells, Gupta and Goldsbrough [125] observed the highest level of PCs after 4 days of their exposure to Cd, which coincided with the peak of cellular Cd concentration (0.6 mM). At this time, there was an eightfold molar excess of PC over Cd. However, the PCs could not be detected after 12 days and the cellular concentration of Cd was still 0.2 mM (the intracellular concentration of Cd decreased as a result of increase in the cell mass). This led them to suggest that PCs possibly function as transport carriers for Cd into the vacuole, where the acidic pH favors dissociation of the Cd–PC complexes, followed by breakdown of the PCs and possible sequestration of the metal in some other form, in agreement with the model proposed by Vogeli–Lange and Wagner [124].

Later, while working on Cd-tolerant and Cd-sensitive plants of *Silene vulgaris*, De Knecht et al. [103] observed that, in response to a range of Cd concentrations, the root tips of Cd-tolerant plants exhibited a lower rate of PCs production accompanied by a lower rate of larger chain PC synthesis than those of Cd-sensitive plants, although both the plants (root tips) accumulated nearly similar levels of Cd at a particular metal-exposure concentration. Second, the tolerant plants reached the same PC concentration as the sensitive plants only after exposure to high Cd concentrations, and at an equal PC concentration the composition of PC and the amount of sulfide incorporated per unit PC-thiol were the same in both the populations.

The authors concluded that the lower concentration of PCs in the Cd-tolerant plants than in the Cd-sensitive plants could be because of greater transport of Cd–PC complexes in vacuoles in the former and, as suggested by Vogeli–Lange and Wagner [124] and Gupta and Goldsbrough [125], the PC–Cd complexes might be getting dissociated in the vacuole because of its acidic pH, followed by breakdown of the PCs or their reshuttling into the cytoplasm. Thus, the observed lower PC concentration in Cd-tolerant plants might be a result of a lower Cd concentration in the cytoplasm caused by (1) a faster transport of the metal into the vacuole when compared to that in the Cd-sensitive plants; and (2) return of the dissociated PCs (in the vacuole) into the cytoplasm, obviating the need of their fresh synthesis for the additional Cd uptake.

Because the enzymes involved in PC synthesis are present in cytoplasm but PCs are also found in the vacuole, a transport mechanism must be involved, and an insight into this comes from the work on *S. pombe*. A Cd-hypersensitive mutant, deficient in producing HMW complex, was observed [92]. This was found to be as a result of mutation within the *hmt1* (heavy metal tolerance 1) gene encoding an ATP-binding cassette (ABC)-type protein associated with vacuolar membrane [126]. ABC-type proteins represent one of the largest known families of membrane transporters. They can mediate tolerance to a wide diversity of cytosolic agents. The presence of HMT1 protein in the vacuolar membrane suggests the possibility of an ABC-type transporter-mediated resistance to Cd, by its sequestration in the vacuole [127].

The yeast *hmt1⁻* mutant harboring *hmt1*-expressing multicopy plasmid (pDH35) exhibited enhanced resistance to Cd compared to the wild-strain (*hmt1⁻* mutant) and accumulated more Cd with HMW complex formation [127]. The vacuolar vesicle derived from the *hmt1⁻* mutant complemented with *hmt1* cDNA (*hmt1⁻*/pDH35), i.e., HMT1 hyperproducer exhibited ATP-dependent uptake of LMW apophytochelatin and LMW–Cd complexes, but that from the *hmt1⁻* mutant did not show any such activity. HMW-Cd complex was not an effective substrate for the transporter proteins. The vacuolar uptake of Cd^{2+}, which was ATP dependent, was also observed, but was not attributable to HMT1. The electrochemical potential generated by vacuolar ATPase did not drive transport of peptides or complexes.

The observation of Ortiz et al. [127] is also supported by work on oat tonoplast vesicles [128,129]. Tonoplast vesicles from oat roots have a Cd^{2+}/H^+ antiporter [129]. The vesicles also show MgATP-dependent transport of PCs and Cd–PC complex [128], and the peptide transport is not driven by electrochemical potential generated by the vacuolar ATPase. Based on the information available, Rauser [18] proposed a model, somewhat similar to that proposed by Ortiz et al. [127], for the transport of Cd and Cd-binding complexes across the tonoplast. PCs synthesized in the cytosol combine with Cd to form LMW complex that is moved across the tonoplast by ABC-type transporters. Apo–PCs are also transported by them. The energy required for the transport is derived from ATP.

Once inside the vacuole, more Cd, transported by Cd^{2+}/H^+ antiporter, is added to the LMW, along with Apo–PCs and sulfide complexes, to produce HMW complexes. Genetic and biochemical analyses suggest that the formation of sulfide moiety in the HMW PC-Cd-S^{2-} complex involves purine metabolism, which serves as the source of sulfide [130,131]. The sulfide-rich HMW complex is more stable in the acidic environment of the vacuole and has a higher Cd-binding capacity than the LMW complex. The LMW complex functions as a cytosolic carrier and the vacuolar HMW complex is the major storage form of cellular Cd. Whether LMW and HMW complexes in plants are compartmentalized as depicted in the model and are of the same peptide composition, however, awaits direct evaluation. Nevertheless, the studies [127,128] do indicate a central role of vacuole in sequestration and detoxification of Cd, and maybe heavy metals in general, and that tolerance to metals could also be due to increased ability of plants to transport them into the vacuole (see Figure 15.2 in Chapter 15).

16.12 ROLE OF PCS IN DETOXIFICATION OF HEAVY METALS OTHER THAN CADMIUM

As stated earlier, synthesis of PCs is induced by most heavy metals, including the multiatomic anions [74,76], in most of the higher plants [73,83]. It has also been observed that the enzyme involved in its synthesis, PC synthetase, needs the presence of heavy metals for its activation; a crude preparation of the enzyme from *S. vulgaris* was activated best by Cd^{2+}, and by Ag^+, Bi^{3+}, Pb^{2+}, Zn^{2+}, Cu^{2+}, and Au^+ in decreasing order [95]. No activation of the enzyme was detected by the metals of the hard-acceptor category including Al^{3+}, Ca^{2+}, Fe^{3+}, Mg^{2+}, Mn^{2+}, Na^+, and K^+. The trend of activation observed by Grill et al. [95], however, was not observed for PC synthetase from tobacco cells, except that Cd was the most effective activator, followed by Ag^+. The activation by Cu^{2+} was next to Ag^+, and Pb^{2+}, Zn^{2+}, and Hg^{2+} produced only weak stimulation of the enzyme activity [96]. Thus, although the enzyme has a rather nonselective domain for binding with metals, it is mostly activated by heavy metals.

This strongly suggests that intracellular metabolism of heavy metals, other than Cd as well, might be largely mediated through PCs. The view also stems from the fact that the heavy metal ions that activate PC synthetase *in vitro* are also able to induce PC synthesis *in vivo* [96], with one exception: Ni^{2+} induced PC synthesis *in vivo*, but did not activate PC synthetase activity *in vitro*. Furthermore, the indication of possible involvement of PC in heavy metal tolerance also comes from genetic evidence; in addition to being sensitive to Cd^{2+}, phytochelatin-deficient *cad1* mutants of *Arabidopsis* are also sensitive to Hg^{2+} [114], and GSH-deficient strains of *S. pombe* show reduced tolerance to Pb^{2+} as well as showing no tolerance to Cd^{2+} [112].

PC synthetase activity is detected mostly in roots, but not in leaves or fruits [61,101]. The constitutive presence of PC synthetase in roots suggests an important role of PCs in metal detoxification. Because plants assimilate various metal ions from soil, the first organ exposed to these ions is the root. Localization of PC synthetase to roots and stems probably provides an effective means of restricting the heavy metals to these organs by chelation in the form of Cd–PC complexes. It has been demonstrated that PCs are able to protect enzymes from heavy metal poisoning *in vitro* [65,132]; many metal-sensitive plant enzymes (rubisco, nitrate reductase, alcohol dehydrogenase, glycerol-3-phosphate dehydrogenase, and urease) were more tolerant to Cd in the form of a Cd–PC complex compared with the free metal ion. Free PCs could reactivate the metal-poisoned enzymes (nitrate reductase poisoned by Cd–acetate) *in vitro* more effectively than other chelators such as GSH or citrate [65].

Recognition of PCs as the chelators of heavy metals in general and protectors of plants against their toxic effect, however, requires careful consideration. For instance, in tobacco cells not selected for metal tolerance, BSO increased the toxicity of Cd but not of Zn or Cu, as if the control of sequestration differed between the metals [133]. This may, of course, be true, but has not been properly demonstrated. Second, although most of the heavy metals are able to induce synthesis of PCs in plants, only a few of them (Cd, Cu, and Ag) form complexes with the peptides [76]. Recently, As has been reported to form complexes with PCs in arsenate-tolerant *Holcus lanatus* [134]. In fact, PC–metal complex formation has been reported mostly for Cd. A few reports of PC forming complexes with Cu are also available [74,75,135], and formation of PC–Zn complexes has been observed in cells (of *Rauvolfia*) grown in micronutrient concentration of Zn [65]. PCs have also been reported to form complexes with Hg and Pb *in vitro* [136,137].

Nevertheless, genetic evidence is that PCs are involved in tolerance to these metals; PC-deficient *cad1* mutants of *Arabidopsis* are also sensitive to Hg^{2+} [114], and GSH-deficient strains of *S. pombe* show reduced tolerance to Pb^{2+} [112]. Thus, although the involvement of PCs in making plants resistant to heavy metals other than Cd cannot be overlooked, more information is required on their induction by individual heavy metals in different plant species. Also, information is required on the formation of PC–metal complexes and cellular localization of the metals (individual) and Apo–PC and metal–PC complexes before the functional significance of PCs known for Cd can be generalized for all heavy metals.

16.13 COMPLEX FORMATION WITH ORGANIC ACIDS

Organic acids are the other group of biomolecules that can function as chelators of heavy metals inside the cell, converting the metals to almost inactive and nontoxic forms. With regard to Al, at least two organic acids are known to function as chelators. One is citric acid [138]; nearly two-thirds of Al in hydrangea leaves remain present in the cell sap in soluble form as Al–citrate complex at a 1:1 molar ratio of Al to citrate, a nontoxic form of Al.

Another acid that has been reported to form intracellular complex with Al is oxalic acid [139]. About 90% of Al in buckwheat remains present as soluble oxalate Al complex in the symplasm, and the intracellular concentration of Al detected is as high as 2 mM. The complex occurs in molar ratio of 1:3, Al:oxalate. Oxalic acid can form three species of complexes with Al at an Al:oxalic acid molar ratio of 1:1, 1:2, and 1:3. The 1:3 Al–oxalate complex is the most stable, with a stability constant of 12.4 [140]. This stability constant is much higher than that of Al–citrate (8.1) or Al:ATP (10.9), meaning that formation of 1:3 Al–oxalate complex can prevent binding of Al to cellular components, thereby detoxifying Al very effectively. The report is in contrast to the order of stability constant for Al–organic acid complexes: Al–citrate > Al–oxalate > Al–malate [141]. It is not known, however, whether the Al complexes of citrate or oxalate remain located in cytoplasm or in the vacuole.

Among the heavy metals reported to be chelated by organic acids inside the cells are Zn and Ni. After exposure to high concentrations of various heavy metals, vacuoles of the Zn- and Ni-tolerant plants, as well as those of the nontolerant plants, often contain high concentrations of zinc and nickel [142,143], as well as some Cu and Pb [144] and Cd [145–147]. The results of the studies on Zn- and Ni-tolerant plants suggested that organic acids could be involved in their sequestration in the vacuole; the Zn-tolerant plants, including *Silene vulgaris*, exhibited enhanced accumulation of malate [148,149] and the Ni-tolerant plants showed accumulation of malate, malonate, or citrate [148,150] upon their exposure to Zn and Ni, respectively. The details of their transportation and sequestration inside the vacuole and the roles of the organic acids in the process are not available.

In one of the models for the transport of Zn into vacuole, it has been postulated that malic acid would bind Zn in the cytosol, thereby detoxifying it, and the Zn–malate complex would be transported over the tonoplast into the vacuole where it would dissociate [39]. After this, malate would be retransported into the cytosol. Vacuolar Zn would remain bound to stronger chelators, such as citrate, oxalate, etc., when present. Brune et al. [151] reported that barley mesophyll cell vacuoles contain appreciable concentration of phosphate (30 to >100 mol m^{-3}); malate (>10 mol m^{-3}); sulphate (>4 mol m^{-3}); citrate (~1 mol m^{-3}); and amino acids (>10 mol m^{-3}) when grown in hydroponic culture. They hypothesized that these organic and inorganic salts interact with the divalent cations, thereby buffering the vacuolar free Zn concentration to low values even in the presence of high Zn levels (292 mmol m^{-3}) in the vacuolar space.

According to Wang et al. [152], citrate is the most efficient ligand for metal complexation in the vacuole at vacuolar pH values of 6 to 6.5. The results of Brune et al. [151] demonstrate the importance of compartmentalization and transport as homeostatic mechanisms within leaves to handle possibly toxic zinc levels in shoots. The dependence of plants on organic acids for detoxification of Zn could be the reason for poor induction of PCs by the metal [74,76].

The mechanism of detoxification adapted by plant probably varies from metal to metal and, for a metal, from species to species, and it is difficult to reconcile the idea of tolerance by means of any single mechanism. For example, Zn-tolerant *Agrostis capillaries* and *Silene vulgaris*, which exhibit increases in malate levels [153], are only slightly Ni tolerant [43]; Ni-tolerant *Alyssum bertolonii*, which is very rich in malate [150], is nontolerant to Zn [39]. Similarly, as stated earlier, BSO increases the toxicity of Cd to the tobacco cells not selected for Cd tolerance, but not of Zn or Cu. Again, for Al detoxification, plants follow several strategies.

16.14 ANTIOXIDATIVE SYSTEM IN METAL TOLERANCE

It is generally considered that virtually all the biochemical effects of heavy metals may ultimately lead to damage of cells and tissues [1]. Thus, arguments are made that heavy metal tolerance could also be linked, to some extent, with reactive oxygen scavenging capability of a plant species [154]. However, little direct evidence supports this hypothesis, although indirect evidence does suggest such a relationship.

De Vos et al. [109] observed that 20 µM Cu reduces the GSH pool in the roots of Cu-sensitive *Silene cucubalus*, and this is accompanied by enhanced MDA accumulation, indicating oxidative damage of membrane. However, even at 30-µM concentration, Cd caused less decrease in the GSH pool than copper and no lipid peroxidation despite inducing nearly 13-fold increase in PC levels, compared to only 6-fold increase induced by Cu. In the Cu-tolerant plants, the decrease of GSH level was significantly less than in the Cu-sensitive plants. Because GSH is an important antioxidant, it may be concluded that tolerance to copper in Cu-tolerant *S. cucubalus* could be because of its more efficient antioxidative system than that of Cu-sensitive plants. Gallego et al. [155] also correlated oxidative damage in sunflower cotyledons induced by Cd with decrease in the GSH pool.

Further indirect evidences come from the study of responses of the antioxidative enzymes to metal treatment. Cakmak and Horst [156] observed significant increase in peroxidase activity in soybean root in response to Al treatment with concomitant increase in MDA content, indicating the induction of oxidative stress with the plant responding by increasing the level of one of its antioxidative enzymes. Subsequently, Ezaki et al. [157] reported Al stress induced appearance of two cationic peroxidases and two moderately anionic peroxidases in tobacco cells. They also produced evidence that at least one of the isoenzymes is produced by enhanced expression of *pAL201* gene and opined the possibility of the isoenzyme having some function in Al resistance. Significant enhancement in the activity of peroxidase in response to heavy metals has also been reported [158–162]. Hendry et al. [158] observed enhancement in the activity of peroxidase in the Cd-sensitive, but not in the Cd-tolerant, plants of *Holcus lanatus* in response to Cd.

The activity of ascorbate peroxidase has also been reported to increase in plants in response to heavy metals [161–163]. Working on *Phaseolus aureus*, Shaw and Rout [161] observed metal-specific (Hg and Cd) differences in the response of the enzyme by the older seedlings when compared to the younger ones, with Hg inducing the activity of the enzyme while Cd had not. This was accompanied by death of the Cd-treated older seedlings after exposure for more than 36 h, suggesting that the enhanced synthesis of ascorbate peroxidase in response to Hg could be protecting the plant against the oxidative stress induced by the metal; this was not possible in the case of Cd treatment. Like peroxidase, ascorbate peroxidase is also induced more in the metal-sensitive plant (*Alssum maritinum*) than in the tolerant plant (*A. argentums*, a nickel hyperaccumulator) [163].

Reports are also available showing enhancement in the activity of catalase, probably the main H_2O_2 scavenging enzyme, in response to heavy metal treatment [1], suggesting possible involvement of the enzyme in heavy metal tolerance. Furthermore, the perimedullar tuber tissue of potato cultivar resistant to Cd showed a higher constitutive level of catalase and also significantly greater increase in the activity of the enzyme in response to exposure to Cd when compared to the Cd-sensitive cultivar [164]. The observation is in contrast to that observed for peroxidase and ascorbate perox-idase and suggests that the enzyme could be of greater importance than peroxidase and ascorbate peroxidase in reducing the metal-induced increase in H_2O_2 level, leading to tolerance of the plant (Cd-tolerant cultivar) to the metal (Cd).

Superoxide dismutase (SOD) has been considered to play the most significant role in active oxygen species scavenging because its action prevents the accumulation of $O_2^{\bullet-}$ radical, which could lead to generation of toxic $HO^{\bullet}$ [1]. It is also the most widely studied enzyme in the context of environmental stresses. In fact, its involvement in tolerance to environmental stresses — partic-ularly drought and frost, which lead to oxidative stress — is well established. This is on the basis of observation of its enhanced synthesis in response to environmental stresses or observation of increase in threshold of tolerance to environmental stresses in the organisms manipulated for enhanced expression of the enzyme [165–168]. The enzyme is also considered to be involved actively in heavy metal tolerance in plants. This is because of observation of significant increase in the activity of enzyme in plants exposed to various heavy metals [156,169,170]. More importantly, like catalase, the activity of SOD in response to heavy metals is increased more in the metal-tolerant plants than in the sensitive ones [158]; this speaks further in favor of an important role of the enzyme in heavy metal tolerance.

Although the observations of various workers presented here suggest an active involvement of the antioxidative components in heavy metal tolerance in addition to the involvement of other processes, it must be kept in mind that contradictory observations have also been reported [1]. Furthermore, the database in support of the involvement of antioxidative machinery in metal tolerance is very limited, particularly the observation from the studies involving metal-tolerant and metal-sensitive varieties of a species. Therefore, at this stage, it would be premature to draw a definite conclusion in favor of the involvement of the antioxidative system in metal tolerance in plants; it would be wise at present to treat the idea as a supposition. Nevertheless, it is worth mentioning that *Arabidopsis* transgenic line, AtPox(4-1), showing enhanced expression of peroxi-dase shows significantly less lipid peroxidation upon exposure to Al and greater tolerance to the metal than the nontransgenic plant [171].

16.15 METAL–METAL INTERACTIONS

Toxicological evidence of plant and nonessential metal interactions comes from a variety of experiments with variable doses and durations of exposure; this leaves a dearth of information on

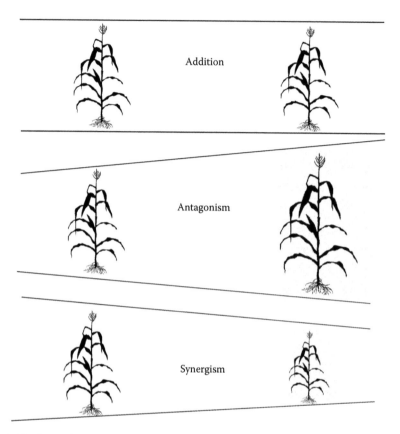

FIGURE 16.6 Metal–metal interactions on model plant systems.

the outcome of interaction of essential (Cu, Zn) and nonessential (Pb, Hg) metals. Therefore, Rauser [172] felt that it might be prudent to begin checking crucial laboratory responses using plants in various levels and durations of exposure of pollution using essential and nonessential metals. Heavily polluted soils and waters often contain mixtures of metals that may have antagonistic, synergistic, or no effects (additive) on plants. Therefore, metal–metal interactions on model plant systems bear exploring. Symeonidis and Karataglis [173] divided plant responses to combinations of metals in the growth medium into three groups (Figure 16.6):

- *Additive:* relative growth under conditions of multiple metal stress is equal to the product of the relative growth produced by the individual metals in isolation (e.g., Cu–Co).
- *Antagonistic:* relative growth under conditions of multiple metal stress is greater than that of the product of the relative growth produced by the individual metals in isolation (e.g., Cu–Cd, Ca–Cd).
- *Synergistic:* relative growth under growth conditions of multiple metal stress is less than that of the product of the relative growth produced by individual metals in isolation (e.g., Cu–Zn).

Cd and Zn belong to group IIB transition elements with similar electronic configuration and valence state; both have affinity to sulfur, nitrogen, and oxygen ligands. Thus, both these elements have similar geochemical and environmental properties [174]. Most of the ores are mixtures of metals in which potentially toxic metals (As, Cd, and Hg) other than the sought-after elements may also be present. Following extraction, which varies in efficiency but is never complete, the contaminant metals are also released into the environment freely.

Ore extraction of Zn from mines and nonferrous metal production processes in smelters with subsequent release of zinc effluents to the environment is normally accompanied by cadmium environmental pollution because zinc ore (ZnS) generally contains 0.1 to 5% and sometimes even higher cadmium [175]. Similarly tyres containing ZnO and sewage sludges applied to agricultural soils as fertilizers also contain Cd as a major contaminant. Thus, this association of Cd and Zn in the environment, their chemical similarity, and the interactive functions are of considerable importance [176]. Factors regulating essential and nonessential metal accumulation at the organismal and cellular level is vital for understanding phytotoxicity [177].

Cadmium has even been described as an antimetabolite of Zn by scientists because of observed Zn deficiency in most of the Cd-treated systems. It has been hypothesized that elements whose physical and chemical properties are similar will act antagonistically to each other biologically [176]. In recent years, a number of workers have documented responses of plants to combinations of Zn and Cd in soil, as well as in solution culture in soil-crop systems under actual field conditions [174].

Aravind and Prasad [178–180] conclusively demonstrated that Zn showed an antagonistic interaction with Cd and alleviated Cd toxicity in *Ceratophyllum demersum*, a freshwater macrophyte. The possible mechanisms identified include:

- Zn inhibited Cd uptake directly by competition and indirectly by controlling the H+ATPase, leading to a reduced intracellular concentration of Cd.
- Zn reduced Cd-increased peroxidation, membrane leakage, and lipolysis.
- Zn inhibited the formation of toxic reactive oxygen species triggered by Cd-influenced NADPH oxidase.
- Zn substantially increased activities of the antioxidant enzymes like SOD, CAT, POD, and APX, significantly quenching the formation of ROS.
- Zn restored and enhanced the functioning of carbonic anhydrase in Cd-exposed *C. demersum* by competitive substitution.

16.16 CONCLUSION

Thus, it can be seen that metals, including heavy metals, are nature's gift to mankind, and modern civilization would not have developed without bringing them into use. However, at the same time, metals — particularly the heavy metals — are very toxic to the living organism; thus, suitable measures must be taken to prevent excessive exposure to them and to immobilize them in the areas of their "hot spots." It is increasingly realized that this can be achieved by the use of plants. Plants can remediate metal pollutants in mainly two ways:

- Phytostabilization: plants convert pollutants to less bioavailable forms and/or prevent pollutants' dispersal by wind erosion or leaching.
- Phytoextraction: plants accumulate pollutants in their harvestable tissues, thus decreasing the concentration of the pollutants in the soil.

With regard to mercury, another concept under phytoremediation is phytovolatilization. Metal-contaminated soil can also be remediated by the use of proper soil amendment practices, which can decrease or increase the metal uptake by plants. When crop plants are grown, the amendment should be such as to decrease the metal availability to the plants, which should also preferably be metal excluders.

Metal resistance in plants could be a result of its exclusion or uptake and proper sequestration inside the cells. This varies depending upon the metals as well as on the plant species. For Al, resistance is mostly due to (1) its exclusion mediated by secretion of organic acids by the roots, which form complexes with the metal, making it unavailable to the plants; and/or (2) efflux of H^+, which increases the rhizosphere pH due to which Al^{3+} species in proximity with the root gets converted to less toxic and less available forms. Resistance to Al due to intracellular complex formation with oxalic acid has also been reported.

For the heavy metals in general, the resistance is achieved by intracellular sequestration inside the vacuoles. This is believed to be mostly mediated through the formation of complexes with phytochelatins, the nontranslationally synthesized low molecular weight polypeptides. However, the evidence for this is mostly available from work on Cd. Although most of the heavy metals are known to induce synthesis of phytochelatins, only a few, like Cd, Cu, and Ag, are reported to form stable complexes with them. In fact, more studies are needed involving different plant species and their ecotypes tolerant and sensitive to various heavy metals before phytochelatins can be recognized as a detoxifier of heavy metals in general.

Moreover, some reports indicate that the accumulation of the heavy metals like Zn and Ni is accompanied by enhanced accumulation of organic acids, although the details of their chelation and sequestration are not available. Also, a totally different mechanism is used for detoxification of Hg by certain bacteria in which the organic and inorganic forms of the element are converted to volatile nontoxic form. Furthermore, the involvement of antioxidative machinery in heavy metal tolerance is increasingly advocated, although not sufficiently substantiated. The information available thus far gives an indication that, in plants, heavy metal tolerance, or metal tolerance in general, could be a result of integrated functioning of more than one mechanism rather than a result of a singular process (Table 16.3). It is hoped that future research will lead to further understanding of the various processes involved in heavy metal tolerance, or metal tolerance in general, and throw light on the nature of the interaction between them.

TABLE 16.3
Adaptive Plant Ecophysiological, Molecular Basis of Metal Tolerance and Detoxification Processes

Mechanism	Key ref.
Plasma membrane (passive uptake and active efflux)	177, 182
Ferritins, metallothioneins, glutathione derived peptides (phytochelatins)	18, 20, 102, 134, 172, 183–190
Over expression of glutathione (precursor for metal sequestration)	111, 191, 192
Low molecular weight organic acids and amino acids	14, 24, 181, 193–203
Heat shock proteins	204–206
Vacuolar compartmentation	177, 182, 183, 207
Metal transporters = cation efflux family (formerly cation diffusion family)	12, 208–214
Hairy root cultures, rhizofiltration and metal complexation	215–218
Rhizosphere biotechnology and physiology and biochemistry of metal tolerance	23, 219–221
Genetic and transgenic strategies for metal hyperaccumulation	177, 222–231
Mycorrhizae	232–240
Naturally occurring metal accumulators	241, 242
Metal–metal interactions (antagonism type) e.g., zinc prevents cadmium toxicity; ferritin prevents HM toxicity	178–180, 186

Source: Modified from Prasad, M.N.V. (Ed.) *Heavy Metal Stress in Plants: from Biomolecules to Ecosystems*, 2nd ed. Springer–Verlag, Heidelberg, 2004, 462; xiv.

REFERENCES

1. Shaw, B.P., Sahu, S.K., and Mishra R.K., Heavy metal induced oxidative damage in terrestrial plants, in *Heavy Metal Stress in Plants — from Biomolecules to Ecosystems*, Prasad, M.N.V., Ed., Springer–Verlag, Heidelberg, 2004, chap. 4.
2. Manahan, S.E., *Environmental Chemistry*, Lewis Publishers, Boston, 1990, chap. 17.
3. Ochiai, E.-I., *Bioinorganic Chemistry, an Introduction*, Allyn and Bacon, Inc., Boston, 1977, chap. 1.
4. Fergusson, J.E., *The Heavy Elements: Chemistry, Environmental Impact and Health Effects*, Pergamon Press, New York, 1990, chap. 2.
5. Evans, A.M., Ore, mineral economics and mineral exploration, in *Introduction to Mineral Exploration*, Evans, A.M., Ed., Blackwell Science, Oxford, 1995, chap. 1.
6. Chaterjee, K.K., *An Introduction to Mineral Economics*, Wiley Eastern Limited, Bombay, 1993, chap. 6.
7. Wedepohl, K.H., The composition of the upper Earth's atmosphere and the natural cycles of selected metals. Metals in natural raw materials. Natural resources, in *Metals and Their Compounds in the Environment: Occurrence, Analysis and Biological Relevance*, Merian, E., Ed., John Wiley & Sons, Inc., New York, 2000, chap. 1.
8. Forstner, U. and Wittmann, G., *Metal Pollution in the Aquatic Environment*, Springer–Verlag, New York, 1979, chap. B.
9. Sheehan, P.J., Functional changes in the ecosystem, in *Effects of Pollutants at the Ecosystem Level*, Sheehan, P.J., Miller, D.R., Butler, G.C., and Bourdeau, P., Eds., John Wiley & Sons, New York, 1984, chap. 6.
10. Martin, M.H. and Coughtrey, P.J., *Biological Monitoring of Heavy Metal Pollution*, Applied Science Publishers, London, 1982, chap. 1.
11. Baker, A.J.M., Metal tolerance, *New Phytol*, 106, 93, 1987.
12. Prasad, M.N.V., Metallothioneins, metal binding complexes and metal sequestration in plants, in *Heavy Metal Stress in Plants: from Biomolecules to Ecosystems,* 2nd ed., Prasad, M.N.V., Ed., Springer–Verlag, Heidelberg, 2004, chap.3.
13. Lasat, M.M., Baker, A.J.M., and Kochian, L.V., Altered zinc compartmentation in the root symplasm and stimulated Zn^{2+} absorption in to the leaf as mechanisms involved in zinc hyperaccumulation in *Thlaspi caerulescens*, *Plant Physiol.*, 118, 875, 1998.
14. Kramer, U., CotterñHowells, J.D., Charnock, J.M., Baker, A.J.M., and Smith, J.A.C., Free histidine as a metal chelator in plants that accumulate nickel, *Nature*, 373, 635, 1996.
15. Salt, D.E., Prince, R.C., Pickering, I.J., and Raskin, I., Mechanism of cadmium mobility and accumulation in Indian mustard, *Plant Physiol.*, 109, 1427, 1995.
16. De Miranda, J.R., Thomas, M.A., Thurman, D.A., and Tomsett, A.B., Metallothionein genes from the flowering plant *Mimulus guttatus*, *FEBS Lett.*, 260, 277, 1990.
17. Evans, I.M., Gatehouse, L.N., Gatehouse, J.A., Robinson, N.J., and Coy, R.R.D., A gene from pea (*Pisum sativum*) and its homology to metallothionein genes, *FEBS Lett.*, 262, 29, 1990.
18. Rauser, W.E., Phytochelatins and related peptides: structure, biosynthesis, and function, *Plant Physiol.*, 109, 1141, 1995.
19. Ow, D.W., Heavy metal-tolerant genes: prospective tools for bioremediation, *Res. Conserv. Recycling*, 18, 135, 1996.
20. Schmöger, M.E.V., Oven, M., and Grill, E., Detoxification of arsenic by phytochelatins in plants, *Plant Physiol.*, 122, 793, 2000.
21. Prasad, M.N.V. (Ed.) *Metals in the Environment: Analysis by Biodiversity*, Marcel Dekker Inc., New York, 2001, 504.
22. Pulford, I.D. and Watson, C., Phytoremediation of heavy metal-contaminated land by trees — a review, *Environ. Int.*, 29, 529, 2003.
23. Prasad, M.N.V. and Strzalka, K. (Eds.) *Physiology and Biochemistry of Metal Toxicity and Tolerance in Plants*, Kluwer Academic Publishers, Dordrecht, 2002, 432.
24. Kerkeb, L. and Kramer, U., The role of free histidine in xylem loading of nickel in *Alyssum lesbiacum* and *Brassica juncea*, *Plant Physiol.*, 1312, 716, 2003.

25. Köhl, K.I. and Lösch, R., Experimental characterization of metal tolerance, in *Heavy Metal Stress in Plants — from Biomolecules to Ecosystems*, M.N.V. Prasad, Ed., Springer–Verlag, Heidelberg, 2004, chap. 17.

26. Misra, S.G., and Mani, D., *Soil Pollution*, Ashish Publishing House, New Delhi, 1991, chap. 5.

27. Dean, J.G., Bosqui, F.L., and Lanouette, V.H., Removing heavy metals from waste water, *Environ. Sci. Technol.*, 6, 518, 1972.

28. Valerio, F., Brescianini, C., Lastraioli, S., and Coccia, S., Metals in leaves as indicators of atmospheric pollution in urban areas, *Int. J. Environ. Anal. Chem.*, 37, 245, 1989.

29. Lagerwerff, J.V. and Specht, A.W., Contamination of roadside soil and vegetation with cadmium, lead and zinc, *Envrion. Sci. Technol.*, 4, 583, 1970.

30. Iskandar, I.K. and Kirkham, M.B., *Trace Elements in Soil — Bioavailability, Flux and Transfer*, CRC Press, Boca Raton, FL, 2001, 304.

31. Chamberlain, A.C., Fallout of lead and uptake by crops, *Atmos. Environ.*, 17, 693, 1983.

32. Bertine, K.K. and Goldber, E.D., History of heavy metal pollution in southern California coastal zone — reprise, *Environ. Sci. Technol.*, 11, 297, 1977.

33. Ruch, R.R., Gluskoter, H.J., and Shimp, N.F., Occurrence and distribution of potentially volatile trace elements in coal, *Environ. Geol. Notes*, 61, 1, 1973.

34. Haug, A., Molecular aspects of aluminum toxicity, *Crit. Rev. Plant Sci.*, 1, 345, 1984.

35. Kochian, L.V., Cellular mechanisms of aluminum toxicity and resistance in plants, *Annu. Rev. Plant Physiol. Plant Mol. Biol.*, 46, 237, 1995.

36. Cunningham, S.D. and Ow, D.W., Promises and prospects of phytoremediation, *Plant Physiol.*, 110, 715, 1996.

37. Bradshaw, A.D., Populations of *Agrostis tenuis* resistant to lead and zinc poisoning, *Nature*, 169, 1098, 1952.

38. Jowett, D., Populations of *Agrostis* spp. tolerant to heavy metals, *Nature*, 182, 816, 1958.

39. Ernst, W.H.O., Verkleij, J.A.C., and Schat, H., Metal tolerance in plants, *Acta Bot. Neerl.*, 41, 229, 1992.

40. Peterson, P.J., The distribution of zinc-65 in *Agrostis tenuis* Sibth. and *A. stolonifera* L. tissues, *J. Exp. Bot.*, 20, 863, 1969.

41. Reilly, C., Accumulation of copper by some Zambian plants, *Nature*, 215, 667, 1967.

42. Strange, J. and Macnair, M.R., Evidence for a role of cell membrane in copper tolerance of *Mimulus guttatus* Fischer ex DC, *New Phytol.*, 119, 383, 1991.

43. Schat, H. and Ten Bookum, W.M., Genetic control of copper tolerance in *Silene vulgaris*, *Heredity*, 68, 219, 1992.

44. Singh, S.P. and Pandey, A.K., Cadmium toxicity in a cyanobacterium: effect of modifying factors, *Environ. Exp. Bot.*, 24, 257, 1981.

45. Gruen, L.C., Interaction of amino acids with silver (I) ions, *Biochem. Biophys. Acta*, 386, 270, 1975.

46. Singh, S.P. and Yadav, V., Cadmium uptake in *Anacystis nidulans*: effect of modifying factors, *J. Gen. Appl. Microbiol.*, 31, 39, 1985.

47. McKnight, D.M. and Morel, F.M.M., Release of weak and strong copper-complexing agents by algae, *Limnol. Oceanogr.*, 24, 823, 1979.

48. Fogg, G.E. and Westlake, D.F., The importance of extracellular products of algae in freshwater, *Int. Assoc. Theroret. Appl. Limnol.*, 12, 219, 1955.

49. Rauser, W.E., Structure and function of metal chelators produced by plants: the case for organic acids, amino acids, phytin and metallothioneins, *Cell Biochem. Biophys.*, 31, 19, 1999.

50. Wilhelm, S.W. and Trick, C.G., Iron-limited growth of cyanobacteria: multiple siderophore production is a common response, *Limnol. Oceanogr.*, 39, 1979, 1994.

51. Olafson, R.W., Loya, S., and Sim, R.G., Physiological parameters of prokaryotic metallothionein production, *Biochem. Biophys. Res. Commun.*, 95, 1495, 1980.

52. Moffett, J.W. and Brand, L.E., Production of strong extracellular Cu chelators by marine cyanobacteria in response to Cu stress, *Limnol. Oceanogr.*, 41, 388, 1996.

53. Olson, R.J., Chisholm, S.W., Zettler, E.R., Altabet, M.A., and Dusenberry, J.A., Spatial and temporal distributions of prochlorophyte picoplankton in the North Atlantic Ocean, *Deep-Sea Res.*, 37, 1033, 1990.

54. Vymazal, J., Toxicity and accumulation of lead with respect to algae and cyanobacteria: a review, *Acta Hydrochim. Hydrobiol.*, 18, 531, 1990.

55. Arduini, I., Godbold, D.L., and Onnis, A., Cadmium and copper uptake and distribution in Mediterranean tree seedlings, *Physiol. Plant.*, 97, 111, 1996.

56. Kramer, U., Smith, R.D., Wenzel, W.W., Raskin, I., and Salt, D.E., The role of metal transport and tolerance in nickel hyperaccumulation by *Thlaspi goesingense* Halacsy, *Plant Physiol.*, 115, 1641, 1997.

57. Guerinot, M.L. and Salt, D.E., Fortified foods and phytoremediation. Two sides of the same coin, *Plant Physiol.*, 125, 164, 2001.

58. Baker, A.J.M. and Brooks, R.R., Terrestrial higher plants which hyperaccumulate metallic elements — a review of their distribution, ecology and phytochemistry, *Biorecovery*, 1, 81, 1989.

59. Reeves, R.D., The hyperaccumulation of nickel by serpentine plants, in *The Vegetation of Ultramafic (Serpentine) Soils*, Baker, A.J.M., Proctor, J., and Reeves, R.D., Eds., Intercept, Andover Hampshire U.K., 1992, 253.

60. Hamer, D.H., Metallothionein, *Ann. Rev. Biochem.*, 55, 913, 1986.

61. Rauser, W.E., Phytochelatins, *Ann. Rev. Biochem.*, 59, 61, 1990.

62. Steffens, J.C., The heavy metal-binding peptides of plants, *Annu. Rev. Plant Physiol. Plant Mol. Biol.*, 41, 553, 1990.

63. Murasugi, A., Wada, C., and Hayashi, Y., Cadmium-binding peptide induced in fission yeast *Schizosaccharomyces pombe*, *J. Biochem.*, 90, 1561, 1981.

64. Murasugi, A., Wada, C., and Hayashi, Y., Occurrence of acid labile sulfide in cadmium-binding peptide 1 from fission yeast, *J. Biochem.*, 93, 661, 1983.

65. Kneer, R. and Zenk, M.H., Phytochelatins protect plant enzymes from heavy metal poisoning, *Phytochemistry*, 31, 2663, 1992.

66. Howden, R., Goldsbrough, P.B., Anderson, C.R., and Cobbett, C.S., Cadmium-sensitive, *cad1* mutants of *Arabidopsis thaliana* are phytochelatin deficient, *Plant Physiol.*, 107, 1059, 1995.

67. Rauser, W.E. and Meuwly, P., Retention of cadmium roots of maize seedlings. Role of complexation by phytochelatins and related thiol peptides, *Plant Physiol.*, 109, 95, 1995.

68. Rauser, W.E. and Curvetto, N.R., Metallothionein occurs in roots of *Agrostis* tolerant to copper, *Nature*, 287, 563, 1980.

69. Weigel, H.J. and Jager, H.J., Subcellular distribution and chemical form of cadmium in bean plants, *Plant Physiol.*, 65, 480, 1980.

70. Murasugi, A., Wada, C., and Hayashi, Y., Purification and unique properties in UV and CD spectra of Cd-binding peptide 1 from *Schizosaccharomyces pombe*, *Biochem. Biophys. Res. Commun.*, 103, 1021, 1981b.

71. Grill, E., Winnacker, E.-L., and Zenk, M.H., Phytochelatins: the principal heavy-metal complexing peptides of higher plants, *Science*, 230, 674, 1985.

72. Gekeler, W., Grill, E., Winnacker, E.-L., and Zenk, M.H., Algae sequester heavy metals via synthesis of phytochelatin complexes, *Arch. Microbiol.*, 150, 197, 1988.

73. Gekeler, W., Grill, E., Winnacker, E.-L., and Zenk, M.H., Survey of the plant kingdom for the ability to bind heavy metals through phytochelatins, *Z. Naturforsch.*, 44c, 361, 1989.

74. Grill, E., Winnacker, E.-L., and Zenk, M.H., Phytochelatins, a class of heavy-metal-binding peptides from plants, are functionally analogous to metallothioneins, *Proc. Natl. Acad. Sci. USA*, 84, 439, 1987.

75. Ahner, B.A. and Morel, F.M.M., Phytochelatin production in marine algae. 2. Induction by various metals, *Limnol. Oceanogr.*, 40, 658, 1995a.

76. Maitani, T., Kubota, H., Sato, K., and Yamada, T., The composition of metals bound to class II metallothionein (phytochelatin and its desglycyl peptide) induced by various metals in root cultures of *Rubia tinctorum*, *Plant Physiol.*, 110, 1145, 1996.

77. Ahner, B.A., Kong, S., and Morel, F.M.M., Phytochelatin production in marine algae. 1. An interspecies comparison, *Limnol. Oceanogr.*, 40, 649, 1995b.

78. Kondo, N., Isobe, M., Imai, K., Goto, T., Murasugi, A., and Hayashi, Y., Structure of cadystin, the unit-peptide of cadmium-binding peptides induces in a fission yeast, *Schizosaccharomyces pombe*, *Tetrahedron Lett.*, 24, 925, 1983.

79. Kondo, N., Imai, K., Isobe, M., Goto, T., Murasugi, A., Wada–Nakagawa, C., and Hayashi, Y., Cadystin A and B, major unit peptides comprising cadmium binding peptides induced in fission yeast separation: revision of structures and synthesis, *Tetrahedron Lett.*, 25, 3869, 1984.

80. Reese, R.N. and Winge, D.R., Sulfide stabilization of the cadmium γ-glutamyl peptide complex of *Schizosaccharomyces pombe*, *J. Biol. Chem.*, 263, 12832, 1988.

81. Jackson, P.J., Unkefer, C.J., Doolen, J.A., Watt, K., and Robinson, N.J., Poly (γ-glutamylcysteinyl) glycine: its role in cadmium resistance in plant cells, *Proc. Natl. Acad. Sci. USA*, 84, 6619, 1987.

82. Steffens, J.C., Hunt, D.F., and Williams, B.G., Accumulation of nonprotein metal-binding polypeptides (γ-glutamyl-cysteinyl)$_n$-glycine in selected cadmium-resistant tomato cells, *J. Biol. Chem.*, 261, 13879, 1986.

83. Grill, E., Gekeler, W., Winnacker, E.-L., and Zenk, M.H., Homo-phytochelatins are heavy metal-binding peptides of homo-glutathione containing Fabales, *FEBS Lett.*, 205, 47, 1986.

84. Klapheck, S., Fliegner, W., and Zimmer, I., Hydroxymethyl-phytochelatins [(γ-glutamylcysteine)$_n$-serine] are metal-induced peptides of the Poaceae, *Plant Physiol.*, 104, 1325, 1994.

85. Meuwly, P., Thibault, P., Schwan, A.L., and Rauser, W.E., Three families of thiol peptides are induced by cadmium in maize, *Plant J.*, 7, 391, 1995.

86. Scheller, H.V., Huang, B., Hatch, E., and Goldsbrough, P.B., Phytochelatin synthesis and glutathione levels in response to heavy metals in tomato cells, *Plant Physiol.*, 85, 1031, 1987.

87. Delhaize, E., Jackson, P.J., Lujan, L.D., and Robinson, N.J., Poly(γ-glutamylcysteinyl)Glycine synthesis in *Datura innoxia* and binding with cadmium, *Plant Physiol.*, 89, 700, 1989.

88. Ruegsegger, A., Schmutz, D., and Brunold, C., Regulation of glutathione synthesis by cadmium in *Pisum sativum* L., *Plant Physiol.*, 93, 1579, 1990.

89. Tukendorf, A. and Rauser, W.E., Changes in glutathione and phytochelatins in roots of maize seedlings exposed to cadmium, *Plant Sci.*, 70, 155, 1990.

90. Rauser, W.E., Schupp, R., and Rennenberg, H., Cysteine, γ-glutamylcysteine, and glutathione levels in maize seedlings, *Plant Physiol.*, 97, 128, 1991.

91. Ruegsegger, A. and Brunold, C., Effect of cadmium on γ-glutamylcysteine synthesis in maize seedlings, *Plant Physiol.*, 99, 428, 1992.

92. Mutoh, N. and Hayashi, Y., Isolation of mutants of *Schizosaccharomyces pombe* unable to synthesize cadystin, small cadmium-binding peptides, *Biochem. Biophys. Res. Commun.*, 151, 32, 1988.

93. Berger, J.M., Jackson, P.J., Robinson, N.J., Lujan, L.D., and Delhaize, E., Precursor–product relationships of poly(γ-glutamylsysteinyl)glycine biosynthesis in *Datura innoxia*, *Plant Cell Rep.*, 7, 632, 1989.

94. Mendum, L.M., Gupta, S.C., and Goldsbrough, P.B., Effect of glutathione on phytochelatin synthesis in tomato cells, *Plant Physiol.*, 93, 484, 1990.

95. Grill, E., Loffler, S., Winnacker, E.-L., and Zenk, M.H., Phytochelatins, the heavy-metal-binding peptides of plants, are synthesized from glutathione by a specific γ-glutamyldipeptidyl transpeptidase (phytochelatin synthase), *Proc. Natl. Acad. Sci. USA*, 86, 6838, 1989.

96. Chen, J., Zhou, J., and Goldsbrough, P.B., Characterization of phytochelatin synthase from tomato, *Physiol. Plant.*, 101, 165, 1997.

97. Klapheck, S., Schlunz, S., and Bergmann, L., Synthesis of phytochelatins and homo-phytochelatins in *Pisum sativum* L., *Plant Physiol.*, 107, 515, 1995.

98. Hayashi, Y., Nakagawa, C.W., Mutoh, N., Isobe, M., and Goto, T., Two pathways in the biosynthesis of cadystins (γ-EC)$_n$G in the cell-free system of the fission yeast, *Biochem. Cell. Biol.*, 69, 115, 1991.

99. Wurz, H., Tanaka, A., and Fruton, J.S., Polymerizaion of *dipeptide* amides by cathepsin C, *Biochemistry*, 1, 19, 1962.

100. Howden, R., Anderson, C.R., Goldsbrough, P.B., and Cobbett, C.S., A cadmium-sensitive, glutathione-deficient mutant of *Arabidopsis thaliana*, *Plant Physiol.* 107, 1067, 1995.

101. Steffens, J.C., The heavy metal-binding peptides of plants, *Annu. Rev. Plant Physiol. Plant Mol. Biol.*, 41, 553, 1990.

102. Reddy, G.N. and Prasad, M.N.V., Heavy metal binding proteins/*peptide*: occurrence structure, synthesis and functions — a review, *Environ. Exp. Bot.*, 30, 251, 1990.

103. De Knecht, J.A., van Dillen, M., Koevoets, P.L.M., Schat, H., Verkleij, J.A.C., and Ernst, W.H.O., Phytochelatins in cadmium-sensitive and cadmium-tolerant *Silene vulgaris*. Chain length distribution and sulfide incorporation, *Plant Physiol.*, 104, 255, 1994.

104. Vogeli-Lange, R. and Wagner, G.J., Relationship between cadmium, glutathione and cadmium-binding peptides (phytochelatins) in leaves of intact tobacco seedlings, *Plant Sci.*, 114, 11, 1996.

105. Huang, B., Hatch, E., and Goldsbrough, P.B., Selection and characterization of cadmium tolerant cells in tomato, *Plant Sci.*, 52, 211, 1987.

106. Huang, B. and Goldsborough, P.B., Cadmium tolerance in tobacco cell culture and its relevance to temperature stress, *Plant Cell Rep.*, 7, 119, 1988.

107. Jackson, P.J., Roth, E.J., McClure, P.R., and Naranjo, C.M., Selection, isolation, and characterization of cadmium-resistant *Datura innoxia* suspension cultures, *Plant Physiol.*, 75, 914, 1984.

108. Gupta, S.C. and Goldsbrough, P.B., Phytochelatin accumulation and cadmium tolerance in selected tomato cell lines, *Plant Physiol.*, 97, 306, 1991.

109. De Vos, C.H.R., Vonk, M.J., Vooijs, R., and Schat, H., Glutathione depletion due to copper-induced phytochelatin synthesis causes oxidative stress in *Silene cucubalus*, *Plant Physiol.*, 98, 853, 1992.

110. Chen, J. and Goldsbrough, P.B., Increased activity of γ-glutamylcysteine synthetase in tomato cells selected for cadmium tolerance, *Plant Physiol.* 106, 233, 1994.

111. Zhu, Y.L., Pilon-Smits, A.H., Jouanin, L., and Terry, N., Overexpression of glutathione synthetase in Indian mustard enhances cadmium accumulation and tolerance, *Plant Physiol.*, 119, 73, 1999.

112. Glaeser, H., Coblenz, A., Kruczek, R., Ruttke, I., Ebert–Jung, A., and Wolf, K., Glutathione metabolism and heavy metal detoxification in *Schizosccharomyces pombe*: isolation and characterization of glutathione-deficient, cadmium-sensitive mutants, *Curr. Genet.*, 19, 207, 1991.

113. Schat, H. and Kalff, M.M.A., Are phytochelatins involved in differential metal tolerance or do they merely reflect metal-imposed strain? *Plant Physiol.*, 99, 1475, 1992.

114. Howden, R. and Cobbett, C.S., Cadmium-sensitive mutants of *Arabidopsis thaliana*, *Plant Physiol.*, 99, 100, 1992.

115. Loeffler, S., Hocheberger, A., Grill, E., Winnacker, E.-L., and Zenk, M.H., Termination of the phytochelatin synthase reaction through sequestration of heavy metals by the reaction product, *FEBS Lett.*, 258, 42, 1989.

116. Verkleij, J.A.C., Koevoets, P., Van't Reit, J., Bank, R., Nijdam, Y., and Ernst, W.H.O., Poly-gamma-glutamycysteinylglycines or phytochelatins and their role in cadmium tolerance of *Silene vulgaris*, *Plant Cell Environ.*, 13, 913, 1990.

117. Reese, N.R., White, C.A., and Winge, D.R., Cadmium-sulfide crystallites in Cd-(γEC)$_n$G peptide complexes from tomato, *Plant Physiol.*, 98, 225, 1992.

118. Speiser, D.M., Abrahamsom, S.L., Banuelos, G., and Ow, D.W., *Brassica junceae* produces a phytochelatin-cadmium-sulfide complex, *Plant Physiol.*, 99, 817, 1992.

119. Rauser, W.E. and Meuwly, P., Retention of cadmium roots of maize seedlings. Role of complexation by phytochelatins and related thiol peptides, *Plant Physiol.*, 109, 195, 1995.

120. Dameron, C.T., Smith, B.R., and Winge, D.R., Glutathione-coated cadmium-sulfide crystallites in *Candida glabrata*, *J. Biol. Chem.*, 264, 17355, 1989.

121. Barbas, J., Santhanagopalan, V., Blaszczynski, M., Ellis, W.R., Jr, and Winge, D.R., Conversion in the peptide coating cadmium: sulfide crystallites in *Candida glabrata*, *J. Inorg. Biochem.*, 48, 95, 1992.

122. Strasdeit, H., Duhme, A.-K., Kneer, R., Zenk, M.H., Hermes, C., and Nolting, H.-F., Evidence for discrete Cd(Scys)$_4$ units in cadmium phytochelatin complexes from EXAFS spectroscopy, *J. Chem. Soc. Chem. Commun.*, 16, 1129, 1991.

123. Hayashi, Y., Nakagawa, C.W., Uyakul, D., Imai, K., Isobe, M., and Goto, T., The change of cadystin components in Cd-binding peptides from the fission yeast during their induction by cadmium, *Biochem. Cell Biol.*, 66, 288, 1988.

124. Vogeli-Lange, R. and Wagner, G.J., Subcellular localization of cadmium and cadmium-binding peptides in tobacco leaves. Implication of a transport function for cadmium-binding peptides, *Plant Physiol.*, 92, 1086, 1990.

125. Gupta, S.C. and Goldsbrough, P.B., Phytochelatin accumulation and stress tolerance in tomato cells exposed to cadmium, *Plant Cell Rep.*, 9, 466, 1990.

126. Ortiz, D.F., Krepppel, L., Speiser, D.M., Scheel, G., McDonald, G., and Ow, D.W., Heavy metal tolerance in the fission yeast required an ATP-binding cassette-type vacuolar membrane transporter, *Embo J.*, 11, 3491, 1992.

127. Ortiz, D.F., Ruscitti, T., McCue, K.F., and Ow, D.W., Transport of metal-binding peptides by HMT1, a fission yeast ABC-type vacuolar membrane protein, *J. Biol. Chem.*, 270, 4721, 1995.

128. Salt, D.E. and Rauser, W.E., MgATP-dependent transport of phytochelatins across the tonoplast of oat roots, *Plant Physiol.*, 107, 1293, 1995.

129. Salt, D.E. and Wagner, G.J., Cadmium transport across tonoplast of vesicles from oat roots. Evidence for Cd^{2+}/H^+ antiport activity, *J. Biol. Chem.*, 268, 12297, 1993.

130. Speiser, D.M., Ortiz, D.F., Kreppel, L., Scheel, G., McDonald, G., and Ow, D.W., Purine biosynthetic genes are required for cadmium tolerance in *Schizosaccharomyces pombe*, *Mol. Cell Biol.*, 12, 5301, 1992.

131. Juang, R.-H., McCue, K.F., and Ow, D.W., Two purine biosynthetic enzymes that are required for Cd tolerance in *Schizosacromyces pombe* utilize cysteine sulfinate *in vitro*, *Arch. Biochem. Biophys.*, 304, 392, 1993.

132. Thumann, J., Grill, E., Einnackder, E.-L., and Zenk, M.H., Reactivation of metal-requiring apoenzymes by phytochelatin-metal complexes, *FEBS Lett.*, 282, 66, 1991.

133. Reese, R.N. and Wagner, G.J., Effects of buthionine sulfoximineon Cd-binding peptide levels in suspension-cultured tobacco cells treated with Cd, Zn, or Cu, *Plant Physiol.*, 84, 574, 1987.

134. Hartley–Whitaker, J., Ainsworth, G., Vooijs, R., Ten Bookum, W., Schat, H., and Meharg, A.A., Phytochelatins are involved in differential arsenate tolerance in *Holcus lanatus*, *Plant Physiol.*, 126, 299, 2001.

135. Reese, R.N., Mehra, R.K., Tarbet, E.B., and Winge, D.R., Studies on the γ-glutamyl Cu-binding peptide from *Schizosaccharomyces pombe*, *J. Biol. Chem.*, 263, 4186, 1988.

136. Mehra, R.K., Kodati, R., and Abdullah, R., Chain length-dependent Pb(II)-coordination in phytochelatins, *Biochem. Biophys. Res. Commun.*, 215, 730, 1995.

137. Mehra, R.K., Miclat, J., Kodati, R., Abdullah, R., Hunter, T.C., and Mulchandani, P., Optical spectroscopic and reverse-phase HPLC analyses of Hg(II) binding to phytochelatins, *Biochem. J.*, 314, 73, 1996.

138. Ma, J.F., Kiradate, S., Nomoto, K., Iwashita, T., and Matsumoto, H., Internal detoxification mechanism of Al in hydrangea. Identification of Al form in the leaves, *Plant Physiol.*, 113, 1033, 1997.

139. Ma, J.F., Hiradate, S., and Matsumoto, H., High aluminum resistance in buckwheat. II. Oxalic acid detoxifies aluminum internally, *Plant Physiol.*, 117, 753, 1998.

140. Nordstrom, D.K. and May, H.M., Aqueous equilibrium data for mononuclear aluminum species, in *The Environment Chemistry of Aluminum*. Sposito, G. Ed., CRC Press, Boca Raton, FL, 1996, 39.

141. Zheng, S.J., Ma, J.F., and Matsumoto, H., Continuous secretion of organic acids is related to aluminum resistance during relatively long-term exposure to aluminum stress, *Physiol. Plant.*, 103, 209, 1998.

142. Ernst, W.H.O., Ecophysiological studies on heavy metal plants in South Central Africa, *Kirkia*, 8, 125, 1972.

143. Brookes, A., Collins, J.C., and Thurman, D.A., The mechanism of zinc tolerance in grasses, *J. Plant Nutr.*, 3, 695, 1981.

144. Mullins, M., Hardwick, K., and Thurman, D.A., Heavy metal location by analytical electron microscopy in conventionally fixed and freeze-substituted roots of metal tolerant and nontolerant ecotypes, in *Heavy Metals in Environment (Athens 1985)*, Lekkas, T.D., Ed., CEP Consultants, Edinburgh, 1985, 43.

145. Heuillet, E., Moreau, A., Halpern, S., Jeanne, N., and Puiseux-Dao, S., Cadmium binding to a thiol-molecular in vacuoles of *Dunaliella bioculata* contaminated with $CdCl_2$: electron probe analysis, *Biol. Cell*, 58, 79, 1986.

146. Rauser, W.E. and Ackerley, C.A., Localization of cadmium in granules within differentiating and mature root cells, *Can. J. Bot.*, 65, 643, 1987.

147. Krotz, R.M., Evangelou, B.P., and Wagner, G.J., Relationship between cadmium, zinc, Cd-peptide, and organic acid in tobacco suspension cells, *Plant Physiol.*, 91, 780, 1989.

148. Brooks, R.R., Shaw, S., and Asensi–Marfil, A., The chemical form and physiological function of nickel in some Iberian *Alyssum* species, *Physiol. Plant.*, 51, 167, 1981.

149. Godbold, D.L., Horst, W.J., Collins, J.C., Thurman, D.A., and Marschner, H., Accumulation of zinc and organic acids in roots of zinc tolerant and nontolerant ecotypes of *Deschampsi caespitosa*, *J. Plant. Physiol.*, 116, 59, 1984.

150. Pelosi, P., Fiorentini, R., and Galoppini, C., On the nature of nickel compounds in *Alyssum bertolonii* Desv., *Agric. Biol. Chem.*, 40, 1641, 1976.

151. Brune, A., Urback, W., and Dietz, K.-J., Compartmentation and transport of zinc in barley primary leaves as basic mechanisms involved in zinc tolerance, *Plant Cell Environ.*, 17, 153, 1994.

152. Wang, J., Evangelou, B.P., Nielsen, M.T., and Wagner, G.J., Computer-simulated evaluation of possible mechanisms for sequestering metal ion activity in plant vacuoles, *Plant Physiol.*, 99, 621, 1992.

153. Ernst, W.H.O., Physiological and biochemical aspects of metal tolerance, in *Effects of Air Pollutants in Plants*, Mansfield, T.A., Ed., Cambridge University Press, Cambridge, 1976, 115.

154. Stroinski, A., Some physiological and biochemical aspects of plant resistance to cadmium effect. I. Antioxidative system, *Acta Physiol. Plant.*, 21, 175, 1999.

155. Gallego, S.M., Benavides, M.P., Tomaro, M.L., Effect of cadmium ions on antioxidant defense system in sunflower cotyledons, *Biol. Plant.*, 42, 49, 1999.

156. Cakmak, I. and Horst, W.J., Effect of aluminum on lipid peroxidation, superoxide dismutase, catalase, and peroxidase activities in root tips of soybean (*Glycine max*), *Physiol. Plant.*, 83, 463, 1991.

157. Ezaki, B., Tsugita, S., and Matsumoto, H., Expression of a moderately anionic peroxidase is induced by aluminum treatment in tobacco cells, *Physiol. Plant.*, 96, 21, 1996.

158. Hendry, G.A.F., Baker, A.J.M., and Ewart, C.F., Cadmium tolerance and toxicity, oxygen radical processes and molecular damage in cadmium-tolerant and cadmium-sensitive clones of *Holcus lanatus* L., *Acta Bot. Neerl.*, 41, 271, 1992.

159. Subhadra, A.V., Nanda, A.K., Behera, P.K., Panda, B.B., Acceleration of catalase and peroxidase activities in *Lemna minor* L. and *Allium cepa* L. in response to low levels of aquatic mercury, *Environ. Pollut.*, 69, 169, 1991.

160. Mazhoudi, S., Chaoui, A., Ghorbal, M.H., and Ferjani, E.E., Response of antioxidant enzymes to excess copper in tomato (*Lycopersicon esculentum*, Mill.), *Plant Sci.*, 127, 129, 1997.

161. Shaw, B.P. and Rout, N.P., Age-dependent responses of *Phaseolus aureus* Roxb. to inorganic salts of mercury and cadmium, *Acta Physiol. Plant.*, 20, 85, 1998.

162. Teisseire, H. and Guy, V., Copper-induced changes in antioxidant enzymes activities in fronds of duckweed (*Lemna minor*), *Plant Sci.*, 153, 65, 2000.

163. Schicker, H. and Caspi, H., Response of antioxidative enzymes to nickel and cadmium stress in hyperaccumulator plants of the genus *Alyssum*, *Plant Physiol.*, 105, 39, 1999.

164. Stroinski, A. and Kozlowska, M., Cadmium-induced oxidative stress in potato tuber, *Acta Soc. Bot. Pol.*, 66, 189, 1997.

165. Breusegem, F.V., Slooten, L., Stassart, J.-M., Moens, T., Botterman, J., Montagu, M.V., and Inze, D., Overproduction of *Arabidopsis thaliana* FeSOD confers oxidative stress tolerance to transgenic maize, *Plant Cell Physiol.*, 40, 515, 1999.

166. Loggini, B., Scartazza, A., Brugnoli, E., and Navari–Izzo, F., Antioxidative defense system, pigment composition, and photosynthetic efficiency in two wheat cultivars subjected to drought, *Plant Physiol.*, 119, 1091, 1999.

167. Thomas, D.J., Thomas, J.B., Prier, S.D., Nasso, N.E., and Herbert, S.K., Iron superoxide dismutase protects against chilling damage in the cyanobacterium *Synechococcus* species PCC7942, *Plant Physiol.*, 120, 275, 1999.

168. Scandalios, J.G., Molecular genetics of superoxide dismutases in plants, in *Oxidative stress and the molecular biology of antioxidant defenses*, Scaldalios, J.G., Ed., Cold Spring Harbor Laboratory Press, Cold Spring Harbor, New York, 1997, 527.

169. Chongpraditium, P., Mori, S., and Chino, M., Excess copper induces a cytosolic Cu, Zn-superoxide dismutase in soybean root, *Plant Cell Physiol.*, 33, 239, 1992.

170. Przymusinski, R., Rucinska, R., and Gwozdz, E., The stress-stimulated 16-kDa polypeptide from lupine roots has properties of cytosolic Cu:Zn-superoxide dismutase, *Environ. Exp. Bot.*, 35, 485, 1995.

171. Ezaki, B., Katsuhara, M., Kawamura, M., and Matsumoto, H., Different mechanisms of four aluminum (Al)-resistant transgenes for Al toxicity in *Arabidopsis*, *Plant Physiol.*, 127, 918, 2001.

172. Rauser, W.E., Phytochelatin-based complexes bind various amounts of cadmium in maize seedlings depending on the time of exposure, the concentration of cadmium and the tissue, *New Phytol.*, 158, 269, 2003.

173. Symeonidis, L. and Karataglis, S., Interactive effects of Cd, Pb, and Zn on root growth of two metal tolerant genotypes of *Holcus lanatus* L., *Bio Metals*, 5, 173, 1992.

174. Nan, Z., Li, J., Zhang, J., and Cheng, G., Cadmium and Zinc interactions and their transfer in soil-crop system under actual field conditions, *Sci. Total Environ.*, 285, 187, 2002.

175. Adriano, D.C., *Trace Elements in Terrestrial Environments: Biogeochemistry, Bioavailability, and Risks of Metals*, 2nd ed., Springer–Verlag, New York, 2001, 866.

176. Das, P., Samantaray, S., and Rout, G.R., Rout, Studies on cadmium toxicity in plants: a review, *Environ. Pollut.*, 98, 29, 1997.

177. Clemens, S., Palmgren, M.G., and Krämer, U., A long way ahead understanding and engineering plant metal accumulation, *Trends Plant Sci.*, 7, 309, 2002.

178. Aravind, P. and Prasad, M.N.V., Zinc alleviates cadmium induced toxicity in *Ceratophyllum demersum*, a fresh water macrophyte, *Plant. Physiol. Biochem.*, 41, 391, 2003.

179. Aravind, P. and Prasad, M.N.V., Carbonic anhydrase impairment in cadmium-treated *Ceratophyllum demersum* L. (free floating freshwater macrophyte): toxicity reversal by zinc, *J. Anal. Atomic Spectrophotometry*, 19, 52, 2004.

180. Aavind, P. and Prasad, M.N.V., Zinc protects chloroplasts and associated photochemical functions in cadmium exposed *Ceratophyllum demersum* L., a freshwater macrophyte, *Plant Sci.*, 166, 1321, 2004.

181. Prasad, M.N.V. (Ed.), *Heavy Metal Stress in Plants: from Biomolecules to Ecosystems*, 2nd ed. Springer–Verlag, Heidelberg, 2004, 462; xiv.

182. Hall, J.L., Cellular mechanisms for heavy metal detoxification and tolerance, *J. Exp. Bot.*, 53, 1, 2002.

183. Prasad, M.N.V., Cadmium toxicity and tolerance to vascular plants, *Envir. Exp. Bot*, 35, 525, 1995.

184. Prasad, M.N.V., Trace metals, in *Plant Ecophysiology*, Prasad, M.N.V., Ed., John Wiley & Sons, New York, 1997, 207–249.

185. Briat, J.F. and Lebrun, M., Plant responses to metal toxicity, *CR Acad. Sci. Paris/Life Sci.*, 322, 43, 1999.

186. Cobbett, C. and Goldsbrough, P.B., Phytochelatins and metallothioneins: roles in heavy metal detoxification and homeostasis, *Annu. Rev. Plant Biol*, 531, 159, 2002.

187. Rama Kumar, T. and Prasad, M.N.V., Metal-binding properties of ferritin in *Vigna mungo* (L.) Hepper (black gram): possible role in heavy metal detoxification, *Bull. Environ. Contam. Toxicol.*, 62, 502, 1999.

188. Mejáre, M. and Bülow, L., Metal-binding proteins and peptides in bioremediation and phytoremediation of heavy metals, *Trends Biotechnol.*, 19, 67, 2000.

189. Vatamaniuk, J.O.K., Mari, S., Lu, Y.P., and Rea, P.A., Mechanism of heavy metal ion activation of phytochelatin PC synthase blocked thiols are sufficient for PC synthase-catalyzed transpeptidation of glutathione and related thiol peptides, *J. Biol. Chem.*, 27540, 31451, 2002.

190. Sanità Di Toppi, L., Prasad, M.N.V., and Ottonello, S., Metal chelating peptides and proteins in plants, in *Physiology and Biochemistry of Metal Toxicity and Tolerance in Plants*, Prasad, M.N.V. and Strzalka, K., Eds., Kluwer Academic Publishers, Dordrecht, 2002, 59.

191. Zhu, Y.L., Pilon–Smits, E.A.H., Tarun, A.S., Weber, S.U., Jouanin, L., and Terry, N., Cadmium tolerance and accumulation in indian mustard is enhanced by over expressing gamma–glutamyl cysteine synthetase; *Plant Physiol.*, 121, 1169, 1999.

192. Xiang, C., Werner, B.L., Christensen, E.M., and Olver, D.J., The biological functions of glutathione revisited in *Arabidopsis* transgenic plants with altered glutathione levels, *Plant Physiol.*, 126, 564, 2001.

193. Banks, M.K., Waters, C.Y., and Schwab, A.P., Influence of organic acids on leaching of heavy metals from contaminated mine tailings, *J. Environ. Health A*, 29, 1045, 1994.

194. Barcelo, J. and Poschenrieder, C., Fast root growth responses, root exudates, and internal detoxification as clues to the mechanisms of aluminum toxicity and resistance: a review, *Environ. Exp. Bot.*, 48, 75, 2002.

195. Deiana, S., Gessa, C., Palma, A., Premoli, P., and Senette, C., Influence of organic acids exuded by plants on the interaction of copper with the polysaccharidic components of the root mucilages, *Organic Geochem.*, 34, 651, 2003.

196. Delhaize, E., Ryan, P.R., and Randall, P.J., Aluminum tolerance in wheat (*Triticum aestivum* L.). II. Aluminum-stimulated excretion of malic acid from root apices, *Plant Physiol.*, 103, 695, 1993.

197. Harmens, H., Koevoets, P.L.M., Verkleij, J.A.C., and Ernst, W.H.O., The role of low molecular weight organic acids in the mechanism of increased zinc tolerance in *Silene vulgaris* (Moench) Garcke, *New Phytol.*, 126, 615, 1994.

198. Huang, J.W., Blaylock, M.J., Kapulnik, Y., and Ensley, B.D., Phytoremediation of uranium-contaminated soils: role of organic acids in triggering uranium hyper accumulation in plants, *Environ. Sci. Technol.*, 32, 2004, 1998.

199. Ma, L.Q., Komar, K.M.M., Tu, C., Zhang, W., Cai, Y., and Kennelley, E.D., A fern that hyperaccumulates arsenic. A hardy, versatile, fast-growing plant helps to remove arsenic from contaminated soils, *Nature*, 409, 579, 2001.

200. Larsen, P.B., Degenhardt, J., Tai, C.Y., Stenzler, L., Howell, S.H., and Kochian, L.V., Aluminum-resistant *Arabidopsis* mutants that exhibit altered patterns of Al accumulation and organic acid release from roots, *Plant Physiol.*, 117, 9, 1998.

201. Pigna, M., Del Gaudio, S.D., and Violante, A., Role of inorganic and organic ligands on the mobility of arsenate in soil environments. SYM01 arsenic in soil and groundwater, environment: biogeochemical interactions, *Proc. 7th Int. Conf. Biogeochem. Trace Elements*, Uppsala '03, 2, 24–25, 2003.

202. Sanità Di Toppi, L., Gremigni, P., Pawlik Skowronska, B., Prasad, M.N.V., and Cobbett, C.S., Responses to heavy metals in plants — molecular approach, in *Abiotic Stresses in Plants*, Sanità Di Toppi, L. and Pawlik Skowronska, B., Eds., Kluwer Academic Publishers, Dordrecht, 2003, 133.

203. Wu, L.J., Luo, Y.M., Christie, P., and Wong, M.H., Effects of EDTA and low molecular weight organic acids on soil solution properties of a heavy metal-polluted soil, *Chemosphere*, 50, 819, 2003.

204. Reddy, G.N. and Prasad, M.N.V., Tyrosine is not phosphorylated in cadmium-induced HSP cognate in maize (*Zea mays* L.) seedlings: role in chaperone function? *Biochem. Arch.*, 9, 25, 1993.

205. Neumann, D., Lichtenberg, O., Günther, D., Tschiersch, K., and Nover, L., Heat-shock protein-induced heavy metal tolerance in higher plants, *Planta*, 194, 360, 1994.

206. Reddy, G.N. and Prasad, M.N.V., Cadmium-induced proteins and protein phosphorylation changes in rice (*Oryza sativa* L.), *J. Plant Physiol.*, 145, 67, 1995.

207. Prasad, M. N. V., Metallothioneins and metal binding complexes in plants, in *Heavy Metal Stress in Plants: from Molecules to Ecosystems*, Prasad, M.N.V. and Hagemeyer, J., Eds., Springer–Verlag, Heidelberg, 1999, 51–72.

208. Assuncão, A.G.L., Martins, P.D.C., Folter, S.D., Vooijs, R., Schat, H., and Aatrs, M.G.M. Elevated expression of metal transporter genes in three accessions of the metal hyperaccumulator *Thlaspi caerulescens*, *Plant Cell Environ.*, 24, 217, 2001.

209. Lasat, M.M., Pence, N.S., Garvin, D.F., Ebbs, S.D., and Kochian, L.V., Molecular physiology of zinc transport in the Zn hyperaccumulator *Thlaspi caerulescens*, *J. Exp. Bot.*, 51, 71, 2000.

210. Persans, M.W., Salt, D.E., Kim, S.A., and Guerinot, M.L., Phylogenetic relationships within cation transporter families of *Arabidopsis*, *Plant Physiol.*, 126, 1646, 2001.

211. Pence, N.S., Larsen, P.B., Ebbs, S.D., Lasat, M.M., Letham, D.L.D., Garvin, D.F., Eide, D., and Kochian, L.V., The molecular basis for heavy metal hyper accumulation in *Thlaspi caerulescens*, *Proc. Natl. Acad. Sci. USA*, 97, 4956, 2000.

212. Guerinot, M.L., The *ZIP* family of metal transporters, *Biochim. Biophys. Acta*, 1465, 190, 2000.

213. Prasad, M.N.V., Phytoremediation of metal-polluted ecosystems: hype for commercialization, *Russ. J. Plant Physiol.*, 50, 686, 2003.

214. McGrath, S.P. and Zhao, F., Phytoextraction of metals and metalloids from contaminated soils, *Curr. Opin. Biotechnol.*, 14, 1, 2003.

215. Brewer, E.P., Saunders, J.A., Angle, J.S., Chaney, R.L., and McIntosh, M.S., Somatic hybridization between heavy metal hyperaccumulating *Thlaspi caerulescens* and canola, *Agron Abstr.*, 154, 1997.

216. Nedelkoska, T.V. and Doran, P.M., Hyperaccumulation of cadmium by hairy roots of *Thlaspi caerulescens*, *Biotechnol. Bioeng.*, 67, 607, 2000.

217. Shanks, J.V. and Morgan, J., Plant "hairy root" culture, *Curr. Opin. Biotechnol.*, 10, 151, 1999.

218. Palmer, C.E., Warwick, S., and Keller, W., Brassicaceae (Cruciferae) family — plant biotechnology and phytoremediation, *Int. J. Phytoremed.*, 3, 245, 2001.

219. Wenzel, W.W., Lombi, E., and Adriano, D.C., Biogeochemical processes in the rhizosphere: role in phytoremediation of metal-polluted soils, in *Heavy Metal Stress in Plants — from Molecules to Ecosystems*, Prasad, M.N.V. and Hagemeyer, J., Eds., Springer–Verlag, Heidelberg, 1999, 271–303.

220. Wenzel, W. W., Bunkowski, M., Puschenreiter, M., and Horak, O., Rhizosphere characteristics of indigenously growing nickel hyperaccumulator and excluder plants on serpentine soil, *Environ. Pollut.*, 123, 131, 2003.

221. Gobran, G.R., Wenzel, W.W., and Lombi, E., *Trace Elements in the Rhizosphere*, CRC Press, Boca Raton, FL, 2001, 344.

222. Clemens, S., Molecular mechanisms of plant metal tolerance and homeostasis, *Planta*, 212, 475, 2001.

223. Karenlampi, S., Schat, H., Vangronsveld, J., Verkleij, J.A.C., Van Der Lelie, C., Mergeay, M., and Tervahauta, A.I., Genetic engineering in the improvement of plants for phytoremediation of metal polluted soils, *Environ. Pollut.*, 107, 225, 2000.

224. Maywald, F. and Weigel, H.J., Biochemistry and molecular biology of heavy metal accumulation in higher plants, *Landbauforschung Volkenrode*, 47, 103, 1997.

225. Misra, S. and Gedamu, L., Heavy metal tolerant transgenic *Brassica napus* L. and *Nicotiana tabacum* L. plants, *Theor. Appl. Genet.*, 78, 161, 1989.

226. Pollard, A.J., Powell, K.D., Harper, F.A., and Smith, J.A.C., The genetic basis of metal hyperaccumulation in plants, *Crit. Rev. Plant Sci.*, 21, 539, 2002.

227. Raskin, I., Plant genetic engineering may help with environmnetal cleanup, *Proc. Natl. Acad. Sci. USA*, 93, 3164, 1996.

228. Rugh, C.L., Wilde, H.D., Stacks, N.M., Thompson, D.M., Summers, A.O., and Meagher, R.B., Mercuric ion reduction and resistance in transgenic Arabidopsis thaliana plants expressing a modified bacterial merA gene, *Proc. Natl. Acad. Sci. USA*, 93, 3182, 1996.

229. Peña, L. and Séguin, A., Recent advances in the genetic transformation of trees, *Trends Biotechnol.*, 19, 500, 2001.

230. Gisbert, C., Ros, R., Haro, A.D., Walker, D.J., Bernal, M.P., Serrano, R., and Navarro-Aviñó, J.N., A plant genetically modified that accumulates Pb is especially promising for phytoremediation, *Biochem. Biophys. Res. Commun.*, 303, 440, 2003.

231. Kramer, U. and Chardonnens, A.N., The use of transgenic plants in the bioremediation of soils contaminated with trace elements, *Microbiol. Biotechnol.*, 55, 661, 2001.

232. Bi, Y.L., Li, X.L., Christie, P., Hu, Z.Q., and Wong, M.H., Growth and nutrient uptake of arbuscular mycorrhizal maize in different depths of soil overlying coal fly ash, *Chemosphere*, 50, 863, 2003.

233. Entry, J.A., Watrud, L.S., and Reeves, M., Accumulation of ^{137}Cs and ^{90}Sr from contaminated soil by three grass species inoculated with mycorrhizal fungi, *Environ. Pollut.*, 104, 449, 1999.

234. Colpaert, J.V., and Vandenkoornhuyse, P., Mycorrhizal fungi, in *Metals in the Environment: Analysis by biodiversity*, Prasad, M.N.V., Ed., Marcel Dekker Inc., New York, 2001, 37–58.

235. Khan, A.G., Relationships between chromium biomagnification ratio, accumulation factor, and mycorrhizae in plants growing on tannery effluent-polluted soil, *Environ. Int.*, 26, 417, 2001.

236. Khan, A.G., Kuek, C., Chaudhry, T.M., Khoo, C.S., and Hayes, W.J., Role of plants, mycorrhizae and phytochelators in heavy metal contaminated land remediation, *Chemosphere*, 41, 197, 2000.

237. Riesen, T.K. and Brunner, I., Effect of ectomycorrhizae and ammonium on ^{134}Cs and ^{85}Sr uptake into *Picea abies* seedlings, *Environ. Pollut.*, 93, 1, 1996.

238. Shetty, K.G., Hetrick, B.A.D., Hoobler, D., and Schwab, A.P., Effects of mycorrhizae and other soil microbes on revegetation of heavy metal contaminated mine spoil, *Environ. Pollut.*, 86, 181, 1994.

239. Shetty, K.G., Hetrick, B.A.D, and Schwab, A.P., Effects of mycorrhizae and fertilizer amendments on zinc tolerance of plants, *Environ. Pollut.*, 88, 307, 1995.

240. El Kherbawy, M., Angle, J.S., Heggo, A., and Chaney, R.L., Soil pH, rhizobia, and vesicular-arbuscular mycorrhizae inoculation effects on growth and heavy metal uptake of alfalfa (*Medicago sativa* L.), *Biol. Fertil. Soil*, 8, 61, 1998.

241. Brooks, R.R. (Ed.) *Plants that Hyperaccumulate Heavy Metals*. CAB International, Wallingford, U.K., 1998.

242. Prasad, M.N.V. and Freitas, H., Metal hyperaccumulation in plants — biodiversity prospecting for phytoremediation technology, *Electron. J. Biotechnol.*, 6, 275, 2003.

17 Bacterial Biosorption of Trace Elements

Kyoung-Woong Kim and So-Young Kang

CONTENTS

17.1 INTRODUCTION

Metals are ubiquitous in environments and are essential to organisms, e.g., K, Na, Mg, Ca, Mn, Fe, Co, Ni, Cu, Zn, and Mo; other metals have no known essential biological functions, e.g., Al, Ag, Cd, Sn, Au, Sr, Hg, and Pb [1]. All these elements can interact with microorganisms and be accumulated by physicochemical mechanisms and transport systems associated with cell growth and metabolism [2,3]. Heavy metals are not biodegradable and tend to be readily accumulated in living organisms, causing various diseases [4–6]. Virtually all metals can exhibit toxicity to above certain threshold concentrations whether essential or nonessential [1,7].

Heavy metal contamination has been increased in aqueous environments near many industrial facilities, such as metal plating facilities, mining operations, and tanneries. The soils in the vicinity of many military bases are also reported to be contaminated and pose a risk of groundwater and surface water contamination with heavy metals [8].

Physicochemical methods such as chemical precipitation, membrane filtration, ion exchange, and activated carbon adsorption have been developed for the removal of heavy metals from wastewater [9,10]. However, the practical application of such processes is sometimes restricted due to technical or economical constraints. The biological removal of metals through biosorption has distinct advantages over conventional methods: the process rarely produces undesirable or deleterious chemical byproducts and it is highly selective, efficient, easy to operate, and cost effective in the treatment of large volumes of wastewater containing toxic heavy metals [11–13]. Algae, fungi, yeast, and bacteria can remove heavy metals from aqueous solutions by binding the cationic metals onto negatively charged functional groups distributed on their cell walls, such as carboxyl and phosphoryl groups [7,14,15].

According to Beveridge [16], bacteria can be used as an excellent biosorbent for metal sorption because it has high surface-to-volume ratios related with the active sorption site in bacterial cell walls. Particularly, the pure microbial strains have shown extremely high capacities for the selective uptake of metals from dilute metal-bearing solutions [17,18]. The advantages of bacteria as a sorbent for metal removal from wastewater have been well documented and reviewed [19–21]. Although bacterial biosorption can be applied as a useful strategy for metal removal in soil and preventing contamination of groundwater, the funding is insufficient to study the microbial remediation of heavy metals in pilot scale applications comparing with the bioremediation of organic contaminants.

This chapter investigates the characteristics and mechanisms of bacterial biosorption and provides the potential applicability of bacterial biosorption as advanced technology in contaminated environment sites.

17.2 BACTERIAL BIOSORPTION

17.2.1 BACTERIA AS A SORBENT

The bacterial biomass has been successful for the selective removal of heavy metals and radionuclides as sorbents [22–25]. The use of bacteria offers a potential feasibility with cost effectiveness and high removal efficiency because microbial biomass is made up of abundant natural materials and can be grown extremely fast.

17.2.1.1 Structure of Bacterial Cells

Bacteria generally can be divided into two kinds of bacteria: Gram positive and Gram negative by Gram straining. The walls of Gram-positive bacteria consist of three primary components: cytoplasm mixed with a peptidoglycan, to which teichoic acids are covalently bound [26,27]. Each of these polymers is an amphoteric group, but with a net negative charge [16,28]. The envelope of Gram-negative bacteria is more complex than that of Gram-negative bacteria. It consists of two membrane bilayers (the outer and plasma membrane) that are chemically and functionally distinct from one another and sandwich a thin peptidoglycan layer between them [27].

The cell surfaces of Gram-positive and Gram-negative bacteria, whether living or nonliving, possess abundant functional groups that bind metal ions in EPS. These also include phosphate, carboxyl, hydroxyl, and amino functional groups, among others [29]. In Gram-negative cell, the EPS is composed of polysaccharides and protein, which are less firmly bound to the cell surface. External polysaccharides of Gram-negative bacteria offer many functional groups such as carboxylate, hydroxyl, sulphate, phosphate, and amino that can coordinately interact with heavy metals ions. In Gram-positive cells, teichoic acids as well as polysaccharides and proteins that are not anchored in the cell wall contribute to the EPS [7,28]. The Gram-positive bacteria, therefore, can accumulate more heavy metal ions than Gram-negative bacteria.

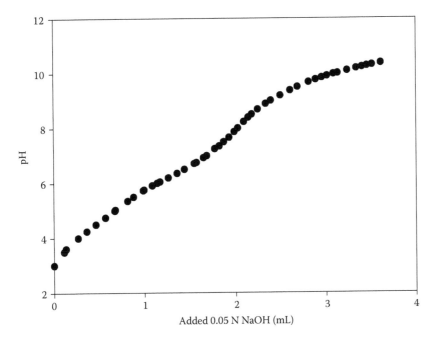

FIGURE 17.1 Biomass potentiomatric titration: 1 g of *P. aeruginosa* was washed by distilled ionized water and then potentiometrically titrated with 0.05 *N* NaOH.

17.2.1.2 Case Study: Identification of Functional Groups

Several investigations have shown that *Pseudomonas aeruginosa*, which is a Gram-negative bacterium commonly distributed in environmental sources such as soil, water, and plant surfaces, has high efficiency for metal uptake [30]. For instance, Chang and Hong [31] have found that the amount of mercury adsorbed on *P. aeruginosa* biomass was higher than that bound to a cation-exchanger resin (AG 50W-X8 resin) with 180 mg Hg/g dry cells and 100 mg Hg/g dry resin, respectively. Hu et al. [32] have identified that *P. aeruginosa* strain CSU showed the highest affinity and maximal capacity for uranium (100 mg U/g dry weight) and was also competitive with commercial cation-exchange resins.

In this part, the study was performed to obtain more information of functional groups of *P. aeruginosa* by potentiometric titration of an aqueous cellular suspension and IR analysis of the lyophilized biomass in solid phase. The washed protonated biomass was potentiometrically titrated with 0.05 *N* NaOH. The biomass titration data are shown in Figure 17.1. It would seem that the *P. aeruginosa* cell wall has two main functions. To distinguish the different groups of binding sites, IR analysis of the lyophilized biomass was used.

In Figure 17.2, the pattern for biosorbent revealed a complex and additive impact of chemical texture. The amines group presents relative transmittance at 3500 to 3300 cm⁻¹ (N–H stretching) and at 1140 to 1080 cm⁻¹ (C–N stretching). The N–H stretching peak lies in a spectrum region occupied by a broad and strong band (3600 to 3300 cm⁻¹). This may be due to hydroxyl groups that are hydrogen bonded to various degrees. The presence of intense OH peaks in the spectrum could be due to the great water content of lyophilized biomass and to the real presence of hydroxyl groups in the biomass.

The C–N stretching peak is also covered by another strong band (1125 to 1090 cm⁻¹) that can be attributed to an alcoholic C–O stretching. The amide group presents relative transmittance at 1490 to 1440 cm⁻¹ (N–H stretching) and at 1661 cm⁻¹ (C–O stretching) at 1550 cm⁻¹ (C–N–H stretching). Again, the C–N stretching peak is covered, while a peak at 1414 cm⁻¹ can be assigned

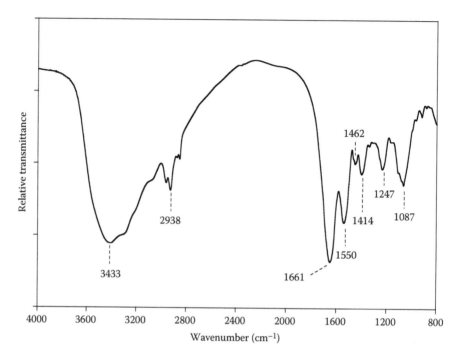

FIGURE 17.2 IR spectrum in solid phase of the lyophilized *P. aeruginosa* in KBr disk.

to O–H stretching of the acidic group. The carboxyl group presents some relative transmittance peaks (O–H stretching at 3100 to 2900cm⁻¹ and at 1414 cm⁻¹; C–O stretching at 1320 to 1211 cm⁻¹).

17.2.2 MECHANISMS OF BIOSORPTION

According to the location where metal biosorption occurs, the mechanisms of biosorption are classified as the following processes:

- Intracellular interaction
- Cell surface interaction
- Extracellular interaction

Figure 17.3 schematically summarizes alternative process pathways to remove heavy metals in trace elemental environments. The biosorption mechanisms of heavy metals show that the heavy metals are adsorbed by physicochemical interactions between metal ions and the bacterial surface. Biosorption is mainly the passive interaction of metals independent of metabolisms. Nutrients are not supported for continuing the bacterial activity and dead cells as well as living cells are effectively used in the removal of heavy metals.

17.2.2.1 Intracellular Interaction

Active transport of the metal across the cell membrane yields intracellular accumulation, which depends on the bacterial metabolism [33]. Essential metals are actively taken up by specialized uptake systems because they are needed, but other, nonessential metals may also be taken up because they are mistaken for an essential metal [34]. In high concentration of toxic metals, microorganisms actively take up the metal ions to detoxify the surrounding environment. Actually, a variety of bacteria is capable of converting metal and metalloid ions to organometallic and organometalloid compounds by intracellular ligands such as metallothioneins [35]. The pathway

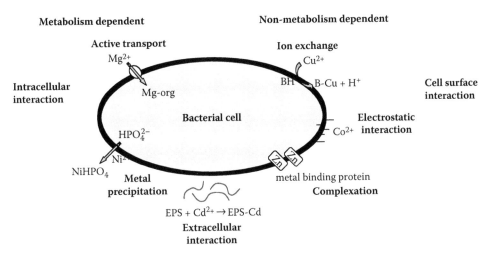

FIGURE 17.3 Schematic diagram showing the mechanisms of bacterial biosorption.

of the formation of organometallic and organometalloid compounds has been comprehensively reviewed [36,37].

The microorganism also has a metabolically sponsored process such as bioprecipitation to enhance the metal uptake [38,39]. The insoluble metal compound precipitates as metal ions combined with anoic species produced by cell metabolism [37]. For instance, Basnakova and Macaskie [40] reported that *Citrobacter* sp. could accumulate high levels of uranium, nickel, and zirconium through the formation of metal phosphate precipitates.

17.2.2.2 Cell Surface Interaction

In the case of physicochemical interaction based on physical adsorption, ion exchange, and complexation between the metal and functional groups of the cell surface, metal uptake does not depend on the metabolism [1,41]. The mechanisms by which metal binds onto the cell surface most likely include electrostatic interactions, van der Waals forces, covalent bonding, or some combination of these processes [27,28]. These passive parts showed a rapid initial uptake and surface-mediated mechanism.

Electrostatic interactions have been demonstrated to be responsible for cobalt biosorption by main algae by Kuyucak and Volesky [42]. The negatively charged groups, such as carboxyl, hydroxyl, and phosphoryl groups of the bacterial cell wall, adsorb metal cations by electrostatic forces. The ion exchange is related to cellular metal accumulation because cell walls of microorganisms contain polysaccharides as basic building [33]. Cell-based on their displacement by ion exchange, the following ascending order of light metal affinity toward biomass was observed: $Na^+<K^+ < Mg^{2+} < Ca^{2+}$ [43].

The metal sorption may also take place through complex formation on the cell surface by interaction between metals and metal-binding proteins of the organism [19]. Uranium and thorium biosorption onto *Rhzopus arrhizus* has a mechanism based not only on physical adsorption but also on complexation on the cell wall network [44].

17.2.2.3 Extracellular Interaction

As has already been pointed out, some bacteria can produce large quantities of extracellular polymeric substances (EPS), including negatively functional groups [22,45]. The EPS can bind and accumulate cation heavy metals such as magnesium of cadmium. Recent studies from Loaëc et al. [46] showed that the polymer from *Alteromonas macleodii* possessed affinity for lead, cadmium, and zinc. Lead was preferentially absorbed, but between zinc and cadmium competed for the same binding site. In a study using extracted EPS from activated sludge, Liu et al. [47] demonstrated that the metal capacity had 1.48 mg of Zn^{2+}; 1.12 mg of Cu^{2+}; 0.83 mg of Cr^{3+}; 0.90 mg of Cd^{2+}; 1.10 mg of Co^{2+}; and 0.25 mg each of Ni^{2+} and CrO_4^{2-} by polymers.

17.3 METAL REMOVAL BY BACTERIAL BIOSORPTION

17.3.1 BIOSORPTION ISOTHERMS

Sorption phenomena can be quantified and evaluated by fitting experimental data to one of several sorption isotherms. The adsorption isotherms describe the relation between the activity or equilibrium concentration of the adsorptive and the quantity of adsorbate on the surface of adsorbent at constant temperature. The most widely used sorption models are the Langmuir isotherm and the Freundlich isotherm. Although they are rather simplistic when applied to biological systems, mathematical sorption isotherms can be used to understand the surface behavior of biosorbent, mechanisms of biosorption, and distribution of metal ions between the liquid and solid phases [12,14].

The Langmuir adsorption isotherm equation is shown in the following equation [48]:

$$\Gamma = \Gamma_{max} \frac{K_{ads}C_e}{1 + K_{ads}C_e} \tag{17.1}$$

where

Γ is the amount of adsorbed metal ion per wet mass of resin (μmol/g)
Γ_{max} is the maximum adsorption capacity of metal ion (μmol/g)
C_e is the equilibrium concentration of adsorbate in solution (μmol/L)
K_{ads} is the equilibrium adsorption constant (L/μmol)

The parameter represents the uptake capacity when the surface is completely covered with metal ions and is an indication of the biosorbent maximum uptake capacity, Γ_{max}. Constant K_{ads} is related to adsorption energy, reflecting quantitatively the affinity between the biosorbent and the metal ion [49].

The Freundlich isotherm suggests a concentration-dependent increase of metal sorption onto the adsorbent. It is based on a heterogeneous distribution of active sites as well as on the interaction between sorbed metals.[50]. One of the major disadvantages of the Freundlich equation is that it does not predict an adsorption maximum.

The Freundlich isotherm can be explained by the following equation [48]:

$$\Gamma = K_d C_e^{1/n} \tag{17.2}$$

where

K_d is the distribution constant
n is the Freundlich exponent known as adsorption intensity

To evaluate the sorption capacity and to understand the pattern of metal biosorption on *P. aeruginosa*, the experimental metal biosorption isotherms were obtained (Figure 17.4). Metal sorption studies carried out at varying initial metal concentrations revealed that specific metal uptake increased with increase in initial metal concentration. Equilibrium sorption isotherm studies showed

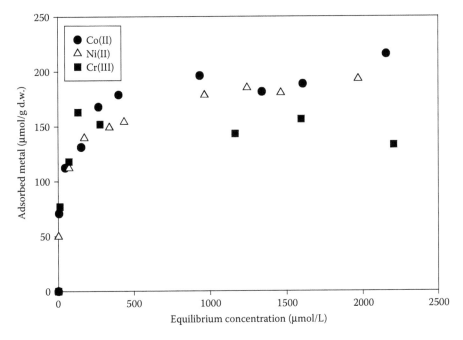

FIGURE 17.4 Biosorption isotherms of Co^{2+}, Ni^{2+}, and Cr^{3+} onto *P. aeruginosa*. The biomass was contacted with metal solution for 10 h at 25°C and 180 rpm in shaking incubator.

that metal uptake by *P. aeruginosa* was a chemically equilibrated and saturated mechanism. When experimental data were applied to adsorption models such as Langmuir and Freundlich, the data were found to fit in the Langmuir isotherm model reasonably. From the result, the maximum amounts of metals taken up by *P. aeruginosa* were 188.7 μmol Co^{2+}/g dry weight; 166.7 μmol Ni^{2+}/g dry weight; and 149.3 μmol Cr^{3+}/g dry weight.

17.3.2 COMPARING METAL CAPACITY

Uptake of toxic metal ions may contribute to the detoxification of polluted environments. Therefore, the investigation of the metal capacity of bacteria is fundamental for the field application of biosorption because it gives information about the removal efficiency of metal ions in the process [33]. As a necessary factor for the design of equipment, the metal capacity of bacteria is usually used by the parameter Γ_{max}. Table 17.1 shows the maximum metal capacities reported in literature by bacteria.

17.3.3 INFLUENCE OF ENVIRONMENTAL CONDITIONS

In the biosorption process, many environmental factors can influence the metal capacity of bacteria due to the change of bacterial surface properties and the characteristics of metal-bearing streams such as a variable pH and competing ions.

17.3.3.1 Bacterial Growth Phase

The physiological changes between the exponential and stationary phases are often reported to be significant so that cells in the stationary phase have distinct characteristics. When *Arthrobacter*, for example, reached the stationary phase, t its factor changed from rod to coccoid [26]. Thus, the cell growth phase can be an important factor that affects metal sorption.

TABLE 17.1
Comparison of Maximum Metal Capacities (Γ_{max}) by Bacterial Biosorption[a]

		Metal biosorption		
Metal	**Bacteria**	**Γ_{max}**	**pH**	**Ref.**
Ag	*Streptomyces noursei*	38.6	6	Mattuschaka and Straube, 1993
Al	*Chryseomonas luteola*	55.3	5	Ozdemir and Baysal, 2004
Cu	*Arthrobacter* sp.	148	6	Veglió et al., 1997
	Arthrobacter sp.	6.6	4.5	Pagnanelli et al., 2000
	Brevibacterium sp.	34.3	6.3	Vecchio et al., 1998
	Pseudomonas aeruginosa	79.5	5.5	Chang et al., 1997
	Pseudomonas putida	6.6	6	Pardo et al., 2003
	Thiobacillus ferrooxidans	23.12	5	Ruiz-Manriquez et al., 1998
	Zoogloea ramigera	270	5.5	Norberg and Persson, 1984
	Zoogloea ramigera	29	4	Aksu et al., 1992
	Zoogloea ramigera	34.05	4	Sag and Kutsal, 1995
	Streptomyces noursei	9	5.5	Mattuschaka and Straube, 1993
Cd	*Arthrobacter* sp.	13.4	6	Pagnanelli et al., 2000
	Bacillus laterosporus	159.5	7	Zouboulis et al., 2004
	Bacillus simplex	1.8	6	Valentine et al., 1996
	Brevibacterium sp.	15.7	6.3	Vecchio et al., 1998
	Pseudomonas aeruginosa	42.4	6	Chang et al., 1997
	Pseudomonas cepacia	130	7.4	Savvaidis et al., 1992
	Pseudomonas putida	8	6	Pardo et al., 2003
	Streptomyces noursei	3.4	6	Mattuschaka and Straube, 1993
Cr(III)	*Pseudomonas aeruginosa*	7.7	4	In this study
	Streptomyces noursei	10.6	5.5	Mattuschaka and Straube, 1993
Cr(VI)	*Bacillus laterosporus*	72.6	7	Zouboulis et al., 2004
	Bacillus licheniformis	62	7	Zouboulis et al., 2004
	Bacillus simplex	0.4	6	Valentine et al., 1996
	Chryseomonas luteola	3	5	Ozdemir and Baysal, 2004
Co	*Pseudomonas aeruginosa*	11.1	4	In this study
Hg	*Pseudomonas aeruginosa*	180	7.4	Chang and Hong, 1994
Mg	*Arthrobacter* sp.	406	5.5	Veglió et al., 1997
Ni	*Arthrobacter* sp.	12.7	6	Veglió et al., 1997
	Bacillus simplex	0.8	6	Valentine et al., 1996
	Pseudomonas aeruginosa	9.8	4	In this study
	Zoogloea ramigera	57.43	4.5	Sa and Kutsal, 1995
Pb	*Arthrobacter* sp.	130	5	Veglió et al., 1997
	Streptomyces noursei	36.5	6.1	Mattuschka and Straube, 1993
	Zoogloea ramigera	81.23	4.5	Sa and Kutsal, 1995
	Brevibacterium sp.	74.6	6.3	Vecchio et al., 1998
	Pseudomonas putida	56.2	6.5	Pardo et al., 2003
	Pseudomonas aeruginosa	79.5	5.5	Chang et al., 1997
Zn	*Pseudomonas cepacia*	200	7.4	Savvaidis et al., 1992
	Streptomyces noursei	1.6	5.8	Mattuschaka and Straube, 1993
	Thiobacillus ferrooxidans	9.7	4.5	Celaya et al., 2000
	Pseudomonas putida	6.9	7	Pardo et al., 2003

[a]Milligrams per gram of dry weight sorbent.

In the case of *P. aeruginosa*, the amount of cobalt did not change substantially with the cell growth phase. The quantities of cobalt taken up by the cells were 180.3 μmol/g dry cells at midexponential phase and 178.9 μmol/g dry cells at stationary phase, respectively (no data shown). From the result, the sorption of cobalt by *P. aeruginosa* appeared to be independent of the cell growth phase. Chang et al. [51] reported an increase in sorption of lead and cadmium by *P. aeruginosa* with increasing culture age; however, copper uptake was also found to be independent of growth phase.

17.3.3.2 pH

Numerous studies show that the biosorption of heavy metal from aqueous solution depends strongly on pH. Rao et al. [52] studied Cu^{2+} biosorption by *G. lucidum* and *A. niger* at initial copper concentration of 0.5 m*M* and found that the metal binding had an increasing trend from pH 2 to 6, with the maximum occurring between pH 5 and 6.

On the other hand, for some biosorbents, pH plays a different role in biosorption when the initial ion concentrations are different. Ke et al. [53] reported results for the biosorption of Ag^+ by using *Datura* cells. The binding was pH independent when the initial concentration was 0.1 m*M*, but it became strongly pH dependent when the initial concentration increased to 1 m*M*. These investigations suggested that at least two binding sites are involved: one site is pH independent and displays a greater affinity and lower availability than the other site, which is pH dependent.

Figure 17.5 shows that the adsorption of metal ions by the *P. aeruginosa* depended highly on solution pH. The adsorption percentage of metals in the lower pH levels (e.g., pH 2 or 3) was significantly low due to competition with the H^+ ions for binding sites on the surface of bacteria; the increase in pH favored metal sorption mainly because of the elevated levels of negatively charged groups on the cell surface.

The negatively charged groups existing on the surface of the microbial cell wall may undergo protonation at low pH, leading to an increase in the positive charge density on the cell surface. Such a process results in a competition between H^+ and cationic ions for the same binding sites, as suggested by Hu et al. [32]. These coworkers noted that an increased H^+ concentration resulted in suppressed uranium uptake by *P. aeruginosa* CSU biomass. The metal biosorption capacities declined substantially under lower pH condition. They concluded that the reduction in uranium-loading capacities with decreasing pH may be due to protonation of the cell wall, high solubility of uranyl ions, and cell structure damage at very acidic conditions.

17.3.3.3 Competing Cations

Most studies on biosorption using microorganisms have involved the removal of only one kind of metal ion from aqueous solutions. However, the presence of only one kind of heavy metal is a rare situation in nature or in wastewater.

To study the competitive biosorption behaviors between trivalent chromium and bivalent cobalt and nickel ions using *P. aeruginosa*, mixed metal solution was used in a batch sorption system (Figure 17.6). In a ternary system, cobalt (−93%) and nickel (−91%) uptake was strongly affected by the presence of chromium when compared with uptake in a single system. The decrease in uptake of chromium ion was negligible in the presence of bivalent ions.

At pH 4 of mixture solutions, cobalt, nickel, and chromium ions are present in the positively charged forms. In this state, they can compete with each other for negatively charged surfaces of the biomass. It is well documented that the ionic charge and ionic radii of cations affect the ion exchange as well as adsorption phenomena [42]. Because monovalent or bivalent ionic species are sorbed to a lesser extent than polyvalent ones, cobalt and nickel do not suppress the prevailing affinity of chromium to the binding sites of *P. aeruginosa*.

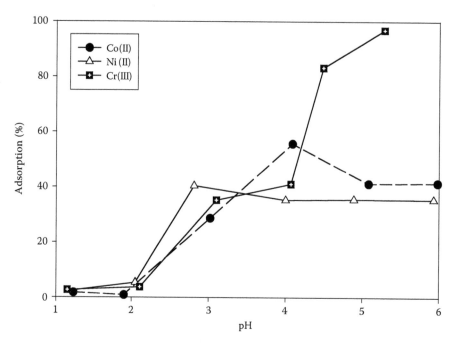

FIGURE 17.5 Effect of initial pH on biosorption of Co(II), Ni(II), and Cr(III) by *P. aeruginosa*. The effects of solution pH on heavy metal sorption were studied by adjusting the initial pH of the metal solution (50 mg/L) over the range of pH 1 to 6. Solution pH was adjusted with NaOH or HNO_3.

17.3.3.4 Ionic Strength and Organics

Ionic strength also plays an important role in the metal biosorption. Chang and Hong [31] reported that the mercury uptake by *P. aeruginosa* decreased with increasing ionic strength. Cho et al. [54] showed that no significant decrease in the binding of Cd^{2+} and Zn^{2+} occurred up to the ionic strength of 10^{-3} *M*.

In natural systems, the metal removal can be affected by the presence of other organics. The presence of organic and inorganic ligands that act as chelators is of particular concern [55]. Such ligands may compete with the microorganisms for heavy metals, and once metal–ligand complexes are formed, they may not be adsorbed by the cells. Metal uptake will be most affected if the binding constants for the ligands are greater than those for the bacteria. In studies using *Thiothrix* species strain to adsorb copper, nickel, and zinc, the bacteria did not bind copper chelated to ethylenediaminetetraacetic acid and nitrilotriacetic acid [56].

17.4 APPLICATION AND POTENTIAL BENEFITS IN METAL-CONTAMINATED ENVIRONMENTS

The main target of the biosorption process is to remove heavy metal, which can be quite toxic even at low concentrations. Biosorption is particularly suited as a polishing step in wastewater with a low to medium initial metal concentration from a few to about 100 ppm [57]. In high initial metal concentrations, biosorptive treatment of wastewaters may be economically applied by combining the pretreatment using other technologies such as precipitation, which is currently used for 90% of heavy metal removal or electrolyte recovery. As a result of the biosorptive process, it offers high effluent quality and avoids the generation of toxic sludge. Figure 17.7 shows several designs utilizing biosorption in the contaminated sites.

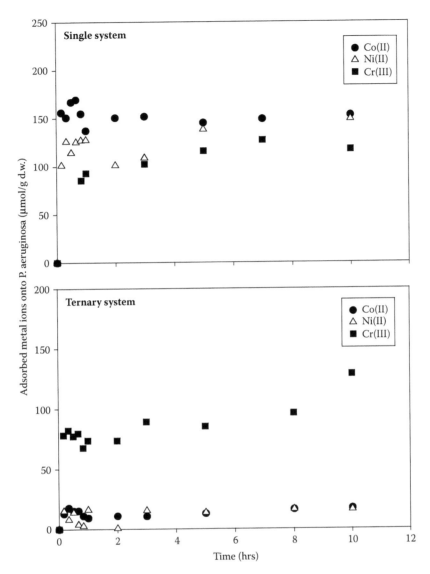

FIGURE 17.6 Adsorption equilibrium for each metal ion in synthesized wastewaters. The solution pH was adjusted to pH 4.

17.4.1 WASTEWATER TREATMENT

Recent research demonstrated that bacteria can enable recovery of valuable elements or further containment of highly toxic or radioactive species [15,20,58]. It is clear that some microbiological methods for the treatment of metal-containing wastes offer potentially efficient and cost-effective alternatives [59]. Actually, the high removal efficiency (almost 99%) of heavy metals could be obtained by using bacteria in batch system containing mixture metal solution (Figure 17.8). To increase the removal efficiency of heavy metal, the sequent bioreactor can be conducted in bacterial biosorption or adjunct to existing treatment technologies.

Contacting large volumes of metal-bearing aqueous solutions with microbial biomass in conventional unit processing operation is not typical because of the solid and liquid separation problems. Recent cell immobilization technology is therefore often studied for its potential to improve

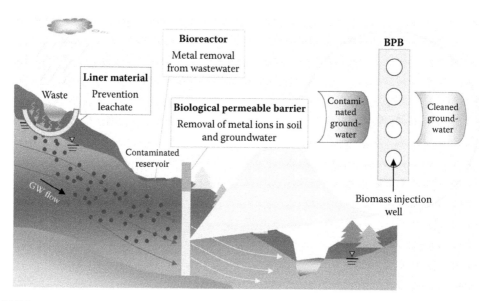

FIGURE 17.7 Bioremediation in contaminated environmental sites using biosorption.

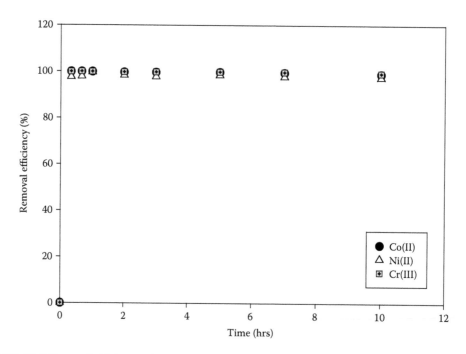

FIGURE 17.8 Removal efficiency of each metal ion in ternary system at pH 4. The initial concentrations of Co, Ni, and Cr were 1 mg/L.

fermentation productivity [21]. The bacterial immobilization can provide remarkable stability and prevent the loss of bacteria. A packed bed or fluidized bed reactor containing the immobilized biomass also can effectively remove contaminants using a much smaller reactor than biosorption processes containing free cells.

17.4.2 Groundwater Treatment

The biosorption process can be applied *in situ* without the expense of pumping out the contaminated groundwater or excavating the soil. This technique provides low-cost, easy operating, and safe treatment of contaminants in groundwater. The immobilized microbial stratum may be placed in an engineered trench across the flow path of a contaminated plume to create a BPB (biological permeable barrier). Contaminated groundwater enters the BPB, to which electron donor and nutrients may be supplied through the groundwater gradient, while the remediated groundwater exits the BPB. Selective removal has the potential and flexibility to treat a wide range of wastewater contaminants beyond nitrates. The most relevant work on true bacterial biosorption has been done by Brierley [60]. He found that the biosorption process was useful for cost-effective treatment of high-volume, low-concentration wastes such as mine runoff using artificial wetlands with undefined biota.

17.4.3 Protection from Pollutant Plumes

The migration of contaminants from a hazardous site is a concern for the protection of downstream resources. Biobarriers serve as an alternative technology for controlling the migration of contaminants from hazardous waste sites. The biobarrier can be applied in the field by injecting starved bacteria and then nutrients into a series of injection wells. The pore space is sealed by bacterial growth and EPS production and then a biobarrier is formed in soil [61]. The biobarrier has applicability as an alternative liner material in landfills to the contaminated sites. It is able to immobilize heavy metals *in situ*, thus protecting environments from the hazardous lechate.

ACKNOWLEDGMENT

This research was supported by the Gwangju Institute of Science and Technology (GIST) Research Fund and National Research Laboratory Project (Arsenic Geoenvironment Lab.) to KWKIM.

REFERENCES

1. Gadd, G.M., Metals and microorganisms, *FEMS Microbiol. Lett.,* 100, 197, 1992.
2. Levi, P. and Linkletter, A., Metals, microorganisms and biotechnology, in *Metals and Microorganisms*, Hughes, M.N. and Poole, R.K., Ed., Chapman & Hall Ltd, New York, 1989, 303.
3. Gadd, G.M. and Sayer, J.A., Influence of fungi on the environmental mobility of metals and metalloids, in *Environmental Microbe–Metal Interaction*, ASM Press, Washington, D.C., 2000, 237.
4. Clarkson, T.W., Metal toxicity in the central nervous system, *Environ. Health Perspect.,* 75, 59, 1987.
5. Adriano, D.C., *Trace Elements in Terrestrial Environments: Biogeochemistry, Bioavailability, and Risks of Metals*, Springer–Verlag, New York, 2001.
6. Rajapaksha, R.M., Tobor–Kaplon, M.A., and Baath, E., Metal toxicity affects fungal and bacterial activities in soil differently, *Appl. Environ. Microbiol.,* 70, 2966, 2004.
7. Ehrlich, H.L., Microbes and metals, *Appl. Microbiol. Biotechnol.,* 48, 687, 1997.
8. USEPA, Sediment-based remediation of hazardous substances at a contaminated military base, no. R825512C014, 1996.
9. Cheremisinoff, P.N., *Handbook of Water and Wastewater Treatment Technology*, Marcel Dekker, Inc., New York, 1995.
10. Reynolds, T.D. and Richards, P.A., *Unit Operations and Processes in Environmental Engineering*, PWS Pub. Co., Boston, 1996.
11. Mullen, M.D. et al., Bacterial sorption of heavy metals, *Appl. Environ. Microbiol.,* 55, 3143, 1989.
12. Volesky, B., Removal and recovery of heavy metals by biosorption, in *Biosorption of Heavy Metals*, Volesky, B., Ed., CRC Press, Boca Raton, FL, 1990, 7.
13. Lehmann, M., Zouboulis, A.I., and Matis, K.A., Removal of metal ions from dilute aqueous solutions: a comparative study of inorganic sorbent materials, *Chemosphere,* 39, 881, 1999.

14. Gadd, G.M., Biosorption, *J. Chem. Technol. Biotechnol.*, 55, 302, 1992.

15. Volesky, B. and Holan, Z.R., Biosorption of heavy metals, *Biotechnol. Prog.*, 11, 235, 1995.

16. Beveridge, T.J., Role of cellular design in bacterial metal accumulation and mineralization, *Annu. Rev. Microbiol.*, 43, 147, 1989.

17. Nourbakhsh, M.N. et al., Biosorption of Cr^{6+}, Pb^{2+}, and Cu^{2+} ions in industrial waste water on *Bacillus* sp., *Chem. Eng. J.*, 85, 351, 2002.

18. Kang, S.Y., Lee, J.U., and Kim, K.W., Selective biosorption of chromium(III) from wastewater by *Pseudomonas aeruginosa*, the 227th American Chemical Society National Meeting, Anaheim Division of Environmental Chemistry, 2004, ENVR 91.

19. White, C., Wilkinson, S.C., and Gadd, G.M., The role of microorganisms in biosorption of toxic metals and radionuclides, *Int. Biodeterior. Biodegr.*, 35, 17, 1995.

20. Lovley, D.R. and Lloyd, R., Microbes with a mettle for bioremediation, *Nat. Biotechnol.*, 18, 600, 2000.

21. Schiewer, S. and Volesky, B., Biosorption process for heavy metal removal, in *Environmental Microbe–Metal Interactions*, Lovley, D.R., Ed., ASM Press, Washington, D.C., 2000, 329.

22. Marquès, A.M. et al., Uranium accumulation by *Pseudomonas* sp. EPS-5028, *Appl. Microbiol. Biotechnol.*, 35, 406, 1991.

23. Valentine, N.B. et al., Biosorption of cadmium, cobalt, nickel, and strontium by *Bacillus simplex* strain isolated from the vadose zone, *J. Ind. Microbiol.*, 16, 189, 1996.

24. Texier, A.C., Andrès, Y., and Le Cloirec, P., Selecive biosorption of lanthanide (La, Eu, Yb) ions by *Pseudomonas aeruginosa*, *Environ. Sci. Technol.*, 33, 489, 1999.

25. Lloyd, R. and Lovley, D.R. Microbial detoxification of metals and radionuclides, *Environ. Biotechnol.*, 12, 248, 2001.

26. Atlas, R.M., *Principles of Microbiology*, Mosby-Year Book, Inc., 1995.

27. Flemming, H.C., Sorption sites in biofilms, *Water Sci. Tech.*, 32, 27, 1995.

28. Beveridge, T.J. and Fyfe, W.S., Metal fixation by bacterial cell walls, *Can. J. Earth Sci.*, 22, 1893, 1985.

29. Wingender, J., Neu, T.R., and Flemming, H.C., What are bacterial extracellular polymeric substances? in *Microbial Extracellular Polymeric Substances*, Wingender, J., Neu, T.R., and Flemming, H.-C., Eds., Springer–Verlag, Berlin, Germany, 1999, 1.

30. Langley, S. and Beveridge, T.J., Metal binding by *Pseudomonas aeruginosa* PAO1 is influenced by growth of the cells as a biofilm, *Can. J. Microbiol.*, 45, 616, 1999.

31. Chang, J.-S. and Hong, J., Biosorption of mercury by the inactivated cells of *Pseudomonas aeruginosa* PU21(Rip64), *Biotechnol. Bioeng.*, 44, 999, 1994.

32. Hu, M.Z.-C. et al., Biosorption of uranium by *Pseudomonas aeruginosa* strain CSU: characterization and comparison studies, *Biotechnol. Bioeng.*, 51, 237, 1996.

33. Vegliò, F. and Beolchini, F., Removal of metals by biosorption: a review, *Hydrometallurgy*, 44, 301, 1997.

34. Ledin, M., Accumulation of metals by microorganisms — processes and importance for soil systems, *Earth Sci. Rev.*, 51, 1, 2000.

35. Robinson, N.J., Whitehall, S.K., and Cavet, J.S., Microbial metallothioneins, *Adv. Microbiol. Physiol.*, 44, 183, 2001.

36. White, C., Sharmanl, A.K., and Gadd, G.M., An integrated microbial process for the bioremediation of soil contaminated with toxic metals, *Nat. Biotechnol.*, 16, 572, 1998.

37. Kotrba, P. and Ruml, T., Bioremediation of heavy metal pollution exploiting constituents, metabolites and metabolic pathways of livings, *Collect. Czech. Chem. Commun.*, 65, 1205, 2000.

38. Gadd, G.M., Bioremedial potential of microbial mechanisms of metal mobilization and immobilization, *Curr. Opin. Biotechnol.*, 11, 271, 2000.

39. Valls, M. and de Lorenzo, V., Exploiting the genetic and biochemical capacities of bacteria for the remediation of heavy metal pollution, *FEMS Microbiol. Rev.*, 26, 327, 2002.

40. Basnakova, G. and Macaskie, L.E., Accumulation of zirconium and nickel by *Citrobacter* sp., *J. Chem. Technol. Biotechnol.*, 74, 509, 1999.

41. Volesky, B., Detoxification of metal-bearing effluents: biosorption for the next century, *Hydrometallurgy*, 59, 203, 2001.

42. Kuyucak, N. and Volesky, B., Accumulation of cobalt by marine alga, *Biotechnol. Bioeng.*, 33, 809, 1989.

43. Lee, H.S. and Volesky, B., Interaction of light metals and protons with seaweed biosorbent, *Water Res.*, 31, 3082, 1997.

44. Tsezos, M. and Volesky, B., Biosorption of uranium and thorium, *Biotechnol. Bioeng.*, 23, 583, 1981.

45. Christensen, B.E., The role of extracellular polysaccharides in biofims, *J. Biotechnol.*, 10, 181, 1989.

46. Loaëc, M., Olier, R., and Guezennec, J.G., Uptake of lead, cadmium and zinc by a novel bacterial exopolysaccharide, *Water Res.*, 31, 1171, 1997.

47. Liu, Y., Lam, M.C., and Fang, H.H., Adsorption of heavy metals by EPS of activated sludge, *Water Sci. Technol.*, 43, 59, 2001.

48. Stumm, W. and Morgan J.J., *Aquatic Chemistry: Chemical Equilibria and Rates in Natural Waters*, Wiley Interscience Publication, New York, 1996, 521.

49. Holan, Z.R. and Volesky, B., Biosorption of lead and nickel by biomass of marine algae, *Biotechnol. Bioeng.*, 43, 1001, 1994.

50. Stumm, W., *Chemistry of the Solid–Water Interface*, John Wiley & Sons, Inc., 1996, 87.

51. Chang, J.-S., Law, R., and Chang, C.-C., Biosorption of lead, copper and cadmium by biomass of *Pseudomonas aeruginosa* PU21, *Water Res.*, 31, 1651, 1997.

52. Rao, C.R.N., Iyengar, L., and Venkobachar, C., Sorption of copper(II) from aqueous phase by waste biomass, *J. Environ. Eng.*, 119, 369, 1993.

53. Ke, H.Y. et al., Luminescence studies of metal ion-binding sites on *Datura innoxia* biomaterial, *Environ. Sci. Technol.*, 28, 586, 1994.

54. Cho, D.-Y. et al., Studies on the biosorption of heavy metals onto *Chlorella vulgaris*, *J. Environ. Sci. Health*, 29, 389, 1994.

55. Wang, L. et al., A combined bioprocess for integrated removal of copper and organic pollutant from copper-containing municipal wastewater, *J. Environ. Sci. Health A*, 39, 223, 2004.

56. Shuttleworth, K.L. and Unz, R.F., Sorption of heavy metals to the filamentous bacterium *Thiothrix* strain A1, *Appl. Environ. Microbiol.*, 59, 1274, 1993.

57. Volesky, B., Detoxification of metal-bearing effluents: biosorption for the next century, *Hydrometallurgy*, 59, 203, 2001.

58. Kang, S.Y., Lee, J.U., and Kim K.W., A study of the biosorption characteristics of Co^{2+} in wastewater using *Pseudomonas aeruginosa*, *Key Eng. Mate.s*, in press, 2004.

59. Mattuschaka, B. and Straube, G., Biosorption of metals by a waste biomass, *J. Chem. Tech. Biotechnol.*, 58, 57, 1993.

60. Ozdemir, G. and Baysal, S.H., Chromium and aluminum biosorption on *Chryseomonas luteola* TEM05, *Appl. Microbiol. Biotechnol.*, 64, 599, 2004.

61. Pagnanelli, F. et al., Biosorption of metal ions on *Arthrobacter sp.*: Biomass characterization and biosorption modeling, *Environ. Sci. Technol.*, 34, 2773, 2000.

62. Vecchio, A. et al., Heavy metal biosorption by bacterial cells, *Fresenius J. Anal. Chem.* 361, 338, 1998.

63. Pardo, R. et al., Biosorption of cadmium, copper, lead, and zinc by inactive biomass of *Pseudomonas putida*, *Anal. Bioanal. Chem.*, 376, 26, 2003.

64. Ruiz-Manriquez, A. et al., Biosorption of Cu by *Thiobacillu ferroxidans*, *Bioproc. Eng.*, 18, 113, 1998.

65. Norberg, A.B. and Persson, G., Accumulation of heavy metal ions by zoogloea ramigera, *Biotechnol. Bioeng.*, 26, 239, 1984.

66. Aksu, Z., Sag, Y., and Kutsal, T., The biosorption of copper(II) by *C. vularis* and *Z. ramigera*, *Environ. Technol.* 13, 579, 1992.

67. Sag, Y., and Kutsal, T., Biosorption of heavy metals by Zoogloea ramigera use of adsorption isotherms and a comparison of biosorption characteristics, *Chem. Eng. J.*, 60, 181, 1995.

68. Zouboulis, A.I., Loukidou, M.X., and Matis, K.A., Biosorption of toxic metals from aqueous solutions by bacteria strains isolated from metal-polluted soils, *Proc. Biochem.*, 39, 909, 2004.

69. Savvaidis, I. Hughes, M., and Poole, R., Differential pulse polarography: a method of directly measuring uptake of metal ions by live bacteria without separation of biomass and medium, *FEMS Microbiol. Lett.*, 92, 181, 1992.

70. Veglio, F., Beolchini, F., and Gasbarro, A., Biosorption of toxic metals: an equilibrium study using free cells of *Arthrobacter sp.*, *Process Biochem.*, 32, 99, 1997.

71. Celaya, R.J., et al., Biosorption of Zn(II) by *Thiobacillus ferroxidans*, *Bioproc. Eng.*, 22, 539, 2000.

72. White, C., Wilkinson, S.C., and Gadd, G.M., The role of microorganisms in biosorption of toxic metals and radionuclides, *Int. Biodeterior. Biodegr.*, 35, 17, 1995.

73. Brierley, C.L., Bioremediation of metal contaminated surfaces and groundwaters, *Geomicrobiol. J.*, 8, 201, 1990.
74. Costerton, J.W., Isolation of pollutants using a biobarrier technology, *20th Annu. RREL Res. Symp.*, EPA, Washington D.C., 1994.

18 Electroremediation of Heavy Metal-Contaminated Soils —Processes and Applications

Alexandra B. Ribeiro and Jose M. Rodríguez–Maroto

CONTENTS

18.1 GENERAL PRINCIPLE

Electrokinetic remediation uses an electric current density of the order of milliamps per square centimeter applied to the cross-sectional area of a soil mass between the electrodes, producing electric potential drops of the order of volts per centimeter. The main interest in electrokinetic soil remediation in environmental cleanup operations lies in an attempt to concentrate and confine contaminants close to an electrode and remove them if possible. Several authors [1–6] have critically reviewed the state of knowledge of this cleanup technology.

The general principle of the electrokinetic process is shown in Figure 18.1. Due to the electric field present when a low-level direct current (DC) is passed between a pair of electrodes placed in a system containing charged particles (e.g., moist contaminated soil), the pollutant species are driven towards one of the electrodes, from where they may be removed. Three main mechanisms are responsible for this movement: electromigration, electroosmosis, and electrophoresis. On inert electrodes, the electrode reactions can be controlled to produce H^+ ions at the anode and OH^- ions at the cathode. If pH in the soil is not controlled, an acid front will propagate into the soil pores from the anode, and an alkaline front will move out from the cathode. In Figure 18.1, the contaminated soil volume is separated from the electrolyte solutions by passive membranes to prevent the soil mixing with the electrolytes.

18.2 DEFINITIONS AND SOME ASPECTS RELATED TO THE TRANSPORT OF SPECIES

Electrokinetic extraction of contaminants from soil involves the simultaneous flows of fluid, electricity, and chemicals under the combined influences of hydraulic, electrical, and chemical gradients. Thus, coupled flows that occur as flows of one type are induced by driving forces or gradients of another type [7]. The net transport of contaminants depends on several complex interactions. Electrokinetic soil remediation invokes three main mass-transport phenomena: electromigration (migration of ionic and/or polar species); electroosmosis; and electrophoresis. Other types of mass transport, such as diffusion and hydraulic convection, are usually present in some extension.

18.2.1 ELECTROMIGRATION

Electromigration is the movement of ions under an applied electric field. It is the predominant transport mechanism in soils under an electric potential gradient [8], particularly when dealing with soluble charged species like the heavy metal cations Pb^{2+}, Cd^{2+}, Cu^{2+}, or Zn^{2+}. Here, the soil zeta

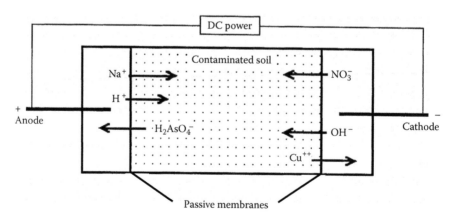

FIGURE 18.1 Schematic principle of the electrokinetic process.

potential may be absent or small [9]. Positive ions are driven towards the cathode and negative ions towards the anode. The electromigration transport is given by:

$$J_m = - u * c \, \Phi_e \tag{18.1}$$

where J_m is the migrational flux; $u*$ and c the ionic mobility and concentration of species; and Φ_e the gradient of electric potential.

The current efficiency of electromigration of a specific ionic species is expressed as the proportion of electrical charge carried by the species of interest, relative to the amount of charge carried by all charged species in solution [10].

18.2.2 ELECTROOSMOSIS

Electroosmosis describes the mass flux of pore fluid relative to soil particles under the influence of an imposed electric potential gradient. When an electrical field applies across a wet soil mass, cations are driven towards the cathode and anions towards the anode. As the ions migrate, they carry their water of hydration and they exert a viscous drag on the pore fluid around them [7].

Electromigration of the species within the diffuse double layer will induce an electroosmotic pore fluid transport towards the electrode polarized with a charge opposite that in the double layer. This transport is generally from the anode to the cathode because the species in the diffuse double layer are often positively charged [8]. However, it may reverse under certain circumstances; when the electrolyte concentration is high and the pH of the pore fluid is low, it is possible to reverse the polarity of the surface charge and initiate a reversed electroosmotic flow, from cathode to anode [4,11–13].

Electroosmosis is the major mechanism of removal of uncharged and/or weakly dissociated organic contaminants (phenols, for instance), where the soil has a finite zeta potential [14,15]. The electroosmotic flux, J_{eo}, is usually obtained from expressions such as:

$$J_{eo} = - k_e \, c \, \Phi_e \tag{18.2}$$

where k_e is the electroosmotic permeability of soil.

The electroosmotic component of transport will almost disappear in coarse sands and high-plasticity clays at low water contents (electromigration will dominate). In fine sands, silts, and low-activity clays at high water contents and low conductivities, electroosmotic transport would be as significant a transport mechanism as electromigration [8].

18.2.3 ELECTROPHORESIS

Electrophoresis is the movement of a charged colloid under an applied electric field. The charged particles are attracted to one of the electrodes and repelled from the other [7]. Negatively charged clay particles move towards the anode. Electrophoresis involves discrete particle transporting through a liquid, which, as a whole is at rest. This movement can be neglected — usually in systems where the solid phase is stationary, like soil systems [16]. However, in unconsolidated soils, the electrophoresis may play a role in the cementation of the soil; if the electric current is applied to slurry, the role of electrophoresis is significant [2].

18.2.4 DIFFUSION

Diffusion is the movement of species under a chemical concentration gradient. In free solutions, it is usually expressed by Fick's law and in porous media as well. However, in the latter case, the effective diffusion coefficient must be obtained correcting the diffusive coefficient, to take into account the porosity and the tortuosity effects, which can decrease this transport more than one order of magnitude. The diffusive flux in soils, J_d, can be calculated by:

$$J_d = -D^* \nabla c \tag{18.3}$$

where D^* is the effective diffusion coefficient and ∇c is the concentration gradient.

In general, in the usual conditions for the electrokinetic treatments, this is a secondary transport and can be important only in some areas of soil where gradients are especially high. The areas where acid and basic fronts, or metal cations and hydroxyl ions, are to meet can be an example.

18.2.5 ADVECTION BY HYDRAULIC GRADIENTS

In general, the transport under fluid advection generated by hydraulic gradients is not an important contribution to global transport. However, one of the applications of electrokinetic treatment is to act as a reactive barrier to avoid the advance of contamination into groundwater. Particularly in this case, the hydraulic gradients convert in an important driving force to the movement of water and, therefore, of the solved contaminants across the electrical barriers.

In any case, the mass flux from hydraulic gradients, J_h, can be calculated by:

$$J_h = -k_h c \nabla h \tag{18.4}$$

where k_h is the hydraulic conductivity of the soil and ∇h is the hydraulic gradient. Yeung [17] has analyzed the importance of coupling effects between hydraulic and electrical gradients of fluid flow quantity in fine-grained soils.

18.3 REACTIONS IN THE ELECTRODE COMPARTMENTS

The applied electric current normally leads to electrolysis of water at the electrodes, generating an acidic medium at the anode and an alkaline medium at the cathode. The water electrolysis reactions can be described by:

$$\text{Anode: } 2H_2O \rightarrow O_2\uparrow + 4H^+ + 4e^- \tag{18.5}$$

$$\text{Cathode: } 4H_2O + 4e^- \rightarrow 2H_2\uparrow + 4OH^- \tag{18.6}$$

However, when the concentration of ions in the electrolytes is increased, other electrode processes may take place at the surface of inert electrodes. Equation 18.7 and Equation 18.8 are examples in which the Me^{n+} is a metal ion with n positive charges.

$$\text{Cathode: } Me^{n+} + ne^- \rightarrow Me \tag{18.7}$$

$$\text{Cathode: } Me(OH)_n + ne^- \rightarrow Me + nOH^- \tag{18.8}$$

Equation 18.7 represents the deposition of the metal at the cathode surface. Metals can also precipitate as oxides, hydroxides, carbonates, and phosphates, depending on pH, the chemical constituents of the pore water, and current flow [18].

If chlorides occur in the solution, chlorine gas can be produced:

$$\text{Anode: } 2Cl^- \longrightarrow Cl_2\uparrow + 2e^- \tag{18.9}$$

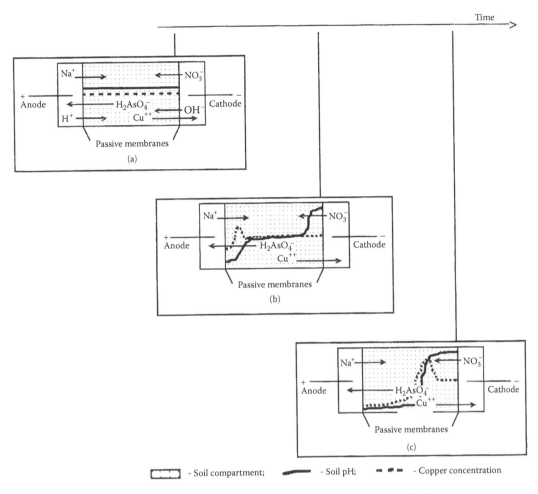

FIGURE 18.2 Soil pH and Cu^{2+} concentration profiles during electrokinetic remediation.

The reactions in Equation 18.5 and Equation 18.6 are the primary causes of the pH changes in the soil system chemistry during electrokinetics, when the technique is applied without conditioning the process fluid at the electrodes (unenhanced electrokinetic remediation). The development of the acid and the alkaline fronts can have a significant effect on the magnitude of electroosmosis, as well as on solubility, ionic state and charge, and level of adsorption of the contaminants [19].

Figure 18.2 shows schematically the soil pH and the Cu^{2+} concentration ($[Cu^{2+}]$) profiles that tend to be created during the electrokinetic soil remediation process [10]. In this figure:

- Before applying the current, the soil pH and the $[Cu^{2+}]$ are constant through the whole soil compartment. Then, when a dc current starts to pass, the soluble positive ions start to be driven towards the cathode and the negative ions towards the anode. At the electrodes, H^+ ions or OH^- ions will be created and will also move towards the opposite electrode. If $NaNO_3$ is used as electrolyte solution, Na^+ and NO_3^- will be present and will cross the barriers (passive membranes), starting their movement in the direction of one of the two electrode compartments.
- As time passes, the H^+ ions produced at the anode induce the propagation of an acid front in the soil, and the OH^- ions at the cathode induce an alkaline front. Due to the decrease in the soil pH near the anode compartment, a solubilization of Cu ions is promoted. These ions will move towards the cathode. A decrease in $[Cu^{2+}]$ takes place

near the passive membrane, which separates the soil from the anode compartment, with a concomitant increase of $[Cu^{2+}]$ a bit further into the soil. Due to the high soil pH near the cathode compartment, a precipitation of Cu will also happen in the soil, and Cu^{2+} does not move into this electrode compartment.

- Soil pH and $[Cu^{2+}]$ profiles reflect the propagation of the H^+ ions from anode compartment into the soil and of the OH^- ions from the cathode compartment. Because the ionic mobility of H^+ is 1.8 times higher than the mobility of OH^-, the pH fronts meet in the soil at approximately two-thirds of the distance from the anode compartment. Until this distance, the $[Cu^{2+}]$ is low, reflecting the solubilization/desorption and movement to which the Cu ions were submitted. A peak of $[Cu^{2+}]$ develops in the soil, due to the soil pH change, which may induce precipitation of the Cu ions when their hydroxide solubility is exceeded.

18.4 OTHER TYPES OF PHENOMENA RELATED TO MOBILITY OF SPECIES

Contaminants can undergo several types of phenomena, which can be divided into two classes [16]: those that would occur in the absence of the applied electric field, such as

- Ion exchange
- Adsorption–desorption of the contaminant from soil surfaces
- Precipitation–dissolution reactions
- Interactions between soluble chemical constituents in the pore water

The second class comprises those that occur because of the electric field, such as (1) plating of a metal on the electrode(s); and (2) electrolysis of water in the electrode compartments.

In the case of the second class, plating of a metal can be minimized by proper selection of the electrode. The nature of electrode reactions during electrokinetic treatment of soil depends on the material of the electrodes and the type and concentration of chemical species in the fluid around the electrodes [3]. For electrokinetic remediation purposes, inert electrodes (i.e., electrode material such as carbon, platinum, or titanium, which do not take part in the reaction) are mostly used (see Section 18.12).

18.4.1 ION EXCHANGE AND SORPTION

The soil particles and specialty clays have an active surface, which can interact with contaminants to their sorption or ion exchange. The large surface of some types of clays such as montmorillonite or illite increases this capacity of retention and the negative charges present usually in their structures attract cations, making their movement difficult. In this sense, the presence of other nontoxic ions (H^+, Na^+, etc.) competing for the same fixation sites, promoting their desorption, can help to increase the mobility of the contaminants.

On the other hand, when sorption/desorption is evaluated in soils, equilibrium isotherms are commonly used, assuming that the contaminants sorbed reach instantaneous equilibrium at all times. However, this assumption is only valid if movement of species is slow enough, but usually, in electrokinetic remediation, the migration velocity of species is high and therefore the kinetics of the sorption process is relevant to evaluate its effects on remediation.

18.4.2 PRECIPITATION AND DISSOLUTION

A great variety of natural species is in a solid phase or in pore aqueous solution (e.g., CO_3^{2-}, SO_4^{2-}, S^{2-}) in the natural soils and, if the contaminants react with them, precipitation can occur. Moreover, the acid or basic fronts or other ions generated or introduced in the electrode compartments can also produce precipitation or dissolution of the contaminants. Thus, the natural buffer capacity of the soils to remediate, as well as the ions proceeding of their own electrokinetic process, must be considered in order to know the effect on dissolution/precipitation and therefore on the contaminants' mobility and soil electric conductivity.

18.4.3. MOVEMENT OF AN ACID FRONT AND AN ALKALINE FRONT IN THE SOIL COMPARTMENT

The acid (H^+ ions) generated at the anode advances through the soil towards the cathode by [11,20]:

- Ion migration due to electrical gradient
- Pore fluid advection due to electroosmotic flow
- Pore fluid flow due to any externally applied or internally generated hydraulic potential difference
- Diffusion due to the chemical gradients developed

The alkaline medium developed at the cathode first advances towards the anode by ionic migration and diffusion. However, the transport of OH^- in the soil is overshadowed by any electroosmotic advection and neutralization by the H^+ ions transported to this zone (generation of water takes place within this zone) [4,11].

According to Acar and Alshawabkeh [20], the extent of the "meeting zone" is controlled by the dissolution/precipitation chemistry of the species available in the soil; their interaction with the soil and with the hydrogen/hydroxide ions; and the electrochemistry of the species produced and/or injected in the electrode compartments. In fact, the acid–alkaline fronts that meet close to the cathode compartment within the soil, coupled with precipitation of hydroxide species in the high pH zone, result in the development of a low conductivity region close to the cathode.

18.4.4 MOVEMENT OF CONTAMINANTS

As a consequence of different phenomena previously explained, acidification of the soil facilitates desorption of the contaminants. The driving mechanisms for the transport of the species in the soil are the same as those for the acid/base transport. As a result, cations tend to accumulate at the cathode and anions at the anode, and transfer of H^+ and OH^- ions across the medium is continuous. Extraction and removal are accomplished by electrodeposition, precipitation, or ion exchange at the electrodes or in an external extraction system placed in a unit cycling the processing fluid [11,21,22].

Extraction of contaminants by electrokinetic methods is based on the assumption that the contaminant is in the liquid phase in the soil pores [23] or in exchangeable positions in the solid phase, in contact with the previous one. Electroosmotic advection should be able to transport nonionic as well as ionic species through the soil towards the cathode. This is, perhaps, best achieved when the state of the material (dissolved, suspended, emulsified, etc.) is suitable for the flowing water to carry it through the tight pores of soil without causing a plug of concentrated material to accumulate at some point in the soil [23].

Polar organic molecules, ionic micelles, and colloidal electrolytes should migrate under the influence of an electric field, as well as being transported by the water. The size of these molecules or micelles and their tendency to agglomerate or be adsorbed onto soil surfaces are probably the main factors that control their removal from soil pores by electrokinetics [23]. Removal of cationic species, on the other hand, should be enhanced by the electroosmotic flow of water as they migrate towards the cathode electrode compartment by the applied potential [23]. In noncompact soils, concurrent colloidal transport may be significant in altering the microstructure of the porous medium and in facilitating the transport of adsorbed metal ions and organics [24].

The concentration of the metal on the soil strongly influences the retention energy and the operative removal mechanism. At low concentrations of metals, removal is most likely due to desorption. Numerous sorption sites on soils have a wide range of binding energies. At low metal concentrations, the high energy-binding sites are occupied first. At higher metal concentrations, the high energy-binding sites are completely occupied and the lower energy-binding sites begin to fill, resulting in a decrease in the average metal–soil binding energy. As a result, it is easier to remove a given percentage of the metal from the more highly contaminated soil. However, even for the highly contaminated soil, it becomes progressively harder to remove the remaining metals (i.e., metals associated with the high energy-binding sites). On the other hand, the formation of insoluble compounds on the soil surface or in the pore liquid can occur if the concentration of the metal is high; if the soil pH is alkaline; or if anions such as Cl^- and SO_4^{2-} are present in sufficient quantity [16].

The relative magnitude of contribution of either process (electroosmosis and electromigration) to decontaminate under a given set of initial and boundary conditions is determined by the soil and contaminant type, as well as their interaction, is yet unclear. At low concentrations of cations, electroosmotic water flow may contribute a significant percentage to, if not be totally responsible for, the overall decontamination process. At high concentrations of ionic species, electrolytic migration and intensity of electrochemical reactions may play more important roles than electroosmosis in the decontamination process [23].

Detailed descriptions of the coupled transport processes of fluid, electricity, and contaminants during electrokinetics, and associated complicated features generated by various electrochemical reactions, have been given by several authors [3,7,9,12,25,26].

18.5 RELEVANCE OF THE TECHNIQUE

An electrical gradient is a much more effective driving force than a hydraulic gradient for moving a fluid, particularly through fine-grained soils of low hydraulic conductivity [3]. In fact, the values of hydraulic conductivity of different soils can differ by orders of magnitude (from 1×10^{-9} to 1×10^{-4} cm/s because it varies with the square of some effective pore size). However, the values of the coefficients of electroosmotic conductivity range generally from 1×10^{-5} to 10×10^{-5} cm²/V.s and are relatively independent of the soil type [7,25].

Electrokinetic remediation appears to be one of the most effective techniques for restoration of polluted sites. The obtained results, at different scales, show that electrokinetic remediation of soils can be an effective technology in heterogeneous and fine textured soils (clayey to fine sandy soils) because, in gross texture soils, other technologies can present good enough cleanup rates and better economic results. According to Yeung et al. [13], the technology is particularly effective and economical in fine-grained soils where pump-and-treat technology is impractical, if not impossible. The complementary nature of the electromigrational component of mass flux to electroosmotic mass flux is the reason why electrokinetic remediation can be a technically feasible and cost-effective means of extracting soluble and predominant species from all types of soils. Electromigrational flux will transport the species even when electroosmotic flux ceases or does not develop [4].

Soils of high water content (preferably saturated) and low physical and chemical fixation capacity are the most favorable to electrokinetic application. However, Sandia National Laboratories [27] in a demonstration into the program SITE, removed hexavalent chromium with acceptable efficiencies from an unsaturated soil. An optimal moisture content was determined to be in the range 10 to 12% w/w, corresponding to approximately 25% saturation.

The soils with a high buffer capacity are difficult to be electroremediated and it is close to obligatory to enhance the process by means of chemical reagents [28,29]. Thus, in soils with high content in illite, montmorillonite, carbonates, hematite, or other metallic oxides, the electrokinetics may need special operative conditions to be effective.

Electrokinetic technology has been applied to multiple types of contaminants. In general terms, it is possible to affirm that electrokinetics is not a selective technology transporting species because the electric potential is the driven force for electromigration of all (natural and contaminants) ions. Furthermore, the electroosmotic flow transports all the solutes in pore aqueous phase without distinction. Therefore, there is no reason to limit the type of contaminant to be removed, although naturally, its speciation in soil is a determining factor to the efficiency. A chemical element can be presented inside the soil as many different contaminant species; a precipitated solid (carbonate, sulphide, hydroxide, etc.); sorbed phase on mineral or organic matter; solved in the aqueous phase; and others with very different mobility, depending on environmental conditions.

18.5.1 *In Situ* vs. *ex Situ* Processes

A trend favors developing natural *in situ* rather than *ex situ* processes. The *in situ* processes are expected to have minimal construction and mobilization costs compared with alternatives for excavating or mixing contaminated soils for use in other *in situ* or *ex situ* technologies. The costs of remediation may be increased further when using disturbed soil in *ex situ* technologies by inadvertent contamination of uncontaminated soils [30]. In some cases, excavation of soil for *ex situ* treatment is not feasible (e.g., under buildings that must not be demolished) and electromigration provides a means to recover these contaminants *in situ* in a cost-effective manner [31].

Applied as the cleaning agent to a contaminated soil, the electric current promotes a redistribution of contaminants within the soil from fractions more tightly bound into the soluble and the exchangeable fractions, and those most weakly bound to soil particles or nonspecifically adsorbed. During the electrokinetic process, electromigration is one of the most important mechanisms by which the pollutant species are driven towards one of the electrodes, where they may be removed. Because electromigration requires the ions to be in solution and in exchangeable forms, the referred redistribution of contaminants shows that electrokinetics is very promising for the remediation of soils. However, care must be taken in order not to interrupt the decontamination process (by interrupting the electric current) at a stage at which the removal is still happening because increased mobility of heavy elements would endanger other ecosystem sections — groundwater, for example [32–34].

The fact that the technique requires a conducting pore fluid in a soil mass may be a shortcoming if concerns about introducing an external fluid in the soil [4] are present. *Ex situ* use of the technology minimizes or excludes this problem. Due to the previously presented, *ex situ* (on-site or off-site) treatment may be most desirable because it allows the joined actions [10]:

- Possibility to judge easily whether the soil is in a condition assuring a successful project; whether its contamination occurs in concretions or in solid forms (as paint residues); or whether the soil contains metal or isolating objects, which make a pretreatment compulsory/necessary
- Pretreatment such as sieving and/or incubation with an acid or with a complexing agent
- Making soil sorting and pollution uniform by their types, possibly to submit to different treatment processes (e.g., a certain coarser or finer soil fraction, considered more or less polluted, by a contaminant, sorted out and submitted to other soil cleaning processes in a nearby available plant)
- Soil homogenization (e.g., wood, stones, and metal pieces can easily be taken out when the soil is excavated and clay lumps can be reduced to pieces) contributing to increasing the efficiency of the process
- Confining the soil volume, thus allowing a much easier monitoring of the process without exposing the environment to possible risks

- Consideration of the possibility of a downstream, controlled phytoremediation scheme (e.g., using heavy metal plant accumulators) to profit from the soluble contaminants, taking advantage of the plant uptake and without endangering groundwater
- Climatic considerations (cover or not cover the processing units)

A lot of contaminated areas currently submitted to remediation correspond to places that became dear due to urban expansion. A common target consists of building at the place and the necessity to remove the soil, in any case. Excavation and transport are, then, already considered a part of the plan, without taking into consideration the best remediation practices. Advantages to an *ex situ* remediation approach can be taken from this procedure.

18.5.2 DISADVANTAGES THAT INTERFERE WITH EFFICIENCY OF THE PROCESS

The electrokinetic soil remediation previously described serves to concentrate and confine contaminants close to an electrode. However, when the media that separate the electrode compartments from the soil are passive membranes, some disadvantages can be depicted. The ions from the electrolytes are able to move across the passive membranes into the soil and out on the other side of the soil compartment because of their electric charge. Due to the soil cation exchange capacity, some of the heavy metal ions may be exchanged with other cations, which are not soil contaminants. In this case, it is preferable to have a transport of cations into the soil. However, if the transport of these cations is more rapid than the cation exchange process, it results in a transport of harmless cations from the anode to the cathode. This raises the costs of the method and results in inefficient remediation if the electrolytes are more mobile than the heavy metals in the soil in the electric field.

Care must be taken when considering the electrolyte solutions to be used in the process because they may create additional problems to the soil. For example, in the case of using $NaNO_3$ as electrolyte solution, Na^+ ions will pass from the anode compartment into the soil, where they may promote deflocculation of the clay soil fraction during their transport through the soil into the cathode compartment.

When the technique is applied without conditioning (unenhanced electrokinetic remediation), water electrolysis reactions at the electrode compartments (Equation 18.5 and Equation 18.6) promote the development of an acid and an alkaline front in the soil. The rate at which the hydrogen and hydroxide ions are produced is fixed by the current and occasionally also by competition of other types of electrode processes. The mobility of H^+ is higher than the mobility of OH^-, so an accumulation of contaminants (e.g., heavy metals) appears close to the cathode compartment. Excavation of the soil volume in the accumulation zone may be a way of attaining the soil-cleaning target concentrations. However, the excavated soil volume must be submitted to proper treatment and/or disposal, so a transfer of the contaminants always occurs at least point-wise.

These limitations may be overcome by the use of ion exchange membranes instead of passive membranes or using enhancing chemical reagents.

18.6 PROCESS ENHANCEMENT WITH CHEMICAL REAGENTS

The physicochemical soil-contaminant interactions occurring during the electrorestoration have a high impact on the efficiency of cleanup. Along with the electrokinetic processes, many kinds of independent or related reactions are simultaneously or alternatively occurring; this is a very good chance for electrokinetic remediation of soils to improve its results by means of stimulating the favorable and diminishing the unfavorable reactions. Thus, in a natural way, the Faradic reactions in the electrodes generate acid and basic fronts that can be accelerated or restrained to promote high or low necessary pH in the soil pore water. This is why different enhancement methods have been basically proposed to control the high pH.

These methods include adding chemical enhancing reagents to improve metal solubility in the soil [35–38], using the CEHIXM process [39], and conditioning cathode and anode pH [40–44]. In some studies, water with added caustic or lime was circulated through the anode compartment in order to control the pH, favoring the electroosmotic flow (permeability 0.5 to 2.5 10⁻⁵ cm²/V.s) and increasing the efficiency of the process. However, it is important that extremely high pH does not occur in the anode because an increased degree of undesirable soil swelling can be produced in this case.

The enhancing agents will further complicate the chemistry of the soil-contaminant reactions and must be carefully considered. In this sense, the enhancement by pH conditioning of electrodes has been modeled [45–47], allowing calculation of the pH and contaminant profiles in the soil and select operative conditions to get an optimized enhancement.

When no external ions are added in the electrolytes, the electrical current is expected to be at a low level [42] because metal hydroxides and others are precipitated near the cathode. The precipitation of species decreases concentration of ionic species in the soil pore fluid, decreases electrolyte strength, and renders a zone of low conductivity. The formation of this high-resistance zone causes a significant increase in the voltage drop across the soil if a constant current is operated, increasing the energy expenditure. Thus, the use of pH conditioning solutions to control the catholyte pH can stop or diminish the generation of the basic front and limit the precipitation of metals, resulting in an increase in conductivity and decreasing the voltage drop.

The use of solutions to a partial or total neutralization of OH⁻ in the catholyte renders acid soils that usually can generate electroosmotic flow towards the anode, producing positive or negative effects on the removal efficacy [38,44,48]. In addition, when agents to complex metals are used, it is necessary to take into account the opposite movement of the metal positive ion and the negative charged complex metal agent. This last phenomenon, with metal simultaneously moving towards anode and cathode, probably will produce a high accumulation of metal in the middle part of the soil. A list of very important characteristics of the enhancing reagents for a good performance has been reported (literally from Alshawabkeh et al. [49]):

- They should not form insoluble salts with the contaminant within the range of pH values expected to develop during the process.
- They should form soluble complexes with the contaminant that can electromigrate efficiently under DC electric field.
- They should be chemically stable over a wide range of pH values.
- They should have higher affinity for the contaminant than the soil particle surface.
- They and the resulting complexes should not have a strong affinity for the soil particle surface.
- They should not generate toxic residue in the treated soil.
- They should not generate an excessive quantity of wastewater or the end products of the treatment process should be amenable to concentration and precipitation after use.
- They should be cost effective, including reagent cost and treatment costs for the waste collected and/or wastewater generated.
- They should not induce excessive solubilization of soils or minerals or increase the concentrations of any species in the soil pore fluid.
- If possible, they should complex with the target species selectively.

Also, the electrokinetic process permits the supply to the soil of other external reagents by electromigration and/or electroosmotic flow. Moreover, when the soil receives electrical energy under certain conditions, it can become a network of microelectrolytic cells generating *in situ* the necessary environment to bring out reduction or oxidation reactions permitting the enhancement of the electroremediation. The latter, as electroosmotic flow, is a result of the double diffuse layer that, under electrical current, acts as a condenser, due to the low conductivity of soil solid particles

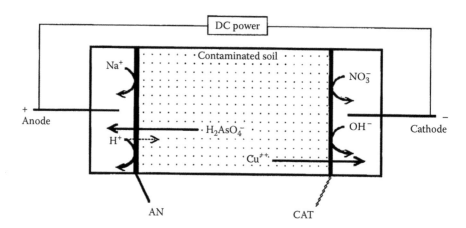

FIGURE 18.3 Schematic principle of the electrodialytic process. AN = anion exchange membrane; CAT = cation exchange membrane.

and the high conductivity of the surrounding pore aqueous phase. In this sense, a successful reduction of Cr(VI) inside the soil by means of supply of ferrous iron and electrical energy has been recently reported [50].

18.7 ELECTRODIALYTIC SOIL REMEDIATION

Electrodialytic remediation is a technique for removal of contaminants from polluted substrates. The method is based on a combination of the electrokinetic movement of ions (e.g., in a contaminated soil) with the principle of electrodialysis [10]. The general principle of the electrodialytic process is presented in Figure 18.3. It is similar to the electrokinetic remediation process (depicted in Figure 18.1), apart from the ion exchange membranes (instead of passive membranes), which are used to separate the contaminated soil volume (central cell compartment) from the two electrode compartments at the extremes.

An anion exchange membrane, AN, interposed between the anode compartment and the soil, will prevent cations from passing from this electrode compartment into the soil (apart from H^+, which can be an exception) and will allow anions to pass from the soil into this compartment. Similarly, a cation exchange membrane, CAT, interposed between the soil and the cathode compartment, will prevent the passage of negatively charged ions into the soil and will permit the cations (e.g., heavy metals and natural cations present in the soil) to pass from the soil into the cathode compartment (Figure 18.3).

The method has mostly been applied on a bench scale to heavy metal-contaminated soils [2,31–33,52–59]. The electrodialytic concept [51] has been the research topic of the Environmental Electrochemistry Group at the Technical University of Denmark (EEG/DTU). Laboratory electrodialytic cells are under continuous development to improve the transport processes; in addition, the concept has been scaled up into a pilot plant scale remediation cell, in collaboration with a commercial company (Soilrem Holding A/S, previously the Danish company A·S Bioteknisk Jordrens). The Technical University of Denmark is also running a pilot plant called "MegaLab" at the DTU Campus.

Due to the electric field present when a DC is passed between a pair of electrodes and placed in a system having charged particles (e.g., contaminated, moist soil), the pollutant species (and natural ions in the soil) are driven towards one of the electrodes, from where they may be removed. The three main mass-transport phenomena invoked during electrokinetic soil remediation (electromigration, electroosmosis, and electrophoresis) continue to act during the electrodialytic process. Electrodialysis provides some improvements over the electrokinetic concept [10]:

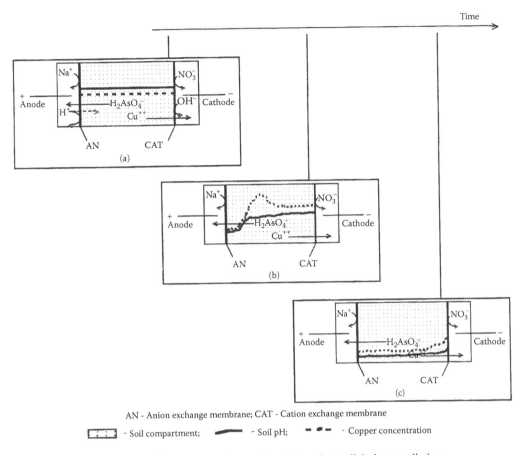

AN - Anion exchange membrane; CAT - Cation exchange membrane

$\boxed{\;\vdots\;}$ - Soil compartment;   ▬▬ - Soil pH;   ▬ ◖▬ - Copper concentration

FIGURE 18.4 Soil pH and Cu^{2+} concentration profiles during electrodialytic remediation.

- CAT acts as a "rectifier," only allowing cations to pass, which should lead to higher efficiencies for the heavy metal transport to cathode. AN is not a complete rectifier because it may let H^+ pass. Although this is known to happen [60], it has not been fully explained and its extension is unknown.
- There will be no flow of electrolytes between the two electrode compartments (assuming that the membranes are ideally permselective).
- The soil is continuously emptied of anions and cations until no more metal ions are available to be transported (indicated by a substantial rise in the resistance over the soil volume).
- The resistance will presumably increase until a certain level and become constant.

Figure 18.4 shows schematically the soil pH and the copper concentration ($[Cu^{2+}]$) profiles that will be developed during the electrodialytic remediation process [10]. A comprehensive overview on the study of practical and theoretical aspects concerning the use of ion exchange membranes in electrodialytic soil remediation can be found in Hansen [56]. In Figure 18.4 [10]:

- First, before applying the current, the soil pH and the $[Cu^{2+}]$ are constant through the whole soil compartment. Then, when a dc current starts to pass, in the soil compartment the soluble positive ions start to be driven towards the cathode and the negative ions towards the anode. At the inert electrodes, H^+ ions and OH^- ions will be formed. However, due to the CAT, the OH^- ions will not be able to pass from the cathode compartment into the soil. The AN will retard but not prevent the passage of H^+ ions from the anode compartment into the soil; this means that an acid front will start to propagate from this electrode compartment into the soil. If $NaNO_3$ is used as electrolyte solutions, Na^+ and NO_3^- will be present in the electrode compartments. However, due to AN and CAT, respectively, these ions will stay in the corresponding electrode compartments.

- As time passes, soluble cations and anions present in the soil compartment continue to move towards the electrode of opposite charge. Due to the usual cation exchange capacity of the soil, there are more "free" cations to move than anions. If the current density used in the process is high compared with the ion velocity in the soil solution, the result is that not enough mobile anions will be available in the soil compartment to cross the AN into the anode compartment. At the interface of AN/soil, the water present will dissociate into H^+ ions and OH^- ions and the OH^- ions will cross the membrane towards the anode; the H^+ ions will add to the electromigration, moving towards the cathode compartment. An acid front will be formed and propagate into the soil from the vicinity of the AN in the direction of the cathode. These H^+ ions join with the H^+ ions, which could pass from the anode compartment, through the membrane, into the soil. The promoted acid front and the concomitant decrease in the soil pH will facilitate the desorption/dissolution of copper in the soil. A decrease in $[Cu^{2+}]$ happens near AN, with an increase in $[Cu^{2+}]$ a bit further into the soil.

- If enough time is given for the process, the acid front will sweep through the whole soil volume. The decrease in the pH will facilitate the redistribution of Cu from forms more tightly bound to soil particles into forms that can electromigrate. The $[Cu^{2+}]$ will decrease in the soil because Cu^{2+} ions will pass into the cathode compartment. Eventually, the whole soil volume will attain copper levels that can be considered acceptable from a remediation point of view.

18.7.1 CHARACTERISTICS OF ION EXCHANGE MEMBRANES FOR ELECTRODYALITIC REMEDIATION

Since 1948, ion exchange membranes have been applied to a variety of processes:

- To concentrate or dilute ions, as in demineralization of potable water, desalination of brackish water to drinking water, and concentration of seawater for salt production [61]
- To LiCl battery chemistry [62] and waste water treatment [63]
- To exclude OH^- ions generated in the cathode to migrate from the cathode compartment towards soil [64,65]
- For the electrodialytic remediation previously discussed

In the electrodialytic remediation, the use of the selective membranes increases the efficiency of the removal process. In an electrodialytic cell for soil remediation purposes, the ion exchange membranes should be characterized by [10,56]:

- Good conducting properties (high conductivity and thereby low electric resistance) in order to have a low power consumption
- Good chemical stability over a wide pH range and in the presence of oxidizing agents
- Low degree of swelling so that, during the remediation process, no problems will develop due to changes in shape and size
- Good mechanical stability and the possibility that, after regeneration the membranes can be reused in the cell in reproducible conditions

- A lower electroosmotic transport of water than the soil volume to maintain a wet interface between the soil and the membrane, in order to ensure contact between the soil and the membrane
- A high exchange capacity in order to maximize the fluxes of counter-ions and minimize electrical resistance
- High selectivity for opposite charged ions and a high permeability (high permselectivity), in order not to waste electricity in the transport of co-ions through the membranes and into the soil volume

The selective membrane bounding the cathode compartment must be resistant to strong bases; meanwhile, the anion exchange membrane near the anode must resist strong oxidant and acidic conditions.

18.7.2 ELECTRODIALYTIC VS. ELECTROKINETIC TECHNIQUES: EXPECTED IMPROVEMENTS

In addition to the previous advantages, the presence of the ion exchange membranes does not allow the ions of the electrolyte solutions (in Figure 18.4, Na^+ and NO_3^- ions from the exemplified $NaNO_3$ electrolytes) to pass into the soil compartment. Therefore, the presence of the membranes allows the use of any kinds of electrodes, inert or reacting ones. Without the membranes, the use of reacting electrodes would be out of the question because their electrode processes could possibly create new polluting ions, which could be transported into the soil. Even in the case of the formation of harmless ions (due to the electrode processes), current would be wasted in moving them, contributing to make the whole process more expensive [10].

The ion exchange membranes also prevent the removed species from the soil compartment from passing again into the soil. These species will stay in the electrode compartment to which they were transported, although some may change polarity in certain medium conditions. The Cu^{2+} transported to the cathode compartment is an example because it might form $HCuO_2^-$ at very high pH [66]; however, due to the presence of the cation exchange membrane, these ions will stay inside the cathode compartment.

Although the soil acidification favors remediation, when the increase in the hydrogen ion concentration is considered in conjunction with electromigration of a species of interest, a substantial increase in the hydrogen ion transference number may hinder transport of other species. This means that the acidification should be controlled to a limit at which the desorption will occur but, at the same time, the acidification should be kept at a minimum so that the transference number of H^+ ions is not higher than needed. Otherwise, energy would be wasted in carrying the H^+ ions from the anode to the cathode, and the soil would end with a low pH, incompatible with any further use after treatment [2]. The desired control of the pH is possible in the electrokinetic (EK) and in the electrodialytic (ED) soil remediation cells; however, the latter case is easier to control, due to the use of the ion exchange membranes.

18.8 GEOMETRY OF ELECTROKINETIC SYSTEMS

In order to analyze the effect of the geometry of the system in soil electroremediation, it is convenient to conduct a separated study dependent on the scale application: laboratory and larger scales, pilot plant and field tests. However, common in both cases is a clear lack of experimental and theoretical studies on two-dimensional systems and electrodes that can be installed horizontally or vertically.

18.8.1 LABORATORY EXPERIMENTS

Most of the laboratory experiments have been carried out in one-dimensional arrangements, although with some differences among them. One kind of laboratory experiment employs different planar electrodes for cathode and anode; others employ cylindrical electrodes.

Thus, for example, in electroosmotic remediation, Schultz [67] used a panel consisting of galvanized steel screen and a piece of corrugated plastic wrapped in a geotextile as a cathode. The cathode compartment was a cavity filled with coke particle or a panel consisting of iridium-oxide coated expanded titanium mesh sandwiches between two pieces of perforated corrugated plastic, then wrapped in a geotextile. In this way, the soil and the electrode were separated and the corrugated plastic easily extracted the electrode fluids. Recently, a one-dimensional cell, 10 cm in length and diameter with planar electrodes, was used in bench tests [68] for electrokinetic removal of chromium and cadmium.

Other authors use two cylinders of graphite [44] or platinum wires [50] as electrodes; the electrokinetic removal of copper from Chinese red soil 600 g inside a PVC column with a length of 10 cm and 6.5 cm in diameter [44] has been recently studied. A two-dimensional modeling and laboratory arrangement for experimental validation has also been recently reported using cylindrical graphite electrodes [69].

18.8.2 FIELD APPLICATIONS

In field applications and pilot plants one-dimensional and two-dimensional systems have been used. In the simplest configuration, a group of sheet electrodes, electrode trenches, or lines of rod electrodes placed in boreholes produce a one-dimensional approach (Figure 18.5). In some cases, the field system is designed in a similar way to the laboratory installation but with larger electrode panels. Thus, Schultz [67] inserted electrode panels about 50 cm wide into the soil to an average depth of 4.5 m for a system 3 m × 10 m. The 10-m row of anode panels consisted of 16 panels with an average gap of 12 cm between them. The 10-m row of cathode panels consisted of 15 panels separated by 15 cm each. The final arrangement of the electrodes was similar to pile sheets (Figure 18.5a). DC power was supplied by a rectifier capable of delivering 65 A at 120 V.

Sandia National Laboratories [27] also essayed a one-dimensional configuration at a chromium-contaminated site of the U.S. Department of Energy in Albuquerque, New Mexico. The electrode system consisted of an anode row and four rows of cathodes parallel to the anode row, two at each side; all were 6 m depth and separated 1 m each one of the immediate (Figure 18.5b). In this way, overall hexavalent chromium efficiency rates varied from 36 mg/kW-h up to 136 mg/kW-h, operating with a limited maximum current of 15 A per electrode.

Another, more recent field scale demo study [68] was conducted on (9.14 m × 4.57 m × 3 m) soil at the Naval Air Weapon Station (NASW), Point Mogu, California. The electrokinetic system included a matrix of electrodes. Three cathodes were centered between six anodes (Figure 18.5c). The anode–cathode distance was about 4.5 m and the depth of the electrode wells was 3 m. The spacing between same polarities' electrodes was 1.5 m. The anode consisted of 0.9 m long, 2.54 cm diameter titanium hollow tubes (Eltech) with an iridium oxide coating. The cathode electrodes were constructed of 3.2 mm × 5.2 cm × 3 m long (shortened to 1.2 m after 118 days) stainless steel mesh. The anode and cathode wells were capped with PVC couplings for easy removal. Electrical power was applied to the electrode array via three 10-kW power supplies.

Also, commercial applications have used one-dimensional configurations. Thus, in 1987, the first electrokinetic commercial remediation project was conducted in a former paint factory contaminated with lead and copper in Groningen, the Netherlands [70]. The system consisted of a vertical array of alternating anodes and cathodes spaced on 3-m intervals (Figure 18.5d). After 430 hours of operation, lead content was reduced 70% and copper 80%.

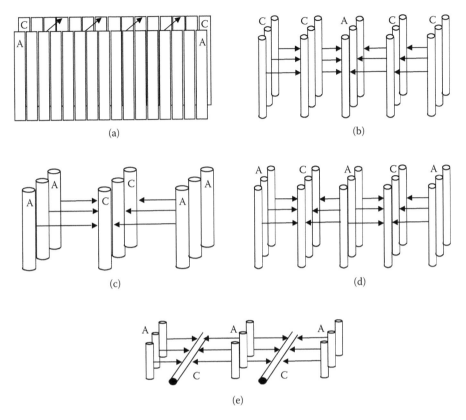

FIGURE 18.5 One-dimensional configurations for electrokinetic remediation at field scale.

One of the larger commercial electrokinetic remediations was carried out in a Dutch Royal Air Force base in 1992 with 2600 m^3 of soil contaminated with cadmium and other heavy metals [70]. This operated a hybrid vertical–horizontal system consisting of large tubular cathodes and short vertical anodes (Figure 18.5e). The tubular anodes were separated at 1.5-m intervals between the cathodes. The electrokinetics reduced the cadmium from 7300 to 47 ppm in 580 days.

These one-dimensional arrangements usually have the advantage of being the simplest and most cost-effective configuration. The frequent existence of areas of inactivity of electric fields or dead zones between electrodes of the same polarity are disadvantages. For this reason, the efficiency of the operations carried out with one-dimensional systems depends especially on spacing between electrodes of the same polarity (anode–anode or cathode–cathode) determining the number of electrodes required. If spacing between same-polarity electrodes is decreased, dead area inside the interest dominiom is minimized, improving the efficiency. Naturally, this also results in an increase of the cost of the equipment.

If two-dimensional systems are employed, the performance is very similar, although when two-dimensional configurations are employed, usually the principal objective is the generation of a radial or axial-symmetrical flow from the peripheral electrodes to the central electrode. If the selected contaminants to be removed are positive ions of heavy metals, the peripheral electrodes will be anodes and the center will be occupied by the cathode favoring the apparition of high concentrations of cations around the central zone and their easier and faster removal. If anions are to be removed, the polarity of electrodes reverses this position.

The peripheral electrodes can be ordered according to different geometrical distributions: hexagonal, square, and triangular, or, ideally, in any way close to a circle (Figure 18.6). The number

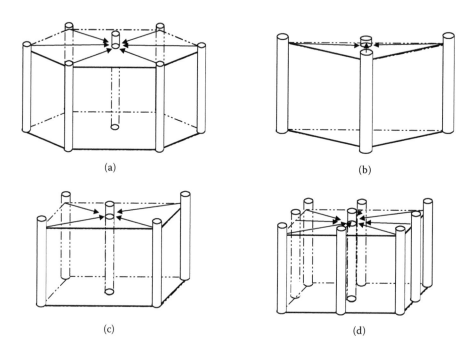

(a)   (b)

(c)   (d)

FIGURE 18.6 Two-dimensional configurations for electrokinetic remediation at field scale.

of outer electrodes corresponding to an inner electrode depends not only on the geometrical configuration but also on additional technical decisions.

Thus, the central electrode is surrounded by six peripheral electrodes in the hexagonal distribution (Figure 18.6a) and in the triangular by three (Figure 18.6b). However, if a square distribution is used, the center can be surrounded by four (Figure 18.6c), eight (Figure 18.6d), or even more electrodes with opposite charge to the central electrode. Also, these two-dimensional configurations generate the corresponding dead zones, without electrical activity. Obviously, the extension and situation of these areas strongly depend on the selected geometrical configuration and the number of electrodes in the system; however, in general terms, these dead zones in two-dimensional systems are smaller than in one-dimensional arrangements.

Electrical field spatial distribution indicates [71] that dead area in all the cells has a shape of a curvilinear triangle whose base is the distance between same-polarity electrodes (Figure 18.7). The height of this triangular area depends on processing time, electrode spacing, and alignment. The height of this triangle is expected to be larger in the case of one-dimensional compared to two-dimensional configurations due to the electrode alignment. An approximate practical method for comparing the efficiency between configurations assumed that this height is half the length of the triangle base for one-dimensional and a quarter of the length of the base for two-dimensional applications [71]. Figure 18.7 shows approximate distributions of the resulting inactive spots for selected configurations.

Another important difference between one- and two-dimensional configurations is that the current density is constant between the different polarity electrodes in the former and increases from external to center electrodes in the latter. Also, as another geometrical consequence, the soil volume extension swept for the desired acid or basic front proceeding from the outer electrodes is favored with respect to this in the one-dimensional installations.

In both types of configurations, the highest is the total number of electrodes in the installation, the highest is the cost, the more cost of immobilized capital and less time to do the cleanup. This produces two partial costs going in opposite directions and, therefore, an optimization of them is recommended to reach a minimum total final cost.

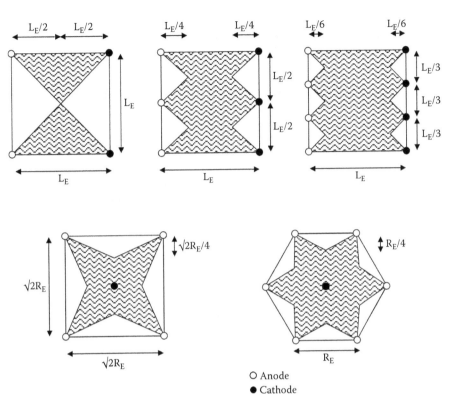

FIGURE 18.7 Approximate evaluation of ineffective areas for one and two-dimensional electrode configurations. (Modified from Alshawabkeh, A.K. et al., *J. Soil Contamination*, 8, 617, 1999. With permission.)

18.8.3 NUMBER OF ELECTRODES

A comparison of the number of electrodes required per unit surface area for one-dimensional, two-dimensional, hexagonal, and square configurations has been published [71]. In the one-dimensional configuration, three cases were provided in which the spacing among same-polarity electrodes equals: (1) spacing among opposite-polarity electrodes; (2) one-half of opposite-polarity spacing; and (3) only one-third of opposite-polarity spacing. Then, the number of electrodes is calculated based on unit of surface area. Considering a unit cell, the number of electrodes per unit area is

$$N = \left[\frac{F_1}{L_E^2} \right]_{1D\,flow} = \left[\frac{F_1}{\pi R_E^2} \right]_{2D\,flow} \tag{18.10}$$

where

N (L^{-2}) is the number of electrodes per unit surface area of site to be treated

L_E (L) and R_E (L) are one- and two-dimensional opposite-polarity electrode spacing, respectively

F_1 (dimensionless) is a shape factor depending on geometrical configuration of the system [71]

F_1 is calculated by addition of the fractions of outer electrodes corresponding to one central electrode because, usually, the outer electrodes are serving to several unit cells.

Table 18.1 shows the values of the F_1 for selected configurations [71]. Approximate calculations of the percentage of ineffective area for each configuration are also summarized. For one-

TABLE 18.1

Impact of Electrode Configuration on Electrode Requirements and Size of Ineffective Areas

Configuration	Electrode spacing		No. electrodes per cell	Area of cell	No. electrodes per unit area		Ineffective area	
	Opposite charge	Same charge			N	Percent increase	$A_{ineff.}$	Percent area
1-D	L_E	L_E	1	L_E^2	$1/L_E^2$	0	$L_E^2/2$	50
1-D	L_E	$L_E/2$	2	L_E^2	$2/L_E^2$	100	$L_E^2/4$	25
1-D	L_E	$L_E/3$	3	L_E^2	$3/L_E^2$	200	$L_E^2/6$	17
Square	R_E	$R_E \sqrt{2}$	2	$2 R_E^2$	$1/R_E^2$	0	R_E^2	50
Hexagonal	R_E	R_E	3	$1.5R_E^2\sqrt{3}$	$2/R_E^2\sqrt{3}$	15.5	$3R_E^2/4$	29

Source: Alshawabkeh, A.K. et al., *J. Soil Contamination*, 8, 617, 1999. With permission.

dimensional configurations, the ineffective area is half of total area, when the spacing of the electrodes of the same polarity is equal to the opposite-charge electrode spacing. Thus, it is not practical to use such a scheme unless remediation is implemented in two stages in which the electrode polarity can be changed.

This is also the case for the two-dimensional square arrangement. The results show that one-dimensional configurations with the same-polarity electrode spacing of one-half and one-third of opposite-polarity require an increase of 100 and 200% in number of electrodes when compared with the same of equal electrode spacing. With respect to two-dimensional configurations, the hexagonal configuration needs only 15% more electrodes than square configuration.

18.9 ELECTRIC OPERATIVE CONDITIONS

The most usual operation procedure drives a constant current through the soil during the cleanup with usual values about a few amperes per square meter (0.1 to 10 A/m^{-2}) of current density [67] and potential gradients depending on soil resistance (0.02 S/m^{-1}) from 100 to 400 V/m^{-1}. Obviously, if a higher current intensity is used, the process is faster, but consumes a much more important quantity of energy because the power consumption is proportional to the square of current intensity. If the soil has a high conductivity, it is necessary to operate with higher values of electric intensity than if it has a high resistance because the current must be distributed to move not only contaminants ions but also the natural ions present in the soil.

Other authors [44,72,73] prefer to operate in a laboratory, applying constant DC potential gradients in the range 30 to 400 V/m^{-1}. In this operative mode, when the resistance of the soil increases, the electrical current naturally goes down. Obviously, in this case, the conductivity control of soil is determining for cleaning time because the capacity of the electromigration to move ions basically depends on intensity through the soil. If the process operates to constant voltage and is not controlled [72,73], at the beginning of operation, a sharp increase in the protons and hydroxyl concentrations is produced in the anode and cathode compartments, respectively, and is responsible for the corresponding increment of electric current and cleanup rate. Afterwards, the depletion of free ions due to extraction and neutralization results in a high resistance in a specific zone of the soil and a high useless drop potential. This effect produces a reduction of the most important electrokinetic processes (electromigration or electroosmotic flow). It has been calculated [45,46] that, operating with a higher applied voltage, the meeting time between acid and basic fronts and the subsequent neutralization becomes shorter, usually resulting into lower energy efficiency.

In some cases, the authors [68] prefer to operate using a constant current density (5 A/m^{-2}) at laboratory scale, but use constant voltage (45 to 60 V 10 to 13 V/m^{-1}) at field scale, with a resultant electric current from soil resistance (10 to 17 A/m^{-2}). Electrokinetics, Inc. [70] reported that the addition of Cadex™ as an enhancement agent at the cathode removed 98% of the total cadmium, 40% of total lead, and 35% of total chromium at the same site (Point Mugu), with an energy expenditure of 3500 kw/h/m³ of soil for 1200 h of treatment. A conclusion of this comparison is that if no extremely longer periods of cleanup are desired, a higher current density in the field (10 A/m⁻²) is to be recommended to achieve better removal rates.

A secondary effect of current application to soil during electrokinetics is an increase in temperature because some electrical energy is transformed into thermal energy. This affects the cleanup in a different way, depending on whether the system is operated under constant current or constant voltage conditions. If under constant current, an increase in the pore water temperature will not affect the electromigration, but the electroosmotic flow will decrease. On the other hand, if the system operates at constant voltage, temperature increase will produce an increment of the electromigration [74]. In any case, the optimal electrical conditions of operation are directly related with the design of the system (geometry and electrode spacing) and all of the variables of the system must be jointly analyzed.

18.9.1 Current Efficiency: Critical Current Density through Membranes

The current efficiency is important when an electric field is applied over an ion exchange membrane. The counter-ion transport number normally decreases with increasing electrolyte concentration. In the electrodialytic process, the progressing electrolyte accumulation in one compartment and depletion in the other cause electrolyte diffusion; this counteracts the selective transference and reduces the current efficiency. The effects of transference and diffusion are not additive, so the description of the transport processes becomes complex.

The critical current density is another important aspect to consider when an electric field is applied over an ion exchange membrane. If the current density over the membrane is increased continuously, at a certain value, the current efficiency will drop sharply and the potential drop across the membrane will rise sharply. This is because, in the membrane, the current is carried almost exclusively by the counter-ions, whereas in the solution it is carried by the counter-ions and co-ions. Therefore, the transference of counter-ions away through the membrane is higher than the transference in the solution up to the membrane surface. The difference must be compensated by convection and diffusion. The resistance in the membrane will increase until water dissociation at the surface becomes a further source of ions [75]. In the case of a cation exchange membrane, the H^+ ions will assist in transporting the electrolyte cations through the membrane; however, the OH^- ions are transferred back to the solution [76] and the opposite in an anion exchange membrane.

18.10 REMEDIATION TIME REQUIREMENTS

In different configurations, the time required for remediation is a function of contaminant transport rates and electrode spacing. Some authors [67,71] have provided practical methods to estimate it in one-dimensional systems. Thus, neglecting the hydrodynamic dispersion and assuming, as usual, homogeneous soil and linear electric field, the linear velocity of cations through the soil with typical electroosmotic flow is:

$$v = \frac{\left(u^* + k_e\right)}{R_t}\Phi_e \tag{18.11}$$

where

u^* ($L^2V^{-1}T^{-1}$) is the effective ionic mobility of the species in soil
k_e ($L^2V^{-1}T^{-1}$) is the coefficient of electroosmotic permeability

Φ_e (VL^{-1}) is the voltage gradient

R_t (dimensionless) is a time-delay factor to account for the time required for contaminant desorption and dissolution

The value of R_t depends on soil type, pH, and type of contaminant. Sorption retardation factor (R_d) can be used to estimate R_t (R_t =1 for nonreactive contaminants). However, the values of R_t might be different from R_d because R_t should account for time delay due to all chemical reactions (solubilization, complexation, and desorption), but R_d values account only for sorption.

Time required to remediation T_R (T) is calculated using this velocity and opposite-polarity electrode spacing, L_E, as:

$$T_R = \frac{1}{\beta} \frac{L_E}{\sigma^* \Phi_e} \tag{18.12}$$

where σ^* (siemens L^{-1}) is the effective conductivity in the soil medium and β (L^3 C^{-1}) is a lumped property of the contaminant and the soil that measures the rate of reactive transport of a species relative to the electric conductivity of a medium, given by:

$$\beta = \frac{\left(u^* + k_e\right)/R_t}{\sigma^*} \tag{18.13}$$

Typical values of β for contaminated fine-grained soils are estimated to be in the range of 10^{-8} to 10^{-6} m^3 C^{-1} [48]. If time is to be calculated using current density, Equation 18.12 becomes

$$T_R = \frac{1}{\beta} \frac{L_E}{I_d} \tag{18.14}$$

where I_d (amps L^{-2}) is electric current density.

In treatments employing two-dimensional systems, it is assumed that radial electrical distribution, R_w, is the radius of central well (usually cathode), R_E, is the distance between central and peripheral electrodes, and Z is the depth of the site. The difference between this case and the one-dimensional case is that the current density in radial flow is a function of the radial distance (r); however, in both cases, it assumes operating with a constant total current. The electric current per unit depth for the radial transport is given by:

$$I_z = \left(2\pi r\right)\sigma * \Phi_r \tag{18.15}$$

where I_z (amps L^{-1}) is the current per unit depth and Φ_r (VL^{-1}) is the radial voltage gradient.

Contaminant transport rate depends on the voltage gradient, which is a nonlinear function of the radial distance. Ignoring dispersion and accounting for ion migration and electroosmosis, the radial velocity of ions transport is given by

$$v(r) = \frac{\left(u^* + k_e\right)}{R_t} \Phi_r \tag{18.16}$$

where v(r) (LT^{-1}) is the radial velocity of reactive species transport. Substituting Φ_r from Equation 18.15 into Equation 18.16 yields:

$$v(r) = \beta \frac{I_z}{2\pi r} \tag{18.17}$$

The velocity of contaminant transport is a nonlinear function of the radial distance even if the soil is homogeneous and isotropic. The time required for the contaminants to be transported from the outside electrodes (anodes) to the center electrode (cathode) is calculated integrating along the radius $dt = dr / v(r)$ from R_w to R_E, leading to:

$$T_R = \frac{\pi}{\beta} \frac{\left(R_E^2 - R_w^2\right)}{I_z} \tag{18.18}$$

In order to provide time evaluation as a function of the voltage, a transformed voltage expression, Φ_R (V), which is constant along the radius, is given by:

$$\Phi_R = \frac{I_z}{\pi \sigma *} \tag{18.19}$$

Substituting the value of I_z from Equation 18.19 and simplifying because $(R_W)^2 << (R_E)^2$, Equation 18.18 yields:

$$T_R = \frac{1}{\beta} \frac{R_E^2}{\sigma * \Phi_R} \tag{18.20}$$

The form of Equation 18.20 for radial transport is similar to the form of Equation 18.12 for one-dimensional transport. In both, β and $\sigma*$ are soil properties and Φ_e and Φ_R are constants if the first ones are. However, the comparison of the two equations shows that, although T_R is a function of the linear distance between the electrodes for the one-dimensional case, it is a function of square of the radial distance for two-dimensional configurations. This is important for selection of electrode spacing. Selection for radial spacing in radial systems is much more critical than for one-dimensional systems because time and cost remediation will significantly increase when radial spacing increases.

18.11 ENERGY CONSUMPTION

If a large area is to be electrokinetically remediated, alternating rows of anodes and cathodes, the power input per unit volume of uniform soil, assuming steady state operation, can be calculated for an anode–cathode pair of rows such as:

$$\frac{Power}{Soil\ volume} = \frac{I \Delta V}{v_s} = \frac{(\Delta V)^2 / R}{d\ Y\ L_E} = \frac{\sigma * (\Delta V)^2}{L_E^2} \tag{18.21}$$

where

I is the current applied
ΔV is the applied electrical potential
v_s is the soil volume
R is the electrical resistance between the anode row and the cathode row

σ^* is the soil electrical conductivity
d is the treated depth
Y is the length of the electrode row
L_E is the anode–cathode separation distance

The electrical field energy per soil volume (E) is the product of the preceding expression and the duration, T_R, of the remediation project:

$$E = \frac{Power.\ Time}{Soil\ volume} = \frac{I(\Delta V)T_R}{v_s} = \frac{\sigma^* T_R (\Delta V)^2}{L_E^2} \tag{18.22}$$

However, as explained previously, time depends on applied potential and anode–cathode distance and can be substituted in the expression by a function of these ones.

As an example, if only electroosmotic remediation is operating on soil, the necessary time to the clean up equals the ratio between necessary water to purge from the soil ($N\omega v_s$) and electroosmotic flow rate $Q_{EO} = k_e\ d\ Y\ (\Delta V)/L_E$, where N is the number of pore volumes of soil to be clean (directly related to retention factor, R_t); ω is the soil porosity; and k_e is the electroosmotic permeability.

Combining both, the necessary time to remediation is:

$$T_R = \frac{N\omega L_E^2}{k_e(\Delta V)} \tag{18.23}$$

This may be rearranged to express the required applied potential to complete the remediation project within the desired time, T_R:

$$(\Delta V) = \frac{N\omega L_E^2}{k_e T_R} \tag{18.24}$$

The substitution of this potential applied in the equation for energy consumption renders an equation for energy consumption as a function of the cleanup time and characteristics of the system, but not of the electrical operation conditions:

$$E = \frac{Power.\ Time}{Soil\ volume} = \frac{\sigma^* N^2 \omega^2}{k_e^2}\left(\frac{L_E^2}{T_R}\right) \tag{18.25}$$

This indicates that a soil maintaining its physical and chemical characteristics from laboratory to field, only maintains its energetic consumption per volume unit if also it maintains the ratio (L_E^2/T_R) in both scales.

In conclusion, several factors affect energy requirements for electrokinetic remediation at a specific site, including the properties of soil and contaminants, electrode configuration, and time of operation. As previously explained, the current or potential is changing along the process because soil resistance is doing it; the energy consumption does it as well. However, these kinds of approximate procedures using average values permit an estimative calculation of the energy expenditure per unit volume of treated soil for electrokinetic remediation with one- and two-dimensional configurations of electrodes:

$$\frac{Energy}{Soil\ volume} = \frac{I(\Delta V)T_R}{V_s} = \left(\frac{I_d(\Delta V)T_R}{L_E}\right)_{1-D} = \left(\frac{I_Z(\Delta V)T_R}{\pi\left(R_E^2 - R_W^2\right)}\right)_{2-D} \qquad (18.26)$$

18.12 ELECTRODE REQUIREMENTS

The electrode should be electrically conductive, chemically inert, porous, and hollow. The anode operates under highly corrosive conditions and needs to be very resistant. Chemically inert and electrically conducting materials such as graphite, coated titanium, or platinum have been used as anodes, to prevent dissolution of electrodes and generation of undesirable corrosion products in an acidic–oxidant environment.

The requirements for the cathode are much smaller; thus, any conductive material that does not corrode in the basic environment can be used. The hollow in the electrode is to act as outlet of solutions from the subsurface or as inlet for enhancing reagents. Special electrodes have also been used combining the electrokinetic and lysimeter technologies, to allow the operation under unsaturated conditions without significantly altering the soil moisture content [27]. In these, the ceramic of anodes was treated with hydrochloric acid and a surfactant to alter its negative zeta potential and to prevent electroosmotic flow out of the anodes.

REFERENCES

1. Pamukcu, S. and Wittle, J.K., Electrokinetic removal of selected heavy metals from soil, *Environ. Progress*, 11, 241, 1992.
2. Ottosen, L.M., Electrokinetic remediation. Application to soils polluted from wood preservation. Ph.D. thesis, Technical University of Denmark, 1995.
3. Yeung, A.T. and Datla, S., Fundamental formulation of electrokinetic extraction of contaminants from soil, *Can. Geotechnical J.*, 32, 569, 1995.
4. Acar, Y.B. et al., Electrokinetic remediation: basics and technology status, *J. Hazardous Mater.*, 40, 117, 1995.
5. Page, M.M. and Page, C.L., Electroremediation in contaminated soils, *J. Env. Eng. ASCE,* 128, 208, 2002.
6. Virkutyte, J., Sillanpaa, M. and Latostenmaa, P., Electrokinetic soil remediation-critical overview, *Sci. Total Environ.*, 289, 97, 2002.
7. Mitchell, J.K., Conduction phenomena, in *Fundamentals of Soil Behavior*, 2nd ed., John Wiley & Sons, New York, 1993, chap. 12.
8. Acar, Y.B. et al., Enhance soil bioremediation with electric fields, *Chemtech*, April, 40, 1996.
9. Denisov, G., Hicks, R.E. and Probstein, R.F., On the kinetics of charged contaminant removal from soils using electric fields, *J. Colloid Interface Sci.*, 178, 309, 1996.
10. Ribeiro, A.B., Use of electrodialytic remediation technique for removal of selected heavy metals and metalloids from soils., Ph.D. thesis, Technical University of Denmark, 1998.
11. Acar, Y.B. and Alshawabkeh, A., Principles of electrokinetic remediation, *Environ. Sci. Technol.*, 27, 2638, 1993.
12. Eykholt, G.R. and Daniel, D.E., Impact of system chemistry on electroosmosis in contaminated soil, *J. Geotech. Eng.*, 120, 797, 1994.
13. Yeung, A.T., Hsu, C. and Menon, R.M., EDTA-enhanced electrokinetic extraction of lead, *J. Geotech. Eng.*, 122, 666, 1996.
14. Acar, Y.B., Li, H. and Gale, R.J., Phenol removal from kaolinite by electrokinetics, *J. Geotech. Eng.*, 118, 1837, 1992.
15. Shapiro, A.P. and Probstein, R.F., Removal of contaminants from saturated clay by electroosmosis, *Environ. Sci. Technol.*, 27, 283, 1993.
16. Reed, B.E. et al., Electronic (EK) remediation of a contaminated soil at several Pb concentrations and applied voltages, *J. Soil Contamination*, 5, 95, 1996.

17. Yeung, A.T., Effects of electrokinetic coupling on the measurement of hydraulic conductivity, in *Hydraulic Conductivity and Waste Contaminant Transport in Soils* , ASTM, Philadelphia, PA., 1994, 569.

18. Segall, B.A. and Bruell, C.J., Electroosmotic contaminant-removal processes, *J. Environ. Eng.,* 118, 84, 1992.

19. Probstein, R.F and Hicks, R.E., Removal of contaminants from soils by electric fields, *Science*, 260, 498, 1993.

20. Acar, Y.B. and Alshawabkeh, A.N., Electrokinetic remediation. I: Pilot-scale tests with lead-spiked kaolinite, *J. Geotech. Eng.,* 122, 173, 1996.

21. Acar, Y.B., Alshawabkeh, A.N. and Gale, R. J., Fundamentals of extracting species from soils by electrokinetics, *Waste Manage.*, 13, 141, 1993.

22. Acar, Y.B. et al., Removal of cadmium (II) from saturated kaolinite by the application of electrical current, *Geotechnique*, 44, 239, 1994.

23. Pamukau S., and Wittle, J., Electrokinetically enhanced *in situ* soil decontamination, in *Remediation of Hazardous Waste Contaminated Soils*, D.L. Wise and D.J. Trantolo (Eds.), Marcel Dekker, Inc., New York, 1994.

24. Kuo, C.C. and Papadopoulos, K.D., Electrokinetic movement of settled spherical particles in fine capillaries, *Environ. Sci. Technol.*, 30, 1176, 1996.

25. Mitchell, J.K., Conduction phenomena: from theory to geotechnical practice, *Géotechnique*, 41, 299, 1991.

26. Alshawabkeh, A.N. and Acar, Y.B., Electrokinetic remediation. II: Theoretical model, *J. Geotech. Eng.*, 122, 186, 1996.

27. EPA/540/R-97/509 Sandia National Laboratories *in situ* electrokinetic extraction technology, USEPA, 1999.

28. Yeung, A.T., Hsu, C. and Menon, R.M., Physicochemical soil–contaminant interactions during electrokinetic extraction, *J. Hazardous Mater.*, 55, 221, 1997.

29. Puppala, S. et al., Enhanced electrokinetic remediation of high sorption capacity soils, *J. Hazardous Mater.*, 55, 203, 1997.

30. Marks, R.E., Acar, Y.B. and Gale, R.J., *In situ* remediation of contaminated soils containing hazardous mixed wastes by bioelectrokinetic remediation and other competitive technologies, in *Remediation of Hazardous Waste Contaminated Soils*, D.L. Wise and D.J. Trantolo (Eds.), Marcel Dekker, Inc., New York, 1994.

31. Hemmings, R.L. et al., Removal of hazardous and radioactive metals from soil using electrochemical migration, in *Proc. Symp. Waste Manage. 92*. Tucson AZ, 2, 1279, 1992.

32. Ribeiro, A. et al., Electrokinetic removal of copper from a polluted soil, in *Proc. 15th World Congr. Soil Sci.*, Acapulco, 3b, 210, 1994.

33. Ribeiro, A.B. and Mexia, J.T., A dynamic model for the electrokinetic removal of copper from a polluted soil, *J. Hazardous Mater.*, 56, 257, 1997.

34. Ribeiro, A.B. et al., Redistribution of Cu, Cr and As in a Portuguese polluted soil during electrokinetic remediation, in *Contaminated Soils. 3rd International Conference on the Biogeochemistry of Trace Elements*, R. Prost (Ed.), Colloques 85, INRA editions, ISBN 2-7380-0775-9, Paris, 1997, 11.

35. Reed, B.E. et al., Chemical conditioning of electrode reservoir during electrokinetic soil flushing of Pb-contaminated silt loam, *J. Environ. Eng.*, 121, 805, 1995.

36. Wong, J.S., Hicks, R.E., and Probstein, R.F., EDTA-enhanced electroremediation of metal contaminated soils, *J. Hazardous Mater.*, 55, 61, 1997.

37. Sah, J.G. and Chen, J.Y., Study of the electrokinetic process on Cd- and Pb spiked soils. *J. Hazardous Mater.*, 58, 301, 1998.

38. Yang, G.C. and Lin, S., Removal of lead from silt loam soil by electrokinetic remediation, *J. Hazardous Mater.*, 58, 285, 1998.

39. Karim, M.A. and Khan, L.I., Removal of heavy metals from sandy soil using CEHIXM process, *J. Hazardous Mater.*, 81, 83, 2001.

40. Hicks, R.E. and Tondorf, S., Electrorestoration of metal contaminated soils, *Environ. Sci. Technol.*, 28, 2203, 1994.

41. Lee, H. and Yang, J., A new method to control electrolytes pH by circulation system in electrokinetic soil remediation, *J. Hazardous Mater.*, 85, 195, 2000.

42. Saichek, R.E. and Reddy, K.R., Effect of pH control at the anode from the electrokinetic removal of phenantrene from kaolin soil, *Chemosphere,* 51, 273, 2003.
43. Zhou, D.M., Zorn, R. and Czurda, K., Electrochemical remediation of copper-contaminated kaolinite by conditioning anolyte and catholyte pH simultaneously, *J. Environ. Sci.,* 15, 396, 2003.
44. Zhou, D.M., Deng, C.-F. and Cang, L., Electrokinetic remediation of a Cu-contaminated red soil by conditioning catholyte pH with different enhancing chemical reagents, *Chemosphere,* 56, 265, 2004.
45. Wilson, D.J., Rodríguez–Maroto, J.M. and Gómez–Lahoz, C., Electrokinetic remediation I. Modeling of simple systems., *Sep. Sci. Technol.,* 30, 2937, 1995.
46. Wilson, D.J., Rodríguez–Maroto, J.M. and Gómez–Lahoz, C., Electrokinetic remediation II. Amphoteric metals and enhancement with a weak acid., *Sep. Sci. Technol.,* 30, 3111, 1995.
47. Park, J.S. et al., Numerical analysis for electrokinetic soil processing enhanced by chemical conditioning of the electrode reservoirs, *J. Hazardous Mater.,* 99, 71, 2003.
48. Kim, S.O., Kim, K.W. and Stueben, D., Evaluation of electrokinetic removal of heavy metals from tailing soils, *J. Environ. Eng.,* 128, 705, 2002.
49. Alshawabkeh, A.K., Yeung, A. and Bricka, R.M., Practical aspects of *in situ* electrokinetic remediation, *ASCE, J. Environ. Eng.,* 125, 27, 1999.
50. Pamukcu, S., Weeks, A. and Wittle, J. K., Enhanced reduction of Cr(VI) by direct electric current in a contaminated clay, *Environ. Sci. Technol.,* 38, 1236, 2004.
51. Ottosen, L.M. and Hansen, H.K., Electrokinetic cleaning of heavy metal polluted soil, in *Internal Report, Department of Physical Chemistry & Department of Geology and Geotechnical Engineering, Technical University of Denmark,* Denmark, 1992, 9.
52. Jensen, I., Jensen, J.B. and Sloth, P., Application of a modified electrokinetic cell for the removal of toxic heavy metals from polluted soils, in *Internal Report, Department of Physical Chemistry, Technical University of Denmark,* 1994, 32.
53. Jensen, J.B., Kubes, V. and Kubal, M., Electrokinetic remediation of soils polluted with heavy metals. Removal of zinc and copper using a new concept, *Environ. Technol.* 15, 1077, 1994.
54. Kliem, B.K. et al., Electrodialytic remediation of heavy metal polluted soil and waste-soil contaminated with mercury, in *5th International Symposium on the Reclamation Treatment and Utilization of Coal Mining Wastes and 3rd Conference on Environmental and Mineral Processing,* Ostrava, 1996, 7.
55. Czediwoda, A., Stichnothe, H. and Schönbucher, A., Electrokinetic removal of phenantrene from fine-grained soil, in *3rd International Symposium on Environmental Geotechnology,* 1996.
56. Hansen, H.K. Practical and theoretical aspects concerning the use of ion exchange membranes and resins in electrokinetic soil remediation. Ph.D. dissertation, Department of Physical Chemistry, The Technical University of Denmark, 216 pp., 1995.
57. Hansen, H.K. et al., Electrochemical analysis of ion exchange membranes with respect to a possible use in electrodialytic decontamination of soil polluted with heavy metals, *Separation Sci. Technol.,* 32, 2425, 1997.
58. Ribeiro, A.B. et al., Electrodialytic removal of Cu, Cr and As from chromated copper arsenate-treated timber waste, *Environ. Sci. Technol.,* 34, 784, 2000.
59. Ottosen, L.M et al., Removal of Cu, Pb and Zn in an applied electric field in calcareous and noncalcareous soils, *J. Hazardous Mater.,* B85, 291, 2001.
60. Elmidaoui, A. et al., Transfer of H_2SO_4, Na_2SO_4 and $ZnSO_4$ by dialysis through an anion exchange membrane, *Desalination,* 101, 39, 1995.
61. McRae, W., Electrodialysis, in *Process Technologies for Water Treatment.* S. Stucki (Ed.), Plenum Press, New York, 1988, 45–63.
62. Tilak, B.V. and Fritts, S.D., Comments on the estimation of caustic current efficiency of ion exchange membrane chlor-alkali cells, *J. Appl. Electrochem.,* 22, 675, 1992.
63. Fane, A.G. et al., Metal recovery from wastewater using membranes, *Water Sci. Technol.,* 25, 5, 1992.
64. Li, Z.H. and Neretnieks, I., Electroremediation: removal of heavy metals from soils by using cation-selective membranes, *Environ. Sci. Technol.,* 32, 394, 1998.
65. Ribeiro, A. et al., Looking at each step of a sequential extraction procedure applied to a contaminated soil before and after an electrodialytic remediation experiment, in *16th World Congress of Soil Science,* ISSS, Montpellier, 1998.

66. Pourbaix, M., *Atlas of Electrochemical Equilibria in Aqueous Solutions*, Pergamon Press, New York, 1966.
67. Schultz, D.S., Electroosmosis technology for soil remediation: laboratory results, field trial and economic modeling, *J. Hazardous Mater.,* 55, 81, 1997.
68. Gent, D.B. et al., Bench and field-scale evaluation of chromium and cadmium extraction by electrokinetics, *J. Hazardous Mater.,* 110, 53, 2004.
69. Vereda–Alonso, C. et al., Two-dimensional model for soil electrokinetic remediation of heavy metals. Application to a copper spiked kaolin, *Chemosphere*, 54, 895, 2004.
70. EPA 402-R-97-006. Resource Guide for electrokinetics laboratory and field processes applicable to radioactive and hazardous mixed wastes in soil and groundwater from 1992 to 1997, USEPA, 1997.
71. Alshawabkeh, A.K. et al., Optimization of 2-D electrode configuration for electrokinetic remediation, *J. Soil Contamination*, 8, 617, 1999.
72. Manna, M., Sanjay, K. and Shekhar, R., Electrochemical cleaning of soil contaminated with dichromate lixiviant, *Int. J. Miner. Process.* 72, 401, 2003.
73. Sanjay, K., Arora, A., Shekhar, R. and Das, R.P., Electroremediation of Cr(VI) contaminated soils: kinetics and energy efficiency, *Colloids Surfaces A: Physicochem. Eng. Aspects,* 222, 253, 2003.
74. Kristensen, I.V., The effect of soil temperature on electrodyalitic remediation, in *2nd Symposium Heavy Metals in the Environment and Electromigration Applied to Soil Remediation*, Lyngby, 1999, 91.
75. Ottosen, L.M., Hansen, H.K. and Hansen, C.B. Water Splitting at ion-exchange membranes and potential differences in soil during electrodialytic soil remediation, *J. Appl. Electrochem.*, 30, 1197, 2000.
76. Helfferich, F., *Ion Exchange*, McGraw–Hill Book Company, New York, 1962.

19 Application of Novel Nanoporous Sorbents for the Removal of Heavy Metals, Metalloids, and Radionuclides

Shas V. Mattigod, Glen E. Fryxell, Kent E. Parker, and Yuehe Lin

CONTENTS

ABSTRACT

A new class of hybrid nanoporous materials for removing toxic heavy metals, oxyanions, and radionuclides from aqueous waste streams has been developed at the Pacific Northwest National Laboratory. These novel materials consist of functional molecules such as thiols, ethylenediamine-complexed copper, and carbamoylphosphonates that are self-assembled as monolayers within the nanopores of a synthetic silica-based material. Tests indicated that these sorbents (self-assembled monolayers on mesoporous silica — SAMMS) can achieve very high sorbate loadings (~6 meq/g) very rapidly with relatively high specificity (K_d: 1×10^8 ml/g). Because of the specifically tunable nature of the functionalities, these nanoporous sorbents can be targeted to remove a selected category of contaminants such as heavy metals (Ag, Cd, Cu, Hg, and Pb), oxyanions (As and Cr), and radionuclides (^{137}Cs, [129]I, [237]Np, and isotopes of Pu, Th, and U) from waste streams.

19.1 INTRODUCTION

Successful synthesis of silica-based nanoporous materials using liquid crystal templating was achieved about a decade ago [1,2]. Since then, use of these nanoporous materials in diverse applications such as catalysis, sensor technology, and sorbents has proved to be feasible. In sorbent technology applications, the nanoporous materials offer a significant advantage over conventional sorbents, due to their high surface areas (~500 to 1000 m²/g). However, the pore surfaces of these novel materials need to be activated before they can be deployed as effective sorbents.

Typically, the nanoporous materials are synthesized through a combination of oxide precursors and surfactant molecules in solution reacted under mild hydrothermal conditions. Under these conditions, the surfactant molecules form hexagonally ordered rod-like micelles, and the oxide materials precipitate on these micellar surfaces to replicate the organic templates formed by the rod-like micelles. Subsequent calcination at 500°C removes the surfactant templates and leaves a high surface area nanoporous ceramic substrate. The pore size of these ceramic substrates can be controlled by using surfactants of different chain lengths.

The authors have developed a method to activate the pore surfaces of silica-based nanoporous materials so that these materials can be used as effective sorbents. This process consists of synthesizing, within pores, self-assembled monolayers of adsorptive functional groups selected to adsorb specific groups of contaminants. Molecular self-assembly is a unique phenomenon in which functional molecules aggregate on an active surface, resulting in an organized assembly that has order and orientation.

In this approach, bifunctional molecules containing a hydrophilic head group and a hydrophobic tail group adsorb onto a substrate or an interface as closely packed monolayers (Figure 19.1). The driving forces for the self-assembly are the intermolecular interactions between the functional molecules (such as van der Waals forces). The tail group and the head group can be chemically modified to contain certain functional groups to promote covalent bonding between the functional organic molecules and the substrate on one end, and the molecular bonding between the organic molecules and the metals on the other. For instance, populating the head group with alkylthiols (which are well known to have a high affinity for various soft heavy metals, including mercury) results in a functional monolayer that specifically adsorbs heavy metals such as Ag, Cd, Cu, Hg, and Pb.

If the head group consists of Cu-ethylenediamine complex, the monolayer will sorb oxyanions (As, Cr, Se, Mo) with high specificity. Additional monolayers with head groups designed by the authors include acetamide and propinamide phosphonates for binding actinides (Am, Pu, U, Th); Hg- and Ag-thiol for sorbing radioiodine; and a ferricyanide Cu-EDA complex for selectively bonding radiocesium. The functionalized monolayer and substrate composite (Figure 19.1) was designated as SAMMS. Various self-assembled monolayer functionalities and the contaminants that they were designed to target are shown in Figure 19.2.

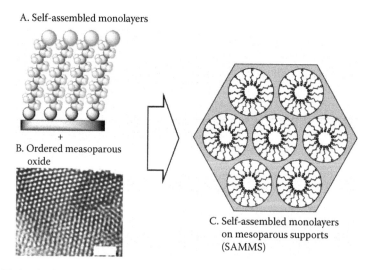

A. Self-assembled monolayers

+

B. Ordered measoparous oxide

C. Self-assembled monolayers on mesoparous supports (SAMMS)

FIGURE 19.1 Technological basis of novel nanoporous sorbents.

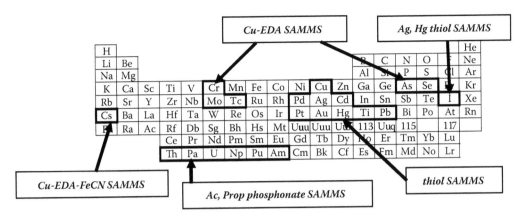

FIGURE 19.2 SAMMS technology and the targeted contaminant groups.

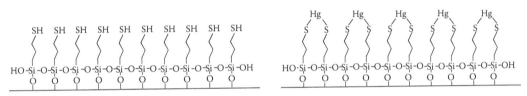

FIGURE 19.3 Schematic conformation of thiol-functionalized monolayers (left) and the bidentate bonding of mercury to the thiol moieties (right).

19.2 MATERIALS AND METHODS

19.2.1 SYNTHESIS OF SELF-ASSEMBLED MONOLAYERS

The nanoporous silica substrate was prepared using cetyltrimethylammonium chloride/hydroxide as the template and silicate and mesilyletene solutions as the reactants. Following hydrothermal reaction at 105°C for 1 week, the product was washed, dried, and calcinated at 580°C for 12 h to remove the template. The resulting nanoporous material had a surface area of 900 m²/g with an average pore size of 5.5 nm in diameter. The details of the substrate synthesis are provided by Feng et al. [3].

The thiol-functionalized SAMMS was synthesized by using trismethoxymercaptopropylsilane and allowing the molecular self-assembly to occur on the pore surface of the silica substrate in a toluene suspension [3]. The resulting monolayer had cross-linked silanes bonded to the pore surface with exposed thiol head groups (Figure 19.3) with a functional surface density of 6.2 silanes/nm². Similarly, Cu-EDA SAMMS was generated by a monolayer of EDA-terminated silane [2-amino-ethyl-3-aminopropyl trimethoxysilane], which was saturated with copper [4] (Figure 19.4). The surface population of this functionality was calculated to be 4.9 silanes/nm².

For radiocesium removal, the Cu-EDA FeCN SAMMS was synthesized from the Cu EDA SAMMS form by saturating the Cu sites with ferricyanide ions [5] (Figure 19.5). Actinide-adsorbing SAMMS was prepared by monolayer grafting of carbamoylphosphonate (CMPO) silanes (acetamide and propionaminde phosphonate silanes) [6] (Figure 19.6). The resulting SAMMS materials (APH SAMMS and PPH SAMMS, respectively) had functional densities of 2.2 to 2.4 silanes/nm² as measured by solid state ²⁹Si nuclear magnetic resonance (NMR) spectrometry [7].

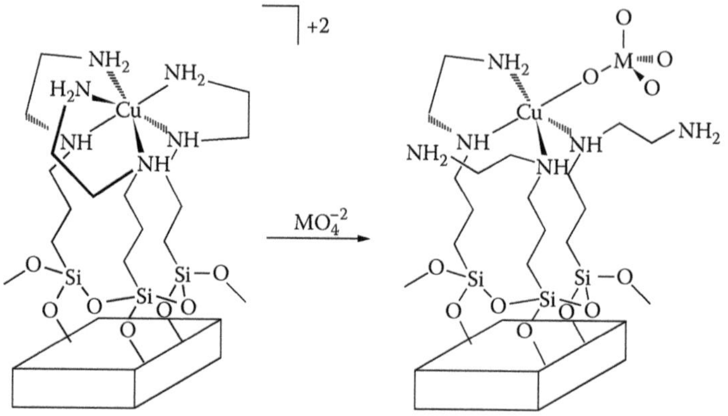

FIGURE 19.4 Schematic of copper ethylenediamine functionality (left) and the tetrahedral oxyanion binding mechanism (right).

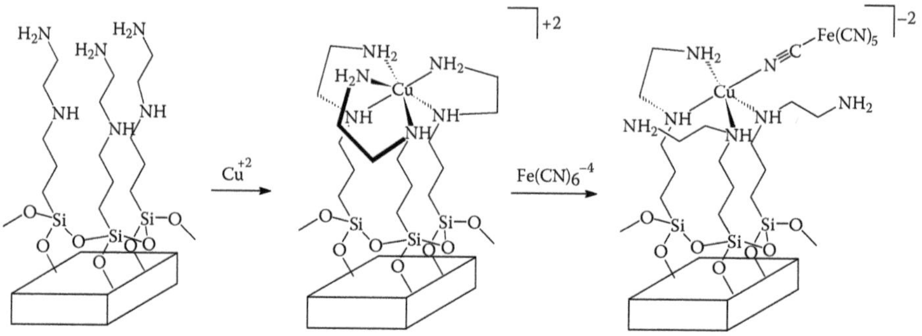

FIGURE 19.5 Sequence of synthesis of ferricyanide Cu-EDA SAMMS (right) starting with EDA SAMMS (left) and Cu-EDA SAMMS (middle).

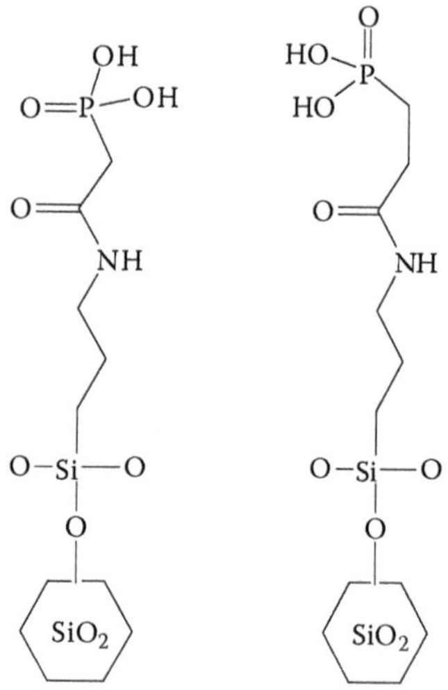

FIGURE 19.6 Schematic structures of acetamide and propinamide phosphonate SAMMS.

19.3 ADSORPTION EXPERIMENTS

Heavy metal adsorptive properties of thiol-SAMMS was tested by contacting known quantities of sorbent with a fixed volume of 0.1 M $NaNO_3$ solution containing the metal of interest (Ag, Cd, Cu, Hg, and Pb). The initial concentrations of these metals ranged from 0.05 to 12.5 meq/L and the solution-to-sorbent ratio in these experiments ranged from ~200 to 5000 ml/g. The suspensions were continually shaken and allowed to react under ambient temperature conditions (~25°C) for approximately 8 h. Next, the sorbent and the contact solutions were separated by filtration and the residual metal concentrations in aliquots were measured by using inductively coupled plasma mass spectrometry (ICP-MS).

Kinetics of adsorption of mercury by thiol-SAMMS was tested by contacting 200 mg of sorbent with 500 mL of 0.1 M $NaNO_3$ solution spiked with 0.1 meq/L concentration of mercury. The mixture was stirred constantly, and periodically aliquots of solution were drawn to monitor the residual mercury concentration. For comparison, mercury adsorption kinetic performance of a resin (GT-73) was also studied.

Experiments to evaluate the radioiodine adsorption performance of Hg-thiol and Ag-thiol SAMMS were conducted by equilibrating sorbent samples with aliquots of groundwater spiked with 3.65×10^7 bq/L of ^{125}I (Table 19.1). Solution-to-sorbent ratios ranging from 100 to 10,000 mL/g were used to evaluate the degree of ^{125}I loading on these materials. The mixture was gently agitated for ~20 h at $25 \pm 3°C$ and portions of equilibrated solutions were filtered and counted for residual ^{125}I activity. Analysis of ^{125}I in liquid samples was conducted by gamma-ray spectrometry, using a calibrated Wallac® 1480 Wizard™ 3-in. NaI detector with built-in software.

Oxyanion adsorption characteristic of Cu-EDA SAMMS was evaluated by contacting the sorbent with 3 meq/L of Na_2SO_4 solution containing either chromate (~0.02 to 18 meq/L) or arsenate (~0.01 to 29 meq/L) ions. Solution to sorbent ratio in these experiments ranged from 100 to 500 ml/g. After 12 h of contact, filtered aliquots of solution were analyzed by inductively-coupled plasma atomic emission spectrometry (ICP-AES). Adsorption kinetics experiment was conducted by contacting 200 mg of sorbent with 500 mL of groundwater spiked with 0.005 meq/L concentration of arsenate. The mixture was stirred constantly, and periodically aliquots of solution were drawn to monitor the residual arsenate concentration.

The cesium adsorption performance of Cu-EDA FeCN SAMMS was evaluated by contacting 50 mg of the sorbent with 10 mL of solution containing 0.5 to 8.4 meq/L of cesium. After 2 h of reaction, the solution was separated by filtration and the residual concentration of cesium was measured by using ICP-MS. A cesium adsorption kinetics experiment was conducted by contacting

TABLE 19.1
Composition of Groundwater Sample

Constituent	mg/L	Constituent	mg/L
Al	0.01	$Alk(CO_3^{2-})$	54.1
Ca	49.5	Cl	7.8
Fe	0.07	Br	0.10
K	1.7	F	0.17
Mg	14.6	NO_2	0.68
Mn	0.17	NO_3	27.2
Na	13.2	SO_4	82.5
Si	16.5	pH (SU)	8.1

50 mg of sorbent with 10 mL of cesium solution (0.015 meq/L). Periodically, aliquots of filtered solution were analyzed for cesium concentration by using ICP-MS.

The actinide-specific APH- and PPH-SAMMS were tested by contacting 100 mg quantities of each of these sorbents with 10-mL portions of solutions containing 2×10^6 counts per minute(CPM)/L of Pu(IV) in a matrix of acidified (pH = 1) 1 M NaNO$_3$ with separately spiked (0.01 M) Pu-complexing ligands such as phosphate, sulfate, ethyelenediaminetetraacetate (EDTA), and citrate. After 1 to 4 h of reaction, the solution was separated by filtration, mixed with Ultima Gold™ scintillation cocktail, and the residual alpha activity of Pu(IV) was measured by using a liquid scintillation counter (2550 TR/AB Packard Instruments, Meriden, Connecticut).

19.4 RESULTS AND DISCUSSION

The data from the adsorption experiments (Figure 19.7 and Table 19.2) indicated that thiol-SAMMS adsorbed the heavy metals with significant affinity. The predicted adsorption maxima were 0.56, 0.72, 1.27, 4.11, and 6.37 meq/g for Cu, Pb, Cd, Ag, and Hg, respectively. The calculated distribution coefficients were 4.6×10^1 to 1.8×10^5; 2.2×10^2 to 8.6×10^3; 2.2×10^2 to 1.9×10^4; 1.2×10^3 to 8.7×10^5; and 1×10^3 to 3.5×10^8 ml/g for Cu, Pb, Cd, Ag, and Hg, respectively.

Such selectivity and affinity in binding these heavy metals by thiol-SAMMS can be explained on the basis of the hard and soft acid base principle (HSAB) [8–10], which predicts that the degree of cation softness directly correlates with the observed strength of interaction with soft base

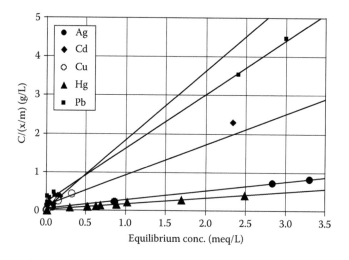

FIGURE 19.7 Langmuir adsorption isotherms for heavy metal adsorption by thiol-SAMMS.

TABLE 19.2
Heavy Metal Adsorption Characteristics of Thiol–SAMMS

Heavy metal	Adsorption maximum (meq/g)	K_d (ml/g)
Ag	4.11	$1.2 \times 10^3 – 8.7 \times 10^5$
Cd	1.27	$2.2 \times 10^2 – 1.9 \times 10^4$
Cu	0.56	$4.6 \times 10^1 – 1.8 \times 10^5$
Hg	6.37	$1.0 \times 10^3 – 3.5 \times 10^8$
Pb	0.72	$2.2 \times 10^2 – 8.6 \times 10^3$

functionalities such as thiols (–SH groups). According to the HSAB principle, soft cations and anions possess relatively large ionic size, low electronegativity, and high polarizability (highly deformable bonding electron orbitals); therefore, they mutually form strong covalent bonds. The order of adsorption maxima observed in this experiment appears to reflect the order of softness calculated by Misono et al. [11] for these heavy metals.

The kinetics data indicated that thiol-SAMMS adsorbed ~99% of the dissolved mercury within the first 5 min of reaction (Figure 19.8). Comparatively, the resin (GT-73) adsorbed only ~18% of the dissolved mercury during the initial 5 min. These data showed that thiol-SAMMS substrate adsorbs mercury about two to three orders of magnitude faster than the commercial GT-73 ion-exchange resin. After 8 h of reaction, thiol-SAMMS reduced the residual concentration of mercury to ~0.04 mg/L; the resin material was not capable of reducing mercury concentration below 1mg/L. Calculated distribution coefficients (K_d) indicated that thiol-SAMMs adsorbed mercury at about one to three orders of magnitude higher selectivity ($7 \times 10^3 - 3.6 \times 10^5$ ml/g) than the resin material ($4.5 \times 10^2 - 1.9 \times 10^3$ ml/g).

Results from the radioiodine adsorption experiments indicated that Hg-thiol, and Ag-thiol SAMMS very effectively adsorbed ^{125}I from the groundwater matrix (Table 19.3). Both forms of SAMMS exhibited very high distribution coefficients (K_d: 2.9×10^4 to 1.2×10^5 ml/g), indicating that radioiodine was sorbed with high specificity even in the presence of anions in the groundwater that were present in significantly higher concentrations than radioiodine. Such selectivity and

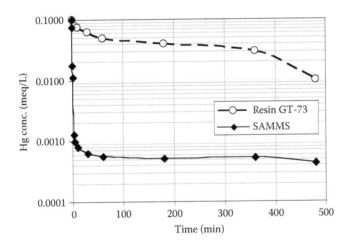

FIGURE 19.8 Kinetics of mercury adsorption by thiol-SAMMS and an ion exchange resin.

TABLE 19.3
Adsorption of Radioiodine (^{125}I) by Ag and Hg Thiol–SAMMS

Eq. activity (Bq/mL)	Ads. density (Bq/g)	K_d (ml/g)	Eq. activity (Bq/mL)	Ads. density (Bq/g)	K_d (ml/g)
	Ag-thiol–SAMMS			Hg-thiol–SAMMS	
108	3.58×10^6	3.31×10^4	611	1.76×10^7	2.88×10^4
409	1.79×10^7	4.38×10^4	942	3.59×10^7	3.81×10^4
848	3.64×10^7	4.29×10^4	1911	1.82×10^8	9.52×10^4
2299	1.73×10^8	7.53×10^4	2801	3.24×10^8	1.16×10^5
2961	3.50×10^8	1.18×10^5			

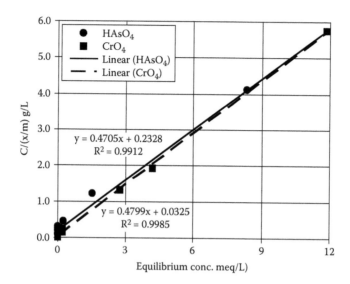

FIGURE 19.9 Langmuir adsorption isotherms for arsenate and chromate by Cu-EDA SAMMS.

affinity in binding of ^{125}I is because iodine is a much softer base than other anions and would bond preferentially to very soft cations — namely, Ag and Hg.

Results of the arsenate and chromate adsorption tests indicated that Cu-EDA SAMMS very effectively adsorbed both these oxyanions from Na_2SO_4 solution. The predicted adsorption maxima from the Langmuirian fit to the data were 2.13 and 2.08 meq/g for arsenate ($HAsO_4$) and chromate (CrO_4), respectively (Figure 19.9). The calculated distribution coefficients were 2.4×10^2 to 1.0×10^4 and 1.7×10^2 to 1.0×10^6 ml/g for arsenate and chromate, respectively. The bonding mechanism of these oxyanions was studied by Kelly et al. [12,13] using x-ray adsorption fine spectroscopy (XAFS), which indicated that bonding was monodentate in nature. The copper ion was bonded directly to the oxyanion by a shared oxygen. The adsorption process changed the coordination of copper from octahedreal to trigonal bipyramidal geometry. The bonding did not alter the tetrahedral symmetry of $HAsO_4$ ion, but the symmetry of the CrO_4 ion was distorted with two short and two long Cr–O bond distances.

The data from the kinetic experiment (Figure 19.10) showed that, within the first 2 min, about 98% of the arsenate was removed by Cu-EDA SAMMS. The adsorption was remarkably fast and the reaction reached a steady state in about 60 min. The residual concentration of arsenate was 2.03×10^{-5} meq/L (~0.8 ppb) — well below the EPA-proposed new regulatory limit (10 ppb) for arsenic in drinking water.

The data from the adsorption experiments indicated that Cu-EDA FeCN SAMMS adsorbed cesium with significant affinity (Figure 19.11). The predicted adsorption maximum was 1.34 meq/g compared to the measured value of 1.33 meq/g. Both these loading values are close to the theoretical maximum adsorption capacity of 1.5 meq/g calculated for this form of SAMMS material. Although the initial solution contained one to two orders of magnitude excess of Na than Cs, the calculated distribution coefficients ranging from 7.7×10^2 to 4.8×10^4 ml/g indicated that Cs bonded preferentially to the FeCN sites. The observed selectivity of ferrocyanide ligand for Cs confirms the previous observations of preferred bonding of this ligand with Cs than with other alkali earth cations [14,15].

The data from the adsorption kinetics experiment indicated very rapid binding of cesium by the Cu-EDA FeCN SAMMS (Figure 19.12). About 99.8% of the cesium present in solution was adsorbed, reducing the residual dissolved concentration to $<3 \times E-5$ meg/L within 1 min of contact

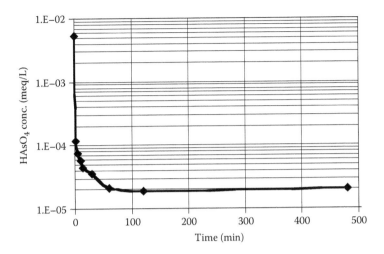

FIGURE 19.10 Kinetics of arsenate adsorption by Cu-EDA-SAMMS.

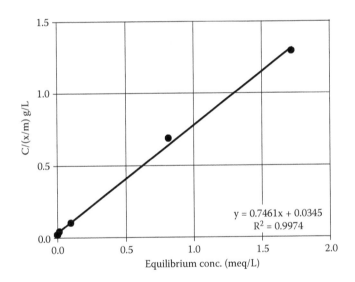

FIGURE 19.11 Langmuir adsorption isotherms for cesium by Cu-EDA FeCN SAMMS.

time. Bulk of the Cs adsorption (99.96%) had occurred within 20 min, resulting in very low residual concentrations of ~6.4×10^{-6} meq/L (0.85 ppb).

Plutonium adsorption data (Table 19.4) indicated that APH and PPH SAMMS adsorbed this actinide with high specificity (K_d: 1.7×10^4 to 2.1×10^4 ml/g). However, on average, APH SAMMS performed slightly better in adsorbing Pu(IV) from solution. In this experiment, the presence of complexants with differing chelating strengths did not significantly affect Pu(IV) adsorption by these nanoporous sorbents. Considering that the complexation constants of Pu(IV) with these ligands vary in the order EDTA > citrate > phosphate >> sulfate > nitrate [16], very high adsorption affinity shown by these sorbents indicates that the CMPO ligand (APH and PPH) functionality-based adsorption substrates are capable of chelating Pu(IV) much more strongly than these ligands. Additionally, the APH SAMMS has been shown to adsorb Pu(IV) very rapidly with bulk of the sorbate removed from solution in under 1 min [17].

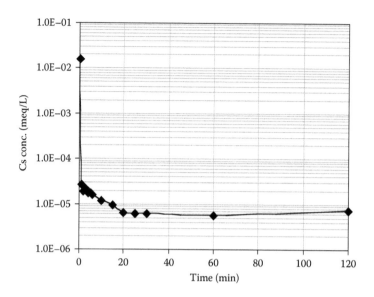

FIGURE 19.12 Kinetics of cesium adsorption by Cu-EDA FeCN SAMMS.

TABLE 19.4
Adsorption of Pu(IV) by APH– and
PPH–SAMMS

Ligand	APH SAMMS	PPH SAMMS
	K_d (ml/g)	
Nitrate	2.12×10^4	1.58×10^4
Nitrate + phosphate	2.04×10^4	1.75×10^4
Nitrate + sulfate	1.98×10^4	1.82×10^4
Nitrate + EDTA	2.05×10^4	1.56×10^4
Nitrate + citrate	2.31×10^4	1.87×10^4

19.5 SUMMARY AND CONCLUSIONS

A new class of hybrid nanoporous materials has been developed for removing toxic heavy metals, oxyanions, and radionuclides from aqueous waste streams. Tests showed that thiol-SAMMS designed for heavy metal adsorption showed significant loading (0.56 to 6.37 meq/g) and high selectivity (K_d: 4.6×10^1 to 3.5×10^8 ml/g) for contaminants such as Cu, Pb, Cd, Ag, and Hg.

Results of the arsenate and chromate adsorption tests indicated that Cu-EDA SAMMS very effectively adsorbed both these oxyanions with predicted adsorption maxima of ~2.1 meq/g, with distribution coefficient ranging from 1.7×10^2 to 1.0×10^6 ml/g. Adsorption experiments conducted using Ag- and Hg-capped thiol SAMMS sorbents very effectively adsorbed ^{125}I from a groundwater matrix. Both forms of thiol-SAMMS exhibited very high distribution coefficients (K_d: 2.9×10^4 to 1.2×10^5 ml/g), indicating that radioiodine was sorbed with high specificity even in the presence of anions in the groundwater that were present in significantly higher concentrations than radioiodine.

Another sorbent, designed to adsorb radiocesium (Cu-EDA FeCN SAMMS) specifically, when tested exhibited loading as high as 1.33 meq/g and distribution coefficients ranging from

7.7×10^2 to 4.8×10^4 ml/g. Tests conducted using carbamoylphosphonate-functionalized nanoporous substrates (APH and PPH SAMMS) showed that these sorbents very effectively adsorbed Pu(IV) from solutions containing complexing ligands, such as EDTA, citrate, phosphate, sulfate, and nitrate. Distribution coefficients as high as 2×10^4 ml/g confirmed that the CMPO-based functionalities assembled on nanoporous substrates are very effective scavengers for actinide ions such as Pu.

Self-assembled monolayers of selected functionalities on nanoporous silica substrates can achieve very high sorbate loadings very rapidly with relatively high specificities. These novel classes of sorbent materials therefore will be very effective in removing a wide range of targeted contaminants from waste streams.

ACKNOWLEDGMENTS

This study was supported by the Office of Science/Office of Biological and Environmental Research of the U.S. Department of Energy and the IR&D funds from Battelle. Pacific Northwest National Laboratory is operated for the U.S. Department of Energy by Battelle under contract DE-AC06-76RLO 1830.

REFERENCES

1. Beck, J.S., J.C. Vartuli, W.J. Roth, M.E. Leonowicz, C.T. Kresge, K.D. Schmitt, C.T-W. Chu, D.H. Olson, E.W. Sheppard, S.B. McCullen, J.B. Higgins, and J.L. Schlenker. A new family of mesoporous molecular sieves prepared with liquid crystal templates. *J. Am. Chem. Soc.* 114, 10834–10843, 1992.
2. Kresge, C.T., M.E. Leonowicz, W.J. Roth, J.C. Vartuli, and J.S. Beck. Ordered mesoporous molecular sieves synthesized by a liquid crystal template mechanism. *Nature* 359, 710–712, 1992.
3. Feng, X., G.E. Fryxell, L.Q. Wang, A.Y. Kim, J. Liu, and K.M. Kemner. Functionalized monolayers on ordered mesoporous supports. *Science.* 276, 865, 1997.
4. Fryxell, G.E., J. Liu, A.A. Hauser, Z. Nie, K.F. Ferris, S.V. Mattigod, M. Gong, and R.T. Hallen. Design and synthesis of selective mesoporous anion traps, *Chem. Mater.* 11, 2148–2154, 1999.
5. Lin Y., G.E. Fryxell, H. Wu, and M. Engelhard. Selective sorption of cesium using self-assembled monolayers on mesoporous supports. *Env. Sci. Technol.* 35, 3962–3966, 2001.
6. Birnbaum, J.C., B. Busche, Y. Lin, W.J. Shaw, and G.E. Fryxell. Synthesis of carbamoyl-phosphonate silanes for the selective sequestration of actinides. *Chem. Commun.* 1374–1375, 2002.
7. Yantasee, W.Y. Lin, Y., G.E. Fryxell, B.J. Busche, and J.C. Birnbaum. Removal of heavy metals from aqueous solution using novel nanoengineered sorbents: self-assembled carbamoylphosphonic acids on mesoporous silica. *Sep. Sci. Tech.* 15, 3809–3825, 2003.
8. Pearson, R.G., Hard and soft acids and bases part 1, *J. Chem. Educ.* 45, 581–587, 1968.
9. Pearson, R.G. Hard and soft acids and bases part 2, *J. Chem. Educ.* 45, 643–648, 1968.
10. Hancock, R.D. and A.E. Martell. Hard and soft acid and base behavior in aqueous solution: steric effects make some metal ions hard: a quantitative scale of hardness-softness for acids and bases. *J. Chem. Educ.* 74, 644, 1996.
11. Misono, M., E. Ochiai, Y. Saito, and Y. Yoneda. A dual parameter scale for the strength of Lewis acids and bases with the evaluation of their softness. *J. Inorg. Nucl. Chem.* 29, 2685–2691, 1967.
12. Kelly, S., K. Kemner, G.S. Fryxell, J. Liu, S.V. Mattigod, and K.F. Ferris. An x-ray absorption fine structure spectroscopy determination of the binding mechanisms of tetrahedral anions to self-assembled monolayers on mesoporous supports. *J. Synchrotron Rad.* 8, 922–924, 2001.
13. Kelly, S., K. Kemner, G.E. Fryxell, J. Liu, S.V. Mattigod, and K.F. Ferris. X-ray absorption fine structure spectroscopy study of the interactions between contaminant tetrahedral anions to self-assembled monolayers on mesoporous supports. *J. Phys Chem.* 105, 6337–6346, 2001.
14. Mekhail, F.M. and K. Benyamin. Sorption of cesium on zinc hexacyanoferrate(III), zinc hexacyanoferrate(III) and hexacyanocobaltate(III). *Radiochim. Acta.* 55, 95–99, 1991.

15. Ayers, J.B. and W. H. Waggoner. Synthesis and properties of two series of heavy metal hexacyanof-errates. *J. Inorg Nucl Chem.* 33, 721, 1971.
16. Cleveland, J.M. *The Chemistry of Plutonium,* American Nuclear Society, LaGrange Park, IL, 1979.
17. Fryxell, G.E., Y. Lin, and H. Wu. Environmental applications of self-assembled monolayers on mesoporous supports (SAMMS). *Stud. Surf. Sci. Cat.* 141, 583–589, 2002.

Section IV

Bioremediation

20 Phytoremediation Technologies Using Trees

I.D. Pulford and N.M. Dickinson

CONTENTS

20.1 INTRODUCTION

Regeneration of industrially contaminated land (brownfield sites) is now a priority in many countries due to risk to human health and the requirement for development land. In the U.K., for example, the government has set a target of 60% of new housing to be built on brownfield sites [1,2]. This high level of activity, however, can incur excessive cleanup costs. Traditional methods for dealing with heavy metal-contaminated soil have relied heavily upon engineering-based techniques such as:

- Excavation and removal to landfill (dig and dump)
- Encapsulation or containment *in situ*
- Separation methods, such as soil washing
- Chemical stabilization

These technologies tend to be expensive [3] and may be prohibitive, with estimates of the amount of contaminated land needing treatment increasing (estimates of up to 100,000 sites covering

300,000 ha in the U.K.). These methods also lack environmental sensitivity. Regeneration of urban brownfield land requires innovative, low-cost, ecologically sensitive and effective techniques.

Current problems associated with reclamation of brownfield land fall into two categories: (1) disposal of existing contaminated soil cover; and (2) import of new topsoil. It is often more cost effective to work *in situ* with existing contaminated soil and to attempt to restore a healthy soil rather than to import expensive topsoil. For these reasons, natural attenuation is often favored for low-value brownfield land — sometimes combined with use of appropriate additives or amendments [4,5]. This chapter evaluates the current status of knowledge as to whether phytoremediation through tree planting may also be suited to the task [6,7].

20.1.1 THE POTENTIAL BENEFITS OF TREES

Phytoremediation, the use of plants to remove contaminants (phytoextraction) or to stabilize the soil (phytostabilization), offers a low-cost alternative that retains the integrity of the soil and is visually unobtrusive. Trees have been suggested as appropriate plants for phytoremediation of heavy metal-contaminated land because they provide a number of beneficial attributes:

- Large biomass
  - The idea that plants could take up large amounts of heavy metals into their aerial tissues and thereby clean up contaminated soil originated from the discovery and study of hyperaccumulator plants [8,9]. Such plants take up and tolerate very high concentrations of heavy metals. Their main disadvantage is their low productivity because total offtake is the product of tissue concentration and biomass yield; thus, low biomass limits their effectiveness as metal extractors. Trees, on the other hand, include many high yielding biomass species that would need to accumulate only moderate amounts of metal to be effective. Harvested trees also have more ready afteruses.
- Genetic variability
  - Because species of many fast growing, short-rotation trees, such as *Salix* and *Populus*, are genetically diverse, the opportunity exists to select genotypes with traits for resistance to high metal concentrations, or for high or low metal uptake. In Europe, there are already extensive breeding programs for *Salix*, particularly to select traits of high biomass and disease resistance [10,11]. It is evident from field observation that certain trees have no difficulty growing on contaminated sites [7]. This may be due to acclimation [12], avoidance of contaminated soil by roots [13], or accumulation in the root and limited translocation into aerial tissues [14].
- Established management practices
  - Agronomic practices for the cultivation and harvest of woody plants for biomass crops are well established, but this is not the case for hyperaccumulator plants. *Salix* (willows and osiers) and *Populus* (poplars or cottonwoods) are commonly grown in a short-rotation coppice (SRC) system, with harvesting on a 3- to 5-year cycle over a coppice life cycle of up to 30 years. Well-established forestry practices of planting, management, and harvesting have been developed with coppice on arable land, and they can readily be adapted for use on contaminated land [15–17]. Trees are typically grown at a density of 10,000 to 20,000 per hectare, producing dry matter yields of 15 odt ha^{-1} annum^{-1} or higher [18].
- Economic value
  - Development of SRC systems for *Salix* has been driven by the various end uses of the trees. This has predominantly been as a biomass fuel by burning of the chipped stems, but other uses are for the production of chipboard, paper, and charcoal, and for specialist uses such as basket weaving. Progress is also being made towards more

efficient conversion of biomass to energy fuels using anaerobic digestion, fermentation, thermochemical conversion, and improved combustion techniques [19].

- Public acceptability
  - Growing trees on contaminated land is a use that has a high degree of public acceptability and fits well into urban or rural landscapes. Urban regeneration in the U.K. and elsewhere is transforming postindustrial and derelict landscapes using multipurpose forestry to increase lowland forest cover in urban and urban fringe areas. Economic and political changes are driving the remediation of brownfield land and are starting to address the legacy of contamination from an earlier industrial history. "Low-cost" restoration to "soft end-uses" is becoming increasingly prioritized.
- Site stability
  - Trees act as a canopy protecting the soil surface from dispersal by wind or erosion by water. The roots help to stabilize the substrate. Uptake of water, and transpiration through the leaves, may help to limit the downward migration of heavy metals by leaching and thus help to protect ground- and surface waters. Leaf fall contributes a significant amount of organic material to the soil surface.

Use of trees on heavy metal-contaminated sites also has possible disadvantages. Metals may be transferred into more mobile and bioavailable forms by deposition on the soil surface following leaf fall or deeper in the soil due to root or microbial action. In general, trees tend to acidify soil, which again would cause increased bioavailability of metals. As a result of these processes, metal uptake by plants and transmission up the food chain are risks, as well as metals leaching to ground- or surface waters.

The potential use of trees in phytoremediation builds on a considerable knowledge base of metal tolerance in plants [20,21], impacts of metal contamination on trees and woodlands [22,23], and experimental studies on the responses of trees to metal contamination [24–30].

20.2 FACTORS AFFECTING UPTAKE OF HEAVY METALS BY TREES

Phytoextraction is the subset of phytoremediation that aims to maximize the uptake of heavy metals from soil into the plant. Thus, measurements of metal concentrations in plant tissue have been commonly carried out in order to assess the effectiveness of various tree species. Much, but not all, of this work has focused on the growth of *Salix* spp. in SRC [18,31–34]. *Populus*, the other genus within the same family as willows (the Salicaceae), has also been studied [35–39]. Pine [40,41], sycamore [24,26,42], birch [43–48], and oak [30] have also attracted interest. In Mediterranean climates, *Acacia retinoides* and *Eucalyptus torquata* growing on Cu mine wastes have been found to take up Cu and Pb from soil [48]. Species of *Acacia* and *Leucaena leucocephala* are potential candidates for tropical soils [49,50].

Concentrations of metals in tree tissues are affected by a number of factors; this frequently makes comparison between studies difficult. This is caused by at least five common but inherent variables, leading to errors and deficiencies that are becoming increasingly apparent.

20.2.1 SUBSTRATE VARIABILITY

Field and pot experiments have been carried out on a wide range of contaminated substrates (Table 20.1). Variations in pH, redox potential, particle size, and organic matter content, as well as the forms and concentrations of the heavy metals, mean that bioavailability of metals will be considerably different in the various substrates.

There are major deficiencies in the reliability of estimates generally used to quantify the contamination of land. These deficiencies are caused by insufficient consideration of heterogeneity of contamination within a given sampling location and thus uncertainties in the measurements. Soil

TABLE 20.1
Various Substrates in Reported Studies of Heavy Metal Uptake by Trees

Plant growth substrate	Ref.
Soil marginally elevated in Cd	34, 90, 134–137
Sewage sludge-treated sites	31, 33, 39, 41, 42, 56, 60, 138, 139
Airborne industrial pollution	33, 83, 89, 140
Industrial/mining wastes	24, 43, 45, 58, 59, 141, 142
Landfill sites	35–38, 143, 144
Artificial contamination	46, 59

sampling is the main source of uncertainty in measurements of contamination [51] and bias is created by spatial variation of metal concentrations across the plot [52]. Standard sampling protocols potentially provide very unreliable estimates of contamination, due to problems with precision (random error) and bias (systematic error) [53]. Using bulked or composite samples clearly masks the heterogeneity of contaminant concentrations across the site, but using a number of separate sampling points for analysis of variability also has major problems. Spatial mapping of contamination and appropriate use of statistics in the design of sampling are essential before it is possible to assess whether planting trees alters metal concentrations in soil.

Using pot experiments in which soils are thoroughly mixed prior to planting is one way to evaluate the effects of tree roots on bioavailability of metals in soil, but extrapolation to the field may not be valid. In pots, the roots are forced to exploit the whole volume of soil, but in the field, plant roots may avoid hotspots of metals and proliferate in less contaminated zones [24]. In nonwoody plants, this can double the bioavailability and the bioconcentration factor (BCF) [54].

Different metals and their bioavailability also raise questions of risk. For example, 700 mg kg^{-1} of antimony (Sb) in surface soils associated with mining of this metal was found to be immobile and biologically unavailable over a wide pH range; thus, much lower concentrations of co-contaminants (As and Cu) pose a more significant risk [55].

20.2.2 TEMPORAL VARIATIONS IN TISSUE CONCENTRATIONS OF METALS

In short-rotation coppice systems using *Salix*, a decline in tissue metal concentrations has been noted after the first year of growth [31,33]. Morin [56] found a similar decline over 3 years in a number of different tree species (ash, birch, cottonwood, maple, and pine) growing on sludged mine spoil. Such decreases in metal concentrations may be partly attributed to dilution due to growth, but may also be due to relatively rapid rates of uptake from soil during the first year, when trees are growing quickly and roots are becoming established.

In addition to this year-to-year variability, many workers have measured variations in the metal contents of various tissues in trees within a single growing season [18,32,44,57–60]. On highly contaminated sites, an increase in metal concentrations in leaves has been interpreted as a survival trait whereby toxic metals can be lost from the tree at leaf fall. Redistribution of metals from leaves and bark to wood at the end of the growing season has also been recorded in trees growing on moderately contaminated sites [32].

20.2.3 VARIABILITY BETWEEN COMPONENT PARTS OF TREES

Many studies have noted variation in metal concentrations between different tissues and organs of trees [14,18,24,31,39,42,56,59–64]. In general, the order of metal concentrations is roots > leaves > bark > wood, although different elements show different patterns of distribution. Lead, Hg, and

Cr, especially, are immobilized in the roots, whereas Cd, Cu, Zn, Co, Mo, B, and Ni tend to be found more distributed to the other tissues, reflecting their greater mobility within the tree.

Knowledge relating to uptake patterns of heavy metals in trees is still very limited, but it is known that metals tend to remain located in the roots rather than being transported to aerial tissues [43,65]. Heavy metals in soils may or may not enter the root tissues and then move to the xylem [66,67]. Onward mobility and translocation through the plant also differ between metals and between plants. Accurate modeling of uptake and fate of trace elements in plants has not yet been achieved, but clearly depends on solubility and mobility.

Uptake (and toxicity) depends on ability of trace elements to cross membranes and to be transported. Ionic status and organic complexation are particularly important. For example, the inner lumen walls of xylem elements are lined with fixed negative charges that bind cations, removing them from the transpiration stream as they move up the xylem. Monovalent ions (e.g., K^+) move at the same rate as the transpiration stream, but divalent ions (e.g., Ca^{2+}) move slowly. Divalent ions bind preferentially to the fixed negative charges, displacing the monovalent cations (which move back into solution). However, some metals (e.g., Cu^{2+}, Ni^{2+}, and Fe^{3+}) move at the same rate as the transpiration stream because it contains dissolved organic molecules (amino acids, organic acids, sugars). Changes occur in speciation as they form complexes with organic molecules (and may become negatively charged); for example, Cu and Ni bind with amino acids.

20.2.4. TOXICITY OF TRACE ELEMENTS TO PLANTS

Background concentrations of metals available for uptake in soil are variable, as are the plant tissue concentrations associated with toxicity. Experimental evidence for minimum toxic concentrations of heavy metals known to affect trees was reviewed by Pahlsson [68]; unfortunately, most of it relates to short-term studies of seedlings grown in artificial culture. The results vary greatly according to whether metal salts were supplied in a growth medium of solution, sand, or soil, and between species and soil types. For example, in one study of boron toxicity, eight tree species were grown in sand receiving nutrient solution containing elevated boron concentrations from 0.2 to 20 mg l^{-1}; however, despite some foliar symptoms of toxicity at 5 mg l^{-1} and above, growth was not severely retarded even at 20 mg l^{-1} [69]. Otherwise, no comprehensive study of boron toxicity in woody species exists in the literature; neither soil nor plant analysis can be used to predict precisely the growth of plants on high B soils [70].

Mobility varies between species and in relation to different metals, but some generalizations can be made. There are two basic groups in relation to mobility in plants: (1) trace elements immobilized in roots, e.g., Al, Cr, Hg, Pb, Sn, and V; and (2) mobile trace elements, e.g., As, B, Cd, Cu, Mn, Mo, Ni, Se, and Zn. The mobile elements may be phytotoxic, giving reduced growth (As, B, Cu Mn, Ni, and Zn) or may accumulate with no discernible effect on plants (Cd, Mo, Se, Be, and Co). This suggests that the most scope is for targeting the latter trace elements for phytoextraction.

20.2.5 PLASTICITY OF PLANT RESPONSE TO METALS

In the field, it is increasingly apparent that trees can survive and grow in contaminated soil with exceptionally high levels of multiple metals [22,23,31,71–73]. Seedlings may be more sensitive than saplings or mature trees, and metal concentrations in seedlings are often higher than in mature plants [74]. Tree roots can actively forage towards less contaminated zones of soil [25] and, even with highly reduced growth, they can "sit and wait" for favorable growth conditions [75]. Trees are able to grow and survive by being resilient and, undoubtedly, they have a large capacity for phenotypic adjustment to metal stress [4,27,29,76].

Metal tolerance to zinc and lead has been identified in populations of birch and willow [5,77,78]. Eltrop et al. [45] found *Salix caprea* (goat willow) growing in soil with total concentrations of

17,000 mg Pb kg^{-1} (with NH$_4$Oac extractable concentrations of 4000 mg Pb kg^{-1}) and *Betula pendula* (silver birch) in soil with total concentrations of 29,000 mg Pb kg^{-1} (extractable concentrations of 7000 mg Pb kg^{-1}). Established woodlands have been shown to remain healthy following industrial fallout, which has raised total soil concentrations above 10,000 mg Cu kg^{-1} and 100 mg Cd kg^{-1} [22,79].

Despite this ability to survive contamination, results from other experimental studies have been extrapolated and it has been concluded that, in some soils, heavy metals may sufficiently inhibit growth of trees to be of significance to commercial forestry [80]. The best direct evidence is from situations in which trees have been planted on reclaimed metalliferous mining spoils [43]. In many soils, it is likely that elevated concentrations of available metals could be ameliorated prior to planting to avoid such effects; for example, liming and fertilization have been found to affect the heavy metal concentrations in *Pinus sylvestris* (Scots pine) [81]. Substantial variation exists, even within clones of the same species [12,82].

20.3 FACTORS AFFECTING METAL OFFTAKE DURING HARVEST

Further recent advances in knowledge also help to evaluate the likelihood of using trees for phytoremediation.

20.3.1 RECYCLING METALS TO THE SURFACE OF SOIL

Tree foliage can contain high concentrations of heavy metals, especially Cd and Zn. On highly contaminated sites, these may be at their peak values just before leaf senescence, appearing not to be translocated and redistributed within the plant prior to leaf fall. If these metals return to the rhizosphere, they represent a significant pool of potentially bioavailable metal. Under such circumstances, phytoremediation will fail in its aim of cleaning up the soil. It does, however, raise the question of improving offtake of metal by harvesting the trees prior to leaf fall. This is not normal practice in SRC systems in which the stems without leaves are harvested over the winter period. With long-term tree covers that are not regularly harvested (and in the years intervening in the usual 3-year SRC harvest), this recycling process of metals to surface soil may be significant.

20.3.2 METALS LOCATED IN TREE ROOTS

Roots usually contain the highest concentrations of all metals in trees; depending on the size of the root biomass, this could be a significant amount. Death and decomposition of roots during normal growth processes or following harvest of above-ground biomass probably release heavy metals back into the rhizosphere. Because the roots can contain a significant pool of metals, it may be that the root bole should be harvested at the end of the SRC cycle (after 25 to 30 years) or when a mature tree is cut down.

Root density and the depth of rooting are particularly significant in the context of phytoremediation [83]. Studies on 246 coppice stools of five *Salix* and five *Populus* clones in four different soil types at seven sites in the U.K., with a stool age of 3 to 9 years, found rooting to a depth of more than 1.3 m; however, 75 to 95% of roots were in the top 36 cm. Wetter soils had shallower root systems [84]. It has also been found that *Salix cinerea* has a much reduced uptake of Cd and Zn into leaves and bark when grown in wetland compared to that in drier soils [85].

20.3.3 METALS LOCATED IN STEMS

Although the two are often reported together simply as "stem" tissue, tree bark usually contains a higher concentration of metals than wood does [14,86–88]. The higher biomass of wood means that it contains a greater pool of metal than the bark does. Wood and bark (stem) tissue contains a significant pool of heavy metals (concentration × biomass yield), one that increases with the age of the tree. This pool of metals is much less bioavailable than those in the roots and leaves. Regular cutting of trees in the SRC system is designed to increase stem biomass, but also increases the ratio of bark to wood and thus may increase this pool of immobilized metal.

20.3.4 MODELING METAL OFFTAKE

Uptake data are frequently converted to BCFs that allows a better comparison to be made of the metal uptake abilities of trees grown under different conditions.

$$BCF = \frac{\text{concentration in plant tissue (mg kg}^{-1})}{\text{concentration in soil (mg kg}^{-1})}$$

Care must be taken when collating such data because of the different strengths and types of acids and extraction processes that are routinely used to measure soil concentrations of heavy metals (see Section 20.4.1). Table 20.2 shows recently published BCFs for uptake of Cd, Cu, Ni, Pb, and Zn by trees. BCF values > 1 indicate that metal is accumulated in the tree relative to the soil.

This approach suggests potential for enhanced uptake of Cd by *Salix*, especially when only slightly elevated concentration of Cd is in the substrate. Zinc uptake may also be significant but, in view of limited zootoxicity of this metal, is of less concern on brownfield land. In general, transfer of Cu, Ni, and Pb from soil to plant is very poor.

Although BCF values can give a guide to the uptake of metals from soil, the concentration of a metal in plant tissue does not alone give any information about the absolute amount of metal removed from the soil. Knowledge of the biomass yield is also required in order to calculate metal offtake (concentration × yield). Some estimates of metal offtake obtained from field trials using *Salix* are given in Table 20.3. Some of the trials measured offtake over a number of years; these have been corrected to an annual figure to allow comparison between trials of differing duration. As pointed out earlier, yearly values may not be equal and, with coppiced systems especially, the effects of tree growth and harvest may be important.

Such figures do, however, show some consistent features. Cadmium, Cu, Ni, and Pb are removed at the rate of the order of tens of grams per hectare per year, whereas the value for Zn is about 100 times higher. When compared with estimates of the metal content of the soil to rooting depth, such offtake figures generally give estimates of hundreds to tens of thousands of years to reduce the soil metal contents to acceptable values. Only the case of Cd, in which the amount taken out by the trees can be a significant percentage of the soil metal, offers hope of cleanup within a realistic time scale (Table 20.4).

20.4 OUTSTANDING ISSUES

20.4.1 IMPORTANCE OF BIOAVAILABILITY OF HEAVY METALS IN SOIL

Soil contamination guidelines tend to be based on some measure of acid extractable metal. Measurement of a true total metal content in soil requires dissolution of the whole of the solid soil matrix, which, because the predominant mineral type in most soils is silicate, involves the use of hydrofluoric acid (HF). For safety reasons, there is a tendency not to use this acid and thus a "pseudototal" value is often obtained using less aggressive reagents. This means that the "total" values reported in the literature have been obtained by various methods of differing extracting power (Table 20.5). Therefore, care must be taken when these data are interpreted.

One measure of the success of phytoremediation is the decrease in soil metal content resulting from tree growth. It could be argued that total (or pseudototal) metal values are meaningless in

TABLE 20.2

Bioconcentration Factors Reported in the Literature for Trees Growing on Contaminated Land

	Cd	Cu	Ni	Pb	Zn	Ref.
Alnus incana						143
Leaves	—	0.05	—	—	0.07	
Twigs	—	0.03	—	—	0.06	
Betula pendula						143
Leaves	0.06	0.02	—	—	0.37	
Twigs	0.11	0.03	—	—	0.32	
Fraxinus excelsior						143
Leaves	—	0.01	—	—	0.02	
Twigs	—	0.04	—	—	0.05	
Salix viminalis						143
Leaves	0.83	0.01	—	—	0.37	
Twigs	0.72	0.03	—	—	0.28	
Sorbus mongeotii						143
Leaves	—	0.01	—	—	0.03	
Twigs	—	0.02	—	—	0.05	
S. viminalis "Orm"						145
Roots	1.1	0.2	—	0.1	0.5	
Wood	1.2	0.1	—	0.1	0.3	
Leaves	1.4	0.1	—	0.02	0.8	
S. viminalis clone 78183						134
Stem	2.9–16.8	—	—	—	—	
Leaves	4.8–27.9	—	—	—	—	
S. viminalis clone 78198						33
Calcereous soil — stem	0.65–1.57	—	—	—	0.11–0.31	
Calcereous soil — leaves	1.61–2.74	—	—	—	0.59–1.71	
Acid soil — stem	0.79–0.93	—	—	—	0.06–0.08	
Acid soil — leaves	1.29–1.82	—	—	—	0.23–0.36	
20 *Salix* clones						31
Stem	0.0–0.45	0.01–0.03	0.01–0.06	—	0.04–0.15	

this context and that some measure of change in the bioavailability of the metal should be used. There have been two broad approaches to this. First, single selective extractants, which have been used extensively in the study of soil chemistry, are employed to give an assessment of bioavailability. This approach assumes the model of a large pool of non-, or partly, bioavailable metal and a smaller pool of bioavailable metal, upon which the plant roots draw. Because this more available metal is removed by plant uptake (or by leaching), it may be slowly replenished by weathering of the larger, nonavailable pool (Figure 20.1).

Because the bioavailable pool contains the environmentally active metal, it could be argued that phytoremediation is successful if it depletes this pool. This argument has two uncertainties: the rate at which metal moves from the nonavailable to the available pool, and the return of metal

TABLE 20.3
Offtake Estimates by *Salix*[a]

	Cd	Cu	Ni	Pb	Zn	Ref.
Dredged river sediment	12	38	—	57	795	145
Coppice soils (estimates based on 25 years)	2.6–16.5	—	—	—	—	134
Calcareous soil pH 7.3 (average over 5 years)	34		—	—	2680	33
Acid soil pH 5.2 (average over 2 years)	24		—	—	7250	33
Calcareous soil, 7.3 (3-year-old trees)	44	187	—	—	3851	83
Sewage sludge soil, pH 6.3 (maximum value after 1 year's growth)	62	59	37	—	822	31
Assuming biomass production of 15 odt/ha/yr — stems	33	—	—	—	—	136
If leaves are also harvested	61.6	—	—	—	—	136
Assuming 10 odt/ha/yr	34.3–76.7	58.6–81.5	9.3–13.8	—	954–1560	18

[a] Grams of metal per hectare per year.

TABLE 20.4
Number of Years to Reduce Soil Cd Concentrations by 5 mg kg^{-1} by Planting *Salix* to Different Soil Depths[a]

	Plant tissue concentration (mg kg^{-1})			
Depth (cm)	10	25	50	100
10	33	13	7	3
20	67	27	14	7
40	133	53	28	14

Note: Calculated according to rates of metal uptake into above-ground plant tissues. Calculation assumes (1) yields of 15 ODT ha^{-1}; (2) bulk density of 1; and (3) consistent Cd uptake throughout the period. Fertilization and irrigation may be required to achieve these yields.

TABLE 20.5
Extracting Procedures Used to Measure Total or Pseudototal Metal Concentrations in Contaminated Soil

Acid extraction scheme	Ref.
2M HNO3	33, 83, 90, 143
7M HNO3	34, 134–136
HCl + HNO3 (3:1)	31, 45, 146
HNO3 + HCl (3:1)	147
HNO3 + HCl + HClO4 (3:3:1)	46
HNO3 + HF + HClO4	89
HNO3 + HCl + HF microwave digestion	145

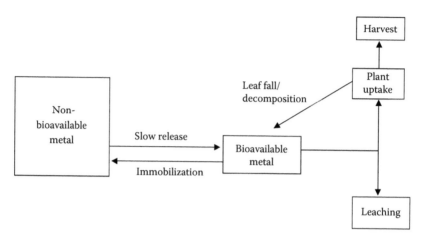

FIGURE 20.1 Model of bioavailability of metals in soil.

to the soil by leaf fall or plant decomposition. Measures of the bioavailable pool have been made by three broad approaches:

- Measurement of metal in soil solution, which is the source of all metal taken up by plant roots
- Extraction by simple salt solutions, e.g., 0.01 M CaCl$_2$, 0.1 M NaNO$_3$, 1 M NH$_4$NO$_3$
- Extraction by a metal complexing reagent, e.g., 0.05 M EDTA, 0.005 M DTPA

The second way in which the changes in soil metal chemistry are studied is by use of sequential extraction. This assumes that metals are held in soil in a series of pools of decreasing bioavailability and uses increasingly aggressive reagents to remove successive pools of metal. This approach allows study of changes in the distribution of metal held in different ways.

Published results of bioavailability of metals measured by single extractants have shown mixed results. Hammer and Keller [89] used three separate extractants (0.1 M NaNO$_3$; 0.005 M DTPA/0.1 M TEA in 0.01 M CaCl$_2$, pH 7.3; 0.02 M EDTA/0.5 M NH$_4$OAc, pH 4.65) to assess the bioavailability of Cd, Cu, and Zn in two contaminated Swiss soils; one was calcareous (pH 7.78) and one was acidic (pH 5.27). Under *Salix viminalis,* they measured a significant decrease in NaNO$_3$ extractable Cd and Zn in both soils, but DTPA–Cd only in the acidic soil, compared with the unplanted soil. Conversely, they found an increase in DTPA–Zn and Cu in the acid soil.

Eriksson and Ledin [90] used 1 M NH$_4$NO$_3$ and 0.01 M CaCl$_2$ to measure the bioavailability of Cd under *Salix viminalis* in Swedish soils and found a decrease in the planted soils. Pulford et al. [31] used 0.05 M NH$_4$EDTA to extract an organic (LOI 22%; pH 6.3) sewage sludge-treated soil following coppicing of the trees. In plots where the most productive clone (*Salix aurita* x *cinerea* x *viminalis,* Rosewarne White) had grown, extractable Cd, Cu, Ni, and Zn were higher than in adjacent unplanted areas.

Watson [32] used 1 M NH$_4$OAc (pH 6) and 0.025 M NH$_4$EDTA (pH 4.6) to measure bioavailability of metals under five *Salix* clones growing on a sewage disposal site (LOI 9.9%, pH 6.2) compared to unplanted areas. No significant effect on the NH$_4$OAc extractable metals was found, but the EDTA extractable metals were lower under the trees. On another sewage sludge site (pH 5.4 to 5.8), Watson [32] measured the following increases in the concentrations of metals in the soil solution on planted sites compared to unplanted: Cd: +133%; Cr: +9%; Cu: +50%; Ni: +51%; Pb: +58%; and Zn: +102%.

The second approach, sequential extraction, is useful for allocating metal fractions in soil, but the process is operationally defined, so interpreting these in terms of bonding to specific soil components should be avoided. A large number of extraction schemes of varying complexity have been used for natural soils and some have been applied to contaminated soils to assess the effects of tree growth on metals. Hammer and Keller [89] used a six-step extraction scheme based on that of Benitez and Dubois [91]:

1. $0.1 M$ $NaNO_3$
2. $1 M$ NaOAc, pH 5
3. $0.1 M$ $Na_4P_2O_7$
4. $0.25 M$ $NH_2OH.HCl$ + $0.05 M$ HCl
5. $1 M$ $NH_2OH.HCl$ + 25% HAc
6. HF, HNO_3, $HClO_4$

They found that Cd was extracted predominantly in step 2 for the calcareous soil and step 3 for the acid soil; Cu was predominantly removed in step 3 for both soils. Zinc was more evenly distributed over a number of steps: 2, 3, 4, and 5 for the calcareous soil; and 3 and 5 for the acid soil. Only very small changes in the distribution of metals among the six steps were found after growth of *Salix*. Slight, but significant, decreases were measured for Cd step 3 and Cu step 4 in the acid soil; Zn step 4 in the calcareous soil; and steps 1, 3, and 4 in the acid soil.

Watson [32] used the four-stage BCR sequential extraction scheme of Davidson et al. [92] on a calcareous sewage sludge disposal site (pH 7.1 to 7.8):

1. $0.11 M$ HAc
2. $0.1 M$ $NH_2OH.HCl$, pH 2
3. $8.8 M$ H_2O_2/1 M NH_4OAc
4. Aqua regia

Cadmium and zinc were extracted mainly in steps 1 and 3; Cu and Pb mainly in step 3; and Ni and Cr in steps 3 and 4. Step 2 extracted only very small amounts of all six metals.

In the case of single and sequential extraction, care must be taken in the interpretation of any changes in metal extractability. For example, the increases in EDTA-extractable Cd, Cu, Ni, and Zn reported by Pulford et al. [31] were measured in soil after coppicing of willow. Cutting of the trees could trigger changes in the rhizosphere (for example, due to root death) that could cause the release of low molecular weight organic compounds, which could complex the metals resulting in greater extractability. Changes measured under growing trees tend to show a decrease in EDTA-extractable metals, which may be due to plant uptake.

Overall, the few studies of this type reported do tend to suggest that tree growth causes changes in the bioavailable pool of metal, shown particularly by the increase in soil solution metal concentrations measured by Watson [32]. Whether an increase or a decrease in bioavailability is measured depends on the stage in the growth cycle when the measurements are made, i.e., the balance between plant uptake and retention of metal in the rhizosphere and the multiple additional factors described earlier.

20.4.2 ROOT FORAGING AND THE RHIZOSPHERE

Soil heterogeneity has a large influence on the location of root activity and growth in soil. Plant roots selectively forage favorable regions in the soil and, similarly, may avoid unfavorable or toxic zones [24]; this has been found to have a profound effect on metal uptake by plants [54]. In low-level or moderately contaminated soils, of course, some elements (e.g., Cd) may be undetectable by plant roots. Furthermore, root processes may alter the speciation and mobility of metals. Phytoavailability of metals is largely related to the aqueous speciation and free ion activity of that metal. The free metal ion is in rapid equilibrium with cell-surface binding sites, and is thus available for uptake; however, other metal species (e.g., colloidal metals and metals complexed to strongly binding organic ligands) are not available for plant uptake (Figure 20.1) [93].

Clearly, there is much scope for transformation of metals between labile and nonlabile pools and for changes between aqueous- and solid-phase metals. The role of rhizosphere processes in metal mobility and availability is poorly understood, but root exudates are likely to have a major influence on speciation and metal bioavailability. For example, investigations of Ni hyperaccumulator plants in rhizotrons have shown that exudation of organic ligands from roots forms Ni–organic complexes that enhance the solubility of the metal in the rhizosphere and increase Ni uptake by the plant [94].

Mycorrhizae are associated with the roots of most of the higher plants, and long-lived woody species are almost invariably mycorrhizal [95]. Over 150 taxa of mycorrhizae have been found to be associated with *Salix* alone in Britain [96]. Mycorrhizal fungi are known to be influential in affecting establishment of trees on nutrient-deficient and severely contaminated soils [97] and in limiting uptake of metals by tree roots from soil. It has been argued that mycorrhizae provide an effective way for tree roots to survive in metal-contaminated soils [98]. Much shorter life cycles offer much more opportunity for selection of metal resistance in mycorrhizae, avoiding the necessity for an adaptive response of tree roots and allowing their survival in otherwise toxic soils.

There is little doubt of the protective role of mycorrhizae in metal-contaminated and otherwise environmentally stressed soils; they provide benefits of enhanced nutrient uptake while offering protection against uptake of toxic metals. However, the role of mycorrhizae is less certain in soils that contain elevated concentrations of metals that are not necessarily toxic or detrimental to growth of the plant or fungus. Ericoid mycorrhizae are known to improve the ability of plants to grow on metal-contaminated soils. In unpolluted soils, the fungus increases the solubility and mobility of Zn using extracellular organic acids. However, in Zn-polluted soil, the fungus loses this ability and protects the plant from excessive Zn uptake [99]. Thus, the fungus appears to maintain homeostasis of metals under different soil conditions. Uptake of Cd, Ni, and Pb from contaminated soils has been shown to be lower in *Quercus rubra* (red oak) seedlings infected with the ectomycorrhiza *Suillus luteus* [100].

Soil processes that influence pollutants can be complex. One example is from evidence that *Pinus sylvestris* (Scots pine) roots and associated mycorrhizal infections increase soil bacterial communities that, in turn, enhance the degradation of mineral oils and petroleum hydrocarbons (PHC) in contaminated soil [101]. Whether mycorrhizae increase or decrease uptake of metals by plants is uncertain, but this is of obvious relevance to remediation of contaminated soils [102]. Ectomycorrhizae have been shown to increase and decrease uptake of Cd by Norway spruce (*Picea abies*) seedlings, depending on the concentration of the metal to which the seedling is exposed [103]. Practical application of this knowledge by inoculating trees roots is only in the rudimentary stages, although mycorrhizae may be less abundant and of less importance on fertile soils [98,103–107].

20.4.3 OTHER ECOLOGICAL PROCESSES IN SOIL

The role of soil fauna, such as earthworms, in altering the mobility of metals in soils in relation to tree growth is equally little understood, although their presence in soil can markedly increase tree growth [108]. Earthworms are intricately involved with the activity of living roots through decomposition, soil microbial activity, and mineralization [109]. It has long been suspected that earthworms are likely to alter the solubility of heavy metals in soil, but this is largely unexplored in the literature.

Ireland [110] found that *Dendrobaena rubida* altered the solubility of Pb, Zn, and Ca in reclaimed Welsh mining spoils, reporting a 50% increase in availability of water-soluble Pb in feces. Devliegher and Verstraete [111,112] found that availability of copper, chromium, and cobalt was raised by between 6 and 30% in the soil by *Lumbricus terrestris*. Ma et al. [113,114] found that species of *Pheretima* increased the mobility of metals in Pb/Zn mine tailing diluted with uncontaminated soil. They suggested that earthworms may benefit attempts to use plants for phytoextraction by increasing the amount of metal in the soil available for plant uptake. The presence of *L. terrestris* in microcosms increased the concentration of Cd, Cu, and Zn into roots and shoots (and of Pb and Fe into roots) of *Lolium perenne* seedlings [115,116].

Experimental work has shown that earthworm activity enhances tree seedling growth associated with enhanced soil organic matter, improved nutrition (including NO_{3-}, NH_{4+}, and Ca^{2+}), and increased mycorrhizal colonization [117]. Yield of a tropical leguminous woody shrub, *Leucaena leucocephala*, in amended Pb–Zn mine tailings has been found to be increased by 10 to 30% in the presence of burrowing earthworms (*Pheretima* spp.) [113,114]. The earthworms increased available forms of N and P in soil, increased metal bioavailability, and raised metal uptake into plants by 16 to 53%.

Some evidence indicates that earthworms increase metal bioavailability in relatively low-level metal-contaminated soils with higher organic matter contents [114,118]. This agrees with results of experiments in which the addition of exogenous humic acid to soil has been shown to increase plant-available metals [119]. This appears to be due to the formation of metal–humic complexes that prevent transformation to insoluble species. If this is correct, clearly earthworms may be of considerable benefit to phytoextraction. The current state of knowledge certainly suggests that phytoremediation should be viewed in an ecological context that includes a consideration of the plant–soil–animal interactions that influence metal mobility.

20.5 PHYTOEXTRACTION VS. PHYTOSTABILIZATION

From the evidence of the BCFs listed in Table 20.2, only Cd and Zn will accumulate in sufficiently high concentrations in the above-ground tissues of trees (BCF > 1) to be candidates for phytoremediation. When the amounts accumulated are compared with the amounts of metal in the soil, only removal of Cd from marginally contaminated soil is possible within a realistic timescale. Zinc is usually present in high concentrations, of which the amount of metal in the plant tissues represents a very small proportion. The other metals commonly studied (Cr, Cu, Ni, Pb) are poorly bioavailable in the soil or are poorly translocated out of the root.

Willows accumulate Zn and Cd more readily than other trees and woody plants do and it has been suggested that this may pose a threat of transfer of metals to the food web [120,121]. It has also been demonstrated that Zn can be transferred to aphids via plant uptake, resulting in Zn concentrations in aphids four times greater than in the soil [122]. Cadmium concentrations in aphids reflected those in plants, but neither metal appeared to be transferred to predatory ladybirds. High Cd concentrations in tissues of small mammals have been recorded at similar sites where willow is grown on dredged sediments [123].

Alternatives to phytoremediation include *ex situ* washing methods with strong chelating agents such as EDTA [124] or acids [125]. Attempts to use the same chemicals *in situ* to enhance metal uptake into plants have had limited success [126], are often prohibitively expensive, and may be destructive to fertility and soil biota; also, residual concentrations in soil pore water may pose a risk of ground water contamination. Nevertheless, BCFs can vary by an order of magnitude over a relatively short pH range from 5.5 to 7.0 in leaf and root vegetables [127], and Cd uptake into *Salix* has been found to be highly pH dependent in field stands [90]. It may be possible to enhance Cd uptake with low-cost organic or inorganic acid soil amendments.

Considerable potential benefits can be gained from the growth of trees on contaminated land, resulting from the stabilization of the soil and/or the contaminant. The protection afforded simply by the presence of the large above-ground biomass of trees can result in a decrease in wind- and water erosion of the soil [128]. Leaf litter can accumulate on the soil surface, forming a barrier over the contaminated soil, which can also help its physical stabilization. Tree roots form a significant below-ground biomass that can effectively bind the soil [129]. They also take up large amounts of water lost from the leaf surface in the transpiration stream. This represents a significant upward movement of water from the soil, via the plant, to the atmosphere, which decreases the potential for metals leaching from the soil. It was estimated in one study that leaching under a tree cover was about 16% less than under grass [130].

In addition to the physical stabilization, chemical stabilization of certain elements may occur. For example, it has been shown that Cr is strongly held in the roots, regardless of the form supplied to the plant (CrIII or CrVI). CrIII is highly insoluble and thus poorly bioavailable, and it has been suggested that CrVI can be reduced to CrIII in the rhizosphere [131]. Lead is another element that is strongly bound in root tissues, possibly due to the formation of lead phosphate [132].

In addition to the use of various soil amendments, including liming, zeolites, beringite, and organic matter, tree cover provides a potentially sustainable vegetation that may allow the contaminants to remain permanently immobilized in soil or woody biomass. At the present time, evidence demonstrating that this happens is insufficient. It is known that metals can become vertically mobile in soil profiles under mature woodlands [22,79] and, clearly, this may threaten underground aquifers in the longer term. Other studies of highly contaminated mature woodlands [133] have demonstrated very low metal mobility and relatively steady-state conditions. Further studies are required to evaluate the long-term feasibility of using trees to stabilize metal contaminants in soil.

20.6 CONCLUSIONS

Urban regeneration, remediation of brownfield land, and cleanup of contaminated soils require innovative, low-cost, ecologically sound, and effective techniques for removal of heavy metals from soil. Phytoremediation using trees provides a potential opportunity to extract or stabilize metals. Phytoextraction involves the use of high yielding plants that readily transport targeted metals from soil to vegetation, allowing removal of metals by harvesting the plants, without damaging the soil or requiring its disposal to landfill. The process takes longer, but meanwhile allows greening of the land, and harvested plants can be used as bioenergy crops.

We provide gathering evidence that selected naturally occurring clones of *Salix* (willows and osiers) are likely to meet this requirement for cadmium, resolving risks associated with its widespread and ubiquitous contamination of soils and restoring soil health within a realistic time. When the concentration of Cd in a soil is just above guideline values or annual inputs are small, phytoextraction by *Salix* may provide an efficient and cost-effective method of cleanup. However, despite many claims to the contrary, field evidence is lacking and phytoextraction using trees remains an unproven technology for reasons discussed. There is probably very limited scope for other metals on brownfield land, although opportunity to reduce phytotoxic concentrations of Zn from agricultural soils may exist.

Tree planting provides aesthetic and ecological improvement of derelict and underused land. The likelihood of the success of phytostabilization is increasing and a good range of field-based evidence already exists. However, outstanding gaps in knowledge relate to the long-term success of ensuring no risk to human health (from re-entrainment of particulates or increased downward movement of metals to water bodies) or to the wider environment (e.g., to food chains).

The probable outcome of current research efforts will be that both phytoremediation strategies using trees will form part of an integrated solution to brownfield land, alongside natural attenuation, soil washing, and chemical and biological additives to contaminated soils.

REFERENCES

1. DETR, Contaminated land: implementation of part IIa of the Environmental Protection Act 1990 in England. Draft circular, London, Department of Environment, Transport and Regions, 2000.
2. Kirby, P. and Elliott, S., The municipal engineer in regeneration and the "brownfield debate," in *Land Reclamation: Extending the Boundaries*, Moore, H.M., Fox, H.R. and Elliott, S. (Eds.) Swets & Zeitlinger, Lisse, 2003, 3.
3. BIO-WISE, *Contaminated Land Remediation: A Review of Biological Technology*, London, DTI, 2000
4. Vangronsveld, J. and Cunningham, S.D., Metal-contaminated soils: *in situ* inactivation and phytorestoration, R.G. Landes Company, Georgetown, TX, 1998.
5. Vangronsveld, J., Ruttens, A., Mench, M., Boisson, J., Lepp, N. W., Edwards, R., Penny, C., and van der Lelie, D., *In situ* inactivation and phytoremediation of metal- and metalloid-contaminated soils: field experiments, in *Bioremediation of Contaminated Soils*, Wise, D.L. and Trantolo, D.J. (Eds.) Marcel Dekker, Inc., New York, 2000, 859–885.
6. Pulford, I.D. and Watson, C., Phytoremediation of heavy metal-contaminated land by trees - a review. *Environ. Int.*, 29, 529, 2003.
7. Dickinson, N.M., Strategies for sustainable woodlands on contaminated soils, *Chemosphere*, 41, 259–263, 2000.
8. Chaney, R.L., Malik, M., Li, Y.M., Brown, S.L., Brewer, E.P., Angle, J.S., and Baker, A.J.M., Phytoremediation of soil metals, *Curr. Opin. Biotechnol.*, 8, 279, 1997.
9. Brooks, R.R., *Plants that Hyperaccumulate Heavy Metals*, CAB International, Wallingford, MA, 1998, 380.
10. Lindegaard, K.N. and Barker, J.H.A., Breeding willows for biomass, *Asp. Appl. Biol.* 49, 155, 1997.
11. Larsson, S., New willow clones for short rotation coppice from Svalof Weibull AB, in *Willow Vegetation Filters for Municipal Wastewaters and Sludges. A Biological Purification System*, Aronsson, P. and Perttu, K. (Eds.), Swedish University of Agricultural Sciences., Uppsala, 1994, 193.
12. Punshon, T. and Dickinson, N.M., Acclimation of *Salix* to metal stress, *New Phytol.*, 137, 303, 1997.
13. Johnson, M.S., McNeilly, T., and Putwain, P.D., Revegetation of metalliferous mine spoil contaminated by lead and zinc, *Environ. Pollut.* 12, 261, 1977.
14. Dickinson, N.M. and Lepp, N.W., Metals and trees: impacts, responses to exposure and exploitation of resistance traits, in *Contaminated Soils*, Prost, R. (Ed.), INRA, Paris, 1997, 247.
15. Mitchell, R.J., Marrs, R.H., Le Duc, M.G., and Auld, M.H.D., A study of the restoration of heathland on successional sites: changes in vegetation and soil chemistry properties, *J. Appl. Ecol.*, 36, 770, 1999.
16. Tabbush, P. and Parfitt, R., Poplar and willow varieties for short-rotation coppice, Edinburgh, Forestry Commission Information Note FCIN17, 1999.
17. Hutchings, T., The opportunities for woodland on contaminated land, Edinburgh, Forestry Commission, Information Note FCIN44 2002 Information Note, FCIN44.
18. Riddell–Black, D., Heavy metal uptake by fast growing willow species, in *Willow Vegetation Filters for Municipal Wastewaters and Sludges: a Biological Purification System*, Aronsson, P. and Perttu, K. (Eds.), Swedish University of Agricultural Sciences, Uppsala, 1994, 145.
19. Raiko, M.O., Gronfors, T.H.A., and Haukka, P., Development and optimization of power plant concepts for local wet fuels, *Biomass Bioenerg.*, 24, 27, 2003.
20. Bradshaw, A.D. and McNeilly, T., *Evolution and Pollution*, Edward Arnold, London, 1982.
21. Bradshaw, A.D., Adaptations of plants to soils containing toxic metals — a test for conceit, in *Origins and Development of Adaptation*, 102 ed., Evered, D. and Collins, G.M., Pitman Publishing, London, 1984, 4.
22. Dickinson, N.M., Watmough, S.A., and Turner, A.P., Ecological impact of 100 years of metal processing at Prescot, Northwest England, *Env. Rev.*, 4, 8, 1996.
23. Watmough, S.A. and Dickinson, N.M., Dispersal and mobility of heavy metals in relation to tree survival in an aerially-contaminated woodland ecosystem, *Environ. Pollut.*, 90, 139, 1995.
24. Turner, A.P. and Dickinson, N.M., Survival of *Acer pseudoplatanus* L. (sycamore) seedlings on metalliferous soils, *New Phytol.*, 123, 509, 1993.
25. Turner, A.P. and Dickinson, N.M., Copper tolerance of *Acer pseudoplatanus* L. (sycamore) in tissue culture, *New Phytol.*, 123, 523, 1993.

26. Watmough, S.A. and Dickinson, N.M., Variability of metal resistance in *Acer pseudoplatanus* L. (sycamore) callus tissue of different origins, *Environ. Exp. Bot.*, 36(3), 293–302, 1996.

27. Watmough, S.A., Gallivan, C.C., and Dickinson, N.M., Induction of zinc and nickel resistance in *Acer pseudoplatanus* L. (sycamore) callus cell lines, *Environ. Exp. Bot.*, 35, 465, 1995.

28. Dickinson, N.M., Metal resistance in trees, in *Heavy Metals and Trees*, Glimmerveen, I., Ed., Institute of Chartered Foresters (ICF), Edinburgh, 1996, 85.

29. Dickinson, N.M., Turner, A.P., and Lepp, N.W., How do trees and other long-lived plants survive in polluted environments? *Funct. Ecol.*, 5, 5, 1991.

30. Wisniewsi, L. and Dickinson, N.M., Toxicity of copper to *Quercus robur* (English oak) seedlings from a copper-rich soil, *Environ. Exp. Bot.*, 50, 99, 2003.

31. Pulford, I.D., Riddell–Black, D., and Stewart, C., Heavy metal uptake by willow clones from sewage sludge-treated soil: the potential for phytoremediation, *Int. J. Phytoremediation*, 4, 59, 2002.

32. Watson, C., The phytoremediation potential of *Salix*: studies of the interaction of heavy metals and willows, Ph.D. thesis, University of Glasgow, 2002.

33. Hammer, D., Kayser, A., and Keller, C., Phytoextraction of Cd and Zn with *Salix viminalis* in field trials, *Soil Use Manage.*, 19, 187, 2003.

34. Greger, M. and Landberg, T., Use of willows in phytoextraction, *Int. J. Phytoremediation*, 1, 115, 1999.

35. Laureysens, I., Blust, R., De Temmerman, L., Lemmens, L., and Ceulemans, R., Clonal variation in heavy metal accumulation and biomass production in a poplar coppice culture. I. Seasonal variation in leaf, wood and bark concentrations, *Environ. Pollut.*, 131, 485, 2004.

36. Laureysens, I., Bogaert, J., Blust, R., and Ceulemans, R., Biomass production of 17 poplar clones in a short-rotation coppice culture on a waste disposal site and its relation to soil characteristics, *Forest Ecol. Manage.*, 187, 295, 2004.

37. Laureysens, I., De Temmerman, L., Hastir, T., Van Gysel, M., and Ceulemans, R., Clonal variation in heavy metal accumulation and biomass production in a poplar coppice culture. II. Vertical distribution and phytoextraction potential, *Environ. Pollut.*, 133, 541, 2005.

38. Laureysens, I., Deeraedt, W., Indeherberge, T., and Ceulemans, R., Population dynamics in a 6-year old coppice culture of poplar. I. Clonal differences in stool mortality, shoot dynamics and shoot diameter distribution in relation to biomass production, *Biomass Bioenerg.*, 24, 81, 2003.

39. Drew, A.P., Guth, R.L., and Greatbatch, W., Poplar culture to the year 2000, in *Proc. Poplar Councils, USA and Canada Joint Meeting*, Syracuse/Cornwall (Ontario), 1987, 109.

40. Ekvall, L. and Greger, M., Effects of environmental biomass-producing factors on Cd uptake in two Swedish ecotypes of *Pinus sylvestris*, *Environ. Pollut.*, 121, 401, 2003.

41. Berry, C., Growth and heavy metal accumulation in pine seedlings grown with sewage sludge, *J. Environ. Qual.*, 14, 415, 1985.

42. Lepp, N.W. and Eardley, G., Growth and trace metal content of European sycamore seedlings grown in soil amended with sewage sludge, *J. Environ. Qual.*, 7, 614, 1978.

43. Borgegård, S.O. and Rydin, H., Biomass, root penetration and heavy metal uptake in birch, in a soil cover over copper tailings, *J. Appl. Ecol.*, 26, 585, 1989.

44. Ehlin, P.O., Seasonal variations in metal contents of birch, *Geologiska Foreningens i Stockholm Fohandkingar*, 104, 63, 1982.

45. Eltrop, L., Brown, G., Joachim, O., and Brinkmann, K., Lead tolerance of *Betula* and *Salix* in the mining area of Mechernich/Germany, *Plant Soil*, 131, 275, 1991.

46. Kopponen, P., Utriainen, M., Lukkari, K., Suntioinen, S., Karenlampi, L., and Karenlampi, S., Clonal differences in copper and zinc tolerance of birch in metal-supplemented soils, *Environ. Pollut.*, 112, 89, 2001.

47. Utriainen, M.A., Karenlampi, L.V., Karenlampi, S.O., and Schat, H., Differential tolerance to copper and zinc of micropropagated birches tested in hydroponics, *New Phytol.*, 137, 543, 1997.

48. Pyatt, F.B., Copper and lead bioaccumulation by *Acacia retinoides* and *Eucalyptus torquata* in sites contaminated as a consequence of extensive ancient mining activities in Cyprus, *Ecotox. Environ. Safe.*, 50, 60, 2001.

49. Rout, G.R., Samantaray, S., and Das, P., Chromium, nickel and zinc tolerance in *Leucaena leucocephalla* (K8), *Silvae Genet.*, 48, 151–157, 1999.

50. Cheung, K.C., Wong, J.P.K., Zhang, Z.Q., Wong, J.W.C., and Wong, M.H., Revegetation of lagoon ash using the legume species *Acacia auriculiformis* and *Leucaena leucocephala*, *Environ. Pollut.*, 109, 75, 2000.

51. Ramsey, M.H. and Argyraki, A., Estimation of measurement uncertainty from field sampling: implications for the classification of contaminated land, *Sci. Total Environ.*, 198, 243, 1997.

52. Arrouays, D., Martin, S., Lepretre, A., and Bourennane, H., Short-range spatial variability of metal contents in soil on a one hectare agricultural plot, *Commun. Soil Sci. Plan*, 31, 387, 2000.

53. CLAIRE Technical Bulletin TB7. Improving the reliability of contaminated land assessment using statistical methods: part 1 — basic principles and concepts, London, 2004.

54. Millis, P.R., Ramsey, M.H., and John, E.A., Heterogeneity of cadmium concentration in soil as a source of uncertainty in plant uptake and its implications for human health risk assessment, *Sci. Total Environ.*, 2003.

55. Flynn, H.C., Meharg, A.A., Bowyer, P.K., and Paton, G.I., Antimony bioavailability in mine soils, *Environ. Pollut.*, 124, 93, 2003.

56. Morin, M.D., Heavy metal concentrations in 3-year old trees grown on sludge-amended surface mine spoil, in *Proceedings of a Symposium on Surface Mining, Hydrology, Sedimentology and Reclamation*, Groves, D.H. (Ed.), OES Press, Lexington, KY, 1981, 297.

57. Ross, S.M., Toxic metals: fate and distribution in contaminated ecosystems, in *Toxic Metals in Soil–Plant Systems*, Ross, S.M. (Ed.), John Wiley & Sons, Chichester, U.K., 1994, 189.

58. Dinelli, E. and Lombini, A., Metal distributions in plants growing on copper mine spoils in Northern Apennines, Italy: the evaluation of seasonal variations, *Appl. Geochem.*, 11, 375, 1996.

59. McGregor, S.D., The uptake and extraction of heavy metals contaminated soil by coppice woodland, Ph.D. thesis, University of Glasgow, 1999.

60. Hasselgren, K., Utilization of sewage sludge in short-rotation energy forestry: a pilot study, *Waste Manage. Res.*, 17, 251, 1999.

61. Lepp, N.W., Uptake, mobility and loci of concentrations of heavy metals in trees, in *Heavy Metals and Trees*. Proceedings of a Discussion Meeting, Glasgow, Glimmerveen, I. Institute of Chartered Foresters, Edinburgh, 1996, 68.

62. Punshon, T., Dickinson, N.M., and Lepp, N.W., The potential of *Salix* clones for bioremediating metal polluted soil, in *Heavy Metals and Trees*. Proceedings of a Discussion Meeting, Glasgow, Glimmerveen, I. Institute of Chartered Foresters, Edinburgh, 1996, 93.

63. Sander, M.L. and Ericsson, T., Vertical distributions of plant nutrients and heavy metals in *Salix viminalis* stems and their implications for sampling, *Biomass Bioenerg.*, 14, 57, 1998.

64. Pulford, I.D., Watson, C., and McGregor, S.D., Uptake of chromium by trees: prospects for phytoremediation, *Environ. Geochem. Health*, 23, 307, 2001.

65. Arduini, I., Godbold, D.L., and Onnis, A., Cadmium and copper change root growth and morphology of *Pinus pinea* and *Pinus pinaster* seedlings, *Physiol. Plantarum*, 92, 675, 1994.

66. Flowers, T.J. and Yeo, A.R., *Solute Transport in Plants*, Blackie, London, 1992.

67. Marschner, H., *Mineral Nutrition of Higher Plants*, 2nd ed., Academic Press, 1988.

68. Pahlsson, A.-M.B., Toxicity of heavy metals (Zn, Cu, Cd, Pb) to vascular plants: a literature review, *Water, Air Soil Pollut.*, 47, 287, 1989.

69. Hodgson, D.R. and Buckley, G.P., A practical approach towards the establishment of trees and shrubs on pulverised fuel ash, in *The Ecology of Resource Degradation and Renewal* (5th Symposium of The British Ecological Society), Chadwick, M.J. and Goodman, G.T. (Eds.), Blackwell Scientific Publications, Oxford, 1975, 305.

70. Nable, F.H., Bañuelos, G.S., and Paull, J.G., Boron toxicity, *Plant Soil*, 193, 181, 1997.

71. Dickinson, N.M. and Lepp, N.W., A model of retention and loss of fungicide-derived copper in different-aged stands of coffee in Kenya, *Agr. Ecosyst. Environ.*, 14, 15, 1985.

72. Dickinson, N.M., Turner, A.P., and Lepp, N.W., Survival of trees in a metal-contaminated environment, *Water Air Soil Pollut.*, 57/8, 627, 1991.

73. Glimmerveen, I., *Heavy Metals and Trees*, Institute of Chartered Foresters, Edinburgh, 1996, 206.

74. Lehn, H. and Bopp, M., Prediction of heavy-metal concentrations in mature plants by chemical analysis of seedlings, *Plant Soil*, 101, 9, 1987.

75. Watmough, S.A., Adaptation to pollution stress in trees: metal tolerance traits, Ph.D. thesis, Liverpool John Moore University, 1994.

76. Dickinson, N.M., Turner, A.P., Watmough, S.A., and Lepp, N.W., Acclimation of trees to pollution stress: cellular metal tolerance traits, *Ann. Bot.*, 70, 569, 1992.

77. Brown, M.T. and Wilkins, D.A., Zinc tolerance of mycorrhizal *Betula* spp., *New Phytol.*, 99, 101, 1985.

78. Denny, H.J. and Wilkins, D.A., Zinc tolerance in *Betula* species. I: Effect of external concentration of zinc on growth and uptake, *New Phytol.*, 106, 517, 1987.

79. Martin, M.H. and Coughtrey, P.J., Cycling and fate of heavy metals in a contaminated woodland ecosystem, in *Pollutant Transport and Fate in Ecosystems*, 6th ed., Coughtrey, P.J., Martin, M.H., and Unsworth, M.H. (Eds.), Blackwell, London, 1987, 319.

80. Burton, K.W. and Morgan, E., The influence of heavy metals upon the growth of sitka spruce in South Wales forest. 1. Upper critical and foliar concentrations, *Plant Soil*, 73, 327, 1983.

81. Derome, J. and Saarsalmi, A., The effect of liming and correct fertilization on heavy metal and macronutrient concentrations in soil solution in heavy-metal polluted Scots pine, *Environ. Pollut.*, 104, 249, 1999.

82. Punshon, T., Lepp, N.W., and Dickinson, N.M., Resistance to copper toxicity in some British willows, *J. Geochem. Explor.*, 52, 259, 1994.

83. Keller, C., Hammer, D., Kayser, A., Richner, W., Brodbeck, M., and Sennhauser, M., Root development and heavy metal phytoextraction efficiency: comparison of different plant species in the field, *Plant Soil*, 249, 67, 2003.

84. Crow, P. and Houston, T.J., The influence of soil and coppice cycle on the rooting habit of short rotation poplar and willow coppice, *Biomass Bioenerg.*, 26, 497, 2004.

85. Vandecasteele, B., Quataert, P., and Tack, F.M.G., The effect of hydrological regime on the bioavailability of Cd, Mn and Zn for the wetland plant species *Salix cinerea*, *Environ. Pollut.*, 135, 303, 2005.

86. Lepp, N.W., Hartley, J., Toti, M., and Dickinson, N.M., Patterns of soil copper contamination and temporal changes in vegetation found in the vicinity of a copper rod-rolling factory, in *Contaminated Soils*, Prost, R. (Ed.), INRA, Paris, 1997, CD ROM (059.PDF).

87. Lepp, N.W. and Dickinson, N.M., Biological interactions: the role of woody plants in phytorestoration, in *Metal-Contaminated Soils: in Situ Inactivation and Phytorestoration*, Vangronsveld, J. and Cunningham, S.D., R.G. Landes, Georgetown, 1998.

88. Dickinson, N.M., Lepp, N.W., and Ormand, K.L., Copper contamination of a 68-year old *Coffea arabica* L. plantation, *Environ. Pollut. B.*, 7, 223, 1984.

89. Hammer, D. and Keller, C., Changes in the rhizosphere of metal-accumulating plants evidenced by chemical extractants, *J. Environ. Qual.*, 31, 1561, 2002.

90. Eriksson, J. and Ledin, S., Changes in phytoavailability and concentration of cadmium in soil following long term *Salix* cropping, *Water Air Soil Pollut.*, 114, 171, 1999.

91. Benitez, N.L. and Dubois, J.-P., Evaluation of the selectivity of sequential extraction procedures applied to the speciation of cadmium in some soils of the Swiss Jura, *Int. J. Environ. Anal. Chem.*, 74, 289, 1998.

92. Davidson, C.M., Duncan, A.L., Littlejohn, D., Ure, A.M., and Garden, L.M., A critical evaluation of the three-stage BCR sequential extraction procedure to assess the potential mobility and toxicity of heavy metals in industrially-contaminated land, *Anal. Chim. Acta*, 363, 45, 1998.

93. Nolan, A.L., Lombi, E., and McLaughlin, M.J., Metal bioaccumulation and toxicity in soils — why bother with speciation? *Aust. J. Chem.*, 56, 77, 2003.

94. Wenzel, W.W., Bunkowski, M., Puschenreiter, M., and Horak, O., Rhizosphere characteristics of indigenously growing nickel hyperaccumulator and excluder plants on serpentine soil, *Environ. Pollut.*, 123, 131, 2003.

95. Peat, H.J. and Fitter, A.H., The distribution of arbuscular mycorrhizas in the British flora, *New Phytol.*, 125, 845, 1993.

96. Watling, R., Macrofungi associated with British willows, *Proc. R. Soc. Edinburgh*, 96B, 133, 1992.

97. Kahle, H., Response of roots of trees to heavy metals, *Environ. Exp. Bot.*, 33, 99, 1993.

98. Wilkinson, D.M. and Dickinson, N.M., Metal resistance in trees: the role of mycorrhizae, *Oikos*, 72, 298, 1994.

99. Martino, E., Perotto, S., Parsons, R., and Gadd, G.M., Solubilization of insoluble inorganic zinc compounds by ericoid mycorrhizal fungi derived from heavy metal polluted sites, *Soil Biol. Biochem.*, 35, 133, 2003.

100. Dixon, R., Response of ectomycorrhizal *Quercus rubra* to soil cadmium, nickel and lead, *Soil Biol. Biochem.*, 20, 555, 1988.

101. Heinonsalo, J., Jorgensen, K.S., Haahtela, K., and Sen, R., Effects of *Pinus sylvestris* root growth and mycorrhizosphere development on bacterial carbon source utilization and hydrocarbon oxidation in forest and petroleum-contaminated soils, *Can. J. Microbiol.*, 46, 451, 2000.

102. Leyval, C., Turnau, K., and Haselwandter, K., Effect of heavy metal pollution on mycorrhizal colonization and function: physiological, ecological and applied aspects, *Mycorrhiza*, 7, 139, 1997.

103. Jentschke, G., Winter, S., and Godbold, D.L., Ectomycorrhizas and cadmium toxicity in Norway spruce seedlings, *Tree Physiol.*, 19, 23, 1999.

104. Jones, M.D. and Hutchinson, T.C., The effects of nickel and copper on the axenic growth of ectomycorrhizal fungi, *Can. J. Bot.*, 66, 119, 1986.

105. Stenstrom, E. and Ek, M., Field growth of *Pinus sylvetris* following nursery innoculation with mycorrhizal fungi, *Can. J. Forest Res.*, 20, 914, 1990.

106. Wilkins, D.A., The influence of sheathing (ecto-) mycorrhizas of trees on the uptake and toxicity of metals, *Agr. Ecosyst. Environ.*, 35, 245, 1991.

107. Colpaert, J.V. and Van Assche, J.A., The effects of cadmium on ectomycorrhizal *Pinus sylvestris* L., *New Phytol.*, 123, 325, 1993.

108. Haimi, J. and Einbork, M., Effects of endogeic earthworms on soil processes and plant growth in coniferous forest soil, *Biol. Fert. Soils*, 13, 6, 1991.

109. Uyl, A., Didden, W., and Marinissen, J., Earthworm activity and decomposition of 14C-labeled grass root systems, *Biol. Fert. Soils*, 36, 447, 2002.

110. Ireland, M.P., The effect of the earthworm *Dendrobaena rubida* on the solubility of lead, zinc and calcium in heavy metal contaminated soil in Wales, *J. Soil Sci.*, 26, 313, 1975.

111. Devliegher, W. and Verstraete, W., *Lumbricus terrestris* in a soil core experiment: effects of nutrient enrichment processes (NEP) and gut-associated processes (GAP) on the availability of plant nutrients and heavy metals, *Soil Biol. Biochem.*, 28, 489–496, 1996.

112. Devliegher, W. and Verstraete, W., The effect of *Lumbricus terrestris* on soil in relation to plant growth: effects of nutrient-enrichment processes (NEP) and associated processes (GAP), *Soil Biol. Biochem.*, 29, 341, 1997.

113. Ma, Y., Dickinson, N.M., and Wong, M.H., Toxicity of Pb/Zn mine tailings to the earthworm *Pheretima* and the effects of burrowing on metal availability, *Biol. Fertil. Soils*, 36, 79, 2002.

114. Ma, Y., Dickinson, N.M., and Wong, M.H., Interactions between earthworms, trees, soil nutrition and metal mobility in amended Pb/Zn mine tailings from Guangdong, China, *Soil Biol. Biochem.*, 35, 1369, 2003.

115. Rida, A.M.M.A., Trace-element concentrations and growth of earthworms and plants in soils with and without cadmium, copper, iron, lead and zinc contamination: interactions of soil and earthworms, *Soil Biol. Biochem.*, 28, 1029, 1996.

116. Rida, A.M.M.A., Trace-element concentrations and growth of earthworms and plants in soils with and without cadmium, copper, iron, lead and zinc contamination: interactions of plants, soil and earthworms, *Soil Biol. Biochem.*, 28, 1037, 1996.

117. Welke, S.E. and Parkinson, D., Effect of *Aporrectodea trapezoides* activity on seedling growth of *Pseudotsuga menziesii*, nutrient dynamics and microbial activity in different forest soils, *Forest Ecol. Manage.*, 173, 169, 2003.

118. Cheng, J. and Wong, M.H., Effects of earthworms on Zn fractionation in soils, *Biol. Fert. Soils*, 36, 72, 2002.

119. Halim, M., Conte, P., and Piccolo, A., Potential availability of heavy metals to phytoextraction from contaminated soils induced by exogenous humic substances, *Chemosphere*, 52, 265, 2003.

120. Vandecasteele, B., De Vos, B., and Tack, F.M.G., Cadmium and Zinc uptake by volunteer willow species and elder rooting in polluted dredged sediment disposal sites, *Sci. Total Environ.*, 299, 191, 2002.

121. Vandecasteele, B., Lauriks, R., De Vos, B., and Tack, F.M.G., Cd and Zn concentration in hybrid poplar foliage and leaf beetles grown on polluted sediment-derived soils, *Environ. Monit. Assess.*, 89, 263, 2003.

122. Green, I.D., Merrington, G., and Tibbett, M., Transfer of cadmium and zinc from sewage sludge amended soil through a plant-aphid system to newly emerged adult ladbirds (*Coccinella septempunctata*), *Agr. Ecosyst. Environ.*, 99, 171, 2003.

123. Mertens, J., Luyssaert, S., Verbeeren, S., Vervaeke, P., and Lust, N., Cd and Zn concentrations in small mammals and willow leaves on disposal facilities for dredged material, *Environ. Pollut.*, 115, 17, 2001.

124. Sun, B., Zhao, F.J., Lombi, E., and McGrath, S.P., Leaching of heavy metals from contaminated soils using EDTA, *Environ. Pollut.*, 113, 111, 2001.

125. Chen, Y.X., Lin, Q., Luo, Y.M., He, Y.F., Zhen, S.J., Yu, Y.L., Tian, G.M., and Wong, M.H., The role of citric acid on the phytoremediation of heavy metal contaminated soil, *Chemosphere*, 50, 807, 2003.

126. Lombi, E., Zhao, F.J., Dunham, S.J., and McGrath, S.P., Phytoremediation of heavy metal-contaminated soils: natural hyperaccumulation vs. chemically enhanced phytoextraction, *J. Environ. Qual.*, 30, 1919, 2001.

127. DEFRA, Soil guideline values for cadmium contamination, Environment Agency, Bristol, 2002.

128. Johnson, M.S., Cooke, J.A., and Stevenson, J.K.W., Revegetation of metalliferous wastes and land after mining., in *Mining and its Environmental Impacts*, Harrison, R.M. (Ed.), Royal Society of Chemistry, London, 1992, 31.

129. Stomp, A.M., Han, K.H., Wilbert, S., and Gordon, M.P., Genetic improvement of tree species for remediation of hazardous wastes, In Vitro *Cell Dev.*, B, 29, 227, 1993.

130. Garten, C.T., Modeling the potential role of a forest ecosystem in phytostabilization and phytoextraction of 90Sr at a contaminated watershed., *J. Environ. Radioact.*, 43, 305, 1999.

131. James, B.R., Remediation-by-reduction strategies for chromate-contaminated soils., *Environ. Geochem. Health*, 23, 175, 2001.

132. Cotter–Howells, J.D., Champness, P.E., Charnock, J.M., and Pattrick, R.A.D., Identification of pyromorphite in mine-waste soils by ATEM and EXAFS, *Eur. J. Soil Sci.*, 45, 393, 1994.

133. Lepp, N.W. and Dickinson, N.M., Natural bioremediation of metal-polluted soils — a case history from the U.K., in *Risk Assessment and Sustainable Land Management Using Plants in Trace Element-Contaminated Soils*, Mench, M. and Mocquut, B. (Ed.), INRA, Bordeaux, France, 2003.

134. Klang–Westin, E. and Eriksson, J., Potential of *Salix* as phytoextractor for Cd on moderately contaminated soils, *Plant Soil*, 249, 127, 2003.

135. Klang–Westin, E. and Perrtu, K., Effects of nutrient supply and soil cadmium contamination on cadmium removal by willow, *Biomass Bioenerg.*, 23, 415, 2002.

136. Greger, M. and Landberg, T., Use of willow clones with high Cd accumulating properties in phytoremediation of agricultural soils with elevated Cd levels, in *Contaminated Soils: 3rd International Conference on the Biogeochemistry of Trace Elements*, Prost, R. (Ed.), INRA, Paris, 1997, 505.

137. Ostman, G., Cadmium in *Salix* — a study of the capacity of *Salix* to remove cadmium from arable soils, in *Wilow Vegetation Filters for Municipal Waste Waters and Sludges: a Biological Purification System*, Aronnsson, P. and Perttu, K. (Eds.), Swedish University of Agricultural Sciences, Uppsala, 1994, 153.

138. Labrecque, M., Teodorescu, T.I., and Daigle, S., Effect of wastewater sludge on growth and heavy-metal bioaccumulation of two *Salix* species, *Plant Soil*, 171, 303, 1995.

139. Labrecque, M., Teodorescu, T.I., and Daigle, S., Early performance and nutrition of two willow species in short-rotation intensive culture fertilized with wastewater sludge and impact on soil characteristics, *Can. J. Forest Res.*, 28, 1621, 1998.

140. Landberg, T. and Greger, M., Differences in uptake and tolerance to heavy metals in *Salix* from unpolluted and polluted areas, *Appl. Geochem.*, 11, 175, 1996.

141. Salt, C.A., Hipkin, J.A., and Davidson, B., Phytoremediation — a feasible option at Lanarkshire Steelworks? in *Heavy Metals and Trees*. Proceedings of a Discussion Meeting, Glasgow, Glimmerveen, I. Institute of Chartered Foresters, Edinburgh, 1996, 51.

142. Fernandes, J.C. and Henriques, F.S., Metal contamination in leaves and fruits of holm-oak (*Quercus rotundifolia* LAM) trees growing in a pyrites mining area at Aljustrel, Portugal, *Water Air Soil Pollut.*, 48, 409, 1989.

143. Rosselli, W., Keller, C., and Boschi, K., Phytoextraction capacity of trees growing on a metal-contaminated soil, *Plant Soil*, 256, 265, 2003.

144. Ettala, M.O., Short-rotation tree plantations at sanitary landfills, *Waste Manage. Res.*, 6, 291, 1988.

145. Vervaeke, P., Luyssaert, S., Mertens, J., Meers, E., Tack, F.M.G., and Lust, N., Phytoremediation prospects of willow stands on a contaminated sediment: a field trial, *Environ. Pollut.*, 126, 275, 2003.

146. Vervaeke, P., Luyssaert, S., Mertens, J., De Vos, B., Speleers, L., and Lust, N., Dredged sediment as a substitute for biomass production of willow trees established using the SALIMAT technique, *Biomass Bioenerg.*, 21, 81, 2001.

147. Vangronsveld, J., Colpaert, J.V., and Van Tichelen, K.K., Reclamation of a bare industrial area contaminated by nonferrous metals: physicochemical and biological evaluation of the durability of soil treatment and revegetation., *Environ. Pollut.*, 94, 131, 1996.

21 Stabilization, Remediation, and Integrated Management of Metal-Contaminated Ecosystems by Grasses (Poaceae)

M.N.V. Prasad

CONTENTS

21.1 INTRODUCTION

Heavy metals are being released into the environment by anthropogenic activities, primarily associated with industrial processes, manufacturing, and disposal of industrial and domestic refuse and waste materials [1]. Mining activities produce a large quantity of waste materials (such as tailings), which frequently contain excessive concentrations of heavy metals. These mining activities and waste materials create pollution and land degradation without vegetation coverage. Phytoremediation of metalliferous mine tailings is necessary for long-term stability of the land surface or removal of toxic metals. Soils and water contaminated with heavy metals pose a major environmental and human health problem that needs an effective and affordable technological solution. Phytostabilization is an important subset of phytoremediation technology in which plants stabilize the pollutants in soils, thus rendering them harmless [2].

Metal-tolerant grasses are preferred for phytostabilization, and heavy metal hyperaccumulators are the best choice for phytoextraction. Plants that can adapt to wetland conditions are useful for phytofiltration. The success of reclamation schemes depends upon the choice of plant species and their methods of [3,4]. There are some important considerations when selecting plants for phytostabilization. Plants should be tolerant of the soil metal levels as well as the other inherent site conditions (e.g., soil pH, salinity, soil structure, water content, lack of major nutrients, and organic materials). Plants chosen for phytostabilization should also be poor translocators of metal contaminants to above-ground plant tissues that could be consumed by humans or animals. Additionally, the plants must grow quickly to establish ground cover, have dense rooting systems and canopies.

The heavy metal acquisition in the members of Poaceae (grasses) is by strategy II, which is much simpler than strategy I. Roots of grasses release to soil specific chelating substances called phytosiderophores, which are mainly derivatives of mugineic acid synthesized from nicotianamine — a nonprotein aminoacid. They extract and chelate ferric ions from soil sorption complex; the whole complex with oxidized Fe ions is taken up by plant. Ferric ions release and their immediate reduction takes place inside root cells. Strategy II is more effective and more resistant to unfavorable environmental factors than strategy I is because two important steps (Fe release from chelate and Fe reduction) are shifted from the rhizosphere into homeostatic conditions inside living root cells. The tolerance of monocots to heavy metals is generally higher than that of dicots and requires rather a long exposure time to heavy metals.

The tolerance of *Agrostis tenuis* Sibth. populations growing in mining areas was investigated by exposing tillers to different concentrations of metals in nutrient solutions. An index of tolerance was calculated by comparing root length with specimens grown in solutions without trace metals [5]. Species of *Agrostis* (Poaceae) have different mechanisms of tolerance to trace metals. Relationships between different species and metal complexation processes in roots and plant tissues have been characterized [6–18] (Table 21.1).

21.2 VETIVER GRASS FOR PHYTOSTABILIZATION OF METALLIFEROUS ECOSYSTEMS

The vetiver grass (*Vetiveria zizanioides L.*) has rather unique morphological and physiological characteristics. It has the ability to control erosion and sedimentation, to withstand extreme soil and climatic variations, and to tolerate elevated heavy metal concentrations in water and soil [107,108]. Vetiver is sterile and noninvasive; it does not compete with native vegetation and, as a nurse plant, it fosters the voluntary return of native plants. Most importantly, however, vetiver is proven technology: its effectiveness as an environmental protection tool has been demonstrated around the world and in some parts of the U.S. [109]. Due to its unique character, vetiver grass has been widely known for its effectiveness in erosion and sediment control [110,111].

The most conspicuous characteristics of vetiver grass include its fast growth, large biomass, strong root system, and high level of metal tolerance; therefore, it is an important candidate for stabilization of metal-contaminated soils. Results from glasshouse studies show that, when adequately supplied with nitrogen and phosphorus fertilizers, vetiver can grow in soils with very high levels of acidity, aluminum, and manganese. Vetiver growth was not affected and no obvious symptoms were observed when soil pH was as low as 3.3 and the extractable manganese reached 578 mg kg^{-1}, and plant manganese was as high as 890 mg kg^{-1}. Bermuda grass (*Cynodon dactylon*), which has been recommended as a suitable species for acid mine rehabilitation, has 314 mg kg^{-1} of manganese in plant tops when growing in mine wastes containing 106 mg kg^{-1} of manganese [112]. Vetiver also produced excellent growth at a very high level of soil aluminum saturation percentage (68%), but it did not survive an aluminum saturation level of 90% at soil pH of 2.0. The toxic level of aluminum for vetiver would be between 68 and 90% [113]. It also has been observed that vetiver thrives well on highly acidic soil with aluminum saturation percentage as high as 87%.

In Australia, *V. zizanioides* has been successfully used to stabilize mining overburden and highly saline, sodic, magnesic, and alkaline (pH 9.5) tailings of coal mines, as well as highly acidic (pH 2.7) arsenic tailings of gold mines [114]. In China, it has been demonstrated that *V. zizanioides* is one of the best choices for revegetation of Pb/Zn mine tailings due to its high metal tolerance [115,116]; furthermore, this grass can also be used for phytoextraction because of its large biomass.

TABLE 21.1
Adaptive Physiological Investigations of Metal Toxicity and Tolerance Using Grasses

Grasses' interaction with metals and associated functions	Ref.
Agrostis capillaries	
Phosphate uptake and transport in L.: effects of nontoxic levels of aluminum and the significance of P and Al speciation	19
Cd–thiolate protein	14
Quantification of metalothionein in small root samples exposed to Cd	9
Isolation and partial purification of Cd-binding protein from roots	10
The amount of Cd associated with Cd-binding protein in roots	11
HPLC characterization of Cd-binding protein	15
Relationship between tolerance to heavy metals	7
A. gigantea	
Differential tolerance of three cultivars to Cd, Cu, Pb, Ni, and Zn	16
Agrostis tenuis	
Pb- and Zn-tolerant populations	6
Combined tolerance to Cu, Zn, and Pb	5
Subcellular distribution of Zn and Cu in the roots of metal-tolerant clones	17
ATPase from the roots: effect of pH, Mg, Zn, and Cu	18
Agrostis tenuis and *A. capillaris* L.	
Relationship between tolerance to heavy metals	7
Agrostis sp.	
Heavy metal-tolerant populations and other grasses	8
Agrostis tennis	
Evolutionary processes in copper-tolerant populations	20
Agrostis stolonifera	
The rapid evolution of copper tolerance	21
The uptake of copper and its effect upon respiratory processes of roots of copper-tolerant and -nontolerant clones	22
Populations resistant to lead and zinc toxicity	6
Heavy metal tolerance in populations of sibth. and other grasses	23
Combined tolerance to copper, zinc, and lead	5
Variability between and within *Agrostis tenuis* populations regarding heavy metal tolerance	24
Selective adaptation to copper in population studies	25
Occurence of metalothioenein in roots tolerant to Cu	12
Avena sativa	
Oxidative damage by excess of Cu	26
Growth and some photosynthetic characteristics of field grown under copper and lead stress	27
Deschampsia cespitosa	
Metal cotolerance	28
Multiple and cotolerance to metals: adaptation, preadaptation and "cost"	29
Role of metallothionein like proteins in Cu tolerance	30
Festuca rubra	
Selective adaptation to copper in population studies	25

TABLE 21.1
Adaptive Physiological Investigations of Metal Toxicity and Tolerance Using Grasses
(continued)

Grasses' interaction with metals and associated functions	Ref.
Holcus lanatus	
Environment-induced Cd tolerance	31
Induction and loss of metal tolerance to Pb, Zn, and Cd factorial combination	32
Tolerance to lead, zinc, and cadmium in factorial combination	33
Cadmium tolerance and toxicity, oxygen radical process, and molecular damage in cadmium-tolerant and cadmium-sensitive clones	34
Interactive effects of Cd, Pb, and Zn on root growth of two metal-tolerant genotypes	35
Hordeum vulgare	
Accumulation of cadmium, molybdenum, nickel, and zinc by roots and leaves	36
Compartmentalization and transport of zinc in barley primary leaves is the basic mechanism of zinc tolerance	37
Zinc stress induces changes in apoplasmic protein content and polypeptide composition of barley primary leaves	37
Comparison of Cd-, Mo-,Ni-, and Zn-stress is partially related to the loss of preferential extraplasmic compartmentalization	38
Critical levels of 20 potentially toxic elements	39
Comparative analysis of element composition of roots and leaves of seedlings grown in the presence of Cd, MO, Ni, and Zn	36
Effects of aluminum on the growth and distribution of calcium in roots of an aluminum-sensitive cultivar of barley	40
Oryza sativa	
Isolation of Cd-binding protein from Cd-treated plant	41
Chemical form of Cd and other trace metals	42
Limiting steps on photosynthesis Cu-treated plants	43
Cd content in rice and its daily intake in various nations	44
Cd content in rice and rice field soils in China, Indonesia, and Japan with difference to soil type and daily intake from rice	45
Cd-induced peroxidase activity and isozymes	46
Cd and Ni effects on mineral nutrition and interaction with ABA and GA	47
Heavy metalñhormone interactions in rice plants: effects on growth, net photosynthesis, and carbohydrate distribution	48
Sorghum bicolor	
Detailed characterization of heat shock protein synthesis and induced thermotolerance in seedlings	49
Heat shock proteins in ethanol, sodium arsenite, sodium malonate, and the development of thermotolerance	50
Arsenate-inducible heat shock proteins	50
Triticum aestivum	
Physiological and ultrstructural effects of cadmium	51
Aluminum effects on photosynthesis and elemental uptake in an aluminum-tolerant and -nontolerant wheat cultivars	52
The effect of crop rotations and tillage practices on cadmium concentration in grains	53
Physiological and ultrastructural effects of cadmium in wheat leaves	54

TABLE 21.1
Adaptive Physiological Investigations of Metal Toxicity and Tolerance Using Grasses
(continued)

Grasses' interaction with metals and associated functions	Ref.
Citrate reverses the inhibition of wheat root growth caused by aluminum; water relations of wheat cultivars grown with cadmium	55
Differential transport of Zn, Mn, and sucrose along the longitudinal axis of developing grains	56
Transport of zinc and manganese to developing wheat grains	57
Uptake and distribution of ^{65}Zn and ^{54}Mn in wheat grown at sufficient and deficient levels of Zn and Mn, I. during vegetative growth.	59
Uptake and distribution of ^{65}Zn and ^{54}Mn in wheat grown at sufficient and deficient levels of Zn and Mn, II. during grain development	59
Induction of microsomal membrane proteins in roots of aluminum-resistant cultivars of L. under conditions of aluminum stress	60
Induction of microsomal membrane proteins in roots of an Al-resistant cultivars under conditions of Al stress	60
Residual Cd in soil and accumulation in grain	61
Differential aluminum tolerance in a high yielding, early maturing wheat	62
Effect of Cu on active and posssibe Rb influx in roots	63
Influence of Se on uptake and toxicity of Cu and Cd	64
Chemical form of Cd and other heavy metals	42
Mn effects on photosynthesis and chlorophyll	65
Mn produces organic acids in Mn-tolerant and Mn-sensitive cultivars	66
Mn tolerance	67
Chlorophyll and leaf growth as a measure of Mn tolerance	68
Heat shock and protection against metal toxicity	69
A protein similar to pathogenesis-related (PR) proteins is elicited by metal toxicity in roots	70
Chemical form of cadmium and other heavy metals in rice and wheat plants	42
Effect of heavy metals on isoperoxidases of wheat	71
Influence of toxic metals on the multiple forms of esterases	72
Influence of selenium on uptake and toxicity of copper and cadmium in wheat	64
Aluminum induces rapid changes in cytosolic pH and free calcium and potassium concentrations in root protoplasts	73
Excess manganese effect on photosynthetic rate and concentration of chlorophyll grown in solution culture	65
Effects of excess manganese on production of organic acids in Mn-tolerant and Mn-sensitive cultivars of *Triticum aestivum* (wheat)	66
Chlorophyll content and leaf elongation rate as a measure of managanese tolerance	67
Pedigree analysis of manganese tolerance in Canadian spring wheat	68
Influence of aluminum on photosynthetic function in phosphate-deficient	74
Increased nonphotochemical quenching in leaves of aluminum-stressed plants is due to Al $^{3+}$-induced elemental loss	75
Metal content, growth responses, and some photosynthetic measurements on field-cultivated in wheat growing on ore bodies enriched in Cu	76
Citrate reverses the inhibition of root growth caused by Al	55
Crop rotation and tillage practices on Cd accumulationin grain	53
Distribution of Cd, Cu, and Zn in fruit	77
Mechanisms of Al tolerance — role in nitrogen nutrition and differential pH	78–81
Al tolerance is independent of rhizosphere pH	82, 83

TABLE 21.1
Adaptive Physiological Investigations of Metal Toxicity and Tolerance Using Grasses
(continued)

Grasses' interaction with metal and associated function	Ref.
Differential uptake and toxicity of ionic and chelated Cu cultivars	78, 79
Kinetics of Al uptake by excised roots ofAl-tolerant and Al-sensitive cultivars	84, 85
Interactive effects of Cd, Cu, Mn, Ni, and Zn on root growth in solution culture	86
Effects of biological inhibitors on kinetics of Al uptake by excised roots and purified cell wall material of Al-tolerant and Al-sensitive cultivars	87
Vetiveria ziznioides	
Integrated management of metal-contaminated environment	88
Zea mays	
Measurements of Pb, Cu, and Cd bindings with mucilage exudates	89
Compartmentalization of cadmium, copper, lead, and zinc in seedlings of maize and induction of metallothionein	90
Silicon amelioration of aluminum toxicity	91
Inhibition of photosynthesis by lead	92
Changes in photosynthesis and transpiration of excised leaves by Cd and Pb	93, 94
The effect of heavy metals on plants, II net changes in photosynthesis and transpiration by Pb, Cd, Ni, and Tl	95
Differing sensitivity of photosynthesis and transpiration to Pb contamination	93
Effect of pH on Cd distribution in maize inbred lines	96, 97
PC concentration and binding state of Cd in roots of maize genotypes differing in shoot and root Cd partitioning	96
Cd distribution in maize inbred lines and evaluation of structural physiological characteristics	97
Zn and Cd in corn plants	98
X-ray microanalytical study and distribution of Cd in roots	99
Compartmentalization of Cd, Cu, Pb, and Zn in seedlings and MT induction	90
Localization of Pb	100
Mobilization of Cd and other trace metals from soil by root exudates	101, 102
Metal-binding properties of high molecular weight soluble exudates	89
Measurements of Pb, Cu, and Cd bindings with mucilage exudates from roots	103
Regulation of assimilatory sulfate reduction by Cd	10
Quantification of metalothionein in small root samples exposed to Cd	104
Cd binding proteins in roots	105
Changes in seedling glutathione exposed to Cd	11, 106

Recent research also suggests that *V. zizanioides* has higher tolerance to acid mine drainage (AMD) from a Pb/Zn mine, and wetland microcosms planted with this grass can effectively adjust pH and remove SO_4^{2-}, Cu, Cd, Pb, Zn, and Mn from AMD [117]. All of these demonstrate that *V. zizanioides* has great potential in phytoremediation of heavy metal-contaminated soils and water, and an integrated vetiver technique can be developed for remediation of metal pollution, especially in mining areas.

Vetiver grass is tolerant to a wide variety of environmentally adverse conditions and tolerates elevated concentrations of trace elements; its stems are stiff and erect, forming dense hedges [109]. Acid sulfate soils (ASS) are highly erodible and difficult to stabilize and rehabilitate due to extremely acidic conditions, with pH between 3 and 4. Eroded sediment and leachate from ASS are also extremely acidic. The leachate from ASS has led to disease and death of fish in several coastal zones of eastern Australia. Vetiver has been successfully used to stabilize and rehabilitate highly erodible ASS on the coastal plain in tropical Australia, where actual soil pH is around 3.5 and oxidized pH is as low as 2.8 [118].

A series of glasshouse trials was carried out to determine the tolerance of vetiver to high soil levels of heavy metals. A literature search indicated that most vascular plants are highly sensitive to heavy metal toxicity and most plants were also reported to have very low threshold levels for arsenic, cadmium, chromium, copper, and nickel in the soil. Table 21.3 shows results that demonstrate that vetiver is highly tolerant to these heavy metals. Research on wastewater treatment in Thailand found that vetiver grass could absorb substantial quantities of Pb, Hg, and Cd in waste water [119].

21.3 VETIVER IN COMBINATION WITH GREEN MANURE LEGUMES ON LEAD/ZINC MINE TAILINGS

Shu and Xia [120] conducted a field trial to compare growth performance, metal accumulation of vetiver (*Vetiveria zizanioides*), and two legume species (*Sesbania rostrata* and *Sesbania sesban*) grown on tailings amended with domestic refuse and/or fertilizer. The lead (Pb)/zinc (Zn) tailings in China contained high concentrations of heavy metals (total Pb, Zn, Cu, and Cd concentrations of 4164, 4377, 35, and 32 mg kg^{-1}, respectively), and low contents of major nutrient elements (N, P, and K) and organic matter. It was revealed that domestic refuse alone and the combination of domestic refuse and artificial fertilizer significantly improved the survival rates and growth of *V. zizanioides* and two *Sesbania* species, especially the combination.

However, artificial fertilizer alone did not improve the survival rate and growth performance of the plants grown on tailings. Roots of these species accumulated similar levels of heavy metals, but the shoots of two *Sesbania* species accumulated higher (three- to fourfold) concentrations of Pb, Zn, Cu, and Cd than shoots of *V. zizanioides* did. Most of the heavy metals in *V. zizanioides* were accumulated in roots, and the translocation of metals from roots to shoots was restricted. Intercropping of *V. zizanioides* and *S. rostrata* did not show any beneficial effect on individual plant species in terms of height, biomass, survival rate, and metal accumulation, possibly due to the rather short experimental period of 5 months.

One developing alternative remediation technique for metal-contaminated sites is phytostabilization, also called "in-place inactivation" or "phytorestoration." It is a type of phytoremediation technique that involves stabilizing heavy metals with plants in contaminated soils. To be a potentially cost-effective remediation technique, plants selected must be able to tolerate high concentrations of heavy metals and to stabilize heavy metals in soils by roots of plants with some organic or inorganic amendments, such as domestic refuse, fertilizer, and others. Revegetation of mining wastes is one of the longest practiced and well documented approaches for stabilization of heavy metals in mining wastes [3].

Shu and Hanping conducted field experiments to assess the role of vetiver grass in phytostabilization of metal-contaminated sites at Guangdong Province, South China. They compared the growth of four grasses (*Vetiveria zizanioides*, *Paspalum notatum*, *Cynodon dactylon*, and *Imperata cylindraca* var. *major*) on Lechang Pb/Zn mine tailings with different amendments, for screening the most useful grass and the most effective measure for revegetation of tailings. They also investigated the abilities of heavy metal accumulation in the four tested plants for assessing their different roles in phytoremediation.

They observed that the concentrations of Pb, Zn, and Cu in shoots and roots of *V. zizanioides* were significantly less than those of the other three species and the shoot/root metal concentration quotients (MS/MR for Pb, Zn, and Cu) in *V. zizanioides* were also lower than those of other three species, which indicated that *V. zizanioides* was an excluder of heavy metals. First, roots of the species accumulated low levels of metals by avoiding or restricting uptake. Second, shoots of the species accumulated much lower concentrations of metals by restricting transport. In general, metal tolerance and metal uptake were functionally related; exclusion was one of the basic strategies of metal uptake by tolerant species [121]. Judging from the metal contents in plant tissues, *V. zizanioides* was more suitable for phytostabilization of toxic mined lands than *P. notatum* and *C. dactylon*, which accumulated relatively high levels of metals in their shoots and roots. It was also noted that *I. cylindrica* accumulated lower amounts of Pb, Zn, and Cu than *C. dactylon* and *P. notatum* did and could also be considered for phytostabilization of tailings.

21.4 TOLERANCE OF VETIVER GRASS TO SUBMERGENCE

These grasses are ideal for biological or ecological measures protecting or stabilizing inner slopes of rivers, reservoirs, and lakes, because plants established on the inner slopes are almost all drowned by the elevated water level in the rainy season. Thus, the key is to screen out plant species strongly tolerant to submergence in order to stabilize and vegetate the "wet" slopes effectively (Table 21.2).

The selected eight grasses were all excellent plant species for soil and water conservation in southern China. The common features of the eight species are high erosion control and high resistance to adverse conditions. They all have been applied widely in erosion control and slope stabilization in southern regions of China. Among them, vetiver is the sole high-stalk type; the other seven species are procumbent types. These plants were raised in pots first and then put into a cement tank filled with water to investigate their tolerance to complete submergence. In an experiment conducted for 3 years, Xia et al. [122] have observed that vetiver and Bermuda grass could tolerate the longest time of submergence — at least up to 100 days — and probably more than that.

21.5 REHABILITATION OF GOLD MINE TAILINGS IN AUSTRALIA

The environment of fresh and old gold tailings is highly hostile to plant growth. The fresh tailings are typically alkaline (pH = 8 to 9); low in plant nutrients; and very high in free sulphate (830 mg kg^{-1}), sodium, and total sulfur (1 to 4%). Vetiver established and grew very well on these tailings without fertilizers, but growth was improved by the application of 500 kg ha^{-1} of di-ammonium phosphate. Due to high sulfur content, old gold tailings are often extremely acidic (pH = 2.5 to 3.5), high in heavy metals, and low in plant nutrients. Revegetation of these tailings is very difficult and often very expensive and the bare soil surface is highly erodible. These tailings are often the source of above-ground and underground contaminants to the local environment. When vetiver was

TABLE 21.2
A Comparative Study on the Tolerance of Eight Grasses to Submergence

Scientific name of the grass	Common name of the grass	Duration of submergence
Vetiveria zizanioides Nash	Vetiver grass	100
Cynodon dactylon Pars	Bermuda grass	100
Paspalum notatum Flugge	Bahia grass	70
Axonopus compressus Beauv	Carpet grass	30–40
Chrysopogon aciculatus Trin	Aciculate chrysopogon	25–30
Paspalum conjugatum Bergius	Sour paspalum	25–30
Stenotaphrum secundatum Kuntze	St. Augustine	20–30
Eremochloa ophiuroides Hack.	Common centipede grass	10

Source: Xia et al., *Proc. 3rd Int. Vetiver Conf.*, Guangzhou, China, 532.

adequately supplied with nitrogen and phosphorus fertilizers, excellent growth was obtained on sites with pH ranging from 2.7 to 3.6, high in sulphate (0.37 to 0.85%), total sulfur (1.31 to 3.75%), and low in plant nutrients. Liming was not needed on sites with higher pH (3.5), but the addition of 20 t ha^{-1} of agricultural lime significantly improved vetiver growth on sites with pH of 2.7.

21.6 REHABILITATION OF MINE TAILINGS IN SOUTH AFRICA AND CHINA

Rehabilitation trials conducted by De Beers on slime dams at several sites have found that vetiver possess the necessary attributes for self-sustainable growth on kimberlite spoils. Vetiver grew vigorously on the alkaline kimberlite and contained runoff, arrested erosion, and created an ideal microhabitat for the establishment of indigenous grass species [123]. Vetiver has also been used successfully in the rehabilitation diamond mines at Premier and Koffiefonteine and slime dams at the Anglo–American platinum mine at Rastenburg, and the Velkom, President Brand gold mine (Tantum, pers. com.). In China, vetiver produced biomass more than twice that of *Paspalum notatum, Cynodondactylon,* and *Imperata cylindrica* in the rehabilitation of the Lechang Pb and Zn mine; tailings there contain very high levels of heavy metals (Pb at 3231 mg kg^{-1}, Zn at 3418 mg kg^{-1}, Cu at 174 mg kg^{-1}, and Cd at 22 mg Kg^{-1}) [124].

Vetiver grass has also been successfully used to rehabilitate coal mine tailings that were saline and highly sodic, with high levels of soluble sulfur, magnesium, and calcium and extremely low nitrogen and phosphorus levels [125]. Bentonite tailings are extremely erodible because they are highly sodic (with exchangeable sodium percentage (ESP) values ranging from 35 to 48%), high in sulphate, and extremely low in plant nutrients. Vetiver established readily on these tailings and effectively controlled erosion, conserved moisture, and improved seedbed conditions for the establishment of indigenous species. Residue of bauxite processing known as red mud is highly caustic, with pH levels as high as 12. Vetiver established successfully on the red mud when its pH was raised to 9.0.

Vetiver has been used very successfully for this purpose in Australia to contain leachate runoff from landfills. For this application, high-density planting is required at strategic locations such as the toes of major slopes of highly contaminated areas.

The vetiver system is proven technology; its effectiveness as an environmental protection tool has been demonstrated around the world. It is a very cost-effective, environmentally friendly, and practical phytoremedial tool for the control and attenuation of heavy metal pollution when appropriately applied.

Jarvis and Leung [126] observed *Chamaecytisus palmensis* plants growing in hydroponic culture exposed to Pb(NO3)2, with and without the addition of the chelating agents H-EDTA and EDTA, for 1 week. The unchelated lead accumulated predominantly in root tissue; lead chelated with H-EDTA or EDTA was taken up principally by the shoots. With transmission electron microscopy, ultrastructural observations were carried out on ultrathin sections derived from lead-treated *C. palmensis* tissues. Unchelated lead was found in cell walls, bacteroids and mitochondria in root nodule tissue, and in middle lamellae and intercellular spaces in root tissues. In roots, chelated lead was found in mitochondria, and in shoot tissues it was found in chloroplasts, pit membranes, and plasmodesmata.

21.7 EXPRESSION OF STRESS PROTEINS BY THE MEMBERS OF POACEAE

The relationship between some plant hormones and heavy metal stress in *Oryza sativa*; trace element stress and expressionof signalling/metabolic pathways in some Poaceae are detailed in Tables 21.3 and 21.4 respectively.

TABLE 21.3

Relationships between Some Plant Hormones and Heavy Metal Toxicity in *Oryza sativa*

Phytohormone	Function	Ref.
GA$_3$	Partially reverses the effect of Cd or Ni	48
ABA	Enhances plant growth inhibition by Cd or Ni; does not affect the influence of Al on growth	48; 130

TABLE 21.4

Trace Element Induced Stress and Expression of Signaling/Metabolic Pathways in Some Poaceae

Signaling/metabolic function	Metal	Plant	Ref.
[Ca^{2+}]$_c$ concentration (Ca-dependent signaling):			
Increased			
	Al	*Hordeum vulgare*	40
		Triticum aestivum	
Decreased	Al	*Triticum aestivum*	73
Expression of stress-responsive gene:			
Induced	Cu, Al, Cd, Co	*Hordeum vulgare*	131
Not induced	Cd, Cu, Zn, Al		
Ethylene content increased	Cd, Cu,Zn, Al	*Triticum aestivum*	132
Induction of Octadecenoic pathway	Fe^{2+}	*Oryza sativa*	70
	Cu	*Oryza sativa*	133

Cd induced 70-kDa protein in roots of maize (*Zea mays* L.), which was precipitated by hsp70 antibodies. *In vitro* phosphorylation assay showed that the hsc70 (heat shock proteins induced by factors other than heat were termed "hsp cognates" [hsc]) is a phosphoprotein. Phosphoamino acid analysis of immunoprecipitated hsc70 by paper chromatography showed that serine and tyrosine are phosphorylated in control seedlings. Serine (but not tyrosine) is phosphorylated in Cd-treated seedlings in spite of increase in hsc70 amount [127]. This could possibly be due to inhibition of a tyrosine kinase.

This change in hsp70 cognate phosphorylation might be playing the role of a chaperone. Possibly, the hsc70 in maize might be acting to limit and rescue the damage to proteins caused by environmental stress like other chaperones [127]. Cd induced a number of stress proteins ranging in molecular weight from 10 to 70 kDa. In *Oryza sativa*, Cd-induced proteins are 70-, 42-, 40-, 26-, 23-, 15-, and 11-kDa proteins [128]. Similar results were also noticed in Cd-treated maize [127–129].

Barley and maize seedlings exhibited retardation in shoot and root growth after exposure to Cu, Cd, and Pb. The Zn ions practically did not influence these characteristics. The total protein content of barley and maize roots declined with an increase in trace metal concentration [134,135]. Total glutathione content was decreased when exposed to Cd (more so in roots than in shoots). Robinson and Jackson [136] suggested the involvement of glutathione in the biosynthesis of PCs. Rauser and Curvetto [13] have isolated Cu-binding complexes from *Agrostis gigantea*. In oat (*Avena sativa*) roots, Cd transport from cytosol to vacuole across the tonoplast is demonstrated through Cd^{2+}/H$^+$ antiport activity [137]. Compartmentalization, complexation, and transport are the primary and basic mechanisms involved in Zn tolerance in barley (*Hordeum vulgare*) [37]. In wheat, transport of Rb and Sr to the ear in mature excised shoots was affected by temperature and stem length and also, perhaps, the ligands in the xylem [138].

Barley (*Hordeum vulgare* L.) was used as a bioassay system to determine the mobility of heavy metals in biosolids applied to the surface of soil columns with or without plants. Three weeks after barley was planted, all columns were irrigated with the disodium salt of the chelating agent, EDTA (ethylenediamine tetraacetic acid). Drainage water, soil, and plants were analyzed for heavy metals (Cd, Cu, Fe, Mn, Ni, Pb, and Zn). Total concentrations of the heavy metals in all columns at the end of the experiment generally were lower in the topsoil with EDTA than in that without EDTA. The chelate increased concentrations of heavy metals in shoots.

With or without plants, the EDTA mobilized Cd, Fe, Mn, Ni, Pb, and Zn, which leached to drainage water. Drainage water from columns without EDTA had concentrations of these heavy metals below detection limits. Only Cu did not leach in the presence of EDTA. Even though roots retarded the movement of Cd, Fe, Mn, Ni, Pb, and Zn through the EDTA-treated soil, the drainage water from columns with EDTA had concentrations of Cd, Fe, Mn, and Pb that exceeded drinking water standards by 1.3, 500, 620, and 8.6 times, respectively. Because the chelate rendered Cd, Fe, Mn, Ni, Pb, and Zn mobile, it is suggested that the theory for leaching of soluble salts, put forward by Nielsen and associates in 1965, could be applied to control movement of the heavy metals for maximum uptake during chelate-assisted phytoremediation using grasses [139] (Figure 21.1).

Chen et al. [140] used a modified glass bead compartment cultivation system in which glass beads were used in the hyphal compartment but replaced by coarse river sand in the compartments for host plant roots and mycorrhizal hyphae. Arbuscular mycorrhizal (AM) associations were established using maize (*Zea mays* L.) and two AM fungi, *Glomus mosseae* and *G. versiforme*. When the standard and modified cultivation systems were compared, the new method yielded much more fungal tissue in the hyphal compartment. Using *G. versiforme* as the fungal symbiont, up to 30 mg of fungal dry matter (DM) was recovered from the hyphal compartment of mycorrhizal maize and about 6 mg from red clover.

Multielement analysis was conducted on samples of host plant roots and shoots and on harvested fungal biomass. Concentrations of P, Cu, and Zn were much higher in the fungal biomass than in the roots or shoots of the host plants but fungal concentrations of K, Ca, Mg, Fe, and Mn were similar to or lower than those in the plants. Nutrient concentrations were also significantly different between the two AM fungi; these may be related to differences in their proportions of extraradical mycelium to spores. The high affinity of the fungal mycelium for Zn was very striking and is discussed in relation to the potential use of arbuscular mycorrhiza in the phytoremediation of Zn-polluted soils.

Vetiver grass is a tall perennial tussock grass from Asia that has been used in a variety of soil conservation applications in that region. Interest in this grass outside Asia is increasing, but its application is handicapped by a lack of quantitative knowledge of its flow-retarding and sediment-trapping capability [88]. Vetiver grass hedge is being considered for control of soil erosion on a cultivated flood plain of low slope subject to overland flow [88].

Rengel [142] for the first time reported the coexistence of traits for tolerance to Zn toxicity and Zn deficiency in a single plant genotype (i.e. *Holcus lanatus*). Genotypes tolerant to zinc (Zn) toxicity accumulated Zn in their roots better than Zn-sensitive genotypes, even in Zn-deficient soil. *Holcus lanatus* L. ecotypes differing in tolerance to Zn toxicity were grown in Zn-deficient Laffer soil, which was amended with Zn to create a range of conditions from Zn deficiency to Zn toxicity. Increasing Zn additions to the soil, up to the sufficiency level, improved growth of all ecotypes. At toxic levels of added Zn, the Zn-sensitive ecotype suffered a greater decrease in growth than the Zn-tolerant ecotypes. All ecotypes accumulated more Zn in roots than in shoots, with root concentrations exceeding 8g Zn kg^{-1} dry weight in extreme cases.

When grown in Zn-deficient or Zn-sufficient soil (up to 0.5mg Zn kg^{-1} soil added), ecotypes tolerant to Zn toxicity took up more Zn, grew better, and had greater root and shoot Zn concentration than the control (Zn-sensitive ecotype). Zn-tolerant ecotypes transported more Zn, copper (Cu), and iron (Fe) from roots to shoots in comparison with the Zn-sensitive ecotype. The average Zn uptake rate from Zn-deficient soil (no Zn added) was greater in the Zn-tolerant ecotypes than in the Zn-sensitive ecotype. In conclusion, ecotypes of *H. lanatus* tolerant to Zn toxicity also tolerate

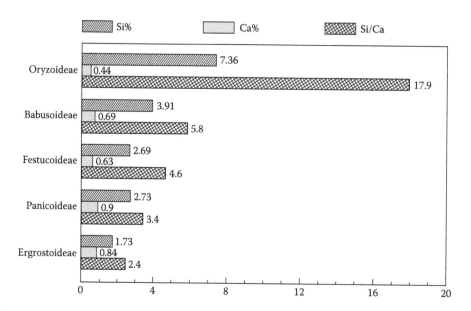

FIGURE 21.1 Ranking of silicon, calcium percentage, and Si/Ca ratio in the subfamilies of Poaceae.

Zn deficiency better than the Zn-sensitive ecotype because of their greater capacity for taking up Zn from Zn-deficient soil.

Zinc tolerance in *Setaria italica* L. *in vitro* was achieved by Samantaray et al. [142] through callus culture derived from leaf base and mesocotyl explants on Murashige and Skoog's medium, supplemented with 0.5 mg/l kinetin and 3.0 mg/l 2,4-D + 0.24 m*M* zinc. Tolerant calli showed vigorous growth in medium containing 0.24 m*M* zinc compared to the non-tolerant calli. Biochemical studies on the basis of activities of peroxidase and catalase as well as estimation of protein and chlorophyll were more in tolerant calli than non-tolerant ones. The accumulation of zinc in the callus was increased significantly with the increase in zinc concentrations in the medium. Plant regeneration via somatic embryogenesis was achieved in tolerant and non-tolerant calli on MS medium containing 1.0 mg/l BA, 1.0 mg/l Kn, and 0.5 mg/l 2,4-D. The somatic embryo-derived plantlets were tested in MS liquid medium with 0.24 m*M* zinc for selection of tolerant clones. This study may help in the selection and characterization of metal-tolerant lines of *Setaria italica* for breeding programs [142].

Using a pot experiment, Chen et al. [143] investigated the uptake of cadmium (Cd) by wetland rice and wheat grown in a red soil contaminated with Cd. The phytoremediation of heavy metal-contaminated soil with vetiver grass was also studied in a field plot experiment. Results showed that chemical amendment with calcium carbonate (CC), steel sludge (SS), and furnace slag (FS) decreased Cd uptake in wetland rice and wheat by 23 to 95% compared with the unamended control. Among the three amendments, FS was the most efficient at suppressing Cd uptake by the plants, probably due to its higher content of available silicon (Si). The concentrations of zinc (Zn), lead (Pb), and Cd in the shoots of vetiver grass were 42 to 67%, 500 to 1200% and 120 to 260% higher in contaminated plots than in control, respectively. Cadmium accumulation by vetiver shoots was 218 g Cd/ha at a soil Cd concentration of 0.33 mg Cd/kg. These results suggest that heavy metal-contaminated soil could be remediated with a combination of chemical treatments using rapidly growing grasses.

The U.S. introduced automobile exhaust catalysts in 1974, followed later by Japan and Europe. This resulted in enormous emissions of platinum (Pt) into the environment, far greater than had been expected. For Pt speciation *Lolium multiflorum* (forage plant) was found to be a reliable bioindicator, in particular for heavy metals [144]. It was shown that Pt was bound to a protein in

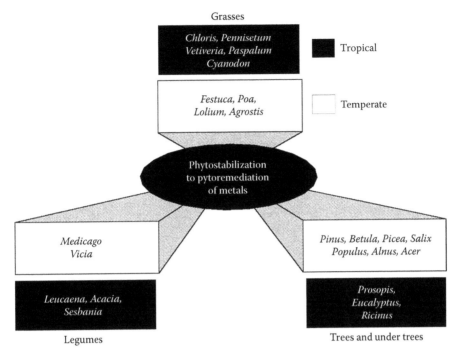

FIGURE 21.2 A community of grass, legume, and tree-based phytoremediation systems would be advantageous.

the high molecular mass fraction in this grass species (160 to 200 kDa) [145]. In Pt-treated cultures, about 90% of the total Pt was observed in a fraction of low molecular mass (<10 kDa) [145]. The Pt-binding ligands could be identified only tentatively; some Pt coeluted with phytochelatin fractions and some with polygalacturonic acids [146].

21.8 SIGNIFICANCE OF SILICON-ACCUMULATING GRASSES FOR INTEGRATED MANAGEMENT AND REMEDIATION OF METALLIFEROUS SOILS

Silica is the second most abundant element in the Earth's crust, so it would be useful for amelioration of aluminum (Al) toxicity in acidic soils. Ranking of silicon accumulation in the subfamilies of Poaceae is shown in Figure 21.2. Subsoil acidification is a serious global environmental concern. Acid soils occupy nearly 30% (3.95 ha) of the arable land area in tropical and temperate belts. In addition to the natural processes, farming and management practices such as high use of nitrogen fertilizers, removal of cations by harvested crops, and leaching and runoff of cations resulted in acidification of soils. In many industrialized areas, the atmospheric deposition of sulfur and nitrogen compounds is a major source of proton influx to soils. More than the low pH of the soils, the major problem associated with acid soils is the toxicity of Al and manganese and the deficiency of phosphorus, calcium, magnesium, and potassium.

In addition to these nutritional factors, the acid soils are also characterized by low water-holding capacity due to compaction of soils. Apart from mineral toxicities, soil acidification is also known to change the species spectrum of the forest soils by changing the microbial activity of the soils. Acidification is also known to reduce the degradation of soil organic matter and alters cation and nutrient flow in the ecosystem. In general, most of the acid soils have low exchangeable bases (e.g., Ca, Mg, and K), mainly because of the low cation exchange capacity (CEC) of the soils. The oxides of Al and Fe in wet acidic soils fix large fractions of the phosphate, making it unavailable to plants

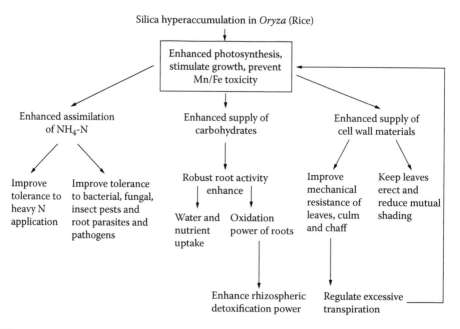

FIGURE 21.3 Silica hyperaccumulation in rice and its functions.

and leading to lower crop yields. Therefore, Al toxicity is the major agronomic problem in acid soils and it has been reported that silica had an ameliorative function [147] (Figure 21.3).

REFERENCES

1. Ross, S.M., Ed. *Toxic Metals in Soil–Plant Systems*. John Wiley & Sons, Chichester, U.K., 469, 1994.
2. Raskin, I. and Ensley, B.D., *Phytoremediation of Toxic Metals*, John Wiley & Sons, Inc., New York, 2000.
3. Bradshaw, A.D. and Chadwick, M.J., *The Restoration of Land: The Ecology and Reclamation of Derelict and Degraded Land*, University of California Press, Berkeley Los Angeles, 302, 1980.
4. Johnson, M.S, Cooke, J.A., and Stevenson, J.K.W., Revegetation of metaliferous wastes and land after metal mining. In: Hester and Harrison R.M. (Eds.), *Mining and its Environmental Impact*, 31, 1994.
5. Karataglis, S.S., Combined tolerance to copper, zinc and lead by populations of *Agrostis tenuis*, *Oikos*, 38, 234, 1982.
6. Bradshaw, A.D., Populations of *Agrostis tenuis* resistant to lead and zinc poisoning, *Nature*, 169, 1098, 1952.
7. Humphreys, M.O. and Nicholls, M.K., Relationship between tolerance to heavy metals in *Agrostis capillaris* L. and *A. tenuis* Sibth, *New Phytol.*, 95, 177, 1984.
8. Jowett, D., Populations of *Agrostis* spp. tolerant of heavy metals, *Nature*, 182, 816, 1958.
9. Rauser, W.E., Isolation and partial purification of cadmium-binding protein from roots of the grass *Agrostis gigantean, Plant Physiol.*, 74, 1025, 1984.
10. Rauser, W.E., Estimating metalothionein in small root samples of *Agrostis gigantea* and *Zea mays* exposed to cadmium, *J. Plant Physiol.*, 116, 253, 1984.
11. Rauser, W.E., The amount of cadmium associated with Cd-binding protein in roots of *Agrostis gigantea*, maize and tomato, *Plant Sci.*, 43, 85, 1986.
12. Rauser, W.E, Metal-binding peptides in plants, in *Sulphur Nutrition and Accumulation in Plants*. De Kok, L.J. et al. (Eds.) SPB Academic Publishing, the Hague, 239, 1993 .
13. Rauser, W.E. and Curvetto, N.R., Metallothionein occurs in roots of *Agrostis* tolerant to excess copper, *Nature*, 287, 563, 1980.

14. Rauser, W.E., Hartmann, H., and Weser, U., Cadmium thiolate protein from the grass *Agrostis gigantean, FEBS Lett.*, 164, 102, 1983.
15. Rauser, W.E., Hunziker, P.E., and Kagi, J.H.R., Reverse phase high-performance liquid chromatography of Cd-binding proteins from the grass *Agrostis gigantea, Plant Sci.*, 45, 105, 1986.
16. Symenoides, L., McNeilly, T., and Bradshaw, A.D., Differential tolerance of three cultivars of *Agrostis capillaris* L. to cadmium, copper, lead, nickel and zinc, *New Phytol.*, 101, 309, 1985.
17. Turner, R.G., The subcellular distribution of zinc and copper within the roots of metal-tolerant clones of *Agrostis tenuis* Sibth, *New Phytol.*, 69, 725, 1970.
18. Veltrup, W., ATPase from the roots of *Agrostis tenuis* Sibth: effect of pH, Mg^{+2}, Zn^{+2}, and Cu^{+2}, *Zeischrift Pflanzenphysiol.*, 108, 457, 1982.
19. Macklon, A.E.S., Lumsdon, D.G., and Sim, A., Phosphate uptake and transport in *Agrostis capillaris* L.: effects of nontoxic levels of aluminum and the significance of P and Al speciation, *J. Exp. Bot.*, 45, 276, 887, 1994.
20. McNeilly, T. and Bradshaw, A.D., Evolutionary processes in populations of copper tolerant *Agrostis tennis* sibth, *Evolution*, 22, 108, 1968.
21. Wu, L., Turman, D., and Bradshaw, A.D., The potential for evolution of heavy metal tolerance in plants. III. The rapid evolution of copper tolerance, *Heredity*, 34, 165, 1975.
22. Wu, L., Turman, D., and Bradshaw, A.D., The uptake of copper and its effect upon respiratory processes of roots of copper-tolerant and nontolerant clones of *Agrostis stolonifera, New Phytol.*, 75, 225, 1975.
23. Gregor, R.P.G. and Bradshaw, A., Heavy metal tolerance in populations of *Agrostis tenuis* sibth. and other grass, *New Phytol.*, 64, 131, 1965.
24. Karataglis, S.S., Variability between and within *Agrostis tenuis* populations regarding heavy metal tolerance, *Phyton*, 20, 23, 1980.
25. Krataglis, S.S., Selective adaptation to copper of populations of *Agrostis tenuis* and *Festuca rubra* (Poaceae) Pl, *Syst. Evol.*, 34, 215, 1980.
26. Luna, C.M., Gonzalez, C.A., and Trippi, V.S., Oxidative damage caused by an excess of copper in oat leaves, *Plant Cell Physiol.*, 35, 11, 1994.
27. Moustakas, M. et al., Growth and some photosynthetic characteristics of field grown *Avena sativa* under copper and lead stress, *Photosynthetica*, 30, 389, 1994.
28. Cox, R.M. and Hutchinson, T.C., Metal cotolerance in the grass *Deschampsia cespitosa, Nature*, 279, 231, 1979.
29. Cox, R.M. and Hutchinson, T.C., Multiple and cotolerance to metals in the grass *Deschampsia cespitosa*, adaptation, preadaptation and cost, *J. Plant Nutr.*, 3, 731, 1981.
30. Schultz, C.L. and Hutchinson, T.C., Evidence against a key role for metalothionein-like protein in the copper tolerance mechanism of *Deschampsia cespitosa* L. Beauv., *New Phytol.*, 110, 163, 1988.
31. Baker, A.J.M., Environmentally induced cadmium tolerance in the grass *Holcus lanatus* L., *Chemosphere*, 13, 585, 1984.
32. Baker, A.J.M. et al., Induction and loss of cadmium tolerance in *Holcus lanatus* L. and other grasses, *New Phytol.*, 102, 575, 1986.
33. Coughtrey, P.J. and Martin, M.H., Tolerance of *Holcus lanatus* to lead, zinc and cadmium in factorial combination, *New Phyto.*, 81, 147, 1978.
34. Hendry, G.A.F., Baker, A.J.M., and Ewart, C.F., Cadmium tolerance and toxicity, oxygen radical process and molecular damage in cadmium-tolerant and cadmium-sensitive clones of *Holcus lanatus* L., *Acta Bot. Neerl.*, 41, 271, 1992.
35. Symeonidis, L. and Karataglis, S., Interactive effects of Cd, Pb, and Zn on root growth of two metal tolerant genotypes of *Holcus lanatus* L., *BioMetals*, 5, 173, 1992.
36. Brune, A. and Dietz, K.J., A comparative analysis of element composition of roots and leaves of barley seedlings grown in the presence of toxic cadmium, molybdenum, nickel and zinc concentratiuons, *J Plant Nutr.*, 18, 853, 1995.
37. Brune, A., Urbach, W., and Dietz, K.J., Compartmentation and transport of zinc in barley primary leaves as basic mechanisms involved in zinc tolerance, *Plant Cell Environ.*, 17, 153, 1994.
38. Brune, A., Urbach, W., and Dietz, K.J., Differential toxicity of heavy metals is partly related to a loss of preferential extraplasmic compartmentation: a comparison of Cd-, Mo-, Ni- and Zn-stress, *New Phytol.*, 129, 403, 1995.

39. Davis, R.D. and Beckett, P.H.T., Critical levels of twenty potentially toxic elements in young spring barley, *Plant Soil*, 68, 835, 1978.

40. Nichol, B.E. and Oliveira, L.A., Effects of aluminum on the growth and distribution of calcium in roots of an aluminum-sensitive cultivar of barley (*Hordeum vulgare*), *Can. J. Bot.*, 73, 1849, 1995.

41. Kaneta, M. et al., Isolation of a cadmium-binding protein from cadmium-treated rice plants, *Agric. Biological Chem.*, 47, 417, 1983.

42. Kaneta, M. et al., Chemical form of cadmium and other heavy metals in rice and wheat plants, *Environ. Health Perspect.*, 65, 33, 1987.

43. Ros, R. et al., Effect of the herbicide MCPA, and the heavy metals, cadmium and nickel on the lipid composition, Mg-ATPase activity and fluidity of plasma membranes from rice, *Oryza sativa* cv. Bahia shoots, *J. Exp. Bot.*, 41, 225, 457, 1990.

44. Rivai, I.F., Koyama, H., and Suzuki, S., Cadmium content in rice and its daily intake in various countries, *Bull. Environ. Contamination Toxicol.*, 44, 910, 1990.

45. Rivai, I.F., Koyama, H., and Suzuki, S., Cadmium contentin rice and rice field soils in China, Indonesia and Japan, with special reference to soil type and daily intake from rice, *Jpn. J. Health Hum. Ecol.*, 56, 168, 1990.

46. Reddy, G.N. and Prasad, M.N.V., Cadmium induced peroxidase activity and isozymes in *Oryza sativa*, *Biochem. Arch.*, 8, 101, 1992.

47. Rubio, M.I. et al., Cadmium and nickel accumulation in rice plants — effects on mineral nutrition and possible interactions of abscisic and gibberellic acids, *Plant Growth Regul.*, 14, 151, 1994.

48. Moya, J.L., Ros, R., and Picazo, I., Metal-hormone interactions in rice plants: effects on growth, net photosynthesis, and carbohydrate distribution, *J. Plant Growth Regul.*, 14, 61, 1995.

49. Howarth, C.J. and Skot, K.P., Detailed characterization of heat shock protein synthesis and induced thermotolerance in seedlings of *Sorghum bicolor* L., *J. Exp. Bot.*, 45, 279, 1353, 1994.

50. Howarth, C.J., Heat shock proteins in sorghum and pearl millet, ethanol, sodium arsenite, sodium malonate and the development of thermotolerance, *J. Exp. Bot.*, 41, 877, 1990.

51. Moustakas, M., Ouzounidou, G., and Eleftheriou, E.P., Physiological and ultrastructural effects of cadmium on wheat (*Triticum aestivum*) leaves. *Proc. 6th Sci.Conf. Hellenic Bot. Soc.*, Paralimni, Cyprus, 332, 1996.

52. Moustakas, M., Ouzounidou, G., and Lannoye, R., Aluminum effects on photosynthesis and elemental uptake in an aluminum-tolerant and nontolerant wheat cultivar, *J. Plant Nutr.*, 18, 669, 1995.

53. Oliver, D.P. et al., The effect of crop rotations and tillage practices on cadmium concentration in wheat grain, *Austr. J. Agric. Res.*, 44, 1221, 1993.

54. Ouzounidou, G., Moustakas, M., and Eleftheriou, E.P., Physiological and ultrastructural effects of cadmium in wheat (*Triticum aestivum* L.) leaves, *Arch. Environ. Contamination Toxicol.*, 32, 154, 1997.

55. Ownby, J.D. and Popham, H.P., Citrate reverses the inhibition of wheat root growth caused by aluminum, *J. Plant Physiol.*, 135, 588, 1989.

56. Pearson, J.N. et al., Differential transport of Zn, Mn and sucrose along the longitudinal axis of developing wheat grains, *Physiol. Plant.*, 97, 332, 1986.

57. Pearson, J.N. et al., Transport of zinc and manganese to developing wheat grains, *Physiol. Plant*, 95, 449, 1995.

58. Pearson, J.N. and Rengel, Z., Uptake and distribution of ^{65}Zn and ^{54}Mn in wheat grown at sufficient and deficient levels of Zn and Mn. I. During vegetative growth, *J. Exp. Bot.*, 46, 833, 1995.

59. Pearson, J.N. and Rengel, Z., Uptake and distribution of ^{65}Zn and ^{54}Mn in wheat growth at sufficient and deficient levels of Zn and Mn. II. During grain development, *J. Exp. Bot.*, 46, 841, 1995.

60. Basu, A., Basu, B., and Taylor, G. J., Induction of microsomal membrane proteins in roots of an aluminum-resistant cultivars of *Triticum aestivum* L. under conditions of aluminum stress, *Plant Physiol.*, 104, 1007, 1994.

61. Brams, E. and Anthony, W., Residual cadmium in a soil profile and accumulation in wheat grain, *Plant Soil*, 109, 3, 1988.

62. Briggs, K.G. et al., Differential aluminum tolerance of high-yielding, early maturing Canadian wheat cultivars and germplasm, *Can. J. Plant Sci.*, 69, 61, 1989.

63. Jensen, P. and Aoalsteinsson, S., Effects of copper on active and passive Rb+ influx in roots of winter wheat, *Physiol. Plantarum*, 75, 195, 1989.

64. Landberg, T. and Greger, M., Influence of selenium on uptake and toxicity of copper and cadmium in pea (*Pisum sativum*) and wheat (*Triticum aestivum*), *Physiol. Plantarum*, 90, 637, 1994.

65. Macfie, S.M. and Taylor, G.J., The effects of excess manganese on photosynthetic rate and concentration of chlorophyll in *Triticum aestivum* grown in solution culture, *Physiol. Plantarum*, 85, 467, 1992.

66. Macfie, S.M., Cossins, E.A., and Taylor, G.J., Effects of excess manganese on production of organic acids in Mn-tolerant and Mn-sensitive cultivars of *Triticum aestivum* (wheat), *J. Plant Physiol.*, 143, 135, 1994.

67. Moroni, J.S., Briggs, K.G., and Taylor, G.J., Chlorophyll content and leaf elongation rate in wheat seedlings as a measure of managanese tolerance, *Plant Soil*, 136, 1, 1991.

68. Moroni, J.S., Briggs, K.G., and Taylor, G.J., Pedigree analysis of the manganese tolerance in Canadian spring wheat (*Triticum aestivum* L.) cultivars, *Euphytica*, 56, 107, 1991.

69. Orzech, K.A. and Burke, J.J., Heat shock and the protection against metal toxicity in wheat leaves, *Plant Cell Environ.*, 11, 711, 1988.

70. CruzñOrtega, R. and Ownby, J.D., A protein similar to PR (pathogenesis-related) proteins is elicited by metal toxicity in wheat roots, *Physiol. Plant*, 89, 211, 1993.

71. Karataglis, K., Moustakas, M., and Symeonides, L., Effect of heavy metals on isoperoxidases of wheat, *Biol. Plantarum*, 33, 3, 1991.

72. Karataglis, S., Symeonidis, L., and Moustakas, M., Effect of toxic metals on the multiple forms of esterases of *Triticum aestivum* cv. Vergina, *J. Agron. Crop Sci.*, 160, 106, 1988.

73. Lindberg, S. and Strid, H., Aluminum induces rapid changes in cytosolic pH and free calcium and potassium concentrations in root protoplasts of wheat (*Triticum aestivum*), *Physiol Plant.*, 99, 405, 1997.

74. Moustakas M., Influence of aluminum on photosynthetic function in phosphate deficient wheat (*Triticum aestivum* L.), *Abs. 10th Congr. Federation Eur. Soc. Plant Physiol.*, Florence, Italy, plant physiology and biochemistry, 268, 1996.

75. Moustakas, M. and Ouzounidou, G., Increased nonphotochemical quenching in leaves of aluminum-stressed wheat plants is due toAl^{3+}-induced elemental loss, *Plant Physiol. Biochem.*, 32, 527, 1994.

76. Lanaras, T. et al., Plant metal content, growth responses and some photosynthetic measurements on field-cultivated wheat growing on ore bodies enriched in Cu, *Physiol. Plantarum*, 88, 307, 1993.

77. Pieczonka, K. and Rosopulo, A., Distribution of cadmium, copper, and zinc in the caryopsis of wheat *Triticum aestivum* L., *Fresenius Anal. Chem.*, 322, 697, 1985.

78. Taylor, G.J. and Foy, C.D., Differential uptake and toxicity of ionic and chelated copper in *Triticum aestivum*, *Can. J. Bot.*, 63, 1271, 1985.

79. Taylor, G.J. and Foy, C.D., Effects of aluminum on the growth and element composition of 20 winter cultivars of *Triticum aestivum* L. (wheat) grown in solution culture, *J. Plant Nutr.*, 8, 811, 1985.

80. Taylor, G.J. and Foy, C.D., Mechanisms of aluminum tolerance in *Triticum aestivum* L. (wheat). I. Diffetnetial pH induced by winter cultivars in nutrient solutions, *Am. J. Bot.*, 72, 695, 1985.

81. Taylor, G.J. and Foy, C.D., Mechanisms of aluminum tolerance in *Triticum aestivum* L. (wheat). II. Differential pH induced by spring cultivars in nutrient solutions, *Am. J. Bot.*, 72, 702, 1985.

82. Taylor, G.J., Aluminum tolerance is independent of rhizosphere pH in *Triticum aestivum* L., *Commun. Soil Sci. Plant Anal.*, 19, 1217, 1988.

83. Taylor, G.J., Mechanisms of aluminum tolerance in *Triticum aestivum* (wheat) V. Nitrogen nutrition, plant-induced and tolerance to aluminum; correlation without causality? *Can. J. Bot.*, 66, 694, 1988.

84. Zhang, G. and Taylor, G.J., Kinetics of aluminum uptake by excised roots of aluminum-tolerant and aluminum-sensitive cultivars of *Triticum aestivum* L., *Plant Physiol.*, 91, 1094, 1989.

85. Zhang, G. and Taylor, G.J., Kinetics of aluminum uptake in *Triticum aestivum* L., *Plant Physiol.*, 94, 577, 1990.

86. Taylor, G.J. and Stadt, K.J., Interactive effects of cadmium, copper, manganese, nickel, and zinc on root growth of wheat (*Triticum aestivum*) in solution culture, in *Plant Nutrition — Physiology and Applications*. (Ed.) M.L. van Beusichem, Kluwer Academic Publishers, 317–322, 1990.

87. Zhang, G. and Taylor, G.J., Effects of biological inhibitors on kinetics of aluminum uptake by excised roots and purified cell wall material of aluminum-tolerant and aluminum-sensitive cultivars of *Triticum aestivum* L., *J. Plant Physiol.*, 138, 533, 1991.

88. http://www.vetiver.com.

89. Morel, J.L., Mench, M., and Guckert, A., Measurements of Pb, Cu and Cd bindings with mucilage exudates from maize *Zea mays* L., *Biol. Fertile Soils*, 2, 29, 1986.

90. Leblova, S., Mucha, A., and Spirhanzova, E., Compartmentation of cadmium, copper, lead and zinc in seedlings of maize (*Zea mays* L.) and induction of metallothionein, *Biologia*, 41, 777, 1986.

91. Barcelo, J., Guevara, P., and Poschenrieder, Ch., Silicon amelioration of aluminum toxicity in teosinte (*Zea mays* L.) subsp. Mexicana, *Plant Soil*, 154, 249, 1993.

92. Bazzaz, F.A., Carlson, R.W., and Rolfe, G.L., Inhibition of corn and sun flower photosynthesis by lead, *Physiol. Plantarum*, 34, 326, 1975.

93. Bazzaz, F.A., Rolfe, G.L., and Carlson, R.W., Effect of cadmium on photosynthesis and transpiration of excised leaves of corn and sun flower, *Plant Physiol.*, 32, 373, 1974.

94. Bazzaz, F.A., Rolfe, G.L., and Carlson, R.W., Differing sensitivity of corn and soybean photosynthesis and transpiration to lead contamination, *J. Environ. Qual.*, 3, 156, 1974.

95. Carlson, R.W., Bazzaz, F.A., and Rolfe, G.L., The effect of heavy metals on plants II. Net photosynthesis and transpiration of whole corn sunflower plants treated with Pb, Cd, Ni and Tl, *Environ. Res.*, 10, 113, 1975.

96. Florijn, P.J., De knecht, J.A., and Van beusichem, M.L., Phytochelatin concentrations and binding state of Cd in roots of maize genotypes differing in shoot/root cd partitioning, *J. Plant Physiol.*, 142, 537, 1993.

97. Florijn, P.J., Nelemans, J.A., and Van Beusichem, M.L., Evaluation of structural and physiological plant characteristics in relation to the distribution of cadmium in maize inbred lines, *Plant Soil*, 154, 103, 1993.

98. Jones, R., Prohaska, K.A., and Burgess, M.S.E., Zinc and cadmium in corn plants growing near electrical transmission towers, *Water Air Soil Pollut.*, 37, 355, 1988.

99. Kahn, D.H. et al., An x-ray microanalytical study of the distribution of cadmium in roots of *Zea mays* L., *Z. Pflanzen Physiol.*, 115, 19, 1984.

100. Malone, C., Koeppe, D.E., and Miller, R.J., Localization of lead accumulated by corn plants, *Plant Physiol.*, 53, 388, 1974.

101. Mench, M. and Martin, E., Mobilization of cadmium and other metals from two soils by root exudates of *Zea mays* L., *Nicotiana tabacum* and *Nicotiana rustica* L., *Plant Soil*, 132, 187, 1991.

102. Mench, M., Morel, J.J., and Guckert, A., Metal binding properties of high molecular weight soluble exudates from maize (*Zea mays* L.) roots, *Biol. Fertil. Soils*, 3, 165, 1978.

103. Nussbaum, S., Schmutz, D., and Brunold, C., Regulation of assimilatory sulfate reduction by cadmium in *Zea mays* L., *Plant Physiol.*, 88, 1407, 1988.

104. Rauser, W.E., and Glover, J., Cadmium-binding protein in roots of maize, *Can. J. Bot.*, 62, 1645, 1984.

105. Rauser, W.E., Changes in glutathione and phytochelatins in roots of maize seedlings exposed to cadmium, *Plant Sci.*, 70, 155, 1990.

106. Rauser, W.E., Schupp, R., and Rennenberg, H., Cysteine, gamma-glutamylcysteine and glutathione level in maize seedlings. Distribution and translocation in normal and cadmium exposed plants, *Plant Physiol.*, 97, 112, 1991.

107. Grimshaw, R.G., Vetiver grass — a world technology and its impact on water, 2003, *Proc. 3rd Int. Vetiver Conf.*, Guangzhou, China, 1.

108. Bevan, O. and Truong, P., Effectiveness of vetiver grass in erosion and sediment control at a bentonite mine in Queensland, Australia, in: *Proceedings of the Second International Conference on Vetiver*, Office of the Royal Development Projects Board, Bangkok, 292, 2002.

109. Truong, P., The global impact of vetiver grass technique on the environment, in *Proc. 2nd Int. Conf. Vetiver*, Office of the Royal Development Projects Board, Bangkok, 46, 2000.

110. National Research Council, *Vetiver Grass — A Thin Green Line against Erosion*, National Academy Press, Washington, D.C., 1993.

111. Truong, P. and Baker, D., Vetiver grass for the stabilization and rehabilitation of acid sulfate soils. *Proc. 2nd Natl. Conf. Acid Sulfate Soils, Coffs Harbour, Australia*, 196, 1996.

112. Taylor, K.W., Ibabuchi, I.O., and Sulford, Growth and accumulation of forage grasses at various clipping dates on acid mine spoils, *J. Environ. Sci. Health*, A24, 195, 1989.

113. Truong, P.N. and Baker, D., Vetiver grass system for environmental protection. Royal development projects protection. Technical bulletin no. 1998/1. Pacific Rim Vetiver Network. Office of the Royal Development Projects Board, Bangkok, Thailand, 1998.

114. Truong, P., Vetiver grass technique for mine tailings rehabilitation, in *Proc. 1st Asia Pacific Conf. Ground Water Bioeng. Erosion Control Slope Stabilization*, Malina, Philippines, 315, 1999.

115. Xia, H.P. and Shu, W.S., Resistance to and uptake of heavy metals by *Vetiveria zizanioides* and *Paspalum notatum* form lead/zinc mine tailings. *Acta Ecol. Sinica*, 21, 7, 1121, 2001.

116. Shu, W.S. et al., Use of vetiver and other three grasses for revegetation of Pb/Zn mine tailings: field experiment, *Int. J. Phytoremediation*, 4, 1, 47, 2002.

117. Shu, W.S., Exploring the potential utilization of vetiver in treating acid mine drainage (AMD), *Proc. 3rd Int. Conf. Vetiver Exhibition, Guangzhou, China*, 2003.

118. Truong, P., Vetiver grass for land rehabilitation, in *Proc. 1st Int. Conf. Vetiver. Bangkok*: Office of the Royal Development Projects Board, 49, 1998.

119. Sripen, K.Y.S., Growth potential of vetiver grass in relation to nutrients in wastewater of Changwat Phetchubari, in *Proc. 1st Int. Vetiver Conf.*, Chiang Rai, Thailand, 1996.

120. Shu, W.S. and Xia, H.P., Developing an integrated vetiver technique for remediation of heavy metal contamination, in *Proc. 3rd Int. Conf. Vetiver Exhibition*, Guangzhou, China, 2003.

121. Baker, A.J.M. and Walker, P.L., Ecophysiology of metal uptake by tolerant plants, in *Heavy Metal Tolerance in Plants: Evolutionary Aspects*, Shaw, A.J. (Ed.), CRC Press, Boca Raton, FL, 155, 1990.

122. Xia, H.P. et al., A preliminary report on tolerance of vetiver to submergence, *Proc. 3rd Int. Vetiver Conf.*, Guangzhou, China, 532.

123. Knoll, C., Rehabilitation with vetiver, *Afr. Mining*, 2(2), 1997.

124. Shu, W.S., Xia, H.P., and Zhang, Z.Q., Use of vetiver and other three grasses for revegetation of a Pb/Zn mine tailings at Lechang, Guangdong: a field experiment. *Proc. 2nd Int. Vetiver Conf.*, Thailand, January, 2000.

125. Radloff, B., Walsh, K., and Melzer, A., Direct revegetation of coal tailings at BHP, Saraji Mine, *Aust. Mining Council Environ. Workshop,* Darwin, Australia, 1995.

126. Jarvis, M.D. and Leung, D.W., Chelated lead transport in *Chamaecytisus proliferus* (L.f.) link ssp. *proliferus* var. *palmensis* (H. Christ): an ultrastructural study, *Plant Sci.*, 161, 433, 2001.

127. Reddy, G.N. and Prasad, M.N.V., Tyrosine is not phosphorylated in cadmium induced hsp70 cognate in maize (*Zea mays* L.) seedlings: role in chaperone function? *Biochem. Arch.* 9, 25, 1993.

128. Reddy, G.N. and Prasad, M.N.V., Cadmium induced protein phopsphorylation changes in rice (*Oryza sativa L.),* *J. Plant Physiol.*, 145, 67, 1995.

129. Neumann, D.O., Lichtenberger, D., Günther, K., Tschiersch, L., and Nover, Heat-shock proteins induce heavy-metal tolerance in higher plants, *Planta*, 194, 360, 1994

130. Bennet, R.J, Breen, C.M., and Feu, M.V., The aluminum signal: new dimensions of aluminum tolerance, *Plant Soil*, 134, 153, 1991.

131. Sasaki, M.Y., Yamamoto, H., and Matsumoto, Putative Ca^{2+} channels of plasma membrane vesicles are not involved in the tolerance mechanism of aluminum in aluminum-tolerant wheat (*Triticum aestivum* L.) cultivar, *Soil Sci. Plant Nutr.*, 40, 709, 1994.

132. Tamás, L., Huttová, J., and Žigová, Z. Accumulation of stress-proteins in intercellular spaces of barley leaves induced by biotic or abiotic factors, *Biol. Plant*, 39, 387, 1997.

133. Yamauchi, M. and Peng, X.X., Iron toxicity and stress-induced ethylene production in rice leaves, *Plant Soil*, 173, 21, 1995.

134. Stiborova, M., Hromadkova, R., and Leblova, S., Effect of ions of heavy metals on the photosynthetic characteristics of maize (*Zea mays* L.), *Biologia* (Bratislava), 41, 1221, 1986.

135. Stiborova, M. et al., Effect of heavy metal ions on growth and biochemical characteristics of photosynthesis of barley, *Hordeum vulgare* L., *Phortosynthetica*, 20, 418, 1986.

136. Robinson, N.J. and Jackson, P.J., Metalothionein-like metal complexes in angiosperms, their structure and function, *Physiol. Plantarum*, 67, 499, 1986.

137. Salt, D.E. and Wagner, G.J., Cadmium transport across tonoplast of vesciles from oat roots. Evidence for a Cd2+/H+ antiport activity, *J. Biol. Chem.*, 268, 12297, 1993.

138. Kuppelwieser, H. and Feller, U., Transport of Rb and Sr to the ear in mature, excised shoots of wheat: effects of temperature and stem length on Rb removal from the xylem, *Plant Soil*, 132, 281, 1991.

139. Madrid, F.M.S., Liphadzi, and Kirkham, M.B., Heavy metal displacement in chelate-irrigated soil during phytoremediation, *J. Hydrol.*, 272, 107, 2003.

140. Chen, B., Christie, P., and Li, X., Modified glass bead compartment cultivation system for studies on nutrient and trace metal uptake by arbuscular mycorrhiza, *Chemosphere*, 42, 185, 2001.

141. Rengel, Z., Ecotypes of *Holcus lanatus* tolerant to zinc toxicity also tolerate zinc deficiency, *Annals of Botany*, 86, 1119, 2000.

142. Samantaray, S., Rout, G.R., Das, P., *In Vitro* selection and regeneration of zinc tolerant calli from *Setaria italica* L., *Plant Science*, 143, 201, 1999.

143. Chen, H.M., Zheng, C.R., Tu, C., and Shen, Z.G., Chemical methods and phytoremediation of soil contaminated with heavy metals, *Chemosphere*, 41, 229, 2000.

144. Messerschmidt, J., Alt, F., and Tölg, G., Detection of platinum species in plant material, *Electrophoresis*, 16, 800, 1995.

145. Messerschmidt, J., Alt, F., and Tölg, G., Platinum species analysis in plant material by gel spermeation chromatography, *Anal. Chim. Acta*, 291, 161, 1994.

146. Oedegard, K.E. and Lund, W., Multielement speciation of tea infusion using cation-exchange separation and size-exclusion chromatography in combination with inductively coupled plasma mass spectrometry, *J. Anal. Spectrum*, 12, 403, 1997.

147. Cocker, K.M., Evans, D.E., and Hodson, M.J., The amelioration of aluminum toxicity by silicon in higher plants: solution chemistry or an in planta mechanisms? *Physiol. Plant*, 104, 608, 1998.

148. Kluepperl et al., 1998.

22 Physiology of Lead Accumulation and Tolerance in a Lead-Accumulating Plant (*Sesbania drummondii*)

Nilesh Sharma and Shivendra Sahi

CONTENTS

ABSTRACT

A prerequisite for phytoremediation of heavy metal-contaminated sites is the identification of metal-accumulating plants. Although a few natural or induced metal accumulators tolerate and thrive under toxic concentrations of lead and other heavy metals, the specific mechanisms governing toxicity tolerance are not well understood. Plants respond to heavy metal toxicity in a variety of ways and have a range of potential mechanisms for the purpose of metal detoxification. *Sesbania drummondii*, a lead accumulator, displays evidence of cellular chelation of the metal with acetate and sulfur-rich compounds under x-ray absorption spectroscopic examinations. On the other hand, this plant uses antioxidative defense mechanisms effectively to counteract the oxidative stress that may be produced as a result of lead application. When grown in $Pb(NO_3)_2$ (500 mg/L), *Sesbania* had many folds enhanced activities of catalase and superoxide dismutase. Furthermore, photosynthetic efficiency (F_v/F_m) and integrity (F_v/F_0) of this plant were largely maintained at this concentration of lead.

22.1 INTRODUCTION

Lead (Pb) contamination in soil is a widespread phenomenon and originates from automobiles, metal-smelting plants, mines, lead-contaminated sewage sludge, industrial wastes, etc.[1]. Lead concentration in roadside fields is sharply elevated due to heavy traffic, particularly in industrialized areas [2,3]. Severe Pb contamination of soil is associated with a variety of environmental problems, including loss of vegetation, ground water contamination, and Pb toxicity to plants, animals, and humans [4]. Thus, there is an urgent need for remediation of contaminated sites using an effective and environmentally friendly technology such as phytoremediation.

Heavy metal accumulation by plants has been utilized in various phytoremediation strategies for decontamination of the environment [5–8]. One of the key goals in phytoextraction is to select plants (hyperaccumulators) that can translocate substantial amounts of toxic metals from their roots to aerial parts, which can be easily harvested. For a plant species to be efficient in lead phytoextraction, it should accumulate metal concentration > 0.1% of shoot dry weight, in addition to having high biomass productivity. A critical balance between metal accumulation and plant biomass productivity will make a plant more efficient for phytoextraction [5]. From this standpoint, plant species such as Indian mustard, pea, and corn were the focus of recent Pb phytoremediation research. These species accumulate high amounts of lead and generate satisfactory biomass [6–8].

Another interesting Pb accumulator is *Sesbania drummondii*, a large, bushy perennial plant with high biomass productivity that grows naturally in seasonally wet places of the southern coastal plains of the U.S. It demonstrates a unique potential of Pb accumulation in aerial parts from an aqueous solution [9]. When grown in Pb-contaminated soil in a greenhouse, this species showed a substantial movement of Pb from roots to shoot with appreciable amounts accumulated in the aerial parts [10]. If its accumulation potential is considered with respect to the enormous biomass (10 to 15 ton/ha/year) that this plant can generate, *Sesbania* should qualify as one of the best known Pb accumulators. Though *Sesbania* and other plant species tolerate and accumulate toxic concentrations of Pb, the specific mechanisms are not well understood. Therefore, understanding the physiological and biochemical basis of metal accumulation will be helpful in chalking out phytoremediation strategy.

Plants have a range of potential mechanisms at the cellular level that might be involved in detoxification and thus tolerance to heavy metal stress. Chelation of metals in the cytosol by high-affinity ligands is potentially a very important mechanism of heavy metal detoxification and tolerance [11]. A number of ligands have now been recognized and their roles have been reviewed. In general, they include amino acids, organic acids, and peptides such as the phytochelatins and the metallothioneins [12]. The phytochelatins (PCs) have been most widely studied in plants, particularly in relation to Cd and Zn [13]. PCs are a family of metal-complexing peptides that have a general structure (y-Glu-Cys)n-Gly, where n = 2 to 11, and are rapidly induced in plants in response to exposure to heavy metals [14,15]. Few reports, however, have appeared demonstrating the production of PC in response to Pb toxicity in plants, particularly those belonging to higher plants [16]. Biochemical evidence indicates that plants exposed to toxic metals may produce ligands for metal detoxification or transport, or both [17]. The internal speciation of Pb in *Sesbania* tissues was examined using x-ray absorption near-edge structure (XANES) and extended x-ray absorption fine structure (EXAFS) [18].

Earlier studies suggest that exposure to Pb and other heavy metals generally interferes with photosynthetic activity, severely affecting health of plants [19–21]. A variety of mechanisms work in tandem to counteract the effects of oxidative stress produced as a result of heavy metal (Pb) accumulation in plant cells [22–24]. Oxidative stress and effects of reactive oxygen species are generally handled by a number of antioxidant enzymes, such as superoxide dismutases, peroxidases, and catalases. Superoxide dismutase (SOD), the first enzyme in the detoxifying process, converts superoxide anion — a powerful reactive oxygen species — to hydrogen peroxide [25]. Hydrogen peroxide (H_2O_2) is scavenged directly by catalase (CAT), which converts it to water and molecular oxygen. Peroxidases also scavenge H_2O_2 indirectly by combining it with antioxidant compounds such as ascorbate and guaiacol [23,24,26]. Measurement of activities of antioxidative enzymes can thus be used to indicate oxidative stress in plants [22]. Thus, accumulation of high amounts of metals in the hyperaccumulator plant suggests the existence of a strong antioxidative defense mechanism to avoid the harmful effects.

This chapter describes

- Uptake and transport of Pb in Pb-accumulating plants
- X-ray absorption spectroscopic analysis to demonstrate biotransformation of accumulated Pb in *Sesbania* tissues
- Use of effective antioxidative defense mechanisms
- Effect of Pb accumulation on photosynthetic activities in *Sesbania* species

22.2 LEAD ACQUISITION AND TRANSPORT

22.2.1 FACTORS GOVERNING LEAD UPTAKE BY PLANTS

Accumulation of Pb by plants depends on several factors, the most important of which are the genetic capability of the plant, as in a hyperaccumulator, and Pb bioavailability. The limiting factor for metal accumulation is the amount of Pb readily available for uptake. The solubility of Pb in soil is limited due to complexation with organic matter, sorption on clays and oxides, and precipitation as carbonates, hydroxides, and phosphates [27]. The application of a synthetic chelator has been used to increase Pb solubility in soil solution and EDTA has been found to be a most effective chelating agent for Pb desorption [7,8,28].

The chemistry and physiology of EDTA–Pb complex is understood better than that of any other Pb–chelate complex. When EDTA is added to Pb-contaminated soil, it complexes with soluble Pb. Due to high affinity of EDTA for Pb, the Pb–EDTA complex formation dominates other metal–EDTA complexation in most soils between pH 5.2 and 7.7 [29]. The soluble soil Pb concentration is not the only factor that influences its uptake in plants. Epstein et al. [30] observed substantially higher Pb accumulations in plants grown in soil containing 4.8 mmol Pb/kg soil and amended with 1 mmol EDTA/kg soil than in plants grown in soil containing 1.5 mmol Pb and 5 mmol EDTA/kg soil. This indicates that a higher ratio of Pb:EDTA in the soil solution will result in enhanced Pb uptake by the plant.

Thus, only a careful consideration of chelate application, taking into account factors like soil type and its total Pb content, can enhance the efficacy of a phytoextraction strategy. EDTA has been shown not only to enhance Pb desorption from the soil components to the soil solution but also to increase its transport into the xylem and its translocation from roots to shoots [7,30–32]. Vassil et al. [32] demonstrated that Pb accumulation in shoots is correlated with the formation of Pb–EDTA in the hydroponic solution and that Pb–EDTA is the major form of Pb taken up and translocated by the plant via xylem stream. The physiological basis of the uptake of Pb–EDTA and, particularly, the possibility of this large molecule to cross the cell membrane are unknown. However, using extended x-ray absorption fine structure spectroscopy, Sarret et al. [16] confirmed the presence of a substantial amount of Pb–EDTA in *Phaseolus vulgaris* leaves, when this plant was grown in a solution of Pb–EDTA.

22.2.2 MODE OF LEAD TRANSPORT

Electron microscopic techniques have mapped out Pb transport via various plant tissues [31,33–35]. Patterns of Pb migration and deposition in tissues differ depending on whether the supply of Pb is in chelated or unchelated form. In *Pinus radiata*, it has been shown in ultrathin sections that Pb grains were exclusively distributed in the outermost layer of the root cell wall and in negligible amounts in the needle when Pb was supplied in unchelated form. Conversely, when Pb was supplied in chelated (EDTA or HEDTA) form, higher intensities of Pb fine grains were detected in needle and the least in the root sections [31]. These ultramicroscopic observations are in conformity with the earlier uptake results in Indian mustard [6], where most lead was accumulated in roots on supply of Pb in the unchelated form.

In *P. radiata* roots, Pb supplied in unchelated form was not found in vacuoles, dictyosomes, or intercellular spaces, but it was often found embedded in or adjacent to cell wall — in many cases, near plasmodesmata [31]. This situation in *P. radiata* suggests that, in the case of Pb deposition between the cell wall and the cell membrane, it is probable that it was transported apoplastically. However, in the case of its aggregation within plasmodesmata, the mode of transport might have been symplastic. Occurrence of two simultaneous modes of Pb transport has also been advocated in other cases [33,36]. Transmission electron microscopic examination coupled with x-ray microanalysis of *Sesbania* root showed Pb deposits along the plasma membrane of the cells. The smaller granules appeared to coat the surface of the plasma membrane while larger deposits, looking like globules, extended deeper into the cell wall [9]. A large deposit was observed in high magnification within the tonoplast of the vacuole [9,18]. This is an indication of symplastic mode of transport in *Sesbania*, though not excluding the possibility of apoplastic migration.

This brings up the question of how *Sesbania* cells capture Pb particles. The fact that it was not detected in the ground cytoplasm but only in the plasma membrane and vacuole may suggest that endocytosis also plays a role in Pb uptake in this species. Reports of pinocytosis or endocytosis as a device for Pb entrapment are also available in the literature [31,35].

Whether following apoplastic or symplastic routes, Pb must cross the Casparian strip-barrier of the root in order to reach the xylem stream; this is difficult for the large Pb particles due to their size and charge characteristics. However, once they have formed a complex with a chelator such as EDTA, their solubility increases and thus the particle size may also decrease. As a result, the complex may become partially invisible to those processes that would normally prevent unrestricted movement, such as precipitation with phosphates or carbonates or binding to cell walls through mechanisms such as cation exchange.

Thus, the general agreement is that synthetic chelates overcome barriers to translocation. Now, how does Pb enter the vessels or tracheids of the xylem having moved symplastically to the parenchyma cells of the vascular cylinder? The mechanism by which transfer of Pb particles from vascular cylinder parenchyma to vessels or tracheids occurs may be a type of highly selective active-carrier transport, as opposed to facilitated diffusion [37]. Translocation of Pb from root to shoot via xylem has been shown in many studies [6,38,39].

22.3 SPECIATION OF ACCUMULATED LEAD IN A LEAD ACCUMULATOR

22.3.1 CHARACTERISTICS OF XANES IN *SESBANIA* SAMPLES

Currently, speciation of Pb in plant tissues (roots and leaves) has been studied by Sharma et al. [18] using L_{III}-edge XANES spectra (Figure 22.1A and B). The fittings of the XANES based on linear combination analysis are depicted in Table 22.1. The root sample grown in presence of Pb primarily consisted of a lead (II) acetate type of structure; over 60% of the spectra was defined by the lead (II) acetate structure (Table 22.1). The remaining part of the lead bound to the roots comprised mainly lead (II) nitrate, lead (II) sulfide, and lead metal (10, 20, and 9%, respectively).

This indicates that *Sesbania* physiologically transforms the starting material, lead (II) nitrate, into organic type lead compound, Pb acetate. The plant leaf sample demonstrated even more changed composition consisting predominantly of lead acetate (52%), lead sulfate (26%), and lead sulfide (14%). The low percentage of the lead nitrate (7.6 %) found in the linear combination

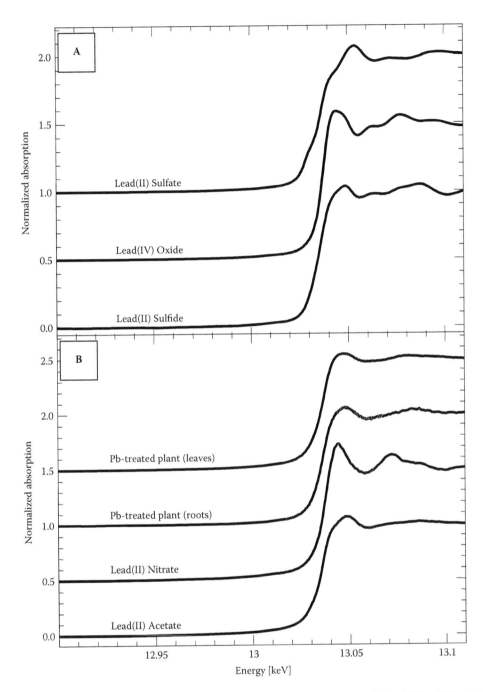

FIGURE 22.1 (A) X-ray absorption near edge structure of lead model compounds: lead (II) sulfide, lead (II) sulfate, and lead (IV) oxide. (B) X-ray absorption near edge structure of lead-treated *Sesbania drummondii* samples, lead (II) nitrate, and lead (II) acetate. (From Sharma, N.C. et al., *Environ. Toxicol. Chem.*, 23, 134, 2004. With permission.)

XANES fitting data is another evidence in favor of biological transformation of lead (II) nitrate into organic types of structures, possibly for storage of the lead into or onto cellular material. It is interesting to note that speciation of Pb into lead acetate does not occur in the starting solution [16].

TABLE 22.1
LC-XANES Fittings of *Sesbania drummondii* Exposed to 500
mg/L Pb(NO$_3$)$_2$ Using Lead (II) and (IV) Model Compounds

Samples	% PbNO$_3$	% PbSO$_4$	% Pbmetal	% PbS	% Pb acetate
Leaves	7.6	25.8	0.0	14.2	52.4
Root	10.1	0.0	8.8	20.2	60.9

Source: Sharma, N.C. et al., *Environ. Toxicol. Chem.*, 23, 134, 2004. With permission.

22.3.2 CHARACTERISTICS OF EXAFS IN *SESBANIA* SAMPLES

EXAFS features (Figure 22.2A and B) and FEFF fittings (Table 22.2 and Table 22.3) shown by Sharma et al. [18] indicate that lead in the root and leaf samples is bound into a chemical moiety with interatomic distances similar to lead (II) acetate. Though the root sample of *Sesbania* displays spectra comparable to the lead (II) acetate type of structure, the presence of a secondary complex complicates the interpretation. The FEFF fittings of root sample, however, indicate a ten-coordinate system that has a mixture of different compounds including lead particles. At the same time, the leaf sample shows an octahedral coordination, making EXAFS interpretation easy. XANES fittings of leaf sample are simple in composition, showing absence of lead particles. Though both samples are complex, they contain compounds of similar structure and composition, as indicated by XANES structure.

Presence of lead acetate in large proportions (>50%) in both the samples (Table 22.1) suggests transport of lead via Pb–organic acid complex. Metal–organic acid complexation has been reported as a phase of metal transport in a variety of plants [17,40,41]. In *Phaseolus vulgaris*, formation of various lead complexes including Pb–salicylate was reported when plants were grown in medium containing Pb and EDTA. Cerussite (lead carbonate) was found to be the predominant Pb species in leaves when *Phaseolus* plants were grown in solution containing Pb(NO$_3$)$_2$ alone.

The evidence for sulfur ligands is a potentially unique aspect of lead speciation in *Sesbania* species. Though sulfur is present in both the samples, the proportion in leaf is double the root content. The nature of sulfur complex also varies. Leaves contain PbSO$_4$ and PbS, but roots have only PbS. These ligand properties are indicative of glutathione (γ-Glu-Cys-Gly) and phytochelatins, small metal-binding polypeptides with the amino acids sequence (γ-Glu-Cys)n- Gly, where $n = 2$ to 14. Phytochelatins (PC$_s$) play an important role in heavy metal homeostasis and detoxification [42] and their induction has been reported as an effective strategy against Cd toxicity [12,43]. A recent study suggested that PC$_s$ might also be involved in arsenic detoxification [44]. Reports of PC induction in response to Pb toxicity are not common in higher plants [16]. However, Pb-induced PC synthesis was confirmed in a common microalga, *Stichococcus bacillaris* [45].

22.4 ANTIOXIDATIVE DEFENSE IN A LEAD ACCUMULATOR

22.4.1 CATALASE ACTIVITY IN *SESBANIA* SEEDLINGS

Catalase activities of *Sesbania drummondii* seedlings grown in the absence or presence of 500 mg/L Pb(NO$_3$)$_2$ were determined by Ruley [10]. The enzyme activities in Pb-treated seedlings increased dramatically, >300% higher than the respective values of controls at week 4 (Figure 22.3). In *Thlaspi caerulescens*, a constitutive Cd hyperaccumulator, an elevated catalase induction in response to Cd stress has been demonstrated [46]. An accumulator of Pb, *Pisum sativum* [7], also demonstrated an elevated activity of CAT in response to 104 or 207 mg/L Pb [47].

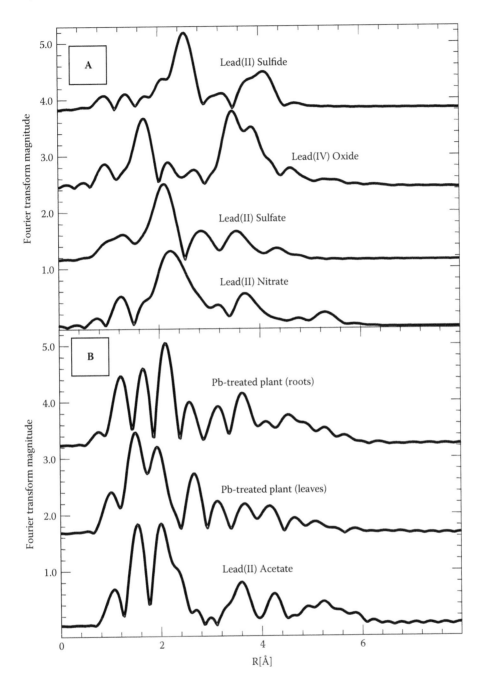

FIGURE 22.2 (A) Extended X-ray absorption fine structure of lead model compounds: lead (II) nitrate, lead (II) sulfide, lead (II) sulfate, and lead (IV) oxide. (B) Extended X-ray absorption fine structure of lead-treated *Sesbania drummondii* samples and lead (II) acetate. (From Sharma, N.C. et al., *Environ. Toxicol. Chem.*, 23, 134, 2004. With permission.)

22.4.2 SOD Activity in *Sesbania* Seedlings

Superoxide dismutase (SOD) activities in the absence and presence of 500 mg/L $Pb(NO_3)_2$ were also reported by Ruley [10]. SOD activities in Pb-treated seedlings gradually increased and were recorded to be >200% higher than the respective values of controls at week 4 (Figure 22.4). Similar

TABLE 22.2
Compilation of Structural Parameters Extracted from Extended X-Ray Absorption Fine Structure Spectra of *Sesbania drummondii* Exposed to 500 mg/L Pb(NO₃)₂

Neighboring atom	Coordination number	R Å[a]	$S^{2(b)}$
Leaves			
Pb-O	2.0	1.90	0.0039
Pb-C	2.0	2.20	0.00086
Pb-S	2.0	2.85	0.0059
Roots			
Pb-O	1.2	1.80	0.000051
Pb-C	1.0	2.01	0.0017
Pb-O	1.2	1.99	0.0016
Pb-O	1.9	2.78	0.0099
Pb-O	1.6	2.75	0.0070
Pb-O	1.6	2.57	0.0099

Distance in angstroms from the metal to the neighboring atom.
[b]Squared standard deviation.

Source: Sharma, N.C. et al., *Environ. Toxicol. Chem.*, 23, 134, 2004. With permission.

to catalase activity, a significant increase in the activity of SOD has been reported in *Pisum sativum* [47]. A hyperaccumulator of Cd, *Alyssum argenteum*, has also been reported to exhibit an elevated SOD activity in response to high concentrations of Cd [48].

22.4.3 GPX Activity in *Sesbania* Seedlings

The guaiacol peroxidase (GPX) activities of *Sesbania drummondii* seedlings grown in 500 mg/L Pb(NO₃)₂ have also been determined [10]. Controls experienced a significant, but temporary, increase in GPX activity at week 2, returning to their original level by week 4 ($p > 0.05$) (Figure 22.5). GPX activities of plants exposed to Pb showed a steady increase during the growth period, registering a level significantly higher than that in control ($p < 0.05$) (Figure 22.5). Increased activities of peroxidase in response to Pb exposure in *Sesbania* are comparable to those reported in a Pb-tolerant species, *Sonchus oleraceus*, in which similar increases in activities of peroxidase were observed [24]. *Phaseolus vulgaris* also demonstrated increased activities of GPX and APX (ascorbate peroxidase) on exposure to 115.8 mg/L Pb–EDTA [22].

22.5 PHOTOSYNTHETIC ACTIVITY IN A LEAD ACCUMULATOR

22.5.1 Efficiency of Photosynthetic Apparatus in *Sesbania*

Seedlings grown in liquid medium contaminated with 500 mg/L Pb (NO₃)₂ were assessed for photosynthetic activity by measuring chlorophyll *a* fluorescence parameters using the Handy-PEA instrument (Hansatech Instruments, U.K.) [10,20]. The following fluorescent parameters were measured: F_o (minimum chlorophyll *a* fluorescence after the dark adaptation) and F_m (maximum fluorescence after the pulse of red light). From these two measurements, the F_v (variable fluores-

TABLE 22.3
Compilation of Structural Parameters Extracted from Extended X-Ray Absorption Spectra of Model Compounds

Neighboring atom	Coordination number	R Å^{a}	$S^{2(b)}$
PbO$_2$			
Pb–O	2.28	2.19	0.00035
Pb–O	3.7	2.33	0.0039
Pb–Pb	2.0	3.36	0.0015
PbSO$_4$			
Pb–O	1.0	2.65	0.0051
Pb–O	1.0	2.57	0.00084
Pb–O	2.0	2.73	0.0010
Pb–O	2.0	2.86	0.0021
Pb–O	2.0	2.98	0.0048
Pb–O	2.0	3.14	0.0044
Pb–O	2.0	3.33	0.0011
PbS			
Pb–S	6.12	2.98	0.0065
Pb–Pb	11.3	4.22	0.0080
Pb(NO$_3$)$_2$			
Pb–O	6.73	2.73	0.0064
Pb–O	5.36	2.87	0.0057
Pb–N	4.89	3.24	0.0039
Pb acetate			
Pb–O	1.0	1.88	0.0025
Pb–C	1.0	2.10	0.0071
Pb–O	1.0	1.89	0.0036
Pb–O	3.0	2.47	0.0046
Pb–O	1.0	3.01	0.0067
Pb–C	1.0	2.82	0.0019
Pb–O	1.0	3.20	0.0058
Pb–C	1.0	1.74	0.0049

[a] Distance in angstroms from the metal to the neighboring atom.
[b] Squared standard deviation.

Source: Sharma, N.C. et al., *Environ. Toxicol. Chem.*, 23, 134, 2004. With permission.

cence calculated as the difference between the minimal and maximal fluorescence), F_v/F_m (ratio of variable to maximal fluorescence), and F_v F_o (ratio of variable to minimal fluorescence) were determined.

F_v/F_m value is an indicator of the metabolic status of plants [49]. An F_v/F_m ratio of 0.8 or above is characteristic of a healthy plant [19]. *Sesbania* seedlings grown in the presence and absence of Pb had F_v/F_m values of 0.8 or higher, except at week 4, when the experimental plants showed lower value (0.78) significantly different from that of the control (Figure 22.6).

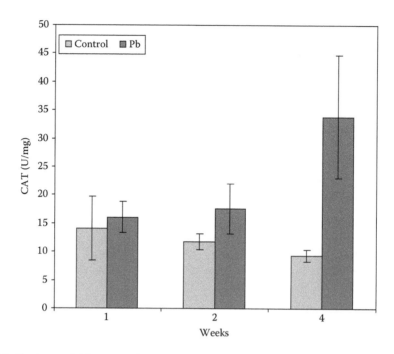

FIGURE 22.3 Catalase activities (U mg^{-1} FW) of *Sesbania drummondii* seedlings grown in the presence of 500 mg/L Pb(NO$_3$)$_2$ for 4 weeks. Values represent means and SE (*n* = 6).

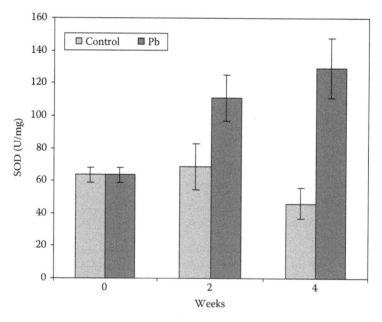

FIGURE 22.4 Superoxide dismutase activities (U mg^{-1} FW) of *Sesbania drummondii* seedlings grown in the presence of 500 mg/L Pb(NO$_3$)$_2$ for 4 weeks. Values represent means and SE (*n* = 6).

22.5.2 Active Photosynthetic Reaction Centers in *Sesbania*

F_v/F_o (ratio of variable to minimal fluorescence) indicates the size and number of active photosynthetic centers in the chloroplast and therefore the photosynthetic strength of the plant. F_v/F_o value of 4.0 or more indicates that the plant is healthy and not suffering photosynthetic stress [19].

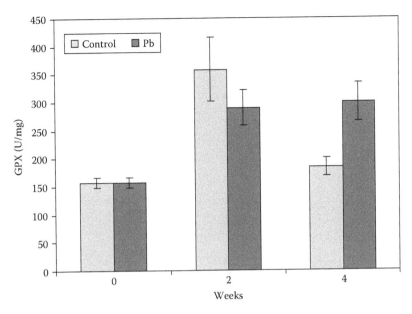

FIGURE 22.5 Guaiacol peroxidase activities (U mg^{-1} FW) of *Sesbania drummondii* seedlings grown in the presence of 500 mg/L Pb(NO$_3$)$_2$ for 4 weeks. Values represent means and SE ($n = 6$).

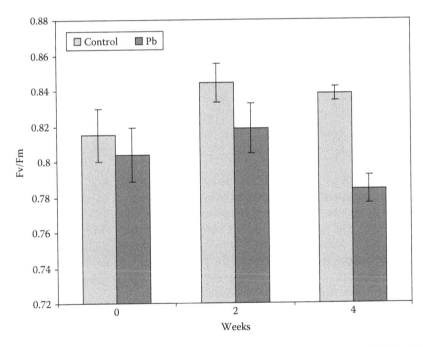

FIGURE 22.6 F_v/F_m values of *Sesbania drummondii* seedlings grown in the presence of 500 mg/L Pb(NO$_3$)$_2$ for 4 weeks. Values represent means and SE ($n = 6$).

Sesbania seedlings grown in the presence and absence of 500 mg/L Pb (as described earlier), displayed F_v/F_o values equal to 4.0 or higher up to week 2. However, in week 4, values dropped off to 3.7 for the experimental group (Figure 22.7) without discernable symptoms.

Though Pb or other heavy metals have been reported to affect photosynthetic activity in plants, *S. drummondii* had little or no effect on photosynthetic activity on exposure to Pb. The photosyn-

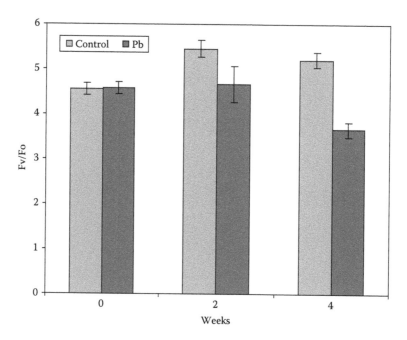

FIGURE 2.7 F_v/F_o values of *Sesbania drummondii* seedlings grown in the presence 500 mg/L Pb(NO$_3$)$_2$ for 4 weeks. Values represent means and SE ($n = 6$).

thetic efficiency of *Sesbania* was more or less comparable to *Pelargonium* sp., where F_v/F_m and F_v/F_o were not significantly affected by Pb accumulation [20]. It is known that Pb interferes directly with δ-aminolevulinic acid dehydratase, an important enzyme in the formation of chlorophyll in plants [16]. This affects chlorophyll synthesis and carbon assimilation, eventually modifying the photolysis of water. The altered H$_2$O photolysis generates reactive oxygen species (ROS) such as superoxide (O$_2^-$) and hydroxyl radicals (·OH) [22–24].

22.6 CONCLUSION

Understanding the physiology of metal accumulation and detoxification mechanisms in plants is crucial to developing strategy for heavy metal remediation. Evidence suggests that *Sesbania drummondii* uses more than one mechanism to counteract lead toxicity. Pb^{2+} in *Sesbania* tissues complexes predominantly with acetate- and sulphur-containing compounds to minimize the toxic effects; however, antioxidative defense mechanisms are also used by this plant effectively to overcome the effects of reactive oxygen species and protect the photosynthetic machinery and efficiency. On the basis of biomass generation capability and physiological backup for tolerance, this plant can be a suitable agent for remediation of lead-contaminated sites.

REFERENCES

1. Zakrzewski, S., *Principles of Environmental Toxicology*, American Chemical Society: Washington, D.C., 1991.
2. Yassoglou, N. et al., Heavy metal contamination of roadside soils in the Greater Athens area, *Environ. Pollut.*, 47, 293, 1987.
3. Ho, Y.B. and Tai, K.M., Elevated levels of lead and other metals in roadside soil and grass and their use to monitor aerial metal depositions in Hong Kong, *Environ. Pollut.*, 49, 37, 1988.

4. Body, P.E., Dolan, P.R., and Mulcahy, D.E., Environmental lead: a review *Crit. Rev. Environ. Control.*, 20, 299, 1991.

5. Huang, J.W. and Cunningham, S.D., Lead phytoextraction: species variation in lead uptake and translocation, *New Phytol.*, 134, 75, 1996.

6. Blaylock, M.J. et al., Enhanced accumulation of Pb in Indian mustard by soil-applied chelating agents, *Environ. Sci. Technol.*, 31, 860, 1997.

7. Huang, J.W. et al., Phytoremediation of lead contaminated soils: role of synthetic chelates in lead phytoextraction, *Environ. Sci. Technol.*, 31, 800, 1997.

8. Begonia, M.F.T. et al., Chelate-assisted phytoextraction of lead from a contaminated soil using wheat (*Triticum aestivum* L.), *Bull. Environ. Conta. Toxicol.*, 68, 705, 2002.

9. Sahi, S.V. et al., Characterization of a lead hyperaccumulator shrub, *Sesbania drummondii*, *Environ. Sci. Technol.*, 36, 4676, 2002.

10. Ruley, A.T., Effects of accumulation of Pb and synthetic chelators on the physiology and biochemistry of *Sesbania drummondii*. M.S. thesis, Western Kentucky University, Bowling Green, KY, 2004.

11. Hall, J.L., Cellular mechanisms for heavy metal detoxification and tolerance, *J. Exp. Bot.*, 53, 1, 2002.

12. Rauser, W.E., Structure and function of metal chelators produced by plants: the case of organic acids, amino acids, phytin and metallothioneins, *Cell Biochem. Biophys.*, 31, 19, 1999.

13. Goldsbrough, P., Metal tolerance in plants: the role of phytochelatins and metallothioneins, in *Phytoremediation of Contaminated Soil and Water,* Terry, N. and Banuelos, G., Eds., CRC Press, Boca Raton, FL, 2000, 221.

14. Rauser, W.E., Phytochelatins and related peptides, structure, biosynthesis and function, *Plant Physiol.*, 109, 1141, 1995.

15. Zenk, J.H., Heavy metal detoxification in higher plants: a review, *Gene*, 179, 21, 1996.

16. Sarret, G. et al., Accumulation of Zn and Pb in *Phaseolus vulgaris* in the presence and absence of EDTA, *Environ. Sci. Technol.*, 35, 2854, 2001.

17. Salt, D.E., Prince, R.C., and Pickering, I.J., Chemical speciation of accumulated metals in plants: evidence from *x*-ray absorption spectroscopy, *Microchem. J.*, 71, 255, 2002.

18. Sharma, N.C. et al., Chemical speciation and cellular deposition of lead in *Sesbania drummondii*, *Environ. Toxicol. Chem.*, 23, 134, 2004.

19. Dan, T.V., KrishnaRaj, S., and Saxena, P.K., Metal tolerance of scented geranium (*Pelargonium* sp. "Freshman"): effects of cadmium and nickel on chlorophyll fluorescence kinetics, *Int. J. Phytorem.* 2, 91, 2000.

20. KrishnaRaj, S., Dan, T.V., and Saxena, P.K., A fragrant solution to soil remediation, *Int. J. Phytorem.,* 2, 117, 2000.

21. MacFarlane, G.R., Chlorophyll: a fluorescence as a potential biomarker of zinc stress in the gray mangrove, *Avicennia marina*, *Bull. Environ. Contam. Toxicol.*, 70, 90, 2003.

22. Geebelen, W. et al., Effects of Pb–EDTA and EDTA on oxidative stress reactions and mineral uptake in *Phaesolus vulgaris*, *Physiol. Plant.*, 115, 377, 2002.

23. Rama Devi, S. and Prasad, M.N.V., Copper toxicity in *Ceratophyllum demersum* (coontail), a free floating macrophyte: response of antioxidant enzymes and antioxidants, *Plant Sci.*, 138, 157,1998.

24. Xiong, Z-T., Bioaccumulation and physiological effects of excess lead in a roadside pioneer species *Sonchus oleraceus*, *Environ. Pollut.*, 97, 275, 1997.

25. Alscher, R.G., Erturk, N., and Heath, L.S., Role of superoxide dismutases (SODs) in controlling oxidative stress in plants, *J. Exp. Bot.*, 53, 1131, 2003.

26. Gupta, A.S. et al., Overexpression of superoxide dismutase protects plants from oxidative stress: induction of ascorbate peroxidase in superoxide dismutase-overexpressing plants, *Plant Physiol.*, 103, 1067, 1993.

27. McBride, M.B., *Environmental Chemistry of Soils,* Oxford University Press, New York, 1994, 406.

28. Kirkham, M.B., EDTA-facilitated phytoremediation of soil with heavy metals from sewage sludge, *Int. J. Phytorem.*, 2, 159, 2000.

29. Sommers, L.E. and Lindsay, W.L., Effect of pH and redox on predicted heavy metal-chelate equilibria in soils, *Soil Sci. Soc. Am. J.*, 43, 39, 1979.

30. Epstein, A.L. et al., EDTA and Pb-EDTA accumulation in *Brassica juncea* grown in Pb-amended soil, *Plant Soil*, 208, 87, 1999.

31. Jarvis, M.D. and Leung, D.W.M., Chelated lead transport in *Pinus radiata*: an ultrastructural study, *Environ. Exp. Bot.*, 48, 21, 2002.

32. Vassil, A.D. et al., The role of EDTA in lead transport and accumulation by Indian mustard, *Plant Physiol.*, 117, 447, 1998.

33. Wierzbicka, M., How lead loses its toxicity to plants, *Acta Soc. Bot. Pollut.*, 64, 81, 1995.

34. Wierzbicka, M., Lead in the apoplast of *Allium cepa* L. root tips — ultrastructural studies, *Plant Sci.*, 133, 105. 1998.

35. Samardakiewicz, S. and Wozny, A., The distribution of lead in duckweed (*Lemna minor* L.) root tip, *Plant Soil,* 226, 107, 2000.

36. Ksiazet, M., Wozny, A., and Mlodzianowski, F., Effect of $Pb(NO_3)_2$ on poplar tissue culture and the ultrastructural localization of lead in cultured cells, *For. Ecol. Manage.*, 8, 95, 1984.

37. Raven, P.H., Evert, R.F., and Eichhorn, S.E., *Biology of Plants*, 6th ed. Worth Publishers, New York, 1999.

38. Salt, D.E. et al., Phytoremediation: a novel strategy for the removal of toxic metals from the environment using plants, *Bio/Technology*, 13, 468, 1995.

39. Wu, J., Hsu, F., and Cunningham, S., Chelate-assisted Pb phytoextraction: Pb availability, uptake, and translocation constraints, *Environ. Sci. Technol.*, 33, 1898, 1999.

40. Erre'calde, O. and Campbell, P.G.C., Cadmium and zinc bioavailability to *Selenastrum capricornutum*(Chlorophyceae): accidental metal uptake and toxicity in the presence of citrate, *J. Phycol.*, 36, 473, 2000.

41. Gardea–Torresdey, J.L. et al., Absorption of copper (II) by creosote bush (*Larrea tridentata*): use of atomic and x-ray absorption spectroscopy, *Environ. Toxicol. Chem.*, 20, 2572, 2001.

42. Sneller, F.E.C., Phytochelatins as a biomarker for heavy metal toxicity in terrestrial plants, Ph.D. thesis, Vrije Universiteit Amsterdam, Print Partners Ipskamp, Enschede/Amsterdam, 1999.

43. Salt, D.E. et al., Mechanisms of cadmium mobility and accumulation in Indian mustard, *Plant Physiol.,* 109, 1427, 1995.

44. Schmoger, M.E.V., Detoxification of arsenic by phytochelatins in plants, *Plant Physiol.*, 122, 793, 2000.

45. Pawlik–Skowronska, B., Correlations between toxic Pb effects and production of Pb-induced thiol peptides in the microalga *Stichococcus bacillaris*, *Environ. Pollut.*, 119, 119, 2002.

46. Boominathan, R. and Doran, P.M., Cadmium tolerance and antioxidative defenses in hairy roots of the cadmium hyperaccumulator, *Thlaspi caerulescens*, *Biotechnol. Bioeng.*, 83, 158, 2003.

47. Malecka, A., Jarmuszkiewicz, W., and Tomaszewska, B., Antioxidant defense to lead stress in subcellular compartments of pea root cells, *Acta Biochimi. Poloni.*, 48, 687, 2001.

48. Schickler, H. and Caspi, H., Response of antioxidative enzymes to nickel and cadmium stress in hyperaccumulator plants of the genus *Alyssum*, *Physiol. Plant.*, 105, 39, 1999.

49. Krause, G.H., Chlorophyll fluorescence and photosynthesis — the basics, *Annu. Rev. Plant Physiol. Plant Mol. Biol.*, 42, 313, 1991.

23 Temperate Weeds in Russia: Sentinels for Monitoring Trace Element Pollution and Possible Application in Phytoremediation

D.I. Bashmakov, A.S. Lukatkin, and M.N.V. Prasad

CONTENTS

23.1 INTRODUCTION

Plant and soil form an integrated system and technogenic contamination of soils is reflected in parts of plants. Soil contamination by heavy metals has several implications, not only for human health but also for many organisms. Economic activities introduce essential changes to the environmental system. These changes are noticed in large cities and urban landscapes by way of pollution. Heavy metals (HMs) represent one of the major environmental pollutants and contaminants of the soil. Evaluation of metal accumulation in soils and plants is of environmental importance due to their health effects on humans and other biota [1]. Green plants are the most relevant components in the decontamination of negative effects of industrial activity. Their resistance and capacity to accumulate different pollutants determine usage of some species in phytomelioration of soils. Phytoremediation is one such important strategy to decontaminate metal-polluted soils that is based primarily on the performance of species selected [2]. In this regard, tolerable herbaceous fast growing plants capable of accumulating HM in shoots are preferred for phytoremediation.

Self-cleaning of soils does not take place or, rather, does so extremely slowly. The toxic metals in topsoil thus are accumulated in plants. Weeds are more adapted to unfavorable soil conditions, such as low moisture and presence of toxic metals, and easily become acclimatized to local situations. One such important exercise is phytoremediation to protect the environment from metal contamination [3–5]. To be successful in phytoremediation, fast growth, easy reproduction, intensive accumulation of heavy metals from soil, and concentration in plant parts are necessary. In this chapter, some weeds of the mean band of Russia (temperate zone) are studied for their capacity to accumulate heavy metals and their possible application in phytoremediation [6–11].

23.2 MATERIALS AND METHODS

Saransk (54° northern latitude; 45° eastern longitude) is the city (254.3 km²) that is the site of this investigation. It is located at the junction of steppes and forests in a central part of Russian flats (on high ground Privolzhskaja). Mean annual air temperature is +3.5 + 4.1°C; in July it is +19.2 +20°C, with a maximum of +39°C. Duration of a vegetation season is 175 to 180 days. The warm season accounts for 70% of the annual precipitation norms. Chernozioms, gray wood, and alluvial soils are present in the terrain of the city. About 30 heavy engineering, electrical engineering, chemical, and food-processing industries are located in Saransk. The water salinity on sewage disposal sites is 900 to 1000 mg/L with pH 6.2 to 8.0. The sewage contains high doses of organics, salts, and Cu, Zn, Pb, Hg, Ni, Cr, Mn, V, W, etc.

Roots, stems, and leaves were harvested in 100-m² plots of six experimental plants: dandelion (*Taraxacum officinale* Wigg.); sagebrush (*Artemisia vulgaris* L. and *Artemisia absinthium* L.); *Amaranthus retroflexus* L.; melilot (*Melilotus officinalis* L. Pall); *Calamagrostis epigeios* (L.) Roth.; and orach (*Chenopodium album* L. *s. l.*). The samples were harvested in the beginning (May 10, 1997), middle (July 3, 1997), and end (September 25, 1997) of growing seasons in regions of industrial and domestic wastes. Plant samples of 0.5 kg fresh weight were made from eight to ten individuals. The collected plants were washed with distilled water, dried, and stored in paper packets. Control samples were harvested from the Smolnij National Park (60 km from Saransk). Heavy metals were determined with the help of spectrophotometric analysis. The method involves the deposition of metal ions from aqueous solution on the residue of zirconium hydroxide at pH 6 to 7, with subsequent filtration and analysis of the residue on an x-ray spectrometer, Spectroskan. Soil samples from the experimental sites were also analyzed for pH and humus contents (Figure 23.1 and Figure 23.2).

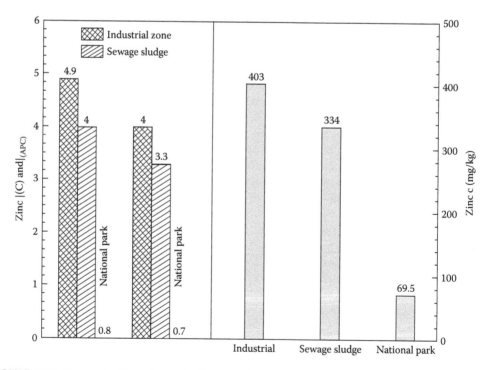

FIGURE 23.1 Zinc content in technogenically contaminated (industrial zone and sewage sludge) and clean soils (national park). Indices *Ic = heavy metal (HM) contents in the investigated sample/HM mean contents in earth crust;** I_{APC} = the HM contents in investigated sample/value of HM approximately permissible concentration (APC).

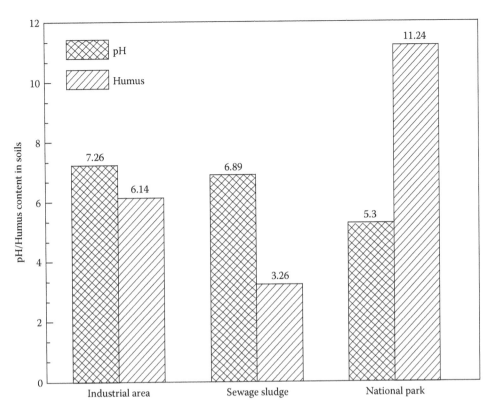

FIGURE 23.2 pH and humus content (% dry matter) in soils of industrial area, sewage sludge, and national park.

The heavy metal content (x) in plant samples (milligrams per kilogram of air-dry weight) was calculated by the formula:

$$x = \frac{(c - c_1) \cdot 100 \cdot 1000}{m \cdot 50}$$

where c is concentration of metal in analyzed sample (mg); c_1 is concentration of metal in a blank sample (mg); and m is the air-dry weight of plant sample taken for the analyzed (g).

23.3 RESULTS AND DISCUSSION

The data obtained indicate that, in the soil samples collected from industrial zone, the total contents of many heavy metals exceed the mean contents of the Earth's crust. There were very big anomalies of Ni (Kc = 10.6), Pb (9.7), and Zn (4.86); anomalies of Cu, Co, and Fe were barely seen. Neutral soil (pH 7.26) and low humus contents (6.14%) promote transition of many heavy metals in soil solution.

Large amounts of heavy metals (Table 23.1) are present in sewage precipitations (SPs) of refining facilities. For many HMs, the values are higher than prescribed limits. The greatest quantities were marked Pb (Kc = 6.4), Ni (5.2), and Zn (4.02); the least was Mn (0.7). SP had neutral pH of 6.89 and contained not enough humus (3.26). It is therefore possible to predict that SPs are characterized by noticeable forms of HMs. The soil samples from Smolnij National Park did not contain HM as compared to the mean contents in the Earth's crust and are characterized

TABLE 23.1
HM Content in Technogenic Contaminated and Clean Soils of a Temperate Area of a Mean Band of Russia

HM	Mean content[a]			I[b]			I$_{APC}$[c]		
	Industrial zone	Refining facilities	National park	Industrial zone	Refining facilities	National park	Industrial zone	Refining facilities	National park
Pb	156.45	102.0	14.0	9.7	6.4	0.8	5.2	3.4	0.5
Zn	403.25	334.0	69.5	4.9	4.0	0.8	4.0	3.3	0.7
Cu	163.3	75.5	17.0	3.4	1.6	0.4	2.9	1.4	0.3
Ni	616.3	305.0	11.2	10.6	5.2	0.2	7.3	3.6	0.1
Co	41.95	31.0	9.0	2.3	1.7	0.5	1.7	1.2	0.4
Fe	55729	48620	20654	1.3	1.2	0.5	1.5	1.3	0.5
Mn	711.35	690.4	503.0	0.7	0.7	0.5	0.5	0.5	0.3
Cr	29.7	122.0	71.6	0.3	1.5	0.8	0.3	1.2	0.7

[a] Milligrams per kilogram.
[b] Ic = the HM contents in investigated sample/HM mean contents in Earth's crust.
[c] I$_{APC}$ = the HM contents in investigated sample/value of HM approximately permissible concentration (APC).

by high humus contents (11.24%) with a pH of 5.3. As a whole, the analysis of the obtained data establishes that the soils of an industrial zone and refining facilities' technogenic oozes fall into the barely contaminated terrains (Table 23.1). Some substances were characterized by the superior soil APC levels designed. It is possible to consider the national park terrain clean.

Plants growing in contaminated terrains have accumulated significant amounts of HM in their organs. On the maiden place, the absolute contents were Fe. Its concentration in roots was from 3.5 up to 33.1 g/kg, and in shoot from 1.5 up to 8.1 g/kg. Plants accumulated Zn, Cu, Mn, Cr, Pb, and Ni (Table 23.2).

The HM absorption from soil by different plant species descended selectively. *Calamagrostis epigeios* was the Pb, Zn, and Cu concentrator. The medicinal dandelion accumulated Zn, Cu, Pb, and Cr. Mugwort stored predominantly Zn, Ni, and Cu and, to a lesser degree, Ni and Mn. *A. absinthium* was a hyperaccumulator of Zn, Cu, and Cr. Orach concentrated Zn, Cu, Zn. The high contents of Zn, Ni, and Cu were marked for a medicinal melilot. Different plants' organs had different levels of HMs (Table 23.2). Note that Zn, Cu, Fe, Pb, and Mn are stored in the roots and leaves and Ni and Cr in the roots and stems.

The high Pb content in the roots and leaves is apparently connected to formation of Pb phosphate complexes in the cell wall. A definite quantity of Pb from the atmosphere goes into plants through leaves. Zn accumulation is connected to its relevant physiological role. Zinc was stored in the roots of dandelion, *Calamagrostis epigeios*, and *Amaranthus retroflexus*. Maximum Cu content was observed in the roots and minimum in leaves. However, for absinthium, orach, and mugwort, the highest Cu content was established in leaves. The Fe and Mn accumulation in the roots and leaves was probably connected with antagonism between Fe and Zn, Mn, and Ni.

The index of biological absorption (I_a = HM concentration in air-dry sample/HM contents in soil) has shown that, on the contaminated soils, a process of Zn, Cu, and Cr accumulation was most intense, and less so for Mn, Fe, and Ni accumulation (Table 23.3). The present research also established that HM availability in soil decreases the intensity of absorption for some plants, but increases it for others. For example, the absinthium shoots in the national park accumulated much less Cr (I_a = 0.55) than those in the industrial zone did (I_a = 16.98).

The Zn accumulation in weeds growing in industrial areas and soils contaminated with sewage sludge is shown in Figures 23.3 and 23.4. For accumulation of other metals, refer to Table 23.1.

TABLE 23.2
HM Content in Shoots and Roots of Some Higher Plants of a Temperate Area of a Mean Band of Russia

Sample location	Plant species	Plant part	Mean HM content[a]						
			Pb	Zn	Cu	Ni	Fe	Mn	Cr
Industrial	*Amaranthus retroflexus*	Leaves	tr.	tr.	176.6	336.4	11,026	60.8	tr.
		Stems	tr.	tr.	tr.	85.9	tr.	tr.	tr.
		Roots	tr.	tr.	11.5	41.4	13,337	6.6	tr.
	Artemisia (absinthium)	Leaves	20.3	12.3	225.4	tr.	11,177	tr.	18.8
		Stems	5.7	92.2	192.0	tr.	6,494	5.7	749.3
		Roots	50.3	tr.	65.6	tr.	14,094	tr.	46.0
	A. retroflexus (mugwort)	Leaves	17.1	115.4	92.2	tr.	8,609	8.2	15.5
		Stems	5.7	tr.	186.7	tr.	3,323	tr.	216.2
		Roots	4.4	416.5	51.6	8.8	4,064	82.9	49.5
	Taraxucum officinale (dandelion)	Leaves	tr.	755.9	100.5	0.9	4,168	9.1	tr.
		Roots	59.1	864.9	309.1	1.1	3,964	16.1	68.8
	Melilotus officinalis	Leaves	tr.	252.0	79.8	tr.	4,225	103.3	tr.
		Stems	26.1	363.2	231.5	44.9	2,299	36.7	24.9
		Roots	tr.	131.6	296.4	26.4	12,159	70.3	12.5
	Calamagrostis epigeios	Leaves	245.2	362.7	17.5	2.3	2,273	143.6	tr.
		Stems	5.4	152.2	265.7	10.4	690	70.4	15.3
		Roots	tr.	746.7	783.7	36.2	22,107	332.6	14.8
	Chenopodium album	Leaves	30.1	1311.0	336.9	tr.	4,463	285.2	6.6
		Stems	12.8	441.2	287.1	tr.	2,048	2.4	8.4
		Roots	94.7	71.5	97.8	tr.	7,241	37.6	16.7
Refining facilities	*A. retroflexus* (mugwort)	Shoots	8.6	1518.7	148.0	73.5	2,071	242.1	tr.
		Roots	157.2	532. 8	405.1	8.8	3,535	63.4	tr.
	Artemisia (absinthium)	Shoots	36.2	843.0	tr.	62.3	2,770	136.7	27.8
		Roots	239.3	1315.7	124.7	34.7	6,663	93.8	tr.
	T. officinale (dandelion)	Shoots	tr.	399.9	23.3	28.6	2,232	56.2	tr.
		Roots	121.9	596.2	108.4	9.5	4,661	5.6	44.1
	M. officinalis (melilot)	Shoots	tr.	1291.7	tr.	230.2	2,492	134.9	40.7
		Roots	90.8	452.2	tr.	73.4	10,942	52.0	8.9
	Calamagrostis epigeios	Shoots	120.8	652.4	28.4	70.2	1,721	149.2	58.1
		Roots	6.2	287.2	tr.	23.6	19,517	15.7	47.1
	Ch. album (orach)	Shoots	tr.	tr.	tr.	tr.	4,143	tr.	63.7
		Roots	102.3	tr.	tr.	tr.	7,796	tr.	4.5
Smolnij National Park	*Ch. album* (orach)	Shoots	tr.	350.7	478.5	0.5	3,076	91.0	66.7
	Artemisia (absinthium)	Shoots	tr.	164.8	41.9	5.7	1,306	141.7	39.9

[a] Milligrams per kilogram of dry sample weight.

TABLE 23.3
The Heavy Metal Biological Absorption Indices[a] (I_a) of Some Higher Plants of a Temperate Area of a Mean Band of Russia

Location of samples	Plant species	Plant part	HM accumulation index						
			Pb	Zn	Cu	Ni	Fe	Mn	Cr
Industrial zone .	T. officinale	Shoots	tr.	1.87	0.62	0.0015	0.08	0.01	tr.
		Roots	0.38	2.14	1.89	0.0018	0.07	0.02	2.32
	Amaranthus retroflexus	Shoots	0.15	tr.	0.30	0.27	0.07	0.03	tr.
		Roots	tr.	tr.	0.07	0.07	0.24	0.01	tr.
	A. absinthium	Shoots	0.06	0.16	1.25	tr.	0.14	0.05	16.98
		Roots	0.32	tr.	0.4	tr.	0.25	tr.	0.15
	Ch. album	Shoots	0.13	1.90	1.88	tr.	0.05	0.16	0.26
		Roots	0.60	0.18	0.59	tr.	0.13	0.05	0.56
	Calamagrostis epigeios	Shoots	0.82	0.65	0.84	0.01	0.02	0.15	0.26
		Roots	tr.	1.85	2.96	0.06	0.39	0.47	0.49
	M. officinalis	Shoots	0.12	0.82	1.16	0.05	0.05	0.08	0.60
		Roots	tr.	0.33	1.81	0.04	0.22	0.09	0.52
Refining facilities	T. officinale	Shoots	tr.	1.20	0.31	0.09	0.05	0.08	tr.
		Roots	1.20	1.70	1.44	0.03	0.10	0.01	0.36
	A. retroflexus	Shoots	0.08	4.55	1.96	0.24	0.04	0.35	tr.
		Roots	1.54	1.60	5.37	0.03	0.07	0.09	tr.
	A. absinthium	Shoots	0.36	2.52	tr.	0.20	0.06	0.20	0.23
		Roots	2.35	3.94	1.65	0.31	0.14	0.14	tr.
	Ch. album	Shoots	tr.	tr.	tr.	tr.	0.09	tr.	0.52
		Roots	1.00	tr.	tr.	tr.	0.16	tr.	0.04
	Calamagrostis epigeios	Shoots	1.18	1.95	0.38	0.23	0.04	0.22	0.48
		Roots	0.06	0.86	tr.	0.08	0.40	0.02	0.39
	M. officinalis	Shoots	tr.	3.87	tr.	0.75	0.05	0.20	0.33
		Roots	0.89	1.35	tr.	0.24	0.23	0.08	0.07
National park	A. retroflexus	Shoots	tr.	5.05	28.14	0.04	0.15	0.18	0.93
	A. absinthium	Shoots	tr.	2.37	2.46	0.51	0.06	0.28	0.56

[a] Index of biological absorption (I_a) = HM contents in a plant organ/HM concentration in soil.

According to the data obtained, Mn and Fe fall into the category of substances of gentle and mean absorption in contaminated conditions; however, in the national park, Zn accumulation was substantial. For example, the Mn absorption by absinthium shoots was more intensive in the national park (I_a = 0.28) than on refining facilities (I_a = 0.20). In the industrial zone, I_a was only 0.05.

Pb falls to a gentle biological absorption substance (Table 23.2). However, On refining facilities, Pb I_a for shoots of *Calamagrostis epigeios* was 1.18. It is established that HM-accumulating capacity of plants depends on the vegetation season. For example, the Cu concentration in an absinthium's above-ground part was maximum in the beginning of the growing season and minimum in July. Pb seasonal dynamics in shoots of *Calamagrostis epigeios* has the following nature: the concentration of this metal increases in the middle of the growing season with a minimum in the spring and autumn (Table 23.4).

High seed production and fast biomass growth was observed in all the plants examined. However, the HM accumulation in an above-ground part was marked only for few species. It was

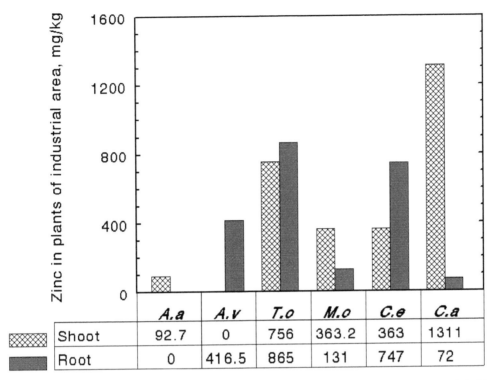

FIGURE 23.3 Zinc in shoots and roots of some weeds growing in industrial area. *A. a = Artemisia absinthium*, (absinthium); *A. v. = Artemisia vulgaris* (sagebrush); *T. o =Taraxacum officinale* (dandelion); *M. o = Melilotus officinalis* (melilot); *C. e = Calamagrostis epigeios*; *C. a = Chenopodium album* (orach).

marked for Zn absorption by mugwort (I_a = 4.55, for the definition off I_a see Table 23.3 on p. 444), *Calamagrostis epigeios* (I_a = 1.95), and medicinal melilot (I_a = 3.87) from refining facilities, and for orach (I_a = 1.9) from the industrial zone; and for Cu absorption by absinthium (I_a = 16.98) from the industrial zone; for Pb absorption by terraneous (I_a = 1.18) from refining facilities (Table 23.3).

23.4 *TARAXACUM OFFICINALE* WIGG. (DANDELION): IDEAL SENTINEL FOR MAPPING METAL POLLUTION

Taraxacum officinale (dandelion) growing on metal-polluted soils has accumulated significant levels of toxic metals. Soil material and plant tissue were collected along transects in two heavily contaminated facilities; a Superfund site revealed that Cd uptake was maximal in *Taraxacum officinale* at 15.4 mg/kg [12–22]. *T. officinale* is the best example for HM accumulation in shoots on alluvial soils and chernozioms at < 7. Zn was absorbed intensively at the beginning of vegetation; Fe in the middle of summer; and Ni, Mn, and Cr in autumn (Figure 23.5 and Figure 23.6). The best time for the harvest of green mass is the end of spring.

At implementation of phytoremediation in the city, the anthropogenic landscape esthetic view is very relevant. The natural tangle of sagebrush is not in harmony with downtown, but this is not true for a dandelion. Furthermore, use of dandelion in phytoremediation has a number of advantages:

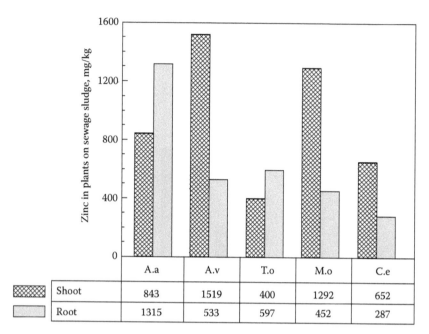

FIGURE 23.4 Zinc in shoots and roots of some weeds growing on sewage sludge. *A. a = Artemisia absinthium,* (absinthium); *A. v. = Artemisia vulgaris* (sagebrush); *T. o =Taraxacum officinale* (dandelion); *M. o = Melilotus officinalis* (melilot); *C. e = Calamagrostis epigeios.*

- The average number of seeds produced is $1.4 \cdot 10^6$ per kilogram; they can be stored at $0°$ for 12 years with viability still maintained.
- Dandelion seeds do not have a dormant period and can be used for cultivation immediately after harvesting.
- Germinating capacity of seeds is 72 to 100%.
- After defoliation, the plant regrows fast.
- The greatest increment of above-ground biomass in the maiden year sowings is August; in the second year, it is from May to June.
- Dandelion is capable of surviving prolonged droughts.

Limitations also exist:

- The plant does not grow well on slopes and inclines.
- Survival of this plant depends on local ecological factors.

Taking the preceding advantages into account, HM accumulation of dandelions from 1 km² of soil was investigated; results are shown in Table 23.5. Thus, *T. officinale* was found to be an ideal sentinel for mapping metal pollution and may be recommended for phytoremediation of metal-contaminated soils [6,7,12–22].

23.5 CONCLUSIONS

On the basis of the data obtained, the following propositions are made: for Cr-contaminated soils, detoxification is possible by cultivating *A. absinthium* and by making use of the biomass in the middle of a growing season. Cu and Zn in soils can be decontaminated by *Chenopodium album* and *A. absinthium*. For better biomass, it is necessary to apply the green manure to sagebrush before the flowering starts. Common garden snail (*Helix aspersa*) feeding on *Taraxacum officinale*

TABLE 23.4
Vegetation Season Influence on HM Accumulation by Plants Collected on Contaminated Terrain (Industrial Zone) of a Temperate Area of a Mean Band of Russia

Plant species	Plant part	Mean HM content[a]						
		Pb	Zn	Cu	Ni	Fe	Mn	Cr
Amaranthus retroflexus	Shoots	tr.	33.7	tr.	tr.	2,078.05	tr.	tr.
	Roots	tr.	271.9	tr.	2.5	6,641	8.9	tr.
	Shoots	tr.	399.2	tr.	57.3	5,058	129.7	19.1
	Roots	tr.	819.1	0.1	96.5	7,932	71.5	2.5
	Shoots	23.2	tr.	48.6	169.3	3,671	20.3	tr.
	Roots	tr.	tr.	11.5	41.4	13,337	6.6	tr.
Taraxacum officinale	Shoots	tr.	289.0	tr.	tr.	4,535	tr.	tr.
	Roots	tr.	369.4	95.5	37.7	1,885	20.3	tr.
	Shoots	2.5	tr.	41.8	11.1	7,132	tr.	50.3
	Roots	tr.	tr.	tr.	tr.	5,969	tr.	tr.
	Shoots	tr.	755.9	100.5	0.9	4,168	8.1	tr.
	Roots	59.1	864.9	309.1	1.1	3,964	16.00	68.8
Calamagrostis epigeios	Shoots	0.5	tr.	tr.	tr.	0.061	tr.	1.1
	Roots	tr.	tr.	tr.	0.01	0.46	tr.	2.8
	Shoots	0.6	tr.	tr.	tr.	0.051	0.01	1.3
	Roots	n.d.	n.d.	n.d.	n.d.	n.d.	n.d.	n.d.
	Shoots	129.9	260.6	137.9	6.2	1,505	105.9	7.4
	Roots	tr.	746.7	483.7	36.2	22,107	332.6	14.8
Artemisia absinthium	Shoots	10.6	65.4	204.5	tr.	8,063	3.8	504.6
	Roots	50.3	tr.	65.6	tr.	14,094	tr.	46.0
	Shoots	0.4	2.6	1.3	tr.	0.12	0.04	tr.
	Roots	0.1	0.2	0.1	0.04	0.36	tr.	6.5

[a] Milligrams per kilogram of dry sample weight.

growing on metal-polluted soils has accumulated significant levels of toxic metals in its gastric system [12–22].

Metal-accumulating plants are of phytoremediation importance. Phytoremediation is one of the strategies to decontaminate toxic levels of metals in the environment and is an upcoming and novel technology. It was observed that some of these temperate weeds are natural hyperaccumulators of toxic metals. Metal accumulation has been investigated in plant roots and shoots of selected weeds: *Calamagrostis epigeios, Taraxacum officinale* (dandelion)*, Artemisia absinthium* (absinthium), *Artemisia vulgaris* (sagebrush), *Amaranthus retroflexus* (mugwort), *Chenopodium album* (orach), and *Melilotus officinalis* (melilot). Heavy metal content in plant parts and soil samples was quantified following x-ray spectrometry. Soil samples were also analyzed for humus content and pH.

It was noted that *Calamagrostis epigeios* was a Pb, Zn, and Cu concentrator. *Taraxacum officinale* accumulated Zn, Cu, Pb, and Cr. *Artemisia absinthium* was a hyperaccumulator of Zn, Cu, and Cr, whereas *Artemisia vulgaris* accumulated predominantly Zn, Ni, and Cu. *Chenopodium album* concentrated Zn, Cu, and Zn; appreciable quantities of Zn, Ni, and Cu were observed in *Melilotus officinalis*. The metal accumulation pattern indicates that Zn, Cu, Fe, Pb, and Mn were stored more often in roots and leaves, and Ni and Cr were stored in roots and stems. It was also observed that the metal accumulation capacity of plants depended on the vegetation season. The data from this investigation may thus be useful for phytoremediation of metal-polluted soils in temperate regions [6,7,12–22].

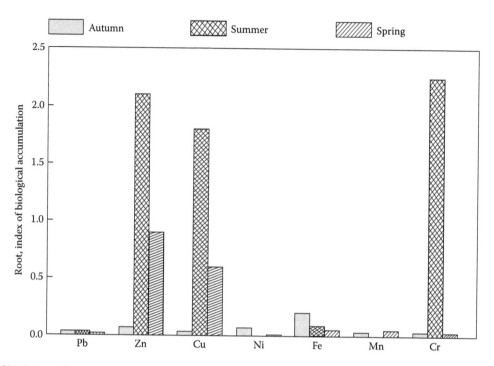

FIGURE 23.5 Seasonal accumulation of heavy metals in shoots of *T. officinale*; index of biological accumulation = contents of a metal in a plant organ (mg/g dry weight)/metal content in the soil (milligrams per kilogram).

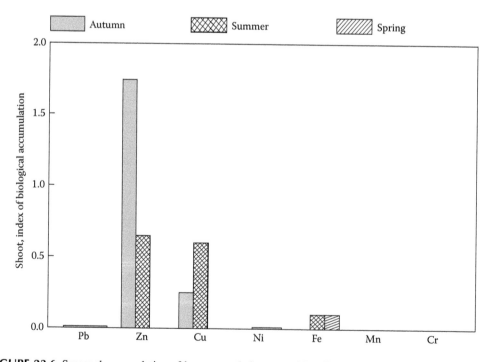

FIGURE 23.6 Seasonal accumulation of heavy metals in roots of *T. officinale*.

TABLE 23.5
Heavy Metal Accumulation by *T. officinale*

Heavy metal	HM accumulation (kg km^{-2})	
	Minimum (from rather clean regions)	Maximum (from the contaminated soils)
Pb	87.53–125.04	108.21–154.59
Zn	123.86–340.37	211.34–3016.20
Cu	35.31–50.44	152.02–217.17
Ni	1.39–1.99	171.99–245.70
Fe	3375.83–4822.61	9974.35–14249.07
Mn	13.79–19.70	525.89–751.27
Cr	30.48–43.55	38.83–55.48

REFERENCES

1. Hadjiliadis, N.D. Cytotoxicity, mutagenic and carcinogenic potential of heavy metals related to human environment, NATO-ASI Series 2, *Environment*, vol. 26, Kluwer, Dordrecht, 629, 1997.
2. Raskin, I. and Ensley, B.D., Eds. *Phytoremediation of Toxic Metals: Using Plants to Clean Up the Environment.* John Wiley & Sons, 352, 2000. Brooks, R.R., Ed. *Plants That Hyperaccumulate Heavy Metals.* CAB International, Wallingford, U.K., 1998.
3. Chaney, R.L., Malik, M., Li, Y.M., Brown, S.L., Brewer, E.P., Angle, J.S., and Baker, A.J.M. Phytoremediation of soil metals. *Curr. Opin. Biotechnol.*, 8, 279, 1997.
4. Djingova, R., Kuleff, I., and Andreev, N. Comparison of the ability of several vascular plants to reflect environmental pollution. *Chemosphere*, 27, 1385, 1993.
5. Glass, D.J. U.S and international markets for phytoremediation, 1999–2000. D.J. Glass Associates Inc. Needham, MA, 266, 1999.
6. Kabata–Pendias, A. and Dudka, S. Trace metal contents of *Taraxacum officinale* dandelion as a convenient environmental indicator. *Environ. Geochem. Health,* 13, 108, 1991.
7. Markert, B. *Plants as Biomonitors: Indicators for Heavy Metals in the Terrestrial Environment.* VCH Publishers, Weinheim, Germany, 1993.
8. Prasad, M.N.V., Ed. *Metals in the Environment — Analysis by Biodiversity.* Marcel Dekker Inc., New York, 504, 2001.
9. Prasad, M.N.V., Ed. *Heavy Metal Stress in Plants — from Biomolecules to Ecosystems.* Springer–Verlag, Heidelberg, 520, 2004.
10. Raskin, I. and Ensley, B.D., Eds. *Phytoremediation of Toxic Metals: Using Plants to Clean Up the Environment.* John Wiley & Sons, New York, 352, 1999.
11. Terry, N. and Bañuelos, G., Eds. *Phytoremediation of Contaminated Soil and Water.* Lewis, U.K., 384, 2000.
12. Marr, K., Fyles, H., and Hendershot, W. Trace metals in Montreal urban soils and the leaves of *Taraxacum officinale. Can. J. Soil Sci.,* 79, 385, 1999.
13. Beeby, A. and Richmond, L. Evaluating *Helix aspersa* as a sentinel for mapping metal pollution. *Ecol. Indicators*, 1, 261, 2002.
14. Cook, C.M. et al. Concentrations of Pb, Zn, and Cu in *Taraxacum* spp. in relation to urban pollution. *Bull. Environ. Contam. Toxicol.,* 53, 204, 1994.
15. Djingova, R. and, Kuleff, I. Bromine, copper, manganese, and lead content of the leaves of *Taraxacum officinale* dandelion. *Sci. Total Environ.,* 50, 197, 1986.
16. Djingova, R. and, Kuleff, I. Monitoring of heavy metal pollution by *Taraxacum officinale*. In: Markert, B., Ed. *Plants as Biomonitors.* Weinheim, Germany: VCH Verlagsgesellschaft, 433–460, 1993.
17. Keane, B. et al., Metal content of dandelion (*Taraxacum officinale*) leaves in relation to soil contamination and airborne particulate matter. *Sci. Total Environ.,* 281, 63, 2001.
18. Kuleff, I. and Djingova, R. The dandelion (*Taraxacum officinale*): a monitor for environmental pollution? *Water Air Soil Pollut.,* 21, 77, 1984.

19. Normandin, L., Kennedy, G., and Zayed, J. Potential of dandelion (*Taraxacum officinale*) as a bioindicator of manganese arising from the use of methylcyclopentadienyl manganese tricarbonyl in unleaded gasoline. *Sci. Total Environ.*, 239, 165, 1999.

20. Robert, E. and Hoagland, R.E., Hydrolysis of 3′,4′-dichloropropionanilide by an aryl acylamidase from *Taraxacum officinale. Phytochemistry*, 14, 383, 1975.

21. Pichtel, J., Kuroiwa, K., and Sawyerr, H.T. Distribution of Pb, Cd and Ba in soils and plants of two contaminated sites. *Environ. Pollut.*, 110, 171, 2000.

22. Simon, L., Martin, H.W., and Adriano, D.C. Chicory *Cichorium intybus* L. and dandelion *Taraxacum officinale* Web. as phytoindicators of cadmium contamination. *Water Air Soil Pollut.*, 91, 351, 1996.

24 Biogeochemical Cycling of Trace Elements by Aquatic and Wetland Plants: Relevance to Phytoremediation

M.N.V. Prasad, Maria Greger, and P. Aravind

CONTENTS

24.1 INTRODUCTION

Aquatic ecosystems (freshwater, marine, and estuarine) act as receptacles for trace elements [1–3]. Several angiospermous families — namely, Cyperaceae, Potamogetonaceae, Ranunculaceae, Haloragaceae, Hydrocharitaceae, Najadaceae, Juncaceae, Pontederiaceae, Zosterophyllaceae, Lemnaceae, Typhaceae, etc. — have aquatic and semiaquatic environments. Aquatic/wetland plants play a crucial role in biogeogenic cycling of trace elements through their active and passive cycling of elements. They act as "pumps for essential and nonessential elements" [3]. Uptake of elements into plant tissue promotes immobilization in plant tissues and thus constructed wetlands are gaining significance for wastewater treatment [4].

Aquatic and wetland plant assemblages occupy specific zones, including the position of aboveground structures and roots in relation to the sediment surface and water table. A typical zonation in tropical aquatic and wetland macrophytes is comprised of:

- Free-floating plants: except for roots, which are situated in the water, most of the plant body is above the water, e.g., *Eichhornia crassipes* (water hyacinth), *Ludwigia* sp. (water primrose), *Pistia stratiotes* (water lettuce), *Lemna* sp., *Wolffia* and *Spirodela* (duckweeds), *Ceratophyllum demersum* (coontail), *Salvinia* sp., and *Azolla* (water ferns).
- Submerged (rooted) plants: these remain submerged in water, e.g., *Egeria densa* (Brazilian elodea), *Elodea canadensis* (elodea), *Hydrilla* and *Vallisneria* (tape grass).
- Emergent (rooted) plants: these plants are rooted to sediments but emergent above the water, e.g., *Alternanthera philoxeroides* (alligator weed), *Typha latifolia* (cattail), and *Phragmites communis* (reed).

Industrialization increased trace element fluxes from terrestrial and atmospheric sources towards the aquatic environment. Industrial effluents, mine discharges, and run-off from agroecosystems often contain metalliferous substrates, which get discharged into nearby aquatic ecosystems. Aquatic systems receive run-offs that contain high levels of contaminants, such as heavy metal effluents from industries, oil and petrol residues, fertilizers, pesticides and animal wastes [5]. Consequently, elevated concentrations of heavy metals, such as lead (Pb), chromium (Cr), cadmium (Cd), copper (Cu), and zinc (Zn), are usually found at high concentrations in the aquatic ecosystem. These metals are progressively added to the aquatic sediments, where they pose threats to the benthic organisms [6].

Sediments formed in the aquatic ecosystems substitute for the role of soil in terrestrial ecosystems. Sediments form biologically important habitats and microenvironments for aquatic life. The metalliferous discharges not only contaminate the interstitial water but also contribute to the metal reservoir pool in sediments.

Many environmental factors are known to modify the availability of metals in water to aquatic plants. Such factors include chemical speciation of the metals, pH, organic chelators, humic substances, particles and complexing agents, and presence of other metals and anions. Rate of influx of these heavy metals into the environment exceeds their removal by natural processes. Therefore, there is enhancement of heavy metals accumulating in the environment.

24.1.1 Aquatic Sediments as Reservoir of Trace Elements

Sediments also act as reservoirs of contaminants, which may enter the water through the desorption process and can be taken up by rooted macrophytes. The availability of metals to the organisms of the upper strata is directly reflected by the sediment characteristics. The major mechanism of accumulation of heavy metals in sediments led to the existence of five categories: exchangeable, bound to carbonates, bound to reducible phases (iron and manganese), bound to organic matter, and residual fraction [6]. These categories have different behaviors with respect to remobilization under changing environmental conditions.

By studying the distribution of metals between the different phases, their availability and toxicity can be ascertained [7,8]. The fraction of metals introduced by human activity pertains to the adsorptive, exchangeable forms bound to carbonates, which are weakly bound and thus equilibrate with the aqueous phase and become more rapidly available [9]. The metal present in the inert fraction, which is of detrital and lattice origin, can be taken as a measure of contribution by natural sources [10]. The fraction of metals bound to organic matter and Fe–Mn oxides is unavailable forms providing a sink for heavy metals; their release from this matrix will be affected only by high redox potential and pH [11]. The criteria of risk assessment code (RAC) indicating sediment that releases in exchangeable and carbonate fractions is shown in Table 24.1 [12].

It is known that metals have shorter retention time in air and water than sediments. However, this depends on the element. For example, Pb has a very short retention time in water and Zn has a longer retention time. Equilibrium is always maintained in the interface between metals in interstitial water and metals in sediment. Therefore, external factors would affect this equilibrium and availability of metals to the aqueous system [13].

Processes responsible for availability of metals from sediments to interstitial waters, biotic and abiotic factors regulating the metal bioavailability in sediments are mentioned below (Figures 24.1 and 24.2):

- Geochemical processes like methylation, metal–metal interactions, and displacement reactions; redox changes; temperature; pH-related adsorption–desorption at sediment particulate surface changes [6,14,15]
- Partitioning of metals between sediment particles and interstitial water, depending on bound forms like sulfides, carbonates, hydroxides, and humic substances [9,11,15–17]
- Speciation of metals in water and "carrier" chelating agents aiding in metal transport. Trace metals in natural waters exist in several different forms (as different oxidation states or forms complexed or bound to inorganic and organic matter). The distribution between these forms is often referred to as the speciation of a metal (Figure 24.3). It has been shown that the toxicity of a metal is related to its speciation; some forms of the

TABLE 24.1
Criteria of Risk Assessment Code

S. no.	Risk assessment code	Criteria (%)
1	No risk	<1
2	Low risk	1–10
3	Medium risk	11–30
4	High risk	31–50
5	Very high risk	>50

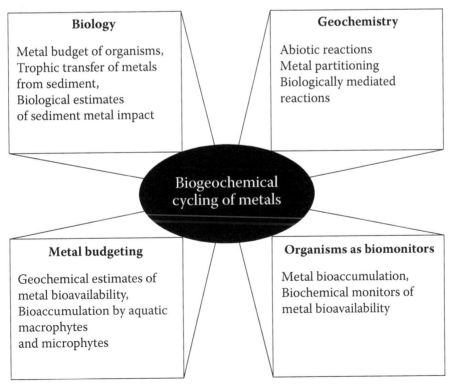

FIGURE 24.1 Processes responsible for biogeochemical cycling of metals from sediments to interstitial waters.

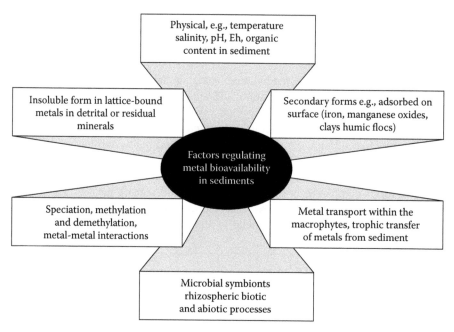

FIGURE 24.2 Biotic and abiotic factors regulating the metal bioavailability in sediments.

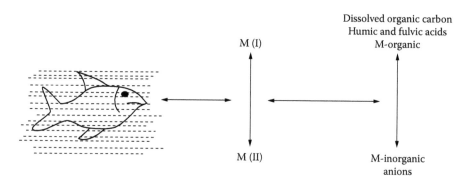

FIGURE 24.3 Trace metals in natural waters exist in different oxidation states or forms complexed or bound to inorganic and organic matter and the distribution of these forms is often referred to as the speciation.

metal are more available and toxic than others. This fact has been recognized in recent environmental legislation relating to surface water quality, e.g., for copper, aluminum, and silver [16]. As a consequence, industry and environmental regulators require an increasing amount of information on metal speciation.

- Microbial activity and animal activity (bioturbation) releasing bound metals [18]; exchange of ions between rhizosphere and metals partitioned in interstitial waters [18] (Figure 24.3 and Figure 24.4)

Once released into the interstitial water, metals become available in the waters of upper surface and thus facilitate the process of bioconcentration. In wetland ecosystems, physicochemical and biological processes operate that provide a suitable situation for removal of metals [19]. It is clear that development of a rational, effective, and economic strategy to remediate contaminated sedi-

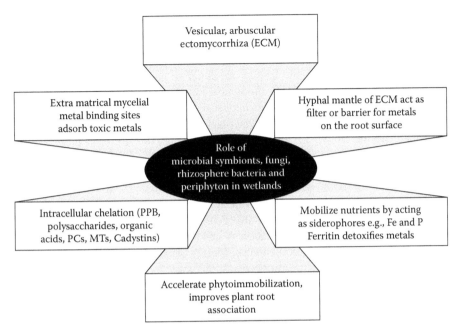

FIGURE 24.4 Role of microbial symbionts, fungi, rhizosphere bacteria, and periphyton in wetlands.

ments concerns understanding of the biogeochemical processes governing metal accumulation in aquatic plants.

24.2 FUNCTIONS OF AQUATIC PLANTS

Aquatic macrophytes are extremely important components of an aquatic ecosystem vital for primary productivity and nutrient cycling [19]. They provide habitat, food, and refuge for a variety of organisms [20]. Therefore, direct effects on macrophytes may lead to indirect effects on the organisms and nutrient cycles dependent on or associated with the plants. Given the ubiquity of submerged plants, their annual growth cycle, and their known capacity to concentrate certain elements in their tissues, it is clear that macrophytes could influence the seasonal storage and cycling of these elements in the aquatic environment. In principle, the submerged plant beds influence the cycling of metal elements by bioconcentration or indirectly by reducing current velocity, thereby favoring sedimentation of suspended particles and thus trace metals.

Depending on the species, a seasonal cycle of submerged plants is characterized by more or less rapid growth in spring, a peak at the end of summer, and more or less rapid decline in autumn in temperate situations, and a decline of growth observed only in summer in tropics [8]. The trend for greater dependence upon roots for heavy metal uptake was in rooted floating leaved taxa with lesser dependence in submerged taxa [21]. The tendency to use shoots as sites of heavy metal uptake instead of roots increases with progression towards submergence and simplicity of root structure.

Submerged rooted plants had some potential for the extraction of metals from water as well as sediments; rootless plants extracted metals rapidly only from water [22]. In submerged plants, leaves are the site of mineral uptake [23]. The foliar absorption of heavy metals is by passive movement through the cuticle, where the negative charges of the pectin and cutin polymers of the thin cuticle and the polygalacturonic acids of the cell walls create a suck inwards. Due to the increase in the charge density inwards, transport of positive metal ions takes place [23,24]. No ions enter stomata and, in submerged plant leaves, no stomata are present.

24.3 TRACE ELEMENT UPTAKE

Different plant species take up different amounts of metals [13,25–33]. In addition, the concentration found in various plant materials varies from site to site due to environmental pollution situations at that site. The variation within the same species can therefore be large. There are also differences in uptake between various metals. Not only the uptake but also the toxicity among different metals varies. Toxicity of different metals to *Elodea canadensis* and *Myriophyllum spicatum* is highest with Cu > Hg > As > Cd > Zn > Pb [34]. Submerged plant leaves have a very thin cuticle. The leaves of submerged plants are therefore very good at taking up metals directly from the water. The foliar uptake of Cd by *Potamogeton pectinatus* was nearly ten times higher than that of *Pisum sativum* [13]. It was shown that lead was accumulated in the shoots via absorption from water [25].

Macrophytes take up heavy metals via roots from the sediment and via shoots directly from the water. Therefore, the integrated amounts of available metals in water and sediment can be indicated by using macrophytes. Plants can also evolve ecotypes fairly soon and thereby be used in unfavorable conditions. Plants are also stationary and long lived and accumulate metals; therefore, they are suitable in monitoring of polluted sites. Metal concentration in plants must be related to the time of the year because the metal concentrations vary by season [35]. Aquatic plants also release metals through their leaves. Plants used as bioindicators must retain the metals in their plant body. *Ceratophyllum demersum, Myriophyllum spicatum, Potamogeton pectinarus, P. perfoliatus,* and *Zannichellia palustris* are proposed as bioindicators [36–42].

The uptake of metals by roots and by leaves increases with increasing metal concentration in the external medium. However, the uptake is not linear in correlation to the concentration increase because the metals are bound in the tissue, causing saturation that is governed by the rate at which the metal is conducted away. The effective uptake (or accumulation factor) is therefore highest at low external concentrations. This is shown for Cd in solution culture and also demonstrated as increased uptake from one and the same metal concentration with increasing root absorption area (root mass) [13]. External factors, such as temperature and light, influence growth and also affect metal uptake [43].

In aquatic macrophytes, the metal removal is accomplished through uptake by roots, chemical precipitation, and ion exchange with or absorption of settled clay and organic compounds [44–46]. Metal uptake is enhanced due to the presence of metal-binding ligands, such as thiols, or synthesis of metal-chelating peptides/proteins (namely, phytochelatins [47–54]) or surfactants such as linear alkyl benzene sulphonate [55]; combinations of metals in wastewaters may exert synergistic or antagonistic influence on metal uptake [56]. Aquatic macrophytes typically have much higher metal contents than nearby terrestrial plants, even when the total metal content of the soils is equivalent. Fritioff and Greger also found that this was the case [57].

However, this is due to the fact that the plant shoot takes up metals from the water, a high shoot biomass and less problematic uptake due to thin cuticle, and that shoots do not need to release metals from colloids or complexes before uptake. The shoot biomass is also much bigger in relation to the root mass in the case for terrestrial plants and, in terrestrial plants, the metals must be translocated a longer distance than in the case of submerged plants. Risk assessment of toxic trace metals in aquatic biota and use of in-built water-renovating strategies on ecologically acceptable principles has gained considerable significance in the field of environmental biotechnology and biotechnological methods of pollution control [58–61].

Aquatic macrophytes such as water hyacinth [*Eichhornia crassipes*] and several duckweeds have attracted the attention of scientists for their ability to accumulate trace metals. Using *E. crassipes*, heavy metal uptake from metal-polluted water, metal speciation, synthesis and characterization of heavy metal-binding complexes in root, and sorption of heavy metals from metal-containing solutions has been extensively investigated [62]. Metals found in nature in more than one valence state are more readily taken up by plants in the reduced form. Often, the metal is reoxidized within the plant tissues [23].

In aquatic ecosystems, the sediments and water interface serve as the habitat for a diverse community in which aquatic macrophytes are prominent. Metals present in surficial sediments in particulate form may exist as constituent elements present in the essentially insoluble products of physical weathering form, i.e., lattice bound (metal in detrital or residual minerals), or in a variety of secondary forms adsorbed on surfaces (iron/manganese oxides, clays, humic flocs) associated with organic matter, with sulphides, etc. The secondary forms are more reactive and more likely to be available.

Many environmental factors are known to modify the availability of metals in water to aquatic plants. As earlier mentioned, such factors include chemical speciation of the metal, pH, organic chelators, humic substances, particles and complexing agents, presence of other metals and anions, ionic strength, temperature, salinity, light intensity, and oxygen level and thus the redox potential [63]. In lakes, pH is important for speciation and thus also for the availability of metals to macrophytes. Most metal concentrations in water increase with decreasing pH, with the highest pH value of about 4. Especially in aquatic systems, the redox potential is important. At low redox potential, metals become bound to sulfides in sediments and are thus immobilized. In water, salinity affects the availability of metals because high salinity causes formation of metal–chloride complexes, which are difficult for plants and other organisms to take up.

External factors can decrease or enhance the efficiency to remove the pollutants from the water. High particle concentration binds elements and organic substances and sediments to the bottom. High cation exchange capacity [CEC] of the particles increases the binding of the positive ions. Outlets at treatment of stormwater, sewage water, and leakage water commonly contain high levels of particles. Plants increase the sedimentation of them, which is one of the mechanisms behind removal of metals from water by plants.

In temperate aquatic ecosystems, the temperature varies with the season. Uptake of metals is influenced by temperature and the uptake increases with increased temperature [64]. The removal efficiency of metals can therefore be season dependent in temperate climate. Another external factor that varies with season is salinity in stormwater. This is due to deicing of roads during winter, using sodium chloride. Salinity decreases the uptake of metals in plants [64]; due to complex formation with chloride, complexes prevent metal uptake. However, in the presence of sediment, sodium can be exchanged with sediment-bound metals and thereby increase the total metal concentration in the water phase, which increases the shoot metal uptake [65] (Figure 24.5). In stormwater treatments, various plants have been tested and it has been found that submerged plants do take up more metals than emergent plant species [57]. *Potamogeton natans* and *Elodea canadensis* were the two species mostly found in the stormwater treatment basins and they took up most metals by their shoots.

24.3.1 AQUATIC MACROPHYTES FOR TRACE ELEMENT BIOMONITORING AND TOXICITY BIOASSAYS

Aquatic plants are represented by a variety of macrophytic species that occur in various types of habitats. Extensive experimentation on various macrophytes such as *Eichhornia* [66–69], *Elodea* [25,33,70–72], *Lemna* [19,73–82], *Myriophyllum* [25,33,83–85], *Potamogeton* [25,33,77,83,86–91], and various other aquatic macrophytes has been conducted and indicates the potential utilization of macrophytes as biomonitor systems for toxic metals (Table 24.2).

Aquatic macrophytes have paramount significance in the monitoring of metals in aquatic ecosystems [129–132]. The use of aquatic plants in water-quality assessment has been common for years as *in situ* biomonitors [133,134]. According to Sawidis et al. [135], the occurrence of aquatic macrophytes is unambiguously related to water chemistry and using these plant species or communities as indicators or biomonitors has been an objective for surveying water quality [136].

In addition, considerable research has been focused on determining the usefulness of macrophytes as biomonitors of polluted environments [60,61,137]. The response of an organism to

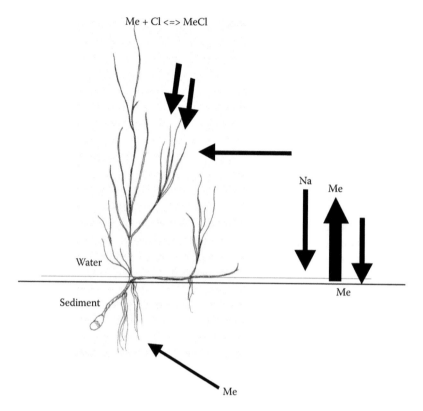

Me + Cl <=> MeCl

Na   Me

Water

Sediment

Me

Me

FIGURE 24.5 Influence of sodium chloride on the metal circulation in plant–water–sediment system.

deficient or excess levels of metal (i.e., bioassays) can be used to estimate metal impact. Such studies done under defined experimental conditions can provide results that can be extrapolated to natural environments. Heavy metals or even lighter metals in excess are often toxic to plants. Even at sublethal concentrations, physiological tests such as changes in pigment composition, photosynthesis, and respiration can reflect stress and predict future plant damage [106,138,139].

Using an aquatic macrophyte as a study material has multifold advantages. Rooted macrophytes, especially, play an important role in metal availability through rhizosphere secretions and exchange processes. This naturally facilitates metal uptake by other floating and emergent forms of macrophytes. The immobile nature of macrophytes makes them a particularly effective bioindicator of metal pollution because they represent real levels present at that site. Data on phytotoxicity studies are considered in the development of water-quality criteria to protect aquatic life and the toxicity evaluation of municipal and industrial effluents [80,140,141]. In addition, aquatic plants have been used to assess the toxicity of contaminated sediment and hazardous waste leachates [142].

Several of the aquatic macrophytes — namely, *Hydrilla verticillata, Certophyllum demersum, Vallisneria Americana*, etc. — detoxify toxic trace elements by inducing phytochelatins [47–55]. In North America, manganese and lead (from methylcyclopentadienyl manganese tricarbonyl [MMT]) uptake by plants growing near highways was much greater in aquatic plants than in terrestrial plants [77]. Plants growing close to the roadway and heavy motor vehicle traffic significantly contributed to these toxic elements [87,143].

Industrial discharge pollutes the bottom sediments with toxic trace elements. Increased density causes turbidity, thereby causing reduced light intensity; thus, the ability for plants to grow on these bottoms will be low. In the sediment, the redox potential is low and most of the trace elements are therefore firmly bound to sulfides. When industrial outlet has been shut down, new, unpolluted

TABLE 24.2
Toxicity Parameters Tested in Various Aquatic Macrophytes

Plant species studied	Concentrations tested	Parameters tested	Contaminants tested	Environment tested	Ref.
Ceratophyllum demersum	Cd: 0.01 to 2 ppm (maximum period of testing: up to 2 weeks) Cu: 0.01 to 64 ppm, Cr: 0.01 to 4.86 μM Fe: 0.01 to 75 μM Pb: 0.01 to 64 ppm Mn: 0.01 to 6.6 μM Zn: 2–64 ppm	Growth; fresh weight; dry weight ratios; metal uptake; mechanism of metal uptake; concentration factors; levels of antioxidant enzymes reflecting metal toxicity; carbonic anhydrase activity; levels of photosynthetic pigments and rate of photosynthesis; electron transport processes; metal–metal interactions between Cd–Zn and Cd–Ca	Cd, Cu, Zn, Cr, Pb, Hg, Fe, Mn, Zn, P, Ni, Co	Erlenmayer flasks, aquaria and samples from natural pond system	37, 54, 84, 92–97, 99–106
Myriophyllum M. spicatum M. alterniflorum M. aquaticum M. exalbescens	Natural metal concentrations existing in field conditions	Metal content, root/shoot metal ratio	Cd, Cu, Zn, Pb, Ni, Cr, Fe, Hg	Natural wetland systems	25, 83, 85
Elodea E. canadensis E. nuttallia E. densa	Cu: 1, 5, 10 ppm	Toxicity symptoms; shoot length variations; dry mass index	Cu, Pb, Cd, Zn, Cr, Ni, Hg, methyl-Hg	Laboratory aquaria	25, 33, 34, 70–72, 108
Potamogeton P. crispus P. perfoliatus P. pectinatus P. filiformis P. orientalis P. lapathifoilum P. attenuatum P. subsessils P. richardsonii	Radiotracers used: 76As, 109Cd, 115Cd, 64Cu, 65Zn, and 69mZn	Biomass and metal content; multispectral remote sensing data; seasonal storage of metals; mass balance calculations; organic selenium, free seleno-amino acids; glutathione-S-transferase activity; bioconcentration factors; metal mobility; single-tracer experiments and double-labeling experiments for estimating metal transport activity	Cu, Pb, Mn, Fe, Cd, Zn, Ni, Cr, Mn, As, Se	Natural lake system; agricultural drainage water medium and laboratory culture media; hydroculture two-compartment system; samples form natural aquatic systems	7, 8, 25, 33, 77, 83, 88, 90, 91, 109, 110
Littorella uniflora			Cu, Pb		25

TABLE 24.2
Toxicity Parameters Tested in Various Aquatic Macrophytes (continued)

Plant species studied	Concentrations tested	Parameters tested	Contaminants tested	Environment tested	Ref.
Isoetes lacustris			Cu. Pb		25
Ruppia maritima	Natural metal concentrations existing in field conditions	Metal content, organic selenium, free seleno-amino acids	Mn, Pb, Cd, Pb, Fe, Se	Agricultural drainage water medium and laboratory culture media	89, 109, 110
Ranunculus *R. baudotii* *R. aquatilis*	Natural metal concentrations existing in field conditions		Cd, Cu, Cr, Zn, Ni, Pb	Samples collected from natural ecosystems	33, 109
Scirpus *S. lacustris* *S. acutes* *S. maritimus*			Cr, Cd, Fe, Pb, Mn		109, 110
Chara spp.					109–111
Lemna *L. minor* *L. gibba* *L. trisulca* *L. paucicostata* *L. valdivinia* *L. polyrrhiza* *L. perpusilla*	Fe: 1, 2, 4, and 8 ppm; Cu: 1, 2, 4, and 8 ppm; Pb: 0–7 ppm; Ni: 0–5 ppm; Cu, Zn, Cr: 5, 10, 15, 20 ppm; Cd: 0.25–10 mM; Ca: 1.1 to 17.6 ppm; As: up to 1 ppm	Growth; fresh weight; fronds number; metal content; metal uptake kinetics; photosynthetic rate; total protein content; vegetative propagation; cadmium content; growth rates; operational PS II quantum yield; nonphotochemical quenching; hydrophobic components bioassay; determination of EC_{50} values	Mn, Pb, Cd, Fe; Mn, Pb, Ba, B, Cd, Cu, Cr, Ni, Se, Zn, Fe	Laboratory cultures; aquaria; natural effluents; sterile cultures; semicontinuous; flow through culture system; sewage stabilization systems	74, 75, 19, 56, 73, 76, 78, 79, 81, 82, 110–116
Najas marina			Cd, Fe, Pb, Mn		109
Distichlis spicata			Cd, Fe, Pb, Mn		109
Naphar *N. lutea* *N. variegatum*		Metal content; uptake potential; glutathione-S-transferase activities	Cu, Ni, Cr, Co, Zn, Mn, Pb, Cd, Hg, Fe	Samples from natural wetlands	96, 117
Phragmites *P. karka* *P. communis*	Natural metal concentrations of the wetland ecosystem	Metal uptake	Ni, Cr, Co, Zn, Mn, Pb, Cd, Cu, Hg, Fe	Natural wetland ecosystem	119, 96, 18
Bacopa monnieri			Cr		119

Species	Treatment/Concentration	Parameters	Metals	Medium/Source	References
Ludwigia; L. palustris; L. natans		Growth; metal content	Hg, Zn, Cu, Fe,	Laboratory aquatic medium; synthetic river water; dechlorinated tap water	71, 85
Mentha aquatica			Cd, Zn, Cu, Fe, Hg	Laboratory aquaria	85, 119
Azolla; A. pinnata	Fe: 1, 2, 4, and 8 ppm; Cu: 1, 2, 4, and 8 ppm	Metal contents; metal uptake kinetics	Cr, Ni, Zn, Fe, Cu, Pb		78, 76
Typha; T. domingensis; T. latifolia			Ni, Cr, Co, Zn, Mn, Pb, Cd, Cu, Hg, Fe	Plants from natural streams	96, 121
Eichhornia crassipes	1, 3, 5, 7, 10, 50, and 100 ppm of the mentioned metals; Choloro complexes of the mentioned metals	Visual symptoms; metal uptake; variation of pH; conductivity and growth media on metal uptake; nitrogen and phosphorus concentrations; metal uptake	Cd, Co, Cr, Cu, Mn, Ni, Pb, Zn; As, Al, Hg, P, Pt, Pd, Os, Ru, Ir, rheffluents	Laboratory stocks; plants collected from natural ponds	66–69, 122
Pistia stratoites	Hg: 1 to 1000 ppb	Metal content; nitrogen and phosphorus concentrations; foliar injury; chlorophyll content; phytomass; leaf injury index; metal content	Cu, Al, Cr, P, Hg	Plants collected from natural ponds	67, 122, 123
Lysimachia nummularia		Growth; metal content	Hg, methyl-Hg	Laboratory aquatic medium; synthetic river water; dechlorinated tap water	71
Hygrophila onogaria		Growth; metal content	Hg, methyl-Hg	Laboratory aquatic medium; synthetic river water; dechlorinated tap water	71
Carex; C. juncella; C. rostrata			Cu, Pb, Zn, Co, Ni, Cr, Mo, U		110, 124
Hydrilla verticillata	Hg: 1–1000 ppb; Fe: 0.025–0.150 ppm	Foliar injury; chlorophyll content; phytomass; leaf injury index; metal content	Hg, Fe, Ni, Hg, Pb		51, 123, 124, 125

TABLE 24.2
Toxicity Parameters Tested in Various Aquatic Macrophytes (continued)

Plant species studied	Concentrations tested	Parameters tested	Contaminants tested	Environment tested	Ref.
Eriocaulon septangulare	Sediment concentrations Hg: 0.007–0.247 µg/g Cd: 0.06–2.53 µg/g Fe: 2.55–34.77 µg/g Pb: 2.8–167.6 µg/g	Metal content	Hg, Pb, Cd, Fe	Naturally growing pond vegetation	32
Salvinia S. molesta S. herzogii S. natans	Hg: 1–1000 ppb Cr: 1, 2, 4, and 6 ppm	Metal uptake; translocation efficiency in different parts; growth rate; chlorophyll content foliar injury; chlorophyll content; phytomass; leaf injury index; metal content	Hg, Cr, Pb	Laboratory aquaria	79, 110
Vallisneria V. spiralis V. americana	Hg: 0.5–20 µM	Metal content; chlorophyll levels; protein content; nitrogen; phosphorus and potassium contents; levels of amino acid-cysteine; nitrate reductase activity; biomass and metal content; multispectral remote sensing data; seasonal storage of metals; mass balance calculations	Cd, Cr, Cu, Ni, Pb, Zn, Hg	Plants from natural river habitat	7, 8, 125
Nymphaea alba			Ni, Cr, Co, Zn, Mn, Pb, Cd, Cu, Hg, Fe		96
Schoenoplectus lacustris			Ni, Cr, Co, Zn, Mn, Pb, Cd, Cu, Hg, Fe		96
Wolffia globosa			Cd, Cr		126
Hydrangea macrophylla	Al: 100 µM	Al content; purification of Al–citrate complex	Al	Hydroponic culture	127

sediment has been deposited on top of the contaminated one and plants are able to colonize the bottom area. Colonization of macrophytes on shallow bottoms brings about increased redox potential in the rhizosphere caused by the photosynthetic oxygen, which has been translocated down to the roots by the lacunar system [144–146] (Figure 24.6). The increased redox potential will increase the availability of heavy metals to the plant roots, thus facilitate the uptake of heavy metals by the roots. Part of the metals taken up will then be translocated to the leaves and thereafter transferred to grazing animals.

Macrophytes can increase metal circulation in the aquatic environment. Thus, several of the macrophytes serve as indicators of metal pollution and also redistribute the metals in the aquatic ecosystem [32,71,136,147–150]. Metals concentrated within macrophytes are available for grazing by fish. These may also be available for epiphytic phytoplankton, herbivorous and detrivorous invertebrates. This may be a major route for incorporating metals in the aquatic food chain [83]. It is therefore of interest to assess the levels of heavy metals in macrophytes due to their importance in ecological processes. The inorganic metal species are, however, not biomagnified and thus do not increase in quantity in higher trophic levels.

In the past, research with macrophytes has been centered mainly on determining effective eradication techniques for nuisance growth of several species, such as *Elodea canadensis, Eichhornia crassipes, Ceratophyllum demersum*, etc. Scientific literature exists for the use of a wide diversity of macrophytes in toxicity tests designed to evaluate the hazard of potential pollutants, but the test species used is quite scattered. Similarly, literature concerning the phytotoxicity tests to be used, test methods, and the value of the result data is scattered. Estuarine and marine plant species are used considerably less than freshwater species in toxicity tests conducted for regulatory reasons [80]. The suitability of a test species is usually based on the specimen availability, sensitivity to toxicant, and reported data.

The sensitivity of various plants to metals was found to be species and chemical specific, differing in the uptake as well as toxicity of metals [151]. Many submerged plants have been used as test species, but no single species is widely used. In a literature survey, only 7% of 528 reported phytotoxicity tests used macrophytic species [152]. Their use in microcosm and mesocosm studies is even rarer and has been highly recommended [153]. Several plant species, such as *Lemna, Myriophyllum*, and *Potamogeton*, have been exhaustively used in phytotoxicity assessment, but

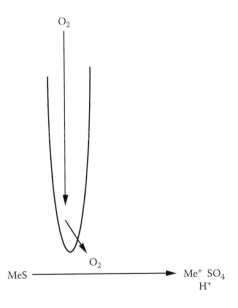

FIGURE 24.6 Changing the redox potential of sediment by roots.

several others have been given less importance as a bioassay tool. Duckweeds have received the greatest attention for toxicity tests because they are relevant to many aquatic environments, including lakes, streams, and effluents. Duckweeds comprise *Spirodela, Wolfiella, Lemna,* and *Wolffia,* of which *Lemna* has been almost exhaustively studied [154].

24.4 REMEDIATION POTENTIAL OF AQUATIC PLANTS

Phytoremediation is the use of green plants to remove or contain environmental contaminants or to render them harmless. Phytoremediation of metals can be divided into the following groups:

- Phytoextraction: metal accumulates in plants that are then harvested and thereby remove the metals from the site
- Phytostabilization: plants reduce the mobility of the metals
- Phyto- and rhizofiltration: the water is cleaned from metals by metal uptake by plant shoot and/or roots and by reducing the water velocity and thereby increasing the sedimentation of metals to the bottom

Aquatic plants have been used frequently to remove suspended solids, nutrients, heavy metals, toxic organics, and bacteria from acid mine drainage, agricultural landfill, and urban stormwater run-off and as bioremediative agents in wastewater treatments [155]

In wetland ecosystems, a wide variety of processes, ranging from physicochemical to biological, operates and can provide a suitable situation for removal of metals. For example, in the case of acidic metal-rich mine drainage, the principal processes include oxidation of dissolved metal ions and subsequent precipitation of metal hydroxides; bacterial reduction of sulphate and precipitation of metal sulphides; the coprecipitation of metals with iron hydroxides; the adsorption of metals onto precipitated hydroxides; the adsorption of metals onto organic or clay substrates; and, finally, metal uptake by growing macrophytes.

The conventional method of removing heavy metals from wastewater has been to mix it with sewage; conventional primary, secondary, and tertiary treatment would then remove them in the site of production of heavy metals. In addition, they provide green space, wildlife habitats, recreational and educational areas. However, secondary and tertiary processes require high input of technology, energy, and chemicals [156]. Such technologies used in the prevention of heavy metal pollution are inadequate or too expensive for some countries. In the past decades, therefore, research efforts have been directed towards wetland plants that comprise rooted, emergent, and surface floating plants as an alternative, low-cost means of removing heavy metals from domestic, commercial, mining, and industrial discharge of wastewater. Macrophytes are cost effective, universally available aquatic plants; with their ability to survive adverse conditions and high colonization rates, they are excellent tools for studies of phytoremediation [157].

24.4.1 FREE-FLOATING AQUATIC PLANTS

Free-floating plants can be used in removal of pollutants from the water phase because they are not in contact with the sediment for e.g., *Lemna* spp., *Eichhornia* spp., and *Azolla* spp. *Eichhornia crassipes* is a species often used in water cleaning. In experiments on metal removal, 0.67 (Cd), 0.57 (Co), 0.18 (Pb), 0.15 (Hg), 0.50 (Ni), and 0.44 (Ag) mg/g DM was accumulated. This gave a removal of 400 (Cd), 340 (Co), 90 (Pb), 110 (Hg), 300 (Ni), and 260 (Ag) g/ha day [158–160].

24.4.2 EMERGENT SPECIES

Aquatic plants are able to take up elements by roots and by shoots or thallus from the water and sediment. Unlike a terrestrial system, inorganics are in equilibrium between the sediment and the

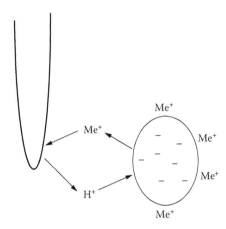

FIGURE 24.7 Release of metals from sediment particles by decreasing the pH.

water phase. Depending on the element, the retention time of the inorganic molecules in water is more or less short. The equilibrium between sediment and water is therefore towards the sediment for inorganic pollutants. Thus, the sediment will be a sink for pollutants. Because plants help to retard the velocity of the bulk flow, they also help in increasing the sedimentation and thus the immobilization of pollutants in wetlands. By influencing the equilibrium, the plants become more or less suitable for phytoextraction or phytostabilization.

Sediment-bound cations may be released from the sediment by roots decreasing the pH in the rhizosphere (Figure 24.7). Also, other mechanisms may occur, such as release of organic acids [161,162] or phytosiderophores [163]; this is the case for some terrestrial plants to be able to make the elements available. Organic substances are also released into the rhizosphere to supply microorganisms with substrate [164] that increases the phytostimulation capacity of bacterial activity in bioremediation. Furthermore, increase in pH of the water by photosynthetic activity and CO_2 uptake [165] of macrophyte shoots will probably change the chemistry of metals in the water. This may increase the precipitation and decrease the retention time of these elements in the water and, thus, phytostabilization.

The redox potential is fairly low in sediment, and most metals are bound to sulphides; thus, hard metal–sulphide complexes are formed. Plants need oxygen for their energy production in the roots. Plants living in such environments have evolved mechanisms to translocate oxygen (photosynthetic or taken up from the air) from shoot to root. Some of this oxygen is then released in the anoxic rhizosphere to protect young root tissue from toxic rhizospheric compounds [166–168]. In addition, this oxygen will change the redox potential of the rhizosphere sediment and thereby release metals from the sulphide complex. Oxygen release rates are highest in the range of -250 mV $<$ Eh < -150 mV [169]. The release of oxygen is species specific under reduced conditions and high rates have been shown for *Typha latifolia* (1.41 mg/h plant), *Phragmites australis* (1 mg/h plant), *Juncus effusus* (0.69 mg/h plant), and *Iris pseudacorus* (0.34 mg/h plant) [169].

24.4.3 SUBMERGENT SPECIES

Macroalgae, plants that mostly need some level of salinity to survive, are also submerged. They take up pollutants only from water due to the absence of root system and a primitive anatomy of the plant body. They are able to accumulate some elements to a high extent (Table 24.3).

TABLE 24.3
Trace Element Contents (μg gDW^{-1}) in Macroalgae and Freshwater Vascular Plants[a] Compared to Reference Terrestrial Plants, as well as Hyperaccumulation Levels[b]

Element	Macroalgae mean	Maximum values for contaminated fresh-water vascular plants	Hyperaccumulating level	Reference plant
Ag	<0.8	67	20	0.2
As	8.2	1200	10	0.1
Au	<4	—	0.1	0.001
Ba	<40	—	4000	40
Br	643	—	400	4
Ce	0.94	—	50	0.5
Cd	—	90		
Co	2.5	350	20	0.2
Cr	2.2	65	150	1.5
Cs	0.11	—	20	0.2
Cu	12	190	1000	10
Hf	0.2	—	5	0.05
Hg	—	1000	—	
I	238	—	300	3
La	0.57	—	20	0.2
Mn	—	8370		
Mo	<0.8	—	50	0.5
Ni	<20	290	150	1.5
Pb	—	1200	100	1
Rb	23	—	5000	50
Sb	0.11	—	10	0.1
Sc	0.49	—	2	0.02
Se	—	21		
Sr	696	—	5000	50
Th	0.06	—	0.5	0.005
U	0.44	1.1	1	0.01
V	5.9	—	50	0.5
Zn	37	7030	5000	50

[a] Markert, B., in Adriano, D.C. et al., Eds., *Biogeochemistry of Trace Elements, Science and Technology Letters*, Northwood, New York, 601, 1994.

[b] According to Dunn, C.E., in Brooks, R.R., Ed. *Plants that Hyperaccumulate Heavy Metals. Their Role in Phytoremediation, Microbiology, Archaeology, Mineral Exploration and Phytomining*, CAB International, Washington, D.C., 119, 1998; Brooks, R.R. and Robinson, B.H., in Brooks, R.R., Ed. *Plants that Hyperaccumulate Heavy Metals. Their Role in Phytoremediation, Microbiology, Archaeology, Mineral Exploration and Phytomining*, CAB International, Washington, D.C., 203–226; and Jones, D.L., *Plant Soil*, 205, 25, 1998.

24.5 WETLANDS

Wetlands are natural or constructed; both types can be used for removal of metals, particles, etc. In constructed wetlands, the flow is constructed as surface flow, subsurface flow or vertical flow, or a mixture [4]. The most important role of plants in wetlands is that they increase theresidence time of water, which means that they reduce the velocity and thereby increase the sedimentation of particles and associated pollutants. Thus, they are indirectly involved in water cleaning. Plants also add oxygen, thus providing a physical site of microbial attachment to the roots and generating

TABLE 24.4
Examples of Plants Used in Treatment of Wetlands [99]

Species	Ref.	Species	Ref.
Emergent plants		**Submerged**	
Scirpus spp.	174	*Ceratophyllum demersum*	174
Typha spp.	174	*Potamogeton* spp.	174
Iris spp.	174	*Elodea canadensis*	174
Phragmites australis	175	*Vallisneria americana*	175
Juncus spp.	174		
Floating		**Rooted floating leaved**	
Spirodela spp.	176	*Nelumbo lutea*	174
Lemna spp.	176	*Nymphoides* spp.	174
Salvinia spp.	177	*Nymphaea* spp.	174

Source: Hammer, D.A., in Constructed Wetland for Wastewater Treatment Conference. Middleton, County Cork, Ireland, 1993.

positive conditions for microbes and bioremediation. For efficient removal of pollutants, a high biomass per volume of water of the submerged plants is necessary. Thus, common and abundant growing plants are probably the best remediators (Table 24.4).

Wetlands serve as sinks for pollutants, reducing contamination of surrounding ecosystems. Although sediments, which tend to be anoxic and reduced, act as sinks, the marsh can become a source of metal contaminants through the activities of the plant species. Plants can oxidize the sediments, making the metals more available. Metals can be taken up by roots and transported upward to above-ground tissues, from which they can be excreted. Decaying litter can accumulate more metals, which may leach or may become available to detritus feeders. Using wetlands for water purification may serve only to delay the process of releasing toxicants to the water. As levels of pollutants increase, the ability of a wetland system to incorporate wastes can be impaired and the wetland can become a source of toxicity.

Uptake of metals in emergent plants only accounts for 5% or less of the total removal capacity in wetlands [170]. Not many studies have been performed on submerged plants. However, higher concentrations of metals in submerged than emerged plants have been found and, in microcosm wetlands, the removal by *Elodea canadensis* and *Potamogeton natans* showed up to 69% removal of Zn [171]. Dushenko et al. [172] found differences in accumulation of As between submerged and emerged plants. When comparing free-floating plants (*Lemna minor*) with emergent plants (*Typha domingensis*), a similar removal of Pb and Cd to about 50% [174] was found. Because *Lemna minor* was the easiest plant to remove the metals, this species is preferred in water cleaning — at least for these two metals.

In salt marshes about 50% of the absorbed metals are retained, and the remaining get transported [178]. Despite the ability of plants for bioconcentrating metals, the overall outcome with regard to biogeochemistry and mobility is that the wetlands generally act as sinks rather than sources for metals. Thus, the mangrove communities also act as effective traps for immobilizing heavy metals, with relatively low export to adjacent ecosystems [180]. However, because many wetland plants, unlike mangroves, are relatively short lived, their ability to stabilize metals may be only for the short term.

Different plant species have different allocation patterns of metals and can have different effects on salt marsh ecosystems. Weis and Weis [180] indicated that the replacement of *S. alterniflora* by invading *P. australis* would be predicted to lead to a reduction in mercury, chromium, and lead availability because of the higher allocation of these metals to leaf tissues in *S. alterniflora*. For a given metal burden, *P. australis* allocates more of the metal pool into below-ground biomass, and recalcitrant tissues (stems, rhizomes, and roots) than *S. alterniflora*. Furthermore, the excretion of metals by leaves is also greater for *S. alterniflora* than for *P. australis*, probably because of the presence of salt glands in the former species.

The movement of metals from below-ground to above-ground tissues and their release from leaf tissue may be important steps in metal flux in marsh ecosystems. Although metals remaining in the roots are generally considered "out of trouble" as far as release to the environment is concerned, studies are needed regarding the turnover of nutritive roots and the potential release of metals from decomposing roots. Decomposing litter of both species becomes highly enriched in metals over time, and evidence indicates that these metals are probably available to detritus feeders

24.5.1 Significance of Metal-Rich Rhizoconcretions, or Plaque, on Roots

A striking feature of roots of some wetland plants is the presence of metal-rich rhizoconcretions, or plaque, on the roots [181–183]. The metals are mobilized from the reduced anoxic estuarine sediments and concentrated in the oxidized microenvironment around the roots. Their concentrations can reach five to ten times the concentrations seen in the surrounding sediments [184]. At higher pH conditions, the presence of plaque enhanced Cu uptake into roots. However, in *T. latifolia* (cattail), the presence of iron plaque did not reduce uptake of toxic metals [184]. Iron plaque increased zinc uptake by rice (*O. sativa*) and movement into shoots [185]. In contrast, Al was not adsorbed onto the iron or the manganese plaque, but rather formed a separate phosphate deposit that resembled the iron and manganese plaques [186].

The discrepancies in effects of plaque on metal uptake need to be resolved by further study. Different metals, environmental conditions, or physiologies may account for these differences. By oxidizing the soil in the immediate vicinity of the rhizosphere, plants can alter the distribution of metals in wetland sediments. Plant activity (metal mobilization by oxidation of the root zone and movement into the rhizosphere) was considered responsible for the increase. The salt marsh metallophyte *S. maritima* roots concentrated trace elements from sediments by producing complex organic compounds and oxidizing the rhizosphere [187].

In macrophytes of Bull Island in Dublin, Ireland, the anaerobic conditions caused iron plaque formation on roots of plants because of oxidation of iron in the rhizosphere. Using six different species of grasses and flowering plants, a comparison has been made on the degree of iron plaque formation [188–190]. It can be added that submerged plants will not survive iron plaque because the plaque prevents photosynthesis by the plant leaves (Nyquist and Greger, unpublished). However, emergent plants with plaque only formed in the rhizosphere and on the part of the shoot situated in the water, survive iron plaque formation.

24.5.2 Influence of Wetland Plants on Weathering of Sulphidic Mine Tailings

Oxygen causes weathering of mine tailings, and if they contain pyrite acid mine drainage (AMD) water containing free metal ions and sulfuric are formed. This can be prevented by covering the tailings with moraine (dry cover technique) or with a high water table (wet cover technique). However, a wetland producing oxygen-consuming organic material on the tailings will enable us to decrease the water table from a couple of meters down to some centimeters. This will prevent accidents in which impoundment walls break due to the high pressure from the water (Figure 24.8).

Wetlands are naturally formed on mine tailings, as long as nutrients are added, because the tailings are nutrient-poor substrate. Twenty years' addition of sewage waters has self-established wetland species like *Carex rostrata*, *Eriophorum angustifolium*, and *Phragmites australis* [191]. However, wetland plants have the ability to take up oxygen from the air or use photosynthetic oxygen and translocate the oxygen to the roots and out into the rhizosphere (Figure 24.9). Thus,

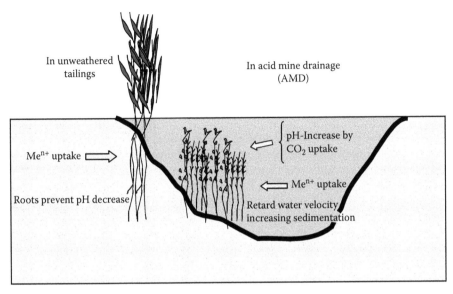

FIGURE 24.8 Remediation of mine tailings by emergent and submerged plants by a) preventing formation of AMD; or b) cleaning AMD from metals and increasing the pH.

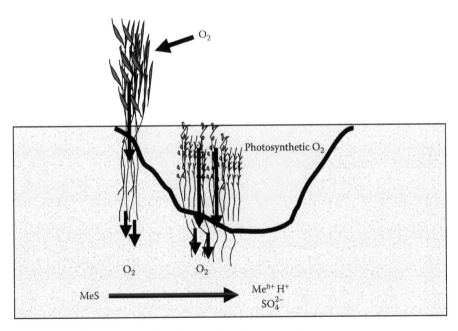

FIGURE 24.9 Weathering of tailings by submerged and emergent plants.

they will increase the redox potential, decrease the pH, and increase the release of metals for uptake. The work by Stoltz and Greger [191], however, showed that established *E. angustifolium* prevented a pH decrease from 6 to 2.6 and decreased the release of As, Cu, Cd, Pb, and Zn up to 99%. In later work [194], it was shown that the signal behind the prevention of pH decrease was to prevent a too high free metal concentration from forming in the tailings. Aquatic plants can tolerate a very low pH, which can be necessary when treating AMD. The mechanisms behind treatment of AMD by aquatic plants are summarized in Figure 24.8.

Eriophorum angustifolium survives in substrates with a wide pH range, from pH 11.0 [192] to about 2.6 [193]. Other wetland plant species that have been found to grow on mine tailings and tolerate a low pH are *Carex rostrata, E. scheuchzeri, Phragmites australis, Typha angustifolia*, and *T. latifolia*; these have been found growing under field conditions in pH as low as 2.1, 4.4, 2.1, 3.0, and 2.5, respectively [193–195].

24.5.3 CONSTRUCTED WETLANDS FOR REMOVAL OF METALS

Constructed wetlands with reed beds and floating-plant systems have been common for the treatment of various types of wastewaters for many years. This strategy is currently gaining importance globally and expanding to address contaminated/polluted soils and water bodies [179, 196–198].

Natural wetland ecosystems are inherently complex. Thus, for the purpose of treatment of metal-contaminated waters, it is advantageous to construct separate tanks within the treatment system, with each tank designated to perform a particular function maximally (occasionally, more than one would be beneficial). The design of wetlands constructed for the treatment of metal-contaminated waters attempts to identify and optimize the key processes that promote the removal of specific targeted metal. Alternatively, this also includes suppression of potentially interfering and competing processes.

Treatment of wastewaters/natural waters containing a single metal such as iron can be achieved using a constructed wetland designated to optimize only one of the possible process. For example, removal of iron involves precipitation of iron hydroxide in an aerobic environment. In contrast, if the water contains a mixture of metals, e.g., iron and zinc in high concentrations, the constructed wetland must adapt different strategies, such as application of aerobic and anaerobic processes. An aerobic environment promotes the precipitation of aluminum and iron hydroxides and coprecipitation of arsenic [1]. An anaerobic situation promotes the reduction of sulphates and the consequent precipitation of sulphides, primarily for copper, cadmium, and zinc [1].

The precipitation of hydroxides is regulated by pH and the availability of oxygen, which can be ensured by

- Construction of shallow wetlands with a maximal depth of about 3 m water
- Organic detritus to be minimized because it demands oxygen for decomposition (it is preferable to use large inorganic substrate)
- Designing the landscaping into ridges and gullies to ensure continual mixing of the water within the system so as to prevent stratification of water into oxygen-rich and oxygen-depleted zones
- Incorporating cascades at the point of influence to promote oxygenation of air
- Utilizing reed beds comprising *Phragmites australis* (common reed), *Typha latifolia* (cattail), etc., which have the ability to transfer oxygen to the root zone [1,4,197]

Glyceria fluitans (floating sweetgrass) is an amphibious plant found growing in the tailings pond of an abandoned lead/zinc mine in Glendalough County, Wicklow, Ireland [190]. Greenhouse experiments demonstrated that *G. fluitans* could grow in sand culture treated with high zinc sulphate solution. Further research confirmed that two populations of *G. fluitans* — one from a metal contaminated and the other from a noncontaminated site — could be grown successfully on mine tailings with a high zinc content. *G. fluitans* and two other wetland plants, *Phragmites australis* and *Typha latifolia*, have since been grown on alkaline and acidic zinc mine tailings in field conditions under fertilized and nonfertilized conditions. Research findings obtained thus far indicate that *G. fluitans* can be easily established on zinc mine tailings. It appears also to have a very low nutrient requirement, thus keeping fertilizer costs to a minimum during rehabilitation of mine tailings.

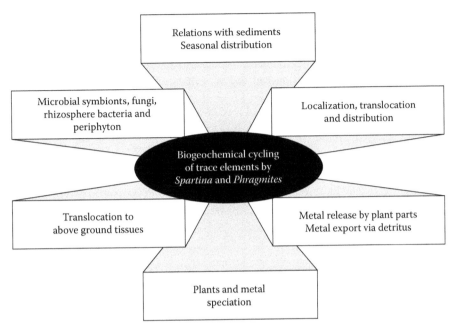

FIGURE 24.10 Role of prominent wetland species in biogeochemical cycling of trace elements.

Wetlands have been constructed in Ireland for the passive treatment of tailing water originating from a lead/zinc mine. Water originating from mine tailings is often characterized by high metal and sulphate concentrations compared to background levels. Conventional methodology of tailing water treatment involves chemical treatment, which is a costly procedure requiring intensive chemical and labor inputs.

Therefore, more recently constructed and natural wetlands have been utilized for metal removal and wastewater quality control. Wetlands with their diversified macrophytes are known to retain substances such as metals from water passing through them. Aquatic macrophytes encompassing many common weeds enable cost-effective treatment and remediation technologies for wastewaters contaminated with inorganics and organics.

Constructed wetlands and its assemblage, anoxic lime stone drains (ALD), and successive alkalinity producing systems (SAPS) have produced promising results [198–201] in restoration of acid mine drainage (AMD) in several real-world ecosystems [202,203].

24.6 BIOGEOGENIC CYCLING OF METALS

Metal uptake, translocation and release by *Spartina alterniflora* and *Phragmites australis* association are implicated in phytoremediation and restoration (Figure 24.10) [180]. Salt marsh metallophytes play a potential role in phytoremediation and restoration of metals. Wetland plants and salt marshes function in a similar fashion with regard to metal uptake patterns and in compartmentalizing them in roots. Some species retain more of their metal burden in below-ground structures than other species, which redistribute a greater proportion of metals into above ground tissues, especially leaves. Storage in roots is most beneficial for phytostabilization of the metal contaminants, which are least available when concentrated below ground.

Wetland/aquatic plants and salt marshes may alter the speciation of metals and may also suffer toxic effects depending upon their bioaccumulation coefficient. In certain salt marsh plants, metals in leaves may be excreted through salt glands and thereby returned to the marsh environment. Metal concentrations of leaf and stem litter may become enriched in metals over time, due in part to cation adsorption or to incorporation of fine particles with adsorbed metals.

Several studies suggest that metals in litter are available to deposit feeders and thus can enter estuarine food webs. Marshes, therefore, can be sources as well as sinks for metal contaminants. *Phragmites australis*, an invasive species in the northeastern U.S., sequesters more metals below ground than the native *Spartina alterniflora*, which also releases more via leaf excretion. This information is important for the silting and use of wetlands for phytoremediation as well as for marsh restoration efforts.

Some aquatic plants are used as food and feed. Water spinach, *Ipomea aquatica*, is commonly used as a vegetable and pig food in Thailand. It is very easy to grow, grows fast, and is present in cultivated water, as well as in industrial areas and big cities. It takes up and accumulates heavy metals to such an extent that a threat to human health has been discussed [31,205]. Recent studies have shown that plants collected in Thailand, near Bangkok, contained up to 530, 350, and 123 $\mu g/g$ DW^{-1} of Pb, Hg, and Cd, respectively [205]. The biggest problem seems to be Hg. According to FAO, the weekly intake of Hg should not exceed 43 μg per week, which means not more than 250 g of plant per day can be eaten. However, the real problem seems to be the property of this plant to accumulate high levels of CH_3–Hg [205]. This plant has also been tested for removal of metals from wastewaters [37].

When this plant is grown, nutrients are necessary for high biomass production, which in turn is positive for those who grow this plant for their own consumption, as well as for selling. Another positive effect with nutrient addition is that the higher the nutrient concentration is, the lower the uptake of Cd, Pb, and Hg [206]. Sobolewski [170] also showed that adding nutrients prevented toxic effects on the plant, likely due to decreased intrinsic concentration of the metals.

24.7 CONCLUDING REMARKS

The aquatic macrophytes listed earlier have several characteristics — namely, hardiness, ability to survive under adverse environmental conditions, and high productivity (Figure 24.11) together with factors controlling bioaccumulation coefficient (Figure 24.12) that would enable them as potential agents of phytotechnology. Furthermore, the anaerobic environment in the water and sediments renders elements in less oxidized forms. This increases solubility and uptake by aquatic plants and may make it possible to a considerable extent in phytoremediation of wetlands. The partial pressure of oxygen in water is only a small fraction of the 21% oxygen in air; the aquatic environment is a strong defense against free radical formation. Unfortunately, low levels of oxygen also greatly reduce the availability of energy liberated by catabolism. This can hinder growth, particularly in stagnant ponds or other situations lacking natural aeration.

The aquatic environment, especially bottom waters and sediments, often has low oxygen content. Metals in reduced form are more easily taken up by plants. Once in the plant, they may be oxidized and become immobile. Thus, aquatic macrophytes generally have a much higher metal content than terrestrial plants and can be usefully employed in phytoremediation. The reason for higher concentration in submerged than in terrestrial plants is that, comparing leaf uptake, the terrestrial plant leaves have cuticle, which prevents the uptake.

Furthermore, most uptake is via roots in terrestrial plants, as well as emergent plants, and the translocation is often low — approximately up to 10% — because the metals bind in the root cell walls due to the negative charge. Submerged plants take up metals via roots to a lesser extent because their roots, at least in most aquatic plants, are few and root biomass is low compared to the shoot part. Aquatic macrophytes serve as accumulators and indicators of pollution and also mitigate metal pollution to a considerable extent [47,93,107,138,207–213].

Macrophytes bioconcentrate metals from water and sediments, resulting in an internal concentration several fold greater than their surroundings. The submerged plants in polluted water bodies are reported to accumulate trace metals to the tune of 10^3 to 10^4 and also reduce the water velocity, facilitating rapid sedimentation of the suspended fine particulates containing trace metals; otherwise,

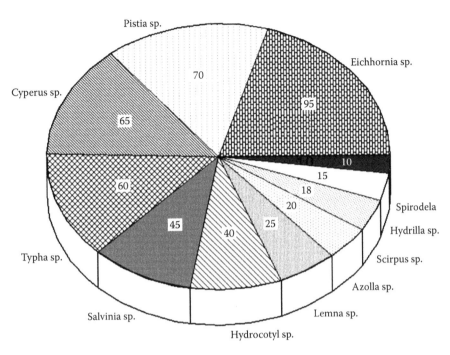

FIGURE 24.11 Bioproductivity (t ha^{-1} yr^{-1}) of selected aquatic and wetland macrophytes under specific conditions enable them to act as potential agents in phytotechnology.

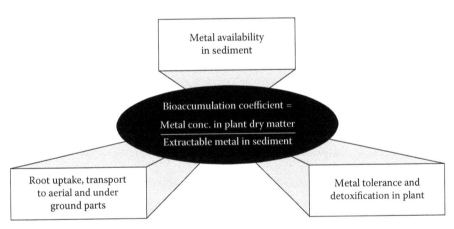

FIGURE 24.12 Factors determining the bioaccumulation coefficient in aquatic/wetland ecosystems.

these are toxic to the biota when present in the interstitial waters in available form [7,8]. Thus, they play a major role in biogeochemical cycling of trace metals.

The reedbed technology developed for bioremediation of organic and inorganic pollutants has attained global significance. Aquatic macrophyte-based phytotechnology for bioremediation has limitations in addition to vast scope. The notable limitations are

- Invasive species could harm the ecosystem.
- Tropical wetlands are rapidly changing.
- Evapotranspiration losses are too high.

Therefore, the future development, potential, and implementation of these systems should be investigated in detail, considering the advancements in biotechnology and molecular biology.

REFERENCES

1. Kadlec, R.H. et al., *Constructed Wetlands for Pollution Control. Control Processes, Performance, Design and Operation*. IWA Publishing, London, 164, 2000.
2. Cacador, I. and Vale, C., Salt marshes, in Prasad, M.N.V. (Ed.) *Metals in the Environment — Analysis by Biodiversity*. Marcel Dekker Inc., New York, 95–116, 2001.
3. Odum, W.E., Comparative ecology of tidal freshwater and salt marshes. *Annu. Rev. Ecolog. Syst.*, 19, 147, 1998.
4. Kadlec, R.H. and Knight, R.I., *Treatment Wetlands*, CRC Press, Boca Raton, FL, 1996.
5. Förstner, U. and Wittmann, G.T.W., Eds., *Metal Pollution in the Aquatic Environment*, Springer, Berlin, 1979.
6. Campbell, P.G.C. et al., Biologically available metals in sediments, National Research Council Canada no. 27694. 298, 1988.
7. St-Cyr, L. and Campbell P.G.C., Trace metals in submerged plants of St. Lawrence River, *Can. J. Bot.*, 72, 429, 1994.
8. St-Cyr, L., Campbell, P.G.C., and Guertin, K., Evaluation of the role of submerged plant beds in the metal budget of fluvial lake, *Hydrobiologia*, 291, 141, 1994.
9. Jain, C.K., Metal fractionation studies on bed sediments of River Yamuna, India, *Water Res.*, 38, 569, 2004.
10. Salomons, W. and Förstner, U., Trace metal plants on polluted sediments, Part II: evaluation of environmental impact, *Environ. Technol. Lett.*, 1, 506, 1980.
11. Mahony, D.J. et al., Partitioning of metals to sediment organic carbon, *Env. Toxicol. Chem.* 15, 2187, 1996.
12. Perin, G. et al., Heavy metal speciation in the sediments of Northern Adriatic sea — a new approach for environmental toxicity determination, in: Lekkas, T.D., Ed., *Heavy Metals in the Environment*, vol 2, C.E.P Consultants, Edinburgh, 1985, 454.
13. Greger, M., Metal availability, uptake, transport and accumulation in plants, in Prasad M.N.V., Ed. *Heavy Metal Stress in Plants: from Biomolecules to Ecosystems*, 2nd ed. Springer–Verlag–Narosa, New Delhi, 1–27, 2004.
14. Nordén, M., Ephraim, J.H., and Allard, B., Europium complexation by an aquatic fulvic acid — effects of competing ions, *Talanta*, 104, 781, 1997.
15. Garcia, A.C. and Prego, R., Chemical speciation of dissolved copper, lead and zinc in a ria coastal system: the role of resuspended sediments, *Anal. Chim. Acta.*, 524, 109, 2004.
16. Gueguen, C. et al., Water toxicity and metal contamination assessment of a polluted river: the upper Vistula River, *Appl. Geochem.*, 19, 153, 2004.
17. Wilkie, J.A. and Herinh, J.G., Adsorption of arsenic onto hydrous ferric oxide: effects of adsorbate/adsorbent ratios and co-occurring solutes, *Colloids and Surfaces: a Physicochemical and Engineering Aspects*, 107, 97, 1996.
18. Peltier, E.F., Webb, S.M., and Gaillard, J.F., Zinc and lead sequestration in an impacted wetland system, *Adv. Environ. Res.*, 8, 103, 2003.
19. Prasad, M.N.V., Greger, M., and Smith, B.N., Aquatic macrophytes, in *Metals in the Environment: Analysis by Biodiversity*, Prasad, M.N.V., Ed., Marcel Dekker Inc., New York, 2001, 259.
20. Chilton, E.W., II, Macroinvertebrate communities associated with three aquatic macrophytes *Ceratophyllum demersum, Myriophyllum spicatum*, and *Vallisneria americana* in Lake Onalaska, Wisconsin, *J. Freshwater Ecol.*, 5, 455, 1990.
21. Denny, P., Solute movement in submerged angiosperms, *Biological Rev.*, 55, 65, 1980.
22. Cowgill, V.M., The hydrogeochemical of Linsley Pond, North Branford. Part 2. The chemical composition of the aquatic macrophytes, *Arch. Hydrobiologie*, 45, 1, 1974.
23. Marschner, H., *Mineral Nutrition in Higher Plants*. Academic Press, New York, 1995.
24. Yamada, Y., Bucovac, M.J., and Wittwer, S.H., Ion binding by surfaces of isolated cuticular membranes, *Plant Physiol.*, 39, 978, 1964.

25. Welsch, R.P.H. and Denny, P., The uptake of lead and copper by submerged aquatic macrophytes in two English lakes, *J. Ecol.*, 68, 443, 1980.

26. Chawla, G., Singh, J., and Viswanathan, P.N., Effect of pH and temperature on the uptake of cadmium by *Lemna minor* L., *Bull. Environ. Contam. Toxicol.*, 47, 84, 1991.

27. Greger, M. and Kautsky, L., Uptake of heavy metals by macrophytes — a comparison between filed samples and controlled experiments, *Proc. 12th Baltic Marine Biol. Symp.*, Eds. E. Bjornestad and K. Jensen. Olsen & Olsen, Fredenborg, 67, 1992.

28. Sinha, S. et al., Chromium and manganese uptake by *Hydrilla verticillata* L.f. Royle: amelioration of chromium toxicity by manganese, *J. Environ. Sci. Health A*, 28, 1545, 1993.

29. Tripathi, R.D. and Chandra, P., Influence of metal chelators and pH on chromium uptake by duckweed *Spirodela polyrrhiza* L. Schleiden. *Bull. Environ. Contam. Toxicol.*, 47, 764, 1991.

30. Zaranyika, M.F., Mutoko, F., and Murahwa, H., Uptake of zinc, cobalt, iron and chromium by water hyacinth *Eichhornia crassipes* in Lake Chiveroh, Zimbabwe, *Sci. Tot. Environ.* 153, 117, 1994.

31. Low, K.S. and Lee, C.K., Copper, zinc, nickel and chromium uptake by "Kangkong air" *Ipomea aquatica* Forsk., *Pertanika*, 4, 16, 1981.

32. Coquery, M. and Welbourn, P.M., Mercury uptake from contaminated water and sediment by the rooted and submerged aqatic macrophytes *Eriocaulon septangulare*, *Arch. Environ. Cont. Toxicol.*, 26, 335, 1994.

33. Greger, M. and Kautsky, L., Effects of Cu, Pb, and Zn on two *Potamogeton* species grown under field conditions, *Vegetatio*, 97, 173, 1991.

34. Brown, B.T. and Rattigan, B.M., Toxicity of soluble copper and other metal ions to *Elodea canadensis*, *Environ. Pollut.*, 20, 303, 1979.

35. Kimball, K.D. and Baker, A.L., Temporal and morphological factors related to mineral composition in *Myriophyllum heterophyllium* Michx, *Aquat. Bot.*, 16, 185, 1983.

36. Greger, M. and Kautsky, L., Use of macrophytes for mapping bioavailable heavy metals in shallow coastal areas, Stockholm, Sweden, *Appl. Geochem.*, suppl. 2, 37, 1993.

37. Suckcharoen, S., *Ceratophyllum demersum* as an indicator of mercury contamination in Thailand and Finland, *Ann. Botan. Fenn.*, 16, 173, 1979.

38. Abo–Rady, M.D.K., Makrophytische Wasserpflanzen als Bioindikatoren für die Schwermetallbelastung der obern Leine, *Archiv. Hydrobiol.*, 89, 387, 1980.

39. Guilizzoni, P., Adams, M.S., and MacGaffet, N., The effect of chromium on growth and photosynthesis of a submersed macrophyte, *Myriophyllum spicatum*, *Ecol. Bull.*, 36, 90, 1984.

40. Sinha, S., Accumulation of Cu, Cd, Cr, Mn, and Pb from artificially contaminated soil by *Bacopa monnieri*, *Environ. Monit. Assess.*, 57, 253, 1999.

41. Zayed, A.M., Gowthaman, S., and Terry, N., Phytoaccumulation of trace elements by wetland plants: I. Duckweed., *J. Environ. Qual.*, 27, 715, 1998.

42. Zhu, Y.L. et al., Phytoaccumulation of trace elements by wetland plants: II. Water hyacinth, *J. Environ. Qual.*, 28, 339, 1999.

43. Ekvall, L. and Greger, M., Effects of environmental biomass-producing factors on Cd uptake in two Swedish ecotypes of *Pinus sylvestris* (L.), *Environ. Qual.*, 121, 401, 2003.

44. Ding, J. et al., Bioconcentration of cadmium in water hyacinth *Eichhornia crassipes* in relation to thiol group content, *Environ. Pollut.*, 84, 93, 1994.

45. Hao, Y.Y., Roach, A.L., and Ramelow, G.J., Uptake of metal ions by nonliving biomass derived from sphagnum moss and water hyacinth roots, *J. Environ. Sci. Health*, 28A, 2333, 1993.

46. Low, K.S., Lee, C.K., and Tai, C.H., Biosorption of copper by water hyacinth roots, *J. Environ. Sci. Health A*, 29, 171, 1994.

47. Rai, U.N. et al., Lead-induced changes in glutathione and phytochelatin in *Hydrilla verticillata* L.f. Royle and *Chara corallina* Wildenow, *J. Env. Sci. Health A*, 30, 537, 1995.

48. Vachon, A. and Campbell, P.G.C., Potential of phytochelatins in aquatic plants as biochemical indicators of exposure to toxic metals — a field study, St. Lawrence River, Quebec, in *Proc. 6th Int. Conf. Heavy Metals Environ.*, Toronto, 2, 45, 1993.

49. Gupta, N. et al., Role of glutathione and phytochelatin in *Hydrilla verticillata* L.f. Royle and *Vallisneria spiralis* L. under mercury stress, *Chemosphere*, 37, 785, 1998.

50. Rai, U.N. et al., Induction of phytochelatins under cadmium stress in water lettuce *Pistia stratiotes*, *J. Environ. Sci. Health A*, 30, 2007, 1995.

51. Gupta, M. et al., Lead-induced changes in glutathione and phytochelatin in *Hydrilla verticillata* L.f. Royle, *Chemosphere*, 33, 2011, 1995.

52. Gupta, M. et al., Lead-induced synthesis of metal-binding peptides phytochelatins in submerged macrophyte *Vallisneria spiralis* L., *Physiol. Mol. Biol. Plants*, 5, 173, 1999.

53. Tripathi, R.D. et al., Induction of phytochelatins in *Hydrilla verticillata* L.f., Royle, *Bull. Environ. Contam. Toxicol.*, 56, 205, 1996.

54. Devi, S.R. and Prasad, M.N.V., Copper toxicity in *Ceratophyllum demersum* L. coontail., a free floating macrophyte: responses of antioxidant enzymes and antioxidants, *Plant Sci.*, 138, 157, 1998.

55. Singh, J. et al., Combined effects of cadmium and linear alkyl benzene, *Ecotoxicology*, 3, 59, 1994.

56. Dirilgen, N. and Inel, Y., Cobalt-copper and cobalt-zinc effects on duckweed growth and metal accumulation, *J. Environ. Sci. Health A*, 29, 63, 1994.

57. Fritioff, Å. and Greger, M., Aquatic and terrestrial plant species with potential to remove heavy metals from stormwater, *Int. J. Phytorem.*, 5, 211, 2003.

58. Abbasi, A.S. and Ramasami, E., *Biotechnological Methods of Pollution Control*. Universities Press. Hyderabad, 168, 1999.

59. Doust, J.L., Schmidt, M., and Doust, L.L., Biological assessment of aquatic pollution: a review, with emphasis on plants as biomonitors, *Biol. Rev. Cam. Phils. Soc.*, 69, 147, 1994.

60. Wang, W., Use of plants for the assessment of environmental contaminants, *Rev. Environ. Contam. Toxicol.*, 126, 87, 1992.

61. Wang, W. and Freemark, K., The use of plants for environmental monitoring and assessment, *Ecotoxicol. Environ. Saf.*, 30, 289, 1995.

62. Fugita, M. and Nakano, K., Metal specificities on induction and binding affinities of heavy metal-binding complexes in water hyacinth root tissue, *Agri. Biochem.*, 52, 2335, 1998.

63. Dean, J.G., Bosqui, F.L., and Lanoutte, V.H., Removing heavy metals from waste water, *Environ. Sci. Technol.*, 6, 518, 1972.

64. Fritioff, Å., Kautsky, L., and Greger, M., Influence of temperature and salinity on heavy-metal accumulation by submersed plants, *Environ. Pollut.*, 133, 265, 2005.

65. Greger, M., Kautsky, L., and Sandberg, T., A tentative model of Cd uptake in *Potamogeton pectinatus* in relation to salinity, *Environ. Exp. Bot.*, 35, 563, 1995.

66. Farago, M.E. and Parsons, P.J., The uptake and accumulation of platinum metals by the water hyacinth *Eichhornia crassipes*, *Inorg. Chim. Acta*, 79, 233, 1983.

67. Klump, A. et al., Variation of nutrient and metal concentrations in aquatic macrophytes along the Rio Cachoeria in Bahia, Brazil, *Environ. Int.*, 28, 165, 2002.

68. Sipauba–Tavares, L.H., Favero, E.G., and Braga, F.M., Utilization of macrophyte biofilter in effluent from aquaculture: I. floating plant, *Braz. J. Biol.*, 62, 713, 2002.

69. Soltan, M.E. and Rashed, M.N., Laboratory study on the survival of water hyacinth under several conditions of heavy metal concentrations, *Adv. Environ. Res.*, 7, 321, 2003.

70. Van der werff, M. and Pruyt, M.J., Long-term effects of heavy metals on aquatic plants, *Chemosphere*, 11, 727, 1982.

71. Ribeyre, F. and Boudou, A., Experimental study of inorganic and methylmercury bioaccumulation by four species of freshwater macrophytes from water and sediment contamination sources, *Ecotoxicol. Environ. Saf.*, 28, 270, 1994.

72. Mal, T.K., Adorjan, P., and Corbett, A.L., Effect of copper on growth of an aquatic macrophyte, *Elodea canadensis. Environ. Pollut.*, 120, 307, 2002.

73. Staves, R.P. and Knaus, R.M., Chromium removal from water by three species of duckweeds, *Aquat. Bot.*, 23, 261, 1985.

74. Wang, W., Toxicity tests for aquatic pollutants by using common duckweed, *Environ. Pollut.*, Ser B., 11, 193, 1986.

75. Wang, W., Site-specific Ba toxicity to common duckweed, *Lemna minor, Aquat. Toxicol.*, 12, 203, 1988.

76. Jain, S.K., Vasudevan, P., and Jha, N.K., Removal of some heavy metals from polluted water by aquatic plants: studies on duckweed and water velvet, *Biol. Wastes*, 28, 115, 1989.

77. Lytle, C.M., Smith, B.N., and McKinnon, C.Z., Manganese accumulation along Utah roadways: a possible indication of motor vehicle exhaust pollution, *Sci. Tot. Environ.*, 162, 105, 1994.

78. Mallick, N., Shardendu, and Rai, L.C., Removal of heavy metals by two free-floating aquatic macrophytes, *Biomed. Environ. Sci.*, 9, 399, 1996.

79. Mohan, B.S. and Hosetti, B.B., Potential phytotoxicity of lead and cadmium to *Lemna minor* grown in sewage stabilization ponds, *Environ. Pollut.*, 98, 233, 1997.

80. Mohan, B.S. and Hosetti, B.B., Aquatic plants for toxicity assessment, *Environ. Res. Sec A.*, 81, 259, 1999.

81. Dirilgen, N. and Dogn, F., Speciation of chromium in the presence of copper and zinc and their combined toxicity, *Ecotoxicol. Environ. Saf.*, 53, 397, 2002.

82. Axtell, N.R, Sternberg, S.P.K., and Claussen, K., Lead and nickel removal using microspora and *Lemna minor*, *Biores. Tech.*, 89, 41, 2003.

83. Cardwell, A.J., Hawker, D.W., and Greenway, M., Metal accumulation in aquatic macrophytes from southeast Queensland, Australia, *Chemosphere*, 48, 653, 2002.

84. Keskinkan, O. et al., Heavy metal adsorption properties of a submerged aquatic plant *Ceratophyllum demersum*, *Biores. Tech.*, 92, 197, 2004.

85. Kamal, M. et al., Phytoaccumulation of heavy metals by aquatic plants, *Environ. Int.*, 29, 1029, 2004.

86. Jones, T.W., Atrazine uptake, photosynthetic inhibition and short-term recovery for the submersed vascular plant, *Potamogeton perfoliatus* L., *Arch. Environ. Contam. Toxicol.*, 15, 277, 1986.

87. Lytle, C.M., Lytle, F.W., and Smith, B.N., Use of XAS to determine the chemical speciation of bioaccumulated manganese in *Potamogeton pectinatus*, *J. Environ. Qual.*, 25, 311, 1996.

88. Wolterbeek, H.T. and Van der Meer, A.J., Transport rate of arsenic, cadmium, copper and zinc in *Potamogeton pectinatus* L.: radiotracer experiments with [76]As, [109, 115]Cd, ^{64}Cu and [65, 69]Zn, *Sci. Total Environ.*, 287, 13, 2002.

89. Wu, L. and Guo, X., Selenium accumulation in submerged aquatic macrophytes *Potamogeton pectinatus* L. and *Ruppia maritima* L. from water with elevated chloride and sulfate salinity, *Ecotox. Environ. Saf.*, 51, 22, 2002.

90. Rai, U.N. et al., Cadmium accumulation and its phytotoxicity in *Potamogeton pectinatus* L. Potamogetonaceae, *Bull Environ. Contam. Toxicol.*, 70, 566, 2003.

91. Tripathi, R.D. et al., Biochemical responses of *Potamogeton pectinatus* L. exposed to higher concentrations of zinc, *Bull. Environ. Contam. Toxicol.*, 71, 255, 2003.

92. Reay, P.F., The accumulation of arsenic from arsenic-rich natural waters by aquatic plants, *J, Appl. Ecol.*, 9, 557, 1972.

93. Ornes, W.H. and Sajwan, K.S., Cadmium accumulation and bioavailability in coontail *Ceratophyllum demersum* L. plants, *Water Air Soil Pollut.*, 69, 291, 1993.

94. Tripathi, R.D. et al., Cadmium transport in submerged macrophyte *Ceratophyllum demersum* L. in presence of various metabolic inhibitors and calcium channel blockers, *Chemosphere*, 31, 3783, 1995.

95. Gupta, P. and Chandra, P., Response of cadmium to *Ceratophyllum demersum* L., a rootless submerged plant, *Waste Manage.*, 16, 335, 1996.

96. Szymanowska, A., Samecka–Cymerman, A., and Kempers, A.J., Heavy metals in three lakes in west Poland, *Ecotoxicol. Environ. Saf.*, 43, 21, 1999.

97. Voeste, D. et al., Pigment composition and concentrations within the plant *Ceratophyllum demersum* L. component of the STS-89 C.E.B.A.S. Minimodule spaceflight experiment, *Adv. Space Res.*, 31, 211, 2003.

98. Garg, P. and Chandra, P., Toxicity and accumulation of chromium in *Ceratophyllum demersum* L., *Bull. Environ. Contam. Toxicol.*, 44, 473, 1991.

99. Dierberg, F.E. et al., Submerged aquatic vegetation-based treatment wetlands for removing phosphorus from agricultural runoff: response to hydraulic and nutrient loading, *Water Res.*, 36, 1409, 2002.

100. Bolsunovskii, A.I.A. et al., Accumulation of industrial radionuclides by the Yenisei River aquatic plants in the area affected by the activity of the mining and chemical plant, *Radiat. Biol. Radioecol.*, 42, 194, 2002.

101. Aravind, P. and Prasad, M.N.V., Zinc alleviates cadmium-induced oxidative stress in *Ceratophyllum demersum* L. — a free-floating freshwater macrophyte, *Plant Physiol. Biochem.*, 41, 391, 2003.

102. Aravind, P. and Prasad, M.N.V., Carbonic anhydrase impairment in cadmium-treated *Ceratophyllum demersum* L. — a free-floating freshwater macrophyte. Toxicity reversal by zinc, *J. Anal. At. Spectrom.*, 19, 52, 2004.

103. Aravind, P. and Prasad, M.N.V., Zinc protects chloroplasts and associated photochemical function in cadmium exposed *Ceratophyllum demersum* L., a freshwater macrophyte, *Plant Sci.,* 166, 1321, 2004.

104. Kumar, G.P. and Prasad, M.N.V., Photosynthetic pigments and gaseous exchange in cadmium-exposed *Ceratophyllum demersum* L. (a freshwater macrophyte) — a model for hormesis, *Bull. Environ. Contam. Toxicol,* 31, 1, 2004.

105. Kumar, G.P. and Prasad, M.N.V., Cadmium toxicity to *Ceratophyllum demersum* L. (a freshwater macrophyte): morphological symptoms, membrane damage and ion leakage, *Bull. Environ. Contam. Toxicol.,* 72, 1038, 2004.

106. Cymerman, A.S. and Kempers, A.J., Bioaccumulation of heavy metals by aquatic macrophytes around Wroclaw, Poland, *Ecotox. Environ. Saf.,* 35, 242, 1996.

107. Stoyanova, D. and Tchakalova, T., Cadmium-induced ultrastructural changes in shoot apical meristem of *Elodea Canadensis* Rich, *Photosynthetica,* 37, 47, 1999.

108. Kupper, H., Kupper, F., and Spiller, M., Environmental relevance of heavy metal-substituted chlorophylls using he examples of water plants, *J. Exp. Bot.,* 47, 259, 1996.

109. Lytle, C.M., Heavy metal bioaccumulation in Great Basin submersed aquatic macrophytes. Ph.D. dissertation, Brigham Young University, Salt Lake City, UT, 1994, 1–200.

110. Mohan, B.S. and Hosetti, B.B., Lead toxicity to *Salvinia natans* grown in macrophyte ponds, *J. Ecotoxicol. Environ. Monit.,* 8, 3, 1998.

111. Davis, J.A., Comparison of static-replacement and flow-through bioassays using duckweed, *Lemna gibba* G-3. US EPA 560/ 81-003. Washington, D.C., 1981.

112. Devi, M. et al., Accumulation and physiological and biochemical effects of cadmium in a simple aquatic food chain, *Ecotox. Environ. Saf.,* 33, 38, 1996.

113. Huebert, D.B. and Shay, J.M. The effect of cadmium and its interaction with external calcium in the submerged aquatic macrophyte *Lemna trisulca* L., *Aquat. Toxicol.,* 20, 57, 1991.

114. Nasu, Y. et al., Comparative studies on the adsorption of cadmium and copper in *Lemna panicicostata*, *Environ. Pollut. Ser A.,* 32, 201, 1983.

115. Huebert, D.B., Dyck, B.S., and Shay, J.M. The effect of EDTA on the assessment of Cu toxicity in the submerged aquatic macrophyte *Lemna trisulca. Aquatic Toxicol.,* 24, 183, 1983.

116. Hutchinson, T.C. and Czyrska, H., Heavy metal toxicity and synergism to floating aquatic weeds, *Vehr. Int. Verein. Theor. Agnew. Limnol.,* 19, 2102, 1975.

117. Campbell, P.G.C. et al., Accumulation of copper and zinc in yellow water lilly *Nuphar veriegatum*: relationships to metal partitioning in adjacent lake sediments, *Can. J. Fish. Aquat. Sci.,* 42, 23, 1985.

118. Chandra, P., Sinha, S., and Rai, U.N., Bioremediation of chromium from water and soil by vascular aquatic plants, *ACS Sump. Ser.,* 664, 274, 1987.

119. Aslan, M. et al., Sorption of cadmium and effects on growth, protein content and photosynthetic pigment composition of *Nasturtium officinale* R. Br. and *Mentha aquatica* L., *Bull. Environ. Contam. Toxicol.,* 71, 323, 2003.

120. Abira, M.A., Ngirigacha, H.W., and Van Bruggen, J.J., Preliminary investigation of the potential of four tropical emergent macrophytes for treatment of pretreated pulp and papermill wastewater in Kenya, *Water Sci.Technol.,* 48, 223, 2003.

121. Satyakala, G. and Jamil, Q., Chromium induced biochemical changes in *Eichhornia crassipes*. Mart. Solms and *Pistia stratoitess, Bull. Environ. Contam. Toxicol.,* 48, 921, 1992.

122. Mhatre, G.N. and Chaphekar, S.B., The effect of mercury on some aquatic plants, *Environ. Pollut. Ser. A Ecological Biological,* 39, 207, 1985.

123. Reid, G.A., Martin, D.F., and Young, K.S., Simulation of *Hydrilla* growth by FeEDTA, *J. Inorg. Nucl. Chem.,* 37, 357, 1975.

124. Sinha, S. and Pandey, K., Nickel induced toxic effects and bioaccumulation in the submerged plant, *Hydrilla verticillata* L.F. Royle under repeated metal exposure, *Bull. Environ. Contam. Toxicol.,* 71, 1175, 2003.

125. Gupta, M. and Chandra, P., Bioaccumulation and toxicity of mercury in rooted-submerged macrophyte *Vallisneria spiralis, Environ. Pollut.,* 103, 327, 1998.

126. Garg, P. and Chandra, P., The duckweed *Wolffia globosa* as an indicator of heavy metal pollution: sensitivity to Cr and Cd, *Environ. Monitor Assess.,* 29, 89, 1993.

127. Ma, J.F. et al., Internal detoxification mechanism of Al in *Hydrangea, Plant. Physiol.,* 113, 1033, 1997.

128. Mortimer, D.C., Freshwater aquatic macrophytes as heavy metal monitors — the Ottawa River experience, *Environ. Monit. Assess.*, 5, 311, 1985.

129. Rai, U.N., Sinha, S., and Chandra, P., Metal biomonitoring in water resources of Eastern Ghats, Koraput Orissa, India, by aquatic plants, *Environ. Monit. Assess.*, 43, 125, 1996.

130. Chandra, P. et al., Monitoring and mitigation of nonpoint source pollution in some water bodies by aquatic plants, *Water Sci.Technol.*, 28, 323, 1993.

131. Ali, M.B. et al., Physico-chemical characteristics and pollution levels of the lake Nainital UP, India: role of macrophytes and phytoplankton in biomonitoring and phytoremediation of toxic metals ions, *Chemosphere*, 39, 2171–2182, 1999.

132. Lovett–Doust, J., Schmidt, M., and Lovett–Doust, L., Biological assessment of aquatic pollution: a review,with emphasis on plants as biomonitors, *Biol. Rev.*, 69, 147, 1994.

133. Mohan, S., Hosetti, B.B., and Tharavathi, N.C., Role of hydrophytes as biological filters: an overview, in *Environmental Impact Assessment and Management*, B.B. Hosetti and A. Kumar, Eds., Daya Publishing House, Delhi, 281, 1998.

134. Lasat, M.M., Phytoextraction of toxic metals: a review of biological mechanisms, *J. Environ. Qual.*, 31, 109, 2002.

135. Sawidis, T., Chettri, M.K., and Zachariadis, G.A., Heavy metals in aquatic plants and sediments from water-system in Macedonia, Greece, *Ecotox. Environ. Saf.*, 32, 73, 1995.

136. Sortjaker, O., Macrophytes and macrophytic communities as test systems in ecotoxicological studies of aquatic systems, *Ecol. Bull. Stockholm*, 36, 75, 1984.

137. Jana, S. and Choudhuri, M.A., Senescence in submerged aquatic angiosperms: effects of heavy metals, *New Phytol.*, 90, 477, 1992.

138. Lösch, R., Plant mitochondrial respiration under the influence of heavy metals, in M.N.V. Prasad, Ed., *Heavy Metal Stress in Plants – from Biomolecules to Ecosystem*, 2nd ed., Springer–Verlag, 181–200, 2004.

139. Weber, C.I. et al., Short-term methods for estimating the chronic toxicity of effluents and receiving waters to freshwater organisms, U.S. Environmental Protection Agency 600.4-89.001. Environmental Monitoring Systems Laboratory, Cincinnati, OH, 1989.

140. Vasseur, P., Ferard, J.F., and Babut, M., The biological aspects of the regulatory control of industrial effluents in France, *Chemosphere*, 22, 119, 1991.

141. Greene, J.C., Se of *Selenastrum capricornutum* to assess the toxicity potential of surface and ground water contamination toxicity caused by chromium waste, *Environ. Toxicol. Chem.*, 7, 35, 1988.

142. Keen, C.L. and Lonnerdal, B., Manganese toxicity in man and experimental animals, in V.L. Schramm and F.C. Wedler, Eds., *Manganese in Metabolism and Enzyme Function*. New York: Academic Press, 35, 1986.

143. Qian, J.H., Zayed, A., and Zhu, X.L., Phytoaccumulation of trace metals by wetland plants III. Uptake and accumulation of ten trace elements by twelve plant species, *J. Environ. Qual.*, 28, 1448, 1999.

144. Wium–Andersen, S. and Andersen, J., Carbon dioxide content of the interstitial water in the sediment of Grane Langso, a Danish lobelia lake, *Limnol. Ocean.*, 17, 943, 1972.

145. Roelofs, J.G.M., Schuurkes, J.A.A.R., and Smits, A.J.M., Impact of acidification and eutrophication on macrophyte communities in soft waters. II. Experimental studies, *Aquatic. Bot.*, 18, 389, 1984.

146. Pip, E. and Stepaniuk, J., Cadmium, copper and lead in sediments and aquatic macrophytes in the lower Nelson River system, Manitoba, Canada. 1. Interspecific differences and macrophyte–sediment relations, *Arch. Hydrobiol.*, 124, 337, 1992.

147. Biernacki, M., Lovett, J., and Doust, J., Laboratory assay of sediment phytotoxicity using macrophyte *Vallisneria Americana*, *Environ. Toxicol. Chem.*, 16, 472, 1997.

148. Sanchezm, J., Marino, N., and Vaquero, M.C., Metal pollution by old lead-zinc mines in Urumea River valley Basque country, Spain. Soil, biota and sediment, *Water Air Soil Pollut.*, 107, 303, 1998.

149. Omote, J., The mobility of trace elements between plants, water and sediments in aquatic ecosystems, in W.W. Wenzel et al., Eds., *Proc. Extended Abstracts. 5th Int. Conf. Biogeochemistry Trace Elements*, 468, 1999.

150. Miller, W.E. et al., Comparative toxicology of laboratory organisms for assessing hazardous waste sites, *J. Environ. Qual.*, 14, 569, 1985.

151. Blanck, H., Wallin, G., and Wangberg, S., Species-dependent variation in algal sensitivity to compounds, *Ecotoxicol. Environ. Saf.*, 8, 339, 1984.

152. Crossland, N.O. et al., Summary and recommendations of the European Workshop on Freshwater Field Tests, Potsdam, June 25–26, 1992.

153. Hillman, W.S. and Culley, D.D., The use of duckweed, *Am. Sci.*, 66, 442, 1978.

154. Tripathi, B.D. and Shukla, S.C., Biological treatment of wastewater by selected aquatic plants, *Environ. Pollut.*, 69, 69, 1991.

155. Brix, H., *Macrophyte-Mediated Oxygen Transfer in Wetlands: Transport Mechanisms and Rates. Constructed Wetland for Water Quality Improvement.* CRC Press, Boca Raton, FL., 1993.

156. Tchnobanoglous, G., Constructed wetlands for waste water treatment engineering considerations, in Copper, P.F. and Findlate, B.C., Eds., *Constructed Wetlands in Water Pollution Control. Advances in Water Pollution Control*, Pergamon Press, Oxford, chap. 11, 1990.

157. Outridge, P.M. and Noller, B.N., Accumulation of toxic trace elements by freshwater vascular plants, *Rev. Environ. Cont. Tox.*, 121, 1, 1991.

158. Markert, B., Plants as biomonitors — potential advances and problems, in D.C. Adriano, Z.S. Chen, and S.S. Yang, Eds., *Biogeochemistry of Trace Elements, Science and Technology Letters*, Northwood, New York, 601, 1994.

159. Dunn, C.E., Seaweeds as hyperaccumulators, in R.R. Brooks, Ed., *Plants that Hyperaccumulate Heavy Metals. Their Role in Phytoremediation, Microbiology, Archaeology, Mineral Exploration and Phytomining*, CAB International, Washington, D.C., 119, 1998.

160. Brooks, R.R. and Robinson, B.H., Aquatic phytoremediation by accumulator plants, in R.R. Brooks, Ed., *Plants that Hyperaccumulate Heavy Metals. Their Role in Phytoremediation, Microbiology, Archaeology, Mineral Exploration and Phytomining*, 203–226, CAB International, Wallingford U.K., 1998.

161. Jones, D.L., Organic acids in rhizosphere — a critical review, *Plant Soil*, 205, 25, 1998.

162. Barber, J.W. and Martin, J.K., The release of organic substances by cereal roots into soil, *New Phytol.*, 76, 69–80, 1976.

163. Römheld, V., The role of phytosiderophores in acquisition of iron and other micronutrients in graminaceous species: an ecological approach, *Plant Soil*, 130, 127, 1991.

164. Greger, M., Johansson, M., and Tillberg, J.E., Aluminium effects on *Scenedesmus obtusiusculus* with different phosphorus status, II. Growth, photosynthesis and pH, *Physiol. Plant.*, 84, 202, 1992.

165. Brix, H., Functions of macrophytes in constructed wetlands, *Water Sci. Tech.*, 29, 71, 1994.

166. Armstrong, W., Armstrong, J., and Beckett, P.M., Measurement and modelling of oxygen release from roots of *Phragmites australis*, in P.F. Cooper and B.C. Findlater, Eds., *Constructed Wetland in Water Pollution Control*. Pergamon Press, London, 41, 1990.

167. Armstrong, W., Brändle, R., and Jackson, M.B., Mechanisms of flood tolerance in plants, *Acta Bot. Neerl.* 43, 307, 1994.

168. Weissner, A. et al., Abilities of helophyte species to release oxygen into rhizospheres with varying redox conditions in laboratory-scale hydroponic systems, *Int. J. Phytorem.*, 4, 1, 2002.

169. Mitch, W.J. and Wise, K.M., Water quality, fate of metals, and predictive model validation of a constructed wetland treating acid mine drainage, *Water Res.*, 32, 1888, 1998.

170. Sobolewski, A., A review of processes responsible for metal removal in wetlands treating contaminated mine drainage, *Int. J. Phytorem.*, 1, 19, 1999.

171. Nyquist, J., Fritioff, Å., and Greger, M., Phytofiltration of stormwater for removal of zinc, in *Phytoremediation Inventory COST Action 837 View*. T. Vanek and J.-P. Schwitzguebel, Eds., 41, Hlavacek-tisk, Prague, Czech Republic, 2003.

172. Dushenko, W.T., Bright, D.A., and Reimer, K.J., Arsenic bioaccumulation and toxicity in Edwards, N.T. Polycyclic aromatic hydrocarbons PAHs in the terrestrial environment — a review, *J. Environ. Qual.*, 12, 427, 1995.

173. Debusk, T.A., Luglin, R.B., and Schwartz, L.N., Retention and compartmentalization of lead and cadmium in wetland microcosms, *Water Res.*, 30, 2707, 1996.

174. Hammer, D.A., Constructed wetland for wastewater treatment: an overview of a low-cost technology, in Constructed Wetland for Wastewater Treatment Conference. Middleton, County Cork, Ireland, 1993.

175. Biddlestone, A.J., Gray, K.R., and Thayanithy, K., A botanical approach to the treatment of wastewaters, *J. Biotech.*, 17, 209, 1991.

176. Koles, S.M., Petrell, R.J., and Bagnall, L.O., Duckweed culture for reduction of ammonia, phosphorus and suspended solids from algal-rich water, In: *Aquatic Plants for Water Treatment and Resource Recovery*. K.R. Reddy and W.H. Smith, Eds., Orlando, FL., Magnolia Publications, 769, 1987.

177. Tanner, C.C., Plants for constructed wetland treatment systems — a comparison of the growth and nutrient uptake of eight emergent species, *Ecol. Eng.*, 7, 59, 1996.

178. Leendertse, P., Scholten, M., and van der Wal, J.T., Fate and effects of nutrients and heavy metals in experimental salt marsh ecosystems, *Environ Pollut.*, 94, 19, 1996.

179. Greenway, M., Suitability of macrophytes for nutrient removal from surface flow constructed wetlands receiving secondary treated sewage effluent in Queensland, *Austr. Water Sci. Technol.*, 48, 121, 2003.

180. Weis, J.S. and Weis, P., Metal uptake, transport and release by wetland plants: implications for phytoremediation and restoration, *Environ. Int.*, 30, 685, 2004.

181. Mendelssohn, I.A. and Postek, M.T., Elemental analysis of deposits on the roots of *Spartina alterniflora* Loisel, *Am. J. Bot.*, 69, 904, 1982.

182. Vale, C. et al., Presence of metal-rich rhizoconcretions on the roots of *Spartina maritima* from the salt marshes of the Tagus Estuary, Portugal, *Sci. Total Environ.*, 97/98, 617, 1990.

183. Sundby, B. et al., Metal-rich concretions on the roots of salt marsh plants: mechanism and rate of formation, *Limnol. Oceanogr.*, 43, 245, 1998.

184. Ye, et al., Zinc, lead, and cadmium accumulation and tolerance in *Typha latifolia* as affected by iron plaque on the root surface, *Aquatic Bot.*, 61, 55, 1998.

185. Zhang, X., Zhang, F., and Mao, D., Effect of iron plaque outside roots on nutrient uptake by rice *Oryza sativa* L. Zinc uptake by Fe-deficient rice, *Plant Soil*, 202, 33, 1998.

186. Batty, L.C. et al., The effect of pH and plaque on the uptake of Cu and Mn in *Phragmites australis* Cav. Trin ex. Steudel, *Ann. Bot.*, 86, 647, 2000.

187. Cacador, I., Vale, C., and Catarino, F., The influence of plants on concentration and fractionation of Zn, Pb, and Cu in salt marsh sediments Tagus Estuary, Portugal, *J. Aquat. Ecosyst. Health.*, 5, 193, 1996.

188. Doyle, M.O. and Otte, M.L., Spatial and temporal variation of soil oxidation by plant roots and burrowing invertabrates and its impact on the distribution of zinc and arsenic in saltmarsh soils, *Biol. Environ.*, 95B, 131, 1995.

189. Doyle, M.O. and Otte, M.L., Organism-induced accumulation of iron, zinc and arsenic in wetland soils. *Environ. Pollut.*, 96, 1, 1997.

190. Wright, D.J. and Otte, M.L., Wetland plant effects on the biogeochemistry of metals beyond the rhizosphere, *Biol. Environ.* 99B, 3, 10, 1999.

191. Stoltz, E. and Greger, M., Accumulation properties of As, Cd, Cu, Pb and Zn by four wetland plant species growing on submersed mine tailings, *Exp. Environ. Bot.*, 47, 271, 2002.

192. Stoltz, E. and Greger, M., Cottongrass effects on trace elements in submersed mine tailings, *J. Environ. Qual.*, 31, 1477, 2002.

193. Nixdorf, B., Fyson, A., and Krumbeck, H., Review: plant life in extremely acidic waters, *Environ. Exp. Bot.*, 46, 203, 2001.

194. Stoltz, Phytostabilization - use of wetland plants to treat mine tailings. Doctoral thesis, Department of Botany, Stockholm University. ISBN: 91-7265-972-6. 2004.

195. Wilkinson, F., Beckett, P.J., and St-Germain, P., Establishment of wetland plants on flooded mine tailings. *Mine, Water Environ.*, IMWA Congress, Sevilla, Spain, 1999.

196. Wood, B. and McAtamney, C., The use of macrophytes in bioremediation, *Biotechnol. Adv.*, 12, 653, 1994.

197. Kadlec, R.L. et al., *Constructed Wetlands for Pollution Control. Control Processes, Performance, Design and Operation*, IWA Publishing, London, 164, 2000.

198. Reddy, K.R. and Smith, W.H., Eds., *Aquatic Plants for Water Treatment and Resource Recovery.* Magnolia Publishing Inc. Orlando, FL, 1032, 1987.

199. Costa–Pierce, B.A., Preliminary investigation of an integrated aquaculture — wetland ecosystem using tertiary treated municipal wastewater in Los Angeles County, California, *Ecol. Eng.*, 10, 341, 1998.

200. Machate, T. et al., Degradation of phenanthrene and hydraulic characteristics in a constructed wetland, *Water Res.*, 31, 554, 1997.

201. McCabe, O.M. and Otte, M.L., Revegetation of mine tailings under wetland conditions, in Brandt J.E. et al., Eds., Proc. 14th Annu. Natl. Meeting — Vision 2000: An Environmental Commitment. AmericanSociety for Surface Mining and Reclamation, Austin, Texas, May 10–16, 99–103, 1997.

202. Crowder, A., Acidification, metals and macrophytes, *Environ. Pollut.*, 71, 171, 1991.

203. Barton, C.D. and Karathanasis, A.D., Renovation of a failed constructed wetland treating acidmine drainage, *Environ. Geol.*, 39, 39, 1999.

204. Karnchanawong, S. and Sanjit, J., Comparative study of domestic wastewater treatment efficiencies between facultative pond and water spinach pond, *Wat. Sci. Technol.*, 32, 263, 1995.

205. Göthberg, M., Greger, M., and Bengtsson, B.E., Accumulation of heavy metals in water spinach *Ipomea aquatica.* cultivated in the Bankok region, Thailand, *Environ. Toxicol.Chem.*, 21, 1934, 2002.

206. Göthberg, A. et al., Uptake and effects of cadmium, mercury and lead in the tropical macrophyte *Ipomea aquatica* in relation to nutrient levels, *J. Environ. Qual.*, 33, 1247, 2004.

207. Beining, B.A. and Otte, M.L., Retention of metals from an abandoned lead-zinc mine by a wetland at Glendalough, Co. Wicklow, Ireland, *Biol. Environ.*, 96, 117, 1996.

208. Vajpayee, P.I. et al., Bioremediation of tannery effluent by aquatic macrophytes, *Bull. Environ. Contam. Toxicol.*, 55, 546, 1995.

209. Mirka, A., Clulow, F.V., and Dave, N.K., Radium[226] in cattails, *Typha latifolia*, and bone of muskrat, *Ondatra zibethica* L., from a watershed with uranium tailings near the city of Elliot Lake, Canada, *Environ. Pollut.*, 91, 41, 1996.

210. Schneider, H. and Rubio, J., Sorption of heavy metal ions by the nonliving biomass of freshwater macrophytes, *Environ. Sci. Technol.*, 33, 2213,1999.

211. Thompson, E.S., Pick, F.R., and Bendell–Young, L.I., The accumulation of cadmium by the yellow pond lily, *Nuphar variegatum*, in Ontario peatlands, *Arch. Environ. Contam Toxicol.*, 32, 161, 1997.

212. Otte, M.L., Kearns, C.C., and Doyle, M.O., Accumulation of arsenic and zinc in the rhizosphere of wetland plants, *Bull. Environ. Contam. Toxicol.*, 55, 154,1995.

213. Siedlecka, A., Some aspects of interactions between heavy metals and plant minerals, *Acta Soc. Bot. Poloniae.*, 64, 265, 1995.

25 Metal-Tolerant Plants: Biodiversity Prospecting for Phytoremediation Technology

M.N.V. Prasad and H. Freitas

CONTENTS

25.1 INTRODUCTION

Metals, radionuclides and other inorganic contaminants are among the most prevalent forms of environmental contaminants, and their remediation in soils and sediments is rather a difficult task [1]. Sources of anthropogenic metal contamination include smelting of metalliferous ore, electroplating, gas exhaust, energy and fuel production, the application of fertilizers and municipal sludges to land, and industrial manufacturing [1,3]. Heavy metal contamination of the biosphere has increased sharply since 1900 [4] and poses major environmental and human health problems worldwide [5]. Unlike many organic contaminants, most metals and radionuclides cannot be eliminated from the environment by chemical or biological transformation [6,7]. Although it may be possible to reduce the toxicity of certain metals by influencing their speciation, they do not degrade and are persistent in the environment [8]. The various conventional remediation technologies used to clean heavy metal-polluted environments are soil *in situ* vitrification, soil incineration, excavation and landfill, soil washing, soil flushing, solidification, and stabilization electrokinetic systems. Each conventional remediation technology has specific benefits and limitations.

All compartments of the biosphere are polluted by a variety of inorganic and organic pollutants as a result of anthropogenic activities and alter the normal biogeochemical cycling. A variety of biological resources have been employed widely in developed and developing nations for cleanup of the metal-polluted sites. These technologies have gained considerable momentum in the last decade and are currently in the process of commercialization [9–19]. The U.S. Environmental Protection Agency's remediation program included phytoremediation of metals and radionuclides as a thrust area during the year 2000.

Plants that hyperaccumulate metals have tremendous potential for application in remediation of metals in the environment. This approach is emerging as an innovative tool with greater potential

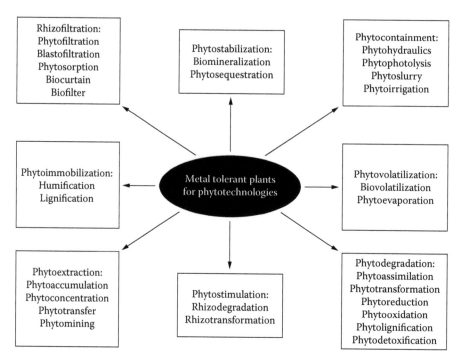

FIGURE 25.1 Various phytotechnologies in which metal-tolerant plants play a key role.

for achieving sustainable development and also to decontaminate metal-polluted air, soil, and water, and for other environmental restoration applications through rhizosphere biotechnology [20] (Figure 25.1 and Figure 25.2). Metal hyperaccumulating plants are thus not only useful in phytoremediation, but also play a significant role in biogeochemical prospecting; they have implications on human health through the food chain and possibly exhibit elemental allelopathy (metallic compounds leached through plant parts of the hyperaccumulator would suppress the growth of other plants growing in the neighborhood) and resistance against fungal pathogens (Figure 25.3) [21]. In order to be realistic about the phytoremediation, focused studies on factors regulating phytoremediation are necessary (Figure 25.4).

The term phytoremediation ("phyto" means plant, and the Latin suffix "remedium" means to clean or restore) actually refers to a diverse collection of plant-based technologies that use naturally occurring or genetically engineered plants for cleaning contaminated environments [1,22]. The primary motivation behind the development of phytoremediative technologies is the potential for low-cost remediation [5].

Although the term "phytoremediation" is a relatively recent invention, the act of phytoremediation is an age-old practice [1]. Research using semiaquatic plants for treating radionuclide-contaminated waters existed in Russia at the dawn of the nuclear era [24,25]. Some plants that grow on metalliferous soils have developed the ability to accumulate massive amounts of indigenous metals in their tissues without exhibiting symptoms of toxicity such plants are termed as "metallophytes" and are of considerable significance for *in situ* decontamination of toxic metals [28,29]. Baker and Brooks [26] and Chaney [29] first suggested using these "hyperaccumulators" for the phytoremediation of metal-polluted sites. However, hyperaccumulators were later believed to have limited potential in this area because of their small size and slow growth, which limit the speed of metal removal [30]. By definition, a hyperaccumulator must accumulate at least 100 $\mu g\ g^{-1}$ (0.01%

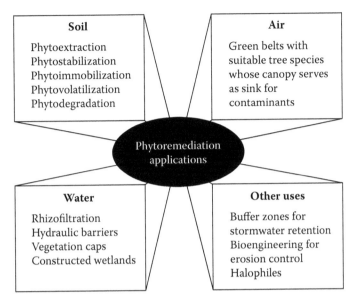

FIGURE 25.2 The important phytoremediation technologies applied are rhizofiltration, phytostabilization, phytovolatilization, and phytoextraction. The term phytoremediation ("phyto" means plant and the Latin suffix "remedium" means to clean or restore) actually refers to a diverse collection of plant-based technologies that use naturally occurring or genetically engineered plants for cleaning contaminated environments. One of the primary objectives behind the development of phytoremediation technologies is its potential for application at a low cost. Although the term phytoremediation is of a relatively recent origin, the practice is not.

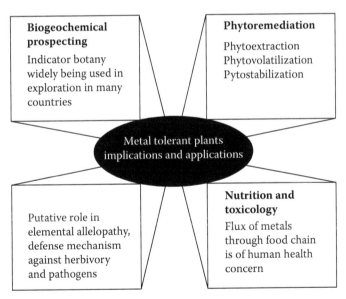

FIGURE 25.3 Metal hyperaccumulators were later believed to have limited potential in the area of phytoremediation, owing to their slow growth and low biomass production, which limit the speed of metal removal. By definition, a hyperaccumulator must accumulate at least 100 µg g^{-1} (0.01% dry wt.); Cd, As, and some other trace metals, 1000 µg g^{-1} (0.1 dry wt.); Co, Cu, Cr, Ni, and Pb and 10,000 µg g^{-1} (1% dry wt.); Mn and Ni. Plants that hyperaccumulate metals have other applications and implications. The most important applications are phytoremediation and biogeochemical prospecting. The other implications are elemental allelopathy and nutrition and toxicology, which are human health-related subjects.

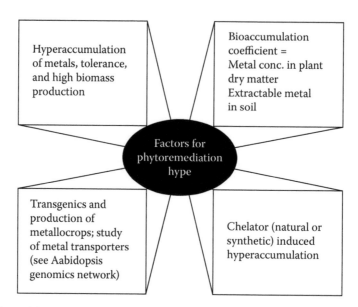

FIGURE 25.4 Several factors would accelerate phytoremediation technology. The prime is genetic engineering and production of transgenics having tolerance and metal accumulation ability for use in phytoremediation, facilitating the factors that would influence the metal bioaccumulation coefficient; in turn, this will depend upon heavy metal availability in the soil, absorption, transport and sequestration, etc., as well as development of low-cost technologies for chelate-induced hyperaccumulation.

dry wt.) of Cd, As, and some other trace metals; 1000 $\mu g\ g^{-1}$ (0.1% dry wt.) of Co, Cu, Cr, Ni, and Pb; and 10,000 $\mu g\ g^{-1}$ (1% dry wt.) of Mn and Ni [19,31].

The following technical terms are connected with phytoremediaton of metals in the environment:

- *Phytoremediation:* use of plants to remediate contaminated soil, water, and air
- *Phytoaccumulation:* the uptake and concentration of contaminants (metals or organics) within the roots or above-ground portion of plants
- *Phytoaccumulation coefficient:* metal concentration in plant dry matter/extractable metal in soil
- *Phytoextraction*: the use of plants at waste sites to accumulate metals into the harvestable, above-ground portion of the plant and, thus, to decontaminate soils
- *Phytoextraction coefficient*: the ratio of extractable metal concentration in the plant tissues (g metal/g dry weight tissue) to the soil concentration of the metal (g metal/g dry weight soil)
- *Phytomining:* use of plants to extract inorganic substances from mine ore
- *Phytostabilization*: plants tolerant to the element in question used to reduce the mobility of elements and thus stabilized in the substrate or roots
- *Phytovolatilization:* the uptake and transpiration of a contaminant by a plant, with release of the contaminant or a modified form of the contaminant to the atmosphere from the plant
- *Rhizofiltration/Phytofiltration*: process in which roots or whole plants of element-accumulating plants absorb the element from polluted effluents and are later harvested to diminish the metals in the effluents
- *Rhizosecretion:* a subset of molecular farming, designed to produce and secrete

25.2 BIODIVERSITY PROSPECTING FOR PHYTOREMEDIATION OF METALS IN THE ENVIRONMENT

Biodiversity prospecting" offers several opportunities, of which the most important is to save as much as possible of the world's immense variety of ecosystems. Biodiversity prospecting would lead to the discovery of a wild plant that could clean polluted environments of the world. This subject is in its infancy, with a great hope of commercial hype. The desire to capitalize on this new idea needs to provide strong incentives for conserving nature [32].

Potentially toxic trace elements are increasing in all compartments of the biosphere, including air, water, and soil, as a result of anthropogenic processes. For example, the metal concentration in river water and sediments has been increased several thousand-fold by effluents from industrial and mining wastes [33]. Aquatic plants in freshwater, marine, and estuarine systems act as receptacles for several metals [34–39]. Published literature indicates that an array of bioresources (biodiversity) has been tested in field and laboratory (Table 25.1). Remediation programs relying on these materials may be successful [40–42,51] (Figure 25.5).

The most successful monitoring methods for metals in the environment are based on gene-based and protein-based bacterial heavy metal biosensors [42]. Mosses, liverworts, and ferns are also capable of growing on metal-enriched substrates. These plants possess anatomical and physiological characteristics enabling them to occupy unique ecological niches in natural metalliferous and manmade environments. For example, groups of specialized bryophytes are found on Cu-enriched substrates — so-called "copper mosses" — and come from widely separated taxonomic groups. Other bryophytes are associated with lead- and zinc-enriched substrates. However, the information about bryophytes growing on serpentine soils is rather scanty. Pteridophytes (ferns) are associated with serpentine substrates in various parts of the world. Brake fern (*Pteris vittata*), a fast growing plant, is reported to tolerate soils contaminated with as much as 1500 ppm arsenic and its fronds concentrate the toxic metal to 22,630 ppm in 6 weeks [43]. Among angiosperms, about 450 metal hyperaccumulators have been identified, which would serve as a reservoir for biotechnological application [44] (Figure 25.6).

25.3 METAL-TOLERANT PLANTS FOR PHYTOREMEDIATION

Mine reclamation and biogeochemical prospecting depend upon the correct selection of plant species and sampling. The selection of heavy metal-tolerant species is a reliable tool to achieve success in phytoremediation. Table 25.2 shows 163 plant taxa belonging to 45 families found to be metal tolerant and capable of growing on elevated concentrations of toxic metals. The use of metal-tolerant species and their metal indicator and accumulation is a function of immense use for biogeochemical prospecting [45–47].

Brassicaceae had the highest number of taxa, i.e., 11 genera and 87 species that are established for hyperaccumulation of metals (Figure 25.7). In Brassicaceae, Ni hyperaccumulation is reported in seven genera and 72 species [48,49], and Zn in three genera and 20 species (Figure 25.8 and Figure 25.9). Different genera of Brassicaceae are known to accumulate metals (Figure 25.10).

Considerable progress had been achieved recently in unraveling the genetic secrets of metal-eating plants. Genes responsible for metal hyperaccumulation in plant tissues have been identified and cloned [50]. These findings are expected to identify new nonconventional crops, *metallocrops*, that can decontaminate metals in the environment [51–53]. The fundamental aspects of microbe/plant stress responses to different doses of metals coupled with breakthrough research innovations in biotechnology would successfully provide answers such as how to apply the biodiversity for advancing phytoremediation technology

TABLE 25.1
Biodiversity Exhibiting Resistance to Metals and with Potential to Clean Up Toxic Metals in all Three Compartments of the Environment (Atmosphere, Hydrosphere, and Lithosphere)

Bacteria	Freshwater algae
Acinetobacter	*Anabaena cylindrica*
Agrobacterium	*A. doliolum*
Alcaligenes eutrophus	*A. inaequalis*
A. faecalis	*A. lutea*
Arthrobacter sp.	*Anacystis nidulans*
Bacillus spp	*Ankistridesmus falcatus*
Citribacter freundii	*Aphanocapsa sp.*
Comamonas sp.	*Asterococcus sp*
Desulfobulbus spp.	*Chlamydomonas acidophila*
Desulfomicrobium	*C. ampla*
Desulfovibrio spp.	*C. bacilus*
Enterobacter colacae	*C. pyrenoidisa*
Leptospirillum sp.	*C. reinhardtii*
Pseudomonas aeruginosa	*C. subglobosus*
P. putida	*C. vulgaris*
P. syringae	*Chamaesiphon minutus*
Photobacterium phosphoreum	*Chara corallina*
Ralstonia eutropha	*Chlamydocapsa bacillus*
R. metallidurans	*Chlamydocapsa cf. petrify*
Salmonella typhimurium	*Chlorella fusca var vacuolata*
S. aureus	*Cladophora glomerata*
Thiobacillus ferrooxidans	*Cosmarium sp.*
Mycorrhizae	*Cyanidium caldarium*
Albatrellus ovinus	*Dictyococcus sp.*
Amanita muscaria	*Dunaliella bioculata*
Chantharellus tubaeformis	*Eisenia bicylis*
C. ciliarius	*Euglena gracilis*
Cortinarius sp.	*E. mutabilis*
Dermocybe sp.	*Eunotia exigua*
Glomus mosseae	*Gleocapsa turfosa*
Gomphidius sp.	*Gleochrysis acoricola*
Hebeloma sp.	*Gleococcus*
Hydnum sp.	*Gleocystis gigas*
Hymenoscyphus ericae	*Gomphenema*
Laccaria laccata	*Hormidium rivulare*
Leccinum spp.	*Hypnomonas*
Oidiodendron maius	*Klebsormidium klebsii*
Paxillus involutus	*K. rivulare*
Pisolithus tinctorius	*Microspora pachyderma*
Russula sp.	*M. stagnorum*
Scleroderma sp.	*M. stagnosum*
Suillus bovinus	*M. strictissimum*
S. luteus	*M. tumidula*
Thelephora terrestris	*M. willeana*
Freshwater algae	*M. floccosa*
Achanthes microcephala	*Microthamnion kutzingianum*
A. minutissima	*Mougeotia*

TABLE 25.1
Biodiversity Exhibiting Resistance to Metals and with Potential to Clean Up Toxic Metals in all Three Compartments of the Environment (Atmosphere, Hydrosphere, and Lithosphere) (continued)

Freshwater algae

Navicula
Nitchia palea
Nostoc calcicola
Oedogonium sp.
O. nephrocytioides
Oocystis elliptica
O. lacustris
Oscillatoria
Phormidium foveolarum
P. luridum
Pinnularia acoricola
Plectonema
Plerococcus
Pseudoanabaena catenata
Pseudococcomyxa adhaerens
Scenedesmusobliqus
S. quadricauda
S. subspicatus
Scenedesmus acutiformis
Schizothrix sp.
Selenastrum capricornutum
Snechococcus sp.
Spartina maritima
Spirogyra sp.
Spirulina platensis
Stegeoclonium sp.
Stegioclonium sp.
Stichococcus sp.
Stichococcus bacillaris
Stigeoclonium tenue
Surirella angustata
Synechocystis aquatilis
Synedra filiformis

Bryophytes

Barbula recurvirostrata
B. acuminatum
B. philonotula
Bryum argenteum
B. rubens
Campylopus bequartii
Cephalozia bicuspidata
Cephaloziella hampeana
C. masalongi
C. nicholsonii
C. rubella
C. stellulifera
C. integerrima

Bryophytes

Ditrichum cornubicum
D. plumbicola
Funaria hygrometrica
Grimmia atrata
Gymnocolea acutiloba
Mielichhoferia macrocarpa
M. elongata
M. nitida
M. mieilichhoferi
Pholia nutans
P. andalusica
P. nutans
Pottia sp.
Scopelophila cataractae
S. ligulata
S. cataractae
S. cataractae
Scapania undulata

Pteridophytes

Asplenium adiantum-nigrum
A. cuneifolium
A. hybrida
A. presolanense
A. ruta-muraria
A. septiontrionale
A. trichomanes
A. viride
Ceratopteris cornuta
Cehaloziella calyculata
Cheilanthes hirta
C. inaequalis var. lanopetiolata
C. inaequalis var. inaequalis
C. hirta Mohria lepigera
Nardia scalaris
Nothalaena marantae
Oligotrichum hercynicum
Ophiglossum lancifolium
Pellea calomelanos
Pteris vittata

Lichens

Bryoria fuscescens
Diploschistes muscorum
Flavoparmelia
Baltimorensis
Hypogyminia physodes
Lobaria pulmonaria
Parmelia caperata

TABLE 25.1
Biodiversity Exhibiting Resistance to Metals and with Potential to Clean Up Toxic Metals in all Three Compartments of the Environment (Atmosphere, Hydrosphere, and Lithosphere) (continued)

Lichens

Peltigera canina
Ramalina duriaei
R. farinacea
R. fastigata

Gymnosperms

Abies chamaecyparis
Chamaecyparis
Cryptomeria
Gingko
Juniperus
Larix picea
Pinus ponderosa
Pseudotsuga
Taxodium

Angiosperms

Acer saccharinum
Aeollanthus biformifolius
Agrostis capillaris
A. gigantea
A. tenuis
Alyssum heldreichii
A. lesbiacum
A. perenne
A. akamasicum
A. alpestre
A. americanum
A. anatolicum
A. argenteum
A. bertlonii
A. bertolonii subsp. scutarinum
A. callichroum
A. carcium
A. cassium
A. chondrogynum
A. cilicium
A. condensatum
A. constellatum
A. corsicum
A. crenulatum
A. cypricum
A. davisianum
A. discolor
A. dubertretii
A. eriophyllum
A. euboeum
A. fallacinum
A. floribundum

Angiosperms

A. giosnanum
A. huber-morathii
A. janchenii
A. malacitanum
A. markgrafii
A. masmenaeum
A. murale
A. obovatum
A. oxycarpum
A. penjwinensis
A. pinifolium
A. pintodasilave
A. pterocarpum
A. robertianum
A. samariferum
A. serpyllifolium
A. singarense
A. smolikanum
A. stolonifera
A. syriacum
A. tenium
A. trapeziforme
A. troodii
A. virgatum
A. wulfenianum
A. montanum
A. serpyllifolium subsp. malacinatum
Amaranthus retroflexus
Anthoxanthum odoratum
Arabidopsis halleri
A. thaliana
Arabis stricta
Armeria maritima subsp. elongata
Arrhenatherum pratensis
Astragalus racemosus
Avenella flexuosa
Berkheya coddi
Betula papyrifera
Bornmuellera glabrescens
B. tymphea
B. baldaccii subsp. baldacci
B. baldaccii subsp. markgrafii
B. baldaccii baldaccii subsp.
Brassica nigra
B. pendula
B. pubescens
B. rapa

TABLE 25.1
Biodiversity Exhibiting Resistance to Metals and with Potential to Clean Up Toxic Metals in all Three Compartments of the Environment (Atmosphere, Hydrosphere, and Lithosphere) (continued)

Angiosperms	Angiosperms
	Mimulus guttatus
B. campestris	*Minuartia hirsuta*
B. hordeaceus	*Nardus stricta*
B. japonica	*Noccaea aptera*
B. juncea	*N. boeotica*
B. napus	*N. eburneosa*
B. narinosa	*N. firmiensis*
B. pekinensis	*N. tymphaea*
B. ramosus	*Pelargonium*
Brachypodium chinensis	*Peltaria dumulosa*
Brachypodium sylvaticum	*P. emarginata*
Calystegia sepium	*Podophyllum peltatum*
Cardamine resedifolia	*Polygonum cuspidatum*
Cardminopsis halleri	*Populus tremula*
Carex echinata	*Pseudosempervium aucheri*
Chrysanthemum morifolium	*Quercus rubra*
Cochlearia aucheri	*Q. ilex*
C. pyrenaica	*Ranunculus baudotti*
C. sempervium	*Rauvolfia serpentina*
Colocasia esculenta	*Ricinus communis*
Cynodon dactylon	*Rumex hydrolapathum*
Danthonia decumbens	*Salix viminalis*
D. linkii	*Sebertia acuminata*
Datura innoxia	*Senecio cornatus*
Deschampsia caespitosa	*Silene cucubalus*
Echinochloa colona	*S. compacta*
Epilobium hirsutum	*S. italica*
Eriophorum angustiflolium	*Solanum nigrum*
Eschscholtzia californica	*Sorghum sudanense subsp. halleri*
Fagopyrum esculentum	*S. sudanens subsp. maritima*
Fagus sylvatica	*Stanleya sp.*
Festuca rubra	*Streptanthus polygaloides*
Fraxinus angustifolia	*Thlaspi alpestre subsp. virens*
Gossypium hirsutum	*T. arvense*
Haumaniastrum katangense	*T. brachypetalum*
Helianthus annuus	*T. bulbosum*
Holcus lanatus	*T. bulbosum*
Hordelymus europaeus	*T. caerulescens*
Hybanthus floribundus	*T. calaminare*
Hydrangea	*T. cepaefolium*
Hydrocotyle umbellata	*T. cepaeifolium subsp.*
Limnobium stoloniferum	*T. cypricum*
Lolium multiflorum	*T. elegans*
L. perenne	*T. epirotum*
Macadamia neurophylla	*T. goesingense*
Medicago sativa	*T. graecum*
Melilotus officinalis	*T. idahoense*

TABLE 25.1
Biodiversity Exhibiting Resistance to Metals and with Potential to Clean Up Toxic Metals in all Three Compartments of the Environment (Atmosphere, Hydrosphere, and Lithosphere) (continued)

Angiosperms	Aquatic macrophytes
T. japonicum	*Lemna minor*
T. jaubertii	*L. trisulca*
T. kovatsii	*Myriophyllum spicatum*
T. liliaceum	*Najas sp.*
T. limosellifolium	*Phragmites australis*
T. magallanicum	*Potamogeton pectinatus*
T. montanum	*P. perfoliatum*
T. montanum var.montanum	*Ruppia sp.*
T. ochroleucum	**Tree crops**
T. oxyceras	*Acer pseudoplatanus*
T. parvifolium	*Betula alleghanensis*
T. praecox	*B. pendula*
T. repens	*B. tauschii*
T. rotundifolium	*Cryptomeria japonica*
T. rotundifolium subsp. cepaefolium	*Eucalyptus camaldulensis*
T. rotundifolium var. corymbosum	*Fagus japonica*
T. stenocarpum	*F. sylvatica*
T. sylvium	*Liriodendron tulipifera*
T. tatraense	*P. deltoides x P. nigra*
T. tymphaeum	*P. maximowiczii*
T. violascens	*P. nigra*
Thinopyrum bessarabicum	*P. taeda*
Trifolium pratense	*P. trichocarpa x P. deltoides*
Viola calaminaria	*Picea abies*
Viola arvensis	*Pinus strobus*
Aquatic macrophytes	*Populus alba*
Arenicola christata	*Prunus virginiana*
A. marina	*Salix arenaria*
Carex sp.	*S. burjatica cv. aquatica*
Ceratophyllum demersum	*S. x caprea*
Glyceria fluitans	*S. viminalis*
Hydrilla verticillata	*S. triandra*
Ipomea aquatica	*S. dasyclados*

Notes: Several of the listed organisms are used for laboratory and field experiments. The results obtained are found to be useful to advance the knowledge of bioremediation and metal monitoring in the environment. (The list is not exhaustive).

Sources: Bargagli, R., *Trace Elements in Terrestrial Plants — an Ecophysiological Approach to Biomonitoring and Biorecovery*, Springer–Verlag, Heidelberg, 1998, 324 pp; Markert, B., *Plants as Biomonitors: Indicators for Heavy Metals in the Terrestrial Environment*, 1993, VCH Publishers; Prasad, M.N.V., in Leeson, A. et al. (Eds.), *Proc. 6th Int.* in Situ *On-Site Bioremediation Symp.*, Battelle Press, Columbus, OH, 2001.

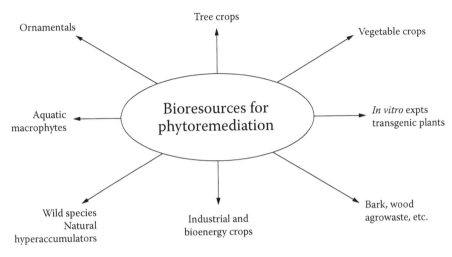

FIGURE 25.5 Biodiversity prospecting for phytoremediation of metals in the environment. Please see the cited references for additional information.

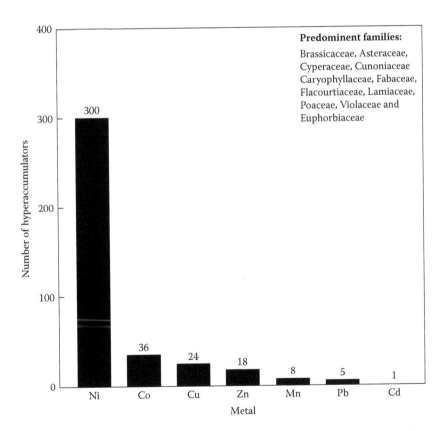

FIGURE 25.6 Taxa of various angiospermous families that hyperaccumulate metals. The families dominating the metal accumulators and hyperaccumulators are Asteraceae, Brassicaceae, Caryophyllaceae, Cyperaceae, Cunoniaceae, Fabaceae, Flacourtiaceae, Lamiaceae, Poaceae, Violaceae, and Euphorbiaceae.

TABLE 25.2
Vascular Plants Growing on Mine Refuse in Portugal

Apiaceae
Daucus crinitus H. Desp.
D. carota L. var. *maritimus* (Lam.) Steud.
Eryngium campestre L.

E. tenue Lam.
Foeniculum vulgare Miller subsp. *piperitum* (Ucria) Big.
Oenanthe crocata L.
Pimpinella villosa Schousb.
Seseli peixotianum Samp.
Aristolochiaceae
Aristolochia longa L.
Aspleniaceae
Asplenium adiantum-nigrum L. var. *corunnense* Christ
Asteraceae
Carlina corymbosa L. subsp. *corymbosa*
Crepis capillaris (L.) Wallr.
Dittrichia viscosa (L.) W. Greuter
Filago lutescens Jordan subsp. *lutescens*
Helichrysum stoechas (L.) Moench.
Hieracium peleteranum Mérat subsp.*ligericum* Zalm
Hispidella hispanica Lam.
Hypochaeris radicata L.
L. viminea (L.) J. & C. Presl. subsp. *viminea*
L. virosa L.
Lactuca viminea (L.) J. & C. Presl
Lapsana communis L. subsp. *communis*
Leontodon taraxacoides (Vill.) Mérat subsp.*longirostris* Finch & P.D. Sell
Logfia gallica (L.) Cosson & Germ.

L. minima (Sm.) Dumort.
Santolina semidentata Hoffmans & Link
Senecio gallicus Vill.
Tolpis barbata (L.) Gaertner
Boraginaceae
Anchusa arvensis (L.) Bieb. subsp. *arvensis*
Echium lusitanicum L. subsp. *lusitanicum*
E. plantagineum L.
Brassicaceae
Alyssum serpyllifolium Desf. subsp. *lusitanicum* Dudley & Pinto da Silva
Lepidium heterophyllum Bentham

Erysimum linifolium (Pers.) Gay subsp. *linifolium*
Campanulaceae
Campanula rapunculus L.
Jasione crispa (Pourret) Samp. subsp. *serpentinitica* P. Silva
Caprifoliaceae
Lonicera periclymenum L. subsp. *periclymenum*
Sambucus nigra L.

Caryophyllaceae
Agrostemma githago L.
Arenaria montana L. subsp. *montana*
A. querioides Pourret ex Willk. subsp. *fontequeri* (P. Silva) R. Afonso
Dianthus laricifolius Boiss. & Reuter subsp. *marizii (Samp.)* Franco
Ortegia hispanica Loefl.
Petrorhagia nantevillii (Burn.) P.W. Ball & Heywood
Saponaria officinalis L.
Silene scabriflora Brot. subsp. *scabriflora*
S. coutinhoi Rothm. & Pinto da Silva
Spergula pentandra L.
S. purpurea (Pers.) G. Don. fil.
Chenopodiaceae
Chenopodium album L. subsp. *album*
Cistaceae
Cistus ladanifer L.
C. salvifolius L.
Tuberaria guttata (L.) Fourr.
Clusiaceae
Hypericum perforatum L.
Convolvulaceae
Convolvulus arvensis L. subsp. *arvensis*
Crassulaceae
Sedum arenarium Brot.
S. forsteranum Sm.
S. tenuifolium Strob.

Dioscoreaceae
Tamus communis L.
Elatinaceae
Elatine macropoda Guss.
Euphorbiaceae
Euphorbia falcata L.
Fagaceae
Castanea sativa Miller
Quercus faginea Lam. subsp. faginea
Q. ilex L. subsp. *ballota* (Desf.) Samp.
Q. pyrenaica Willd.
Gentianaceae

Centaurium erythraea Rafin subsp. *majus* (Hoffmans & Link) Meldéris
Geraniaceae
Geranium purpureum Vill.

Haloragaceae
Myriophyllum alterniflorum DC.
Hypolepidaceae
Pteridium aquilinum (L.) Kuhn subsp. *aquilinum*

TABLE 25.2
Vascular Plants Growing on Mine Refuse in Portugal (continued)

Lamiaceae

Clinopodium vulgare L.

Dorycnium pentaphyllum Scop. subsp. *transmontanum*
Franco

Lavandula stoechas L. subsp. *pedunculata* (Miller) Samp. &
Rozeira

L. stoechas L. subsp. *sampaiana* Rozeira

L. stoechas L. subsp. *stoechas*

Mentha pulegium L.

M. spicata L.

M. suaveolens Ehrh.

Origanum virens Hoffmanns & Link

Phlomis lychnitis L.

Prunella vulgaris L. subsp. *vulgaris*

Salvia verbenaca L.

Teucrium scorodonia L. subsp. *scorodonia*

Thymus mastichina L.

Fabaceae

A. stoechas L.

Acacia dealbata Link

Adenocarpus complicatus (L.) J. Gay

Anthyllis lotoides L.

C. multiflorus (L. Hér.) Sweet

C. striatus (Hill.) Rothm.

Cytisus grandiflorus (Brot.) DC.

G. polyanthos Willk. subsp. *hystrix* (Lange) Franco

Genista triacanthos Brot.

Lotus tenuis Willd.

L. uliginosus Schkuhr.

Lotus corniculatus L. var. *corniculatus*

O. spinosa L. subsp. *antiquorum* (L.) Arcangeli

Ononis cintrana Brot.

Ornithopus compressus L.

Phagnalon saxatile (L.) Cass.

Pisum sativum L. subsp. *elatius* (Bieb.) Ascherson &
Graebner

Pterospartum tridentatum L.

Trifolium arvense L. var. *arvense*

T. glomeratum L.

T. repens L. subsp. *repens*

T. campestre Schreber

Vicia sativa L. subsp. *nigra* (L.) Ehrh.

V. laxiflora Brot.

Liliaceae

Allium vineale L.

A. sphaerocephalos L. subsp. *sphaerocephalos*

Lythraceae

Lythrum hyssopifolia L.

Malvaceae

Malva sylvestris L.

Oleaceae

Fraxinus angustifolia Vahl

Ongagraceae

Epilobium tetragonum L. subsp. *tetragonum*

Orchidaceae

Serapias lingua L.

Papaveraceae

Papaver rhoeas L.

Pinaceae

Pinus pinaster Aiton

Plantaginaceae

Plantago lanceolata L.

P. radicata Hoffmans & Link subsp. *radicata*

Plumbaginaceae

Armeria langei Boiss. subsp. *langei*

Poaceae

Aegilops triuncialis L.

Agrostis curtisii Kerguélen

Arrhenatherum elatius (L.) J. & C. Presl.

Avena sterilis L.

Briza maxima L.

Bromus hordeaceus L. subsp. hordeaceus

Festuca pseudotricophylla Patzke

Holcus lanatus L.

Melica ciliata L. subsp. *ciliata*

Phleum pratense L. subsp. *bertolonii* (DC.) Bornm.

Sanguisorba minor Scop. subsp. *magnolii* (Spach) Coutinho

Setariopsis verticillata Samp.

Trisetaria ovata (Cav.) Paunero

Polygonaceae

Polygonum arenastrum Boreau

P. minus Hudson

Rumex pulcher L.

R. crispus L.

R. acetosella L. subsp. *angiocarpus* (Murb.) Murb.

R. induratus Boiss. & Reuter

Portulacaceae

Portulaca oleraceae L. subsp. Oleraceae

Primulaceae

Anagalis monelli L. var. *linifolia* (L.) Lange

Resedaceae

Reseda virgata Boiss. & Reuter

Rosaceae

Agrimonia procera Wallr.

Crataegus monogyna Jacq.

C. monogyna Jacq. subsp. *brevispina* (G. Kunze) Franco

Filipendula vulgaris Moench

Potentilla erecta (L.) Rauschel

TABLE 25.2
Vascular Plants Growing on Mine Refuse in Portugal (continued)

Rosaceae	Scrophulariaceae
Rosa canina L.	*Anarrhinum bellidifolium* (L.) Willd.
Rubus caesius L.	*Digitalis purpurea* L. subsp. *purpurea*
R. ulmifolius Schott	*Linaria aeruginea* (Gouan) Cav.
Sanguisorba verrucosa (Link) Ces.	*L. spartea* (L.) Willd. subsp. *virgatula* (Brot.) Franco
Rubiaceae	*Odontites tenuifolia* (Pers.) G. Don fil.
Asperula aristata L. fil. subsp. scabra (J. & C. Presl) Nyman	*Scrophularia auriculata*
Galium palustre L.	*Verbascum virgatum* Stokes
Salicaceae	*Digitalis thapsi* L.
Salix salvifolia Brot.	**Thymelaeaceae**
S. triandra L.	*Daphne gnidium* L.
	Valerianaceae
	Centranthus calcitrapae (L.) Dufresne subsp. *calcitrapae*

Sources: Bargagli, R., *Trace Elements in Terrestrial Plants — an Ecophysiological Approach to Biomonitoring and Biore-covery*, Springer–Verlag, Heidelberg, 1998, 324 pp.; Markert, B., *Plants as Biomonitors: Indicators for Heavy Metals in the Terrestrial Environment.* 1993. Vch Publishers; Prasad, M.N.V., in A. Leeson et al. (Eds.), *Phytoremediation, Wetlands, Sediments*, 6(5), 165–172, *Proc. 6th Int.* in Situ *On-Site Bioremediation Symp.*, Battelle Press, Columbus, OH, 2001.

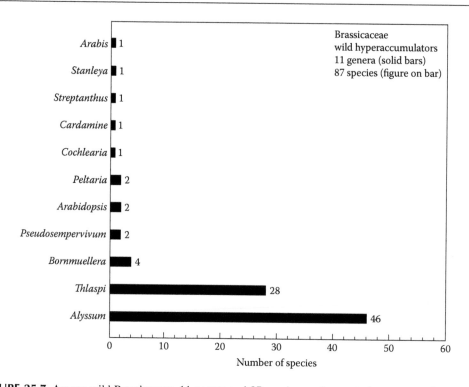

FIGURE 25.7 Among wild Brassicaceae, 11 genera and 87 species are known to hyperaccumulate metals.

25.4 ORNAMENTALS

Nerium oleander leaves collected from urban areas of Portugal accumulated lead up to 78 µg/g dry weight in leaves and are suitable for monitoring lead in air [59]. Canna x generalis is an

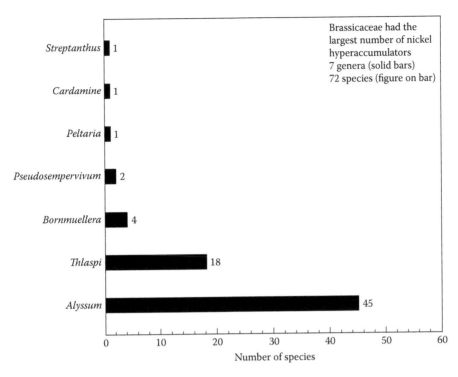

FIGURE 25.8 Brassicaceae has the largest amount of nickel (seven genera and 72 species).

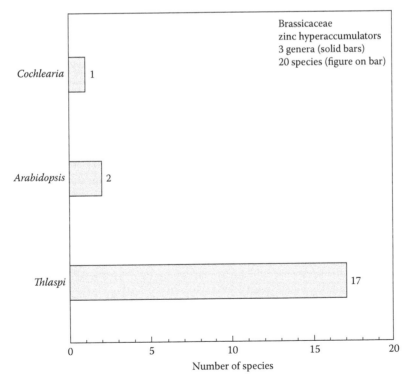

FIGURE 25.9 Brassicaceae has the largest number of zinc hyperacccumulators (three genera and 20 species).

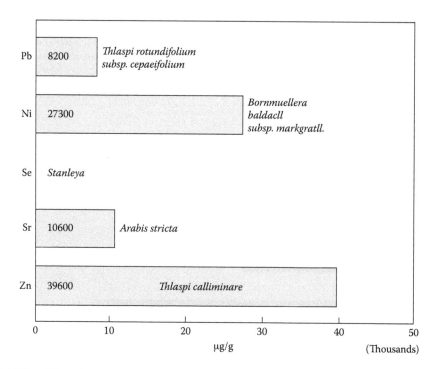

FIGURE 25.10 Selected examples of Brassicaeae that hyperaccumulate lead, nickel, strontium, and zinc.

important ornamental cultivated in urban landscapes. Hydroponic cultures of this plant treated with lead for 1 month suggest that it is suitable for phytoextraction of lead because the plant produces an appreciable quantity of biomass [60]. Pelargonium sp. "Frensham" (scented geranium) was identified as one of the most efficient metal hyperaccumulator plants [61]. In a greenhouse study, young cuttings of scented geranium grown in artificial soil and fed different metal solutions were capable of taking up large amounts of three major heavy metal contaminants (e.g., Pb, Cd, and Ni) in a relatively short time. These plants were capable of extracting from the feeding solution and stocking in their roots amounts of lead, cadmium, and nickel equivalent to 9, 2.7, and 1.9% of their dry weight material, respectively.

With an average root mass of 0.5 to 1.0 g in dry weight, scented geranium cuttings could extract 90 mg of Pb, 27 mg of Cd, and 19 mg of Ni from the feeding solution in 14 days. If these rates of uptake could be maintained under field conditions, scented geranium should be able to clean up heavily contaminated sites in less than 10 years. (Growth and uptake in nutrient solution are extremely different to that in soil, and scientific studies indicate the hydroponic culture is not indicative of a real-world situation, due to ion competition, root impedance, and the fact that plants do not grow root hairs when they are grown in solution.) For example, a phytoremediation lead cleanup program consisting of 16 successive croppings of scented geranium planted at a density of 100 plants m^{-2} over the summer could easily remove up to 72 g of lead m^{-2} yr^{-1}. In the authors' estimates, scented geranium would extract 1000 to 5000 kg of lead per ha^{-1} yr^{-1}. Thus, these reported figures are close to the estimations of metal removal rates of 200 to 1000 kg ha^{-1} yr^{-1} for plants capable of accumulating 1.0 to 2.0% metal [6].

Thus, if scented geranium is planted in soil where the lead contamination is 1000 µg kg^{-1} of soil, which is the acceptable limit for the province of Ontario (Canada), it can clean up the soil completely in 8 years. Scented geranium also has the ability to survive on soils containing one or more metal contaminants (individually or in combination) and on soils contaminated with a mixture of metal and hydrocarbons (up to three metal–hydrocarbon-contaminated soils) > 3% total hydrocarbon in combination with several metal contaminants.

25.5 METAL TOLERANT PLANTS AND CHELATORS MIGHT PROMOTE PHYTOREMEDIATION TECHNOLOGY

Use of soil amendments such as synthetics (ammonium thiocyanate) and natural zeolites has yielded promising results [62–66]. EDTA, NTA, citrate, oxalate, malate, succinate, tartrate, phthalate, salicylate, acetate, etc. have been used as chelators for rapid mobility and uptake of metals from contaminated soils by plants. Use of synthetic chelators significantly increased Pb and Cd uptake and translocation from roots to shoots, facilitating phytoextraction of the metals from low-grade ores. Synthetic cross-linked polyacrylates, hydrogels have protected plant roots from heavy metal toxicity and prevented the entry of toxic metals into roots. Application of low-cost synthetics and natural zeolites on a large scale are applied to the soil through irrigation at specific stages of plant growth; this might be beneficial to accelerate metal accumulation [67].

A major factor influencing the efficiency of phytoextraction is the ability of plants to absorb large quantities of metal in a short period of time. Hyperaccumulators accumulate appreciable quantities of metal in their tissue regardless of the concentration of metal in the soil [68], as long as the metal in question is present. This property is unlike moderate accumulators now used for phytoextraction in which the quantity of absorbed metal is a reflection of the concentration in the soil. Although the total soil metal content may be high, it is the fraction that is readily available in the soil solution that determines the efficiency of metal absorption by plant roots. To enhance the speed and quantity of metal removal by plants, some researchers advocate the use of various chemicals for increasing the quantity of available metal for plant uptake.

Chemicals suggested for this purpose include various acidifying agents [6,20,69], fertilizer salts [71,72], and chelating materials [67,73,74]. These chemicals increase the amount of bioavailable metal in the soil solution by liberating or displacing metal from the solid phase of the soil or by making precipitated metal species more soluble. Research in this area has been moderately successful, but the wisdom of liberating large quantities of toxic metal into soil water is questionable.

Soil pH is a major factor influencing the availability of elements in the soil for plant uptake [75]. Under acidic conditions, H^+ ions displace metal cations from the cation exchange complex (CEC) of soil components and cause metals to be released from sesquioxides and variably charged clays to which they have been chemisorbed (i.e., specific adsorption) [76]. The retention of metals to soil organic matter is also weaker at low pH, resulting in more available metal in the soil solution for root absorption. Many metal cations are more soluble and available in the soil solution at low pH (below 5.5), including Cd, Cu, Hg, Ni, Pb, and Zn [2,76]. It is suggested that the phytoextraction process is enhanced when metal availability to plant roots is facilitated through the addition of acidifying agents to the soil [2,24,70].

Possible amendments for acidification include NH_4-containing fertilizers, organic and inorganic acids, and elemental S. Trelease and Trelease [77] indicated that plant roots acidify hydroponic solutions in response to NH_4 nutrition and cause solutions to become more alkaline in response to NO_3 nutrition. Metal availability in the soil can be manipulated by the proper ratio of NO_3 to NH_4 used for plant fertilization by the effect of these N sources on soil pH, but no phytoremediation research has been conducted on this topic to date.

The acidification of soil with elemental S is a common agronomic practice that can be used to mobilize metal cations in soil. Brown et al. [69] acidified a Cd- and Zn-contaminated soil with elemental S and observed that accumulation of these metals by plants was greater than when the amendment was not used. Acidifying agents are also used to increase the availability of radioactive elements in the soil for plant uptake. Huang et al. [70] reported that the addition of citric acid increases U accumulation in Indian mustard (*B. juncea*) tissues more than nitric or sulfuric acid, although all acids decrease soil pH by the same amount. These authors speculated that citric acid chelates the soil U, thereby enhancing its solubility and availability in the soil solution. The addition of citric acid causes a 1000-fold increase of U in the shoots of *B. juncea* compared to accumulation

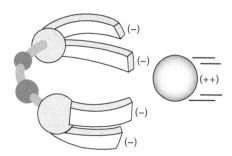

FIGURE 25.11 Citric acid is a naturally occurring chelating agent. The chelation process is water activated. EDTA, NTA, citrate, oxalate, malate, succinate, tartrate, phthalate, salicylate, acetate, etc. have been used for chelate-induced hyperaccumulation. Synthetic soil amendments such as ammonium thiocyanate and natural zeolites have yielded promising results in inducing hyperaccumulation of metals.

in the control (no citric acid addition) (Figure 25.11). Despite the promise of some acidifying agents for use in phytoextraction, little research is reported on this subject.

25.6 CONCLUSIONS

Implementing phytoremediation with the use of biodiversity has certain limitations [6,78,79]. To a considerable extent, these include potential contamination of the vegetation and food chain and the often extreme difficulty in establishing and maintaining vegetation on contaminated sites (e.g., mine tailings with high levels of residual metals). For metal contaminants, plants show the potential for phytoextraction (uptake and recovery of contaminants into above-ground biomass) [62,80,81], filtering metals from water onto root systems (rhizofiltration), or stabilizing waste sites by erosion control and evapotranspiration of large quantities of water (phytostabilization) [82,83].

After the plants have been allowed to grow for some time, they are harvested and then incinerated or composted to recycle the metals. This procedure may be repeated as necessary to bring soil contaminant levels down to allowable limits. If plants are incinerated, the ash must be disposed of in a hazardous waste landfill. Finally, phytoremediation in some countries has limited acceptance by the local government and takes a long time to mitigate the contaminant. Metal hyperaccumulators are generally slow growing with a small biomass and shallow root systems. Plant biomass must be harvested and removed, followed by proper disposal. Plants experience stress due to prevailing high concentrations of metals.

One of the main advantages of phytoextraction is that the plant biomass containing the extracted contaminant can be a resource. For example, biomass that contains selenium (Se), an essential nutrient, has been transported to areas that are deficient in Se and used for animal feed [84,85]; metal hyperaccumulators are of special significance in biogeochemical prospecting of minerals.

Rhizofiltration has the advantage that terrestrial or aquatic plants are used for this purpose. Although terrestrial plants require support, such as a floating platform, they generally remove more contaminants than aquatic plants do. This system can be *in situ* (floating rafts on ponds) or *ex situ* (an engineered tank system). An *ex situ* system can be placed anywhere because the treatment does not need to be at the original location of contamination [86–88].

Rhizofiltration has the following disadvantages:

- The pH of the influent solution may need to be adjusted continually to obtain optimum metal uptake.
- The chemical speciation and interaction of all species in the influent must be understood for proper application.
- A well engineered system is required to control influent concentration and flow rate.

- Plants (especially terrestrial plants) may need to be grown in a greenhouse or nursery and then placed in the rhizofiltration system.
- Periodic harvesting and plant disposal are required.
- Metal immobilization and uptake results from laboratory and greenhouse studies might not be achievable in the field.

In phytovolatilization, the advantages are that contaminants could be transformed to less toxic forms, such as elemental mercury and dimethyl selenite gas, and that contaminants' or metabolites released to the atmosphere might be subject to more effective or rapid natural degradation processes, such as photodegradation [89].

A disadvantage of phytovolatilization is that the contaminant (such as Se) might be released into the atmosphere [84,85,89]. Therefore, adequate planning is needed for phytoremediation-based systems integrated with the environment, e.g., green belts (invaluable ecological niches, particularly in urban industrial areas) and constructed wetlands in which *Eichhornia crassipes* (water hyacinth), *Hydrocotyle umbellata* (pennywort), *Lemma minor* (duckweed), and *Azolla pinnata* (water velvet) are maintained and managed; these can take up Pb, Cu, Cd, Fe, and Hg from aqueous solutions [90].

Nicotianamine (NA), a plant nonproteinogenic amino acid, is an efficient complexing agent for Co^{2+}, Cu^{2+}, Fe^{2+}, Mn^{2+}, Zn^{2+}, and other divalent transition metals. Genetic manipulation of genes involved in the biosynthesis of metal-sequestering compounds and introduction to desirable plant species might attract phytoremediation strategies. It is very advantageous to use commonly cultivated crops such as *Brassica juncea, Armoracia rusticana*, and *Helianthus annuus*, which are reported to accumulate many toxic metals [91]. Plants amenable to genetic manipulation and *in vitro* culture play a significant role in the success of phytoremediation; however, socioecological and regulatory acceptance using genetically modified organisms and their field trials make up a challenging task in the coming days (Figure 25.12 and Figure 25.13).

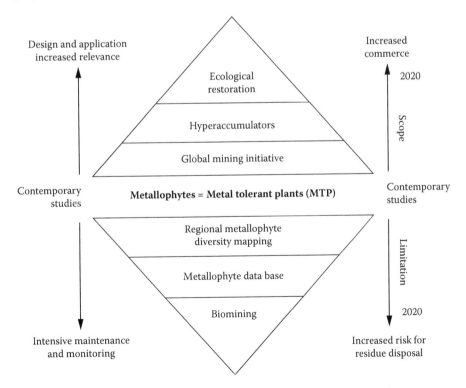

FIGURE 25.12 Advantages and limitations of using metallophytes in phytotechnologies.

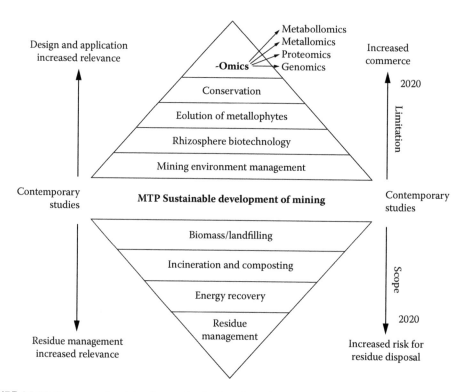

FIGURE 25.13 To move ahead in phytotechnologies, the existing gaps in knowledge must be thoroughly investigated with the available modern tools.

ACKNOWLEDGMENTS

The authors thank Prof. Graciela Munoz, editor in chief, *Electronic Journal of Biotechnology* for according permission to reproduce large portions of text from a previously published article in the online electronic journal http://www. ejbiotechnology. info/content/ vol6/ issue3/index.html.

REFERENCES

1. Cunningham, S.D., Shann, J.R., Crowley, D.E., and Anderson,T.A., Phytoremediation of contaminated water and soil. In E.L. Kruger, T.A. Anderson, and J.R. Coats (Eds.) *Phytoremediation of Soil and Water Contaminants*. ACS symposium series 664. American Chemical Society, Washington, D.C., 2–19, 1997.
2. Blaylock, M.J. and Huang, J.W., Phytoextraction of metals. In *Phytoremediation of Toxic Metals — Using Plants to Clean Up the Environment*, I. Raskin and B.D. Ensley (Eds.), John Wiley & Sons, Inc., New York, 2000, 53–70.
3. Raskin, I., Kumar, P.B.A.N., Dushenkov, S., and Salt, D.E., Bioconcentration of heavy metals by plants. *Curr. Opin. Biotechnol.,* 1994, 5, 285–290.
4. Nriagu, J.O., Global inventory of natural and anthropogenic emissions of trace metals to the atmosphere. *Nature,* 1979, 279, 409–411.
5. Ensley, B.D., Rational for use of phytoremediation. In *Phytoremediation of Toxic Metals — Using Plants to Clean Up the Environment*, I. Raskin and B.D. Ensley (Eds.), John Wiley & Sons, Inc., New York, 2000, 3–12.
6. Cunningham, S.D. and Ow, D.W., Promises and prospects of phytoremediation. *Plant Physiol.,* 110, 715–719.

7. NRC, Challenges of groundwater and soil cleanup. In *Innovations in Groundwater and Soil Cleanup: From Concepts to Commercialization* National Academy Press, Washington, D.C., 1997, 18–41.

8. NRC, Metals and radionuclides: technologies for characterization, remediation, and containment. In *Groundwater and Soil Cleanup: Improving Management of Persistent Contaminants*. Washington, D.C., National Academy Press, 1999, 72–128.

9. Alcantara, E., Barra, R., Benlloch, M., Ginhas, A., Jorrinj, Lopez. J.A., Lora, A., Ojeda, M.A., Pujadas, A., Requejo, R., Romera, J., Sancho, E.D., Shilev, S., and Tena, M., Phytoremediation of a metal-contaminated area in southern Spain. In *Intercost Workshop — Sorrento*. 121–123. 2000.

10. Comis, D., Metal-scavenging plants to cleanse the soil. *Agric. Res.*, 1995, 43, 4–9.

11. Ernst, W.H.O., Revolution of metal hyperaccumulation and phytoremediation hype. *New Phytol.*, 2000, 146, 357–358.

12. Glass, D.J., U.S. and international markets for phytoremediation, 1999–2000. D.J. Glass Associates Inc. Needham, MA, 266, 1999.

13. Glass, D.J., Economic potential of phytoremediation. In *Phytoremediation of Toxic Metals — Using Plants to Clean Up the Environment*, I. Raskin and B.D. Ensley (Eds.), John Wiley & Sons, Inc., New York, 2000, 15–32.

14. Hamlin, R.L., Phytoremediation literature review. Environmental awareness in the United States. http://www.umass.edu/umext/soils and plant. 2002.

15. Prasad, M.N.V. and Freitas, H., Feasible biotechnological and bioremediation strategies for serpentine soils and mine spoils. *Electron. J. Biotechnol.*, 1999, 2(1), 35–50, http://ejb.ucv.cl or http://www.ejb.org.

16. Raskin, I. and Ensley, B.D. (Eds.) *Phytoremediation of Toxic Metals : Using Plants to Clean Up the Environment*, John Wiley & Sons, 352 pp., 2000.

17. Salt, D.E., Smith, R.D., and Raskin, I., Phytoremediation. *Annu. Rev. Plant Physiol. Plant Mol. Biol.*. 1998, 49, 643–668.

18. Vangronsveld, J. and Cunningham, S.D., *Metal-Contaminated Soils: in Situ Activation and Phytorestoration*. Springer–Verlag, Berlin, Heidelberg, 265. 1998.

19. Watanabe, M.E., Phytoremediation on the brink of commercialization. *Environ. Sci. Tech.*, 1997, 31, 182.

20. Wenzel, W.W., Lombi, E., and Adriano, D.C., Biogeochemical processes in the rhizosphere: role in phytoremediation of metal-polluted soils. In *Heavy Metal Stress in Plants — from Molecules to Ecosystems*. Prasad, M.N.V. and Hagemeyer, J. (Eds.), Springer–Verlag, Berlin, 271–303, 1999.

21. Boyd, R.S., Shaw, J.J., and Martens, S.N., Nickel hyperaccumulation in *S. Polygaloids* (Brassicaceae) as a defense against pathogens. *Am. J. Bot.*, 1994, 81, 294–300.

22. Flathman, P.E. and Lanza, G.R., Phytoremediation: current views on an emerging green technology. *J. Soil Contam.*, 1998, 7(4), 415–432.

23. Brooks, R.R. (Ed.). *Plants That Hyperaccumulate Heavy Metals*. CAB International. Wallingford, U.K., 384, 1998.

24. Salt, D.E., Blaylock, M., Kumar, N.P.B.A., Dushenkov, V., Ensley, D., Chet, I., and Raskin, I., Phytoremediation: a novel strategy for the removal of toxic metals from the environment using plants. *Biotechnology*, 1995, 13, 468–474.

25. Timofeev–Resovsky, E.A., Agafonov, B.M., and Timofeev–Resovsky, N.V., Fate of radioisotopes in aquatic environments (in Russian). *Proc. Biol. Inst. USSR Acad. Sci.*, 1962, 22, 49–67.

26. Baker, A.J.M. and Brooks, R.R., Terrestrial higher plants which hyperaccumulate metal elements — a review of their distribution, ecology, and phytochemistry. *Biorecovery*, 1989, 1, 81–126.

27. Baker, A.J.M., Reeves, R.D., and Mcgrath, S.P., *In situ* decontamination of heavy metal polluted soils using crops of metal-accumulating plants — a feasibility study. In R.L. Hinchee and R.F. Olfenbuttel (Ed.), In Situ *Bioreclamation*. Butterworth–Heinemann, Boston, 1991, 600–605.

28. Reeves, R.D. and Brooks, R.R., Hyperaccumulation of lead and zinc by two metallophytes from a mining area of Central Europe. *Environ. Pollut.* Series A., 1983, 31, 277–287.

29. Chaney, R.L., Plant uptake of inorganic waste constitutes. In *Land Treatment of Hazardous Wastes*, J.F. Parr, P.B. Marsh, and J.M. Kla (Eds.), Noyes Data Corp., Park Ridge, NJ, 50–76.

30. Ebbs, S.D., Lasat, M.M., Brandy, D.J., Cornish, J., Gordon, R., and Kochian. L.V., Heavy metals in the environment — phytoextraction of cadmium and zinc from a contaminated soil. *J. Environ. Qual.*, 1997, 26, 1424–1430.

31. Reeves, R.D. and Baker, A.J.M., Metal-accumulating plants. In *Phytoremediation of Toxic Metals — Using Plants to Clean Up the Environment*, Raskin and B.D. Ensley (Eds.), John Wiley & Sons, Inc., New York. 2000, 193–230.

32. Myers, N., The biodiversity challenge: expanded hot-spots analysis. *Environmentalist*, 1990, 10, 243–256.

33. Siegel, F.R., *Environmental Geochemistry of Potentially Toxic Metals*. Springer–Verlag, Heidelberg, 2002, 218.

34. Cole, S., The emergence of treatment wetlands. *Environ. Sci. Technol.*, 1998, 32, 218a–223a.

35. Glass, D.J., *The 2000 Phytoremediation Industry*. D.J. Glass Associates Inc., Needham, MA, 2000, 100.

36. Hansen, D., Duda, P.J., Zayed, A., and Terry, N., Selenium removal by constructed wetlands: role of biological volatilization. *Environ. Sci. Technol.*, 1998, 32, 591–597.

37. Kadlec, R.H., Knight, R.L., Vymazal, J., Brix, H., Cooper, P., and Habert, R., *Constructed Wetlands for Pollution Control. Control Processes, Performance, Design and Operation.* Iwa Publishing, London, 2000.

38. Kaltsikes, P.J., *Phytoremediation — State of the Art in Europe, an International Comparison*. Agricultural University of Athens, Cost Action 837 First Workshop, 2000.

39. Odum, H.T., Woucik, W., and Pritchard, L., *Heavy Metals in the Environment: Using Wetlands for their Removal*. CRC Press, Boca Raton, FL, 2000.

40. Valdes, J.J., *Bioremediation*. Kluwer Academic Publishers, Dordrecht, 20.

41. Wise, D.L., Trantolo, D.J., Cichon, E.J., Inyang, H.I., and Stottmeister, U., *Bioremediation of Contaminated Soils*. Marcel Dekker Inc., New York, 2000.

42. Tsao, D. (Ed.), *Phytoremediation*. Springer–Verlag, Heidelberg, 206, 2003.

43. Ma, L.Q., Komar, K.M., Tu, C., Zhang, W., Cai, Y., and Kennelley, E.D., A fern that hyperaccumulates arsenic. *Nature*, 2001, 409, 579.

44. Brooks, R.R., General introduction. In *Plants That Hyperaccumulate Heavy Metals: Their Role in Phytoremediation, Microbiology, Archaeology, Mineral Exploration and Phytomining*, R.R. Brooks (Ed.). Cab International, New York, 1998, 1–4.

45. Brooks, R.R., *Biological Methods of Prospecting for Minerals*. Wiley-Interscience, New York, 313, 1983.

46. Badri, M. and Springuel, I., Biogeochemical prospecting in the southeastern desert of Egypt. *J. Arid Environ.*, 1994, 28, 257–264.

47. McInnes, B.I.A., Dunn, C., Cameron, E.M., and Kameko, L., Biogeochemical exploration for gold in tropical rain forest regions of Papua, New Guinea. *J. Geochem. Exploration, 1996, 57, 227–243.*

48. Reeves, R.D., Baker, A.J.M., Borhidi, A., and Berazain, R., Nickel hyperaccumulation in the serpentine flora of Cuba. *Ann. Bot.*, 1999, 83, 29–38.

49. Reeves, R.D., Baker, A.J.M., Borhidi, A., and Berazain, R., Nickel-accumulating plants from the ancient serpentine soils of Cuba. *New Phytol.,* 1996, 133, 217–224.

50. Ebbs, S.D. and Kochian, L.V., Phytoextraction of zinc by oat (*Avena Sativa*), barley (*Hordeum Vulgare*), and Indian mustard (*Brassica Juncea*). *Environ. Sci. Technol.*, 1998, 32(6), 802–806.

51. Moffat, A.S., Engineering plants to cope with metals, *Science*, 1999, 285, 369–370.

52. Ebbs, S.D. and Kochian, L.V., Toxicity of zinc and copper to Brassica species: implications for phytoremediation. *J. Environ. Qual.*, 1997, 26, 776–781.

53. Raskin, I., Plant genetic engineering may help with environmnetal cleanup. *Proc. Natl. Acad. Sci. USA*, 1996, 93, 3164–3166.

54. Bargagli, R., *Trace Elements In Terrestrial Plants — an Ecophysiological Approach to Biomonitoring and Biorecovery*. Springer–Verlag, Heidelberg 1998, 324.

55. Markert, B., *Plants as Biomonitors: Indicators for Heavy Metals in the Terrestrial Environment*. VCH Publishers, Weinheim, 1993.

56. Prasad, M.N.V., Bioremediation potential of Amaranthaceae. In *Phytoremediation, Wetlands, and Sediments*, A. Leeson, E.A. Foote, M.K. Banks, and V.S. Magar (Eds.). Proc. 6th Int. *In Situ* On-Site Bioremediation Symp., Battelle Press, Columbus, OH, 2001, 6(5), 165–172.

57. Freitas, H., Prasad, M.N.V., and Pratas, J., Plant community tolerant to trace elements growing on the degraded soils of São Domingos mine in the southeast of Portugal: environmental implications. *Environ. Int.*, 2004, 30, 65–72.

58. Freitas, H., Prasad, M.N.V., and Pratas J., Analysis of serpentinophytes from northeast of Portugal for trace metal accumulation — relevance to the management of mine environment. *Chemosphere*, 2004, 54(11), 1625–1642.

59. Freiats, H., Nabais, V., and Paiva, J., Heavy metal pollution in the urban areas and roads of Portugal using *Nerium Oleander* L. *Proc. Int. Conf. Heavy Metals Environ.*, 1240–1242. Farmer, J.G. (Ed.), Cep Consultants Ltd. Edinburgh.

60. Trampczynska, A., Gawronski, S.W., and Kutrys, S., *Canna X generalis* as a plant for phytoextraction of heavy metals in urbanized area. Zeszyty Naukowe Politechniki Slaskiej, 2001, 45, 71–74.

61. Saxena, P.K., Krishnaraj, S., Dan, T., Perras, M.R., and Vettakkoruma–Kankav, N.N., Phytoremediation of metal contaminated and polluted soils. In *Heavy Metal Stress in Plants — from Molecules to Ecosystems*, M.N.V. Prasad and J. Hagemeyer (Eds.). Springer–Verlag, Heidelberg, 1999, 305–329.

62. Anderson, C.W.N., Brooks, R.R., Stewart, R.B., and Simcock, R., Harvesting a crop of gold in plants. *Nature*, 1998, 395, 553–554.

63. Churchmann, G.J., Slade, P.G., Rengasamy, P., Peter, P., Wright, M., and Naidu, R., Use of fine grained minerals to minimize the bioavaliability of metal contaminants. Environmental Impacts of Metals. Int. Workshop, Tamil Nadu Agricultural University, Coimbatore, India, 1999, 49-50.

64. Huttermann, A., Arduini, I., and Godbold, D.L., Metal pollution and forest decline. In *Heavy Metal Stress in Plants — from Molecules to Ecosystems*. Prasad, M.N.V. and Hagemeyer, J. (Eds.). Springer–Verlag, Berlin, 1999, 253–272.

65. Huang, J.W. and Cunningham, S.D., Lead phytoextraction: species variation in lead uptake and translocation. *New Phytol.*, 1996, 134, 75–84.

66. Zorpas, A.A., Constantinides, T., Vlyssides, A.G., Aralambous, I., and Loizidou, M., Heavy metal uptake by natural zeolite and metals partitioning in sewage sludge compost. *Bioresour. Technol.*, 1999, 71(2), 113–119.

67. Blaylock, M.J., Salt, D.E., Dushenkov, S., Zakharova, O., Gussman, C., Kapulnik, Y., Ensley, B.D., and Raskin, I., Enhanced accumulation of Pb in Indian mustard by soil-applied chelating agents. *Environ. Sci. Technol.*, 1997, 31, 860–865.

68. Baker, A.J.M., Accumulators and excluders — strategies in the response of plants to heavy metals. *J. Plant Nutr.*, 1981, 3, 643–654.

69. Brown, S.L., Chaney, R.L., Angle, J.S., and Baker, A.J.M., Phytoremediation potential of *Thlaspi Caerulescens* and bladder campion for zinc- and cadmium-contaminated soil. *J. Environ. Qual.*, 1994, 23, 1151–1157.

70. Huang, J.W., Blaylock, M.J., Kapulnik, Y., and Ensley, B.D., Phytoremediation of uranium-contaminated soils: role of organic acids in triggering uranium hyperaccumulation in plants. *Environ. Sci. Technol.*, 1998, 32, 2004–2008.

71. Lasat, M.M., Norvell, W.A., and Kochian, L.V., Potential for phytoextraction of ^{137}Cs from a contaminated soil. *Plant Soil*, 1997, 195, 99–106.

72. Lasat, M.M., Fuhrmann, M., Ebbs, S.D., Cornish, J.E., and Kochian, L.V., Phytoremediation of a radiocesium-contaminated soil: evaluation of ^{137}cesium bioaccumulation in the shoots of three plant species. *J. Environ. Qual.*, 1998, 27, 163–169.

73. Huang, J.W., Chen, J., Berti, W.R., and Cunningham, S.D., Phytoremediation of lead contaminated soil: role of synthetic chelates in lead phytoextraction. *Environ. Sci. Technol.*, 1997, 31, 800–805.

74. Huang, J.W., Chen, J., and Cunningham, S.D., Phytoextraction of lead from contaminated soils. In *Phytoremediation of Soil and Water Contaminants*. Kruger, E.L., et al. (Ed.), ACS Symposium Series 664. American Chemical Society, Washington, D.C., 1997, 283–298.

75. Marschner, H., *Mineral Nutrition of Higher Plants*. 2nd ed., Academic Press, New York. 1995.

76. Mcbride, M.B., *Environmental Chemistry of Soils*. Oxford University Press, New York, 1994.

77. Trelease, S.F. and Trelease, H.M., Changes in hydrogen–ion concentration of culture solutions containing nitrate and ammonium nitrogen. *Am. J. Bot.*, 1935, 22, 520–542.

78. Chaney, R.L., Malik, M., Li, Y.M., Brown, S.L., Brewer, E.P., Angle, J.S., and Baker, A.J.M., Phytoremediation of soil metals. *Curr. Opin. Biotechnol.*, 1997, 8(3), 279.

79. Clemens, S., Palmgren, M.G., and Krämer, U., A long way ahead: understanding and engineering plant metal accumulation. *Trends Plant Sci.*, 2002, 7, 309–315.

80. Bañuelos, G.S., Shannon, M.C., Ajwa, H., Draper, J.H., Jordahl, J., and Licht, L., Phytoextraction and accumulation of boron and selenium by poplar (*Populus*) hybrid coles. *Int. J. Phytochem.*, 1999, 1, 81–96.

81. Huang, J.W. and Cunningham, S.D., Lead phytoextraction: species variation in lead uptake and translocation. *New Phytol.*, 1996, 134, 75–84.

82. Terry, N. and Banuelos, G. (Eds.), *Phytoremediation of Contaminated Soil and Water*. Lewis Publishers, Inc., Chelsea, MI, 1999, 408.

83. Heijden, M.G.A. and Sanders, I.R. (Eds.), *Mycorrhizal Ecology*. Springer–Verlag, Heidelberg, 2002, 469.

84. Banuelos, G.S. and Meek, D.W., Selenium accumulation in selected vegetables. *J. Plant Nutr.*, 1989, 12(10), 1255–1272.

85. Bañuelos, G.S., Mead, R.R., and Hoffman, G.J., Accumulation of selenium in wild mustard irrigated with agricultural effluent. *Agric. Ecosyst. Environ.* 1993, 43, 119–126.

86. Dushenkov, S., Vasudev, D., Kapulnik, Y., Gleba, D., Fleisher, D., Ting, K.C., and Ensley, B., Phytoremediation: a novel approach to an old problem. In *Global Environmental Biotechnology*, Wise, D.L. (Ed.). Elsevier Science B.V., Amsterdam, 1997, 563–572.

87. Dushenkov, S., Vasudev, D. Kapulnik, Y., Gleba, D., Fleisher, D., Ting K.C., and Ensley. B., Removal of uranium from water using terrestrial plants. *Environ. Sci. Technol.*, 1997, 31(12), 3468–3474.

88. Dushenkov, V., Kumar, P.B.A.N., Motto, H., and Raskin, I., Rhizofiltration: the use of plants to remove heavy metals from aqueous streams. *Environ. Sci. Technol.*, 1995, 29, 1239–1245.

89. Azaizeh, H.A., Gowthaman, S., and Terry. N., Microbial selenium volatilization in rhizosphere and bulk soils from a constructed wetland. *J. Environ. Qual.*, 1997, 26, 666–672.

90. Carbonell, A.A., Aarabi, M.A., Delaune, R.D., Gambrell, R.P., and Patrick, WH., Jr., Bioavailability and uptake of arsenic by wetland vegetation: effects on plant growth and nutrition. *J. Environ. Sci. Health*, 1998, A33(1), 45–66.

91. Gleba, D., Borisjuk, M.V., Borisjuk, L.G., Kneer, R., Poulev, A., Skarzhinskaya, M., Dushenkov, S., Logendra, S., Gleba, Y.Y., and Raskin, I., Use of plant roots for phytoremediation and molecular farming. *Proc. Natl. Acad. Sci. USA*, 5973–5977, 1999, 96.

92. Comis, D., Green remediation: using plants to clean the soil. *J. Soil Water Conserv.*, 1996, 51, 184–187.

26 Trace Elements in Plants and Soils of Abandoned Mines in Portugal: Significance for Phytomanagement and Biogeochemical Prospecting

M.N.V. Prasad, J. Pratas, and H. Freitas

CONTENTS

26.1 INTRODUCTION

Plant communities that can grow on mine spoils and are capable of accumulating metals in their parts have immense scope for mine reclamation and for biogeochemical exploration. The chemical composition of these plants is usually correlated with the mineral composition of the soil in order to fetch biogeochemical prospecting [1]. Thus, the plant community established on a mine spoil can be useful in mineral exploration as well as in remediation to minimize the impacts of mining [2–4]. Therefore, considering the diversity of plants and their responses in metal-contaminated sites having different levels of metals, it is important to study the composition of the plant communities of abandoned mines or mine spoils, which would serve as a basic approach for initiating steps for mine reclamation and remediation.

Authentic information about plant communities that can grow on metal-contaminated soils is required in large volume to assess this approach for mine reclamation or remediation and also for biogeochemical exploration [5–8]. This study attempts to evaluate plant species and communities that have established on metal laden soils. Plants and soils were analyzed for total Ag, Bi, Cu, Pb, W, and Zn. Patterns of metal accumulation in plant communities were analyzed, indicating their possible application in restoration of mine spoils and tailings. Plants growing in abandoned mine sites usually indicate the mineral composition of the soil. These plant species are tolerant to metals and are able to accumulate or exclude toxic metals. Trace metal-accumulating plants are thus of immense use for biogeochemical prospecting and are invaluable[5,8,9].

Therefore, metal-tolerant perennials and plants with high biomass and bioproductivity are of immense use in phytostabilization and mine restoration [10]. Mining activities leave behind vast

507

amounts of refuse and tailings, which are often very unstable and will become sources of environmental pollution. The direct effects will be loss of cultivated, forest, or grazing land, and the overall loss of production [11]. Air, soil, and water pollution and siltation of rivers are the indirect effects. Both these effects would eventually lead to the loss of biodiversity, amenity, and economic wealth [12].

Reclamation of mine waste with plant communities that are tolerant can fulfill the objectives of stabilization, pollution control, visual improvement, and removal of threats to mankind. The constraints related to plant establishment and amendment of the physical and chemical properties of the toxic metal-mined soils depend upon the choice of appropriate plant species that will be able to grow in such a hostile environment. Thus, the plant community tolerant to trace elements plays a major role in remediation of degraded mine soils. Plants tolerant to toxic levels of trace elements respond by exclusion, indication, or accumulation of metals [13]. Thus, more information about plant communities that can grow on metal-contaminated soils is essential to determine their potential for mine reclamation and biogeochemical exploration [5–8,14,15].

26.2 SITE DESCRIPTION

The study area comprises two major segments, namely, the Palão and Pinheiro mines, which belong to the same geological setting. The Palão mines are located about 2 km north of Penamacor (center east of Portugal), 40° 10′ 50″ N; 7° 11′ 20″ W. The mineralization is in the Schist–Graywake complex and is characterized by a vein network of the stockwork type. These thin veins are made of sphalerite, galena, and traces of other sulphides, within a carbonate gangue, barite, and quartz. The principal mining area corresponds to an open pit mine, currently filled with water, slightly elliptic, and more or less 100×200 m; the vast majority of the veins are less than 3 cm thick. Very scarce are the Veins larger than 50 cm thick are very scarce and are oriented parallel to the major exploitation axis. From the paragenetic point of view, these mines are considered similar to the Pb–Zn vein mineralizations, with identical paragenesis; some of these date from the upper Cretaceous period.

The Pinheiro segment has the same Schist–Graywake complex, located about 3 km east of Penamacor village, 40° 09′ 05″ N; 7° 07′ 55″ W. It is a vein type mineralization composed of arsenopyrite, ferriferous sphalerite, chalcopyrite, pyrite, and wolframite. By its geological and paragenetic emplacement, the mineralization is part of the hercinic mineralization peribatolitic group with W–Sn of the Central Iberian geotectonic zone. The plant community established on a distinct metal vein mineralization (or distinct paragenesis) in a common geological complex of abandoned mines in central Portugal, namely, Palão, Pinheiro, and Mata da Rainha, which were chosen as study material. Plant samples (aerial parts, leaf, needles, fruit, flower, phyllode, root, seed, twigs, whole plant) belonging to 95 species representing 37 families were collected from the mine sites and analyzed for total As, Bi, Cu, Pb, W, and Zn content.

26.3 METHODOLOGY

Several transects were made in the mineralized zones and mine tailings to collect plant samples. On these transects, for every 40 m, soils and plants were collected in a 2-m circle (sampling point) following the methodology suggested by Kovalevskii [7] and Brooks [9].

Mineralized area and mine tailings were chosen as sites for collection of materials. To avoid temporal variations of the different heavy metals by plant species, collection was completed in short duration, i.e., 2 days. Soil samples are always collected from B horizon (not always in the same depth because they are mountain soils with little thickness but variable) to minimize the influence of the organic matter present in superficial horizons and consists of an homogenization of the collection of about 4 points. The soil samples were dried at 80°C and passed through a 100-

mesh sieve. Soil pH was determined in a soil/water suspension with a pH meter. Plant samples were cleaned in abundant freshwater, rinsed with deionized water, and air-dried at room temperature for several days and after crushing. Plants were identified with the help of local herbarium and floras [16,17]. Analytical methods included colorimetry for W [18]; atomic absorption spectrophotometry (Perkin Elmer 2380) for Ag, Co, Cr, Cu, Fe, Mn, Ni, Pb, and Zn; and hydride generation (HGS) for As and Sb [19–21].

Plant samples were cleaned in abundant freshwater and rinsed with distilled water. Dry weights were obtained after oven drying at 60°C for 2 days. Plant samples were analyzed for total Ag, As, Co, Cr, Cu, Ni, Pb, and Zn following the methods stated for soil.

26.4 RESULTS AND DISCUSSION

Six metals were analyzed in all soil samples (Table 26.1). High levels of Pb, Zn, As, and W were recorded in the soils. The Pb in the soil was very high, reaching a maximum of 2497.2 ppm in the collected samples. The maximum Zn content in soils was 3599.6 ppm, a level that can be extremely toxic for plants. Maximum concentration of As in soils was 469.5 ppm. Soil content of W was also high, reaching a maximum of 230.7 ppm. The maximum concentrations of Cu and Bi in soils were 356.7 ppm and 2.74 ppm, respectively.

A total of 95 plant species from 37 families were investigated for their ability to accumulate metals in these mine complexes (Table 26.1). Analysis of metals performed in plant material (aerial parts, leaf, needles, fruit, flower, phyllode, root, seed, and twig) indicated that a variable degree of accumulation of metals was present in different plant parts (Table 26.1). The most representative family is Fabaceae with 11 species, followed by Asteraceae with 9 species, Cistaceae with 7, Lamiaceae with 6, Caryophyllaceae with 5, Poaceae with 5, Rosaceae with 5, Scrophulariaceae with 4, Polygonaceae with 4, Fagaceae with 4, Cyperaceae with 3, and the rest with 1 or 2.

Lead concentration in plants was rather high for some species (Figure 26.1), varying from 0.71 ppm in leaves of *Salix atrocinera* and 1.03 ppm in leaves of *Eucalyptus globulus* to 157.8 ppm in roots and 97 ppm in the aerial parts of *Typha dominguensis*. In the aerial parts of another Poacea, *Briza maxima*, lead content of 97.02 ppm was recorded. *Myriophyllum alterniflorum* also showed high content of Pb in the whole plant, with 93.76 ppm (Figure 26.1). This aquatic species also accumulated Zn in the above-ground tissues, demonstrating its capacity to take up toxic metals from soil (Table 26.1). In *Andryala integrifolia*, a frequent species in other mining areas (unpublished data), a content of 77.02 ppm of lead in the above-ground tissue was recorded. This frequent finding of common species in diverse mining areas with distinct paragenesis is worth pointing out because metals and other environmental stress factors play a critical role. Ecological convergence seems clear in this particularly hard environment.

Other species that showed high content of Pb (Figure 26.1) were *Ononis cintrana* (65.69 ppm), *Lotus corniculatus* L. var. *corniculatus* (62.6), *Elatine macropoda* (62.5), *Daphne gnidium* (59.47), *Digitalis thapsi* (55.89), *Dittrichia viscosa* (37.08), *Pterospartum tridentatum* (38.6), *Ludwigia palustris* (35.1), and *Phlomis lychnitis* (29.8).

With respect to As, the aquatic species *Elatine macropoda* showed a remarkable content of As in the above-ground tissues: 531.4 ppm (Figure 26.2). This species also accumulated high levels of Pb (62.5 ppm), Co (127.8 ppm), and Zn (2258.5 ppm) (Table 26.1). The aquatics *Typha dominguensis* and *Myriophyllum alterniflorum*, with 16.97 and 10 ppm of As, respectively, again were among the plants accumulating more As. These two species also showed high content of zinc in the whole plant: 389.7 ppm in *T. dominguensis* and 507.83 ppm in *M. alterniflorum*. Other species that showed high content of As (Figure 26.2) were *Adenocarpus complicatus* (12.9 ppm), *Asparagus acutifolius* (19.62), *Galium palustre* (11.98), and *Lactuca viminea* (11.64).

TABLE 26.1
Trace Metal Composition of Soil and Plant Parts[a] Collected from the Abandoned Mine Complex

Plants/family Fabaceae	Plant part	As–S	As–P	Bi–S	Bi–P	Cu–S	Cu–P	Pb–S	Pb–P	W–S	W–P	Zn–S	Zn–P
Genista triacanthos Brot.	A	343.8	1.26	2.74	0.003	82.9	2.9	374.1	6.8	<0.1	<0.01	3599.6	39.4
Cytisus striatus (Hill.) Rothm.	A	127.20	0.49	0.82	0.007	62.21	7.74	585.01	3.69	24.77	0.011	497.59	96.45
C. multiflorus (L'Hér.) Sweet	A	79.35	0.27	0.72	0.015	40.5	9.34	288.23	3.54	1.19	0.00	165.13	105.76
C. grandiflorus (Brot.) DC.	A	81.18	0.37	0.53	0.072	49.23	11.31	797.06	3.24	1.12	0	383.91	121.19
Lotus corniculatus L. var. *corniculatus*	A	51.2	0.45	0.64	0.008	28.9	6.5	1631.5	62.6	<0.1	<0.01	589.8	139.3
L. pedunculatus car	A	161.95	0.03	0.48	0.007	36.72	6.33	95.23	5.09	7.26	0.003	102.14	35.87
Hymenocarpos lotoides (L.) Vis.	A	261.2	2.16	2.55	0.009	57.7	9.3	102.5	7.2	142.1	<0.01	145.3	111.7
Ononis cintrana Brot.	A	49.93	0.1	0.58	0.03	33.59	12.14	1361.2	65.69	0.075	0	649.61	397.99
Acacia dealbata Link	L	123.25	0.76	1.26	0.023	44.88	5.88	177.95	7.67	5.50	0.004	983.89	35.85
	T	123.25	0.27	1.26	0.014	44.88	2.93	177.97	3.01	5.50	0.005	983.89	39.49
	S	69.3	0.06	1.56	0.018	30.5	2.3	141.9	0.7	5.5	<0.01	99.3	16.5
Adenocarpus complicatus (L.) J. Gay	L	192.7	12.85	0.81	0.024	139.47	9.17	101.7	7.716	79.84	0.116	160.62	132.68
	T	192.7	0.15	0.81	0.031	139.47	3.84	101.78	1.68	79.84	0.06	160.62	28.75
Pterospartum tridentatum (L.) Willk	A	343.8	0.00	2.74	0.39	82.9	1.5	374.1	38.6	<0.1	<0.01	3599.6	284.0
Asteraceae													
Helichrysum stoechas (L.) Moench.	A	157.8	1.50	0.97	0.05	69.87	10.13	458.68	12.18	31.80	0.04	813.67	120.14
	L	149.84	0.61	0.82	0.06	72.68	10.94	349.85	8.14	38.17	0.046	557.45	120.45
	T	149.84	1.63	0.82	0.06	72.68	10.04	349.9	9.54	38.17	0.046	557.44	89.72
Andryala integrifolia L.	A	265.66	1.65	1.03	0.002	122.80	10.74	337.89	77.02	93.57	0.032	301.24	88.98
Lactuca viminea (L.) J. & C.Presl	A	338	11.64	1.20	0.41	150.97	9.43	86.84	12.48	124.71	0	165.14	59.21
Carlina corymbosa L. subsp. *corymbosa*	A	163.08	0.18	1.10	0.018	41.94	8.24	521.71	11.47	20.5	0.020	769.27	84.656
Tolpis barbata (L.) Gaertner	A	272.25	0.60	1.61		48.11	13.66	99.01	14.50	71.69	0	125.26	170.76
Crepis capillaris (L.) Wallr.	A	263.6	0.39	0.98	0.06	121.94	9.30	88.87	11.48	96.84	0	148.61	114.48

Species													
Chondrilla juncea L.	A	71.7	0.03	1.98	<0.001	130.6	12.3	577.4	6.0	0.1	<0.01	118.4	59.6
Dittrichia viscosa (L.) W. Greuter	A	7.0	1.27	0.16	0.07	32.3	18.6	32.2	11.1	1.9	<0.01	47.6	104.6
	L	104.83	0.74	0.92	0.033	43.54	18.99	436.48	37.08	0.50	0.005	1150.0	179.38
	T	137.56	0.63	1.16	0.004	47.28	6.47	571.2	5.65	0.049	0.002	1517.5	53.40
Phagnalon saxatile (L.) Cass.	A	469.5	7.60	0.40	<0.001	356.7	9.1	62.5	4.6	230.7	0.13	244.9	117.9
Fagaceae													
Quercus coccifera L.	L	9.63	0.21	0.21	0.01	28.82	2.61	33.64	2.61	1.95	0.00	70.03	20.40
	T	9.63	0.11	0.21	0.012	28.81	3.36	33.64	3.19	1.95	0	70.03	15.21
Q. ilex L. subsp. *ballota* (Desf.) Samp.	L	338	0.31	1.20	0.005	150.97	4.785	86.84	2.44	124.71	0.031	165.14	65.83
	T	338	0.32	1.20	0.005	150.97	4.68	86.84	3.62	124.71	0.11	165.14	38.41
Q. pyrenaica Willd.	L	38.32	0.31	0.43	0.005	26.95	14.85	287.89	2.99	0.45	0	322.42	85.58
	T	38.32	0.44	0.43	0.003	26.95	6.57	287.89	3.86	0.45	0	322.42	70.04
Q. suber L.	L	160.93	0.38	0.74	0.019	80.39	5.057	581.02	3.81	42.49	0.021	288.36	57.65
	T	160.93	0.26	0.74	0.06	80.39	6.21	581.02	4.45	42.49	7.32	288.36	65.64
Ericaceae													
Erica australis L.	A	8.73	0.08	0.17	0.007	33.25	3.071	30.74	1.73	3.03	0.001	52.58	7.12
Calluna vulgaris (L.) Hull.	A	10.68	0.36	0.21	0.017	30.78	3.39	33.22	1.29	2.77	0.007	51.74	10.66
Erica umbellata L.	A	11.02	0.23	0.22	0.020	30.64	3.50	33.30	3.43	2.85	0.002	52.11	10.88
E. lusitanica Rudolphi	T	9.3	0.42	0.15	0.007	33.3	38.3	40.8	4.9	1.3	0.03	41.6	125.9
E. lusitanica Rudolphi	A	39.31	0.25	0.85	0.008	31.92	23.45	91.33	10.28	3.39	0.016	70.44	102.6
Cistaceae													
Cistus psilosepalus Sweet	L	116.7	0.5	0.9	0.0	48.0	9.5	623.3	6.9	20.1	0.1	290.1	189.5
	T	116.70	0.42	0.92	0.018	47.98	3.072	623.32	3.79	20.06	0.026	290.09	98.65
Xolantha guttata (L.) Raf.	A	338	1.48	1.20	0.07	150.97	8.99	86.84	10.51	124.71	0.041	165.14	80.98
Cistus ladanifer L.	L	60.34	0.796	0.39	0.02	47.89	5.14	208.85	7.64	13.18	0.03	313.6	134.15
	F	16.1	0.14	0.14	0.009	22.8	5.1	133.9	2.3	<0.1	<0.01	243.9	41.0
	FR	10.78	0.43	0.21	0.039	31.26	5.03	33.59	2.58	2.75	0.005	54.28	32.59
	T	60.34	0.46	0.39	0.018	47.89	2.85	208.85	11.64	13.18	0.033	313.6	72.77
C. salvifolius L.	A	7.0	0.65	0.16	0.027	32.3	3.1	32.2	2.6	1.9	<0.01	47.6	28.9
	L	71.41	0.48	0.43	0.019	30.33	7.98	483.62	10.58	0.98	0.00	229.98	198.00
	T	71.41	0.44	0.43	0.016	30.33	4.066	483.62	17.13	0.98	0.003	229.98	84.21
C. crispus L.	A	9.819	0.42	0.17	0.01	34.54	2.98	34.21	6.26	2.63	0.012	43.13	30.60
	L	109.56	0.93	0.71	0.017	65.02	8.84	550.17	14.16	26.12	0.042	504.13	222.06

TABLE 26.1

Trace Metal Composition of Soil and Plant Parts[a] Collected from the Abandoned Mine Complex (continued)

Plants/family	Plant part	As-S	As-P	Bi-S	Bi-P	Cu-S	Cu-P	Pb-S	Pb-P	W-S	W-P	Zn-S	Zn-P
Cistus monspeliensis L.	L	74.91	0.71	0.46	0	23.44	8.90	154.22	7.09	0	0	133.50	104.11
	T	74.91	0.84	0.46	0.004	23.44	3.17	154.21	5.30	0	0	133.50	52.51
Halimium ocymoides (Lam.) Willk	T	109.56	0.72	0.70	0.011	65.02	4.75	550.17	24.61	26.12	0.012	504.13	105.89
	A	10.36	0.22	0.22	0.005	30.74	3.31	32.84	3.39	2.82	0.006	51.97	30.87
Rosaceae													
Sanguisorba verrucosa (Link) Ces.	A	92.06	1.04	0.78	0.016	31.15	6.54	475.39	8.18	28.41	0.138	280.38	65.70
Pyrus bourgaeana Decne	L	20.2	0.15	0.21	0.027	28.7	22.9	179.8	6.4	<0.1	<0.01	306.4	121.2
	T	20.2	0.09	0.21	0.021	28.7	8.2	179.8	6.3	<0.1	<0.01	306.4	61.1
Rubus ulmifolius Schott	L	110.51	0.32	0.65	0.015	60.29	8.22	435.13	6.03	27.043	0.008	244.91	56.68
Rosa canina L.	T	110.52	0.30	0.65	0.014	60.29	6.52	435.13	3.79	27.043	0.002	244.91	59.10
	L	266.23	0.10	0.44	0.01	140.8	5.13	100.24	4.78	77.3	0	156.2	35.36
	T	266.23	0.048	0.44	0.022	140.85	7.25	100.24	8.18	77.35	0.011	156.22	40.47
Crataegus monogyna Jacq.	L	119.52	0.59	0.64	0.003	60.37	10.02	351.62	7.40	34.25	0.009	296.66	65.71
	T	119.52	0.39	0.64	0.003	60.37	6.02	351.69	4.20	34.25	0.010	296.66	93.51
Hypolepidaceae													
Pteridium aquilinum (L.) Kuhn subsp. *aquilinum*	A	226.90	0.19	1.16	0.002	102.46	7.045	406.78	5.52	75.91	0.003	236.92	285.27
Polygonaceae													
Rumex crispus L.	A	40.6	5.51	0.28	0.009	34.9	4.7	94.9	4.0	13.2	0.02	99.0	25.9
R. acetosella L. subsp. *angiocarpus* Murb.	A	45.91	0.43	0.46	0.054	31.88	6.30	863.20	15.92	6.613	0	344.42	123.7
R. induratus Boiss. & Reuter	A	213.26	1.13	0.80	0.015	102.63	6.19	86.29	14.04	75.86	0.16	138.24	35.05
Polygonum minus Hudson	A	51.2	0.92	0.64	<0.001	28.9	4.9	1631.5	45.8	<0.1	<0.01	589.8	313.8
Caryophyllaceae													
Spergularia purpurea (Pers.) G. Don. fil.	A	135.5	0.45	1.02	0.010	88.35	12.85	2497.2	20.88	2.80	0	846.01	1057.7
Petrorhagia nanteuillii (Burnat) P.W. Ball & Heywood	A	376.4	3.35	0.54	0.092	197.63	2.97	79.02	6.96	116.02	0.028	175.08	27.75

Species														
Silene scabriflora Brot. subsp. *scabriflora*	A	26.19	0.45	0.31	0.02	26.14	6.37	135.86	5.74	0	0	233.49	138.26
Ortegia hispanica Loefl. ex. L.	A	261.2	0.43	2.55	0.003	57.7	13.7	102.5	4.4	142.1	<0.01	145.3	168.6
Saponaria officinalis L.	A	39.3	5.59	0.48	0.009	31.2	5.1	100.9	7.3	3.3	0.01	137.6	41.2
Boraginaceae													
Echium plantagineum L.	A	20.1	0.31	0.21	0.014	20.6	8.8	248.5	6.1	<0.1	<0.01	243.5	158.6
Athyriaceae													
Athyrium felix-femina (L.) Roth.	A	69.3	0.14	1.56	0.002	30.5	7.4	141.9	16.2	5.5	<0.01	99.3	31.7
Aristolochiaceae													
Aristolochia longa L.	A	32.45	0.33	0.39	0.007	26.82	11.64	509.76	10.04	0	0	322.5	204.90
Scrophulariaceae													
Digitalis purpurea L. subsp. *purpurea*	A	42.97	0.94	0.74	0.024	27.45	9.13	163.79	5.32	2.93	0.016	160.16	154.50
D. thapsi L.	A	338	0.49	1.20	0.037	150.97	6.717	86.84	55.89	124.71	0.037	165.14	172.19
Linaria aeruginea (Gouan) Cav.	A	20.24	0.62	0.23	0.055	26.02	6.38	394.19	10.44	0.90	0	243.69	119.71
Anarrhinum bellidifolium (L.) Willd.	A	338	0.90	1.20	0.01	150.97	5.72	86.84	3.93	124.71	0.013	165.14	39.88
Pinaceae													
Pinus pinaster Aiton	N	115.4	1.6	0.6	0.1	69.8	2.5	447.6	2.9	32.2	0.0	524.6	56.7
	T	95.6	0.1	0.5	0.0	63.0	3.4	370.8	4.0	26.8	0.0	436.6	32.3
Apiaceae													
Oenanthe crocata L.	A	131.85	0.14	0.98	0.039	68.03	25.41	255.95	15.47	4.87	0.002	107.57	69.68
Eryngium campestre L.	A	48.7	0.03	0.53	0.004	38.3	4.1	1091.0	4.9	0.1	<0.01	709.5	17.9
Oleaceae													
Olea europaea L.	L	131.76	0.6	0.50	0.023	92.18	5.59	171.16	4.29	46.14	0.02	338.37	41.00
	T	131.76	0.24	0.50	0.012	92.18	3.92	171.16	4.68	46.14	0.003	338.37	48.72
Fraxinus angustifolia Vahl	L	115.6	0.05	1.02	0.00	33.63	7.81	118.57	6.73	6.36	0	100.73	116.87
	T	115.62	0.13	1.02	0.009	33.63	7.20	118.57	7.85	6.36	0.002	100.73	117.63
Valerianceae													
Centranthus calcitrapae (L.) Dufresne	A	338.0	0.3	1.2	0.0	151.0	10.1	86.8	12.8	124.7	0.0	165.1	34.1
Thymelaeaceae													
Daphne gnidium L.	L	134.51	0.50	0.83	0.517	80.53	11.54	668.86	59.47	34.77	0.039	311.86	261.20

TABLE 26.1
Trace Metal Composition of Soil and Plant Parts[a] Collected from the Abandoned Mine Complex (continued)

Plants/family	Plant part	As–S	As–P	Bi–S	Bi–P	Cu–S	Cu–P	Pb–S	Pb–P	W–S	W–P	Zn–S	Zn–P
Dioscoreaceae													
Tamus communis L.	T	134.51	0.36	0.83	0.22	80.53	13.64	668.86	7.38	34.77	0.005	311.86	149.64
	A	29.11	0.49	0.35	0.052	24.88	10.20	113.87	7.56	0	0	197.02	110.3
Lamiaceae													
Mentha pulegium L.	A	43.2	0.52	0.27	0.016	31.08	9.79	118.82	6.11	6.61	0	108.75	34.95
Mentha suaveolens Ehrh.	A	83.08	0.39	0.57	0.007	30.52	11.30	137.42	7.10	3.89	0.024	133.87	324.50
Origanum virens Hoffmanns & Link	A	16.1	0.61	0.14	0.005	22.8	10.3	133.9	5.2	<0.1	<0.01	243.9	62.7
Phlomis lychnitis L.	A	469.5	0.32	0.40	0.098	356.7	14.0	62.5	29.8	230.7	0.37	244.9	710.3
Clinopodium vulgare L.	A	26.12	0.54	0.30	0.036	26.13	10.82	135.86	14.66	0	0	233.49	263.72
Lavandula stoechas L. subsp. *pedunculata* (Miller) Samp. ex. Rozeira	A	159.00	0.73	0.89	0.00	74.70	10.60	704.98	11.97	31.88	0.07	644.8	347.19
Teucrium scorodonia L.	A	124.27	0.18	0.82	0.008	32.59	9.25	139.09	4.35	2.25	0	170.34	112.35
	E	9.99	0.30	0.20	0.018	31.65	4.79	33.19	3.68	2.82	0.034	47.640	45.95
	FL	146.77	0.26	0.82	0.015	68.95	17.47	650.75	9.65	29.43	0.059	595.2	206.98
	T	146.77	0.39	0.82	0.014	68.95	5.47	650.75	8.85	29.43	0.01	595.24	329.4
L. stoechas L. subsp. *stoechas*	A	9.99	0.24	0.20	0.044	31.65	3.07	33.19	2.20	2.827	0.006	47.640	36.72
Origanum virens Hoffmanns & Link	A	16.1	0.61	0.14	0.005	22.8	10.3	133.9	5.2	<0.1	<0.01	243.9	62.7
Cyperaceae													
Eleocharis multicaulis (Sm.) Desv.	A	123.26	0.14	0.41	0.004	33.58	2.99	111.05	3.79	4.84	0.038	107.59	25.46
Cyperus fuscus L.	A	102.27	0.07	0.42	0.010	32.98	4.66	108.58	3.77	4.47	0	115.10	68.70
Scirpus holoschoenus L.	A	93.14	0.30	0.79	0.002	58.99	3.55	1207.1	5.13	3.45	0.006	1020.0	123.18
Juncaceae													
Juncus inflexus L.	A	40.6	1.53	0.28	0.013	34.9	7.0	94.9	6.0	13.2	0.02	99.0	74.7
J. heterophyllus Dufour	A	39.3	1.64	0.48	0.011	31.2	14.1	100.9	5.2	3.3	<0.01	137.6	399.3
Liliaceae													
Ruscus aculeatus L.	P	42.3	0.44	0.57	0.004	26.9	3.0	93.8	3.3	<0.1	<0.01	150.1	134.9
Asparagus acutifolius L.	A	376.4	19.62	0.54	0.05	197.63	3.14	79.02	10.32	116.02	0.038	175.05	37.03

Taxon													
Rutaceae													
Ruta montana L.	A	469.5	0.05	0.40	0.012	356.7	3.6	62.5	5.4	230.7	0.02	244.9	13.6
Haloragaceae													
Myriophyllum alterniflorum DC.	W	46.91	9.94	0.51	0.020	29.06	9.309	876.66	93.76	0.83	0.06	358.95	507.83
Lythraceae													
Lythrum hyssopifolia L.	A	39.95	0.56	0.38	0.023	33.03	11.29	97.93	9.29	8.28	0.045	118.32	879.87
Callitrichaceae													
Callitriche stagnalis (L.) Scop	A	39.3	531.4	0.48	<0.001	31.2	127.8	100.9	62.5	3.3	0.42	137.6	2258.5
Rubiaceae													
Galium palustre L.	A	39.3	11.98	0.48	0.140	31.2	60.7	100.9	12.8	3.3	<0.01	137.6	1726.0
Typhaceae													
Typha dominguensis (Pers.) Steudel	L	51.2	0.27	0.64	0.005	28.9	3.2	1631.5	7.5	<0.1	<0.01	589.8	82.0
	F	51.2	0.12	0.64	0.011	28.9	2.3	1631.5	8.2	<0.1	<0.01	589.8	66.8
	AL	51.2	0.20	0.64	0.013	28.9	5.1	1631.5	4.0	<0.1	<0.01	589.8	39.3
	R	51.2	16.97	0.64	0.021	28.9	7.9	1631.5	157.8	<0.1	0.15	589.8	389.7
Briza maxima L.	A	272.25	0.02	1.61	0.19	48.11	6.46	99.01	97.02	71.69	0.06	125.26	36.21
Arrhenatherum elatius (L.) J. & C. Presl.	A	272.25	0.22	1.61	0.035	48.11	3.71	99.01	8.37	71.69	0	125.26	29.88
Avena sterilis L.	A	272.25	0.20	1.61	0.086	48.11	5.13	99.01	5.60	71.69	0.39	125.26	40.96
Agrostis curtisii Kerguélen	A	10.49	0.61	0.22	0.007	30.30	1.994	39.18	2.03	1.66	0.009	81.12	20.31
Onagraceae													
Ludwigia palustris (L.) Elliott	A	322.6	1.9	1.2	0.1	69.8	53.5	196.5	35.1	4.7	0.0	242.9	1131.0
Dipsacaceae													
Pterocephalus diandrus Lag.	A	261.2	1.08	2.55	0.003	57.7	8.6	102.5	3.9	142.1	<0.01	145.3	138.4
Ranunculaceae													
Ranunculus peltatus Schrank	A	45.9	3.07	0.26	0.016	27.3	5.0	142.7	13.1	<0.1	0.04	118.5	29.0
Myrtaceae													
Eucalyptus globulus Labill	**A**	**42.3**	**0.40**	**0.57**	**0.005**	**26.9**	**1.5**	**93.8**	**1.9**	**<0.1**	**<0.01**	**150.1**	**67.1**
	L	6.27	0.023	0.15	0.013	31.67	2.69	27.25	1.03	3.35	0.02	73.40	2.92

TABLE 26.1
Trace Metal Composition of Soil and Plant Parts[a] Collected from the Abandoned Mine Complex (continued)

Plants/family	Plant part	As–S	As–P	Bi–S	Bi–P	Cu–S	Cu–P	Pb–S	Pb–P	W–S	W–P	Zn–S	Zn–P
Betulaceae													
Alnus glutinosa (L.) Gaertner	L	54.95	0.014	0.92	0.013	32.70	11.90	118.42	6.006	9.34	0.036	99.17	64.71
	T	54.95	0.00	0.92	0.006	32.70	5.55	118.42	1.76	9.34		99.17	38.07
Caprifoliaceae													
Lonicera periclymenum L.	L	20.2	0.96	0.21	0.004	28.7	7.8	179.8	4.8	<0.1	<0.01	306.4	290.5
	T	20.2	0.67	0.21	<0.001	28.7	5.8	179.8	4.2	<0.1	<0.01	306.4	290.4
Salicaceae													
Salix salvifolia Brot.	L	103.60	0.65	0.51	0.014	33.38	5.14	480.72	2.72	4.47	0.005	232.94	117.50
	T	103.60	0.32	0.51	0.0120	33.38	3.87	480.72	1.83	4.47	0.001	232.94	122.25
S. atrocinerea Brot.	L	10.68	0.24	0.21	0.014	30.78	2.61	33.22	0.71	2.77	0.009	51.74	22.71
	T	40.6	0.65	0.28	0.007	34.9	6.2	94.9	3.4	13.2	0.03	99.0	328.5

[a] mg kg[-1].

Notes: A = aerial parts; F = fruit; FL = flower; L = leaf; N = needles; P = phyllaode; R = root; S = seed; T = twig; W = whole plant; P = suffixing element in plant; S = suffixing element in soil.

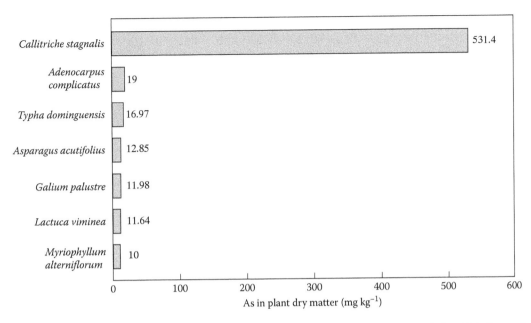

FIGURE 26.1 Accumulation of lead (in mg kg⁻¹ dry weight) in plant species from the Palão, Pinheiro, and Mata da Rainha mine complex.

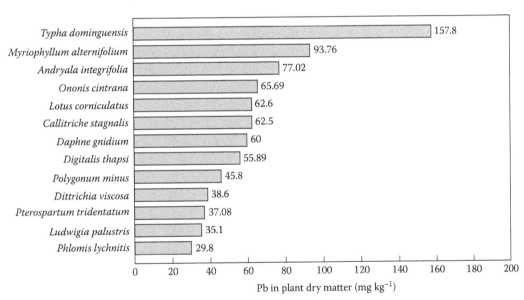

FIGURE 26.2 Accumulation of arsenic (in mg kg⁻¹ dry weight) in plant species of the Palão, Pinheiro, and Mata da Rainha mine complex.

26.5 CONCLUSIONS

It is worth pointing out that the species from the most representative families are typical Mediterranean species, which are well adapted to environmental stress conditions — namely, water stress and very poor and tiny soils, and are represented in the investigated samples. Tolerance to high levels of metals in the soil is the important aspect for application in phytoremediation. It is also advantageous when the plant community can tolerate other environmental stress factors as mentioned earlier. The investigated plant species are present in metal-enriched soils and therefore manage to control the uptake of heavy metals and also cope with very poor nutritional substrates. This study gave important insights for remediation of contaminated soils. The investigated plant community was established on abandoned mines.

The plant diversity observed in this study includes a number of functional groups of plants (semiaquatic, legumes, grasses, and tree species) (Table 26.1) in response to single stress factors or a combination of them. The reclamation of such mines may include metal accumulators belonging to these functional groups to initiate plant successions in such a hostile environment, eventually leading to restoration of the ecosystem [24,25]. The use of legumes may enrich the soil nutrient content and the combined used of perennials and annuals can provide different inputs in terms of organic matter and nutrient recycling, thus contributing in distinct ways to the development of the soil.

Establishing plant communities with tolerance to toxic metals is an approach that would successfully reclaim mines or minimize the hazards of environment [26–31]. The success of any phytoremediation technique depends upon the identification of suitable plant species that hyperaccumulate trace metals and produce a large amount of biomass, using established crop production and management practices. Restoration of abandoned mines by phytostabilization, soil ammendments, and rhizosphere biotechnology facilitates phytoremediation technology and is supported by promising research findings.

In conclusion, studies on plants that accumulate/hyperaccumulate toxic trace elements have significant environmental and biogeochemical implications (Figure 26.3). The emerging phytoremediation technologies aimed at metals in the environment would derive great sources of knowledge and information on this category of plants for the benefit of man and biosphere.

In order to establish the vegetative groundcover on the surface mined sites, the two most important factors influencing species selection are the soil properties and the tolerance levels of the selected plants. Three categories of plants have been noted to possess the reclamation potential

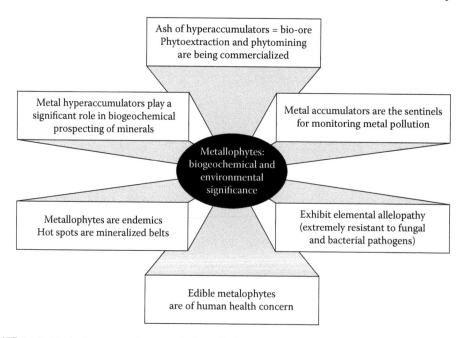

FIGURE 26.3 Plants that accumulate metals (metallophytes) have a major role in mine restoration and also serve as sentinels of mineral exploration (biogeochemical prospecting). The environmental and biogeochemical significance of the metal-tolerant plants is depicted in this figure.

for mine areas. They are grasses, forbs, and trees. Grasses (Poaceae) produce large amounts of biomass and are adapted to initiate regrowth rapidly. Grasses have fibrous root systems that hold soil in place, thereby controlling erosion.

Forbs (herbaceous flowering plants) are generally used in mine revegetation in conjunction with grasses. Forbs usually have broad leaves, flowers, and a branching taproot system. Forbs can be further classified as *legumes* and *non-legumes*. Legumes are especially important for revegetating mined lands because they are capable of using nitrogen from the air to meet their N nutrition requirements, and they can transfer this "fixed" N to other components of the plant/soil system. Nonleguminous forbs are also broad-leaved plants with showy flowers. The establishment of trees and shrubs on mined lands is the final stage of reclamation [22,32,33].

The identified assemblage of plants (Table 26.1) consists of the preceding three categories and is well adapted to the metal toxicity [34]. Phytostabilization of mining sites is a well established environmental compliance using plant species that adapt different strategies, such as metal tolerance, metal accumulation, and metal exclusion [11,23,35–40]. This type of approach requires increasing information about plant communities growing on different kinds of abandoned mine sites in order to take advantage of their potential for phytostabilization of abandoned mines.

The physicochemical properties of the metal-contaminated soils tend to inhibit soil-forming processes and plant growth. In addition to elevated metal concentrations, other adverse factors include absence of topsoil; periodic sheet erosion; drought; surface mobility; compaction; wide temperature fluctuations; absence of soil-forming fine materials; and shortage of essential nutrients [41,42]. The mine-degraded soils usually have low concentrations of important nutrients like K, P, and N [43]. Toxic metals can also adversely affect the number, diversity, and activity of soil organisms, inhibiting soil organic matter decomposition and N mineralization processes. The chemical form of the potential toxic metal, the presence of other chemicals that may aggravate or ameliorate metal toxicity, the prevailing pH, and the poor nutrient status of the trace element-contaminated soil will affect the way in which plants respond to the toxic metal.

Substrate pH affects plant growth mainly through its effect on the solubility of chemicals, including toxic metals and nutrients. Three important factors may affect the bioavailability of metals: (1) soil capacity: pH cation exchange, capacity, organic matter, amount and type of clay, ion interactions, oxides of Fe and Mn, and redox potential; and (2) plant capacity, such as species, cultivar, age of plant part, and 3 plant–metal interaction [44].

Phytoremediation of heavy metal-contaminated soils basically includes phytostabilization and phytoextraction. Some soils are so heavily contaminated that removal of metals using plants would take an unrealistic amount of time. The normal practice is to choose drought-resistant, fast growing crops or fodder that can grow in metal-contaminated and nutrient-deficient soils.

REFERENCES

1. Kabata–Pendias, A., *Trace Elements in Soil and Plants.* CRC Press, Boca Raton, FL, 2001, 432.
2. Ross, S.M., Ed., *Toxic Metals in Soil and Plant Systems.* Wiley, Chichester, U.K., 1994.
3. Siegel, R.R., *Environmental Geochemistry of Potentially Toxic Metals.* Springer–Verlag, Heidelberg, 2002, 218.
4. Sengupta, M., *Environmental Impacts of Mining: Monitoring, Restoration and Control.* Lewis Publishers, Boca Raton, FL, 1993.
5. Badri, M. and Springuel, I., Biogeochemical prospecting in the southeastern desert of Egypt, *J. Arid Environ.,* 28, 257, 1994.
6. Fletcher, W.K., Analytical methods in geochemical prospecting, in *Handbook of Exploration Geochemistry,* Govett, G.J.S., Ed., vol. 1. Elsevier, Amsterdam, 1981, 255, 1.
7. Kovalevskii, A.L., *Biogeochemical Exploration for Mineral Deposits.* Oxonian Press PVT, New Dehli, 1979, 136.

8. McInnes, B.I.A. et al., Biogeochemical exploration for gold in tropical rain forest regions of Papua, New Guinea, *J. Geochem. Exploration,* 57, 227, 1996.

9. Brooks, R.R., *Biological Methods of Prospecting for Minerals.* Wiley-Interscience, New York, 1983, 313.

10. Hooper, D.U. and Vitousek, P.M., The effects of plant composition and diversity on ecosystem processes, *Science,* 277, 1302, 1997.

11. Wong, M.H., Ecological restoration of mine degraded soils, with emphasis on metal contaminated soils, *Chemosphere,* 50, 775, 2003.

12. Bradshaw, A.D., Understanding the fundamentals of succession, in Miles, J. and Walton, D.H. (Eds.), *Primary Succession on Land.* Blackwell, Oxford. 1993.

13. Baker, A.J.M., Accumulators and excluders: strategies in the response of plants to trace metals, *J. Plant Nutr.,* 3, 643, 1981.

14. Pratas, J. et al. Plants growing in abandoned mines of Portugal are useful for biogeochemical exploration of arsenic, antimony, tungsten and mine reclamation, *J. Geochemical Exploration,* 85, 99, 2005.

15. Levinson, A., *Introduction to Exploration Geochemistry.* Applied Publishing Ltd, Calgary, 1994, 612.

16. Franco, J.A., *Nova Flora de Portugal Continente e Açores* Vol. I *Lycopodiaceae — Umbelliferae.* Lisboa, 1971, 555.

17. Franco, J.A., *Nova Flora de Portugal Continente e Açores,* Vol. II *Clethraceae — Compositae.* Lisboa, 659, 1984.

18. Quin, B.F. and Brooks, R.R., The rapid determination of tungsten in soils, rocks and vegetation, *Anal. Chim. Acta,* 58, 301, 1972.

19. Aslin, G.E.M., The determination of arsenic and antimony in geological materials by flameless atomic absorption spectrometry, *J. Geochem. Explor.,* 6, 321, 1976.

20. Van Loon, J.C., *Selected Methods of Trace Metal Analysis: Biological and Environmental Samples.* John Wiley & Sons, London, 1985, 355.

21. Vijan, P.N. et al., A semiautomated method for the determination of arsenic in soil and vegetation by gas-phase sampling and atomic absorption spectrometry, *Anal. Chim. Acta,* 82, 329, 1976.

22. Hooper, D.U. and Vitousek, P.M., Effect of plant composition and diversity on nutrient cycling, *Ecol. Monogr.,* 68, 121, 1998.

23. Tsao, D. (Ed.), *Phytoremediation.* Springer–Verlag, Heidelberg, 206, 2003.

24. Vangronsveld, J. and Cunningham, S.D. *Metal-Contaminated Soils:* in Situ *in Activation and Phytorestoration.* Springer–Verlag, Berlin, 265, 1998.

25. Wenzel, W.W., Lombi, E., and Adriano, D.C., Biogeochemical processes in the rhizosphere: role in phytoremediation of metal-polluted soils, in *Heavy Metal Stress in Plants — from Molecules to Ecosystems,* Prasad, M.N.V. and Hagemeyer, J., Eds. Springer–Verlag, Berlin, 1999, 271.

26. Markert, B., *Plants as Biomonitors: Indicators for Heavy Metals in the Terrestrial Environment.* 1993, VCH Publishers.

27. Palmer, E.F., Warwick, F., and Keller, W., Brassicaceae (Cruciferae) family, plant biotechnology and phytoremediation, *Int. J. Phytoremediation,* 3, 245, 2001.

28. Prasad, M.N.V., *Metals in the Environment — Analysis by Biodiversity.* Marcel Dekker Inc., 2001, New York, 504.

29. Prasad, M.N.V., and K. Strzalka *Physiology and Biochemistry of Metal Toxicity and Tolerance in Plants.* Kluwer Academic Publishers, Dordrecht, 460, 2002.

30. Prasad, M.N.V., Phytoremediation of metals and radionuclides in the environment: the case for natural hyperaccumulators, metal transporters, soil-amending chelators and transgenic plants, in M.N.V. Prasad (Ed.), *Heavy Metal Stress Inlants: from Biomolecules to Ecosystems.* Springer–Verlag, Heidelberg, 2nd ed., 2003, 345–392.

31. Prasad, M.N.V. and Freitas, H., Feasible biotechnological and bioremediation strategies for serpentine soils and mine spoils, *Electron. J. Biotechnol.,* 1999, 2, 35. http://Ejb.Ucv.Cl Or Http://Www.Ejb.Org.

32. Berry, C.R., Growth and heavy metal accumulation in pine seedlings grown with sewage sludge, *J. Environ. Qual.,* 14, 415, 1995.

33. Bending, N.A.D. and Moffat, A.J., Tree performance on mine spoils in the South Wales coalfield, *J. Appl. Ecol.,* 36, 784, 1999.

34. Pulford, I.D. and Watson, C., Phytoremediation of heavy metal-contaminated land by trees — a review, *Environ. Int.,* 29, 529, 2003.

35. Dahmani–Muller, H., van Oort, F., Gélie, B., and Balabane, M., Strategies of heavy metal uptake by three plant species growing near a metal smelter, *Environ. Pollut.*, 109, 231–238, 2000.

36. Shu, W.S. et al., Use of vetiver and other three grasses for revegetation of Pb/Zn mine tailings at Lechang, Guangdong Province: field experiment, in *2nd Int. Vetiver Conf.*, Bangkok, Thailand, 2000.

37. Tang, S.R. et al., The uptake of copper by plants dominantly growing on copper mining spoils along the Yangtze River, the People's Republic of China, *Plant Soil*, 209, 225, 1999.

38. Tang, S.R. et al., Heavy metal uptake by metal-tolerant *Elsholtzia haichowensis* and *Commelina communis* from China, *Commun. Soil Sci. Plant Anal.*, 32, 895, 2001.

39. Yang, Z.Y. et al., Germination, growth and nodulation of *Sesbania rostrata* grown in Pb/Zn mine tailings, *Environ. Manage.*, 21, 617, 1997.

40. Zhang, Z.Q. et al., Soil seed bank as an input of seed source in revegetation of lead/zinc mine tailings, *Restor. Ecol.*, 9, 1, 2001.

41. Wong, J.W.C. et al., Phytostabilization of mimicked cadmium-contaminated soil with lime amendment, in *Proc. 5th Int. Conf. Biogeochem. Trace Elements*, Vienna, 1999.

42. Wong, M.H. et al., Current approaches to managing and remediating metal contaminated soils in China, in *Proc. 5th Int. Conf. Biogeochem. Trace Elements*, Vienna, Austria, 1999.

43. Huenneke, L.F. et al., Effects of soil resources on plant invasion and community structure in Californian serpentine grassland, *Ecology*, 71, 2, 478, 1990.

44. Adriano, D.C., *Trace Elements in Terrestrial Environments: Biogeochemistry Bioavailability and Risks of Metals*. 2nd ed., Springer–Verlag, New York, 866, 2001.

27 Plants That Accumulate and/or Exclude Toxic Trace Elements Play an Important Role in Phytoremediation

M.N.V. Prasad

CONTENTS

27.1 INTRODUCTION

Global industrialization has resulted in the release of large amounts of potentially toxic trace elements into the biosphere — notably, arsenic, cadmium, lead, mercury, and nickel. Cleaning up most of these soils is necessary to minimize the entry of potentially toxic elements into the food chain. Phytoremediation is an environmental cleanup strategy in which green plants are employed to remove or contain environmentally toxic contaminants, or render them harmless [1]. This strategy is rapidly expanding, highlighting the uses of plants beyond food fiber and fuel.

It is estimated that cleanup of toxic metal using conventional technologies will cost at least $200 billion in the U.S. [4]. The sources of metallic contaminants and pollutants are listed in Table 27.1. Lead is one of the most frequently encountered heavy metals in polluted environments. For example, the primary sources of Pb include mining and smelting of metalliferous ores, burning of leaded gasoline, disposal of municipal sewage, and industrial wastes enriched in Pb, as well as use

TABLE 27.1
Sources of Trace Elements

Contaminant	Major source
Aluminum (Al)	Paper coating pretreatment sludge and drinking sludge
Arsenic (As)	Production of pesticides and veterinary pharmaceuticals, and wood preservatives
Cadmium (Cd)	Cd–Ni battery production, pigments for plastics and enamels, fumicides, and electroplating and metal coatings
Chromium (Cr)	Corrosion inhibitor, dyeing and tanning industries, plating operations, alloys, antiseptics, defoliants, and photographic emulsions
Cobalt (Co)	Steel and alloy production, paint and varnish drying agent, and pigment and glass manufacturing
Copper (Cu)	Textile mills, cosmetic manufacturing, and hardboard production sludge
Lead (Pb)	Battery industry, fuel additives, manufacturing of ammunition, caulking compounds, solders, pigments, paints, herbicides, and insecticides
Mercury (Hg)	Electrical apparatus manufacture, electrolytic production of Cl and caustic soda, pharmaceuticals, paints, plastics, paper products, batteries, pesticides, and burning of coal and oil
Nickel (Ni)	Production of stainless steel, alloys, storage batteries, spark plugs, magnets, and machinery
Selenium (Se)	Coal power plant fly ash
Silver (Ag)	Photographic, electroplating, and mirror industries
Tin (Son)	Can production
Zinc (Zn)	Brass and bronze alloy production, galvanized metal production, pesticides, and ink

Note: Sources of trace elements as soil contaminants have been elaborately detailed in Thangavel, P. and Subbhuraam, C.V., *Proc. Indian Natl. Sci. Acad.*, 70, 109, 2004; and Ross, S.M., Ed., *Toxic Metals in Soil Plant Systems*. John Wiley & Sons, Chichester, U.K., 1994, 469.

of Pb-based paint [5]. The threat that heavy metals pose to human and animal health is aggravated by their long-term persistence in the environment. For instance, Pb, one of the more persistent metals, was estimated to have a soil retention time of 150 to 5000 years [6]. Also, the average biological half-life of cadmium has been estimated to be about 18 years [7].

Phytoremediation is an emerging low-cost technology that utilizes plants to remove, transform, or stabilize contaminants located in water, sediments, or soils. Vegetation growing on toxic trace element-contaminated sites is expected to evolve a mechanism of tolerance to withstand the inadequate environment. Therefore, a well adapted flora, tolerant to edaphic climax conditions, is a prerequisite for successful phytoremediation [8,9].

Plants termed phytoremediators are capable of absorbing large amounts of heavy metals from the soil and accumulating these metals in plant tissues [10,11]. Different biochemical studies of heavy metal transport in plants have been conducted [12]. Plants that accumulate and exclude toxic trace elements can play an equally important role in phytoremediation technology. Plants capable of growing on soils contaminated with toxic metals and accumulating extraordinarily high levels of them are shown in Table 27.2. To date, over 450 different hyperaccumulator species have been identified [8]. No one knows why some plants accumulate metals instead of keeping them out. Thus, it is possible to extract and recycle the metals from plants [13].

The ideal phytoextractor should:

- Grow rapidly
- Produce high amount of biomass
- Tolerate and accumulate high concentrations of toxic metals
- Contain substances that deter herbivores from feeding, thus preventing the heavy metal transfer to the food chain

TABLE 27.2
Plants That Hyperaccumulate Trace Elements in Their Tissues

Trace element	Accumulation in plant tissues (mg/g DW)	Number of taxa	Number. of families	Examples
Cd	>0.1	1	1	*Thlaspi caerulescens*
Pb	>1	14	6	*Minuartia verna*
Co	>1	28	11	*Aeollanthus biformifolius*
Cu	>1	37	15	*Aeollanthus biformifolius*
Ni	>1	317	37	*Alyssum bertolonii, Berkheya coddi*
Mn	>10	9	5	*Macadamia neurophylla*
Zn	>10	11	5	*Sedum alfredii Thlaspi caerulescens*
As	>22	2	1	*Pteris vittata, Pityrogramma calomelanos*

Sources: Francesconi, K. et al., *Sci Total Environ.*, 284, 27, 2002; Brooks, R.R., Ed., *Plants That Hyperaccumulate Heavy Metals and Their Role in Phytoremediation, Microbiology, Archeology, Mineral Exploration and Phytomining*. CAB International, New York, 1998; Prasad, M.N.V., Ed., *Heavy Metal Stress in Plants: from Molecules to Ecosystems*. Springer–Verlag, Heidelberg, 2004, 462, xiv; and Hossner, L.R. et al., Amarillo National Resource Center for Plutonium. Amarillo, TX, 1998.

Potentially toxic trace elements are increasing in all compartments of the biosphere, including, air, water, and soil, as a result of anthropogenic processes. For example, the metal concentration in river water and sediments increased several thousand-fold by effluents from industrial and mining wastes [14]. Published literature indicates that an array of bioresources (biodiversity) have been tested in the field and laboratory. Remediation programs relying on these materials may be successful [15–17].

The most successful monitoring methods for metals in the environment are based on bacterial heavy metal biosensors, namely, gene-based biosensors and protein-based biosensors [18]. Mosses, liverworts, and ferns are also capable of growing on metal-enriched substrates. These plants possess anatomical and physiological characteristics enabling them to occupy unique ecological niches in natural metalliferous and manmade environments. For example, groups of specialized bryophytes are found on Cu-enriched substrates — so-called "copper mosses" — and come from widely separated taxonomic groups.

Other bryophytes are associated with lead- and zinc-enriched substrates. *Pteris vittata* (brake fern), a fast growing Pteridophytes (fern) plant, is reported to tolerate soils contaminated with as much as 1500 ppm arsenic and its fronds concentrate the toxic metal to 22,630 ppm in 6 weeks [19]. The fern possesses three key features that are typical of metal/metalloid hyperaccumulator plants: an efficient root uptake, an efficient root-to-shoot translocation, and a much enhanced tolerance to As inside plant cells. After the discovery of this first As hyperaccumulator, several other fern species, including *Pityrogramma calomelanos, Pteris cretica, Pteris longifolia,* and *Pteris umbrosa* [20,21] have recently been added to the list of As hyperaccumulators. The hyperaccumulation trait of these ferns may be potentially exploitable in phytoremediation of As-contaminated soils.

27.2 METAL HYPERACCUMULATORS FOR PHYTOREMEDIATION HYPE

Metal accumulation in higher plants is a complex phenomenon involving (1) transport of metals across the plasma membrane of root cells; (2) xylem loading and translocation; and (3) detoxification and sequestration of metals at the whole plant and cellular levels [22]. Among angiosperms, about 400 metal hyperaccumulators have been identified that would serve as a reservoir for biotechnological application [17,23]. Mine reclamation and biogeochemical prospecting depend upon the correct selection of plant species and sampling. The selection of heavy metal-tolerant species is a reliable tool to achieve success in phytoremediation. Table 27.3 shows plant taxa belonging to 45 families are found to be metal tolerant and are capable of growing on elevated concentrations of toxic metals.

The use of metal-tolerant species and their metal indicator and accumulation is a function of immense use for biogeochemical prospecting [26,27]. Brassicaceae had the highest number of taxa (i.e., 11 genera and 87 species that are established for hyperaccumulation of metals). In Brassicace, Ni hyperaccumulation is reported in seven genera and 72 species [28,29] and Zn in three genera and 20 species [17].

The choice of phytoremediation strategy strongly depends on the risk presented by different metal-polluted soils. For this purpose, instead of chemical analyses, several authors recommend use of plant-based bioassays for risk evaluation. Obviously, metal phytoextraction is not the best approach for industrially polluted soils, where phytostabilization is more appropriate due to a high pollution level suppressing plant growth and productivity. Furthermore, the so-called treatability studies should be conducted before the implementation of phytoextraction technology for the evaluation of the site suitability.

Also, the success of the phytoremediation depends on the nature of the target metal. For example, due to different degrees of soil pollution and solubility, the cases of Cd and Pb are completely different. In many slightly Cd-contaminated agricultural soils, the phytoextraction approach should be more successful than on Pb-polluted soils, which need significant effort to immobilize Pb and to extract much higher metal content. Some crops produce biomass with an added value. For example, crops for industrial products, chemicals, biodiesel and other aromatic compounds.

Certain areas, e.g., rhizosphere biotechnology (and its associated microbes, including mycor-rhizae) deserve a much more exhaustive treatment because this is where trace metals come in contact with plant roots in unsterilized field soil conditions. Sources of heavy metal contaminants in soils (Table 27.1) include [30]:

- Metalliferous mining and smelting sites
- Metallurgical industries
- Sewage sludge applications
- Warfare and military training areas or shooting ranges
- Waste disposal sites
- Agrochemicals
- Electronic industries

Once deposited on the soil, certain metals such as Pb and Cr may be virtually permanent [31].

27.3 MECHANISMS OF METAL UPTAKE BY PLANTS

Following mobilization in the rhizosphere, which is controlled by the soil chemistry, the metals must be taken up by the root cells. Transport proteins and intracellular high-affinity binding sites mediate the uptake of metals across the plasma membrane [32]. Several studies have shown that metal hyperaccumulation of Zn and Cd by *T. caerulescens* involves enhanced metal uptake by the roots [33,34]. Several Zn transporter genes have been cloned recently from *T. caerulescens;* these belong to the ZIP (Zn-regulated transporter/Fe-regulated transporter-like proteins) family [35]. These genes, named *ZNT1* and *ZNT2,* are highly expressed in the roots of *T. caerulescens,* but their expression is not responsive to the Zn status of the plant. Through functional complementation in

TABLE 27.3
Plants That Accumulate Trace Elements

Species	Max. Conc. mg/kg	Ref.
Al		
Hydrangea	3,000	47
Miconia acinodendron	66,100	48
As		
Pterir cretica	3,030	21
P. umbrosa	7,600	21
Pityrogramma calomelanos	8,350	20
Pteris vittata L.	22,630	19
Au		
Brassica juncea	57	49
Cd		
Thlaspi caerulescens	2,130	50
Co		
Hibiscus rhodanthus	1,527	51
Cyanotis longifolia	4,197	52
Haumaniastrum robertii	10,232	51
Cr		
Leptospermum scoparium	20,000	53
Dicoma niccolifera	30,000	54
Sutera fodina	48,000	54
Cu		
Ergrostis recemosa	2,800	55
Vigna dolomitica	3,000	55
Pandiaka metallorum Haumanisatrum	6,270	51
katangense	9,222	51
Ipomea alpina Aeollanthus subcaulis	12,300	1
	13,700	56
Hg		
Lemna minor	25,800	57
Mn		
Eugenia clusioides	10,880	58
Macadamia angustifolia	11,500	58
Alyxia rubricaulis	14,000	58
Maytenus pancheriana	16,370	58
M. sebertiana	22,500	58
M. bureaviana	33,750	58
M. neurophylla	55,200	1
Ni		
Cardamine redisifolia	1,050	59
Alyssum singarense	1,280	60
Thlaspi bulbosum	2,000	50
T. japonicum	2,440	61
T. epirotum	3,000	50
Pseudosempervivum sempervium	3,140	62

TABLE 27.3
Plants That Accumulate Trace Elements (continued)

Species	Max. Conc. mg/kg	Ref.
Ni		
Alyssum tenium	3,420	62
A. fallacinum	3,960	50
Thlaspi ochroleucum	4,000	62
Alyssum alpestre	4,480	62
A. euboeum	4,550	63
A. obovatum	4,590	60
A. condensatum	4,900	60
Thlaspi cypricum	5,120	63
Thlaspi montanum var. montanum	5,530	64
Alyssum virgatum	6,230	60
A. smolikanum	6,600	62
A. murale	7,080	62
A. oxycarpum	7,290	60
A. giosnanum	7,390	60
A. peltarioides subsp.	7,600	60
virgatiforme	7,700	63
A. floribundum	7,860	60
A. penjwinensis	8,170	60
A. anatolicum	9,090	60
A. akamasicum	10,000	60
A. serpylifolium	10,000	65
A. bertolonii subsp.scutarinum	10,200	63
A. syriacum	10,400	60
A. crenulatum	10,900	60
A. callichroum	11,400	60
Bornmulleria sp.	11,500	63
Alyssum eriophyllum	11,700	60
A. discolor	11,800	60
Thlaspi tymphaeum	11,900	50
Alyssum trapeziforme	12,000	66
Thlaspi goesingense	12,400	46
T. graecum	12,500	60
Alyssum heldreichii	12,500	63
A. robertianum	13,400	62
A. bertolonii	13,500	50
A. cilicium	13,500	60
A. huber-morathii	13,600	66
Thlaspi kovatsii	13,600	60
Alyssum markgrafii	13,700	67
Streptanthus polygaloides	14,800	60
Thlaspi caerulescens	16,200	60
Alyssum chondrogynum	16,300	60
A. dubertretii	16,500	60
A. carcium	16,500	61
A. troodii	17,100	60
Pseudosempervium aucheri	17,600	50
Alyssum constellatum	18,100	60
Thlaspi rotundifolium var. corymbosum	—	—
corymbosum	18,300	60

TABLE 27.3
Plants That Accumulate Trace Elements (continued)

Species	Max. Conc. mg/kg	Ref.
Ni		
Peltaria dumulosa	18,900	63
Alyssum samariferum	18,900	68
Bornmuellaria glabrescens	19,200	63
Alyssum davisianum	19,600	60
Alyssum cassium	20,000	60
Thlaspi elegans	20,800	61
T. rotundifolium var.corymbosum	18,300	50
A. samariferum	18,900	60
A. pinifolium	21,100	63
Bornmuellera baldaccii	21,300	60
Alyssum pterocarpum	22,200	60
A. lesbiacum	24,000	60
A. cypricum	23,600	60
A. masmenaeum	24,300	61
Thlaspi jaubertii	26,900	50
T. caerulescens	27,300	60
Alyssum argenteum	29,400	63
Thlaspi sylvium	31,000	50
Bornmuellaria tymphea	31,200	63
Peltaria emarginata	34,400	68
Thlaspi oxyceras	35,600	61
Pb		
Polycarpaea synandra	1,044	69
Acer pseudoplatanus	1,955	70
Thlaspi alpestre	2,740	71
T. rotundifolium	8,200	1
Agrostis tenuis	13,490	72
Minuartia verna	20,000	70
Se		
Acacia cana	1,121	73
Atriplex confertifolia Machaeranthera	1,734	74
glabriuscula	1,800	74
Neptunia amplexicaulis Astragalus	4,334	73
bisulcatus	8,840	74
Astragalus racemosus	14,900	75
Lecythis ollaria	18,200	76
Zn		
Thlaspi idahoense	1,150	64
T. caerulescens	1,400	50
Cochlearia pyrenaica	1,680	77
Thlaspi violascens	2,700	61
T. montanum	3,000	64
T. ochroleucum	3,000	50
T. parvifolium	3,090	61, 64
T. liaceum	3,520	61
T. magellanicum	3,890	50
T. bulbosum	10,500	50

TABLE 27.3
Plants That Accumulate Trace Elements (continued)

Species	Max. Conc. mg/kg	Ref.
Zn		
T. praecox	11,000	61
Arabidopsis thaliana	11,000	50
Thlaspi stenocarpum	16,000	50
T. rotundifolium subsp.cepaeifolium	21,000	50
T. rotundifolium	21,000	78
Thlaspi taraense	25,000	50
Rumex acetosa	26,700	70
Thlaspi alpestre	30,000	50
Arabidopsis halleri	39,600	79
Thlaspi calaminare	39,600	50
Thlaspi caerulescens	51,600	80

Note: The tabulation is based on ascending order of metal accumulation for a given element.

Sources: Thangavel, P. and Subbhuraam, C.V., *Proc. Indian Natl. Sci. Acad.*, 70, 109, 2004; Prasad, M.N.V., Ed., *Heavy Metal Stress in Plants: from Molecules to Ecosystems*. Springer–Verlag, Heidelberg, 2004, 462, xiv; and Palmer, C.E. et al., *Int. J. Phytoremed.*, 3, 245, 2001.

yeast, it was shown that ZNT^1 mediates high-affinity uptake of Zn^{2+} and low-affinity uptake of Cd^{2+} [36]. Specific alterations in Zn-responsive elements, such as transcriptional activators, may play an important role in Zn hyperaccumulation in *T. caerulescens* [36]. However, increased uptake of Cd by *T. caerulescens* cannot be explained by the Zn transport pathway, but may be related to an enhanced expression of the IRT^1 gene, which is essential for Fe uptake [22]. The *IRT1* gene was shown to be able to mediate high-affinity uptake of Cd^{2+} in *A. thaliana* [37,38].

Several classes of proteins have been implicated in transport in plants. These include the metal P-type ATPases that are involved in overall ion homeostasis and tolerance in plants, natural resistance-associated macrophage protein (NRAMP) proteins, and cation diffusion facilitator family proteins [39]. CPx type ATPases have been identified in a wide range of organisms and have been implicated in the transport of potentially toxic metals like Cu, Cd, and Pb across cell membranes [39]. These transporters use ATP to pump a variety of substrates across cell membranes.

Arabidopsis P-type ATPase was the first CPx ATPase reported in plants [40]. Most CPx type ATPases identified so far have been implicated in Cu transport. The physiological role of the metal transporters in higher plants is not clearly known. Because *Arabidopsis* CPx ATPases transport different substrates. They may be present in the membranes and function as efflux pumps. They may also be present at various intracellular membranes and be responsible for the compartmentalization of metals, e.g., sequestration in the vacuoles, golgi, or endoplasmic reticulum.

Because cellular levels of metals must be carefully controlled, transporters represent good candidate for their regulation. How they may be regulated at the transcriptional level or at the translational level as was observed in bacteria and yeast [39] mentioned earlier. In higher plants, three *Nramp* homologues have been identified in rice [41].

In *Arabidopsis*, two genes showing similarity to *Nramps* have also been identified [42]. Initial results suggest that *Arabidopsis Nramp* homologues encode functional metal transporters [43]. Northern analysis indicates that the rice *Nramp* gene *OsNramp1* is primarily expressed in the roots, *OsNramp2* in the leaves, and *OsNramp3* in both tissues of rice [41]. This distinct pattern of expression could mean that they are regulated differently and have distinct functions in different tissues or that they transport distinct but related ions in different parts of the plant.

CDF proteins have been primarily implicated in the transport of Zn, Co, and Cd in bacteria and some eukaryotes. Certain members of CDF have been implicated in Cu or Cd transport. A related Zn transporter (*ZAT1*) from *Arabidopsis* was reported by Van der Zaal et al. [44]. This *ZAT1* transporter may have a role in Zn sequestration in plants. Enhanced Zn resistance was observed in transgenic plants overexpressing *ZAT1*, expressed constitutively throughout the plant. High Zn exposure of these plants led to increased Zn content in the roots. Zinc transporter (ZIP) proteins have been found to be involved in Zn and Fe uptake [45].

In order to enhance metal uptake, the number of uptake sites could be increased, the specificity of the uptake proteins could be altered, and sequestration capacity could be enhanced by increasing the number of intracellular high-affinity binding sites or the rates of transport across organelles. A comprehensive understanding of the metal transport processes in plants is essential for formulating effective strategies to develop genetically engineered plants that can accumulate specific metals.

27.4 PHYTOMASS OF ACCUMULATORS/HYPERACCUMULATORS OF METALS IS A VALUABLE RESOURCE FOR PHYTOEXTRACTION

Plants are selected according to the application and the contaminants of concern. In temperate climates, phreatophytes (e.g., hybrid poplar, willow, cottonwood, aspen) are often selected because of fast growth, a deep rooting ability down to the level of groundwater, large transpiration rates, and the fact that they are natives of most of the countries. The idea of using plants to remove metals from soils came from the discovery of different wild plants, often endemic to naturally mineralized soils that accumulate high concentrations of metals in their foliage [81] (Figure 27.1).

Ideal attributes for metal accumulators are [2]:

- No geographical preference and fast growth
- High bioproductivity
- Capable of producing multiple use products
- Robust and profuse root system
- Ability for metal hyperaccumulation
- High metal tolerance
- Rapid transport to harvestable plant parts
- Capable of accumulating multiple metals with stable properties
- Resistance to disease and pests

Metal hyperaccumulators usually have an antiherbivory function to minimize human health risks through the trophic chain.

27.5 ACCUMULATION OF METALS BY PLANTS

27.5.1 Root Uptake

Roots can reduce soil-bound metal ions by specific plasma membrane bound metal reductases. Plant roots can solubilize heavy metals by acidifying their soil environment with protons extruded from the roots. All of these processes could also be preformed by mycorrhizal fungi or root-colonizing bacteria. Solubilized metal ions can enter the roots via extracellular (apoplastic) or intracellular

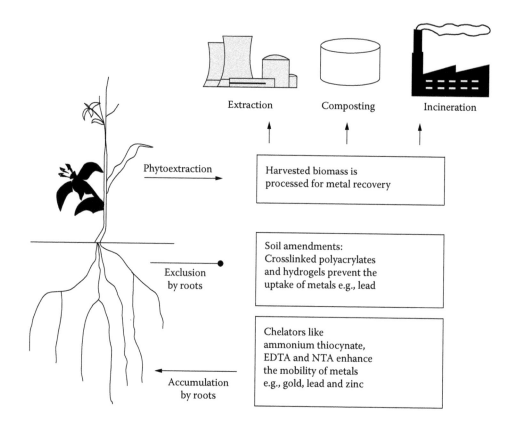

FIGURE 27.1 Phytoextraction consists of (1) planting a species that tends to accumulate and store, transpire, or degrade the target contaminant; (2) letting the crop grow; and (3) harvesting it. Cofiring of phytomass or incineration and composting are concentration methods. Liquid extraction is another separation method. Chelates and soil-amending agents enhance the phytoextraction. The soil–rhizosphere–plant continuum needs critical study for successful phytoremediation [24].

(symplastic) pathways. Nonessential heavy metals may effectively compete for the same transmembrane carriers used by essential heavy metals.

27.5.2 TRANSPORT WITHIN PLANTS

Once in the root, metal ions can be stored or exported to the shoot. Metal transport likely occurs in the xylem, but metals may redistribute in the shoot via the phloem. For metals to enter the xylem,

Shoot system

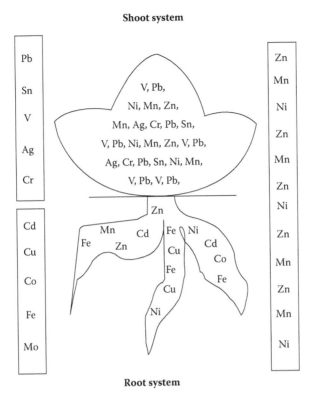

FIGURE 27.2 A generalized pattern of partitioning of metals in the root and shoot system. Silver, chromium, lead, tin, and vanadium accumulate more in shoot (stems and leaves) compared to roots and rhizomes. Cadmium, cobalt, copper, iron, and molybdenum accumulate more in roots and rhizomes than in shoot (stems and leaves). Nickel, manganese, and zinc are distributed more or less uniformly in root and shoot of the plant [82].

they must cross the casparian strip symplastically (intracellular) and this may be the rate-limiting step in metal translocation to the shoot.

27.5.3 MULTIPLE METAL ACCUMULATION

Siedlecka [82] divided the metal-accumulating plants into three categories based on metal accumulation in plant parts (Figure 27.2):

- Accumulate more in roots/rhizomes: Cd, Co, Cu, Fe, and Mo in beetroot, carrot, radish, Jerusalem artichoke, and potato
- Accumulate more in shoots (stems/leaves): Ag, Cr, Pb, Sn, and V in cabbage, cauliflower, tomato, rice, barley, oats, wheat, corn, pigeon pea, chick pea, soybean, peanut, broccoli, lettuce, spinach, and amaranthus
- More or less uniform distribution in roots/shoots: Mn, Ni, and Zn in bush bean, broad bean, mung bean, and cucumber

Soils contaminated with multiple heavy metals can present a difficult challenge for phytoextraction. Although some hyperaccumulators appear to be capable of accumulating elevated concentrations of several heavy metals simultaneously, there is still considerable specificity in metal hyperaccumulation [83]. However, phytoextraction using nontolerant cultivars of *Brassica* sp. is unlikely to succeed in soils contaminated with higher concentrations of Cu, Cd, and Zn, which are usually much more bioavailable, and thus more phytotoxic, than Pb [84]. Generally, monocotyledon species are usually more tolerant to metals than dicotyledon species [85].

The distribution of metals in a contaminated soil is never uniform and, in most agricultural soils, the highest concentrations are usually found near the soil surface. Urban and industrial soils are usually more heterogenous with high metal concentration "hot spots" occurring at depth [86]. Kramer et al. [87] found that, depending on plant species and metal considered, the rates of Ni uptake and root-to-shoot translocation were the same in the Ni hyperaccumulator *Thlaspi goesingense* and the nonaccumulator *Thlaspi arvense*, as long as both species were unaffected by Ni toxicity. Multiple metal hyperaccumulation is particularly advantageous for phytoremediation because soils are often contaminated with various metals such as Cu, Cd, and Zn [88]. Baker et al. [83] reported that, apart from Zn hyperaccumulation, five British populations of *T. caerulescens* also had exceptionally high uptakes of Cd, Co, Mn, and Ni. They suggested that common mechanisms of uptake and translocation existed for several metals in this species.

It is generally accepted that, under natural conditions, a majority of plants have mycorrhizae [89], which have been shown to reduce or enhance metal uptake by plants [85]. Many hyperaccumulators belong to the family Brassicaceae and do not have mycorrhizal associations. It is therefore unlikely that mycorrhizal fungi are directly involved in the enhanced acquisition of metals by Brassicaceous hyperaccumulator plants. In addition, arbuscular mycorrhizal fungi are known to colonize ferns [90], suggesting a possible role of mycorrhizal associations in the recently reported As hyperaccumulation [19]. The effect of mycorrhizal associations on metal root uptake is not clear and appears to be metal and plant specific [91].

The concept of using hyperaccumulator plants to take up and remove heavy metals from contaminated soils was first proposed by Chaney [92]. However, it was not until the early 1990s that field experiments were carried out to test the potential of phytoextraction of metals with hyperaccumulator plants [10]. Hyperaccumulators take up a large quantity of toxic metals through their roots and transport them to the stems or leaves. The word "hyperaccumulator" was coined by the late R.R. Brooks [93] and has been defined as metal accumulation exceeding a threshold value of shoot metal concentration of 1% (Zn, Mn), 0.1% (Ni, Co, Cu, Pb), or 0.1% (Cd) of the dry weight shoot biomass [94].

Compared to nonaccumulator plants, metal concentrations in hyperaccumulator plants are one to three orders of magnitude higher. Apart from these rather arbitrary criteria, hyperaccumulator plants usually have a shoot/root metal concentration ratio greater than one; nonhyperaccumulator plants generally have higher metal concentrations in roots than in shoots [95]. The bioaccumulation factor (shoot/soil concentration ratio) is more important than shoot concentration per se when one considers the potential of phytoextraction for a given species. In metal excluder species, the bioaccumulation factor is typically less than one, and in metal accumulator species, the factor is often greater than one [96].

Several studies indicate the extraction of the various heavy metals (Se, B, Zn, Cd, As, Cu, Co, Ni, Hg, and Cr) used in different industrial processes and other anthropogenic activities from contaminated soils or mine drainage areas using selected hyperaccumulator species [97]. The remediation of other elements (Al, Cs, Sr, and U) from soils by hyperaccumulator crops has not been documented, but is expected to be possible if creative research is applied [98]. The rapidly growing nonaccumulator plants could be engineered so that they achieve some of the properties of hyperaccumulators. Al content of green gram leaves was greater than 1000 mg/kg; therefore, *Vigna radiata* is an Al hyperaccumulator per the definition [99].

Phytoextraction using forestry species in a forestation program is predicted to be a financially attractive option. Trees are potentially the lowest cost plant type used for phytoremediation. Trees have the most massive root systems of all plants; these penetrate the soil for several meters, farther than most herbaceous plants. In some tree species, above-ground biomass can be harvested, and trees will resprout without disturbance of the site. This coppicing shoot system would be valuable if periodic removal of pollutants sequestrated in plant tissue were desirable, as in the case of heavy metals bound to wood often used in chemically enhanced phytoextraction [102]. However, other plant species such as maize and pea (*Pisum sativum* L.) have also been used [103]. The high biomass crop plants, such as Indian mustard, corn, and sunflower, could accumulate significant amounts of Pb when induced through the addition of metal chelates. Simultaneous accumulation of several metals (Pb, Cd, Cu, Ni, Zn) by Indian mustard plants after applying metal chelates has been reported [104]. Metal accumulation efficiency suggested the possibility of using the introduced *Prosopis juliflora* as a metalophyte for the biorecovery of metals from contaminated industrial sites [2,100].

27.6 STRATEGIES FOR ENHANCED UPTAKE OF TRACE ELEMENTS TO FACILITATE PHYTOEXTRACTION

27.6.1 CHELATE-ASSISTED OR CHEMICALLY INDUCED PHYTOEXTRACTION

This strategy of phytoextraction is based on the fact that the application of metal chelates to the soil significantly enhances metal accumulation by plants. The literature to date reports a number of chelates that have been used for chelate-induced hyperaccumulation. These include EDTA, CDTA (*trans*-1,2-diaminocyclohexane-N,N,N′,N′-tetraacetic acid), DTPA (diethylene triaminepentaacetic acid), EGTA [ethyleneglycol-bis(β-aminoethyl ether),N,N,N′,N-tetraacetic acid], EDDHA [ethyl-enediaminedi (o-hydroxyphenylacetic acid)], HEDTA (N-hydroxyethyl enediaminetriacetic acid), HEIDA [N-(2-hydroxyethyl)iminodiacetic acid], and NTA (nitrilo-triacetic acid) [101].

For chemically enhanced phytoextraction, establishment of a high biomass crop is required before chelate application. *Brassica* sp. are to be directly related to the affinity of the applied chelate for the metal [105]. Therefore, it can be concluded that, for efficient phytoextraction to occur, synthetic chelates having a high affinity for the metal of interest should be used: EDTA for lead, EGTA for cadmium [104], possibly citrate for uranium [106], etc. Also, adding ammonium thio-cyanate to the substrate [49] showed that *Brassica juncea* can be induced to accumulate up to 57 mg/kg gold. The mechanisms involved in metal-chelate induced plant uptake and translocation of metals are not well understood. Chemically induced phytoextraction has been described as a two-step process in which plants first accumulate metals in their roots and then, by application of an inducing agent, enhanced transfer of the metals to the shoots occurs [105,107]. This transfer is due to disrupting the plant metabolism that regulates the transport of metal to the shoots. Lombi et al. [84] reported that the application of EDTA alone increases metal mobility in soil and accumulation in roots, but does not substantially increase the transfer of metals to shoots.

Apart from the addition of synthetic chelates, plants secrete to the rhizosphere natural metal-chelating molecules to mobilize soil-bound metals. Thus far, only phytosiderophores, iron-chelating compounds, have been studied in detail. Some of these phytosiderophores include mugeneic and deoxymugeneic acids from barley and corn, and avenic acid from oats [108]. It is also possible that metal-chelating proteins, perhaps related to metalothioneins or phytochelatins, may act as phytosiderophores [81].

Chelate-assisted phytoextraction in field conditions is likely to increase the risk of adverse environmental effects such as ground water pollution due to leaching of metal-laden seepage during extended periods after chelate application. Wenzel et al. [109] hypothesize that free protonated EDTA enters the roots, subsequently forming metal complexes that enhance metal transport to shoots. However, the study of Vassil et al. [110] was conducted in hydroponic conditions, whereas EDTA in soil is expected to form complexes with Ca and other metals. Greman et al. [111] reported ethylenediaminedisuccinate (EDDS) as a promising new chelate for enhanced, environmentally safe phytoextraction of Pb-contaminated soils. It caused only minor leaching of Pb and was significantly less toxic to plants and soil microbes. To avoid possible chelate-metal movement into ground water, the amount, time, and method of chelate application should be carefully controlled.

Due to the severe limitations of chelate-assisted phytoextraction, further efforts should focus on natural, continuous technologies using high biomass perennial plants such as willows or poplar.

27.6.2 RHIZOSPHERE-ASSISTED PROCESSES FOR METAL ACCUMULATION AND EXCLUSION

27.6.2.1 Bioavailability of Metals in Soils

Heavy metal accumulation in soils is highly dependant on the availability of metals for plant uptake. Soils consist of a heterogeneous mixture of different minerals (primary minerals, clay minerals, and hydrous oxides of Al, Fe, and Mn); organic and organo–mineral substances, and other solid components. The binding mechanisms for heavy metals are therefore complex and vary with the composition of the soil, soil acidity, and redox conditions. Heavy metal behavior (e.g., mobility, bioavailability) depends upon several factors (Table 27.2), which can be classified as [31]:

- Geochemical characteristics of a metal
- Plant capacity to take up a metal
- Soil chemical equilibria
- Climatic and other environmental variables
- Agricultural or remedial soil management

Generally, the solubility of metal fractions is in the order [112]:

exchangeable > carbonate specifically adsorbed > Fe–Mn oxide > organic–sulfide > residual

Furthermore, only a fraction of soil metal is readily available (bioavailable) for plant uptake. The bulk of soil metal is commonly found as insoluble compounds unavailable for transport into roots. With the exception of Hg, metal uptake into roots occurs from the aqueous phase. In soil, easily mobile metals such as Zn and Cd occur primarily as soluble or exchangeable, readily bioavailable forms. Cu and Mo predominate inorganically bound and exchangeable fractions. Slightly mobile metals such as Ni and Cr are mainly bound in silicates (residual fraction). Soluble, exchangeable, and chelated species of trace elements are the most mobile in soils and govern their migration and phytoavailability [31]. Others, such as Pb, occur as insoluble precipitates (phosphates, carbonates, and hydroxyoxides), which are largely unavailable for plant uptake [113]. Binding and immobilization within the soil matrix can significantly restrict the potential for metal phytoextraction.

Despite the adverse effect on metal root uptake, soil inactivation with chemical amendments has been proposed as a temporary solution for the remediation of metal-contaminated soils, especially for Pb. Also, the effect of soil amendment on bioavailability is metal specific. Increased mobility of metals can be stimulated by plant roots; this includes changes in pH, reducing capacity, the amount and composition of exudates [114], and use of chelating agents. Soil amendments can increase or decrease biological availability of the contaminant for plant uptake. Bioavailability and metal uptake can often be increased by lowering soil pH, adding chelating agents, using appropriate fertilizers (containing ammonium), altering soil ion composition, soil microorganisms, phytosiderophores, and root exudates [2].

- *Soil pH.* The lower soil pH increases concentration of heavy metals in solution via decreasing their adsorption. Soil pH was adjusted using HNO_3 and $CaCo_3$ to provide a range of pH before planting. Acidified treatments were leached to remove excess nitrate before fertilizers were added. Chaney et al. [97] pointed out that, because soil pH is known to affect plant uptake of most heavy metals from soils, studies needed to be conducted to evaluate the independent effect of soil pH and soil metal concentration on hyperaccumulator yield and metal uptake.

- *Chelate amendments.* Chelate additions (EDTA, HEDTA, DTPA, EGTA, EDDHA, NTA, citrate, and hydroxylamine) are commonly used in soil washing technologies because they cause metal desorption from clay minerals and dissolution of certain precipitates such as Fe and Mn oxides. Artificial chelates, such as EDTA, have been tested to enhance metal phytoavailability and subsequent uptake and translocation in shoots. Two strategies have been proposed regarding the mode of chelate application. The chelates may be added at once a few days before harvest [103] or gradually during the growth period [115]. The type of chelate and its time of application are important considerations. Pierzynski and Schwab [116] investigated the effect of chemical amendments on the potential for phytoextraction of several toxic metals including Cd, Pb, and Zn. They showed that addition of limestone, cattle manure, and poultry litter to soil significantly reduced Zn bioavailability. Experiments indicate that biosurfactants have the potential to enhance metal bioavailability in contaminated soil and sediments [117]. EDTA-amended soils increased Pb availability. Therefore, chelator application might pose a risk to the environment [118]. If the metal availability could be locally improved by increasing reductase activity or the amount of chelating agents — e.g., phytosiderophores [119] — without harmful effects on the environment, hyperaccumulators might be used safely for phytoremediation.
- *Soil fertilizers.* Fertilization with N, P, and K more than doubled annual biomass production without reducing the shoot Ni concentration. This suggested that soil fertility management will be important for commercial phytoextraction [120].
- *Competition for sorption sites.* Using the competition of metal ions in solution for sorption sites may also be a useful tool. For example, addition of phosphate to soil may help to extract Cr, Se, and As on exchange sites by binding to the sites, thereby increasing bioavailability.
- *Soil microorganisms.* The soil microbes have been documented to catalyse redox reactions leading to changes in metal mobility in soils and propensity for uptake into roots. For example, chemolithotrophic bacteria have been shown to enhance environmental mobility of metal contaminants via soil acidification or, in contrast, to decrease their solubility due to precipitation as sulfides [121]. Several strains of *Bacillus* and *Pseudomonas* increased the total amount of Cd accumulated by *Brassica juncea* seedlings [122]. Furthermore, soil microorganisms have been shown to exude organic compounds, which stimulate bioavailability and facilitate root absorption of a variety of metal ions, including Fe^{2+} [123], Mn^{2+} [124], and possibly Cd^{2+} [122]. The microbial activity is stimulated by adding carbon substrates such as agricultural wastes [125], water, and nutrients. Growth of crops also provides these materials to the soil microbiota due to standard farming practices and the process of C loss from roots, called "rhizodeposition." It is interesting that rhizodeposition increases after clipping plants [126]; this could partly explain the enhanced Se removal after cutting treatments, plus the enhanced biomass production.
- *Phytosiderophores.* Plants possess highly specialized mechanisms to stimulate metal bioavailability in the rhizosphere and to enhance uptake into roots [127]. Thus, graminaceous species (grass sp.) have been documented to exude a class of organic compounds termed siderophores (mugineic and avenic acids) capable of enhancing the availability of soil Fe for uptake into roots [128].
- *Root exudates.* It is well established that roots of many plant species release specific metal-chelating or reducing compounds into the rhizosphere to mobilize Fe and, possibly, Zn [85]. For Zn/Cd/Cu/Pb hyperaccumulators, there are no studies on the role of root exudates in metal accumulation to date.

27.6.2.2   Exclusion of Trace Elements to Foster Phytostabilization

The bioavailability of metal ions depends on their solubility in the soil solution, i.e., their general solubility and the stage of equilibrium between the metal cation in its bound form and the free soluble cation. Because the concentration of heavy metal cations in forest soils is usually so low that the solubility behavior of the metal salts will govern their concentration in the soil solution, the dominating factor for the bioavailability of heavy metal cations in soils is their adsorption to soil structures.

The capacity of a given soil to bind given heavy metals depends on the amount and nature of binding sites in the soil structures and the pH of the soil solution. Generally, it can be stated that the lower the pH value is, the more soluble are the metal cations, and the more binding sites that are available in a given soil, the lower will be the solubility of the heavy metals. In the case of cadmium ions, the increase in solubility with decreasing pH values starts at a pH of 6.5. In the case of lead and mercury ions, it starts at a pH value of 4; ions of arsenic, chrome, nickel, and copper start to dissolve at pH values between these two extremes [129]. Thus, the pH value of the soil solution in principle is one of the main factors governing the solubility of heavy metal cations in the soil solution; its influence on plants under heavy metal stress is well established [130–138]. Unfortunately, the acid deposition prevalent all over Europe during recent decades has increased the mobility of heavy metals considerably [139,140].

Thus, increasing the pH could be a measure to reduce the bioavailability of heavy metals. This has been shown by Walendzik [141] for spruce in the Western Sudety Mountains. Liming, however, may not always be a good solution because it may increase the rate of nitrogen mineralization and thus aggravate the NO_3 load in the groundwater [142–144]. The approach of using waste materials such as fuel ash [145] or sewage sludge [146] to improve the growth of trees on mine spoils may not always be successful, as the authors cited previously have shown. Better results using municipal sewage sludge for establishing sagebrush vegetation on copper mine spoils were reported by Sabey et al. [147]. The simple addition of inorganic fertilizer may not work at all [146].

Another way to decrease the bioavailability of heavy metals is to increase the binding sites for heavy metal ions in the soil, e.g., by amendment with humic substances or zeolites [148] or expanded clay and porous ceramic material [149]. When organic substances are added to the soil, it is very important to work with water-insoluble material, which is not available for rapid degradation by microorganisms [150]. The authors found that an addition of hay to a soil contaminated with heavy metals increased the solubility of Cu, Cd, and Zn, but this effect was not observed for Pb. Amendment of the soil with peat had the opposite effect.

Hüttermann and coworkers applied cross-linked polyacrylates, hydrogels, to metal-contaminated soils. When such a compound (Stockosorb K400) was applied to hydrocultures of Scots pine (*Pinus sylvestris*), which contained 1 μM of Pb, two effects were observed: (1) the hydrogel increased the nutrient efficiency of the plants; and (2) the detrimental effect of the heavy metal was completely remediated. Determination of the heavy metal content of the roots revealed that the uptake of the lead was greatly inhibited by the hydrogel. Analysis of the fine roots of 3-year-old spruce grown for one vegetation period in lead-contaminated soil, with and without amendments with the hydrogel, showed that the amendment of the soil with the cross-linked acrylate did indeed prevent the uptake of the lead into the stele of the fine roots. The hydrogel acts as a protective gel that inhibits the entrance of the heavy metal into the plant root [151].

27.6.2.3   Metal Exclusion by Organic Acids

Organic acids are natural products of root exudates, microbial secretions, and plant and animal residue decomposition in soils [152] (Figure 27.3 and Figure 27.4). These biomolecules have been implicated for altering the bioavailabilities and phytoremediation efficiencies of heavy metals in soils. Some researchers showed that amendment of contaminated soils with organic acids reduced

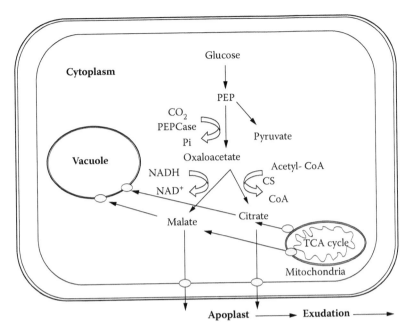

FIGURE 27.3 A comprehensive model to explain the availability of substrates required for the biosynthesis and exudation of organics acids, namely, citrate and malate. These organic acids are present in continuous exchange between mitochondria and the cytosol. Organic acids can be accumulated in the vacuole or excreted into the apoplast by specific carrier proteins, transported towards phloem, and directed to roots for exudation. Plants that exclude toxic trace metals would be the best for photostabilization.

the bioavailability of heavy metals [153]. In contrast, Huang et al. [103] investigated the effect of organic acids amendment of uranium-contaminated soils and found that citric acid significantly increased metal availability and enhanced uranium accumulation many folds in the shoots of selected plants. The contradictory results may be tightly related to the concentration of heavy metal in soil solution and may sequentially be the results of desorption behavior of heavy metal from this soil.

In plants, organic acids may be implicated in detoxification, transport, and compartmentalization of heavy metals. Organic acids are low molecular weight compounds containing carbon, hydrogen, and oxygen and are characterized by one or more carboxylic groups. The number and the dissociation properties of the carboxylic groups determine the negative charges carried by the molecules: the number of metal cations that can be bound in solution or the number of anions that can be displaced from the soil matrix [39]. The most stable ligand–metal complexes have the highest number of carboxyl groups available for binding metal cations. Metal complexes with citrate (tricarboxylate) are more stable than those with malate^{2-}, oxalate^{2-}, or malonate^{2-} (dicarboxylate) and acetate (monocarboxylate) [154].

In several plant species, organic acids participate in the metal exclusion mechanism as metal chelators excreted by the root apex outside the plant and in metal hyperaccumulation as metal chelators inside the plant, with various degrees of metal retention within root and shoot [155,156]. The total concentration of organic acids in the root is generally about 10 to 20 mM, but may vary depending on the degree of cation–anion imbalance because organic acids often provide the negative charges that balance excess cations [157]. Within the plant cell, organic acids are mainly synthesized in mitochondria through the tricarboxylic acid cycle, but the site of preferential storage is the vacuole. Usually, root vacuoles contain two- to tenfold higher concentrations of malate and citrate than cytosol (5 mM) [157] and organometallic chelates can be found in the cell wall, cytoplasm,

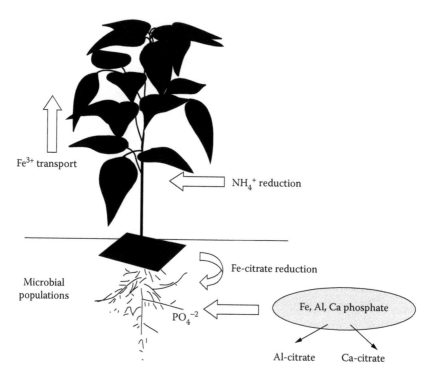

FIGURE 27.4 The ubiquity of organic acids mediating the response of plants to soil stress. Organic acids are strong cation chelators, which act in important adaptive processes in the rhizosphere, such as P and Fe acquisition, Al tolerance, NH4 uptake and reduction, and microbial attraction.

and vacuoles. The composition of root exudates varies greatly, depending on environment, plant species, and age [157–159].

Researchers have found that plant roots exude a variety of organic compounds. Root exudates contain components that play important roles in nutrient solubilization (e.g., organic acids, phytosiderophores, and phenolics), restricting the passage of toxic metals across the root (e.g., citrate, malate, small peptides) and attracting beneficial microorganisms (e.g., phenolics, organic acids and sugars). Often, the excretion of these organic molecules increases in response to soil stress.

27.7 ORGANIC ACIDS PLAY AN IMPORTANT ROLE IN ADAPTIVE PHYSIOLOGY

Krotsky et al. [160] showed that a sorghum cultivar efficiently colonized by N2 fixing free-living bacteria released more malic, fumaric, and succinic acid than a less active cultivar [160,161]. Rennie [161] reported that the addition of sugars and organic acids to maize inoculated with N2 fixing *Azospirillum brasilense* promoted the incorporation of atmosphere-derived nitrogen in the plant. Succinate and malate stimulated nitrogen fixation more than sucrose [162]. More recently, it was found that aluminum-tolerant wheat cultivars that produce high concentrations of low molecular dicarboxylic acids had higher associative nitrogen fixation rates than nontolerant cultivars did [163].

The adequacy of organic acids as carbon and energy sources has been demonstrated with studies on other microbial species. Some *Campylobacter* sp. isolated from roots of *Spartiana alterniflora* were found to metabolize amino and organic acids efficiently [164]. In legume–*Rhizobium* symbiosis, the preferred substrate taken up by the bacteroids from the host is malate, which may be oxidized to oxaloacetate by malate dehydrogenase or may be converted to acetyl CoA by the malic enzyme and pyruvate dehydrogenase. Further oxidation of acetyl CoA in the tricarboxylic acid cycle can generate the large amount of energy required by the nitrogenase reaction.

REFERENCES

1. Cunningham, S.D. and Berti, W.R., The remediation of contaminated soils with green plants: an overview, In Vitro *Cell. Devel. Biol.* 6 *J. Tissue Cult. Assoc.*, 29, 207, 1993.
2. Thangavel, P. and Subbhuraam, C.V., Phytoextraction: role of hyperaccumulatots in metal-contaminated soils, *Proc. Indian Natl. Sci. Acad.*, 70, 109, 2004.
3. Ross, S.M., Ed., *Toxic Metals in Soil Plant Systems*, John Wiley & Sons, Chichester, U.K., 1994, 469.
4. Environmental investments: the cost of a clean environment, U.S. Environmental Protection Agency (EPA). (ed. EPA-230-11-90- 083) 5 (US Government Printing Office, Washington, D.C., 1990).
5. Seaward, M.R.D. and Richardson, D.H.S., Atmospheric sources of metal pollution and effects on vegetation, in: *Heavy Metal Tolerance in Plants: Evolutionary Aspects*, Shaw, A.J., Ed., CRC Press, Boca Raton, FL, 1990, 75.
6. Shaw, B.P., Sahu, S.K. and Mishra R.K., Heavy metal-induced oxidative damage in terrestrial plants, in *Heavy Metal Stress in Plants — from Biomolecules to Ecosystems*, Prasad, M.N.V., Ed., Springer–Verlag, Heidelberg, 2004, chap. 4.
7. Forstner, U., Land contamination by metals: global scope and magnitude of problem, in *Metal Speciation and Contamination of Soil*, Allen, H.E. et al., Eds., CRC Press, Boca Raton, FL, 1995, 1.
8. Baker, A.J.M. and Brooks, R.R., Terrestrial higher plants which hyperaccumulate metallic elements. A review of their distribution, ecology and phytochemistry, *Biorecovery,* 1, 81, 1989.
9. Del Rio, M. et al., Heavy metals and arsenic uptake by wild vegetation in the Guadiamar River area alter the toxic spill of the Aznalcollar mine, *J. Biotechnol.*, 98, 125, 2002.
10. Brown, S.L. et al., Zinc and cadmium uptake by hyperaccumulator *Thlaspi caerulescens* and metal tolerant *Silene vulgaris* grown on sludge-amended soils, *Environ. Sci. Technol.*, 29, 1581, 1995.
11. Lasat, M.M., Baker, A.J.M. and Kochian, L.V., Altered zinc compartmentation in the root symplasm and stimulated Zn^{2+} absorption into the leaf as mechanisms involved in zinc hyperaccumulation in *Thlaspi caerulescens, Plant Physiol.*, 118, 875–883, 1998.
12. Wagner, G.J, Biochemical studies of heavy metal transport in plants, in *Current Topics in Plant Biochemistry, Physiology and Molecular Biology*, Randall, D. et al., Eds., University of Missouri, Columbia, 1995, 21.
13. McGrath, S.P, Phytoextraction for soil reclamation, in *Plants That Hyperaccumulate Heavy Metals. Their Role in Phytoremediation, Microbiology, Archaeology, Mineral Exploration and Phytomining*, Brooks, R.R., Ed., CAB International, Wallingford, Oxon, 1998, 261.
14. Siegel, F.R., *Environmental Geochemistry of Potentially Toxic Metals*, Springer–Verlag, Heidelberg, 2002, 218.
15. Glass, D.J., U.S. and international markets for phytoremediation, 1999-2000. D.J. Glass Associates, Inc. Needham, Massacheusetts, 1999.
16. Glass, D.J., Economic potential of phytoremediation, in *Phytoremediation of Toxic Metals — Using Plants to Clean Up the Environment*, Raskin, I. and Ensley, B.D., Eds., John Wiley & Sons, Inc., New York, 15 2000.
17. Prasad, M.N.V. and Freitas, H., Metal hyperaccumulation in plants — biodiversity prospecting for phytoremediation technology, *Electron. J. Biotechnol.*, 6, 3, 275, 2003. http://www. ejbiotechnology. info/content/ vol6/issue3/index.html electronic journal.
18. Prasad, M.N.V., Ed., *Metals in the Environment — Analysis by Biodiversity*, Marcel Dekker Inc. New York, 2001, 504.
19. Ma, L.Q. et al., A fern that hyperaccumulates arsenic, a hardy, versatile, fast-growing plant helps to remove arsenic from contaminated soils, *Nature*, 409, 579, 2001.
20. Francesconi, K. et al., Arsenic species in an arsenic hyperaccumulating fern, *Pityrogramma calomelanos* a potential phytoremediator of arsenic-contaminated soils, *Sci. Total Environ.*, 284, 27, 2002.
21. Zhao, F.J., Dunham, S.J. and McGrath, S.P., Arsenic hyperaccumulation by different fern species, *New Phytol.*, 156, 27, 2002.

22. Lombi, E. et al., Influence of iron status on calcium and zinc uptake by different ecotypes of the hyperaccumulator *Thlaspi caerulescens*, *Plant Pysiol.*, 128, 1359, 2002.
23. Brooks, R.R., Ed., *Plants That Hyperaccumulate Heavy Metals ó Their Role in Phytoremediation, Microbiology, Archeology, Mineral Exploration and Phytomining*, CAB International, New York, 1998.
24. Prasad, M.N.V., Ed., *Heavy Metal Stress in Plants: from Molecules to Ecosystems*, Springer–Verlag, 2004, 462, xiv.
25. Hossner, L.R., Loeppert, R.H. and Newton, E.J., Literature review: phytoaccumulation of chromium, uranium, and plutonium in plant systems, Amarillo National Resource Center for Plutonium. Amarillo, TX, 1998.
26. Badri, M. and Sringeul, I., Biogeochemical prospecting in the southeastern desert of Egypt, *J. Arid Environ.*, 28, 257, 1994.
27. McInnes, B.I.A. et al., Biogeochemical exploration for gold in tropical rain forest regions of Papua, New Guinea, *J. Geochem. Exploration, 57, 227, 1996.*
28. Reeves, R.D. et al., Nickel-accumulating plants from the ancient serpentine soils of Cuba, *New Phytol.*, 133, 217, 1996.
29. Reeves, R.D. et al., Nickel hyperaccumulation in the serpentine flora of Cuba, *Ann. Bot.*, 83, 29, 1999.
30. Alloway, B.J., *Heavy Metals in Soils*, Blackie and Wiley, New York, 1995.
31. Kabata–Pendias, A., *Trace Elements in the Soil and Plants*, 3rd ed., CRC Press, Boca Raton, FL, 2001.
32. Dutta, R. and Sarkar, D., Biotechnology in phytoremediation of metal-contaminated soils, *Proc. Indian Natl. Sci. Acad.*, 70, 99, 2004.
33. Lasat, M.M., Baker, A.J.M. and Kocjian, L.V., Physiological characterization of root Zn^{2+} absorption and translocation to shoots in Zn hyperaccumulator and nonaccumulator species of *Thlaspi*, *Plant Physiol.*, 112, 1715, 1996.
34. Zhao, F. et al., Characteristics of cadmium uptake in two contrasting ecotypes of the hyperaccumulator *Thlaspi caerulescens*, *J. Exp. Bot.*, 53, 535, 2002.
35. Maser, P. et al., Phylogentic relationships within cation transporter families of *Arabidopsis*, *Plant Physiol.*, 126, 1646, 2001.
36. Pence, N.S. et al., The molecular physiology of heavy metal transporter in the Zn/Cd hyperaccumulator, *Thlaspi caerulescens*, *Proc. Natl. Acad. Sci. USA*, 97, 4956, 2000.
37. Vert, G. et al., ITR1, an *Arabidopsis* transporter essential for iron uptake from the soil and for plant growth, *Plant Cell*, 14, 1233, 2002.
38. Connolly, E.L., Fett, J.P. and Guerinot, M.L., Expression of the IRT1 metal transporter is controlled by metals at the levels of transcript and protein accumulation, *Plant Cell*, 14, 1347, 2002.
39. Williams, L.E., Pittman, J.K. and Hall, J.L., Emerging mechanisms for heavy metal transport in plants, *Biochim. Biophys. Acta*, 1465, 104, 2000.
40. Tabata, K. et al., Cloning of cDNA encoding a putative metal transporting P-type ATPase from *Arabisopsis thaliana*, *Biochim. Biophys. Acta*, 1326, 1, 1997.
41. Belouchi, A., Kwan, T. and Gros, P., Cloning and characterization of the OsNramp family from *Oryza sativa*, a new family of membrane proteins possibly implicated in the transport of metal ions, *Plant Mol. Biol.*, 33, 1085, 1997.
42. Alonso, J.M. et al., EIN2, a bifunctional transducer of ethylene and stress responses in *Arabidopsis*, *Science*, 284, 2148, 1999.
43. Thomine, S. et al., Molecular analysis of metal transport and metal tolerance in plants, Plant Biology Meeting, 1999. Abst 1304.
44. Van der Zaal, B.J. et al., Overexpressin of a novel Arabidopsis gene related to putative zinc-transporter gene from animals can lead to enhanced zinc- resistance and accumulation, *Plant Physiol.*, 119, 1047, 1999.
45. Guerinot, M.L. and Eide, D., Zeroing in on zinc uptake in yeast and plants, *Curr. Opin. Plant Biol.*, 2, 244, 1999.
46. Palmer, C.E., Warwick, S. and Keller, W., Brassicaceae (Cruciferae) family — plant biotechnology and phytoremediation, *Int. J. Phytoremed.*, 3, 245, 2001.
47. Ma, L.Q., Zheng, S.J. and Matsumoto, H., Detoxifying aluminum with buckwheat, *Nature,* 390, 569, 1997.
48. Chenery, E.M., Aluminium in the plant world. I. General survey in dicotyledons, *Kew Bull.*, 173, 1948.
49. Anderson, C.W.N. et al., Harvesting a crop of gold in plants, *Nature,* 395, 553, 1998.

50. Reeves, R.D. and Brooks, R.R., European species of *Thalaspi* L. (Cruciferae) as indicators of nickel and zinc, *J. Geochem. Explor.,* 18, 275, 1983.
51. Brooks, R.R. et al., The elemental content of metallophytes from the copper/cobalt deposits of Central Africa, *Bull. Soc. R. Bot. Belg.,* 119, 179, 1987.
52. Morrison, R.S., Aspects of the accumulation of cobalt, copper and nickel by plants, Ph.D. thesis, Massey University, New Zealand, 1980.
53. Lyon, G.L. et al., Some trace elements in plants for serpentine soils, *Nz. J. Sci.,* 13,133, 1969.
54. Zayed, A.M., and Terry, N. Chromium in the environment: factors affecting biological remediation, *Plant and Soil,* 249, 139, 2003.
55. Malaisse, F. et al., *Aeolanthus biformifolius* De Wild.: a hyperaccumulator of copper from Zaire, *Science,* 199, 887, 1978.
56. Brooks, R.R. et al., Hyperaccumulation of copper and cobalt — a review, *Bull. Soc. R. Bot. Belg.,* 113, 166, 1980.
57. Zayed, A., Gowthaman, S. and Terry, N., Phytoaccumulation of trace elements by wetland plants. 1. Duckweed, *J. Environ.Qual.,* 27, 715, 1998.
58. Jaffre, T., Etude ecologique du peuplement vegetal des Sols Derives de Roches Ultrabsiques en Nouvelle Caledonie. Vol. 124, Travaux et Documents de ORSTOM, Paris, 1980.
59. Vergnano Gambi, O. and Gabrielli, R., Ecophysiological and geochemical aspects of nickel, chromium and cobalt accumulation in the vegetation of some Italian ophiolitic outcrops, *Ofioliti,* 4, 199, 1979.
60. Brooks, R.R. et al., Hyperaccumulation of nickel by *Alyssum linnaeua* (Cruciferae), *Proc. R. Soc. Lond.* B203 387-403. Cannon, H.L., 1960 Botanical prospecting for ore deposits, *Science,* 132, 591, 1979.
61. Reeves, R.D., Nickel and zinc accumulation by species of *Thlaspi* L., *Cochlearia* L., and other genera of the Brassicaceae, *Taxon,* 37, 309, 1988.
62. Brooks, R.R. and Radford, C.C., Nickel accumulation by European species of the genus *Alyssum,* *Proc. R. Soc. Lond. B.,* 200, 217, 1978.
63. Reeves, R.D., Brooks, R.R. and Dudley, T.R., Uptake of nickel by species of *Alyssum, Bornmuellera* and other genera of old world *Tribus alyssaeae, Taxon,* 32, 184, 1983.
64. Reeves, R.D., MacFarlane, R.M. and Brooks, R.R., Accumulation of nickel and zinc by western North American genera containing serpentine-tolerant species, *Am. J. Bot.,* 70, 1297, 1983.
65. Brooks, R.R., Shaw, S. and Asensi Marfil, A., The chemical form and physiological function of nickel in some Iberian *Alyssum* species, *Physiol. Plant.,* 51, 167, 1981.
66. Minguzzi, C. and Vergnano, O., Il cpntenuto di nichel nelle ceneri di *Alyssum bertlonii* Desv Atti, *Soc. Toscana Sci. Natl. Pisa Mem. Ser. A,* 55, 49, 1948.
67. Reeves, R.D., Brooks, R.R. and MacFarlane, R.M., Nickel uptake by California *Streptanthus* and *Caulanthus* with particular reference to the hyperaccumulator *S. polygaloides* Gray (Brassicaceae), *Am. J. Bot.,* 68, 708–712, 1981.
68. Reeves, R.D., Brooks, R.R. and Press, J., Nickel accumulation by species of Peltaria Jacq. (Cruciferae), *Taxon,* 29, 629, 1980.
69. Cole, M.M., Provan, D.M.J. and Tooms, J.S., Geobotany, biogeochemistry and geochemistry in the Bulman–Waimuna Springs area, Northern Territory, Australia, *Trans. Instit. Mining Metallurgy Sect. B,* 77, 81, 1968.
70. Johnston, W.R. and Proctor, J., A comparative study of metal levels in plants from two contrasting lead-mine sites, *Plant Soil,* 4646, 251, 1977.
71. Shimwell, D.W. and Laurie, A.E., Lead and zinc accumulation of vegetation in the Southern Pennines, *Environ. Pollut.,* 3, 291, 1972.
72. Williams, S.T. et al., The decomposition of vegetation growing on metal mine waste, *Soil Biol. Biochem.,* 9, 271, 1977.
73. McCray, C.W.R. and Hurwood, I.S., Selenosis in northwestern Queensland associated with a marine cretaceous formation, *Queensland J. Agric. Sci.,* 20, 475, 1963.
74. Rosenfeld, I. and Beath, O.A., *Selenium — Geobotany, Biochemistry, Toxicity and Nutrition,* Academic Press, New York, 1964.
75. Beath, O.A., Eppsom, H.F. and Gilbert, C.S., Selenium distribution in and seasonal variation of type vegetation occurring on seleniferous soils, *J. Am. Pharmacol. Assoc.,* 26, 394, 1937.

76. Aronow, L. and Kerdel–Vegas, F., Seleno–cystathionine, a pharmacologically active factor in the seeds of *Lecythis ollaria, Nature,* 205, 1185, 1965.

77. Ramaut, J.L., Petit, J. and Maquinay, A., *Cochlearia pyrenacia* — plante calaminaire? *Naturalistes Belg.,* 53, 475, 1972.

78. Linstow, O., Die naturliche Anreicherung von metallsazen und anderen anorganiscgen verbindungen in den pflanzen, *Feddes Rep.,* 31, 1, 1924.

79. Zhao, F.J. et al., Zinc hyperaccumulation and cellular distribution in *Arabidopsis halleri, Plant Cell Environ.,* 23, 507, 2000.

80. Brown, S.L. et al., Phytoremediation potential of *Thlaspi caerulescens* and bladder campion for zinc and cadmium contaminated soil, *J. Environ. Qual.,* 23, 1151, 1994.

81. Raskin, I., Smith, R.D. and Salt, D.E., Phytoremediation of metals: using plants to remove pollutants from the environment, *Curr. Opin. Biotechnol.,* 8, 221, 1997.

82. Siedlecka, A., Some aspects of interactions between heavy metals and plant mineral nutrients, *Acta. Soc. Bot. Pol.,* 64, 265, 1995.

83. Baker, A.J.M. et al., Metal hyperaccumulator plants: a review of the ecology and physiology of a biochemical resource for phytoremediaiton of metal-polluted soils, in *Phytoremediaiton of Contaminated Soil and Water,* Terry, N. and Banuelos, G.S., Lewis Publishers, Boca Raton, FL, 2000, 85.

84. Lombi, E. et al., Phytoremediation of heavy metal-contaminated soils: natural hyperaccumulation vs. chemically enhanced phytoextraction, *J. Environ. Qual.,* 30, 1919, 2001.

85. Marschner, H., *Mineral Nutrition of Higher Plants,* 2nd ed., Academic Press, London, 1995.

86. Robinson, B. et al., Phytoextraction: an assessment of biogeochemical and economic viability, *Plant Soil,* 249, 117, 2003.

87. Kramer, U. et al., The role of metal transport and tolerance in nickel hyperaccumulation by *Thlaspi goesingense* Halacsy, *Plant Physiol.,* 115, 1641, 1997.

88. McGrath, S.P., Zhao, F.J. and Lambi, E., Plant and rhizosphere processes involved in phytoremediation of metal-contaminated soils, *Plant Soil,* 232, 207, 2001.

89. Smith, S.E. and Reed, D.J., *Mycorrhizal Symbiosis,* 2nd ed., Academic Press, London, 1997.

90. Sharma, B.D., Fungal association with Isoetes species, *Am. Fern J.,* 88, 138, 1998.

91. Lasat, M.M. et al., Molecular physiology of zinc transport in the Zn hyperaccumulator *Thlaspi caerulescens, J. Exp. Bot.,* 51, 71, 2000.

92. Chaney, R.L., Plant uptake of inorganic waste constituents, in *Land Treatment of Hazardous Wastes,* Parr, J.F., Marsh, P.B. and Kla, J.S., Eds., Noyes Data Corp., Park Ridge, NJ, 1983, 50.

93. Brooks, R.R. and Wither, E.D., Nickel accumulation by *Rinorea bengalensis* Wall. O.K., *J. Geochem. Explor.,* 7, 295, 1977.

94. McGrath, S.P., Phytoextraction for soil remediation, in *Plants That Hyperaccumulate Heavy Metals,* Brooks, R.R., Ed., CAB International, Wallingford, U.K., 1998.

95. Shen, Z.G., Zhao, F.J. and McGrath, S.P, Uptake and transport of zinc in the hyperaccumulator *Thlaspi cerulescens* and nonhyperaccumulator *Thlaspi ochroleucum, Plant Cell Environ.,* 20, 898, 1997.

96. Baker, A.J.M., Accumulators and excluders — strategies in the response of plants to heavy metals, *J. Plant. Nutri.,* 3, 643, 1981.

97. Chaney, R.L. et al., Improving metal hyperaccumulator wild plants to develop commercial phytoextraction systems: approaches and progress, in *Phytoremediation of Contaminated Soil and Water,* Terry, N., Banuelos, G. and Vangronsveld, J., Eds., CRC Press, Boca Raton, FL, 2000, 129.

98. Chaney, R.L. et al., Potential use of metal hyperaccumulators, *Mining Environ. Manage.,* 3, 9, 1995.

99. Jansen, S. et al., Aluminum hyperaccumulation in angiosperms: a review of its phylogenetic significance, *Bot. Rev.,* 68, 235, 2002.

100. Nagaraju, A. and Prasad, K.S. S. Growth of *Prosopis juliflora* on pegmatite tailings from Nellore Mica Belt, Andhra Pradesh, India, *Envir. Geol.,* 36, 320, 1998.

101. Cooper, E.M. et al., Chelate-assisted phytoextraction of lead from contaminated soils, *J. Environ. Qual.,* 28, 1709, 1999.

102. Blaylock, M.J., Field demonstration of phytoremediation of lead contaminated soils, in *Phytoremediation of Contaminated Soil and Water.* Terry, N., Banuelos, G. and Vangronsveld, J., Eds., CRC Press, Boca Raton, FL, 2000, 1.

103. Huang, J.W. et al., Phytoremediation of lead-contaminated soils: role of synthetic chelates in lead phytoextraction, *Environ. Sci. Technol.,* 31, 800, 1997.

104. Blaylock, M.J. et al., Enhanced accumulation of Pb in Indian mustard by soil-applied chelating agents, *Environ. Sci. Technol.*, 31, 860, 1997.

105. Salt, D.E., Smith, R.D. and Raskin, I., Phytoremediation, *Annu. Rev. Plant Physiol. Plant Mol. Biol.*, 49, 643, 1998.

106. Huang, J.W. et al., Phytoremediation of uranium contaminated soils: role of organic acids in triggering uranium hyperaccumulation in plants, *Environ. Sci. Technol.*, 32, 2004, 1998.

107. Ensley, B.D. et al., Inducing hyperaccumulation of metals in plant shoots, U.S. Patent 5 917 117. Date issued: 29 June, 1999.

108. Welch, R.M. et al., Induction of iron III reduction in pea (*Pisum sativum* L.) roots by Fe and Cu status: does the root-cell plasmalemma Fe III-chelate reductase perform a general role in regulating cation uptake? *Planta*, 190, 555, 1993.

109. Wenzel, W.W. et al., Rhizosphere characteristics of indigenously growing nickel hyperaccumulator and excluder plants on serpentine soil, *Environ. Pollut.*, 123, 131, 2003.

110. Vassil, A.D. et al., The role of EDTA in lead transport and accumulation by Indian mustard, *Plant Physiol.*, 117, 447, 1998.

111. Greman, H. et al., Ethylenediaminedisuccinate as a new chelate for environmentally safe enhanced lead phytoextraction, *J. Environ. Qual.*, 32, 500, 2003.

112. Li, X.D. and Thornton, I. Chemical partitioning of trace and major elements in soils contaminated by mining and smelting activities, *Appl. Geochem.*, 16, 1693, 2001.

113. Pitchel, J., Kuroiwa, K. and Sawyer, H.T., Distribution of Pb, Cd and Ba in soils and plants of two contaminated soils, *Environ. Pollut.*, 110, 171, 1999.

114. Bernal, M.P. and McGrath, S.P., Effects of pH and heavy metal concentrations in solution culture on the proton release, growth and elemental composition of *Alyssum murale* and *Raphanus sativus* L., *Plant Soil*, 166, 83, 1994.

115. Kayser, A., Wenger, K., Keller, A., Attinger, W., Felix, H.R., Gupta, S.K. and Schulin, R., Enhancement of phytoextraction of Zn, Cd and Cu from calcareous soil: the use of NTA and sulfur amendments, *Environ. Sci. Technol.*, 34, 1778–1783, 2000.

116. Pierzynski, G.M. and Schwab, A.P., Bioavailability of zinc, cadmium and lead in metal-contaminated alluvial soil, *J. Environ. Qual.*, 22, 247, 1993.

117. Mulligan, C.N, Yong, R.N. and Gibbs, B.F., Remediation technologies for metal contaminated soil and ground water: an evaluation, *Eng. Geol.*, 60, 193, 2001.

118. McGrath, S.P. and Zhao, F., Phytoextraction of metals and metalloids from contaminated soils, *Curr. Opin. Biotechnol.*, 14, 1, 2003.

119. Briat, J.F. and Lebrun, M., Plant responses to metal toxicity. *CR Acad. Sci. Paris/Life Sci.*, 322, 43–54, 1999

120. Li, Y.M. et al., Development of a technology for commercial phytoextraction of nickel: economic and technical considerations, *Plant Soil*, 249, 107, 2003.

121. Kelley, B.C. and Tuovinen, O.H. Microbial oxidation of minerals in mine tailings, in *Chemistry and Biology of Solid Waste*, Solomons, W. and Foerstner, V., Eds., Springer–Verlag, Berlin, 1988, 33.

122. Salt, D.E., Blaylock, M., Kumar, P.B.A.N., Dushenkov, V., Ensley, B.D., Chet, L. and Raskin, L., Phytoremediation: a novel strategy for the removal of toxic metals from the environment using plants, *Biogeochemistry*, 13, 468–474, 1995.

123. Bural, G.I., Dixon, D.G. and Glick, B.R., Plant growth-promoting bacteria that decrease heavy metal toxicity in plants, *Can. J. Microbiol.*, 46, 237, 2000.

124. Barber, S.A. and Lee, R.B., The effect of microorganisms on the absorption of manganese by plants, *New Phytol.*, 73, 97, 1974.

125. Frankenberger, W.T. and Karlson, U. Dissipation of soil selenium by microbial volatilization, in *Biogeochemistry of Trace Metals,* D.C. Adriano, Ed., Lewis Publishers, Boca Raton, FL, 1992, 365.

126. Terry, N. and Zayed, A.M., Selenium volatilization by plants, in *Selenium in the Environment,* Frankenberger, W.T., Jr. and Benson, S., Eds., Marcel Dekker, New York, 1994, 343.

127. Romheld, V. and Marschner, H., Mobilization of iron in the rhizosphere of different plant species, *Adv. Plant Nutr.*, 2, 155, 1986.

128. Takagi, S., Nomoto, K. and Takemoto, T., Physiological aspects of mugineic acid, a possible phytosiderophores of graminaceous plants, *J. Plant Nutr.*, 7, 469, 1984.

129. Scheffer, F. and Schachtschabel, P., *Lehrbuch der Bodenkunde*, Enke, Stuttgart, 1989.

130. Gerth, J. and Brümmer, G., Quantitäts-Intensitäts-Beziehungen von Cd, Zn und Ni in Böden unterschiedlichen Stoffbestands, *Mitt Dtsch Bodenkund Ges,* 29, 555–566, 1977.

131. Alloway, B.J. and Morgan, H., The behavior and availability of Cd, Ni and Pb in polluted soils, in: Assink, J.W. and Brink, W.J. van den, Eds., *Contaminated Soil,* Nijhoff, Dordrecht, 101–113, 1985.

132. Tyler, G., Berggren, D., Bergkvist, B., Falkengren–Grerup, U., Folkeson, L. and Ruhling, A., Soil acidification and metal solubility in forests of southern Sweden, in Hutchinson, T.C. and Meema, K.M. Eds., *Effects of Atmospheric Pollutants on Forests, Wetlands, and Agricultural-Eco Systems.* Springer–Verlag, Berlin, 347–359, 1987.

133. Neite, H., Zum Einfluß von pH und organischem Kohlenstoffgehalt auf die Löslichkeit von Eisen, Blei, Mangan und Zink in Waldböden, *Z Pflanzenernähr Bodenkd,* 152, 441–445, 1989.

134. Kedziorek, M.A.M. and Bourg, A.C.M., Acidification and solubilization of heavy metals from single and dual-component model solids, *Appl. Geochem.,* 11, 299–304, 1996.

135. Reddy, K.J., Wang, L. and Gloss, S.P., Solubility and mobility of copper, zinc and lead in acidic environments, *Plant Soil,* 171, 53–58, 1995.

136. Chuan, M.C., Shu, G.Y. and Liu, J.C., Solubility of heavy metals in a contaminated soil: effects of redox potential and pH, *Water Air Soil Pollut.,* 90, 543–556, 1996.

137. McBride, M., Sauve, B. and Hendershot, W., Solubility control of Cu, Zn, Cd and Pb in contaminated soils, *Eur. J. Soil Sci.,* 48, 337–346, 1997.

138. Zhu, B. and Alva, A.K., Effect of pH on growth and uptake of copper by single citrumelo seedlings, *J. Plant Nutr.,* 16, 1837–1845, 1993.

139. Wilson, M.J. and Bell, N., Acid deposition and heavy metal mobilization, *Appl. Geochem.,* 11, 133–137, 1996.

140. Mannings, S., Smith, S. and Bell, J.N.B., Effect of acid deposition on soil acidification and metal mobilisation, *Appl. Geochem.,* 11, 139–143, 1996.

141. Walendzik, R.J., Deterioration of forest soils in the Western Sudety Mountains (Poland) and attempts of its limitation, *Sylwan,* 137, 29–38 (in Polish), 1993.

142. Marschner, B., Stahr, K. and Rengen, M., Potential hazards of lime application in a damaged pine forest ecosystem in Berlin, Germany, *Water Air Soil Pollut.,* 48, 45–57, 1989.

143. Kreutzer, K. and Hüttl, R.F., Effects of forest liming on soil processes, *Plant Soil,* 168–169; 447–470, 1995.

144. Schuler, G., Waldkalkung als Bodenschutz, *Allg Forst Zschr.,* 50, 430–433, 1995.

145. Piha, M.I., Vallack, H.W., Michael, N. and Reeler, B.M., A low input approach to vegetation establishment on mine and coal ash wastes in semiarid regions. II. Lagooned pulverized fuel ash in Zimbabwe, *J. Appl. Ecol.,* 32, 382–390, 1995.

146. Borgegard, S-O. and Rydin, H., Utilization of waste products and inorganic fertilizer in the restoration of iron-mine tailings, *J. Appl. Ecol.,* 26, 1083–1088, 1989.

147. Sabey, B.R., Pendleton, R.L. and Webb, B.L., Effect of municipal sewage sludge application on growth of two reclamation shrub species in copper mine spoils, *J. Environ. Qual.,* 19, 580–586, 1990.

148. Baydina, N.L., Inactivation of heavy metals by humus and zeolites in industrially contaminated soil *Euras. Soil Sci.,* 28, 96–105, 1996.

149. Figge, D.A.H., Hetrick, B.A.D. and Wilson, G.W.T., Role of expanded clay and porous ceramic amendments on plant establishment in mine spoils, *Environ. Pollut.,* 88, 161–165, 1995.

150. Herms, U. and Brümmer, G., Einflußgrößen der Schwermetallöslichkeit und bindung in Böden, *Z Pflanzenernähr Bodenkd,* 147, 400–424, 1984.

151. Hüttermann, A., Arduini, I. and Godbold, D.L., Metal pollution and forest decline, in Prasad, M.N.V., Ed., *Heavy Metal Stress in Plants: from Biomolecules to Ecosystems,* Narosa Publishing House. New Delhi, 2nd ed., 295–312, 2004.

152. Khan, A.G., Kuek, C., Chaudhry, T.M., Khoo, C.S. and Hayes, W.J., Role of plants, mycorrhizae and phytochelators in heavy metal contaminated land remediation, *Chemosphere,* 41,1ñ2,197ñ207, 2000.

153. Ryan, P.R., Di Tomaso, J.M. and Kochian, L.V., Aluminum toxicity in roots:an investigation of spatial sensitivity and the role of the root cap, *J. Exp. Bot.,* 44, 437–446, 1993.

154. Kochian, L.V., Cellular mechanisms of aluminum toxicity and resistance in plants, *Annu. Rev. Plant Physiol. Plant Mol. Biol.,* 46, 237–260, 1995.

155. Herrera–Estrella, L., Guevara–Garc, A. and Lopez–Bucio, J., *Heavy Metal Adaptation, Embryonic Encyclopedia of Life Sciences,* Macmillan Publishers, New York, 1999.

156. Jones, D.L., Organic acids in the rhizosphere — a critical review, *Plant Soil*, 205, 25–44, 1998.
157. Gobran, G.R., Wenzel, W.W. and Lombi, E., Eds., *Trace Elements in the Rhizosphere*, CRC Press, Boca Raton, FL, 2000.
158. Sanità Di Toppi, L., Gremigni, P., Pawlik Skowronska, B., Prasad, M.N.V. and Cobbett, C.S., Responses to heavy metals in plants — molecular approach. In *Abiotic Stresses in Plants*. L. Sanità Di Toppi and B. Pawlik Skowronska, Eds., Kluwer Academic Publishers, 133–156, Dordrecht, 2003.
159. Krotzky, A., Berggold, R. and Werner, D., Analysis of factors limiting associative N2 fixation with two cultivars of sorghum mutants, *Soil Biol. Biochem.*, 18, 201–207, 1986.
160. Krotzky, A., Berggold, R. and Werner, D., Plant characteristics limiting associative N2 fixation with two cultivars of sorghum mutants, *Soil Biol. Biochem.*, 20, 157–162, 1988.
161. Rennie, R.J., 15N isotope dilution as a measure of dinitrogen fixation by *Azospirillum brasilense* associated with maize, *Can. J. Bot.*, 58, 21–24, 1980.
162. Christiansen–Weniger, C., Groenman, A.F., van Veen, J.A., Associative N2 fixation and root exudation of organic acids from wheat cultivars of different aluminum tolerance, *Plant Soil*, 139, 167–174, 1992.
163. McClung, C.R. and Patriquin, D.G., Isolation of a nitrogen-fixing Campilobacter species from the roots of *Spartina alterniflora* Loiser, *Can. J. Microbiol.*, 26, 881–886, 1980.
164. Udvardy, M.K. and Day, D.A., Metabolite transport across symbiotic membranes of root nodules, *Annu. Rev. Plant Physiol. Plant Mol. Biol.*, 48, 493–523, 1997.

28 Phytoremediation of Trace Element Contaminated Soil with Cereal Crops: Role of Fertilizers and Bacteria on Bioavailability

Irina Shtangeeva

CONTENTS

28.1 INTRODUCTION

The selection of appropriate plant species is a cornerstone of successful application of phytoremediation methods and probably one of the most important factors affecting the extent of metal removal from contaminated soils. As a general rule, native plant species are preferred to exotic plants that can affect the harmony of the ecosystem [1]. The optimum metal phytoextraction plants should be able to accumulate and tolerate rather large amounts of toxic metals. Combined with a rapid growth rate and the potential to produce large biomass in the field, this can help to remove more metals per planting. Thus, the main goal is to find species able to accumulate large amounts of metals in harvestable plant parts without harmful consequences for the plants.

In the last two decades, the development of phytoremediation was mainly focused on the use of plants as hyperaccumulators [2–7]. Although the metal-accumulating plants have a good potential, so far there is no clear evidence that the plants may be actually used for commercial remediation of contaminated soils. Most hyperaccumulators are relatively small and grow slowly and often can accumulate only one metal; however, in many cases, metal pollution of soils is a multielement problem. Therefore, one of the alternative ways is to identify plants with a large biomass production capable of mobilizing several metals in the rhizosphere, assimilating the metals, and tolerating high metal contents in their tissues.

Recently, phytoremediation has been modified to include certain native grasses as potential candidates [8,9]. Plant species with large biomass are easily available around the world and, though they cannot be classified as hyperaccumulators, these plants can compensate the lower metal concentrations in their aboveground parts by greater biomass production and higher transpiration rate. These factors are critical in proving the successful field applicability of phytoremediation techniques. Comparing a large mass of soil with a mass of the plants to be used for soil cleaning, it becomes apparent that only large biomass species able to take up great amounts of metals may be used for phytoremediation of metal-contaminated soils.

Moreover, as a "soft" technique, phytoextraction may be successfully used probably only for remediation of slightly, and sometimes moderately, contaminated soils. In this case, an ability of certain plant species with high root and shoot biomass to affect chemical situation in the rhizosphere (by producing specific organic exudates capable of transferring metals to more soluble and more available to plant forms) may be expected to result in actual metal removal from contaminated soils. The attempt to use phytoextraction for cleaning highly contaminated soils is probably the main problem of phytoremediation philosophy. Each technique has certain limits where it may be the most suitable. It seems that phytoextraction procedures may be successfully applied mainly for remediation of slightly contaminated soils. Furthermore, even in this case we will have to "help" plants: first, by enhancing plant yields; second, by stimulating an increased uptake and translocation of several metals; and, last, by improving the ability of plants to tolerate high contents of metals in their tissues.

To identify the plant species best adapted for site-specific conditions (different levels and types of metal contamination, physicochemical parameters of soil), it is necessary to screen various natural and cultural species and evaluate these species in controlled greenhouse and field conditions. This will help to increase the number of potential candidate plants by expanding the type, range, and adaptability of the species perspective for metal phytoextraction and phytoaccumulation.

28.2 APPLICATION OF CEREAL CROPS IN PHYTOREMEDIATION STUDIES

The most important group of cultivated plants is cereal crops, which include eight genuses: wheat, rice, rye, oats, barley, corn (maize), sorghum, and some of the millets. Wheat, oats, barley, and rye can grow under relatively cold conditions (10 to 12°C). Rice, maize, and sorghum are warm-season crops; rice is unique in that it can germinate and thrive in water. Proso millet is a short-season crop grown on a small acreage in the areas of spring-sown grains [10].

In spite of wide diversity, cereal crops have many common morphological characters. The cereals have fibrous roots and the main root biomass is located in the arable layer. The depths of root penetration differ among different cereal crops: more than 100 cm for wheat, 120 to 200 cm for rye, and up to 200 cm for corn. Notice that plant species with fibrous roots and relatively large proportion of fine roots seem to have additional advantages for increased nutrient and metal uptakes.

- Wheat is planted on more acreage than any other crop is. It is grown most effectively in grassland climates, which have the appropriate level of rainfall and include a cold season. Wheat is classified in several ways, based on botany of the plant. The literature describes 18 species of wheat, but only two are of commercial importance: durum wheat and bread wheat. These two types of wheat are grown extensively throughout the world and account for 90% of world production [11].
- Rice is one of the most important crop plants and feeds more people than any other plant species on Earth. Cultivated rice belongs to two species, *Oryza sativa* and *O. glaberrima*. Of the two, *O. sativa* is by far the more widely utilized. A big difference between rice and other cereal crops is that this species often grows in water and thus the conditions of metal uptake can differ significantly between rice and other cultural plants (water-soluble metals may be more easily taken by plants) [10].
- Barley occurs in many types, but the most commonly cultivated type is *Hordeum vulgare* L. The best soils for growing barley are loam and clay loam. This plant is the most tolerant of cereal species to alkaline soils (pH 6 to 8) and the least tolerant of acid soils (below pH 5) of the cereal species [12]. As a cultivated crop, oats came after wheat and barley.
- Oat (*Avena*) is a grass; several species exist, though taxonomists do not agree on all the classifications. Of the cultivated species *Avena sativa* (common white oats) is the most important [13].
- Corn, or maize (*Zea mays* L.), was originally produced in the Western Hemisphere by Indians and was then carried to Europe by the early explorers. Today, it is cultivated in many temperate climates [10].
- Rye (*Secale cerealeis* L.) is the second most widely used cereal (after wheat) for bread making, though its gross production is less than 1/15 that of wheat. Rye can be grown on relatively poor soils and it is able to survive more severe winters than most grains.
- Forage sorghum (*Sorghum vulgare*, *S. bicolor* L.) is an important cereal grass native to Africa that ranks fourth after rice, corn, and wheat in terms of the importance for human nutrition. It is especially used as grass in summer because of its great resistance to heat and drought [10].
- Proso millet (*Panicum miliaceum* L.) is a member of the tribe Paniceae of the Panicoideae subfamily of grasses. This tribe of grasses originated in the tropics and most of these crops are grown in the hot, semiarid regions of the world. Proso millet is a warm-season grass and is usually capable of producing seed anywhere from 60 to 100 days after planting. Because of its relatively short growing season, it has a low moisture requirement and is capable of producing food or feed where other grain crops would fail. Proso millet is often planted as an emergency catch crop for situations in which other crops have failed, been hailed out, or were never planted due to unfavorable conditions. Proso is versatile in that it can be successfully grown on many soil types and is probably better adapted than most crops to "poor" land, such as land with soils having low water-holding capacity and low fertility.

28.3 METAL UPTAKE BY CEREAL CROPS

At present, not much of the literature concerns application of cereal crops for phytoremediation of contaminated soils, though experimental data on metal uptake by different cereals may be found in biological and analytical journals. Recent experimental studies showed that certain cereal genotypes are able to tolerate high concentrations of several toxic metals and thus can provide a good basis for successful phytoremediation of metal-contaminated soils [14–20]. Different species, and indeed different cultivars, regulate metal uptake at the soil–root and root–shoot interfaces to varying degrees. Among other cereals, rice and wheat are prospectively the best plants for soil phytoremediation.

Wheat cultivars vary considerably in their ability to grow and yield well in contaminated soils. An ability of wheat to accumulate different metals has been demonstrated in many publications [21–25]. According to recent literature data, different cultivars of wheat may differ in the ability to take up heavy metals [26–28]. Moreover, the experimental data may be very contradictory. It was reported [29] that whole-plant Cd accumulation and Cd translocation to shoots is greater in the bread wheat cultivar than in the durum cultivar. On the other hand, Li et al. [30] found that many durum wheat cultivars accumulate two to three times as much Cd in grain as does bread wheat. Greger and Löfstedt [27] showed that different Cd accumulation in wheat grains grown in nutrient solution was related to variations in the translocation from root to shoot and to the Cd concentration in shoot, flag leaf, and grain coats, but not to the uptake of Cd by roots.

Comparison of hyperaccumulator species *Thlaspi caerulescens* and wheat *Triticum aestevum* showed that root exudates of wheat generally are able to mobilize more metals from soil as compared to hyperaccumulators [31]. Considering that shoot and root biomass of wheat is significantly higher than that of hyperaccumulators, wheat may be a promising plant species for the aims of soil phytoremediation. It is also important to note here that an increase in the degree of metal accumulation in wheat may be accompanied by activation of antioxidant enzymes. It was found that bioaccumulation of metals capable of inducing the antioxidant protection may be fixed in the next generation of wheat seedlings [32].

The author's recent experimental data on wheat *Triticum aestevum* type Umanka indicates that this type can uptake and tolerate large amounts of different metals in its tissues [33]. As an example, Table 28.1 shows concentrations of Th in different parts of wheat seedlings grown in soil artificially contaminated with Th.

It is important that metal content increase not only in roots, but also in leaves of the wheat seedlings. This means that Th was actually transferred from soil to plant. In the course of the short-term (7 days), vegetation test concentration of Th in the contaminated soil decreased ~1.7 times (Figure 28.1). Moreover, Th bioaccumulation had not affected biomass of the plants. After 7 days, lengths of leaves in the experiment and in the control were the same. This indicates that Th was not toxic for the wheat seedlings. However, a rather short-term Th exposure was used for the experiment. The physiological and biochemical consequences of longer radiolitic and chemical Th stress would not appear in the further stages of the plant growth which cannot be excluded.

Rice can accumulate high amounts of different metals, including Cd, Co, Cu, Cr, Ni, Pb, and Zn [34–36]. Successful examples of application of rice to remove Cd from contaminated soils have been reported [37]. Data on using rice to remove Se from drainage water are also available [38].

TABLE 28.1
Mean Concentrations of Th[a] in Wheat Seedlings and in Soil where Plants Were Grown and Ratio of Th Content in Plants Grown in Th-Enriched Soil to That in Control

	Control	Experiment (+Th)	Ratio experiment/control
Leaves	0.10 ± 0.03	1.45 ± 1.25[a]	18.6
Roots	0.71 ± 0.61	43.9 ± 4.6[b]	69.2
Soil	7.0 ± 2.9	39.1 ± 11.6[a]	

[a] mg kg^{-1}.
[b] Differences between control and experiment are significant at $P < 0.01$.
[c] Differences between control and experiment are significant at $P < 0.001$.

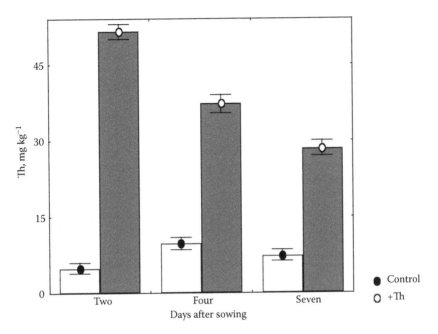

FIGURE 28.1 Variations in Th concentration in clean (control) and artificially contaminated with Th soil within 2, 4, and 7 days after adding Th to the experimental soil.

Application of maize in phytoremediation studies is described in numerous reports in the literature [39–41]. Maize can grow quickly, has high biomass, and has been found to accumulate different metals in its shoots and roots [42]. For example, Kalisova–Spirochova et al. [43] compared maize and sunflower grown in highly contaminated soil (Zn: 75,000 mg kg^{-1}; Pb: 16,000 mg kg^{-1}). The highest values of accumulation of Zn and Pb were found in roots and in leaves of corn (whole plant: 1158 mg kg^{-1} for Zn and 500 mg kg^{-1} for Pb). Sunflower showed considerably lower accumulation ability (whole plant: 47 mg kg^{-1} for Zn and 9 mg kg^{-1} for Pb).

Keltjens and van Beusichem [44] compared an uptake of Cu and Cd by wheat (*Triticum aestevum*) and maize (*Zea mays*) grown in nutrient solution. They found that, at identical external Cd levels, maize accumulated significantly more Cd in the shoots than wheat did, making maize a more pronounced "shoot Cd accumulator" than wheat. Unfortunately, similar metal uptake by these plants when they grow in soil cannot be predicted. In this case, quite opposite effects may be observed; these may be explained by differences that can arise when comparing experiments on metal uptake by plants grown in soil and culture solutions.

Much less information is available on uptake of metals by other cereal crops. Madrid and Kirkham [45] studied an uptake of Cd, Fe, Mn, Ni, and Pb by barley (*Hordeum vulgare* L.) grown in animal-waste lagoon soil. They found that amendment of the soil with EDTA resulted in an increase of uptake of Fe, Mn, Ni, and Pb. Luo and Rimmer [46] studied multielement toxicity at the early stages of barley growth. It was shown that growth of barley was controlled by the amount of plant-available Zn, which depended on the amounts of added Zn and added Cu. The effect of the added Cu was to increase the toxicity of the added Zn. Adams et al. [47] compared metal uptake by wheat and barley and found that, when it was grown under comparable soil conditions, barley could take up much lower amounts of Cd than wheat could.

Cobb et al. [48] compared three plant species (oats, radish, and lettuce) grown in soil contaminated with Cd, Cr, Ni, and Pb. They found that oats were the most tolerant compared to other plant species and were able to accumulate rather high amounts of Cd and Ni.

Knox et al. [49] studied an uptake of Cd by rye, maize, and oats grown in soils with different levels of Cd contamination. It was found that an increase of Cd content in soil up to 20 mg kg^{-1}

resulted in a significant increase of Cd level in all the plant species: oats accumulated 112 mg kg^{-1} Cd in their leaves, and concentrations of Cd in leaves of rye and maize were 43 and 41 mg kg^{-1}, respectively. When concentration of Cd was increased up to 40 mg kg^{-1}, Cd contents in leaves were 80, 98, and 198 mg kg^{-1} in rye, maize, and oats, respectively. Rye was found to be able to accumulate more Cu and Zn in its shoots compared to wheat and oats grown in the same soils [50].

Schumann and Sumner [51] performed experiments with sorghum grown in silty loam soil amended and non-amended with by-products. It was shown that As concentrations in sorghum leaves were linearly correlated (R^2 = 0.895) to total As in soil. An application of coal combustion by-products (fly ash) for crop fertilization led to a significant (four times) increase of As content in leaves that finally reached the level of 120 mg kg^{-1}. Estevez Alvarez et al. [19] studied the uptake of Cr, Cd, Ni, Pb, and Zn by sorghum and toxicity of the metals for this species. They observed a remarkable effect of Zn and Cr on the growth of the plant. Even insignificant increases of Zn and Cr concentrations in soil resulted in a death of sorghum. On the other hand, Cu, Ni, and Pb did not affect the plant yield.

The author's experimental data on oats and barley indicate that these plant species may be used for metal phytoextraction in slightly and moderately contaminated soils [15,16]. Vegetation tests showed that growth of oats and barley in different soils contaminated with various metals and metalloids resulted in a decrease of metal concentrations in soil solutions. As an example, Figure 28.2 illustrates variations in Cu and Zn contents in soil samples treated with 1 M ammonium acetate, pH 4.8 (this soil extraction can characterize sorbed and loosely bound metals) [52]. It is important to note that, after soil cultivation, total amounts of several elements in the soils also decreased (Table 28.2).

Behavior of As in the urban soils was the most interesting. During a 1-month vegetation test, As content in slightly contaminated soils cultivated with oats decreased up to the level of As in the clean soil (Table 28.2). Concentration of As in moderately contaminated soil also decreased significantly. In this case, such an effect was observed as a result of cultivation of soil with both oats and barley.

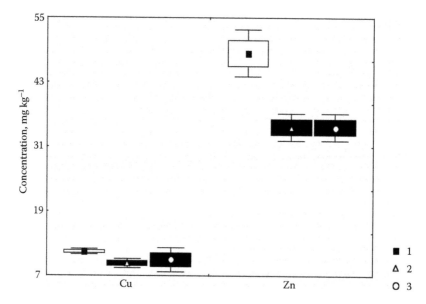

FIGURE 28.2 Concentrations of Cu and Zn in soil leachates before (1) and after growth of oats (2) and barley (3) in the soil.

TABLE 28.2
Total Amounts of Elements[a] in Soils Taken from Different Sites in St. Petersburg before (A) and after (B) Growth of Oats and Barley in Clean (1), Slightly Contaminated (2), and Moderately Contaminated (3) Soils

			Cr	Fe (%)	Co	Zn	As
A		1	20.1 ± 2.8^c	1.29 ± 0.16	2.98 ± 0.38^b	94.7 ± 11.3^c	1.94 ± 0.21^c
		2	27.7 ± 2.9^c	1.57 ± 0.11	8.61 ± 0.47^b	176 ± 19	2.66 ± 0.65^c
		3	126 ± 13	1.57 ± 0.11	5.33 ± 0.61	245 ± 25	9.92 ± 1.04
B	Oats	1	26.3 ± 4.9	1.53 ± 0.16	3.85 ± 0.61	120 ± 17	2.11 ± 0.67
		2	27.7 ± 6.5	1.95 ± 1.22	3.8 ± 0.8^d	140 ± 31	1.79 ± 0.45^d
		3	103 ± 31	1.85 ± 0.45	5.65 ± 0.91	208 ± 59	6.47 ± 0.79^d
	Barley	1	26.4 ± 4.1	1.38 ± 0.31	3.85 ± 0.63	124 ± 25	2.18 ± 0.68
		2	26.7 ± 6.8	1.31 ± 0.24	3.66 ± 0.86^d	148 ± 51	2.17 ± 1.22
		3	116 ± 47	2.2 ± 0.5	6.21 ± 1.43	221 ± 70	6.85 ± 2.01^d

[a]mg kg^{-1}.
[b]Differences between concentration of the elements in soils taken from site 1 and site 3 and soils taken from site 2 and site 3 are significant at $P < 0.01$.
[c]Differences between concentration of the elements in soils taken from site 1 and site 3 and soils taken from site 2 and site 3 are significant at $P < 0.001$.
[d]Differences between concentrations of the elements in cultivated and non-cultivated soils are significant at $P < 0.001$

28.3.1 PHYTOTOXICITY OF SOME ULTRATRACE METALS

Ideally, screening of plant species for phytoremediation purposes should involve a broad range of trace metals and metalloids. The list of elements commonly considered as pollutants is rather short and usually includes Al, As, Cd, Cr, Cu, Hg, Ni, Pb, Se, Zn, and some radionuclides. Meanwhile, many other trace elements may also be toxic if their concentrations in soil exceed ordinary concentrations typical for the soil.

To complete the life cycle, plants must acquire not only macronutrients (N, P, K, S, Ca, and Mg), but also many trace elements [53]. Unfortunately, little or no attention has been paid to the majority of the 90 elements that occur in soil and in different plant species, though it may be assumed that these elements are also involved in specific physiological activities and their biological significance is presently unknown. Until the present time, few investigations have been conducted on phytotoxicity of various trace metals not included in the short list mentioned earlier.

During the last few years, much attention has been paid to toxic effects of the rare earth elements (REE) widely used in agricultural practice in China [54–56]. It was shown that plants can take up rather large amounts of REE and this can affect concentrations of main nutrients in the plants. The experiments performed by the author on Eu effects on plant nutrition showed that germination of wheat seeds in the medium enriched with Eu and growth of wheat seedlings in the soil to which a small amount of Eu was added may result in significant variations in concentrations of macronutrients in the plants. Figure 28.3 shows variations in K content in roots and leaves of 7-day-old wheat seedlings germinated in medium to which 20 µg kg^{-1} Eu was added. Although concentration of K in roots remained unchanged, K content in leaves decreased significantly compared to that in the control plants. Such a decrease of K in generative plant parts may have harmful consequences for future development of the plant.

Many other ultratrace metals — for example, exotic elements such as Sc — were never considered as pollutants. However, even insignificant (1.5 times) increase of Sc content in soil may result in Sc accumulation in plants growing in the soil (concentration of Sc in roots can increase up to 20 times compared to Sc content in the roots of the control plants) and significant variations

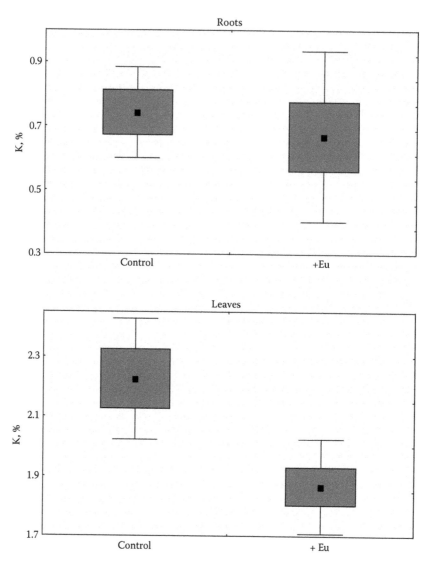

FIGURE 28.3 Variations in K content in roots and leaves of the wheat seedlings grown in the control soil and in the soil where a small amount of Eu was added.

in concentrations of main nutrients in the plants [57]. In particular, an increase of Sc concentration in the plants led to a decrease of K in seeds and increased Na, Ca, and Cs contents in leaves. It is interesting that negative effects of Sc were found to be higher than phytotoxic effects of such radioactive element as Th. At first glance, this seems strange because, in the case of radionuclides, effects of chemical and radiolitic toxicity have been combined. However, vegetation tests performed by the author showed that an increase of Th content in soil can cause only minor variations in elemental composition of wheat seedlings (concentration of Sb in leaves and concentrations of Hf and Cr in roots of the plants grown in Th-contaminated soil became higher ($P < 0.01$) than in the control plants grown in ordinary soil) [33].

Plants have evolved highly specific mechanisms to take up, translocate, and store different elements. The uptake mechanism is selective; plants preferentially acquire some ions over others [1]. For example, mean Au content in soil is very low: $10^{-7}\%$ [58]. It may be assumed that rather high (n_*10^{-6} to $n_*10^{-5}\%$ of dry weight) amounts of Au found in different plant species [59,60] may be attributable to the demand of the plants for these amounts of Au to complete certain biochemical processes.

28.4 METAL DISTRIBUTION BETWEEN ROOTS AND UPPER PLANT PARTS

Uptake of metals by root cells, the point of entry into living tissues, is a step of major importance for the process of metal phytoextraction. However, for phytoextraction to occur, metals must also be transported from the roots to the upper plant parts. It is quite possible that a plant species exhibiting a significant metal accumulation into the roots may have a limited capacity for phytoextraction. For example, many publications have reported that concentrations of trace metals in roots may be several times higher than in shoots [61,62]. Kim et al. [36] found that Pb content in roots may increase with increased concentrations of Pb^{2+} applied to soil, but Pb content in shoots will remain unchanged. In addition, the mechanisms of metal translocation from root to shoot may be different for different metals.

As an example, Table 28.3 shows concentrations of several elements in different parts of three cereal crops (oats, barley, and wheat) grown simultaneously on the same soil. Concentrations of all the elements (except calcium in barley) were higher in roots than in leaves. It is also important to remember that only a part of the total amount of ions associated with the root is absorbed into cells. A significant ion fraction is just physically adsorbed at the extracellular negatively charged sites of the root cell walls [1]. The cell wall-bound fraction cannot be translocated to the shoots and, therefore, cannot be removed by harvesting shoot biomass (phytoextraction).

Binding to the cell wall is not the only plant mechanism responsible for metal immobilization into roots and subsequent inhibition of ion translocation to the shoots. Metals can also be complexed and sequestered in cellular structures (e.g., vacuoles), thus becoming unavailable for translocation to the shoot [63]. In addition, some plants, so-called excluders, possess specialized mechanisms to restrict metal uptake by roots. Movement of metal-containing sap from root to shoot is primarily controlled by two processes: root pressure and leaf transpiration. Following translocation to leaves, part of metals can be reabsorbed from the sap into leaf cells. However, certain amounts of metals may evaporate during transpiration together with water (the water will carry dissolved mineral salts). As has been reported, elemental composition of transpiration solutions correlates well with elemental composition of soil water [64].

TABLE 28.3
Concentrations of Elements[a] in Roots and Leaves of Wheat, Barley, and Oats

Element	Wheat		Barley		Oats	
	Roots	Leaves	Roots	Leaves	Roots	Leaves
Ca	0.30 ± 0.23	0.17 ± 0.10	0.50 ± 0.17	0.50 ± 0.21	0.70 ± 0.34	0.34 ± 0.10
Cr	9.4 ± 1.4	0.82 ± 0.22	10.7 ± 11.2	2.0 ± 0.6	8.9 ± 3.0	1.8 ± 0.4
Fe	686 ± 199	162 ± 30	1319 ± 1373	113 ± 24	1576 ± 683	132 ± 29
Co	0.29 ± 0.08	0.04±0.01	1.1 ± 0.8	0.07 ± 0.02	2.0 ± 1.2	0.07 ± 0.02
Zn	235 ± 90	61.2 ± 11.8	200 ± 80	136 ± 160	165 ± 38	55.0 ± 9.0
As	0.64 ± 0.31	<0.1	0.62 ± 0.30	0.22 ± 0.24	0.79 ± 0.60	0.12 ± 0.06
Sb	0.12 ± 0.07	0.07 ± 0.03	0.17 ± 0.08	0.03 ± 0.01	0.27 ± 0.10	0.05 ± 0.01
Sm	0.16 ± 0.03	0.005 ± 0.002	0.25 ± 0.22	0.03 ± 0.01	0.21 ± 0.02	0.03 ± 0.01
Th	0.56 ± 0.23	0.10 ± 0.03	0.49 ± 0.47	0.10 ± 0.12	0.51 ± 0.13	0.05 ± 0.02
U	0.41 ± 0.05	<0.04	0.30 ± 0.18	0.05 ± 0.03	0.48 ± 0.17	0.05 ± 0.02

[a] mg kg^{-1}.

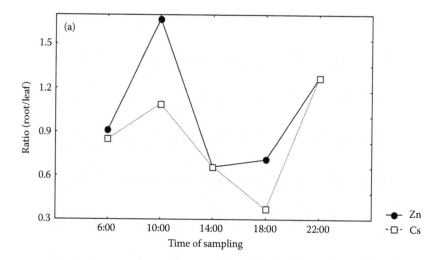

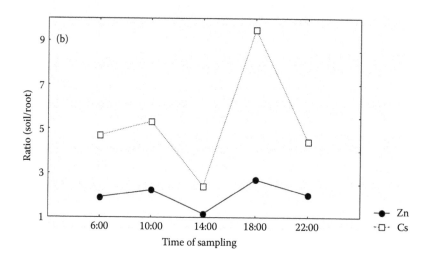

FIGURE 28.4 Daily dynamics of ratio of Cs and Zn concentrations in roots to those in leaves (a) and ratio of Cs and Zn concentrations in soil to those in roots (b) of wheatgrass.

A certain part of metals may also return back to soil. This process is especially active during first phases of plant growth [65]. Figure 28.4 illustrates daily dynamics of ratios of Zn and Cs concentrations in roots to those in leaves of wheatgrass and ratios of Zn and Cs concentrations in roots to those in the soil where the plants grow. During the day, the variations in metal concentrations are quite significant and very similar for different elements.

28.4.1 Effects of Soil Characteristics, Weather Conditions, and Plant Physiological Activity on Metal Uptake

Physicochemical characteristics of contaminated soils are also important for the selection of the remediating plants. Implementation of phytoremediation methods in real environmental conditions requires a careful preliminary assessment of each site and, first of all, detailed analysis of soil matrix. Each site is unique with regard to its chemical and physical characteristics, such as temperature, pH, moisture content, soil texture, nutrients available for microbes and plants, etc. The process of metal phytoextraction is very sensitive to many of these variables.

In particular, it is necessary to take into account a high variability of soil metal contents within plots and between years. The variability between plots may result from microtopographic relief of the area. For example, extreme care is required in the interpretation of phytoremediation experiments in a the case where the experimental site is situated on slightly sloping territory. Under these circumstances, significant variations in concentrations of main cations such as Ca, K, Na, and trace metals along the soil profile that are caused by quite natural reasons can be expected.

Sunlight for photosynthesis, water, and nutrient supply, suitable temperature, and humidity are essential constituents of plant growth [66]. Differences between years may be rather significant because of variations in climatic conditions [67]. This can include plant dry matter yields, nutrient uptake, and distribution of nutrients between different plant parts [50]. As a result, the rates of metal uptake by the same plant species grown in the same place can vary between years depending only on climatic conditions during the vegetation seasons.

Figure 28.5 and Figure 28.6 illustrate distribution of Cr, Ni, and Pb in different parts of wheat (*Triticum aestevum*) grown in 2002 and 2003 in loam soil and in loam sand soil. The two years differed significantly in temperature and mean summer rainfall. The vegetation experiments showed significant variations in concentrations of different elements in the plants. In a more wet and warm season (2002), uptake of metals by the plants was lower compared to that in a more dry and cold season (2003). Additionally, the metal uptake depended on soil texture (it is known that water content and soil texture are generally among the main factors controlling metal uptake [1]). Finally, biomass production, an important parameter of soil phytoremediation, is also determined by climatic conditions during vegetation season [68].

Numerous experiments have been conducted to estimate the adsorption or solubility of trace elements as a function of soil pH and its effect on metal uptake by plants. It was shown that, in general, metal solubility in soil decreases as pH increases [69–71]. However, it is necessary to take into account that plants can survive and produce sufficient biomass only at certain ranges of soil pH: 5.0 to 8.0. Metal phytotoxicity increases with decreasing pH [72] and may result in inhibition of plant growth. An increased metal availability associated with the low soil pH may also be detrimental to rhizosphere microorganisms [40]. For example, Angle and Heckman [73] reported that a change of pH from 6.2 to 5.6 in soil treated with metal-contaminated sewage sludge resulted in a dramatic decline of AM colonization in plant roots. However, it was also reported [67,74] that, even under minimal pH changes, significant variations in metal uptake can often be observed. Therefore, soil pH alone cannot explain these changes in metal bioaccumulation.

Seed germination, planting, water irrigation and fertilization, and time of harvesting are important factors responsible for successful implementation of the phytoremediation method. Many efforts are required to find an optimal combination of all these parameters to provide the most efficient metal phytoextraction. For example, environmental conditions during seed germination and planting have a considerable influence on future plant development, including an ability of the plants to take up nutrients and metals [75]. With respect to cereal crops, the examples of wheat [25], barley, and oats [16] showed that younger plants can take up and translocate higher amounts of metals than older plants can. Laboratory tests clearly indicate that chlorophyll, protein, and many chemical elements in cereal grasses reach their peak concentrations in the period just prior to the jointing stage of the green plant. Although this period lasts for only a few days, cereal grasses should be harvested precisely during this stage of the plant development. These observations were also supported by data on other plant species [76].

Exudation of organic compounds by roots is one of the most important factors influencing the ion solubility and metal uptake through their indirect effects on microbial activity, rhizosphere physical properties, and root growth dynamics and directly through acidification, chelation, precipitation, and oxidation/reduction reactions in the rhizosphere [53,77]. The components of plant

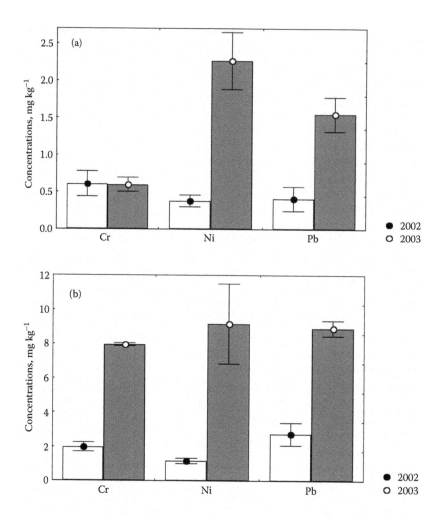

FIGURE 28.5 Concentrations of Cr, Ni, and Pb in leaves (a) and roots (b) of the wheat grown in loam soil in 2002 and 2003.

root exudates are various and complex; they serve as a source of carbon substrate for microbial growth and also contain chemical molecules that promote chemotaxis of soil microbes to the rhizosphere [78]. In addition, root exudates control the nutrient and metal uptake. The root-derived exudates can improve the availability of nutrients in a case when they were released in response to a particular nutrient deficiency.

It was also shown that plants under stress tend to produce large amounts of secondary metabolites, which may influence the availability of nutrients and metals in the rhizosphere [79]. The metal-induced exudation of organic acids might be a common response in plants. It should be remembered that root exudates are species specific and also may vary when the same plant species is growing in different soils [80]. For example, in response to Cu stress, rye can exude two times more organic acids from roots than triticale and six times more than maize [81].

28.4.2 Identification of Soil Amendments Capable of Enhancing Plant Yield and Metal Phytoextraction

The mobilization of metal contaminants in soil and in plant is an important factor influencing the success of phytoremediation. Many metals have low solubility in soils because of adsorption of

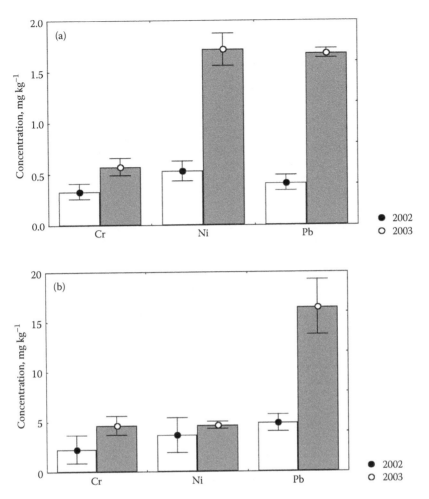

FIGURE 28.6 Concentrations of Cr, Ni, and Pb in leaves (a) and roots (b) of the wheat grown in loam sandy soil in 2002 and 2003.

the metals to soil particles. For example, most clay particles are covered with a thin layer of hydrous Fe, Mn, and Al oxides. These selective sites can maintain metal activity in the soil solution at low levels [82]. Mobility of metals in soil may also be reduced because of metal precipitation as insoluble phosphates, carbonates, and (hydr)oxides [83]. In addition, plants usually do not translocate metals efficiently from roots to shoots.

Commercial fertilizers not only play an important role in the improvement of soil fertility and the increase of crop production, but can also enhance the uptake of metals by plants through improvement of plant growth and root density (in general, the more biomass the plants have, the more metals can be accumulated because the metal uptake is a function of the overall biomass). Organic and inorganic fertilizers can provide a potent means of stimulating the root and shoot biomass. In addition, fertilizers can effectively and specifically increase solubility and, therefore, bioaccumulation of metals because metal bioavailability may often be increased by lowering soil pH and altering soil ion composition. Therefore, phytoremediation effects may be enhanced using the agronomic approaches, and success of the manipulation measures will depend significantly on the agronomic practices applied at the site.

It should be remembered, however, that consequences observed as a result of the soil fertilization may be rather different. The effects will depend on

- Type of soil
- Climatic conditions
- Type and amount of applied fertilizer
- Species of plants growing in the soil
- Character of microbiological processes

In fact, options for increasing agricultural productivity are limited. At present, the phytoremediation practice has several commonly accepted agricultural approaches. They include application of organic materials, sewage sludge, chelates, and inorganic agents such as nitrogen, potassium, calcium, etc.

28.4.3 SEWAGE SLUDGE

Land application of sewage sludge — the solid portion that remains after wastewater treatment — provides a means of disposing unwanted waste products as well as returning valuable nutrients and organic matter to the land [25]. During the past several decades, sludge has been applied in increasing amounts to agricultural lands. Sewage sludge might be an effective and cheap alternative to commercial fertilizers because it usually has high contents of nutrients and organic matter. However, the rather high level of numerous trace metals is the most significant impediment of wider land application of sludge and can negate the benefits of the manipulation measure.

It was assumed that metals in sewage sludge are generally organically bound and therefore less available for plant uptake than more mobile metal salt impurities found in commercial fertilizers [84,85]. However, this conclusion was not supported by experimental data. For example, it was reported that increasing sewage sludge doses induced a linear increase in Cu, Ni, and Zn concentrations in soil solutions [86]. Speir et al. [71] showed that New Plymouth sewage sludge contained 3 to 10 mg kg^{-1} Cd, 180 to 320 mg kg^{-1} Cr, 400 to 570 mg kg^{-1} Cu, 190 to 270 mg kg^{-1} Ni, 64 to 130 mg kg^{-1} Pb, and 1620 to 2400 mg kg^{-1} Zn. Moreover, pH of the tested light-textured sandy soil was markedly reduced after sludge application (up to pH 4 in some samples), presumably as a result of breakdown of the unstable organic matter, nitrification of the NH_4^+–N, and sulphide oxidation. As a consequence, soil solution concentrations of Cu, Ni, and, especially, Zn increased significantly.

Also, sludge is an imbalanced source of plant nutrients. Approximate N:P:K ratios for sewage sludge are 2.5:1.0:0.9 [51]. This ratio implies that P will be in excess for most field and forage crops when they are fertilized with the sludge to supply the total crop N requirement because, for example, recommended N:P:K for maize is 7.5:1.0:4.4, and 9.0:1.0:4.7 for grain sorghum. Finally, the benefits of the sludge application may be reduced with time because of mineralization of the sludge organic fraction [86].

28.4.4 CHELATES

The use of synthetic chelates has been shown to stimulate the potential for Pb accumulation in plants dramatically. These compounds prevent Pb precipitation and keep the metal as soluble chelate–lead complexes available for uptake by roots and transport within a plant. It was reported that an addition of EDTA (ethylene-diamine-tetraacetic acid) at a rate of 10 mmol kg^{-1} soil stimulated Pb accumulation in shoots of maize up to 1.6% and simultaneously altered root/shoot partitioning in a wide variety of crop plants [87]. However, certain differences in Pb accumulation levels between different plant species under equivalent added chelate [88] may be present.

Chemically assisted phytoextraction has also been applied for phytoremediation of soils contaminated with other metals. Anderson et al. [89] reported their finding that plants could be induced to hyperaccumulate Au using thiocyanate and thiosulphate. Uranium concentrations may be strongly increased when citric acid is applied [90]. Thayalakumara et al. [91] showed that EDTA application increased the leaf Cu concentration of grass from 30 to 300 mg kg^{-1}. They found that more Cu was accumulated in the leaves when EDTA was applied in numerous small doses than in just one large dose. The authors also found that, as a result of EDTA applications, about 100 times more Cu was leached than was taken up by grasses. This indicates that a strategy for managing leaching losses must be a part of the EDTA-enhanced phytoremediation procedures.

Chemically assisted phytoextraction can work well when the metal to be extracted has initially a very low bioavailability, and thus is not phytotoxic, allowing the establishment of a large plant biomass before the chelate is applied. In contrast, such metals as Cu, Zn, and Cd are usually more bioavailable in soil and can cause severe phytotoxicity at levels that require remediation, particularly to dicotyledonous species such as *Brassica juncea* [92]. It was shown that *B. juncea* seedlings were able to accumulate 875 mg kg^{-1} Cd with chelating agent vs. 164 mg kg^{-1} without a chelator [93]. This amount of Cd may be very toxic to the plant and result in the inhibition of plant growth, thus limiting the chance of successful chemically assisted phytoextraction on mixed pollution sites. The other issue of concern is the possible leaching of metal–chelate complexes to groundwater, which may represent an environmental hazard [94]. For example, it was reported that application of more than 1 g kg^{-1} EDTA becomes inefficient because Pb concentration in crops is not enhanced and the leaching rate increases [95].

The chelate-induced metal hyperaccumulation may be fatal to the plant. An accumulation of elevated Pb levels is highly toxic and can cause plant death. Because of the toxic effects, it is recommended that chelates should be applied only after a maximum amount of plant biomass has been produced. Prompt harvesting (within 1 week of treatment) is required. Huang et al. [88] suggested the following field application protocol when growing a high biomass crop that is sensitive to chelates. After the crop has become well established and reached a sufficient biomass, a selected chelate may be applied to the root zone to facilitate rapid Pb accumulation. The plants should be harvested shortly after chelate addition to reduce environmental risk.

28.5 ORGANIC MATERIALS

An application of organic materials such as animal manure, poultry litter, and pig slurries can stimulate plant yield and increase aggregate stability, infiltration rate, retention of water and decrease soil bulk density. In many cases, the plant yield may be progressively increased as the rate of the organic materials increases [96]. The enhanced soil fertility and improved soil physical conditions may also be due to the macronutrients contained in the organic materials [68].

However, nutrients are not the only components of organic materials that can cause environmental concern. Heavy metal accumulation in soils fertilized with animal wastes is also possible. The trace metals contained in the organic materials may affect the level of metal concentrations in the soil solution. For example, long-term application of Cu and Zn rich poultry litter was found to increase concentrations of levels of these metals in soil up to toxic [23]. Concentrations of Cu and Zn in animal wastes can vary due to the type of animal and feeding practices [50]. For example, the amounts of Cu and Zn applied to the soil with poultry litter exceeded the annual nutrient requirements of plants, with estimates of 6.6 times more Zn applied than plants require [97]. Thus, the organic fertilizers used for remediation of contaminated soils must be rather clean to exclude the possibility of adding new metals to the soil. At the same time, they must be able to stimulate transfer of certain elements, including toxic metals, to more available to plants forms with a consequent accumulation of the metals in plants.

Application of organic materials is known to affect soil pH; therefore, these materials may influence soil metal mobility through their influence on solubility or dissociation kinetics of metals and changes in the solid/liquid phase equilibrium [98]. Nitrogen, the most important component of organic fertilizers, can highly correlate with P, Cu, and Zn concentrations in above-ground plant parts, suggesting that improvements in N fertility can increase P, Cu, and Zn concentrations in plants [50]. It was shown that Cu is able to form Cu–NH$_4$ complexes that are very mobile in soil

TABLE 28.4
**Concentrations of Metals[a] in Initial Non-fertilized
(Control) and Fertilized Soils**

	Cd	Cu	Pb	Zn
Site 1 — control	0.44 ± 0.06	39.3 ± 0.5	53.1 ± 2.9	128 ± 10
Site 1 + urea	0.43 ± 0.03	42.1 ± 3.7	49.3 ± 5.2	135 ± 9
Site 1 + manure	0.45 ± 0.02	37.6 ± 4	59.5 ± 25	129 ± 11
Site 1 + ispolin	0.45 ± 0.02	38.3 ± 3.3	74.1 ± 44.5	121 ± 10
Site 2 — control	2.4 ± 0.5	159 ± 19	208 ± 31	514 ± 74
Site 2 + urea	2.2 ± 0.1	134 ± 19	195 ± 24	470 ± 50
Site 2 + manure	2.5 ± 1.1	148 ± 12	179 ± 40	499 ± 30
Site 2 + ispolin	2.7 ± 0.2	159 ± 11	181 ± 31	486 ± 43

[a] mg kg^{-1}.

solutions [99]. Thus, an excess of N application may also result in an excess of trace metal contents in plants grown in the fertilized soils. It should also be noted that no direct correlation between N uptake and plant yield has been shown. As reported by Trapeznikov et al. [100], an increase of 24 to 101% (depending on plant species) of N uptake was accompanied by an increase of above-ground biomass of 18 to 52%. However, it is clear that the increased plant biomass resulting from the treatment may be attributed first of all to enhanced soil fertility and not only to N uptake.

An ability of cereal crops to take up large amounts of nutrients (N, P, and K) has been shown in numerous studies [101]. Cereal crops also demonstrated a high tolerance to several metals, so they might use the metal surpluses that may originate from soil contamination. It may be assumed that certain species of cereals, especially in combination with appropriate soil amendments, may be successfully used to improve the processes of metal phytoextraction and phytoremediation of contaminated soils.

28.6 EXPERIMENTAL STUDIES ON THE EFFECTS OF DIFFERENT FERTILIZERS ON METAL REMOVAL FROM CONTAMINATED SOILS USING WHEAT

To optimize and manipulate the process of metal phytoextraction successfully, the author studied the effects of three fertilizers (urea, horse manure, and ispolin — fertilizer on the basis of mixture of organic fertilizers, humic acids, and industrial population of worms) on yields and physiological characteristics of wheat (*Triticum aestivum*) and removal of metals from two Podzol soils with loam (site 1) and sandy loam (site 2) textures. Soil in site 2 was contaminated with several metals (Cd, Cu, Pb, and Zn), and soil in site 1 was relatively clean. Wheat seedlings were grown in the two soils for 36 days. Urea, manure, and ispolin were incorporated into the soils at rates of 10 mg kg^{-1}, 100 mg kg^{-1}, and 100 mg kg^{-1}, respectively.

Metal concentrations in the initial soils before and after application of different fertilizers are presented in Table 28.4. No statistically significant differences were present between concentrations of Cd, Cu, Pb, and Zn in the fertilized and nonfertilized soils. Therefore, soil contamination with the metals by applied fertilizers was negligible.

The amounts of metals taken by plants during the vegetation test are shown in Figure 28.7. Although treatment of soil with all the fertilizers did not automatically result in an increase of metal concentrations in plants, it is seen that application of ispolin led to an increase of Cu and application of urea led to increased uptake of Cd, Cu, Pb, and Zn in the wheat grown in the loam (clean) soil. Amendment of sandy loam (contaminated) soil with manure and ispolin resulted in an

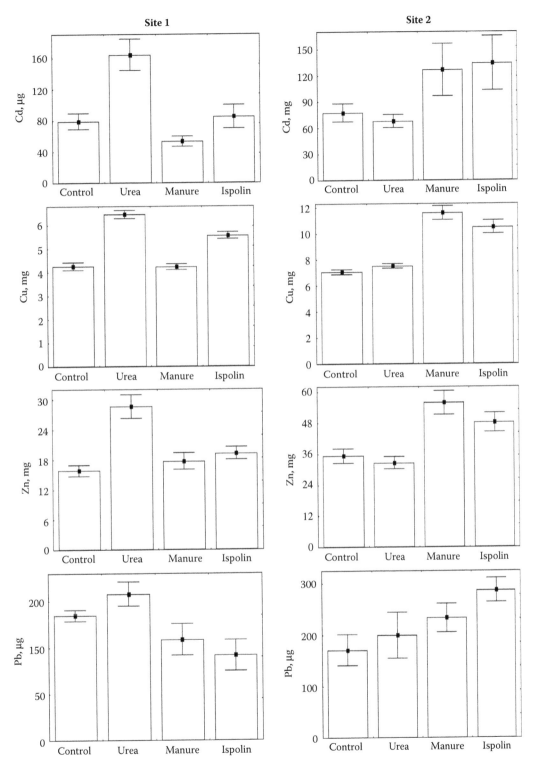

FIGURE 28.7 Variations in amounts of metals taken by wheat grown in clean (site 1) and contaminated (site 2) soils after application of different fertilizers. Control wheat was grown in non-fertilized soil.

TABLE 28.5
Effects of Wheat Growth and Application of Different Fertilizers on Concentrations of Metals[a] in Contaminated Soil

| | Without fertilizers | | Wheat + fertilizers | | | | | |
| | Control | | Urea | | Manure | | Ispolin | |
Element	1a[b]	2a[c]	1b[d]	2b[e]	1b[d]	2b[e]	1b[d]	2b[e]
Cd	2.4 ± 0.5	1.7 ± 0.4[f]	2.2 ± 0.1	1.3 ± 0.4[g]	2.5 ± 1.1	1.6 ± 0.6	2.7 ± 0.2	1.4 ± 0.4[g]
Cu	159 ± 19	164 ± 39	134 ± 19	151 ± 19	148 ± 12	169 ± 15	159 ± 11	143 ± 14[f]
Pb	208 ± 31	170 ± 42	195 ± 24	155 ± 47	179 ± 40	169 ± 33	181 ± 31	143 ± 21[f]
Zn	514 ± 74	507 ± 178	470 ± 50	555 ± 188	499 ± 30	426 ± 79	486 ± 43	404 ± 72[f]

[a] mg kg^{-1}.
[b] Initial nonfertilized soil.
[c] Soil after the end of the experiment (only wheat).
[d] Initial fertilized soil.
[e] Soil after the end of the experiment (wheat + fertilizer).
[f] Differences between metal concentrations in the initial and experimental soils are significant at $P < 0.05$.
[g] Differences between metal concentrations in the initial and experimental soils are significant at $P < 0.01$.

increased metal uptake by the plants grown in the soil. The differences between control and the treatments were significant at $P < 0.05$.

It is known that fertilizers may affect soil pH and therefore mobility of metals in soil [70,71]. The pH of the loam soil amended with urea was the lowest compared to the pH values of the control (without application of fertilizers) and loam soil amended with other fertilizers. In this case, it may be assumed that the decrease in the soil pH was mainly responsible for the variations in metal uptake. On the other hand, pH of the sandy loam soil decreased significantly ($P < 0.05$) only after application of ispolin. Because pH of the soil fertilized with manure did not change compared to that of the control, in this case soil pH cannot be used for explanation of the observed variations in the metal uptake.

The highly variable process of element uptake depends on many factors and pH is not the only parameter determining mobility of trace elements in soil. In particular, these two soils differed significantly in soil texture and content of total carbon (8.4% in loam soil and 3.6% in sandy loam soil). Soil texture and soil organic matter content, like pH, can have a significant influence on metal uptake. Finally, as a function of many factors, metal mobilization, in certain environmental conditions, may be efficient for one or another metal, but may also result in immobilization of others.

The experiments showed that growth of wheat in the nonfertilized, contaminated soil led to a decrease (1.4 times) of Cd content in the soil (Table 28.5). Amendment of the soil with urea enhanced the effect: the decrease of soil Cd was more significant. The best effect was demonstrated after application of ispolin: over a short period (36 days) concentrations of Cd, Cu, Pb, and Zn in the contaminated soil decreased 1.2 to 1.4 times compared with those in the initial soil (the differences were statistically significant). It should be noted that calculations made by McGrath et al. [102] showed that it would take 9 years to reduce Zn concentration in soil from 440 mg kg^{-1} to 300 mg kg^{-1} using plant hyperaccumulators. Application of wheat together with ispolin for remediation of metal-contaminated soil allowed them to affect simultaneously the uptake of several metals and, more importantly, to reduce time required for the soil cleanup. In particular, if the decrease of Zn content in this soil continued with approximately the same rate, the level of 300 mg kg^{-1} may be reached within several months.

To estimate the effect of the soil type and amendment of soils with ispolin on element behavior in plants, a principal component analysis of the plant samples was performed (Figure 28.8). Roots and leaves of the plants grown in the two soils were separated from each other. The PC1 was

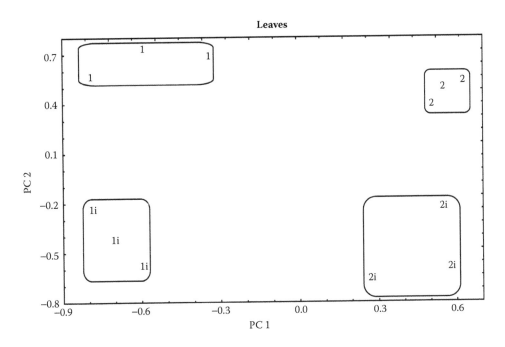

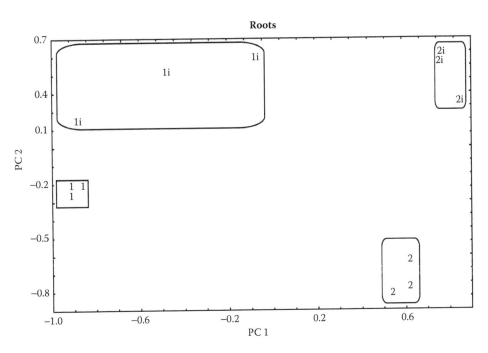

FIGURE 28.8 Score plot of the PC1 and PC2 for leaves and roots of the wheat seedlings grown in clean (1) and contaminated (2) soils. 1 and 2: plants were grown in nonfertilized soils; 1i and 2i: plants were grown in soil fertilized with ispolin.

responsible for the separation. In roots, Na, Cu, S, Pb, Zn, Sb, and P were highly correlated with the first PC. In leaves S, Sb, Mo, Na, Cu, Ca, As, and K provided the main contribution to the PC1. As one might expect, with small exceptions (P in roots), concentrations of these elements were different in plants grown in the two soils. Although the list of the elements responsible for the separation of the plant samples into different groups was similar for roots and for leaves, certain differences between different plant parts were observed. This might reflect different functions of roots and leaves in a plant and, as a result, different behavior of certain elements in the two plant parts.

The second PC was responsible for separation of the plants grown in the control and fertilized soils. Highly correlated with the PC2 elements were K, Rb, Mo, P, Pb, B, and Sb in roots and Rb, K, Mo, As, Mg, Sr, and Cd in leaves. In both cases, K, Rb, and Mo provided the most significant contribution to separation of the plant samples. In ispolin, potassium and molybdenum were presented in a form that may be easily taken by a plant. An application of ispolin resulted in an increase of the level of potassium in soil. This may be explained by rather high (5 to 8%) concentration of potassium in ispolin. Rb is a chemical analogue of K and similarity of physico-chemical parameters of K^+ and Rb^+ might result in a similar behavior of the two ions in the plant. In particular, it is known that, under K^+ deficiency, plants can take up higher amounts of Rb [103]. Other elements were rather different for different plant parts. As it was in the previous case, concentrations of many of these elements in plants grown in fertilized and nonfertilized soils differed significantly, though an addition of ispolin to both soils did not result in an increase of concentrations of these elements in the soils.

28.6.1 EFFECTS OF MINERAL ELEMENTS ON PLANT BIOMASS AND METAL UPTAKE

Deficiency and excess of mineral elements in soil can significantly affect plant yield. It is known that nonsufficient nutrient availability can reduce growth of leaf biomass [104]. In addition, metal mobility in soil solution and, as a result, the rate and extent of metal phytoextraction will also depend on the presence and amounts of various inorganic chemicals in the rhizosphere. For example, phosphorus is a major nutrient, and plants respond favorably to the application of P fertilizer by increasing biomass production. The application of P fertilizer, however, can inhibit the uptake of some major metals due to metal precipitation as pyromorphite and chloropyromorphite [105].

Potassium has been found to enhance yielding capacity, resistance to stresses and diseases and crop quality, though it was also reported that potassium fertilizer depending on the K/Mg ratio in the soil can increase or decrease the plant yield. (When this ratio is near 1, the yield increases, but when it is over 3, the yield can decrease [106].) High bioavailable potassium content in soil may also result in Mg and Ca deficiency in plants. This may be illustrated by data presented in Table 28.6 and Figure 28.9.

An application of ispolin, fertilizer, which has a very high content of available potassium, resulted in a significant increase of K concentration in soil. Total amount of K and content of exchangeable K were higher in the soils amended with this fertilizer. As a result, concentration of K in the plants grown in the fertilized soils was higher than in the control plants. Although Ca and Mg concentrations in both soils remained unchanged after application of ispolin, concentrations of these two nutrients in plants decreased. The most significant decrease was observed in leaves. When the K/Mg ratio in soil was the highest, a statistically significant decrease of Mg concentration in roots was also observed. Figure 28.9 shows variations in plant biomass resulting from application of ispolin. The variations were soil dependent.

In loam soil, an application of ispolin stimulated plant yield, but fertilization of sandy loam soil did not affect the growth of plant biomass (only in this case the K/Mg ratio in the soil was the highest: 6.2). Other characteristics of the physiological state of the plants — for example, chlorophyll index — demonstrated significant effects from ispolin application in the plants grown in loam soil and an absence of such an effect in the plants grown in sandy loam soil (Figure 28.9).

Zhao et al. [24] studied the effects of soil amendments with KNO_3, KCl, and K_2SO_4 on plant yield and metal uptake. They found that shoot and root dry weight of wheat was reduced significantly by the addition of K fertilizer in KCl and K_2SO_4 forms, but that only minimal changes occurred when soil was treated by KNO_3. This experiment showed that anions such as Cl^- and SO_4^{2-} increased

TABLE 28.6
Total Amounts and Concentrations of Exchangeable Mg, K, and Ca[a]
in Fertilized and Nonfertilized Soils and in Plants Grown in the Soils

Element	Loam soil		Sandy loam soil	
	Control	+ Ispolin	Control	+ Ispolin
Mg total	5465 ± 140	5460 ± 358	4077 ± 793	4057 ± 575
Mg fraction	315 ± 10	333 ± 11	54 ± 4	72 ± 8[b]
Mg roots	2437 ± 138	2873 ± 698	2237 ± 107	1903 ± 134[b]
Mg leaves	3747 ± 165	3113 ± 230[b]	3560 ± 79	2443 ± 253[b]
K total	3883 ± 156	4170 ± 150[b]	1400 ± 125	1720 ± 149[b]
K fraction	333 ± 10	702 ± 51[b]	80 ± 3	447 ± 83[b]
K roots	17,700 ± 1300	34,700 ± 6447[b]	11,300 ± 600	29,100 ± 500[b]
K leaves	80,400 ± 5500	98,500 ± 3051[b]	59,600 ± 1900	81,200 ± 7200[b]
Ca total	35,200 ± 1600	33,500 ± 3400	9265 ± 1525	9170 ± 1378
Ca fraction	5110 ± 192	5123 ± 125	962 ± 69	992 ± 65
Ca roots	5660 ± 175	6633 ± 1731	4403 ± 286	3897 ± 231
Ca leaves	6633 ± 439	4593 ± 153[b]	4063 ± 301	2370 ± 160[b]
K/Mg fraction	1.1	2.1	1.5	6.2

[a] mg kg^{-1}.
[b] Differences between control and ispolin-fertilized samples are significant at $P < 0.05$.

Cd uptake by plants; this may be explained as a result of Cl$^-$ and SO$_4^{2-}$ complexing with Cd^{2+} and thereby increasing the bioavailability of Cd^{2+} in soil. Similar results were reported by Norvell et al. [30], who showed that accumulation of Cd in wheat may be enhanced by presence of Cl$^-$ ions in soil. Although the mechanism of the ion interactions was not clear, the authors suggested that it is likely to involve increased solubility or availability of soil Cd resulting from the formation of chlorocomplexes in soil solution.

Potassium also has an effect on plant uptake of Cd. In particular, Cd concentrations in shoots increased two times with increasing K addition [24]. Lorenz et al. [107] reported that an excess addition of such cations as K$^+$ and NH$_4^+$ to soil caused a substantial increase of major and heavy metal ions in soil solution as well as their uptake by plants. This may be explained by the fact that NH$_4^+$ ions have an acidifying effect in soil, due to the nitrification processes or the release of H$^+$ ions to soil solution during plant uptake of NH$_4^+$ ions.

An application of amendments such as lime, apatite, and zeolite will increase Ca content in soil and also increase soil pH [49]. Because the increase of pH may lead to decrease of metal mobility in soil, these Ca fertilizers may be used for metal stabilization in contaminated soils. It is also known that many enzymatic and physiological processes in plant cells are under the regulatory control of Ca^{2+} [108]. In particular, externally supplied Ca^{2+} can affect the toxicity of some elements, perhaps by competing with metal uptake [104].

Kim et al. [36] studied the effects of the presence of K$^+$, Ca^{2+}, or Mg^{2+} in soil on transport and toxicity of Pb^{2+} and Cd^{2+} in rice. In contrast to the data cited previously, in this experiment K$^+$ had little effect on uptake or toxicity of Pb^{2+} and Cd^{2+}. Ca^{2+} and Mg^{2+} blocked Cd^{2+} transport into rice roots and Cd^{2+} toxicity for the root growth, which suggested that detoxification effect of these ions is directly related to their blocking the entry of the metal. Similarly, Ca^{2+} blocked Pb^{2+} transport into the roots and Pb^{2+} toxicity for the root growth. The authors suggested that the protective effect of Ca^{2+} on Pb^{2+} toxicity may be related to inhibition of the metal accumulation in the root tip, a potential target site of Pb^{2+} toxicity. Mg^{2+} did not ameliorate the Pb^{2+} toxicity for the root growth as much as Ca^{2+} did, though it decreased Pb^{2+} uptake by roots in the same manner as Ca^{2+}. These results suggest that the protective effect of Ca^{2+} on Pb^{2+} toxicity may involve multiple mechanisms,

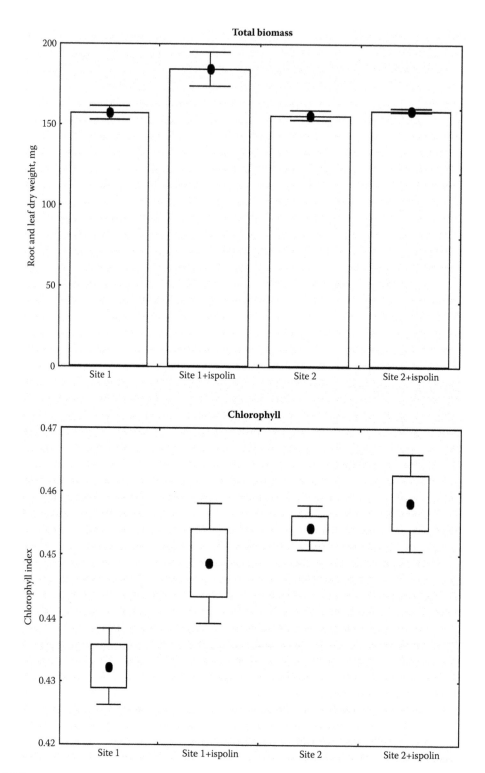

FIGURE 28.9 Variations in root and leaf biomass and chlorophyll index in wheat seedlings grown in fertilized and nonfertilized loam soil (site 1) and sandy loam soil (site 2).

including competition at the entry level, and that Pb^{2+} and Cd^{2+} may compete with divalent cations during transport of the metals into roots of rice plants.

Thus, soil amendments capable of enhancing the transfer of metals from soil solid phase to soil solution and increasing the metal uptake by plants can help to enhance metal phytoextraction. Application of fertilizers is very sensitive to soil parameters and plants growing in the soils; thus, the success of soil fertilization for enhancement of plant production and metal phytoextraction will depend on the selection of appropriate combination of the plant species and soil amendment.

28.7 SOIL BIOTA AS A PROMISING MEANS TO AFFECT METAL PHYTOEXTRACTION

Another possible way to stimulate yield of plants and increase plant metal uptake is an appropriate manipulation of plant—microorganism interactions. From 20 to 50% of plant biomass is below the ground [109]; from 30 to 60% of assimilatory products of photosynthesis are in the root zone, where they are used for root growth, root respiration, ion uptake, and release of organic exudates into the rhizosphere [53]. Unfortunately, due to the high complexity of the soil ecosystem, many aspects of the interactions between plants and soil microorganisms are still poorly understood. Numerous studies describing uptake of metals by plants often ignore the impact of soil biota on the processes of plant growth and metal uptake.

The structure of the soil microorganism community depends significantly on the physicochemical parameters of soil. For example, it was reported that the highest biomass of microorganisms in Podzol soil is presented by mycelium of fungi; the contribution of bacteria is less than 2.3% of the total amount of all microorganisms and the amount of actinomycetes may be very small [110]. Chernozem is enriched with bacteria (up to 90% of the total amount of soil microorganisms), the part of actinomycetes is 2 to 5%, and contribution of fungi may be less that 1%. The microbial colonization is also specific to certain plant species.

Elevated levels of trace metals can affect qualitative and quantitative structure of microbial communities. Many reports have demonstrated that heavy metals adversely influenced microorganism growth, morphology, and biochemical activities resulting in the decrease of microbial biomass and diversity [111,112]. In particular, it was shown that such trace metals as Cu, Ni, Pb, and Zn inhibited the extent of mycorrhizal colonization of cereal plants [113]. Cd, Cu, and Ni also significantly inhibited the *Azotobacter* population [21].

Moreover, it seems that any, even insignificant, disturbance in metal concentrations in the growth medium can negatively affect the soil microbial population. Pot experiments performed by the author with wheat seedlings grown in soil to which a small amount of nontoxic Cr^{3+} was added demonstrated considerable variations in the diversity of the rhizosphere microorganisms [114]. In particular, a large part of ordinary microbial strains died within the first hours after Cr^{3+} was added to soil; the number of the microorganisms that were previously presented in negligible amounts increased significantly. It is interesting that total amounts of the rhizosphere microorganisms remained at approximately the same level as in the control. It would be reasonable to suggest that, as a result of the variations in the microbial community, certain changes in chemical situation in the rhizosphere may be expected.

Microbial processes play an important role in defining the speciation and mobility of toxic metals and radionuclides in soil [115]. It has been reported that soil microorganisms, primarily bacteria and fungi, can significantly affect metal uptake by plants [116,117]. Rhizosphere biota can interact symbiotically with roots to enhance the potential for metal uptake and directly influence metal solubility by altering their chemical properties. In particular, plants and associated microorganisms can change soil pH, thus affecting the availability of main nutrients and trace metals [118].

Superior uptake properties to facilitate the movement of mineral elements into plants were demonstrated for different microbial strains. Preliminary evidence indicates that soil microbes are also important in the metal translocation from roots to upper plant parts [40,119]. Therefore, it is quite possible to manipulate the soil environment in such a way that metal uptake would be increased and, as a result, the process of metal removal from contaminated soils would be enhanced. In addition, it is clear that an improvement of nutrient availability as a result of the physiological activity of the rhizosphere microorganisms will increase plant growth, with a consequent removal of more metals from soil.

28.7.1 ARBUSCULAR MYCORRHIZAL FUNGI

Despite the important role that rhizosphere microorganisms play in plant interactions with the soil environment in general, and toxic metals in particular, relatively few studies have focused on the effects of these microorganisms on metal remediation efforts [120]. Among the rhizosphere organisms used in the phytoremediation studies, the arbuscular mycorrhizal fungi (AMF) have received special attention.

Data on metal uptake by mycorrhizal plants and the role of AMF in the metal uptake by host plants are rather contradictory. It is assumed that, in general, plant–AMF symbiosis may play an important role by increasing metal stress tolerance [121]. It was reported that fungi may constitute a biological barrier against transfer of toxic metals to the shoot [122]. In some cases, AMF can reduce excess plant uptake of elements like Zn, Cd, and Mn. The mechanisms by which AMF provide protection against metal toxicity to themselves and to plants are still relatively unclear. A number of mechanisms have been postulated; however, in some cases, data have been ambiguous and circumstantial [123]. In other cases, AMF have been shown to enhance or have no effect on the uptake of Zn, Pb, Ni, and Cu [113].

Inconsistent results on the effects of AMF on metal uptake may be due to a range of factors that are not controlled or accounted for, such as an inherent metal uptake capacity of plants, corresponding fungal properties, and soil adsorption/desorption characteristics [121].

28.7.2 BACTERIA

Bacteria have the ability to produce metabolites that, in suitable plant hosts, show activity similar to auxins, giberellins, cytokinins, and ethylene [124]. The role of bacteria in the rhizosphere chemistry is of particular interest. The specific exudates that root-colonizing bacteria excrete into surrounding soil catalyze the oxidation-reduction reactions that can alter mobility of metal ions [125]. Therefore, bacteria potentially may be used to facilitate movement of metals that would otherwise be unavailable to plants. It was shown that, in some cases, bacteria are perhaps more proficient than root exudates at solubilizing and absorbing certain metals [126]. For example, Zhou et al. [127] used bacterial extracellular polymers (BEP) to affect fate of uptake and translocation of inorganic pollutants in terrestrial ecosystems. An addition of BEP to metal-contaminated soil resulted in 1.6- to 12.8-fold higher soil-soluble Cu concentration over the control. Consequently, the Cu uptake by the ryegrass grown in the soil was increased by 31%.

However, the question remains whether the rhizosphere population can be altered in such a way that metal uptake by host plants would be increased. Preliminary studies with different bacterial systems showed that it is very difficult to overcome the competitiveness of the indigenous population and that inoculation is rarely successful [128]. Meanwhile, it may be assumed that the problem lies in the way by which interactions between bacteria and plant will be affected. Why it is desirable to change species populations in the rhizosphere and the consequences of such variations must be considered.

It seems that modification of the soil microbial biomass by conventional agricultural practices would be easier than attempts to modify it by means of direct addition of microorganisms to soil. In particular, in many experimental studies, the new population of microorganisms was just added to soil. If the experimental design is changed slightly — for example, if seeds are treated before sowing with certain strains of microorganisms — the proposed results may show considerable promise as a means to enhance the metal uptake. Seed germination is an important stage of plant growth. The conditions during the germination may have a crucial significance for further life of the plant.

28.8 EFFECTS OF *CELLULOMONAS* AND *MYCOBACTERIUM* STRAINS ON METAL PHYTOEXTRACTION BY CEREAL CROPS

In the author's experiments, seeds of wheat, oats, and barley were germinated on a moist filter paper to which culture of phosphate-mobilizing bacteria *Cellulomonas* sp. 32 SPBTI or *Mycobacterium* sp. 12 was added. It was shown that the bacteria are able to affect the uptake of different macro- and trace elements [119]. As an example, Figure 28.10 illustrates transfer of potassium from soil solution to leaves of oats treated and not treated with *Cellulomonas*. Potassium content in leaves of the plants increased and concentration of exchangeable K^+ in the soil solution decreased as a result of the bacterial treatment. Thus, *Cellulomonas* was favorable for supply of the plants with potassium. Similar variations in K content were observed after treatment of wheat seeds with *Mycobacterium*. However, treatment of barley seeds did not significantly change concentrations of K in soil or in different plant parts.

The ability of microorganisms to accumulate large amounts of potassium is well known [129]. Growth rate of bacteria also depends upon K^+ [130]. As a consequence of the metabolic activity of plants and microorganisms, concentration of K^+ in soil solution near the root surface can decrease, thereby inducing desorption of K^+ held on the external surface of soil particles. Therefore, the increase of K content in the plants and decrease of K concentration in soil are quite appreciable.

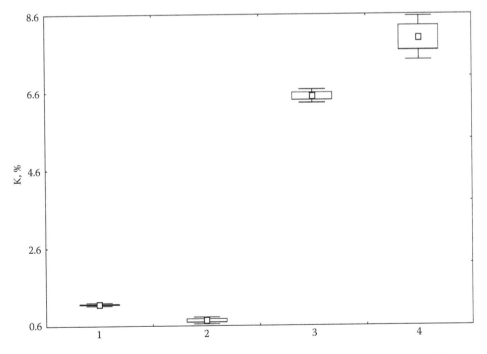

FIGURE 28.10 Concentration of exchangeable K^+ in soil solutions after cultivation of soil with nontreated (1) and treated with *Cellulomonas* (2) oats and K content in leaves of nontreated (3) and treated with *Cellulomonas* (4) plants.

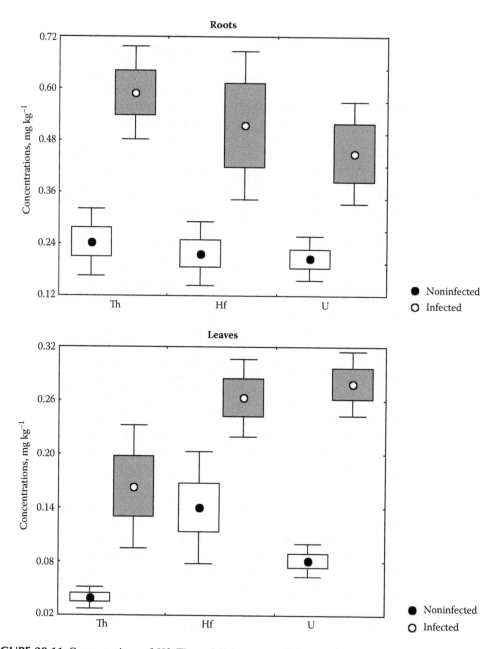

FIGURE 28.11 Concentrations of Hf, Th, and U in roots and leaves of the wheat seedlings infected and noninfected with *Cellulomonas*.

It was also found that presowing treatment of wheat seeds by bacteria may result in significant variations in uptake of different elements and their translocation from roots to leaves. Concentrations of many elements in roots and leaves of the plants treated by bacteria were much higher than those in the control plants. In particular, concentrations of Cr, Rb, Ag, Sb, Cs, Hf, Th, and U increased in roots and in leaves of the wheat seedlings treated with *Cellulomonas* compared with those in the control. Figure 28.11 illustrates such variations in the example of three elements: Hf, Th, and U. On the other hand, concentrations of many rare earth elements (La, Ce, Sm, Yb, and Lu) and

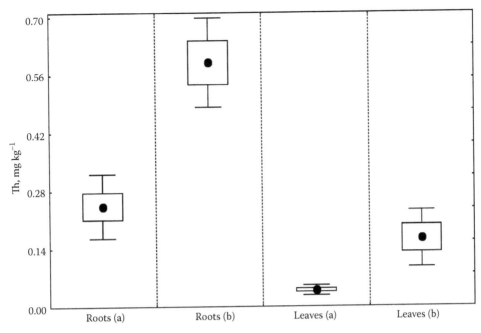

FIGURE 28.12 Thorium in roots and leaves of wheat seedlings noninfected (a) and infected (b) with *Myco-bacterium.*

Fe increased only in roots, and contents of K, Sc, Zn, and Eu were higher only in leaves, of the plants treated by the bacteria.

Similar variations in element uptake were observed after treatment of seeds with *Mycobacterium.* Figure 28.12 shows concentrations of Th in roots and leaves of wheat (*Triticum astevum*) infected and non-infected with *Mycobacterium.* The differences in Th uptake by treated and nontreated with *Mycobacterium* plants are statistically significant. Figure 28.13 illustrates the dynamics of Th in soil artificially contaminated with this metal after growth in the soil wheat seedlings inoculated with *Mycobacterium.*

It is known that, because of their high complexing ability, all radionuclides can be bound tightly to solid surfaces [131]. Therefore, one might expect that an addition of Th to the experimental soil may result in a quite rapid absorption of Th by soil organic matter and Th adsorption on the surface of soil particles. The observed variations in Th content in the soil might be caused solely by physiological activity of the experimental plants and microorganisms. Thus, it may be assumed that the changes in species diversity in the rhizosphere and, as a result, variations in the organic exudates into the soil can stimulate transfer of certain elements to more bioavailable forms. As a consequence, roots will be able to take up more metals and the rate of metal translocation to leaves may also be increased.

28.9 CONCLUSIONS

Cereal crops in combination with certain soil/plant treatments were shown as perspective "tools" to improve metal phytoextraction and removal of metals from contaminated soils. In conclusion, it may be said that a greenhouse experiment is the key stage in identifying the optimal procedures required to increase metal bioavailability and to stimulate metal phytoextraction, though only during a field trial, when the results of the previous model experiments are tested in real environmental conditions, can the feasibility of the manipulation measures finally be estimated. At this stage,

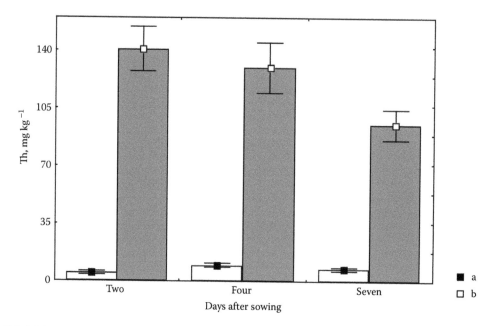

FIGURE 28.13 Dynamics of Th concentration in clean (a) and artificially contaminated with Th (b) soils after growing in the soils of wheat seedlings infected with *Mycobacterium*.

necessary modifications and corrections may be made, taking into account the local situation. The plant species and soil treatments found to be promising under batch experiment and/or greenhouse conditions may be less effective in the field.

REFERENCES

1. Lasat, M.M., Phytoextraction of toxic metals. A review of biological mechanisms, *J. Environ. Qual.*, 31, 109, 2002.
2. Baker, A.J.M. and Brooks, R.R., Terrestrial higher plants which hyperaccumulate metal elements — a review of their distribution, ecology, and phytochemistry, *Biorecovery*, 81, N1, 81-126, 1989.
3. Reeves, R.D., The hyperaccumulation of nickel by serpentine plants, in *The Vegetation of Ultramafic (Serpentine) Soils*, Baker, A.J.M., Proctor, J., and Reeves, R.D., Eds., Intercept Ltd., Andover, Hampshire, U.K., 253, 1992.
4. Baker, A.J.M. et al., The possibility of *in situ* heavy metal decontamination of polluted soils using crops of metal-accumulating plants, *Resour. Conserv. Recycl.*, 11, 41, 1994.
5. Brooks, R.R., *Plants That Hyperaccumulate Heavy Metals*, CAB International, Wallington, Oxon, U.K., 1998.
6. McGrath, S.P., Dunham, S.J., and Correll, R.L. Potential for phytoextraction of zinc and cadmium from soils using hyperaccumulator plants, in *Phytoremediation of Contaminated Soil and Water*, Terry, N. and Banuelos, G.S., Eds., Lewis Publishers, Boca Raton, FL, 2000, 109.
7. Gardea–Torresdey, J.L. et al., Bioaccumulation of cadmium, chromium and copper by *Convolvulus arvensis* L.: impact on plant growth and uptake of nutritional elements, *Bioresour. Technol.*, 92, 229, 2004.
8. Del Rio, M. et al., Heavy metals and arsenic uptake by wild vegetation in the Guadiamar River area after the toxic spill of Aznalcollar Mine, *J. Biotechnol.*, 98, 125, 2002.
9. Visoottiviseth, P., Francesconi, K., and Sridochan, W., The potential of Thai indigenous species for phytoremediation of arsenic contaminated land, *Environ. Pollut.*, 118, 452, 2002.
10. Norton, P.B., Ed. *The New Encyclopaedia Britannia*, 15th ed., Chicago, 1994, 37.

11. Shellenberger, J.A. Processing and utilization: wheat, in *Processing and Utilization in Agriculture*, vol. 2, Wolff, I.A., Ed., CRC Press Inc., Boca Raton, FL, 2000, 91.
12. Burger, W.C. Barley, in *Processing and Utilization in Agriculture*, vol. 2, Wolff, I.A., Ed., CRC Press Inc., Boca Raton, FL, 2000, 187.
13. Young, V.L., Processing and utilization: oats, in *Processing and Utilization in Agriculture*, vol. 2, Wolff, I.A., Ed., CRC Press Inc., Boca Raton, FL, 2000, 223.
14. Husted, S. et al., Elemental fingerprint analysis of barley (*Hordeum vulgare*) using inductively coupled plasma mass spectrometry, isotope ratio mass spectrometry and multivariate statistics, *Anal. Bioanal. Chem.* 378, 171, 2004.
15. Shtangeeva, I. et al., Combination of ICP-AES and instrumental neutron activation analysis as effective methods for studying of distribution of elements in soil and plants, *J. Geostandards Geoanal.*, 25, 299, 2001.
16. Shtangeeva, I. et al., Decontamination of polluted urban soils by plants. Our possibilities to enhance the uptake of heavy metals, *J. Radioanal. Nucl. Chem.*, 249, 369, 2001.
17. Kashem, M.A. and Singh, B.R., Metal availability in contaminated soils: II. Uptake of Cd, Ni and Zn in rice plants grown under flooded culture with organic matter addition, *Nutr. Cycling Agroecosyst.*, 61, 257, 2001.
18. Bjerre, G.K. and Schierup, H.H., Uptake of six heavy metals by oat as influenced by soil type and addition of cadmium, lead, zinc and copper, *Plant Soil*, 88, 57, 1985.
19. Estevez Alvarez, J.R. et al., Nuclear and related analytical methods applied to the determination of Cr, Ni, Cu, Zn, Cd and Pb in red ferralitic soil and *Sorghum* samples, *J. Radioanal. Nucl. Chem.*, 247, 479, 2001.
20. Joner, E.J. and Leyval, C., Time-course of heavy metal uptake in maize and clover as affected by root density and different mycorrhizal inoculation regimes, *Biol. Fertil. Soils*, 33, 351, 2001.
21. Athar, R. and Ahmad, M., Heavy metal toxicity: effect on plant growth and metal uptake by wheat, and free living *Azobacter*, *Water Air Soil Pollut.*, 138, 165, 2002.
22. Ryan, M.H. and Angus, J.F., Arbuscular mycorrhizae in wheat and field pea crops on a low P soil: increased Zn-uptake but no increase in P-uptake or yield, *Plant Soil*, 250, 225, 2003.
23. Narwal, R.P. and Singh, B.R., Effect of organic materials on partitioning, extractability and plant uptake of metals in alum shale soil, *Water Air Soil Pollut.*, 103, 405, 1998.
24. Zhao, Z.Q. et al., Effects of forms and rates of potassium fertilizers on cadmium uptake by two cultivars of spring wheat (*Triticum aestivum* L.), *Environ. Int.*, 29, 973, 2004.
25. Frost, H.L. and Ketchum, H.L., Trace metal concentration in durum wheat from application of sewage sludge and commercial fertilizer, *Adv. Environ. Res.*, 4, 347, 2000.
26. Berkelaar, E. and Beverley Hale, B., The relationship between root morphology and cadmium accumulation in seedlings of two durum wheat cultivars, *Can. J. Bot.*, 78, 381, 2000.
27. Greger, M. and Löfstedt, M., Comparison of uptake and distribution of cadmium in different cultivars of bread and durum wheat, *Crop Sci.*, 44, 501, 2004.
28. Hacisalihoglu, G., Hart, J.J., and Kochian, L.V., High- and low-affinity zinc transport systems and their possible role in zinc efficiency in bread wheat, *Plant Physiol.*, 125, 456, 2001.
29. Hart, J.J et al., Characterization of cadmium binding, uptake, and translocation in intact seedlings of bread and durum wheat cultivars, *Plant Physiol.*, 116, 1413, 1998.
30. Norvell, W.A. et al., Association of cadmium in durum wheat grain with soil chloride and chelate-extractable soil cadmium, *Soil Sci. Soc. Am.*, 64, 2162, 2000.
31. Zhao, F.J., Hamon, R.E., and McLaughlin, M.J., Root exudates of the hyperaccumulator *Thlaspi caerulescens* do not enhance metal mobilization, *New Phytol.*, 151, 613, 2001.
32. Murzaeva, S.V., Effect of heavy metals on wheat seedlings: activation of antioxidant enzymes, *Appl. Biochem. Microbiol.*, 40, 98, 2004.
33. Shtangeeva, I. and Ayrault, S., Phytoextraction of thorium from soil and water media, *Water, Air Soil Pollut.*, 154, 19, 2004.
34. Kashem, M.A. and Singh, B.R., Metal availability in contaminated soils: II. Uptake of Cd, Ni and Zn in rice plants grown under flooded culture with organic matter addition, *Nutr. Cycling Agroecosyst.*, 61, 257, 2001.
35. Fazeli, M.S. et al., Enrichment of heavy metals in paddy crops irrigated by paper mill effluents near Nanjangud, Mysore District, Karnatake, India, *Environ. Geol.*, 34, 297, 1998.

36. Kim, Y.-Y, Yang, Y.-Y, and Lee, Y., Pb and Cd uptake in rice roots, *Physiol. Plantarum*, 116, 368, 2002.

37. Tu, C. et al., Chemical methods and phytoremediation of soil contaminated with heavy metals, *Chemosphere*, 41, 229, 2000.

38. Zhang, Y. and Frankenberger, W.T., Removal of selenate in simulated agricultural drainage water by a rice straw bioreactor channel system, *J. Environ. Qual.*, 32, 1650, 2003.

39. Ali, N.A., Bernal, M.P., and Ater, M., Tolerance and bioaccumulation of copper in *Phragmites australis* and *Zea Mays*, *Plant Soil*, 239, 103, 2002.

40. Pawlowska, T.E. et al., Effects of metal phytoextraction practices on the indigenous community of arbuscular mycorrhizal fungi at a metal-contaminated landfill, *Appl. Environ. Microbiol.*, 6, 2526, 2000.

41. Singh, S.P., Tack, F.M., and Verloo, M.G., Uptake of metals in maize as affected by irrigation water quality and clay content of the soil, *Agrochimica*, 41, 27, 1997.

42. Huang, J.W. and Cunningham, S.D., Lead phytoextraction: species variation in lead uptake and translocation, *New Phytol.*, 134, 75, 1996.

43. Kalisova–Spirochova, I. et al., Accumulation of heavy metals by *in vitro* cultures of plants, *Water Air Soil Pollut.*, 3, 269, 2003.

44. Keltjens, W.G. and van Beusichem, M.L., Phytochelatins as biomarkers for heavy metal stress in maize (*Zea mays* L.) and wheat (*Triticum aestivum* L.): combined effects of copper and cadmium, *Plant Soil*, 203, 119, 1998.

45. Madrid, F. and Kirkham, M.B., Heavy metal uptake by barley and sunflower grown in abandoned animal lagoon soil, in *Proc. 17th WCSS*, Thailand, 2002, 401-1.

46. Luo, Y. and Rimmer, D.L., Zinc-copper interaction affecting plant growth on a metal-contaminated soil, *Environ. Pollut.*, 88, 79, 1995.

47. Adams, M.L. et al., Predicting cadmium concentrations in wheat and barley grain using soil properties, *J. Environ. Qual.*, 33, 532, 2004.

48. Cobb, G.P. et al., Accumulation of heavy metals by vegetables grown in mine wastes, *Environ. Toxicol. Chem.*, 19, 600, 2000.

49. Knox, A.S. et al., Remediation of metal and radionuclides contaminated soils by *in situ* stabilization techniques, in *Environmental Restoration of Metals Contaminated Soils*, Iskandar, I.K., Ed., CRC Press, Inc., Boca Raton, FL, 2000, 21.

50. Pederson, G.A., Brink, G.E., and Fairbrother, T.E., Nutrient uptake in plant parts of 16 forages fertilized with poultry litter. Nitrogen, phosphorus, potassium, copper, and zinc, *Agron. J.*, 94, 895, 2002.

51. Schumann, A.W. and Sumner, M.E., Formulation of environmentally sound waste mixtures for land application, *Water Air Soil Pollut.*, 152, 195, 2004.

52. Wyatt, P.H., A technique for determining the acid neutralization capacity of till and other surficial sediments, *Geol. Surv. Can.*, 84-1A, 597, 1984.

53. Marschner, H., *Mineral Nutrition of Higher Plants*, Academic Press, London, 1995.

54. Diatloff, E., Smith, F.W., and Asher, X.J., Rare earth elements. Effects of lanthanum and cerium on root elongation of corn and mungbean, *J. Plant Nutr.*, 18, 1963, 1995.

55. Todorovsky, D.S., Minkova, L., and Dariena, P.B., Effect of the application of superphosphate on rare earth element content in the soil, *Sci. Total Environ.*, 203, 13, 1997.

56. Wang, Z. et al., Accumulation of rare earth elements in corn after agricultural application, *J. Environ. Qual.*, 30, 37, 2001.

57. Shtangeeva, I. et al., Improvement of phytoremediation effects with help of different fertilizers, *Soil Sci. Plant Nutr.*, 2004, 50, 885-889.

58. Kabata–Pendias, A. and Pendias, H., *Trace Elements in Soils and Plants*, CRC Press, Boca Raton, FL, 2001.

59. Shtangeeva, I.V., Variations of the elemental composition of plants and soils, *J. Radioanal. Nucl. Chem.*, 177, 381, 1994.

60. Dunn, C.E. Biogeochemical prospecting for metals, in *Biogeochemical Systems in Mineral Exploration and Processing*, Brooks, R.R. and Dunn, C.E., Eds., Hall, GEM, Ellis Horwood, Hemel Hemstead, 1995, 371.

61. Shtangeeva, I.V., Heydorn, K., and Ovchinnikova, N., About differences in uptake of elements by plants grown in water and soil, in *Mengen und Spurenelemente*, Anke, M., Ed., 20, Friedrich–Schiller–Universitat, Jena, 64, 2000.

62. Takas, T. and Voros, I., Effect of metal nonadapted arbuscular mycorrhizal fungi on Cd, Ni and Zn uptake by ryegrass, *Acta Agronom Hungarica*, 51, 347, 2003.

63. Lasat, M.M., Baker, A.J.M., and Kochian, L.V., Altered Zn compartmentation in the root symplasm and stimulated Zn absorption into the leaf as mechanisms involved in Zn hyperaccumulation in *Thlaspi caerulescens*, *Plant Physiol.*, 118, 875, 1998.

64. Kolotov, B.A., Demidov, V.V., and Volkov, S.N., Chlophyll composition as a fundamental evidence of environmental degradation under contamination with heavy metals, *DAN*, 393, 567, 2003.

65. Shtangeeva, I.V., Behavior of chemical elements in plants and soils, *Chem. Ecol.*, 11, 85, 1995.

66. Adriano, D.C., *Trace Elements in Terrestrial Environments; Biogeochemistry, Bioavailability and Risk of Metals*, 2nd ed., Springer, New York, 2001.

67. Keller, C. et al., Root development and heavy metal phytoextraction efficiency: comparison of different plant species in the field, *Plant Soil*, 249, 67, 2003.

68. Matsi, T., Lithourgidis, A.S., and Gagianas, A.A., Effects of injected liquid cattle manure on growth and yield of winter wheat and soil characteristics, *Agron. J.*, 95, 592, 2003.

69. Arduini, I. et al., pH influence on root growth and nutrient uptake of *Pinus pinaster* seedlings, *Chemosphere*, 36, 733, 1998.

70. Wang, H.F., Takematsu, N., and Ambe, S., Effects of soil acidity on the uptake of trace elements in soybean and tomato plants, *Appl. Radiation Isotopes*, 52, 803, 2000.

71. Speir, T.W. et al., Heavy metal in soil, plants and groundwater following high-rate sludge application to land, *Water Air Soil Pollut.*, 150, 319, 2003.

72. Weng, L. et al., Phytotoxicity and bioavailability of nickel: chemical speciation and bioaccumulation, *Environ. Toxicol. Chem.*, 22, 2180, 2003.

73. Angle, J.S. and Heckman, J.R., Effect of soil pH and sewage sludge on VA mycorrhizal infection of soybeans, *Plant Soil*, 93, 437, 1986.

74. Kashem, M.A. and Singh, B.R., The effect of fertilizer additions on the solubility and plant-availability of Cd, Ni and Zn in soil, *Nutr. Cycling Agroecosyst.*, 62, 287, 2002.

75. Bajji, M., Kinet, J.-M., and Luts, S., Osmotic and ionic effects of NaCl on germination, early seedling growth, and ion content of *Atriplex halimus* (Chenopodiaceae), *Can. J. Bot.*, 80, 297, 2002.

76. Pezzarossa, B. et al., Phosphatic fertilizers as a source of heavy metals in protected cultivation, *Commun. Soil Sci. Plant Anal.*, 21, 737, 1990.

77. Uren, N.C. and Raisenauer, H.M., The role of root exudates in nutrient acquisition, *Adv. Pl. Nutr.*, 3, 79, 1988.

78. Dakora, F.D. and Phillip, D.A., Root exudates as mediators of mineral acquisition in low-nutrient environments, *Plant Soil*, 245, 35, 2002.

79. Robinson, D., Hodge, A., and Fitter, A., Constraints in the form and function of root systems, in *Root Ecology*, de Kroon, H. and Visser, E.J.W., Eds., Springer, Berlin, 2003, 1.

80. Ciezlinski, G. et al., Low-molecular-weight organic acids in rhizosphere soils of durum wheat and their effect on cadmium bioaccumulation, *Plant Soil*, 203, 109, 1998.

81. Nian, H. et al., A comparative study on the aluminum- and copper-induced organic acid exudation from wheat roots, *Physiol. Plantarum*, 116, 328, 2002.

82. Chaney, R.L., Metal speciation and interactions among elements affect trace element transfer in agricultural and environmental food-chains, in *Metal Speciation: Theory, Analysis and Applications*, Kramer, J.R. and Allen, H.E., Eds., Lewis Publishers Inc, Chelsea, MI, 1988, 218.

83. Blaylock, M.J. and Huang, J.W., Phytoextraction of metals, in *Phytoremediation of Toxic Metals: Using Plants to Clean up the Environment*, Raskin, I. and Ensley, B.D., Eds., John Wiley & Sons Inc, New York, 1999, 53.

84. Hendrickson, L.L. and Corey, R.B., Effect of equilibrium metal concentrations on apparent selectivity coefficients soil complexes, *Soil Sci.*, 131, 163, 1981.

85. Bell, P.F., James, B.R., and Chaney., R.L., Heavy metal extractability in long-term sewage sludge and metal salt amended soils, *J. Environ. Qual.*, 20, 481, 1991.

86. Martins, A.L.C., Bataglia, O.L., and de Camargo, O.A., Copper, nickel and zinc phytoavailability in an oxisol amended with sewage sludge and liming, *Scintia Agricola*, 60, 747, 2003.

87. Blaylock, M.J. et al., Enhanced accumulation of Pb in Indian mustard by soil-applied chelating agents, *Environ. Sci. Technol.*, 31, 860, 1997.

88. Huang, J.W. et al., Phytoremediation of lead-contaminated soils: role of synthetic chelates in lead phytoextraction, *Environ. Sci. Technol.*, 31, 800, 1997.

89. Anderson, C.W..N. et al., Harvesting a crop of gold in plants, *Nature*, 395, 553, 1998.

90. Huang, J.W. et al., Phytoremediation of uranium contaminated soils: role of organic acids in triggering uranium hyperaccumulation in plants, *Environ. Sci. Technol.*, 32, 2004, 1998.

91. Thayalakumara, T. et al., Plant uptake and leaching of copper during EDTA-enhanced phytoremediation of repacked and undisturbed soil, *Plant Soil*, 254, 415, 2003.

92. Ebbs, S.D. and Kochian, L.V., Phytoextraction of zinc by oat (*Avena sativa*), barley (*Hordeum vulgare*), and Indian mustard (*Brassica juncea*), *Environ. Sci. Technol.*, 32, 802, 1998.

93. Zhu, Y.L. et al., Cadmium tolerance and accumulation in Indian mustard is enhanced by overexpressing gamma-glutamylcysteine synthetase, *Plant Physiol.*, 12, 1169, 1999.

94. McGrath, S.P., Zhao, F.J., and Lombi, E. Plant and rhizosphere processes involved in phytoremediation of metal-contaminated soils, *Plant Soil*, 232, 207, 2001.

95. Schmidt, U., The effect of chemical soil manipulation on mobility, plant accumulation, and leaching of heavy metals, *J. Environ. Qual.*, 32, 1939, 2003.

96. Barzegar, A.R., Yousefi, A., and Daryashenas, A., The effect of addition of different amounts and types of organic materials on soil physical properties and yield of wheat, *Plant Soil*, 247, 295, 2002.

97. Mohanna, C. and Nys, Y., Effect of dietary zinc content and sources on the growth, body zinc deposition and retention, zinc excretion and immune response in chickens, *Br. Poult. Sci.*, 40, 108, 1999.

98. van der Watt, H.H., Sumner, M.E., and Cabrera, M.L., Bioavailability of copper, manganese, and zinc in poultry litter, *J. Environ. Qual.*, 23, 43, 1994.

99. Panin, M.S. and Gul'kina, T.I., The sorption of copper by the main soil types in the Irtysh Semipalatinsk region, *Agrochemistry*, 1, 75, 2004.

100. Trapeznikov, V.K. et al., Mineral nutrition of spring wheat cultivars and species under different fertilization regimes, *Agrochemistry*, 1, 51, 2004.

101. *Encyclopeadia Americana*, international. ed., vol. 6, American Corporation, International Headquarters, New York, 1966, 223.

102. McGrath, S.P. et al., The potential for the use of metal-accumulating plants for the *in situ* decontamination of metal-polluted soils, in *Integrated Soil and Sediment Research: a Basis for Proper Prediction*, Eijsackers, H.J.P. and Hamers, T., Eds., Kluwer Academic Publishers, Dordrecht, 1993, 673.

103. Tylor, G., Influence of acidity and potassium saturation on plant uptake of indigenous soil rubidium, *Environ. Exp. Bot.*, 38, 181, 1997.

104. Baker, D.A., Uptake of cations and their transport within the plants, in *Metals and Micronutrients*, Robb, D.A. and Pierpoint, W.S., Eds., Academic Press, New York, 1983, 3.

105. Chaney, R.L. et el., Progress in risk assessment for soil metals, and *in situ* remediation and phytoextraction of metals from hazardous contaminated soils, in *Proc. U.S-EPA Phytoremediation: State of Science*, Boston, MA, 2000.

106. Nemeskeri, E. and Nagy, L., Influence of growth factors on the yield and quality of dry beans, *Acta Agronomica Hungarica*, 51, 307, 2003.

107. Lorenz, S.E. et al., Application of fertilizer cations affect cadmium and zinc concentrations in soil solutions and uptake by plants, *Eur. J. Soil Sci.*, 45, 159, 1994.

108. Broadley, M.R. et al., Variation in the shoot calcium content of angiosperms, *J. Exp. Bot.*, 54, 1431, 2003.

109. Pucheta, E. et al., Below-ground biomass and productivity of a grazed site and a neighboring ungrazed exclosure in a grassland in central Argentina, *Aust. Ecol.*, 29, 201, 2004.

110. Balgoveshchenskaya, G.G. and Duhonina, T.M., Soil microbial communities and their functioning under chemization conditions, *Agrochemistry*, 2, 80, 2004.

111. Roane, T.M. and Kellog, S.T. Characterization of bacterial communities in heavy metal-contaminated soils, *Can. J. Microbiol.*, 42, 593, 1996.

112. Konopka, A. et al., Microbial biomass and activity in lead-contaminated soil, *Appl. Environ. Microbiol.*, 65, 2256, 1999.

113. Jamal, A. et al., Arbuscular mycorrhizal fungi enhance zinc and nickel uptake from contaminated soil by soybean and lentil, *Int. J. Phytoremediation*, 4, 205, 2002.

114. Shtangeeva, I.V. et al., Cr^{+3} as toxic element for plants. Role of physiological activity of microorganisms and plant roots, in *Metal Ions in Biology and Medicine*, 5, Collery, Ph. et al., Eds., John Libbey Eurotext, Paris, 1998, 349.

115. Banaszan, J.E., Riitman, B.E., and Reed, D.T. Subsurface interactions of actinide species and microorganisms: implications for the bioremediation of actinide-organic mixtures, *J. Radioanal. Nucl. Chem.*, 241, 385, 1999.

116. Del Val, C., Barea, J.M., and Azcon–Aguilar, C., Diversity of arbuscular mycorrhizal fungus populations in heavy-metal-contaminated soils, *Appl. Environ. Microbiol.*, 65, 718, 1999.

117. Andres, Y. et al., Contribution of biosorption to the behavior of radionuclides in the environment, *J. Radioanal. Nucl. Chem.*, 247,(1), 89-93, 2001.

118. Dington, J., Nutrient supply to plants, in *Nutrient Cycling in Terrestrial Ecosystems*, Harrison, A.F., Ineson, P., and Heal, O.W., Eds., Elsevier Applied Science, London, 1990, 441.

119. Shtangeeva, I., Heydorn, K., and Lissitskaia, T., The potential effect of microbial activity on the uptake of elements studied by exploratory multivariate analysis of data obtained by INAA, *J. Radioanal. Nucl. Chem.*, 249, 375, 2001.

120. Lovley, D.R. and Coates, J.D., Bioremediation of metal contamination, *Curr. Opin. Biotechnol.*, 8, 285, 1997.

121. Leyval, C., Turnau, K., and Haselwandter, K., Effect of heavy metal pollution on mycorrhizal colonization and function: physiological, ecological and applied aspects, *Mycorrhiza*, 7, 139, 1997.

122. Joner, E.J. and Leyval, C., Uptake of [109]Cd by roots and hyphae of a *Glomus mosseae/Trifolium subterraneum* mycorrhiza from soil amended with high and low concentrations of cadmium, *New Phytol.*, 135, 353, 1997.

123. Hartley, J. et al., Do ectomycorrhizal fungi exhibit adaptive tolerance to potentially toxic metals in the environment? *Plant Soil*, 189, 303, 1997.

124. Lynch, J.M., Interactions between bacteria and plants in the root environment, in *Bacteria and Plants*, Rhodes M.E. and Skinnen F.A., Eds., London, 1982, 1.

125. Wielinga, B. et al., Microbial and geochemical characterization of fluvially deposited sulfidic mine tailing, *Appl. Environ. Microbiol.*, 65, 1548, 1999.

126. Jackson, W.R., Humic, fulvic and microbial balance: organic soil conditioning, in *An Agricultural Text and Reference Book*, W.R. Jackson, Ed., Umi Research Pr, 1993, 34.

127. Zhou, L.X., Zhou, S.G., and Zhan, X.H., Sorption and biodegradability of sludge bacterial extracellular polymers in soil and their influence on soil copper behavior, *J. Environ. Qual.*, 33, 154, 2004.

128. Scott, A.J. et al., Role of soil microorganisms in uptake of heavy metals by hyperaccumulators, *Proc. 6th Int. Conf. Biogeochem. Trace Elements*, Evand, L. et al., Eds., Guelph, 2001, 288.

129. Leonard, R.T., Potassium transport and the plasma membrane-ATPase in plants, in *Metals and Micronutrients: Uptake and Utilization by Plants*, Robb, D.A. and Pierpont, W.S., Eds., Academic Press, New York, 1983, 71.

130. Hughes, M.N., *The Inorganic Chemistry of Biological Processes*, John Wiley & Sons, London, 1974.

131. Mortvedt, J.J., Plant and soil relationships of uranium and thorium decay series radio-nuclides — a review, *J. Environ. Qual.*, 23, 643, 1994.

29 Phytomanagement of Radioactively Contaminated Sites

H. Vandenhove

CONTENTS

29.1 INTRODUCTION

The application of nuclear energy and the use of radionuclides for industrial, medical, and research purposes have caused significant contamination of certain sites and their environments, which could result in health problems for several centuries if nothing is undertaken to remedy these situations. Except for the close environment of the facility, where decontamination activities may be feasible and affordable, the contamination often extends over a vast area and decontamination would result in considerable amounts of waste. Therefore, a search should be conducted for more realistic, yet efficient, remediation options; phytomanagement is among the potential options.

Major sources of radioactive contamination and major radionuclides released with long-term impact are [1]:

- Nuclear weapon testing (release of mainly ^{14}C, ^{137}Cs, ^{90}Sr, and ^{95}Zr)
- Nuclear weapon production (release of mainly ^{137}Cs, ^{106}Ru, and ^{95}Zr)
- Nuclear power production
    - During the mining operation, the main radionuclide discharged is ^{222}Rn; the environment of the U mining and milling sites is contaminated through dispersion of ^{238}U (and daughters — e.g., ^{226}Ra, ^{210}Pb, and ^{210}Po) and ^{232}Th.
    - During the operational phase, small amounts of radionuclides are routinely released, mainly ^{14}C.
- Nuclear accidents can involve only small local contamination, such as in the Goiania accident involving a discarded ^{137}Cs medical source; other accidents have been of much greater significance (Chernobyl, Three Mile Island, Kyshtym, and Windscale). Although the cocktail of radionuclides released is case dependent, in general, elements released with a scale impact were ^{137}Cs, ^{90}Sr, and ^{131}I (^{210}Po, ^{95}Zr, and ^{144}Ce).
- Natural sources of contamination comprise those of terrestrial linked with ^{40}K, and the ^{238}U, ^{235}U, and ^{232}Th decay chains. World weight averaged concentrations are 420 Bq kg^{-1} for ^{40}K and, respectively, 33 and 45 Bq kg^{-1} for ^{238}U and ^{232}Th series radionuclides. In most minerals, natural levels of radionuclides are very low. In others, e.g., zircon and rare earths, the concentration of ^{238}U and ^{232}Th may be considerably elevated (e.g., for apatite, 2 kBq kg^{-1}; for zircon, 10 kBq kg^{-1}; for U ore, ~200 kBq kg^{-1}; and for monazite, 350 kBq kg^{-1}. During recent years, increasing interest has been paid to contamination linked with industries handling materials containing elevated levels of natural radionuclides (NORs). Ore extraction and processing may lead to enhanced levels of naturally occurring radionuclides (NORs) in products, by-products, and waste and in surroundings and installations of the facility. The most contaminating industries are uranium mining and milling, metal mining and smelting, and the phosphate industry [2,3].

In this chapter, the potential role of different phytomanagement options for the remediation of sites contaminated with radionuclides is discussed and illustrated with some examples. The phytomanagement options considered are: phytoextraction (including rhizofiltration), alternative land uses and phytostabilization. The radionuclides considered are the fission products ^{137}Cs and ^{90}Sr and the natural radionuclides uranium and, to some extent, radium and thorium.

29.2 POSSIBLE ROLE OF PHYTOMANAGEMENT

29.2.1 Phytoextraction

Phytoextraction is an approach to the cleanup of contaminated soils using accumulator plants. This process requires that the target metal (radionuclide) be available for the plant root, absorbed by the roots, and translocated from the root to the shoot and biomass production should be substantial. The metal (radionuclide) is removed from the site by harvesting of the biomass; after that, it is processed to recover the metals or further concentrate the metal (thermal, microbial, chemical treatment) to facilitate disposal.

To maximize the metal content in the biomass, it is necessary to use a combination of improved soil management inputs (e.g., optimized soil pH and mineral nutrition, addition of agents that increase the availability of metals in soils), improved genotypes with optimized metal uptake, translocation and tolerance, and improved biomass yield. The economics of the operation not only depend on the phytoextraction efficiency but also on costs associated with crop management (soil management, crop establishment costs [yearly returning for annual crops], crop harvest); postharvest biomass transport; and biomass treatment and potential disposal.

Contrary to the research on heavy metals, the phytoextraction approach for radionuclides (RN) is rather new. Furthermore, most experiments conducted to test the phytoremediation approach for RN have been done in hydroponics systems [4,5].

29.2.2 THE POTENTIAL FOR PHYTOEXTRACTION

The soil-to-plant transfer of metal contaminants is a major process that must be considered in the management of metal-contaminated sites. These fluxes should be minimized to reduce exposure of soil pollutants in the food chain or maximized in an attempt to remediate a soil by phytomining or phytoextraction. This section will give an overview of the soil–plant fluxes of metal contaminants. Examples will be given for three radiocontaminants that are considered important due to their toxicity and/or ubiquity. These elements are the natural radionuclide uranium, one of the predominant contaminants in the uranium mining and milling industry; the phosphate industry and other NORM industries; and the long-lived fission products radiocaesium (^{137}Cs) and radiostrontium ^{90}Sr.

The flux of an element from soil to plant is often referred to as the crop off-take or removal of that element, i.e., the removal of the element from soil with the harvested part of the crop. The removal of a contaminant from the soil with the harvested biomass (in Bq ha^{-1}) is the product of the concentration in the plant (C_{plant} in Bq kg^{-1}) and the yield of the harvested biomass (kg ha^{-1}):

$$\text{Crop removal} = \text{yield} \times C_{plant} \qquad (29.1)$$

The concentration of an element in the plant depends on its concentration in the soil, type of soil, plant type, etc. It has been demonstrated that crop concentrations of nonessential trace elements, U, ^{137}Cs, and ^{90}Sr, are proportional to their concentration in soil, for the same crop, soil type, etc. The proportionality constant is defined as the transfer factor (TF, dimensionless)

$$TF = C_{plant}/C_{soil} \qquad (29.2)$$

where C_{plant} is the concentration of the radiocontaminant in the plant (Bq/g) and C_{soil} is the concentration of the contaminant in the soil. The transfer factor is thus an important parameter determining the potential of phytoextraction. Table 29.1 gives a summary of transfer factors.

TABLE 29.1
TFs for Cs, Sr, U and Ra: Total Ranges, Ranges of Recommended Values, and Upper Limits (Several Sources)

	Total range	Range of recommended values	Upper limit	Comment on conditions for upper limit
Cs	0.00025–7.5	0.0038–0.29	7.5	Brassica, organic soil
Sr	0.0051–22	0.017–3.2	22	Green vegetables, sandy soil
U	0.000006–21.13	0.00075–0.02	21.13	Tubers, sandy soil
Ra	0.00048–0.66	0.00097–0.12	0.66	Ryegrass, organic soil
Th	0.000021–0.270	0.00004–0.058	0.27	Grass, U mining area

With Equation 29.1 and Equation 29.2, the soil-to-plant transfer factor (TF), the percentage yearly reduction in soil activity can be calculated as follows:

$$\text{Annual removal } (\%) = \frac{\text{TF} \times \text{Yield}}{W_{soil}} \times 100 \qquad (29.3)$$

where W_{soil} is the weight of the contaminated soil layer (kg ha^{-1}). As made clear by Equation 29.3, the annual removal percentage increases with yield and TF. However, TF and yield values are not independent: high yield is often associated with lower TFs because of growth dilution effects.

Phytoextraction requires several years and the future trend in radionuclide concentration in the soil can be calculated according to:

$$C_{soil,t} = C_{soil_{t=0}} \exp\left\{-\left(\frac{\text{TF} \times \text{yield}}{W_{soil}} + \frac{0.69}{t_{1/2}}\right) \times t\right\} \qquad (29.4)$$

The second term in the exponent of Equation 29.4 was included to account for radioactive decay (t_f is the half-life of the radionuclide). Given the half life of ^{238}U (4.5 10^9 years), this component will not affect the phytoextraction potential. For ^{137}Cs and ^{90}Sr, with half-lives of 30 years, the phytoextraction potential will be affected (2.33% yearly loss in activity merely through radioactive decay). Equation 29.4 assumes a constant bioavailability of the contaminant, i.e., a constant TF.

For a soil depth of 10 cm and a soil density of 1.5 kg dm^{-3}, soil weight is 1,500,000 kg. Table 29.2 shows the percent of annual removal for different crop yields and TFs. Yields of more than 20 ton ha^{-1} and TFs higher than 0.1 (Table 29.1) may be regarded as average values or upper limits, except for Sr. This would result in an annual reduction percentage of 0.1% (excluding decay). In the case of a TF of 1, annual reduction is about 1%.

Rearranging Equation 29.4 allows calculation of the number of years needed to attain the required reduction factor as a function of annual removal percentage. Table 29.3 tabulates the years required to attain a reduction of the contaminant concentration up to a factor of 100, given an

---

TABLE 29.2
Percentage Yearly Reduction of Soil Contamination due to Phytoextraction (and Radioactive Decay [t_f 30 years])

Yield (t ha^{-1})	Annual reduction % due to phytoextraction					Annual reduction % due to phytoextraction and radioactive decay				
	5	10	15	20	30	5	10	15	20	30
TF (g g^{-1})										
0.01	0.0033	0.0067	0.01	0.013	0.02	2.333	2.337	2.34	2.343	2.35
0.1	0.033	0.067	0.1	0.133	0.2	2.363	2.397	2.43	2.463	2.53
1	0.33	0.67	1.00	1.33	2.00	2.66	3	3.33	3.66	4.33
2	0.67	1.33	2.00	2.67	4.00	3	3.66	4.33	5	6.33
5	1.67	3.33	5.00	6.67	10.00	4	5.66	7.33	9	12.33
10	3.33	6.67	10.00	13.33	20.00	5.66	9	12.33	15.66	22.33

Notes: Soil depth 10 cm; soil density 1.5 kg dm^{-3}.

Source: Vandenhove, H. et al., *Int. J. Phytorem.*, 3(3), 301, 2001.

TABLE 29.3
Years Required to Decontaminate a Soil Given a Required (Desired)
Reduction Factor and a Given Annual Removal Percentage

Desired reduction factor	% activity remaining $C_{soil,t}/C_{soil,t=0}$	Annual removal, % y^{-1}							
		20	15	10	5	3	2	1	0.1
1.5	66	2	2.5	4	8	13	20	40	400
2	50	3	4	7	14	23	34	69	693
3	33	5	7	11	22	36	55	110	1108
4	25	6	9	13	27	46	69	138	1609
5	20	7	10	15	31	53	80	160	1650
10	10	10	14	22	45	76	114	229	2301
20	5	13	18	28	58	98	148	298	2994
50	2	18	27	37	76	128	194	389	3910
100	1	21	28	44	90	151	228	458	4603

Notes: Soil depth 10 cm; soil density 1.5 kg dm^{-3}.

annual extraction percentage or percentage reduction in radionuclide activity varying between 0.1 and 20%. With an annual removal of 0.1%, it would take more than 2000 years to decontaminate a soil to 10% of its initial activity. With an annual removal of 1%, more than 200 years are required. Thus, it is clear that measures should be taken to increase the annual removal efficiency through crop selection, increasing the bioavailability with soil additives, and technical actions (e.g., decreased soil depth).

Decreasing the soil depth increases the removal percentage, according to Equation 29.3, and may intensify root–soil contact and result in an increased TF. The soil can be spread on geomembranes, which impede roots from penetrating to deeper layers. These membranes will also limit contamination of the underlying clean soil, but a substantial area may be needed for treatment. In most cases of soil contamination, control on the depth of the contamination is limited, though it may be feasible and advantageous to excavate and spread out the soil over the desired soil depth for phytodecontamination purposes. Therefore, considering a soil depth of 10 cm may be a realistic assumption.

Decreasing the soil depth is not specific to radionuclide and plant. The other factors influencing radionuclide bioavailability (crop selection and increasing the bioavailability of the radionuclide of concern) will be discussed per radioelement. For each element, a short description on the behavior of the element in soil will be given.

29.2.2.1 Uranium

29.2.2.1.1 Behavior of U in Soil

Natural uranium exists as three isotopes — ^{238}U, ^{235}U, and ^{234}U — with a relative abundance of 99.27, 0.720, and 0.0055%, respectively. The world average ^{238}U activity in soil is 40 Bq kg^{-1} [6]. Uranium behavior is similar to that of other heavy metals and its physiological toxicity mimics that of lead. Uranium is chemically toxic to kidneys and insoluble U-compounds are carcinogenic [7].

The complex, pH-dependent speciation of U in soils makes the study of U uptake by plants difficult. Uranium is present in soil primarily (80 to 90%) in the +VI oxidation state as the uranyl (UO_2^{+2}) cation [8]. Under acidic conditions, UO_2^{+2} is the predominant U species in the soil. Hydroxide complexes form under near-neutral conditions, and carbonate complexes predominate under alkaline conditions. Generally, negligible amounts of UO_2^{+2} are present in available forms, due to the high solid–liquid distribution coefficient of uranium (K_D = ratio of the radionuclide in the solid soil fraction to the radionuclide concentration in the soil solution: Bq kg^{-1}/Bq dm^{-3}) (620 dm^3 kg^{-1}, [9]). Though extensive work has been done on U solubility in soils, comparatively little information is available regarding the uptake and translocation of U by plants as affected by soil properties. Work has focused primarily on the U content of native plants growing in U-contaminated environments [11,12] or U uptake by field and garden crops of importance to animals and humans [13–15].

29.2.2.1.2 Phytoextraction Potential of U

The U soil-to-plant TF varies with plant compartment and plant species. Roots incorporate much more uranium than stems, leaves, and seeds do [16–18].

Leafy vegetables generally show higher U TFs, followed by root, fruit, and grain crops (Table 29.4). TF values for the plants that have been studied rarely exceed a value of 0.1, except in plants grown on highly contaminated, often acidic U-mining sites. A study of U accumulation by mixed grass species, forbs, and big sagebrush (*Artemesia* spp.) growing near a U mine/mill complex indicated that plants growing on exposed tailings (low pH) had a TF value that approached unity; plants grown in proximity to the tailings pile had TF values ranging from <0.1 to approximately 0.4, depending on the distance from the tailings and the wind direction [12]. The plant with the highest TF for natural conditions is pH sensitive and does not grow at a pH below 4.

Ebbs et al. [18] showed that free UO$_2$$^{+2}$ is the U species most readily taken up and translocated by plants. Because this species is only present at a pH of 5.5 or less, acidification of U-contaminated soils may be necessary for phytoextraction (Figure 29.1). The uranyl cation also binds to the soil solids and organic matter, reducing the extent of plant uptake [21]. Therefore, in addition to acidification, soil amendments that increase the availability of U by complexation may also be required.

Ebbs et al. [18] and Huang et al. [20] studied the role of acidification and chelating agents on the solubilization of uranium. Of the organic acids tested (acetic acid, citric acid, oxalic acid, malic acid), citric acid was the most effective for increasing U in the soil solution. The results also indicated that citric acid solubilized more U than simple acidification (HCl, S, HNO$_3$). Chelating agents like EDTA and DTPA did not increase uranium solubility significantly. Compared to potassium citrate, citric acid was much more effective for increasing U solubility and thus accumulation in plants. With the addition of 20 to 25 mmol citric acid per kilogram of soil, soil pH decreased 0.5 to 1.0 pH units, depending on initial soil pH.

Following citric acid treatment (20 mmol kg^{-1}) the U accumulation in Indian mustard (*Brassica juncea*) was increased 1000-fold [20]. Ebbs et al. [18] and Ebbs [8] observed an increase in U accumulation in beet (*Beta vulgaris*) of a factor of ten after citric acid treatment (25 or 20 mmol kg^{-1}). Citric acid was always applied 1 week before harvest by spraying a solution over the soil surface. The citric acid-triggered U accumulation in plants was observed within 24 h after application and a maximum was reached after 3 days. Thus, plants would maintain very low U concentrations for most of their growth and start accumulating rapidly following citric acid application, after which they could be quickly harvested. Citric acid is rapidly biodegradable, so no problems with residual levels of citric acid in soil should occur.

The potential to phytoextract uranium (U) from a sandy soil contaminated at low levels was tested in the greenhouse by Vandenhove et al. [22]. Two soils were tested: a control soil (317 Bq ^{238}U kg^{-1}) and the same soil washed with bicarbonate (69 Bq ^{238}U kg^{-1}). Ryegrass (*Lolium perenne* cv. *Melvina*), Indian mustard (*Brassica juncea* cv. *Vitasso*) and redroot pigweed (*Amarathus retroflexus*) were used as test plants. Plants were selected on the basis of their reportedly high transfer factors and relative important dry-weight production. Citric acid addition resulted in a decreased dry weight production (all plants tested) and crop regrowth (ryegrass) (Figure 29.2).

The annual removal of the soil activity with the biomass was less than 0.1% for the control scenario. The addition of citric acid (25 mmol kg^{-1}) 1 week before the harvest increased U uptake up to 500-fold (Figure 29.2). With a ryegrass and mustard yield of 15,000 kg ha^{-1} and 10,000 kg

TABLE 29.4
Some Specific U-TF to Natural Vegetation and Garden Crops

Plant/crop/vegetation type		TF	Experimental conditions
Natural vegetation at U-milling site[a,b]			
Sagebrush		0.12	Background
		0.90	Edge of tailing pond
Mixed grasses		0.07	Background
		0.07	Edge of tailing pond
		0.69	Exposed tailings
Forbs		0.16	Background
		0.08	Edge of tailing pond
		1.1	Exposed tailings
Grasses[c]		0.003–0.18	Sandy and loamy soil
Vegetables			
Leaf vegetables	Lettuce[c]	0.025	Loam, pH 7
	Spinach[c]	0.033	Loam, pH 7
	Turnip green[a]	0.0058	Lake sediments, sandy, pH 4, 1.5% OM
	Indian mustard[e,f]	0.007	Loam, OM 4.2%, pH 7.7
Root crops	Carrots[c]	0.0057	Sandy loam, pH 7
	Potato flesh[c]	0.019	Loam, pH 7.3
	Potato flesh[c]	0.002	Sand, pH 4.9
	Potato[d]	0.009-0.0009	Sand, pH 8.1 (TF for fresh weight)
	Sugar beet[c]	0.01–0.06	Sand or sandy loam, pH 7
	Red beet[c]	0.0024	Loam, pH 5.1–7.0
	Turnip root[a]	0.00099	Lake sediments, sandy, pH 4, 1.5% OM
Fruit vegetables	Tomato[c]	0.0005	Sandy loam, pH 7
	Cucumber[c]	0.0009	Sandy loam, pH 7
Cereals	Bush bean[a]	0.0018	Lake sediments, sandy, pH 4, 1.5% OM
	Rice[c]	0.0005	Loam, pH 7.3
	Barley[c]	0.0021	Loam, pH 7.3
	Corn[c]	0.00021	Loam, pH 7.3
	Corn[a]	0.00021	Lake sediments, pH 4, 1.5% OM
	Corn[e]	0.006–0.01	Not specified
Amaranthus[g]		~0.007	

[a] Whicker, F.W. and Ibrahim, SA., Radioecological investigations of uranium mill tailings systems. Fort Collins, Colorado State University, 1984, 48.
[b] Ibrahim, S.A. and Whicker, F.W., *J. Radioanal. Nucl. Chem.,* 156(2), 253, 1992.
[c] Frissel, M.J. and van Bergeijck, K.E., Fifth Report of the IUR working group soil-to-plant transfer factors: report of the working group meeting in Guttannen Switzerland, Bilthoven, RIVM, 1989, 240.
[d] Lakshmanan, A.R. and Venkateswarlu, V.S., *Water, Air Soil Pollut.,* 38, 151, 1988.
[e] Mortvedt, J.J., *J. Environ. Qual,* 23, 643, 1994.
[f] Ebbs, S.D. et al., *J. Environ. Qual.,* 27, 1486, 1998.
[g] Huang, J.W. et al., *Environ. Sci. Technol.,* 32, 2004, 1998.

ha^{-1}, respectively, up to 3.5 and 4.6% of the soil activity could annually be removed with the biomass (Table 29.5). With the required activity reduction level of 1.5 and 5 for the bicarbonate-washed and control soil, respectively, it would take 10 to 50 years to attain the release limit.

A linear relationship was found between the plant ^{238}U concentration and the ^{238}U concentration in the soil solution of the control, bicarbonate-washed, or citric acid-treated soil. This points to the importance of the soil solution activity concentration in determining U uptake and thus to the importance of solubilizing agents to increase plant uptake.

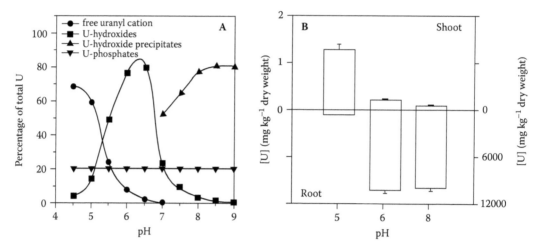

FIGURE 29.1 Dependence of U speciation and U uptake on pH. (After Ebbs, S.D. et al., *J. Environ. Qual.*, 27, 1486, 1998.)

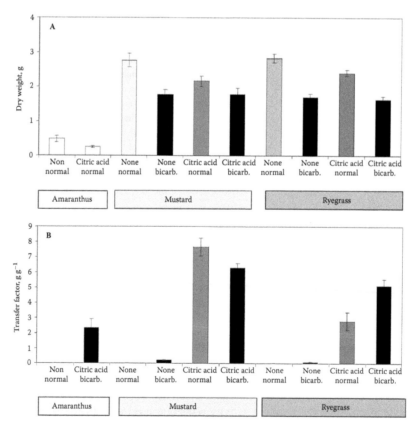

FIGURE 29.2 Effect of citric acid addition 1 week before harvest on the dry weight production and uranium transfer factor of ryegrass and mustard and redroot pigweed.

TABLE 29.5
Phytoextraction Potential[a] of Untreated and Citric Acid-Treated Soils (First Harvest)

Plant	Soil	Citric acid	Annual extraction %
Amaranthus	Control	No	0.0009 ± 0.0003
	Control	Yes	0.178 ± 0.058
Mustard	Control	No	0.010 ± 0.002
	Control	Yes	4.618 ± 0.384
	Bicarb.	No	0.103 ± 0.030
	Bicarb.	Yes	3.284 ± 0.250
Ryegrass	Control	No	0.007 ± 0.004
	Control	Yes	2.810 ± 0.689
	Bicarb.	No	0.052 ± 0.008
	Bicarb.	Yes	3.477 ± 0.474

[a] Annual removal, %.

Notes: Soil depth 10 cm; soil density 1.5 kg dm^{-3}.

Addition of citric acid is required to attain extraction levels that may make phytoextraction a feasible remediation option. Important uncertainties affect the phytoextraction potential:

- How does plant growth stage affect the efficiency of citric acid addition and what is the expected yield at the growth stage of maximal accumulation?
- What is the optimal level of citric acid addition to attain the highest increase in U-TF and the lowest impact on yield?
- What is the effect of continuous citric acid treatment on the soil?
- What is the global cost of phytoextraction compared to other remediation options?

Given the deleterious effect of citric acid on plant growth, one should possibly revert to other soil additives to increase the phytoavailability. However, several authors [18,20] have screened a number of additives, and only citric acid was found to be effective in increasing the TF. Though the experimental conditions of the latter authors were similar to the experimental conditions used by Vandenhove et al. [22], they did not mention a significant decrease in dry weight production or plants' dying off after citric acid treatment. When equivalent amounts of citric acid at pH 7 (adjusted with KOH) were added, ryegrass regrowth was also hampered (Vandenhove and Van Hees, results not shown). This implies that plants' dying off following citric acid addition is not predominantly due to a low pH, but rather to an increase in ion strength of the soil pore water when nutrients are solubilized under the action of citric acid.

Apart from application of soil additives to increase U export with the plant biomass, plant selection may also be important for improving the phytoextraction potential. However, the expected effects are small compared to the increase in uptake following soil amendments like citric acid addition. Huang et al. [20] recorded a difference of a factor of two in TF for four varieties of Indian mustard (*Brassica juncea*). They also demonstrated that plant species and cultivars differed significantly in response to citric acid treatment. Of the species tested, four showed significant potential in citric acid-triggered U accumulation: Indian mustard *(B. juncea),* Chinese mustard *(B. narinosa),* Chinese cabbage *(B. chinensis)* and redroot pigweed *(A. retroflexus).* A twofold difference between cultivars was also observed.

29.2.2.2 Radium

29.2.2.2.1 Behavior of Ra in Soil and Soil Factors Affecting Ra Availability

Radium is one of the prominent potential contaminants linked with industries extracting or processing material containing naturally occurring radionuclides. It (^{226}Ra) is a natural decay product of ^{238}U, an alpha emitter, and is present in the soil at concentrations of 37 to 370 Bq kg^{-1}. The radium concentration may reach critical levels — orders of magnitudes higher — in tailings from uranium mills, phosphate mines, and processing sites, as well as in various waste products from the NORM-industry.

Radium is the last member of the alkaline earth metals, a group of metals whose lighter members (Ca and Mg) play a very important role in plant growth and nutrition. Because radium is highly electropositive, it reacts readily with many agents, most of which are insoluble. It also coprecipitates with barium and strontium to form insoluble sulphates. Due to its basic character, it is not easily complexed. The most stable complexes are formed with EDTA [23].

Radium has a high affinity for the regular exchange sites of the soil. According to Simon and Ibrahim [24], organic matter adsorbs about ten times as much radium as clay, which is more adsorptive than other soil minerals. Increased exchangeable Ca [25], increasing pH [19], and high soil-sulphate content [12] are reported to decrease the radium transfer factor. Vasconcellos et al. [26], however, did not find a relation between soil Ca concentration and Ra TF.

29.2.2.2.2 Phytoextraction Potential of Ra

The transfer factor depends on soil characteristics, plant type, the part of the plant concerned, climate conditions, and the physicochemical form of radium. Compared to many other radionuclides, little information is available on the uptake of radium from soil. Among reported values, large discrepancies are noticeable for the Ra-transfer factors (TF: concentration in plants to concentration in soil). Conducting a field study in a high natural background area testing seven field and garden crops, Linsalata et al. [34] reported values ranging from 0.3 10^{-4} to 0.02 g g^{-1} (lower value for corn; higher value for carrot). TFs on loamy, sandy, and organic soils ranged from 0.05 to 0.44 g g^{-1} for clover and from 0.14 to 0.62 g g^{-1} for ryegrass (lowest value on loam and highest on organic soil) [27]. IUR [28] reported 95% confidence limits ranging from 2.9 10^{-4} to 0.12 g g^{-1} considering a large series of crops and soil types (Table 29.6).

Thus far, no information was found by the author on studies dedicated to the phytoextraction of radium or methods to promote radium availability in the soil–plant environment. Addition of EDTA or decrease in pH could be expected to decrease the radium transfer factor. Highest TFs for radium are around 0.1 g g^{-1}. Considering a crop yield of 20 t ha^{-1}, only 0.13% (see Table 29.2) of the radium could be annually extracted with plants.

29.2.2.3 Thorium

29.2.2.3.1 Behavior of Th in Soil and Soil Factors Affecting Th Availability

More than 99% of natural thorium exists in the form of ^{232}Th ($t_{\frac{1}{2}}$ 4.5 × 10^9 a). Thorium occurs naturally in the Earth's crust at an average background concentration of 8 to 12 mg kg^{-1}. The typical concentration range of Th in soils is 2 to 12 mg kg^{-1} with an average value of 6 mg kg^{-1} [29]. More than 120 minerals contain thorium and thorium compounds [30]. The major source of thorium is monazite, which is a component of different granites and other igneous rocks. The geochemical behavior of Th is very similar to the behavior of rare earth elements (especially cerium), zirconium, and uranium.

The geochemistry of thorium is simplified by the existence of just one valent state: +4. It has been shown [31] that the Th ion is largely hydrolyzed at pH above 3.2, and the hydroxyl complexes are involved in the sorption process. The adsorption of Th on clays, oxides, and organic matter increases with increasing pH and is completed at pH 6.5. On the other hand, it has been suggested

TABLE 29.6
Some IUR Recommended Values for Ra-TF for Different Crops and Soil Types and 95% Confidence Intervals[a]

Crop group	Soil Type	Recommended TF value	95% Confidence interval[b]
Cereals (maize)	Sand	0.00104	0.00048–0.00223
Tubers (potatoes)	Sand	0.00097	0.00077–0.00123
	Clay	0.00358	0.00029–0.04362
Brassicas (beans)	Sand	0.00458	0.00223–0.01057
	Clay	0.0210	0.0092–0.0479
Root vegetables (carrots)	Sand	0.00676	0.00321–0.0142
	Clay	0.0379	0.0123–0.1161
	Organic		
Grass	Sand	0.12	0.0692–0.2083
	Loam	0.0617	0.0410–0.0928
	Clay	0.0330	0.0243–0.0449

[a] TF: Bq kg^{-1} dry mass plant per Bq kg^{-1} dry mass soil.
[b] 95% confidence intervals of the mean (i.e., geometric mean $\pm$ t_n standard errors) (t_n ~2 if $n > 10$).

Source: IUR, Sixth report of the IUR working group soil-to-plant transfer factors: report of the working group meeting in Guttannen, Switzerland, 198924-26, Bilthoven, RIVM: 240, 1989.

[32] that mobility of Th in soil may be less affected by soil pH than by soil organic matter. Tetravalent thorium may be strongly complexed with soil organic matter, thus increasing the mobility of Th in soil.

29.2.2.3.2 Phytoextraction Potential of Th

Information on Th-TF is rather scant. Shtangeeva et al. [33] report TFs for Couch grass ranging from ~0.01 to 0.05 and for plantain from 0.008 to 0.03 g g^{-1} — the higher values for the shoots and the lower values for the roots. For garden crops cultivated on high background areas, Linsalata et al. [34] found values ranging from 10^{-5} to 10^{-3} g g^{-1}. Whicker et al. [35] measured values ranging from 2×10^{-5} to 2×10^{-3} g g^{-1} for crops grown on a contaminated lake bed. Only Vera Tome et al. [36] reported rather high Th-TF, ranging from 0.013 to 0.270 g g^{-1} for grass pasture samples collected in granitic and alluvial soils around a disused uranium mine.

As was said for radium, no information was found by the author on studies dedicated to the phytoextraction of thorium or methods to promote radium availability in the soil–plant environment. Because the highest TF for thorium observed is around 0.1 g g^{-1} (yet most TFs observed are a factor of 100 to 10,000 lower), the maximal amount annually extracted by plants is of the order of 0.1%.

29.2.2.4 Caesium

29.2.2.4.1 Behavior of Cs in Soil and Soil Factors Affecting Cs Availability

After the Chernobyl accident, the study of the fate of radiocaesium in the environment was of particular importance, given its relatively long half-life, its widespread contamination, and its similarity with K favoring its uptake by plants [37–40].

The soil K status affects Cs availability: in general, the higher the soil K is, the lower is the TF. This K effect is explained by the increased solution concentration of a cation, which competes with radiocaesium for uptake. The consensus now is that the solid/liquid partitioning of radiocaesium in the soils, and thus its soil bioavailability, is governed by the presence of micaceous, potassium-bearing clay minerals [41–43]. The process relates to the action of a small number of

TABLE 29.7
Annual Removal of ^{137}Cs for Some Agricultural Crops[a]

Crop	Yield (ton dm ha⁻¹)	TF (g g⁻¹)	Crop off-take (% of total in soil)
Cereals (grain)	5–7	0.0004–0.25	0.0005–0.06
Potato tuber	6–10	0.003–0.89	0.0006–0.3
Leafy vegetables	5–10	0.008–1.7	0.001–0.6
Grassland	10–15	0.01–1.0	0.007–1.0

[a] Expressed as fraction of total content in the plough layer (arable crops) or in the 0- to 12.5-cm layer (grassland).

Note: The range in dry weight-based transfer factors (TF) represents a typical range and is derived from Nisbet A.F. and Woodman R.F.M., *Health Phys.*, 78(3), 279, 2000.

Source: IUR, Eighth report of the working group soil-to-plant transfer factors, report of the Working Group Meeting in Madrid, Spain, IUR Pub R-9212-02, IUR Technical Secretariat, Balen, Belgium, June 1–3, 1992.

very selective sites — the so-called frayed edge sites (FES) — located at the edges of illite particles. In general, the higher the soil clay content is, the lower is the caesium availability.

In studies following the Chernobyl accident, lime was applied to reduce the radiocaesium TF through an increase in pH (extent of increase not reported) and transfer reductions of a factor of two were obtained [38,44]. Plant growth may, however, be reduced at lower pH and in reducing conditions (high water content). For example, the Cs-TF may be much higher on an organic soil, but yield may also be substantially reduced.

Enhanced uptake of caesium by plants is reported in the presence of increased amounts of NH_4 [7,45,46] or when K is depleted [47–49]. NH_4 additions increased the TF up to twofold and also enhanced biomass production; both factors favored the phytoextraction potential [50]. Extremely low soil fertility with regard to potassium may increase the radiocaesium TF 10- to 100-fold, but will also decrease plant growth. The rather high TF for ryegrass recorded by IUR [28], 3.3 g g⁻¹, may have been obtained for soil with low fertility.

Given the number of parameters that may possibly enhance the radiocaesium TF (low pH, high ammonium, low K), it may be concluded that increasing the TFs reported in literature two- to threefold is a possibility worthy of testing.

29.2.2.4.2 Phytoextraction Potential of Cs

In normal agricultural systems, the annual Cs flux is small compared to that present in soil (Table 29.7). The ^{137}Cs off-take values are all less than 1%; the highest off-take is found for grassland. The high sorption of ^{137}Cs in soil and the typical K levels in soil required for optimal plant growth limit high off-take values.

Table 29.8 presents radiocaesium TFs for sandy soils (generally higher TF) for some crops and natural vegetation. Most crops have a TF $\ll$ 1 g/g, resulting in annual removal percentages of around 3%, decay included (Table 29.3).

Many crops may have a wide range of TFs, even for a specific soil type. For example, the range of TFs observed for ryegrass grown on sandy soil extends from 0.05 to 3.3 [52]; this means that TFs substantially higher than 1 are within reach. Potential ryegrass yield is between 15 and 20 t/ha as long as fertilization is adequate. This would mean a phytoextraction ranging between 5.3 and 6.3%, decay included (Table 29.2).

Using sandy Cs-contaminated soil (5 Bq g⁻¹) at a waste processing facility, Vandenhove [50] tested the effect of ammonium addition on the phytoextraction potential of two crops with reportedly high TF and DW production in a greenhouse experiment. Target was 1 Bq g⁻¹ Cs. Ammonium

TABLE 29.8
Average Radiocaesium Soil-to-Plant Transfer Factors

Crop	Transfer factor (gg⁻¹)
Winter wheat (leaves + grain)[a]	0.03
Summer wheat (leaves)[a]	0.05
Ryegrass[a]	0.03
Potato (tuber)[a]	0.09
Lettuce[a]	0.24
Yellow lupine (seeds)[b]	1.64
Yellow lupine (straw)[b]	1.02
Sunflower (seeds)[b]	0.15
Sunflower (straw)[b]	0.59
Sunflower (roots)[b]	2.88
Ryegrass[c]	1.2 (0.05–3.3)
Ryegrass[d]	1.87 (0.92–2.82)
Flax (leaves)[e]	0.66 (0.57–0.84)
Indian mustard[f]	0.4–0.5
Red root pigweed (amaranthus species)[f]	2.2–3.2
Tepary bean[f]	0.2–0.3
Different amaranthus species[g]	0.53–2.03

[a] Lembrechts, J., *Sci. Tot. Environ.*, 137, 81, 1993.
[b] Gopa, Belarus: study on alternative biodiesel sources in relation with soil decontamination. Project N° TACIS/REG93, 1996.
[c] IUR, Fifth report of the IUR working group soil-to-plant transfer factors: report of the working group meeting in Guttannen, Switzerland, 198924-26, Bilthoven, RIVM: 240, 1989.
[d] Vandenhove, H. et al., *Sci. Tot. Environ.*, 187, 237, 1996.
[e] Van Hees, M. and Vandenhove, H., Unpublished data, 1998.
[f] Lasat, M.M. et al., *J. Environ. Qual.*, 27, 165, 1998.
[g] Sorochinsky, B., in *Proc. Chernobyl Phytoremediation Biomass Energy Conversion Workshop.*, Slavutych, Ukraine, 1998, 229.

addition increased DW production with 20% and the TF with 80%, resulting in a TF of 0.8 g g⁻¹. With a realistic yield of 20 t ha⁻¹ under field conditions, this would result in an annual reduction of 3.3% (decay included). This would imply 50 years of continued phytoextraction to reach the target of 1 Bq g⁻¹. Lasat et al. [55] observed that adding ammonium did not significantly affect the Cs-TF for three species (amaranthus, tepary bean, Indian mustard) (Figure 29.3).

Amaranthus species screened by Lasat et al. [55] and Sorochinsky [56] have TFs as high as 3.2 g g⁻¹ with a yield potential projected at 30 t ha⁻¹ a⁻¹ (based on two harvests per year, which is very improbable). The TFs reported by Lasat et al.[55] were derived for a sandy loam from Hanford with contamination levels of 15 Bq g⁻¹ soil; thus, the annual removal, including decay, would be 8.3% (Table 29.3). Their target was a fourfold reduction in soil activity (to 4 Bq g⁻¹), which would require 16 years. In the more likely event of a yield of 15 t ha⁻¹, the time required would be 28 years.

29.2.2.5 Strontium

29.2.2.5.1 Factors Affecting the Strontium TF

As radiocaesium, radiostrontium is among the most abundant radionuclides in the suite of nuclear fission products that are routinely or accidentally released. Its relatively long half-life (~30 years)

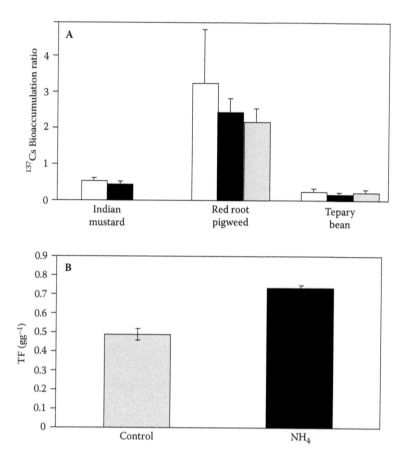

FIGURE 29.3 Effect of ammonium addition on the Cs-TF of Indian mustard, redroot pigweed, and tepary bean. (Sorochinsky, B., in *Proc. Chernobyl Phytoremediation Biomass Energy Conversion Workshop*, Slavutych, Ukraine, 1998, 229) and ryegrass (Vandenhove, H., Internal SCK•CEN report R-3407, 1999).

and its metabolic similarity with Ca favor its uptake in plants. For the Sr-TFs, the TFs for each crop tend to be highest for sand and lowest for clay and organic soil types; the maximal differences between soil types for any one crop tend to be around one order of magnitude. The soil-exchangeable Ca content seems to affect the Sr-TF, but not significantly. Sauras et al. [58] found that lower Sr-TF was observed with increased soil solution Ca + Mg and increased CEC.

29.2.2.5.2 Phytoextraction Potential of Sr

The off-take of ^{90}Sr is higher than that of ^{137}Cs because the Sr availability is typically tenfold above that of Cs. The TFs of ^{90}Sr in green vegetables and *Brassica* are typically around unity and upper levels are around 10 (Table 29.9). With a yield of 10 ton/ha for leafy vegetables and a TF of 10, 9% could be annually removed, considering that contamination is restricted to 10 cm and accounting for decay. Under those conditions, a fivefold reduction of the soil contamination would require 15 years. Phytoextraction of ^{90}Sr has not yet been investigated at field scale. The high off-take values in agricultural crops (Table 29.9) suggest that this phytoextraction should be explored for this element.

29.2.3 Conclusions for the Potential of Phytoextraction

Except for radiostrontium, TFs are generally too low to allow phytoextraction to be efficient without soil additives that increase bioavailability. Following citric acid addition, the uranium TF may

TABLE 29.9
Annual Crop Off-Take of ^{90}Sr for Some Agricultural Crops[a]

Crop	Yield (ton dm/ha)	TF (g/g)	Crop off-take (% of total in soil)
Cereals (grain)	5–7	0.02–0.94	0.0037–0.22
Potato tuber	6-10	0.03–1.4	0.006–0.5
Leafy vegetables	5–10	0.45–9.1	0.07–3.0

[a] Expressed as a fraction of total content in the plough layer (arable crops) or in the 0- to 12.5-cm layer (grassland).

Note: The range in dry weight based transfer factors (TF) represents a typical range. The TFs of ^{90}Sr have been derived from Nisbet A.F. and Woodman R.F.M., *Health Phys.,* 78(3), 279, 2000.

Source: IUR, Eighth report of the working group soil-to-plant transfer factors, report of the Working Group Meeting in Madrid, Spain, IUR Pub R-9212-02, IUR Technical Secretariat, Balen, Belgium, June 1–3, 1992.

increase with a factor of 500 resulting in annual activity reductions up to 3%; however, even then, it would still require more than 50 years to reduce the soil contamination level with only a factor of 5. Moreover, the effect of continuous citric acid treatment on the soil and plant growth and TF is not clear. For Cs, the same remark can be made. TFs are higher than the U-TF by a factor of 10 to 100. However, additives will only increase the availability with maximally a factor of two, without really affecting the DW production.

Given the low TFs observed for Ra and Th and the actual absence of an adequate method to increase the TFs, the feasibility of phytoextraction of Ra- and Th-contaminated soils is strongly questioned. The high off-take values in agricultural crops for Sr suggest that phytoextraction should be explored for this element. The question about long-term effectiveness still remains: will TF remain constant or will it decrease while radionuclide concentration decreases and when ageing processes occur?

There is the question of cost of waste treatment and site monitoring. With respect to this last issue, phytoextraction involves costs at different stages in the process. The soil must be prepared for crop establishment and the crop must be well maintained. During crop establishment or before the harvest, the soil may need to be treated with radionuclide-specific amendments to improve crop yield. The treatment of 1 m^3 of contaminated soil (10 m^2 if 1 dm soil layer) will annually result in about 10 to 20 kg biomass (~2 to 4 kg ashes), which must be harvested, transported, and treated; this entails considerable costs. This scheme of action must be repeated on a yearly basis for several years.

29.2.4 RHIZOFILTRATION

Rhizofiltration is the use of plants to accumulate compounds from aqueous solutions through adsorption on the roots or assimilation through the roots and eventual translocation to the aerial biomass (phytoextraction). Rhizofiltration is being investigated for the removal of radionuclides from aqueous waste streams, including groundwater and wastewater.

The removal of a radionuclide from an aqueous waste stream is governed by the plant dry weight production and the concentration factor, CF (ratio of Bq/g plant to Bq/ml water). Because absorption in (waste) water per volume is lower than in soil, the CF is higher than the TF. This becomes clear when considering the relationship between TF and CF:

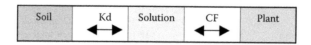

$$TF = CF/K_D \qquad\qquad (29.5)$$

where CF is the concentration factor (i.e., the ratio of the radionuclide activity concentration in plant shoots to that in soil solution) and K_D is the solid–liquid distribution coefficient of the radionuclide ($dm^3\ kg^{-1}$) (i.e., the ratio of radionuclide activity concentration in the solid phase to that in the soil solution).

The K_D for Cs in soils ranges from 10 to 10^5 ml g^{-1} [59]; for Sr, from 1 to 100; for U, the geometric mean K_D per soil group ranges from 35 to 1600 (full range from 0.03 to 395,100) [60]; and, for Ra, the average K_D per soil group ranges from 490 to 36,000 ml g^{-1} (full range from 1 to 1.810^7 [61]). The K_D for all radionuclides studied is generally substantially higher than 1, so it is clear that the CF exceeds the TF with the same factor and that rhizofiltration is generally more effective than soil phytoextraction.

Vasudev et al. [4] commented upon the removal of radionuclides from contaminated waste water and ground water. They found a 95% reduction in the U concentration in water through U absorption by sunflowers (*Helianthus annuus*) after a 24-h contact time. In a comparable experiment using Chernobyl-contaminated pond water, the Cs and Sr levels decreased 90 and 80%, respectively, after a 12-day contact time with 8-week-old sunflower plants replaced every 48 h. Sunflowers showed higher removal rates than timothy, meadow foxtail, Indian mustard, and peas (Cs-CF to shoots ranging from 400 to 2300 L kg^{-1}, Cs–CF to shoot from 860 to 8600 L kg^{-1}; for Sr, respectively, 650 to 2000 and 110 to 980 L kg^{-1}).

Dushenkov et al. [62] tested a number of species for the removal of uranium from water in a greenhouse experiment. Beans and mustard were less effective than sunflower in U removal. The latter removed 95% of the U in the water in less than 24 h. Practically all U was concentrated in the roots, with almost none in the shoots (Figure 29.4).

Concentration of uranium in the water did not affect the CF, but pH did: an average bioaccumulation coefficient for U in sunflower roots was 6624 and 3370 L/kg, respectively, at pH 5 and pH 7. As mentioned before, U accumulation is affected by U speciation. The U species most readily taken up by the plant is the uranyl form. This form predominates at pH below 5.5 [18]. Uranium was removed much faster from the contaminated water compared to Cs and Sr.

In natural waters, uranium is usually complexed with carbonate, hydroxide, sulphate, and phosphate. These complexes increase the solubility of U and make uranium precipitation more challenging. For this reason, it seems that U is one of the best candidates for a biological removal process.

However, Dushenkov et al. [62] demonstrated that the rhizofiltration method has its limits. Setting up an experiment with rather highly contaminated waste water (1000 µg L^{-1} U; 20% above the upper limit of local processing waste water) and high flow rate (1.05 L min^{-1}), 95% of the U was removed by 6-week-old sunflowers grown for 2 weeks in the waste water. This resulted in effluent concentrations of 40 to 70 µg/L^{-1}, above the 20 µg L^{-1} drinking water limit.

(*Disclaimer*: This section is certainly not exhaustive. No attempt has been made to perform a literature search for CFs for the radionuclides considered. Rather, a few figures are highlighted and the potential of rhizofiltration is discussed to some extent.)

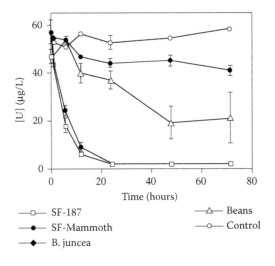

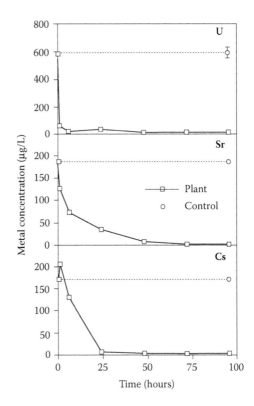

FIGURE 29.4 Removal of uranium by different sunflowers cultivars (left) and removal of Cs, Sr, and U by sunflowers (right) in a hydroponics system. (Dushenkov, S. et al., *Environ. Sci. Technol.,* 31, 3468, 1997.)

29.3 ALTERNATIVE LAND USE: NONFOOD CROP PRODUCTION IN CONTAMINATED AREAS

29.3.1 Introduction

The question of how to manage large territories contaminated with a broad spectrum of radionuclides still remains 15 years after the Chernobyl accident. An area of 29,200 km² in Belarus, Ukraine, and Russia, was contaminated with levels exceeding 185 kBq ^{137}Cs m^{-2} [63] and as much as 4300 km² of agricultural land had to be excluded from use. The Sr contamination occurs mainly within a 70-km radius from the reactor, although some significant contamination (37 to 74 kBq m^{-2}) can be found in the area northeast of Gomel [64]. Belli and Tikhomirov [65] reported that, 9 years after the accident, the radiocaesium concentrations in plants grown in forests and on meadows did not significantly decline. In the more contaminated territories of Belarus, Ukraine, and Russia, no lifting of the restrictions on land use is likely in the foreseeable future.

When an appropriate countermeasure for a specific area is selected, apart from radiological criteria, the optimal solution to a given problem will depend as much on economic, social, and political factors as on sound scientific considerations [66]. Many studies have targeted possible agricultural countermeasures in response to concentration levels in food and agricultural crops that are too high. Most studies have been conducted to test the effect of different physical and chemical countermeasures.

In contrast to this, information on long-term effects of countermeasures and, especially, the change to nonfood crops is still limited. Countermeasures can also be based on the selection of crops that exhibit smaller radionuclide uptake, on food processing, or choosing for nonfood crops such that the products from the land are radiologically acceptable [65,67,68]. Impact on dose to people and on the ecology and economy of the affected area may vary enormously: change in crop variety will have a much smaller impact than more radical changes such as substitution of vegetables by cereals or changing from an arable or cattle system to forestry.

- When an alternative crop or land use is advocated, the principle questions to be asked are:
- What is the fate of the radionuclide in the cultivation system and conversion routes and what is the expected radionuclide concentration in the end-products?
- How does the radionuclide behave during the biomass processing?
- What is the exposure during biomass cultivation and processing?
- How well are the crops adapted to the climate and soil conditions prevailing in the contaminated area?
- What are the conclusions with regard to economic feasibility for the production and use of these alternative crops?
- What prospective land use for large contaminated surfaces do these various alternative crops offer?

With respect to the fate of radionuclides in the cultivation system and conversion routes and the concentration to be expected in end- and waste products, one should have an idea of the entry radionuclide flux. This depends on the deposition level, the crop accumulation factor (which depends on plant and soil characteristics), and the radionuclide accumulation in the source material (e.g., wood, rape seeds, root beet, etc.). Whether the source material is safe for conversion (e.g., burning of wood or straw), the products are acceptable for use, and the waste products to be disposed of as radwaste or not depends on the radionuclide concentration and the exemption limits prevailing in each country.

29.3.2 Liquid Biofuels

Crops used for liquid biofuel production, e.g., rapeseed, wheat, sugar beets, barley, potatoes, and winter rye, may also be considered as suitable alternative crops. TFs to the useable product (rapeseed, wheat grains, beet root, barley, potatoes, and winter rye) are low (Table 29.10) and the liquid biofuels are almost free from activity. Also, radiocaesium levels in the waste products are generally of no concern.

Caesium levels in oil cake from oil seed rape (~2000 t ha^{-1}) and the pulp and vines from sugar beet (~4000 t ha^{-1}) may be too high for use as animal fodder and for incineration; they may need to be disposed of as radwaste. Yet, this is only in case of high contamination levels because only about 3% of the total radiocaesium and 6% of the total radiostrontium taken up will be potentially involved in the soil–plant–fodder–animal–man chain. Moreover, usually, fodder constitutes only 10% of oil cake and pulp and thus animal fodder can generally be used without restriction. The cost of liquid biofuels is actually (in general) about a few times to several hundred times the cost of fossil fuels, so a price subsidy is needed [53,71].

On the other hand, the production of rapeseed and the processing to edible rapeseed oil are profitable technologies. Furthermore, levels of caesium and strontium in the rapeseed oil after three filtrations and bleaching were below detection limit.

29.3.3 Willow Short-Rotation Coppice for Energy Production

The feasibility of willow short-rotation coppice (SRC) for energy production for revaluation of contaminated land was studied by Vandenhove et al. [71–74]. Coppicing is a method of vegetative forest regeneration by cutting trees at the base of the trunk at regular time intervals. Fast-growing species of the *Salix* genus (willows) are frequently used in a coppice system because of the ease

TABLE 29.10
Cs Transfer Factors[a] to Different Plant Parts of Some Potential Biofuel Crops

Crop	Plant component	Cs-TF, 10^{-3} m² kg⁻¹
Spring wheat	Leaves	0.23–0.42
	Straw	0.23–0.36
	Grain	0.08–0.16
Cereals	Grain	0.0048/26 (2.6–260)[b]
	Grain (peaty)	83(8.3–830)
Winter wheat	Straw	0.27–0.44
	Seeds	0.08–0.18
Spring rape seed	Stems and leaves	0.12 (0.46)[c]
	Straw	0.017–1.4[c]
	Seeds	0.03–0.04 (0.0006–0.5)[c]
Brassicaceae	Seeds	0.037–3.4[a]
Legumes	Seeds	94(12–750)[a]
Sugar beet	Root	0.43[d]
Root crops	Root	0.025/11 (1.1–110)[a]
Green vegetables	Leaves	0.07–4.86[a]
	Leaves (peaty)	260 (25–2700)[a]

[a] TF, g g⁻¹.
[b] Nisbet, A.F. et al., National Protection Board, NRPB-R304, Didcot, UK, 1999, 50.
[c] Gopa, Belarus: study on alternative biodiesel sources in relation with soil decontamination. Project no.°TACIS/REG93, 1996.
[d] IAEA (International Atomic Energy Authority), *Handbook of Parameter Values for the Prediction of Radionuclide Transfer in Temperate Environments*, Technical Report series no.°364, IAEA, Vienna, 1994.

Source: Data for Belarus from Grebenkov, A., Internal deliverables for RECOVER, 1997.

of the vegetative reproduction and the large biomass production. The harvested biomass is converted into heat or power. As such, this nonfood industrial crop is a potential candidate for the valorization of contaminated land with restricted use.

In the case of reuse of contaminated land, SRC may be preferred to traditional forestry because revenues come sooner and more regularly (every 3 to 5 years) after establishment. SRC yields are also high on good agricultural soils, and it is not a drastic change in land use (easy to apply and easy return to food crops). SRC may also be considered as complementary to forestry, given the different culture requirements of both vegetation systems (forests perform well on sandy soils whereas SRC requires soils with sufficient water retention capacity). SRC has other potential advantages in a contamination scenario. Because it is a perennial crop, dispersion of radionuclides may be reduced. Harvest can be in winter when the soil may be covered by snow, resulting in protection for the people. Finally, SRC cultivation is not labor intensive, which is also advantageous with relation to exposure.

Willow SRC may be a suitable rehabilitation tool for highly contaminated land, but only if

- The radionuclide levels in the wood are below the exemption limits for fuel wood
- The average yearly dose received during coppice cultivation and coppice wood conversion is acceptable
- SRC can be grown successfully in the contaminated territories (soils, climate)

- The cultivation of SRC is technically feasible
- SRC production and conversion are economically profitable

It has been shown [71–74] that, for soils with a medium to high fixation (finer textured soils) and sufficient K, the TF (ratio of concentration in plant biomass to concentration in soil, m^2 kg^{-1}) is $<10^{-5}$ m^2 kg^{-1}, and wood can be safely burned and the ashes disposed of without concern. Only in the case of light-textured soils with a low radiocaesium fixation and low soil K, the TF to wood is around 10^{23} m^2 kg^{-1} and concentrations in wood may be elevated so that prevailing exemption limits are reached. Given that TFs for common forestry and for straw of winter wheat and oil seed rape are comparable, the same applies for burning wood or straw for energy.

Regarding the agrotechnical aspects affecting SRC production and feasibility, this crop has generally a high annual yield of about 12 t ha^{-1}. Sandy soils are only suitable for SRC production if well fertilized and irrigated. Regarding radiation exposure, doses in the vicinity of ash collectors may exceed the acceptable level of 1 mSv a^{-1} for a member of the general public only during the conversion phase and when highly contaminated wood is burned (3000 Bq kg^{-1}). Contributions from other possible exposure pathways are negligible (external exposure during culturing and transport, inhalation dose in the combustion plant and to the public following wood burning).

The economic sustainability of cultivation and conversion of SRC was evaluated by previously mentioned authors [71-74]. Crop yield and capital cost of the conversion units are among the most important parameters affecting system profitability. At the production site, a minimal yield of 6 t ha^{-1} y^{-1} is required for the Belarus production conditions and 12 t ha^{-1} y^{-1} for West European conditions, if all other parameters are optimal. Heat schemes may be a viable option for wood conversion in Belarus; electricity schemes are not. In Europe, subsidies are required to make wood conversion economically feasible. From this study it was also concluded that the existence of a contamination scenario does not necessarily hamper the economic viability of the energy production schemes studied. The cost associated with disposal of contaminated ashes was estimated as less than 1% of the biofuel cost [70,72,73] and would not affect the economic feasibility.

29.3.4 FORESTRY

Forestry can also be considered as an adequate alternative land use. Soil-to-wood TFs to coniferous and deciduous wood are around 10^{-3} m^2 kg^{-1} [75] and thus comparable with the TFs to willow wood observed for soils with low fertility and limited Cs fixation. They are high compared with the TFs observed for willow in finer textured soils and soils with an adequate potassium status. Moreover, annual biomass increase is only 6 t ha^{-1} for forests and may attain 12 t ha^{-1} for SRC grown on soils with an adequate water reserve and fertility status. On these types of soil, SRC may thus be a more promising land-use option than traditional forestry. On soils with low water reserves (e.g., sandy soil), willow yield without irrigation is too low to be economically feasible, so forestry may be the preferred option here [71,72].

29.3.5 FIBER CROPS

Fiber crops are also potentially alternative crops for agricultural land with restricted use. Potentially suitable crops are the annual fiber crops hemp (*Cannabis sativa* L.) and flax *(Linum usitatissimum* L.). Hemp and flax are well known arable crops that have been cultivated for centuries. Ukraine has a legacy of flax and fiber hemp cultivation. In Belarus, there is only some flax production because, in the early 1990s, the acreage for production of flax and hemp declined dramatically in Ukraine. Establishment of fiber crops on contaminated arable land is generally of no radiological concern [76]. The TFs observed in the hemp fibers are higher than the TFs observed in flax by a factor of 4 to 50; cultivation should generally be restricted to areas that are not too contaminated (<1000 kBq m^{-2}). For both crops, contamination levels in the waste products (oil seed cake, chaff, ash after burning of straw) may, however, be high enough so that they should be considered as radioactive waste. The economics of this land use was not investigated.

29.3.6 CONCLUSIONS

Energy production from SRC in a contaminated area is a feasible remediation option on radiological, technicoagricultural, and economic grounds under certain conditions. On sandy soils, it is advisable to install forests or drought-resistant grasses instead of SRC or to apply irrigation because SRC is clearly not adapted to dry sandy soil conditions. For the other energy crops considered, generally no concern existed on radioecological grounds, but conversion to liquid biofuels is not profitable; therefore, the cultivation of crops for liquid fuel production cannot be recommended.

29.3.7 PHYTOSTABILIZATION

Phytostabilization reduces the risk presented by un- or sparsely vegetated contaminated soil by the use of plants and/or soil amendments to establish a stable vegetation cover that may progressively reduce the soil labile metal pool. This technology does not achieve a cleanup of the soil, but changes the mobility of potentially toxic elements by reducing concentrations in the soil water and other freely exchangeable sites within the soil matrix or by reducing re-entrainment of toxic particulates following the development of a stable and permanent vegetation cover. Both processes alter the speciation of soil metals, reducing potential environmental impact. These technologies draw upon fundamental plant and soil chemical processes as well as established agricultural practices.

The development of a stable and self-perpetuating ecosystem as a result of this type of treatment may be a further beneficial process because, in some circumstances, plant root activity may change metal speciation (changes in redox potential, secretion of protons, and chelating agents); the microflora associated with their root systems may produce similar effects. The rainwater infiltration rate can be reduced by increasing the plant-induced evapotranspiration, thus reducing the potential for leaching and acid drainage generating [77–80]. Plants will also physically stabilize topsoils [81]. Physical stabilization will reduce erodability and dust generation, thus reducing the mobility of radionuclides and respective exposure pathways. Some examples of phytostabilization approaches are presented next.

29.3.8 PHYTOMANAGEMENT WITH WILLOW VEGETATION SYSTEMS IN THE CHERNOBYL EXCLUSION ZONE

Remedial actions to control the radionuclide efflux from the Chernobyl Exclusion zone are still being investigated because most radioactive fall-out following the Chernobyl accident was deposited within the Dnieper catchment system, which adjoins the site of the Chernobyl nuclear power plant (ChNPP). This area and adjacent drainage basins form an extensive area from which contaminated water and sediments flow downstream through the Pripyat and Dnieper Rivers across the Ukraine to the Black Sea. Phytostabilization technologies could in this context also be considered as remedial options. Three phytorehabilitation approaches with willow vegetation systems for the Dnieper catchment system in the vicinity of the Chernobyl nuclear power plant were studied [72] (Figure 29.5): the effectiveness of willow vegetation on vertical migration of radionuclides; on the stabilization of the Chernobyl cooling pond sediments; and on horizontal erosion control.

The area of interest to study the vertical migration control by willow was an extremely contaminated zone of 16 km^2 at the left bank of the Pripyat (between 3.7 and 18.5 TBq km^{-2} ^{90}Sr and ^{137}Cs and 0.37 TBq km^{-2} Pu), which is partly protected from the spring floods by a dam. Through modeling exercises, it was shown that, due to their high evapotranspiration rate, willow SRC stands are expected to decrease the groundwater table level by 100 to 200 cm in fertilized stands. Without fertilization, only a decrease of groundwater table level of less than 50 cm was

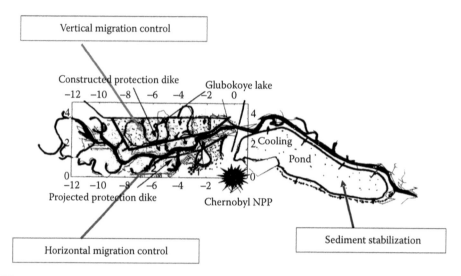

FIGURE 29.5 Phytostabilization approaches at the contaminated area of the Dnieper close to the Chernobyl NPPP.

predicted. Because the immobilization potential of ^{137}Cs and ^{90}Sr in the willow wood is limited, the influence of plant uptake on migration remains low.

Following the effective closure of the ChNPP, the water level of the cooling pond (22.5 km^2; depth between 1.5 and 15 m, with about 111 TBq ^{137}Cs and 37 TBq ^{90}Sr) will drop with 4 to 7 m, and 15 km^2 of the sediments will come to surface and may be in need for stabilization. In this respect, the SALIMAT option was investigated. SALIMATs consist of a roll of willow rods (stems) rolled around a central disposable tube; these are unrolled by dragging them across the lagoon. Small tests have demonstrated that SALIMATs establish well on contaminated pond sediments and produce a full vegetation cover during the second year following establishment. The approximate cost of the phytostabilization option ranges from 0.8 to 1.9 10^6 EUR for the reclamation of 15 km^2 of sediments. This is a low cost compared with the prospective cost of removal of the sediments ($6,000,000, transport and disposal costs not included) or the maintenance of the present water level ($200,000 per year).

The projected area for the horizontal erosion control was the right bank of the Pripyat River, which was significantly less contaminated than the left bank, yet not protected with a dam. After inundation, part of the activity is eroded and transported to the Pripyat with the withdrawing water. It was calculated that, even in case of extremely high flooding, dense willow plantings will effectively decrease the horizontal soil erosion and concomitant transport of radionuclides into the Dnieper River system.

29.3.9 URANIUM MINING TAILINGS AND DEBRIS HEAPS

(Re)vegetation is a commonly employed measure on the capping of engineered waste disposal facilities and on mining residues such as spoil heaps [82,83] or tailings ponds. The final step in closing out an impoundment for uranium mill tailings is the design and placement of a cover that will give long-term stability and control to acceptable levels of radon emanation, gamma radiation, erosion of the cover and tailings, and infiltration and precipitation into the tailings and heaps. Surface vegetation can be effective in protecting tailings or a tailing cover from water and wind erosion.

Factors affecting the effectiveness of surface revegetation on impoundments can be broadly classed into climatological and agrobiological factors. Plants should be chosen to match the local climate conditions. Concerning agrobiological factors, the nature of the ore and the mill processes will largely determine the tailing characteristics from the point of view of sustaining growth. Considerable efforts to correct adverse characteristics, such as low or high pH values and low plant nutrient content, will usually be required before tailings can sustain growth. Depending on the substrate, revegetation requires preparation and amelioration of the topsoil to remove, for instance, acid-generating minerals [84,85]. Techniques and strategies to overcome such difficulties have been developed [86] — for instance, hydroseeding or the use of compost from organic household refuse [87]. The method may be limited to low contaminant concentrations owing to the (root) toxicity of higher concentrations. An adequate soil cover may need to be established.

Water and wind erosion are the primary causes of erosion of tailings or tailing cover material. A vegetation cover may decrease the erosion hazard. However, vegetation surfaces may raise concerns: the vegetation can promote radon emanation by drying out the tailing or tree roots may penetrate the contaminated material and break the cover integrity. Given the increased evapotranspiration rate and interception of precipitation following vegetation establishment, the vegetation cover alters the water household of the tailings and may decrease seepage. The effect of a vegetation cover on the radionuclide dispersion through an alteration of the water balance and also potentially because of the effect of plant roots on the physicochemical characteristics of the tailing material (biologically driven acidification of the tailing material) has not been studied intensively thus far.

For a 35-year-old reclaimed site on a uranium mining dump near Schlema (Saxony, Germany), it was concluded that the biomass could reduce infiltration by 40 to 60% due to interception by the canopy (25 to 40%) and increased transpiration [88]. Of the 165,000 g ha^{-1} U in soil (30 cm depth), only 4 g ha^{-1} was in the above-ground plant parts and 510 g ha^{-1} in the below-ground plant parts. Of the uranium taken up during the growing season, 90% is recycled (returned) with the needles. U-dispersion by uptake through vegetation is thus minimal. It may be concluded from these preliminary results that forest vegetation may reduce infiltration rate and will not favor radionuclide dispersion.

Kistinger et al. [89] evaluated some design criteria for tailing coverage regarding vegetation aspects. Because the plant roots can penetrate the compacted sealing layer (trees have roots up to 3 to 4 m) and because the trees should also have a certain degree of mechanical support in order to minimize probability of uprooting, a vegetation substrate depth of at least 1.5 m is required. The vegetation substrate layer must be such that the critical suction is not exceeded at the top of the clay seal. It must be thick enough for plants to find sufficient water and nutrients not to generate high suction at the seal. Cracks resulting from such suctions become accessible to roots and can be widened as further water is extracted.

29.4 CONCLUSIONS

Different phytomanagement options may be applied for the remediation of radioactive contaminated sites. For vast contaminated surfaces, only the phytostabilization option and alternative land use options seem feasible options. The effect of the vegetation cover on the radionuclide dispersion and the subsequent dose should be thoroughly studied. Similarly, the effectiveness of alternative land use should not be studied on mere radiological grounds only. Socioeconomic factors are equally important.

Phytoextraction for soil cleanup seems only reasonable for limited surfaces and will only be effective under very specific conditions. Side effects (costs, treatment of contaminated biomass, and potential for ground water contamination) should be evaluated in depth.

REFERENCES

1. Bennett, B.G., Worldwide panorama of radioactive residues in the environment, in *Proc. Int. Symp. Restoration Environments Radioactive Resdidues,* IAEA, Vienna, Austria, 2000, 11.

2. Vandenhove, H. et al., Investigation of a possible basis for a common approach with regard to the restoration of areas affected by lasting radiation exposure as a result of past or old practice or work activity — CARE, final report for EC-DG XI-project 96-ET-006, *Radiation Protection* 115, Luxembourg, Office for Official Publication by the European Communities, printed in Belgium, 2000, 238.

3. Vandenhove, H., European sites contaminated by residues from ore extraction and processing industries, in *Int. Symp. Restoration Environments Radioactive Residues*, Arlington, Virginia, USA, IAEA, STI/PUB/1092, Austria, Vienna,1999, 61.

4. Vasudev, D. et al., Removal of radionuclide contamination from water by metal-accumulating terrestrial plants. Prepared for presentation at Spring National Meeting, New Orleans, LA: *in situ* soil and sediment remediation. Unpublished. 1996.

5. Ensley, B.D., Raskin, I., and Salt, D.E., Phytoremediation applications for removing heavy metals contamination from soil and water. *Biotechnol. Sustain. Environ.*, 6, 59, 1997.

6. UNSCEAR, *Sources, Effects and Risks of Ionising Radiation*. United Nations Scientific Committee on the effects of atomic radiation. Report to the General Assembly, United Nations, New York, UNSCEAR, 1993.

7. Hossner, L.R. et al., Literature review: phytoaccumulation of chromium, uranium, and plutonium in plant systems. Springfield, VA, Amarillo National Resource Center for Plutonium, 1998, 51.

8. Ebbs, S.D., Identification of plant species and soil amendments that improve the phytoextraction of zinc and uranium from contaminated soil. Ph.D. thesis, Cornell University, Michigan, 1997, 174.

9. Baes, C.F., Environmental transport and monitoring: prediction of radionuclide Kd values from soil-plant concentration ratios. *Transuran. Am. Nucl. Soc.*, 42, 53, 1982.

10. Whicker, F.W. and Ibrahim, SA., Radioecological investigations of uranium mill tailings systems. Fort Collins, Colorado State University, 1984, 48.

11. Ibrahim, S.A. and Whicker, F.W., Comparative plant uptake and environmental behavior of U-series radionuclides at a uranium mine-mill. *J. Radioanal. Nucl. Chem.*, 156(2), 253, 1992.

12. Ibrahim, S.A. and Whicker, F.W., Comparative uptake of U and Th by native plants at a U production site. *Health Phys.*, 54(4), 413, 1988.

13. Frissel, M.J. and van Bergeijck, K.E., Sixth report of the IUR working group soil-to-plant transfer factors: report of the working group meeting in Guttannen Switzerland, Bilthoven, RIVM, 1989, 240.

14. Sheppard, M.I., Thibault, D.H., and Sheppard, S.C., Concentrations and concentration ratios of U, As and Co in Scots Pine grown in waste site soil and an experimental contaminated soil. *Water, Air Soil Pollut.*, 26, 85, 1985.

15. Sheppard, S.C., Evenden, W.G., and Pollock, R.J., Uptake of natural radionuclides by field and garden crops. *Can. J. Soil Sci.*, 69, 751, 1989.

16. Apps, M.J., Duke, M.J.M., and Stephens–Newsham, L.G., A study of radionuclides in vegetation on abandoned uranium tailings. *J. Radioanal. Nucl. Chem.*, 123(1), 133, 1988.

17. Lakshmanan, A.R. and Venkateswarlu, V.S., Uptake of uranium by vegetables and rice. *Water, Air Soil Pollut.*, 38, 151, 1988.

18. Ebbs, S.D., Norvell, W.A., and Kochian, L.V., The effect of acidification and chelating agents on the solubilisation of uranium from contaminated soil. *J. Environ. Qual.*, 27, 1486, 1998.

19. Mortvedt, J.J., Plant and soil relationship of uranium and thorium decay series radionuclides — a review. *J. Environ. Qual.*, 23, 643, 1994.

20. Huang, J.W. et al., Phytoremediation of uranium contaminated soils: role of organic acids in triggering hyperaccumulation in plants. *Environ. Sci. Technol.*, 32, 2004, 1998.

21. Sheppard, M.I., Sheppard, S.C., and Thibault, D.H., Uptake by plants and migration of uranium and chromium in field lysimeters. *J. Environ. Qual.*, 13(3), 357, 1984.

22. Vandenhove, H., Van Hees, M., and Van Winckel, S., Feasibility of the phytoextraction approach to clean-up low-level uranium contaminated soil. *Int. J. Phytorem.*, 3(3), 301, 2001.

23. Kopp, P., Oestling, O., and Burkart, W., Availability and uptake by plants of radionuclides under different environmental conditions. *Toxicol. Environ. Chem.*, 23, 53, 1989.

24. Simon, S.L. and Ibrahim, S.A., The soil/plant concentration ratio for calcium radium, lead and polonium: evidence for nonlinearity with reference to substrate concentration. *J. Environ. Rad.*, 5, 123, 1987.

25. Menzel, R.G., Competitive uptake by plants of potassium, rubidium, cesium, calcium, strontium and barium from soils. *Soil Sci.*, 77(6), 419, 1954.

26. Vasconcellos, L.M.H. et al., Uptake of ^{226}Ra and ^{210}Pb by food crops cultivated in a region of high natural radioactivity in Brazil. *J. Environ. Rad.*, 5, 287, 1987.

27. Vandenhove, H., Eyckmans, T., and Van Hees, M., Can barium and strontium be used as tracer for radium in soil-plant transfer studies? *J. Environ. Rad.*, 81(2-3), 255-267, 2005.

28. IUR, Sixth report of the IUR working group soil-to-plant transfer factors: report of the working group meeting in Guttannen Switzerland, 198924-26, Bilthoven, RIVM: 240, 1989.

29. Kabata–Pendias, A. and Pendias, H. (Eds.), *Trace Elements in Soil and Plants*, Lewis Publishers, Inc., Chelsea, MI, 2000.

30. Chirkov, I.V., Kaplan, G.E., and Uspenskaya, T.A., *Thorium*, Gosizdat: Moskow, 9, 1961.

31. Syed, H.S., Comparison studies adsorption of thorium and uranium on pure clay minerals and local Malaysian soil sediments. *J. Radioanal. Nucl. Chem.*, 241(1), 11, 1999.

32. Hunsen, R.O. and Huntington, G.L., Thorium movements in morainal soils of the High Sierra, California. *Soil Sci.*, 108, 257, 1969.

33. Shtangeeva, I., Thorium, in *Trace and Ultratrace Elements in Plants and Soil*, 2004.

34. Linsalata, P. et al., An assessment of soil-to-plant concentration ratios for some natural analogues of the transuranic elements. *Health Phys.*, 56(1), 33, 1989.

35. Whicker, F.W. et al., Uptake of natural and anthropogenic ectinides in vegetable crops grown on a contaminated lake bed. *J. Environ. Rad.*, 45, 1, 1999.

36. Vera Tome, F., Blanco Rodríguez, M.P., and Lozano, J.C., Soil-to-plant transfer factors for natural radionuclides and stable elements in a Mediterranean area. *J. Environ. Rad.*, 65, 161. 2003.

37. Lembrechts, J., A review of literature on the effectiveness of chemical amendments in reducing the soil-to-plant transfer of radiostrontium and radiocaesium. *Sci. Tot. Environ.*, 137, 81, 1993.

38. Nisbet, A.F. et al., Application of fertilizers and ameliorants to reduce soil-to-plant transfer of radio-caesium and radiostrontium in the medium to long term — a summary, *Sci. Tot. Environ.*, 137, 173, 1993.

39. Shaw, G., Blockade of fertilizers of caesium and strontium uptake into crops: effects of root uptake process. *Sci. Tot. Environ.*, 137, 119, 1993.

40. Smolders, E. and Merckx, R., Some principles behind the selection of crops to minimize radionuclide uptake from soil. *Sci. Tot. Environ.*, 137, 135, 1993.

41. Evans, D.W., Alberts, J.J., and Clack, R.A., Reversible ion-exchange fixation of cesium-137 leading to mobilization from reservoir sediments. *Geochim. Cosmochim. Acta*, 47, 1041, 1983.

42. Brouwer, E. et al., Caesium and rubidium ion equilibria in illite clays, *J. Phys. Chem.*, 87, 1213, 1983.

43. Cremers, A., Elsen, A., and De Preter, P., Quantitative analysis of radiocaesium retention in soils. *Nature*, 335, 247, 1988.

44. Konoplev, A.V. et al., Influence of agricultural countermeasures on the ratio of different chemical forms of radionuclides in soil and soil solution. *Sci. Tot. Environ.*, 137, 147, 1993.

45. Bondar, P.E. and Dutov, A.I., Parameters of radiocaesium transfer into oats harvest on lime soils in connection with the application of mineral fertilizers and chemical ameliorants, in *Collections of Scientific Works — Problems of Agricultural Radiology*. Kiev, Ukraine,1992, 125.

46. Belli, M. et al., The effect of fertilizer applications on ^{137}Cs uptake by different plant species and vegetation types. *J. Environ. Radioact.*, 27, 75, 1995.

47. Cline, J.F. and Hungate, F.P., Accumulation of potassium, caesium-137 and rubidium-86 in bean plants grown in nutrient solution. *J. Plant Physiol.*, 35, 826, 1960.

48. Shaw, G. et al., Radiocaesium uptake and translocation in wheat with reference to the transfer factor concept and ion competition effects. *J. Environ. Radioact.*, 16, 167, 1992.

49. Smolders, E., Van den Brande, K., and Merckx, R., Concentrations of ^{137}Cs and K in soil solution predict the plant availability of ^{137}Cs in soils. *Environ. Sci. Techn.*, 31, 3432, 1997.

50. Vandenhove, H., Phytoextraction of low-level contaminated soil: study of feasibility of the phyto-extraction approach to cleanup ^{137}Cs contaminated soil from the belgoprocess site; part 2: transfer factor screening test: discussion of results. Internal SCK•CEN report R-3407, 1999.

51. IUR, Eighth report of the working group soil-to-plant transfer factors, report of the Working Group Meeting in Madrid, Spain, IUR Pub R-9212-02, IUR Technical Secretariat, Balen, Belgium, June 1–3, 1992.

52. Gopa, Belarus, Study on alternative biodiesel sources in relation with soil decontamination. Project no.° TACIS/REG93, 1996.

53. Vandenhove, H. et al., Transfer of radiocaesium from podzol to ryegrass as affected by AFCF concentration. *Sci. Tot. Environ.,* 187, 237, 1996.

54. Van Hees, M. and Vandenhove, H., Unpublished data, 1998.

55. Lasat, M.M. et al., Phytoremediation of a radiocaesium contaminated soil: evaluation of caesium-137 accumulation in the shoots of three plant species. *J. Environ. Qual.,* 27, 165, 1998.

56. Sorochinsky, B., Application of phytoremediation technologies in real conditions in the Chernobyl zone, in *Proc. Chernobyl Phytoremediation Biomass Energy Conversion Workshop.* Slavutych, Ukraine, 1998, 229.

57. Nisbet, A.F. and Woodman, R.F.M., Soil-to-plant transfer factors for radiocaesium and strontium in agricultural systems. *Health Phys.,* 78(3), 279, 2000.

58. Sauras, T.Y. et al., ^{137}Cs and ^{90}Sr root uptake prediction under close-to-real controlled conditions. *J. Environ. Rad.,* 45, 191, 1999.

59. Wauters, J., Radiocaesium in aquatic sediments: sorption, remobilization and fixation. Ph.D. dissertation 246, Faculty of Agronomy and Applied Biological Sciences, Kuleuven, Belgium, 1994, 110.

60. Thibault, D.H., Sheppard, M.I., and Smith, P.A., A critical compilation and review of default soil solid/liquid partition coefficients, Kd, for use in environmental assessments. Atomic Energy Canada, AECL,-10125, Whiteshell Nuclear Research Establishment, Manitoba, Canada, 1990.

61. IAEA (International Atomic Energy Authority), *Handbook of Parameter Values for the Prediction of Radionuclide Transfer in Temperate Environments,* Technical Report Series N° 364, IAEA, Vienna, 1994.

62. Dushenkov, S. et al., Removal of uranium from water using terrestrial plants. *Environ. Sci. Technol.,* 31, 3468, 1997.

63. IAEA, One decade after Chernobyl: the basis for decision. *IAEA Bull.,* 38(3), 10, 1996.

64. Rauret, G. and Firsakova, S., Transfer of radionuclides through the terrestrial environment to agricultural products, including the evaluation of agrochemical practices. Final report for the Experimental Collaboration Project no.°2 (ECP-2, EU 16528), 179, 1996.

65. Belli, M. and Tikhomirov, F., Behavior of radionuclides in natural and seminatural environments. ECSC-EC-EAEC 145, 1996.

66. Segal, M.G., Agricultural countermeasures following deposition of radioactivity after a nuclear accident. *Sci. Tot. Environ.,* 137, 31, 1993.

67. Alexakhin, R.M., Countermeasures in agricultural production as an effective means of mitigating the radiological consequences of the Chernobyl accident. *Sci. Tot. Environ.,* 137, 9, 1993.

68. Renaud, P. and Maubert, H., Agricultural countermeasures in the management of the post-accidental situation. *J. Environ. Rad.,* 35, 53, 1997.

69. Grebenkov, A., Internal deliverables for RECOVER (see [71]), 1997.

70. Nisbet, A.F., Woodman, R.F.M., and Haylock, R.G.E., Recommended soil-to-plant transfer factors for radiocaesium and radiostrontium for use in arable systems. National Protection Board, NRPB-R304, Didcot, U.K., 1999, 50.

71. Vandenhove, H. et al., RECOVER — relevancy of short rotation coppice vegetation for the remediation of contaminated areas, final report, EC-DG XII-project FI4-CT0095-0021c, SCK•CEN, BLG 826, 1999.

72. Vandenhove, H. et al., PHYTOR: evaluation of willow plantations for the phytorehabilitation of contaminated arable land and flood plane areas. Final report, EC-DG XII project under contract ERB IC15-CT98 0213, SCK•CEN, BLG 909, 2002.

73. Vandenhove, H. et al., Short rotation coppice for reevaluation of contaminated land. *J. Environ. Rad.,* 56, 157, 2001.

74. Vandenhove, H. et al. Economic viability of short rotation coppice for energy production for refuse of caesium contaminated land in Belarus. *Biomass Bioenergy,* 22, 421, 2002.

75. Zabudko, A.N. et al., Comprehensive radiological investigations of forests in Kaluga, Tula and Or'ol regions, IAEA Technical Meeting on Cleanup Criteria for Forests and Forestry Products Following a Nuclear Accident, unpublished, 1995.

76. Vandenhove, H. and van Hees, M., Fiber crops as alternative land use for radioactively contaminated arable land. *J. Environ. Rad.,* accepted for publication, 2004.

77. Petrisor, I.G., Komnitsas, K., and Lazar, I., Application of a vegetative cover for remediation of sulphidic tailings dumps on a Romanian Black Sea coastal area, Warsaw '98, *Proc. 4th Int. Symp. Exhib. Environ. Contamination Central Eastern Europe*, 15–17 September 1998, Warsaw, Poland, 1998.

78. Van de Vivere, H., Revegetation of industrial sites, in Vandenhove, H. (Ed.), *Topical Days on Phytomanagement of Contaminated Environments*, Mol, Belgium, Report SCK•CEN BLG-844, 2000, 67.

79. Stanley, J., Rehabilitation of mines and other disturbed sites, http://www.hortresearch.co.nz/products/bioremediation/rehab/, 2002.

80. Stanlye, J. et al., Developing optimum strategies for rehabilitating overburden stockpiles at the Grasberg Mine, Irian Jaya, Indonesia, in *Proc. 11th Int.. Peat Congr., Quebec, Canada. Commission V*: 2000, 806.

81. Vandenhove, H., Major sources of radioactive contamination, possible remediation options and role of phytostabilization, in Vandenhove, H. (Ed.), *Topical Days on Phytomanagement of Contaminated Environments*, Mol, Belgium, Report SCK•CEN BLG-844, 2000, 1.

82. Dudel, G.E. et al., Uptake and cycling of natural radionuclides in vegetation developed on uranium mining heaps and tailings, in Vandenhove, H. (Ed.), *Topical Days on Phytomanagement of Contaminaed Environments*, Mol, Belgium, Report SCK•CEN BLG-844, 2000, 49.

83. Brackhage, C. et al., Radionuclide and heavy metal distribution in soil and plants from a 35-year-old reclaimed uranium mining dump site. *Proc. 5th Int. Conf. Biogeochem. Trace Elements,* Vienna, 1999, 1174.

84. Jennings, S.R. and Krueger, J., Clean tailing reclamation: tailing reprocessing for sulfide removal and vegetation establishment, in *Proc. 12th Ann. Conf. Hazardous Waste Res.*, Kansas City (1997) www.engg.ksu.edu/ HSRC/97Proceed/Metals5/clean.html (tested 2003), 1997.

85. Schippers, A., Jozsa, P.-G., and Sand, W., Evaluation of the efficiency of measures for sulfidic mine waste mitigation. *Appl. Microbiol. Biotechnol.*, 49, 698, 1998.

86. Schippers, A., Jozsa, P.-G., and Sand, W., Überprüfung der Effizienz von Maßnahmen zur Sanierung von Bergbaualtlasten, in Przybylski, T., Merkel, B., Althaus, M., and Kurzydlo, H. (Eds.), Rekultivierung und Umweltschutz in Bergbau — Industriegebieten, Band 2, Towarzystwo Przyjaciól Nauk, Legnica, Polen, 307 and 357.

87. McNearny, R.L., Revegetation of a mine tailings impoundment using municipal biosolids in a semiarid environment, *Proc. 1998 Conf. Hazardous Waste Res. — Bridging Gaps Technol. Culture*, Snowbird, Utah 87, 1998 www.engg.ksu.edu/HSRC/ 98Proceed/8McNearny/8mcnearny.pdf (tested 07/08/01).

88. Thiry, Y. et al., Uranium distribution and cycling in Scots pine (*Pinus sylvestris* L.) growing on a revegetated U-mining heap. Accepted by *J. Environ. Rad.*, 2004.

89. Kistinger, S. et al., Evaluation of the long-term durability of engineered dry covers for mining wastes, and consideration of associated design constraints, in *Uranium in the Aquatic Environment*, Merkel, B. et al., Eds., Springer–Verlag, Berlin, 2002, 155.

30 Efficiency and Limitations of Phytoextraction by High Biomass Plants: the Example of Willows

Catherine Keller

CONTENTS

30.1 INTRODUCTION

Phytoextraction with crops and bioenergy plants or trees has been proposed as a suitable alternative to the destructive techniques used thus far to clean up soils contaminated with heavy metals. This is one of the two possible approaches that are followed in phytoextraction; the other one is the use of hyperaccumulating plants. The crop and energy plants have opposite characteristics to hyperaccumulators: they are well known plants with large yields and fast growth, but they are usually not metal specific and they accumulate below the hyperaccumulation thresholds. Plants or seeds are readily available and some have already been tested in the field [1–3]. On the other hand, very few species (if any) of hyperaccumulators are presently available for field application under temperate climate (lack of nurseries or seed producers), so the use of hyperaccumulators might need to be dismissed for remediation of large areas.

In general, the use of plants to remove metals from soils is environmental friendly and its cost is low compared to engineering-based techniques such as soil capping, soil washing, vitrification, land-filling, etc. Additionally, it is an *in situ* and solar-generated technique that could help to rehabilitate large areas of agricultural soils contaminated mostly in the upper layer, by maintaining or even restoring soil fertility. It also produces less waste because the biomass may be recycled for energy and metals. Unlike phytostabilization — its counterpart among the phytoremediation techniques — phytoextraction potentially resolves most of the legislative requirements that ask for metal removal down to a given threshold. Another interest of bioenergy plants or trees — including willows — is that they may also be used in phytostabilization [4].

The general drawbacks of phytoextraction are well known [5] and its optimization still requires a lot of research: phytoextraction is a slow process in comparison to the previously mentioned techniques (10 years is the goal upon which most people agree to recognize phytoextraction as economically acceptable). Its application is limited because of the lack of established methods and successful completed remediation case studies, the lack of recognized economic performance, and the risk of food chain contamination. Additionally, it may not be able to remove 100% of the contaminants and its efficiency has been proved for some contaminants only. Alternatively, the plants might not be able to withstand highly toxic concentrations of the contaminants.

This chapter does not aim to review all the drawbacks in detail; a preliminary comparison with the other available techniques should determine whether these negative aspects are limiting or not for choosing phytoextraction for a given site. In addition, for the reasons mentioned earlier, phytoextraction with hyperaccumulators will not be discussed here.

Once it has been decided to use phytoextraction with crop or bioenergy plants as a cleanup technique, many pitfalls may appear that will limit its efficiency — that is, its ability to reach the goal set in a reasonable time and with limited negative side effects. They may be due to the specificities of the site, soil, plants, and contamination characteristics, as well as to external constraints like legislation. Highlighting these factors will help to define the boundary conditions for the use of phytoextraction and thus the ability to offer an efficient technique to landowners or local authorities as main remediation tool or as part of a "remediation package."

This chapter focuses on phytoextraction with the high biomass willow *Salix viminalis*. It draws examples from field experiments with *S. viminalis* and pot experiments, with additional references to other soils and plants. The field experiments of reference will be presented first. Then, the site characteristics (including the local climatic and edaphic conditions as well as the nature and extent of the contamination), plant characteristics, and aspects related to the legislative background that may be limiting for phytoextraction efficiency will be presented successively. It should be remembered that all these aspects may interact and thus are not truly independent parameters.

30.2 CHARACTERISTICS OF THE EXPERIMENTS

30.2.1 DESCRIPTION OF EXPERIMENTAL SITES

The three sites chosen to illustrate these different aspects and their impacts on the final efficiency of phytoextraction are presented in Table 30.1. Two are agricultural soils and the third is a landfill that has been closed and revegetated with trees. The first two have been tested for phytoextraction [6,7] and the third one was studied in view of a general site rehabilitation [8]. They are all moderately contaminated with metals, mostly Cu, Zn, Cd and Pb with some additional metals in lower concentrations, but no significant organic contamination.

- The Dornach site in northwest Switzerland has been described by Kayser et al. [2]. The source of the heavy metal contamination was a nearby brass smelter emitting

TABLE 30.1
Site Descriptions

	Dornach	Caslano	Les Abattes
Location	Jura edge	Southern Alps	Jura chain
Altitude in meters	300	280	1000
Mean temperature in °C	10	11.7	5.3
Mean rainfall in millimeters	800	1860	1470
Number of samples for analysis 0–0.2 m	20	4	49
pH	7.2	4.9	7.4 (0.1)[a]
% Clay	30[b]	12	43 (13)[a]
% Organic carbon	2.5[b]	5.2	5.6 (2.1)[a]
Total Zn in mg $kg^{-1[c]}$	645 (81)	1158 (216)	245 (179)
Total Cu in mg $kg^{-1[c]}$	525 (62)	264 (43)	188 (158)
Total Cd in mg $kg^{-1[c]}$	2.0 (0.4)	2.8 (0.7)	1.2 (0.7)
Soluble Zn in mg $kg^{-1[c]}$	0.08 (0.03)	7.4 (5.9)	bdl
Soluble Cu in mg $kg^{-1[c]}$	0.7 (0.2)	0.4 (0.1)	bdl
Soluble Cd in mg $kg^{-1[c]}$	2.0 (0.6)	13 (11)	bdl

[a] 11 samples.

[b] 3 samples.

[c] FAC, *Methoden für die Bodenuntersuchungen,* Eidgenössische Forschungsanstalt für Agrikulturchemie und Umwelthygiene, Bern-Liebefeld, Switzerland, 1989.

[d] bdl = below detection limit.

Note: Soil parameters are average values with standard deviations given in parentheses.

Sources: After Kayser, A. et al., *Environ. Sci. Technol.,* 34, 1778, 2000 (Dornach and Caslano); Hammer, D. et al., *Soil Use Manage.,* 19, 187, 2003 (Dornach and Caslano); Keller, C. et al., *Plant Soil,* 249, 67, 2003 (Dornach and Caslano); and Keller, C. et al., unpublished data, 1997–2004 (Les Abattes).

Cu, Zn, and Cd in particulate form until the mid 1980s. The soil has been classified as calcaric regosol [9]. The site is currently included in a crop rotation.

- The Caslano site in southern Switzerland was contaminated with sludges from septic tanks spread on the site for 20 years until 1980. This has led to enrichment of the topsoil of this acidic fluvisol with organic matter and heavy metals (Cd, Cu, and Zn). Currently, the site is used as a private garden and a meadow for fodder production. At both sites, the contaminated layer is between 0.20 and 0.60 m thick.

- The third experimental site was designed for forest rehabilitation and is located on the former landfill "Les Abattes" (Le Locle, Switzerland), in the Swiss Jura mountain chain. The 2-ha landfill was used for inert and uncontaminated waste material deposits [10], but the landfill was capped in 1990 with a final top layer (0.05 to 0.60 m thick) made up by a mixture in varying proportions of contaminated compost originating from contaminated sewage sludge and uncontaminated local calcareous soil and gravel (calcaric anthroposol). Thus, heavy metal concentrations in the final top layer were spatially highly variable as a consequence of the heterogeneous mixing of the components [8].

In addition, for illustration of some specific points, results from other soils are presented. The important characteristics of these soils are given together with the results.

30.2.2 PLANTS FOR PHYTOEXTRACTION

Phytoextraction experiments have been performed on the two agricultural sites with crop plants (tobacco, maize, Indian mustard, sunflower, bioenergy plants [*Salix viminalis*], and hyperaccumulators [*Thlaspi caerulescens, Alyssum montanum*, and *Iberis intermedia*] on miniplots. Only results obtained with the crop plants and *S. viminalis* (Swedish clone n°78980) are presented here. Both experiments have been conducted according to agronomic standards; detailed descriptions of the experiments can be found elsewhere [2,7]. On the landfill, various tree species have been planted in groups of 10 plants of the same species, including pioneer (and bioenergy) trees like *S. viminalis* [8]. In addition to this grid, to study the behavior of *S. viminalis* more precisely, two clones (a local one and the Swedish clone) of *S. viminalis* were planted as cuttings in 1997 or as 1-year-old plants (1 m high) in 1998 in 30 groups of 12 plants with a random distribution of the 30 groups on a 30-m side grid. The whole field was maintained as a typical forest plot (no weed removal but grass around the young plants was cut once per year).

Table 30.2 gives an overview of the main characteristics of the experiments as well as some results obtained with *S. viminalis* that will be presented later. In the case of Les Abattes, only the results of the 1-year-old plants are given. Additionally, the two clones of *S. viminalis* were tested in hydroponics as well as in pots (with various soils) in order to measure their extraction potential in controlled conditions. They were compared to other willow species, *Salix aurita* and *S. irrorata*,

---

TABLE 30.2
Overview of *Salix viminalis* (Clone 78980) Results

	Dornach	Caslano	Les Abattes
Type of experiment	Phytoextraction	Phytoextraction	Forest rehabilitation
Plant density in plants ha^{-1}	12,800 (+20,000)	40,000	10,000
Type of planting	Cuttings	Cuttings	1-year-old plant
Yield after 2 years in t DM ha^{-1}	6.5 (3.8)	15.8 (6.0)	0.9 (0.7)[a]
Yield after 5 years in t DM ha^{-1}	32 (2.5)	24 (11)[b]	—
Yield after 2 years in t DM ha^{-1} normalized to 10,000 plants ha^{-1}	5.08	3.95	0.9
Concentrations after 2 years in mg kg^{-1} DM			
Zn in whole plants	240 (46)[c]	785 (202)	252 (77)
Zn in leaves	665 (50)	1700 (465)	475 (181)
Zn in stems	120 (10)	430 (85)	183 (42)
Cu in whole plants	10 (1)	13 (1)	5.2 (0.7)
Cu in leaves	13 (1)	13 (1)	5.6 (0.7)
Cu in stems	9 (1)	13 (1)	5.1 (0.8)
Cd in whole plants	3.3 (0.5)	2.6 (0.5)	1.3 (0.5)
Cd in leaves	5.1 (0.7)	3.6 (0.5)	1.5 (0.5)
Cd in stems	2.3 (0.1)	1.2 (0.5)	1.2 (0.5)

[a] Plants not cut the first year.
[b] Yield after 4 years in t DM ha^{-1}.
[c] Analyses performed on 3-year-old plants for Zn, Cu, and Cd.

Note: Average yield and concentrations of 4 (Dornach and Caslano) or 30 plots (Les Abattes) are given with standard deviations in brackets.

Sources: After Kayser, A. et al., *Environ. Sci. Technol.*, 34, 1778, 2000 (Dornach and Caslano); Hammer, D. et al., *Soil Use Manage.*, 19, 187, 2003 (Dornach and Caslano); Keller, C. et al., *Plant Soil*, 249, 67, 2003 (Dornach and Caslano); and Keller, C. et al., unpublished data, 1997–2004 (Les Abattes).

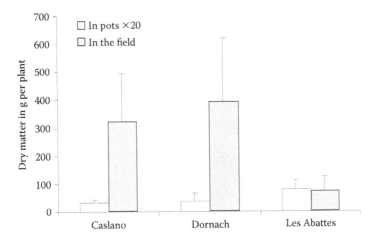

FIGURE 30.1 Comparison of the biomass produced by plants of *Salix viminalis* (clone 78980) grown in three different soils in pots (3-month-old plants) and in the field (second growing season). Ratios between dry biomass produced in pot and in the field are 118, 190, and 18 for Caslano, Dornach, and Les Abattes, respectively. (Modified from Keller, CNATO Series, 2004, in press.)

found to be tolerant to Cd and Zn, and *S. aurita*, which was also able to accumulate Cd and Zn in its shoots (tests performed in hydroponics) [11].

30.3 FACTORS LIMITING THE EFFICIENCY OF PHYTOEXTRACTION

30.3.1 CLIMATE AND SOIL CHARACTERISTICS

Site characteristics include the climatic conditions; the nature of the soil that is contaminated; volume of soil concerned (surface and depth); the nature and the extent of the contamination; the metals to be removed, their concentrations, their availability to plants, and their toxicity; the presence of other inorganic or organic contaminants; and the degree of heterogeneity of the contamination.

30.3.1.1 Climate

Climatic conditions (mostly temperature and rain) play a role directly in biomass production and indirectly in metal concentrations (that are influenced by plant health). The results obtained for *S. viminalis* and presented in Figure 30.1 are typical of the impact of the climatic conditions on the growth of plants. In a pot experiment, no significant difference was measured in the biomass production of *S. viminalis* (clone 78980) grown 3 months in Les Abattes, Dornach, and Caslano soils [12,13]; however, in the field, the yield per plant in Les Abattes was lower due to a lower average temperature and high rainfall (Table 30.1 and Table 30.2). Also, Van Splunger et al. [14] have found that drought increases root length and rooting depth, thus affecting root distribution of Salicaceae and the ability to forage and take up nutrients from soil.

30.3.1.2 Soil Characteristics

Soil characteristics (pH, percent of C, percent of clay, nutrient availability, water table, and compaction) are responsible for the general soil fertility and have a large impact on plant biomass and composition [15] (Figure 30.1, results in pot, and Figure 30.2). In Figure 30.2, the biomass production of four *Salix* clones grown in pots under controlled conditions varied with the type of soil and these variations were not correlated (see, for example, the differences between *S. aurita*

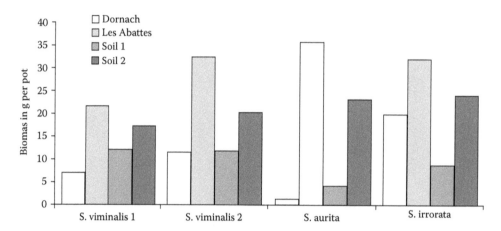

FIGURE 30.2 Biomass production of *S. viminalis* (1 = Swedish clone 78980; 2 = clone Les Abattes), *S. aurita*, and *S. irrorata* grown 8 months in pots in the Dornach and Les Abattes soils and two other agricultural soils —soil 1 and soil 2 — presented for their Zn content in Figure 30.6. (After Keller, C., Raduljica, O., and Hammer, D., unpublished data, 1997–2004.)

TABLE 30.3
Examples of Average Root Depth for Several Tree Species and According to Soil Type and Age of Trees

Soil type	Acer pseudoplatanus		*Populus* spp	Populus tremula	Betula pendula	
	Root depth (cm)	Age (year)	Root depth (cm)	Root depth (cm)	Root depth (cm)	Age (years)
With skeleton, carbonates, average depth	—		—	—	40–60	40–50
Deep, without skeleton, silty clay	50 500	2 20	120–140	90–150	100–150	40–50
Nonhydromorphic sand, sandy soils	110–140	60–70	100–260	30	150–450	10–20
Different soils, slightly hydromorphic	130–140	60–70	110	—	—	
Hydromorphic soils	40–50	50–60	—	—	90–130	90
Presence of groundwater table	40	50-60	80	—	—	

Source: Modified from Polomski, J. and Kuhn, N., *Wurzelsysteme*, Paul Haupt, Bern, Switzerland, 1998.

and *S. irrorata* for Dornach and soil 1). The type of soil also influences the root depth prospecting as shown in Table 30.3, where root depth may vary from 40 to 500 cm for the same species growing in different uncontaminated soils. For willows, it has been shown that root development varies according to soil texture that drives plant-available soil water and the maximum capillary rise. For example, sand retains too little water and silt is usually too compact and oxygen deficient to allow root growth at depth, whereas sandy loam or clay loam allows a deep root system [16].

Like the three sites studied located within the boundaries of towns, many contaminated soils in urban areas are fertile, especially kitchen gardens. Unfortunately, in addition to the contamination, many other contaminated sites are degraded in a broader sense [17]. They are poor in nutrients and have low water retention and more generally heterogeneous biological, chemical, and physical properties unfavorable for plant growth that may induce deficiencies. For example, Bending and Moffat [18] found that the establishment of trees on mine spoils was impaired by acidity, salinity, N deficiency, compaction, and poor water-holding capacity. Agronomic management, irrigation, fertilization, and various additives may need to be used intensively to obtain optimal results; this adds to the costs of phytoextraction. For example, *S. viminalis* has a high nitrogen requirement that must be satisfied to obtain efficient metal removal [19,20]. Iron deficiency may also occur and requires the use of an Fe fertilizer to reduce chlorosis, as observed on *S. viminalis* at Dornach and Caslano [6].

Although the origin of the contamination influences the speciation of the metals present in the soil, soil characteristics have an impact on metal speciation in the soil matrix and soil solution. This is best exemplified in unpolluted environments [21,22], but also occurs in polluted soils [13]. This in turn has an impact on the availability of metals and, ultimately, on metal uptake. For example, at Les Abattes, the large C content combined with a high pH and clay content resulted in undetectable soluble Cd, Zn, and Cu as compared to the Dornach and Caslano soils (Table 30.1).

Kayser [2] has studied the relation between pH and $NaNO_3$-extractable Zn and Cd (defined according to Swiss law [23] as the soluble metal fraction in two soils [one is the Dornach soil and the other is named Rafz] amended with increasing amounts of elemental sulphur in order to decrease soil pH and, consequently, metal bioavailability to plants) [2,24,25]. The increase in $NaNO_3$-extractable Cd and Zn was more pronounced in the Rafz soil (initial pH $CaCl_2$ = 6.8, HNO_3-extractable Zn = 813 mg kg^{-1}, and Cd = 0.9 mg kg^{-1}) than in the Dornach soil (pH 7.4; see Table 30.1) with 26- and 13-fold increases in Rafz for Cd and Zn, respectively, and in Dornach of 15- and 11-fold with the highest S8 addition (400 mmol sulphur kg^{-1} soil). This illustrates that the changes in soil pH (artificially induced or through natural acidification) may affect metal availability in different soils differently.

Additionally, the effect on Cd and Zn concentrations in *S. viminalis* was 4.6-fold for Zn and 1.6-fold for Cd compared to the control plants grown on the non-amended Rafz soil and 2.4- and 1.5-fold for Zn and Cd, respectively, in *S. viminalis* grown on the Dornach soil [25]. This highlights the fact that no strict relationship between plant uptake and metal bioavailability is assessed by chemical extractants. The question of bioavailability in relation to plant uptake is discussed in the next section.

30.3.1.3 Nature and Extent of the Contamination

30.3.1.3.1 Toxicity of the Contaminants

An overall evaluation of the particular contamination issue and site must be made before starting a phytoextraction program. Indeed, sites to be remediated are very seldom contaminated with single metals. Multicontamination with organic and inorganic pollutants (and sometimes with physical constraints, such as compacted soil) is common and may induce toxicity to plants that are not tolerant to all the contaminants present. Plants are potentially sensitive to a broad range of elements if their concentrations in the soil are above a given threshold. However, the most toxic metals for higher plants and certain microorganisms are Hg, Cu, Ni, Pb, Co, Cd, and, possibly, Ag, Be, and Sn [26].

The effect of organic pollutants (toxicity) on plants and their fate in plants is poorly known. Attempts have been made to use plants to reduce their concentrations in soils. However, their uptake by roots, translocation to shoots, volatilization, and/or degradation seem to be highly plant, soil, and component dependent [27–29]. It is also clear that the presence of herbicide residues may reduce or even prevent plant establishment.

In general, critical limits in plant can be quickly reached when a large amount of a metal contaminant is in a form available in the soil for plant uptake, especially when it is combined with a large soil–plant transfer coefficient [30,31] or when the threshold is low. For example, copper is a widespread contaminant and no example of successful extraction of Cu by *Salix* species from contaminated soil has been reported thus far [32]. The remediation efforts are mostly directed towards Cu immobilization through the use of additives [33]. However, its toxicity has been found to reduce phytoextraction efficiency because, for most of the plants, the toxic level is already reached when the Cu concentration in the plant is above 15 to 20 mg kg^{-1} DM [31] or when the concentration in the soil solution reaches 0.02 to 0.06 mg L^{-1} [34]. As a result of Cu toxicity, the biomass is reduced and the metal uptake is impaired [35–37].

In contrast, Punshon et al. [38] have shown that *Salix* species (*S. caprea*, *S. cinerea*, and their hybrids with *S. viminalis*) originating from contaminated sites exhibit Cu tolerance, but that there is a large variability between willows at the species and the population levels. Thus, soils with undesirable contaminants might not be easily decontaminated and a soil pretreatment might be necessary to remove or inhibit the toxic compounds. Alternatively, specific willow clones may need to be selected for these sites. Additionally, metals may have antagonistic or synergic effects on their toxicity and uptake as demonstrated for Cd by Costa and Morel [39].

30.3.1.3.2 Heterogeneity of the Contamination

Contamination is highly heterogeneous at polluted sites; this includes spatial variation in composition and concentration [4,17] as well as variation with depth [40]. An example of surface and at depth heterogeneity is displayed in Figure 30.3 for HNO$_3$-extractable Cd and Cu concentrations at Dornach. The standard deviations (SD) increased with depth, illustrating that the thickness of the contaminated layer varied between 0.2 and 0.7 m within a 400-m^2 area. For Cd, the SD was smaller and similar along the soil profile because, although Cd was brought to the soil by atmospheric deposition along with Cu and Zn, it also had a geogenic origin [41,42]. The high heterogeneity in the thickness of the contaminated layer prevents an optimal root colonization of the contaminated layer (see Table 30.4, third column) because phytoextraction can only be achieved when roots are within the contaminated layer. The question of root colonization will be developed later.

30.3.1.3.3 Limited Availability of Metals

From a different perspective, low availability of a metal to be remediated may reduce the potential for its phytoextraction. Although the metal available fraction in soil may be determined by different methods, it is indeed usually better correlated to plant uptake than total metal content in soil. In Figure 30.4, the low Cd uptake by *S. viminalis* is explained by the low NaNO$_3$-extractable (s) Cd

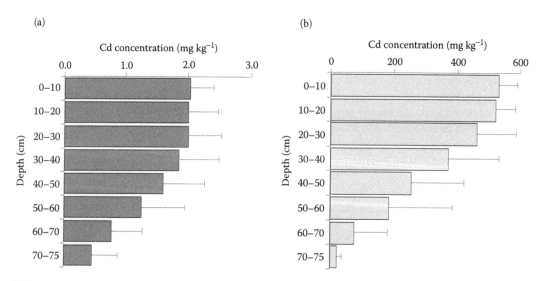

FIGURE 30.3 Profiles of HNO$_3$-extractable Cd and Cu at Dornach. These are means of 15 profiles collected within a 20 × 20 m area with the Humax technique that allows for sampling of undisturbed soil cores. The increasing SD with depth illustrates the variability in the contamination depth.

TABLE 30.4
Cumulative Root Density

	LA (km m⁻²)[a]	LA/shoot biomass[b]	Depth of the contamination		
			True	Shallow	Deep
Indian mustard	2.7 (0.7)	4	0.9 (0.3)	1.9 (0.8)	>0.5
Tobacco	3.7 (0.8)	3	0.5 (0.2)	1.3 (0.3)	0.4 (0.1)
Maize	6.6 (1.8)	4	1.4 (0.6)	>3.3	>0.9
Willows[c]	3.5 (1.8)	3	1.5 (0.7)	>3.2	0.9 (0.2)
Thlaspi (Ganges)	2.6 (1.1)	28	0.4 (0.1)	1.0 (0.0)	0.3 (0.0)

[a] LA = cumulative root density between 0 and 0.2 m (depth) in km m⁻².
[b] LA 0 to 0.2 m/shoot biomass.
[c] 2- and 3-year-old plants.

Notes: Calculated for five crops grown at Dornach and matching of roots and Zn soil contamination expressed as the ratio between the maximal depth with a root length density > 5000 mm dm³ and the soil contamination depth, according to three scenarios: (1) contamination as measured in the soil profile of each plot; (2) hypothetical shallow contamination (0.2 m); and (3) hypothetical deep contamination (0.7 m). (shallow) and (deep) depths are extreme values found in the field experiment. Depth of contamination is calculated after removing 150 mg kg⁻¹ Zn (Swiss guide value) from the total soil Zn concentration. ">" means that the deepest root sampling gave a root length density above 5000 mm dm³. Standard deviations are in brackets.

Sources: Swiss guide value: OIS (ordinance relating to impacts on the soil), Swiss Federal Legislation, SR 814.12, 1998; modified from Keller, C. et al., *Plant Soil*, 249, 67, 2003.

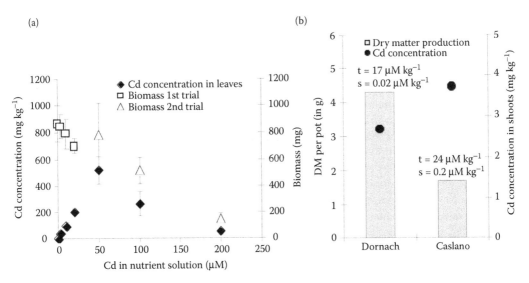

FIGURE 30.4 Cadmium concentration and biomass production: comparison between a) *Salix viminalis* (clone 78980) grown in hydroponics with increasing Cd concentrations and b) the same clone grown in pots in the Dornach and Caslano soils. s = NaNO₃-extractable Cd (soluble) and t = HNO₃-extractable (pseudototal). (After Hammer, D. and Keller, C., *J. Environ. Qual.*, 31, 1561, 2002 and Cosio, C. et al., unpublished data, 1998–2002.)

content in the soil compared to hydroponics that show the uptake potential by this plant. Similar Cd concentrations in solution in both cases give similar Cd concentrations in the shoots of *S. viminalis* (clone 78980).

A low metal uptake due to low metal availability increases the number of years necessary for metal removal. It may lead to the use of additives to increase bioavailability [43–47]. EDTA, NTA, sulphur, and other chelating agents have been tested on willows [2,48], as well as on other crop plants, but results have not been conclusive thus far. EDTA is often preferred because it enables large metal mobilization, especially for Pb phytoextraction [47,49]. However, the main drawbacks are its persistence, the risk of groundwater contamination through uncontrolled leaching [50,51], and its high cost. Its use should therefore be limited to a situation in which leaching can be controlled or prevented and thus it may not be applicable at any site.

In contrast, Fisher et al. [52] tested different depths of incorporation of lime and organic matter on metal uptake and biomass production of Geyer willow (*Salix geyeriana*). They found that these additives improved the growth performance without decreasing Cd uptake, but increased the overall Cd extraction. However, it decreased Zn uptake and may not be suitable at Zn-contaminated sites. This approach needs to be investigated and tested carefully prior to application, but may provide an alternative to the addition of mobilizing agents.

30.4 FACTORS LIMITING THE EFFICIENCY OF PHYTOEXTRACTION SPECIFICITY OF WILLOWS

Ideally, the plant chosen for phytoextraction should be

- Contaminant tolerant
- Adapted to the site conditions
- Able to produce a large biomass quickly with high metal concentrations in the shoots
- Easy to manage, dispose of, and/or recycle
- Able to be grown continually or repeatedly

No plants that possess all these characteristics are available; however, crop plants and bioenergy plants usually produce a large biomass in a short time. Crop plants are usually annual but tree species — including willows — are perennial and thus do not require annual planting. In addition, their agronomic requirements are well known and their harvest is mechanized. They accumulate metals in their tissues moderately, but could be used for multiple contamination because they may be able to take up more than one metal. In the end, the choice will depend on the final total extraction obtained multiplying the yield with the metal concentration in the shoots. This will require preliminary studies to estimate these two values.

A large genetic variability exists in willows and active breeding programs for the bioenergy production have already produced a broad range of clones, especially of *S. viminalis* [53]. They have been selected for rather rich and fertilized agricultural soils and may thus not adapt easily to contaminated sites (see earlier comments). However, clones also exist that are adapted to extreme environments and grow spontaneously at contaminated sites [54–56]. They may have developed and enhanced metal tolerance to the metal present in their original soil [38,57] and may also be able to stand a limited nutrient availability. Vandecasteele et al. [56] proposed to use *Salix* species as bioindicators because the foliar content in Zn and Cd of *Salix* spp. grown on contaminated dredged sediments reflected the soil metal content, indicating that metal accumulation is probably a common trait to many *Salix* species. However, they also stressed the risk of metal transfer to the food chain. This is a recognized risk of phytoextraction, but is probably less problematic with willows than with hyperaccumulators.

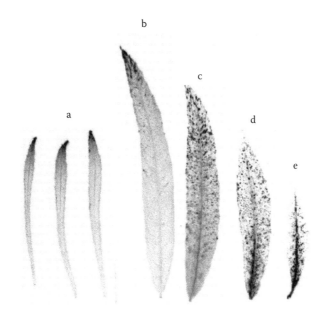

FIGURE 30.5 Cadmium visualization in leaves of *S. viminalis* (clone 78980): autoradiographs of *S. viminalis* leaves grown in hydroponics and after 30 days exposure to 10 m*M* Cd spiked with 2.2 10^{-3} m*M* of 109 Cd; (a) leaves n°1–3; (b) leaf n°11; (c) leaf n°16; (d) leaf n°21; and (e) leaf n°27. Leaves were counted from the top to the base of the stem; leaf n°1 is the first identifiable leaf. The analyzed plant had a total of 29 identifiable leaves. (After Cosio, C., Ph.D. thesis n°2937, Swiss Federal Institute of Technology, Lausanne, Switzerland, 2004.)

Thus, because of the wide range of wild species that can be collected at contaminated sites, as well as the existing nurseries for biomass production, a large choice of species and clones can be tested according to site specifications and metal tolerance.

30.4.1　Concentrations in Plant Parts

As already mentioned, high biomass plants may be more or less tolerant to a range of contaminants, depending on the contaminant, the species, and even the cultivars or clones. *Salix* spp. presents a large plasticity in Cd tolerance and accumulation [58]. For this reason, Cd extraction seems to be the most likely application of phytoextraction using willows. However, some critical limits may also be quickly reached for other elements, such as boron, which has a narrow deficiency/toxicity range varying widely between plant species [26].

In addition to having various accumulation capabilities, plants do not react in the same way to increasing concentrations in the soil [59]. In general, their efficiency will decrease with increasing soil concentrations [60].

Another point is the distribution of metals inside the plants: metal concentration in the plant varies with the organ and the age of this organ [31,61–63]. Figure 30.5 shows Cd distribution within leaves of *S. viminalis* grown in hydroponics: Cd is located at the edges and tips of the younger leaves but is mostly at the base of the older leaves [64]. In addition, concentration varies with the species [65], the metal [66,67], and the soil characteristics. In general, Zn, Cd, and Ni tend to accumulate in shoots, whereas Cu and Pb remain in roots. This has consequences for the time of the harvest and the parts to be harvested.

As illustration, in Figure 30.6, Zn concentrations in shoots of *S. viminalis* and *S. aurita* grown in two different soils are shown: in soil 1 they were larger in leaves than in stems or roots. In addition, the ratio of concentration between leaves and roots was smaller for *S. aurita* than for the two *S. viminalis* clones. Conversely, in soil 2, Zn concentrations were larger in roots. This means

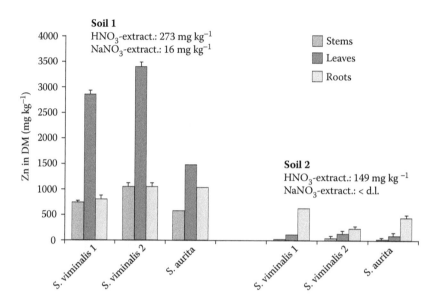

FIGURE 30.6 Zn concentrations in different parts of two *Salix viminalis* clones (Swedish clone 78980 and local clone from Les Abattes) and *Salix aurita* grown during 3 months in pots in two different soils. (After Keller, C. et al., in *ICOBTE: 5th Int. Conf. Biogeochem. Trace Elements*, Vienna, Austria, 1999, 518.) According to the Swiss legislation (OIS: Ordinance relating to impacts on the soil), Swiss Federal Legislation, SR 814.12, 1998), soil 1 is above the guide value for the HNO_3-extractable Zn and above the cleanup value for the $NaNO_3$-extractable Zn. In soil 2, HNO_3- and $NaNO_3$-extractable Zn concentrations are below the guide values.

that extrapolation of given results to different situations (different metals, soils, and clones) is difficult and that leaves and stems must be collected if *Salix* is used for phytoextraction because concentrations are the highest in leaves. In general, pot experiments may thus be necessary prior to applying full-scale phytoextraction.

30.4.2 ROOT PROSPECTING

As demonstrated earlier, the extent of the root systems varies with the site conditions and the age of the plants (Table 30.3). However, for a given soil, plants develop very different root systems [20] as well as different root:shoot ratios. This is true for willows known to present clonal variation in rooting strategy [68]. Such characteristics have been measured at Dornach for various species and the results are presented in Table 30.4. Maize had the largest root length density and thus was able to prospect the whole soil profile efficiently; *T. caerulescens* had the largest root:shoot ratio (together with a relative larger proportion of fine roots), expressing an ability for increased element uptake [40]. Willow gave intermediate results.

Root colonization of the soil is mostly driven by plant type and soil characteristics and may not match the metal distribution properly. Figure 30.7 displays several of the possible situations that lead to optimal or suboptimal metal extraction. Because of the localization of the contaminated layer and the size of their root system, as observed for various species, plants may [69,71]:

- Not colonize the whole layer (case III)
- Not be able to reach the contaminated layer if it is at depth (IV)
- Grow deeper than the contaminated layer (V)
- Avoid the contaminated spots (VI)

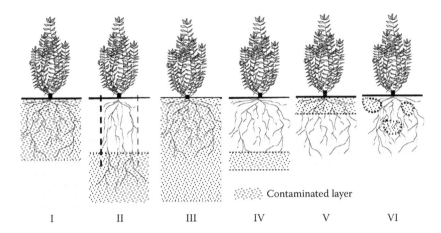

FIGURE 30.7 Various possible combinations of root distribution and contaminated layers. *Salix viminalis* is taken as an example. I: optimal situation where the root system develops only within the contaminated layer; II: situation where the roots are forced to reach the contaminated layer; III to VI: situations where the root system distribution does not match the contamination.

On the other hand, root systems may be manipulated to a certain extent to force the root to reach the contaminated layer [72].

Practically, this may result in a decrease in extraction efficiency as shown in Table 30.4: for each miniplot, soil cores were taken down to 0.75 m and metal concentrations of the soil were measured every 0.10 m, as well as total root length for the same layer. A coefficient taking into account the true depth of contamination and the maximal root depth (with root length density > 5000 mm dm³) was calculated. The same calculation was made for a hypothetical steady shallow (0.20 m, case III of Figure 30.7) or deep (0.70 m, case V of Figure 30.7) contamination. In general, maize and willows were more suitable when the contaminated layer was thick. However, in the specific case of Dornach, none of the species was optimal, so a mixed culture or a rotation with plants with various root colonization would be best. These results can probably be extended to other contaminated sites and emphasize the necessity of an appropriate and site-specific plant management.

Additionally, total root length in the contaminated layer (and thus the extent of the root prospecting) and the root length/shoot biomass ratio may have an impact on metal concentrations in the biomass. For example, Van Noordwijk et al. [73] found a positive correlation between Cd uptake by maize and root length.

30.4.3 Plant Management

As discussed with Figure 30.6, the field management applied for the bioenergy production may not be efficient for phytoextraction with willows: for biomass production, *Salix* spp. are left 20 to 30 years on site and cut in winter every 3 to 4 years [16]. From Figure 30.6 and data obtained at Dornach and Caslano, it was found that Cd and Zn concentrations were the highest in leaves. At Dornach, leaves accounted for approximately 20 and 15% of the biomass produced after 3 and 4 years, respectively, but concentrations were between twice (for Cd) and five times (Zn) larger than in stems [6]. Thus, leaves must be harvested in order to reach the maximal extraction efficiency during phytoextraction and to prevent metal accumulation in the litter and the topsoil up to toxic levels. Because leaves must be harvested, the chipping process (chopping stems into small pieces) used in biomass production is not suitable; thus, other harvesting machines like adapted stick harvesters may need to be used. This problem may increase the price of the harvesting process and

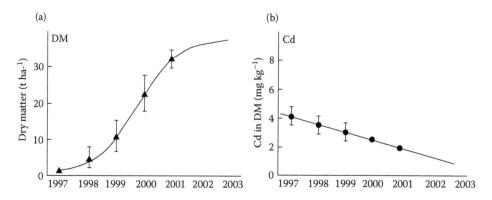

FIGURE 30.8 Biomass production and Cd concentrations measured in *S. viminalis* grown at Dornach between 1997 and 2001. (Modified from Hammer, D. et al., *Soil Use Manage.,* 19, 187, 2003.)

thus phytoextraction drastically. In addition, many studies are necessary to assess the decrease in plant vigor that may arise from a systematic leaf harvest.

In addition, Cd and Zn concentrations measured in leaves and stems of *S. viminalis* decreased with time (Figure 30.8) — most probably because roots had progressively extended downwards outside the contaminated layer [40], thus reducing the extraction efficiency with time. Similar results were obtained at Caslano after 4 years (Keller et al., unpublished data) and are attributed to the same reason. Therefore, it may be necessary to remove the plants every 2 or 3 years to keep an optimal extraction.

On the other hand, whereas Cd and Zn concentrations in shoots were lower at Les Abattes than those recorded at Dornach and Caslano, they increased with time (from 62 in 1998 to 175 in 1999) and 252 (2000) Zn mg kg^{-1} DM). This reflects the progressive establishment of the plants on the site and also the fact that the roots probably remained within the contaminated layer because the underlying layer was composed of marl and lime and compacted, and thus did not facilitate root colonization (Figure 30.9). Again, these contrasting results emphasize the need for a careful study of the site to be remediated and for monitoring plant concentrations to assess metal extraction in the long term.

The choice of the planting technique may have also a long-term effect on phytoextraction efficiency. At Les Abattes, the field management was conducted as a forest management with limited weed control, and cuttings and 1-year-old plants (1 m high) were planted. At Dornach and Caslano, only cuttings were used and weeds were systematically removed. In the first case, cuttings did not survive, but in the second, a good survival percentage was obtained and maintained. Weed competition and subsequent neglect may thus influence plant survival and the result of phytoextraction [74].

Another point to take into account before starting phytoextraction is an assessment of the plant disposal that may differ according to the site location, the volume of "waste" produced, the time of the year and duration of the biomass production, the accessibility of disposal facilities, and the legislation (these parameters may have an impact on the plant species to be chosen).

Finally, the choice of the species (one or several) needs a thorough study of the plant's potential and its suitability for the site. Willows have a high potential for use in phytoextraction, but other alternatives must not be discarded before thorough evaluation. As it has been shown, the results are not easily predictable and pre-experiments in pots or even in miniplots will be necessary before applying phytoextraction to the site.

30.5 LEGISLATION AND TIME REQUIRED

Models have been proposed to evaluate the time needed to decontaminate a given soil with a given plant [2,3,75]. This time varies with the target value and the depth of contamination, as well as

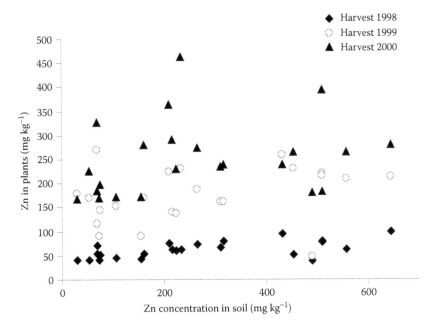

FIGURE 30.9 Zn concentrations in shoots of *S. viminalis* clone 78980 grown in the field at Les Abattes (planted as 1-year-old plants) and sampled after 6, 18, and 30 months after planting.

with the type of function (linear or exponential) used to describe the decrease of metal concentrations in soil. Indeed, concentrations in the plants and yields may not be constant with time, and efficiency may thus decrease because the available pools decrease and also because the roots are going deeper or because perennial plants need a certain time to establish themselves. Unfortunately, not enough long-term data sets are available to validate either of these models.

The target values are also highly variable and depend mostly on legislation and the perception of a clean (or uncontaminated) soil. Some countries have already developed legislation that gives a framework for soil remediation; others have not [76]. If the target value in a given country is very low, it will take more time to clean up a soil below this value than it will in a country where the target value is high. In the second case, phytoextraction may be an option, but it might take too long in the first one. Using phytoextraction as a stripping technique to remove the bioavailable pool only has also been proposed. The extent of the replenishment of the metal available pool is difficult to assess in the first place, and it appears also that, in some countries like Switzerland [23], total and soluble contents must be reduced below a given threshold; the BSM approach may not be suitable for this reason. Finally, as mentioned earlier, the decrease of metal concentration in soil may not be linear and it remains to be tested whether the target values can effectively be reached.

In addition to the total or soluble metal content, target values could theoretically be based on the sum of metals, functional criteria and toxicological aspects, exposure pathways, or the idea of restoring soil multifunctionality. This will lead to a large range of concentrations that may have an impact on the technical and economical feasibility of phytoextraction whatever the option chosen is.

Thus far, calculations made from preliminary experiments or, at most, from a few years of field experiments, have led to various results (results obtained with addition of large amounts of chelating agents like EDTA are not taken into account), predicting remediation within few years, decades, or centuries [1–3,6,75]. Obviously, all the factors presented previously added to obtain such a large range of results. However, Cd seems to be the element that is the most readily taken up by willows and for which the best results in the shortest time would be expected. This emphasizes again the necessity to choose the sites to be remediated by phytoextraction, as well as the planning of the procedure to be applied, carefully.

30.6 CONCLUSION

Phytoextraction has a great potential for cleaning soils contaminated with heavy metals, especially in cases in which conventional technologies are not efficient, not possible, or too expensive. From the early results obtained in the lab and the field, it is clear that phytoextraction could be applied as the main remediation tool or, more likely, as a "polishing" technique combined with other conventional and "bio" techniques. Some evidence also indicates that, under temperate climate, some sites are unlikely to be cleaned by high biomass plants or hyperaccumulators; elements such as Pb, Cr, and Cu are not easily translocated to the plant shoots, whereas Cd is more suitable for extraction with willows [32].

There is obviously no single phytoextraction technique and each site will need a tailor-made scheme of the willows that may be part of it. The main difficulty in doing so is the optimization of the efficiency. Indeed, it is difficult to spot the limiting factors and even more to try to find a hierarchy among them because of their large number and their interdependency. To help practitioners choose or discard phytoextraction technology, decision trees have been proposed (see, for example, the US-EPA phytoremediation decision trees at: www.clu-in.org/techdrct/techpubs.asp [77]) that include some of these factors. However, once phytoextraction has been selected as the appropriate technique, much effort must be devoted to reach a full control of the technique and to obtain a maximal efficiency — in particular, in the long term. Research has a key position on that aspect and is responsible for the successful development of phytoextraction.

ACKNOWLEDGMENT

This research was part of the project n°ENV4-CT97-0598/n°OFES97.0362 funded by the Swiss Federal Agency for Teaching and Research and was also financed by the Swiss Federal Institute of Technology, Zürich. I thank all the people involved in these researches, including D. Hammer (EPFL), C. Cosio (EPFL), D. Vainstein (EPFL), and A. Kayser (ETHZ) for their work on the projects; U. Zimmermann, A. Grünwald, J.-D. Teuscher, E. Rossel, and K. Truninger (EPFL and ETHZ) for the help in plant and soil analyses; W. Attinger (ETHZ) for the Dornach field management; E. Frossard (ETHZ) for the greenhouse facilities; R. Schulin (ETHZ) for the scientific support and for allowing the work in Dornach; and V. Barbezat and W. Rosselli (Snow, Forest and Landscape Institute) for the management of the Les Abattes field experiment. Maria Greger, University of Stockholm, Sweden, provided the *Salix viminalis* cuttings (clone 78980).

REFERENCES

1. Felix, H.R., Field trials for *in situ* decontamination of heavy metal-polluted soils using crops of metal-accumulating plants, *Z. Pflanzenernähr. Bodenk.*, 160, 525, 1997.
2. Kayser, A., Wenger, K., Keller, A., Attinger, W., Felix, H.R., Gupta, S.K., and Schulin, R., Enhancement of phytoextraction of Zn, Cd and Cu from calcareous soil: the use of NTA and sulfur amendments, *Environ. Sci. Technol.*, 34, 1778, 2000.
3. Klang–Westin, E. and Eriksson, J., Potential of *Salix* as phytoextractor for Cd on moderately contaminated soils, *Plant Soil*, 249, 127, 2003.
4. Dickinson, N.M., Phytoremediation of industrially contaminated sites using trees, Series, N. Eds. 2004, in press.

5. Vangronsveld, J. and Cunningham, S.D., Introduction to concepts, in *Metal-Contaminated Soils: in Situ Inactivation and Phytorestoration*, Vangronsveld, J. and Cunningham, S.D. Eds., R.G. Landes Company, Georgetown, TX, 1998, 1.

6. Hammer, D., Kayser, A., and Keller, C., Phytoextraction of Cd and Zn with *Salix viminalis* in field trials, *Soil Use Manage.*, 19, 187, 2003.

7. Hammer, D. and Keller, C., Phytoextraction of Cd and Zn with *Thlaspi caerulescens* in field trials, *Soil Use Manage.*, 19, 144, 2003.

8. Rosselli, W., Keller, C., and Boschi, K., Phytoextraction capacity of trees growing on a metal-contaminated soil, *Plant Soil*, 256, 265, 2003.

9. Geiger, G., Federer, P., and Sticher, H., Reclamation of heavy metal-contaminated soils: field studies and germination experiments, *J. Environ. Qual.*, 22, 201, 1993.

10. Dubois, J.-P., Analysis report concerning the metal concentration in the soil of the landfill "Les Abattes" (City of Le Locle) (in French), Swiss Federal Institute of Technology, Lausanne, Switzerland, 1991.

11. Guadagnini, M., personal communication, 1997.

12. Hammer, D., Application of phytoextraction to contaminated soils: factors which influence heavy metal phytoavailability, Ph.D thesis no.°2576, Swiss Federal Institute of Technology, Lausanne, Switzerland, 2002.

13. Hammer, D. and Keller, C., Changes in the rhizosphere of metal-accumulating plants evidenced by chemical extractants, *J. Environ. Qual.*, 31, 1561, 2002.

14. Van Splunger, I., Voesenek, L.A.C.J., Coops, H., De Vries, X.J.A., and Blom, C.W.P.M., Morphological responses of seedlings of four species of Salicaceae to drought, *Can. J. Bot.*, 74, 1988, 1996.

15. Li, Y.M., Chaney, R.L., Brewer, E.P., Roseberg, R., Angle, J.S., Baker, A.J.M., Reeves, R., and Nelkin, J., Development of a technology for commercial phytoextraction of nickel: economic and technical considerations, *Plant Soil*, 249, 107, 2003.

16. Danfors, B., Ledin, S., and Rosenqvist, H., *Short-Rotation Willow Coppice — Growers' Manual*, Swedish Institute of Agricultural Engineering, Uppsala, 1998.

17. Genske, D.D., *Urban Land: Degradation — Investigation — Remediation*, Springer–Verlag, Berlin, Germany, 2003.

18. Bending, N.A.D. and Moffat, A.J., Tree performance on mine spoils in the South Wales coalfield, *J. Appl. Ecol.*, 36, 784, 1999.

19. Ericsson, E., Effect of varied nitrogen stress on growth and nutrition in three *Salix* clones, *Physiol. Plant.*, 51, 423, 1981.

20. Polomski, J. and Kuhn, N., *Wurzelsysteme*, Paul Haupt, Bern, Switzerland, 1998.

21. Keller, C. and Védy, J.-C., Distribution of Cu and Cd fractions in two forested soils, *J. Environ. Qual.*, 23, 987, 1994.

22. Iyengar, S.S., Martens, D.C., and Miller, W.P., Distribution and plant availability of soil zinc, *Soil Sci. Soc. Am. J.*, 45, 735, 1981.

23. OIS (ordinance relating to impacts on the soil), Swiss Federal Legislation, SR 814.12, 1998.

24. Kayser, A., Evaluation and enhancement of phytoextraction of heavy metals from contaminated soils, Ph.D. thesis no.°13563, Swiss Federal Institute of Technology, Zürich, Switzerland, 2000.

25. Kayser, A., Schröder, T.J., Grünwald, A., and Schulin, R., Solubilization and plant uptake of zinc and cadmium from soils treated with elemental sulfur, *Int. J. Phytorem.*, 3, 381, 2001.

26. Kabata–Pendias, A., *Trace Elements in Soils and Plants*, 3rd ed., CRC Press, London, U.K., 2001.

27. Siciliano, S.D. and Germida, J.J., Degradation of chlorinated benzoic acid mixtures by plant–bacteria associations, *Environ. Toxicol. Chem.*, 17, 728, 1998.

28. Bhadra, R., Spanggord, R.J., Wayment, D.G., Hughes, J.B., and Shanks, J.V., Characterization of oxidation products of TNT metabolism in aquatic phytoremediation systems of *Myriophyllum aquaticum*, *Environ. Sci. Technol.*, 33, 3354, 1999.

29. Burken, J.G. and Schnoor, J.L., Predictive relationships for uptake of organic contaminants by hybrid poplar trees, *Environ. Sci. Technol.*, 32, 3379, 1998.

30. Baker, A.J.M. and Walker, P.L., Ecophysiology of metal uptake by tolerant plants, in *Heavy Metal Tolerance in Plants: Evolutionary Aspects*, Shaw, A.J., Ed., CRC Press Inc., Boca Raton, FL, 1990, 155.

31. Sauerbeck, D., Der Transfer von Schwermetallen in die Pflanze, in *Beurteilung von Schwermetallkontaminationen im Boden*, DECHEMA Fachgespräche Umweltschutz, Stuttgart am Mainz, 1989, 281.

32. Pulford, I.D. and Watson, C., Phytoremediation of heavy metal-contaminated land by trees — a review, *Environ. Int.*, 29, 529, 2003.

33. McBride, M.B. and Martinez, C.E., Copper phytotoxicity in a contaminated soil: remediation tests with adsorptive materials, *Environ. Sci. Technol.*, 34, 4386, 2000.

34. Pahlsson, A.-M.B., Toxicity of heavy metals (Zn, Cu, Cd, Pb) to vascular plants, *Water Air Soil Pollut.*, 47, 287, 1989.

35. Ebbs, S.D. and Kochian, L.V., Toxicity of zinc and copper to *Brassica* species: implications for phytoremediation, *J. Environ. Qual.*, 26, 776, 1997.

36. Ouzounidou, G., Ilias, I., Tranopoulou, H., and Karataglis, S., Amelioration of copper toxicity by iron on spinach physiology, *J. Plant Nutr.*, 21, 2089, 1998.

37. Hewitt, E.J., Metal interrelationships in plant nutrition. I. Effects of some metal toxicities on sugar beet, tomato, oat, potato, and marrowstem kale grown in sand culture, *J. Exp. Bot.*, 4, 59, 1953.

38. Punshon, T., Lepp, N.W., and Dickinson, N.M., Resistance to copper toxicity in some British willows, *J. Geochem. Explor.*, 52, 259, 1995.

39. Costa, G. and Morel, J.-L., Cadmium uptake by *Lupinus albus* (L) — cadmium excretion, a possible mechanism of cadmium tolerance, *J. Plant Nutr. Soil Sci.*, 16, 1921, 1993.

40. Keller, C., Hammer, D., Kayser, A., Richner, W., Brodbeck, M., and Sennhauser, M., Root development and heavy metal phytoextraction efficiency: comparison of different plant species in the field, *Plant Soil*, 249, 67, 2003.

41. Baize, D. and Sterckeman, T., Of the necessity of knowledge of the natural pseudo-geochemical background content in the evaluation of the contamination of soils by trace elements, *Sci. Tot. Environ.*, 264, 127, 2001.

42. Dubois, J.-P., Okopnik, F., Benitez, N., and Védy, J.-C., Origin and spatial variability of cadmium in some soils of the Swiss Jura, in *Proceedings of the 16th World Congress on Soil Science*, Montpellier, France, 1998.

43. Cooper, E.M., Sims, J.T., Cunningham, S.D., Huang, J.W., and Berti, W.R., Chelate-assisted phytoextraction of lead from contaminated soils, *J. Environ. Qual.*, 28, 1709, 1999.

44. Elkhatib, E.A., Thabet, A.G., and Mahdy, A.M., Phytoremediation of cadmium contaminated soils: role of organic complexing agents in cadmium phytoextraction, *Land Contamin. Reclam.*, 9, 359, 2001.

45. Lombi, E., Zhao, F.J., Dunham, S.J., and McGrath, S.P., Phytoremediation of heavy-metal-contaminated soils: natural hyperaccumulation vs. chemically enhanced phytoextraction, *J. Environ. Qual.*, 30, 1919, 2001.

46. Wenger, K., Kayser, A., Gupta, S.K., Furrer, G., and Schulin, R., Comparison of NTA and elemental sulfur as potential soil amendments in phytoremediation, *Soil Sediment Contam.*, 11, 655, 2002.

47. Liphadzi, M.S., Kirkham, M.B., Mankin, K.R., and Paulsen, G.M., EDTA-assisted heavy-metal uptake by poplar and sunflower grown at a long-term sewage-sludge farm, *Plant Soil*, 257, 171, 2003.

48. Robinson, B.H., Mills, T.S., Petit, D., Fung, L.E., Green, S.R., and Clothier, B.E., Natural and induced cadmium accumulation in poplar and willow: implications for phytoremediation, *Plant Soil*, 227, 301, 2000.

49. Huang, J.W., Chen, J., and Cunningham, S.D., Phytoextraction of lead from contaminated soils, in *Phytoremediation of Soil and Water Contaminants*, Kruger, E.L. et al., Eds., American Chemical Society, Washington, D.C., 1997, 283.

50. Wenzel, W.W., Unterbrunner, R., Sommer, P., and Sacco, P., Chelate-assisted phytoextraction using canola (*Brassica napus* L.) in outdoors pot and lysimeter experiments, *Plant Soil*, 249, 83, 2003.

51. Wu, L.H., Luo, Y.M., Xing, X.R., and Christie, P., EDTA-enhanced phytoremediation of heavy metal contaminated soil with Indian mustard and associated potential leaching risk, *Agr. Ecosyst. Environ.*, 102, 307, 2004.

52. Fisher, K.T., Brummer, J.E., Leininger, W.C., and Heil, D.M., Interactive effects of soil amendments and depth of incorporation on Geyer willow, *J. Environ. Qual.*, 29, 1786, 2000.

53. Lindegaard, K.N. and Barker, J.H.A., Breeding willows for biomass, *Asp. Appl. Biol.*, 49, 155, 1997.

54. Dickinson, N.M., Punshon, T., Hodkinson, R.B., and Lepp, N.W., Metal tolerance and accumulation in willows, in *Willow Vegetation Filters for Municipal Wastewaters and Sludges. A Biological Purification System*, Aronsson, P. and Perttu, K., Eds., Swedish University of Agricultural Sciences, Uppsala, Sweden, 1994, 121.

55. Álvarez, E., Fernández Marcos, M.L.C.V., and Fernández–Sanjurjo, M.J., Heavy metals in the dump of an abandoned mine in Galicia (NW Spain) and in the spontaneously occurring vegetation, *Sci.Tot. Environ.*, 313, 185, 2003.

56. Vandecasteele, B., De Vos, B., and Tack, F.M.G., Cadmium and zinc uptake by volunteer willow species and elder rooting in polluted dredged sediment disposal sites, *Sci. Tot. Environ.*, 299, 191, 2002.

57. Kopponen, P., Utrianen, M., Lukkari, K., Suntioinen, S., Kärenlampi, L., and Kärenlampi, S., Clonal differences in copper and zinc tolerance of birch in metal-supplemented soils, *Environ. Pollut.*, 112, 89, 2001.

58. Landberg, T. and Greger, M., Differences in uptake and tolerance to heavy metals in *Salix* from unpolluted and polluted areas, *Appl. Geochem.*, 11, 175, 1996.

59. Baker, A.J.M., Accumulators and excluders — strategies in the response of plants to heavy metals, *J. Plant Nutr.*, 3, 643, 1981.

60. Keller, C., Kayser, A., Keller, A., and Schulin, R., Heavy metal uptake by agricultural crops from sewage-sludge treated soils of the Upper Swiss Rhine valley and the effect of time, in *Environmental Restoration of Metal-Contaminated Soils*, Iskandar, I.K., Ed., CRC Press LLC, Boca Raton, FL, 2000, 273.

61. Dinelli, E. and Lombini, A., Metal distribution in plant growing on copper mine spoils in Northern Apennines, Italy: the evaluation of seasonal variations, *Appl. Geochem.*, 11, 375, 1996.

62. Luyssaert, S., Van Meirvenne, M., and Lust, N., Cadmium variability in leaves of a *Salix fragilis*: simulation and implications for leaf sampling, *Can. J. For. Res.*, 31, 313, 2001.

63. Sander, M.-L. and Ericsson, T., Vertical distributions of plant nutrients and heavy metals in *Salix viminalis* stems and their implications for sampling, *Biomass Bioenerg.*, 14, 57, 1998.

64. Cosio, C., Phytoextraction of heavy metal by hyperaccumulating and nonhyperaccumulating plants: comparison of cadmium uptake and storage mechanisms in the plants, Ph.D. thesis no.°2937, Swiss Federal Institute of Technology, Lausanne, Switzerland, 2004.

65. Nissen, L.R. and Lepp, N.W., Baseline concentrations of copper and zinc in shoot tissues of a range of *Salix* species, *Biomass Bioenerg.*, 12, 115, 1997.

66. McGregor, S.D., Duncan, H.J., Pulford, I.D., and Wheeler, C.T., Uptake of heavy metals from contaminated soil by trees, in *Heavy Metals and Trees, Proceedings of a Discussion Meeting*, Glimmerveen, Institute of Chartered Foresters, Edinburgh, U.K., 1996, 68.

67. Hasselgren, K., Utilization of sewage sludge in short-rotation energy forestry: a pilot study, *Waste Manage. Res.*, 17, 251, 1999.

68. Good, J.E.G., Bellis, J.A., and Munro, R.C., Clonal variation in rooting of softwood cutting of woody perennials occurring naturally on derelict land, *Int. Plant Propag. Soc. Comb.*, 1978, 192.

69. Breckle, S.-W. and Kahle, H., Effect of toxic heavy metals (Cd, Pb) on growth and mineral nutrition of beech (*Fagus sylvatica* L.), *Vegetatio*, 101, 43, 1992.

70. Turner, A.P. and Dickinson, N.M., Survival of *Acer pseudoplatanus* L. (sycamore) seedlings on metalliferous soils, *New Phytol.*, 123, 509, 1993.

71. Schnoor, J.L., Licht, L.A., McCutcheon, S.C., Wolfe, N.L., and Carreira, L.H., Phytoremediation of organic and nutrient contaminants, *Environ. Sci. Technol.*, 29, 318, 1995.

72. Chappell, J., *Phytoremediation of TCE in Groundwater Using Populus*, Technology Innovation Office, Office of Solid Waste and Emergency Response, Washington, D.C., 1998.

73. Van Noordwijk, M., Van Driel, W., Brouwer, G., and Schuurmans, W., Heavy-metal uptake by crops from polluted river sediments covered by nonpolluted topsoil, *Plant Soil*, 175, 105, 1995.

74. Rees, S., Pulford, I.D., and Duncan, H.J., Utilizing contaminated and disturbed industrial land for short rotation coppice, in *Contaminated Soil '98* Thomas Telford, London, 1998, 1225.

75. Robinson, B.H., Fernandez, J.-E., Madejon, P., Maranon, T., Murillo, J.M., Green, S., and Clothier, B.E., Phytoextraction: an assessment of biogeochemical and economic viability, *Plant Soil*, 249, 117, 2003.

76. Adriano, D.C., *Trace Elements in Terrestrial Environments*, 2nd ed., Springer–Verlag, New York, 2001.

77. ITCR (Interstate Technology and Regulatory Cooperation Work Group), Phytoremediation decision tree, *http://www.clu-in.org/download/partner/phytotree.pdf*, 1999.

78. Keller, C., Factors limiting efficiency of phytoextraction at multimetal contaminated sites, NATO Series, 2004, in press.

79. Keller, C., Raduljica, O., and Hammer, D., unpublished data, 1997–2004.

80. Cosio, C., Vainstein, D., Hammer, D., and Keller, C., unpublished data, 1998–2002.

81. Keller, C., Hammer, D., Kayser, A., and Schulin, R., Zinc availability in contaminated soils as a function of plant (Willows) growth and additive (NH4Cl), in *ICOBTE: 5th Int. Conf. Biogeochem. Trace Elements*, Vienna, Austria, 1999, 518.

82. FAC, *Methoden für die Bodenuntersuchungen,* Eidgenössische Forschungsanstalt für Agrikulturchemie und Umwelthygiene, Bern-Liebefeld, Switzerland, 1989.

Section V

Risk Assessment

31 Risk Assessment, Pathways, and Trace Element Toxicity of Sewage Sludge-Amended Agroforestry and Soils

K. Chandra Sekhar and M.N.V. Prasad

CONTENTS

31.1 INTRODUCTION

Heavy metal accumulation in farming soil depends mainly on the heavy metal concentration in fertilizers or amendments and the application rate of each. Use of municipal sewage sludge as soil amendment is of current interest because of environmental and economic concerns. Application of sludges, as sources of plant nutrients, to soil at rates consistent with the nutrient requirements of a crop are believed to be most beneficial. It has been observed that sludge application reduces the surface runoff and gives some protection against soil erosion [1]. The composition of sewage sludge varies greatly depending on the source of sewage (domestic, industrial, etc.) [2]. The undesirable consequence of sludge application arises from the heavy metal content of the sludge, which may be taken up by the plants [3] and will have indirect impacts on animal health, human health, and/or the environment.

When considering the toxicity of heavy meals, the route by which the smallest amount of an element can cause harm is used as the limiting concentration. For most heavy trace elements, this limiting route of exposure falls into one of the three categories: plant growth, animal health, or human health [4,5]. When elements such as Cu and Zn are applied to soil in excess, they can accumulate in plant tissues and interfere with plant growth [4]; different soil types and soil pH will affect how plants react to these heavy trace elements in the soil. Additionally, many plants have different element uptake rates and heavy metal tolerances. All these factors will affect the potential crop effects of heavy trace elements in sewage sludge. Decrease in crop yield can occur and this could lead to crop loss if heavy trace elements are applied in excess to heavy metal-intolerant plants [4,6].

During the last decade, concern about the hazards of toxic trace elements for human health and the environment has increased worldwide. In many countries, legislative and administrative measures have been taken to reduce environmental contamination and to prevent adverse effects resulting from environmental exposure to metal pollutants [7,8]. Because the consensus is that trace elements in soil may be taken up by plants and thus enter the food chain, value limits for maximum tolerable metal concentration in agricultural soils were set in various countries [1,9]. Sewage sludge can be the most important localized source of heavy metal increase in agricultural soils, so the existing limits have been developed to regulate sludge application in agriculture [10,11] (Table 31.1).

In India, treated or untreated sludge is dumped indiscriminately on land or directly applied to agricultural land as manure. In drought-prone areas of India, it has been a common practice to cultivate food crops in sludge-amended soils. Land application of municipal sewage sludge is one of the most commercial alternative methods for solving the waste disposal problem and increasing the organic matter content of the soil in India. However, little work has been carried out on sludge-amended soils and the risks associated with them.

As a result of their nonbiodegradable nature, heavy trace elements present in soil accumulate by organisms and undergo biochemical cycles in the environment during which they are transformed with various chemical species [12]. Knowledge of the chemical state of trace metal in the biogeochemical cycle or biological fluids is important for undertaking their reactivity, transport, and toxicity [13,14]. The potential toxicity of various trace elements is controlled to a large extent by their physicochemical forms. Biochemical and toxicological investigations have shown that, for living organisms, the chemical form of a specific element or the oxidation state in which the element is introduced into the environment, as well as the quantities, is crucial [15,16].

To get information on the activity of specific elements in the environment — particularly for those in contact with living organisms — it is necessary to determine the total content of the element and to gain an indication of its individual chemical and physical forms [16]. Thus, the chemical form and species of the metal present in sludge-amended soils (SAS) in which vegetables are grown becomes an important parameter. In all the studies carried out on the application of sludge and its possible entry into food chain, the total metal concentration is reported, rather than the forms in

TABLE 31.1
Permissible Limits of Trace Elements in Biosolids

| Trace elements | King County biosolids for 2000 | | EPA standards | |
	South (mg/kg)	West Point (mg/kg)	Exceptional quality (mg/kg)	Maximum limit (mg/kg)
Arsenic	7.3	7.1	41	75
Cadmium	4.4	3.7	39	85
Chromium	36.5	45.3	1200	3000
Copper	642.1	529.1	1500	4300
Lead	51.0	141.3	300	840
Mercury	2.7	2.8	17	57
Molybdenum	14.6	11.1	Na	75
Nickel	21.9	35.0	420	420
Selenium	6.4	6.0	36	100
Zinc	714.9	803.8	2800	7500

which metals are present. This chapter reviews application of sludge as manure and health risks associated with the trace elements present in the sludge, as well as factors affecting metal mobility.

31.2 SEWAGE SLUDGE

Sewage sludge (often referred to as biosolids) is the mud-like material that remains after the treatment of household and industrial wastewater that flows into sewage treatment plants. This malodorous mess contains volatiles, organic solids, nutrients, disease-causing pathogenic organisms, heavy trace elements, etc. The chemical composition and physical characteristics of the sewage sludge depend upon the nature of the sewage water treated in the sewage plant and the treatment process used. The options for dealing with sewage sludge include its application to agricultural land, incineration, land reclamation, landfill, forestry, and sea disposal. The benefits of recycling biosolids onto agricultural land include providing essential nutrients for crop needs and organic matter for improving the tilth, water-holding capacity, soil aeration, and as an energy source for earthworms [17] and beneficial microorganisms [18].

Crop yield on land amendments with biosolids can be as great as or greater than that on land fertilized with only commercial synthetic fertilizers (Table 31.2) [19]. Organic nitrates constitute a significant fraction of the organic solids. When applied to the land for agricultural or horticultural purpose, nitrates are gradually mineralized through a biological decomposition process. They then became nutrients feeding plants. The inorganic fraction contains mineral salts; adding minerals to soil to improve plant growth has been recognized for centuries. Sludges contain most of the 13 mineral elements considered essential for plant growth that are often not economically viable with chemical fertilizers [20]. Many plant species respond to biosolids favorably because they release nitrogen and also contain phosphorus and potassium [21]. The crops use the organic nitrogen found in biosolids very efficiently; because it is released slowly throughout the growing seasons, the crop can take it up as it grows [22]. Biosolids may also correct crops' micronutrient deficiencies [23].

31.3 TRACE ELEMENT TOXICITY

Despite these positive results of soil amendment with biosolids, the dark side of the sludge is evident in the health risks. Sewage sludges also contain components considered to be harmful to the environment. Sewage sludge contains three constituents of environmental concern: (1) heavy metals; (2) organic pollution; (c) pathogenic organism [24]. Generally, sludges contain heavy trace elements such as Cu, As, Cd, Ni, Zn, and Pb in diverse concentration; these originate from a number of different sources, such as industry, commerce, business, domestic household waste, corroding pipes, and runoff from roads and roofs [5].

TABLE 31.2
Comparison of Main Nutrient Loads between Various Fertilizers Used

	Nutrient load in 1000 t		
	N	P	K
Manure	128	20.5	162
Mineral fertilizer	53	7.4	27
Sewage sludge	3.7	2.2	0.25
Compost	2.9	0.74	1.8
Other organic water	1.5	0.57	1.5
Total	189	31.4	192

Some heavy trace elements (Cu, Zn, etc.) are micronutrients essential for plant growth and therefore are beneficial to the crops; however, in excessive amounts they can reduce growth or can be toxic to plants [25]. Other heavy trace elements present in sewage sludges (As, Cd, Hg, Pb, and Se) are not essential for plants, but are toxic above certain defined levels [26–28]. Similarly, animals will have different tolerance levels for heavy trace elements, e.g., cattle are more susceptible to Se poisoning than sheep are [4]. Many trace elements present in sewage sludge form stable complexes, when applied to soil, with biomolecules and their presence in even small amounts can be detrimental to plants and animals [29,30].

In the soil environment, a heavy metal ion can undergo a number of processes and will be distributed among its different chemical forms and physical phases (water-soluble mobile phase, organic matter bound, oxide bound, etc.). It was found that trace elements that form stable complexes with ligands tend to be more toxic and remain in the soil for much smaller periods of time [15,31,32]; this can result in phytotoxicity and increased movement of trace elements into the food chain. Figure 31.1 illustrates the speciation of heavy trace elements in a soil–water system [33,34].

The processes that determine the speciation of trace elements in a soil environment are [33–35]:

- Precipitation and dissolution
- Sorption and desorption
- Complexation with organic compounds
- Complexation with inorganic compounds

Which of these processes occurs depends on the chemical characteristics of the trace elements and properties of the soil environment.

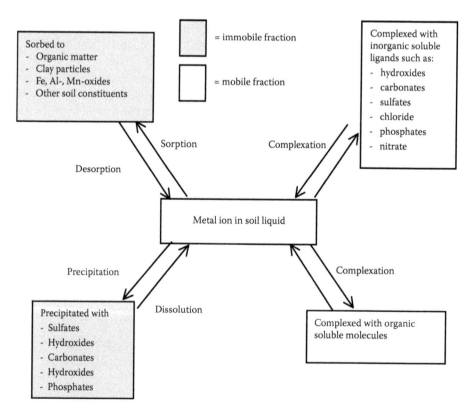

FIGURE 31.1 Fractionation of trace elements in soil–water system.

Depending on their nature, heavy trace elements are associated in a variable manner with different phases making up the sludge. The distribution of heavy trace elements in the different forms and phases in which they occur in soil and sludge can be determined using sequential extraction procedures [36,37]. These procedures provide information about the differentiation of the relative binding strength of the metal on various solid phases and about their potential reactivity under different physicochemical environmental conditions.

31.4 PROPERTIES AFFECTING TRACE ELEMENT MOBILITY

Once they have entered the soil, heavy trace elements can be removed from a given soil volume if they enter the soil solution because this is the mobile part of the topsoil system. The more metal that can be found in solution, the more of it is available. The soil properties determining the speciation of heavy trace elements are as follows:

- pH. Sorption of heavy trace elements depends strongly on pH; in general, sorption increases with increasing pH. That is, the lower the pH value is, the more metal can be found in solution and thus the more metal is mobilized. When pH falls to below 5, mobility is enhanced as a result of the increased proton concentration. At pH values above 7, some heavy trace elements tend to form hydroxyl complexes, which will increase the solubility of the metal in question [38,39]. A soil that has high pH and high cation exchange capacity will be able to immobilize the trace elements added to it via the sewage sludge. An alkaline soil may not always remain that way; thus, if the pH drops at a later stage (i.e., becomes acidic), the trace elements may get released [40].
- Organic matter. The organic matter makes trace elements soil most strongly onto the soil constituents [41,42]. The chemical composition of soil water determines which ligands are available for complexation, whether precipitation is likely to occur, and whether competitive sorption exists. For example, the presence of chloride ions in the soil solution has been shown to enhance the mobility of some heavy trace elements by the formation of more soluble complexes [43].
- Redox potential. The oxidation state of a given metal is determined by the redox potential. Different oxidation forms of a metal behave differently chemically. For example, chromium in the oxidation state (+6) is much more soluble and toxic than chromium in the oxidation state (+3) [43,44].
- Cation exchange capacity (CEC). The CEC of a soil is a measure of the negative charge density of a soil as function of the soil's ability to adsorb positively charged ions, cations. Thus, a high CEC reflects a soil with a high sorption capacity. Quebec researchers [45] suggest that the maximum heavy metal concentration in soils be based on the CEC of the soil, which is a measure of soil's ability to retain heavy metal ions. CEC increases with increasing clay content of the soil. Thus, they recommend maximum heavy metal concentration for fine textured soils, such as clay and clay loam, rather than for coarse textured soils, such as sand. Table 31.3 shows the recommended maximum concentration (ppm) of heavy trace elements in soils based on their cation exchange capacities [45].
- Soil texture. Texture reflects the particle size distribution of the soil and thus the content of fine particles like oxides and clay. These compounds are important adsorption media for heavy trace elements in soils.
- Temperature. Several chemical reactions are temperature dependent in the way that they proceed at a higher rate when temperature increases.

It is now well established that determination of the individual physicochemical forms and phase of metal are required for understanding the role of trace elements in the environment or human health [4,31,47]. One of the main interests has been the state of elements found in the soil and

TABLE 31.3
Recommended Maximum Concentrations (ppm) of Trace Elements in Soils Based on Cation Exchange Capacities (CEC)

CEC[a]	Cu	Co	Hg	Cd	Cr	Zn	Pb	Ni
CEC > 15	50	34	0.14	2.4	120	160	70	60
CEC < 15	25	>17	0.07	1.2	60	80	35	30

[a] Measured as milliequivalents (meq) per 100 g.

their effect on living organisms. Speciation in environmental, toxicological and biochemical aspects usually refer to very low concentrations; these may be even two orders of magnitude lower than the total concentration of the element, which is mostly in the range of extreme trace analysis [48–51].

In most soil environments, sorption is the dominating speciation process; thus, the largest fraction of heavy metal in a soil is associated with the solid phase of that soil [52]. Pollution problems arise when heavy trace elements are mobilized into the soil solution and taken up by plants or transported to the surface or ground waters. Therefore, the properties of the soil are very important in the attenuation of heavy trace elements in the environment. The U.S. Environmental Protection Agency (USEPA) has issued numeric standards for ten trace elements (As, Cd, Cr, Cu, Pb, Hg, Mo, Ni, Se, and Zn) [53]. However, the movement of trace elements from soils into ground water, surface water, plants, and wildlife — and of the hundreds of other toxins in sludge, which EPA chooses not to regulate — is poorly understood [54]. Their movement depends upon at least the following factors: plant species, soil type, soil moisture, soil acidity or alkalinity, sludge application rate, slope, drainage, and the specific chemistry of the toxins and of the sludge [6,55].

In creating its regulations, the EPA assumed that sludge-treated land would be under the perpetual care of a farmer who would lime the soil to keep it alkaline and prevent the trace elements from moving dangerously. For this reason, a build-up of toxic heavy trace elements in soils is often dismissed as irrelevant. However, in the real world, farmers go out of business while acid precipitation keeps soaking soils with dilute acid year after year [56]. Soil acidity seems to be the key factor in promoting or retarding movement of toxic trace elements into ground water, wildlife, and crops [50,57].

31.5 BIOAVAILABILITY OF TRACE ELEMENTS

The trace elements present in sewage sludge may be divided into three categories on the basis of their availability to plants (Figure 31.2) [58–60]:

- Unavailable forms, such as insoluble compounds (oxides or sulphides)
- Potentially available forms, such as insoluble complexes, trace elements linked to ligands, or forms attached to clay and organic matter
- Mobile and available forms, such as hydrated ions or soluble complexes

The fraction of the total metal content in a soil available for plant uptake, i.e., the bioavailable fraction, is usually considered to be the sum of water-soluble and exchangeable metal. Factors affecting transference of the trace elements to the plants are temperature, total metal contents of the soil, and time [3,61]. Temperature has been shown to have an effect on the resulting metal concentration in plants grown on sewage sludge-amended soils. When the temperature rises, the metal activity in the soil solution may increase and the plant roots may be more active and have

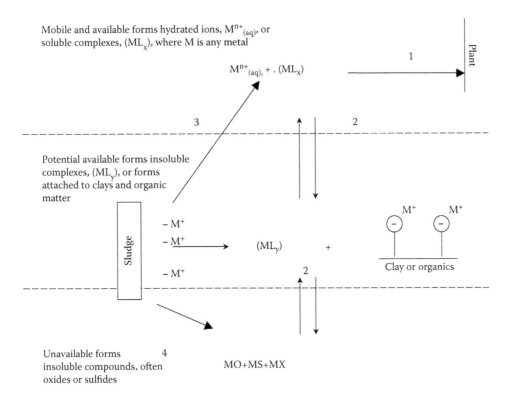

Mobile and available forms hydrated ions, $M^{n+}_{(aq)}$, or soluble complexes, (ML_x), where M is any metal

$M^{n+}_{(aq)}, + . (ML_x)$

Potential available forms insoluble complexes, (ML_y), or forms attached to clays and organic matter

Unavailable forms insoluble compounds, often oxides or sulfides

Key

1. Uptake by plant
2. Changes in form due to action of microorganisms or changes in pH or Eh
3. Soluble forms passed directly into solution
4. Formation of insoluble compounds

FIGURE 31.2 Availability of trace elements from sewage sludge. (From Ashwathanarayana, U., *Soil Resources and the Environment*, Oxford and IBH Publishers, New Delhi, 1999.)

faster absorption rates. Furthermore, the absorption rate of the roots may be increased as a result of higher evapotranspiration from the plant [3].

The concentration levels in plants also depend on the concentration levels in the soil. Several studies have established that metal concentrations in plants increase with increasing metal content of a specific soil [3,61–63]. Time after application of the sludge — the so-called residual time — may affect the mobility and thus the metal content in the plants grown on it [64]. When the organic matter applied with the sludge is decomposed, sorption sites are lost and the accompanying pH decrease also affects desorption of trace elements [65]. It has been shown that the metal content increased in successively harvested plants grown on a sewage sludge-amended soil. Thus, the availability of trace elements in the amended soil may change over time [43,66].

The large amount of organic matter in the sludge has been shown to present excessive plant uptake on a short term basis [43,67]. It has also been argued that a fraction of the organic matter resists degradation and will provide protection against plant uptake and leachability of the trace elements on a long-term basis. However, this depends on soil and sludge chemistry [67,68]. Repeated application of trace element-contaminated sewage sludge can result in an accumulation of toxic trace elements in the soil [69]. Once accumulated, trace elements are highly persistent in the topsoil [15,70,71] and can cause potential problems such as phytotoxicity [72], injury to soil microorganisms [73], or elevated transfer to the food chain [74–76].

Current debates on the bioavailability of sludge-borne trace elements center on two hypotheses. The first concerns how the bioavailability of trace elements is likely to change following the termination of sludge application. Some advocate the "sludge protection hypothesis," which states that sludge-borne trace elements are maintained in chemical forms of low bioavailability by inorganic components of the sludge and that the specific metal absorption capacity added in sludge will persist in the soil. According to this hypothesis, sludge-borne trace elements should become less bioavailable with time as surface adsorbed trace elements become occluded.

In contrast, others believe that sludge-derived organic matter contributes significantly to the metal adsorption capacity and that slow mineralization of this organic matter could release trace elements into more soluble, more phytoavailable forms. This is often termed the "sludge time bomb hypothesis." Because the decomposition of sludge organic matter is often associated with an acidification of soil, if this is uncorrected, further increase in the bioavailability/phytoavailability of the sludge-borne trace elements would be expected [31,77,78].

From an ecotoxicological point of view, any addition of trace elements will have some impact on organisms in the soil, vegetation, and food chain [79]. If a metal enrichment policy of environmental protection is followed, adopting the approach maintaining the metal balance in soils can diminish the potentially adverse effects of sewage sludge application to soil. If an accumulation of trace elements in sewage-treated soils is accepted, the maximum permissible metal concentrations in the soil depend upon the organisms that regulations are intended to protect, assuming that the cause–effect relationships are known [80].

Transfer coefficients (concentration of metal in the aerial portion of a plant relative to total concentration in the soil) are a convenient way of quantifying the relative differences in bioavailability of trace elements to plants. Alloway and Ayres [30] and Kloke et al. [76] have given generalized transfer coefficients for soils and plants (Table 31.4). However, soil pH, soil organic matter content, and plant genotype can have marked effects on metal uptake. The transfer coefficients are based on root uptake of trace elements, but plants can accumulate relatively large amounts of trace elements by foliar absorption of atmosphere deposits on plant leaves.

Table 31.4 shows that Cd, Tl, and Zn have the highest transfer coefficients — a reflection of their poor sorption in the soil [29]. In contrast, trace elements such as Cu, Co, Cr, and Pb have low coefficients because they are usually strongly bound to the soil colloids. The most toxic trace elements for high plants and several microorganisms are Hg, Cu, Ni, Pb, Co, Cd, and, possibly, Hg, Be, and Sn [81]. Although the occurrence and toxicity will depend on soil factors such as pH, plant genotype, and source of trace elements, a general indication of the toxic levels of some trace elements is given in Table 31.5 [82].

TABLE 31.4
Transfer Coefficient of Trace Elements in the Soil–Plant System

Element	Transfer coefficient	Element	Transfer coefficient
As	0.01–0.1	Ni	0.1–1
Be	0.01–0.1	Pb	0.01–0.1
Cd	1–10	Se	0.1–10
Co	0.01–0.1	Sn	0.01–0.1
Cr	0.01–0.1	Tl	1.0–10
Cu	0.01–0.1	Zn	1.0–10
Hg	0.1–1.0		

Source: From Kloke, A. et al., in *Health*, Springer–Verlag, Berlin, 1984. With permission.

TABLE 31.5
Normal and Phytotoxic Metal Concentrations
Generally Found in Plant Leaves[a]

	Concentration in leaves (μg/g)	
Element	Normal range	Toxicity
Ag	0.01–0.8	1–4
As(III)	0.02–7	5–20
Cd	0.1–2.4	5–30
Cu	5–20	20–100
Cr	0.03–14	5–30
Hg	0.005–0.17	1–3
Ni	0.02–5	10–100
Pb	5–10	30–300
Sb	0.0001–2	1–2
V	0.001–1.5	5–10
Zn	1–400	100–400

[a] Alloway, B.J., in *Heavy Metals in Soils*, Alloway, B.J. (Ed.), Blackie and Sons, Glasgow, U.K. 1990; based largely on Bowen, H.J.M., *The Environmental Chemistry of the Elements*, Academic Press, London, 1979. With permission.

Usually, lime application reduces uptake of Zn and Ni more than Cd [83–85]; no decrease in availability is noticed in the case of Mo and Se [85]. Metal uptake in response to liming may also vary among plant species [86,87]. Liming decreased more Cd uptake by lettuce and carrot than by potatoes and peanuts [74]. Some evidence indicates that pH of about 6 is high enough to regulate metal uptake [88]. Increase of pH to 6.5, recommended for controlling trace elements in the food chain, does not seem to be necessary [89]. This finding is of practical importance because liming of acidic soils to pH 6.5 is often costly and can require a considerable amount of time.

The effect of other soil properties on element uptake by plants is less evident than that of pH, and the results are often conflicting. Hinesly et al. [91] conducted a study to determine the effect of CEC on Cd uptake by corn. The soil CEC inversely affected Cd uptake by corn when the metal was applied as a soluble salt, but not when Cd was supplied as a constituent of sewage sludge. These experiments were further confirmed in greenhouse studies conducted by Korcak and Fanning [92].

Some trace elements exhibit affinity for soil organic matter (OM), which has the CEC property and chelating ability. Therefore, addition of sewage sludge, peat, or plant residues can bind trace elements in soil. On the other hand, because of chelating ability, OM is viewed as a source of soluble complexing agents for trace elements. The binding ability of organic matter is not permanent; it is generally agreed that the OM level in soil must eventually return to a value not much greater than that of the original soil [38].

Trends in metal availability as a function of metal content in soils can be described by three models: (1) linear (constant partitioning model); (2) plateau (saturation model); and (3) the Langmuir sorption model (Figure 31.3) [10]. Usually, uptake of trace elements by plant tops does not occur in linear response to concentrations of the metal in soils, except at a low range of concentrations [10,38,92]. Uptake of trace elements by plants becomes less efficient at higher metal loading in soil, and the plateau relationship is used to describe this saturation effect [93]. Soils have a finite capacity to immobilize trace elements by adsorption or precipitation. When this protective potential is exceeded, a Langmuir type of relationship is expected (Figure 31.3). This relationship is found sometimes for trace elements added to soil in soluble salt forms [94]. Under more realistic field conditions, when trace elements are introduced to soil with sewage sludges or with industrial wastes,

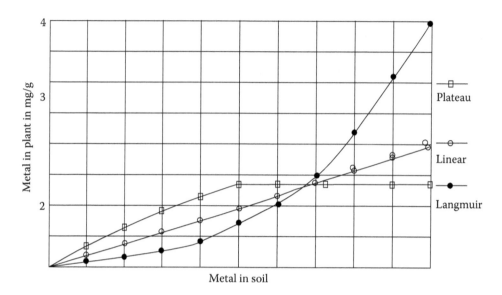

FIGURE 31.3 Metal availability as a function of metal content of soils. (From Dudka, S. and Miller, W.P., *J. Environ. Sci. Health B*, 34(4): 681–708, 1999.

the plateau model prevails because it best describes the plant response to increased metal concentrations in the soil [10].

The simple soil–plant relationship of plant element uptake is often modified by environmental, plant, and soil factors [95]. As a result, only a small proportion of elemental variability in plants can be explained by element concentrations in soils. Plants can accumulate trace elements, in their tissues due to their ability to adapt to various environmental conditions [96–98]. Plant uptake of elements from the soil solution requires positional availability to the plant root. The element must be moved to the root through diffusion or mass flow or the root must grow to the element [99]. This transfer requires that the element move through a solution phase. Therefore, water solubility and a variety of complexation, chelation, and other chemical reactions controlled by pH become important in regulating element availability [100].

Except for a few special cases, plant tissue concentrations do not positively correlate with total element content of untreated soils [101]. The element concentrations in plant tissues vary greatly, depending on species and cultivate, plant organ, age, and environmental conditions [102–107]. The extent of increase in trace element concentration above control for crops grown on metal-contaminated soils is strongly affected by crop species [108]. Studies conducted in England report [109] the response of many crop species grown in the same experiment on two soils contaminated with long-term sludge application. The relative crop uptake removes factors other than crop species (or cultivate). Leafy vegetables and, interestingly, wheat grains had the highest relative increased uptake of Cd [110,111].

31.6 ENVIRONMENTAL PATHWAYS AND HEALTH RISK ASSESSMENT

Trace elements' accumulation in soils and their entry into the food chain associated with land application of municipal sewage sludge has been reported [112]. Accumulation of trace elements in the tissue of edible vegetables grown on sludge-treated soils is well established [74,113]. Increased levels of Cd, Zn, and Ni were demonstrated in beets and chard grown on soils receiving sludge for many years [114,115]. Although these studies contain valuable information, they accept the site "as is," with little or no information on the amounts and characteristics of applied sludge, or soil changes that occurred after initiation of sludge application [116].

Other researchers investigated the extent of sludge-born metal accumulation in the edible tissues of seven vegetable crops [117]. According to these workers, metal accumulation was maximum in leaf tissue and the Zn and Cd content of edible fruits, tubes, and root tissues increased significantly; the Cu and Pb levels remain unchanged. Higher metal levels in vegetation against storage tissues for different vegetable species [118,119] has also been reported. Dowdy et al. also carried out extensive work on the growth and metal uptake of snap beans grown on sewage sludge-amended soil [121]; performance of goats and lambs fed on corn silage produced on sludge-amended soil [121]; and trace metal and mineral composition of milk and blood from goats fed silage produced on sludge-amended soil [122,123].

There is good reason to believe that livestock grazing on plants treated with sewage sludge will ingest the pollutants through the grazed plants or by eating sewage sludge along with the plants [124]. Sheep eating cabbage grown on sludge developed lesions of the liver and thyroid gland. Pigs grown on corn treated with sludge had elevated levels of cadmium in their tissues [125]. Cows, goats, and sheep are also likely to eat sludge directly. In grazing, these animals may pull up plants by the roots and thus ingest substantial quantities of soil [56]. A cow may ingest as much as 500 kg (1100 lb) of soil each year [126].

Small mammals have been shown to accumulate trace elements after sewage sludge was applied to forest lands. Shrews, shrew moles, and deer mice absorbed trace elements from sludge [127]. Insects in the soil absorb toxic trace elements, which then accumulate in birds [128]. It has been shown that sewage sludge applied to soils can increase the metal intake of humans eating beef (cow or buffalo milk) produced from those soils [129,130]. Studies have indicated that sludge application to soil increases the metal content in plants in this order [3,131]:

$$Zn > Cd > Ni > Cu > Pb = Hg = Cr$$

The three basic methods for formulating limits to metal additions in soils receiving sewage sludges are: (1) analysis of pathways of metal transfers; (2) limits consistent with the no observed adverse effect level (NOAEL); and (3) metal balance approach [1,101]. The governing principle of the pathway approach [132] is that cumulative pollutant loading limits, which were derived from the analyses of various exposure pathways, will not be exceeded. To protect public health and the environment against the impact of high-metal sewage sludges, this regulation also defines maximum concentrations for metal present in the sewage sludge beyond which land application of the sludge is not permitted. The upper limit was calculated maximum concentration allowed according to the cumulative pollutant loading limits assuming sludge is applied at the annual rate of 10 tons per hectare for 100 years.

In adopting the environmental exposure pathway approach, one must select a target organism for each pathway. The process of selecting the contaminant-loading limit for each pathway is also based on the NOAEL. From the long-term perspective, it is very difficult to foresee whether the approach to setting limits for trace elements in soils allowing substantial increase in soil metal concentrations will be effective in protecting environment, food chain, and human health from adverse effects of potentially toxic trace elements. Limits consistent with the NOAEL are based on actual cases of effects due to trace elements, but not necessarily derived from studies that involved land application of sewage sludge.

The Commission of the European Communities (CEC) issued a directive to limit the inputs of potentially toxic trace elements to soil from sewage sludge so as to protect plant growth, crop quality, and human and animal health [133]. The directive has three types of metal limits:

- Concentrations allowed in sludges used for agriculture
- Maximum concentrations permitted in sludge-treated soils
- The 10-year average annual rate of addition of trace elements in sludge

The third category of defining rules for trace elements in soils, the metal balance approach, is concerned with the fact that, in industrial countries, the metal inputs to soil minus losses through crop removal, leaching, and erosion is positive for all trace elements studied [134,135]. These observations have led to a very cautious approach to the intentional addition of trace elements to soils. Quality standards and limits for pollutants in biosolids were developed from extensive environmental risk assessments conducted by USEPA and the U.S. Department of Agriculture.

The EPA used a rigorously reviewed methodology developed specifically for conducting the risk assessment [136,137]. The goal of the risk assessment was to protect a person, animal, or plant that is highly and continuously exposed to pollutants in biosolids. The rationale for this goal is that the general population would be protected if the regulations were developed to protect highly exposed individuals [19]. The risk assessment process was the most comprehensive analysis of its kind ever undertaken by the EPA. The approach has since been applied to other materials, such as municipal solid waste compost. The resultant part 503 rule was designed to provide "reasonable worst case," not absolute, protection to human health and the environment [19].

The initial task of the 10-year risk assessment process was to establish a range of concentrations for trace elements and organic compounds that had the greatest potential for harm based on known human, animal, and plant toxicities. Maximum safe accumulations for the chemical constituents in soil were established from the most limiting 14 pathways of exposure, which included risk posed to human health, plant toxicity and uptake, effects on livestock or wildlife, and water quality impacts [138]. The EPA screened a total of 200 chemical constituents and 50 of these were selected for further evaluation. The 503 rule was then limited to ten trace elements (arsenic, cadmium, chromium, copper, lead, mercury, molybdenum, nickel, selenium, and zinc). Chromium was subsequently dropped on a court challenge because the risk assessment had shown a very low risk level for this metal. The most limiting pathway for each of the nine regulated trace elements was used to develop pollutant concentration limits and a lifetime loading rate standard [139].

Under part 503, the cumulative loading limits established by EPA for eight trace elements would allow the concentrations of these elements to increase to levels 10 to 100 times the normal background concentrations in soil. The time for each of the eight elements to reach its cumulative loading limit when biosolid with typical trace element concentrations was applied annually at the rate of 5 dry tons per acre [19,140]. The cumulative loading limits were developed to ensure that the soil trace elements never reached harmful levels. Further applications of biosolids to the site would be prohibited if the cumulative loading limit for any of the eight trace elements was reached.

Although supporters of the 503 regulations believe that sludge-amended soils will maintain an ability to immobilize toxic trace elements in unavailable form, it is impossible to predict long-term consequences of high metal loading of soils. Predictions of no adverse effect on crops and food chain employ the sludge protection hypothesis, which states that the specific adsorption capacity added with sludge will persist in soil and will be effective in metal immobilization as long as trace elements are present in soils [92]. The experimental data, however, suggest that immobilization of trace elements in sludge-amended soils is due in some part to sludge organic matter. Therefore, the sludge time bomb hypothesis points out that slow mineralization of organic matter added with sludge could release trace elements into the more soluble forms, which eventually will adversely affect crop productivity and food chain quality [38].

In composts and sewage, as in wastes, trace elements are present in various chemical forms, which differ in respect to solubility in water, bioavailability, and stability [37]. The utilization of composts as fertilizers depends not only on the total content of trace elements but also on the chemical forms and phases in which they are present in compost and sewage [141]. The form in which a metal exists significantly affects its bioavailability, toxicity, and mobility in the soil to

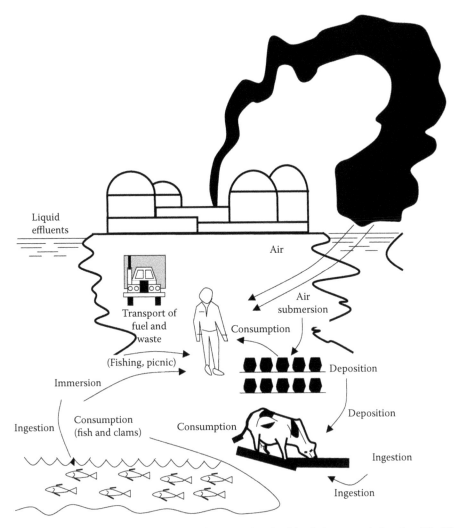

FIGURE 31.4 Pollutant transport and health risk associated with sludge-amended soils. (Modified and redrawn from USAEC, U.S. Atomic Energy Commission, WASH-1209, Washington, D.C., 1973.)

plant and, further, to human [142,143]. Possible distribution pathways of trace elements from sludge-amended soils are well illustrated in Figure 31.4 [144], which shows the movement of trace elements and the pathways that help determine whether individuals have been, are being, or will be exposed to trace elements contaminants.

Environmental pathways can be completed or potential. A completed pathway indicates that human exposure to contaminants has occurred, is occurring, or will occur. A potential exposure pathway indicates that human exposure to contaminants could have occurred, could be occurring, or could occur in the future. The movement, mobility, and transport of trace elements from soil to human can be well understand by the speciation analysis of trace elements [37,145]. Biological cycling includes the bioaccumulation, bioconcentration, bioavailability, and toxicity; geochemical cycling involves the transport, adsorption, and precipitation of the element in the water–soil system.

It is now well established that no meaningful interpretation of biological or geochemical cycling can be made without speciation information [146,147]. When the sewage sludge amendment of soils is judged by the potential health effects of trace elements, it should be emphasized that sewage sludge is not the only source of trace element contamination of soils. The knowledge of how trace elements are distributed in the environment and which processes affect the distribution can be used to evaluate the potential of sludge amendment of a certain soil with respect to trace element migration. Amending an acidic sandy soil with sewage sludge seems to be a bad idea. Such a soil may have no retaining capacity and, when the sludge starts to decompose, trace elements are susceptible to migrate at a high rate.

The movement, mobility, and transport of trace elements from soil to human can be well understood by the speciation of trace elements, which essentially include four main steps:

- Evaluation of the quantity of trace elements in the sewage sludge
- Evaluation of the levels of trace elements in the soil, water, vegetation, milk, etc.
- Identification of the exposure pathways and determination of ambient levels that result from the sources
- Identification of the population exposed to the trace element pollutants by carrying out epidemiological studies

31.7 INDIAN SCENARIO

Treated and untreated sewage and sewage sludge are extensively used for cultivation in the suburban areas of India and food crops and vegetables are grown in sludge-amended soils. Olaniya et al. [149] conducted a case study on trace element pollution in agricultural soil and vegetation due to the application of municipal solid waste in the Dhapa disposal site near Kolkata. At this site, fresh municipal solid waste results in accumulation of toxic trace elements in the soil; when edibles are grown on such soils, toxic trace elements enter into the food chain and pose risks to human health [149]. To determine the accumulation of trace elements, Srikanth et al. [150] conducted studies on forage grass (Guinea grass) cultivated in urban sewage sludge along the bank of River Musi, which receives sewage from the city of Hyderabad.

The same authors reported high concentrations of Pd, Cd, and Cr in various vegetables grown in urban sewage sludge, indicating possible health hazard for consumers [151]. The risk of toxic trace elements like Cd among the farmers working on sludge farms was reported by carrying out urine analysis on male sewage sludge farmers [152]. Ayyadurai et al. [153] studied the concentration of Pb and Cd in milk of cows and buffalos fed with fodder grown on wastewater sludge-irrigated soils. They reported that buffalo milk contained higher levels of Pb in comparison to that of cow milk in all the samples analyzed [153].

Much work has been carried out on the effects of trace elements on soil and vegetation, and their entry into human food chain [154]. Recently, a case study was conducted by the authors on the speciation and accumulation of trace elements in vegetation grown on SAS and their profound transfer to food chain. The results revealed that Ni, Cr, Pb, and Zn were associated often with the mobilizable fraction in the SAS; this was clearly indicated by the high concentration of these trace elements in the vegetables. The concentration factor (CF) for these trace elements in vegetables was calculated and it was found that Ni had maximum CF. The epidemiological studies showed that the trace element content — particularly, Ni, Pb, and Zn — was very high in people consuming these vegetables grown on SAS, indicating the metal transfer to food chain [156].

Forage grass grown on SAS was also analyzed for trace element content and Ni, Zn, and Pb concentrations were found to be high. Corresponding milk samples were collected from cows and buffalos fed on these forage grasses and high concentrations of Ni and Zn were found in milk compared to controls [156]. It is well known that the movement, mobility, and transport of trace elements from soil to human can be well understood by the speciation of the trace elements. Very little work has been carried out in India on these lines for assessing the mobility, phytoavailability, and transfer of trace elements to the human food chain.

31.8 INTERNATIONAL SCENARIO

31.8.1 SHORT-ROTATION FORESTRY USING SEWAGE SLUDGE AND BIOSOLIDS — IMPLICATIONS

Short-rotation forestry (SRF) is advocated as a means of phytoremediation and for disposing of sewage sludge [157]. Fast growing trees such as *Salix* (willow), *Populus* (poplar), *Alnus* (alder), and *Eucalyptus* (eucalypts) are ideal trees for SRF and have gained importance in mitigating CO_2, industrial effluents, and other air pollutants. Large quantities of fertilizer and water are required to sustain the intensive management of SRF. This demand prompted the application of sewage sludge, with the additional benefit of providing a viable alternative for the waste disposal [157–160].

SRF has been successful in Sweden for the production of biofuel for the past 20 years. SRF takes its name from the short rotation time typical of poplar and willows (4 to 6 years) compared to conifers (60 to 120 years) or broadleaved trees (35 to 50 years). Under optimum conditions (i.e., intensive management strategies), SRF can yield between 12,000 and 20,000 kg dry matter (DM) ha^{-1} yr^{-1}, although in general, lower yields of between 6000 and 9000 kg DM ha^{-1} yr^{-1} are more likely. If SRF is used for the generation of energy, approximately 4.5 MWh of energy can be produced from burning 1 ton of dry matter (50% moisture) [158].

A substantial body of data now indicates that biomass plantations are irrigated with sewage effluents; this helps in recycling effluents, biosolids, and other wastes. Current research on the use of sludge as a fertilizer for SRF indicates strongly that willow plantations may be able to perform the majority of the required cleanup steps well. SRF could be used as "vegetation filters" to utilize the nutrients in municipal sewage sludge, wastewater, leakage water, and bioash (wood ash), although dredged sediments containing organic chemicals and pathogenic bacteria have also been effectively "purified" following application onto biomass plantations [160, 161].

The tree species used in biomass forestry vary between areas in which they are grown. Many of the SRF species commonly used, especially *Populus* spp., have also shown considerable potential in the remediation of groundwater contaminated with inorganic chemicals — specifically, the *P. deltoides* × *nigra* and *P. trichocarpa* × *deltoides* clones. SRF fulfills important environmental and ecological factors: there is no net CO_2 contribution to the atmosphere compared to energy production using fossil fuel; in fact, there is a small net uptake. Pesticide use is lower than on conventional agricultural crops, and the growth of willows for SRF improves the condition of the soil as well as biodiversity

Primary requirements of SRF are a rich supply of nitrogen and water. The ability of fast growing trees to remove nutrients and water can also be applied to wastewater treatment; excessive concentrations of nutrient chemicals pose environmental problems when present in excess, such as eutrophication of streams and lakes, causing algal blooms, and a progressive reduction in potability. Standards usually monitor the biological oxygen demand (BOD), total N, ammonia N, and total P in water and often these strict regulatory limits cannot be met by conventional treatment methods alone.

The elimination of P from wastewater generally involves the use of large quantities of chemical agents, which are difficult to clean up once the process has been completed. Obarska–Pempkowiak [162] found that *Salix viminalis* and *S. arenaria* could significantly reduce the nitrogen concentration of municipal wastewater; in particular, *S. viminalis* reduced the total N concentration of wastewater from over 35 g m^{-2} to less than 5 g m^{-2} over a period of 270 days. Typical N and P contents of applied sludges are approximately 4.5 kg N t^{-1} and 2.8 kg P t^{-1} in the raw cake, of which 1.5 and 1.4 kg t^{-1} N and P respectively are available in the first cropping season. Willow and poplars irrigated with sewage sludge showed an increased yield and that the trees took up almost 95% of the N and P load of the effluent, simultaneously lowering the BOD of the effluent by an average of 96%.

Studies have emphasized that SRF can be used for "collecting" metals from the soil — in effect, using the trees as biological filters. Goransson and Philippot [163] found that almost all of the Cd applied during sludge application was taken up by *Betula pendula* and that, theoretically, the trees could remove 1.5 kg Cd ha^{-1} yr^{-1} when watered with sludge. Landberg and Greger [164] found that net transport of heavy metals to the shoots varied widely between 1 and 72%; therefore, certain clones were able to accumulate more metals than others, indicating the need for clone selection for resistant trees. The efficiency with which willows can remove metals from sewage sludge-fertilized soils has been demonstrated in several key studies.

Some evidence suggests that fast growing trees may be able to play a major role in reducing the concentration of organic chemicals in contaminated sludges, soils, and groundwater (see next section). Using land-farming techniques, Harmsen et al. used willows to remove PAHs from dredged harbor sediments and concluded that the high yield of biomass was instrumental to reducing the costs. Willows and poplar trees were also used in a multispecies wetland system to remove bacteria and other pathogens from municipal wastewater.

31.8.2 SLUDGE USAGE — INTERNATIONAL REGULATIONS

The residual product generated from sewage treatment is termed sludge. Sewage sludge contains heavy metals, organic compounds, and pathogens, in addition to substantial amounts of nutrients. Sludge may be disposed of by depositing, burning, or dumping into the sea or it may be used in forestry and agriculture. On the one hand, spreading sludge onto the land is desirable because the nutrients of sludge would participate in the biogeochemical cycle in ecosystems. However, due to the inherent nature of its composition, sludge involves the risk of harming the environment because the substances may be accumulated in the soil or damage the ecosystem. Thus, health considerations must be considered by environmental regulatory agencies with respect to the use of sludge because it contains pathogens in the form of bacteria, viruses, and parasite eggs. This content can be reduced by stabilization or disinfection of the sludge.

To ensure that the environment and man are fully safeguarded against the harmful effects arising from the uncontrolled use of sludge, it is necessary to regulate the agricultural use of sewage sludge. Therefore, most countries have legislation regulating the use of municipal sewage sludge in forestry and agriculture [165]. These regulations set maximum limits for the content of heavy metals in sludge (except in the U.K.). Limitations are set on the amount of sludge that can be spread on a given area, based on total amount of sludge (calculated as the amount of dry solids in the sludge), heavy metals, or nutrients added to the soil. It is especially with respect to this point that different national regulations vary [165]. In addition, there are restrictions in relation to certain crops and land uses, (it is normally not permissible to apply sludge to land to be used for growing vegetables). Germany prohibits spreading sludge on forest land, but Denmark and the U.K. permit this. Before application of sludge, chemical composition of the soil must be taken into account (pH, texture, and concentration of heavy metals).

Sludge producers are required to make analyses of the sludge. Typically, the frequency of the analysis is determined by the size of the treatment works and is intensified if the composition of the sludge is variable. The frequency varies from a daily sludge sampling to one analysis every second or fourth year; sometimes, with respect to certain substances, analysis is not required at all. Usually, sludge is analyzed for its content of heavy metals, nitrogen, and phosphorus, but identification of other substances may also be required. In Norway and the U.S., samples must also be taken for the control of pathogen content. Furthermore, in the U.S., it must be controlled whether the vector attraction reduction requirements of the sludge are met or not, so that the sludge exerts only a limited attraction to animals, which can transmit infection to other animals as well as to man. The frequency of the analyses and the substances for analysis vary among the different countries, so the use of sludge may be indirectly limited in some countries by the high cost of analysis.

For the purpose of controlling the legislation, the sludge producers must register the production and the agricultural use of their sludge. Therefore, the controls must be drawn up differently in the different countries. In all countries, except the U.K., maximum limits are set for the content of heavy metals in sludge. In addition, some countries have set value limits for arsenic, selenium, and molybdenum. Germany also limits the content of the following organic compounds: polychlorated biphenyles (PCB), polychlorated dibenzoefuranes and polychlorated dibenzoedioxines (PCBP, PCDF, PCDD), and organic halogens (AOX). The German value limits for the content of organic compounds in sludge are not set on the basis of qualified toxicological tests, but they are set out of prudence [165]. In Germany, the present content of PCB in sludge is unlikely to limit use of sludge, but the content of PCDD/PCDB may exceed the fixed limits. Furthermore, the analyses are complicated and costly [165].

In general, heavy metals and persistent organic compounds cause damage to the environment by accumulating, while the introduction of nutrients may harm the environment through the leaching of nutrients to the aquatic environment. The extent to which organic compounds are accumulated has not yet been clarified. The effects of heavy metals and organic compounds on the soil fauna have not been sufficiently investigated. One of the most important sources of pollution from persistent organic compounds is pesticides; the pollution caused by heavy metals arises from industrial deposition and fertilizers.

When comparing the regulations, remarkably large differences appear concerning the restrictions on heavy metals with respect to their concentrations in sludge and soil and to the quantities that may be spread. The general tendency appears to be that countries that permit the highest concentrations of heavy metals in sludge also permit the largest quantities of sludge to be spread annually and permit the highest concentrations of heavy metals in soil. Countries that have the most stringent requirements for the heavy metal content in sludge also have a tendency to have the lowest value limits for the permissible heavy metal content in soils.

The great variation between the countries with respect to the amount of heavy metals that can be applied to soils and the lesser variation with respect to the permitted heavy metal concentrations in soils where sludge can be applied give rise to very different time horizons for the use of sludge on individual areas. Therefore, the value limits for heavy metal concentrations in soils can be regarded as an "emergency brake" on the use of sludge. However, this emergency brake is only effective if the control of heavy metal concentrations in soils is effective in the countries that risk exceeding the value limits after a few years' use of sludge.

31.9 CONCLUSIONS

Biosolids can provide essential plant nutrients, improve soil structure and tilth, add organic matter, enhance moisture retention, and reduce soil erosion. Sewage sludge contains organic matter and plant nutrients beneficial to soil health and crop production. Applying sewage sludge to agricultural lands fits well with the current concepts of resource recycling and sustainable agriculture. However, because sewage sludge contains contaminants, such as trace elements, that may decrease the quality of agricultural soils, careful control of contaminant levels in soils receiving sludge is needed.

Contamination of soil by trace elements is a concern because they are persistent and may affect plant, animal, and human health. Application to agricultural soils is a beneficial method of managing sewage sludge, but adding contaminants to soil in this, and other, waste materials must be controlled. The trace element content of sewage sludge is the main factor that still restricts its agricultural use.

However, steps can be taken to decrease the availability of trace elements in sludge-amended soils. Trace element content in plants is strongly influenced by pH and sludge treatment. Increasing the pH of a sewage-amended soil or heat-treating it can reduce the plant metal uptake. Manipulating the soil pH is the most effective and rapid method of controlling the availability of trace elements in sludged soil. Liming the soil to a pH of 6.5 to 7 can reduce the mobile fraction of most of the trace elements in soil.

The trace elements in sludge are present in various chemical forms that differ in respect to solubility in water, bioavailability, and stability [166,167]. The utilization of sewage sludge as fertilizer depends not only on the total content of trace elements but also on the chemical forms in which they are present. The form in which a metal exists significantly affects its biological activity availability, toxicity, and mobility in the soil environment. It is well known that vegetation grown on sludge-amended soils poses a direct entry for trace elements into human food chain. India has no administration or legislative measures to control the continued application of sludge as a manure.

Most regulations in India concerning trace elements in food, occupational health, and environment are based on the total element contents and are frequently given as maximum limits or guideline levels. In contrast, few regulations pay attention to the molecular species (speciation) in which the elements are bound. There is a great need for the development of more species-specific analytical and toxicological data and modified regulations are necessary if India wants to enter the global market in the area of food, agriculture, and environment. Also, there is a need for creating a different speciation scheme to study the toxicity, mobility, and phytoavailability of these trace elements for the development of a model to predict the soil-to-cell movement of metal pollutants. The model representing the pathways of trace elements in the environment and prediction about the fate of these pollutants based on the speciation studies will help in reducing the risk due to the application of sewage sludge as manure.

ACKNOWLEDGMENTS

The authors are grateful to the director of the Indian Institute of Chemical Technology (IICT) and Dr. M. Vairamani, head of the Analytical Chemistry Division, IICT, for their encouragement and support.

REFERENCES

1. Mcgrath, S.P, Chang, A.C, Page, A.L., and Witter, E. Land application of sewage sludge: scientific perspectives of heavy metal loading limits in Europe and United States. Environ. Rev. 1994, 2(1): 108–118.
2. Ashwathanarayana, U. *Soil Resources and the Environment*, Oxford and IBH Publishers, New Delhi, 1999.
3. Sims, J.T. and Kline, J.S. Chemical fractionation and plant uptake of heavy metals in soils amended with co-composted sewage sludge. *J. Environ. Qual.* 1991, 20(2): 387.
4. Adriano, D.C. *Trace Elements in the Terrestrial Environment*, Springer–Verlag, New York, 1986.
5. Rutger's cooperative extension (RCE) fact sheet FS 955. Land application of sewage biosolids #5: heavy metals. New Brunswick, NJ, 2000.
6. U.S. Environmental Protection Agency. Clean Water Act. Section 503, vol. 58, No.32 USEPA, Washington, D.C., 1993.
7. Schmidt, J.P. Understanding phytotoxicity threshold for trace elements in land applied sewage sludge. *J. Environ. Qual.* 1997, 26: 4–10.
8. Towers, W., Paterson, E., and Coull, M.C. Potential impact of draft proposals to revise the EC directive controlling sewage sludge application to land. *J. Chart. Inst. Water E.* 2002, 16(1): 65–71.
9. Heinze, J.E. Science vs. politics in the environmental regulatory process. *Chim. Oggi.* 1999, 17(1–2): 14–19.
10. Mbila, M.O., Thompson, M.L., Mbagwu, J.S.C. et al., Distribution and movement of sludge derived trace metals in selected Nigerian soils. *J. Environ Qual.* 2001, 30(5): 1667–1674.
11. Dudka, S. and Miller, W.P. Accumulation of potentially toxic elements in plants and their transfer to human food chain. *J. Environ. Sci. Health B.* 1999, 34(4): 681–708.

12. Bengtsson, M. and Tillman, A.M. Actors and interpretations in an environmental controversy: the Swedish debate on sewage sludge use in agriculture. *Resour. Conserv. Recycling* 42, 2004, 65–82.
13. Bowen, H.J.M. *The Environmental Chemistry of the Elements*, Academic Press, London, 1979.
14. Davis, D.J.A. and Bennett. B.G. *Exposure Commitment. Assessments of Environmental Pollutants*, vol. 3. Monitoring and Assessment Research Center, London, 1983.
15. Department of the Environment Interdepartmental Committee for the Reclamation of Contaminated Land List of Triggers. *Concentration for Contamination*, DOE, London, 1987.
16. Alloway, B.J. and Jackson, A.P. The behavior of heavy metal in sewage sludge-amended soil. *Sci. Total Environ.* 1991, 100: 151.
17. Kumaresan, M. and Riyezuddin, P. Chemical speciation of trace metals — a review. *Res. J. Chem. Environ.* 1999, 3(4): 59–79.
18. Barrera, I., Andres, P., and Alcaniz, J.M. Sewage sludge application on soil: effect of two earthworm species. *Water Air Soil Pollut.* 2001, 129(1–4): 319–332.
19. Obbard, J.P. Ecotoxicological assessment of heavy metals in sewage sludge amended soils. *Appl. Geochem.* 2001, 16(11–12): 1405–1411.
20. Evanyles, G.K. Agricultural land application of biosolids in Virginia — risks and concerns. Virginia Co-Operative Extension Publications, Virginia State, Petersburg. 1999, 482–504.
21. What do you do with the sludge? http://www.bettertechnology/sewagesludge/pelletization. 2000.
22. United States Environmental Protection Agency (USEPA). Region and Biosolid Management Program http://www.cpa.gov/regions/water/npde/biosolids, 1997.
23. Pennsylvania Department of Environmental Protection. http://www. dep.state.pa.vs/dep/biosolids, 1999.
24. Elliot, H.A. Land application of municipal sewage sludge. *J. Soil Water Conserv.* 1986, 41: 5–10.
25. Questions on sludge. Cornell Waste Management Institute, videoconference on land application of sewage sludges, New York, 1996.
26. Heckmann, J.R. and Kluchinski, D. Chemical composition of municipal leaf waste and hand-collected leaf litter. *J. Environ. Qual.* 1996, 25: 355–362.
27. Wong, C.W.J., Li, K., Fang, M., and Su, D.C. Toxicity evaluation of sewage sludges in Hong Kong. *Environ. Int.* 2001, 27(5): 373–380.
28. RCE fact sheet FS 956. Land application of sewage sludge (biosolids) #6: soil amendments and heavy metals, 2000.
29. USEPA technical support documented for land application of sewage sludge, vol. 1, report no. EPA/822/R-93-00/a, 1992.
30. Alloway, B.J. and Ayres, D.C. *Chemical Principles of Environmental Pollution*, Blackie Academic and Professionals, Glasgow, Scotland 1993, 109.
31. Qureshi, S., Richards, B.K., Steenhuis, T.S., McBride, M.B., Baveye, P., and Dousset, S. Microbial acidification and pH effects on trace element release from sewage sludge. *Environ. Pollut.*, 132, 61-71, 2004.
32. Fergusson, J.E. *Inorganic Chemistry and the Earth*, Pergamon Press, Oxford, 1982.
33. Greenwood, N.N. and Earnshaw, A. *Chemistry: the Elements*, Pergamon Press, Oxford, 1984.
34. Essington, M.E. and Mattigod, S.V. Element partitioning in size and density fractionated sewage sludge and sludge amended soil. *Soil Sci. Soc. Am. J.* 1990, 54(2): 385–394.
35. Essington, M.E. and Malttigod, S.V. Trace element solid phase associations in sewage sludge and sludge-amended soils. *Soil Sci. Soc. Am. J.* 1991, 55(5): 350–356.
36. Keefer, R.F. and Singh, R.N. Correlation of metal-organic fractions with soil properties in sewage sludge amended soils. *Soil Sci.* 1986, 142(1): 20–26.
37. Obrador, A., Rico, M.I., Alvarez, J.M., and Novillo, J. Influence of thermal treatment on sequential extraction and leaching behaviour of trace metals in a contaminated sewage sludge. *Biores. Technol.* 2001, 76(3): 259–264.
38. Legreton, M. Speciation of heavy metals in sewage sludge and sludge-amended soils, *Int. J. Environ. Anal. Chem.* 1993, 51: 161–165.
39. McBride, M.B. Toxic metal accumulation from agricultural use of sludge; are USEPA regulations protective? *J. Environ. Qual.* 1995, 24: 5–18.
40. Stennes, E. in Cadmium Alloway, B.J. (Ed.) *Heavy Metals in Soils*, Blackie Publishers, Glasgow, chap. 11, 1990.

41. Nemeth, T. Mobility of some heavy metals in soil — plant systems studied on soil monoliths. *Water. Sci. Technol.* 1993, 28(3): 389.

42. Antoniadis, V. and Alloway, B.J. The role of dissolved organic carbon in the mobility of Cd, Ni and Zn in sewage sludge-amended soils. *Environ. Pollut.* 2002, 117(3): 515–521.

43. Ashwortha, D.J. and Allowaya, B.J. Soil mobility of sewage sludge-derived dissolved organic matter, copper, nickel and zinc. *Environ. Pollut.*, 2004, 127: 137–144.

44. Saverbeck, D.R. Plant, element and soil properties governing plant uptake and availability of heavy metals derived from sewage sludge. *Water Air Soil Pollut.* 1991, 57–58: 227.

45. Milacic, R. and Stupar, J. Fractionation and oxidation of chromium in tannery waste and sewage sludge-amended soils. *Environ. Sci. Technol.* 1995, 29(2): 506.

46. Webber, M.D. and Singh, S.S. Contamination of agricultural soils, Canada SIS publications, Eastern Cereal and Oilseed Research Center (ECORC), Canada, 2000.

47. Fergusson, J.E. *The Heavy Elements. Chemistry, Environmental Impact and Health Effects*, Pergamon Press, Oxford, 1990.

48. Langenbech, T. Heavy metals in sludge from a sewage treatment plant of Rio de Tanners. *Environ. Technol.* 1994, 15(1): 1997.

49. Harte, J., Holden, C., Schnieder, R., and Shirley, C. *Toxic A to Z*, University of California Press, Berkeley and Los Angeles, 1991.

50. Landgard, S. in *Metals in the Environment*, Waldron, H.A. (Ed.), Academic Press, London, chap. 4 1980.

51. Brallier, S. et al. Liming effects on availability of Cd, Cu, Ni and Zn in soil-amended with sewage sludge 16 years previously. *Water Air Soil Pollut.* 1996, 86: 195–206.

52. Vulkan, R., Mingelgrin, U., and Ben–Asher, P., Cu and Zn speciation in the solution of a soil sludge mixture. *J. Environ. Qual.* 2002, 31(1): 193–203.

53. Fuentes, A., Llorems, M., Saez, J., Soler, A., Aguilar, I.M., Ortuno, J.F., and Meseguer, V.F. Simple and sequential extraction of heavy metals from different sewage sludges. *Chemosphere* 2004, 54(8): 1039–1047.

54. National Technical Information Service (NTIS). The Part 503 sewage sludge regulations are available on diskette from the, telephone 1-800-553-6847; purchase item No.PB93-5004781 NC; price $60.00. 1996.

55. Welch, J.E. and Lund L.J. Zinc movement in sewage sludge-treated soils as influenced by soil properties, irrigation water quality and soil moisture level. *Soil Sci.* 1989, 147(3): 208–214.

56. Haag, E. Just say no to sludge. *Dairy Today* 1992, 3: 82–83.

57. New U.S. Waste policy pt-2 sewage sludge. *Rachel's Environ. Health Wkly.* Annapolis, MD, 2001.

58. Sanely, S.P. et al. Cadmium uptake by crops and implications for human dietary intake. *Environ. Pollut.* 1994; 86: 5–13.

59. O'Neill, P. Arsenic, in *Heavy Metals in Soils*, Allowey, B.J. (Ed.), Blackie and Sons, Glasgow, U.K., chap. 5, 1990.

60. Nyamangara, J. Use of sequential extraction to evaluate Zn and Cu in a soil amended with sewage sludge and inorganic metal salts. *Agri. Ecos. Environ.* 1998, 69(2): 135–141.

61. Qiao, X.L., Luo, Y.M., Christie, P., and Wong, M.H. Chemical speciation and extractability of Zn, Cu and Cd in two contrasting biosolids amended clay soils. *Chemosphere* 2003, 50(6): 823–829.

62. Moreno, J.L. Transference of heavy metals from a calcareous soil amended with sewage sludge compost of Berley Plant. *Bioresour. Technol.* 1996, 55(3): 251.

63. Stahl, R.S. and James, B.R. Zinc sorption by B horizon soils as a function of pH. *Soil. Sci. Soc. Am. J.* 1991, 55(6): 1592–1597.

64. Fuentes, A., Lloréns, M., Sáez, J., Aguilar, M.I., Ortuño, J.F., and Meseguer, V.F. Phytotoxicity and heavy metals speciation of stabilized sewage sludges. *J. Hazardous Mater.* 2004, A108 2004, 161–169.

65. Udom, B.E., Mbagwu, J.S.C., Adesodun, J.K., and Agbim, N.N. Distributions of zinc, copper, cadmium and lead in a tropical ultisol after long-term disposal of sewage sludge. *Environ. Int.* 2004, 30: 467–470.

66. Su, D.C. and Wong, J.W.C. Chemical speciation and phytoavailability of Zn, Cu, Ni and Cd in soils amended with fly ash-stabilized sewage sludge. *Environ. Int.* 2004, 29(7): 895–900.

67. Paulsrud, B. and Nedland, T.K. Strategy for land application of sewage sludge in Norway. *Water Sci. Technol.* 1997, 36(11): 283–290.

68. Weissenhorn, I., Mench, M., and Leyval, C. Bioavailability of heavy metals and asbuscular mycorrhiza in a sewage sludge-amended sandy soil. *Soil. Biol. Biochem.* 1995, 27(3): 287–296.
69. Manahan, S.E. *Environmental Chemistry* (5th ed.), Lewis Publishers, Chelsea, MI, 1991.
70. Hooda, P.S., McNulty, D., and Alloway, B.J. Plant availability of heavy metals in soils previously amended with heavy application of sewage sludge. *J. Sci. Food Agric.* 1997, 73(4): 446–454.
71. Chang, A.C., Page, A.L., Foster, K.W., and Jones, T.E. A comparision of Cd and Zn accumulation by four cultivates of barley grown in sludge-amended soil. *J. Environ. Qual.* 1982, 11: 409–412.
72. McGrath, S.P. and Lane, P.W. An explanation for apparent losses in metals in long-term experiment with sewage sludge. *Environ. Pollut.* 1989, 60: 235–256.
73. Berti, W.R. and Jacobs, L.W. Chemistry and phytotoxicity of soil trace elements from repeated sewage sludge application. *J. Environ. Qual.* 1996, 25: 1025–1032.
74. McGrath, S.P., Chaudri, A.M., and Giller, K.E. Long-term effects of metals in sewage sludge on soils, microorganisms and plants. *J. Indust. Microbiol.* 1995, 14: 94–104.
75. Chaney, L.R., Bruins, R.J.F., Baker, D.E., Korcak, R.F., Smith, J.E., and Cole, D. Transfer of sludge — applied trace elements to the food chain. In *Land Application of Sewage Sludge. Food Chain Implications*. Page, A.L., Logan, T.J., and Ryan, J.A. (Eds.), Lewis Publishers, Inc. Chelsea, MI, 1987, 66–99.
76. Kloke, A., Sauerbeck, D.R., and Vetter, H. Changing metal cycles and human health, in Nriagu, J.O. (Ed.), *The Contamination of Plants and Soils with Heavy Metals and Transport of Metals in Terrestrial Food Chains*, 1984, 113–141, Springer-Verlag, Germany.
77. Page, A.L., Logan, T.J., and Ryan, J.A., *Land Application of Sludge-Food Chain Implication*, Lewis Publishers, Chelsea, MI, 1987.
78. Hapke, H.J. Metal accumulation in the food chain and load of feed and food. In *Metals and Their Compounds in the Environment*, Merian, E. (Ed.), VCH, New York, 1991, 469–489.
79. Davies, R.D. and Carlton Smith, C. Crops as indicators of the significance of contamination of soil by heavy metals technical report, TR.140, Water Research Center, Stevenage, U.K., 1980, 44.
80. Nriagu, J.O. A silent epidemic of environmental metal poisoning? *Environ. Pollut.* 1988, 50: 139–161.
81. Nriagu, J.O. Global metal pollution: poisoning the biosphere? *Environment* 1990, 32(7): 7–33.
82. Kabata–Pendias, A. and Pendias, H. *Trace Elements in Soils and Plants*. CRC Press, Boca Raton, FL, 1984.
83. Alloway, B.J. In *Heavy Metals in Soils*, Alloway, B.J. (Ed.), Blackie and Sons, Glasgow, U.K., chap. 3, 1990.
84. Singh, B.R. and Narwal, R.P. Plant availability of heavy metals in a sludge-treated soil. 1. Effect of sewage sludge and soil pH and the yield and chemical composition of rape. J. *Environ. Qual.*, 1983, 12(3): 358–365.
85. Singh, B.R. and Narwal, R.P. Plant availability of heavy metals in a sludge-treated soil. 2. Metal extractability compared with plant metal uptake. *J. Environ. Qual.* 1984, 13: 344–349.
86. McBride, M.B., Richards, B.K., Steenhuis, T.S., Russo, J.J., and Suave, S. Mobility and solubility of toxic metals an nutrients in soil — 15 years after sludge application. *Soil. Sci.* 1997, 162: 487–500.
87. Krebs, R., Gupta, S.K., and Furrer, G. Solubility and plant uptake of metals with and without liming of sludge-amended soil. *J. Environ. Qual.* 1998, 27(1): 18–23.
88. Hsiau, P.C. and Lo, S.L. Effects of lime treatment on fractionation and extractabilities of heavy metals in sewage sludge. *J. Environ. Sci. Health A.* 1997, 32(9–10): 2521–2536.
89. Vlamis, J., Williams, D.E., Corey, J.E., Page, A.L., and Ganje, J.J. Zinc and cadmium uptake by barley in field plots fertilized 7 years with urban and suburban sludge. *Soil Sci.* 1985, 139: 81–87.
90. Hooda, P.S. and Alloway, B.J. The effect of liming on heavy metal concentrations in wheat, carrots and spinach grown on previously sludge-applied soils. *J. Agri. Sci.* 1996, 127(3): 289–294.
91. Hinesly, T.D., Alexander, D.E., Redborg, K.E., and Zielar, E.L. Differential accumulation of Cd and Zn by corn hybrids grown on soil amended with sewage sludge. *Agron. J.* 1982, 74: 469–474.
92. Korcak, R.F. and Fanning, D.S. Availability of applied heavy metals as a function of type of soil materials and metal source. *Soil Sci.* 1985, 140: 23–34.
93. Chaney, R.L. and Ryan, J.A. Heavy metal and toxic organic pollutants in MSW — composts. Research results on phytoavailability, bioavailability, fate, etc. In *Science and Engineering of Composting: Design, Environmental, Microbiological and Utilization Aspects.* Hointive, H.A.J. and Keeners, H.M. (Eds.), Worthington, OH, Renaissance Publications, 1993, 451–506.

94. Dudka, S., Piotrowska, M., and Chlopecka, A. Effect of elevated concentration of Cd and Zn in soil on spring wheat yield and the metal contents of the plant. *Water Air Soil Pollut.* 1994, 76: 333–341.

95. Hendrickson, L.L. and Corey, R.B. Effect of equilibrium metal concentration on apparent selectivity coefficient of soil complexes. *Soil Sci.* 1981, 131: 163–171.

96. Miner, G.S., Gutierrez, R., and King, L. Soil factors affecting plant concentrations of Cd, Cu and Zn on sludge amended soils. *J. Environ. Qual.* 1997, 26(4): 989–994.

97. Oehme, F.W. *Toxicity of Heavy Metals in the Environment.* Marcel Decker Inc., New York, 1978, 199–260.

98. Page, A.L., Bingham, F.T., and Chang, A.C. Cadmium in effect of heavy metal pollution on plants, in *Effects of Trace Elements on Plant Functions*, vol. 1, Lepp, N.W. (Ed.), Applied Science Publishers, Englewood, NJ, 1982, 77–110.

99. Perez–Espinosa, A., Moreno, C.J., and Moral, R. Effect of sewage sludge and cobalt treatment on tomato fruit yield, weight and quality. *J. Plant Nutr.* 1999, 22(2): 379–385.

100. Tepper, L.B. Beryllium, In *Metals in the Environment*, Waldron, H.A. (Ed.), Academic Press, London, 1980.

101. Wong, C.W.J., Xiang, L., Gu, X.Y., and Zhou, L.X. Bioleaching of heavy metals from anaerobically digested sewage sludge using FeS2 as an energy source. *Chemosphere* 2004, 55(1): 101–107.

102. Lund, L.J., Betty, E.E., Page, A.L., and Elliot, R.A. Occurrence of naturally high Cd levels in soils and its accumulation by vegetables. *J. Environ. Qual.* 1981, 10: 551–556.

103. Chang, A.C., Hinsley, J.D., Bates, T.E., Dones, H.E., Dowdy, R.H., and Ryan, J.A. Effect of long term sludge application on accumulation of trace elements by crops. In *Land Application of Sludge: Food Chain Implications*, Page, A.L., Logan, T.J., and Ryan, J.A. (Eds.), Lewis Publishers, Chelsea, MI, 1987, 53–66.

104. Dudka, S., Piotrowska, M., and Terelak, H. Transfer of Cd, Zn and Pb from industrially contaminated soil to crop plants: a field study. *Environ. Pollut.* 1997, 94(2): 181–188.

105. Kubota, J., Welch, R.M., and Van Campen, D.R. Partitioning of Cd, Cu, Pb and Zn among above-ground parts of seed and grain crops grown in selected locations in the U.S.A. *Environ. Geochem. Health* 1992, 14(3): 91–100.

106. Wolnik, K.A., Fricke, F.L., Capar, S.G., Brandel, G.L., Mayer, M.W,. and Satzger, R.D. Bonnin elements in major raw agricultural crops in the United States. 1. Cadmium and lead in lettuce, peanuts, potatoes, soybeans, sweet corn, and wheat. *J. Agric. Food. Chem.* 1983, 31: 1240–1244.

107. Wolnik, K.A. et al. Elements in major raw agricultural crops in the United States. 2. Other elements in lettuce, peanuts, potatoes, soybeans, sweet corn, and wheat. *J. Agric. Food Chem.* 1983, 31: 1244–1249.

108. Wolnik, K.A. et al. Elements in major raw agricultural crops in the United States. 3. Cadmium, lead, and eleven other elements in carrots, field corn, onions, rice, spinach, and tomatoes. *J. Agric. Food Chem.* 1985, 33: 807–811.

109. Hirsch, M.P. Availability of sludge borne silver to agricultural crops. *Environ. Toxicol. Chem.* 1998, 17(4): 610–616.

110. Carlton Smith, E.H. and Davies, R.D. Comparative uptake of heavy metals by crops grown on sludge treated soil. In *Heavy Metals in the Environment.* Proc. Int. Conf. Vol. I. Herdelberg, 1983 CEP consultants, Ltd, Edinburgh, U.K., 1983: 393–396.

111. Council for Agricultural Science and Technology (CAST) effects of sewage sludge and on the cadmium and zinc content of crops. Ames, IA, 1980.

112. Bingham, F.T., Page, A.L., Mahler, R.J., and Ganje, T.J. Growth and Cd accumulation of plants grown on a soil treated with a Cd-enriched sewage sludge. *J. Environ. Qual.* 1975, 4: 207–211.

113. Brandel, G.L. Jelenik, C.F., and Cornelicusin, P. In *Proc. 2nd Natl. Conf. Municipal Sludge Range. Inf. Translocations.* Inc. Rukeville 1975, 214–217.

114. Sally, L.B., Chaney, R.L., Cheryl Lloyd, A., Scott Angel, J., and Rayan, J.A. Relative uptake of cadmium by garden vegetables and fruits grown on long term biosolid amended soils. *Environ. Sci Technol.* 1996, 30: 3508–3511.

115. Chaney, R., Hornick, L., and Simon, P.W. Heavy metal relationships during land utilization of sewage sludge in the Northeast, in: *Land as a Waste Management Alternative*, Loehr, R.C. (Ed.), Ann Arbor Science, Ann Arbor, MI, 1977: 283–314.

116. Le Riche, H.H. Metal contamination of soil in the Woburn Market Garden experiment resulting from the application of sewage sludge. *J. Agric. Sci.* 1968, 71: 205–208.
117. Cole, J.L., McCracken, I.D., Foster, N.G., and Aitken, M.N. Using Collembola to assess the risk of applying metal rich sewage sludge to agricultural land in Western Scotland. *Agri. Ecos. Environ.* 2001, 83(1–2): 177–189.
118. Dowdy, R.H. and Larson, W.E. The availability of sludge-borne metals to various vegetable crops. *J. Environ. Qual.* 1975, 4: 278–282.
119. Giordano, F.M., Mertvedt, J.J., and Mays, D.A. Effect of municipal waste on crop yields and uptake of heavy metals. *J. Environ. Qual.* 1975, 4: 394–399.
120. Giordino, P.M. and Mays, D.A. Yield and heavy metal content of several vegetables species grown in soil amended with sewage sludge. In *Biological Implications of Heavy Metals in Environment.* ERDA Rep. Conf 750929, Simp Ser., 1976, 42: 417–425.
121. Dowdy, R.H., Larsen, W.E., Titred, J.M., and Latterell, J.J. Growth and metal uptake of snap beans grown of sewage sludge amended soils — a 4-year study. *J. Environ. Qual.* 1978, 7: 252–257.
122. Dowdy, R.H., Bray, B.J., Goodrich, R.N., Marten, G.C., Pamp, D.E., and Larsen, W.E. Performance of goats and lambs fed corn silage produced on sludge-amended soils. *J. Environ. Qual.* 1983, 12(4): 467–472.
123. Dowdy, R.H., Bray, B.J., and Goodrich, R.N. Trace metals and mineral composition of milk and blood from goats fed silage produced on sludge-amended soil. *J. Environ. Qual.* 1983, 12: 473–478.
124. Bray, B.J., Dowdy, R.H., Goodrich, R., and Pamp, D.E. Trace metal accumulated in tissue, of goats fed silage produced on sewage sludge-amended soil. *J. Environ. Qual.* 1985, 14: 115–118.
125. Hill, J., Stark, B.A., and Wilkinson, J.M. Accumulation of potentially toxic elements by sheep given diets containing soil and sewage sludge 1. Effect of type of soil and levels of sewage sludge in the diet. *Anim. Sci.* 1998, 67: 73–86.
126. See, D.J. Risk and other toxicological studies with swine fed corn grown on municipal sewage sludge-amended soil. *J. Anim. Sci.* 1982, 55(3): 613–619.
127. Varma, M.M. and Wade Talbot, W. Organic pollutants in municipal sludge — health risk. *J. Environ. Syst.* 1986–1987, 16(4): 295–308.
128. Hegstrom, L.J. and West, S.D. Heavy metal accumulation in small mammals following sewage sludge application to forest. *J. Environ. Qual.* 1989, 18: 345–349.
129. David, T.S. et al. Uptake of polychlorobipheny's present in trace amounts from dried sewage sludge. *Bull. Env. Cont. Toxicol.* 1981, 27: 689–694.
130. Wild, S.R. et al. The influence of sewage sludge application to agricultural land on human exposure to polychlorinated dibenzo-P-dioxins, furnace and metals. *Environ. Pollut.* 1994, 83: 357–369.
131. Mchachlan, M.S. et al., A study of the influence of sewage sludge fertilization on the concentration of PCDD/F and PCB in soil and milk. *Environ. Pollut.* 1994, 85: 337–343.
132. Chaney, R.L., Sterrett, S.B., Morella, M.C., and Lloyd, C.A. Effect of sludge quality and rate, soil pH, and time on heavy metal residues in leafy vegetables. University of Wisconsin — Madison Conf. Appl. Res. Practical Municipal Industrial Waste Proc. University of Wisconsin, Madison, 1982, 6: 444–458.
133. USEPA. Technical support documented for land application of sewage sludge, vol. 1, report no. EPA/822/R-93-00/a, 1992.
134. CEC. Council directive (86/278/EEC) on the protection of the environment and in particular of the soil, when sewage sludge is used in agriculture. *Off. J. Eur. Common.* no. L181, annex 1986, 1A: 6–12.
135. Anderson, A. Trace elements in agricultural soils, fluxes, balances and background values. Swedish EPA. Rep 4077 Solna, Sweden, 1992.
136. Driel, V.W. and Smilde, K.W. Micronutrients and heavy metals in Dutch agriculture. *Fertilizer Res.* 1990, 25: 115–126.
137. National Academy of Sciences (USA). Risk assessment and management. Framework for decision making. Washington, D.C., 1983.
138. USEPA guidelines for carcinogen assessment. Guidelines for estimating exposure. Guidelines for mutagenicity risk assessment. Guidelines for health assessment of suspect development toxicants and guidelines for health assessment of chemical mixtures FR. 1986, 51: 185.
139. USEPA. A guide to the biosolids risk assessments for the EPA part 503 rule; EPA/833211, B-93-005; USEPA, Office of Wastewater Management U.S. Government Printing Office; Washington, DC, 1995.

140. McBridge, M.B. Growing food crops on sludge amended soils: problems with the U.S., Environmental Protection Agency. Method of estimating toxic metal transfer. *Environ. Toxicol. Chem.* 1998, 17(11): 2274–2281.

141. USEPA. National sewage sludge survey. Availability of information and data and anticipated impacts on proposal regulations. *Fed. Regul.* 1990, 55: 218.

142. Purves, D. and Mackenzie, E.J. Effect of application of municipal compost on uptake of Cu, Zn and B by garden vegetables, *Plant Soil* 1973, 39: 361–371.

143. McGrath, S.P., Zha, F.J., Danhem, S.J., Crosland, A.R., and Coleman, R. Long term change in the extractability and bioavailability of Zn and Cd after sludge application. *Environ. Qual.* 2000, 29: 875–883.

144. Brain, K., Richards, T., Steenhuis, S., Peverly, J.H., and Murray McBride, B. Metal mobility at an old heavily loaded sludge application site. *Environ. Pollut.* 1998, 99: 365–377.

145. USAEC, U.S. Atomic Energy Commission, The potential radiological implications of nuclear facilities in a large region in the USA in the year 2000 (WASH-1209), Washington, D.C., 1973.

146. McBridge, M.B, Brain, K., Richards, T., Steenhuis, S., and Spiecs, G. Molybdenum uptake by forage crops grown on sewage sludge amended soils in the field and greenhouse. *J. Environ. Qual.* 2000, 29: 848–854.

147. Caroli, S., *Element Speciation in Bioinorganic Chemistry.*, John Wiley & Sons, New York, 1996.

148. Quevanvillee, Ph., Ure, A., Mantau, H., and Griepink, B. Improvements of analytical measurements within the BCR program: case of soil and sediment speciation analyses. *Int. J. Environ. Anal. Chem.* 1993, 51: 129–134.

149. Olaniya, M.S., Sui, M.S., Bhinde, A.R., and Swarnakar, S.N. Heavy metal pollution of agricultural soil and vegetation due to the application of municipal solid waste — a case study. *Ind. Environ. Health* 1998, 40(2): 160–168.

150. Olaniya, M.S. Studies on heavy metal content in city refuse compost and sewage sludge and their impact on environment. Ph.D. thesis, Osmania University, 1990.

151. Srikanth, R., Shravan, K.C.H., and Khanam, A. Heavy metal content in forage grass grown in urban sewage sludge. *Int. J. Env. Health* 1992, 34: 103–107.

152. Srikanth, R. and Raja Papi Reddy, S. Lead, cadmium and chromium levels in vegetables grown in urban sewage sludge — Hyderabad, India. *Food Chem.* 1991, 40: 229–234.

153. Srikanth, R., Khanam, A., and Rao, V. Cadmium levels in the urine of male sewage sludge faunes of Hyderabad, India. *J. Toxicol. Environ. Health* 1994, 43: 1–6.

154. Ayyadurai, K.M., Eswarmurthy, N., Jebarathinam, S., Swaminathan, C.S., and Krishnaswamy, V. Studies on the concentration of lead and cadmium in milk of cow and buffalo. *Ind. J. Environ. Health* 1998, 40: 367–371.

155. Bansal, O.P. Heavy metal pollution of soil and plants due to sewage irrigation. *Ind. J. Environ. Health* 1998, 40: 51–57.

156. Chandra Sekhar, K., Rajni Supriya, K., Kamala, C.T., Chary, N.S., Nageswara Rao, T., and Anjaneyulu, Y. Speciation, accumulation of heavy metals in vegetation grown on sludge amended soils and their transfer to human food chain — a case study. *Toxicol. Environ. Chem.* 2001, 82: 33–43.

157. Chandra Sekhar, K., Sastry, G.S.R., Rajni Supriya, K., Kamala, C.T., and Chary, N.S. Speciation studies of heavy metals in sludge-amended soils for measuring their phytoavailability. *Proc. Int. Congr. Chem. Environ.*, Indore, 2002, 55.

158. Punshon, T. Tree crops, in *Metals in the Environment — Analysis by Biodiversity*, Marcel Dekker, New York, 2001, 321–351.

159. Bhati, M. and Singh, G. Growth and mineral accumulation in *Eucalyptus camaldulensis* seedlings irrigated with mixed industrial effluents. *Bioresour. Technol.* 2003, 88: 221–228.

160. Aronsson, P. and Perttu, K. Willow vegetation filters for municipal wastewaters and sludges. A biological purification system. *Proc. Willow Vegetations Filters*, Uppsala, Sweden, 1994.

161. Sennerby–Forsse, L. The Swedish Energy Forestry Program Proceedings of Willow Vegetation Filters for Municipal Waste Waters and Sludges: a Biological Purification System. Uppsala, Sweden, 1994, 19–22.

162. Obarska–Pempkowiak, H. Removal of nitrogen and phosphorus from municipal wastewater by willow — a laboratory approach. *Proc. Willow Vegetation Filters Municipal Wastewaters Sludges: Biological Purification Syst.*, Uppsala, Sweden, 1994, 83–90.

163. Goransson, A. and Philippot, S. The use of fast growing trees as "metal collectors." *Proc. Willow Vegetation Filters Municipal Wastewaters Sludges: Biological Purification Syst.* Sweden, 1994, 129–132.

164. Landberg, T. and Greger, M. Can heavy metal tolerant clones of *Salix* be used as vegetation filters on heavy metal contaminated land? *Proc. Willow Vegetation Filters Municipal Wastewaters Sludges: Biological Purification Syst.* Sweden, 1994, 133–144.

165. Morsing, M. The use of sludge in forestry and agriculture. A compararison of the legislation in different countries. *Forskningsserien*, no 5. Danish Forest and Landscape Research Institute, Lyngby, Denmark, 1994, 661, introduction and summary.

166. Horswell, J., Speir, T.W., and Van Schaik, A.P. Bioindicators to assess impact of heavy metals in land applied sewage sludge. *Soil Biol. Biochem.* 2003, 35(11): 1501–1505.

167. Merrington, G., Oliver, I., Smernik, R.J., and McLaughlin, M.J. The influence of sewage sludge properties on sludge borne metal availability. *Adv. Environ. Res.* 2003, 8(1): 21–36.

32 Trophic Transfer of Trace Elements and Associated Human Health Effects

Amit Love and C.R. Babu

CONTENTS

32.1 INTRODUCTION

Trace elements (metals and metalloids) are ubiquitous in their presence in water, air, and soil. These elements are introduced into the environment by natural and anthropogenic processes [1]. It has been shown that anthropogenic processes have become the most important factor in the global biogeochemical cycling of some trace elements [2]. Anthropogenic activities such as mining, metal smelting, incineration and disposal of wastes, and combustion of fossil fuels have greatly influenced the mobilization of these elements in the biosphere [3]. Elevated concentrations of different elements in environmental resources such as water, air, and soil are of great concern because these elements are infinitely persistent and immutable in nature [4]. Persistence of trace elements lends them the potential to bioaccumulate in living organisms — directly, through exposure to contaminated water, air, and soil, and indirectly, by trophic transfer through the food chain.

Society is concerned for the increased accumulation of these elements in biota because of the imminent probability of humans being exposed to these elements through trophic transfer (exposure via food chains) [5]. Trophic transfer of trace elements to humans is an exposure pathway generally characterized by chronic, sublethal exposure to trace elements, which can result in the manifestations of various disease states. It is acknowledged that growth, health, fertility, and well-being of humans is influenced by the amount and proportion of various trace elements from food and environment to which they are exposed [6]. Trace elements such as Cd, Cr, and As have adverse health effects and are known carcinogens and mutagens in humans [7–9]; additionally, some trace elements are also involved in generation of reactive oxygen species [10] that have been implicated in many human diseases. This chapter elucidates and outlines current understanding of the process of bioaccumulation and trophic transfer of trace elements in biota, and the associated human health effects.

32.2 BIOACCUMULATION AND TROPHIC TRANSFER OF TRACE ELEMENTS

Living organisms bioaccumulate trace elements due to direct exposure to contaminated water, air, and soil or indirectly through contaminated food. The relative contribution of each of these sources to the total body burden of the organism varies depending upon the specific environmental conditions, the organism, and the interaction between them. It is possible that, in a particular environmental setting, direct exposure due to contaminated water, air, and soil individually or collectively may contribute more to the trace element burden of a living organism than exposure through food or vice versa. For example, the potential human health risk associated with inorganic arsenic exposure is predominantly due to intake of As-contaminated water; however, in the case of methyl mercury exposure, health risk is associated with the consumption of contaminated food like fish and shell fish [11,12].

Bioaccumulation of trace elements in receptor biota and humans through food is essentially related to trophic transfer of trace elements in terrestrial and aquatic ecosystems. Increasingly, sophisticated analytical methods such as stable isotope analysis are being used to resolve food chain structure and define trophic position of biota in the food web [13,14]. Trophic transfer of trace elements within the food web has been demonstrated by relating the metal levels in the dietary components with those assimilated by animals [15]; indices such as bioaccumulation, bioconcentration, and biomagnification factors have been routinely used to estimate trophic transfer in the food chains [16].

Exposure resulting from terrestrial and aquatic food chains to humans and biota must be analyzed separately due to the fundamental differences between them. The number of trophic levels in the terrestrial food chain is greatly reduced compared to the aquatic food chain. Moreover, the terrestrial herbivores demonstrate low efficiency of converting plant biomass into protein due to a large percentage of indigestible structural material containing high levels of cellulose and lignin in terrestrial plants. However, the low energy efficiency causes higher consumption rates of food stuffs in herbivores, which can effectively intensify metal uptake in food webs within contaminated areas [17].

This chapter focuses on trophic transfer of trace elements as an exposure pathway for biota and humans. Trophic transfer of trace elements chronically exposes humans to sublethal concentration of elements, which can build overtime to reach toxic levels. The most important trace elements in terms of their trophic transfer potential through the food chain are Cd, Hg, Pb, As, and Se [8]. Trophic transfer of trace elements and their bioaccumulation in biota have associated ecological and human health effects.

TABLE 32.1
Trace Element Content in a Typical International Diet

Meal	As (µg)	Cd (µg)	Hg (µg)	Pb (µg)
Breakfast				
Milk (200 ml)				0.7
Fruits (200 g)		2		1.6
Cereals (100g)		0.2		3.2
Lunch				
Spinach (100 g)		6		3.1
Broccoli (50 g)		0.5		1.6
Tomatoes (50 g)		0.7		1.6
Egg (80 g)		0.8		1.6
Fish (100 g)	100	1.5	0.6	
Water (300 ml)	3	0.03		0.6
Dinner				
Meat (200 g)		2		6.3
Carrots (50 g)		0.9		
Cauliflower (50 g)		0.5		
Rice (80 g)		0.4		1.3
Potato (100 g)		2		1.3
Wine (125 ml)	2.5			0.6
Total	105.5	17.53	0.6	23.5

Source: Adapted from Rojas, E., Herrera, L.A. et al., *Mut. Res.*, 443, 157, 1999.

Humans are necessarily exposed to trace elements through the consumption of contaminated food items in the absence of any specific environmental and occupational exposure. An average human diet includes an assortment of products derived from plants and animals that would include food items like fruits, vegetables, grains and cereals, milk, eggs, meat, and fish (Table 32.1).

As is evident, these items are sourced from different plant and animal species. Predominantly, all the plant products are derived form terrestrial agricultural food chains; while animal products may come from terrestrial or aquatic animal species. To estimate human exposure due to trophic transfer of trace elements and how it relates to human health reasonably, it is imperative to understand the processes associated with the trace element content in different tissues of plant and animal species.

32.2.1 Terrestrial Ecosystems

Trace elements enter the terrestrial food chains predominantly through plants that mobilize these elements from contaminated soils. It is necessary to differentiate the various steps through which a trace element passes as it progresses from soil through the food chain to the human receptor. As trace elements are transferred from the soil to the plant and then further to primary and secondary consumers, different processes influence the amount of trace element available at each trophic level and what will eventually reach the human endpoint [5]. Exposure of different animal species and humans to trace elements due to consumption of plant products is influenced not only by the trace element content in the consumed plant part, but also by the plant-mediated changes in trace element speciation that may alter the rate and extent of gastrointestinal absorption after ingestion of plant tissue by animals [18]. Furthermore, trophic transfer and bioaccumulation of trace elements by animals is influenced by different ecological host factors [19]. In terrestrial ecosystems, trophic transfer of trace elements has two distinct components keeping in view humans as the ultimate receptor: bioaccumulation of trace elements by the plants that mobilize these elements into the food chain and trophic transfer and bioaccumulation in animals.

32.2.1.1 Bioaccumulation of Trace Elements in Plants

32.2.1.1.1 Bioavailability and Uptake of Trace Elements

Plants represent an important pathway for the movement of potentially toxic trace elements from contaminated soils to humans [18]. Elevated levels of trace elements in soil may lead to their uptake by plants, but, generally, a direct relationship between the total trace element content of soil and trace element content in different tissues of plants has not been conclusively established [20,21]. The trace elements in soils are associated with different chemical fractions of the soil, which affect their solubility and have a direct bearing on their mobility, bioavailability, and plant tissue levels [22]. The upper limit of trace element availability to plants depends upon the composition of the soil solution because root uptake requires the presence of soluble chemical species [23,24].

Plants' uptake of trace elements from the soil is influenced by physicochemical characteristics of the rhizosphere, the rhizosphere processes, and the chemical form of the trace element in the soil [22,25]. Plant roots and their associated microorganisms are known to modify the composition of soil, resulting in altered chemical mobility in the soil and acquisition of trace elements by the roots [26]. Rhizosphere soil may differ considerably from bulk soil with respect to modified pH, Eh, and concentration gradients of many minerals and organic ligands. The possible consequences of such changes on speciation of trace elements could explain to some extent the species-specific differences in soil to plant transfer factors [27].

The rhizosphere processes influence the rates of solubilization and the equilibrium concentration of individual ions associated with the soil solution, subsequently affecting the availability of essential and contaminant ions for absorption by the plants [23]. One of the hypotheses to explain the large uptake of certain trace metals by hyperaccumulators is that they solubilize the target metal in their rhizosphere [28–31].

Soil microorganisms affect trace element mobility and availability to plants by production of metal chelators, siderophores, alteration of soil pH, and solubilization of metal phosphates; they also influence root parameters such as root morphology, growth, and root exudation, leading to increased uptake trace metals [32]. Abou-Shanab et al. [32] showed that rhizobacteria play an important role in increasing the Ni availability in soil by releasing Ni from the nonlabile phase in the soil, thus enhancing the availability to the plants. In contrast, vesicular–arbuscular fungi in the roots of the plants reduced the uptake of trace elements, leading to increased plant biomass due to decreased uptake of heavy metals [33–35]. Furthermore, Leyval and Joner [36] also concluded that ecto- and AM-mycorrhizae tended to reduce metal concentration in the shoots of nonhyperaccumulator plants. Certain plant growth-promoting bacteria also reduce the metal stress on plants. Burd et al. [37] found that *Kluyvera ascorbata* SUD 165, a plant growth-promoting bacteria, decreased toxicity of Ni to seedlings by lowering the level of stress ethylene induced by Ni in canola, rather than by reducing uptake by seedling.

Soils are often contaminated with more than one trace element. These trace elements interact in various ways to increase or decrease the uptake and transfer of each other from the soil to plants. Podar et al. [38] showed that the relative uptake of cadmium in the presence of zinc is reduced by the plants, suggesting that human health risk from consuming parts of plants grown on Cd-contaminated soils is relatively low when Zn is also present. In addition to the variation in the bioavailability of different elements in the soils, the uptake and translocation characteristics of metals vary between different plant species, and even between population and cultivars [39]. Baker [40] found that plant species differed considerably in their metal uptake characteristics and that, for any given species, uptake may vary between different metals. For example, the genus *Salix* is known for its high uptake of Cd [41]. A study of native plants in the forest ecosystem of Norway indicated that interspecific differences in the uptake were more evident for nonessential elements like Cd and Pb, compared to essential ones such as Zn and Cu [42].

32.2.1.1.2 Partitioning of Trace Elements in Different Plant Tissues

Plants partition the trace elements into different tissues based upon the physiological mechanisms of the plant species and also the trace metal elements under consideration. Shoot metal content is determined by root uptake and sequestration within root vacuoles, translocation in the xylem and phloem, and dilution within the shoot through growth [43]. Nan et al. [44] studied the transfer of Cd and Zn from calcareous soils near a nonferrous mining and smelting bases to spring wheat (*Triticum aestivum*) and corn (*Zea mays*) tissues under field conditions. They found that Zn and Cd were present in highest concentrations in the roots; the lowest concentration was found in the seeds. This pattern of tissue distribution of trace elements suggests the presence of a physiological mechanism that prevents the excessive movements of Zn and Cd in the edible parts of the plants.

Torres and Johnson [15] found that roots of *Scirpus robustus* had elevated concentration of Pb and Ni compared to seeds, which was consistent with pattern of bioaccumulation in plants. Lubben and Sauerbeck [45] also found that seeds had lower trace element content compared to the other plant parts. Generally, nonessential trace elements become immobilized in the roots and undergo limited translocation to the above-ground structures such as leaves and seeds. Root-to-shoot transfer may not be constant for a given species; soil factors and probably also the age of the plant play an equally significant role [27].

Broadley et al. [43] found that plant species of families Chenopodiaceae, Polygonaceae, and Amaranthaceae belonging to order Caryophyllales have the general ability to accumulate trace elements in their shoots. On the other hand, crop plants such as cereals and alliums have low concentrations of shoot metal. It seems that certain plant groups have evolved their own spectrum of ecological and physiological adaptations. For instance, grasses have developed effective regulatory mechanisms at root level. In grass shoots, the element content, even on ore soils, tends to be much lower than broad-leaved herbs.

On the other hand, some plants termed as hyperaccumulators have extreme adaptation to accumulate high concentration of trace elements in their shoots. Brooks et al. [46] introduced the term "hyperaccumulator" to describe plants capable of accumulating more than 1000 µg of Ni/g on a dry leaf weight basis in their natural habitats. The same criteria were also applied to elements such as Co, Cu, and Pb; whereas for Cd and Zn, the respective thresholds are 100 and 10,000 µg/g dry leaves [47,48]. Metal hyperaccumulation is a rare phenomenon in terrestrial higher plants. To date about 400 plant species have been identified as metal hyperaccumulators, representing <0.2% of all angiosperms [47,48].

Generally, non-hyperaccumulator plants have higher metal concentration in roots than in shoots [49,50]. The differential expression of transport or metal chelating proteins may account for phylogenetic differences in shoot metal concentrations between angiosperm clades and also for association between subsets of metals independent of phylogeny [43]. This kind of information on differentiation in shoot element content based on phylogenetic differentiation may be useful in estimating transfer coefficients when experimental data are not available and in improving predictions relative to soil-to-plant metal transfer in general [43]. On a relative scale, species-specific differences exist with respect to the concentration of metals in the roots and shoots of different plant species. The adaptive strategy adopted by the plant species with respect to the distribution of trace elements in different tissues has a profound impact on the trophic transfer of trace elements in the terrestrial food chain.

32.2.1.1.3 Chemical Modification of Trace Elements in Plants

The chemical modification of inorganic ions by plants can significantly influence their bioavailability to consuming animals. The nature and extent of this influence depends upon the element, and plant transport and deposition processes. Therefore, prediction of metal transfer in food chains must be partially based on an understanding of the processes governing chemical form and speciation in specific plant systems [18].

32.2.1.2 Bioaccumulation of Trace Elements in Animals and Trophic Transfer

Increasingly, animals are used as sentinels of human exposure and as bioindicators of environmental contamination [51,52]. Different groups of animals differ with respect to the magnitude and type of exposure to trace elements. Several workers have advocated the use of free ranging animals such as mammals and birds for various reasons [53–55]. Small mammals depend exclusively on the quality of food, water, and air in a particular habitat and locality; they are physiologically similar enough to show comparable biological and physiological effects as humans [56,57]. Depending upon the group considered, birds are conspicuous, may occur over a wide geographical range, and show adverse effects at low concentrations [58]. Moreover, these animals can also provide an indication of the threat to other animal species and ecosystem health of the area. In this discussion, the focus is on small mammals and birds to understand the factors that influence bioaccumulation, trophic transfer of trace elements, and tissue distribution of the assimilated trace elements. Bioaccumulation patterns of trace elements in birds and small mammals can be affected due to exposure differences, individual differences, and species differences [19].

32.2.1.2.1 Exposure Differences

Differences in the exposure of animals to trace elements arise due to differences in habitat; home range; temporal patterns of a species; behavioral pattern, particularly foraging behavior; food chain effects; and spatial distribution of animal populations. Because habitat characteristics affect the home range of animal species, it becomes an important factor in determining the amount of exposure to trace elements. Attuquayefio et al. [59] found that small mammals inhabiting unproductive ecosystems had significantly larger home ranges than those inhabiting productive and species-rich habitats. It is well known that heavy metal-contaminated habitats have low primary productivity, which could probably result in animals foraging in adjacent areas where contamination is low and the abundance of food is greater. The consequential dilution of trace metals intake can lead to lack of correlation between dietary metal intake and the body/tissue concentration.

Marinussen and Vanderzee [60] found that a reduction in the cadmium concentration of tissues occurs due to increase in the home range of the animals. Larison et al. [61] found that foraging behavior of ptarmigans inhabiting the Colorado Rocky Mountains significantly influenced Cd exposure rates of the birds. They found that ptarmigans fed predominantly on willow (*Salix* spp.), which had elevated levels of Cd during winters and spring. However, during summers when other food items such as *Trifolium* spp., *Bistorta bistorta*, and *Geum rossii* become available, ptarmigans switched over to less contaminated food items and, in the process, consumed far less Cd. This expansion of diet of ptarmigans reduced the overall Cd exposure of the animal during summer months.

Exposure to trace elements such as Cd and Pb in free ranging wildlife like small mammals and birds depends on the dietary preferences of the animal species. Metcheva et al. [62] estimated the metal content in tissues of six species of small mammals from different habitats and found that the heavy metal loading in small mammals is not only due to specific accumulation features of physiological origins or preferable bioaccumulation of elements in respective organs and tissues, but also due to ecological and food characteristics. Hunter and Johnson [63] analyzed the total body burden and tissue distribution of heavy metals in small mammals. They found that *Apodemus sylvaticus*, which fed on fruits and seeds of ground cover vegetation and canopy species, had the lowest metal burden due to minimal translocation of metals to the reproductive structures of plants parts. On the other hand, *Sorex araneus* fed predominantly on detritivorous soil invertebrates and had the highest concentration of heavy metals.

Cooke and Johnson [64] also found high Cd concentration in the liver of *S. araneus*, most which was probably related to the food chain position of shrews, which accumulate higher Cd levels from their insectivorous diets. Mertens et al. [65] found that elevated concentration of Cd and Zn were present in the tissues of small mammals inhabiting disposal facilities for dredged materials; elevated tissue concentration of trace elements was related to the amount of Cd and Zn present in dietary items of the animals. Trace element content in the tissue of shrews was the highest as compared to all other small mammals due to high bioavailability of elements in the insectivorous diets. Ma [66] found that concentration of Pb in different tissues of small mammals was also very much dependent upon the diet.

Temporal changes associated with seasonal differences in the trace element content of the vegetation of an area can significantly affect the metal loading of the consuming animal because the trace element content in forage or food items can increase or decrease according to the season and plant species. Hunter et al. [67] reported seasonal variation in the concentration of Pb in vegetation growing on contaminated sites; lowest Pb concentration in vegetation was present in summer due to rapid increase in plant biomass resulting in growth dilution. These seasonal fluctuations may have implications for interpretation of concentration of lead in animal tissues because the metal content in diet reduces in summer. This indicates that the summer season may represent lower dietary exposure conditions for consumers than those experienced at other times of the year.

The diets of adult and juvenile small mammals also differ; in the case of wood mice, juveniles eat less seed and animal food compared to adults [68]. Such a marked shift in diet can lead to differential exposure of different age classes in the same species. However, Erry et al. [69] found no statistically significant difference in the stomach content of juveniles and adults with respect to arsenic content, suggesting that the degree of arsenic contamination may be similar in a wide range of animal forage. They also suggested that dietary overlap between different species and age classes is more pronounced on contaminated than uncontaminated sites, perhaps because diversity of food is limited and leads to similar dietary intake.

Another confounding factor that can obscure many ecotoxicological model predictions is the ingestion of soil by small mammals. The relative proportion of soil in the diet of small mammals may be less [70], but it can contribute immensely to the total trace element exposure. Erry et al. [69] found that soil contributed approximately 15% of total arsenic exposure to the small mammals. Shore [71] found significant correlation between liver and kidney levels of Cd in field mice, common shrews, and short-tailed voles and the levels of metals in the soils. Sharma and Shupe [72] also found significant relationship between Cd levels in rock squirrel (*Spermophilus variegates*) livers and the soil concentration; however, the same pattern was not found for As and Pb.

Shift in dietary preferences due to change in habitat type also has a profound influence on the exposure of trace elements. Hunter et al. [67] found that voles fed primarily on leaves and seeds of trees in woodlands; however, in the absence of trees, the bank voles fed opportunistically on any available plant and animal material. This leaf- and invertebrate-based diet resulted in higher exposure for the animals due to high trace metal content in metallophyte forbs [40] growing on contaminated sites. Furthermore, to obtain the energy equivalent of the fruit- and seed-rich diet, voles consumed increased amounts of leaf and animal material (soil invertebrates), leading to increased exposure to trace elements [73].

Some animal species or families are more vulnerable not only because of their foraging behavior and habitat characteristics but also because of their characteristic breeding biology. Seals that leave the water to breed on the beaches near industrial regions are exposed to higher levels of contaminants than those present near more pristine beaches. Similarly, birds that nest near urban estuaries and bays are more exposed to trace elements than those nesting in remote places [19].

32.2.1.2.2 Individual and Species Differences

Individual differences in bioaccumulation of trace elements may be related to age, gender, size, weight, nutrition, and genetic factors [19]. The studies on age-related accumulation in animals have more or less focused on non-essential trace elements because of efficient regulatory homeostatic mechanisms that exist for essential elements. Age-related bioaccumulation of trace elements in animals has largely been found in the case of Cd [61,69,74–77]. Hunter et al. [76] found significant correlation between total body weight and Cd in liver and kidney of common shrew, reflecting the effect of increasing age on accumulation of trace metals in tissues. Pedersen and Myklebust [78] suggested that age-dependent accumulation of Cd in willow ptarmigan liver and kidney reaches a threshold value in birds in the first year of life. However, in mammals like cervides, a linear age-dependent accumulation of Cd seems to exist [79]. Larison et al. [61] found that Cd in the body tissues of ptarmigan chicks was negligible compared to older birds, which accumulated substantial amount of Cd over time.

Milton et al. [80] did not find any age accumulation of Pb in the bone tissue of small mammals across all age groups from subadults to adults. They suggested that the absence of a clear trend of age accumulation of Pb in bones of small mammals was due to exposure of unweaned pups and young animals to high dietary lead (via milk) and the fact that maximum bone deposition rate of Pb occurs in young animals [81]. This led to a rapid increase in the concentration of Pb in bone tissue of young animals, but in the case of adults and subadults, reduction in the rate of skeletal development could have produced a steady state Pb concentration. Erry et al. [69] also found that concentration of As in the stomach contents and body tissues of *Apodemus sylvaticus* and *Clethrionomys glareolus* did not vary with age class.

Gender differences in the bioaccumulation potential of trace elements have been observed in small mammals and birds. These differences have been at least partly attributed to the difference between the behavior of male and female individuals of a species. Larison et al. [61] found differences in the wintering distribution of male and female ptarmigans. They found that female ptarmigans overwinter in areas heavily contaminated with Cd, resulting in greater exposure of female ptarmigans and leading to higher accumulation of Cd in females compared to males.

Erry et al. [69] observed significant variation in arsenic intake between sexes in bank voles in which stomach arsenic concentration was higher in females than in males. Arsenic residues in the liver and, to a lesser extent, in the kidney were also higher in females than in males. The gender difference between male and female bank voles could not be explained on the basis of dietary preferences of the male and female voles because male and female voles had similar diets. It was suggested that the difference could have arisen due to the larger home range of male voles, which might have led to the dilution of the metal content. The same argument could not be used to explain the absence of any difference with respect to trace element content of wood mice male and female individuals, even when males of the species have a larger home range. This shows that gender differences in trace element content are arising due to factors other than behavioral and need further investigation.

Species-specific differences with respect to retention and accumulation of nonessential trace elements in body tissues are determined by species dietary uptake as well as by physiological and behavioral characteristics [82]. Lock and Janssen [83] found that some species differences may relate to differences in internal toxicodynamics, leading to species differences even in closely related organisms. The risk of heavy metals accumulating to a toxic dose level in small mammals varies considerably between Rodentia and Insectivora from the same polluted area [77,84]. This could be due to differences in the diet and foraging behavior of the species of the two orders. Few studies have brought out the physiological basis of species-specific variation in mammalian species.

32.2.1.2.3 Tissue Distribution of Trace Elements

Distribution patterns of assimilated trace elements in tissues of animals highlight important differences between metabolically essential and non-essential elements [85]. Concentration of essential trace elements in animal tissues is effectively regulated through homeostatic mechanisms; these elements are widely distributed in different body tissues without restriction to any specific body organs or tissue. On the other hand, nonessential elements like Pb, As, and Cd have very specific target organs where they preferentially accumulate [74–76,85,86]. Even though, essential trace element content in the body is generally regulated by the homeostatic mechanisms. These mechanisms may prove to be ineffective for animals exposed to very high concentrations of these elements, such as in cases in which animals inhabit highly polluted habitats [87].

Accumulation and distribution of trace metal elements in different body tissues are influenced by many factors, which include total heavy metal intake, Ca, phosphate and vitamin status in the diet, and metal–metal interaction [88,89]. Tissue accumulation of trace elements is also influenced by the presence of other competing elements. For example, Zn and Ca intake are known to influence the extent to which ingested Cd is absorbed and retained [90]. It may be noted that low Ca levels in the body can exacerbate Cd uptake [91]. Flick et al. [89] found that increased dietary Zn intake ameliorates the potential toxicity of administered Cd, presumably by competing with the latter element for specific enzymes and metabolic centers. Such competitive inhibition, which has been established mainly in lab studies, may have considerable significance in field studies because the two metals are naturally associated in the biosphere [92].

Hunter et al. [63].found that, regardless of the degree of environmental contamination, the retention of Cu in animal species was regulated by some kind of absorption–excretion equilibrium that adapts to maintain a homeostatic control. No retention control mechanism was evident in the case of Cd in small mammals whose distribution is characterized by preferential accumulation in the liver and kidney in mammals. Generally, trace elements concentrate in liver and kidney of animals; in contrast, the trace element content in the muscle tissue remains constant [74]. Kidney has greater affinity for Cd accumulation compared to liver in young animals and during low dietary exposure, although on a total organ basis, liver usually contains a greater percentage of the total body burden [76]. Age accumulation of Cd in renal tissue is a well established phenomenon and is associated with the formation of stable Cd–metallothionein complexes in liver and kidney [63,92,93]. On chronic exposure to high doses of Cd, kidneys become saturated with Cd and subsequent accumulation of Cd continues in the liver [76]. Erry et al. [69] found that arsenic concentrates preferentially in liver and kidney of mammals, but bone tissue bioaccumulates a maximum amount of Pb followed by kidney, liver, and muscle in small mammals [57,85].

Other workers have also found the same pattern of accumulation of lead [73,94]. Differential deposition of Pb in bone is because Pb mimics Ca during metabolism [57,95,96]. Bone tissue constitutes a physiologically important compartment into which Pb is absorbed, with a slow rate of equilibrium and half-life of over 10 years in humans [97]. In the animal body, one of the major repositories of lead other than bone tissue is kidney [80,86]. Even though liver is the first organ to encounter absorbed dietary lead distributed via bloodstream, Pb concentration in liver did not exceed 15 mg/kg even when kidney values in the same individual were as high as 60 mg/kg [98].

Seasonal variation in tissue concentration of trace elements has been reported for different organisms. Wlostowski et al. [99] found that low ambient temperature (+5°C) decreases the Cd accumulation in liver and kidney of bank voles compared to when voles were exposed to Cd at 20°C. They suggested that Cd intake may be affected through changes in Cu metabolism due to low temperature.

32.2.2 Aquatic Ecosystems

Exposure of humans to trace elements through the aquatic food chain is predominantly due to consumption of fish and aquatic invertebrates such as prawn and lobsters. Fish and aquatic invertebrates occupy different trophic levels depending upon their respective feeding ecology and foraging behavior. The bioaccumulation of trace elements varies greatly between different aquatic species and also between different trace elements. Some animal species, such as zooplanktons, accumulate high levels of metals, but others, such as fish, try to regulate the concentration of metals or sequester them with metallothioneins [100]. Due to the contrasting behavior and the fact that planktonic organisms and aquatic invertebrates form the base of many aquatic food chains, bioaccumulation patterns in aquatic invertebrates (including planktonic organisms) and fishes are dealt with separately.

32.2.2.1 Bioaccumulation in Planktonic Organisms and Aquatic Invertebrates

Planktonic organisms and aquatic invertebrates form the base of many aquatic food chains and thus have a crucial role in transfer of trace elements in the aquatic food web. It has been increasingly recognized that trophic transfer of trace elements in aquatic food chains is an important source for trace element accumulation in higher trophic-level organisms [101–103]. Trace element concentration in aquatic biota at any trophic level results from a combined effect of uptake from water and the diet; the uptake process is peculiar to each metal and taxa [104].

The uptake and bioaccumulation of trace elements in aquatic organisms is controlled by numerous geochemical, ecological, and physiological factors [105]. Aquatic invertebrates are exposed to trace elements from the particulate and the dissolved phases of water [106]. Trace elements dissolved in water are directly adsorbed onto the body surfaces, in contrast to trace elements present in the particulate phase, which are accumulated by animals following ingestion and digestion of food [106]. In copepods, trace elements accumulated from the dissolved phase are often associated with the exoskeleton of the organism [107]; consequently, they are less available to higher level consumers such as fish than trace elements, which are bioaccumulated in internal tissues of the organism [108].

Bioaccumulation of trace elements in aquatic invertebrates is strongly related to assimilation efficiency, particularly for trace elements accumulated from ingested food particles [107,109]. Trophic transfer of elements from zooplankton and other aquatic invertebrates to higher trophic levels is controlled by assimilation efficiency, efflux rate, and feeding and growth rates of the organism [110]. Elements with low assimilation efficiency in aquatic biota near the base of the food chain (zooplanktons) are unlikely to be transferred up the food chain. For example, in the case of methyl mercury, Mason et al. [111] suggested that its biomagnification in marine systems is mainly due to high assimilation by marine organisms.

All aquatic invertebrates accumulate trace metals in their tissues whether or not these are essential to metabolism. The difference in the accumulation pattern and intertaxon variability between aquatic invertebrates depends, among other things, upon the physiological mechanisms present in the organism with respect to trace metal elements. The physiological mechanisms for regulation of essential elements in different aquatic invertebrates vary from (1) regulation of body element concentration; (2) accumulation without excretion; and (3) accumulation with some excretion. For non-essential metals, the physiological mechanisms are (1) accumulation without excretion; and (2) accumulation with excretion [112].

Food chain transfer of trace elements and potential bioaccumulation at higher trophic levels is controlled by the quantity of element accumulated in the prey species at the lower trophic level as well as the form of detoxified storage of the accumulated element in the prey species [110,112]. Nott and Nicolaidou [113] have shown that the bioavailability to neogastropod mollusk predators of metals in detoxified metalliferous granules in prey varies between metals and type of granules. Similarly, the physicochemical form of accumulated Cd in Oligochaete worm, *Limnodrilus hoffmiesteri*, is critical in the assimilation of Cd by the predator *Palaemonetes pugio* [114,115].

Different groups of aquatic invertebrates behave differently with respect to essential and nonessential metals. Decapods usually regulate their body concentrations of essential elements such as Zn, Cu, and Fe to approximately constant levels [116,117], but they are net accumulators of nonessential elements Cd and Pb [116]. On the other hand, amphipods and barnacles are net accumulators (to different degrees) of Zn, Cu, Fe, Cd, and Pb [117].

Decapods such as prawns, which are an important food source inhabiting metal-contaminated sites, have elevated trace element concentrations due to high uptake rates of individual organisms

[118]. Differences in trace metal contents in tissues of different prawn species have been attributed to habitat differences (estuarine, marine, etc.) and also to contaminant levels and dietary shifts due to change in habitat types [118–121]. Changes in the trace element content in body tissues of aquatic biota have also been attributed to seasonal differences. Rapid growth of aquatic invertebrates at relatively high temperature leads to low concentration of non-essential metals such as Cd and Pb due to growth dilution. In penaid prawn (*Metapenaeopsis palmensis*) found in tropical coastal waters of Hong Kong, growth dilutes the Cd body concentration [122]. In contrast, typical Cd concentration in temperate caridean species is higher due to slower growth rates [123].

Bottom-dwelling aquatic fauna includes a wide diversity of organisms, including aquatic insects, benthic invertebrates, and bottom-dwelling fish species [124]. Sediments in aquatic ecosystems act as a source and sink for pollutants. Trace element concentration in the sediments is three to five orders of magnitude higher compared to that in the overlying water column [125]. The high levels of trace elements and their relatively high availability for uptake indicate that elements associated with sediments can pose a direct risk to benthic organisms [126]. The uptake of sediment-associated trace elements is largely a function of bioavailability in combination with physiological factors such as age and sexual condition [127]. Bioaccumulation of trace elements in benthic invertebrates such as lobsters has a great significance with respect to human health because lobster muscles are consumed by humans —thus, the indirect risk of trophic transfer of trace elements from benthic invertebrates.

Bioaccumulation of sediment-associated trace elements is highly species dependent because of diversity in feeding ecology and living habitats of benthic organisms [128]. Mode of feeding can significantly influence trace element bioaccumulation from sediments [129]. For example, deposit-feeding invertebrates may accumulate contaminants to a greater extent than filter feeders. Also, benthic organisms may preferentially select higher organic carbon sediments, thus influencing the exposure to sediment-associated contaminants [130]. Trace element bioaccumulation from sediments is influenced by metal speciation, transformation, metal–metal interactions, sediment chemistry, presence of metal–binding sites, and binding to dissolved organic matter [125].

It may be noted that physicochemical parameters of the water and sediments affect the bioavailability of trace elements to aquatic organisms. In particular, pH has a strong control on metal speciation because it determines the degree of metal hydrolysis, polymerization, aggregation, precipitation, and proton competition for available metal-binding sites or ligands [131]. Water hardness may influence metal accumulation by metal speciation and form insoluble organic and inorganic complexes that precipitate and become unavailable for bioconcentration [132,133]. Water hardness may also influence metal uptake independent of effect of metal speciation by inhibition of metal absorption and reduction in membrane permeability [134]. Metals can exist as free ions and inorganic complexes; this also influences their bioavailability.

32.2.2.2 Bioaccumulation in Fish and Trophic Transfer

Fish are ideal indicators of heavy metal contamination in aquatic ecosystems because they occupy different trophic levels [135]. Fish species bioaccumulate trace elements in different tissues, predominantly from their diet [136,137]. Many fish adapt to a wide variety of food sources and often switch from one food source to another as environmental and food supply conditions change [124]. Due to this capacity of trophic adaptability, trace element concentration in water and tissue trace element content do not have a consistent relationship. For example, fish from lakes with trace element levels below detection in water can carry trace element burdens that present a human health hazard. Yet, elevated concentration in water has also been measured in lakes where fish have a relatively low burden [138].

Trophic level differences in different fish species with respect to elemental content have been reported by a number of workers [139–142], although much of the work is with reference to mercury. In general, carnivorous species have higher levels than herbivorous, omnivorous, and planktivorous species [143] and larger carnivores have higher levels than smaller carnivores. However, bottom-dwelling fish can sometimes have higher levels than carnivorous fish, particularly when they are ingesting sediments [144]. For example, Campbell [137] found that bottom-feeding redear sunfish (*Lepomis microlophus*) had higher levels than bass and bluegill sunfish, which are predominantly water-column feeders. Villar et al. [145] found a difference in the metal content in the tissues of the two fish species, *Prochilodus lineatus* (detritivore) and *Pterodoras granulosus* (omnivore). *P. lineatus* had a high concentration of metals in the tissues because it primarily fed on the metal-rich detritus. Thus, it is essential to understand the feeding location as well as the trophic level to understand contaminant levels.

Burger et al. [135] estimated trace element levels (As, Cd, Cr, Cu, Pb, Mn, Hg, and Sr) in 11 species of fish occupying different trophic levels with varied dietary habits. They found that species-specific differences existed in all the species studied with respect to all the trace elements. Bowfin and channel fish (both piscivores) had the highest level of all the elements except Mn and Sr. They also found that trophic relationships alone were not able to account for the elemental concentration in different fish species and suggested that metal levels in fish may also reflect age; older and larger fish have higher levels of trace elements. Such a correlation has been seen for mercury [146].

Stewart et al. [147] demonstrated the role of food web structure and physiology of trace element accumulation in the prey species in the differential bioaccumulation of trace elements by fish species; this resulted in some species of fish having very high concentrations of Se while others did not in the San Francisco Bay. Two dominant food webs present in the estuary region of the bay were based upon bivalves and crustacean zooplanktons. The dominant bivalve, *Potamocorbula amurensis,* had a tenfold slower rate of loss of Se from the body compared to that of the crustacean zooplanktons. This resulted in higher tissue Se concentration in the bivalves, which was then reflected in the higher tissue levels of Se in the predatory species of the bivalves. The tissue concentration far exceeded the threshold levels at which Se acted as teratogen and carcinogen, whereas concentration of Se in the predators of the zooplankton was less than the threshold value.

In this case, concentration of Se in water and sediments was not high ($<1\mu g/lt$), but in some of the top consumers like white sturgeon, the tissue levels were as high as >10 $\mu g/g$. This case is an example in which basic physiological and ecological processes can drive wide differences in exposure and effects among different species. These processes are rarely considered in traditional risk assessment studies of contaminant impacts. Besser et al. [148] also found that metal bioavailability to higher order consumers such as trout can be substantially modified by processing metals in the stream food web. Substantial variation in diets of higher consumers such as fish with respect to trace metal content exist due to differential accumulation of metals among invertebrate taxa and differences in taxonomic composition among different locations [149,150].

The differences in the bioaccumulation and trophic transfer potential of different trace elements may be related to their bioavailability in aquatic environments, chemical characteristics of the element, and food web processing of the element. Chen and Folt [138] found that, although As and Pb bioaccumulate in aquatic biota, the concentrations of both these elements biodiminish with increasing trophic levels. The elemental content in fish species was 10 to 20 times lower than what was present in the zooplanktons. Higher levels of As were present in planktivorous fish compared to the omnivores and piscivores. Kay [151] and Handy [152] concluded that Cd also did not biomagnify in the aquatic food web; however, methyl mercury had the capability to bioaccumulate and biomagnify in aquatic biota [135,153]. Besser et al. [148] found that Cd had a higher bioaccumlation factor compared to Pb, suggesting that Cd was highly available in the stream food web.

32.3 HUMAN EXPOSURE TO TRACE ELEMENTS

Exposure of the human population to trace elements occurs from multiple media (water, air, soil) and food. Estimation of exposure to trace elements from food is extremely complex due to varied dietary habits of human populations. Factors that influence dietary intake include age, sex, race, residential region, ethnicity, and personal preferences [154]. Moreover, in metropolitan cities, the problem is further compounded by the fact that many food items comprising the diet of an average human are sourced from different places and not from a specific locality. The most important elements in terms of trophic transfer via the food chains to humans are As, Cd, Hg, and Pb [8].

32.3.1 CADMIUM

Cd is a non-essential heavy metal occurring naturally in Zn and Pb ores [9]. Industrial uses of the metal and agricultural activities have led to widespread dispersion of this element in the environment and human food items. Studies have shown that the average concentration of trace elements in the general agricultural environment has increased over the years. Kjellstrom et al. [155] reported that the Cd content of Swedish wheat increased about threefold between 1900 and 1980. On similar lines, Jones et al. [156] found that the Cd content in the herbage grown in a semirural, undisturbed site in England almost doubled between 1860 and 1990.

One of the primary concerns with respect to cadmium is its transfer from agricultural produce to humans [157]. It is widely acknowledged that vegetable foods contribute to >70% of Cd intake of humans [158]. Accumulation and translocation of trace elements via the agricultural food chain depends upon soil and climatic factors, plant species (variety cultivated), and agronomic management practices [7,8,159]. As has already been discussed, the bioaccumulation of the trace elements can vary with the species and variety.

It is possible that, in some plant species, the edible portion might have low trace element content; however, dietary exposure is not only a function of the elemental content of the plant part consumed and also of the amount of the tissue consumed. For example, even though reproductive structures of flowering plants, such as grains, have relatively low elemental content compared to the other plant tissues, they are consumed in relatively large amounts [157] and therefore contribute largely to the dietary exposure [154]. Cd levels in most vegetables, including bulbs, roots, and tubers, are usually below 0.05 mg/kg, although leafy vegetables such as spinach and lettuce may have considerably higher levels [160]. Most plant-based food stuffs contain higher Cd concentration of approximately 25 µg/kg fresh weight, which represents food items such as cereals, root tubers, and vegetables [7].

Cadmium content in muscle meats is relatively low. In contrast, cadmium content in visceral meats (kidney, liver, and pancreas) is high because these organs preferentially accumulate Cd and other non-essential elements in the animal body [161]. The Cd levels in muscle meats are of the order of 0.01 mg/kg for slaughter animals [7]; cadmium content in the liver and kidney of calves, pigs, and poultry ranges from 0.02 to 0.2 mg/kg and 0.05 to 0.5 mg/kg, respectively.

Aquatic food species (fish, crab, oysters, etc.) bioconcentrate Cd and therefore can have high Cd concentrations [157]. Typical Cd concentration reported in fish muscle is about 0.02 mg/kg, although higher levels may also be found in some fish species. Certain other sea foods may also accumulate Cd from contaminated waters. Certain shell fish can have Cd content in excess of 50 – 100 µg/g fresh weight [7]. Although milk has moderate Cd concentration it is a major source of Cd for infants and children, contributing about 50 and 25% of Cd intake for bottle-fed and breast-fed infants, respectively [158,162]. A number of studies have been carried out to estimate the exposure of the human newborn to Cd via breast feeding [162–164].

Mean daily intake of Cd from food in developed countries has been estimated to be in the range of 16 to 60 µg/day [165]. The provisional tolerable weekly intake (PTWI) of Cd is 7 µg/kg of body weight per week or 1 µg/kg of body weight per day [166]. It may be noted that the safety margin between the exposure in normal diet and the level of exposure that produces deleterious effects is very low.

People who habitually consume a diet high in Cd — for example, due to high consumption of shellfish and sea food — are of particular concern. Bioavailability of Cd from various foods is not well understood. Factors that may influence bioavailability include the chemical form of Cd-consumed tissue, content of competing ions and ligands in the diet, effects of food preparation methods, and the nutritional state of the consuming animal [157].

Database estimation of dietary Cd intake probably provides a reasonable method for estimating relative contributions of different food groups to dietary Cd intake, but its use in health risk assessment is doubtful. Its use in health risk assessment should be validated by analysis of duplicated diets. In a study in the U.S., the database calculation gave an estimate of Cd intake to about 24 µg/day; however, an actual analysis of duplicate diets estimated intake at 56 µg/day [167].

32.3.2 MERCURY

Mercury is ubiquitous in the environment and derives from natural and anthropogenic sources. In nature, it is present in three different forms: elemental mercury (Hg0), inorganic mercury compounds (I-Hg), and organic mercury (primarily methyl mercury, MeHg) [12]. Speciation of Hg is critical in determining the toxicity of different forms of Hg, of which methyl mercury is the most toxic and also the most important with respect to dietary exposure to humans [168]. Inorganic Hg deposits in aquatic environments and is converted through methylation to MeHg by microorganisms [169]. Methyl mercury is readily bioaccumulated and transferred in the aquatic food web with a tendency for biomagnification resulting in high concentration in predatory fish and aquatic wildlife [12].

Fish are the primary source of methyl mercury in the human diet [170]. Nearly all of the mercury present in the fish muscle is in the form of methyl mercury [171]. The dietary uptake of methyl mercury in fish is influenced by their size, diet, and trophic position [172,173]. Estimated concentration of mercury in 3-year-old large-mouth bass collected from 53 lakes in Florida in the U.S. varies form 0.04 to 1.53 µg/g wet weight [174]. Methyl mercury content was highest in long-lived, larger fish that feed on other fish, such as tile fish, king mackerel, sword fish, and shark (Table 32.2).

In humans, 90 to 100% of Me Hg is absorbed through the gastrointestinal tract, where it easily enters the blood stream and distributes through the body [12]. The reference dose set by the U.S. Environmental Protection Agency for ingested methyl mercury exposure is 0.1 µg/kg body weight per day. It has been suggested that a food preparation factor should be used in risk assessment because, when the fish is cooked (especially deep-frying), the concentration of Hg in micrograms per kilogram increases, although the cooked fish retains the same amount of Hg as was present in the raw fish [175,176].

TABLE 32.2
Mean and Range of Mercury (Hg) in Different Seafood Species

Species	Mean (ppm)	Range (ppm)
Northern lobster (American)	0.31	0.05–1.31
King crab	0.09	0.02–0.63
Tuna (fresh or frozen)	0.32	ND–1.30
Shrimp	ND	ND
Catfish	0.07	ND–0.31
Tilefish	1.45	0.65–3.73
Sword fish	1.00	0.10–3.22
King mackerel	0.73	0.30–1.67
Shark	0.96	0.05–4.54

Notes: ND: not detectable. Data from U.S. Food and Drug Administration, Center for Food Safety and Applied Nutrition, Office of Seafood (May 2001).

Source: Adapted from Counter, S.A. and Buchanan, L.H., *Toxicol. Appl. Pharmacol.*, 198, 209, 2004.

In the U.S., 41 states have issued advisories on limiting fish intake, especially for pregnant women and women who may become pregnant [177]. It may be noted that methyl mercury exposure during fetal development and breast feeding is strongly related to the maternal Hg burden [178]. Maternal constitution factors that affect Hg secretion into breast milk are maternal age and lactation stage [179].

Mercury concentration in food crops is generally low, with most of the dietary intake of Hg deriving from consumption of seafood. Exposure of mercury through food items other than fish contaminated with methyl mercury is much less via trophic transfer. Concentration of mercury in most other food items in the average human diet is below the detection limit (20 µg/kg fresh weight) [180,181]. Levels of Hg in most field crops are sufficiently low to cause any concern from the human health viewpoint [182]. This is because mercury is strongly sorbed to soil constituents and as Hg^{2+} or hydrolyzed species is rather immobile in soils [159]. Mercury poisoning on account of consumption of agricultural crops reported in Iraq was due to the consumption of seeds treated with mercury fungicides and not trophic transfer [183].

32.3.3 LEAD AND ARSENIC

Industrial and vehicular pollution are primary sources of contamination of air, water, soil, and food [184]. The major source of lead for nonoccupationally exposed humans is through food and water [7,168]. The amount of lead in food crops depends upon the uptake of lead from the contaminated soils by the plants. Lead is strongly retained by most soils, resulting in very low soil solution Pb concentration compared to other elements such as Cd [185]. Lead content is more in the roots than in the stems and leaves; seeds and fruits have the lowest concentration. Pb concentration in crops is generally well within guidelines or regulatory levels set by respective countries (Table 32.3) [159].

Particulate lead can deposit on the leaf surfaces due to atmospheric fallout of the element. Lead content in dairy products ranges from 0.003 to 0.083 µg/g; vegetables have 0.005 to 0.65 µg/g and meat, fish, and poultry products have 0.002 to 0.16 µg/g [7]. Lead exposure can be further enhanced by domestic food processing, such as food canning, serving of food in glazed pottery, and delivery of water from lead pipes [186]. Because of detrimental effects of lead on the developing infant central nervous system [187], studies have been carried out on the lead levels in human breast milk [163,168]. The WHO considers 2 to 5 ng/g of Pb in human milk "normal" [188]. Diet is the main

TABLE 32.3
Lead Concentrations in Some Field Crops

Crop/country	Mean (mg/kg fresh range weight)		Ref.
Wheat:			
The Netherlands	0.16	0.03–0.65	182
U.S. (all states)	0.037	<0.001–0.716	245
Carrots:			
The Netherlands	0.05	0.011–0.21	182
U.S. (all states)	0.009	0.001–0.125	245
Onions:			
The Netherlands	0.02	0.009–0.05	182
U.S. (all states)	0.005	<0.002–0.054	245
Spinach:			
The Netherlands	0.09	0.01–0.29	182
U.S. (all states)	0.045	0.016–0.17	245

Source: Adapted from McLaughlin, M.J. et al., *Field Crops Res.*, 60, 143, 1999.

source of maternal exposure to Pb [189]. Moreover, during pregnancy and lactation, Pb is mobilized from the bones and is likely to result in increased concentration of lead in blood and breast milk, with potentially toxic effects on the fetus and the mother [190].

Arsenic occurs in food, air, water, and soil, and practically all human populations are exposed to As in one form or another. The major sources of exposure to As are food and water [9]. The principal cause of elevated soil As is the widespread use of As compounds as insecticides, herbicides, and defoliants for agriculture [159]. Arsenic also enters the soil by mining and smelting of non-ferrous metals, application of phosphate fertilizers, fossil fuel combustion, and application of municipal sewage sludge [191]. All these sources include inorganic salts and various organic compounds of arsenite As (III) and arsenate As (V).

The degree of toxicity and the resulting pathologic states depend upon the chemical form of arsenic present in food items [192]. Food contains organic and inorganic forms of arsenic [159], whereas drinking water predominantly has inorganic arsenic [193]. In terms of dietary exposure of As to humans, organo–As compounds found in seafood dominate. Consequently, the total human intake of arsenic depends upon the quantity of seafood consumed [194]. Marine animals have a limited ability to bioconcentrate inorganic As from sea water, but they can bioaccumulate organo–arsenic compounds. Falconer et al. [195] detected arsenic concentrations ranging from 12 to 216 µg/g in *Pleuronectes platessa* from the North Sea. Food crops have highly variable and intermediate percentages of such compounds [159]. Organo–As compounds are absorbed readily from the gastrointestinal tract, but are not metabolized in the body and readily excreted out. It is inorganic As (III) and As (V) that pose a greater health risk.

32.4 HUMAN HEALTH EFFECTS

Human health effects associated with trophic transfer are of diseased states associated with chronic rather than acute exposure to trace elements. Research on the health effects of chronic exposure to trace elements suggests that physiological alterations may occur at levels that were formerly considered to be safe [8]. Neurological and neurophysiological effects, nephrotoxicity, reproductive toxicity, teratogenicity, and carcinogenicity remain at the forefront of research on the health effects of trace elements [196,197].

32.4.1 CADMIUM

Cadmium can cause irreversible renal tubular injury, nonhypertrophic emphysema, osteoporosis, anemia, eosinophelia, anosmia, and chronic rhinitis [157,158,198]. The sentinel sign of Cd adverse effect is renal tubular dysfunction, which is characterized by low molecular weight proteinuria [199] and can occur in concert with anemia [200,201] or bone mineral loss [202,203]. Renal tubular dysfunction caused due to Cd exposure is generally irreversible and can constitute a significant health effect [92,158]. In a study of nearly 1700 subjects aged between 20 and 80 years, sampled randomly from the general population of four areas of Belgium having various levels of exposure to industrially derived Cd pollution, it was concluded that 10% of the Belgian population absorb sufficient Cd to cause renal dysfunction [204].

Recently, an association between cadmium exposure and chronic renal failure (end stage renal disease, ESRD) was shown [205]. Elevated levels of Cd in the diet and drinking water were concluded to be causative factors in the 1964 occurrence of Itai Itai disease in Toyoma prefecture in Japan [206]. The source of the Cd was acid drainage from a Pb–Zn–Cd mine into the Jintsu River. Water from the river was used for drinking and irrigation of the rice paddies. Cd intake from food and water was estimated to be 300 to 600 µg/day [158]. The disease was characterized by severe pain, bone fractures, proteinuria, and severe osteomalacia, which appeared mainly among women. Exposure to low doses of Cd has led to decreases in the density of bone tissue [207,208].

Cadmium exposure also enhances susceptibility to bacterial, protozoal, and viral infections and results in impaired and humoral and cell-mediated immune response [209].

Several groups of individuals have been identified as "at risk" from excess Cd exposure, including: persons having severe nutritional deficiencies (Fe, Ca, Zn, protein, vitamin D) that are aggravated by Cd; persons consuming more than normal levels of visceral meats, fish, and shellfish; pregnant and lactating females with a negative Ca balance; nursing infants; persons with kidney ailments; and multiparous, postmenopausal women [158,210,211].

The International Agency for Research on Cancer (IARC) has classified cadmium as a human carcinogen (group I) on the basis of sufficient evidence in humans and experimental animals [212]. Cadmium has been associated with cancer of the lung, prostrate, pancreas, and kidney. It is predominantly a nongenetic carcinogen. Many indirect mechanisms are implicated in carcinogenicity of Cd, such as modification of gene expression and signal transduction, interference with enzymes of the cellular antioxidant system and generation of reactive oxygen species, inhibition of DNA repair, DNA methylation, role in apoptopsis, and disruption of E-cadherin-mediated cell–cell adhesion [198].

32.4.2 MERCURY

Health effects of mercury are highly dependent on the different forms mercury. Methyl mercury is the most toxic form because of its neurotoxic properties [213]. In humans, the main target for mercury, especially methyl mercury, is the central nervous system; exposure to methyl mercury can cause serious brain damage, including psychological disturbances, impaired hearing, loss of sight, ataxia, loss of motor control, and general debilitation [214]. In humans, 90 to 100% of methyl mercury is absorbed through the gastrointestinal tract from which it enters the blood stream and is distributed throughout the body [12]. It is transported across the blood–brain barrier by an amino acid carrier and readily accumulates in the brain [215].

Although methyl mercury is distributed across the body, its most serious effects are on the developing brain. In adult brain, methyl mercury damage is focal — for example, involving loss of neurons in the visual cortex and loss of granule neurons in the cerebellum. In the developing brain, the damage is more diffuse and extensive [12]. Methyl mercury affects the formation of microtubules and thus neuronal migration and cell division [170,213,216]. The earliest symptoms are parestias and numbness in hands and feet. Later, coordination difficulties and concentric constriction of visual fields may develop with auditory symptoms. At high exposure levels, methyl mercury may result in a loss of neurons in each lobe of the brain, and the developmental effects may include hyperactive reflexes, deafness, blindness, cerebral palsy, mental retardation, and general paralysis [183,217].

At low exposure levels, the neurodevelopmental effects may be subtle and include deficits in language, learning, attention, and, to a lesser degree, fine motor and visual–spatial organizational impairments. Several possible molecular targets of methyl mercury exposure in the nervous system include the blood–brain barrier, cytoskeleton, axonal transport, neurotransmitter production, secretion, uptake and metabolism cell signaling, protein, DNA and RNA synthesis, and respiratory and energy-generating systems [216].

The Minamata catastrophe in Japan in the 1950s was caused by methyl mercury poisoning from fish contaminated by mercury discharges to the surrounding sea. In the early 1970s, more than 10,000 persons in Iraq were poisoned by eating bread baked from mercury-polluted grain [9]. Gender-related susceptibility of Me Hg neurotoxicity has been extensively studied, but only some evidence indicates that women are more affected than men when exposure occurs in adulthood [218]. Males seem to be more affected by exposure during stages of early development [219,220]. Sakamota et al. [221] found that a declining male birth ratio was associated with increased male fetal death due to Me Hg exposure in Minamata, Japan. Methyl mercury and inorganic Hg can also induce DNA strand breaks in cells [222] and inhibit DNA repair. Although, some evidence indicates that inorganic and methyl mercury can cause renal tumors in rodents, evidence for carcinogenicity of Hg in humans is inadequate [223].

32.4.3 LEAD

Occupational and environmental chronic Pb exposure can damage the central nervous system, kidneys, and cardiovascular, reproductive, and hematological systems [224]. Gastrointestinal absorption of Pb varies with age, diet, and nutritional status. Age is a critical variable in absorption levels, with adults absorbing 7 to 15% from dietary sources; in infants and children, absorption levels can reach 40 to 50% [225]. Chronic exposure of Pb follows a prolonged disease progression. Long-term exposure can lead to distal motor neuropathy, possible seizures, and coma [8]. Infants and young children have long been known to be at risk to toxicity of Pb because of higher Pb intake relative to body size and greater absorption from the gastrointestinal tract. The central nervous system of the developing fetus may be at even greater risk because of immaturity of the blood–brain barrier [226].

A number of cross-sectional studies and prospective epidemiological studies have shown impairment of cognitive behavioral development in children [224]. In acute toxicity, lead can also induce encephalopathy in children, with symptoms of headache, confusion, stupor, coma, and seizures [226]. It is suggested that immature endothelial cells forming the capillaries in the developing brain are less resistant to the effects of Pb than capillaries from mature brain are. These cells permit fluid and cations, including Pb, to reach newly formed components of the brain, particularly astrocytes and neurons [227]. Pb also produces deficits in neurotransmission through inhibition of cholinergic function, possibly by reduction of extracellular Ca [226].

Overt effects of lead on the kidney in man and experimental animals, particularly rat and mouse, begin with acute morphological changes consisting of nuclear inclusion bodies or lead–protein complexes and ultrasructural changes in organelles, particularly mitochondria. Progression of acute nephropathy to chronic irreversible nephropathy occurs slowly, over months or years, and only after years of heavy exposure [226]. Experimental studies suggest a possible threshold for lead nephrotoxicity. Mortada et al. [228] found that long-term Pb exposure may also give rise to kidney damage. It was found that blood lead level of 60 µg/dL is the threshold for proximal tubular cell injury from lead [229].

Buchet et al. [231] found that workers who did not have lead levels of over 62 to 63 µg/dL for up to 12 years did not have Pb nephropathy. It has also been recognized that severe Pb intoxication is associated with sterility, abortion, still births, neonatal morbidity, and mortality from exposure *in utero* [230]. Effects of Pb on the hematological system have also been known for a long time. Anemia is a well known symptom of Pb poisoning; Pb inhibits activity of γ-aminolevulinic acid dehydratase (ALAD) and ferrochetalase, which are involved in heme synthesis, and also leads to changes in RBC morphology and survival [231].

More than 90 % of the body burden of Pb is localized in the bone, with an average half-life of about 10 years [224]. The accumulated Pb in bone tissue follows the same general physiology as that of bone Ca metabolism [232]. During periods like pregnancy and lactation, bone Pb stores may be mobilized even long after cessation of exposure [233,234]. Endogenous exposure of Pb may occur during the critical period of organ development in the fetus and nursing child. Roussow et al. [235] found that exposure to Pb *in utero* resulted in sixfold increase in brain accumulation as compared to 3.5- to 2-fold increase during other periods. Some evidence indicates that certain genetic and environmental factors can increase the detrimental effects of Pb on neural development, thereby rendering certain children more vulnerable to Pb neurotoxicity [236]. Pb has direct and indirect effects on bone turnover. Indirect effects of Pb are via kidney dysfunction and inhibition of 1,25-dihydroxy vitamin D3; direct effects are on osteoblast and osteoclast functions and inhibition of synthesis of bone matrix components [226,232].

32.4.4 ARSENIC

The toxicity of As varies with its chemical state, ranging from virtually nontoxic forms of organic and pure elemental arsenic to acutely toxic trivalent arsenic trioxide. Organo–As compounds, which have a potential for bioaccumulation and trophic transfer, have relatively reduced human health risk associated with them. Predominantly, health risk is associated with exposure to inorganic As (II) and As (V) through drinking water. The existing drinking water limit for As is 50 µg/lt; some evidence indicates that the permissible limit must be lowered because carcinogenicity has been observed at As levels less than an order of magnitude from the drinking water limit [11].

Epidemiological data have shown that chronic exposure of humans to inorganic arsenical compounds is associated with liver injury, peripheral neuropathy, and increased incidence of cancer of lung, skin, bladder, and liver [9,237]. Arsenic can also cross the placenta and can cause fetotoxicity, decreased birth weight, and congenital malformation [196]. Several mechanisms have been implicated in arsenic-induced genotoxicity, which includes oxidative stress [238,239]. DNA repair inhibition [240,241] and direct mutagenesis [242] have been reported. In humans, an increased percentage of apoptosis was found in the buccal epithelial cells from individuals chronically exposed to arsenic in China [243]. Inorganic arsenic is a known human carcinogen causing lung cancer by inhalation and skin cancer via ingestion [244].

32.5 CONCLUSIONS

Anthropogenic activities such as mining, smelting, and combustion of fossil fuels have altered the biogeochemical cycles of many trace elements dramatically. This has resulted in increased trace element content in environmental resources (land, air, and water) and biota. Bioaccumulation of trace elements by living organisms is influenced by the chemical attributes of the trace element, physicochemical characteristics of the ambient environment, physiological make up of the organism, and ecological host factors. The increasing level of trace elements in the tissues of plants and animals due to bioaccumulation and trophic transfer has adverse effects on ecological and human health.

Human beings are top consumers of many terrestrial and aquatic food chains; this results in exposure to trace elements due to consumption of contaminated plant and animal products sourced form biota with elevated trace element content. Dietary exposure is affected by dietary preferences (choice of food items), age, sex, residential region, and ethnicity, among other things; the assimilation of the ingested trace elements is influenced by nutritional status, metal–metal interactions, and the chemical form of the element in the ingested food. Trace elements vary with respect to their potential for trophic transfer to humans via food; elements such as cadmium and methyl mercury are probably the most important with respect to their ability for trophic transfer and potential effects on human health.

Current advances in key areas, such as bioavailability; uptake; assimilation and tissue distribution; detoxification mechanisms of trace elements at the organismal level; and the role played by the ecological characteristics such as food chain length, foraging behavior, and feeding ecology, have led to a better understanding of the variability in the host and environmental factors affecting trace element cycling. These advances are critical for ecological risk assessment and also provide information on variability of trace elements in different constituents of the human diet.

Recent data have indicated that the adverse health effects related to trace element exposure occur at lower levels than previously expected. In this regard, it is essential to identify the risk factors associated with the trace element exposure and to identify risk groups in human populations in order to achieve a reliable risk assessment. Finally, better ways should be developed to put the advances in the body of knowledge with respect to trace elements into risk assessment and regulatory practice to minimize the ecological and health risks associated with trace element exposure.

ACKNOWLEDGMENT

A.L. received a research fellowship from the Council of Scientific and Industrial Research, India.

REFERENCES

1. Brown, G.E., Jr., Foster, A.L., and Ostergren, J.D., Mineral surfaces and bioavailability of heavy metals: a molecular-scale perspective, *Proc. Nat. Acad. Sci. USA.*, 96, 3388, 1999.
2. Nriagu, J.O. and Pacyna, J.M., Quantitative assessment of worldwide contamination of air, water and soils by trace metals, *Nature*, 333, 134, 1988.
3. McGrath, S.P., Zhao, F.J., and Lombi, E., Plant and rhizosphere processes involved in phytoremediation of metal-contaminated soils, *Plant Soil*, 232, 207, 2001.
4. Meagher, R.B., Phytoremediation of toxic elemental and organic pollutants, *Curr. Opin. Plant Biol.*, 3, 153, 2000.
5. Pierzynski, G.M., Sims, J.T., and Vance, G.J., *Soils and Environmental Quality*, 2nd ed., CRC Press, Boca Raton, FL, 2000, 19.
6. Underwood, E.J., Trace elements and health: an overview, *Phil. Trans. R. Soc. Lond. B.*, 288, 5, 1979.
7. Rojas, E., Herrera, L.A., Poirier, L.A., and Ostrosky–Wegman, P., Are metals dietary carcinogens? *Mut. Res.*, 443, 157, 1999.
8. Robson, M., Methodologies for assessing exposures to metals: human host factors, *Ecotox. Environ. Safety*, 56, 104, 2003.
9. Jarup, L., Hazards of heavy metal contamination, *Brit. Med. Bull.*, 68, 167, 2003.
10. Ercal, N., Gurer–Orham, H., and Aykin–Burns, N., Toxic metals and oxidative stress part I: mechanisms involved in metal-induced oxidative damage, *Curr. Topics Med. Chem.*, 1, 529, 2001.
11. Schoen, A., Beck, B., Sharma, R., and Dube, E., Arsenic toxicity at low doses: epidemiological and mode of action considerations, *Toxicol. Appl. Phamacol.*, 198, 253, 2004.
12. Counter, S.A. and Buchanan, L.H., Mercury exposure in children: a review, *Toxicol. Appl. Pharmacol.*, 198, 209, 2004.
13. Peterson, B.J. and Fry, B., Stable isotopes in ecosystem studies, *Ann. Rev. Ecol. Sys.*, 18, 293, 1987.
14. Cabana, G. and Rasmussen, J.B., Modeling food chain structure and contaminant bioaccumulation using stable nitrogen isotopes, *Nature*, 372, 255, 1994.
15. Torres, K.C. and Johnson, M.L., Bioaccumulation of metals in plants, arthropods, and mice at a seasonal wetland, *Environ. Toxicol. Chem.*, 20, 2617, 2001.
16. Newman, M.C., *Fundamentals of Ecotoxicology*, Ann Arbor Press, Chelsea, MI, USA, 1998, 71.
17. Gnamus, A., Byrne, A.R., and Horvat, M., Mecury in the soil-plant-deer-predator food chain of a temperature forest in Solvenia, *Environ. Sci. Technol.*, 34, 3337, 2000.
18. Cataldo, D.A., Wildung, R.E., and Garland, T.R., Speciation of trace inorganic contaminants in plants and bioavailability to animals: an overview, *J. Environ. Qual.*, 16, 289, 1987.
19. Peakall, D. and Burger, J., Methodologies for assessing exposure to metals: speciation, bioavailability of metals, and ecological host factors, *Ecotox. Environ. Safety*, 56, 110, 2003.
20. John, M.K., van Laerhoven, C.J., and Chuah, H.H., Factors affecting plant uptake and phytotoxicity of cadmium added to soils, *Environ. Sci. Technol.*, 6, 1005, 1972.
21. Kuboi, T., Noguchi, A., and Yazaki, J., Family-dependent cadmium accumulation characteristics in higher plants, *Plant Soil*, 92, 405, 1986.
22. Xian, X., Effect of chemical forms of cadmium, zinc, and lead in polluted soils on their uptake by cabbage plants, *Plant Soil*, 113, 257, 1989.
23. Cataldo, D.A. and Wildung, R.E., The role of soil and plant metabolic processes in controlling trace element behavior and bioavailability to animals, *Sci. Total Env.*, 28, 159, 1983.
24. Darrah, P.R. and Stauton, S., A mathematical model of root uptake incorporating root turnover, distribution within the plant and recycling of absorbed species, *Eur. J. Soil Sci.*, 51, 643, 2000.
25. Waisel,Y., Eshel, A., and Kafkafi, U., *Plant Roots: The Hidden Half*, 3rd ed., Marcel Dekker Inc., New York, 2002, 617.

26. Hinsinger, P., Bioavailability of trace elements as related to root-induced chemical changes in the rhizosphere, in *Trace Elements in the Rhizosphere*, Gobran, G.R., Wenzel, W.W., and Lombi, E., Eds., CRC Press, Boca Raton, FL, 2001, 25.

27. Pinel, F., Leclerc–Cessac, E., and Staunton, S., Relative contributions of soil chemistry, plant physiology and rhizosphere induced changes in speciation on Ni accumulation in plant shoots, *Plant Soil*, 255, 619, 2003.

28. Knight, B., Zhao, F.J., McGrath, S.P., and Shen, Z.G., Zinc and cadmium uptake by the hyperaccumulator *Thlaspi caerulescens* in contaminated soils and its effects on the concentration and chemical speciation of metals in soil solution, *Plant Soil*, 197, 71, 1997.

29. McGrath, S.P., Shen, Z.G., and Zhao, F.J., Heavy metal uptake and chemical changes in the rhizosphere of *Thlaspi caerulescens* and *Thlaspi ochroleucum* grown in contaminated soils, *Plant Soil*, 188, 153, 1997.

30. Whiting, S.N., de Souza, M.P., and Terry, N., Rhizosphere bacteria mobilize Zn for hyperaccumulation by *Thlaspi caerulescens*, *Environ. Sci. Technol.*, 35, 3144, 2001.

31. Whiting, S.N., Leake, J.R., McGrath, S.P., and Baker, A.J.M., Zinc accumulation by Thlaspi caerulescens from soils with different Zn availability: a pot study, *Plant Soil*, 236, 11, 2001.

32. Abou–Shanab, R.A., Angle, J.S., Delorme, T.A., Chaney, R.L., van Berkum, P., Mowad, H., Ghanem, K., and Gozlan, H.A., Rhizobacterial effects on nickel extraction from soil and uptake by *Alyssum murale*, *New Phytol.*, 158, 219, 2003.

33. Killham, K. and Firestaone, M.K., Vesicular-arbuscular mycorrhizal mediation of grass response to acidic and heavy metal deposition, *Plant Soil*, 72, 39, 1983.

34. Heggo, A., Angle, J.S., and Chaney, R.L., Effect of vesicular–arbuscular mycorrhizae fungi on heavy metal uptake by soybeans, *Soil Biol. Biochem.*, 22, 865, 1990.

35. Tam, P.C.F., Heavy metal tolerance by ectomycorrhizal fungi and metal amelioration by *Pisolithus tinctorium*, *Mycorrhiza*, 5, 181, 1995.

36. Leyval, C. and Joner, E.J., Bioavailability of heavy metals in the mycorrhizosphere, in *Trace Elements in the Rhizosphere*, Gobran, G.R., Wenzel, W.W., and Lombi, E., Eds., CRC Press, Boca Raton, FL, 2001, 165.

37. Burd, G.I., Dixon, D.G., and Glick, B.R., A plant growth-promoting bacterium that decreases nickel toxicity in seedlings, *Appl. Environ. Microbiol.*, 3663, 64, 1998.

38. Podar, D., Ramsey, M.H., and Hutchings, M.J., Effect of cadmium, zinc and substrate heterogeneity on yield, shoot metal concentration and metal uptake by *Brassica juncea*: implications for human health risk assessment and phytoremediation, *New Phytol.*, 163, 313, 2004.

39. John, M.K. and van Laerhoven, C.J., Differential effects of cadmium on lettuce, *Environ. Pollut.*, 10, 163, 1976.

40. Baker, A.J.M., Accumulators and excluders — strategies in the response of plants to heavy metals, *J. Plant Nutr.*, 3, 643, 1981.

41. Landberg, T. and Greger, M., Can heavy metal-tolerant clones of *Salix* be used as vegetation filters on heavy metal-contaminated land? In *Willow Vegetation Filters for Municipal Wastewater and Sludges: a Biological Purification System*, Aronsson, P. and Perttu, K., Eds., Swedish University of Agricultural Sciences, Uppsala, 1994, 133.

42. Berthelsen, B.O., Steinnes, E., Solberg, W., and Jingsen, L., Heavy metal concentrations in plants in relation to atmospheric heavy metal deposition, *J. Environ. Qual.*, 24, 1018, 1995.

43. Broadley, M.R., Willey, N.J., Wilkins, J.C., Baker, A.J.M., and White, P.J., Phylogenetic variation in heavy metal accumulation in angiosperms, *New Phytol.*, 152, 9, 2001.

44. Nan, Z., Li, J., Zhang, J., and Cheng, G., Cadmium and zinc interactions and their transfer in soil-crop system under actual field conditions, *Sci. Total Environ.*, 285, 187, 2002.

45. Lubben, S. and Sauerbeck, D., The uptake and distribution of heavy metals in spring wheat, *Water Air Soil Pollut.*, 57–58, 239, 1991.

46. Brooks, R.R., Lee, J., Reeves, R.D., and Jaffre, T., Detection of nickeliferous rocks by analysis of herbarium specimens of indicator plants, *J. Geochem. Explor.*, 7, 49, 1977.

47. Brooks, R.R., Geobotany and hyperaccumulators, in *Plants that Hyperaccumulate Heavy Metals*, Brooks, R.R., Ed., CAB International, Wallingford, U.K., 1998, 55.

48. Baker, A.J.M., McGrath, S.P., Reeves, R.D., and Smith J.A.C., Metal hyperaccumulator plants: a review of the ecology and physiology of a biochemical resource for phytoremediation of metal-polluted soils, in *Phytoremediation of Contaminated Soil and Water*, Terry, N., Banuelos, G., and Vangronsveld, J., Eds., Lewis Publishers, Boca Raton, FL, 2000, 85.

49. Baker, A.J.M., Reeves, R.D., and Hajar, A.S.M., Heavy metal accunulation and tolaerance in British population of the mettalophyte *Thlaspi caerulescens* J. and C. Presl (Brassicaceae), *New Phytol.*, 127, 61, 1994.

50. Shen, Z.G., Zhao, F.J., and McGrath, S.P., Uptake and transport of Zn in the hyperaccumulator *Thlaspi caerulescens* and the nonhyperaccumulator *Thlaspi ochroleucuns*, *Plant Cell Environ.*, 20, 898, 1997.

51. Rapport, D.J., What constitutes ecosystem health? *Perspect. Biol. Med.* 33, 120, 1989.

52. Peakall, D.B., *Animal Markers as Pollution Indicators*, Chapman & Hall, London, 1992.

53. Burger, J. and Gochfeld, M., Heavy metal and selenium concentrations in eggs of Herring gulls (*Larus argentatus*): temporal differences from 1989 to 1994, *Arch. Environ. Contam. Toxicol.*, 29, 192, 1995.

54. O'Brien, D.J., Kaneene, J.B., and Poppenga, R.H., The use of mammals as sentinels for human exposure to toxic contaminants in the environment, *Environ. Health Perspect.*, 99, 351, 1993.

55. Talmage, S. and Walton, B.T., Small mammals as monitors of environmental contaminants, *Rev. Environ. Contam. Toxicol.*, 119, 47, 1991.

56. McBee, K. and Bickham., J.W., Mammals as bioindicators of environmental toxicity, in *Current Mammology*, Geoways, H.H., Ed., vol. 2, Plenum Press, New York, 1990, 37.

57. Reinecke, A.J., Reinecke, S.A., Musilbono, D.E., and Chapman, A., The transfer of lead from earthworms to shrews (*Myosorex varius*), *Arch. Environ. Contam. Toxicol.*, 39, 392, 2000.

58. Burger, J., Metals in avian feathers: bioindicators of environmental pollution, *Rev. Environ. Toxicol.*, 5, 203, 1994.

59. Attuquayefio, D.K., Gorman, M.L., and Wolton, R.J., Home range sizes in wood mouse *Apodemus sylavticus* habitat, sex and seasonal differences, *J. Zool.*, 210, 45, 1986.

60. Marinussen, M.P.J.C. and Vanderzee, S.E.A.T.M., Conceptual approach to estimating the effect of home range on the exposure of organism to spatially variable soil contamination, *Ecol. Model.*, 87, 83, 1996.

61. Larison, J.R., Likens, G.E., Fitzpatrick, J.W., and Crock, J.G., Cadmium toxicity among wildlife in Colorado Rocky Mountains, *Nature*, 406, 181, 2000.

62. Metcheva, R., Teodorova, S., and Topashka–Ancheva, M., A comparative analysis of the heavy metal loading of small mammals in different regions of Bulgaria I: monitoring points and bioaccumulation features, *Ecotox. Environ. Saf.*, 54, 176, 2003.

63. Hunter, B.A. and Johnson, M.S., Food chain relationships of copper and cadmium in contaminated grassland ecosystems, *Oikos*, 38, 108, 1982.

64. Cooke, J.A. and Johnson, M.S., Cadmium in small mammals, in *Environmental Contaminants in Wildlife: Interpreting Tissue Concentration*, Beyer, W.N., Heinz, G., and Redmond–Norwood, A., Eds., Lewis Publishers, Boca Raton, FL, 1996, 377.

65. Mertens, J., Luyssaert, S., Verbeeren, S., Vervaeke, P., and Lust, N., Cd and Zn concentrations in small mammals and willow leaves on disposal facilities for dredged material, *Environ. Pollut.*, 115, 17, 2001.

66. Ma, W., Lead in mammals, in *Environmental Contaminants in Wildlife: Interpreting Tissue Concentration*, Beyer, W.N., Heinz, G., and Redmond–Norwood, A., Eds., Lewis Publishers, Boca Raton, FL, 1996, 281.

67. Hunter, B.A., Johnson, M.S., and Thompson, D.J., Ecotoxicology of copper and cadmium in a contaminated grassland ecosystem: small mammals, *J. App. Ecol.*, 24, 601, 1987.

68. Watts, C.H.S., The foods eaten by wood mice (*Apodemus sylvaticus*) and bank voles (*Clethrionomys glareolus*) in Wytham Woods, Berkshire, *J. Animal Ecol.*, 37, 25, 1968.

69. Erry, B.V. et al., Arsenic contamination in wood mice (*Apodemus sylvaticus*) and bank voles (*Clethrionomys glareolus*) on abandoned mine sites in southwest Britain, *Environ. Pollut.*, 110, 179, 2000.

70. Beyer, W.N., Connor, E.E., and Gerould, S., Estimates of soil ingestion by wildlife, *J. Wildlife Manage.*, 58, 375, 1994.

71. Shore, R.F., Predicting cadmium, lead and fluoride levels in small mammals from soil residues and by species–species extrapolation, *Environ. Pollut.*, 88, 333, 1995.

72. Sharma, R.P. and Shupe, J.L., Lead, cadmium and arsenic residues in animal tissues in relation to those in their surrounding habitats, *Sci. Total Environ.*, 7, 53, 1977.

73. Purcell, P.W., Hynes, M.J., and Fairley, J.S., Lead levels in small Irish rodents (*Apodemus sylvaticus* and *Clethrionomys glareolus*) from around a tailing pond and along motorway verges, *Proc. Royal Irish Acad.*, 92B, 79, 1992.

74. Lodenius, M., Soltanpour–Gargari, A., Tulisalo, E., and Henttonen, H., Effects of ash application on cadmium concentration in small mammals, *J. Environ. Qual.*, 31, 188, 2002.

75. Dodds–Smith, M.E., Johnson, M.S., and Thompson, D.J., Trace metal accumulation by the shrew *Sorex araneus*. III. Tissue distribution in kidney and liver, *Ecotox. Environ. Safety*, 24, 118, 1992.

76. Hunter, B.A., Johnson, M.S., and Thompson, D.J., Ecotoxicology of copper and cadmium in a contaminated grassland ecosystem: tissue distribution and age accumulation in small mammals, *J. App. Ecol.*, 26, 89, 1989.

77. Ma, W., Methodological principles of using small mammals for ecological hazard assessment of chemical soil pollution, with examples on cadmium and lead, in *Ecotoxicology of Soil Organisms*, Donker, M.H., Eijackers, H., and Heinbach, F., Eds., SETAC, CRC Press, Boca Raton, FL, 1994, 357.

78. Pedersen, H.C. and Myklebust, I., Age-dependent accumulation of cadmium and zinc in the liver and kidneys of Norwegian willow ptarmigan, *Bull. Environ. Contam. Toxicol.*, 51, 381, 1993.

79. Froslie, A., Haugen, A., Holt, G., and Norheim, G., Levels of cadmium in liver and kidneys from Norwegian cervides, *Bull. Environ. Contam. Toxicol.*, 37, 453, 1986.

80. Milton, A., Johnson, M.S., and Cooke, J.A., Lead within ecosystems on metalliferous mine tailings in Wales and Ireland, *Sci. Total Environ.*, 299, 177, 2002.

81. Vega, M.M., Solorzano, J.C., Medina, A.A., Luna, C.H., and Salvinus, J.V.C., Lead: intestinal absorption and mobilisation during lactation, *Human Exp. Toxicol.*, 15, 872, 1996.

82. Jefferies, D.J. and French, M.C., Lead concentration in small mammals trapped on roadside verges and field sites, *Environ. Pollut.*, 3, 147, 1972.

83. Lock, K. and Janssen, C.R., Zinc and cadmium body burdens in terrestrial oligochaetes: use and significance in environmental risk assessment, *Environ. Toxicol. Chem.*, 20, 2067, 2001.

84. Ma, W., Denneman, W., and Faber, J., Hazardous exposure of ground-living small mammals to cadmium and lead in contaminated terrestrial ecosystems, *Arch. Environ. Contam. Toxicol.*, 20, 266, 1991.

85. Milton, A., Cooke, J.A., and Johnson, M.S., Accumulation of lead, zinc, and cadmium in a wild population of *Clethrionomys glareolus* from an abandoned lead mine, *Arch. Environ. Contam. Toxicol.*, 44, 405, 2003.

86. Myklebust, I. and Pedersen, H.C., Accumulation and distribution of cadmium in willow ptarmigan, *Ecotoxicology*, 8, 457, 1999.

87. Laurinolli, M. and Bendell–Young L.I., Copper, zinc, and cadmium concentrations in *Peromyscus maniculatus* sampled near an abandoned mine, *Arch. Environ. Contam. Toxicol.*, 30, 481, 1996.

88. Stewart, A.K. and Magee, A.C., Effects of zinc toxicity on copper, phosphate and magnesium metabolism in young rats, *J. Nutr.*, 287, 82, 1964.

89. Flick, D.F., Kraybill, H.F., and Dimitroff, J.M., Toxic effects of cadmium: a review, *Environ. Res.*, 4, 71, 1971.

90. Powell, G.W., Miller, W.J., Morton, J.D., and Clifton, C.M., Influence of dietary Cd level and supplemental Zn on Cd toxicity in the bovine, *J. Nutr.*, 84, 205, 1964.

91. Scheuhammer, A.M., Influence of reduced dietary calcium on the accumulation of lead, cadmium, and aluminum in birds, *Environ. Pollut.*, 94, 337, 1996.

92. Friberg, L., Piscator, M., Nordeberg, G.F., and Kjellstorm, T., *Cadmium in the Environment*, CRC Press, Cleveland, OH, 1974.

93. Suzuki, K.T., Studies of cadmium uptake and metabolism by the kidney, *Environ. Health Perspect.*, 54, 21, 1984.

94. Roberts, R.D., and Johnson, M.S., Dispersal of heavy metals from abandoned mine workings and their transference through the terrestrial food chains, *Environ. Pollut.*, 16, 293, 1978.

95. Robinson, C.J., Hall, S., and Beshir, S.O., Hormonal modulation of mineral metabolism in reproduction, *Proc. Nutr. Soc.*, 42, 169, 1983.

96. Fullmer, C.S., Intestinal calcium and lead absorption: effects of dietary lead and calcium, *Environ. Res.*, 54, 159, 1991.

97. Mautino, M., Lead and zinc intoxication in zoological medicine: a review, *Zoo. Wildlife Med.*, 28, 28, 1997.

98. Shore, R.F. and Douben, P.E.T., The ecotoxicological significance of Cd intake and residues in terrestrial small mammals, *Ecotox. Environ. Safety*, 29, 101, 1994.

99. Wlostowski, T.A., Krasowska, A., and Dworakowski, W., Low ambient temperature decreases cadmium accumulation in the liver and kidneys of the bank vole (*Clethrionomys glareolus*), *Biometals*, 9, 363, 1996.

100. Barron, M.G., Bioaccumulation and bioconcentration in aquatic organisms, in *Handbook of Ecotoxicology*, Hoffman, D.J., Rattner, B.A., Burton, G.A., Jr., and Cairns, J., Jr., Eds., Lewis Publishers, Boca Raton, FL, 2003, 877.

101. Suedel, B.C., Boraczek, J.A., Peddicord, R.K., Clifford, P.A., and Dillon, T.M., Trophic transfer and biomagnification potential of contaminants in aquatic ecosystems, *Rev. Environ. Contam. Toxicol.*, 136, 21, 1994.

102. Fisher, N.S. and Wang, W-X., The trophic transfer of silver in marine herbivores: a review of recent studies, *Environ. Toxicol. Chem.*, 17, 562, 1998.

103. Reinfelder, J.R., Fisher, N.S., Luoma, S.N., Nichols, J.W., and Wang, W-X., Trace element trophic transfer in aquatic organisms: a critique of kinetic model approach, *Sci. Total Environ.*, 219, 117, 1998.

104. Chen, C.Y., Stemberger, R.S., Klane, B., Blum, J.D., Pickhardt, P.C., and Folt, C.L., Accumulation of heavy metals in food web components across a gradient of lakes, *Limnol. Oceanogr.*, 45, 1525, 2000.

105. Shi, D. and Wang, W-X., Understanding the differences in Cd and Zn bioaccumulation and subcellular storage among different population of marine clams, *Environ. Sci. Tech.*, 38, 449, 2004.

106. Wang, W.-X. and Fisher, N.S., Delineating metal accumulation pathways for marine invertebrates, *Sci. Total. Environ.*, 237/238, 459, 1999.

107. Wang, W-X. and Fisher, N.S., Accumulation of trace elements in a marine copepod, *Limnol. Oceanogr.*, 43, 273, 1998.

108. Reinfelder, J.R. and Fisher, N.S., Retention of elements absorbed by juvenile fish (*Menidia menidia, Menidia beryllina*) from zooplankton prey, *Limnol. Oceanogr.*, 39, 1783, 1994.

109. Fisher, N.S., Teyssie, J.L., Fowler, S.W., and Wang, W-X., The accumulation and retention of metals in mussels from food and water: a comparison under field and laboratory condition, *Environ. Sci. Tech.*, 30, 3232, 1996.

110. Wang, W.-X. and Fisher, N.S., Assimilation efficiencies of chemical contaminants in aquatic invertebrates: a synthesis, *Environ. Toxicol. Chem.*, 18, 2034, 1999.

111. Mason, R.P., Reinfelder, J.R., and Morel, F.M.M., Uptake, toxicity, and trophic transfer of mercury in a coastal diatom, *Environ. Sci. Technol.*, 30, 1835, 1996.

112. Rainbow, P.S., Trace metal concentrations in aquatic invertebrates: why and so what? *Environ. Pollut.*, 120, 497, 2002.

113. Nott, J.A. and Nicolaidou, A., Transfer of metal detoxification along marine food chains, *J. Mar. Biol. Assoc (U.K.)*, 70, 905, 1990.

114. Wallace, W.G. and Lopez, G.R., Relationship between subcellular cadmium distribution in prey and cadmium, *Estuaries*, 19, 923, 1996.

115. Wallace, W.G., Lopez, G.R., and Levington, J.S., Cadmium resistance in an oligochaete and its effect on cadmium trophic transfer to an omnivorous shrimp, *Mar. Ecol. Prog. Ser.*, 172, 225, 1998.

116. Rainbow, P.S., Phylogeny of trace metal accumulation in crustaceans, in *Metal Metabolism in Aquatic Environments*, Langston, W.J. and Bebianno, M., Eds., Chapman & Hall, London, 1998, 285.

117. Rainbow, P.S. and White, S.L., Comparative strategies of heavy metal accumulation by crustaceans: zinc, copper, and cadmium in a decapod, and amphipod and a barnacle, *Hydrobiologia*, 174, 245, 1989.

118. Abdennour, C., Smith, B.D., Boulakoud, M.S., Samraoui, B., and Rainbow, P.S., Trace metals in marine, brackish and freshwater prawns (Crustacea, Deacpoda) from northeast Algeria, *Hydrobiologia*, 432, 217, 2000.

119. Johanson, P., Hansen, M.M., Asmund, G., and Neilsen, P.B., Marine organisms as indicators of heavy metal pollution exposure from 16 years of monitoring at a lead–zinc mine in Greenland, *Chem. Ecol.*, 5, 35, 1991.

120. Ridout, P.S., Willcocks, A.D., Morris, R.J., White, S.L., and Rainbow, P.S., Concentration of Mn, Fe, Cu, Zn and Cd in the mesoplagic decapod *Systellaspis debilis* from the east Atlantic Ocean, *Ocean. Mar. Biol.*, 87, 285, 1985.

121. Paez–Osuma, F. and Ruiz–Fernandez, C., Trace metals in the Mexican shrimp *Peneaus vannamei* from estuarine and marine environments, *Environ. Pollut.*, 87, 243, 1995.

122. Rainbow, P.S., Trace metal concentration in a Hong Kong penaid prawn, Metapenaeopsis palmensis (Haswell), in *Proc. 2nd Int. Marine Bio. Workshop: The Marine Flora and Fauna of Hongkong and Southern China*, Morton, B., Ed., Hong Kong University Press, 1986, 1221.

123. Rainbow, P.S., Heavy metals in invertebrates, in *Heavy Metals in Marine Environment*, Furness, R.W., and Rainbow, P.S., Eds., CRC Press, Boca Raton, FL, 1990, 67.

124. Wetzel, R.G., *Limnology: Lake and River Ecosystems*, 3rd ed., Academic Press, San Diego, CA, 2001, 396.

125. Bryan, G.W. and Langston, W.S., Bioavailability, accumulation and effects of heavy metals in sediments with special reference to United Kingdom estuaries: a review, *Environ. Pollut.* 79, 89, 1992.

126. Morales–Hernandez, F., Soto–Jimnez, M.F., and Paez–Osuna, F., Heavy metals in sediments and lobster (*Panulirus gracilis*) from the discharge area of the submarine sewage outfall in Mazatlan Bay (SE Gulf of California), *Arch. Environ. Contam. Toxicol.*, 46, 485, 2004.

127. Tessier, A. and Campbell, P.G.C., Partitioning of trace metals in sediments: relationships with bioavailability, *Hydrobiologia*, 149, 43, 1987.

128. Watling, L., The sedimentary milieu and its consequences for resident organisms, *Am. Zool.*, 31, 789, 1991.

129. Knezovich, J.P., Harrison, F.L., and Wilhelm, R.G., The bioavailability of sediment-sorbed organic chemicals: a review, *Water Air Soil Pollut.*, 32, 233, 1987.

130. Schrap, S.M., Bioavailability of organic chemicals in aquatic environments, *Comp. Biochem. Physiol.*, 100C, 13, 1991.

131. Smith, D.S., Bell, R.A., and Kramer, J.R., Metal speciation in natural waters with emphasis on reduced sulfur groups as strong metal binding sites, *Comp. Biochem. Physiol. C*, 133, 65, 2002.

132. Brown, P.L. and Markich, S.J., Evaluation of the free ion activity model of metal–organisms interaction: extension of the conceptual model, *Aquat. Toxicol.*, 51, 177, 2000.

133. Hodson, P.V., The effect of metal metabolism on uptake, disposition and toxicity in fish, *Aquat. Toxicol.*, 11, 3, 1988.

134. Barron, M.G. and Albeke, S., Calcium control of zinc uptake in rainbow trout, *Aquat. Toxicol.*, 50, 257, 2000.

135. Burger, J., Gaines, K.F., Boring, C.S., Stephens, W.L., Snodgrass, J., Dixon, C., McMohan, M., Shukla, S., Shukla, T., and Gochfeld, M., Metal levels in fish from the Savannah River: potential hazards to fish and other receptors, *Environ. Res A.*, 89, 85, 2002.

136. Langston, W.J. and Spence, S.K., Biological factors involved in metal concentrations observed in aquatic organisms, in *Metal Speciation and Bioavailability in Aquatic Systems*, Tessier, A. and Turner, D.R., Eds., John Wiley & Sons, New York, 1995, 407.

137. Campbell, K.R., Concentrations of heavy metal associated with urban runoff in fish living in storm-water treatment ponds, *Arch. Environ. Contam. Toxicol.*, 27, 352, 1994.

138. Chen, C.Y. and Folt, C.L., Bioaccumulation and dimunition of arsenic and lead in a fresh water food web, *Environ. Sci. Tech.*, 37, 3878, 2000.

139. Lemly, D.A., Guidelines for evaluating selenium data from aquatic monitoring and assessment studies, *Environ. Monit. Assess.*, 28, 83, 1993.

140. Lemly, D.A., Metabolic stress during winter increases the toxicity of selenium to fish, *Aquat. Toxicol.*, 27, 133, 1993.

141. Barron, M.G., Bioaccumulation and bioconcentration in aquatic organisms, in *Handbook of Ecotoxicology*, Hoffman, D.J., Rattner, B.A., Burton, G.A., Jr., and Cairns, J., Jr., Eds., Lewis Publishers, Boca Raton, FL, 1995, 652.

142. Sydeman, W.J. and Jarman, W.M., Trace metals in seabirds, stellar sea lion, and forage fish and zooplankton from central California, *Mar. Pollut. Bull.*, 36, 828, 1998.

143. Phillips, G.R., Lenhart, T.E., and Gregory, R.W., Relations between trophic position and mercury accumulation among fishes from the Tongue River Reservoir, Montana, *Environ. Res.*, 22, 73, 1980.

144. Tayel, F.T.R. and Shriadah, M.M.A., Fe, Cu, Mn, Pb, and Cd in some fish species from western harbor of Alexandria, Egypt, *Bull. Natl. Inst. Oceanogr. Fish.*, 22, 85, 1996.

145. Villar, C., Stripeikis, J., Colautti, D., D'Huicque, L., Tudino, M., and Bonetto, C., Metal contents in two fishes of different feeding behavior in lower Parana River and Rio de la Plata Estuary, *Hydrobiologia*, 457, 225, 2001.

146. Lange, T.R., Royals, H.E., and Connor, L.L., Mercury accumulation in largemouth bass (*Micropterus salmoides*) in a Florida lake, *Arch. Environ. Contam. Toxicol.*, 27, 466, 1994.

147. Stewart, A.R., Luoma, S.N., Schlekat, C.E., Doblin, M.A. and Hieb, K.A, Food web pathway determines how selenium affects aquatic ecosystems: a San Francisco Bay case study, *Environ. Sci. Tech.*, 38, 4519, 2004.

148. Besser, J.M., Brumbaugh, W.G., May, T.W., Church, S.E., and Kimball, B.A., Bioavailability of metals in stream food webs and hazards to brook trout (*Salvelinus fontinalis*) in the upper Animas River watershed, Colorado, *Arch. Environ. Contam. Toxicol.*, 40, 48, 2001.

149. Moore, J.W., Luoma, S.N., and Peters, D., Downstream effects of mine effluent on an intermontane riparian system, *Can. J. Fish Aquat. Sci.*, 48, 222, 1991.

150. Clements, W.H. and Kiffney, P.M., Integrated laboratory and field approach for assessing impacts of heavy metals at the Arkansas River, Colorado, *Environ. Toxicol. Chem.*, 13, 397, 1994.

151. Kay, S.H., Cadmium in aquatic food webs, *Res. Rev.*, 96, 13, 1985.

152. Handy, R.D., Dietary exposure to toxic metals in fish, in *Toxicology of Aquatic Pollution*, Taylor, E.W., Ed., University Press, Cambridge, 1996, 29.

153. Castilhos, Z.C. and Bidone, E.D., Hg biomagnification in ichthyofauna of the Tapajos River region, Amazonia, *Bull. Environ. Contam. Toxicol.*, 64, 693, 2000.

154. Yost, K.J., Miles, L.J., and Parsons, T.W., A method for estimating dietary intake of environmental trace contaminants: cadmium-a case study, *Environ. Int.*, 3, 473, 1980.

155. Kjellstrom, T., Lind, B., Linnman, L., and Elinder, C., Variation in Cd concentration in Swedish wheat and barley, *Arch. Environ. Health*, 30, 321, 1975.

156. Jones, K.C., Jackson, A., and Johnston, A.E., Evidence for an increase in the Cd content of herbage since the 1860s, *Environ. Sci. Technol.*, 834, 26, 1992.

157. Wagner, G.J., Accumulation of cadmium in crop plants and its consequences to human health, *Adv. Agron.*, 51, 173, 1993.

158. Ryan, J.A., Pahren, H.R., and Lucas, J.B., Controlling cadmium in the human chain: review and rational based on health effects, *Environ. Res.*, 28, 251, 1982.

159. McLaughlin, M.J., Parker, D.R., and Clarke, J.M., Metals and micronutrients — food safety issues, *Field Crops Res.*, 60, 143, 1999.

160. Satarug, S., Haswell–Elkins, M.R., and Moore, M.R., Safe levels of cadmium intake to prevent renal toxicity in human subjects, *Br. J. Nutr.*, 84, 791, 2000.

161. van Bruwaene, R., Kirchmann, R., and Impens, R., Cadmium contamination in agriculture and zoo technology, *Experentia*, 40, 43, 1984.

162. Honda, R., Tawara, K., Nishijo, M., Nakegawa, H., Tanebe, K., and Saito, S., Cadmium exposure and trace elements in human breast milk, *Toxicology*, 186, 255, 2003.

163. Frkovic, A., Kras, M., and Alebic–Juretic, A., Lead and cadmium content in human milk from northern Adriatic area of Croatia, *Bull. Environ. Contam. Toxicol.*, 58, 16, 1999.

164. Sonawane, B.R., Chemical contaminants in human milk: an overview, *Environ. Health Perspect.*, 103, 197, 1995.

165. Sherlock, J.C., Cadmium in foods and diet, *Experientia*, 40, 152, 1984.

166. Walker, R. and Herrman, J.L., Summary and conclusions of the joint FAO/WHO Expert Committee on Food Additives, Report 55, World Health Organization, Geneva, 2000.

167. Reeves, P.G. and Vanderpool, R.A., Cadmium burden of men and women who report regular consumption of confectionary sunflower kernels containing a natural abundance of Cd, *Environ. Health Perspect.*, 105, 1098, 1997.

168. Dorea, J.G., Mecury and lead during breast feeding, *Br. J. Nutr.*, 92, 21, 2004.

169. Jensen, S. and Jernelov, A., Biological methylation of mercury in aquatic organisms, *Nature*, 223, 753, 1969.

170. Clarkson, T.W., Mercury: major issues in environmental health, *Environ. Health Perspect.*, 100, 31, 1993.

171. Francesconi, K.A. and Lenanton, R.C.J., Mercury contamination in a semienclosed marine embayment: organic and inorganic mercury content of biota, and factors influencing mercury levels in fish, *Mar. Environ. Res.*, 33, 189, 1992.

172. MacCrimmon, H.R., Wren, C.D., and Gots, B.L., Mercury uptake by lake trout, *Salvelinus namaycush*, relative to age, growth, and diet in Tadenac Lake with comparative data from other Precambrian shield lakes, *Can. J. Fish Aquat. Sci.*, 40, 114, 1983.

173. Cabana, G., Trembley, A., Klaff, J., and Rasmussen, J.B., Pelagic food chain structure in Ontario lakes: a determinant of mercury levels in lake trout (*Salvelinus namaycush*), *Can. J. Fish Aquat. Sci.*, 51, 381, 1994.

174. Lange, T.R., Royals, H.E., and Connor, L.L., Influence of water chemistry on mercury concentration in large mouth bass from Florida lakes, *Trans. Am. Fish Soc.*, 74, 122, 1993.

175. Burger, J., Dixon, C., Boring, C.S., and Gochfeld, M., Effects of deep frying fish on risk from mercury, *J. Toxicol. Environ. Health, Part A*, 66, 817, 2003.

176. Morgan, J., Berry, M.R., and Graves, R.L., Effects of commonly used cooking practices on total mercury concentration in fish and their impact on exposure assessments, *J. Exposure Anal. Environ. Epidemiol.*, 7, 119, 1997.

177. Mahaffey, K.R., Methyl mercury exposure and neurotoxicity, *J. Am. Med. Assoc.*, 280, 737, 1998.

178. Barbosa, A.C. and Dorea J.G., Indices of mercury contamination during breast feeding in the Amazon basin, *Environ. Toxicol. Pharmacol.*, 6, 71, 1998.

179. Drexler, H. and Schaller, K.H., The mercury concentration in breast milk resulting from amalgam fillings and dietary habits, *Environ. Res.*, 77, 124, 1998.

180. WHO, Mercury, IPCS, *Environmental Health Criteria 1*, World Health Organization, Geneva, 1976, 131.

181. Inskip, M.J., and Piotrowski, J.K., Review of the health effects of methylmercury, *J. Appl. Toxicol.*, 5, 113, 1985.

182. Wiersma, D., van Goor, B.J., and van der Veen, N.G., Cadmium, lead, mercury and arsenic concentrations in crops and corresponding soils in the Netherlands, *J. Agric. Food Chem.*, 34, 1067, 1986.

183. Bakir, F., Dalmuji, S., Amin–Zaki, L., Murtadha, L., Khalidi, A., Al-Rawi, N., Tikriti, S., Dhakir, H., Clarkson, T.W., Smith, J., and Doherty, R., Methylmercury poisoning in Iraq. An interuniversity report, *Science*, 181, 230, 1973.

184. Wade, M.J., Davis, B.K., Carlisle, J.S., Klien, A.K., and Valoppi, L.M., Environmental transformation of toxic metals, *Occup. Med.*, 8, 574, 1993.

185. Elliot, H.A., Liberali, M.R., and Huang, C.P, Competitive absorption of heavy metals by soils, *J. Environ. Qual.*, 15, 214, 1986.

186. Needleman, H.L. and Bellinger, D., The health effects of low level exposure to lead, *Ann. Rev. Public Health*, 12, 111, 1991.

187. Mendola, P., Selevan, S.G., Gutter, S., and Rice, D., Environmental factors associated with a spectrum of neurodevelopmental deficits, *Ment. Retard. Dev. Disability Res. Rev.*, 8, 188, 2002.

188. WHO, Minor and trace elements in human milk, *Report of a Joint WHO/IEAE Collaborative Study*, World Health Organization, Geneva, 1989.

189. Chamberlain, A.C., Prediction of response of blood lead to airborne and dietary lead from volunteer experiments with lead isotopes, *Proc. R. Soc. Lond.*, 224B, 149, 1985.

190. Silbergeld, E.K., Lead in bone: implications for toxicology during pregnancy and lactation, *Environ. Health Perspect.*, 91, 63, 1991.

191. O'Neill, P., Arsenic, in *Heavy Metals in Soils*, Alloway, B.J., Ed., John Wiley & Sons, New York, 1990, 83.

192. Hughes, M.F., Arsenic toxicity and potential mechanism of action, *Toxicol. Lett.*, 133, 1, 2002.

193. NRC, *Arsenic in Drinking Water*, Goyer, R.A., Contor, K.P., Eaton, D.L., Henderson, R.F., Kosnett, M.J., Ryan, L.M., Thompson, K.M., Vahter, M.E., and Walker, B., Jr., Subcommittee on arsenic in drinking water, National Research Council, National Academy of Sciences, National Academy Press, Washington, D.C. 2001.

194. Suner, M.A., Deseva, V., Munoz, O., Lopez, F., Montoro, R., Aeias, A.M., and Blanco, J., Total and inorganic arsenic in fauna of the Guadalquivir Estuary: environmental and human health implications, *Sci. Total Environ.*, 242, 261, 1999.

195. Falconer, C.R., Shepherd, R.J., Pirie, J.M., and Topping, G., Arsenic levels in fish and shellfish from the North Sea, *Exp. Mar. Biol. Ecol.*, 71, 193, 1983.

196. Lewis, R., Metals, in *Occupational and Environmental Medicine*, LaDou, J., Ed., McGraw–Hill, New York, 1997, 405.

197. Gochfeld, M., Factors influencing susceptibility to metals, *Environ. Health Perspect.*, 105 (suppl. 4), 817, 1997.

198. Waisberg, M., Joseph, P., Hale, B., and Beyersmann, D., Molecular and cellular mechanisms of cadmium carcinogenesis, *Toxicology*, 192, 95, 2003.

199. Nogawa, K., Kido, T., Yamada, Y., Tsuritani, I., Honda,R., Ishizaki, M., and Terahata, K., α_1-microglobulin in urine as an indicator of renal tubular damage caused by environmental cadmium exposure, *Toxicol. Lett.*, 22, 63, 1984.

200. Horiguchi, H., Teranishi, H., Niiya, K., Aoshima, K., Katoh, T., Sakuragawa, N., and Kasuya, M., Hypoproduction of erythropoieten contributes to anemia in chronic cadmium intoxication: clinical study on Itai-Itai disease in Japan, *Arch. Toxicol.*,68, 632, 1994.

201. Horiguchi, H., Sato, M., Konno, N., and Fukushima, M., Long-term cadmium exposure induces anemia in rats through hypoproduction of erythropoietin in kidneys, *Arch. Toxicol.*, 71, 11, 1996.

202. Aoshima, K., Fan, J., Cai, Y., Katoh, T., Teranishi, H., and Kasuya, M., Assessment of bone metabolism in cadmium-induced renal tubular dysfunction by measurement of biochemical markers, *Toxicol. Lett.*, 136, 183, 2003.

203. Goyer, R.A., Epstein, S., Bhattacharya, M., Korach, K.S., and Pounds, J., Environmental risk factors for osteoporosis, *Environ. Health Perspect.*, 102, 390, 1994.

204. Buchet, J.P., Lauwerys, R., Roels, H., Brenard, A., Bruaux, P., Claeys, F., Ducoffre, G., DePlaen, P., Staessen, J., Amery, A., Lijnen, P., Thijs, L., Rondia, D., Sartor, F., Saint Remy, A., and Nick, L., Renal effects of cadmium body burden of the general population, *Lancet*, 336, 699, 1990.

205. Hellstorm, L., Elinder, C.G., Dahlberg, B., Lundberg, M., Jarup, L., Persson, B., and Axelson, O., Cadmium exposure and end-stage renal disease, *Am. J. Kidney Dis.*, 38, 1001, 2001.

206. Hallenback, W.H., Human health effects of exposure to cadmium, *Experientia*, 70, 136, 1984.

207. Alfven, T., Elinder, C.G., Carlesson, M.D., Grubb, A., Hellstorm, L., Persson, B., Petersson, C., Spang, G., Schutz, A., and Jarup, L., Low-level cadmium exposure and osteoporosis, *J. Bone Miner. Res.*, 15, 1579, 2000.

208. Honda, R., Tsuritani, I., Ishida, M., Noborisaka, Y., and Yamada, Y., Relationship between ultrasonic calcaneal bone measurements and urinary cadmium excretion in women living in non-Cd-polluted area, *J. Bone Miner. Res.*, 12 (suppl 1), S1, 1997.

209. Blakley, B.R. and Rajpal, S.T., The effect of cadmium on antibody responses to antigens with different cellular requirements, *Int. J. Immunopharmacol.*, 8(8), 1009, 1986.

210. Babich, H. and Davis, D.L., Food tolerances and action levels: do they adequately protect children? *Bioscience*, 31, 429, 1981.

211. Spivey Fox, M.R., Nutritional factors that may influence bioavailability of Cd, *J. Environ. Qual.*, 17, 175, 1988.

212. IARC, Cadmium and cadmium compounds, in *Beryllium, Cadmium, Mercury and Exposure in the Glass Manufacturing Industry, IARC Monographs on the Evaluation of Carcinogenic Risks to Humans*, vol. 58, International Agency for Research on Cancer, Lyon, 1993, 119.

213. WHO, Methylmercury, *Environmental Health Criteria*, vol. 101, World Health Organization, Geneva, 1990.

214. Dixon, R. and Jones, B., Mercury concentrations in stomach contents and muscle of five species from northeast coast of England, *Mar. Pollut. Bull.*, 28, 741, 1994.

215. Kerper, L.E., Ballatori, N., and Clarkson, T.W., Methylmercury transport across the blood–brain barrier by an amino acid carrier, *Am. J. Physiol.*, 262, R761, 1992.

216. Castoldi, A.F., Coccini, T., Ceccatelli, S., and Manzo, L., Neurotoxicity and molecular effects of methylmercury, *Brain Res. Bull.*, 55, 197, 2001.

217. Amin–Zaki, L., Elhassani, S., Majeed, M.A., Clarkson, T.W., Doherty, R.A., and Greenwood, M., Intrauterine methyl mercury poisoning in Iraq, *Pediatrics*, 54, 587, 1974.

218. Magos, L., Peristianis, G.C., Clarkson, T.W., Brown, A., Preston, S., and Snowden, R.T., Comparative study of sensitivity of male and female rats to methylmercury, *Arch. Toxicol.*, 48, 11, 1981.

219. Grandjean, P., Weihe, P., White, R.F., and Debes, F., Cognitive performance of children prenatally exposed to "safe" levels of methyl mercury, *Environ. Res. A.*, 77, 165, 1998.
220. Gimenez–Llort, E., Ahlbom, E., Dare, E., Vahter, M., Ogren, S.-O., and Ceccateeli, S., Prenatal exposure to methylmercury changes dopamine-modulated motor activity during early ontogeny: age and gender effects, *Environ. Toxicol. Pharmacol.*, 9, 61, 2001.
221. Sakamoto, M., Nakano, A., and Akagi, H., Declining Minamata male birth ratio associated with increased male fetal death due to heavy methylmercury pollution, *Environ. Res.*, 87, 92, 2001.
222. Costa, M., Zhitkovich, A., Gargas, M., Pautenbach, D., Finley, B., Kuykendau, J., Billing, R., Clarkson, T.J., Wettehahn, J., Xu, J., Patierno, S., and Bogdanffy, M., Interlaboratory validation of a new assay for DNA–protein crosslinks, *Mut. Res.*, 369, 13, 1996.
223. IARC, Mercury, in beryllium, cadmium, *Beryllium, Cadmium, Mercury and Exposure in the Glass Manufacturing Industry, IARC Monographs on the Evaluation of Carcinogenic Risks to Humans*, vol. 58, International Agency for Research on Cancer, Lyon, 1993, 119.
224. WHO, Environmental health criteria 165: inorganic lead, World Health Organization, Geneva, 1995.
225. NEPI, Assessing the bioavailability of metals in soil for use in human health risk assessments, Bioavailability Policy Project Phase II: Metals Task Force Report, National Environmental Policy Institute, Washington, D.C., 2000.
226. Goyer, R.A., Lead toxicity: current concerns, *Environ. Health Perspect.* 100, 177, 1993.
227. Goldsein, G.W., Lead poisoning and brain cell function, *Environ. Health Perspect.*, 89, 91, 1990.
228. Mortada, W.I., Sobh, M.A., El-Defrawy, M.M., and Farahat, S.E., Study of lead exposure from automobile exhaust as a risk for nephrotocixity among traffic policemen, *Am. J. Nephrol.*, 21, 274, 2001.
229. Buchet, J.P., Roels, H., Bernard, A., and Lauwerys, R., Assessment of renal function of workers exposed to inorganic lead, cadmium or mercury vapor, *J. Occup. Med.*, 22, 741, 1980.
230. Goyer, R.A. and Rhyne, B.C., Pathological effects of lead, *Int. Rev. Exp. Pathol.* 12, 1, 1973.
231. Gurer, H. and Ercal, N., Can antioxidants be beneficial in the treatment of lead poisoning? *Free Radic. Biol. Med.*, 29, 927, 2000.
232. Pounds, J.G., Long, G.J., and Rosen, J.F., Cellular and molecular toxicity of lead in bone, *Environ. Health Perspect.*, 91,17, 1991.
233. Gulson, B.L., Mahaffey, K.R., Vidal, M., Jameson, C.W., Lew, A.J., Mizon, K.J., Smith, A.J., and Korsch, M.J., Dietary lead intake for mother/child pairs and relevance to pharmokinetic models, *Environ. Health Perspect.*, 105, 1334, 1997.
234. Gulson, B.L., Jameson, C.W., Mahaffey, K.R., Mizon, K.J., Patison, N., Law, A.J., Korsch, M.J., and Salter, M.A., Relationships of lead in breast milk to lead in blood, urine, and diet of the infant and mother, *Environ. Health Perspect.*, 106, 667, 1998.
235. Rossouw, J., Offermeier, J., and van Rooyen, J.M., Apparent central neurotransmitter receptor changes induced by low-level lead exposure during different development phases in the rat, *Toxicol. Appl. Pharmacol.*, 91, 132, 1987.
236. Lidsky, T.I. and Scheider, J.S., Lead neurotoxicity in children: basic mechanism and clinical correlates, *Brain*, 126, 5, 2003.
237. Leonard, A. and Lauwerys, R.R., Carcinogenicity, teratogenicity and mutagenicity of arsenic, *Mut. Res./Rev. Genetic. Toxicol.*, 75, 49, 1980.
238. Lynn, S., Gurr, J.R., Lai, H.T., and Jan, K.Y., NADH oxidase activation is involved in arsenite-induced oxidative DNA damage in human vascular smooth muscle cells, *Circ. Res.*, 86, 514, 2000.
239. Ramos, O., Carrizales, L., Yanez, L., Mejia, J., Batres, L., Ortiz, D., and Diaz–Barriga, F., Arsenic increased lipid peroxidation in rat tissues by a mechanism independent of glutathione levels, *Environ. Health Perspect.*, 103 (suppl. 1), 85, 1995.
240. Hartmann, A. and Speit, G., Effect of arsenic and cadmium on the persistence of mutagen-induced DNA lesions in human cells, *Environ. Mol. Mutagen.*, 27, 98, 1996.
241. Hartwig, A., Groblinghoff, U.D., Beyersmann, D., Natrajan, A.T., Filon, R., and Mullenders, L.H., Interaction of arsenic(III) with nucleotide excision repair in UV-irradiated human fibroblasts, *Carcinogenesis*, 18, 399, 1997.
242. Wiencke, J.K., Yager, J.W., Varkonyi, A., Hultner, M., and Lutze, L.H., Study of arsenic mutagenesis using the plasmid shuttle vector pZ189 propagated in DNA repair proficient human cells, *Mut. Res.*, 386, 335, 1997.

243. Feng, Z., Xia, Y., Tian, D., Wo, K., Schmitt, M., Kwok, R.K., and Mumford, J.L., DNA damage in buccal epithelial cells from individuals chronically exposed to arsenic via drinking water in Inner Mongolia, China, *Anticancer Res.*, 21(1A), 51, 2001.
244. HopenhaynRich, C., Biggs, M., Fuchs, A., Bergoglio, R., Tello, E., Nicolli, H., and Smith, A., Bladder cancer mortality associated with arsenic in drinking water in Argentina, *Epidemiology*, 7, 117, 1996.
245. Wolnik, K.A., Fricke, F.L., Capar, S.G., Meyer, M.W., Satzger, R.D., Bonnin, E., and Gaston, C.M., Elements in major raw agricultural crops in the United States. 3. Cadmium, lead and 11 other elements in carrots, field corn, onions, rice, spinach and tomatoes, *J. Agri. Food Chem.*, 33, 807, 1985.

33 Trace Metal Accumulation, Movement, and Remediation in Soils Receiving Animal Manure

Karamat R. Sistani and Jeffrey M. Novak

CONTENTS

ABSTRACT

Trace metals are important components in an animal's daily food rations because they are essential for maintaining various physiological processes. Because some animal feeds contain inadequate supplies, trace metals are added into feeds to ensure an optimal supply and to minimize health disorders. Unfortunately, trace metal levels in feeds are frequently fortified to amounts greater than recommended to ensure adequate uptake. Trace metals not retained by the animal are excreted in manure. The return of manure to the land completes a natural recycling process. However, trace metals from manure are also known to be a potential source of pollution to the environment.

The application of animal manures on agricultural lands as a fertilizer has the benefit of resource recycling and waste minimization from livestock production facilities. Research from around the world has shown, however, that continuous manure applications can cause their accumulation in soils because crop uptake of trace metals is small. Although trace metals are accumulating in manure-treated soils, there are few reports of levels reaching phytotoxic threshold concentrations. High levels of trace metals in soils, however, can potentially pose a water quality risk when runoff or leaching transports trace metals into surface water sources. Research has shown that the risk of trace metals causing environmental problems depends upon the element's solubility and binding affinity to organic matter and minerals.

Countermeasures to reduce trace metal concentrations in soils include phytoremediation and additions of soil chemical amendments. It is recommended that trace metal accumulation in soils may be avoided and/or reduced by optimizing their concentration in feed stocks, improving animal trace metal uptake efficiencies, and reducing application rates onto the same fields.

33.1 INTRODUCTION

The well-being of humankind is inflexibly linked to the stock of nutrient elements, including trace metals in the biogeosphere and their capacity for cycling and manipulation. The capacity to produce usable plant biomass depends upon the adequacy and balance of nutrient elements. The uptake of nutrients by plants has been well-researched for over three centuries by European scientists. Early colonial farmers in America knew that some soils were more fertile than others and that the soil nutrients' status depended upon the chemical composition of the geologic strata.

Soils are composed of various rocks and minerals that, after weathering by physical and chemical agents, will release metals into the soil solution, mostly as cations and anions [1,2]. Metals released by these solid phases can be used by plants as nutrients. The quantity of metals required by plants depends on the metabolic functions of the metal in question [3]. Because the amount of metal assimilated by plants varies widely, their essentiality in the plant nutrient cycle has been conveniently classified into groups called macro- and micronutrients (Table 33.1). Macronutrients are required in larger amounts and are utilized for the formation of amino acids, cell membrane components, maintaining electrochemical balances, and formation of enzymes (Table 33.1). Although micronutrients are required in smaller quantities, they still have essential roles in plant physiological reactions (Table 33.1). In this chapter, micronutrients will be referred to as trace metals.

TABLE 33.1
Grouping of Elements into Macro- and Micronutrient Categories and Physiological Functofunctions of Each Trace Metal in Plant Growth

Element	Physiological function in plant
Macro	
N	Structural component of cell wall, amino acid formation
P	Transfer of energy, formation of genetic material
K	Electrochemical balance, nutrient translocation
Ca	Structural component of cell wall
Mg	Core molecule in chlorophyll
S	Component of amino acids, enzymes
Micro	
B	Component of enzymes
Co	Nitrogen fixation
Cl	Electrochemical balance
Cu	Coenzyme, component of chlorophyll
Fe	Chlorophyll component, some enzyme functions
Mn	Chlorophyll component, some enzyme functions
Mo	Nitrogen fixation, conversion of N into amines
Zn	Protein formation, chlorophyll production

Source: Thompson, L.M. and Troeh, F.R., *Soils and Soil Fertility.* 4th ed. McGraw–Hill Publishing Co., New York.

Trace metal concentrations in soils can vary widely because of chemical compositional differences in soil parent materials and geologic deposits. Literature has shown that soils formed over serpentine geologic deposits are high in Cr and Ni [4] and rock deposits containing cinnabar contribute to a high level of soil Hg [5]. Anthropogenic factors are, however, frequently more responsible for high soil trace metal concentrations. For instance, excessively high trace metal concentrations in soils can also occur as a result of additions of fertilizers, sewage sludges, animal manure, irrigation waters, and mine spoils. In localized areas, such as those contaminated by mining or industrial wastes, other non-nutrient elements like Cd, Pb, and Hg may also occur in soils.

Application of animal manures to fields is another recognized pathway for the introduction of trace metals to soils. Cattle, poultry, and swine feeds are frequently fortified with various levels of trace metals in order to maintain various physiological processes and to avoid animal health disorders [6–8]. In some cases, trace metals are supplemented in amounts that exceed the daily recommended allowances because some metals are not easily sorbed by the animal's digestive tract. Trace metals that pass through the animal's gut are excreted in the manure. Literature often reports a close relationship between trace metal intake by livestock and the manure's trace metal concentrations [9–11].

Optimum use of trace metal-enriched animal manures as a beneficial fertilizer requires that application rates be balanced with crop nutritional uptake rates. Agronomic crops assimilate much smaller amounts of trace metals compared to macronutrients [12]. Consequently, if trace metal application rates exceed plant removal rates, it is highly probable that they will accumulate in soil. Studies in Asia, Europe, and the U.S. [10,13–16] have reported trace metal accumulation in soils that receive continuous long-term applications of animal manure.

Trace element accumulation in soil poses serious animal health, environmental, and agronomic issues [17]. Alonso et al. [18] investigated the effects of Cu-enriched pig slurry additions to cattle grazing pastures in Galicia, Spain. They reported that more than 20% of the cattle grazing in the field had liver Cu concentrations that exceeded the potential Cu toxic concentration of 150 mg/kg fresh weight. Cu toxicity has also been reported in sheep and cattle that have been grazing on pastures contaminated with pig slurry [19–21]. Soils with high trace metal concentrations can be a point source of contamination for surface water bodies [22,23] and can cause occurrences of crop phytotoxicity [24]. Trace metals can be transported offsite by runoff [25] and by leaching [26,27].

Soil chemical and physical features have a large influence on the solubility and thus offsite transport potential of trace metals. For example, trace metal bioavailability, mobility, and toxicity are regulated by soil pH, texture, and the kinetic distribution between solution and solid phases [28]. Soil pH is probably the most important intrinsic soil property influencing trace metal mobility [28]. Additionally, binding of trace metals to organic matter is known to control their mobility and leaching potential [29–32].

Remediation strategies to reduce trace metals concentrations in soils and sediments are available. Conventional treatment strategies like soil washing and incineration are destructive to the background matrix. Destruction of key soil quality factors (e.g., organic matter, cation exchange capacity, etc.) in the soil matrix lowers the potential reuse of the material. On the other hand, in some examples, nondestructive remediation technologies can reduce trace metal concentrations or create compounds that render trace metals immobile.

Phytoremediation is a plant-based technology that can reduce trace metal concentrations in contaminated soils and sediments [33–35]. In phytoremediation, specific plants are used to phytoextract target trace metals and then translocate the metal to above-ground plant biomass [34]. After harvesting, the target trace metal is removed from the system. A disadvantage of using phytoremediation is that hyperaccumulator plant growth is slow and production of above-ground biomass is limited. This means that considerable time may be required for plants to reduce trace metal concentrations. Time estimates to remove trace metals from soils have been modeled by examining relationships between target metal concentrations and the quantity of metal removed in the harvested crop. These projected time estimates can vary from a few to several hundred years depending on the reduction level required, yield, bioavailability, leaching potential, and plant uptake [36–38].

Because trace metal accumulation in soils will probably have a long residence time, it is important to understand reasons for the accumulation and to determine soil factors controlling mobility. An understanding of these factors is critical for the development of physical or chemical remediation strategies or adjustments in manure management practices to reduce trace metal accumulation. This review has been designed initially to determine sources and distributions of metals in various animal manure types. Then, an analysis between metal uptake rates by plants vs. manure application rates is examined to determine potential imbalances and accumulation to phytotoxic levels. Pathways that influence trace metal transport offsite into water bodies are then examined. Finally, the chapter concludes with the plant- and chemical-based remediation strategies available as countermeasures to reduce soil trace metal concentrations.

33.2 SOURCES AND DISTRIBUTIONS OF TRACE METALS IN MANURE

Trace metal assimilation by animals is important for the proper functioning of many physiological processes, including reproduction, enzymatic catalysis, and electron transport reactions (Table 33.2) [39,40]. If the supply of trace metals is inadequate in the livestock's diet, then growth and/or health of the animal species can be adversely affected [39].

Animals ingest trace metals through consuming plant materials; however, dietary metal requirements may not be met because of low metal concentrations in the feed sources. In many cases, the absorption and bioavailability of the trace metals by livestock are low; consequently, trace minerals have traditionally been supplemented to animal diets to ensure an adequate supply. Unfortunately, trace metals are frequently supplemented in amounts that exceed the daily recommended intake amount [41]. For example, daily dietary concentrations of 150 to 250 mg/kg Cu_2SO_4 and 2500 to 3000 mg/kg $ZnSO_4$ (more than 25 times the minimum requirement) have been reported to stimulate swine growth [42,43]. On an international scale, the U.S. National Research Council recommends between 3.5 and 6.0 mg Cu/kg in the daily diet of swine (piglets to finishing pigs). However, Europe's current recommendation through Directive 70/534/EEC allows maximum Cu levels between 35 and 175 mg/kg [41]. Current European regulations for the maximum daily Zn content in swine diets are also higher than NRC recommendations.

The reduction of Cu and Zn in dietary supplies is one of the avenues available to limit trace metal accumulations. Jondreville et al. [41] reported that annual Cu and Zn accumulation could be reduced by 35% if Cu and Zn concentrations were reduced in swine feed rations from 100

TABLE 33.2
Physiological Role of Selected Trace Metals in Animal Diets

Trace element	Physiological role in animals
B	Increase bone strength and feed efficiency in swine
Co	Vitamin formation
Cu	Enzyme formation, reproductive processes
Fe	Cytochrome functions and electron transport process
Mn	Enzyme formation, reproductive process
Mo	Enzyme formation
Zn	Enzyme formation, fetus development

Sources: Hostetler, C.E. et al., *Vet. J.*, 166, 125, 2003; Bolan, N.S. and Adriano, D.C., *Crit. Rev. Environ. Sci. Technol.* 34(3): 291-338, 2004.

TABLE 33.3
Trace Metal Concentration Ranges in Cattle, Poultry, and Swine Manure

Trace element	Cattle manures	Poultry manures	Swine manure
		mg/kg dry weight	
B	0.3 to 24	—	—
Co	0.3 to 24	1.2 to 5.0	11
Cu	2 to 62	4.4 to 31	13
Mn	30 to 550	166 to 242	168
Fe	—	80 to 560	—
Mo	0.05 to 49	5 to 42	34
Zn	15 to 250	36 to 158	198

Sources: Kabata–Pendias, A. and Pendias, H., *Trace Elements in Soils and Plants*, CRC Press, Boca Raton, FL, 1984; Edwards, D.R. and Daniels, T.C., *Bioresour. Technol.*, 41, 9, 1992; Arora, C.L. et al., *Indian J. Agric. Sci.*, 45, 80, 1975; Adriano, D.C., Trace Elements in the Terrestrial Environment. Springer-Verlag, Berlin, 1986.

to 20 mg/kg and from 250 to 100 mg/kg, respectively. Spears et al. [44] evaluated the effects of lowering Cu and Zn supplementation from 15 to 5 mg/kg and from 100 to 25 mg/kg, respectively, on trace metal concentrations in swine feces. Results demonstrated that lower Cu and Zn concentrations in dietary intake reduced Cu and Zn excretions in feces by 40% without negatively affecting carcass characteristics or carcass value. Van Heugten et al. [8] demonstrated similar results. Reducing trace metal intake by animals is a possibility, but the lack of relevant information on possible health side-effects (e.g., impairment of immune system [45]) could limit acceptance of this practice.

Not all of the trace metals consumed by animals are absorbed by the animal's digestive tract; consequently, the manure is often trace metal enriched (Table 33.3). For example, swine can excrete approximately 80 to 95% of the total daily Cu and Zn contained in dietary supplements [19,42,46]. Kunkle et al. [9] found that Cu concentrations in poultry litter were linearly related to that in feed and that the litter Cu concentration was about 3.25-fold higher than values measured in the feed. Edwards and Daniels [47] reported that poultry manures were enriched in trace metals like Mn, Fe, Cu, and Zn. From numerous poultry manure compositional investigations, they summarized that the mean Mn, Fe, Cu, and Zn concentrations were 304, 320, 53, and 354 mg/kg, respectively [47]. Barker and Zublena [48] conducted a Cu and Zn concentration assessment in swine manure measured from several swine operations in North Carolina and reported that the mean concentrations were 15 and 62 mg/kg, respectively. These results suggest that animal manures are a significant source of several trace metals and that lack of ruminant assimilation and subsequent deposition of excreta in fields allows these trace metals to accumulate in soils.

Trace metal concentrations in animal manures vary greatly (Table 33.3). This should not be unexpected, considering the large number of management variables associated with manure management systems [49,50]. As shown in Table 33.5, trace metal concentrations like Cu, Fe, Mn, and Zn will vary depending upon source (broiler cake vs. poultry litter vs. feed). The large variation in trace metal concentration is probably a result of animal age, type of feed source, trace metal supplements, bedding differences, and manure storage practices [40]. Because of this large variation in trace metal concentrations, it is difficult to imply that manures from certain livestock types will have a specific trace metal compositional signature.

33.3 TRACE METAL ACCUMULATION IN MANURE-TREATED SOILS

Because some soils can have fertility levels that are out of balance (Table 33.6), animal manures have historically been applied to soils as a fertilizer and to improve the soil's physicochemical properties [51,52]. A large portion of the approximately 10 million Mg of broiler litter (a mixture of manure, wasted feed, feather, and bedding materials such as wood shavings) produced annually in the southeastern U.S. is applied to hay field, pasture, and row crops [53]. After applications of poultry litter to low-fertility soils (Table 33.7), monthly monitoring of trace metal concentrations in topsoils shows that levels do not change dramatically within 1 year after application. Although manures contain essential micronutrients and organic matter, their high trace metal concentration is recognized as a significant source for trace metal accumulation in soils [7,10,14,26].

Brink et al. [54] reported that application of swine effluent (annual mean of 10 ha cm) to Bermuda grass pasture added annual averages of 0.6 kg ha^{-1} Cu, 2.2 kg ha^{-1} Fe, 0.3 kg ha^{-1} Mn, and 0.86 kg ha^{-1} Zn to the soil. Evers [55] also reported application of 9 Mg ha^{-1} broiler litter added averages of 5.85 kg ha^{-1} Zn, 5.0 kg ha^{-1} Fe, 9.4 kg ha^{-1} Cu, and 6.55 kg ha^{-1} Mn to soil. In another study, it was reported [56] that high Cd levels in soils and vegetables across Sydney, Australia, were due to repeated applications of poultry manures. Animal manure applications to agricultural lands in England and Wales accounted for 25 to 45% of the total annual Cu and Zn inputs [7].

Kingery et al. [26] examined long-term application (15 to 28 years) of poultry litter on nutrient levels in several Ultisols in Alabama. They reported that topsoil Cu and Zn concentrations (2.5 and 10 mg kg^{-1}) in poultry litter-treated soils were higher than topsoil Cu and Zn concentrations (0.75 and 2.2 mg kg^{-1}, respectively) measured in control soils (no history of manure application). Nicholson et al. [7] modeled potential annual trace metal loadings in soil treated with 250 kg N ha^{-1} yr^{-1} from various manure sources. The quantities of trace metals were based on the manure dry matter and typical metal concentrations in manures. Significant quantities of Cu and Zn were added to soil through the application of swine (2.2 and 1.6 kg ha^{-1}), poultry (1.1 and 0.5 kg ha^{-1}), and cattle (1.0 and 0.3 kg ha^{-1}) manures, respectively. These researchers noted that the modeled trace metal loading rates may be higher in areas where Cu- and Zn-enriched swine and poultry manures have been applied for many years [7].

Many studies have suggested additional reasons for trace metal accumulation in manure-treated soils. These reasons include repeated applications of manure to the same field and a low trace metal crop nutrient requirement. Repeated manure applications to the same fields occur because manure transportation costs are high and land availability for manure disposal is limited. Frequently, animal manures are not transported very far, and they are often applied close to the production facility. Land available for manure application continues to decline because of high real estate prices, encroachment of urban sprawl, zoning restriction, and exhaustion of the soil's nutrition assimilation capacity. Expecting some animal producers to locate more land available for manure application may not be economically feasible. Unless directed to cease by a restrictive nutrient management plan, some fields will probably continue to receive trace metal-enriched animal manures.

The low trace metal requirement by crops is also an important reason for their accumulation. Researchers have conducted experiments to determine the sufficiency level of trace metals utilized by agronomic crops (Table 33.4). Data in this table exemplify that the sufficiency range of trace metals in crop tissue is much smaller than for macronutrients (Table 33.4). For example, the nutrient sufficiency range for trace metals like Co and Mo are orders of magnitude smaller than for macronutrients like N and K. Through field experiments, researchers have measured the amount of trace metals removed by harvesting crops. For example, Bermuda grass (*Cynodon dactylon* L.), legumes, and other temperate annual and perennials grasses are commonly used in southern and southeastern U.S. pasture systems to assimilate nutrients contained in animal manures [51,54,55,57,58]. The grass can remove 0.009 and 0.218 kg of Cu and Zn, respectively, per 7.3 Mg/ha yield of above-ground biomass [59].

TABLE 33.4
Sufficiency Range of Macro- and Micronutrients Assimilated by Agronomic Crops

Element	Sufficiency range (mg/kg, dry weight)	Plant part measured in
Macro		
N	1000 to 6000	Leaf tissue
P	200 to 500	Mature leaves
K	1500 to 4000	Recent and mature leaves
Ca	500 to 1500	Mature leaves
Mg	150 to 400	Dry matter
S	150 to 500	Leaf tissue
Micro		
B	20	Dry matter
Co	0.021 to 0.55	Dry matter
Cl	50 to 200	Dry matter
Cu	2 to 20	Leaf tissue
Fe	50 to 75	Leaf tissue
Mn	10 to 200	Leaf tissue
Mo	0.15 to 0.30	Dry matter
Zn	15 to 20	Leaf tissue

Source: Jones Jr., J.B., *Agronomic Handbook, Management of Crops, Soils and their Fertility.* CRC Press, Boca Raton, FL, 2003.

TABLE 33.5
Calcium, Magnesium, Potassium, Copper, Iron, Manganese, and Zinc Content of Broiler Cake, Litter, and Feed

Variables	Ca	Mg	K	Cu	Fe	Mn	Zn
	g kg^{-1}			mg kg^{-1}			
Broiler cake	99.1 (8.7)a	24.4 (2.1)a	125.7 (8.9)a	2763 (72)a	3818 (119)a	2307 (102)a	1848 (92)a
Broiler litter	26.2 (2.4)b	6.1 (0.9)b	30.3 (1.5)b	662 (43)b	1055 (57)b	556 (23)b	36 (27)b
Broiler feed	10.1 (1.3)	1.8 (0.3)	9.0 (1.1)	210 (19)	202 (18)	169 (14)	139 (11)

Notes: Standard error in parentheses; means followed by the same letter in each column (excluding feed) were not significantly different at a 0.05 probability level according to Tukey's test.

Source: Sistani, K.R. et al., *Bioresour. Technol.*, 90(1), 27, 2003.

Novak and Watts [38] examined the effects of applying swine manure effluent to a 1-ha field in North Carolina cropped with coastal Bermuda grass. They estimated that approximately 10.5 and 15.6 kg of Cu and Zn, respectively, were applied annually after spraying 4.32×10^6 L ha^{-1} of swine manure effluent. Performing a mass balance revealed that, after grass Cu and Zn removal with harvested above-ground biomass, approximately 10.49 and 15.38 kg of Cu and Zn, respectively, would remain in the 1-ha field. This shows that the annual rate of Cu and Zn additions wase much higher than the grasses' Cu and Zn removal rate.

The investigation by Novak and Watts [38] is an important demonstration of why trace metals accumulate in soils. If trace metals accumulate to phytotoxic levels, sensitive crops can potentially suffer serious yield declines [24,60,61]. Although agronomic crops vary in their sensitivity to trace

TABLE 33.6
Initial Soil Chemical Analyses[a] and Nutrient Composition of Broiler Litter Applied in 2000 and 2001

	pH	N	P	K	Ca	Mg	Cu	Fe	Mn	Zn
	g kg^{-1}				mg kg^{-1}					
Soil										
0–5 cm depth	6.2	1.59	0.74	0.30	1.09	0.13	29	283	160	35
5–10 cm depth	6.4	0.56	0.35	0.17	0.59	0.09	10	231	192	7
Litter										
May (2000)	7.7	34.9	22.5	29.6	30.1	6.1	687	837	631	416
July (2000)	7.8	32.7	19.8	28.7	29.8	6.5	632	795	703	456
May (2001)	7.4	32.5	20.8	29.1	30.9	6.4	541	1040	657	455
July (2001)	7.6	30.7	18.1	28.3	26.2	5.8	457	702	559	380

[a] Mehlich 3, except for pH and N.

Source: Sistani, K.R. et al., *Agron. J.*, 96, 525, 2004.

TABLE 33.7
Soil Concentrations of Copper, Iron, Manganese, and Zinc at Different Sampling Times in 2000 and 2001

Sampling dates	Soil depth (cm)	Cu	Fe	Mn	Zn
		Mg/kg			
Jan 2000	0–10	19.8 ± 7.2[a]	257.7 ± 36.3	176.5 ± 30.5	21.2 ± 15.3
Feb 2000	0–10	20.7 ± 7.3	236.7 ± 42.3	172.6 ± 29.6	26.8 ± 18.2
Mar 2000	0–10	32.5 ± 6.9	287.1 ± 10.4	153.4 ± 16.1	40.9 ± 11.1
Apr 2000	0–10	43.6 ± 13.2	511.8 ± 70.5	284.5 ± 54.7	57.6 ± 19.5
May 2000	0–10	36.3 ± 8.6	459.2 ± 43.1	250.9 ± 46.4	46.6 ± 16.8
Jun 2000	0–10	50.3 ± 13.6	483.2 ± 29.8	260.6 ± 32.8	56.4 ± 19.3
Jul 2000	0–10	38.2 ± 7.6	321.4 ± 27.2	160.6 ± 23.8	43.0 ± 12.6
Aug 2000	0–10	21.9 ± 6.9	499.5 ± 66.6	239.4 ± 41.1	37.3 ± 10.4
Sep 2000	0–10	27.3 ± 7.7	301.9 ± 31.3	132.9 ± 20.2	31.7 ± 12.5
Oct 2000	0–10	34.7 ± 7.7	423.5 ± 26.2	172.4 ± 21.9	35.7 ± 10.7
Nov 2000	0–10	33.3 ± 7.4	372.4 ± 23.3	156.4 ± 25.8	34.9 ± 10.6
Dec 2000	0–10	25.8 ± 8.9	118.8 ± 1.6	182.2 ± 30.8	39.6 ± 14.2
Mar 2001	0–10	31.0 ± 6.3	421.0 ± 46.8	145.4 ± 17.9	41.0 ± 12.1
Jun 2001	0–10	31.3 ± 6.2	109.6 ± 3.0	125.3 ± 12.5	32.6 ± 10.1
Sep 2001	0–10	37.2 ± 7.2	302.8 ± 36.7	108.1 ± 14.5	47.2 ± 15.8
Dec 2001	0–10	32.8 ± 6.1	220.3 ± 14.9	85.0 ± 11.9	46.0 ± 14.0

[a] Data points are averages of 24 values, ± standard deviations, and extracted by Mehlich 3.

Source: Sistani, K.R. et al., *Bioresour. Technol.*, 90(1), 27, 2003.

metal phytotoxicity [3,62], critical plant available (Mehlich 3 extractable) concentrations in North Carolina soils for sensitive plants have been reported as >120 mg kg^{-1} for Zn and >60 mg kg^{-1} for Cu [62]. Some crops, however, are sensitive to soil Zn concentrations far below 120 mg kg^{-1}. For instance, peanuts (*Arachis hypogaea* L.) are sensitive to Zn concentrations as low as 12 mg kg^{-1} [63]. Because peanuts are sensitive to low soil Zn concentrations, fields used for their production have a Zn concentration limit of 20 mg kg^{-1} [62].

The concern about possible trace metal phytotoxicity and a subsequent decline in crop yields due to manure applications has been previously investigated. These studies have shown that agronomic crops grown in soils treated with animal manure can cause an increase in trace metal concentration in pasture grasses [64], in corn leaves (*Zea mays* L.) [65,66], and in corn ear leaf tissue [67]. In the study by Batey et al. [64], pasture grasses treated with Cu-enriched pig slurry had elevated Cu concentrations, but not to a phytotoxic level. Much of the increase in Cu concentration was attributed to direct foliar contamination during manure application [64]. Korneagy et al. [66] applied approximately 103 kg Cu ha^{-1} as Cu-enriched pig manure over a 3-year period to a corn crop; this resulted in a doubling of the Cu level in corn roots and slight increases in the ear leaves, but did not influence grain yield or Cu concentration in leaves.

Mullins et al. [68] reported similar Cu concentration results in corn ear leaves following application of Cu-enriched swine manure to three Virginia soils. In this study, corn yields did not decline in spite of the high Cu uptake concentrations. Sutton et al. [69] reported that four annual applications of Cu-enriched swine manure (15.3 kg Cu ha^{-1}) neither decreased corn grain yield nor increased Cu concentrations in corn tissues. These studies show that although Cu-enriched manure applied to soils can increase crop tissue Cu concentrations, the increase in Cu uptake is insufficient to influence yields adversely. Except for the sensitivity of peanuts to soil Zn concentrations, examination of the literature suggests a minimal number of reports of decreased crop yields caused by trace metal accumulations in manure-treated soils.

33.4 PATHWAYS FOR OFFSITE METAL TRANSPORT

There are many physically and chemically based pathways for offsite trace metal transport from manure-treated soils. This review will focus on two common pathways: trace metal movement in runoff and leaching through soils. A few field runoff simulation studies have shown high Cu, Fe, and Zn concentrations in runoff from poultry litter-treated soils. Less information exists in the literature documenting runoff metal losses from swine and cattle manure-treated soils, however. With regards to leaching risks, some long-term field studies show limited migration of metals in the soil profile because of strong metal binding by topsoil organic matter.

In contrast to these reports, other field sites treated with manures have revealed that Cu and Zn can leach to subsoil depths. Leaching of trace metals in manure-treated soils is facilitated by dissolved organic carbon compounds [40]. It is important that the ability of offsite trace metal transport into water bodies and aquatic ecosystems be known because studies have reported that metal enrichment can pose a significant threat to water quality [23] and can cause detrimental impacts on aquatic plants, shellfish, and other wildlife [70].

33.4.1 RUNOFF

The few investigations of trace metal movement via runoff from manure-treated soils have reported that runoff losses from fields treated with poultry litter can contain measurable concentrations of trace metals such as Cu, Fe, and Zn [25,71]. Moore et al. [25] reported that runoff collected from plots in Arkansas treated with 2 to 9 Mg ha^{-1} of poultry litter contained an average soluble Cu concentration of 0.93 mg L^{-1} and that runoff from control plots averaged 0.01 mg Cu L^{-1}. In a similar study, runoff samples from plots in Arkansas treated with 5 Mg ha^{-1} of poultry litter had average concentrations of Fe (0.2 mg L^{-1}) and Zn (0.06 mg L^{-1}) that were not high enough to be a direct threat to aquatic life [71]. Concentrations above 0.3 mg L^{-1} and 0.18 mg L^{-1} of Fe [72] and Zn [73], respectively, are judged to be potentially harmful to freshwater fish.

On the other hand, Cu concentrations (mean of 0.45 mg L^{-1}) in runoff samples were high enough to be of environmental concern [71]. Elevated Cu concentrations in streams can be toxic to algae [70] and maximum concentrations of 0.02 mg L^{-1} are recommended to protect freshwater fish. These studies show that runoff samples collected from plots treated with poultry litter can contain elevated Cu concentrations that, if not diluted, can be a concern for fish in freshwater surface water systems. It can also be perceived that the risk of trace metal contamination of water bodies might not be as great as considered, despite elevated concentrations at the soil surface, because interaction of trace metals with topsoil organic matter may limit transport of trace metals across the landscape. Further studies are needed to evaluate trace metal movement in runoff, particularly using several different manure types (e.g., swine, cattle, etc.).

33.4.2 Leaching

Leaching of trace metals in manure-treated fields is commonly investigated by sampling topsoil and subsoil depths and extracting for available trace metals using various reagents (e.g., double acids, Mehlich 3, chelating reagents, etc.). This approach has been used in several studies in which soils were treated with poultry litter [26], swine manure [13,38,64], and cattle manure [74,75]. These investigations have reported contrasting results concerning trace metal leaching to subsoil depths.

Trace metals are generally regarded as relatively immobile in most soils because they are bound to soil organic matter [30,76] and are sorbed by Fe–Mn oxides and clays, and precipitated as carbonates, hydroxides, and phosphates [77]. Batey et al. [64] applied up to 12.1 kg ha^{-1} of Cu-enriched hog manure slurry onto a field containing a mixture of perennial ryegrass (*Lolium perenne* L.) and white clover (*Trifolium repens* L.). Analyses of soil profile samples taken from this field showed that trace metal accumulation was limited to the top several centimeters of the soil profile [64]. Lack of leaching to subsoil depths supports the contention that Cu will mostly accumulate in topsoil by forming strong bonds with various soil organic and inorganic components. Other manure application studies related to Cu accumulation or leaching in soil reported similar results [67,74,75].

Few investigations have reported trace metal leaching to subsoil depths in fields treated with swine manure [13,38] and poultry litter [26]. King et al.[13] treated two North Carolina Ultisols with swine lagoon effluent (1340 kg N ha^{-1} yr^{-1}) and reported significantly different Cu concentrations between control and treated soils in samples collected between 15 and 30 cm deep. In this study, swine manure treatment on topsoil and subsoil Zn and Mn concentrations had no significant effects.

Novak and Watts [38] collected topsoil (0 to 15 cm) and subsoil samples (15 to 45 cm deep) in North Carolina Ultisols treated for 10 years with swine manure effluent. Nearby soils without a history of swine manure effluent application were collected in a similar manner to that used for controls. Copper and Zn were extracted from the soils using Mehlich 3 extractant that contains the chelating agent ethylenediaminetetraacetic acid (EDTA) [78]. After 10 years of swine effluent application, topsoil and subsoil mean Cu concentrations in the treated soils were similar to those of controls (Table 33.8). Novak and Watts [38] found higher mean Zn concentrations in treated topsoils than in those of the controls. In subsoil depths (15 to 45 cm), however, the Mehlich 3 extractable mean Cu and Zn concentrations in treated vs. control soils showed no differences, indicating very minimal leaching (Table 33.8).

In a similar study, Kingery et al. [26] examined the effects of long-term (15 to 28 years) poultry litter applications (6 and 22 Mg/ha/yr) to some Ultisols in the Sand Mountain region of Alabama. They collected topsoils and subsoils and extracted for available Cu and Zn, using a double acid extractant [79]. These workers found the highest Cu concentrations in topsoils (0 to 15 cm deep) treated with litter compared to samples without litter application. Additionally, litter-treated soils collected at 30 cm had higher mean Cu concentrations than in samples without litter application. Below 45 cm, the mean Cu concentrations showed no differences among treatments [26]. Similarly, the greatest Zn concentrations were measured in topsoils treated with poultry litter. There were minimal differences in mean Zn concentrations between treatments measured in subsoil samples

TABLE 33.8
Mean Cu and Zn Concentrations Extracted Using Mehlich 3 Reagent from Control Soils (0 years) and from Soils after 10 years of Swine Manure Effluent Application

Metal	Depth (cm)	Control soils (0 yrs)[a] (mg/kg)	Soils treated for 10 years (mg/kg)
Cu	0 to 15	0.48 (0.30)a	3.59 (2.77)a
	15 to 45	0.27 (0.10)a	0.86 (0.46)a
Zn	0 to 15	1.69 (1.39)b	7.87 (4.67)c
	15 to 45	1.39 (0.78)d	1.33 (0.41)d

[a] Mean Cu and Zn concentrations after 0 vs. 10 years' effluent application were tested by depth using a *t*-test and means followed by a different letter were significantly different at a $P < 0.05$ level of rejection.

Source: Novak, J.M. and Watts, D.W., *Trans. ASAE* (in review), 2004.

(30 to 45 cm deep). In work done with cattle manure applications, Chang et al. [75] found Zn accumulations to a depth of 30 cm.

The leaching of Cu and Zn to subsoil depths has been explained by metal chelation with soluble organic compounds from the litter [80,81]. It appears that the highest trace metal concentrations in manure-treated fields are measured in topsoil. By extracting subsoils and finding an increase in Cu and Zn concentrations, scientists have suggested that some downward leaching of Cu and Zn can occur in manure-treated soils. Measured subsoil concentration increases are, however, small. Nevertheless, lack of substantial Cu and Zn concentrations in subsoil depths cannot exclude the fact that soluble forms of trace metals can leach through the soil profile and could cause enrichment in shallow ground water. Trace metal enrichment of shallow ground water as a result of leaching has not been sufficiently documented in soils treated with animal manures. Additional studies examining trace metal concentrations in shallow ground water beneath manure-treated fields should be conducted to determine whether trace metal leaching is a significant water quality issue.

33.5   REMEDIATION STRATEGIES

When trace metals accumulate in soils to threshold concentrations that pose a detrimental agronomic or environmental effect, a number of remediation strategies are available to reduce or immobilize their concentrations. This review will focus on two nondestructive remediation strategies: phytoremediation and *in situ* chemical immobilization using chemical amendments. Conventional remediation strategies like incineration, soil washing, solidification, or removal of contaminated soil are costly and often harmful to desirable soil properties like organic matter, cation exchange capacity, and texture. More recently, nondestructive remediation technologies have been employed that do not require offsite transport of contaminated materials; instead, metals are removed by growing crops or immobilized as insoluble metal forms by adding a chemical amendment.

33.5.1   PHYTOREMEDIATION

The development of phytoremediation, employing a plant-based approach to remediate metals, has previously been reviewed [33–35,37]. Crops utilized in phytoremediation can extract the target metal from soil and then hyperaccumulate the metal in above-ground plant biomass [35]. The metal-enriched plant biomass can be harvested and removed from the contaminated site, thus avoiding extensive excavation and disposal costs, and loss of topsoil through removal [82]. Plants capable of hyperaccumulating metals are usually specific to one target metal [83]. Because of this trait, maximizing metal removal through phytoremediation probably will require growth and cultivation of several sequential plants to reduce soil metal concentrations to acceptable levels [33].

The success of the phytoremediation process depends on adequate plant yields, high metal concentration in above-ground plant biomass, and harvesting of plant biomass to ensure metal removal from the contaminated site [35]. Hyperaccumulator plants possess an ability to take up abnormally high amounts of metals in their shoots [35]. Baker and Brooks [84] have reported a number of plant species capable of accumulating exceptionally high concentrations (usually > 0.1% in leaf dry matter) of Ni, Zn, Cu, Co, and Pb in their above-ground plant parts. These metal hyperaccumulator plant species have been recognized on the basis of dry leaf matter metal concentrations, often growing in mine waste areas with high metal concentrations.

Baker et al. [83] reported that a few plants in the genera *Thlaspi* and *Cardaminopsis* in the Brassicaceae family are capable of accumulating Zn to 1.0% (dry weight). Several other *Thlaspi* species isolated from lead/zinc-mineralized soils in central European countries have been reported with up to 2% Zn in leaf tissue [85]. High Cu concentrations (up to 13,700 $\mu g\ g^{-1}$) have been measured in leaf tissue of *Aeollanthus biformifolius* De Wild. (Lamiaceae), a dwarf perennial herb endemic to the southern part of Zaire [86]. Baker and Brooks [84] have identified 26 hyperaccumulator plants for Co (> 1000 $\mu g\ g^{-1}$), all of which are native to Zaire. The majority are slow growing herbs in a range of families including Lamiaceae, Scrophulariaceae, Asteraceae, and Fabaceae [83]. Baker et al. [83] reported that the highest Co concentration measured in a hyperaccumulator was 10, 200 $\mu g\ g^{-1}$ in *Haumaniastrum robertii* (Robyns) Duvign. Et Plancke (Lamiaceae).

Hyperaccumulator plant species are slow growing with low biomass production, so their utilization without some form of trait modification is limited. Therefore, using plant species with high biomass production, such as pea (*Pisum sativum* L.), oat (*Avena sativa* L.), canola, (*Brassica napus* L.), and Indian mustard (*Brassica juncea* L. Czern.), has been suggested to enhance metal uptake and removal from the contaminated site [82,87]. These plants can produce more above-ground biomass; however, they have other limitations, such as their lower metal uptake amount and need for high amounts of macronutrients like N and P.

Another geochemical factor that limits the success of phytoremediation is the metal's bioavailability [88,89]. Metal bioavailability can be increased by adding a chelating agent like EDTA to the metal-contaminated soils. This approach was employed by Shen et al. [90], who treated Pb-contaminated soil with EDTA. They reported that cabbage (*Brassica rapa* L.) shoots had elevated Pb concentrations between 4620 and 5010 mg/kg dry matter compared to controls with only 126 and 101 mg/kg at days 7 and 14, respectively, after treatment. Other investigations have added chelating agents such as HEDTA (hydroethylenediaminetriacetic acid) and NTA (nitrilotriacetic acid) to enhance metal uptake by plants in pot and field experiments [82,91].

Although it has been recognized since the 1990s that metal uptake by hyperaccumulator plants could be a useful technology for metal remediation, slow growth coupled with some soil ediphatic conditions has limited its successful application. Phytoremediation is limited by soil ediphatic conditions that inherently limit plant growth, e.g., lack of water, low/high soil pH values, and availability of nutrients [35]. If growth conditions are improved by adjusting soil pH, optimizing soil moisture, and supplying micronutrients, phytoremediation may be a viable technology for trace metal reduction. To date, the utilization of phytoremediation has not proved to be an economically viable solution to reduce soil trace metal concentrations.

33.5.2 IMMOBILIZATION USING CHEMICAL AMENDMENTS

It has been mentioned previously that some conventional trace metal remediation technologies can reduce metal concentrations, but can also be destructive to the background media. These forms of metal remediation technologies may be unsuitable in situations in which the background material may be used to support subsequent plant growth for recreational purposes. Recent studies have demonstrated that *in situ* contaminant immobilization may be the preferred approach to remediate metals in soils and shallow sediments [92,93].

In situ immobilization methods typically reduce the mobility and bioavailability of the target metal by redirecting solid-phase speciation in favor of less labile phases through sorption or precipitation modes [94,95]. This type of technology requires the addition of a chemical amendment to the contaminated media that has a higher affinity for the metal than the background matrix. The amendment, however, should not impair ediphatic processes responsible for maintaining good soil quality. Otherwise, other chemical and physical problems may be created by adding chemical amendments to contaminated soil or sediments.

Many additives have been screened for their potential to immobilize heavy metals in soils. Many of these additives are alkaline materials such as lime [96,97], zeolites [80], incinerator ashes [98], Fe-rich byproducts from TiO_2 pigment production [99], and hydroxyapatite [100]. In addition to the amendment fixing the metal, additions of alkaline amendments will increase soil pH values, thus causing other exchange sites (present on clay surfaces, iron oxides, and organic matter) to be more reactive to metal binding [80]. Applications of chemical amendments have been shown to bind significant amounts of Co, Cu, and Zn in contaminated soils and sediments. For example, apatite was effective at binding Co, Cu, and Zn in contaminated media [92,93]. Additionally, significant binding of Zn in contaminated materials by Fe-oxides [99], and by zeolites [80,92] has been reported.

Some problems have been noted with the use of chemical amendments to immobilize metals. The amendment may contain trace metal impurities that may exceed environmental limits if applied at sufficiently high concentrations. Additionally, because the same metal sorption mechanisms will also bind nontarget macronutrients such as P and Fe, amendment may reduce the concentrations of essential plant nutrients [92].

33.6 CONCLUSIONS

Trace elements are inorganic components in feeds and are part of the chemical makeup of the animal's body. They are essential for maintaining various physiological processes and are required in sufficient amounts to avoid health disorders, reproductive problems, or disruption in the production of animal products for consumers (milk, wool, eggs, etc.). Consequently, to reduce the incidence of animal health problems and disruption in product generation, trace metals are supplemented in the animal's daily rations. Unfortunately, in some cases, trace elements are supplemented in amounts that exceed the daily recommendations. Under these scenarios, trace metals not assimilated by the animal digestive system resulted in the production of trace metal-rich manures. Crops do not require large amounts of trace metals, and the scientific literature has shown that continual application of trace metal-enriched manures to the same fields results in their accumulation.

Trace metal accumulation in soils is not an isolated incident, but rather is a worldwide issue because it has been reported in Europe, Asia, and the U.S. Agronomic and environmental issues concerning trace metal accumulation have been discussed in this review. Few studies have shown a phytotoxic effect of trace metals causing a serious crop yield decline. One study showed that high soil Zn contents can be phytotoxic to sensitive crops like peanuts. Except for peanuts, trace metal accumulation in manure-treated soils has not caused widespread crop yield declines.

As a matter of fact, some investigations have shown that Cu-enriched manure applications to corn have not resulted in elevated Cu concentrations in tissue samples. Nevertheless, low trace metal uptake by agronomic crops implies that metals will continue to accumulate in manure-treated soils. If current manure management protocols do not change for the next 100 years, some fields in Europe may contain soils with sufficient trace metal concentrations for crop growth of sensitive crops to be adversely affected [10].

The buildup of trace metals in soils probably should be a concern because it may contribute to future environmental issues. For instance, studies have shown that runoff from manure-treated fields had elevated Cu concentrations that, if undiluted prior to entry into a water body, could be toxic to fish. Data on the long-term consequences of trace metal accumulation in soils with respect to ground- and surface water enrichment are lacking. Long-term monitoring of Cu and Zn concentrations in ground- and surface water systems in watersheds high in animal production may be needed to address potential issues of water quality impairment by trace metals.

Countermeasures to reduce trace metal concentrations in soils include plant removal (phytoremediation) and immobilizing the metal using chemical amendments. Both of these technologies have been shown to be effective at reducing metal concentrations, but they have inherent limitations. Areas where phytoremediation could be used may be inhospitable for plant growth, thereby requiring additional management inputs to produce a harvestable crop. Chemical amendments added to soils may bind nontarget plant nutrients and can contain trace metals as contaminants.

Perhaps the most significant countermeasure to reduce trace metal accumulation is simply by a reduction in the trace metal supplementation in feedstuffs [41]. A possible mechanism to achieve this is to lower the maximum permitted trace metal levels in feedstuffs and feed additives. A problem with this approach is that some animals may suffer health problems because of a reduction in trace metal uptake. Further measures, such as improved trace metal assimilation through reformulating the organic/inorganic compounds that carry the trace elements or the addition of phytase, can complement the reduction or even make the reduction feasible.

REFERENCES

1. Kabata–Pendias, A. and Pendias, H., *Trace Elements in Soils and Plants*, CRC Press, Boca Raton, FL, 1984.
2. Pierzynski, G.M., Sims, J.T., and Vance, G.F., *Soils and Environmental Quality*. Lewis Publ., Boca Raton, FL, 1994.
3. Marschner, H., *Mineral Nutrition of Higher Plants*, 2nd ed., Academic Press, New York, 1995.
4. Proctor, J. and Woodell, S.R.J., The ecology of serpentine soils. *Adv. Ecol. Res.*, 9, 255, 1975.
5. Harsh, J.B. and Doner, H.E., Characterization of mercury in a river-wash soil. *J. Environ. Qual.*, 10, 333, 1981.
6. Capar, S.G., Tanner, Friedman, and Boyer, Multielement analysis of animal feed, animal wastes, and sewage sludge. *Environ.Sci. Technol.*, 7, 785, 1978.
7. Nicholson, F.A., Chambers, Williams, and Unwin, Heavy metal contents of livestock feeds and animal manures in England and Wales. *Bioresources Technology.*, 23, 23, 1999.
8. Van Heugten, E., O'Quinn, Funderburke, Flowers, and Spears, Effects of supplemental trace mineral levels on growth performance, carcass characteristics, and fecal mineral excretion in growing-finishing swine. Available at: (http://mark.asci.ncsu.edu/SwineReports/2002/vanheugten1.htm), 2002.
9. Kunkle, W.E., Carr, Carter, and Bossard, Effects of flock and floor type on the levels of nutrient and heavy metals in broiler litter. *Poultry Sci.*, 60, 1160, 1981.
10. Van Driel, W. and Smilde, K.W., Micronutrients and heavy metals in Dutch agriculture. *Fertilizer Res.*, 25, 115, 1990.
11. Miller, R.E., Lei, X., and Ullrey, D.E., Trace elements in animal nutrition. In: *Micronutrients in Agriculture*, 2nd ed., J.J. Mortvedt (Ed.), Soil Science Society America, Madison, WI, 1991, 593.
12. Jones Jr., J.B., *Agronomic Handbook, Management of Crops, Soils and their Fertility.* CRC Press, Boca Raton, FL, 2003.
13. King, L.D. et al., Westerman, Cummings, Overcash, and Burns, Swine lagoon effluent applied to "coastal" Bermuda grass: II. Effects on soil. *J. Environ. Qual.*, 14, 14, 1985.
14. Eneji, A.E., Honna, T., and Yamamoto, S., Manuring effect on rice grain yield and extractable trace elements in soils. *J. Plant Nutr.*, 24, 967, 2001.
15. Hansen, M.N., Risk of heavy metal accumulation in agricultural soil when livestock manure and organic waste is used for fertilizer. In *Proc. Global Perspective in Livestock Waste Management, 4th International Livestock Waste Management Symposium and Technology Expo.*, Penang, Malaysia, 2002, 269.

16. Franzluebbers, A.J., Wilkinson, S.R., and Stuedemann, J.A., Bermuda grass management in the southern Piedmont, USA: IX. Trace elements in soil with broiler litter application. *J. Environ. Qual.*, 33, 778, 2004.
17. Senesi, G.S., Baldassarre, Senesi, and Radina, Trace element inputs into soils by anthropogenic activities and implications for human health. *Chemosphere,* 39, 343, 1999.
18. Alonso, M.L., Benedito, Miranda, Castillo, Hernandez, and Shore, The effect of pig farming on copper and zinc accumulation in cattle in Galicia (Northwestern Spain). *Vet. J.*, 160, 259, 2000.
19. Parkinson, R.J. and Yells, R., Copper content in soil and herbage following pig slurry application to grassland. *J. Agric. Sci.*, 105, 183, 1985.
20. Christie, P. and Beattie, J.M., Grassland soil microbial biomass and accumulation of potentially toxic metals from long-term slurry application. *J. Appl. Ecol.*, 26,597, 1989.
21. Poole, D.B., McGrath, Fleming, and Moore, Effects of applying copper-rich pig slurry to grassland. *Irish J. Agric. Res.*, 29, 34, 1990.
22. Pierzynski, G.M., Sims, J.T., and Vance, G.F., *Soils and Environmental Quality.* 2nd ed. CRC Press, Boca Raton, FL, 1990.
23. Xue, W., Nhat, Gachter, and Hooda, The transport of Cu and Zn from agricultural soils to surface water in small catchment. *Adv. Environ. Res.*, 8, 69, 2003.
24. Wong, M.H. and Bradshaw A.D., A comparison of the toxicity of heavy metals, using root elongation of rye grass (*Lolium prenne.*). *New Phytol.*, 91, 255, 1982.
25. Moore, P.A., Jr., Daniels, Gilmour, Shreve, Edwards, and Wood, Decreasing metal runoff from poultry litter with aluminum sulfate. *J. Environ. Qual.*, 27, 92, 1998.
26. Kingery, W.L., Wood, Delaney, Williams, and Mullins, Impact of long-term land application of broiler litter on environmentally related soil properties. *J. Environ. Qual.*, 23, 139, 1994.
27. Li, Z. and Shuman, L.M., Mobility of Zn, Cd, and Pb in soils as affected by poultry litter extract I. leaching in soil columns. *Environ. Pollut.*, 95, 219, 1997.
28. Hesterberg, D., Biogeochemical cycles and processes leading to changes in mobility of chemicals in soils. *Agric, Ecosyst. Environ.,* 67, 121, 1998.
29. del Castilho, P., Chardon, W.J., and Salomons, W., Influence of cattle-manure slurry application on the solubility of cadmium, copper, and zinc in manured acidic, loamy-sand soil. *J. Environ. Qual.*, 22, 689, 1993.
30. Stevenson, F.J., *Humus Chemistry.* 2nd ed. John Wiley & Sons, New York, 1994.
31. Chowdhury, A.K., McLaren, Cameron, and Swift, Fractionation of zinc in some New Zealand soils. *Commun. Soil Sci. Plant. Anal.*, 28, 301, 1997.
32. Yin, Y., Impellitteri, You, and Allen, The importance of organic matter distribution and extract soil:solution ratio on the desorption of heavy metals from soils. *Sci. Total Environ.*, 287, 107, 2002.
33. Raskin, I., Kumar, Dushenkov, and Salt, Bioconcentration of heavy metals by plants. *Curr. Opin. Biotechnol.*, 5, 285, 1994.
34. Salt. D.E., Baylock, Kumar, Dushenkov, Ensley, Chet, and Raskin, Phytoremediation: a novel strategy for the removal of toxic metals from the environment using plants. *Biotechnology,* 13, 468, 1995.
35. Wenzel, W.W., Adriano, Salt, and Smith, Phytoremediation: a plant-microbe-based remediation system, In *Bioremediation of Contaminated Soils.* D.C. Adriano, J.M. Bollag, W.R. Frankenburg, and R.C. Sims (Eds.), American Society of Agronomy Monograph 37. ASA, Madison, WI, 1999, 457.
36. Alloway, B.J., *Heavy Metals in Soils.* 2nd ed. Blackie, Glasgow, Scotland, 1995.
37. McGrath, S.P., Dunham, S.J., and Correll, R.L., Potential for phytoextraction of zinc and cadmium form soils using hyperaccumulator plants. In *Phytoremediation of Contaminated Soil and Water.* N. Terry and G.S. Banuelos (Eds.), CRC Press, Boca Raton, FL, 2000, 110.
38. Novak, J.M. and Watts, D.W., Copper and zinc accumulation, profile distribution, and crop removal in coastal plain soils receiving long-term, intensive applications of swine manure. *Trans. ASAE* (in review), 2004.
39. Hostetler, C.E., Kincaid, R.L., and Mirando, M.A., The role of essential trace elements in embryonic and fetal development in livestock. *Vet. J.*, 166, 125, 2003.
40. Bolan, N.S. and Adriano, D.C., Distribution and bioavailability of trace elements in livestock and poultry manure byproducts. *Crit. Rev. Environ. Sci. Technol.* 34(3): 291–338, 2004.
41. Jondreville, C., Revy, P.S., and Dourmad, J.Y., Dietary means to better control the environmental impact of copper and zinc by pigs from weaning to slaughter. *Livestock Prod. Sci.*, 84, 147, 2003.

42. Brumm, M.C., Sources of manure: swine. In *Animal Waste Utilization, Effective Use of Manure as a Soil Resource*. J.L. Hatfield and B.A. Stewart (Eds.), Ann Arbor Press, MI, 1998, 49.

43. Poulsen, H.D., Zinc and copper as feed additives, growth factors or unwanted environmental factors. *J. Anim. Feed Sci.*, 7, 135, 1998.

44. Spears, J.W., Creech, B.A., and Flowers, W.L., in *Proc. Reducing Copper and Zinc in Swine Waste through Dietary Manipulation, Anim. Waste Manage. Symp.* G.B. Havenstein (Ed.), 1999, 179.

45. Dorton, K.L., Engle, Hamar, Sicilano, and Yemm, Effects of copper source and concentration on copper status and immune functions in growing and finishing steers. *Anim. Feed Sci. Tech.*, 110, 31, 2003.

46. Unwin, R.J., Copper in pig slurry: some effects and consequences of spreading on grassland. In *Inorganic Pollution in Agriculture*, MAFE reference book no. 326. London, 1977, 306.

47. Edwards, D.R. and Daniels, T.C., Environmental impacts of on-farm poultry waste disposal — a review. *Bioresour. Technol.*, 41, 9, 1992.

48. Barker, J.C. and Zublena, J.P., Livestock manure nutrient assessment in North Carolina, In *Proc. 7th International Symposium Agricultural Wastes*. American Society of Agricultural Engineers, St. Joseph, MO, 1995, 98.

49. Sistani, K.R., Adeli, Brink, Tewolde, and Rowe, Seasonal and management impact on broiler cake nutrient composition. *J. Sustainable Agric.*, 24(1): 27–37, 2003.

50. Sistani, K.R., Brink, McGowen, Rowe, and Oldham, A change of management practice among broiler producers: broiler cake vs. broiler litter. *Bioresour. Technol.*, 90(1), 27, 2003.

51. Sistani, K.R., Pederson, Brink, and Rowe, Nutrient uptake by ryegrass cultivars and crabgrass from a highly phosphorus-enriched soil. *J. Plant Nutr.*, 26(12), 2521, 2003.

52. Tewolde, H., Sistani, K.R., and Rowe, D.E, Broiler litter as a complete nutrient source for cotton. In *Proceedings of the Beltwide Cotton Conferences*, D.A. Richter (Ed.), San Antonio, 2004.

53. Bagley, C.P., Evans, R.R., and Burdine, W.B., Jr., Broiler litter as a fertilizer or livestock feed. *J. Prod. Agric.*, 9, 342, 1996.

54. Brink, G.E., Rowe, Sistani, and Adeli, Bermudagrass cultivar response to swine effluent application. *Agron. J.*, 95(3), 597, 2003.

55. Evers, G.W., Ryegrass-Bermuda grass production and nutrient uptake when combining nitrogen fertilizer with broiler litter. *Agron. J.*, 94, 905, 2002.

56. Jinadasa, K.B., Milham, Hawkins, Cornish, Williams, Kaldor, and Conroy, Survey of cadmium in vegetables and soils of greater Sydney, Australia. *J. Environ. Qual.*, 26, 924, 1997.

57. Brink, G.E., Pederson, Sistani, and Fairbrother, Uptake of selected nutrients by temperate grasses and legumes. *Agron. J.*, 93, 887, 2001.

58. Brink, G.E., Rowe, D.E., and Sistani, K.R. Broiler litter application effects on yield and nutrient uptake of "Alicia" Bermuda grass. *Agron. J.*, 94, 911, 2002.

59. Zublena, J.P., Soil facts: nutrient removed by crops in North Carolina. *North Carolina Cooperative Extension Service Bulletin. AG-439-16*. Available at: (http://www.soils.ncsu.edu/publications/soil-facts/AG-439-16/), 1991

60. Brady, N.C., *The Nature and Properties of Soils*. 9th ed. Macmillian, New York, 1984.

61. Whitehead, D.C., *Nutrient Elements in Grassland: Soil–Plant Relationships*. CABI Publishers, New York, 2000.

62. Tucker, M.R., Stokes, D.H., and Stokes, C.E., Heavy metals in North Carolina soils: occurrence and significance. North Carolina Department of Agriculture and Consumer Services, Raleigh, NC. Available at; (http://www.ncagr.com/agronomi/pdffiles/hflyer.pdf), 2003.

63. Kiesling, T.C., Lauer, Walker, and Henning, Visual, tissue, and soil factors associated with Zn toxicity of peanuts. *Agron. J.*, 69, 765, 1977.

64. Batey, T.E., Berryman, C. and Line, C., The disposal of copper-enriched pig manure on grasslands. *J. Br. Grassland Soc.*, 27, 139, 1972.

65. Wallingford, G.W., Murphy, Powers, and Manges, Effects of beef-feedlot manure and lagoon water on iron, zinc, manganese and copper content in corn and in DTPA soil extracts. *Soil Sci. Soc. Am. Proc.*, 39, 482, 1975.

66. Kornegay, E.T., Hedges, Martens, and Kramer, Effect of soil and plant mineral levels following application of manures of different copper levels. *Plant Soil*, 45, 151, 1976.

67. Payne, G.G., Martens, Kornegay, and Lindemann, Availability and form of copper in three soils following eight annual applications of copper-enriched swine manure. *J. Environ. Qual.,* 17, 740, 1988.

68. Mullins, G.L., Martens, Gettier, and Miller, Form and availability of copper and zinc in a Rhodic Paleudult following long-term $CuSO_4$ and $ZnPO_4$ applications. *J. Environ. Qual.,* 11, 573, 1982.

69. Sutton, A.L., Nelson, Mayrose, and Kelly, Effect of copper levels in swine manure on corn and soil. *J. Environ. Qual.,* 12, 198, 1983.

70. Manahan, S.E., *Environmental Chemistry,* 5th ed., Lewis Publishers, Chelsea, MI, 1991.

71. Edwards, D.R., Moore Jr., Daniels, Srivastava, and Nichols, Vegetative filter strip removal of metals in runoff from poultry litter-amended fescue grass plots. *Trans. Am. Soc. Agric. Eng.,* 40, 121, 1997.

72. National Academy of Sciences, National Academy of Engineering (NAS/NAE), Water quality criteria, Washington, D.C., U.S. Government Printing Office, 1972.

73. U.S. Environmental Protection Agency (USEPA), Water quality criteria. *Fed. Reg.* Washington, D.C., 1980.

74. Kuo, S., Effects of drainage and long-term manure application on nitrogen, copper, zinc, and salt distribution and availability in soils. *J. Environ. Qual.,* 19, 365, 1981.

75. Chang, C., Sommerfeldt, T.G., and Entz, T., Soil chemistry after eleven annual applications of cattle feedlot manure. *J. Environ. Chem.,* 20, 475, 1991.

76. McLaren, R.G. and Crawford, D.V., Studies on soil copper. II. The specific adsorption of copper by soils. *J. Soil Sci.,* 24, 443, 1973.

77. McBride, M., *Environmental Chemistry of Soils.* Oxford University Press, New York, 1994.

78. Mehlich, A., Mehlich 3 soil test extractant: a modification of Mehlich 2 extractant. *Commun. Soil Sci. Plant Anal.,* 18, 1, 1984.

79. Southern Cooperative Series, Reference soil test methods for southern region of the United States. *Southern Cooperative Service Bulletin 289.* Georgia Agricultural Experimentation Station, Athens, GA, 1983.

80. Oste, L.A., Lexmond, T.M., and Van Riemsdijk, W.H., Metal immobilization in soils using synthetic zeolites. *J. Environ. Qual.,* 31, 813, 2002.

81. Bolan, N., Adriano, Mani, and Khan, Adsorption, complexation, and phytoavailability of copper as influenced by organic manure. *Environ. Toxicol. Chem.,* 22, 450, 2003.

82. Blaylock, M.J., Salt, Dushenkov, Zakharova, Gussman, Kapulnik, Ensley, and Raskin, Enhanced accumulation of Pb in Indian mustard by soil-applied chelating agents. *Environ. Sci. Technol.,* 31, 860, 1997.

83. Baker, A.J., McGrath, Reeves, and Smith, Metal accumulator plants: a review of the ecology and physiology of biological resources for phytoremediation of metal-polluted soils, In: N. Terry (Ed.), *Phytoremediation.* Ann Arbor Press, MI, 1989, 86.

84. Baker, A.J. and Brooks, R.R., Terrestrial higher plants which hyperaccumulate metallic elements — a review of their distribution, ecology and phytochemistry. *Biorecovery,* 1, 81, 1989.

85. Reeves, R.D. and Brooks, R.R., Hyperaccumulation of lead and zinc by two metallophytes from mining areas in Central Europe. *Environ. Pollut.,* 31, 277, 1983.

86. Malaisse, F., Gregoire, Brooks, Morrison, and Reeves, *Aeolanthus biformifolius* De Wild.: a hyper-accumulator of copper from Zaire. *Science,* 199, 887, 1978.

87. Banuelos, G.S., Ajwa, Mackey, Wu, Cook, Akohoue, and Zambruzuski, Evaluation of different plant species for phytoremediation of high soil selenium. *J. Environ. Qual.,* 26, 639, 1997.

88. Davis, A., Drexler, Ruby, and Nicholson, Micromineralogy of mine wastes in relation to lead bio-availability, Butte Montana. *Environ. Sci. Technol.,* 27, 1415, 1993.

89. Zhang, M., Alva, Li, and Calvert, Chemical association of Cu, Zn, Mn, and Pb in selected sandy citrus soils. *Soil Sci.,* 162, 181, 1997.

90. Shen, Z.G., Li, Wang, Chen, and Chua, Lead phytoextraction from contaminated soil with high-biomass plant species. *J. Environ. Qual.,* 31, 1893, 2002.

91. Kayser, A., Wenger, Keller, Attinger, Felix, Gupta, and Schulin, Enhancement of phytoextraction of Zn, Cd, and Cu from calcareous soils: the use of NTA and sulfur amendments. *Environ. Sci. Technol.,* 34, 1778, 2000.

92. Knox, A., Seaman, Mench, and Vangronsveld, Remediation of metal- and radionuclide-contaminated soils by *in situ* stabilization techniques, in I.K. Iskander (Ed). *Environmental Restoration of Metal Contaminated Soils.* Ann Arbor Press, Chelsea, MI, 2000, 21.

93. Seaman, J., Meehan, T., and Bertsch, P., Immobilization of ^{137}Cs and U in contaminated sediments using soil amendments. *J. Environ. Qual.,* 30, 1206, 2001.

94. Knox, A., Kaplan, Adriano, Hinton, and Wilson, Apatitie and phillipsite as sequestering agents for metals and radionuclides. *J. Environ. Qual.,* 32, 515, 2003.

95. Seaman, J.C., Hutchinson, Jackson, and Vulava, *In situ* treatment of metals in contaminated soils with phytate. *J. Environ. Qual.,* 32, 153, 2003.

96. Hooda, P.S. and Alloway, B.J., The effects of liming on heavy metal concentrations in wheat, carrots and spinach grown on previously sludge-applied soils. *J. Agric. Sci.,* 127, 289, 1996.

97. Singh, B.R. and Myhr, K., Cadmium uptake by barley as affected by Cd sources and pH levels. *Geoderma,* 84, 185, 1998.

98. Vangronsveld, J., Colpaert, J., and Van Tichelsen, K., Reclamation of a bare industrial area contaminated by nonferrous metals: physicochemical and biological evaluation of the durability of soil treatment and revegetation. *Environ. Pollut.,* 94, 131, 1996.

99. Chlopecka, A. and Adriano, D.C., Mimicked *in situ* stabilization of metals in a cropped soil: bioavailability and chemical form of zinc. *Environ. Sci. Technol.,* 30, 3294, 1996.

100. Boisson, J., Ruttens, Mench, and Vangronsveld, Evaluation of hydroxyapatite as a metal immobilization soil additive for the remediation of polluted soils. Part 1. Influences of hydroxapatite on metal exchangeability in soil, plant growth and plant metal accumulation. *Environ Pollut.,* 104, 225, 1999.

101. Sistani, K.R., Brink, Adeli, Tewolde, and Rowe, Year-round soil nutrient dynamics from broiler litter application to three Bermuda grass cultivars. *Agron. J.,* 96, 525, 2004.

102. Thompson, L.M. and Troeh, F.R., *Soils and Soil Fertility.* 4th ed. McGraw–Hill, New York, 1993.

103. Arora, C.L., Nayyar, and Randhawa, Note on secondary and microelement contents of fertilizers and manures. *Indian J. Agric. Sci.,* 45, 80, 1975.

104. Adriano, D.C. *Trace Elements in the Terrestrial Environment.* Springer–Verlag, Berlin, 1986.

Subject Index

Biodiversity Index

A

H

I

J

K

L